Wireless Networks

Series Editor

Xuemin Sherman Shen, University of Waterloo, Waterloo, ON, Canada

The purpose of Springer's Wireless Networks book series is to establish the state of the art and set the course for future research and development in wireless communication networks. The scope of this series includes not only all aspects of wireless networks (including cellular networks, WiFi, sensor networks, and vehicular networks), but related areas such as cloud computing and big data. The series serves as a central source of references for wireless networks research and development. It aims to publish thorough and cohesive overviews on specific topics in wireless networks, as well as works that are larger in scope than survey articles and that contain more detailed background information. The series also provides coverage of advanced and timely topics worthy of monographs, contributed volumes, textbooks and handbooks.

Zuxuan Wu • Yu-Gang Jiang

Deep Learning for Video Understanding

Springer

Zuxuan Wu
Fudan University
Shanghai, China

Yu-Gang Jiang
Fudan University
Shanghai, China

ISSN 2366-1186 ISSN 2366-1445 (electronic)
Wireless Networks
ISBN 978-3-031-57678-2 ISBN 978-3-031-57679-9 (eBook)
https://doi.org/10.1007/978-3-031-57679-9

This Springer imprint is published by the registered company Springer Nature Switzerland AG
The registered company address is: Gewerbestrasse 11, 6330 Cham, Switzerland

Preface

Video understanding is a critical technique which aims to recognize the objects and activities in videos and further analyze their evolution over time. In an era dominated by video data, the significance of video understanding is increasing day by day since it benefits applications like online education, telemedicine, etc. This book is dedicated to offering a comprehensive overview for the field of video analytics, and an introduction to the cutting-edge techniques and methodologies anchored in deep learning.

Spatial and temporal dimensions of video bring complex patterns and diverse contents, and thus, there is a wide range of tasks in video understanding that aims to comprehend information in videos from different aspects and levels. Chapter 1 provides a panoramic overview of video understanding, delineating its significance in the broader context of artificial intelligence and providing a timeline for the development of this dynamic field. Chapter 2 lays the foundational groundwork by delving into the basics of deep learning, from fundamental neural network architectures to key concepts like convolution and pooling. This chapter equips readers with the knowledge to embark on a deeper comprehension of video-centric techniques. Chapters 3–5 focuses on classic tasks in video understanding, i.e., action recognition, video localization, and captioning. These sections introduce prevalent solutions in different video understanding tasks and showcase the power of deep learning in constantly refreshing the performance records on popular benchmarks. Other than this, two emerging research directions, unsupervised video feature learning and efficient video understanding, are separately discussed in Chaps. 6 and 7, offering readers insights on how to enable machines to autonomously extract meaningful representations from unlabeled video data or in an efficient manner. Finally, in the concluding Chap. 8, we summarize the content in this book and point out several possible future directions in this field.

We hope this book could serve as a guiding beacon for researchers, practitioners, and enthusiasts by leading them through the ocean of video understanding and sparks their desire to investigate this captivating and ever-evolving field. We would also like to express our appreciation to the students in Fudan Vision and Learning Lab for their contributions to this work.

Shanghai, China
December 30, 2023

Zuxuan Wu
Yu-Gang Jiang

Contents

Chapter 1
Overview of Video Understanding

1.1 Video and Video Understanding

Videos are becoming mainstream communication channels. According to the 2023 Global Internet Phenomena Report from Sandvine [1], video data account for more than 65% of all the Internet traffic and are at an annual growth rate of approximately 24%. As a result, such a trend has promoted the development of various video-related applications including online education, remote healthcare, online advertising, virtual reality (VR), etc.

A fundamental technique that contributes to the success of the aforementioned applications is *video understanding*, which refers to the analysis and interpretation of video contents with advanced algorithms. Traditional video understanding approaches [2–5] typically rely on interest point detectors [6, 7] or local descriptors [8] to extract visual features of video clips. However, designing robust and scalable hand-crafted features is challenging, particularly when confronted with complicated appearances and movements.

Recently, the boom of deep learning has revolutionized the field of computer vision, enabling machines to perform diversified vision tasks with superior performance. Starting with AlexNet [9], the winner of the 2012 ImageNet Large-Scale Visual Recognition Challenge, a series of deep visual models [10–14] have continuously set new accuracy records for popular benchmarks [15, 16]. Along with this, the prominent success of deep learning in the image domain also inspired researchers to investigate its application in the realm of video understanding. One straightforward idea is to apply the top-performing image models to videos in a frame-by-frame manner. Early works [17, 18] have indeed implemented this idea and achieved decent results on several video benchmarks, such as Sports-1M [18]. However, neglecting the temporal dimension in modeling poses significant challenges for these methods when analyzing videos that exhibit strong temporal correlations. For example, while image models can readily recognize the presence of a door in a video sequence, they often struggle to accurately predict whether a

Z. Wu, Y.-G. Jiang, *Deep Learning for Video Understanding*, Wireless Networks,
https://doi.org/10.1007/978-3-031-57679-9_1

person is opening or closing the door. Additionally, analyzing videos frame by frame will undoubtedly result in a significant increase in computational cost, making it inefficient and impractical for real-world applications.

These challenges have motivated researchers in the field to develop innovative and efficient solutions for various video tasks. The pursuit of such solutions confers immense value and fascination to the study of video understanding.

1.2 Video Understanding Tasks and Definitions

Due to the prevalence of video applications in people's daily lives, video understanding has become an umbrella term for a broad range of video tasks. In this subsection, we will briefly introduce classic video understanding tasks and their corresponding benchmark datasets.

Video action recognition [19] aims to recognize human actions in a video sequence. As one of the most representative video tasks, video action recognition relies on computer vision and machine learning techniques to first extract representative features from video frames, followed by predicting the action categories, such as running, swimming, or eating. The video action recognition benchmarks commonly used by researchers have iterated over two generations. The first generation of datasets are overall smaller in size and shorter in video length, including: (1) UCF101 [20] that has a total of 13,320 videos with 101 action categories and (2) HMDB [21] that consists of 6766 videos with 51 classes. By contrast, the second generation of datasets are more challenging due to their larger size and longer duration, which include: (1) Kinetics-400 [22], a large-scale dataset that contains approximately 240,000 video clips with 400 action classes, and (2) Something-something V2 [23], a fine-grained human activity dataset that covers 174 action categories.

Temporal action localization is the task of localizing and identifying actions within a video over time. It involves detecting not only what action is being performed, but also when it starts and ends. Unlike video action recognition that takes short video clips as inputs, temporal action localization focuses on the untrimmed long videos in the wild. The popular benchmarks include: (1) THUMOS14 [24] that contains 413 untrimmed videos with 20 categories of actions, (2) ActivityNet-1.3 [25] that includes 200 activity classes and around 20,000 videos of more than 600 hours, and (3) EPIC-Kitchens 100 [26] that is the largest cooking dataset in egocentric vision.

Video captioning refers to the process of automatically producing a description of the video that accurately captures its main elements, such as the objects, people, actions, and events depicted in it. The most widely used benchmarks are: (1) MSVD [27] that contains 1970 short video clips with paired human-generated captions, (2) MSR-VTT [28] that has 10,000 video clips, each paired with 20 human-generated captions, (3) ActivityNet Captions [29] that is a large-scale benchmark includes over 20,000 videos and more than 100,000 human-generated captions, and (4) YouCook2 [30] that focuses on cooking videos.

Efficient video understanding is devoted to analyzing and interpreting video data in a computationally efficient manner. Compared to static images with a single spatial dimension, the extra temporal dimension in video data undoubtedly brings about expensive training costs and inference overhead for deep models, preventing their deployment in the real-world applications. To address this issue, efficient video understanding aims at reducing the memory and time consumption by optimizing the network architecture and training process.

Unsupervised video representation learning learns the meaningful patterns and structures in the visual content of the videos, such as object motion, scene dynamics, and temporal dependencies, without the need of manual labels. Unsupervised video representation learning has the potential to unlock new insights and applications in video analysis and understanding, especially in scenarios where labeled data are scarce or expensive to obtain.

1.3 The Timeline of Video Understanding Techniques

Deep-learning-based video understanding techniques have witnessed superior advancements in the past decade. Here is a rough timeline of some significant developments, as shown in Fig. 1.1.

- **2014**: A two-stream CNN [17] was introduced for video action recognition, where two sub-networks are adopted to process spatial and temporal information in video data, respectively.
- **2015**: The 3D CNNs were introduced in [31], which can directly extract the spatial–temporal features in videos through 3D convolutions, leading to significant improvements in video classification accuracy. In addition, [32] proposed a hybrid neural network architecture by combining the CNNs and LSTMs.
- **2016**: TSN [33] utilized a combination of temporal segment sampling and consensus pooling, which achieved state-of-the-art results on multiple video action recognition datasets.

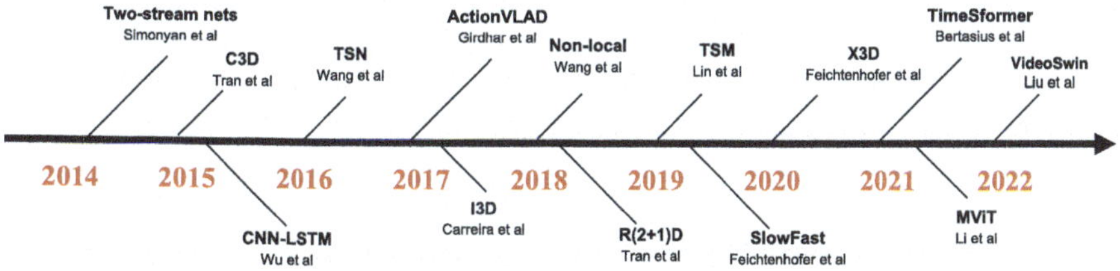

Fig. 1.1 The timeline of deep-learning-based video understanding techniques

- **2017**: ActionVLAD [34] presented a novel spatial–temporal aggregation method called VLAD (vector of locally aggregated descriptors) for the better fusion of deep features. Additionally, the first large-scale video action recognition dataset, Kinetics-400, was introduced in [35].
- **2018**: R(2+1)D [36] investigated the effectiveness of spatial–temporal convolutions in neural networks for recognizing human actions in videos and proposed a novel architecture that combines the benefits of two-stream CNNs and 3D CNNs. Non-local [37] modeled the long-range dependencies in video data by allowing each output position to attend to all input positions.
- **2019**: TSM [38] improved the efficiency and accuracy of video action recognition by incorporating temporal shift operations into convolutional neural networks. Slowfast [39] involved a slow pathway and a fast pathway to operate the video inputs at different frame rates, where both spatial semantics and temporal dynamics could be combined for video understanding.
- **2020**: X3D [40] was a lightweight video model with superior performance, which leveraged the factorized spatial–temporal convolution, depthwise separable convolution, and squeeze-and-excitation modules to vanilla 3D CNNs.
- **2021-**: TimeSformer [41] was a pioneering work that applied self-attention mechanism to the video tasks. Following this, MViT [42] and VideoSwin [43] further explored the hierarchical design and window attention for video action recognition.

The timeline of deep-learning-based video understanding techniques highlights some of the significant milestones in this field. From the early days of deep learning, when Convolutional Neural Networks were first applied to video classification tasks, to the more recent developments in Video Transformers, the progress in this area has been impressive. However, the potential for video understanding is far from exhausted. As deep learning continues to advance, we can expect further developments in video understanding, leading to novel applications in diverse areas such as online shopping, autonomous driving, and video recommendation. These applications have the potential to revolutionize industries and improve our daily lives in many ways, such as enabling personalized shopping experiences, enhancing road safety, and providing more relevant video content. Therefore, understanding the current state-of-the-art deep-learning-based video understanding techniques and keeping up with the latest advancements are crucial for researchers and practitioners in this field.

The remainder of this book will explore the advances in deep learning for video understanding, starting with a concise overview of classic network architectures such as CNN, RNN, and Transformer. Throughout the book, we will examine how these techniques have been utilized in various real-world video applications, including action recognition, temporal action localization, video captioning, and many others. We will end the book by summarizing the main points and offering insights on future research directions. This book covers a range of topics related to deep learning for video understanding, from fundamental concepts to advanced techniques for efficient video analysis. Chapter 2 provides an introduction to deep learning basics for video understanding, including convolutional neural networks (CNNs), recurrent neural networks (RNNs), and attention mechanisms. Chapters 3–4 focus on two essential video understanding tasks: action recognition and action localization, respectively. Chapter 5 explores deep learning techniques for video captioning, which involves generating natural language descriptions of video content. Chapter 6 introduces unsupervised feature learning approaches for video representation learning, including self-prediction pretext tasks, contrastive learning and masked video modeling, and Chap. 7 delves into efficient deep learning models and training strategies for practical video understanding applications. Finally, Chap. 8 reflects on the current status of video understanding research and outlines several promising future directions. By reading this book, readers are expected to gain a comprehensive understanding of how deep learning can be utilized to extract valuable information from video data, as well as the tools and techniques needed to implement their own video understanding systems.

Chapter 2
Deep Learning Basics for Video Understanding

2.1 Convolutional Neural Networks (CNNs)

Convolutional Neural Networks (CNNs) was first introduced in 1989 by Yann LeCun primarily used for handwritten character recognition [44]. CNNs require fewer parameters than fully connected networks, making them superior to their peer deep networks with easier parallel computation on GPUs [9]. Over time, CNNs have dominated a wide range of tasks such as image classification, object detection, semantic segmentation, and instance segmentation. In recent years, researchers extend CNNs' ability to applications including natural language processing, recommendation systems, etc. [45].

This section is organized as below: First, we will explain convolutions in neural networks, followed by an introduction of pooling, which is a common operation in CNNs; lastly, we will present background and detailed architectures of five classic CNNs in a chronological order.

2.1.1 Convolution

Convolution is a fundamental mathematical operation used extensively in signal processing and image processing. In image processing, a two-dimensional convolution is widely used since images are two-dimensional structures. It involves convolving a filter kernel with an input image to produce a feature map. The output is computed by summing the element-wise product of the filter kernel and a region of the input image. Given an image $\boldsymbol{X} \in \mathbb{R}^{M \times N}$ and a filter $\boldsymbol{W} \in \mathbb{R}^{U \times V}$, generally $U << M, V << N$, the convolution output y_{ij} is defined as following:

Z. Wu, Y.-G. Jiang, *Deep Learning for Video Understanding*, Wireless Networks,
https://doi.org/10.1007/978-3-031-57679-9_2

$$y_{ij} = \sum_{u=1}^{U} \sum_{v=1}^{V} w_{uv} x_{i-u+1, j-v+1}. \tag{2.1}$$

The convolution operation in the input image $\boldsymbol{X}$ and the filter $\boldsymbol{W}$ is defined as

$$\boldsymbol{Y} = \boldsymbol{W} * \boldsymbol{X}, \tag{2.2}$$

where $\boldsymbol{X}$ is the input and $\boldsymbol{W}$ is the kernel function. The output Y is also referred to as the feature map, and the convolution operation is denoted by $*$.

To improve the flexibility of feature extraction and increase the diversity of convolution, the standard convolution operation can be supplemented with a sliding stride for the convolution kernel and a zero padding operation. This additional step allows for a wider range of features to be extracted and enhances the effectiveness of the convolution operation.

Finally, the size of the feature map can be calculated using a formula that takes into account the input size, filter size, stride, and padding:

$$o = \left\lfloor \frac{n + 2p - f}{s} \right\rfloor + 1, \tag{2.3}$$

where o is the height/width of the feature map, f is the height/width of the filter conducting the convolution operation, n denotes the input's height/width, and stride and padding sizes are denoted as s and p, respectively.

2.1.2 Pooling

The pooling function explores the statistical properties of neighboring regions of a specific location and uses the derived statistics to represent that all elements in that region. For instance, the max pooling function [46] selects the highest value in a neighboring rectangular region. Other frequently used pooling functions involve the average value of adjacent rectangular areas, the L2 norm, and the weighted average function based on the distance from the center pixel [47]. Convolution operations are effective in decreasing the number of connections in a network, but they do not significantly reduce the number of units in the feature map. Directly attaching a classifier to the feature map can result in dense weight matrix, making the model prone to overfitting. To solve this problem, a pooling function can be added after the convolution operations, so as to further reduce feature dimension and thus avoid overfitting.

2.1.3 Classic Convolutional Neural Networks

In this subsection, we discuss five classic CNNs that have played a significant role in the history of computer vision. These models serve as the foundation for modern

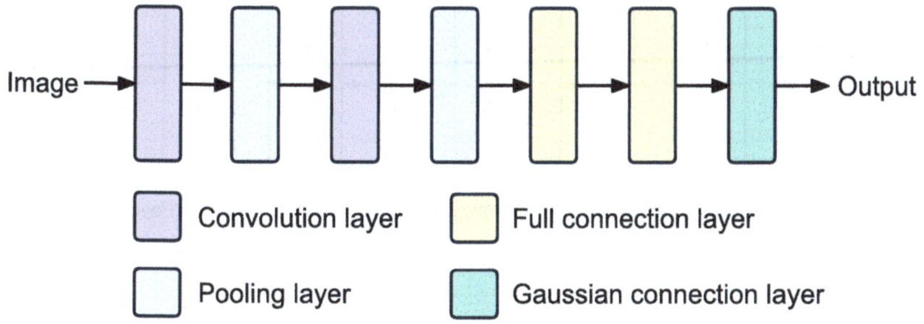

Fig. 2.1 The structure of the Lenet-5

deep learning architectures and are still used as benchmarks for performance comparisons:

LeNet-5 is one of the first published CNNs that has gained widespread recognition for its exceptional performance in computer vision tasks. This model was proposed in 1989 by Yann LeCun [44] in order to recognize handwritten digits in images. At that time, Yann LeCun published the first study on successfully training a convolutional neural network via backpropagation. However, the performance of LeNet-5 and its variants [48] is restricted by the scarcity of labeled training data and computational resources, resulting in unsatisfactory outcomes on complex computer vision tasks. The structure of the LeNet-5 is illustrated in Fig. 2.1. LeNet-5 has a total of 7 layers. The size of the input image is $32 \times 32 = 1024$, and the digits of the output correspond to 10 categories. In general, Lenet-5 is composed of two parts: convolutional blocks and fully connected blocks. The basic units in each convolutional block are a convolutional layer, a sigmoid activation function, and an average pooling layer. Each convolutional layer employs both a convolution kernel and a sigmoid activation function. These functional layers map inputs to multiple 2D feature representations, often increasing the number of channels simultaneously. The first convolutional layer has 6 output channels, while the second convolutional layer has 16 output channels. Each pooling operation reduces the dimensionality by a factor of 4 through spatial downsampling with stride 2. The output shape of the convolution is determined by the batch size, the number of channels, height, and width.

AlexNet was the first modern deep convolutional network model. It was proposed by Krizhevsky et al. in 2012 [9] and won the 2012 ImageNet image classification competition. The overall structure of the AlexNet is shown in Fig. 2.2. AlexNet consists of five convolutional layers, three pooling layers, and three fully connected layers, where the last layer is the output layer using the Softmax function. Compared with LeNet-5, AlexNet uses Rectified Linear Units (ReLU) as the non-linear activation function to make the training process several times faster and Dropout to prevent overfitting. In addition, AlexNet also conducts parallel training in GPUs to speed up the training process and utilizes data

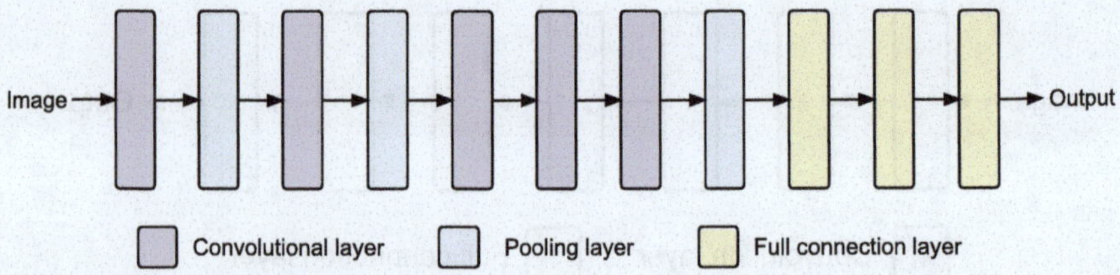

Fig. 2.2 The structure of the AlexNet

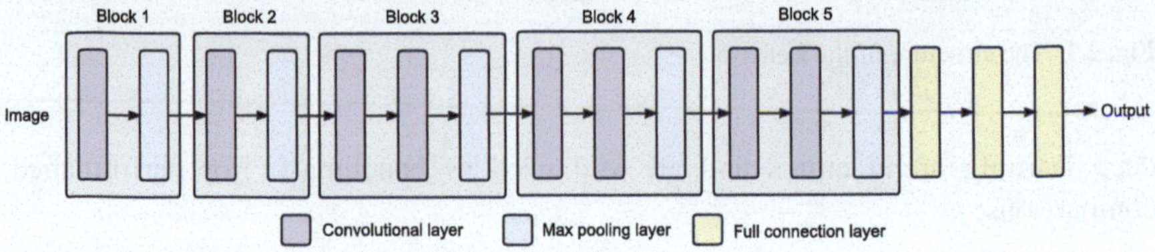

Fig. 2.3 The structure of the VGG-11

augmentation to improve accuracy. AlexNet takes 224 × 224 × 3 images as input and outputs conditional probabilities as indicators of 1000 classes. Because the network size exceeded the memory limits of a single GPU at one time, AlexNet split the network in half and put them on two GPUs, with GPUs communicating only at certain layers.

VGGNet [10] provides a typical template to guide succeeding researchers in designing new networks. Neural network architecture design is becoming more coarse-grained and abstract, from individual neurons, to layers and eventually to blocks. The concept of using blocks was first introduced in VGGNet, where blocks were used to expand the network's depth from the 8 layers in AlexNet to 16–19 layers, thereby improving the discriminative power of CNNs. Additionally, VGGNet's utilization of (3 × 3) convolutional filters enables it to capture fine details in input images. The basic building block of a classical CNN is displayed in the following sequence: convolutional layers with padding to preserve resolution, non-linear activation functions (e.g., ReLU), and pooling layers (e.g., max pooling layer). A VGG block is similar to the aforementioned basic building block, consisting of a series of convolutional layers followed by a max pooling layer that performs spatial downsampling. A typical VGG network has 5 VGG blocks, with the first two blocks consisting of one convolutional layer each and the last three blocks consisting of two convolutional layers each. The structure of the VGG network is illustrated in Fig. 2.3. The first module in the network produces 64 output channels, with the number of output channels doubling in each subsequent module until it reaches 512. Because it has 8 convolutional layers and 3 fully connected layers, it is commonly referred to as VGG-11.

GoogLeNet [11] shined in the 2014 ImageNet image classification competition. GoogLeNet addresses the problem of determining the best convolution kernel

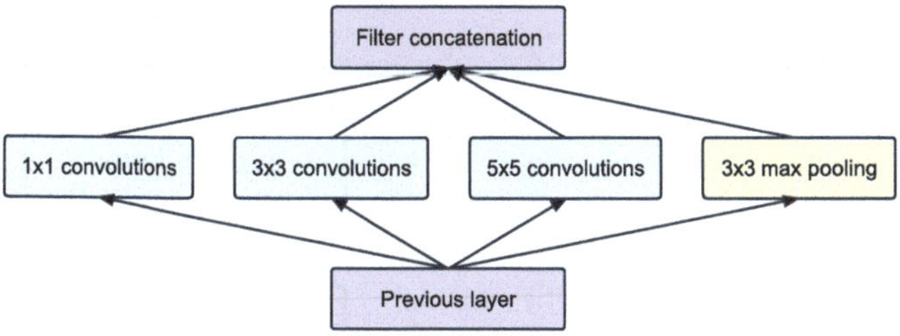

Fig. 2.4 The structure of an Inception block

size. Previous popular CNNs use kernels that vary from size 1×1 to 11×11, without a conclusion about the best kernel size. GoogLeNet proposed a novel architecture Inception block as illustrated in Fig. 2.4. With four parallel paths, the Inception block enlarges the capacity of CNNs while maintaining an affordable computational cost. In this way, GoogLeNet takes the advantage of several different-sized kernels in one model. Specifically, the Inception block has four parallel paths: The first three paths apply convolutions with filter sizes of 1×1, 3×3, and 5×5 to capture information at different spatial scales; the two middle paths use 1×1 convolutions to reduce the number of channels and thus the model's computational complexity; and the fourth path uses max pooling followed by a 1×1 convolution to change the number of channels. All four paths use appropriate padding to match input and output sizes, and their outputs are concatenated along the channel dimension to form the block output. In practice, the number of output channels per layer in the Inception block is a key hyperparameter that is often tuned to achieve the best performance.

ResNet [12] improves the efficiency of information propagation by adding shortcut connections (also known as residual connections) to the non-linear convolutional layer. Residual connections led to ResNet's championship in the 2015 ImageNet image classification competition. A deep network can be regarded as a combination of shallow networks with some identity mappings, and the function of a certain layer in the network is similar to a mapping from x to $H(x)$. It is expected that deep networks outperform their shallow counterparts, but deep models are extremely difficult to train with gradient descent. To solve this problem, ResNet no longer expects a certain layer of the network to obtain a mapping from x to $H(x)$ and decides to let the hidden layers learn the mapping from x to $F(x)$ in which $F(x) = H(x) - x$. The mapping from x to $F(x)$ is called a residual mapping that is often easier to optimize in reality. An illustration of the residual block structure can be found in Fig. 2.5. ResNet follows the convolutional block design of VGG and consists of residual blocks, each starting with two convolutional layers with the same number of output channels. Batch normalization and ReLU activation functions follow each convolutional layer. The two convolutional layers are then skipped via the cross-layer shortcut, and the input is added directly before the final ReLU activation function, producing the output $x + F(x)$. To enable this addition, the outputs of the two convolutional layers must be of the same shape as the inputs.

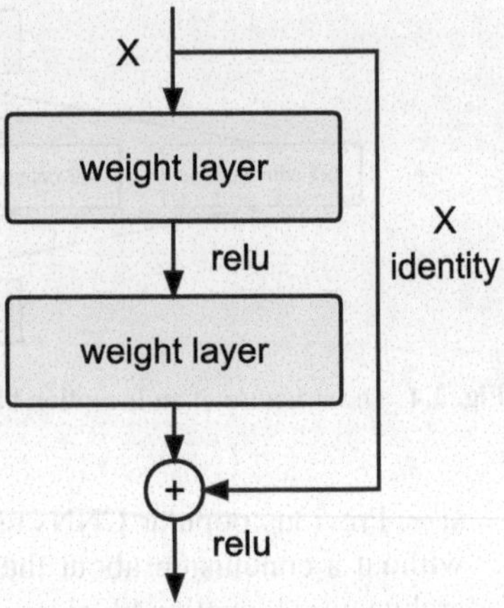

Fig. 2.5 The structure of a residual block

One obvious trend is that networks are becoming deeper. As the number of network layers increases, convolutional models turn to smaller convolution kernels of size 1 × 1 and 3 × 3. The network structure gradually transforms to fully convolutional networks (FCNs) [49], lessening the aggregation layer and the fully connected layer. Please refer to [50, 51] for more detailed reviews.

2.2 Recurrent Neural Networks (RNNs)

The CNN architecture discussed above exploits convolution for efficient spatial (2D) information extraction, such as images. Considering domains that involve sequential data, CNNs might not be the optimal choice. In tasks such as video understanding, RNNs are the preferred models. In particular, RNNs are designed to capture the dynamic correlations of sequential data via recurrent connections, which use state variables to memorize previous information and determine current output in combination with current input.

RNNs perform differently from a fully connected network. At the time step t, a typical block in RNNs will take as input an external input vector $\boldsymbol{x}^{(t)} \in \mathbb{R}^n$ and a hidden state $\boldsymbol{h}^{(t-1)} \in \mathbb{R}^r$ generated at the previous time step $t-1$ to calculate current hidden state $\boldsymbol{h}^{(t)} \in \mathbb{R}^r$. Then the current hidden state will be used to generate the output vector $z^{(t)} \in \mathbb{R}^m$:

$$\boldsymbol{h}^{(t)} = \sigma\left(W_x \boldsymbol{x}^{(t)} + W_h \boldsymbol{h}^{(t-1)} + \boldsymbol{b}_h\right) \tag{2.4}$$

$$z^{(t)} = \text{softmax}\left(W_z \boldsymbol{h}^{(t)} + \boldsymbol{b}_z\right),$$

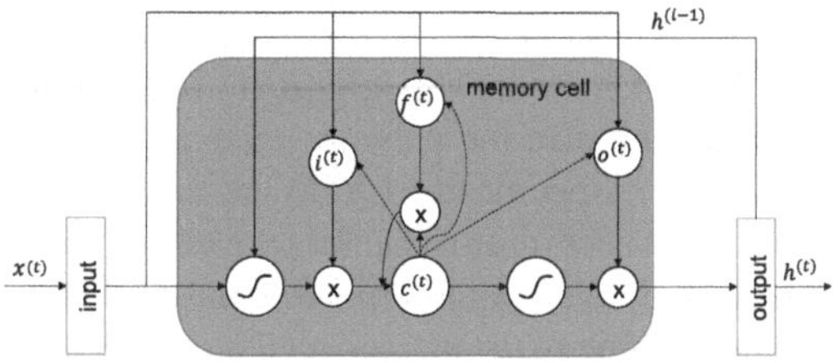

Fig. 2.6 The structure of an LSTM unit

where $W_x \in \mathbb{R}^{r \times n}$, $W_h \in \mathbb{R}^{r \times r}$, and $W_z \in \mathbb{R}^{m \times r}$ are weight matrices, while $\boldsymbol{b}_h$ and $\boldsymbol{b}_z$ denote biases. σ defines the sigmoid function $\sigma(x) = \frac{1}{1+e^{-x}}$, whereas softmax $(\cdot)$ represents the softmax function.

Traditional RNNs endure two major problems: One is their limitation in modeling long-term dependencies and the other is storage limitation [52], hindering RNNs from approximating the sequence of arbitrary complexity. In addition, *vanishing* and *exploding* gradients further impede RNNs' real-world applications. The *vanishing* gradient means as gradients propagate back through time, they might be shrunk exponentially during the process. And the *exploding* gradient problem occurs when gradients become too large during training, making it difficult to train the network effectively. To tackle the barriers above, Long Short-Term Memory models are proposed as one of the solutions.

Long Short-Term Memory (LSTM) is a type of Recurrent Neural Network (RNN) that addresses the challenge of storing and accessing information over long time sequences. LSTM introduces non-linear multiplicative gates and a memory cell. The non-linear gates, including the input, output, and forget gates, control the flow of information into and out of the memory cell.

As illustrated in Fig. 2.6, an LSTM unit takes an external vector $\boldsymbol{x}^{(t)} \in \mathbb{R}^n$ as input and maps it to an output vector $z^{(t)} \in \mathbb{R}^m$ by computing the activations of its units with the following equations recursively from $t = 1$ to $t = T$:

$$
\begin{aligned}
\boldsymbol{i}^{(t)} &= \sigma\left(W_{xi}\boldsymbol{x}^{(t)} + W_{hi}\boldsymbol{h}^{(t-1)} + W_{ci}\boldsymbol{c}^{(t)} + \boldsymbol{b}_i\right), \\
\boldsymbol{f}^{(t)} &= \sigma\left(W_{xf}\boldsymbol{x}^{(t)} + W_{hf}\boldsymbol{h}^{(t)} + W_{cf}\boldsymbol{c}^{(t)} + \boldsymbol{b}_f\right), \\
\boldsymbol{c}^{(t)} &= \boldsymbol{f}^{(t)}\boldsymbol{c}^{(t-1)} + \boldsymbol{i}_t \tanh\left(W_{xc}\boldsymbol{x}^{(t)} + W_{hc}\boldsymbol{h}^{(t-1)} + \boldsymbol{b}_c\right), \\
\boldsymbol{o}^{(t)} &= \sigma\left(W_{xo}\boldsymbol{x}^{(t)} + W_{ho}\boldsymbol{h}^{(t-1)} + W_{co}\boldsymbol{c}^{(t)} + \boldsymbol{b}_o\right), \\
\boldsymbol{h}^{(t)} &= \boldsymbol{o}^{(t)} \tanh\left(\boldsymbol{c}^{(t)}\right),
\end{aligned}
\tag{2.5}
$$

where $\boldsymbol{x}^{(t)}$ and $\boldsymbol{h}^{(t)}$ represent the input and hidden vectors at time step t, respectively. The activation vectors for the input gate, forget gate, memory cell, and output gate are denoted by $\boldsymbol{i}^{(t)}$, $\boldsymbol{f}^{(t)}$, $\boldsymbol{c}^{(t)}$, and $\boldsymbol{o}^{(t)}$, respectively. The weight matrix between two vectors α and β is represented by $W_{\alpha\beta}$. For instance, W_{xi} corresponds to the weight matrix from input $\boldsymbol{x}^{(t)}$ to the input gate $\boldsymbol{i}^{(t)}$.

At each time step t, an LSTM unit takes the input $\boldsymbol{x}^{(t)}$ and the previous hidden state $\boldsymbol{h}^{(t-1)}$ as input. The information stored in the memory cell is determined by the information in the previous memory cell unit $\boldsymbol{c}^{(t-1)}$ and the activation of the input gate $\boldsymbol{i}_t$. Specifically, LSTM has three main steps: (1) to forget previous state's information according to the forget gate; (2) to update the information of the current state; and (3) to output information. In step 1, $\boldsymbol{f}^{(t)}$ serves as the forget gate. In step 2, the current state of the memory cell is obtained by summing up previous information and the current input signal. The output gate controls the information received by hidden state variable $\boldsymbol{h}^{(t)}$. To summarize, LSTM is capable of long-term temporal memory with carefully designed memory units and gates. As LSTM discards some long-term information and is required to memorize only partial information, it avoids the issues of vanishing/exploding gradients.

2.3 Transformer

2.3.1 *Self-attention*

Attention function is used for correlating, which allows the similarity computations between samples. For example, in natural language processing (NLP), the positions of words in a sentence contribute to their translation. Attention mechanism can be described as mapping a query (Q) and a set of key–value pairs $\{K_i, V_i \mid i = 1, 2, \ldots, m\}$ to an output, where the query, keys, values are all vectors. Weighted sums of the values are computed based on compatibility between the query and the relevant key, where the weight for each value is assigned. The steps are as follows:

1. Computing the similarity between Q and K:

$$f(Q, K_i), i = 1, 2, \ldots, m \tag{2.6}$$

2. Normalizing the obtained similarity:

$$\alpha_i = \frac{e^{f(Q,K_i)}}{\sum_{j=1}^{m} f\left(Q, K_j\right)}, i = 1, 2, \ldots, m \tag{2.7}$$

3. Computing the weighted sum of V with the weight α_i:

$$Attention(Q, K, V) = \sum_{i=1}^{m} \alpha_i V_i \tag{2.8}$$

The similarity between Q and K can be computed in various ways:

1. Dot product

$$f(Q, K_i) = Q \cdot K_i \tag{2.9}$$

2. Cosine

$$f(Q, K_i) = \frac{Q \cdot K_i}{\|Q\| \cdot \|K_i\|} \tag{2.10}$$

3. Perception

$$f(Q, K_i) = MLP(Q, K_i) \tag{2.11}$$

Self-attention is a variant of attention, which improves the modeling of internal correlation of data, reducing the dependence on external information. Text-related tasks usually employ self-attention to get the correlations between words and then solve the long-distance dependency problem. Self-attention exploits a similar equation as in attention. The self-attention, similar to the attention, maps a query (Q) and a set of key–value pairs $\{K_i, V_i \mid i = 1, 2, \ldots, m\}$ to an output as well, with same query (Q), key (K), and value (V) are the same instead.

2.3.2 *Transformer*

Transformer [53] is a machine learning model that only relies on attention mechanisms to learn global dependencies among inputs without the need of recurrence. The model architecture is illustrated in Fig. 2.7.

Positional Encoding The inputs of the encoder and decoder are first converted into vectors of dimension d_{model}, followed by summing with a positional encoding vector that provides absolute and relative positional information using sine and cosine functions of different frequencies, given by

$$\begin{aligned} PE_{(pos,2i)} &= \sin\left(pos/10000^{2i/d_{\text{model}}}\right) \\ PE_{(pos,2i+1)} &= \cos\left(pos/10000^{2i/d_{\text{model}}}\right), \end{aligned} \tag{2.12}$$

where pos is the position and i is the dimension.

Attention While multiple attention mechanisms exist, the one used in Transformer is called "Scaled Dot-Product Attention," a self-attention mechanism that computes a scaled dot product of the queries with all keys, where the scale, $\sqrt{d_k}$, is given by the square root of the dimension of keys. The final attention score is further derived

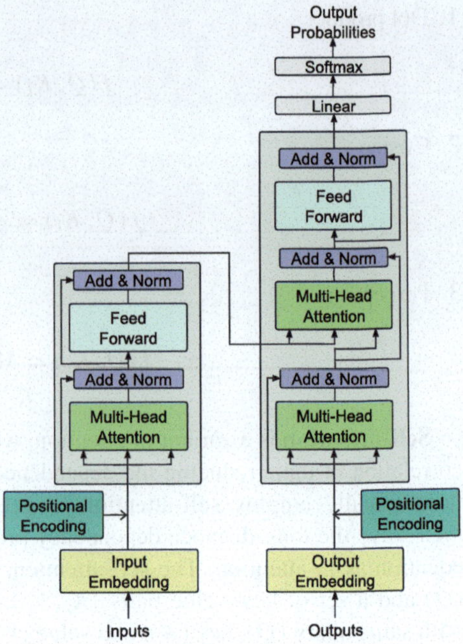

Fig. 2.7 The model architecture of the Transformer

by applying a softmax function to the above stated result. Formally, the scaled dot-product attention is given by

$$\text{Attention}(Q, K, V) = \text{softmax}\left(\frac{QK^T}{\sqrt{d_k}}\right)V, \tag{2.13}$$

where Q, K, and V are queries, keys, and values matrices.

Multi-head Attention Rather than relying on a single attention function, the Transformer architecture employs a technique known as "multi-head attention." This approach involves using multiple self-attention functions, which are combined to produce the final output. Specifically, the queries, keys, and values are projected to d_k, d_k, d_v dimensions h times with h attention functions performed in parallel. The obtained h outputs are then concatenated and projected back to the original dimension as the final output. The formula for multi-head attention is given by

$$\begin{aligned} \text{MultiHead}(Q, K, V) &= \text{Concat}\left(\text{head}_1, \ldots, \text{head}_{\text{h}}\right) W^O \\ \text{where } \text{head}_i &= \text{Attention}\left(QW_i^Q, KW_i^K, VW_i^V\right), \end{aligned} \tag{2.14}$$

where $W_i^Q \in \mathbb{R}^{d_{\text{model}} \times d_k}$, $W_i^K \in \mathbb{R}^{d_{\text{model}} \times d_k}$, $W_i^V \in \mathbb{R}^{d_{\text{model}} \times d_v}$, and $W_i^O \in \mathbb{R}^{hd_v \times d_{\text{model}}}$ are parameter matrices for projections.

Feed-Forward Layer The feed-forward layer consists of two linear transformations followed by a non-linear activation function, typically a rectified linear unit (ReLU) function. Given an input $\mathbf{x} \in \mathbb{R}^{d_{\text{model}}}$, the feed-forward layer computes

$$\text{FFN}(\mathbf{x}) = \max(0, \mathbf{x}\mathbf{W}_1 + \mathbf{b}_1)\mathbf{W}_2 + \mathbf{b}_2,$$

where $\mathbf{W}_1 \in \mathbb{R}^{d\text{model} \times d_{\text{ff}}}$ and $\mathbf{W}_2 \in \mathbb{R}^{d\text{ff} \times d_{\text{model}}}$ are weight matrices, $\mathbf{b}_1 \in \mathbb{R}^{d_{\text{ff}}}$ and $\mathbf{b}_2 \in \mathbb{R}^{d_{\text{model}}}$ are bias vectors, and d_{model} and d_{ff} are hyperparameters that represent the dimension of the input and output feature spaces, respectively.

The feed-forward layer is typically applied to each input embedding in the Transformer model, providing them with a non-linear transformation of features. This allows the Transformer model to capture complex interactions between features and provides robust representations for downstream tasks.

Layer Normalization Layer normalization works by normalizing the inputs with respect to the layer dimension such that each feature has zero mean and unit variance across training examples. This helps to reduce the impact of the internal covariate shift and improves the stability and convergence of the network. In the Transformer model, layer normalization is applied before and after the feed-forward and multi-head attention layers.

Encoder The Transformer encoder is comprised of several layers, each containing two sub-layers. The first sub-layer is a multi-head self-attention mechanism that calculates the relevance of the input token embedding. The second sub-layer is a position-wise fully connected feed-forward network that applies a non-linear transformation to the embedding. To ensure the stability of the model, a residual connection and layer normalization are applied to the output of each sub-layer. By processing the input tokens, the encoder captures their information and generates a high-dimensional representation that can be used by the decoder to generate the output sequence.

Decoder In contrast to the encoder, each decoder layer has three sub-layers. The first sub-layer is a masked multi-head self-attention mechanism, which is similar to the self-attention mechanism in the encoder, but it prevents the decoder from attending to future tokens. The second sub-layer is a multi-head attention mechanism that queries the encoder outputs to incorporate the encoder's information into the decoder. The third sub-layer is a position-wise fully connected feed-forward network that applies a non-lincar transformation to the decoder's inputs. Similar to the encoder, each sub-layer in the decoder is followed by a residual connection and layer normalization. Finally, the decoder applies a linear layer and a softmax function to generate a probability distribution over the vocabulary for the next output token.

The Transformer model has shown remarkable success in various natural language processing tasks, such as machine translation, language modeling, and text generation. More recently, researchers are exploring the potential of Transformer models in video understanding, and Transformer models show promising results.

More specifically, researchers have demonstrated that the Transformer architectures can effectively capture temporal and spatial relationships in videos, enabling more accurate and efficient video understanding. The success of Transformer model in video understanding motivates further research and attention in this field, with aims of creating more advanced video understanding tools and models.

In summary, Transformer is an architecture based entirely on self-attention to obtain feature representations, addressing the limitations of recurrent and convolutional models. The model uses a positional encoding to incorporate absolute and relative positional information to the input and applies multi-head attention to make the model focus on information from various representation subspaces. It also uses the concepts of residual connections and layer normalization to improve data modeling. Transformer has achieved state-of-the-art results on a variety of tasks, making it one of the most influential machine learning models in both academia and industry.

2.3.3 *Vision Transformer*

The Vision Transformer (ViT) is a state-of-the-art deep learning model in computer vision, which utilizes the Transformer architecture originally developed for natural language processing tasks. With few modifications, ViT applies the standard Transformer model to process image data and achieves superior performance compared with conventional CNN models. The ViT model preprocesses an image by dividing it into several patches and arranging them into linear embedding sequences, which are fed into a Transformer encoder. Finally, an MLP head generates the results for various computer vision tasks.

To handle 2D images, ViT first reshapes an image $\mathbf{x} \in \mathbb{R}^{H \times W \times C}$ into a sequence of flattened 2D patches $\mathbf{x}_p \in \mathbb{R}^{N \times (P^2 \cdot C)}$, where $N = HW/P^2$ is the number of patches and (P, P) is the resolution of each patch. The embedding operation consists of patch embedding and position embedding:

$$\mathbf{z}_0 = \left[\mathbf{x}_{\text{class}}\, ; \mathbf{x}_p^1\mathbf{E}; \mathbf{x}_p^2\mathbf{E}; \cdots ; \mathbf{x}_p^N\mathbf{E}\right] + \mathbf{E}_{\text{pos}}\,, \tag{2.15}$$

where $\mathbf{z}_0$ serves as the input of the Transformer encoder, and $\mathbf{x}_{\text{class}}$ is a learnable embedding used for classification. The patch embedding is mapped to the dimensions of D with a trainable linear projection. To retain positional information, ViT adds 1D learnable position embeddings to patch embeddings. The model encodes the distance in the image to measure the similarity of position embeddings, which means adjacent patches have more similar position embeddings.

The Transformer encoder of ViT consists of multi-head self-attention (MSA) and MLP blocks. Residual connections are used between every two blocks, and the architecture is illustrated in Fig. 2.8.

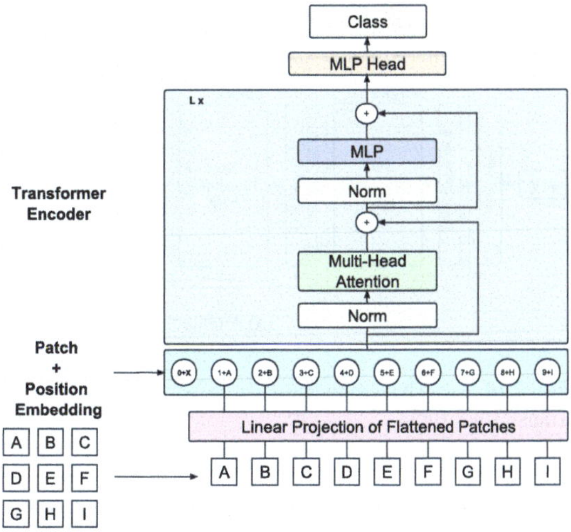

Fig. 2.8 The model architecture of ViT

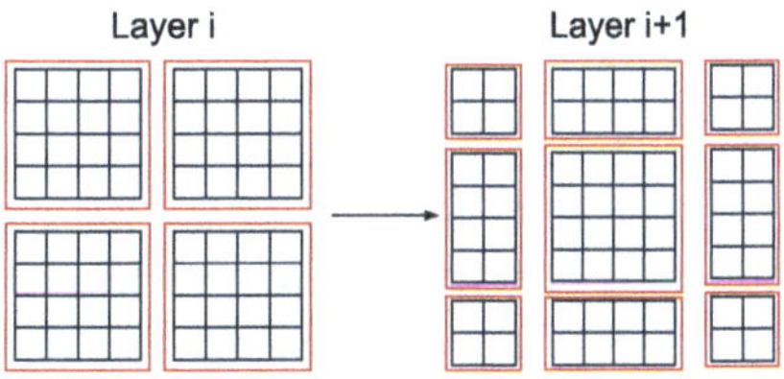

Fig. 2.9 The illustration of shifted windows

2.3.4 Swin Transformer

The success of the Transformer in natural language processing has inspired its application to various computer vision tasks. The Vision Transformer (ViT) has shown remarkable performance in image classification but lacks inductive bias in CNNs for modeling images. The Swin Transformer thus explores seminal designs that work well for CNNs in Transformers.

Shifted Window The Swin Transformer [54] employs a unique approach called the shifted window mechanism in each Transformer block, which refers to dividing the image into small, overlapping windows. These shifted windows ensure that different patches can interact with each other, creating cross-window connections. Figure 2.9 shows an illustration of the shifted windows.

Patch Merging Similar to pooling, patch merging aggregates values from the same position in small windows and forms a new patch. All patches are then concatenated to reduce the resolution of the feature map. This process is repeated in each stage, and the resulting feature maps contain a mixture of information from both local and global scales. By merging patches across the image, the Swin Transformer can efficiently capture long-range dependencies and spatial information while maintaining high-resolution features in the earlier stages of the model.

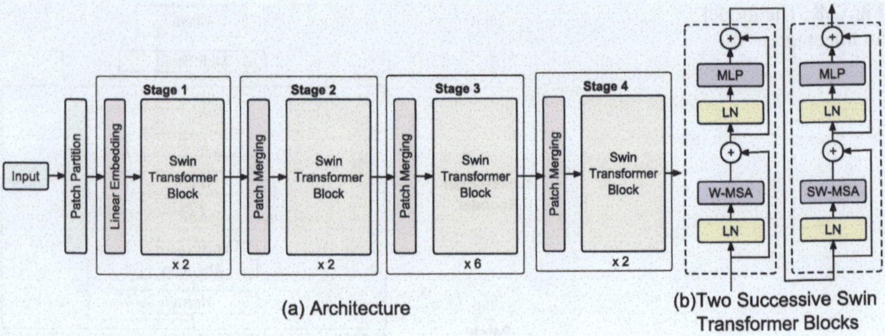

Fig. 2.10 (**a**) The architecture of a Swin Transformer (Swin-T) (**b**) Two Successive Swin Transformer Blocks

Swin Transformer Block The Swin Transformer block is a modified version of the standard Transformer block that replaces the multi-head self-attention (MSA) module with a shifted window-based MSA module while keeping the other layers unchanged. Figure 2.10b illustrates the structure of a Swin Transformer block, which includes a LayerNorm (LN) layer, a shifted window-based MSA module, a 2-layer MLP with GELU non-linearity, and a residual connection. The LN layer normalizes the inputs before each MSA module and MLP, and the shifted window-based MSA module captures the dependencies between different regions of an image using a set of shifted windows.

As Fig. 2.10a shows, the overall structure of a tiny version of the Swin Transformer (Swin-T) resembles the convolutional neural network (CNN) architecture. The model consists of 4 stages, and every stage is made up of Patch Merging (linear embedding in Stage 1) and Swin Transformer blocks.

Swin Transformer is a powerful computer vision model that performs attention on small windows of images to enable multi-scale feature extraction. The shifted windows and patch merging mechanisms make Swin Transformer more effective for dense recognition tasks compared to ViT.

2.4 Summary

This chapter provides an introduction to the essential techniques of deep learning for video understanding, covering Convolutional Neural Networks (CNNs), Recurrent Neural Networks (RNNs), and Transformer. The chapter starts by explaining how CNNs can be utilized to extract spatial information, followed by RNNs, which can capture dynamic correlations in sequential data. Moreover, we explore the self-attention mechanism, which enables the model to extract data correlations, and introduce the popular Transformer model. Finally, we introduce the Vision Transformer (ViT) model and the Swin Transformer, which extend the ability of Transformer model to the field of computer vision and significantly enhance the performance of various vision tasks.

Chapter 3
Deep Learning for Action Recognition

3.1 Action Recognition with Convolutional Neural Networks

The outstanding performance of CNNs in image analysis tasks [55, 56] has inspired their application in video classification. The basic approach to apply CNNs in video classification involves treating each video clip as a set of frames and extracting their features by pre-trained CNN models, such as AlexNet [9], VGGNet [10], GoogleNet [11], and ResNet [12]. The derived frame-level features are then aggregated to form video-level representations, which serve as the inputs of the final classifier. To quantify the capability of CNNs as static feature extractors in video action recognition tasks, Zha et al. [57] studied the classification performance of the model with features from different layers of deep models and with multiple SVM kernels. They demonstrated that using static CNN models pre-trained on image domain can also obtain decent performance for video classification.

Although utilizing off-the-shelf CNNs as feature extractors achieves satisfactory performances in video action recognition tasks, it fails to capture the underlying dynamic information and relationships among frames in video clips. Since CNNs are able to act as an end-to-end pipeline to learn features from raw data [11, 49, 58], researchers start to develop end-to-end CNN models for video action recognition, with the aim of learning hidden spatial–temporal patterns from the video data. One of the earliest works in this area was [59], where the authors designed a 3D CNN model with 3D convolutional kernels. The 3D convolution kernels empower CNN models to better capture motion information in video data by analyzing the information between frames. Karparthy et al. [18] explored three temporal connectivity patterns, namely late fusion, early fusion, and slow fusion, to learn the spatial–temporal features for video action recognition. They designed several variants of 3D CNN-architecture to explore better ways to extend the capability of original 2D CNN models to 3D CNN models. Their experiments suggested that the models with a stack of frames as inputs perform similarly to the models with a single frame as input. Tran et al. [60] further extended the use of 3D convolutions

Z. Wu, Y.-G. Jiang, *Deep Learning for Video Understanding*, Wireless Networks,
https://doi.org/10.1007/978-3-031-57679-9_3

in modern deep architectures [61] and showed that 3D CNNs exhibit more powerful spatial–temporal feature learning capabilities compared to 2D CNNs with deeper and more complex architectures.

However, the great potential of 3D CNNs to capture the dynamic temporal information does not bring much performance gain when recognizing actions, which is due to the insufficient training videos and the complexity of spatial–temporal patterns. 3D CNNs face the difficulty of capturing the real significant information in video data. Besides, time and computational consumption surges for training CNNs with 3D volumes. To handle 3D signals more effectively, many studies have started decoupling spatial–temporal modeling for advanced video content understanding. Sun et al. introduced the factorized spatial–temporal convolutional network (FSTCN), which decouples the original 3D convolutional kernels into a combination of 2D convolutional kernels and 1D temporal convolutional kernels, with the construction manner of using 2D convolution kernels in the shallow layers of the network and 1D convolutional kernels in the deeper layer of the network [62]. Motivated by the fact that both spatial and temporal components in videos provide evidence for recognition, Simonyan and Zisserman [17] proposed a two-stream pipeline (see Fig. 3.1) and break down the video representation learning into separate spatial and temporal feature learning. In particular, the authors applied the CNN to multiple-frame dense optical flow for capturing temporal clues among adjacent frames. In the testing stage, the model will analyze 25 uniformly sampled optical flow frames for each video clip and average the prediction scores, while the final prediction is generated by a weighted sum of the scores of the spatial and temporal streams. Based on the idea of two-stream networks, Ye et al. [63] designed experiments for different parameters and explored the impacts of different experimental settings, including various dropout ratios, network architectures, etc.

Intrigued by the effective two-stream architecture, many extensions of the two-steam approach emerged. An important line of research is to design feature fusion modules to aggregate features of spatial and temporal information. Among the

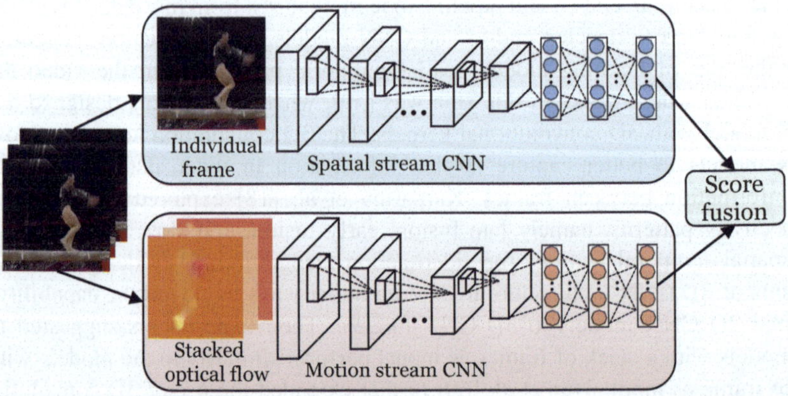

Fig. 3.1 Two-stream CNN framework

works, Wang et al. [64] introduced the point trajectories from the improved dense trajectories for better two-stream convolutional feature maps pooling [65], while Feichtenhofer et al. [66] chose to fuse the two-stream features through the convolutional layers to enhance the model spatial–temporal modeling. Feichtenhofer et al. [66] also explored the impacts of fusing the two-stream features from different layers in a network and found that it is a better choice to fuse spatial and temporal features from the convolution layer rather than softmax layer of the model. Utilizing features from the convolution layer will significantly reduce the number of parameters but offers the same performance, compared with using features from the softmax layer. In their following works, Feichtenhofer et al. [67] designed a spatial–temporal residual network where the two streams interacted with residual connections and then proposed the multiplicative residual gating [68] between the two streams for better learning the spatial–temporal patterns in videos.

Following the two-stream network paradigm, another research focus is to improve the network efficiency. Computationally expensive and storage-demanding optical flow pre-computing is a barrier to swift inferences. Zhang et al. [69] replaced optical flow images with motion vectors as a means of achieving real-time action recognition. Zhu et al. [70] proposed to capture motion information between adjacent frames implicitly using CNN architectures, thus avoiding the computation of traditional optical flow and greatly reducing computation and storage costs. Weng et al. [71] relied on a 3D CNN to extract the motion information in videos and designed an adaptive selection strategy conditioned on the input video. Computational costs are reduced by selecting important time steps. As aforementioned, 3D convolutions in modern deep architectures [61] overtake 2D CNNs in feature extraction but still suffer from poor convergence and high time consumption, leading to limited results in classification tasks. To this end, many 3D CNNs variants emerge to address the challenges. Inflated 3D Convolutional Network (I3D) [35] is a 3D extension of 2D CNN with exceptional performance in image recognition tasks. The key idea behind I3D is to inflate 2D CNN pre-trained weights and filters to a 3D architecture, so that it can process sequences of frames while keeping knowledge learned from the image domain and alleviating the problem of training from scratch on insufficient 3D training data. Experiments demonstrate noteworthy advancements in performance on their large-scale video action recognition dataset, Kinetics-400 [22]. Taking both high performance and low computational complexity into account, different kinds of 3D CNN factorization methods are proposed. P3D [72] and R2+1D [36] factorized the 3D convolution to two separate operations: a 2D spatial convolution and a 1D temporal convolution. Xie et al. [73] found that a top-heavy network with 2D kernels replacing the bottom 3D convolutions results in higher recognition accuracy. They proposed the Separable 3D CNN (S3D) network architecture where the bottom 3D convolutions are replaced by the 2D convolutions and the top 3D convolutions are replaced by the factorized 3D convolutions to reduce the model size to another level. Furthermore, Tran et al. [60] introduced the group convolution to factorize the 3D convolutions. They empirically demonstrated that the amount of channel interactions significantly affected the classification performance of 3D group convolutional networks. Their approach not only achieved state-of-the-art performance but also accomplished it at 2–3 times the speed of prior best methods. See Chap. 7 for a complete review of efficient action recognition methods.

Optical flow allows two-stream networks to analyze the short-term motion between adjacent video frames. However, these networks are limited when it comes to capturing long-range temporal information. To address this challenge, Wang et al. [33] introduced a the Temporal Segment Network (TSN) which divided a whole video into several segments uniformly along the time dimension. TSN considers the consistency of predictions between segments with segmental consensus operators like average pooling, max pooling, etc. The global sparse sampling strategy of TSN gives the model long-range temporal modeling abilities while lowering the training costs over long video sequences. In contrast to the simple segmental consensus operation in TSN, such as max or average poolings, Diba et al. [74] proposed a temporal linear encoding (TLE) layer via end-to-end learning for feature encoding among segments throughout entire videos. In addition, the Temporal Relation Network (TRN) method, created by Zhou et al. [75], involves training a model on multi-scale temporal segments and inferring temporal dependencies among video frames across multiple time scales.

Although 3D convolution kernels can jointly model the spatial and temporal relationships, it suffers from high computational costs, intensive memory requirements and optimization difficulties. Thus, it is necessary to set each kernel with limited temporal resolution. Stacking multiple short temporal 3D convolutions for temporal information extraction is a common model building strategy, however, this design approach still tends to cause temporal information loss, especially for frames that are far apart. As a result, different temporal modeling methods have been proposed. Diba et al. [76] adopted the densely connected structure in their T3D model for improved temporal information modeling. Wang et al. [37] realized that a single short 3D convolution operation can only see limited temporal information, so they proposed a plug-and-play module named non-local module which completes the aggregation of global information among frames through the attention mechanism along temporal and spatial axes. Feichtenhofer et al. designed the SlowFast [39] model which is composed of a Slow pathway and a Fast pathway. The Fast pathway corresponds to processing high-rate frames aiming to capture motions, while the Slow pathway focuses more on learning spatial semantics. By combining the two pathways, SlowFast further improves the video recognition performance of the convolutional neural network.

3.2 Feature Aggregation for Long-Range Temporal Modeling

The aforementioned convolutional-based methods mainly focus on classifying short video clips and are limited in handling excessively long untrimmed videos, which poses a challenge for real-world applications that require classification of events in long videos. Many of these works have to sample many frames in different time steps and apply average pooling or maximum pooling on pre-extracted segmental features to solve the challenge of long videos. However, it could often lead to sub-optimal performance in downstream tasks due to a lack of proper long-term temporal modeling. To address this issue, the exploration of advanced long-range temporal feature aggregation techniques is needed. Feature aggregation is the crucial step in the analysis of video data, where multiple features are combined into a compact representation to reduce feature redundancy. In the context of video action recognition, proper aggregation of the long-range temporal features plays a key role in capturing the dynamics of actions. As such, there has been a growing interest in the development of long-range temporal feature aggregation methods to capture the temporal continuity of actions or events in videos. In this section, we delve into the area of feature aggregation methods specifically designed for temporal information modeling, which is a fundamental component of video analysis. Two types of temporal aggregation methods will be introduced, namely the bag of visual words (BoVW) method and the Recurrent Neural Network-based method.

3.2.1 Bag of Visual Words and its Variants

The bag-of-words (BOW) method is initially used in natural language processing (NLP) and then transferred to the field of computer vision. When the BOW is applied to the field of computer vision, it is also known as the bag-of-visual-words (BOVW) or bag-of-features (BOF). Similar to the process of BOW that converts text into word frequency vectors by accumulating word frequency information in NLP, the most basic application of BOW in vision tasks is to quantify each local feature descriptor, (e.g., SIFT features [77]) into a word and accumulate each word frequency, aiming to produce global compact visual features for downstream tasks [78].

With respect to videos, video frame features extracted by deep learning networks are considered local [79, 80] and the BOVW method can be used to aggregate these frame features into a compact global video representation for each video. Although both spatial and temporal relationships are required and aggregated for video classification tasks, BOVW focuses on temporal information aggregation. Figure 3.2 shows the basic pipeline of BOVW for video action recognition with the following stages: (1) feature extraction and preprocessing, (2) codebook generalization, and (3) feature encoding and pooling.

In the natural language processing domain, texts consist of discrete words and all words form a codebook. Each word corresponds to an index according to the codebook. However, in visual tasks, the feature space is a high-dimensional continuous space. If we would like to treat each image patch or each frame as a "word," we need to build a codebook for image features first. The process of constructing a codebook in vision tasks typically involves clustering all the feature descriptors using cluster algorithms such as K-Means [81] to partition the feature space into regions, resulting in K word vectors denoted as $C = \{\mathbf{c_i}\}^K$. Then, when given a set of input features, the traditional BOVW method will categorize each

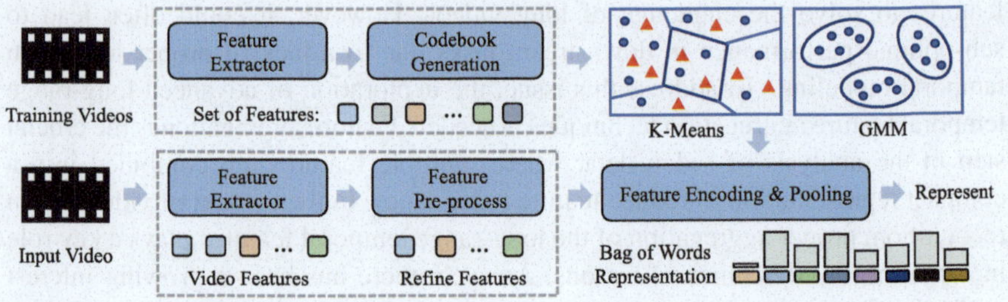

Fig. 3.2 Overview of the pipeline for obtaining the BOVW representation for video action recognition

feature descriptor by computing its distance from the K visual word vectors through the nearest-neighbor strategy. And the set of feature descriptors $\mathbf{x} = \{\mathbf{x}_i\}^N$ can be encoded into a K-dimension vector $\mathbf{V}$ calculated by

$$\mathbf{V}_k = \sum_{i}^{N} a_k(\mathbf{x_i}), \tag{3.1}$$

where $a_k(\mathbf{x})$ is the indicator function and it takes the value 1 when $\mathbf{x}$ is closest to the kth visual word and 0 otherwise. However, such representations of individual features result in great information loss. To address this problem, James et al. [82] improved the "nearest-neighbor strategy to quantify features" by proposing a modification from 0-1 absolute categorization to distance-weighted soft-assignment.

$$a_k(\mathbf{x_i}) = \exp\left(-\frac{\|\mathbf{x_i} - \mathbf{c_k}\|_2}{2\sigma}\right), \tag{3.2}$$

where σ is the scaling factor. Using multiple distance information among different word centers can better quantify a feature descriptor, improving the problem of quantization accuracy loss to some extent and enhancing the performance of the BOW model in various tasks.

Besides accumulating the word frequency, Vector of Locally Aggregated Descriptors (VLAD) [83] and Fisher Vector (FV) [84] are higher dimensional feature encoding methods by aggregating high order statistics. VLAD is the method that accumulates the sum of residuals between the features and the nearest-neighboring words, formulated as below:

$$\mathbf{V}_k = \sum_{i=1}^{N} a_k(\mathbf{x_i})(\mathbf{x_i} - \mathbf{c_k}). \tag{3.3}$$

VLAD performs normalization on each cumulative residual vector $\mathbf{V}_k$ and then concatenates all the residual vectors as the final feature aggregation results. FV is another feature aggregation method derived from the fisher kernel [85]. It consists of three stages: Gaussian Mixture Model (GMM) estimation, encoding, and pooling. GMM [81] is widely used to capture the probability distribution of features, which relates to the codebook construction stage. Then, given a set of features, FV calculates the gradients of the likelihood function with respect to the weight, mean, and variance, then normalizes and concatenates them to obtain the global feature representation.

The advent of VLAD and FV has prompted attempts to apply them in video tasks. Xu et al. [79] applied VLAD and FV, respectively, when pooling the CNN descriptors, improving video retrieval performance on TRECVID datasets [86, 87]. This study demonstrates the strength of VLAD and FV in video tasks and highlights their capabilities in temporal feature modeling.

However, traditional aggregation methods, such as the VLAD method, can be challenging to be implemented in an end-to-end training process due to the need for offline clustering and the calculation of non-derivable form of $a_k(\mathbf{x_i})$ and c_k. To address this, NetVLAD [88] was introduced with modified $a_k(\mathbf{x_i})$ as a soft-assignment format and visual words matrix as a learnable parameter. The K-dimension vector codebook is formulated as below:

$$\mathbf{V}_k = \sum_{i=1}^{N} \frac{e^{w_k^T \mathbf{x_i} + b_k}}{\sum_{k'} e^{w_{k'}^T \mathbf{x_i} + b_{k'}}} (\mathbf{x_i} - \mathbf{c_k}). \tag{3.4}$$

The improved NetVLAD is integrated into the network as a deep learning module for end-to-end training. On top of that, ActionVLAD [34] applies NetVLAD to video action recognition by extending a temporal dimension to jointly aggregate the features across space and time as below.

$$\mathbf{V}_k = \sum_{t=1}^{T} \sum_{i=1}^{N} \frac{e^{w_k^T \mathbf{x_{it}} + b_k}}{\sum_{k'} e^{w_{k'}^T \mathbf{x_{it}} + b_{k'}}} (\mathbf{x_{it}} - \mathbf{c_k}), \tag{3.5}$$

where x_i refers to the feature of the ith patch in the tth frame. The study by Girdhar et al. [34] demonstrates that the NetVLAD end-to-end training paradigm yields performance improvements by aggregating features along both the spatial and temporal dimensions. Miech et al. [80] improve BOF and FV as end-to-end training methods named soft-DBOF and NetFV by borrowing the training paradigm from NetVLAD. Then, they apply NetVLAD, soft-DBOF, and NetFV methods, respectively, to aggregate a sequence of video frame features and achieve the best performance in the first YouTube-8M Challenge [89], demonstrating the effectiveness of end-to-end pipe lines in aggregating video features for temporal modeling.

Overall, the BOVW method and its variants, including VLAD/FV and their end-to-end variants, play an important role in aggregating video temporal features for advanced video content understanding. The BOVW-based methods can effectively aggregate feature sequences to obtain the distribution information of temporal features, enhancing video content understanding. However, it is worth mentioning that the BOVW-based methods suffer from being less sensitive to time. When aggregating a sequence of video features, even shuffling the order between video segments/frames will not affect the output of the model. Therefore, when applying BOVW-based methods to video action recognition, it is essential for the feature extraction network to capture temporal ordering information in video segments to ensure accurate recognition.

3.2.2 *Temporal Aggregation with Recurrent Neural Networks*

Recurrent Neural Networks (RNNs) have become one of the mainstream approaches in the field of video recognition since the RNN-based methods contain powerful abilities to process sequential data and handle long-term temporal dependencies [80, 90, 91]. Long Short-Term Memory (LSTM) [92] and Gate Recurrent Unit (GRU) [93] are widely used RNN models to model the temporal dynamics in videos due to their ability to avoid the severe "vanishing gradient" effect. Various approaches have been explored by researchers for leveraging RNNs in temporal information modeling for action recognition. Donahue et al. [94] designed a two-layer LSTM networks as depicted in Fig. 3.3 and demonstrated that the joint finetuning of the CNN models and LSTMs resulted in a slight improvement in performance compared to using only the LSTM model. Wu et al. [32] combined the outputs of CNN and LSTM models to jointly model spatial–temporal cues, finding them to be highly complementary. In addition, Wu et al. [91] further employed a CNN to extract features from the spectrogram from the video soundtrack to enhance the visual cues captured by the CNN and LSTM. In a multi-granular spatial–temporal architecture [95], LSTMs were employed to model the temporal information of frame, motion, and clip streams. Finally, Miech et al. [80] further verified that using LSTM and GRU as temporal aggregation modules can achieve good video classification performance on the large-scale video understanding dataset.

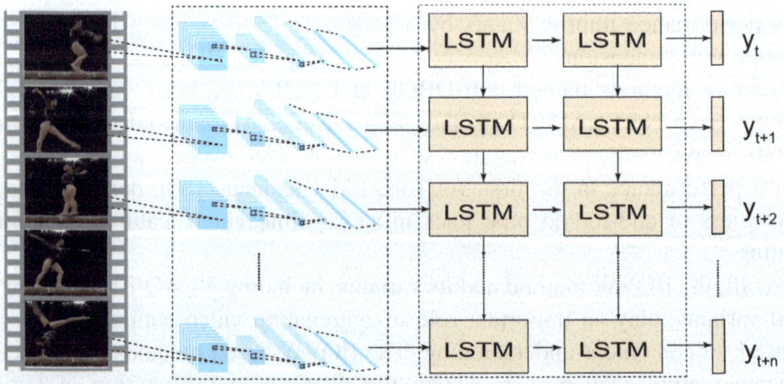

Fig. 3.3 Utilizing LSTMs to explore temporal dynamics in videos with CNN features as inputs

3.3 Action Recognition with Transformer Networks

Transformer [53] is originally designed for natural language translation tasks and replaces the function of LSTM in sequence modeling. Transformers show powerful abilities in a variety of downstream tasks for natural language processing.

Recently, Vision Transformer [14] first uses Transformer in image recognition. Concretely, it divides an image into uniform patches and considers these patches as a sequence sending to the Transformer block, which achieves great success in 2D image recognition for its long-range interactions and feature modeling abilities. After that, many improved versions of Vision Transformers [54] and training paradigms [96] gradually emerged, replacing the convolutional-based methods and reaching new state-of-the-art in image recognition.

Inspired by Vision Transformer, researchers in video action recognition develop video transformers hoping to reproduce the success of Transformers in the image recognition field. The time dimension in video data poses challenges in spatial–temporal information modeling for Transformers. We will introduce the research and development of the video Transformer in a chronological order in the following subsections.

3.3.1 Temporal Modeling Transformer

Video Transformer Network (VTN) [97] is the first to introduce Transformer in video action recognition. As shown in Fig. 3.4, VTN is composed of three modules, a 2D spatial feature extractor, a temporal-attention-based encoder, and an MLP

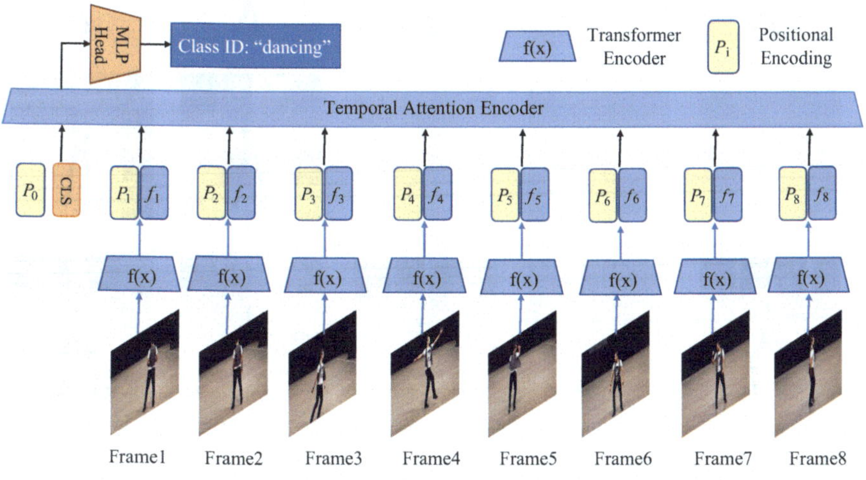

Fig. 3.4 Video Transformer Network architecture. Reproduced from VTN [97]

classification head. The spatial extraction module could be any 2D feature extraction network, such as conv-based ResNet [12] or Vision Transformer [14], and the weights of the feature extraction module can be pre-trained and fixed or trained from scratch. To tackle the challenge of modeling the long sequence of video data, VTN utilizes Longformer [98] as the temporal attention module to deal with long videos. Following BERT [99], a classification ([CLS]) token is added to the spatial feature sequence to act as the ultimate video representation, which is then utilized by a MLP classification head. Before being supplied to the classification head, the representation of the CLS token undergoes a Layer normalization. The MLP head consists of two linear layers incorporating a GELU non-linearity and a Dropout layer sandwiched between them.

3.3.2 *Vanilla Vision Transformer Variants*

TimeSformer [41] is a convolution-free transformer network for video action recognition. Besides, it proposes five different space–time self-attention schemes as shown in Fig. 3.5. In contrast to VTN [97] with a divided temporal attention module on the spatial feature level, TimeSformer proposes different self-attention modules at the patch level. It first divides a video clip $X \in R^{H \times W \times 3 \times L}$ into N non-overlapping patches, and each patch size is $P \times P$, resulting in $N = HW/P^2 \times L$. Then the five self-attention schemes are considered. Space Attention

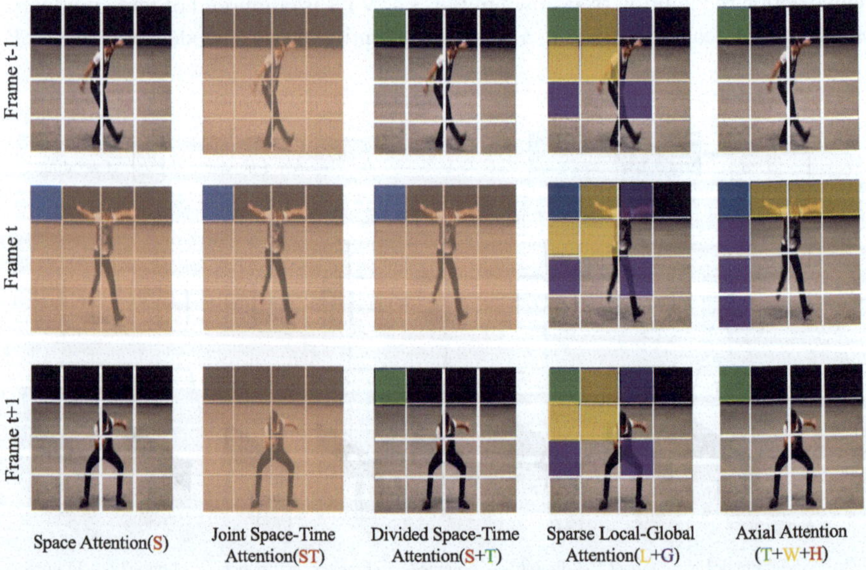

Fig. 3.5 Different space–time self-attention schemes. Reproduced from TimeSformer [41]

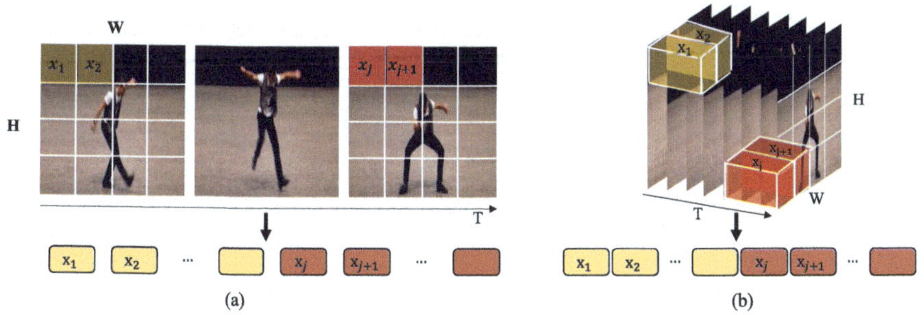

Fig. 3.6 (**a**) Uniform frame sampling. (**b**) Tubelet embedding. Reproduced from ViViT [100]

refers to performing self-attention only on each individual frame and averaging the token of each frame for action recognition. Joint Space–Time Attention treats each patch equally and performs global attention on these patches. Divided Space–Time Attention, Sparse Local Global Attention, and Axial Attention are designed to relieve the attention computational cost. Considering the efficiency and precision, TimeSformer adopts the Divided Space–Time Attention as default to perform self-attention on both spatial and temporal domains.

ViViT [100] first proposes tube embedding—it divides an video clip $X \in R^{H \times W \times 3 \times L}$ into non-overlapping, spatial–temporal "tubes" with $N = (HWL)/(TP^2)$ tokens as shown in Fig. 3.6. To leverage ImageNet pre-trained weights, ViViT initiates the 3D convolution filters from 2D filters for video classification to "inflate" them through replication along the temporal dimension. Alternatively, the filters can be initialized with zeros for all temporal positions, except for those central frames. After the tokenization process, a global attention is utilized to model spatial–temporal clues. With the tube embedding, ViViT reduces the number of tokens and achieves decent results on different benchmarks.

3.3.3 Video Transformer with Convolutional Advantages

Because Vision Transformers inherently lack inductive bias, i.e., translation equivariance, and locality, researchers redesign the network architecture and develop plug-and-play modules to enhance Vision Transformers. Video Swin Transformer (VideoSwin) [43] and Multi-scale Transformer (MViT, MViTv2) [42, 101] study the hierarchical architecture and introduce an inductive locality bias into video transformers. VideoSwin (Fig. 3.7) follows the hierarchical architecture of origin Swin Transformer [54] and introduces a 3D shifted window attention module and 3D relative position bias. It inflates the 3D convolution with weights from the 2D Swin network, and designs four different scales of VideoSwin (i.e., Swin-T, Swin-S, Swin-B, Swin-L). MViT and MViTv2 design several scale stages, which reduce the spatial resolution and expand the channel dimension as convolution-

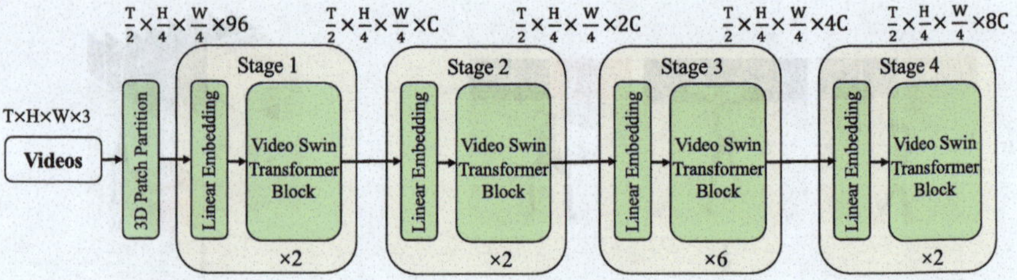

Fig. 3.7 Overall architecture of Video Swin. Reproduced from VideoSwin [43]

(a) Frames before patch shift (b) Highlight patches for patch shift (c) Frame after patch shift

Fig. 3.8 An example of temporal patch shift. Reproduced from PST [102]

based methods. Reshaping the network allows shallow network blocks to focus on high-resolution spatial features and deep network blocks to focus more on complex high-dimensional semantic features. Most of the previous Video Transformers [41, 43, 100] rely on 2D transformers that are pre-trained on large-scale image datasets such as ImageNet [15] and then they expand the 2D Transformers to the video domain through inflation or interpolation. Instead, owing to the multi-scale design, MViT and MViTv2 could be trained from scratch and outperform the state-of-the-art convolution-based methods without image-based pre-training.

3.3.4 *Lightweight Video Transformer*

Without a doubt, Video Transformer achieves great advancements in action recognition tasks. Because of the quadratic computational complexity of self-attention for processing a large number of patches, directly utilizing the spatial–temporal Transformer for video data brings huge computing costs and memory consumption (Fig. 3.8). To effectively and efficiently build Video Transformers, Patch Shift Transformer (PST) [102] presents a method called temporal patch shift (TPS) to enable efficient 3D self-attention modeling of video data. TPS achieves this by shifting patches in a mosaic pattern along the temporal dimension, resulting in the conversion of spatial self-attention into a spatial–temporal one with minimal computational cost. The temporal patch shifting module reduces the original

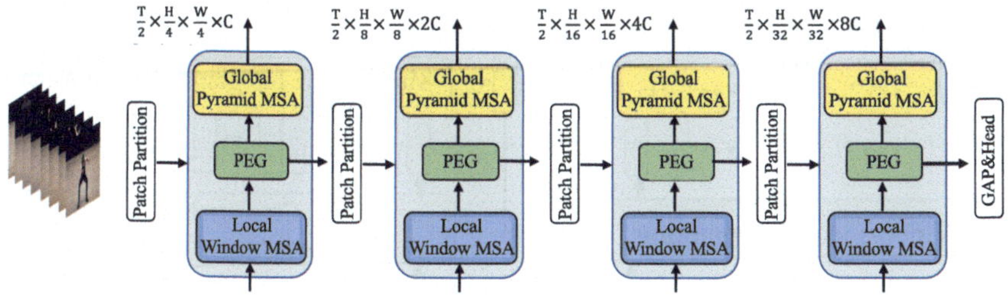

Fig. 3.9 Overview of Dualformer. Reproduced from Dualformer [103]

$O(N^2T^2)$ complexity to $O(N^2T)$, which greatly saves both the computation cost and memory consumption (Fig. 3.9). Furthermore, Dualformer [103] incorporates a dual-attention mechanism that effectively models both the short-range interactions by applying the local window attention and the long-range interactions by applying a global pyramid attention. With this, Dualformer effectively reduces the computation cost associated with attention and outperforms the vanilla vision Transformer in terms of accuracy.

3.3.5 CLIP for Video Transformer

Most Transformer-based methods for action recognition rely on ImageNet [15] pre-training. With pre-trained spatial attention, Video Transformer can converge quickly during the training process. More recently, several works that rely on pre-trained CLIP models show competitive performance. X-CLIP [104] adopted the CLIP as its backbone to recognize video actions. Unlike most previous works that extract video features from the backbone network and recognize action categories by a classification head, X-CLIP learns to align video representations and their textual labels by contrastive losses. Video encoders and text encoders are jointly optimized during training. The video encoder in X-CLIP [104] consists of two transformers that operate in a cascade manner: a cross-frame communication attention and a multi-frame integration attention. The former takes the original frame patches as inputs and utilizes self-attention to extract the representation of each frame. The latter performs attention on different frames to model the entire video representation. The text encoder is identical to the CLIP text encoder but with video-specific prompting that has been designed by a cross-attention module. During the process, the text representation serves as a query, while the video representation acts as key and value, and the contrastive loss is optimized (Fig. 3.10). With the use of the pre-trained CLIP model and a lightweight attention module, X-CLIP outperforms state-of-the-art ImageNet pre-trained models while requiring fewer floating point operations per second (FLOPs).

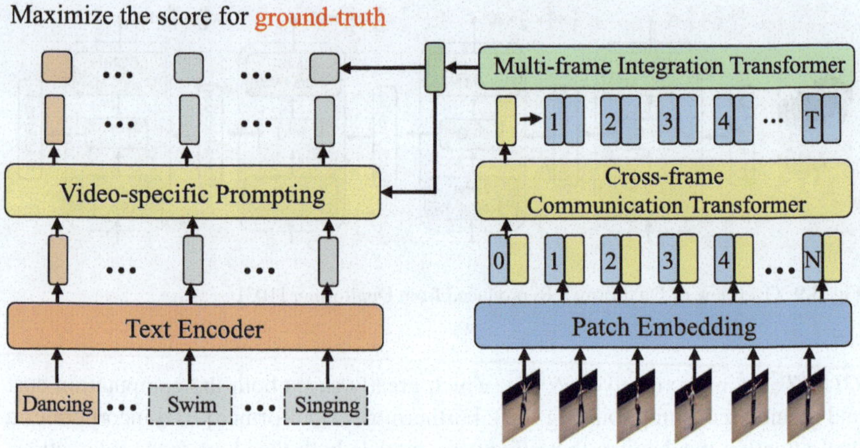

Fig. 3.10 Overview of X-CLIP. Reproduced from X-CLIP [104]

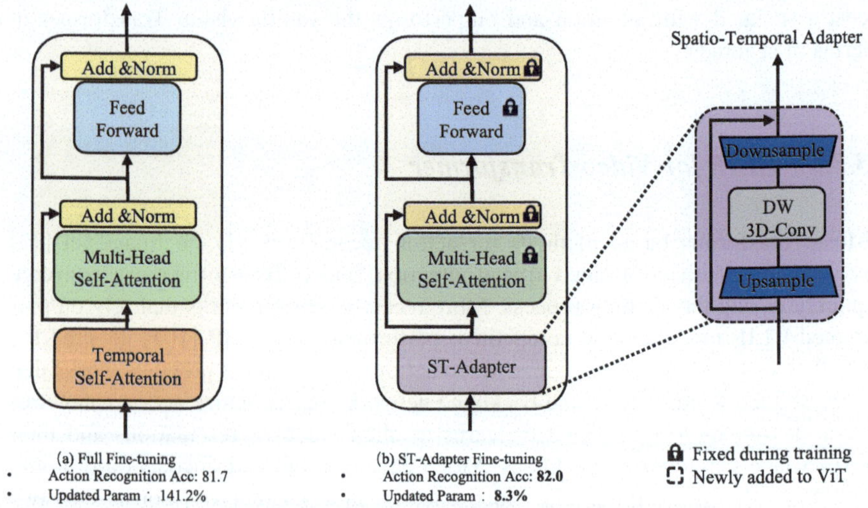

Fig. 3.11 Image-to-video transfer learning strategies of ST-Adapter. Reproduced from ST-Adapter [105]

ST-Adapter [105] is designed to transfer the CLIP model for video recognition with the fewest parameters finetuned. As shown in Fig. 3.11a, traditional video transformers (i.e., TimeSformer [41]) add temporal self-attention in vanilla Transformers and finetune the whole model parameters with video data. However, ST-Adapter adds a lightweight spatial–temporal adapter to transfer the image pre-trained model to video recognition. The adapter is composed of a down projection layer, a depth-wise 3D convolution and an up projection layer. Only the adapter parameters are updated during training. Besides, ST-Adapter demonstrates the superiority of CLIP pre-trained models compared to the ImageNet pre-trained models. Though ST-Adapter [105] is simple and straightforward, it is the first work that investigates parameter-efficient image-to-video transfer learning scenarios and establishes a baseline for subsequent research.

3.4 Datasets

In this section, we will introduce some popular video datasets. As we know, the development of video recognition has progressed with the publication of new datasets. We display and compare popular public video datasets in Table 3.1 with published year, modality, FPS, Action categories, number of videos, and specific tasks/applications.

Table 3.1 Human action video datasets

Datasets	Years	Modality	FPS	Actions	Videos	Application
KTH [106]	2004	Grey	25	6	600	Human action recognition in real outdoor conditions
Hollywood [107]	2008	RGB	30	8	397	Realistic actions recognition from movies
HMDB51 [21]	2011	RGB	30	51	6766	Human activity recognition
UCF101 [20]	2013	RGB	25	101	13,320	Human Sports activity recognition
YouTube Sports 1M [18]	2013	RGB	–	487	1,133,158	Human Sports activity recognition
ActivityNet [25]	2015	RGB	–	203	27,801	Human activity understanding
FCVID [108]	2015	RGB	–	239	91,223	Human activity understanding
Charades [109]	2016	RGB	–	157	6800	Daily activity understanding
YouTube 8M [89]	2016	RGB	–	4716	8,264,650	Human activity recognition
Kinetics-400 [22]	2017	RGB	30	400	306,245	Human activity recognition
SS-V1 [110]	2017	RGB	12	174	108,499	Human–object interaction
Kinetics-600 [111]	2018	RGB	30	600	495,547	Human activity recognition
SS-V2 [112]	2018	RGB	12	174	220,847	Human–object interaction
AVA	2018	RGB	–		268k	Spatial-temporally Localization
Epic-Kitchens	2018	RGB	–	149	39,596	Human–object interaction
Kinetics-700 [113]	2019	RGB	30	700	650,317	Human activity recognition
VideoLT	2021	RGB	–	1004	256k	Long-Tail video recognition

KTH [106] The KTH dataset was proposed by Ivan Laptev et al. at http://www.nada.kth.se/cvap/actions in 2004. The dataset consists of 25 actors and each actor with 6 activities, including "walking," "jogging," "running," "boxing," "handclapping," and "handwaving." Each of the datasets is collected indoors and outdoors with changing conditions of four scenarios.

Hollywood [107] Ivan Laptev et al. introduced the dataset, available at (https://www.di.ens.fr/~laptev/download.html), in 2008 to facilitate the recognition of natural human actions in various realistic video environments. The dataset comprises videos collected from 32 Hollywood movies and includes eight classes of actions: "AnswerPhone," "GetOutCar," "HandShake," "HugPerson," "Kiss," "SitDown," "SitUp," and "StandUp." The test set includes 397 manually labeled action samples.

HMDB [21] It was created by H. Kuehne et al. at https://serre-lab.clps.brown.edu/resource/hmdb-a-large-human-motion-database/ in 2011. The HMDB dataset contains 51 categories and each category with at least 101 clips, for a total of 6766 clips. The action categories can be divided into 5 groups: (i) General facial actions, (ii) Facial actions with object manipulation, (iii) General body movements, (iv) Body movements with object interaction, and (v) Body movements for human interaction.

UCF101 [20] In 2013, Khurram Soomro et al. introduced the UCF101 dataset, which is an extension of UCF50 [114] and contains 101 categories of human actions collected from YouTube. The action categories can be categorized into five types, namely Human-Object Interaction, Body-Motion Only, Human-Human Interaction, Playing Musical Instruments, and Sports. The dataset comprises 13,320 clips with a frame rate of 25 and is currently one of the most popular datasets for human activity recognition.

YouTube Sports 1M [18] It was proposed by Karpathy, Andrej et al. at https://github.com/gtoderici/sports-1m-dataset/blob/wiki/ProjectHome.md in 2013. The Sports-1M dataset consists of 1 million YouTube videos annotated with 487 categories. The categories can be divided in 6 types namely, "Aquatic Sports," "Team Sports," "Winter Sports," "Ball Sports," "Combat Sports," and "Sports with Animals." The datasets are split to train, valid, and test set by 70%, 10%, and 20%.

ActivityNet [25] The ActivityNet dataset, made available by Fabian Caba Heilbron et al. at http://activity-net.org/ in 2015, represents a significant benchmark for the task of human activity understanding in video analysis. With 203 activity classes represented across an average of 137 untrimmed videos per class and 1.41 activity instances per video, ActivityNet comprises 849 video hours in total. This large-scale dataset is not only useful for action recognition but is also crucial for developing and evaluating effective methods for temporal action detection and grounding.

FCVID [108] In 2015, Jiang et al. introduced a dataset available at http://bigvid.fudan.edu.cn/FCVID. This dataset includes 91,223 Internet videos manually annotated across 239 categories, covering diverse topics such as procedural events ("making hotdog"), objects ("giraffe"), scenes ("forest"), *etc*. FCVID is a comprehensive dataset for video categorization, boasting some of the most accurate manual annotations among the largest available datasets.

Charades [109] The charades dataset was proposed by Sigurdsson1et al. at https://allenai.org/data/charades in 2016. There are 157 categories collected in this dataset. Each video contains an average of about 6.8 actions, which is also suitable for temporal action localization. The datasets focus on indoor homes, covering 15 types of rooms to represent typical homes. It contains human-object interactions, which could aid robots equipped with computer vision in the future.

YouTube 8M dataset [89] The YouTube 8M dataset was proposed by Abu-El-Haijaet al. at https://research.google.com/youtube8m/ in 2016. It is the largest multi-label video classification dataset, with up to 8 million videos and 500K hours with 4800 annotated visual entities. The dataset groups 4800 entities into 24 top-level categories to measure statistics and illustrate diversity.

Kinetics [22, 111, 113] The Kinetics datasets are composed of Kinetics-400 [22], Kinetics-600 [111], and Kinetics-700 [113], which contains 400, 600, and 700 categories, respectively. Each category has more than 400 samples and the samples last about 10s averagely. The Kinetics datasets are proposed by Google at https://www.deepmind.com/open-source/kinetics. Since the release, it became the most popular benchmark for video recognition, especially for Transformer-based architecture.

Something-Something [110, 112] Something-Something datasets have two versions, Something-Something V1 [110] was proposed at https://20bn.com/datasets/something-something in 2017. Previous video datasets focus on the specific scenes but ignore the moving object, while Something is collected by some moving object and human-object interaction recognition. It contains 174 categories with more than 100,000 samples. Something-Something V2 [112] enriched the datasets with more than 200,000 samples at https://developer.qualcomm.com/software/ai-datasets/something-something in 2018. Its categories are the same as Something-Something V1.

AVA [115] The AVA dataset was proposed by Chunhui Gu et al. at https://research.google.com/ava/ in 2018. It contains 14 pose classes, 49 person-object interaction classes, and 17 person-person interaction classes in the vocabulary. The dataset is proposed for spatial–temporal localization of human actions with 211K training and 57K validation video segments.

Epic-Kitchens [116] Damen1 et al. proposed the datasets at https://epic-kitchens.github.io/2023 in 2018. The dataset is a large-scale egocentric video benchmark recorded by 32 participants in their native kitchen environments. It contains 149 categories with their start and end timestamps, which is widely used in temporal localization. The datasets densely label a total of 39.6K action segments and 454.3K object bounding boxes.

VideoLT [117] In 2021, Zhang et al. introduced VideoLT, a large-scale long-tailed video recognition dataset available at https://github.com/17Skye17/VideoLT. With

256,218 untrimmed videos annotated across 1004 classes featuring a long-tailed distribution, VideoLT represents a crucial step toward realizing effective video recognition systems that can better operate in real-world settings.

3.5 Summary

This chapter provides an overview of the literature on deep learning for video action recognition. We review the evolution of CNN-based and Transformer-based deep learning algorithms for video classification, together with long-range temporal feature aggregation methods. We also present an overview of commonly used video action recognition datasets to demonstrate the increasingly challenging datasets inspire the advances and innovation of deep learning for video action recognition.

Chapter 4
Deep Learning for Video Localization

4.1 Action Localization

4.1.1 Introduction of Action Localization Task

Background With the exponential growth of web videos, how to accurately and efficiently detect actions and events in videos that users are interested in has become a challenging problem, therefore, action localization has attracted much attention from academia and industry. Action localization can be used in versatile real-world applications. For instance, in sports video records, action localization can automatically locate athletes' actions to facilitate referees to make correct decisions.

In the past few years, video recognition [35, 37–40, 42, 43, 72, 100, 118] has been extensively explored with the rapid development of extensive video datasets and growing computing power. Although recognizing actions in trimmed videos has achieved decent performance, localizing potential video segments that contain actions in untrimmed videos remains a challenging problem. This may result from the substantial amount of background frames in untrimmed videos, i.e., frames that are irrelevant to video labels. To cope with untrimmed videos, models are required to distinguish between foreground and background frames, i.e., localizing foreground frames that are related to video labels. Typically, action localization (AL) contains two mainstream sub-tasks, i.e., temporal action localization (TAL) and spatial–temporal action localization (STAL). TAL refers to localizing the action instances and predicting their classes in a video, while STAL extends TAL to spatial–temporal dimension and requires localizing both the bounding boxes of actors and the time positions of action instances among continuous video frames.

Generally, TAL is categorized into two-stage methods [31, 119–123], single-stage methods [53, 124–126], class-specific attention methods [127–130], class-agnostic attention methods [131–135], etc. In the following, action localization methods will be delivered in accordance with the taxonomy illustrated in Fig. 4.1. Note that our main emphasis is placed on the introduction of TAL.

Z. Wu, Y.-G. Jiang, *Deep Learning for Video Understanding*, Wireless Networks ,
https://doi.org/10.1007/978-3-031-57679-9_4

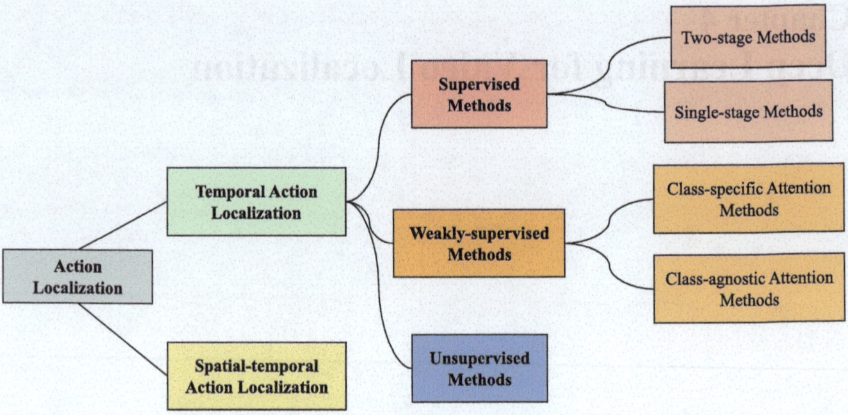

Fig. 4.1 The taxonomy of action localization methods

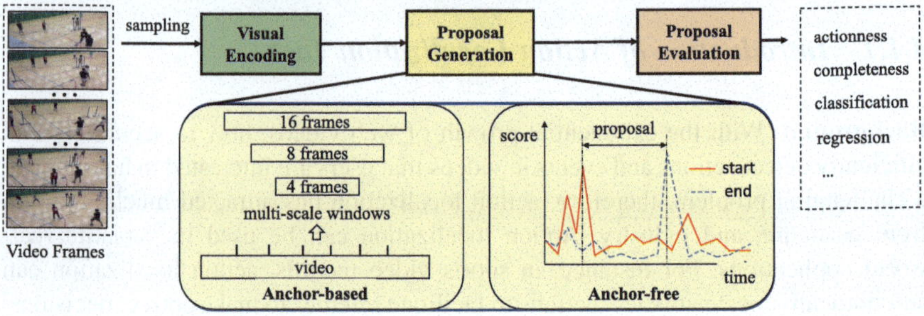

Fig. 4.2 The pipeline of TAL

Definition & Pipeline The formal definition of TAL is demonstrated as follows. Given an untrimmed video V which contains N action instances, the start and end time positions of action instances are denoted as $(t_{s,n}, t_{e,n})$, where $n \in N$. Each action instance n has single or multiple labels l_n from C action classes. TAL aims to localize potential action instances at the temporal dimension and predict the corresponding action classes. Given a model M, the TAL task can be formulated as $\{(t_{s,n}, t_{e,n}, l_n)\}_{n=1}^{N} \leftarrow M(V)$.

As shown in Fig. 4.2, the common pipeline of TAL includes three components, i.e., visual encoding, proposal generation, and proposal evaluation. Particularly, in visual encoding, several snippets are first sampled from the video and then fed into backbone networks which are pre-trained on large video understanding datasets, e.g., I3D [35] and C3D [31]. Lastly, the extracted features are concatenated to represent the video. Once the features are ready, an anchor-based or anchor-free proposal generation strategy is followed to generate candidate proposals. The anchor-based strategy utilizes multi-scale temporal windows around a set of pre-defined anchors as initial proposals whose boundaries are refined during training, while the anchor-free strategy predicts the boundaries of proposals or actionness scores directly. Afterward, for proposal evaluation, sub-networks are used to evaluate the actionness score, completeness score, classification score, and boundary precision of these proposals.

Evaluation Metric For temporal action localization, mean Average Precision (mAP), which is computed by averaging the AP over all activity categories, is a widely used evaluation metric. In addition, the threshold of temporal intersection over union (tIoU) is also employed to classify the positive or negative detection results. Specifically, the official tIoU of ActivityNet is set to $\{0.5, 0.75, 0.95\}$ while that of THUMOS'14 is set to $\{0.3, 0.4, 0.5, 0.6, 0.7\}$.

In the meantime, spatial–temporal action localization is evaluated by two major metrics, i.e., frame-AP and video-AP. As indicated in [115], the frame-AP is computed based on 2D IoU for each frame between the ground truth and predictions, with a threshold of 0.5. On the other hand, Video-AP is computed based on 3D IoUs between the predicted and ground-truth tubes, with a threshold of 0.5.

4.1.2 Supervised Action Localization

4.1.2.1 Two-Stage Methods

Generally, two-stage methods follow the "proposal-and-classification" paradigm: first extracting temporal action proposals and then performing classification for each of the proposals. A temporal action proposal gives a class-agnostic indication of where an action instance could potentially exist, including its starting and ending boundaries. Thanks to the substantial advancements in action recognition, models are now capable of accurately identifying "what" action is occurring in a trimmed video segment, whereas detecting "when" the action takes place remains a significant obstacle. Therefore, many researchers study how to generate precise action proposals that can greatly facilitate the subsequent classification. Depending on how proposals are generated, two-stage methods can be further categorized into the following four types, namely, sliding window-based methods [119–121], anchor-based methods [136–139], temporal actionness grouping methods [122, 140] and boundary-based methods [123, 141, 142].

(i) **Sliding window-based** methods generate action proposals by moving multi-scale fixed-size windows along the temporal dimension of the video (Fig. 4.3). The length and density of proposals can be controlled by manipulating the window sizes and moving strides, respectively. Segment-CNN (S-CNN) [119] makes the first attempt to apply sliding windows in generating proposals in TAL. Specifically, S-CNN adopts six sliding windows of different sizes to acquire video segments, then processes these segments by three C3D-based sub-networks. First, a proposal generation network is designed to predict the probability of an action being present in the current segment. Subsequently, substantial background segments of low probability are removed. The second sub-network undertakes the classification

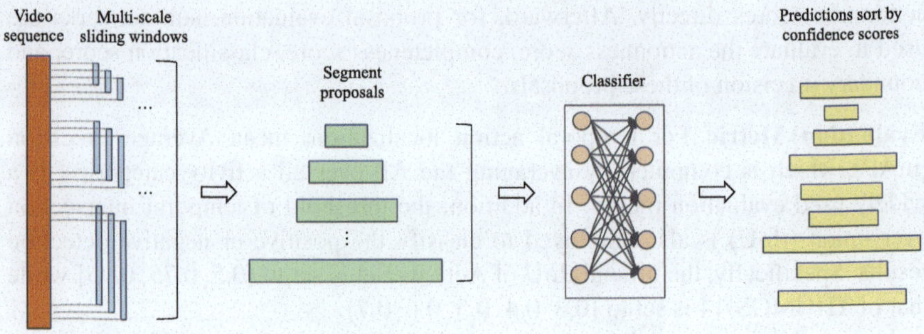

Fig. 4.3 Sliding window-based temporal proposal generation. Reproduced from Shen et al. [143]

by classifying video segments into $C + 1$ classes. Next, the third one performs localization by predicting not only the action category of the segment but also the overlap (measured by temporal IoU) between the current segment and the ground-truth action instance.

Nevertheless, action boundaries of S-CNN are retained after proposal generation and are therefore confined to pre-defined boundaries of a specific set of temporal windows. In order to overcome such drawback, Shou et al. proposed Convolutional-De-Convolutional (CDC) network [120], refining temporal boundaries from proposal segments. Inspired by the success of de-convolution operations in semantic segmentation, CDC filters perform convolution in space and apply de-convolution in time at the same time. The down-sampling convolution summarizes spatial–temporal patterns from raw videos into high-level semantics while the up-sampling de-convolution enables frame-level prediction of action class scores. The CDC filter manages to outperform separate calculations in positioning accuracy and computational efficiency.

Yang et al. designed a temporal preservation convolutional network (TPC-Net) [121] to further improve the CDC networks. Yang et al. believe that the temporal information lost for downsampling cannot be recovered by the following upsampling operation in CDC networks [121], which is damaging to precise boundary localization. Thus, they adopt a no-downsampling structure that maintains the full temporal resolution throughout the whole network, with only spatial dimensions reduced. Inspired by atrous convolution in 2D semantic segmentation, the proposed TPC filter simply pads the original feature map with zero values to keep the temporal resolution. In this way, TPC-Net can take advantage of pre-trained C3D weight as initialization, while fulfilling frame-level action localization with less temporal information loss.

Despite their simplicity, sliding window-based methods are plagued by their intrinsic shortcomings. Firstly, denser sampling of proposals naturally leads to higher recall rates but also results in heavier computational costs due to the need of recomputing a larger number of overlapped areas. Secondly, boundaries are restricted to several fixed-size temporal windows, which lacks flexibility. Thirdly,

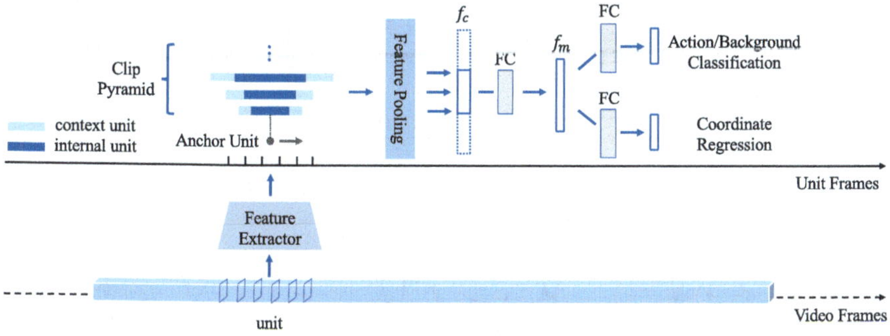

Fig. 4.4 The architecture of TURN. Reproduced from Gao et al. [136]

these methods are often sensitive to negative samples, which limits their localization effectiveness.

(ii) **Anchor-based** methods, whose efficacy has been proven in object detection, are employed in TAL to perform accurate boundary localization as well as reduce computation costs. The idea is to generate multi-scale windows (i.e., anchor boxes) at each temporal position as a prior, then refine the prediction by regressing the offset between ground-truth segments and anchor boxes. Temporal Unit Regression Network (TURN) [136] (Fig. 4.4) is the first method that applies anchor mechanism and boundary regression in temporal action proposal generation. TURN first extracts clip-level features using C3D [31]. Each clip is considered a video *unit* that serves as a minimum processing block of the network as well as an anchor point. A clip pyramid is constructed at each anchor unit by building temporal windows of multiple scales. Clip-level features are created by aggregating adjacent unit features within the windows and are utilized to generate two predictions: (1) a confidence score suggesting the likelihood of action instance existing within the input clip, (2) temporal coordinate offsets generated by temporal boundary regression. TURN is much more computationally efficient than sliding window-based methods.

Cascaded Boundary Regression (CBR) [137] model adopts a structure similar to TURN, but with a cascaded refinement process for boundary regression, where regressed temporal windows are input into the network iteratively for improvement. By employing a cascaded structure, the model is able to progressively refine the predicted boundaries by observing distinct visual information in each step, leading to more accurate boundary localization.

Xu et al. proposed a model named R-C3D [138] that closely follows the two-staged anchor-based pipeline of Faster R-CNN [144]. It mainly contains three parts: (1) a feature extraction network that extracts video features using pre-trained 3D ConvNets; (2) a temporal proposal network that predicts a sparse collection of temporal proposals based on anchor windows, together with the confidence scores of these proposals; (3) a classification network that performs classification and boundary regression based on fixed-size RoI pooled features of proposals. The

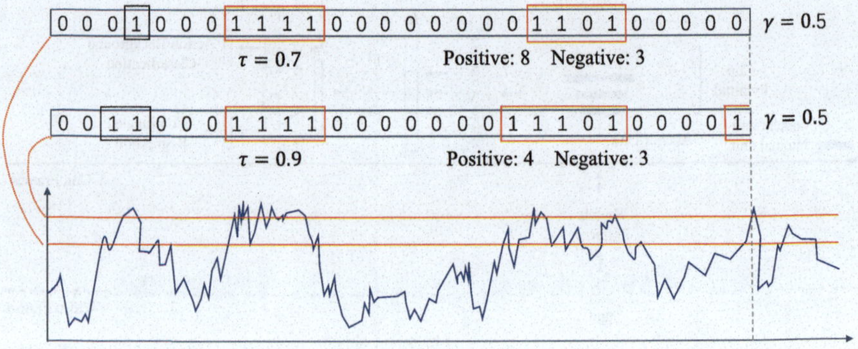

Fig. 4.5 The procedure of temporal actionness grouping. Reproduced from Xiong et al. [122]

temporal proposal network filters out most of the background segments, thus saving computational cost and improving performance.

Furthermore, Chao et al. [139] argue that there exist disadvantages to directly adopt the Faster R-CNN structure to the TAL task and introduce TAL-Net with three architectural changes (i.e., receptive field alignment, context feature extraction, and late feature integration) so as to adapt the seminal detection framework better for TAL.

(iii) **Temporal actionness grouping** method is first proposed by Xiong et al. [122] to efficiently detect activities of various lengths with accurate temporal boundaries. Actionness is a score ranging from 0 to 1, defined as a class-agnostic measure of the likelihood of any activity instance residing in a snippet. The authors observe that there is significant variation in the duration of actions, with some lasting only a second while others last a few minutes, thus posing substantial difficulties for sliding window-based methods. To generate proposals that are both reliable and flexible, they first learn a binary classifier based on TSN [33] to evaluate the actionness score for each snippet in the video. They then conduct the proposed temporal actionness grouping (TAG, Fig. 4.5) to group consecutive snippets with high actionness scores into proposals. The complete process of TAG is (1) get a number of action fragments with actionness scores above a certain threshold τ; (2) use these fragments as starting points and expand them recursively by absorbing succeeding fragments, until the portion of low-actionness snippets goes beyond a tolerance threshold γ.

Structured segment network (SSN) [140] also uses TAG to generate temporal proposals, while adding a series of mechanisms to learn action completeness. The authors posit that each complete action instance can be split into three stages, i.e., starting, course, and ending stages. Previous methods learn imprecise boundaries because they fail in distinguishing incomplete action fragments from proposals. To tackle this issue, SSN extends a temporal proposal $p_i = [s_i, e_i]$ obtained by TAG into $p'_i = [s_i - d_i/2, e_i + d_i/2]$, where $d_i = e_i - s_i$. Then the visual feature of the extended proposal is extracted via a structured temporal pyramid pooling layer and

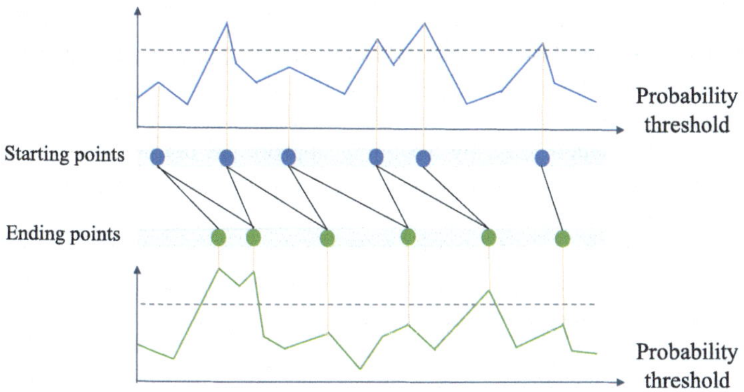

Fig. 4.6 Boundary-based temporal action proposal generation. Reproduced from Lin et al. [123]

is used to predict the classification scores, completeness scores, and offsets, which guide the refinement of temporal intervals of proposals. With this explicit modeling of action completeness, the model learns from complete action instances only at the training stage and is able to confirm the integrity of actions at the inference stage.

(iv) **Boundary-based** methods follow a "bottom-up" fashion that directly assesses the probability of action's existence at each temporal position to generate flexible and precise action boundaries (Fig. 4.6). Boundary sensitive network (BSN) [123] first adopts the idea of temporal action proposal generation. The proposed boundary sensitive network handles a sequence of snippet-level features extracted by a two-stream network. BSN includes three carefully designed modules: a temporal evaluation module that directly evaluates starting, ending, and actionness probabilities of each location on the temporal axis; a proposal generation module that produces proposals by combining candidate starting or ending locations with high probabilities, then constructs a boundary sensitive feature for each proposal by incorporating boundary features; and a proposal evaluation module that regresses the overlap ratios between proposals and ground-truth action segments as confidence scores. Finally, confidence scores of redundant proposals are suppressed through soft-NMS.

BSN generates and evaluates each proposal separately, resulting in inefficient computation and temporal context capturing. To address these shortcomings, Lin et al. proposed BMN [141] with a boundary matching mechanism to predict reliable confidence scores based on densely distributed proposals. BMN models the video's global context by gathering all proposal features and assessing them simultaneously through a 2D BM confidence map. Each point in this 2D map represents the confidence score of a proposal $\phi_{i,j}$ with starting time t_j and temporal duration $d = t_i$. The confidence scores are fused with boundary probabilities and are used for proposal selection during inference. BSN++ [142] further improves boundary-based methods by integrating a complementary boundary generator that captures abundant context information.

While the aforementioned methods are primarily based on CNNs, some other works try to exploit the Graph Convolutional Networks (GCNs) [145] in temporal action localization. The main distinction among these methods lies in how they utilize the graph structure to represent video sequences and model the relationships of elements in the video, e.g., by assigning appropriate definitions to nodes and edges of the graph. P-GCN [146] proposed to create a proposal graph for modeling inter-proposal relations using GCN. The nodes and edges of the graph represent temporal action proposals and relations between these proposals, respectively. Two separate GCNs on the same proposal graph with distinct features are used for class prediction and boundary regression. G-TAD [147] introduced by Xu et al. utilizes GCN to capture both temporal and semantic context from videos, in which each node represents a fixed-length video snippet, each edge represents the dependency between a pair of nodes, and each sub-graph denotes an action proposal associated with the context. By using this definition, G-TAD converts TAL into a sub-graph localization problem. In BC-GNN [148] proposed by Bai et al., proposal boundaries and proposal contents serve as nodes and edges, respectively, in a GCN. The Graph Construction Module (GCM) constructs a graph structure that explicitly models the relationships between proposal boundaries and proposals' contents. The Graph Reasoning Module (GRM) performs both edge update and node update on the graph.

4.1.2.2 Single-Stage Methods

Single-stage methods [124–126, 149–151] tackle proposal generation and classification simultaneously. Proponents of this paradigm argue that an independent proposal generation step is unnecessary because it has the following weaknesses: (1) proposal generation and action classification are performed by two separate sub-networks, which are typically not end-to-end trainable; (2) action boundaries are fixed and imprecise in the case of sliding window-based proposal generation; (3) the independent proposal generation stage introduces additional computational overhead and increases processing time.

Lin et al. proposed a Single Shot Action Detector (SSAD) [124] model that combines characteristics of single-stage detection methods YOLO [152, 153] and SSD [154]. SSAD first extracts clip-level feature sequences using a pre-trained C3D or two-stream network, then directly performs action classification as well as boundary offset regression following YOLO. The model also combines features of multiple convolutional layers for prediction following SSD in order to address action instances with different lengths. SS-TAD [149] proposed by Buch et al. utilize a modified gated recurrent unit (GRU) network to accumulate evidence across time and directly predicts at each time step. Long et al. introduced a Gaussian Temporal Awareness Network (GTAN) [150], which uses a collection of mixed Gaussian kernels to represent action proposals with varying lengths. Lin et al. proposed the first fully anchor-free TAL method [151] which incorporates a novel saliency-based refinement network to gather discriminative boundary features based on coarse predictions and refines boundary prediction using the aggregated feature.

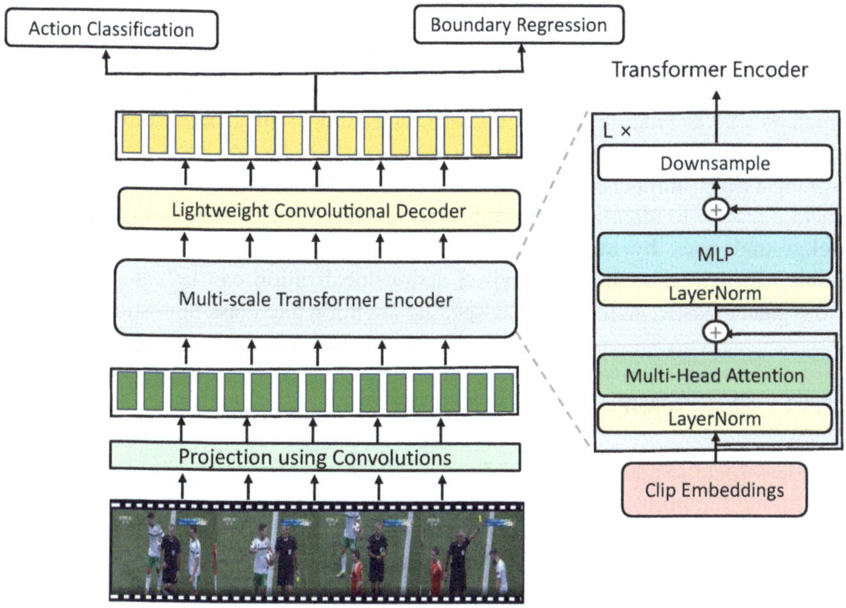

Fig. 4.7 The architecture of Actionformer. Reproduced from Zhang et al. [126]

Due to the great success of the self-attention mechanism in natural language processing and image understanding, recent literature adopts the Transformer [53] architecture in temporal action localization. Moreover, due to its superior capability in sequence modeling, Transformer is rather suitable for video processing. Liu et al. propose TadTR [125] that tackles TAL as a Transformer-based set prediction problem. TadTR uses the temporal deformable attention (TDA) network to capture long-term context in video sequences, then generates action predictions through a Transformer decoder by taking learnable queries as input. Bipartite matching is used to assign predictions to each ground-truth location during training. TadTR eliminates the necessity for many handcrafted components in prior TAL methods, such as NMS or anchor mechanism that explicitly relies on prior knowledge. Zhang et al. proposed a simple yet powerful model named Actionformer [126] that combines a temporal feature pyramid with local self-attention (Fig. 4.7). The model adopts a single-stage anchor-free manner that achieves high localization performance without relying on a first-stage proposal generator or pre-defined anchor windows. Computational costs are reduced significantly thanks to the local self-attention mechanism that constrains the computation within a small-size temporal window. The temporal FPN structure predicts action segments at various resolutions, while ensuring that the receptive fields of each position on the downsampled feature maps are large enough even within these small temporal windows.

4.1.3 Weakly Supervised Action Localization

Sun et al. [155] made the first attempt in weakly supervised action localization, which only takes video-level action labels for training, with ground-truth boundaries of action instances removed. Later, works also use order and frequency of actions as well as the number of events [156–161] as weak supervision. The *de facto* practice resorts to attention mechanism for distinguishing foreground frames from background ones by attention scores, hence localizing the regions containing actions. Basically, weakly supervised action localization can be categorized into two typical classes, including class-specific attention and class-agnostic attention.

4.1.3.1 Class-Specific Attention Methods

The class-specific attention methods compute an attention map between all action classes and temporal positions. Given the attention map $A \in \mathbb{R}^{C \times T}$, with C and T indicating the number of classes and temporal positions, respectively. The class-specific attention score can be denoted as:

$$S(c,t) = \frac{exp(A(c,t))}{\sum_{t=1}^{T} exp(A(c,t))}, \tag{4.1}$$

which represents the score distribution of class c over temporal position t. Then localized segments are predicted according to class-specific attention scores, and a segment is regarded as an action region if its score exceeds the threshold. To optimize attention scores, different strategies of training objectives are proposed, including multi-instance learning (MIL), co-activity similarity loss (CASL), and center loss.

MIL-based methods, e.g., Action Graphs [127], UNet [128], Actionbytes [129] and W-TALC [130], view the action instances in a video as the concept of bags in multi-instance learning and regard the video-level label as the label of a bag. Particularly, the training loss is formulated as:

$$\mathcal{L}_{MIL} = \frac{1}{n} \sum_{i=1}^{n} \sum_{c=1}^{C} -y_i^c \log\left(p_i^c\right), \tag{4.2}$$

where n is the number of videos and y_i^c is a binary label indicating whether video i contains class c. More importantly, the predicted probability score p_i^c for video i of class c is obtained by

$$p_i^c = \frac{exp(s_i^c)}{\sum_{c=1}^{C} exp(s_i^c)}, \quad s^c = \frac{1}{k} \sum \text{Top-}k_t A(c,t). \tag{4.3}$$

s_i^c refers to the confidence score of class c for video i, which is in practice computed by averaging the Top-k activation scores of temporal positions for each class. Namely, $\text{Top}_t^k A(c,t)$ stands for the highest k activation scores for class c when temporal position t ranges from $(1, T)$, where A is the attention map as described in Eq. (4.1). Thereafter, the probability score p_i^c of video i is computed by applying softmax function on all class-wise scores.

CASL, introduced by WTALC [130] and widely used afterwards in Deep Metric [162], WOAD [163], Actionbytes [129], etc., shares similar spirit with contrastive learning. Given a pair of videos, CASL targets minimizing the distance between their foreground features and maximizing the distance between foreground and background features from both videos. The loss is defined as:

$$\begin{aligned}\mathcal{L}_{ij}^c = \frac{1}{2}\Big\{&\max\left(0,\ d\left(F_i^c, F_j^c\right) - d\left(F_i^c, B_j^c\right) + \delta\right)\\ &+ \max\left(0,\ d\left(F_i^c, F_j^c\right) - d\left(B_i^c, F_j^c\right) + \delta\right)\Big\},\end{aligned} \tag{4.4}$$

where i and j are two sampled videos from class c, F_i^c and F_j^c are foreground features, B_i^c and B_j^c are background features. d is the measurement of similarity between two features and δ is a parameter of margin.

Center loss was first proposed by Wen et al. [164] to solve deep face recognition and was later applied to temporal action localization in 3C-Net [160]. The rationale behind center loss is to reduce intra-class variance by learning the feature center for every class and penalizing the distance between samples and their class centers. Considering original center loss is only available for single-label instances, therefore, the authors employ class-specific attention to make center loss adapted to multi-label action localization. The center loss calculates the L_2-norm distance between foreground features and cluster center features of class c as follows:

$$\mathcal{L}_C = \frac{1}{n}\sum_{i=1}^{n}\sum_{c\in y^i} \left\| F_i^c - z_c \right\|_2^2, \tag{4.5}$$

where F_i^c is the foreground feature of video i, z_c is the cluster center feature which is updated based on mini-batch to save computational cost, the centers are updated by taking the mean value of corresponding class features.

4.1.3.2 Class-Agnostic Attention Methods

Class-agnostic attention is used in RPN [131], BG modeling [132], AutoLoc [133], STPN [134], MAAN [135], etc. The common pipeline is to split untrimmed videos into segments, then obtain embedded feature sequences of them which are fed to fully connected layers or convolutional layers followed by an activation function (e.g., sigmoid or softmax) to scale weights to [0, 1].

RPN [131] designs a relational prototypical network that takes label features as input and outputs relational prototypes, then computes the similarity between features and prototypes as the distance measurement. The class-agnostic attention module computes snippet-wise scores which are aggregated into video-level scores through global weighted average pooling. The classification loss is represented by cross-entropy loss between the predicted score and ground-truth video-level label. To solve the action-background separation challenge, a clustering loss is used to guide the separation and improve the integrity of actions.

BG modeling [132] posits that background frames are also important for weakly supervised temporal action localization, similar to previous methods, it computes the classification loss for both foreground and background frames. Besides, a self-guided attention loss is constructed via bottom-up and top-down attention maps. The bottom-up class-agnostic attention weights estimate whether a video frame belongs to foreground or not while the top-down attention maps are used to supervise the refining process of bottom-up class-agnostic attention weights. The self-guided loss is formulated as

$$\mathcal{L}_G = \frac{1}{T}\sum_t |\lambda_t - \hat{\lambda}_t^{fg}| + |\lambda_t - \hat{\lambda}_t^{bg}|, \tag{4.6}$$

where λ_t is the bottom-up class-agnostic attention map, $\hat{\lambda}_t^{fg}$ is the top-down attention map of foreground classes y, and $\hat{\lambda}_t^{bg}$ is the top-down attention map of background classes. To encourage foreground features to be distinct from background ones, a foreground-background clustering loss is employed. By denoting z_f and z_b as class-agnostic foreground features and background features, respectively, the loss expression equals:

$$\mathcal{L}_C = -\log z_f - \log z_b. \tag{4.7}$$

4.1.4 *Unsupervised Action Localization*

Unsupervised action localization aims to the scenario where only the number of actions is known while ground-truth boundaries and classes are excluded from training. In order to detect and segment actions in videos without manual annotations, early attempts have been made to discover activities in instructional videos. For instance, Sener et al. [165] propose the first fully unsupervised method for understanding video activities in the temporal dimension. In particular, visual representation is discriminatively optimized by closing the distances between features from the same actions and widening the gaps between features of different actions. On the flip side, temporal modeling is instantiated by the Generalized Mallows Model (GMM). Kukleva et al. [166] propose to learn temporal-aware frame feature embeddings, then use the Viterbi algorithm to assign frames to video segments considering their temporal order. Later, Gong et al. [167] take the first attempt to explore strict unsupervised settings for TAL by a procedure of clustering and localization. It generates pseudo labels through clustering and then performs the localization step with temporal co-attention methods to refine clustering results.

4.1.5 Spatial–Temporal Action Localization

Spatial–temporal Action Localization (STAL) extends TAL to the spatial–temporal dimension, localizing not only action instances in time but also the actors in space. This gives a more accurate reflection of real-world scenarios than TAL, as it better mimics how humans recognize actions by integrating clues from actors, objects, and scenes. For example, considering a composite action *"making French fries"* which contains multiple atomic actions, such as *"cutting potato," "boil potato,"* and *"fry potato,"* TAL methods may fail by dividing the composite action into segmented actions because the atomic actions are very different. However, by attending to the same object, e.g., potato, one can easily identify that these atomic actions are subtasks of *"making French fries."*

In a word, the spatial relationships between actors and objects are essential for differentiating actions with similar objects and improving the continuity of predicted proposals, thus exploring human-object interaction and spatial–temporal context information is the key for STAL methods. As indicated in [168], existing spatial–temporal action localization methods can be categorized into frame-level methods, clip-level methods, and spatial–temporal dependent methods.

(i) Frame-level methods mainly contain two stages. Action proposals are generated by region proposal algorithms in the first stage and are then used for action localization and recognition in the second stage. Jain et al. [169] adapt the selective search sampling strategy [170] for STAL by generating spatial–temporal bounding boxes, denoted as tubelets, based on super-voxels. On the other hand, they perform merging relying on features computed by independent motion evidences (IMEs), which serve as the measure of how points in videos contribute to global motions.

Some works [171–175] employed optical flow data to extract action feature, which also used linking strategy to connect the frame candidates into human action tubes in spatial and temporal dimensions. To solve the action detection problem from frame-level, [174] proposed a new model, named Cascade Proposal and Location Anticipation (CPLA). The model designed a region-based proposal network which predicted the action proposals from an input video. Then, the proposals were fed into a detection network to further regress more accurate temporal segments and classify their categories. These results can be temporally linked and combined with a spatial–temporal action tube. Ye et al. [175] fused the static and kinematic information to address STAL. They employed the point matching strategy to solve the tracking question. The advantage of the method is that it can gather the action area into a continuous temporal-spatial action tube. To

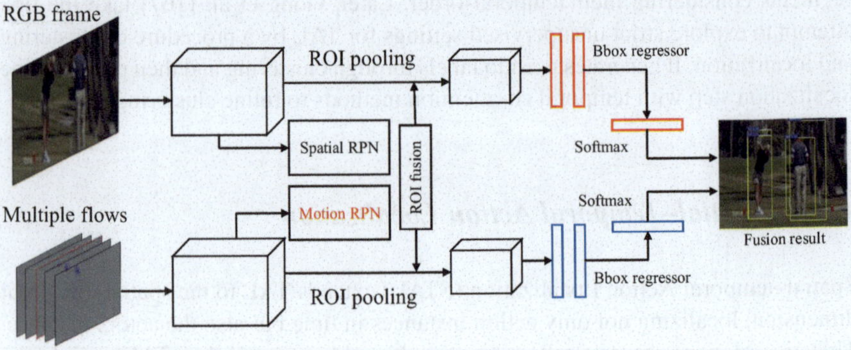

Fig. 4.8 Overview of the two-stream faster R-CNN framework. It takes RGB images and multiple flows as input, then applies RoI pooling and RPN network to detect objects in videos frames. Reproduced from Peng et al. [176]

get the candidates, the framework also designed an human action tracking strategy, which is built based on the spatial information.

In light of object detection approaches, Peng et al. [176] proposed a multi-region R-CNN method for STAL. As Fig. 4.8 shows, it starts from frame-level object detection based on Faster R-CNN. The framework took RGB frame image and optical flow data as input. The network contains several independent convolutional and max-pooling layers, and the last convolutional layer is employed to deal with appearance and motion feature via region of interest pooling. Experimental results show stacking optical flow improves frame-level localization significantly.

Although frame-level methods achieved decent performance at the early stages of STAL, the context information among video frames, which is very important for STAL, was under-explored. For example, actions like *"sit down"* and *"stand up"* can only be recognized if we observe the temporal order of frames; complex actions like *"sit down then stand up"* require further reasoning along the temporal dimension.

(ii) Clip-level methods overcome the limitations of frame-level methods by leveraging the temporal information across a sequence of frames to predict action categories and regress action tubelets. Carreira et al. extracted spatial–temporal information from videos using Inflated 3D ConvNet (I3D) which is inflated from a 2D ConvNet. Due to the great performance, the network is also widely used as feature extractor. Instead of operating at frame-level, Kalogeiton et al. [177] applied an action tubelet detector (ACT-detector) to model temporal continuity of videos. In specific, ACT-detector adapts single shot multibox detector (SSD) for video and obtains scored action class as well as anchor cuboids from frame features stacked along temporal dimension. Gu et al. [115] contributed a new dataset about spatial–temporal action localization. The dataset contains million labels to localize per person's action. It also includes long-term videos, such as over 15 minutes. Due to the abundant labels and the challenge task, the proposed dataset is widely evaluated, as shown in Fig. 4.9.

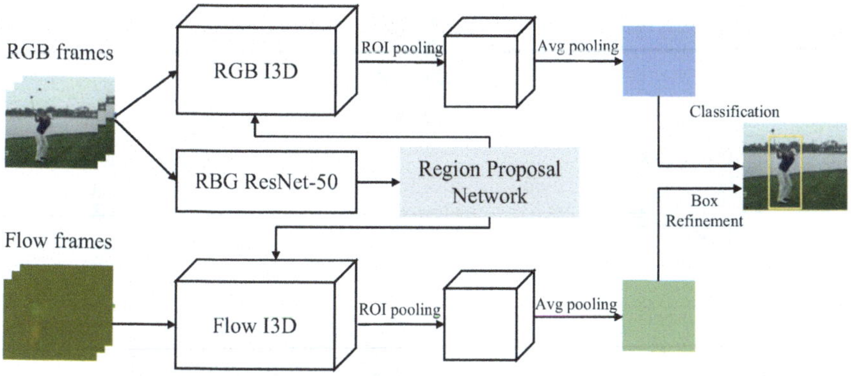

Fig. 4.9 The pipeline of the spatial–temporal action localization approach. Region proposals are detected and regressed with Faster-RCNN for both RGB frames and optical flow. Reproduced from Gu et al. [115]

Unlike previous works, Yang et al. [178] proposed a progressive-learning framework for STAL. It defines a handful of coarse-scale proposal cuboids that are further refined to approach target actions through a few steps. Compared with prior works, such a progressive-learning paradigm can deal with the spatial moving action within the tubes and achieve effective spatial–temporal modeling. Rather than focusing on short-range videos, Wu et al. [179] noticed that the long-term temporal in video data is hard to employ. To better understand the vision task, they proposed to obtain more supportive information extracted from the long video and maintained a feature bank to store the rich video representation.

(iii) Spatial–temporal dependent methods not only consider temporal modeling but also the interactions between actors and objects. For example, Sun et al. [180] developed a human action detection framework that finds the spatial–temporal connections among humans, objects, and environments. To achieve such an ability, they used an off-the-shelf network to compute pairwise relationships.

A mainstream approach involves utilizing graphs to model actor-object interactions. Zhang et al. [181] conduct STAL training with the aid of domain knowledge from a graph convolutional network. As shown in Fig. 4.10, they proposed to enhance the standard I3D network for action detection in videos. The proposed method models long-term human behavior, which can represent the relationships between humans and objects. Tomei et al. [182] developed a plugin module that does not require end-to-end training for STAL. The graph structure is used to encode relationships between actors and objects, and self-attention operations are used for spatial–temporal modeling. Ji et al. [183] designed a novel model to learn the representation between the actions and humans. A spatial–temporal graph is proposed to capture the person, objects, and their relationships. With such representation, they researched spatial–temporal scene graph prediction. Pan et al. proposed a method [184] which focused on modeling second-order relations to represent the relation between actors. The method directly uses spatial–temporal

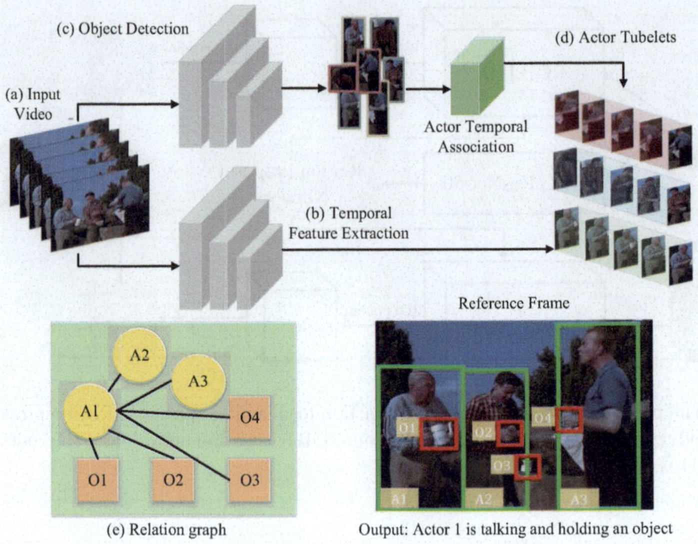

Fig. 4.10 Overview of the STAL training method with graph convolution network. Reproduced from Zhang et al. [181]

grid feature maps to get context features. They demonstrated that the context does not use any other detection methods and makes the whole model more simpler and flexible.

Some other approaches aim to improve network design using self-attention mechanism. Girdar et al. [185] introduce Action Transformer which takes a modified Transformer as a head to gather the contextual information regarding the people of interest and the objects surrounding them. Besides, the attention mechanism learns to attend to body parts such as hands and faces, which can aid in action recognition. MViT [42] is built for vision understanding, especial for recognition. The method employed transformer models to explore the connection of multi-scale features. MViT contains multi-scale channels to represent different resolutions. This method produces feature maps of high spatial resolution in early layers to capture lower-level patterns, and the maps at deeper layers are used to capture high-level semantic information. Compared with CNN-based methods [39], MViT achieves higher mAP with better efficiency and fewer parameters.

4.1.6 *Datasets*

To evaluate the efficacy of the proposed methods, several datasets for action localization have been introduced. Table 4.1 summarizes benchmark TAL datasets.

Table 4.1 Benchmark datasets for TAL, *Avg. Duration* refers to the average length of videos in seconds, *Avg. Instances* refers to the average number of action instances in a video

Dataset	Domain	#Videos	#Classes	Avg. Duration	Avg. Instances
THUMOS [186]	Sports	413	20	212s	15.5
ActivityNet [25]	Open	19,994	200	115s	1.54
Charades [109]	Indoor	9848	157	30s	6.75
Breakfast [187]	Cooking	1712	48	162s	6
COIN [188]	Open	11,827	180	142s	3.9
AVA [115]	Movies	437	80	900s	3361.5

THUMOS [186] Jiang et al. released the THUMOS'14 dataset which includes 1574 videos in the training set and 1010 videos in the validation set, categorized into 101 action classes. For action localization task, it splits 220 and 213 videos for train and test, annotated into 20 classes.

ActivityNet [25] The ActivityNet v1.2 has 9682 videos annotated into 100 classes and ActivityNet v1.3 has 19,994 videos annotated into 200 classes which is the most popular version for temporal action localization task; the average duration of a video is 115 seconds.

Charades [109] The Charades dataset is an indoor activity dataset that has rich annotations for multi-label classification, temporal action localization, and temporal video grounding. It contains 9848 videos in total with an average duration of 30 seconds, annotated into 157 classes.

Breakfast [187] The Breakfast dataset includes 48 fine-grained cooking actions and 52 participants in total, each performing 10 cooking activities in daily-life kitchens. Each video lasts for 162 seconds and contains 6 action instances on average.

COIN [188] The COIN dataset is collected from YouTube and built for comprehensive instructional video analysis, which contains 11,827 daily-life videos categorized into 180 tasks. It has a hierarchical structure of annotation levels including domain, task, and step.

AVA [115] The AVA dataset is made public in 2018 and then became the most popular one for spatial–temporal action localization. It is densely annotated into 80 atomic visual actions which contain spatial bounding boxes and temporal timestamps, leading to 1.59M action labels in total.

4.1.7 Summary

This section provides a comprehensive study of action localization which aims to localize potential action instances and predict action categories at temporal (TAL)

or spatial–temporal (STAL) dimensions. Action localization has a wide range of applications in real-life scenarios, such as sports video analysis, medical video analysis, automatic video clipping, video highlight generation, etc. Supervised TAL can be categorized into two basic types, i.e., two-stage methods and single-stage methods, considering the expensive annotation cost of supervised TAL, weakly supervised and unsupervised are attracting more and more attention in the near years. Besides, STAL aims to locate the temporal interval of the action and the spatial bounding box of the actor, which is a more challenging task yet more closely related to realistic requirements. At last, popular benchmark datasets for TAL and STAL are introduced and we expect there to be more efforts to address the challenges of action localization in the future.

4.2 Temporal Video Grounding

4.2.1 Introduction of Temporal Video Grounding Task

Background

With the continuous development of multimedia technology on terminal devices, the amount of user-uploaded videos is increasing rapidly, which results in a new challenge of effective and efficient retrieval of numerous videos. Compared with text-image retrieval, video retrieval or video moment retrieval (i.e., temporal video grounding) is more challenging, because videos contain rich dynamics and semantics which require models to be capable of temporal reasoning. Also, huge amounts of videos are hard to manage since they occupy ample storage (e.g., YouTube-8M [89] has 6.1M videos, HowTo100M [189] has 1.2M videos). Therefore, it is important to develop efficient retrieval techniques of specific video clips from an untrimmed video, as it allows users to easily locate the moments of interest.

Temporal video grounding (TVG), which is also known as temporal sentence/language video grounding or single video moment retrieval, was first proposed in 2017 [190, 191]. It aims to predict the start and end boundary points of the events described by the given natural language query. Figure 4.11 demonstrates the targeted segment of this task given the natural language query input *"A dog wags his tail as he tries to help."* While videos can be considered as a collection of consecutive frames and queries can be considered as a sequence of words, the temporal video grounding task requires an integral understanding of both sentence semantics and video contents, which needs a joint modeling of visual and textual features.

By taking a closer look at TVG, we can see TVG shares similarities with classic tasks, such as temporal action localization, video grounding, and reading comprehension in NLP. The closest task to TVG is temporal action localization (TAL), which aims to detect a video moment that is relevant to a specific action. Compared with TAL, TVG is more challenging since it not only needs to model

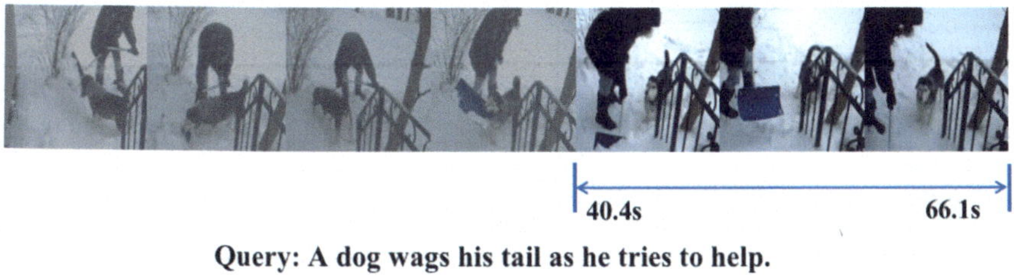

Fig. 4.11 An example of the temporal video grounding task, i.e., to determine the start and end timestamps of the target video segment corresponding to the given sentence query

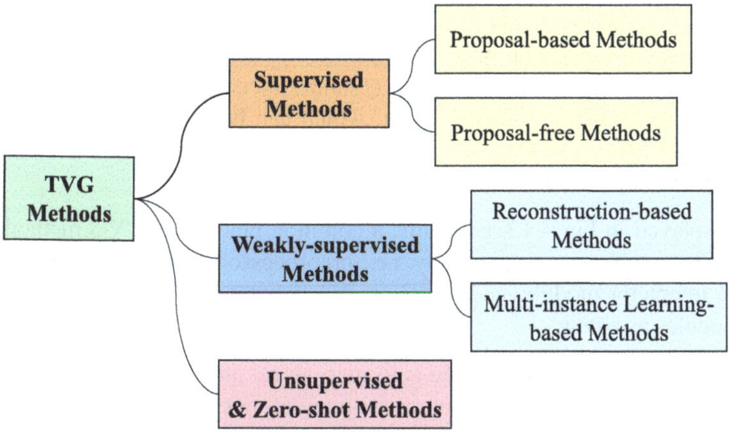

Fig. 4.12 A taxonomy of methods for TVG

sentences and videos but also exploits the fine-grained interactions between these two modalities. Besides, language queries are not restricted by close-set action classes in TVG and are more complex and diverse, which makes this problem more difficult.

Due to its great potential in industrial applications, TVG has drawn significant attention from researchers and numerous relevant algorithms have been proposed in the past few years. As indicated by Zhang et al. [192], existing approaches can be divided into four categories, including proposal-based [190, 191, 193–195], proposal-free [196–198], reconstruction-based methods [199–203], and multi-instance learning [202, 204–207] methods. Although there are various types of TVG methods, most of them share a common pipeline. Therefore, we will first demonstrate the general pipeline, then take an overview of TVG methods from an aspect of supervised, weakly supervised, and unsupervised settings in the next section, see Fig. 4.12.

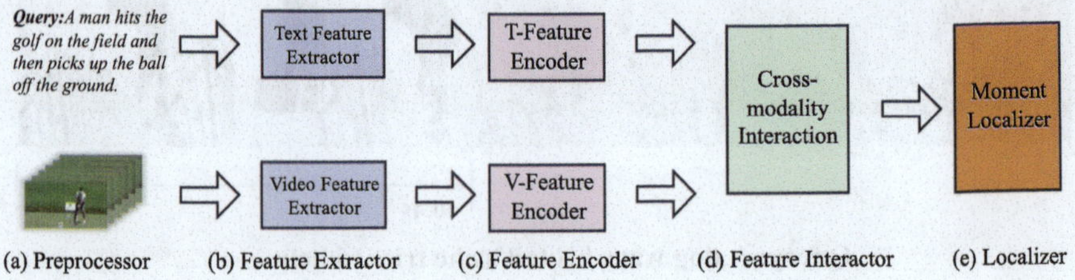

Fig. 4.13 The common pipeline of TVG models

Formulation We denote an untrimmed video as $V = \{f_t\}_{t=1}^T$ and a language query as $Q = \{q_j\}_{j=1}^L$, where T and L are the number of frames and words, respectively. The ground-truth timestamp of a video moment is denoted as $[\tau_s, \tau_e]$, where $1 \leq \tau_s < \tau_e \leq T$. Given a video V, query Q, and a model M, the goal of TVG is to retrieve the ground-truth video moment, which can be written mathematically as $[\tau_s^{'}, \tau_e^{'}] \longleftarrow M_{TVG}(V, Q)$. Therefore, model M is optimized by maximizing the overlap between prediction $[\tau_s^{'}, \tau_e^{'}]$ and ground truth $[\tau_s, \tau_e]$.

As displayed in Fig. 4.13, generally, a common pipeline of TVG methods consists of five components sequentially, i.e., preprocessor, feature extractor, encoder module, interactor module, and localizer. We elaborate on the designs of the last three modules as below.

(i) **The feature encoder module** is to map visual and text features to the same dimension and transform features by integrating contextual information. The mapping of query features and video features can be defined as $Q \in \mathbb{R}^{n \times d_q} \mapsto Q' \in \mathbb{R}^{n \times d}$ and $V \in \mathbb{R}^{n \times d_v} \mapsto V' \in \mathbb{R}^{n \times d}$, respectively.
(ii) **The feature interactor module** is used to model the cross-modality interaction between visual and textual clues, which is an essential component in TVG. The cross-modality attention mechanism is usually used to fuse the text and video representations in a cross-modality interaction module. Through carefully designed interaction modules, the segment which is most relevant to the query will be highlighted. And the different design of feature interactors highly affects TVG performance.
(iii) **The localizer module** mainly generates candidate predictions based on the learned feature representations, and different prediction modules will result in different prediction results. Proposal-based and proposal-free methods predict the start-end boundaries of target moments differently. For proposal-based methods, the localization predictor calculates a score for each proposal and ranks the proposals by the score, while the proposal with a higher score should be closer to the ground-truth moment ideally. For proposal-free methods, a regressor is usually used to directly predict the temporal boundaries of the target segment.

Evaluation Metrics In TVG, Intersection over Union (IoU) in the range of [0.0, 1.0] is widely used for measuring the similarity between two temporal moments, larger IoU indicates better matching of two video moments. The most commonly used evaluation metrics for TVG are mean IoU (mIoU) and $< R@k, tIoU@m >$. mIoU is the average of all temporal IoUs (tIoU) of the test set, $< R@k, tIoU@m >$ is the percentage of samples that have at least one moment that has $tIoU > m$ among the top k results, which is computed as:

$$R@k, IoU@m = \frac{1}{N}\sum_{i=1}^{N} r(k, m, q_i), \quad (4.8)$$

where $r(k, m, q_i)$ is positive sample for query q_i, the widely used k and m are $k \in \{1, 5, 10\}$ and $m \in \{0.3, 0.5, 0.7\}$.

The rest of this section is organized as follows: (1) Sect. 4.2.2 introduces supervised temporal video grounding methods; (2) Sect. 4.2.3 introduces weakly supervised temporal video grounding methods; (3) Sect. 4.2.4 introduces unsupervised and zero-shot temporal video grounding methods; (4) Sect. 4.2.5 introduces datasets and discussions about potential applications and future works; (5) Sect. 4.2.6 summarizes this section.

4.2.2 *Supervised Temporal Video Grounding Methods*

The supervised temporal video grounding methods can be categorized into two parts including proposal-based methods and proposal-free methods.

4.2.2.1 Proposal-Based Methods

Proposal-based methods usually cut the video into multiple candidate segments via a sliding window strategy, a proposal generation module, or pre-defined anchors. Then the proposal selection task is treated as a cross-modal matching task by ranking the candidate proposals produced by the matching module. Representative works include CTRL [190], MCN [191], ROLE [208], MCF [195], ACRN [194], TCMN [209], ASST [210], ACL [193], SLTA [211], MMRG [212], QSPN [213], SAP [214], BPNet [215], APGN [216], MAN [217], SCDM [198], TMN [218], 2D-TAN [219], etc.

(i) **Sliding-window methods.** CTRL [190] is the first work to generate proposals via sliding windows. As Fig. 4.14 displays, the features of generated candidate proposals and queries are extracted by a visual feature extractor and a textual feature extractor, respectively. The features are then fused by a multimodal fusion module followed by two fully connected layers, which aim to supervise proposal-query alignment and generate proposal boundaries, respectively.

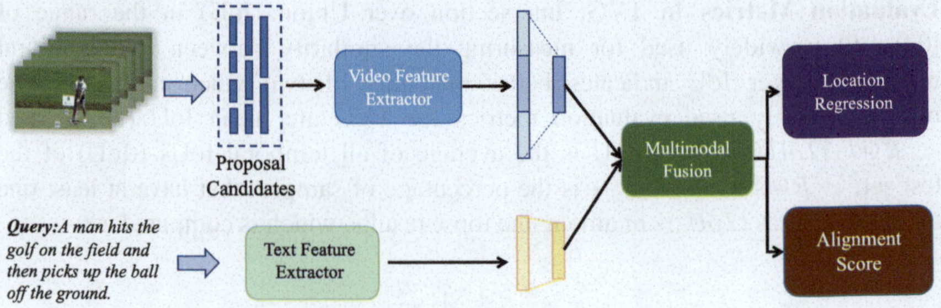

Fig. 4.14 The architecture of CTRL. Reproduced from Gao et al. [190]

Hendricks et al. [191] propose MCN, which adopts sliding windows of multi-scale to generate candidate proposals. Then the video and text representations are projected into a common embedding space, and the distance between them is minimized to supervise model learning. However, a simple fusion strategy and the coarse video-query input without temporal modeling may result in weak performance on TVG, since both videos and sentences have rich temporal information which is important for temporal reasoning. Some works intend to investigate the fine-grained integration of video and query or introduce knowledge to refine the proposals [193–195, 208–212].

(ii) **Proposal generation methods.**

Although sliding-window-based approaches are simple and effective, they require a dense sampling of windows and hence lead to a heavy computation burden. Researchers have devoted themselves to improving efficiency by learning to generate fewer proposals. For instance, QSPN [213] alleviates the computation burden by using a pre-trained segment proposal network (SPN) [138] and refines proposing candidate segments by interacting video features with corresponding queries.

Due to the abundant computation of overlapped proposals, the two-stage propose-and-rank pipeline of QSPN is still time-consuming, and the individual moment query matching may also ignore the contextual information. To overcome these drawbacks, BPNet [215] and APGN [216] jointly train a proposal-free module that only generates a few proposals. Besides, the models take a whole video as input so that contextual information of all proposals in a video is considered.

(iii) **Anchor-based methods.** The sliding window and proposal generation methods split a video into several proposal candidates at input space, thus it suffers from costly computation and it also lacks contextual information. To mitigate this issue, researchers explore anchor-based methods.

TGN [220] is the first anchor-based method, which can localize the temporal segment in one stage without handling redundant pre-defined segment candidates. As Fig. 4.15 shows, TGN models frame-by-word interactions across two modalities using a sequential LSTM module. In contrast, Yuan et al. [198] propose SCDM, which uses a hierarchical temporal convolutional module to locate the

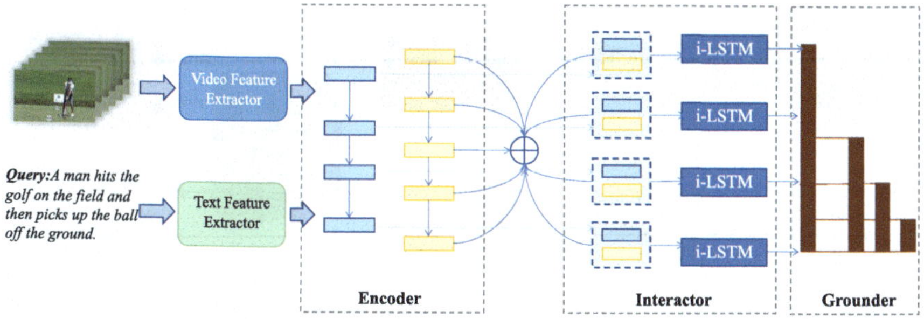

Fig. 4.15 An overview of TGN. Reproduced from Chen et al. [220]

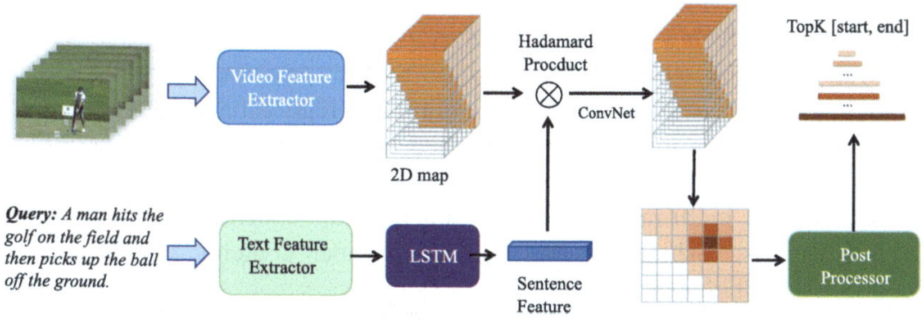

Fig. 4.16 An overview of 2D-TAN. Reproduced from Zhang et al. [219]

moment and combines it with a semantics-conditioned dynamic module to make full use of sentence semantics. Instead of considering one-dimensional temporal features, 2D-TAN [219] (Fig. 4.16) models the temporal relations of video proposals using a 2D map and achieves better performance. Wang et al. [221] propose a structured multi-level interaction network (SMIN), which adds multi-levels of content-boundary-moment interaction on the 2D candidate proposal feature map. SMIN takes advantage of the inherent structure of different candidate proposals by exploiting the structured multi-modality interaction and intra-moment interaction.

4.2.2.2 Proposal-Free Methods

There also exists end-to-end methods straightforwardly regressing the start and end boundaries of the target moment based on the multimodal features of video and query. These methods are named proposal-free methods and do not depend on any candidate proposal generation process.

(i) **Regression-based methods.** Regression-based methods directly predict timestamps (i.e., start and end point) of a video, and representative works include ABLR [197], ExCL [222], DEBUG [223], GDP [224], DRN [225], HLGT [226], HVTG [227], PMI [196], DRFT [228], etc. Yuan et al. propose

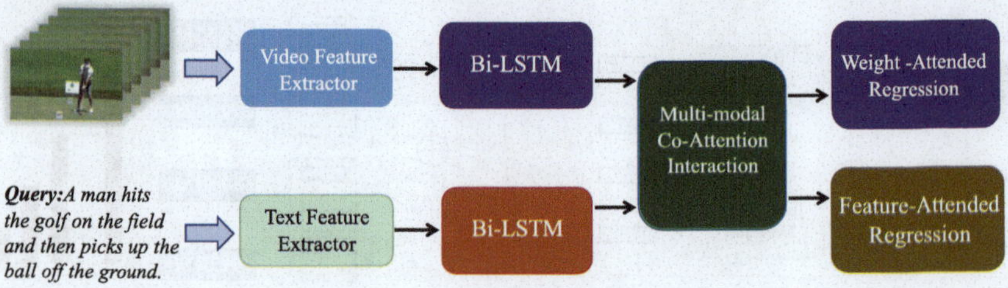

Fig. 4.17 The architecture of ABLR. Reproduced from Yuan et al. [197]

ABLR [197] (Fig. 4.17), which regresses the temporal coordinates directly without producing any anchors. Specifically, a cross-modality co-attention strategy is used to model the correlation between two modalities in ABLR, and two types of regressors, attention weight-based regression and attended feature-based regression modules, are designed to predict the temporal coordinates. Pairwise modality interaction (PMI) proposed by Chen et al. [196] learned fine-grained multi-modality interaction in both sequence and channel levels in a pairwise fashion to better understand video contents. PMI also directly predicts the boundaries.

(ii) **Span-based methods.** Span-based methods usually predict the probability of being the start or end point for a video frame-by-frame, and representative works include L-Net [229], VSLNet [230], SeqPAN [231], CI-MHA [232], PPT [233], etc. L-Net [229] introduces span prediction in TVG for the first time. It contains a cross-gated attended recurrent network to build fine-grained interactions between video and query, a self-interactor to dynamically encode and aggregate vision-language information, and a segment localizer that predicts the start-end position of a video moment. VSLNet [230] treats TVG from a perspective of QA where videos are treated as text passages, and a VSLNet is added to a span-based QA network. A query-guided highlighting strategy is proposed to narrow the differences between TVG and QA, which improves video-query matching by helping the model distinguish minor differences between different frames.

4.2.3 Weakly Supervised Temporal Video Grounding Methods

Obtaining extensive labeled data for TVG is time-consuming and labor-intensive. To avoid the expensive labor of annotating video segments and captions, researchers study TVG without access to the timestamps of the video corresponding to a query—this is often known as weakly supervised temporal video grounding. Generally, existing studies fall into two groups, i.e., (1) methods based on reconstruction [199–203], which find video proposals that match the given query first

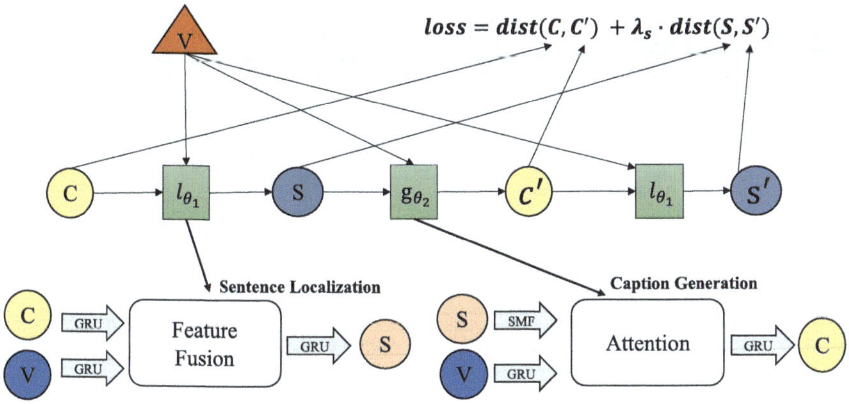

Fig. 4.18 Architecture of WS-DEC. Reproduced from Duan et al. [199]

and then update models by reconstructing text queries, using predictions chosen from intermediate proposals, (2) methods using multi-instance learning (MIL) [204–207, 234, 235], which enhances video–text pairing by treating videos as a bag of instances and constructing precise alignments between videos and queries.

4.2.3.1 Reconstruction-Based Methods

Early methods generate timestamps and captions of segments through cyclic training. Duan et al. [199] proposed the first reconstruction-based approach by decoupling the task into moment localization and caption generation (see Fig. 4.18). Moment localization is performed by retrieving a video moment S given the caption C first. The caption $C^{'}$ is then generated according to the retrieved video moment S and the generated caption $C^{'}$ is used to retrieve a video moment $S^{'}$ at last. Therefore, the training objective is to simultaneously minimize the distance of $(C, C^{'})$ and $(S, S^{'})$.

One follow-up work is SCN [200] which first generates and scores candidate proposals through a proposal generation module, then selects top-N proposals considering exploitation and exploration. Next, a semantic completion module computes the similarity between proposals and queries, providing feedback to the proposal generation module to refine the proposals' scores.

To simplify and align the proposal scoring process for training and inference, [201] presents a mechanism for reconstructing attention that uses proposal-level attention to directly evaluate all proposed candidates. The idea is to construct a multi-level framework capable of dealing with multi-scale proposals and generating more discriminative features.

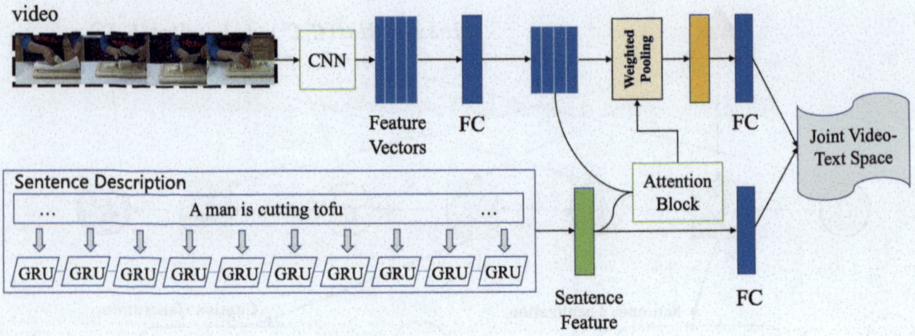

Fig. 4.19 Architecture of TGA. Reproduced from Mithun et al. [204]

4.2.3.2 Multi-Instance Learning (MIL) Methods

Multi-instance learning (MIL) [236] considers a video as a collection of instances groups, i.e., each video consists of multiple instances but only has a video-level label. TGA [204] (Text-Guided Attention) solves weakly supervised TVG (See Fig. 4.19) via MIL. TGA aligns the video frames and text implicitly by highlighting frames that are corresponded to the text, then the video features are pooled along time dimension based on the attention weights to obtain the text-dependent global feature. The goal of training is to reduce the distance between the video features contingent on the text and the textual features.

Another stream of solutions builds MIL upon contrastive learning. With carefully designed positive and negative samples, the iterative training process learns fine-grained alignment between video and query. CCL [207] and VCA [237] implement contrastive learning to solve weakly supervised TVG.

CCL [207] builds a counterfactual contrastive learning paradigm to tackle counterfactual positive and negative results. The counterfactual transformations contain robust counterfactual transformations that damage the visual information of inessential proposals and generate robust positive results, and destructive counterfactual transformations that destroy text-related visual contents and retain unnecessary content to achieve effective contrastive learning.

VCA [237] proposes to align video moments from co-occurred videos in a mini-batch to improve the robustness and discrimination of video features, which is motivated by the observation that similar video moments from various videos can be described by similar sentences, therefore, a VCA-NCE loss is introduced to mining hard positives from video moments in co-occurred videos.

LCNet [202] applies cycle consistent loss as the learning objective (See Fig. 4.20). Considering existing methods' lack of video–text local correspondence modeling, LCNet utilizes a hierarchical feature representation module to model different local correspondences.

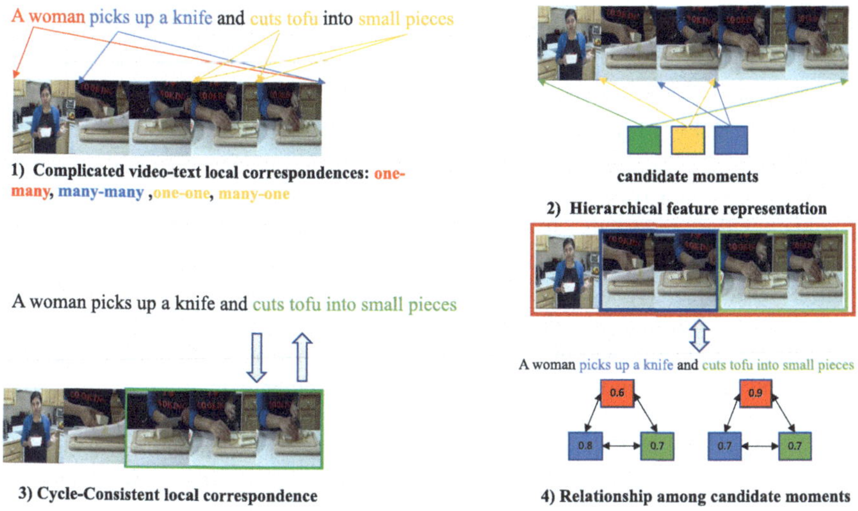

Fig. 4.20 Model structure of LCNet. Reproduced from Yang et al. [202]

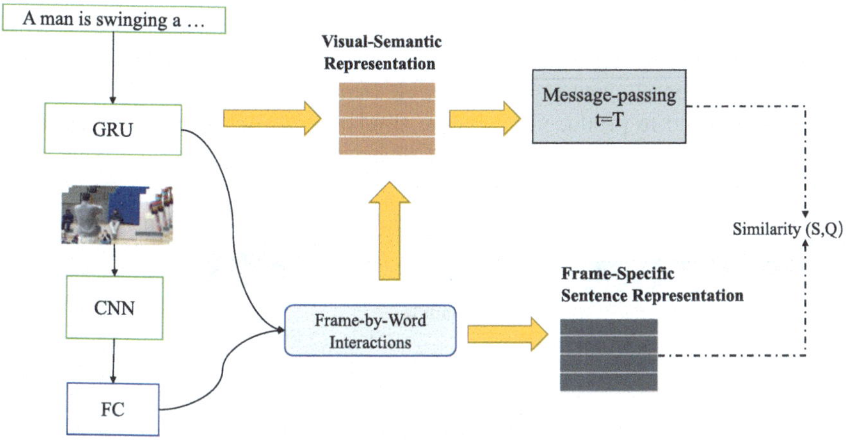

Fig. 4.21 Model structure of LoGAN. Reproduced from Tan et al. [206]

LoGAN [206] enhances fine-grained alignment between video moment and texts with a triplet loss, which takes a video with its query as positive video-sentence pair and a video with queries from other videos as negative video-sentence pair (See Fig. 4.21). Furthermore, video contextual information is aggregated through graph propagation and the frame-by-word interactions are performed for precise semantic retrieval.

4.2.4 Unsupervised and Zero-shot Temporal Video Grounding Methods

Unsupervised TVG is defined by Liu et al. [238] that removes any paired supervision. Liu et al. [238] proposes a Deep Semantic Clustering Network (DSCNet) that exploits comprehensive semantic data from the entire query collection to generate candidate video proposals. In the first step, an encoder–decoder is employed to acquire the semantic characteristics of all queries and categorize them into various clusters to represent diverse semantics. A two-branch video component is then used, with one branch using attention to aggregate query semantics and match video segments, and the other branch using foreground attention to identify foreground frames. Pseudo labels derived from deep semantic features are employed to supervise the training.

Zero-shot TVG is first explored by Nam et al. [239], assuming that only unlabeled videos are available for target datasets. The main challenge is to generate suitable pseudo-supervision for explicitly optimizing models, which consists of two key processes: (1) finding meaningful temporal regions that are likely to be queried; (2) generating queries for the temporal regions.

Nam et al. [239] propose Pseudo-Supervised Video Localization (PSVL) (See Fig. 4.22) to address zero-shot TVG. In order to obtain meaningful proposal segments, a temporal event proposal module that takes K-means as the clustering strategy is applied to produce frame-wise similarity matrix. Then a pseudo-query generation module takes an off-the-shelf object detector to generate nouns and a

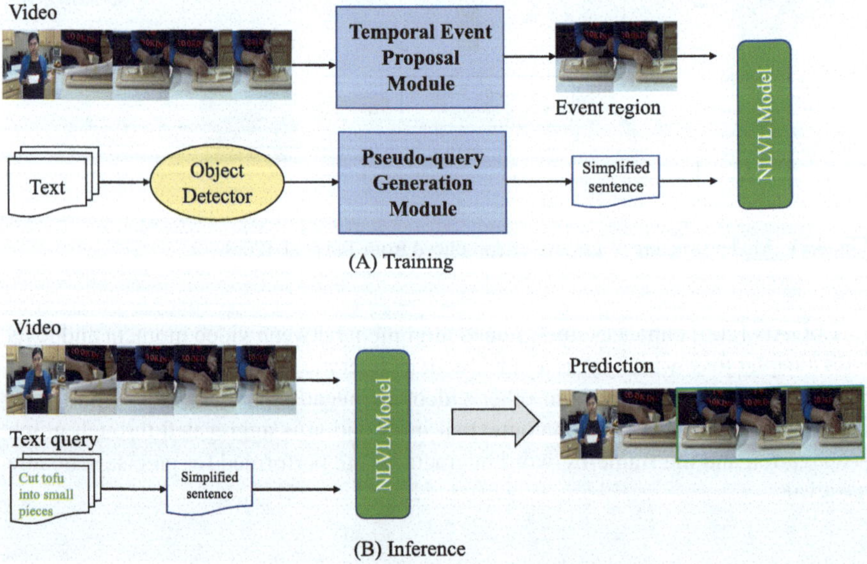

Fig. 4.22 Model structure of PSVL. Reproduced from Nam et al. [239]

Table 4.2 Popular TVG benchmark datasets

Dataset	Source	Domain	# Videos	# Moments	# Queries	Vocab. Size
ActivityNet Captions [29]	YouTube	Open	14,926	71,953	71,953	15,505
Charades-STA [190]	Homes	Indoor	6672	11,767	16,124	1303
TACoS [242]	Lab Kitchen	Cooking	127	3290	18,818	2344
DiDeMo [191]	Flickr	Open	10,464	26,892	40,543	7785
MAD [191]	Movie	Open	650	–	384,600	61,400

VerbBERT model to generate verbs, respectively. Next, the nouns and verbs are concatenated to construct a simplified sentence as the query of the proposal segment. Once the pseudo labels are prepared, any TVG model can adapt to zero-shot setup by training on pseudo data in the same way as supervised TVG.

Another similar work is [240], which proposes Prompt-based Zero-Shot Video Moment Retrieval (PZVMR) method. PZVMR is different from PSVL in two aspects: firstly it replaces K-means with sliding windows to generate proposal segments; secondly it generates verbs' representations through prompt learning on a pre-trained CLIP [241] model.

4.2.5 *Datasets*

This section introduces the most popular datasets for TVG. Table 4.2 lists the TVG datasets.

ActivityNet Captions [29] Krishna et al. extend ActivityNet [25] dataset for dense video captioning. It contains 20k videos with a total duration of 849 video hours and 100k descriptions. Each video contains 3.65 query sentences, constructing a total of 100k sentences for the dataset. Considering the test set is unavailable to the public, Table 4.2 only reports statistics on train and validation set.

Charades-STA [190] Charades-STA dataset is extended from Charades dataset [109] which contains 9848 videos with an average duration of 30 seconds, and rich annotations of 12,404 clip-query pairs for training set and 3720 clip-query pairs for testing set.

TACoS [242] TACoS dataset is built from the MPII Cooking Composite Activities dataset [243]. There are 127 videos in total with an average duration of 4.5 minutes, annotated into 18,818 clip-query pairs.

DiDeMo [191] DiDeMo dataset is sourced from YFCC100M [244] dataset which covers various human activities. It is divided into train, validation, and test sets, with 8395, 1065, and 1004 videos, respectively. There are 10,464 videos in total annotated into 26,892 moments and 40,543 queries, the average number of annotations for each video is 3.87.

MAD [191] MAD dataset is collected from movies and is different from existing TVG datasets due to unbiased annotations. It contains 650 movies with an average duration of 110 minutes, annotated into 384,600 queries.

4.2.6 Summary

This section gives a systematic introduction to the temporal video grounding (TVG) task. TVG aims to localize a moment from a video given a sentence which is essential for realistic applications such as fine-grained video retrieval. The taxonomy of TVG methods consists of supervised methods, weakly supervised methods and unsupervised methods. To be specific, supervised methods mainly consist of proposal-based and proposal-free methods, and weakly supervised methods are divided into reconstruction-based and MIL methods. At last, we present benchmark datasets used for evaluation.

Chapter 5
Deep Learning for Video Captioning

5.1 Introduction of Video Captioning Task

5.1.1 Problem Formulation

Given an input video $V = \{f_1, \dots, f_{N_v}\}$ with N_v frames (or clips), the goal of video captioning is to generate a natural language sentence $S = \{w_1, \dots, w_{N_s}\}$ consisting of N_s words to describe the video's content. Specifically, let $\mathbf{V} = \{\mathbf{v}_1, \mathbf{v}_2, \dots, \mathbf{v}_{N_v}\} \in \mathbb{R}^{D_v \times N_v}$ and $\mathbf{W} = \{\mathbf{w}_1, \mathbf{w}_2, \dots, \mathbf{w}_{N_s}\} \in \mathbb{R}^{D_w \times N_s}$ denote the visual features of a video V and the textual features of a sentence S, where $\mathbf{v}_m$ and $\mathbf{w}_n$ denote the mth frame (or clip) feature vector and the nth word vector, respectively. Therefore, given the video representations $\mathbf{V}$, video captioning aims to compute the conditional probability of the generated word sequence, i.e.,

$$P(\mathbf{w}_1, \mathbf{w}_2, \dots, \mathbf{w}_{N_s} | \mathbf{V}). \tag{5.1}$$

Since there exist contextual relations between the words in a sentence, it is natural to generate words sequentially with the chain rule. In other words, we aim to estimate the joint probability over the generated words, which is computed by the sum of the log probabilities as follows:

$$\log P(\mathbf{W}|\mathbf{V}) = \sum_{t=1}^{N_s} \log P\left(\mathbf{w}_t | \mathbf{V}, \mathbf{w}_1, \dots, \mathbf{w}_{t-1}\right). \tag{5.2}$$

During training, we input the begin-of-sentence symbol $\langle BOS \rangle$ into a decoder (e.g., an LSTM), which indicates the beginning of the sentence generation process. Then we optimize the model parameters θ to maximize the *log* probability of the ground-truth video description conditioned on video representations, which can be formulated as follows:

Z. Wu, Y.-G. Jiang, *Deep Learning for Video Understanding*, Wireless Networks,
https://doi.org/10.1007/978-3-031-57679-9_5

$$\theta^* = \underset{\theta}{argmax} \sum_{t=1}^{N_s} \log P\left(\mathbf{w}_t | \mathbf{V}, \mathbf{w}_1, \ldots, \mathbf{w}_{t-1}; \theta\right). \tag{5.3}$$

The optimization is usually performed over a training dataset containing video–text pairs using stochastic gradient descent. The description generation process continues until the end-of-sentence symbol $\langle EOS \rangle$ is predicted. At inference time, we select the word with the maximum probability at each time step and set it as the input for the next time step until the end-of-sentence symbol is generated.

5.1.2 A Common Encoder–Decoder Framework to Video Captioning

As described in Sec. 5.1.1, video captioning is often formulated as a sequence-to-sequence learning problem. Therefore, existing video captioning models follow an encoder–decoder framework, where the encoder extracts video features and the decoder generates a sentence word by word depending on the extracted representation.

Let Φ denote the encoder network and Ψ denote the decoder network. The encoder Φ converts the input video $V = \{f_1, \ldots, f_{N_v}\}$ into a continuous-space representation, which may be a series of continuous vectors with varying sizes:

$$\Phi(V) = \mathbf{V} = \mathbf{v_1}, \ldots, v_{N_v}\}. \tag{5.4}$$

The feature representation $\mathbf{V}$ is then used to generate sentence S by the decoder network Ψ. The architecture of the decoder must be selected in accordance with the form of output. We use RNN as an example as it is the most commonly used choice when the output is a sentence of a natural language, as in the case of automatic video description.

The output is sequentially processed by the decoder RNN. The RNN changes its internal state $\mathbf{h}_t$ to produce an output y_t at each step t based on its previous internal state $\mathbf{h}_{t-1}$, the previous output y_{t-1}, and the encoder representation $\mathbf{V}$. It then outputs the symbol y_t:

$$\begin{bmatrix} y_t \\ \mathbf{h}_t \end{bmatrix} = \Psi\left(\mathbf{h}_{t-1}, y_{t-1}, \mathbf{V}\right). \tag{5.5}$$

The RNN runs recursively until the end-of-sequence symbol is produced.

To illustrate the frameworks, Fig. 5.1 shows a general view of the encoder–decoder architecture. On the encoder side, different feature extractors are used to obtain video features, like the 2D [10] and/or 3D CNNs [61]. Additionally, an attention mechanism is utilized to capture the temporal and spatial clues. Other

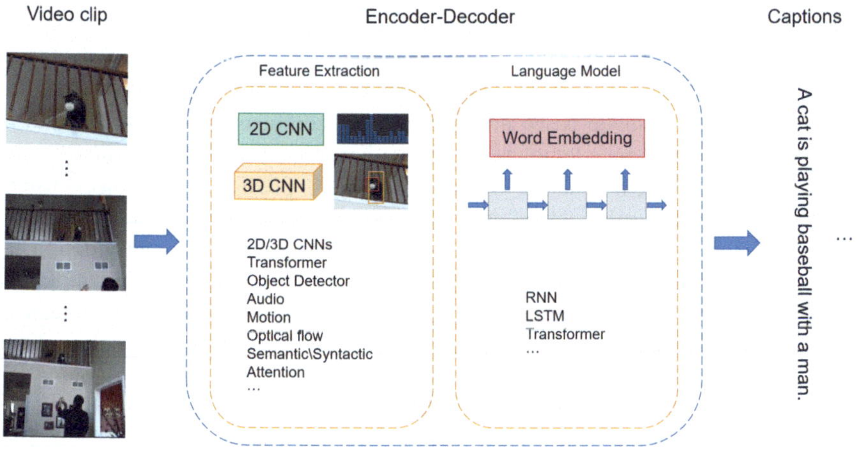

Fig. 5.1 A common encoder–decoder framework of video captioning

features are also taken into consideration, such as audio, optical flow, motion, semantic, etc. On the decoder side, RNNs such as LSTM [92] are studied to generate video sentences based on the extracted features. Recently, Transformer [53] has been introduced to video captioning tasks for both encoder and decoder designs.

5.2 Video Captioning Methods

Compared with image captioning, it is more complex to describe videos with natural language sentences, as there are two additional challenges of video captioning. In terms of the encoder, videos have richer information than images, such as temporal and spatial information and multimodal information. How to use such rich information to generate more accurate and fine-grained video representation is an important aspect for further work. In terms of the decoder, language generation needs to ensure semantic and syntactic correctness, and some methods have been proposed for this purpose.

Broadly speaking, there exist two types of methods for video captioning: template-based language method and sequence learning method. Early studies mostly belonged to the former, which highly depend on sentence templates and regularly generated sentences with syntactic structures. The latter is the focus of present research, which is often designed as an encoder–decoder framework to extract video features and then generate natural language sentences. In this section, we briefly introduce some of the early template-based language studies followed by more detailed discussion on current sequence learning research.

5.2.1 *Template-Based Language Methods*

Inspired by the success of image recognition and activity recognition, a straightforward way for template-based approaches is to combine detected words into a sentence using a sentence template. Template-based methods typically first divide sentences into different syntactic units (e.g., subject, predicate, and object) using specific grammatical rules, where each syntactic unit has a certain association with those detected words (e.g., objects, actions, and attributes) from the video. The syntactic units are then composed to a sentence conditioned on provided templates. As a result, the captioning quality is to some extent decided by the sentence templates, and the sentences are regularly generated with a syntactic structure.

Kojima et al. [246] is one of the earliest works that apply the concept hierarchy of actions to extract semantic meanings of video frames by verifying correspondences between semantic features of human actions and natural language concepts. Tan et al. [247] proposed to make use of audio–visual concept classifiers and pre-defined concept-based templates for video event recounting. Rohrbach et al. [248] worked on learning an intermediate semantic representation for each video through a conditional random field model and then translating it to natural sentences by using statistical machine translation techniques. Furthermore, by integrating semantic unaries, hand-centric features, and enforcing consistent topics, Rohrbach et al. [249] attempted to detailedly generate coherent multi-sentence video descriptions. In [250], Guadarrama built semantic hierarchies to predict nodes (a set of labels for sentence generation) by trading off specificity with semantic similarity. Thomason et al. [251] introduced a framework to generate simple descriptions for short videos, which uses a probabilistic graphical model to combine visual detection results with additional linguistic information and select the best subjects, verbs, objects, and places for a language template. Additionally, Xu et al. [245] proposed a unified framework that jointly updates a deep video model and a compositional language model to minimize the distance of their outputs in a video–text embedding space, as shown in Fig. 5.2.

Although these template-based language methods can generate complete sentences with grammatical correctness, generated descriptions are very inflexible. Meanwhile, the vocabulary is usually limited to a specific domain. When encountering a domain with a rich vocabulary, it becomes impractical and too expensive to design grammatical rules and sentence templates in advance.

5.2.2 *Sequence Learning Methods*

In contrast to the aforementioned template-based language method, sequence learning methods learn the probability distribution of each word in a sentence with a Markov chain model and thus are able to produce sentences with more flexible syntactical structures. Most sequence learning methods regard video captioning as

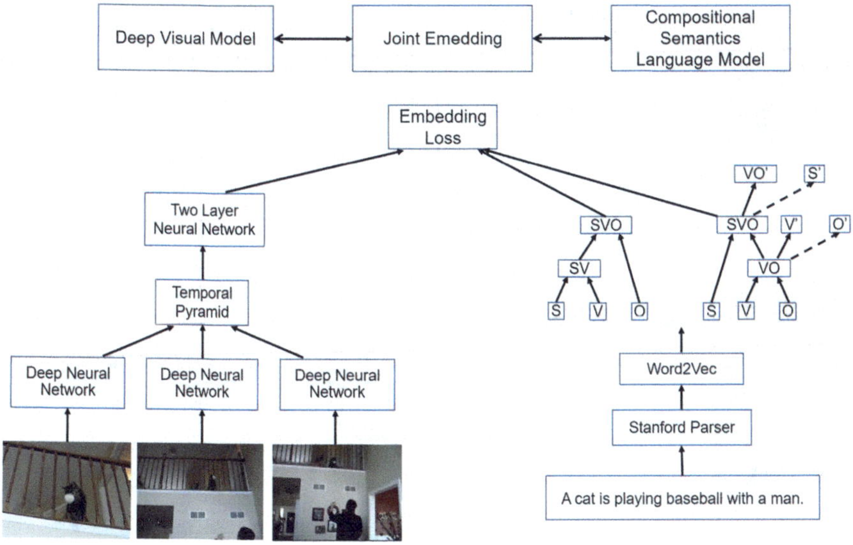

Fig. 5.2 Unified framework for joint video-language modeling. Reproduced from Xu et al. [245]

a sequence-to-sequence generation (or translation) task and design an encoder–decoder framework that encodes a video into a feature representation and then decodes it into a sentence in a word-by-word manner. In fact, the encoder–decoder framework was first introduced in machine translation and then applied in captioning.

Normally, a CNN is used as the encoder to extract features from video frames, and the decoder is often a sequence model such as RNN models. Current methods also resort to attention mechanisms to assign non-uniform weights to different frame features, learning better dependencies among frames.

Based on this architecture, we divide sequence learning methods into the following categories: multimodal fusion-based methods, semantic/syntax guidance-based methods, spatial/temporal structure-based methods, and others. Note that some approaches belong to more than one category. We therefore mention their corresponding modules in all related sections.

5.2.2.1 Multimodal Fusion-Based Methods

There are a number of sequence learning methods only using features from a single modality to generate captions. For instance, Donahue et al. [94] built upon traditional activity and video recognition approaches and leveraged a conditional random field to estimate subjects, predicates, and objects in the video. These representations were then concatenated and input to an LSTM for sentence generation.

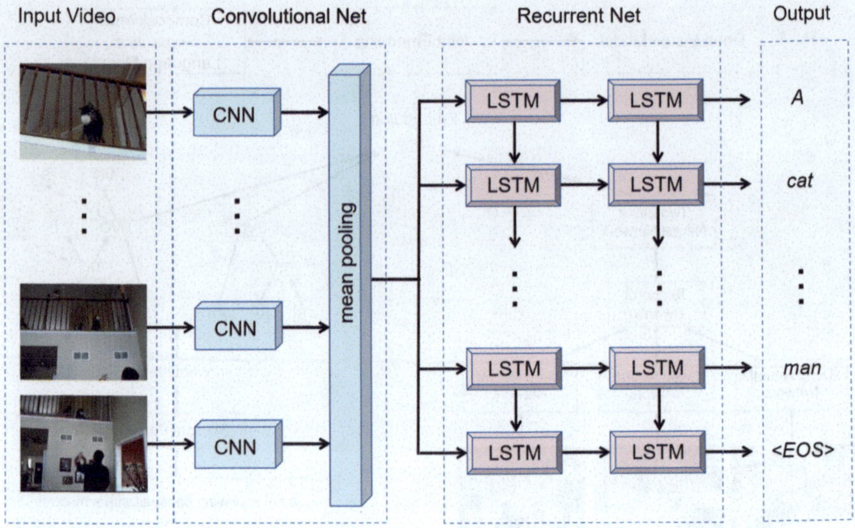

Fig. 5.3 LSTM-YT architecture. Reproduced from Venugopalan et al. [252]

As videos are multimodal, most video captioning approaches typically combine different modalities such as visual and acoustic clues for improved results, and a key component in these approaches lies in "how to fuse." From this perspective, we further classify multimodal fusion-based methods into several categories as follows.

Early Fusion Early fusion, or feature-level fusion, is a relatively simple fusion of the multiple input modalities. In [252], Venugopalan et al. proposed to extract features for each frame and then apply mean pooling across the entire video, before inputting the derived global video features to a two-layer LSTM, as shown in Fig. 5.3. The LSTM generates words sequentially based on the video features and the previous word (or the begin-of-sentence symbol) until it outputs the end-of-sentence symbol. The model was then extended to S2VT [253] by inputting both frames and optical flow images into the model. Specifically, the generation process can be divided into two stages: an encoding stage and a decoding stage. In the encoding stage, S2VT discards the mean pooling; instead, it inputs a frame feature vector successively at every time step to the first LSTM. Zero padding vectors are used as input for the second LSTM since it has not started to generate a sentence. In the decoding stage, the second LSTM receives the begin-of-sentence tag and begins to generate a sentence word by word. Again, zero padding vectors are input to the first LSTM after the visual frames are exhausted. Furthermore, Ramanishka et al. [254] combined visual, audio, and semantic clues and extended S2VT to a multimodal framework. In [255], Jin et al. proposed a multimodal fusion encoder to fuse almost all available modalities including image, video, aural, speech, and meta modalities.

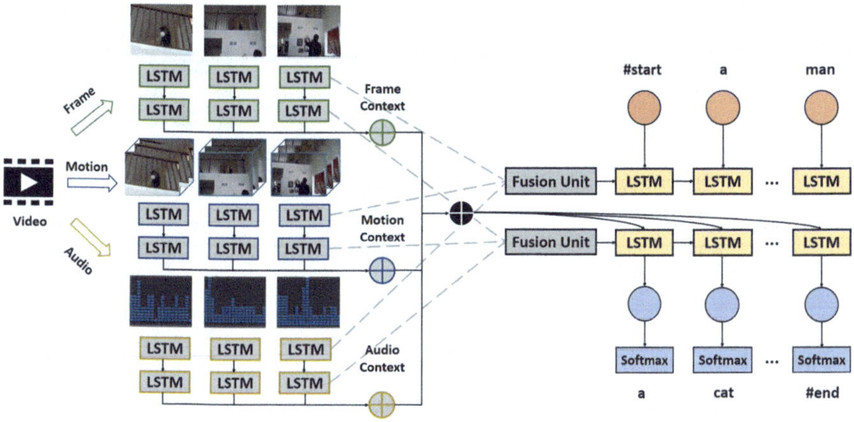

Fig. 5.4 MA-LSTM architecture. Reproduced from Xu et al. [256]

Attention Fusion Attention fusion, or soft attention, is to dynamically assign weights to each modality and then combine them. In [257], Hori et al. introduced a multimodal attention framework that can selectively make use of features from different modalities. The modality-dependent attention mechanism could attend to both specific modalities and specific times. Xu et al. [256] proposed MA-LSTM, which consists of three LSTMs encoding features of different modalities, a child-sum fusion unit dynamically combining complementary video representations from different modal streams, and a multi-level attention mechanism exploring additional clues, as illustrated in Fig. 5.4. Long et al. [258] presented a multi-faceted attention mechanism to first selectively pick up important information from temporal, motion, and semantic features. Then through a simple but effective feature scaling operation, it is easier to learn relationships between those representations. Wang et al. [259] introduced a hierarchically aligned cross-modal attention (HACA) module to adaptively learn the attentive representations and combine both global and local temporal information from multiple modalities.

Memory Mechanisms Several methods use memory mechanisms to organize multimodal features. Motivated by the fact that memory modeling can help solve long-term sequential problems, Wang et al. [260] proposed a multimodal memory model (M^3) that constructs a visual and textual shared memory and applies reading/writing operations on the memory structure to model long-term visual–textual dependencies, as illustrated in Fig. 5.5. The constantly updated content from the multimodal memory guides visual attention to learn effective video representations and enhance visual–textual alignment. A content-based memory addressing mechanism is further designed for better convergence during training. In [261], Pei et al. proposed a Memory-Attended Recurrent Network (MARN), in which a memory structure is designed to store the description information for each word in the vocabulary. The memory is used to build full relationships between a

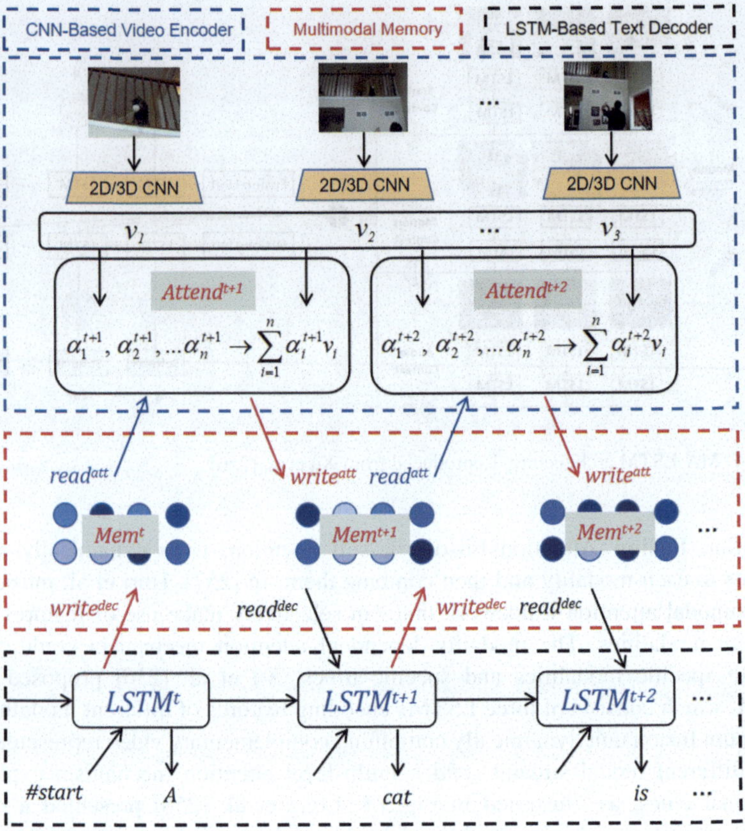

Fig. 5.5 M^3 architecture. Reproduced from Wang et al. [260]

word and its relevant video content, since a word may appear multiple times across similar video scenes. Specifically, the memory exists as a mapping structure, which maps each word to its corresponding description. In a word, the memory enables MARN to model consistencies between adjacent words explicitly and generate more precise captions.

Other Fusion Strategies There are also methods that attempt to design other novel multimodal fusion strategies. Zhang et al. [262] proposed a task-driven dynamic visual fusion (TDDF) model, which can effectively utilize task status to select appropriate fusion patterns. The TDDF unit fuses motion features, appearance features, and model status information and generates a dynamic visual input for each LSTM decoder. Liu et al. [263] presented a Sibling Convolutional Encoder (SibNet), which consists of two branches for processing visual content and semantic information, respectively. Concretely, the content branch captures detailed visual content with an autoencoder, while the semantic branch, on the other hand, encodes high-level semantic information via visual–semantic joint embedding.

5.2.2.2 Spatial/Temporal Structure-Based Methods

Extensive approaches also explore the spatial/temporal structures of videos, such as adopting spatial/temporal attention mechanisms or modeling spatial–temporal interactions among objects through object detection.

Spatial/Temporal Attention Given a sequence of features, temporal attention is a mechanism that dynamically assigns different weights to each feature such that the decoder can attentively focus on important features when generating sentence words. Similarly, different regions of the video frames also play different roles in the sentence generation process, e.g., objects often bring more important information than background. Thus, spatial attention methods focus on the importance of different regions. In [265], Yao et al. pointed out that describing videos requires properly modeling the dynamic temporal structure of the static video frames. Thus they proposed a novel architecture that captures both local and global temporal information of videos. First, a 3D CNN trained on video action recognition tasks is designed to extract representations with local fine-grained motion information. Second, a temporal attention mechanism is used to further exploit the global temporal structures. In order to fully explore the spatial and temporal clues in a video, Yan et al. [266] proposed the use of a spatial–temporal attention mechanism (STAT) using an encoder–decoder architecture. The encoder extracts global features, local features, and motion features with a 2D CNN, an R-CNN, and a 3D CNN, respectively. Then the decoder fuses these three kinds of features via a two-stage attention mechanism (e.g., spatial attention and temporal attention) to generate sentences with dynamic temporal structures. Motivated by the findings of human perception that salient motion can attract attention, Chen et al. [267] suggested to guide spatial attention by optical flow images and presented a novel attention framework called Motion Guided Spatial Attention (MGSA). Specifically, a CNN makes use of the input dense optical flow to compute a rough spatial attention map. To further relate the derived attention maps, a Gated Attention Recurrent Unit (GARU) is designed to adaptively incorporate attention information from previous maps when generating the next one. In addition, Chen et al. [264] designed a spatial information extraction and integration method (MGRMP) with no external object detectors needed. As illustrated in Fig. 5.6, a Recurrent Region Attention module is designed to compute multiple region-wise representations for each frame. Then conditioned on motion information, temporal connections among the regions are established. Finally, the Adjusted Temporal Graph Decoder generates captions with the help of the graph network to dynamically adjust temporal features during decoding.

Utilizing Detected Objects Several methods adopt pre-trained object detectors and design novel frameworks to better model the spatial–temporal relationships among detected objects. To learn effective spatial–temporal representations, Zhang et al. [269] proposed object-aware aggregation with a bidirectional temporal graph. To derive temporal clues for objects, the bidirectional temporal graph constructs a forward and a backward graph according to the similarities among object instances

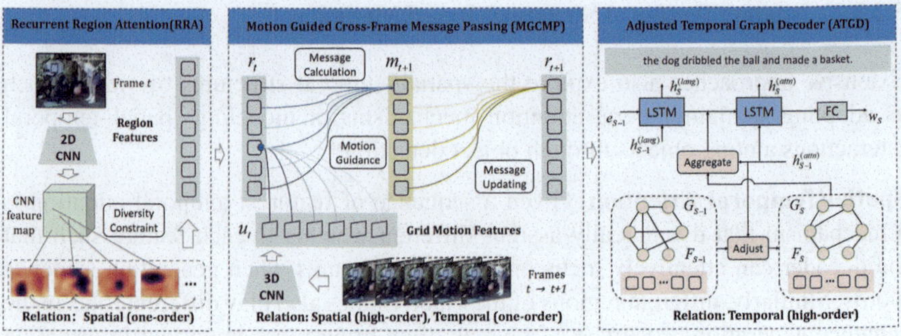

Fig. 5.6 MGRMP architecture. Reproduced from Chen et al. [264]

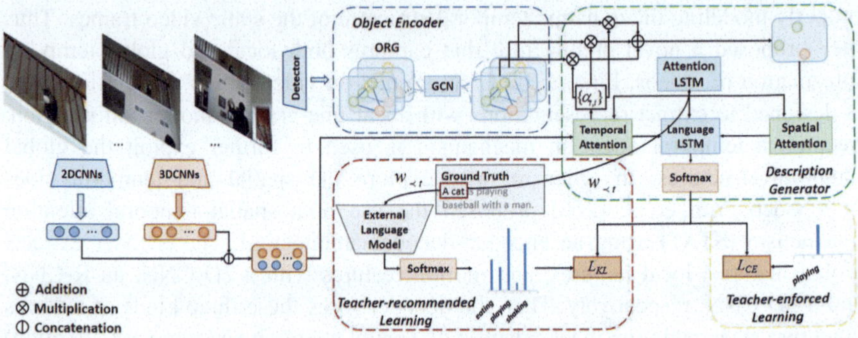

Fig. 5.7 ORG-TRL architecture. Reproduced from Zhang et al. [268]

or object regions. The aggregation module then learns spatial–temporal correlations for objects and global context. To deal with the problem of inadequate object interaction and insufficient training, Zhang et al. [268] introduced a model with an effective training strategy. The caption model, ORG-TRL, learns detailed interaction features across detected objects with an object relational graph (ORG) encoder, as shown in Fig. 5.7. A teacher-recommended learning (TRL) strategy further uses the external language model (ELM) to inject extra knowledge into the model, which helps to generate more potential words and alleviate the long-tailed problem of the caption training data. In [270], Pan et al. proposed a spatial–temporal graph model that explicitly captures object relationships in space and time. Specifically, spatial graphs and temporal graphs are first constructed to reconstruct the input video as a spatial–temporal graph. Then graph convolutions are performed to update the graph representation. A separate scene branch is also introduced to provide additional global context information. To alleviate the noise in feature learning because of the variable number of objects, an object-aware knowledge distillation mechanism is introduced, which uses local information as a soft regularization.

5.2.2.3 Semantic/Syntactic Guidance-Based Methods

The semantic and syntactic correctness of the generated sentences are also important. Some methods leverage additional semantic or syntactic information to guide description generation.

Semantic Guidance Several methods build upon semantic guidance to help generate semantically correct descriptions. Considering that the connections between sentence semantics and video content remain inadequately exploited, Pan et al. [272] proposed a unified framework—LSTM-E, which simultaneously bridges contextual relationships between video content and natural language. Specifically, a visual–semantic embedding space is jointly learned by minimizing a relevance loss and a coherence loss. Coherence measures how words in a sentence are contextually related to one another, while relevance measures the extent to which the meaning of a sentence corresponds to the video content. To take advantage of complementary semantic information between images and videos, Pan et al. [271] presented a video multiple instance learning model (LSTM-TSA) to learn high-level attributes from both images and videos, as shown in Fig. 5.8. To further transfer the learned semantic attributes into sentences, a transfer unit is designed that takes into account both static attributes and dynamic context. SibNet [263] introduced a visual–semantic joint embedding branch that maps videos and captions into a shared semantic space, as well as a bidirectional ranking loss to facilitate the learning of high-level semantic information. Sun et al. [273] proposed a Multimodal Semantic Attention Network (MSAN) framework that produces multimodal semantic attributes by adopting a multi-label classification approach with an auxiliary classification loss. Moreover,

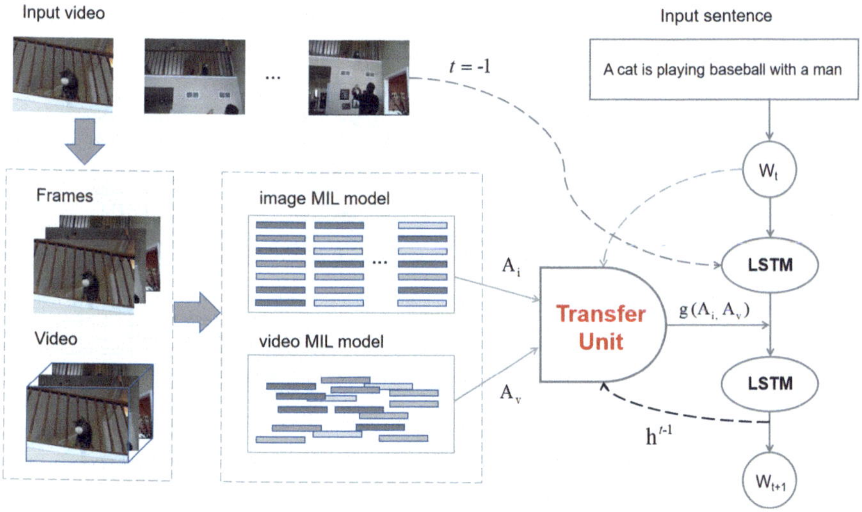

Fig. 5.8 LSTM-TSA architecture. Reproduced from Pan et al. [271]

the MSAN framework incorporates an attention-based fusion strategy that automatically prioritizes different semantic information as the caption is generated. In [274], Ryu et al. proposed a Semantic Grouping Network (SGN) that encodes videos into semantic groups containing filtered phrases and aligned frames. This approach enables the model to develop a contextual understanding when obtaining video representations, as each of the constructed semantic groups has a unique meaning and is utilized to predict the next word. To deal with the inadequacy of semantic representations and the limitations in expressiveness of generated captions, Perez-Martin et al. [275] presented an Attentive Visual–Semantic Specialized Network (AVSSN). This approach utilizes redesigned LSTM layers with adaptive attention gates, which can dynamically select relevant information for generating each word of the caption. In order to fully exploit linguistic semantics at different granularities, Ye et al. [276] proposed a Hierarchical Modular Network (HMN) to relate video representations with linguistic semantics from three levels, e.g., entity, predicate, and sentence, as shown in Fig. 5.9. Specifically, the entity module is capable of identifying key objects that contain significant semantic information, without the need to process all video objects. Then the predicate module attentively summarizes the object features to obtain the context-related action features. Finally, the sentence module computes the global video features based on initial video context features, object features, and action features.

Syntactic Guidance Similarly, some methods focus on syntactic information to guide the description generation and adjust the grammar of the resulting sentence. One typical example is the part-of-speech (POS) tagging, which assigns each word in the input text to a part-of-speech category. For instance, Wang et al. [277] proposed a POS sentence generator that predicts the overall syntactic POS

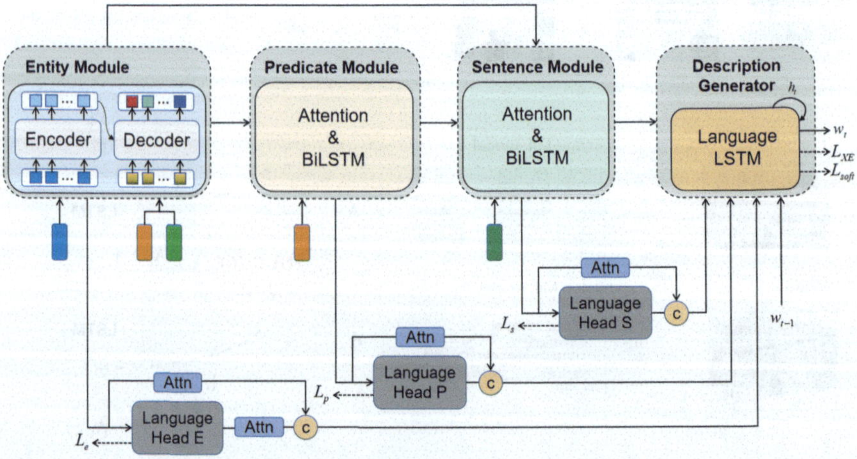

Fig. 5.9 HMN architecture. Reproduced from Ye et al. [276]

information of the generated sentence. The information is then utilized to guide the decoder in generating more precise descriptions concerning both syntax and semantics. Hou et al. [278] took into account both syntax representation learning and visual cue translation to generate accurate descriptions. Specifically, the syntax representation learning module is responsible for processing extracted video features and generating POS tagging sequences. The visual cue translation module, on the other hand, translates the visual cues into lexical words based on the POS tagging sequences. Considering that the predicate in a description sentence highly relates to the static scene and the dynamic motions in a video, Zheng et al. [279] proposed a Syntax-Aware Action Targeting (SAAT) module. This module calculates scene representations through self-attention of the bounding boxes obtained by an object detector and then predicts actions to guide the action-guided captioner, as illustrated in Fig. 5.10. Instead of predicting the POS tag for each word, their approach focuses solely on the most correlated visual components to the target action. Deng et al. [280] proposed a Syntax-guided Hierarchical Attention Network (SHAN), which utilizes visual features and sentence-context features to help generate visual words and non-visual words, respectively. To further enable the adaptive selection of word-related features, a hierarchical attention mechanism guided by syntax is introduced, which consists of content attention and cross-modal syntax attention. Content attention focuses on semantically related time intervals for each feature, while cross-modal syntax attention models the significance of different features with respect to the target word. Perez-Martin et al. [281] presented a Visual–Semantic–Syntactic Aligned Network (SemSynAN), which learns to integrate visual, semantic, and syntactic representations of the input video. Specifically, the encoder consists of a global representation extractor, a concept detector, and a syntax detector. The concept detector learns a semantic representation of the video by generating possible keywords, while the syntax detector learns a syntactic representation of descriptions by solving a POS template retrieval problem. The decoder has three specialized LSTM layers and a fusion gate, among which the three LSTM layers operate on different combinations of visual, semantic, and syntactic information, and the fusion gate further leverages those information dynamically.

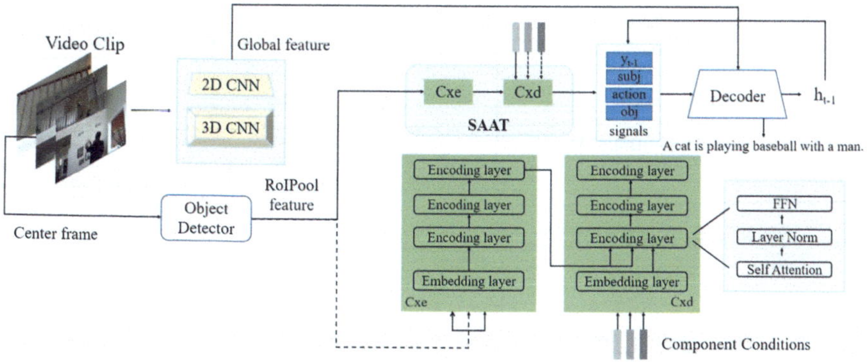

Fig. 5.10 SAAT involved architecture. Reproduced from Zheng et al. [279]

5.2.2.4 Other Methods

Apart from the categories mentioned previously, researchers also have explored other types of formulations for video captioning under different settings.

Reinforcement Learning Methods It was discovered that Reinforcement Learning (RL) is a reliable method for generating precise sentences. Pasunuru et al. [283] presented a mixed-loss policy gradient approach based on reinforcement learning, which directly optimizes sentence-level task-based metrics instead of word-level cross-entropy loss. Besides, an entailment corrected reward (CIDEnt) is used to ensure the generated caption is logically matched with the ground truth. To enhance the fine-grained generation of video descriptions with a broader range of activities, Wang et al. [282] introduced a hierarchical framework (HRL), where a manager produces semantically meaningful goals, a worker receives the goals and generates the corresponding words, and the internal critic helps the manager to update the goals according to whether the worker has finished them, as shown in Fig. 5.11. Considering that uniformly sampling from input videos can result in redundant visual information and unnecessary computation cost, Chen et al. [284] designed a plug-and-play reinforcement-learning-based module, i.e. PickNet, which can select informative frames from input videos. The reward for each frame picking action consists of two components: a language reward and a visual diversity reward. Li et al. [285] proposed a novel multi-task reinforcement learning approach to train an end-to-end video captioning model, which is in essence a deepened version of the S2VT model. The model is trained progressively in three steps, including the standard S2VT training, reinforcement training, and multi-task training. The reinforcement training is introduced to enhance generalization performance of the

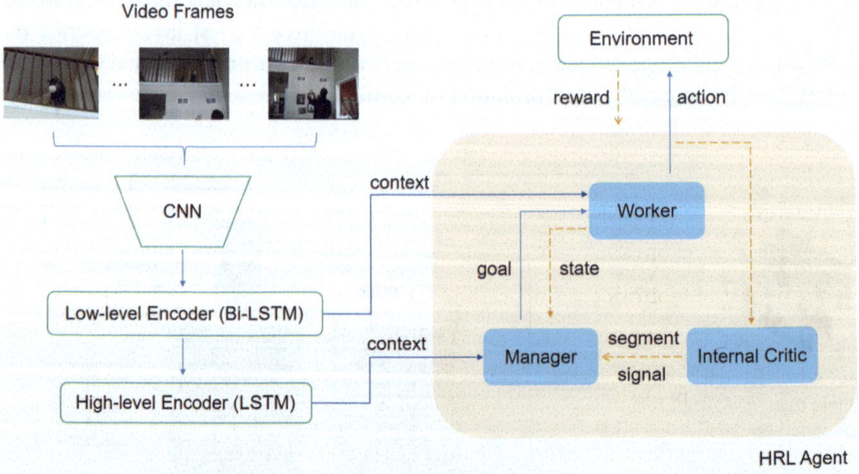

Fig. 5.11 HRL architecture. Reproduced from Wang et al. [282]

model, while the multi-task training is designed to mine effective attributes from the videos and regulate the search space.

Non-autoregressive Methods As discussed previously, mainstream video captioning methods are autoregressive, which means future values are predicted based on past values. Due to the sequential manner of the decoding paradigm, autoregressive methods suffer from a slow inference speed. Moreover, similar descriptions are probably generated because of insufficient training and inadequate decoding. Recently, several non-autoregressive methods have been proposed to mitigate the abovementioned difficulties. Yang et al. [286] proposed a Non-autoregressive Coarse-to-Fine (NACF) model as shown in Fig. 5.12. The NACF model adopts a coarse-to-fine captioning procedure, which initially generates visual words in parallel to form a coarse-grained sentence template. Subsequently, specific decoding algorithms are applied to generate a fine-grained description. Liu et al. [287] proposed an Object-Oriented Non-autoregressive method (O2NA) to enable video captioning in a controllable way by transmitting strong control signals conditioned on selected objects. First, an object predictor detects all objects in the input video. Then an object generator locates focused objects for the caption generation process. These focused objects could also be the pre-defined concerned objects given by the user for controllability. Based on the focused objects, the caption generator produces related words to obtain a draft caption. An iterative approach is further introduced to improve the draft caption. Zhong et al. [288] proposed a non-autoregressive network (DSA-LST) that reconstructs linguistic skeleton-level tags to establish semantic dependencies among visual word sequences. Specifically, the network aligns the visual words group and the video frame to identify similar videos and their corresponding captions, which is supervised by skeleton tags.

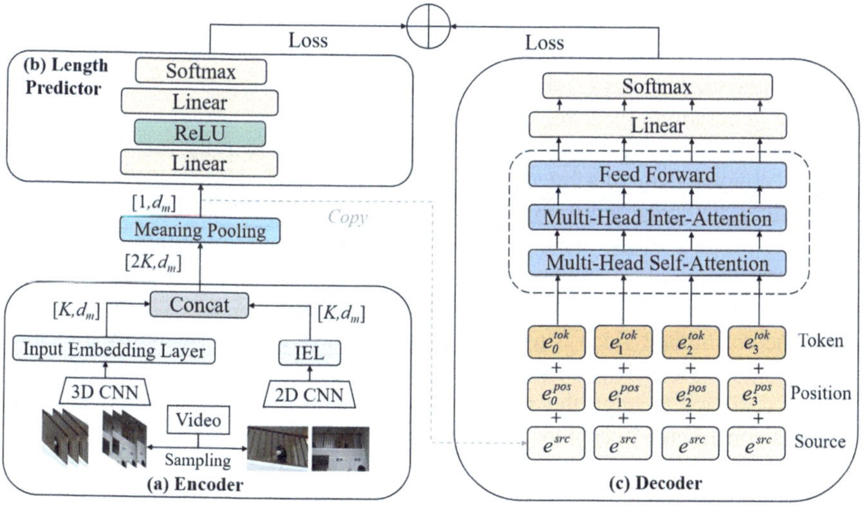

Fig. 5.12 NACF architecture. Reproduced from Yang et al. [286]

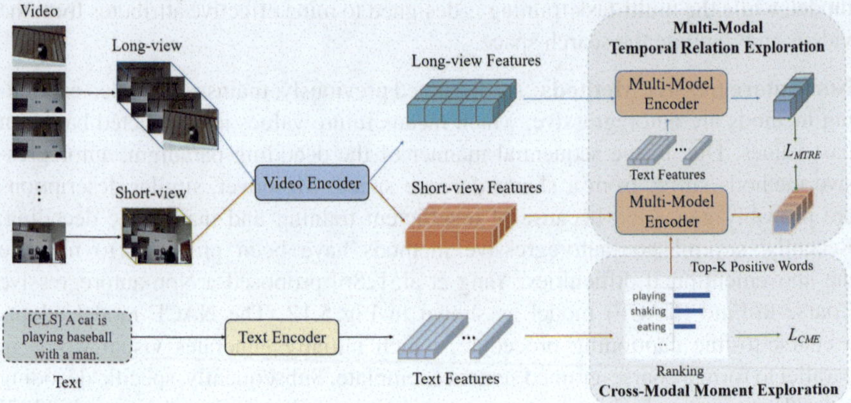

Fig. 5.13 HiTeA architecture. Reproduced from Ye et al. [289]

Pre-training Methods Since video-language pre-training has advanced the performance of various downstream video-language tasks recently, researchers also attempted to design novel video-language pre-training frameworks for video captioning. The CLIP4Caption framework proposed by Tang et al. [290] builds upon the success of the CLIP4clip [291] model, which is extended from CLIP [241] and shows promising performance in video–text retrieval. The CLIP4Caption framework pre-trains a video–text matching network using CLIP4Clip and learns highly correlated video and text features during the pre-training process. In addition, a novel metric-based voting strategy is designed to ensemble multiple captioning results for a more powerful caption result. In Seo et al. [292], a pre-training framework, Multimodal Video Generative Pre-training (MV-GPT), is introduced, for multimodal video captioning. A multimodal video encoder and a sentence decoder are jointly trained in MV-GPT without the need for manually annotated captions. This framework leverages unlabeled video data by training with a bidirectional utterance generation objective, aiming to generate a future/current utterance given the current video context and current/future utterances. To fully make use of the temporal characteristic of a video, Ye et al. [289] proposed a Hierarchical Temporal-Aware video-language pre-training framework (HiTeA) with two novel temporal-aware pre-training tasks, named as cross-modal moment exploration (CME) and multimodal temporal relation exploration (MTRE). As illustrated in Fig. 5.13, two different temporal views for the input video are generated by a video encoder first, where the untrimmed video is regarded as a long view and the randomly truncated video segment as a short view. To further explore the information of these two views, CME is designed to align the short-view video with fine-grained semantics, and MTRE is constructed to capture temporal relations between moments and events.

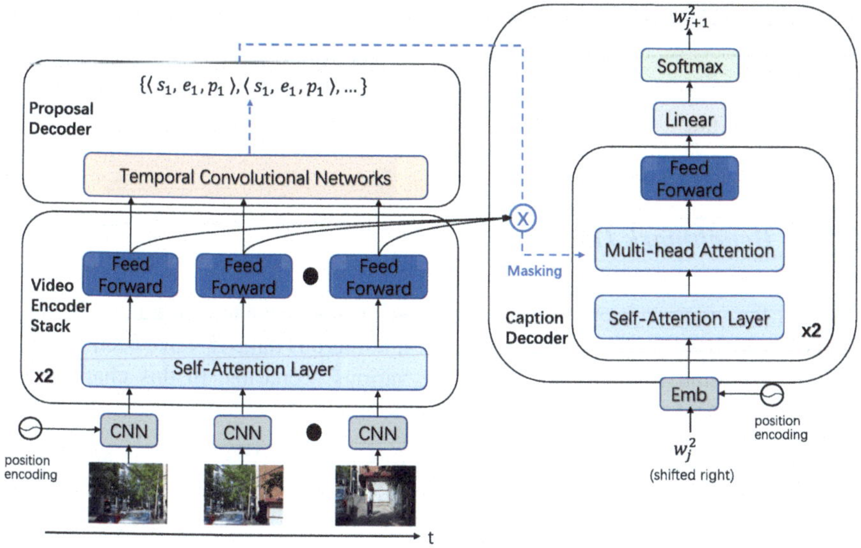

Fig. 5.14 End-to-end dense video captioning with masked transformer architecture. Reproduced from Zhou et al. [293]

Dense Video Captioning Dense video captioning is a task introduced by Krishna et al. [29] that aims to describe multiple events in a long video. This task can be decomposed into two sub-problems: first localizing the events and then generating descriptions for each localized segment. Most of the previous methods treated these two sub-problems separately, which reduces the impact of the language descriptions for the event proposal. Zhou et al. [293] proposed an end-to-end transformer-based model and included a differentiable masking scheme that guarantees consistent event proposals and language descriptions, as shown in Fig. 5.14. The proposal decoder produces video event proposals with explicit anchors, and the caption decoder generates words based on the event proposals and visual features. Li et al. [294] proposed an integrated end-to-end architecture that combines the generation of temporal event proposals and sentences. The temporal event proposals are predicted and refined with event/background classification and temporal coordinate regression. The descriptiveness regression is further designed to compute descriptive complexity for each event. Video paragraph captioning is another attempt similar to dense video captioning, which intends to generate a paragraph of language descriptions (usually more than one sentence) for an untrimmed video. However, it is relatively different from dense video captioning since video paragraph captioning ignores the localization of events but focuses on the coherence of the description.

5.3 Datasets and Measures

In this section, we review standard benchmark datasets and evaluation metrics for video captioning.

5.3.1 Benchmark Datasets

Datasets for video captioning usually consist of human-annotated pairs of videos and corresponding sentences. In recent times, numerous datasets have been created, each with unique characteristics to cover various scenarios. In this chapter, we highlight the most commonly employed datasets.

The Microsoft Research Video Description Corpus (MSVD) Chen and Dolan [27] contains 2089 YouTube video clips and 85K English descriptions, which were collected by having annotators watch a short video segment and describe the main event in one sentence. Noted that the original corpus also provides descriptions in other languages, but only the English annotations are commonly used for video captioning tasks. Besides, some of the videos were removed from YouTube before archived, leaving 1970 video clips available on the official website. The standard split of MSVD (also called Youtube2text) has 1300 videos for training and validation and 670 for testing, as in [250].

YouCook Dataset Das et al. [295] is composed of 88 cooking videos collected from YouTube. These videos were roughly evenly divided into six cooking categories, such as baking and grilling. The videos were filmed in various kitchen settings, from a third-person perspective with dynamic camera movements. Each annotator was required to write a minimum of three sentences for each video, and each sentence has at least 15 words. As a result, a total of 2668 sentences were generated for all of the videos.

YouCook2 Dataset Zhou et al. [30] contains 2000 cooking videos collected from YouTube. The videos were shot from a third-person viewpoint, covering 89 cooking recipes. Each video was annotated with imperative English sentences and their corresponding temporal boundaries. On average, the duration and the number of segments per video are 5.27 minutes and 7.7, respectively.

TACoS Multi-level Corpus (TACoS-ML) Rohrbach et al. [249] was created from the MPII Cooking Activities dataset 2.0 [296]. TACoS-ML contains a total of 185 lengthy videos, which regularly capture various activities that occur in a cooking environment. Each annotator wrote sentences along time intervals of a long video and matched each interval with a single concise sentence. In total, 14,105 distinct intervals and 52,593 sentences were collected.

Montreal Video Annotation Dataset (M-VAD) Torabi et al. [297] comprises 48,986 video clips taken from 92 DVDs, with an average length of 6.2 seconds.

Most of the clips have only one annotation sentence. The annotations were obtained by narration isolation and video alignment using Descriptive Video Service (DVS). Due to the considerable diversity of visual and textual content in movies and the rareness of annotation sentences for each movie clip, video captioning tasks on the M-VAD dataset are known to be highly challenging.

MPII Movie Description Corpus (MPII-MD) Rohrbach et al. [298] consists of over 68K video clips extracted from 94 HD movies. Similar to M-VAD, most of the clips have only one annotation sentence. The annotations were obtained by transcribing audio descriptions and then temporally aligning them with the full movies.

MSR Video to Text (MSR-VTT-10K) Xu et al. [28] contains 10,000 web video clips and a total of 200K clip–sentence pairs. The video clips cover a variety of scenarios and categories, resulting in a diverse and challenging dataset. Each clip was annotated with roughly 20 natural sentences. MSR-VTT-10K also provides the category information for each clip, including music, sports, news and TV shows, etc. The standard split of MSR-VTT-10K has 6513 clips for training, 2990 for validation, and 497 for testing.

Large-Scale Movie Description Challenge (LSMDC) Rohrbach et al. [299] was developed using the M-VAD and MPII-MD datasets, with the inclusion of additional movies. LSMDC 2016 contains 101,079 training clips, 7408 validation clips, and 19,031 testing clips. The most recent version of the dataset is LSMDC 2021.

T-GIF Li et al. [300] was collected for the purpose of advancing the development of animated GIF understanding, which contains 100K animated GIFs and 120K natural language descriptions. The standard split of TGIF has 80,000 animated GIFs for training, 10,708 for validation, and 11,360 for testing. There is one sentence per animated GIF for the training and validation splits and three sentences for the test split.

Charades Sigurdsson et al. [109] consists of 9848 videos of daily indoor activities, with an average length of 30 seconds. Charades provides a total of 27,847 video descriptions, 66,500 temporal action annotations, and 41,104 object labels. Thus it can also be used for action classification and localization tasks. The standard split of Charades has 7985 videos for training and 1863 for testing.

ActivityNet Captions Krishna et al. [29] is introduced for dense-captioning events, which contains 20K videos extracted from ActivityNet [25] with 100K descriptions in total. Most of the videos have more than 3 annotation sentences with their corresponding start/end time. The standard split of ActivityNet Captions has 10,044 videos for training, 4926 for validation, and 5044 for testing.

Vatex Wang et al. [301] is a recently introduced multilingual video description dataset. It contains 41,269 videos and 825,380 captions in both English and Chinese. Among these captions, there are more than 206,000 English–Chinese description

Table 5.1 Comparison of video captioning benchmark datasets

Dataset	Context	Sentence source	#Video	#Clip	#Sentence	#Word
MSVD	Multi-category	AMT workers	–	1970	70,028	607,339
ActyNet-Cap	Multi-category	Humans	20,000	100,000	100,000	1,350,000
Vatex	Multi-category	AMT workers	41,269	–	825,380	–
TGIF	Multi-category	Crowdsourcing	102,068	–	125,782	–
YouCook	Cooking	AMT workers	88	–	2668	42,457
YouCook2	Cooking	Humans	2000	–	13,829	121,695
TACoS-ML	Cooking	AMT workers	273	14,105	52,593	–
M-VAD	Movie	DVS	92	48,986	55,905	519,933
MPII-MD	Movie	Script+DVS	94	68,337	68,375	653,467
LSMDC 2016	Movie	Script+DVS	200	128,085	128,118	1,157,155
Charades	Indoor	AMT workers	9848	–	27,847	–
MSR-VTT-10K	20 categories	AMT workers	7180	10,000	200,000	1,856,523

pairs. The standard split of Vatex has 25,991 videos for training, 3000 for validation, and 12,278 for testing.

Table 5.1 summarizes key statistics and comparisons of popular datasets for video captioning. Figure 5.15 illustrates a few examples from some of the datasets.

5.3.2 *Evaluation Metrics*

Video captioning is generally evaluated by comparing the performance of the generated results (candidate sentences) against human descriptions (reference sentences). The widely used measures include BLEU, METEOR, ROUGE-L, and CIDEr, as described in the following section. Such automatic evaluation metrics are based on N-gram (e.g., a sequence of N words or tokens) so that they measure similarities of sentences by lexical or grammatical matching, while ignoring semantics. Subsequently, the following section also touches upon SPICE, a common evaluation metric in image captioning that takes into account semantic information, despite its relatively low popularity in video captioning.

Bilingual Evaluation Understudy (BLEU) Papineni et al. [302] analyzes co-occurrences of N-grams (up to 4-gram) between the candidate and the reference sentences. First, BLEU modifies N-gram precision by dividing the number of the clipped N-gram with the number of candidate N-grams in the test corpus. To take different sequence lengths of N-grams into consideration, BLEU uses the weighted geometric mean. Additionally, a penalty factor is adopted to filter out excessively short candidates. The overall formula to calculate BLEU is

Fig. 5.15 Examples from (**a**) MSVD, (**b**) M-VAD, (**c**) MPII-MD, and (**d**) MSR-VTT-10K datasets

$$\text{BLEU} = BP * exp\left(\sum_{n=1}^{N} \omega_n log p_n\right),$$

where BP is the penalty factor, N is the maximum N-gram length, ω_n are the weights assigned to different N-gram lengths, and p_n is the modified precision score.

Metric for Evaluation of Translation with Explicit ORdering (METEOR) Banerjee and Lavie [303] was designed to explicitly address and improve the weaknesses in BLEU. In particular, METEOR computes F_{mean} by combining the precision (P) and recall (R) for unigram matches via a harmonic mean that places most of the weight on recall. The resulting formula used is

$$F_{mean} = \frac{10PR}{R + 9P}.$$

Then, METEOR introduces a *Penalty* to consider longer matches, which is computed through the following formula:

$$Penalty = 0.5 * \left(\frac{\#chunks}{\#unigrams\_matched} \right)^3,$$

where *chunks* is a group of mapped unigrams and "#" refers to quantity. The resulting METEOR score is computed as follows:

$$\text{METEOR} = F_{mean} * (1 - Penalty).$$

Recall-Oriented Understudy for Gisting Evaluation (ROUGE) Lin [304] is a metric evaluation package containing four different measures: ROUGE-N, ROUGE-L, ROUGE-W, and ROUGE-S. And ROUGE-L is often used for evaluation of automatic video captioning, which estimates the similarity of two sentences based on the longest common subsequence. The first step is to calculate the Recall (R_{lcs}) and Precision (P_{lcs}) regarding to LCS as follows:

$$R_{lcs} = \frac{LCS(s_r, s_p)}{len(s_r)},$$

$$P_{lcs} = \frac{LCS(s_r, s_p)}{len(s_p)}.$$

Specifically, R_{lcs} is the ratio of LCS of a reference sentence (s_r) and a predicted sentence (s_p) to the length of the reference sentence. P_{lcs} is similar to R_{lcs} except that it is divided by the length of the predicted sentence. The resulting ROUGE-L score is computed as follows:

$$\text{ROUGE} - \text{L} = \frac{(1 + \beta^2) R_{lcs} P_{lcs}}{R_{lcs} + \beta^2 P_{lcs}},$$

where β is a constant number commonly set to be large, which means ROUGE-L almost only considers R_{lcs}.

Consensus-based Image Description Evaluation (CIDEr) Vedantam et al. [305] measures the similarity between a generated sentence and the ground-truth sentences. Specifically, the goal of CIDEr is to evaluate how much a candidate sequence c_i agrees to the consensus of a collection of reference descriptions

$S_i = \{s_{i1}, \ldots, s_{im}\}$. First, all words in the sentences, including both candidate and reference descriptions, are transformed into their root forms. Each sentence is described by a set of N-grams ω_k present in it. Subsequently, the TF-IDF (Term Frequency Inverse Document Frequency) weighting $g_k(s_{ij})$ for each w_k is computed. Refer to [306] for details of TF-IDF. In addition, the CIDEr$_N$ score for N-grams of size N is computed by taking the mean cosine similarity between the candidate description and the reference descriptions. It can be expressed as follows:

$$\text{CIDEr}_N(c_i, S_i) = \frac{1}{m} \sum_j \frac{\mathbf{g}^N(c_i) \times \mathbf{g}^N(s_{ij})}{\|\mathbf{g}^N(c_i)\| \|\mathbf{g}^N(s_{ij})\|},$$

where $\mathbf{g}^N(c_i)$ is a vector formed by $g_k(c_i)$ in correspondence with all the N-grams of size N and $\|\cdot\|$ is the magnitude operator. At the end, the CIDEr score combines the values from N-grams of different lengths as follows:

$$\text{CIDEr}(c_i, S_i) = \sum_{N=1}^{N_{max}} \omega_N \text{CIDEr}_N(c_i, S_i).$$

Several modifications have been implemented to enhance the robustness of the basic CIDEr metric against "gaming," resulting in the CIDEr-D. Refer to details of CIDEr-D in [305].

Semantic Propositional Image Caption Evaluation (SPICE) Anderson et al. [307] is a metric that estimates the caption quality based on scene graph, which is a general structure denoting the logical and spatial representation of a graphical scene. First, SPICE applies semantic parsing to convert both candidate and reference captions into semantic representations, which explicitly capture the objects, attributes, and relations in captions. Specifically, let c and $S = \{s_1, \ldots, s_m\}$ denote a candidate caption and a set of reference captions, respectively, and SPICE constructs their scene graphs $G(c)$ and $G(S)$, where $G(S)$ is formed as the union of scene graphs $G(s_i)$ for each $s_i \in S$. The semantic parsing function G can be formulated as follows:

$$G(c) = \langle O(c), E(c), K(c) \rangle,$$

where $O(c)$ is the set of objects mentioned in c, $E(c)$ denotes the collections of hyper-edges representing object relations, and $K(c)$ denotes the set of attributes of objects. Second, the semantic propositions in the scene graph are formed as logical tuples to compute the similarity between candidate and reference scene graphs. Each tuple contains a combination of objects, attributes, and relations. The tuple generating function T is defined as follows:

$$T(G(c)) \triangleq O(c) \cup E(c) \cup K(c).$$

Finally, precision P, recall R, and $SPICE$ are calculated as follows:

$$P(c, S) = \frac{|T(G(c)) \otimes T(G(S))|}{|T(G(c))|},$$

$$R(c, S) = \frac{|T(G(c)) \otimes T(G(S))|}{|T(G(S))|},$$

$$SPICE(c, S) = F_1(c, S) = \frac{2 \cdot P(c, S) \cdot R(c, S)}{P(c, S) + R(c, S)},$$

where the binary matching operator $\otimes$ is a function that identifies and returns matching tuples in two scene graphs.

5.4 Summary

In this chapter we present a review and categorization of the deep learning methods for video captioning developed in recent years. In addition, most commonly used benchmark datasets and evaluation metrics are summarized.

Chapter 6
Unsupervised Feature Learning for Video Understanding

6.1 Learning with Unlabeled Videos

Researchers have been leveraging unsupervised visual understanding with videos throughout the evolution of computer vision. We seek to present readers with a comprehensive understanding of deep-learning-based SSVL techniques, especially ones with state-of-the-art DNNs for video understanding (e.g., ResNet [12], ViT [14] and etc). To do that, it is necessary to introduce the infancy progress in unsupervised spatial–temporal learning and preliminaries, including an overview of SSVL category and downstream tasks before the formal discussion.

Early attempts of extracting video information in an unsupervised manner can be traced back to the phase of statistical machine learning when those methods aim to overcome the inferiority of hand-crafted features. Both primal and present-day approaches were motivated by spatial and temporal consistency. For instance, the earliest works went beyond traditional hand-crafted features via applying linear independent component analysis (ICA) [308] on patches from frames of unlabeled videos [309, 310]. Later in 2011, with independent subspace analysis (ISA) [311], the refined model of ICA, Le et al. [312] exploited stacked convolutions to capture high-level motion information. Another stream of works resorted to slow feature analysis (SFA) [313, 314], which is based on the hypothesis that features of neighboring frames change slowly and smoothly, e.g., Zou et al. [315] encoded invariance in consecutive frames from a fixed location of videos by SFA. On the other hand, Taylor et al. employed Restricted Boltzmann Machines (RBMs) [316] as early generative models [317]. Echoing up-to-date video prediction pretext tasks, as early as in 2014, Ranzatol et al. [318] proposed to train Recurrent Neural Networks (RNNs) [319] by predicting or extrapolating future frames. Motivated by word2vec [320], a sequence modeling technique in NLP, Ramanathan et al. [321] introduced a strategy to model video frame embeddings via capturing the semantics

Z. Wu, Y.-G. Jiang, *Deep Learning for Video Understanding*, Wireless Networks,
https://doi.org/10.1007/978-3-031-57679-9_6

in context, i.e., averaging the features within a window involving both past and future frames.

Later, the advent of AlexNet [9], GoogLeNet [11], and VGGNet [10] led to an overwhelming boost to the performance of visual understanding. Thereafter, researchers have been devoted to exploiting unlabeled videos for sophisticated visual representation learning. Pretext tasks based on capturing spatial–temporal consistency between video frames were designed. In particular, unsupervised video tracking was exploited to aid visual understanding of objects [322–324] for 2D object detection. Wang and Gupta [322] utilized a Siamese triplet network to close the distance between representations of tracked frames from videos. Similarly, Pathak et al. [323] trained a CNN on motion segments generated without supervision.

Downstream Tasks To validate the efficacy of self-supervised video learning, models are evaluated by transferring them to a variety of downstream tasks. The core task of video understanding refers to action recognition (see Chap. 3), which has also been known as video classification. Efforts have also been made to tackle tasks such as action localization (see Chap. 4). Besides, a small portion of works exploit speech (text) information in videos in training and hence can also be applied to tasks with video–text collaboration, e.g., text-to-video retrieval, video captioning (see Chap. 5), etc.

6.2 Pretext Tasks Based on Self-prediction

Self-prediction pretext tasks in videos have advanced the development of SSVL by learning representations via solving pre-designed tasks aimed at predicting variation, information, or characteristics within the video. Supervisory signals of self-prediction tasks come from different sources, i.e., intrinsic characteristics of the RGB domain and cross-modal correspondences. In particular, RGB-signal-based methods either capture the spatial and temporal transformations or exploit self-targeted generation.

6.2.1 Predicting Spatial–Temporal Transformations

Researchers have made numerous successful attempts to leverage inherent spatial and temporal context relations in massive videos for video understanding. Generally, the surrogate signals are extracted from spatial and temporal transformations on the video, respectively, or jointly.

Spatial Transformation Jing et al. [325] managed to conduct SSVL with spatial information only by adapting imaged-based **rotation transformation** for video.

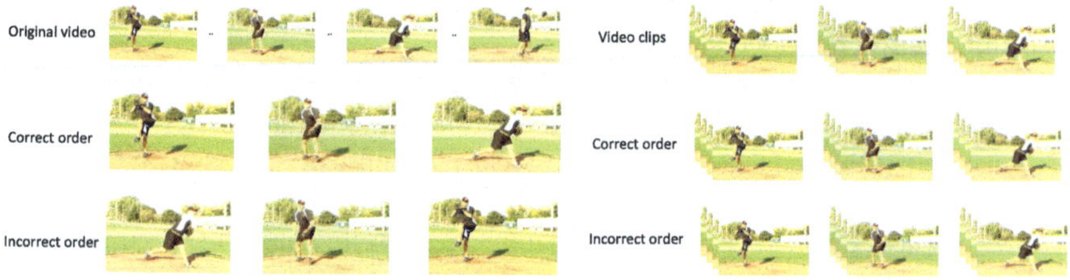

Fig. 6.1 Shuffle the temporal ordering of input video

The proposed 3DRotNet exploits a pretext task to predict the rotations applied to videos. Following [326], a classification task is formulated in practical because all the videos are rotated with four different degrees (0°, 90°, 180°, and 270°) during training. By recognizing the rotations of videos, the 3DRotNet can actually capture both spatial and temporal information within videos.

Temporal Transformation To utilize the characteristic temporal information videos contain, versatile pretext tasks exploring temporal transformations are designed. In general, we divide them into three categories, i.e., temporal ordering, temporal skipping, and temporal coherence.

(A) Temporal ordering of frames and clips is one of the common supervisory signals. Through the verification and validation of the shuffled input and the prediction of the correct order, temporal ordering provides supervisory signals to video models. Figure 6.1 shows how to shuffle the input frames/clips and define the positive/negative samples:

(i) **Frame-based tasks with 2D CNNs.** Shuffle-and-Learn [327] was proposed to shuffle a sequence of frames from a video and verify its temporal validation. The frame order verification problem is usually implemented as a binary classification task. O3N [328] suggested an odd-one-out task in which one disordered clip out of three needs to be identified. The odd-one-out learning framework can be applied to a variety of questions in different domains. Since multi-class classification problem is harder and more complicated than binary verification, to further explore the use of spatial–temporal signals in the video, OPN [329] adopted a frame order prediction method as a proxy task to learn richer representations. In order to better capture the concept of ordering, OPN performed pairwise feature extraction on extracted features from frames in a tuple, which means the concatenated pairwise features are finally used to predict the order. Considering that the direction of some actions may be ambiguous, they adopted the forward–backward grouping strategy when computing permutations, which reduces the number of possible permutations to half. However, early attempts including the approaches mentioned above are limited by the 2D CNN architectures that are focusing on learning spatial features of every single frame, rendering the incapability of capturing temporal information for video understanding.

(ii) Clip-based tasks with 3D CNNs. Since 3D CNN-based methods are capable of learning both spatial and temporal information simultaneously, many pretext tasks are proposed to take advantage of the capability of 3D CNNs. While OPN only uses frames and 2D-CNNs and they group the bidirectional orders as the same class, Clip Order [330] instead directly uses clips and 3D-CNNs. Besides, non-overlapping clips are uniformly sampled and shuffled with the number of sampled clips restricted between 2 and 5 in case the prediction task is too hard to solve. They utilize clip features extracted by a 3D-CNN to predict the order of clips and also find that pairwise concatenated clip features perform better and the clip order prediction method works well when using C3D [61], R3D, R(2+1)D [36] networks as the feature extractors.

(iii) Playing direction. Reversing the video can be seen as a special case of the temporal-ordering-based method. Specifically, AoT [331] trained a CNN to tell the playing direction of the given video using optical flow as input instead of RGB appearance of the video.

(B) Temporal Skipping Video frames can be sampled by skipping a number of frames according to a wide variety of schemes. Changing playback speed by using a uniform frame-skipping strategy is a special case of frame skipping, which relies on the video speed as another type of supervisory signal. SpeedNet [332] was trained to tell whether or not the video is sped up with the target of learning to determine whether the motion of an object is natural, which is also the reason why they did not use a regression objective. By solving the simple binary classification task, SpeedNet can be further used to predict arbitrary speeds of videos and the learned space–time representations can be used for video understanding tasks.

Jenni et al. [333] proposed two novel temporal transformations called "periodic" motions and "warping." In periodic motion transformations, a midpoint s will be randomly selected to determine where to switch the playback direction with the selected speed instead of reversing the whole playback. For warping transformations, they pick a set of frames with a random number of skipped frames. As illustrated in Fig. 6.2, the authors combined the proposed periodic motions and temporal warp changes with speed changes as well as random temporal permutations to train a 3D CNN to predict both the transformation type and the speed type with raw RGB frames as input. The results showed that motion statistics can be implicitly learned and the pre-trained model transfers well to the action recognition task.

(C) Temporal Coherence Based on the assumption that the adjacent frames in the same video tend to contain semantically similar information, many works adopted CNNs to learn slow temporally coherent features. Since representations should be temporally smooth, Goroshin et al. [334] utilized the slowness prior on autoencoders. The slowness prior can be integrated into the model by minimizing the norm of the feature difference between adjacent frames. They put forward a reconstruction term to penalize the reconstruction error instead of the contrastive

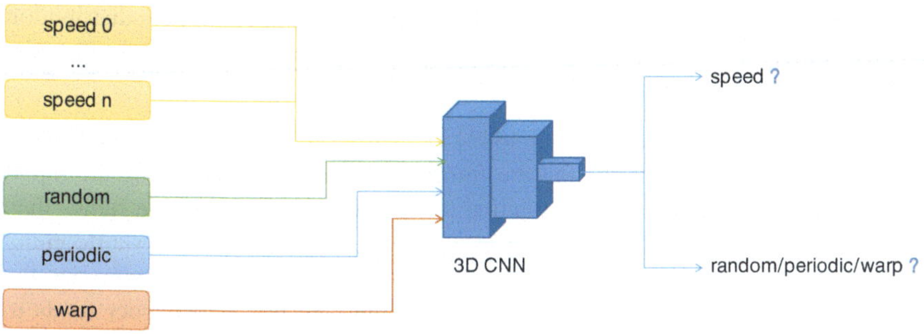

Fig. 6.2 The 3D-CNN to distinguish temporal transformation on both motion and speed. Reproduced from [333]

term, which only pushes away frames that are not adjacent and hence cannot guarantee the learning of informative and discriminating features.

In addition to slow feature learning that encourages features of nearby frames to be close, steady feature learning [335] encourages consistent feature evolution using a regularizer based on contrastive loss. Unlike slow-feature-based approaches, Agrawal et al. [336] first proposed to use egomotion as supervision, using a Siamese-like network to predict camera transformation through pairs of images.

Spatial–Temporal Transformation Buchler et al. [337] proposed to combine both spatial and temporal ordering tasks. By training two complementary permutation prediction tasks simultaneously, general and meaningful feature representations can be learned. They suggested that permutations should be conditioned on the different stages of the network rather than generated randomly and thereby employed a reinforcement learning approach to find a suitable permutation. 3D ST-puzzle [338] proposed a space–time cubic puzzles solving task to predict the original arrangement of a randomly permuted sequence of 3D spatial–temporal crops, extracted from a video clip. Besides, Video Jigsaw [339] is also an extension of the 2D puzzles.

6.2.2 Predicting by Generation

Colorization In the age of the Internet, colored videos dominate digital media. It is a promising way to make use of the abundant unlabeled videos in full color. Inspired by the colorization-based pretext tasks on 2D images [340, 341], researchers have devoted themselves to extending the approaches to the video domain. Based on the assumption that color is temporally stable in videos, color coherence can also serve as a supervisory signal for video feature learning. For example, Vondrick et al. [342] put forth a color prediction method that combines the colors of all pixels in the reference frame using a linear approach to predict the target color y_i. More specifically, a categorical cross-entropy loss is employed to predict the distribution of possible colors for each pixel, which is more appropriate for the multimodal nature of colorization task [341].

Frame Prediction Srivastava et al. [343] introduced a composite LSTM encoder–decoder model for both predicting the future frames and reconstructing the input. This model was designed to be attentive to the whole input sequence rather than just the last few frames so as to reconstruct it, which can alleviate the shortcoming of the future predictor model. Besides, the authors tried not to just memorize the input by learning trivial representations as a high-capacity autoencoder tends to do. Specifically, a sequence of frames is encoded by the Encoder LSTM into a representation that is then fed to the two different Decoder LSTMs. On the other hand, instead of predicting low-level pixels, Vondrick et al. [344] used unlabeled videos to predict future visual representations for higher semantic level anticipating. For the uncertain future, they trained a mixture of K networks to predict different possible outputs.

DynamoNet [345] conducted self-supervised feature learning by predicting and reconstructing frames as well but designed a different prediction method, i.e., the frames are generated from the sample-specific dynamic motion filters produced by the conv–deconv layers rather than directly from these layers. When incorporated into the classification model, the dynamic motion filters can learn motion representations to improve the classification.

VideoGPT [346] adapted VQ-VAE [347] and Transformer-based image generation models to the video domain. In VideoGPT, a GPT-like architecture models the discrete latent representations autoregressively using spatial–temporal position encodings, while VQ-VAE learns downsampled discrete latent representations by employing 3D CNNs and the axial self-attention. A VQ-VAE is first trained to learn a set of discrete latent codes across space–time, then a GPT architecture autoregressive prior network is learned over these codes, and finally, the generated latent representations from the autoregressive prior are then decoded to videos of the original resolution using the VQ-VAE decoder to reconstruct video frames.

GAN-Based Video Generation Mathieu et al. [348] first applied the generative adversarial training method, i.e., GAN, in video prediction. The generative model G is trained to produce a frame conditioned on the input frames, while the discriminative model D learns to predict the probability that the last frame of the sequence is generated by G. Both the G and D are designed to be multi-scale by gradually making higher resolution predictions for the purpose of preserving long-range dependencies as well as resolution. Additionally, they introduced the Image Gradient Difference Loss to sharpen the image prediction. VGAN [349] can generate videos from scratch rather than based on past frames using a two-stream architecture. As illustrated in Fig. 6.3, VGAN generates static background and moving foreground independently with two separate streams and then associates the two-way outputs for final video generation outputs.

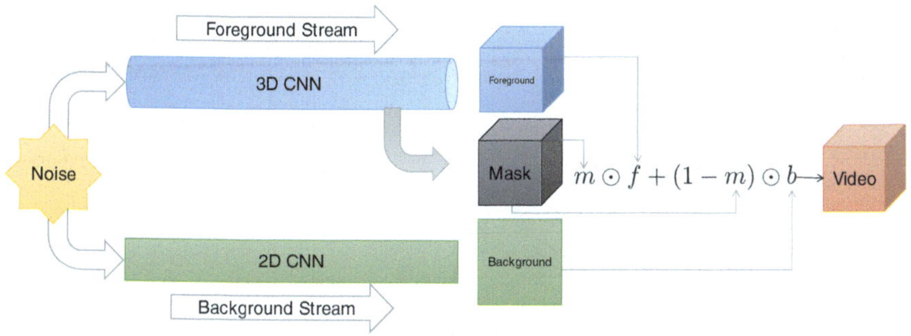

Fig. 6.3 Two streams of the VGAN generator network. Reproduced from [349]

Tulyakov et al. thought it is unnecessary to map a video clip to a point in the latent space, and maybe a latent space of images is enough because a video clip about objects performing actions can be generated by traversing the points in the image latent space. Therefore, they proposed MoCoGAN [350], a video generation model where the motion and content of video are decomposed. To be specific, the image latent space is divided into content and motion subspaces—the content vector is sampled once and fixed based on the observation that the content does not change much in a video clip, and the motion embeddings are generated from a series of motion vectors via a recurrent network. In addition to an image generator G_I and an RNN, MoCoGAN uses both image and video discriminators D_I and D_V based on a 2D and a 3D CNN architecture, respectively.

6.2.3 Predicting from Cross-modal Signals

Real-world video samples contain rich information from different modalities. For instance, audio signals are embedded in natural videos; meanwhile, text can be extracted from audio by automatic speech recognition (ASR). Through exploiting consistency and correspondence between different modalities of video data as surrogate signals, self-supervised video learning manages to learn representations with richer information and higher-level semantics.

Optical Flow The motion information provided by the optical flow is quite useful for video feature learning. Some networks either take optical flows as the only input to model motion information (AoT [331]) or incorporate optical flow for frame selection (OPN [329]) as an auxiliary signal.

Dosovitskiy et al. first proposed a CNN-based FlowNet [351] to estimate the optical flow using a pair of images without explicitly providing the matching correlation. Moreover, due to the lack of a large optical flow dataset, they also constructed a synthetic dataset denoted as Flying Chairs, where the samples are generated by using images collected from Flickr as the background and placing moving 3D chair

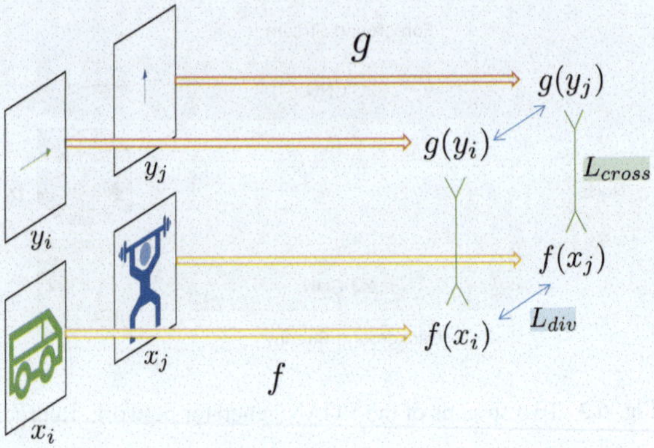

Fig. 6.4 Two-stream architecture of Cross-and-Learn model. Reproduced from [353]

models in the foreground. The representations learned by predicting optical flow using unrealistic data can generalize well to other realistic datasets. Subsequently, FlowNet 2.0 [352] was presented as an extension of FlowNet that can yield smooth flow fields and preserve fine-grained motion details.

Cross-and-Learn [353] trained models with tuples consisting of two paired data (x_i, y_i) and (x_j, y_j), where x and y belong to the RGB domain and optical flow domain, respectively. They enforced both the cross-modal similarity and the diversity of different pairs by minimizing a cross-modal loss (Eq. (6.1)) and a diversity loss (Eq. (6.2)), where d stands for the bounded cosine distance. As illustrated in Fig. 6.4, f and g are trained to decrease the cross-modal distance (vertical) as well as increase the distance between different pairs (horizontal). The corresponding cross-modal loss L_{cross} and the diversity loss L_{div} can be written as

$$L_{cross}(\boldsymbol{x_i}, \boldsymbol{y_i}, \boldsymbol{x_j}, \boldsymbol{y_j}) = \frac{1}{2}[d(f(\boldsymbol{x_i}), f(\boldsymbol{y_i})) + d(g(\boldsymbol{x_j}), g(\boldsymbol{y_j}))], \tag{6.1}$$

$$L_{div}(\boldsymbol{x_i}, \boldsymbol{y_i}, \boldsymbol{x_j}, \boldsymbol{y_j}) = -\frac{1}{2}[d(f(\boldsymbol{x_i}), f(\boldsymbol{x_j})) + d(g(\boldsymbol{y_i}), g(\boldsymbol{y_j}))]. \tag{6.2}$$

PoseFromAction [354] is a triplet network where two appearance networks learn to represent the first and n^{th} RGB frame and a third motion network deals with optical flow maps to learn motion representation. Then the three output features are concatenated to tell whether or not the motion transformation learned from the optical flow can cause the RGB frames appearance change, i.e., whether the pair of frame appearance and optical flow map are sampled from the same video snippet.

Audio Audio (i.e., speech text) from videos can be helpful for understanding the content as well.[1] L^3-Net [355] was trained for the audio–visual correspondence classification task, which could serve as a reference. Owens et al. [356] proposed to train an early-fusion network to predict whether the audio and visual streams of a video are temporally aligned (Fig. 6.5). They fed the network with a clip to represent

[1] Please refer to contrastive learning-based frameworks in Sect. 6.3.5.

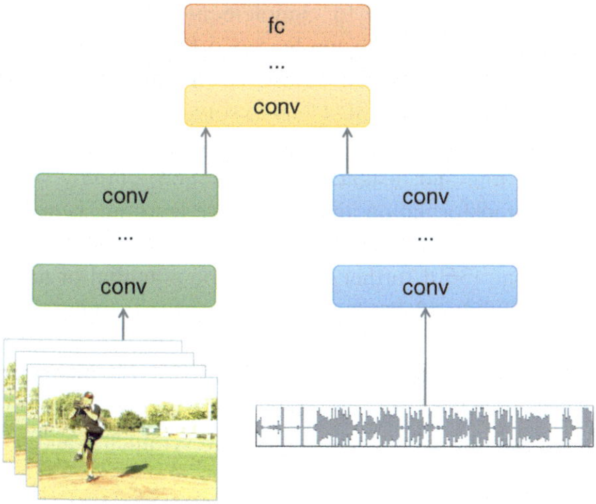

Fig. 6.5 The audio–visual network where two sub-networks are fused early to integrate information across modalities. Reproduced from [356]

a video instead of a single frame used in L^3-Net [355] and created the negative pairs by introducing misalignment from the same video, while L^3-Net randomly paired the video frame and audio clip from the different videos. By solving this misalignment detecting task, a multisensory representation can be learned for video understanding tasks like action recognition even audio–visual applications, e.g., sound source localization, audio–visual action recognition, and on/off-screen audio separation.

6.2.4 *Predicting with Multiple Tasks*

Driven by the thought that a single pretext task may limit the generalizability and robustness of learned representations, many researchers proposed to combine multiple pretext tasks, providing different signals that tend to complement each other to benefit the representation learning.

Inspired by both biological mechanisms and the human visual system, Wang et al. [357] proposed to learn spatial–temporal video representations by regressing both motion and appearance statistics. They first divided video clips into blocks according to three different patterns and then utilized motion boundaries calculated from optical flow components to obtain motion statistical labels, e.g., the largest magnitude location and the orientation. Besides, they generated appearance statistical labels such as the position of largest/smallest color diversity and the corresponding dominant color.

Attempts have also been made to complement self-prediction learning with contrastive learning methods. For example, SeCo [358][2] trained both inter-frame and intra-frame instance discrimination tasks in addition to temporal validation tasks.[3] Kumar et al. [359] trained a 3D CNN to solve multiple tasks simultaneously by using a multi-loss objective. The selected tasks (i.e., rotation prediction, speed prediction, time direction, and instance discrimination) employ signals from both spatial and temporal transformation and contrastive learning, hence yielding better results.

Recently, Jenni et al. [360] further exploited the potential of multitask training by integrating intra-modal and cross-modal contrastive learning. They designed a unified training objective for (a) audio-only temporal prediction including direction and speed classification for short-term feature learning, (b) inter-modal and intra-modal temporal ordering for long-term dynamics capturing, and (c) audio–visual contrastive learning that includes visual–visual and symmetric audio–visual instantiations.

6.3 Contrastive Self-supervised Learning

Contrastive learning has led to revolutionary advances in self-supervised video learning (SSVL) owing to its unique benefits. The core idea behind contrastive learning is to construct a discriminative framework: Similar input pairs attract each other closer, while the dissimilar pairs repel one another further away. This strategy allows the architecture to obtain a better visual representation. Via freely selecting both positive and negative pairs, invariant features in the visual representations are learned and decided. Hence, how to formulate suitable positive and negative pairs matters in contrastive learning.

In the vision area, videos contain a wealth of information in the time domain that serves as another self-supervised signal. This also permits contrastive learning for video representation to separate similar and dissimilar samples from multidimensional ones. In the following sections, we present the current contrastive learning approaches for video representations and organize them based on their dimensions to formulate sample pairs. The mentioned contrastive learning approaches will be categorized into the temporal domain, spatial domain, spatial–temporal association domain, and a specific clustering method.

Before diving into the specific details of contrastive learning methods, we start by introducing the preliminarily used loss function in contrastive learning: Noise Contrastive Estimation (NCE) [361] as well as InfoNCE [362]. NCE transforms the multi-class classification problem into a binary discrimination problem to disguise

[2] More details about SeCo will be illustrated in Sect. 6.3.3.

[3] Temporal validation task is elaborated in Sect. 6.2.1.

between the positive *data sample pairs* and negative *noise sample pairs*. Essentially, the training objective of NCE can be indicated as

$$\mathcal{L}_{\text{NCE}}(i, v) = -E_{P_d}[\log p(D = 1|i, v)] - E_{P_n}[\log(1 - p(D = 1|i, v))], \quad (6.3)$$

where $p(D = 1|i, v)$ refers to the posterior probability that the ith example with a feature vector v belonging to the actual data distribution, P_d refers to the data sample distribution, and P_n refers to the noise sample distribution. Based on NCE, Oord et al. proposed a probabilistic contrastive loss called InfoNCE [362] indicated as

$$\mathcal{L}_{\text{InfoNCE}} = -\log \frac{\exp(q \cdot k_p/\tau)}{\sum_{i=0}^{K} \exp(q \cdot k_i/\tau)}, \quad (6.4)$$

where q and k_p are mutually positive sample pairs, q and remaining samples of $k_i \in K = \{k_1, k_2, \ldots, k_K\}$ are negative sample pairs, and τ adjusts the sensitivity of the loss objective. The aim of InfoNCE is to enhance the similarity between positive representation pairs as large as possible and decrease the similarity between negative representation pairs as small as possible. In most recent contrastive learning works of video understanding, InfoNCE is extensively applied as the basis of the overall loss function.

6.3.1 Spatial Domain

Intuitively, in an attempt to represent the video and complete the downstream video understanding tasks better, some contrastive learning efforts are based on the content of the video, i.e., constructing positive and negative sample pairs in the spatial domain of videos. The typical approach in the spatial field is to model shared features or content from a multi-view perspective. The aim of this multi-view strategy is to explore the invariant features and extract the mutual information between different views.

Here, we categorize this multi-view approach as *Multi-angle Video Capture* and *Multi-visual Enhancement*. The first category investigates the invariant visual contents straightforwardly from varying video sources. The other perspective exploits alternative representations of the same visual content. Though being characterized in various forms, the semantic-preserved components of the same video will be maintained as self-supervised signals. We will move on to highlight some significant methods within the following two categories.

Multi-angle Video Capture Methods in *Multi-angle Video Capture* perspective depict the same scene directly through captured multi-angle videos. One representative work is Time-Contrastive Network (TCN) [363]. TCN adopts a straightforward way to build the contrastive learning framework: simultaneous actions captured by different cameras from distinct angles are considered positive sample pairs, while actions that occurred at a separate time within one video are attributed to negative sample pairs. After operating through the deep network, the learned embeddings are trained with a triplet loss. The trained model, therefore, pays attention to invariant features across varying views while disregarding subtle variants across time.

Multi-visual Enhancement On the other hand, the other form of literature adopts view augmentations to portray the scene. View enhancement accentuates unchanging semantic features and neglects unnecessary visual components. The following methods apply view augmentation from different perspectives: the former emphasizes on the invariant semantic features and the latter ignores unnecessary visual content.

Feichtenhofer et al. conduct a persistent temporal feature learning [364], which extends the augmentation in trendy image-based contrastive learning frameworks to video-based contrastive learning. By directly adding spatial cropping and color augmentation to the clips of videos, invariant features across different views of the video appearance are learned. In this case, different clips in the same videos formulate positive pairs.

Instead of adding artificial visual augmentations, Video Noise Contrastive Estimation (VINCE) [365] performs view augmentation with the aid of intrinsic characteristics of videos. The moving procedure of content in the video provides different views of critical semantic parts, such as the deformation of moving objects, the offset of the recording machine, and surrounding relative static objects. Unlike artificial data augmentations, VINCE expands the selection range of positive sample pairs so that multiple frames originating from the same video are pulled closer.

Background Erasing (BE) [366] highlights the key semantic information by mitigating the impact of video context through contrastive learning. This strategy implants a random frame to the remaining frames of the same clip to generate a distracting clip and pulls this distracting clip closer to the original one. In the meantime, other clips of the same video are repelled from the anchor video clip.

6.3.2 Temporal Domain

The additional temporal domain enables the extension of contrastive learning methods in the video region. Regarding video dynamics, comparable pairs should explore the intrinsic characteristics of the temporal dimensionality in videos. Hence, we demonstrate typical contrastive learning frameworks of the time domain from different perspectives. All these methods exploit the advantages of video dynamics.

Video Tempo Video tempo is an eminent variable in the time domain, and it is available as a self-supervised signal in video contrastive learning. Semantic components of videos are consistent though with different sampling frequencies. Currently, there are two strategies considering the relationship between the time tempo and video content: content-aware and content-agnostic contrastive learning.

The content-aware contrastive framework tends to find invariant video semantics over time. There are two frameworks: *Video Tempo Hierarchical Contrastive Learning (VTHCL)* [367] and *Contrast and Mix (CoMix)* [368], both leveraging the same idea to exploit different speeds of actions in videos to capture the consistency in semantic components of videos. Each video instance v_i is sampled at slow and fast sampling rates, respectively, to generate its slow version $v_{i,s}$ and its fast version $v_{i,f}$. The semantic information of individual videos still holds though with different sampling frequencies. Thus, the slow and fast versions of the i−th video $\{v_{i,s}, v_{i,f}\}$ constitute positive sample pairs. The rest $\{v_{i,s}, v_{j,f}\}$ from varying videos constitute negative sample pairs.

From an alternative perspective, content-agnostic contrastive learning does not consider the content of video clips but uses different video pace transformation approaches instead. Video Pace Contrastive Learning (VPCL) [369] obtains the clips by sampling the videos with varying pace rates. Regardless of the video content, the clips of the same sampling rate constitute the positive pairs, while those clips with different paces constitute the negative pairs. Similarly, Jenni et al. integrate a time-equivariant constraint [370] into contrastive learning. This method aims at digging out the dynamics of videos through temporal augmentations including speeding up the video playback, inversing the playback direction, etc. Clips are then manipulated with multiple temporal transformation measurements. To build a contrastive framework, clips with the same temporal transformation methods are pulled closer, and those with different methods are pushed away. Except for the time-equivariant constraint, instance discrimination on the actual scene is still applied in the contrastive learning framework.

Relative Speed Network (RSPNet) [371] consolidates both content-aware and content-agnostic strategies as two branches to regularize the model simultaneously. As illustrated in Fig. 6.6, in RSP, the clip features are extracted from one video. The features representing the same speed of one video playback are pulled closer and pushed away from the features representing different speeds. In A-VID task, different clip instances are pushed apart and clip instances from the same video are pulled close. The triplet loss and InfoNCE loss constrain the RSP task and A-VID task, respectively.

Sequence Coherence It is a consensus that the salient semantic features do not dramatically change in a short period. Movements in a set of **continuous** sequences decomposed from videos are always considered to contain the same semantic meaning. This precondition permits contrastive learning to use the properties of close frames located on the video timeline. The typical sequence coherence-based methods are described as follows.

Refined from TCN [363],[4] multi-frame Time-Contrastive Network (mfTCN) [372] embeds representations along the video timeline by using the sequential coherence property. To investigate the dynamic motion in the video, mfTCN exploits

[4] The basic version of TCN is illustrated in Sect. 6.3.1.

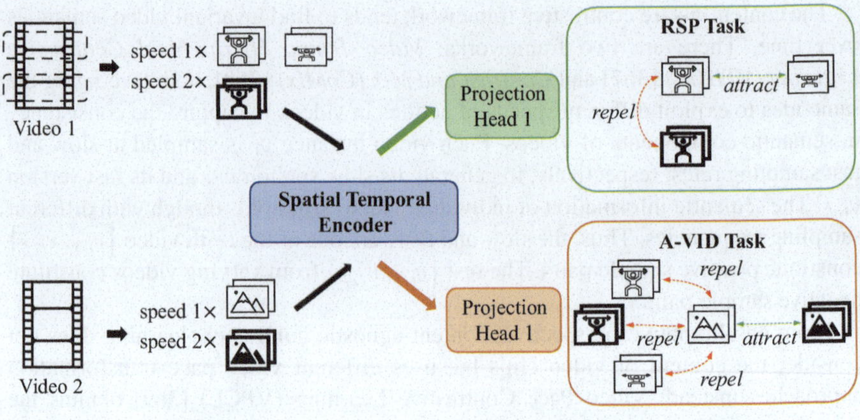

Fig. 6.6 Structure of RSPNet containing RSP task and A-VID task. Reproduced from [371]

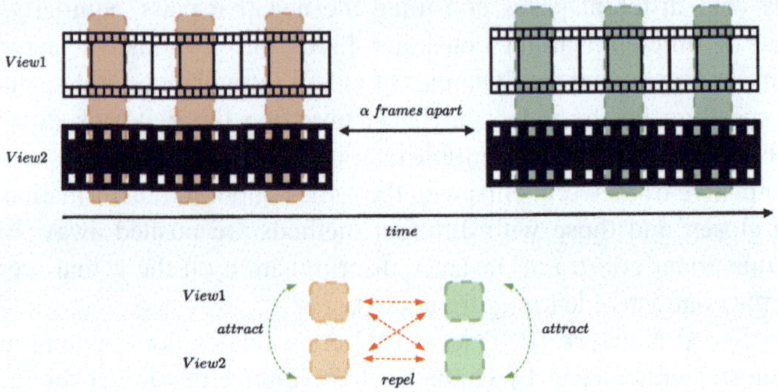

Fig. 6.7 Overview of mfTCN. Videos are sampled at a sufficient distance without overlapping. Clips standing for the same actions in the consecutive sequence window are attracted. Clips over a long time distance are repelled. Reproduced from [372]

consecutive sequences over a short period of time instead of sampling two single frames at separated time steps. As illustrated in Fig. 6.7, the consecutive sequences of varying views are gathered along the timeline, with at least α frames apart and no overlapping. Each consecutive sequence is sampled three times with a stride of three. The sampled frames displaying the same action in the meantime are regarded as positive pairs. Time-separated frames are deemed to be mutually negative sample pairs.

Based on Momentum Contrast (MoCo) [373], VideoMoCo [374] expands the scope into the video domain. VideoMoco adopts a dictionary lookup scheme following MoCo. To set up contrastive representations in the time domain, VideoMoCo applies a temporally robust encoder to generate queries and a temporal decay term to simulate the attenuation procedure of keys, as illustrated in Fig. 6.8. The generator adaptively drops out a quarter of frames along the timeline to generate

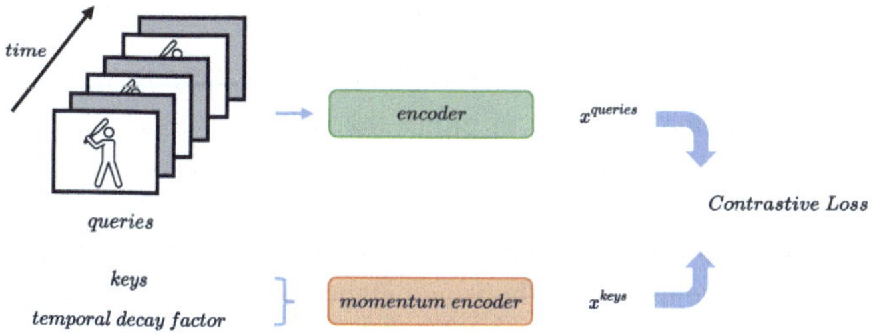

Fig. 6.8 Diagram shows the whole VideoMoCo structure. Reproduced from [374]

queries x^{query}. Apart from that, the temporal decaying factor multiplied by the key highlights the time coherence—the gap between query and key becomes larger as more time passes.

Global–Local Comparison Past knowledge could be beneficial when predicting the future in videos. To capture invariant features comprehensively as time passes, a global–local relationship in the temporal dimension is proposed. Contrastive learning closes the distance between the global context of a long time span and the local features of short-period video segments, hence strengthening understanding of high-level semantics. We explain two representative methods global–local contrastive learning models, CPC [362] and LSTCL [375] in the following.

A prominent work based on global–local relationships to explore the past–future relationship is Contrastive Predictive Coding (CPC) [362]. Aiming to learn generalizable representations, CPC aggregates past message from prior time steps through an RNN as anchors, which are pulled closer to samples in the future time steps and pushed away from samples that are not sequentially related. The mutual information between the archive of past knowledge and the instance of future predictions is then optimized through InfoNCE. In this way, the model obtains the predictive capability through a global–local contrastive learning strategy in the temporal domain.

Long–Short Temporal Contrastive Learning (LSTCL) [375] empowers the model to deduce the context viewed throughout the video from a brief duration of the clip. Equal numbers of samples are extracted from both longer and shorter video clips.

The long–short samples originating from the same video are considered positive pairs and those without the same origin are negative pairs. Meanwhile, long and short video clips can be either overlapped or disjointed because they are collected separately and independently. In this way, local short clips must acquire as much contextual information as possible to match the corresponding global long clips.

6.3.3 Spatial–Temporal Association Domain

Naturally, the underlying potential of either the spatial or temporal domain video contrastive works promotes the associated use of space and time in video contrastive learning. Existing methods use the connections between the two domains of videos in conjunction to exploit the potential of video contrastive learning to the hilt. We will then examine these hybrid methods and elaborate on how they employ spatial–temporal information in the following subsections.

Global–Local Relationship One common way is to assess the *global–local relationships* between spatial–temporal blocks. Built upon CPC,[5] a reputable global–local contrastive loss for the spatial–temporal domain is Dense Predictive Coding (DPC) [376]. As illustrated in Fig. 6.9, DPC splits the video into spatial–temporal blocks and generates their corresponding embeddings. The embeddings over past time are aggregated through an RNN just as they are treated in CPC, while the distinction is that the prediction is over the future spatial–temporal blocks rather than a single future frame. Simultaneously, DPC formulates three types of negative pairs, which include spatial–temporal blocks from distinct videos as *easy negatives*, from the same video while positioned differently in spatial feature maps as *spatial negatives*, and from the same spatial position but within distinct time intervals as *hard negatives*.

Spatial–temporal DeepInfomax (ST-DIM) [377] leverages the global–local relationship in another way. This approach extends the image-based DeepInfomax (DIM) [378] into the video to maximize the mutual information over spatial and temporal axes. The contrast is performed on consecutive frames x_t and x_{t+1}. In ST-DIM, the anchor is the global fully accessed features on x_t. Given the anchor, the positive sample is the local partial spatial location in the feature map on the next

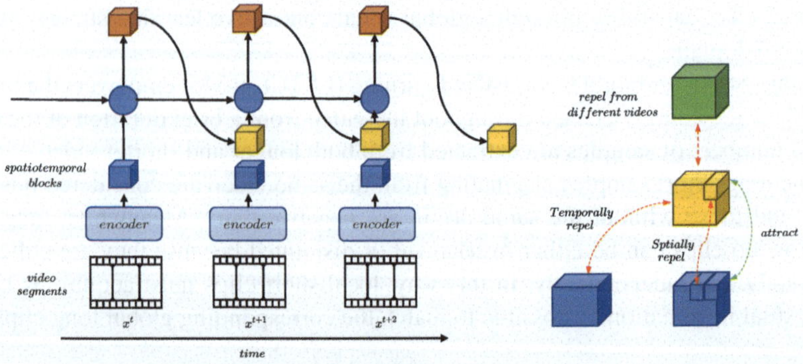

Fig. 6.9 Diagram of DPC. Reproduced from [376]

[5] CPC is illustrated in Sect. 6.3.2.

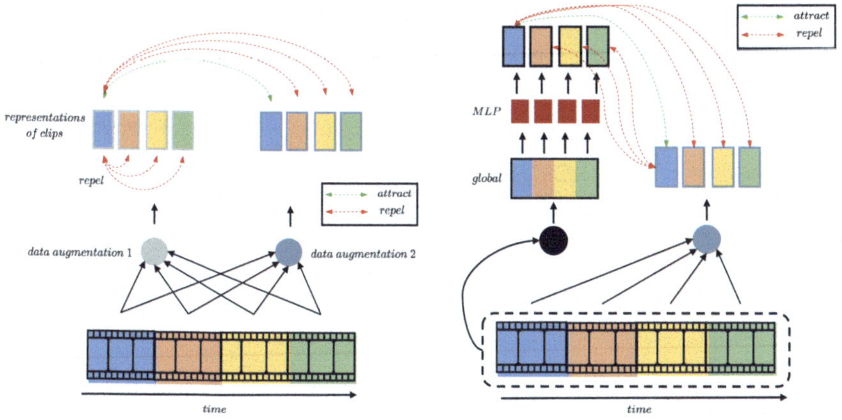

(a) The local-local contrastive part of TCLR (b) The global-local contrastive part of TCLR

Fig. 6.10 Diagrams indicate the local–local and global–local parts of TCLR. Reproduced from [379]

x_{t+1} frame. The negative sample is the partial location in the feature map but on a randomly sampled x^* frame that is non-consecutive.

Not limited to the global–local relationship, Temporal Contrastive Learning for Video Representation (TCLR) [379] combined instance discrimination, the local–local contrastive learning as well as global–local temporal contrastive learning. As illustrated in Fig. 6.10, the left and right diagrams show the local–local contrastive procedure and the global–local contrastive procedure, respectively. In the local–local pattern, non-overlapping clips sampled from the same video through different view transformations are treated as positive pairs, and those sampled at different time steps are treated as negative samples. In the global–local pattern, the feature map of the global clip obtained by temporal pooling is split into several segments. Each segment from the global features forms a positive pair with features from the corresponding local clip; and it forms negative fairs with features from irrelavant clips.

The above methods explore the global–local relationship of the spatial–temporal features, mainly relying on the time axes. Contextualized Spatial–Temporal Contrastive Learning (ConST-CL) [380] leverages this relationship between the regional and globally pooled features. As illustrated in Fig. 6.11, given two spatial–temporal distant clips randomly sampled from one video, the context features F_c' are obtained from a network $f(x)$, which also serve as *key* and *value* to the transformer. The region features $\{h_j', h_i\}$ are generated by spatial–temporal ROIAlign, where h_i is used as the *query* to the transformer. The transformer network aims to reconstruct h_j' conditioned on h_i and context features.

Exploring Spatial–Temporal Features in Parallel Another way is to exploit the inherent features of videos *in parallel*. This genre starts with building contrastive

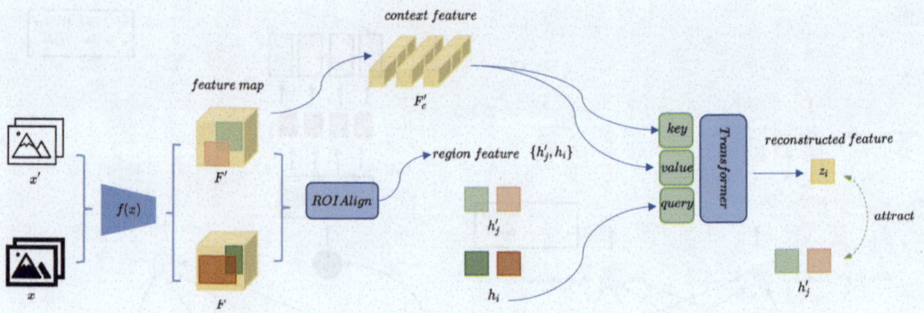

Fig. 6.11 Diagram of ConST-CL structure. Reproduced from [380]

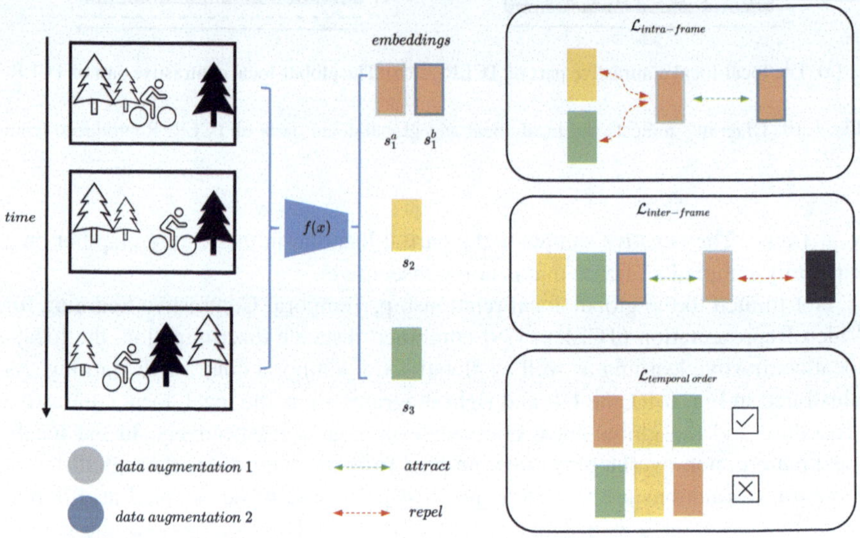

Fig. 6.12 Diagram indicating the components and procedure of SeCo. Reproduced from [358]

tasks in the temporal domain and spatial domain, respectively, and then combines these contrastive training losses into a single training objective.

Sequence Contrastive Learning (SeCo) [358] consists of two contrastive tasks: intra-frame temporal instance discrimination task and inter-frame spatial instance discrimination task together with temporal order validation task in the temporal domain. As illustrated in Fig. 6.12, three frames from one video are sampled randomly in a time sequence, and their embeddings are denoted as $\{s_1, s_2, s_3\}$. s_1^a and s_1^b are embeddings of s_1 through different view augmentations. The intra-frame task identifies whether embeddings are from one frame: The anchor s_1^a constitutes the positive pairs with s_1^b while constitutes the negative pairs with $\{s_2, s_3\}$. The inter-frame task determines whether embeddings are from one video: the anchor s_1^a constitutes the positive pairs with $\{s_1^b, s_2, s_3\}$ while constitutes the negative pairs with embeddings from the other videos. The overall training objective is written as Eq. (6.5).

$$\mathcal{L}_{SeCo} = \mathcal{L}_{intra-frame} + \mathcal{L}_{inter-frame} + \mathcal{L}_{temporal-order}. \tag{6.5}$$

Following the formulation in SeCo, the video-level contrastive learning framework (VCLR) [381] contrasts video segments over a long time duration. The video is split equally into K segments, and each segment is sampled randomly twice to produce an anchor collection $s^a = \{s_1^a, s_2^a, \ldots, s_K^a\}$ and the corresponding multi-view frames $s^b = \{s_1^b, s_2^b, \ldots, s_K^b\}$, forming positive pairs. Negative pairs are sampled from the other videos.

Instead of isolating the spatial and temporal contrastive structures, Macro-to-Micro Semantic Correspondence (MaMiCo) [382] builds a temporally nested pyramid structure, combining Spatial-MaMiCo and Temporal-MaMiCo simultaneously while linking them via a macro-to-micro strategy. T-MaMiCo assists in capturing the temporal context: It maximizes the similarity between the video-level samples and clip-level samples of a given video while minimizing the similarity of samples from the other videos. S-MaMiCo assists in capturing spatial clues. In S-MaMiCo, samples are generated through a dense projection head: The dense feature pairs between clip level and frame level with the highest similarity are pulled together, and features from the other clips are pushed away.

Exploring Spatial–Temporal Features in Sequence Apart from the above parallel approaches, methods such as Contrastive Video Representation Learning (CVRL) [383] exploit spatial–temporal features of videos *in a consecutive manner*. CVRL starts with employing a specific temporal sampling manner: Previous works apply a random sampling strategy to extract frames uniformly over the temporal domain, while this strategy simulates a motion-evolving procedure. The time interval t is sampled from a monotonically decreasing distribution, and two clips are extracted with an interval t. The extracted temporal clips are then fed into a temporally consistent spatial augmentation block so that two clips sampled from the same video are each subjected to a unique data enhancement. After a sequential operation from both temporal and spatial domains, the final augmented clips with the same origin are regarded as positive pairs; meanwhile, clips with different origins are regarded as negative pairs.

Derived from CVRL, a recent work named Shuffled-CVRL (SCVRL) [384] integrates an extra shuffled contrastive framework into CVRL to promote the awareness of motion over a long time duration. The sampled clips with different spatial augmentations are regarded as positive pairs. Then one of the augmented clips is shuffled to generate another feature map, which is viewed as the negative sample. This shuffled contrastive loss is then added to the original loss of CVRL to build the overall contrastive loss. With the aid of this additional shuffled contrastive loss, the performance of CVRL is further improved.

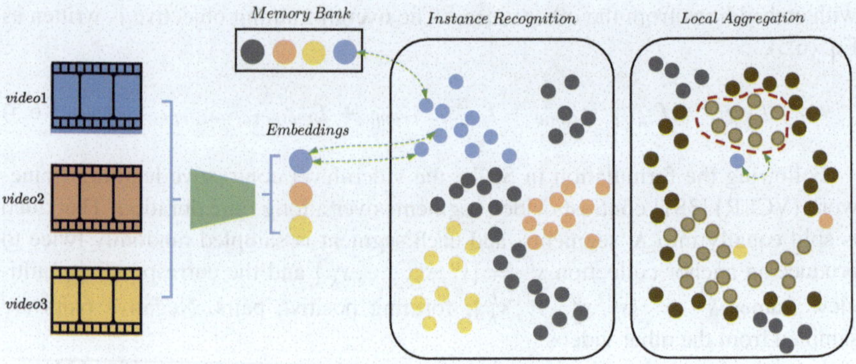

Fig. 6.13 Diagram of VIE structure. Reproduced from [385]

6.3.4 Clustering

Clustering is also a way of learning feature representations. Unlike typical approaches that discriminate instance pairs, clustering aims at aggregating similar samples into groups for feature learning. Below, we present cluster-based contrastive learning methods.

One type of cluster-based contrastive learning architecture adopts a clustering algorithm to group embeddings of videos (Fig. 6.13). Video Instance Embedding (VIE) [385] framework adopts *Instance Recognition* (IR) and *Local Aggregation* (LA) tasks together to form the clusters with an aim to group similar samples closer and push away dissimilar samples at the same time. Similarly, with the aid of Improved Dense Trajectories (IDT) descriptor, Tokmakov et al. [386] starts by aggregating embeddings of IDT descriptors via k-means clustering and then applies the LA loss for improved results.

Another type of approaches first groups embeddings into varying clusters and using these clusters as cues to sample positives and negatives. Clustering-based Contrastive Learning (CCL) [387] leverages a parameter-free FINCH clustering algorithm [388] to encode video features into different clusters as indicators to sample positives and negatives. Similar to CCL, Self-supervised Learning with Iterative Clustering (SLIC) [389] also uses the FINCH clustering method to generate sampling cues. Additionally, SLIC integrates an instance triplet loss and a temporal discrimination loss into the overall training objective. To discriminate instances, positives are sampled from multi-views that include typical RGB and optical flow embeddings of the anchor cluster, while negatives are sampled from the other clusters. For temporal discrimination, negatives are non-overlapping clips regardless of which cluster they fall into.

Clustering-based techniques are also extensively used for multimodal contrastive learning. Multimodal Clustering Network (MCN) [390] jointly groups features of distinct modalities into a vector to form clusters. By combining a pairwise contrastive loss together with a clustering loss, the joint embeddings of cross-domain information are better represented. Multi-level Feature Learning for Contrastive Multi-view Clustering (MFLVC) [391] aggregates cross-modal features of the same semantic meaning into one group and matches these clusters with the corresponding semantic labels.

6.3.5 Multimodal Self-supervision

Besides supervisory signals extracted from the spatial–temporal domain (i.e., RGB domain), contrastive learning leverages information from other modalities as well. Generally speaking, signals from the same video yet different domains serve as the positive pairs, and multimodal signals from different videos act as negative pairs. Below, we present the multimodal-based frameworks in four folds, i.e., audio-based, text-based, optical-flow-based, and multiple-domain-based.

Audio Domain Real-world videos are abundant with audio information, which can also serve as surrogate supervision. To build upon contrastive learning, efforts have been made to harness video–audio pairs for improved video representation.

AVTS [392], denoted audio–visual temporal synchronization, leverages semantic correspondence between video and audio clues to facilitate feature learning. With video clips encoded by an MCx network [36] and audio samples modeled by a VGGNet [10], training is optimized by a contrastive loss (i.e., NCE in Eq. (6.3)) to tackle the task of syncing video clips with audio ones. In particular, networks are fed with three types of video–audio pairs, i.e., positive ones, hard negative ones, and super hard negative ones. In particular, hard negative pairs are relatively "hard" since they are drawn from the same video yet sliced from different locations. Meanwhile, the difficulty is further escalated when the video and audio segments overlap, as demonstrated in Fig. 6.14. The hard-core contrasting task manages to facilitate high-level semantic understanding.

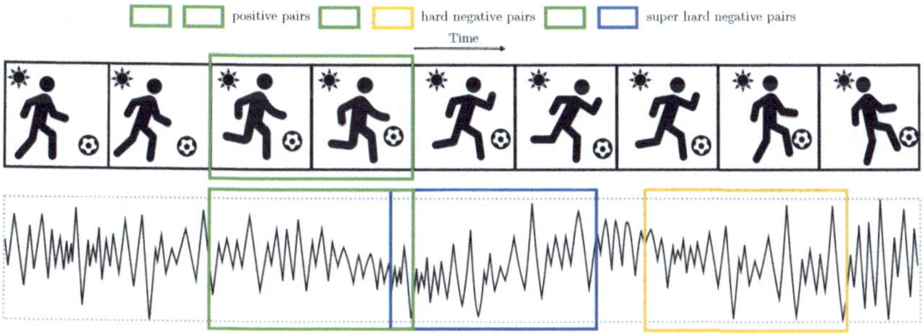

Fig. 6.14 Video–audio pairs in AVTS [392]

AVID [393] refers to the audio–visual instance discrimination task, which aims to model video–audio discrimination with the aid of contrastive learning. For network instantiation, a R(2+1)D [36] serves as the video encoder, and a 9-layer 2D CNN is used for audio. Besides, the positive set is acquired by the cross-modal agreement (CMA) methods. Typically, for video clips v_i, v_j and the corresponding audio clips a_i, a_j, CMA rates their agreement scores $s_{i,j}$ via the equation:

$$s_{i,j} = \min(v_i^T v_j . a_i^T a_j). \tag{6.6}$$

Therefore, high scores indicate high similarity both in the video feature space and in the audio feature space. Based on the CMA policy, positive sets $\mathcal{P}_i$ and negative sets $\mathcal{N}_i$ for the ith sample can be formulated by

$$\mathcal{P}_i = \text{TopK}_j(s_{i,j}), \quad \mathcal{N}_i = \overline{\mathcal{P}_i}. \tag{6.7}$$

On top of that, authors perform NCE-based contrastive loss (Eq. (6.3)) in three different scenarios: (1) Self-AVID is applied within the same modality, i.e., video–video pairs and audio–audio pairs; (2) Cross-AVID indicates contrasting with samples from the other modality, i.e., video–audio pairs and audio–video pairs; (3) Joint-AVID stands for the association of self-AVID and cross-AVID.

Inter–Intra Contrastive Learning(IICL) [394] resorts to applying clustering-based contrastive learning on different modalities. As illustrated in Fig. 6.15, it contains both instance-level contrastive blocks and cluster-level contrastive blocks. The cluster features Y_v and Y_a of the video and audio are generated separately and contrasted with each other over the corresponding cluster.

Text Domain We refer to the text as one acquired without human supervision, which is typically generated by automatic speech recognition (ASR). In particular, instructional videos are exploited for SSVL, whose features are computed by the video encoder associated with the text encoder. Thereafter, contrastive learning is conducted on pairs of spatial–temporal embeddings and text embeddings. While real-world videos usually consist of tedious and futile sounds that hardly can be

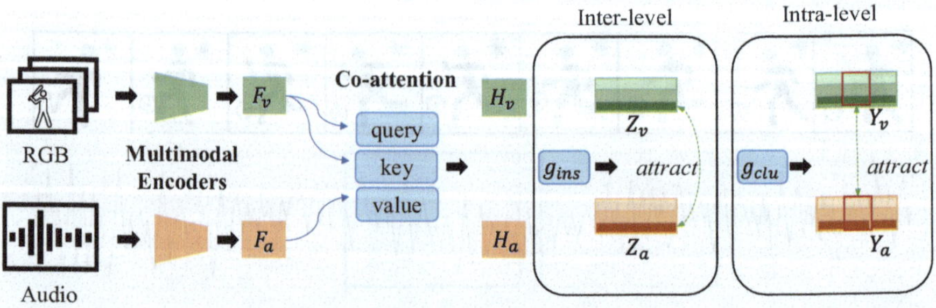

Fig. 6.15 Schematic of IICL framework. Reproduced from [394]

interpreted as coherent text, instructional videos are narrated by meaningful dubs and hence produce enriched text information. Therefore, the following introduced framework is targeted at instructional videos with narration text.

MIL [395] proposes an enhanced contrastive learning paradigm for video–text alignment, which is denoted as multiple instance learning and noise contrastive estimation (MIL-NCE), in order to support contrasting with multiple positive pairs instead of a single one. In consequence, MIL-NCE alleviates the common deviation brought about by misalignment between narration and video segments, particularly, the loss can be formulated as

$$\mathcal{L}_{\text{MIL-NCE}} = -\mathbb{E}[\log \frac{\sum_{p \in \mathcal{P}_i} \exp(x_i \cdot y_p)}{\sum_{p \in \mathcal{P}_i} \exp(x_i \cdot y_p) + \sum_{n \in \mathcal{N}_i} \exp(x_i \cdot y_n)}], \tag{6.8}$$

where x_i denotes the ith video clip and y refers to the corresponding text. Besides, $\mathcal{P}_i$ refers to the positive sets of video–text pairs for x_i, and $\mathcal{N}_i$ corresponds to the negative one. In the specific implementation, $\mathcal{P}_i$ is composed of K narrations that temporally approach clip x_i, encouraging at least one correct matching between the clip and the narration. Also, an asymmetric sampling for negative sets is employed, which not only obtains negative narrations based on the given clip but also includes negative videos based on the given narration.

Optical Flow Domain The second category employs optical flow that stands for the motion of objects as supplementary visual signals. We illustrate two approaches that leverage the optical flow in producing positive and negative sample pairs: Contrastive Multi-view Coding (CMC) [396] and Co-training Contrastive Learning of visual Representation (CoCLR) [397]. These two approaches both apply multimodal visual information, but with different tactics.

CMC [396] deals with simultaneous RGB frame and optical flow. It composes of two streams: the ventral stream that directly handles the connection of visual content between each frame and the dorsal stream that handles the inter-frame optical flow. CMC constructs positive pairs from the same video clip: The anchor RGB frame attracts both RGB frames and optical flow that originated from the same clip. In the ventral stream, the current RGB frame repels individual frames from different clips; in the dorsal stream, the current RGB frame repels flow information from the other clips. The concatenated contrastive loss consists of separate training objectives built within both streams.

CoCLR [397] is a self-supervised co-training scheme on both RGB streams and optical flows. Two video clips depicting the same action might contain dissimilar RGB features and similar optical flow representation. However, InfoNCE may push them away owing to different origins. CoCLR allows semantically similar samples to be treated as positive sample pairs without the requirement of sample labels. After pre-training both RGB and optical flow streams, the optical features that are most comparable to the RGB features are extracted and fed into the RGB stream as positive sample pairs for continuous training. CoCLR adopts a MIL-NCE for training, as displayed in Eq. (6.8).

Multiple Domains Video bears the natural strength of abundance in multimodal signals. Researchers have been acutely aware of such advantages and have designed extensive related pretext tasks. More importantly, more than one modality can be incorporated with vanilla RGB visual information in SSVL, contributing to stronger consistency captured, better representation learned, and more extensive downstream tasks to be applied. Moreover, contrastive learning alleviates the intricacy of designing elaborate self-prediction-based pretext tasks on multiple modalities.

Multimodal versatile network (MMV) [398] takes advantage of abundant information from different modalities in videos by collaborating vision with audio and language in contrastive learning. In particular, they hypothesize that the relationship between these multimodal features constitutes fine and coarse spaces (FAC), where visual and audio representation x_v, x_a belongs to the fine-grained space yet text embedding x_t falls into a coarse-grained space. To propel interaction within FAC, contrastive loss for training is formulated as

$$\mathcal{L}(x) = \lambda_{va}\mathcal{L}_{\mathrm{NCE}}(x_v, x_a) + \lambda_{vt}\mathcal{L}_{\mathrm{MIL-NCE}}(x_v, x_t). \tag{6.9}$$

Specifically, negative pairs are sampled from different videos, while positive ones consist of multimodal signals from the same location of the same video.

Video–Audio–Text Transformer (VaTT) [399] associates unlabeled videos with the corresponding audio and text samples in SSVL with Vision Transformer. Similar to MVM [398], a contrastive loss of VATT is computed between video–audio pairs and video–text pairs via NCE (Eq. (6.3)) and MIL-NCE (Eq. (6.8)), respectively. On top of using a respective transformer for each modality, VATT performs joint training of multimodal data on an identical weight-sharing backbone and manages to achieve encouraging performance on supervised downstream tasks. A strategy denoted DropToken is also proposed for efficient training, where the high-resolution input is kept, whereas a certain portion of tokens will be dropped randomly. Overall, VATT demonstrates the promising future for omnipotent foundation models.

6.4 Masked Video Modeling on ViTs

There has been a proliferation of large-scale and advanced NLP models after full-attention-based Transformer [53] was introduced. In the meantime, masked language modeling on Transformers, i.e., GPT (2018, generative pre-training) [400] and BERT (2019, bidirectional encoder) [99], have dominated self-supervised pre-training in NLP ever since. However, contemporary vision models remained CNN-based. Besides, the mainstream of self-supervised visual learning falls into techniques based on contrastive learning.[6] There has existed a discrepancy between the modeling of language and vision until the Vision Transformer (ViT) [14] was

[6] Please refer contrastive learning in Sect. 6.3.

introduced. In essence, the design of ViT borrows substantially from the language counterpart, which involves dealing with images and videos as sequences and aggregating global visual information with self-attention.

Thereafter, ViTs have made giant strides in a wide range of visual tasks. Especially, the wake of ViTs witnessed unprecedented progress in self-supervised learning via Masked Visual Modeling,[7] which can also be regarded as an attempt to align language and vision learning. Specifically, in 2021, BEiT [401] and MAE [402] achieve performance superior to contrastive learning-based methods. Since then, masked image modeling with ViTs has become all the rage. Subsequently, research interests are aroused in extending masked modeling from 2D images to 3D videos, i.e., Masked Video Modeling (MVM). In this section, we begin with the preliminary overview of MVM and then enumerate MVM-based frameworks built upon ViTs.

6.4.1 Preliminary

To put it simply, masked video modeling drives the training by a mask-then-predict strategy, i.e. masking inputs with a certain strategy, forcing deep neural networks to learn by encoding a portion of visible inputs, and then decoding prediction targets correlated with the masked portion. Concerning masked modeling applied to video Vision Transformers (ViTs),[8] it is customary to utilize patches as the fundamental units at which masks are applied. Moreover, we refer to the set of masked patches as P_M and the set of visible ones as P_V. Typically, for the universal MVM framework, an encoder–decoder architecture is adopted, where the encoder Ψ_{en} is instantiated with ViT, while the decoder Ψ_{de} differs in specific designs and may be absent in some scenarios. On top of that, Ψ_{en} learns video representation from unmasked patches, and the decoder aims to predict given targets of masked patches with a loss $\mathcal{L}$. Finally, the optimization for masked video modeling can be formulated as

$$\mathcal{L} = \frac{1}{|P_M|} \sum_{(t,i,j)\in P_M} L_{\text{pred}}(\tilde{X}_{(t,i,j)}, X_{(t,i,j)}), \tag{6.10}$$

where $\tilde{X}$ denotes predicted targets and L_{pred} can be realized by a given loss function, e.g., L_1 loss, Mean Squared Error (MSE, which can be equivalently termed as L_2 loss), etc. Generally speaking, MVM frameworks differ in masking strategies, encoder/decoder architectures, and prediction targets. To begin with, we introduce different masking strategies in MVM as below. Subsequently, by roughly dividing

---

[7] Note that Masked Visual Modeling can also be termed as Masked Autoencoder.

[8] Note that masked video modeling can also be conducted on backbones other than ViTs, as the implementation of Image-GPT [403], S-DAE [404] and Context Encoder [405]. However, we refer to MVM in this chapter to specific frameworks with ViTs as the default encoder backbone.

MVM methods into 3 categories based on prediction targets: (a) tokens, (b) features, (c) pixels, we deliver the corresponding frameworks in detail.

Masking Strategies Given a training video $X \in \mathbb{R}^{T \times H \times W}$ of MVM, before entering transformer blocks of the encoder, it is first broken down into 3D patches P size of $T_p \times H_p \times W_p$ and projected into feature space as "tokens" $Z \in \mathbb{R}^{\frac{T}{T_p} \times \frac{H}{H_p} \times \frac{W}{W_p} \times D}$. Due to this intrinsic character of ViT, the de facto mask on each sample views a patch (or a token) as a unit. The mask can be considered as a binary map $M \in \mathbb{R}^{\frac{T}{T_p} \times \frac{H}{H_p} \times \frac{W}{W_p}}$. Therefore, the masked patch set P_M and the visible masked patch set P_V can be expressed as $P_M = \{P_{(i,j,k)} | M_{(i,j,k)} = 1\}$ and $P_V = \{P_{(i,j,k)} | M_{(i,j,k)} = 0\}$, respectively. Provided a masking ratio μ, versatile measures of computing M are employed in practice, which are also illustrated in Fig. 6.16:

(i) ***Cube masking*** first samples a 2D block-wise map in a random time step and extends the masking to a random number of consecutive frames. Displayed in b) of Fig. 6.16, as a consequence, 3D cubes are randomly sampled until μ percentage of patches are masked. Wei et al. [406] assumed that it is a combination of tube masking and frame masking. Note that ***block masking***, shown as c) of Fig. 6.16, performs in a similar way, whereas only sampling one single 3D cube.

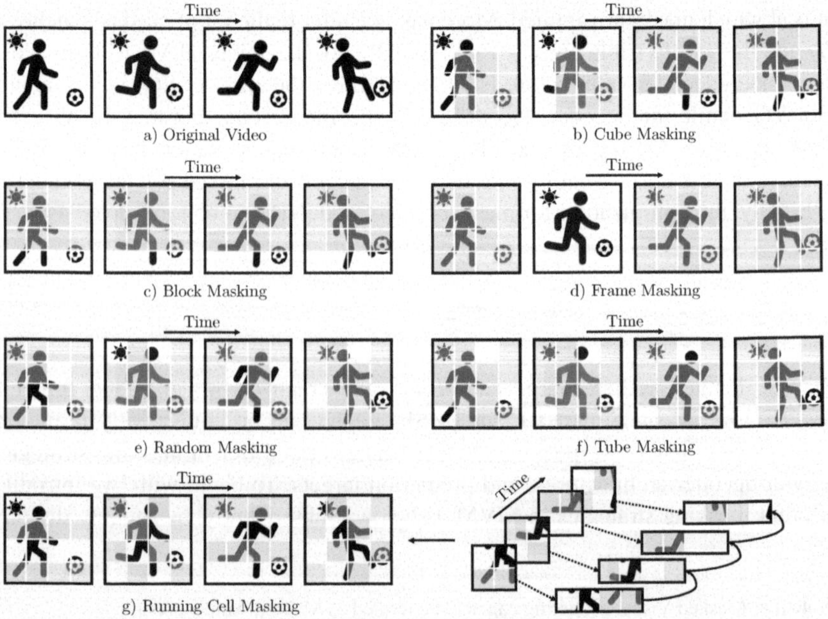

Fig. 6.16 Different masking strategies. Tokens with green shadows are masked out

(ii) ***Frame masking*** randomly casts full-token-masks on μ ratio of randomly chosen time steps, the index set of which is denoted as F. Therefore, the mask M can be written following the rule that if the ith time step belongs to F, $M_{i,j,k}$ equals 1 for identical j and k, on the other hand, $M_{i,j,k}$ is assigned to 0. Please refer to d of Fig. 6.16 for visualization with $\mu = 1/4$.

(iii) ***Random masking*** conducts masking in a most straightforward manner by uniformly sampling μ ratio of all spatial–temporal patches, as illustrated in e) of Fig. 6.16.

(iv) ***Tube masking***, f in Fig. 6.16, generates masks consistent in the temporal dimension; therefore, the 3D masks are shaped like tubes. In the implementation, first, one can randomly draw μ proportion of patches from one time step with a 2D mask $m_{\text{tube}} \in \mathbb{R}^{\frac{H}{H_p} \times \frac{W}{W_p}}$ and repeat the operation with the same mask on remaining frames. Therefore, the 3D mask M is equivalent to the stacking of corresponding 2D masks m_{tube} copies in temporal dimension by $\frac{T}{T_p}$ times. On the other hand, one can also sample $\frac{\mu}{\sigma}$ portion of spatial patches and extend them randomly to a range of $\sigma \cdot \frac{T}{T_p}$ consecutive frames.

(vi) ***Running cell masking*** generates masks shifting from frame to frame with regular changes within small windows of $w_H \times w_W$ tokens. While each spatial window is denoted as a "cell," the group of spatial windows within a temporal span is denoted as a "running cell." As illustrated in the bottom right of Fig. 6.16, the mask on each time step of the running cell switches iteratively with a period of $w_H \times w_W$. In the consequence, tokens of corresponding spatial location are masked with uniform probability. Thereby, running cell masking enables masked tokens to leverage information leaked by both spatially adjacent and temporally adjacent visible tokens. Since each running cell possesses $w_H \times w_W$ states of masking, all cells of the video cube can be either assigned with masks of arbitrary states or a uniform state (g of Fig. 6.16).

6.4.2 Masked Video Prediction on ViTs

In this section, we enumerate MVM-based SSVL frameworks built upon ViTs by dividing them into three categories based on the prediction target, i.e., (1) discrete tokens, (2) hand-crafted features, and (3) plain pixels. To begin with, in Table 6.1, we outline involved methods with a brief introduction from seven aspects of MVM, i.e., encoder architecture, decoder architecture, masking strategy, masking ratio, prediction target, reconstruction loss, and whether `[MASK]` token is used in the encoder.

Predicting Discrete Tokens To take after BERT-fashioned masked language modeling, early attempts in the vision area focused on representing patches as discrete tokens. Discrete Variational Autoencoder (dVAE) was introduced for discretizing images and videos. Output discrete tokens of corresponding masked patches from

Table 6.1 The comparison between different MVM frameworks. † indicates that pre-training of VIMPAC involves contrastive loss, i.e., InfoNCE. W-ViT* represents the self-designed window-based ViT in MaskViT. V and I denote the video stream and the image stream, respectively. E stands for the encoder, and V indicates the decoder

Framework	Encoder	Decoder	Masking strategy	Masking ratio	[MASK] in E	Pred. target	Rec. loss
BEVT [407]	Swin	Deconv.	V: tube, I: block	50%	✓	Dist. token	NLL
VIMPAC [408]	ViT	Linear	Block	~15%	✓	Dist. token	NLL†
MaskViT [409]	W-ViT*	N/A	Random	50%–100%	✓	Dist. token	NLL
MaskFeat [406]	MViT	Linear	Cube	40%	✓	HOG	MSE
MME [410]	ViT	ViT×1d	Tube	70%	✗	HOG+traj.	MSE
VideoMAE [411]	ViT	ViT×4d	Tube	90%	✗	Pixel	MSE
ST-MAE [412]	ViT	ViT×4d	Random	90%	✗	Pixel	MSE
AdaMAE [413]	ViT	ViT×1d	Adaptive	95%	✗	Pixel	MSE
OmniMAE [414]	ViT	ViT×4d	V: tube, I: rand	V: 95%, I: 90%	✗	Pixel	MSE
VideoMS [415]	ViT	ViT×8d	Motion-centric	75–90% (30%)	✗	Pixel	MSE
VideoMAEv2 [416]	ViT	ViT×4d	E: tube, D: cell	E: 90%, D: 50%	✗	Pixel	MSE

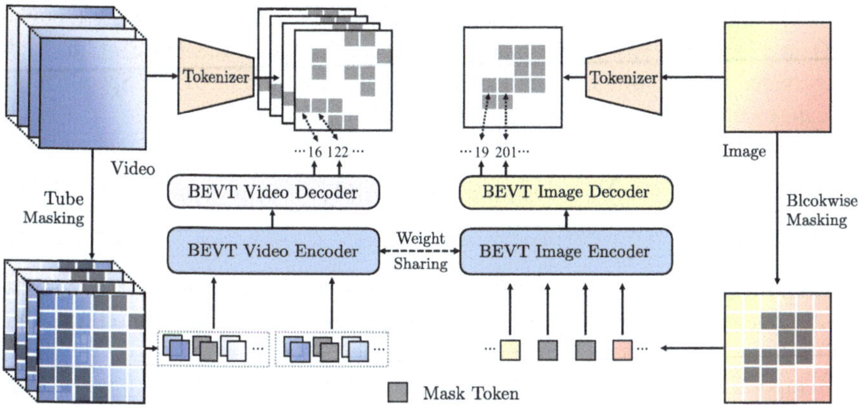

Fig. 6.17 Framework of BEVT. Reproduced from [407]

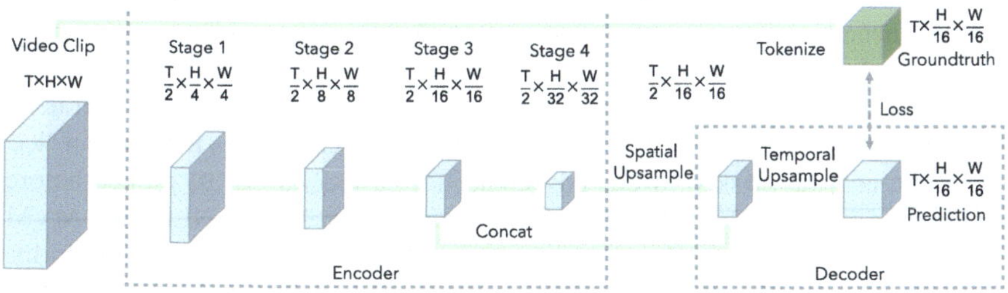

Fig. 6.18 Encoder–decoder architecture of BEVT. Reproduced from [407]

tokenizers serve as the prediction target. In particular, DALL-E [417] and its variants are commonly employed as the tokenizer. In the following paragraph, we introduce three typical MVM frameworks that predict discrete tokens sequentially, i.e., BEVT [407], VIMPAC [408], and MaskViT [409], and the first two frameworks both adopt DALL-E as a pre-trained discretized codebook.

BEVT [407] conducts BERT-style training on both image and video streams with weight-sharing encoders Ψ_{en} instantiated with Swin Transformers, as shown in Fig. 6.17. In order to align with the 3D video stream, for the encoder, the image stream adopts an independent 2D patch embedding followed by a linear projection and reuses 3D positional bias with a relative temporal distance set to zero. A simple decoder acts as the auxiliary prediction head to produce feature maps matched with the ground-truth tokens. As displayed in Fig. 6.18, the hierarchical encoder shifts feature dimensions in 4 stages. Given a video clip X, we denote the output tokens of stage 3 and stage 4 as F_3 and F_4. The decoder fuses the feature maps by first up-sampling F_4 by a transposed convolutional layer (i.e., deconvolution) to obtain $\hat{F}_4$ and processing concatenation of F_3 and $\hat{F}_4$ with a linear head. By proceeding to up-sampling the output in the temporal dimension, the final prediction can be acquired by a `softmax` activation.

For MVM, BEVT takes the initiative to apply a tube masking strategy (shown as f) in Fig. 6.16). Also, a learnable [MASK] token takes the place of masked image/video tokens, which is similar to the implementation of BERT. Specifically, we can rewrite Eq. (6.10) of masked video modeling for BEVT as

$$\mathcal{L}_{\mathrm{MVM}} = -\frac{1}{|P_M|} \sum_{(t,i,j)\in P_M} \log p_{(t,i,j)}(\tilde{X}_{(t,i,j)} | \{X_{(t',i',j')}\}_{(t',i',j')\in P_V}), \tag{6.11}$$

where p denotes the probability value and ground-truth $\tilde{X}$ refers to the corresponding discrete output of the pre-trained discrete codebook. The optimization function for the image stream can be regarded as a particular case of Eq. (6.11) where the temporal dimension contains a single t. In practice, the 2-stream pre-training involves training samples from both image and video datasets and then performs joint optimization by computing gradient of adding up $\mathcal{L}_{\mathrm{MVM}}$ and $\mathcal{L}_{\mathrm{MIM}}$.

VIMPAC [408] gained inspiration from both masked modeling and contrastive learning, hence incorporating two concurrent tasks. Given one video sample v, two clips x_1 and x_2 are independently sampled without spatial augmentations. With regard to the first task (i.e., MVM), the authors exploited a block masking strategy to generate two random masks on x_1 and x_2 separately. Especially, the decoder Ψ_{de} is left out, and the masked tokens are replaced by a specific [MASK] token before entering the encoder. By optimizing the negative log-likelihood (NLL), the MVM loss is analogous to one in Eq. (6.11). On the other hand, the second task performs contrastive learning by viewing the output [CLS] tokens of x_1 and x_2 as a positive pair. Distinct from most other MVM frameworks, VIMPAC adopts the large-scale uncurated HowTo100M [189] as the pre-training dataset.

Moreover, Gupta et al. [409] proposed a MVM-based framework for video prediction, termed as ***MaskViT***, which performs masking on a customized ViT equipped with both spatial window attention and spatial–temporal window attention. MaskViT conducts an efficient and effective generation by an iterative decoding strategy and a dynamic encoder masking ratio ranging from [50%, 100%). A VQ-GAN [418] is used to tokenize video frames into discrete tokens. Accordingly, during pre-training, the learnable [MASK] token is substituted for masked tokens, and MaskViT strives to reconstruct the original discrete tokens from the visible patches.

Predicting Hand-Crafted Features Despite that using discretized tokens as prediction targets leads to promising results, it requires two-stage training for both the tokenizer and the encoder–decoder architecture. On the contrary, nowadays, end-to-end training is more widely appreciated in deep learning, and tokenizers require additional computation overhead. Later effort focuses on exploiting prediction targets for improved efficiency, among which traditional hand-crafted features have been proven to be simple and effective.

Typically, Wei et al. [406] came up with ***MaskFeat*** by investigating different prediction targets in MVM with the same encoder backbone. They observed that predicting Histograms of Oriented Gradients (HOG) [419], a hand-designed feature

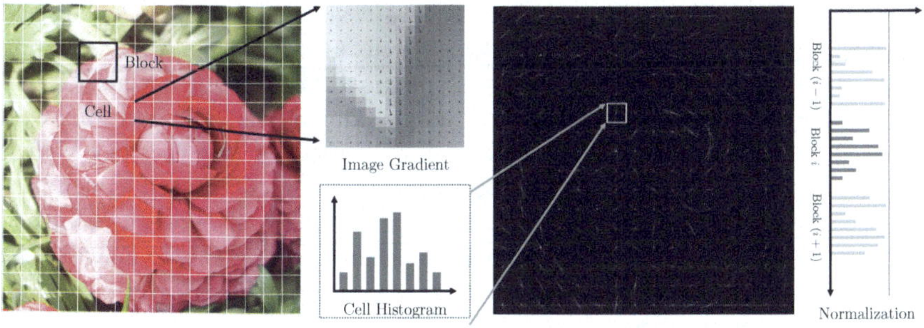

Fig. 6.19 An overview of HOG features extraction

descriptor computed on-the-fly, yields satisfactory performance without heavy computation cost. As displayed in Fig. 6.19, in practice, to compute HOG of a single frame (image), the first step lies in gradients computation between neighboring pixels of each RGB channel both horizontally and vertically, which can be fulfilled by a two-channel convolution. Next, one performs histogramming within $N \times N$ pixel cells by voting gradients into K orientation bins evenly spaced from 0° to 180°. The local contrast normalization of the histogram is conducted among blocks (i.e., $b \times b$ cells). In particular, MaskFeat sets $N = 8$, $K = 9$ and applies l_2 norm. In consequence, the HOG descriptor of each cell contains $b \times b \times K$ channels.

In the implementation, MaskFeat adopts MViT [42] as the backbone of the encoder and a simple linear head as the decoder. During training, 40% of tokens are masked out by cube masking strategy, and a `[MASK]` token is also employed to substitute for tokens covered up by masks. Furthermore, formally, the optimization (Eq. (6.10)) aims to minimize L_2 distance between prediction targets and HOG features.

In a similar spirit, ***MME*** [410] performs masked motion encoding by exploiting manually defined motion trajectory as the prediction target, which incorporates both the position and shape information of long-term motion. MME follows dense trajectories descriptors (DT) [420] to model object motion. Given token sets $Z_{(i,j)} = \{Z_{(0,i,j)}, Z_{(1,i,j)}, \cdots, Z_{(L,i,j)}\}$ within a temporal span of L, the corresponding trajectory $\mathcal{T}_{(i,j)}$ starts with K points densely sampled from the patch of the beginning frame, denoted as $s^0 = \{x_1^0, x_2^0, \cdots, x_k^0\}$. Specifically, $\mathcal{T}_{i,j}$ can be written as dense optical flow.

$$\mathcal{T}_{(i,j)} = (s^0, s^1, \cdots, s^L). \tag{6.12}$$

Accordingly, the prediction target $\tilde{X}_{(i,j)}$ is composed of a concatenation of shape features and motion features. In particular, shape information is represented by a $L \times K$ HOG descriptor with 9 bins, while motion information is extracted by a dense optical flow model. In practice, the trajectory length L is set to 6. Besides, MME instantiates both the encoder and decoder with ViT and trains the network with an MSE loss together with the tube masking strategy.

Predicting Plain Pixels Instead of delicately designing targets of tokens and features, subsequent works design an elegant pretext task by simply reconstructing the pixels in masked patches. He et al. [402] proposed Masked Autoencoder (MAE) for visual representation, which adopts a high masking ratio of 75% and employs a lightweight decoder to reconstruct plain pixels of masked patches. They placed their motivations on a proposition that the successful application of masked modeling in the visual domain is expected to (1) bridge the architecture gap between vision and language by ViTs, (2) apply a high proportion random masking to alleviate negative impacts of visual redundancy, and (3) choose an appropriate decoder design for better latent representations.

Thereafter, Tong et al. [411] and Feichtenhofer et al. [412] concurrently released an adaptation of MAE to the video domain, which we denote as ***VideoMAE*** and spatial–temporal MAE (***ST-MAE***), respectively. These simple yet effective strategies outperform their counterparts above. Basically, they follow the concept of image MAE by adopting a high masking ratio and an asymmetric encoder–decoder architecture for the prediction of masked voxels. Since VideoMAE and ST-MAE share a high degree of similarity, we introduce these two strategies together along with their nuances in implementation (Fig. 6.20):

(A) *Masking strategies.* Video samples suffer from higher redundancy than images for temporal continuity between neighboring frames. Therefore, compared with 2D images, MAEs mask an even higher proportion of videos, i.e., 90%-95% for VideoMAE and 90% for ST-MAE. As illustrated in Fig. 6.16, tube masking, which is also called space-only random masking, is used in VideoMAE, while random masking, which is otherwise denoted as space–time-agnostic masking, is applied in ST-MAE.

(B) *Architecture designs.* For both methods, an encoder Ψ_{en} is instantiated with a vanilla video ViT with space–time joint attention, and a decoder Ψ_{de} is composed of 4-layer ViT with half channels of the encoder. It is worth noting that in MAE pre-training, to improve training efficiency, encoders only process

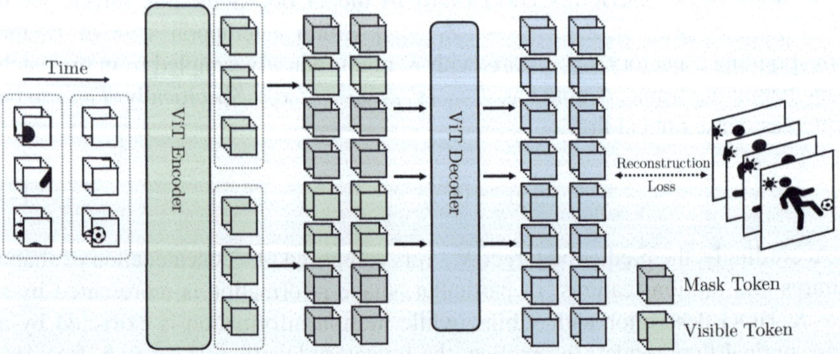

Fig. 6.20 Pipeline of MAE pre-training for videos. Tube masking is adopted in the illustration

unmasked tokens and the decoders are fed with a concatenation of masked tokens and learned unmasked representation. For specification of Eq. (6.10), an MSE loss is adopted as L_{pred}. In particular, pixels within each patch are normalized before computing reconstruction loss in ST-MAE.

Similarly, Girdar et al. came up with ***OmniMAE*** [414], which takes the advantage of both MAE and BEVT by performing a MAE-style joint pre-training of image and video on vanilla ViTs. More importantly, they show that applying an extreme masking ratio of 90% and 95% on image and video stream of OmniMAE, respectively, not only reduces training costs but also acquires satisfactory results.

Moreover, inspired by the remarkable effectiveness of simple pixel-reconstruction tasks, there emerge versatile follow-up works devoted to making further improvements on either efficiency or scalability. For instance, by means of reinforcement learning, ***AdaMAE*** [413] exploits an adaptive sampling policy network $\mathcal{G}_S$, composed sequentially of a multi-head self-attention block, a linear layer, and a `softmax` activation, to assign masks for each token. Specifically, given a training video X, the output token embeddings Z are fed into $\mathcal{G}_S$ to generate a categorical distribution P. Visible tokens are then selected based on P and the masking ratio λ. Authors hypothesize that a higher reconstruction loss implies more significant motion information. Therefore, to optimize $\mathcal{G}_S$ based on policy gradient method [421], they reward tokens leading to more reconstruction error. The corresponding loss can be written as

$$\mathcal{L}(\mathcal{G}_S) = -\sum_{i \in P_M} P^i \cdot \mathcal{L}^i_{MVM}, \tag{6.13}$$

where P_M stands for the set of masked tokens, and $\mathcal{L}^i_{MVM}$ indicates the reconstruction loss for token i. The adaptive masking strategy enables AdaMAE to pre-train with an increased masking ratio yet achieve more competitive results.

On the other hand, ***VideoMAEv2*** [416] aims to accommodate MVM pre-training to both large-scale encoder backbones and enlarged training datasets. To alleviate the major obstacles to scaling up, i.e. training cost and memory overhead, the authors proposed efficient pre-training tactics with the aid of dual masking. In particular, they insert the running cell masking [422] operation before the decoder so as to keep as few as 50% tokens before reconstruction.

VideoMS [415] accelerates pre-training of MVM by performing "mask-then-predict" on $\sigma_{\text{pre}} = 30\%$ portion of pre-selected tokens. The self-defined token selection is based on the proposition that among redundant video patches, those that markedly differ from their temporally neighboring counterparts embody essential motion information and hence play important roles in video representation. Given a patch $P_{(t,i,j)}$, a score I indicating its importance is computed after tokenizing, which can be written as

$$I_{(t,i,j)} = L_2(Z_{(t,i,j)}, Z_{(t-1,i,j)}). \tag{6.14}$$

Tokens with Top-$\sigma_{\text{pre}}\%$ score are viewed as the most motion-centric ones for video modeling. Assume that the masking ratio equals μ, only $(1-\mu)\cdot\sigma_{\text{pre}}$ tokens are fed into the encoder, and as few as σ_{pre} tokens are required for the reconstruction.

Performance We compare the performance of the aforementioned MVM methods when transferring to downstream tasks. Specifically, displayed results cover datasets of Kinetics-400 [22] (Table 6.2) and SomethingSomething-V2 [112] (Table 6.3) for video action recognition. Note that several works, including VideoMAE, VideoMAEv2, and VideoMS, also test their performance on classification datasets of UCF101 [20] and HMDB51 [21]. Generally, MVM-based frameworks demonstrate promising potentials for video understanding, part of which even surpass the supervised counterpart and show state-of-the-art competence.

On top of that, MVM has great potential for adapting to other video downstream tasks, e.g., both MaskFeat, ST-MAE, VideoMAE, and VideoMAEv2 explore the feasibility of transferring to spatial–temporal action localization tasks by achieving the highest mAP of 38.8, 39.0, 39.5, and 42.6 on AVA [115], respectively. Besides, VideoMAEv2 [416] achieves remarkable performance on temporal action detection benchmarks, e.g., 69.9 mAP on THUMOS14 [186] and 18.2 mAP on FineAction [423]. VideoMS succeeds to perform MVM on egocentric first-person video datasets, i.e., Ego4D [424], for improved performance on the object state change classification task. On the other hand, MaskViT also demonstrates promising power in video prediction of robot actions, including datasets of BAIR [425], RoboNet [426] and KITTI [427].

Table 6.2 Results and comparisons of different MVM frameworks on video action recognition benchmark Kinetics-400

Framework	Encoder	Pre-train	Epoch	Frame	GFLOPs	Top-1	Top-5
BEVT	Swin-B	K400+IN-1K	150	32	$282 \times 4 \times 3$	80.6	–
VIMPAC [408]	ViT-L	HowTo100M	100	10	N/A$\times 10 \times 3$	77.4	–
MaskFeat [406]	MViT-L	K400	800	16	$377 \times 10 \times 1$	84.3	96.3
MME [410]	ViT-B	K400	1600	16	$180 \times 7 \times 3$	81.8	–
VideoMAE [411]	ViT-B	K400	1600	16	$180 \times 5 \times 3$	81.5	95.1
ST-MAE [412]	ViT-B	K400	1600	16	$180 \times 7 \times 3$	81.3	94.9
AdaMAE [413]	ViT-B	K400	800	16	$180 \times 5 \times 3$	81.7	95.2
OmniMAE [414]	ViT-B	K400+IN-1K	1600	16	$180 \times 5 \times 3$	80.6	–
VideoMAE [411]	ViT-L	K400	1600	16	$597 \times 5 \times 3$	85.2	96.8
ST-MAE [412]	ViT-L	K400	1600	16	$598 \times 7 \times 3$	84.8	96.2
VideoMAEv2 [416]	ViT-H	K400	1600	16	$1192 \times 5 \times 3$	88.6	97.9

Table 6.3 Results and comparisons of different MVM frameworks on video action recognition benchmark SomethingSomething-v2

Framework	Encoder	Pre-train	Epoch	Frame	GFLOPs	Top-1	Top-5
BEVT [407]	Swin-B	K400+IN-1K	150	32	$321 \times 1 \times 3$	70.6	–
VIMPAC [408]	ViT-L	HowTo100M	100	10	N/A$\times 10 \times 3$	68.1	–
MaskFeat$_{\uparrow 312}$ [406]	MViT-L	800	K400	40	$2828 \times 1 \times 3$	75.0	95.0
MME [410]	ViT-B	K400	1600	16	$180 \times 2 \times 3$	70.5	–
MME [410]	ViT-B	SSv2	1600	16	$180 \times 2 \times 3$	70.0	–
VideoMAE [411]	ViT-B	K400	1600	16	$180 \times 2 \times 3$	69.7	92.3
VideoMAE [411]	ViT-B	SSv2	2400	16	$180 \times 2 \times 3$	70.8	92.4
OmniMAE [414]	ViT-B	SSv2+IN-1K	1600	16	$180 \times 5 \times 3$	69.5	–
AdaMAE [413]	ViT-B	SSv2	800	16	$180 \times 2 \times 3$	70.0	92.7
ST-MAE [412]	ViT-L	K400	1600	16	$598 \times 3 \times 1$	72.1	93.9
ST-MAE [412]	ViT-L	K600	1600	16	$598 \times 3 \times 1$	73.0	94.2
VideoMAE [411]	ViT-L	SSv2	2400	16	$597 \times 2 \times 3$	74.3	94.6
OmniMAE [414]	ViT-L	SSv2+IN-1K	1600	16	$597 \times 5 \times 3$	74.2	–
VideoMAEv2 [416]	ViT-H	SSv2	1600	16	$1192 \times 2 \times 3$	76.8	95.8

6.5 Summary

This section of the book conducts a review of unsupervised feature learning for video understanding, which begins with pretext tasks based on self-prediction and is followed by contrastive self-supervised learning and masked video modeling on ViTs. Above all, unsupervised learning has harvested overwhelming success in video representation and has led to promising performance in video-based downstream tasks, which even surpass the performance of supervised counterparts. In recent literature, efforts have also been made to train foundation models, and large models trained jointly on a vast scale of unlabeled data (e.g., images, text and videos), which have achieved overwhelming performance over a wide spectrum of video-based downstream tasks.

Chapter 7
Efficient Video Understanding

7.1 Design Choices for Compact Neural Networks

As video content becomes increasingly prevalent on the Internet, numerous researchers have focused on the field of video understanding. However, the additional time dimension in video data presents significant challenges for deep neural network models utilized in video understanding. More specifically, methods that rely solely on optical-flow-based approaches [17, 33, 66] for capturing temporal relationships require high computational complexity. Simply incorporating the time dimension into convolution, by transforming the two-dimensional convolution into three-dimensional convolution, also incurs additional parameters and poses challenges for training. Aiming at alleviating the above difficulties, various architectural designs have been proposed one after another.

Convolutional Neural Networks (CNNs) have been extensively studied and widely adopted in various applications. However, the abundance of data in video understanding tasks poses a challenge for basic CNNs that are designed for 2D image domains. As a result, efficient and compact architecture designs are urgently needed for deep learning video models based on CNNs. Over the years, several dominant designs have emerged and gained widespread use in efficient video understanding, including 3D-CNN-based approaches (Sect. 7.1.1), separable convolutions (Sect. 7.1.2), decomposing 3D into spatial and temporal learning (Sect. 7.1.3), and shift-based approaches (Sect. 7.1.4). In the following sections, we will review these approaches in detail.

7.1.1 3D-CNN-Based Approaches

When the research goal transitions from 2D static images to 3D dynamic videos, it seems like a natural idea to upgrade the 2D-CNN to 3D-CNN. However, a common

Z. Wu, Y.-G. Jiang, *Deep Learning for Video Understanding*, Wireless Networks,
https://doi.org/10.1007/978-3-031-57679-9_7

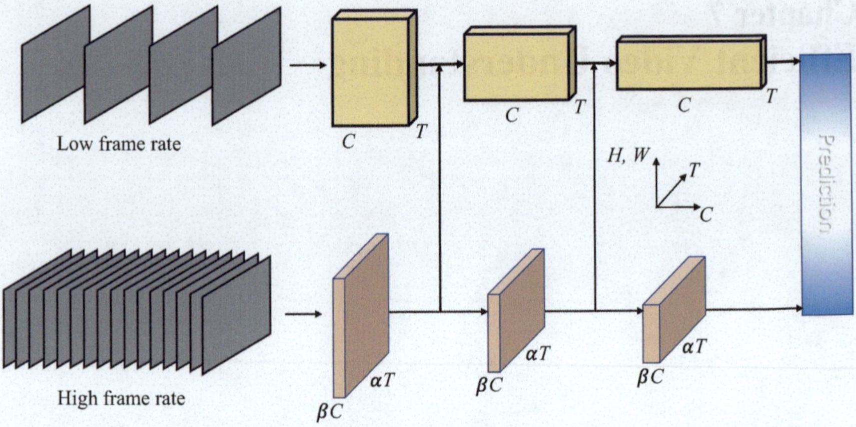

Fig. 7.1 SlowFast has two pathways and the Fast pathway corresponds to a high frame rate, $\alpha\times$ higher temporal resolution, which is lightweight by using a fraction (β, e.g., 1/8) of channels

issue with these models is that they tend to have many more parameters than 2D ConvNets due to the additional time dimension of convolutional kernels, which also makes them more difficult to train.

I3D [35] leverages an ImageNet pre-trained image classification architecture [428] but inflates all the filters and pooling kernels along the temporal dimension to 3D, resulting in a deep and effective spatial–temporal classifier. However, it has been recognized that not all spatial–temporal orientations are equally important. A common intuition is that spatial semantics in vision tend to change slowly, while dynamic motion may develop much faster than its subject. SlowFast [39] builds upon this idea by proposing a two-pathway model. The framework of SlowFast is shown in Fig. 7.1, from which we can see it is composed of two pathways: (1) a slow pathway that takes low-rate inputs to focus on spatial semantics and (2) a fast pathway that, on the other hand, processes high-rate inputs to model the motion information. Note that although the high temporal rate in the fast pathway entails a significant computational overhead, reducing the channel capacity of the fast pathway can make it much more lightweight than expected, achieving an ideal trade-off between computational complexity and temporal modeling ability. Overall, the two pathways are horizontally fused together to form the SlowFast network for video recognition.

Although SlowFast [39] introduced above incorporates a fast pathway with a high frame rate to effectively reduce the computational overhead, the slow pathway is still computationally expensive (Fig. 7.2). In order to design a more lightweight network structure, X3D [40] further explores the parameter savings on more axes, i.e., temporal length, spatial resolution, bottleneck width, and even depth. Specifically, they adopt a simple stepwise extension method to extend a 2D image classification model along one axis at each step to choose the best hyperparameters. Eventually, the desired computational budget could be achieved by an axis-by-axis parameter

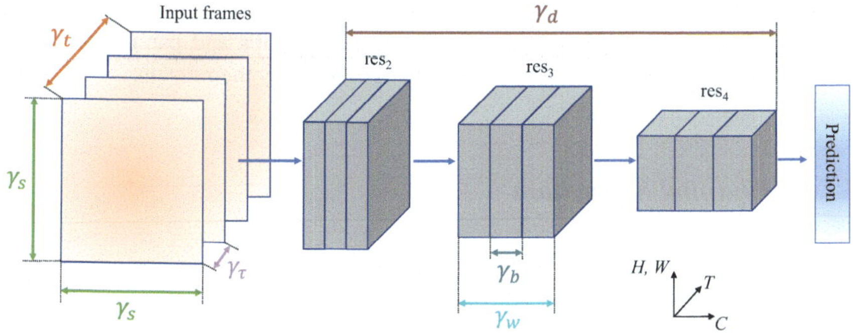

Fig. 7.2 X3D networks. The candidate axes: frame rate γ_τ, temporal duration γ_t, spatial resolution γ_s, depth γ_d, bottleneck width γ_b, and width γ_w

Table 7.1 Comparison to other advanced models on Kinetics-400 dataset. The "test" column indicates the top 1 and top 5 average scores on the Kinetics-400 [35] test set

Model	Top-1	Top-5	Test	GFLOPs × views	Param
Two-Stream I3D [35]	71.6	90.0		216 × N/A	25.0M
R(2+1)D [36]	72.0	90.0		152 × 115	63.6M
Two-Stream R(2+1)D [36]	73.9	90.9		304 × 115	127.2M
ip-CSN-152 [60]	77.8	92.8		109 × 30	32.8M
SlowFast 4 × 16, R50 [39]	75.6	92.1		36.1 × 30	34.4M
SlowFast 8 × 8, R101 [39]	77.9	93.2	84.2	106 × 30	53.7M
SlowFast 8 × 8, R101+NL [39]	78.7	93.5	84.9	116 × 30	59.9M
SlowFast 16 × 8, R101+NL [39]	79.8	93.9	85.7	234 × 30	59.9M
X3D-M	76.0	92.3	82.9	**6.2 × 30**	**3.8M**
X3D-L	77.5	92.9	83.8	24.8 × 30	6.1M
X3D-XL	79.1	93.9	85.3	48.4 × 30	11.0M
X3D-XXL	**80.4**	**94.6**	**86.7**	194.1 × 30	20.3M

search. We compare three variants of X3D [40] with other state-of-the-art models in Table 7.1. The results show that the X3D series outperforms SlowFast [39], while requiring 4.8 and 5.5 × fewer GFLOPs and parameters, respectively.

It is not difficult to find that X3D saves computational complexity at the cost of expensive hyperparameter searching. Each expansion step of X3D requires training 6 models, corresponding to its six axes dimensions. When we need a moderate-size model, let us assume it needs to expand by 10 steps, which means 60 models will be trained independently throughout the process. To address this issue, the adaptive 3D networks (A3D) [429] train different spatial–temporal-width configurations jointly. Through their proposed Spatial–Temporal Distillation (STD), knowledge from each configuration is transferred to one another, enabling better multi-scale spatial and temporal information learning. Each model configuration captures distinct semantic information, making the knowledge sharing process beneficial. During each training iteration, two sub-networks share parameters with a part of the full network. This approach allows the full network to benefit from the training of the sub-networks. A3D [429] achieves higher accuracy with fewer GFLOPs and parameters using lower spatial resolutions and network widths, as shown by their results.

7.1.2 Channel Separation

Separable convolutions are a technique that can be traced back to group convolutions, which were proposed in AlexNet [9] as a solution to overcome computing resource limitations. Group convolutions group input channels and then combine features, with the number of parameters for the grouping convolution being $1/G$ of the total number of parameters, where G is the number of divided groups. Group convolutions have recently been widely used to make 2D image recognition networks lightweight and have been successful in various architectures such as [430–434]. In this section, we will discuss the application of 3D separable convolution in video recognition.

Xception [431] factorizes 2D convolution into channel and spatial dimensions and has shown great results in the image domain. Building upon this idea, the Channel-Separated Convolutional Network (CSN) [60] further proposed to factorize 3D convolution into channel and spatial–temporal dimensions for action recognition. In CSN, all convolutional layers (excluding the first layer) are implemented using either traditional convolutions of size $1 \times 1 \times 1$ for channel interaction or depthwise convolutions of size $k \times k \times k$ for local spatial–temporal interactions, $k = 3$ typically. There are two factorization methods for a 3D bottleneck block called interaction-preserved and interaction-reduced, respectively, and we compare them with the ordinary 3D bottleneck block in Fig. 7.3. The interaction-preserved bottleneck block replaces the original 3D convolution block by a standard $1 \times 1 \times 1$

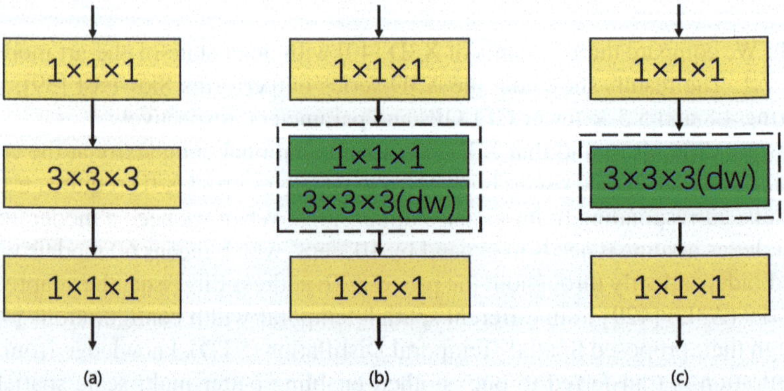

Fig. 7.3 (**a**) A standard ResNet bottleneck block. (**b**) An interaction-preserved bottleneck block. (**c**) An interaction-reduced bottleneck block

Table 7.2 Performance comparison between channel-separated networks (ir-CSN, ip-CSN) and ResNet3D on Kinetics validation

Model	Depth	video@1(%)	GFLOPs	Interactions ($\times 10^9$)	Params(M)
ResNet3D	26	**65.3**	14.3	0.42	20.4
ir-CSN	26	62.4	4.0	0.27	1.7
ip-CSN	26	64.6	5.0	0.42	2.4
ResNet3D	50	69.4	29.5	5.68	46.9
ir-CSN	50	70.3	10.6	5.42	**13.1**
ip-CSN	50	**70.8**	11.9	5.68	**14.3**
ResNet3D	101	70.6	44.7	8.67	85.9
ir-CSN	101	71.3	14.1	8.27	**22.1**
ip-CSN	101	**71.8**	15.9	8.67	**24.5**

convolution and a $3 \times 3 \times 3$ depthwise convolution, which could not only effectively reduce the number of parameters and GFLOPs, but also enable the communication between different channels. On the other hand, in the interaction-reduced bottleneck block, the $3 \times 3 \times 3$ convolution is substituted with a $3 \times 3 \times 3$ depthwise convolution, leading to a complete loss of channel interaction.

Table 7.2 presents the results of interaction-reduced CSN (ir-CSN) and interaction-preserved CSN (ip-CSN) and compares them with ResNet3D baseline under different depths. Note that both ir-CSN and ip-CSN outperform ResNet3D (by approximately 0.9–1.4%) while using far fewer parameters and GFLOPs for deeper network configurations, e.g., 50 and 101 layers. Furthermore, the performance discrepancy between ir-CSN and ip-CSN decreases. It can be inferred that conventional $3 \times 3 \times 3$ convolutions with numerous parameters can be eliminated in the deeper models without any compromise on accuracy.

The Channel-Separated Convolutional Network [60] provides a feasible solution to make 3D convolutional operations more lightweight. Based on the intuition that some feature channels may emphasize static spatial features, while others may focus more on dynamic motion features, Grouped Spatial–Temporal aggregation (GST) [435] further explores asymmetric decomposition by dividing the output channels of a standard 3D convolution layer into a spatial modeling group and a spatial–temporal modeling group, respectively. As shown in Fig. 7.4, the standard 2D convolutions are applied to the spatial path, while 3D spatial–temporal convolutions are used over time.

In order to further save the computational overhead, GST [435] also divides the input channels C_i into spatial C_{i_s} and temporal C_{i_t} groups and applies 2D and 3D convolutions to them, respectively. Figure 7.4c and d illustrates the resulting architecture, where a hyperparameter α determines the proportion of temporal output channels, i.e., $C_{o_t} = \alpha C_o$. The numbers of parameters for the spatial and temporal paths are $(1 - \alpha) HWC_{i_s} C_o$ and $\alpha THWC_{i_t} C_o$, respectively, while the total parameters for the GST module with $C_{i_s} = C_{i_t} = C_i/2$ are $(1 - \alpha + \alpha T) HWC_i C_o/2$, and T, H, and W denote the temporal and spatial dimensions of the input. Additionally, they find that reducing the number of temporal output

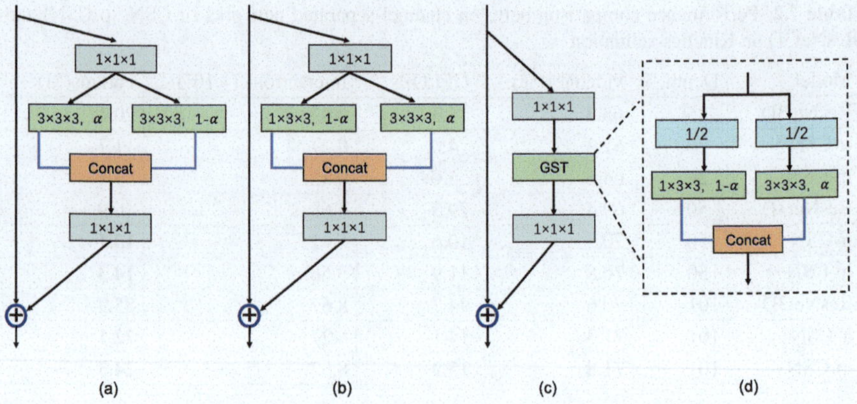

Fig. 7.4 Outline of the proposed method. (**a**) depicts an analogous network to C3D [31]. (**b**) illustrates the substitution of one path with spatial-only convolutions, referred to as GST-Large. (**c**) showcases the method, while (**d**) highlights the proposed GST module

channels leads to a significant reduction in the parameter count without negatively impacting model performance. The spatial–temporal decomposition method [435] is shown to utilize a similar or lower number of parameters compared to 2D models, yet outperform some 3D models [31, 72] in experiments.

7.1.3 Decompose 3D into Spatial and Temporal Learning

In addition to the factorization of 3D convolutions in the feature channel dimension, other approaches [36, 72, 73, 436] have also been proposed to decompose the 3D spatial–temporal representation learning into separate spatial and temporal modeling to reduce the memory demand and computational cost.

A straightforward idea is to decompose a 3D convolutional filter with a size of $3 \times 3 \times 3$ into a $1 \times 3 \times 3$ convolutional filter and a $3 \times 1 \times 1$ convolutional filter. To take advantage of this decomposition, P3D [72] introduces a Pseudo-3D (P3D) block to model the spatial–temporal information in videos. There are three designs of bottleneck building blocks (P3D) presented in Fig. 7.5. P3D-A in Fig. 7.5a develops a cascaded architecture where spatial filters are followed by temporal filters, allowing these two types of filters dependent on each other. The two filters in the second design P3D-B illustrated in Fig. 7.5b hold in a parallel fashion and directly accumulate into the output. P3D-C shown in Fig. 7.5c combines the characteristics of P3D-A and P3D-B by connecting the spatial filters to the final output directly. Experimental results have shown that all P3D ResNet and variants exhibit better performance than prior techniques [31, 33, 65] with significantly smaller model sizes.

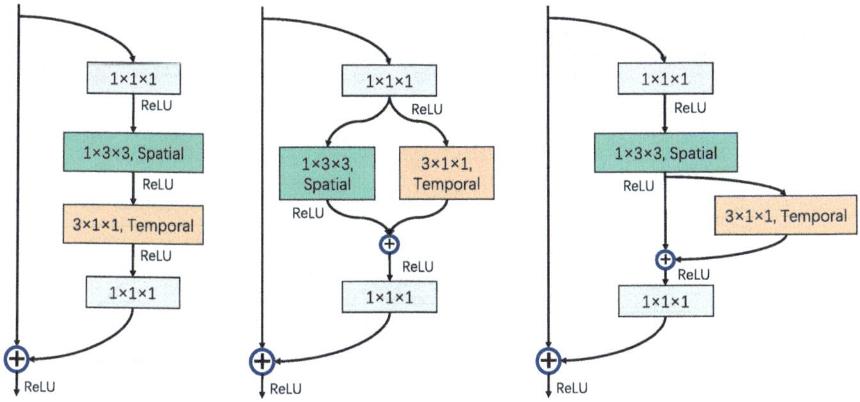

(a) P3D-A (b) P3D-B (c) P3D-C

Fig. 7.5 Bottleneck building blocks of Pseudo-3D. (**a**) Temporal 1D filter follows spatial 2D filter in a cascaded manner. (**b**) Both filters are at different pathways in a parallel fashion. (**c**) Two filters operate in a cascaded manner, and the spatial filter has an extra shortcut connection with the output

R(2+1)D [36] shares the same decomposition technique with P3D [72] but differs in that it employs a uniform spatial–temporal residual block across all layers without bottlenecks. Furthermore, R(2+1)D is intentionally designed to have the same number of parameters as its 3D counterparts, which in turn helps double the number of nonlinearities. The experiments such as architectural design significantly outperforms C3D [31] and P3D [72] on several primary benchmarks [18, 35].

Another interesting work is [73], which presents an efficient approach for video classification by integrating two types of factorized designs: top-heavy architecture that adopts computationally heavy 3D convolutions at the top layers and utilizes 2D convolutions for the lower layers, and bottom-heavy architecture that uses 3D at the bottom layers and 2D in the upper layers. Specifically, they replace 3D convolutions in I3D [35] with two consecutive convolution layers (Sep-Conv on the right side of Fig. 7.6a) and obtain a model referred to as S3D illustrated in Fig. 7.6. By using separable temporal convolution (Sep-Conv) to the network, all branches in an inception block shown in Fig. 7.6b can observe temporal information that contributes to better model performance. The results demonstrate that, compared to I3D [35], S3D achieves a significant model size reduction (12.06M parameters to 8.77M) but maintains a gain in accuracy (71.1% top-1 to 72.2% for Kinetics-Full [22], and 45.8% top-1 to 47.3% for Something-Something V2 [110]).

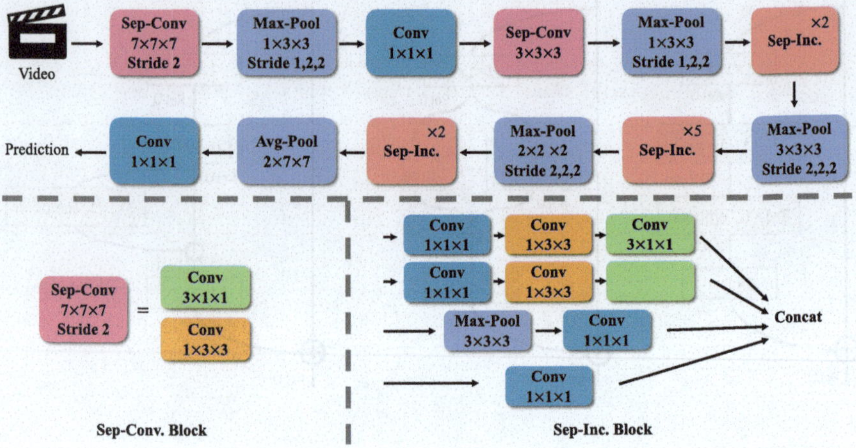

Fig. 7.6 An illustration of the S3D model. Dark red boxes are temporal separable convolutions (Sep-Conv), and pink boxes are 3D temporal separable inception blocks shown in (**b**)

7.1.4 Shift-Based Approaches

The shift operation has been successfully explored in designing compact 2D CNNs for image classification [437, 438]. However, due to the inherent complexity of videos, directly applying shift operations to the temporal dimension of 3D CNNs can lead to high computational and memory costs.

To tackle these challenges, [38] propose a generic Temporal Shift Module (TSM) that enhances the shift strategy for video understanding tasks. In contrast to prior works that shift all channels along the temporal axis, TSM selectively shifts a small number of channels, which alleviates the time complexity caused by excessive data movement. Additionally, the TSM is embedded within the residual branch of a residual block, preserving the original activation and allowing for the extraction of temporal information without damaging the spatial feature information of 2D convolution.

TSM [38] enables efficient temporal modeling in both offline and online video recognition, as shown in Fig. 7.7. In the offline setting, as presented in Fig. 7.7b, TSM shifts a fraction of channels along the temporal dimension, integrating both past and future temporal clues into the current frame without incurring additional overhead. Such a bidirectional shift can be regarded as a replacement for 1D temporal convolutions. In online video understanding, where future frames are not available, only past frames are shifted in a uni-directional fashion (Fig. 7.7c). The results demonstrate the proposed TSM exhibits comparable spatial–temporal modeling capabilities as 3D CNN-based approaches, but with the parameter and computation cost equivalent to that of 2D CNNs.

RubiksNet [439] introduced a learnable 3D-shift operation (RubiksShift) that operates in both spatial and temporal axes, effectively enlarging the receptive

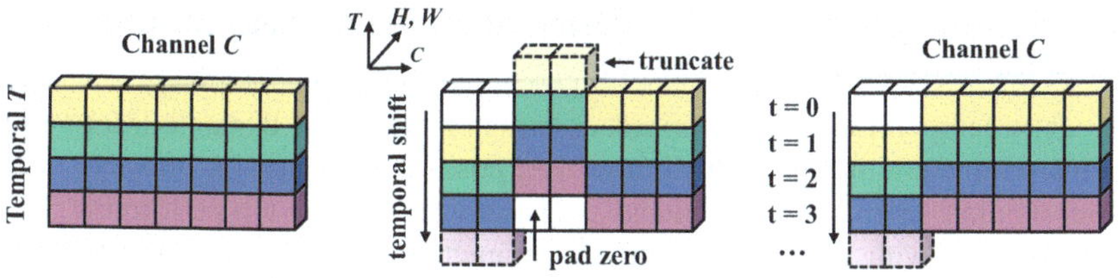

(a) Original tensor (b) Offline shift (c) Online shift

Fig. 7.7 **Temporal Shift Module (TSM)** performs efficient temporal modeling by moving the feature map along the temporal dimension. (**a**) The original tensor without shift. (**b**) Offline temporal shift (bidirection). (**c**) Online temporal shift (uni-direction)

Table 7.3 Comparison with a range of base architectures on Something-Something V2 dataset [110]. (#×) is referred to as efficiency savings relative to a similarly sized TSM model

Method	Size	Input	Top-1	Top-5	**#Param.**	FLOPs/Video
TSN [33]	Large	8	30.0	60.5	24.3M	33G
TRN [75]	Large	8	48.8	77.6	18.3M	16G
bLVNet-TAM [440]	Large	8×2	59.1	86.0	25M	23.8G
TSM [38]	Large	8	58.8	85.6	24.3M	33G
	Medium	8	56.9	84.0	21.4M	29.4G
	Small	8	49.3	77.6	11.3M	14.6G
RubiksNet [439]	Large	8	59.0	85.2	8.5M (2.9×)	15.8G (2.1×)
	Medium	8	58.3	85.0	6.2M (3.5×)	11.2G (2.6×)
	Small	8	57.5	84.3	3.6M (3.1×)	6.8G (2.1×)
	Tiny	8	54.6	82.0	1.9M (5.9×)	3.9G (3.7×)

field of the model. In contrast to prior modules [38] that employed hand-crafted fixed temporal shift throughout the network, RubiksShift allows for efficient and adaptable receptive field enlargement. By integrating spatial shift and temporal shift, a three-dimensional continuous interpolated shift primitive was designed, so that the shift parameters could be optimized through backpropagation. To further speed up the computation, Fan et al. also proposed a quantized variant where all shifts are rounded to the nearest integer during forward propagation. However, this method is difficult to optimize. To address this issue, they introduced attention quantization that gradually learns the integer shift values using annealing and represents them in a one-hot manner. The performance comparison with a range of base architectures on Something-Something V2 dataset [110] is presented in Table 7.3. RubiksNet [439] achieves comparable or superior accuracy compared to previous shift-based video recognition approach [38] while significantly reducing the number of parameters and computation cost.

7.2 Training Strategies for Efficient Video Understanding

The extra temporal dimension in videos brings a wealth of dynamic information compared to static images, which, however, poses a challenge for effective video representation learning concomitantly. Therefore, training top-performing deep video models with tens of millions of parameters [39, 43, 101, 411] typically requires more memory consumption and longer training schedules, which has slowed the progress of video understanding research [441, 442].

In order to mitigate the above issue, a line of work has explored to develop time-efficient [441, 443, 444] and memory-efficient methods [442, 445] to save the overhead of training video models from different aspects. In the following, we review both types of approaches and present representative works.

7.2.1 *Time-Efficient Methods*

Training deep video models has traditionally taken significantly more time than their image counterparts, often by an order of magnitude [441, 444]. This gap is further enlarged as the vision community is witnessing a transfer in backbone models from CNN [9, 12] to Transformer [14, 54]. To achieve faster convergence with competitive performance, various methods have been proposed by initializing weights from image models (Sect. 7.2.1.1) and designing advanced training strategies (Sect. 7.2.1.2).

7.2.1.1 Weight Initialization from Image Models

Pre-training then fine-tuning paradigm is a prevalent paradigm in the vision community, which typically transfers the models pre-trained on large-scale datasets, e.g., ImageNet [15] and CC12M [446], to downstream tasks through cost-effective fine-tuning. Inspired by this, several works [35, 67] leverage image pre-training for further video representation learning to save enormous computational resources.

Initializing from ImageNet pre-trained Models, e.g., ResNet [12], to acquire strong visual representations. Early methods such as ST-ResNet [67] and I3D [35] simply inflate the 2D kernels of pre-trained image models along the temporal dimension to 3D shapes. This improves the efficiency of spatial–temporal feature learning in video and has become a mainstream paradigm for training deep video models [37, 41, 43]. However, the results in recent literatures [39, 40, 60] indicate that the ImageNet pre-training will impair the performance to some extent, which leads to a tough choice between the training time and performance for subsequent research. To overcome this limitation, STS [447] explores the design of better architectures and fine-tuning strategies. They first study the relationship between from-scratch trained 3D kernels [36, 36, 39, 60, 448] and 2D appearance modeling

Table 7.4 Comparisons of training time and Top-1 accuracy on Kinetics-400 dataset with and without ImageNet pre-training and STS convolution. The results are replicated from STS [447]

	w/ Image Pre & STS Conv	Pre-train Time	Fine-Tune Time	Total Time	Speedup	Top-1 Acc
Slowonly50-8f [39]	✗	–	389	389	×1	74.9
	✓	182	76	258	×1.5	75.6
Slowfast50 (4 × 16) [39]	✗	–	696	696	×1	75.6
	✓	237	136	373	×1.9	76.2
Slowfast50 (8 × 8) [39]	✗	–	840	840	×1	76.9
	✓	237	164	401	×2.1	77.2
R2plus1D34-8f [36]	✗	–	778	778	×1	68.7
	✓	230	152	382	×2.0	76.2
CSN50-32f [60]	✗	–	1106	1106	×1	73.6
	✓	176	216	392	×2.8	76.7
X3D-S-13f [448]	✗	–	432	432	×1	73.2
	✓	121	144	265	×1.6	73.5

through the linear-probing experiments on ImageNet. Surprisingly, the spatial–temporal features learned on video data exhibit strong appearance modeling ability.

The above phenomenon implies that image pre-training, with proper adaptation, could benefit the video understanding as the appearance prior. To this end, a 3D Spatial–temporal Separable (STS) convolution [447] is proposed to decompose the vanilla 3D convolution into two groups along the output channel: a static appearance modeling (SAM) group that consists of one 2D convolution and two perpendicular 1D convolutions and a dynamic motion modeling (DMM) group with 3D convolution only. Specifically, the STS convolution could be formulated as

$$\begin{aligned} y_t = \text{concat}(&Conv1D(x_t^{row}; \alpha_0) + Conv2D(x_t; \alpha_1) + Conv1D(x_t^{col}; \alpha_2), \\ &Conv3D(x; \beta)), \end{aligned} \tag{7.1}$$

where $x \in \mathbb{R}^{C\times T\times H\times W}$ is the input video tensor, $x_t^{row} \in \mathbb{R}^{C\times WH}$ and $x_t^{col} \in \mathbb{R}^{C\times HW}$ are reshaped frame tensors, and α and β denote the learnable parameters in SAM and DMM groups, respectively. During training, α_1 and the center kernel of β are initialized with pre-trained weights, while α_0 and α_2 are initialized with zeros. The comparison of training time and performance with and without ImageNet pre-training (as well as STS convolution) is shown in Table 7.4. The results demonstrate that, on average, the proposed method achieves better performance with a 2× speedup compared to the training-from-scratch baseline.

Initializing from counterpart Image Models typically requires the video model and the counterpart image model to be fully isomorphic, which is enabled by the Transformer architecture [14, 53] benefiting from its ability to handle variable-length sequences. Large-scale video pre-training, including video self-supervised learning [408] and video-language pre-training [449, 450], plays a key role in the success of deep neural networks on various downstream tasks recently. Compared with image pre-training,[1] video pre-training is far more challenging due to its dynamic nature. A recent literature [364] indicates that the successful self-supervised learning frameworks in the image domain, e.g., MoCo [373], typically require hundreds of training epochs to achieve competitive performance on video benchmarks (e.g., 800 epochs on Kinetics-400 [35]) if trained from scratch.

To alleviate this problem, BEVT [407] argues that performing image representation learning jointly could make video representation learning more efficient by utilizing the encoded spatial priors. To this end, they propose to learn the spatial–temporal features in video with a two-stream framework: an image stream for spatial modeling and a video stream for temporal modeling. Specifically, the former focuses on spatial representation learning and is pre-trained on image data first by predicting the latent codes of masked image tokens [401], while the latter learns temporal dynamics on video data with the same form of supervision. The two streams share the attention weights during training and work collaboratively for video modeling. Experimental results have shown that such a design enables BEVT to achieve superior results in a significantly shorter training schedule (i.e., 150 epochs on Kinetics-400). OmniVL [451] inherits this spirit and further proposes a decoupled joint pre-training paradigm for vision-language pre-training. They first pre-train on image language to learn the alignment between the appearance features and language modality and then do joint pre-training with video language together to learn the motion-language alignment incrementally. The results on a wide variety of downstream tasks demonstrate that learning from spatial to temporal could not only improve the efficiency of vision-language training but also enjoy the complementarity of data from different sources.

7.2.1.2 Advanced Training Strategies

End-to-end training with video clips is a popular approach to train deep video models nowadays [39, 41, 42, 101]. Considering the diversity of video data and numerous parameters in video models, fitting a competitive-performing model on a large-scale video dataset, such as Kinetics-400 [35], oftentimes consumes a considerable amount of time [441]. To address this problem, a few methods have been proposed to accelerate the training of video models by designing and training strategies.

One of the most representative work is Multigrid [441]. Memory limitations and temporal redundancy of videos make it impractical, and unnecessary, to take all the frames in a video as inputs to train deep models. Instead, existing methods typically sample a video clip of fixed shape, e.g., $T \times 224 \times 224$, to represent the whole video [39, 41, 43, 67]. However, a trade-off between the convergence speed and accuracy

[1] Here we refer to self-supervised image pre-training and image-language pre-training collectively as image pre-training.

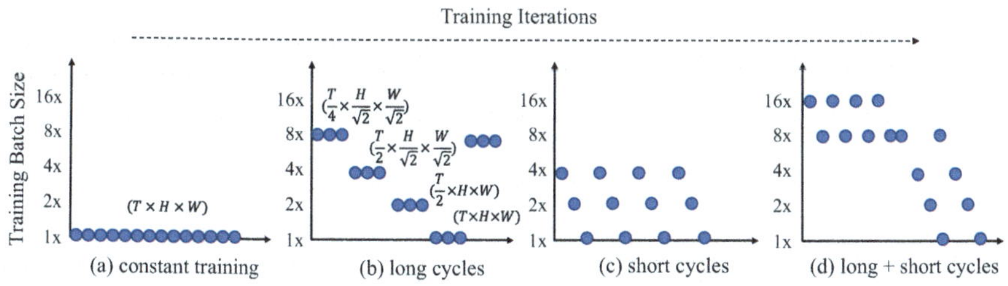

Fig. 7.8 Illustration of different training schedules: (**a**) Constant training method adopts a fixed shape of batch data during training. (**b**) The long cycle method gradually increases the spatial and temporal resolution of batch data (while reducing the batch size) over a long period. (**c**) The short cycle method moves through a variety of spatial–temporal shapes in a short period. (**d**) The proposed long + short cycle method changes the shape at two frequencies simultaneously

could be observed by adjusting the shape of sampled clips: High-resolution models tend to perform well with longer training time, while low-resolution models train faster at the sacrifice of accuracy [36, 37, 39]. In order to balance both aspects, Multigrid [441] introduces a multigrid training strategy that yields a significant training speedup without a loss in performance.

Multigrid methods in numerical analysis [452] reveal that the optimization problems could be resolved by alternating coarse and fine grids. Inspired by this, Multigrid [441] proposes to dynamically adjust the spatial–temporal resolution of the batched data during training by resampling the video clips on different grids. Note that in order to control the overall computation for an iteration, the batch size and clip shape satisfy the following requirements:

$$b = B\frac{T}{t}\frac{H}{h}\frac{W}{w}, \tag{7.2}$$

where B is the batch size, and T, H, W are the number of frames and spatial resolution, respectively. The proposed method defines a set of sampling grids on the concrete input signals and carefully designs a schedule to specify which grid to use in each iteration. Following the coarse-to-fine curriculum learning, they start with large batch size and small resolution and gradually move to the opposite scale. With a fixed number of training epochs, increasing the batch size on average could therefore save the wall-clock time.

Multigrid [441] implements three types of multigrid training schedules for comparison: constant training (Fig. 7.8a) that uses a fixed mini-batch shape through training; long cycles (Fig. 7.8b) that update the sampling grid at a smaller frequency (every several epochs); short cycles (Fig. 7.8c) that rapidly traverse from small resolutions to large resolutions (every iteration); long + short cycles (Fig. 7.8d) that combine the above long and short cycles and move through different shapes at both frequencies simultaneously. The training time and Top-1 accuracy on the validation set of Kinetics-400 are shown in Fig. 7.9. We can observe a performance degradation for constant training with $\geq 1.5\times$ epochs. By contrast, the results of three variants

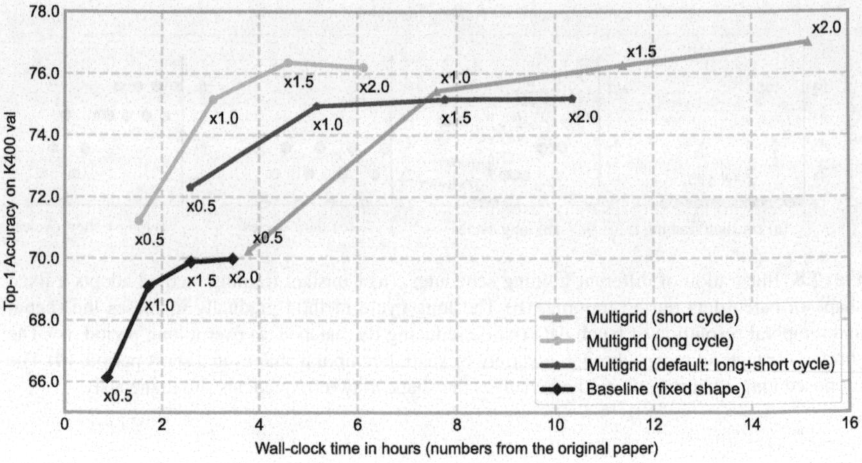

Fig. 7.9 Comparison of training time and validation accuracy between multigrid and other baselines. The numbers on top of each point denote the training epochs relative to the baseline 1.0× schedule

with multigrid training could be continuously improved up to 2.0× epochs without overfitting, and the multigrid training with long and short cycles achieves the most competitive trade-off.

Drawing inspiration from multigrid training, MultiVSR [444] further proposes a multi-stage training strategy to accelerate the training of video super-resolution models. Video super-resolution (VSR) [453] aims to recover high-resolution (HR) videos from low-resolution (LR) inputs. This task is challenging due to the high cost of training complex spatial–temporal information. Existing VSR methods typically fix the spatial–temporal shape of the output video during training, which is verified by MultiVSR [444] to be unnecessary and even increase the difficulty of spatial–temporal reconstruction learning. To address this issue, they propose to learn the video super-resolution in an easy-to-hard manner by gradually increasing the output resolution of the VSR models with two cycles: a spatial cycle that varies the spatial size from small to large and a temporal cycle that moves the temporal length from short to long.

While appealing, the experimental results demonstrate that simply applying the above multi-stage strategy will lead to an obvious performance drop, which results from the "continuously decaying" learning rate schedule adopted in most VSR models [454, 455]. MultiVSR [444] argues that a relatively large learning rate is required at the beginning of each stage to adapt to the switch of different output resolutions. To this end, they introduce a dynamic learning rate scheduler, which re-starts the learning rate with large values when the spatial or temporal size changes. Quantitative comparison between the multi-stage training and several baselines on a

Table 7.5 Quantitative comparison on REDS4 with BasicVSR and EDVR-M

Method	Batchsize	Multi	Speedup	PSNR	SSIM
BasicVSR [455]	32	–	–	31.42	0.8909
BasicVSR [455]	64	–	1.9×	31.62	0.8943
Multi-S	64	S	2.4×	31.56	0.8932
Multi-T	64	T	2.5×	31.61	0.8941
Multi-ST	64	S&T	**3.1**×	31.54	0.8925
EDVR-M [454]	32	–	–	30.46	0.8684
EDVR-M [454]	64	–	1.9×	30.46	0.8689
Multi-ST	64	S	**2.3**×	30.44	0.8685

"S" and "T" denote spatial and temporal, respectively

standard video super-resolution dataset REDS [456] is shown in Table 7.5. We can see that the proposed method could accelerate the training of VSR methods while achieving superior performance.

7.2.2 *Memory-Efficient Methods*

In addition to reducing training time, the development of memory-efficient training methods can also be of great value in reducing the bar for training video models, especially in resource-limited scenarios.

7.2.2.1 CLIP Transfer Learning

In recent years, vision-language pre-training (VLP) [241, 457] has gained considerable attention by means of distinguished transferability. By learning cross-modal alignments in a contrastive manner over large numbers of image–text pairs, VLP models, i.e., CLIP [241], develop powerful visual understanding capability and thus achieve good empirical performance on various visual and cross-modal tasks. As a result, there is a growing interest in using frozen CLIP features for efficient video understanding tasks.

Efficient prompting [458] is a pioneering work that directly transfers the frozen CLIP features to a series of video downstream tasks. Taking advantage of the strong generalization capability of CLIP, they fine-tune a set of learnable vectors, i.e., prompts, on each downstream task to learn the task-specific knowledge adaptively, which saves prominent computational costs compared with training the complete model in an end-to-end manner. Specifically, efficient prompting [458] first obtains the per-frame visual features with the frozen CLIP image encoder, on top of which a lightweight Transformer block is built to model the temporal interaction between different frames. For the textual representation, they additionally append several

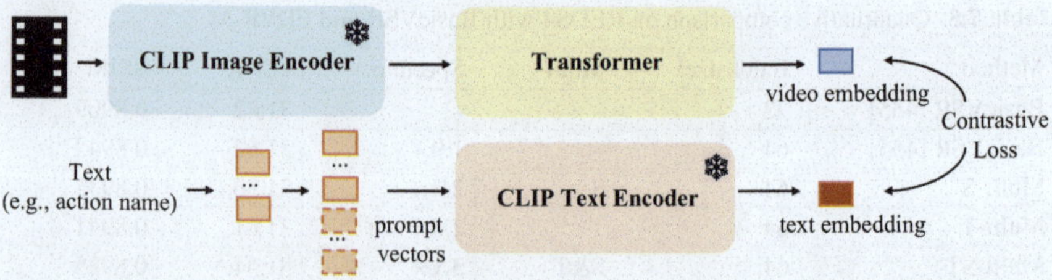

Fig. 7.10 Overview of the efficient prompting [458] and its deployment on various video understanding tasks including action recognition, text–video retrieval, and action localization

prompt vectors to the input word embeddings, which are jointly input to the CLIP text encoder to get the text embedding. With the above unimodal embeddings, efficient prompting reformulates different video tasks into a unified contrastive paradigm for fine-tuning with minimal effort:

- Action recognition (AR) classifies the category of video clips. The category names are taken as text input to learn a joint space with the corresponding videos.
- Action localization (AL) localizes action in untrimmed long videos. Following the two-stage practice that first detects action proposals and then performs proposal-level classification, efficient prompting inputs the detected action proposals, instead of the whole video, to the visual encoder to align with the corresponding action descriptions.
- Text-to-Video Retrieval (TVR) retrieves a relevant video from a database given a descriptive sentence. The cross-modal nature makes TVR easily compatible with CLIP training.

The architecture of efficient prompting [458] and its adaption on three classic video understanding tasks are illustrated in Fig. 7.10. Extensive experiments are conducted on 10 popular action recognition, action localization, and text–video retrieval benchmarks. The results demonstrate that with prompt vectors trainable only, efficient prompting [458] could achieve competitive performance under close-set, few-shot, and even zero-shot scenarios.

Efficient Video Learner (EVL) [459], on the other hand, abandons the text encoder of CLIP and only reserves the image encoder for visual representation modeling. Similar to efficient prompting [458], they first adopt the frozen CLIP as the backbone to extract per-frame features $X_{1..N} = \{X_i\}_{i=1}^{N}$ independently, where N is the number of input frames. Considering CLIP [241] inherently lacks the temporal modeling ability that lays the core of video understanding, EVL [459] further introduces a temporal decoder block to fuse the temporal information dynamically:

$$\hat{X}_{1..N} = \mathrm{Tconv}(X_{1..N}) + \mathrm{CFAttn}(X_{1..N}), \tag{7.3}$$

where Tconv and CFAttn denote the temporal depthwise convolution and cross-frame attention, respectively. After that, a set of learnable temporal positional

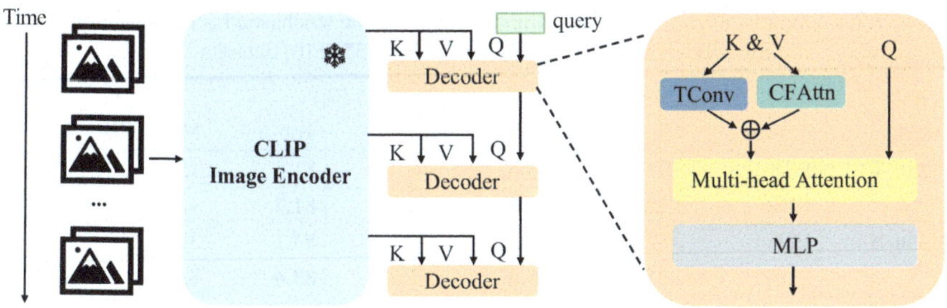

Fig. 7.11 Architecture of the EVL [459]. The frozen CLIP [241] takes the video frames as inputs and generates per-frame features, which are further processed by several motion-enhanced Transformer decoder blocks for temporal modeling

Table 7.6 Comparisons of trainable parameters and Top-1 accuracy on Kinetics-400 with other CLIP-based methods

Method	New Params (M)	#Frames×#Views	Acc. (%)
Efficient prompting [458] (A5)	9.43*	16 × 5	76.9
ActionCLIP [460]	105.15	8 × 1	81.1
		16 × 1	81.7
		32 × 1	82.3
		16 × 3	82.6
		32 × 3	83.8
EVL ViT-B/16 (1 Layer)	7.41	8 × 3	81.1
EVL ViT-B/16 (4 Layers)	28.70	8 × 3	82.9
EVL ViT-B/16 (4 Layers)	28.78	32 × 3	**84.2**

embeddings are added to the enhanced frame features $\hat{X}_{1..N}$ to inject positional information in the temporal dimension. EVL also learns a query token to gather the frame-level representations through cross-attention, the output of which will be used for category prediction. We illustrate the architecture of EVL in Fig. 7.11.

As previously mentioned, EVL [459] freezes the CLIP backbone during training and only back-propagates gradients to several lightweight decoders. This approach significantly reduces both memory consumption and training time. Table 7.6 compares EVL to other CLIP-based video recognition models, and the results show that EVL achieves superior accuracy with a relatively small number of parameters.

7.2.2.2 Efficient Gradient Backpropagation

A large proportion of the memory consumption in model training comes from storing the activation values (*forward path*) and gradients (*backward path*) for backpropagation, which inspires some work to explore efficient gradient backpropagation algorithms to save training overhead. In the video domain, the remarkable redundancy in the temporal dimension offers more opportunities for improvement.

Table 7.7 Comparisons of end-to-end training (E2E) and the stochastic backpropagation (SBP) using Swin-T and Swin-B as backbones on K400 and Epic-55 [116] datasets

		K400 (%)		Epic-55 (%)		
Model	Training	Top-1	Top-5	Top-1	Top-5	Mem (GB/gpu)
Swin-T	E2E	78.8	93.6	46.0	81.5	15.2
	SBP	**78.2**	**93.1**	**45.2**	**81.7**	4.4
Swin-B	E2E	82.7	95.5	49.4	83.2	32.5
	SBP	**81.9**	**95.2**	**47.3**	**83.6**	8.6

SBP [445] observes the activation maps, and gradients of different frames are similar at the shallow layers but differ significantly at the deep layers. Based on this, they argue that all the feed-forward computations should be reserved to provide temporally diversified representations in the deeper layers, but some backward calculations, on the other hand, can be discarded.

Taking the video Transformers as an example. We term the representation of a single frame as a temporal node, which is densely connected with other nodes for temporal modeling during the forward process, while for the backward, SBP randomly samples several nodes per layer and only propagates the gradients through the sampled nodes. With gradients pruned layer by layer, an unavoidable problem is that the gradients of the nodes in shallow layers will be inaccurate due to the error accumulation. To address this problem, SBP simply keeps the sampling location constant across different layers to ensure a consistent gradient delivery. The quantitative comparisons of action recognition accuracy and training memory usage between end-to-end training and stochastic backpropagation are shown in Table 7.7. With around 1% performance degradation, SBP reduces the memory consumption for training by 71% and 74% on Kinetics-400 (K400) and EPIC-Kitchens55 (Epic-55) [116] datasets, respectively.

In contrast to performing gradient pruning during backpropagation, Full Video Action Recognition (FVAR) [461] approximates the individual frame gradients with the gradients of temporally aggregated frame activations to make the gradient computation more efficient. They argue that for the videos with significant motion, randomly sampling frames, which is the popular training strategy for existing deep video models [38, 39, 67], tends to miss the salient frames and thus produce incorrect predictions. Instead, training on full video could effectively avoid the information loss at the input end, which, however, is computationally infeasible due to the unaffordable memory consumption.

Taking a closer look at the memory usage for training video models, it is not difficult to spot that the main bottleneck lies in the storage of activation values and gradients for backpropagation: As the final classification loss depends on all the input frames, it is necessary to store all activations for each frame in each intermediate layer. Fortunately, the prominent temporal redundancy in videos offers the possibility of optimizing the gradient calculation. Based on the assumption that similar-looking frames in a video should also share similar activation values, and furthermore similar gradients, FVAR [461] chooses TSM as the backbone network

Table 7.8 Comparison between TSM [38] (baseline model) and FVAR (with different clustering algorithms) on fully controlled dataset Move4MNIST

Model	#Frames	#Clusters	FLOPs /Video	Runtime Mem./Video	Top-1
TSM [38]	8	–	14.56G	1.04 GB	90.13 ± 0.38
	16	–	29.12G	1.72 GB	93.78 ± 0.33
	all	–	58.24G	3.15 GB	**98.83 ± 0.16**
FVAR-uniform	all	8	28.61G	1.56 GB	90.25 ± 0.28
FVAR-slope	all	8	28.61G	1.56 GB	93.33 ± 0.19
FVAR-cumulative	all	8	28.61G	1.56 GB	**94.08 ± 0.25**
FVAR-uniform	all	16	38.51G	1.79 GB	92.73 ± 0.25
FVAR-slope	all	16	38.51G	1.79 GB	94.06 ± 0.18
FVAR-cumulative	all	16	38.51G	1.79 GB	**95.27 ± 0.21**

and adopts ReLU as the activation function to make the network approximately linear. With this, the following transformation can be derived:

$$\sum_i \nabla_{\mathbf{w}} L(h(\mathbf{x}_i \mathbf{w})) = \nabla_{\mathbf{w}} L\left(\sum_i h(\mathbf{x}_i \mathbf{w})\right), \tag{7.4}$$

where $\mathbf{x}$ denotes the activation values of different frames, $\mathbf{w}$ is the trainable weights, $h(\cdot)$ is an activation function, and $L(\cdot)$ is the loss function. Therefore, adding the activations of all frames first to calculate the gradients could significantly reduce memory consumption.

As the above Eq. (7.4) only holds if the activations of similar frames are approximately linear, the next question is how to temporally cluster the frames in a video. FVAR [461] simply binarizes the activation values of different frames with the sign function and computes a fast Hamming distance to measure their similarity:

$$\mathcal{C}_N(\mathbf{x}) = \sum_{i=1}^{N-1} H(\mathbf{x}_i, \mathbf{x}_{i+1}), \tag{7.5}$$

where $\mathbf{x}_i$ is the binarized activation of frame i, $H(\cdot, \cdot)$ is the Hamming distance, and N is the total number of frames. FVAR [461] implements two clustering algorithms for comparison: cumulative clustering that divides the maximum cumulative Hamming distance into several segments evenly and slope clustering that defines the segment boundaries as frame indexes with the top cumulative distance increasing slopes.

In order to evaluate the effectiveness of the proposed method, FVAR [461] creates a fully controlled dataset Move4MNIST that includes four action classes up, down, left, right. The comparisons of runtime memory usage and top-1 accuracy between the baseline methods TSM and FVAR are shown in Table 7.8. The results in the line 1–3 demonstrate that for TSM using more frames could improve the

Mask%	Recon%	GFLOPs	Acc
0	-	180.6	94.8
50	50	99.3	93.7
75	75	57.6	91.8
75	25	45.9	91.6
90	90	35.2	89.7
90	10	18.3	89.6

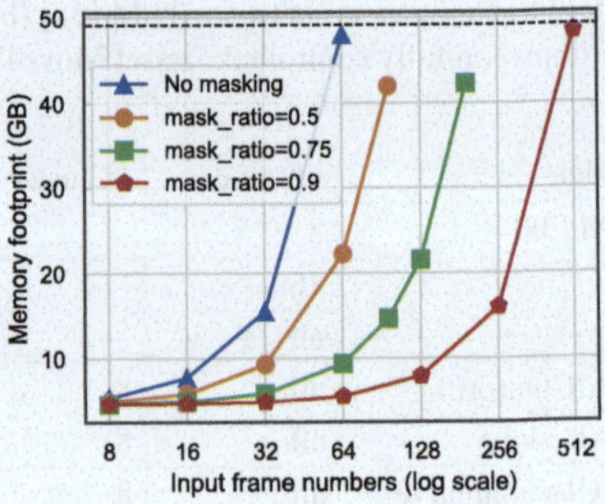

Fig. 7.12 **Left**: The trade-off between GFLOPs and classification accuracy on UCF101 with different masking ratios. **Right**: Memory footprint of training one video with different input frames, replicated from Turbo [442]

recognition performance significantly, which however will result in more memory consumption. Comparatively, FVAR could effectively limit resource costs with acceptable accuracy loss.

7.2.2.3 Input Pruning

Input pruning is widely explored in the video domain to improve the inference efficiency [462, 463], but limited efforts have been paid to applying it for efficient training. Recently, the distinct success of masked autoencoders (MAE) [402, 411] reveals that the redundancy in visual data enables a small proportion of patches ($\sim$25% for image and $\sim$10% for video) enough to reconstruct the pristine inputs. This motivates Turbo [442] to sparsely sample visual tokens for efficient video training. Specifically, they randomly mask the input tokens at a large ratio ($\geq$ 50%) and only input the visible ones to a visual encoder. After that, a lightweight decoder takes the updated representations as input to reconstruct the invisible tokens. Turbo [442] training is a general training framework that could be applied to various video understanding tasks, where both the reconstruction loss and the default objective should be combined for joint optimization. The computational costs (left) and memory footprints (right) with and without Turbo training on UCF101 [20] are compared below. We can observe that at the cost of a negligible loss of accuracy, dropping $\geq$50% tokens could bring remarkable computational overhead and memory savings (Fig. 7.12).

7.3 Dynamic Inference for Video Understanding

Besides compact network designing (Sect. 7.1) and efficient training strategies (Sect. 7.2), there exists another line of work focusing on reducing the expensive overhead at inference time—dynamic inference. Dynamic inference is agnostic to specific network architectures and training processes, aiming at adaptively allocating computation to make inference efficient but also keeps the accuracy.

Due to the inherent data redundancy, information in videos is not evenly distributed, i.e., certain segments of videos contain more important information than others. By allocating more computation to more informative parts and less or even no computation to unimportant parts, we can make inference more efficient without sacrificing accuracy (Sect. 7.3.1). Furthermore, current one-size-fits-all models, using the same number of parameters for all input samples, are also not optimal, ignoring the fact input samples are of varying difficulty to be correctly recognized. Thus, network configurations can also be adaptively adjusted on a per-input basis (Sect. 7.3.2).

7.3.1 Reducing Input Redundancy

As videos contain spatial appearance information and temporal motion information, we can reduce the redundancy along the temporal dimension (Sect. 7.3.1.1) and the spatial dimension (Sect. 7.3.1.2) separately. We can also consider these two dimensions jointly and remove spatial–temporal redundancy cooperatively (Sect. 7.3.1.3). Besides, videos are naturally multimodal, containing visual, text, and audio clues, but not all modalities are created equal. In a multimodal recognition pipeline, removing redundant modalities can also achieve faster inference (Sect. 7.3.1.4).

7.3.1.1 Temporal Redundancy Reducing

Typically, state-of-the-art video recognition frameworks uniformly sample frames across the temporal dimension and use aggregation for final predictions. However, information is not uniformly distributed along the temporal dimension, and the number and indices of frames required for video recognition differ across videos. An efficient video recognizer should adaptively sample frames based on the characteristics of each input video. Figure 7.13 illustrates that for a harder instance, more frames are selected to assure an accurate prediction. To reduce temporal redundancy, [462, 464–467] sample most salient frames, neglecting other frames been. Ghodrati et al. [468] and Fan et al. [469] process frames sequentially, and the inference will be terminated in advance, once a reliable prediction is produced.

Salient frame sampling aims to derive a frame selection strategy that achieves accurate predictions while using as few frames as possible. However, this is a nontrivial problem since videos are weakly labeled for classification, and there are no supervised labels to indicate the importance of each frame.

AdaFrame [462] addresses this problem by introducing a memory-augmented LSTM that effectively explores temporal saliency. Specifically, the memory-augmented LSTM acts as an agent that interacts with the video frame sequence. At

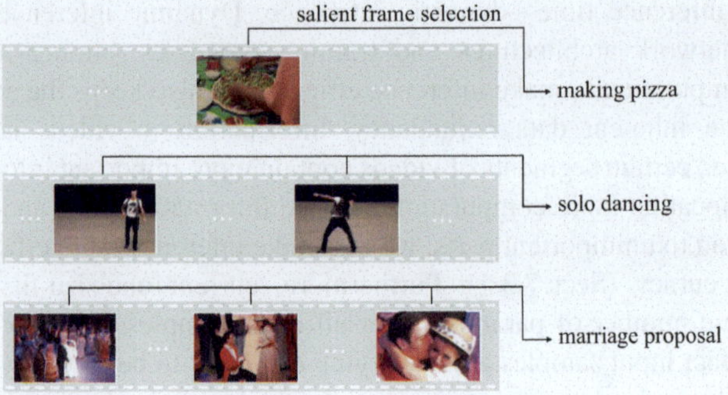

Fig. 7.13 Frame usage depends on the difficulty of the video content

each time step, it takes the features of the current frame, the hidden state of previous frames, and a global context vector as input and outputs the current hidden state, as shown in Eq. (7.6):

$$\boldsymbol{h}_t, \boldsymbol{c}_t = \text{LSTM}([\boldsymbol{v}_t, \boldsymbol{u}_t], \boldsymbol{h}_{t-1}, \boldsymbol{c}_{t-1}), \tag{7.6}$$

where $\boldsymbol{h}_t, \boldsymbol{c}_t, \boldsymbol{h}_{t-1}$ and $\boldsymbol{c}_{t-1}$ represent the hidden states and cell outputs of the current and previous steps, respectively. The features of current frame $\boldsymbol{v}_t$ and a global context vector $\boldsymbol{u}_t$ are concatenated, and the resulting hidden state $\boldsymbol{h}_t$ is then input to a prediction network f_p, a selection network f_s, and a utility network f_u.

The global memory maintains downsampled representations of video frames, both spatially and temporally. This memory serves as a source of global context for the current frame, which is computed as a weighted average of all the stored representations. A cross-attention layer is utilized to generate the weight for the global context computation. This layer takes the hidden state $\boldsymbol{h}_t$ as a query and the representations stored in the memory as keys. The prediction network, denoted as $f_p(\boldsymbol{h}_t; \mathbf{W}_p)$, is a fully connected layer that maps the hidden state $\boldsymbol{h}_t$ to $\boldsymbol{s}_t \in \mathbb{R}^C$, which represents the probability scores for each of the C categories. The score $\boldsymbol{s}_t$ is then used to generate a reward r_t that measures the gain of the transition from the last time step (i.e., whether observing one more frame provides more information to make a more accurate prediction). The selection network f_s defines the policy as a Gaussian distribution with fixed variance to decide which frame to look at next. Specifically, f_s transforms the hidden state $\boldsymbol{h}_t$ to a scalar $a_t = \text{sigmoid}(\mathbf{W}_s^T \boldsymbol{h}_t)$, which is then used as the mean of the Gaussian distribution. During training, AdaFrame samples from the distribution to generate an index of the frame to jump to next. The utility network consists of a single fully connected layer, which generates an output $\hat{V}_t = \mathbf{W}_u^T \boldsymbol{h}_t$, and it will be served as a critic to estimate the advantage of seeing more frames in the future. With all its components, AdaFrame is equipped with the ability to adaptively look ahead during inference, meaning that it only processes a subset of informative frames selected on a per-input basis.

Besides AdaFrame [462], there are several other works that share the idea of salient frame sampling. SCSampler [464] highlights that a dense and uniform frame sampling strategy for long, untrimmed videos can be computationally expensive and may include irrelevant frames, leading to poor accuracy. To address these issues, SCSampler proposed an extremely lightweight network to identify salient video clips. It then samples a reduced set of these salient clips for classification, resulting in both running time savings and improved accuracy.

To address the frame sampling problem, MARL [465] adopts a Markov decision process framework and leverages multi-agent reinforcement learning techniques for its solution. Specifically, it trains N parallel agents, each of which learns to select an informative frame or clip. To enable contextual reasoning and account for historical states, an RNN-based observation network is designed context-aware and shared by all agents. Once all agents have completed the frame selection process, a classification network aggregates all the selected frames and produces a final video-level prediction.

Gao et al. [466] argued that untrimmed videos contain redundant information in both the long term and short term. To tackle the problem of short-term redundancy within a clip, they used a two-stream framework with only a single frame and its accompanying audio to capture the appearance and dynamic information, respectively. Since only using a single image and audio to hallucinate a video descriptor, this framework is referred to as IMGAUD2VID. To reduce long-term redundancy, they introduce an LSTM-based IMGAUD-SKIMMING network that selects a subset of informative and task-relevant image–audio pairs for the final recognition.

TSQNet [467] proposes a novel approach to measure frame saliency based on the assumption that salient patterns are closely associated with category-specific semantics. Motivated by this insight, Xia et al. [467] formulated saliency measurement as a query-response task. To measure frame saliency over all categories, a Temporal Saliency Query mechanism was proposed. More specifically, the approach constructs prototypes (i.e., common patterns) for all categories and uses them as queries. The similarities between the prototypes and the frames are considered as the response, which is then utilized to calculate the saliency specific to each class.

Conditional early termination is another stream of works, which executes inference dynamically. Existing salient frames selection [462, 464–467] relies on a learning-based policy to sample informative frames, and thus the policy network may introduce additional computational costs. Instead of learning a sampling policy, conditional early termination learns to dynamically determine how many frames will be used for processing a video based on its difficulty. As illustrated in Fig. 7.14, frames are processed in sequence, and once the classifier is confident enough to make a reliable prediction, the inference is terminated in advance. This enables the model to allocate the computation adaptively on a per-input basis: The more complex the video is, the more frames are computed to make an accurate prediction.

FrameExit [468] uses a series of cascaded gates to adaptively determine where to exit, once a reliable prediction is made. The architecture consists of a simple and deterministic sample policy, a shared feature extractor, an accumulated feature pooling module, and a cascade of classifiers accompanied by their early exiting gates.

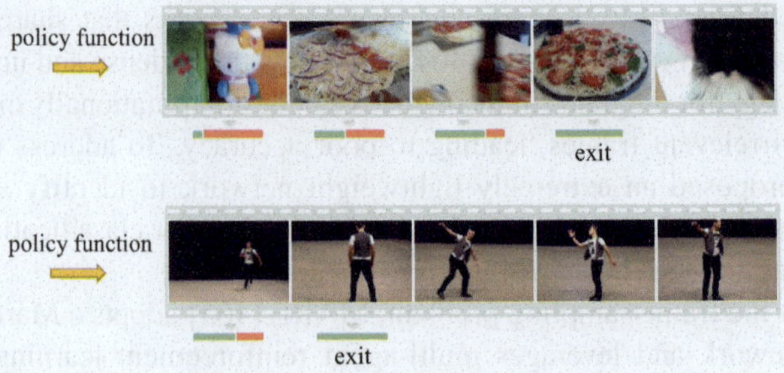

Fig. 7.14 Illustration of conditional early termination, the usage of frames is determined by the difficulty, a difficult video requires more frames to ensure a reliable prediction

The frame sampling policy π is a parameter-free and deterministic function, which samples frames from a coarse-to-fine manner. The feature extractor $\Phi(x_i)$ is a 2D image model that extracts the representation of frame x_i. Since the 2D model can only extract static appearance information, a temporal aggregation module is introduced to model the temporal dynamics. The accumulated feature pooling module aggregates the previous representation z_{t-1} and feature of the current frame $\Phi(x_i)$ to produce the latest clip-level representation z_t as shown in Eq. (7.7), where $\Psi(\cdot)$ is the temporal aggregation function that can be instantiated as an average/max pooling, LSTM, or self-attention module. Each gate in the cascade is a binary function $g_t(z_{t-1}, z_t) \rightarrow \{0, 1\}$, and the input is the fused representation of z_{t-1} and z_t. These two representations are first input to a shared MLP independently, and the output features are then concatenated and transformed by a linear layer and a sigmoid activation function, producing a scalar ranging between 0 and 1. This scalar is used as an indicator of whether the model is confident enough to make an accurate prediction. Furthermore, each classifier f_t receives the current clip-level representation z_t and makes a prediction for the whole video.

$$z_t = \Psi(z_{t-1}, \Phi(x_i)). \tag{7.7}$$

During inference, FrameExit begins by selecting a single frame and subsequently increases the number of frames according to the deterministic sampling policy. Such a process is repeated until an early exiting signal is emitted. In particular, each frame is processed independently by a shared feature extractor to extract its features. In order to capture temporal cues, the current feature representation is then combined with the previous states by an accumulated feature pooling module. Using the merged feature as inputs, the gate module determines whether to terminate the inference process, while the classifier makes a prediction at each step. If a gate module predicts a halt signal, the inference process is terminated, and the prediction made by the current classifier is treated as the video-level prediction.

Until now, we have explored methods to reduce temporal redundancy, by either selecting salient frames or employing conditional early termination. Nevertheless, given that videos contain not only temporal clues but also spatial information, redundancy also exists in the spatial dimension. In the next section, we will delve into techniques for enhancing inference speed by reducing spatial redundancy.

7.3.1.2 Spatial Redundancy Reducing

Not all regions in a frame contain an equal amount of information, and often the most informative region is just a small patch. This suggests that spatial redundancy can be reduced by localizing the salient region in a frame [463, 470]. Another approach is to allocate higher resolutions to frames that contain finer information, as not all frames require the same level of resolution. Using adaptive resolution across frames can also help mitigate spatial redundancy [471, 472].

Salient region localization argues that most of the regions in a frame belong to the background and are irrelevant to the recognition task. As shown in Fig. 7.15, the task-relevant region is usually a small patch that shifts smoothly across frames. By attending only to the relevant region and ignoring the background, the inference process may become more efficient and accurate.

AdaFocus [463] formulates the problem of salient region localization as a sequential decision task and models it using a reinforcement learning algorithm. Figure 7.16 demonstrates an overall illustration of the entire framework. At each iteration, the current frame is first processed by a lightweight network f_G to obtain

Fig. 7.15 Informative region is a small patch and shifts across frames

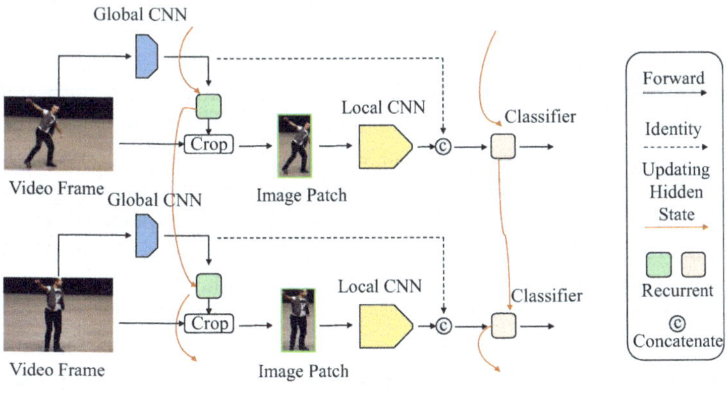

Fig. 7.16 Overview of AdaFocus [463] framework

a quick glimpse. Based on this coarse feature, a policy network π determines which region the frame should be focused on. Once the salient region is localized, a high-capacity CNN f_L is activated to extract a fine-grained feature from the selected region. Finally, features from all time steps are combined to make a final prediction.

f_G is a global CNN that is responsible for processing the entire frame, and it is designed to be lightweight to extract a coarse feature $e_t^G = f_G(v_t)$ for each frame v_t. The policy π is a recurrent network that takes in the coarse feature e_t^G and hidden state h_{t-1}^{π} of the previous time step and chooses the salient region $\tilde{v_t}$ from multiple candidates according to Eq. (7.8). Unlike the global CNN, f_L is a high-capacity local CNN that only focuses on the selected subregion, enabling it to extract a discriminative and fine-grained feature e_t^L for the current frame v_t. The aforementioned features e_t^G and e_t^L are first processed by global average pooling and then concatenated into a vector $[\overline{e}_t^G, \overline{e}_t^L]$. This concatenated vector is then fed into the recurrent classifier f_C along with the previous hidden state to produce the prediction p_t at the tth frame.

$$\tilde{v_t} \sim \pi\left(\cdot | e_t^G, h_{t-1}^{\pi}\right). \tag{7.8}$$

The training process of AdaFocus is divided into three stages. At stage one, only f_G, f_L and f_C are trained, while the policy network π is left out. During this stage, the subregion $\tilde{v_t}$ is randomly sampled from the entire frame. The aim of this stage is to help the model learn to extract task-relevant information and prepare for policy learning. In the second stage of training, reinforcement learning is used to train the policy π, while f_G, f_L, and f_C are kept fixed. The selection of the salient region $\tilde{v_t}$ is rewarded to measure its informativeness. The PPO algorithm [473] is utilized to maximize the discounted expected return. In the last stage of training, f_L and f_C are further fine-tuned using the learned policy from the previous stage to achieve better performance.

To simplify the training process and eliminate the need for a three-stage pipeline in AdaFocus [463], the subsequent AdaFocus V2 [470] introduces an end-to-end training strategy. It also replaces the reinforcement learning-based patch selection operation with a differentiable interpolation-based method, as the former can be susceptible to convergence issues.

Adaptive resolution is another line of work that mitigates spatial redundancy by dynamically adjusting the resolution on a per-frame basis. This genre of works [471, 472] argue that frames in a video require different resolutions, as different frames contain different levels of details. The use of a one-resolution-fits-all approach, while widely adopted, is computationally expensive. As shown in Fig. 7.17, the resolution of each frame should be adaptive and dynamically changed conditioned on inputs, i.e., the less important frame should be assigned a lower resolution and even skipped.

AR-Net [471] utilizes a policy that dynamically determines the most suitable resolution for each frame. This lightweight policy network samples the per-frame resolution from a discrete distribution. The policy is made trainable using the

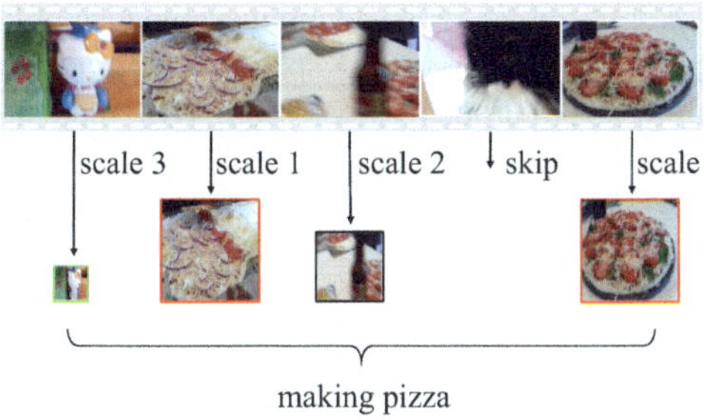

Fig. 7.17 A conceptual overview of adaptive resolution scenario

Gumbel Softmax [474] method, which handles the non-differentiability of the discrete policy. AR-Net is composed of a policy network that includes an LSTM unit and several backbone networks designed for various resolution inputs. The policy network is implemented as a lightweight feature extractor to make the overhead negligible. The LSTM module can aggregate information across time to make better decisions. Additionally, the backbone networks are adopted in different capabilities to process inputs with different resolutions. High-resolution frames demand a more high-capacity network to extract more detailed representation. The policy network and backbone networks are trained together, utilizing the Gumbel Softmax [474]. During inference, the frames of a video are processed sequentially. First, each frame is passed through the policy network to determine the optimal resolution. Next, the corresponding backbone network is used to recognize the resized frame and generate a frame-level prediction. Finally, the frame-level predictions are aggregated to obtain a prediction for the video as a whole.

ResKD [472] improves the accuracy of low-resolution models through cross-resolution knowledge distillation. This approach suggests that the decrease in accuracy is not solely caused by information loss during downsampling, but rather by a mismatch between the input scale and the architecture.

Up till now, we have explored how to make inference more efficient from the perspective of temporal (Sect. 7.3.1.1) and spatial (Sect. 7.3.1.2) redundancy. However, these two approaches have been studied separately from each other. In the following subsection, we will introduce techniques for jointly modeling spatial–temporal redundancy.

7.3.1.3 Unified Spatial–Temporal Redundancy Reduction

Most of the previous works focused on exploring either spatial [463, 470–472] and temporal [462, 465–468] redundancy reduction separately. These two lines are orthogonal to each other and can be combined to get further improvements. Recently proposed works [475, 476] argue that considering spatial–temporal redundancy simultaneously can achieve a better balance between accuracy and efficiency.

AdaFocus V3 [475] generalizes the idea of localizing important temporal frames and spatial regions to 3D spatial–temporal cubes. While the overall framework is similar to AdaFocus V1/V2 [463, 470], there are some key differences to note:

- The lightweight global encoder f_G receives all the frames at the same time to capture coarse global features, while AdaFocus V1/V2 receives inputs frame by frame.
- The policy network π determines the location of salient 3D cubes, instead of 2D patches in AdaFocus V1/V2. Additionally, π receives the global features of the entire video, eliminating the need for a recurrent design.
- The cubes selected by the policy π are organized in a descending order and are processed progressively by the local encoder f_L and the final classifier f_C. In this way, the computation is first allocated to the most informative cubes.
- AdaFocus V3 enables adaptive inference: Once the prediction is sufficiently reliable, the inference can be early terminated.

Extensive studies investigated efficient inference for CNN-based models, leaving the popular video transformer underexplored. Wang et al. [476] propose a plug-in Spatial–Temporal-Token Selection (STTS) module that chooses informative tokens from both spatial and temporal dimensions. Sampling spatial–temporal tokens directly can lead to fragmented tokens that may reduce contextual information. To address this issue, STTS first identifies the relevant frames and subsequently focuses on the significant regions within those frames. The task of choosing tokens is approached as a ranking problem, and tokens with higher ranking scores are utilized for recognition, while those with lower scores are discarded. Specifically, there are a temporal token selector and a spatial token selector in STTS, and each of them is implemented as a lightweight MLP network. First, the temporal selector scores each frame, selecting informative frames. These selected frames are then fed into the spatial selector, which selects patches with high-importance scores. However, this token selection operation is non-differentiable. To enable end-to-end training, [476] employ a differentiable Top-K selection algorithm.

We have discussed how to mitigate both spatial and temporal redundancy (Sects. 7.3.1.1–7.3.1.3). However, we have only considered the visual modality and neglected the multimodal nature of video input. In the next subsection, we will explore input redundancy in a multimodal scenario.

7.3.1.4 Modality Redundancy Reducing

Videos are multimodal that contain static appearance information, dynamic motion clues, as well as other supplementary modalities such as text and audio. However, most of the multimodal models [17, 477] simply treat all the modalities equally and fuse all modalities to obtain final predictions. Despite their excellent performance, these models suffer from expensive computational costs and are thus limited in

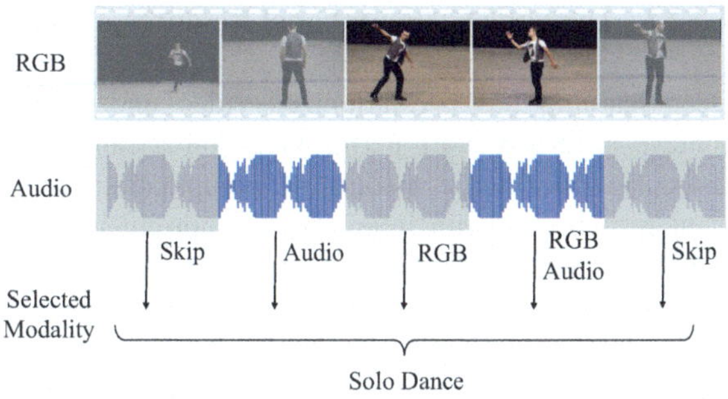

Fig. 7.18 An illustration of adaptive modality selection for efficient action recognition

resource-intensive applications. Some recent works [71, 478] suggest that utilizing information from all modalities equally may hinder overall performance. As demonstrated in Fig. 7.18, for a complex action, only a few inputs are necessary to make an accurate prediction, and different parts may require different modalities on a per-segment basis.

Panda et al. [478] argue that the informative modalities may be overwhelmed by uninformative ones, and using all modalities equally may be counterproductive. The authors propose AdaMML to learn a policy to select input modalities for efficient video recognition. AdaMML includes a lightweight policy network that determines the appropriate modality to use for each segment, and the selected modalities for each segment are then routed to corresponding high-capacity recognition models to obtain a fused segment-level prediction. Once all segments have been processed, the final prediction for the video is generated by computing the average of the predictions from all segments. Specifically, the selection decisions are sampled from a Gumbel Softmax distribution [474] to make it end-to-end trainable.

Thus far, our discussion is focused on redundancy in input data, which is agnostic to the model architecture (i.e., the input is dynamically sampled while keeping the model's computational graph static). In the next section, we will discuss how to eliminate network redundancy through adaptive network pruning.

7.3.2 Adaptive Network Pruning

Using a one-size-fits-all model that applies the same amount of computation to all videos can produce decent recognition results but at the cost of significant computational costs. To improve efficiency without sacrificing accuracy on a per-input basis, we can dynamically adjust the capacity of the recognition network based on the complexity of the input video. In other words, an adaptive network pruning approach can be used to optimize efficiency and accuracy simultaneously. In this section, we will provide a review of the most representative adaptive network pruning techniques for efficient video understanding.

7.3.2.1 Dynamic Computational Resource Allocation

Deep learning has revolutionized video understanding by enabling models to achieve state-of-the-art performance on a range of challenging video analysis tasks. However, these models are computationally expensive and require significant resources to run inference. One promising approach to address this issue is to dynamically allocate computational resources during inference to improve efficiency without sacrificing accuracy. By adjusting the computational budget according to the complexity of the input video, dynamic inference can significantly reduce the computational cost of deep video models and enable them to be applied in real-world scenarios.

LiteEval [479] is an efficient video recognition framework that allocates computational resources dynamically on a per-frame basis. It adopts a coarse-to-fine manner by using a lightweight coarse network by default and activating a more resource-intensive fine network only when necessary. LiteEval includes a coarse CNN/LSTM pair, a conditional gating module, and a fine CNN/LSTM pair. Since the expensive CNN model is only activated when there is a need for fine-grained features, it can be regarded as dynamically pruned from the overall network.

The coarse CNN is lightweight and operates on downsampled frames to provide a coarse feature v_t^c of the current frame. Then, the coarse LSTM takes in the current coarse feature v_t^c, the previous hidden state h_{t-1}^c, and cell output c_{t-1}^c to produce the current hidden state h_t^c and cell output c_t^c. The conditional gating module is a one-layer MLP that decides whether to extract a finer feature from the current frame. It takes in the concatenation of the current coarse feature v_t^t, the previous hidden state h_{t-1}^f, and cell output c_{t-1}^f of the fine LSTM as input and outputs the probability b_t of whether to activate the fine LSTM. If the fine LSTM is activated, it updates its state following Eq. (7.9), where $[\cdot, \cdot]$ represents the concatenate operation. Otherwise, it reuses the previous state h_{t-1}^f and c_{t-1}^f. The classifier takes in the fine hidden state h_t^f and makes a prediction $p_t = \text{softmax}(\mathbf{W}_p^T h_t^f)$, where $\mathbf{W_p}$ is the classifier's parameter.

$$h_t^f, c_t^f = \text{fLSTM}\left(\left[v_t^c, v_t^f\right], h_{t-1}^f, c_{t-1}^f\right). \tag{7.9}$$

During inference, the coarse CNN first computes coarse features for the current frame and updates the coarse LSTM. Based on these coarse features and LSTM states, the gating module then determines whether to compute finer features for obtaining more detailed information. If necessary, the fine CNN is employed to compute these fine features.

Li et al. [480] argue that using 3D convolution throughout the network is unnecessary, and the usage of 3D convolutions should be selected dynamically. To achieve this, they proposed Ada3D, a framework that learns to adaptively choose 3D convolutions on a per-clip basis. It decides on-the-fly how many frames are needed, and it further determines how many 3D convolutions are to be activated.

The network architecture is dynamically configured to make the inference process more efficient.

7.3.2.2 Adaptive Network Quantization

Dynamic quantization is a type of network pruning that can improve the efficiency of video recognition. According to [481], the bit-width of network weights and activations is another crucial factor that affects the efficiency of video recognition. The authors suggested that the network should not be fixed at 32-bit precision, and instead the precision should be dynamically selected based on how informative the input content is. They introduce VideoIQ, a dynamic quantization framework that can select the network precision on a per-frame basis. Figure 7.19 (bottom) gives an overview of VideoIQ, which adaptively selects the optimal precision, i.e., higher precision when the content is more informative.

While the idea of dynamic precision selection in deep neural networks is promising for improving efficiency, it presents two significant challenges. The first is how to determine the optimal network precision for each instance without incurring additional computational costs. To tackle this, [481] propose training a lightweight policy network to sample quantization from a discrete distribution, with the Gumbel Softmax strategy [474] to make the process trainable. The second challenge is how to quantize the network in a flexible manner without adding storage or computational overhead. Sun et al. [481] address this by truncating the least significant bits to adjust the numerical precision. Extensive experiments show that VideoIQ achieves remarkable reductions in computational and memory usage while surpassing the performance of existing state-of-the-art methods.

In this subsection, we have discussed how to make inference more efficient, including removing the redundancy in the input data (Sect. 7.3.1) and pruning the network architecture (Sect. 7.3.2). The first kind posits that information is not distributed uniformly in input data and certain parts may be more informative than others. Therefore, computational resources should be adaptively allocated depending on the importance of input. The other line of work argues that the

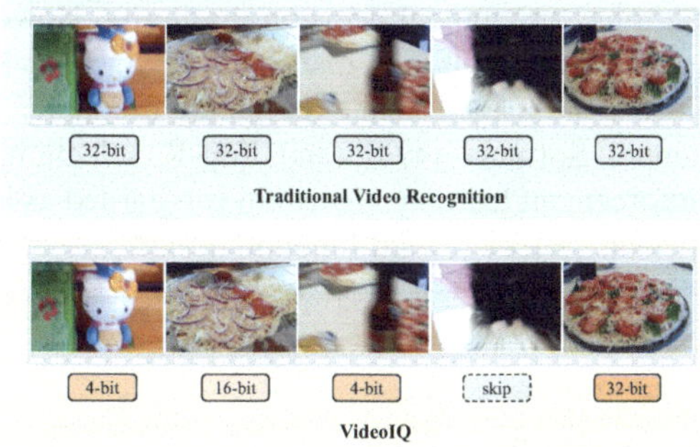

Fig. 7.19 A conceptual overview of VideoIQ [481] framework

Table 7.9 A summary of all aforementioned methods for an overall comparison, all results are copied from the source papers

	FCVID		ActivityNet		Mini-kinetics	
Method	mAP	GFLOPs	mAP	GFLOPs	Top-1	GFLOPs
Temporal redundancy reduction						
AdaFrame [462]	80.2%	75.13	71.5%	78.97	–	–
SCSampler [464]	81.0%	42.0	72.9%	42.0	70.8%	42.0
Listen2Look [466]	–	–	72.3%	81.36	–	–
TSQNet [467]	83.5%	262	76.6%	26.1	73.2%	19.7
FrameExit [468]	–	–	76.1%	26.1	72.8%	19.7
Spatial redundancy reduction						
AdaFocus [463]	83.4%	26.6	75.0%	26.6	72.2%	26.6
AdaFocus V2 [470]	85.0%	27.0	79.0%	27.0	75.4%	27.0
AR-Net [471]	81.3%	35.1	73.8%	33.5	71.7%	32.0
ResKD [472]	84.4%	17.4	80.0%	17.4	75.4%	8.7
Spatial–temporal redundancy reduction						
AdaFocus V3 [475]	85.9%	26.8	79.5%	21.7	75.0%	17.5
Modality redundancy reduction						
AdaMML [478]	85.8%	93.7	73.9%	94.01	–	–
HCMS [71]	88.8%	377.4	87.5%	576.7	–	–
Adaptive network pruning						
LiteEval [479]	80.0%	94.3	72.7%	95.1	61.0%	99.0
Ada3D [480]	81.9%	35.6	82.6%	42.2	78.9%	42.4
VideoIQ [481]	82.7%	27.0	74.8%	28.1	72.3%	20.4

network configuration should not remain static, i.e., the model's architecture and capacity should be dynamically configured based on the input sample. These two lines of work are based on the assumption that one-size-fits-all scheme will lead to redundancy and inefficiency. To have an overall comparison of accuracy–efficiency trade-off for all aforementioned methods, we summarize their results on three frequently used datasets FCVID [108], ActivityNet [25] and Mini-kinetics in Table 7.9.

To make the inference more adaptive, most of the methods introduce a policy to determine the optimal decision on-the-fly. However, the policy network is usually sampled from a discrete distribution, which makes it non-differentiable. A variety of tricks are used to make the policy trainable, e.g., reinforcement learning [473], Gumbel Softmax [474], and Top-K selection [482]. Among these tricks, the reinforcement learning algorithm is regarded as hard to train, since the performance relies on a tricky reward function design.

7.4 Summary

This section of the book provides an overview of the literature on efficient video understanding. We start by discussing design options for compact neural networks. Then, we explore several carefully crafted strategies for training deep video models with reduced time and memory requirements. Finally, we review a diverse range of methods for enhancing the inference efficiency of video models through spatial–temporal input selection and network pruning.

Chapter 8
Conclusion and Future Directions

8.1 Concluding Remarks

In this book, we have sketched a grand picture of **Deep Learning for Video Understanding** for readers from diverse aspects. Starting with a foundational overview of video understanding and the basics of deep learning, we navigated through specialized domains such as action recognition, video localization, and temporal grounding, with each chapter unfolding the corresponding task definitions, benchmarks, and prevalent solutions. Beyond this, our exploration also extended to the unsupervised learning of video representations and efficient video analytics.

8.2 Future Research Directions

The booming development of deep learning has stimulated the wide exploration in video understanding. However, there still exist many open problems in this field, which are summarized as below:

1. Large-Scale Video Foundation Models For a long period of time, the research in video understanding has followed a task-specific paradigm, i.e., designing specific prediction heads and even model architectures to cater to the settings of different video tasks. Despite the satisfactory results achieved on established benchmarks, such a paradigm not only hinders the knowledge sharing between video tasks but also leads to the inefficient utilization of training resources. To address this issue, large-scale video foundation models that could handle a wide variety of video tasks within a unified framework have attracted more and more attention.

2. The Use of LLMs for Video Understanding Large Language Models (LLMs), such as GPT series [400, 483] and LLaMA [484, 485], have achieved remarkable

Z. Wu, Y.-G. Jiang, *Deep Learning for Video Understanding*, Wireless Networks,
https://doi.org/10.1007/978-3-031-57679-9_8

success in natural language processing (NLP). Inspired by their superior capabilities like instruction-following and reasoning, researchers in the vision community have attempted to adapt or integrate LLMs for enhanced visual understanding. The exploration of LLMs in the video domain is promising.

3. Long-Form Video Understanding Most existing video understanding approaches could only process short video clips but fail on untrimmed long videos due to the GPU memory constraints. In the real world, however, videos can last for tens of minutes or even hours. To overcome this limitation, researchers may make use of the temporal hierarchies within videos for efficient video feature extraction, or memory networks for long-term visual context modeling.

These research directions, including exploring the potential of large-scale deep models or adapting language models for video analysis, as well as addressing the intricate challenges inherent in comprehending long-form video content, mark a pivotal transition toward more advanced and universal video understanding. They collectively represent a frontier where deep learning intersects with the demands of comprehensive and context-aware video understanding.

Reference

1. Sandvine. Global internet phenomena report. https://www.sandvine.com/global-internet-phenomena-report-2023, 2023.
2. Josef Sivic and Andrew Zisserman. Video google: A text retrieval approach to object matching in videos. In *ICCV*, 2003.
3. Jingen Liu, Jiebo Luo, and Mubarak Shah. Recognizing realistic actions from videos "in the wild". In *CVPR*, 2009.
4. Heng Wang, Muhammad Muneeb Ullah, Alexander Klaser, Ivan Laptev, and Cordelia Schmid. Evaluation of local spatio-temporal features for action recognition. In *BMVC*, 2009.
5. Juan Carlos Niebles, Chih-Wei Chen, and Li Fei-Fei. Modeling temporal structure of decomposable motion segments for activity classification. In *ECCV*, 2010.
6. Ivan Laptev and Tony Lindeberg. On space-time interest points. *IJCV*, 2005.
7. Piotr Dollár, Vincent Rabaud, Garrison Cottrell, and Serge Belongie. Behavior recognition via sparse spatio-temporal features. In *VSPETSW*, 2005.
8. Heng Wang, Alexander Kläser, Cordelia Schmid, and Cheng-Lin Liu. Action recognition by dense trajectories. In *CVPR 2011*, 2011.
9. Alex Krizhevsky, Ilya Sutskever, and Geoffrey E Hinton. Imagenet classification with deep convolutional neural networks. In *NeurIPS*, 2012.
10. Karen Simonyan and Andrew Zisserman. Very deep convolutional networks for large-scale image recognition. In *ICLR*, 2015.
11. Christian Szegedy, Wei Liu, Yangqing Jia, Pierre Sermanet, Scott Reed, Dragomir Anguelov, Dumitru Erhan, Vincent Vanhoucke, and Andrew Rabinovich. Going deeper with convolutions. In *CVPR*, 2015.
12. Kaiming He, Xiangyu Zhang, Shaoqing Ren, and Jian Sun. Deep residual learning for image recognition. In *CVPR*, 2016.
13. Gao Huang, Zhuang Liu, Geoff Pleiss, Laurens Van Der Maaten, and Kilian Q Weinberger. Convolutional networks with dense connectivity. *TPAMI*, 2019.
14. Alexey Dosovitskiy, Lucas Beyer, Alexander Kolesnikov, Dirk Weissenborn, Xiaohua Zhai, Thomas Unterthiner, Mostafa Dehghani, Matthias Minderer, Georg Heigold, Sylvain Gelly, Jakob Uszkoreit, and Neil Houlsby. An image is worth 16×16 words: Transformers for image recognition at scale. In *ICLR*, 2021.
15. Jia Deng, Wei Dong, Richard Socher, Li-Jia Li, Kai Li, and Li Fei-Fei. Imagenet: A large-scale hierarchical image database. In *CVPR*, 2009.
16. Tsung-Yi Lin, Michael Maire, Serge Belongie, James Hays, Pietro Perona, Deva Ramanan, Piotr Dollár, and C Lawrence Zitnick. Microsoft coco: Common objects in context. In *ECCV*, 2014.

Z. Wu, Y.-G. Jiang, *Deep Learning for Video Understanding*, Wireless Networks,
https://doi.org/10.1007/978-3-031-57679-9

17. Karen Simonyan and Andrew Zisserman. Two-stream convolutional networks for action recognition in videos. In *NeurIPS*, 2014.
18. Andrej Karpathy, George Toderici, Sanketh Shetty, Thomas Leung, Rahul Sukthankar, and Li Fei-Fei. Large-scale video classification with convolutional neural networks. In *CVPR*, 2014.
19. Yi Zhu, Xinyu Li, Chunhui Liu, Mohammadreza Zolfaghari, Yuanjun Xiong, Chongruo Wu, Zhi Zhang, Joseph Tighe, R Manmatha, and Mu Li. A comprehensive study of deep video action recognition. *arXiv preprint arXiv:2012.06567*, 2020.
20. Khurram Soomro, Amir Roshan Zamir, and Mubarak Shah. Ucf101: A dataset of 101 human actions classes from videos in the wild. *arXiv preprint arXiv:1212.0402*, 2012.
21. Hildegard Kuehne, Hueihan Jhuang, Estíbaliz Garrote, Tomaso Poggio, and Thomas Serre. HMDB: a large video database for human motion recognition. In *ICCV*, 2011.
22. Will Kay, Joao Carreira, Karen Simonyan, Brian Zhang, Chloe Hillier, Sudheendra Vijayanarasimhan, Fabio Viola, Tim Green, Trevor Back, Paul Natsev, et al. The kinetics human action video dataset. *arXiv preprint arXiv:1705.06950*, 2017.
23. Raghav Goyal, Samira Ebrahimi Kahou, Vincent Michalski, Joanna Materzynska, Susanne Westphal, Heuna Kim, Valentin Haenel, Ingo Fruend, Peter Yianilos, Moritz Mueller-Freitag, et al. The" something something" video database for learning and evaluating visual common sense. In *ICCV*, 2017.
24. Haroon Idrees, Amir R Zamir, Yu-Gang Jiang, Alex Gorban, Ivan Laptev, Rahul Sukthankar, and Mubarak Shah. The thumos challenge on action recognition for videos "in the wild". *CVIU*, 2017.
25. Fabian Caba Heilbron, Victor Escorcia, Bernard Ghanem, and Juan Carlos Niebles. Activitynet: A large-scale video benchmark for human activity understanding. In *CVPR*, 2015.
26. Dima Damen, Hazel Doughty, Giovanni Maria Farinella, Antonino Furnari, Evangelos Kazakos, Jian Ma, Davide Moltisanti, Jonathan Munro, Toby Perrett, Will Price, et al. Rescaling egocentric vision. *arXiv preprint arXiv:2006.13256*, 2020.
27. David L Chen and William B Dolan. Collecting highly parallel data for paraphrase evaluation. In *ACL*, 2011.
28. Jun Xu, Tao Mei, Ting Yao, and Yong Rui. Msr-vtt: A large video description dataset for bridging video and language. In *CVPR*, 2016.
29. Ranjay Krishna, Kenji Hata, Frederic Ren, Li Fei-Fei, and Juan Carlos Niebles. Dense-captioning events in videos. In *ICCV*, 2017.
30. Luowei Zhou, Chenliang Xu, and Jason J Corso. Towards automatic learning of procedures from web instructional videos. In *AAAI*, 2018.
31. Du Tran, Lubomir Bourdev, Rob Fergus, Lorenzo Torresani, and Manohar Paluri. Learning spatiotemporal features with 3d convolutional networks. In *ICCV*, 2015.
32. Zuxuan Wu, Xi Wang, Yu-Gang Jiang, Hao Ye, and Xiangyang Xue. Modeling spatial-temporal clues in a hybrid deep learning framework for video classification. In *ACM MM*, 2015.
33. Limin Wang, Yuanjun Xiong, Zhe Wang, Yu Qiao, Dahua Lin, Xiaoou Tang, and Luc Van Gool. Temporal segment networks: Towards good practices for deep action recognition. In *ECCV*, 2016.
34. Rohit Girdhar, Deva Ramanan, Abhinav Gupta, Josef Sivic, and Bryan Russell. Actionvlad: Learning spatio-temporal aggregation for action classification. In *CVPR*, 2017.
35. Joao Carreira and Andrew Zisserman. Quo vadis, action recognition? a new model and the kinetics dataset. In *CVPR*, 2017.
36. Du Tran, Heng Wang, Lorenzo Torresani, Jamie Ray, Yann LeCun, and Manohar Paluri. A closer look at spatiotemporal convolutions for action recognition. In *CVPR*, 2018.
37. Xiaolong Wang, Ross Girshick, Abhinav Gupta, and Kaiming He. Non-local neural networks. In *CVPR*, 2018.
38. Ji Lin, Chuang Gan, and Song Han. Tsm: Temporal shift module for efficient video understanding. In *CVPR*, 2019.

39. Christoph Feichtenhofer, Haoqi Fan, Jitendra Malik, and Kaiming He. Slowfast networks for video recognition. In *ICCV*, 2019.
40. Christoph Feichtenhofer. X3d: Expanding architectures for efficient video recognition. In *CVPR*, 2020.
41. Gedas Bertasius, Heng Wang, and Lorenzo Torresani. Is space-time attention all you need for video understanding? In *ICML*, 2021.
42. Haoqi Fan, Bo Xiong, Karttikeya Mangalam, Yanghao Li, Zhicheng Yan, Jitendra Malik, and Christoph Feichtenhofer. Multiscale vision transformers. In *ICCV*, 2021.
43. Ze Liu, Jia Ning, Yue Cao, Yixuan Wei, Zheng Zhang, Stephen Lin, and Han Hu. Video swin transformer. In *CVPR*, 2022.
44. B. B. LeCun, J. S. Denker, D. Henderson, R. E. Howard, W. Hubbard, and L. D. Jackel. Handwritten digit recognition with a backpropagation network. In *NeurIPS*, 1990.
45. Yann LeCun, Yoshua Bengio, and Geoffrey Hinton. Deep learning. *nature*, 521(7553):436–444, 2015.
46. Yi-Tong Zhou and Rama Chellappa. Computation of optical flow using a neural network. In *ICNN*, 1988.
47. Matthew D Zeiler and Rob Fergus. Visualizing and understanding convolutional networks. In *Computer Vision–ECCV 2014: 13th European Conference, Zurich, Switzerland, September 6-12, 2014, Proceedings, Part I 13*, pages 818–833. Springer, 2014.
48. Yann LeCun, Léon Bottou, Yoshua Bengio, and Patrick Haffner. Gradient-based learning applied to document recognition. *Proceedings of the IEEE*, 86(11):2278–2324, 1998.
49. Jonathan Long, Evan Shelhamer, and Trevor Darrell. Fully convolutional networks for semantic segmentation. In *CVPR*, 2015.
50. Jiuxiang Gu, Zhenhua Wang, Jason Kuen, Lianyang Ma, Amir Shahroudy, Bing Shuai, Ting Liu, Xingxing Wang, and Gang Wang. Recent advances in convolutional neural networks. In *arXiv preprint arXiv:1512.07108*, 2016.
51. Yoshua Bengio, Aaron Courville, and Pascal Vincent. Representation learning: A review and new perspectives. *TPAMI*, 2013.
52. Yoshua Bengio, Patrice Simard, and Paolo Frasconi. Learning long-term dependencies with gradient descent is difficult. *TNN*, 1994.
53. Ashish Vaswani, Noam Shazeer, Niki Parmar, Jakob Uszkoreit, Llion Jones, Aidan N Gomez, Łukasz Kaiser, and Illia Polosukhin. Attention is all you need. In *NeurIPS*, 2017.
54. Ze Liu, Yutong Lin, Yue Cao, Han Hu, Yixuan Wei, Zheng Zhang, Stephen Lin, and Baining Guo. Swin transformer: Hierarchical vision transformer using shifted windows. In *ICCV*, 2021.
55. Ross Girshick, Jeff Donahue, Trevor Darrell, and Jitendra Malik. Rich feature hierarchies for accurate object detection and semantic segmentation. In *CVPR*, 2014.
56. Ali Sharif Razavian, Hossein Azizpour, Josephine Sullivan, and Stefan Carlsson. CNN features off-the-shelf: an astounding baseline for recognition. In *CVPRW*, 2014.
57. Shengxin Zha, Florian Luisier, Walter Andrews, Nitish Srivastava, and Ruslan Salakhutdinov. Exploiting image-trained CNN architectures for unconstrained video classification. In *BMVC*, 2015.
58. Ross Girshick. Fast r-cnn. In *ICCV*, 2015.
59. Shuiwang Ji, Wei Xu, Ming Yang, and Kai Yu. 3d convolutional neural networks for human action recognition. In *ICML*, 2010.
60. Du Tran, Heng Wang, Lorenzo Torresani, and Matt Feiszli. Video classification with channel-separated convolutional networks. In *ICCV*, 2019.
61. Du Tran, Lubomir D Bourdev, Rob Fergus, Lorenzo Torresani, and Manohar Paluri. C3d: Generic features for video analysis. In *ICCV*, 2015.
62. Lin Sun, Kui Jia, Dit-Yan Yeung, and Bertram E Shi. Human action recognition using factorized spatio-temporal convolutional networks. In *CVPR*, 2015.
63. H. Ye, Z. Wu, R.-W. Zhao, X. Wang, Y.-G. Jiang, and X. Xue. Evaluating two-stream cnn for video classification. In *ICMR*, 2015.

64. Heng Wang and Cordelia Schmid. Action recognition with improved trajectories. In *ICCV*, 2013.
65. Limin Wang, Yu Qiao, and Xiaoou Tang. Action recognition with trajectory-pooled deep-convolutional descriptors. In *CVPR*, 2015.
66. Christoph Feichtenhofer, Axel Pinz, and Andrew Zisserman. Convolutional two-stream network fusion for video action recognition. In *CVPR*, 2016.
67. R Christoph and Feichtenhofer Axel Pinz. Spatiotemporal residual networks for video action recognition. In *NeurIPS*, 2016.
68. Christoph Feichtenhofer, Axel Pinz, and Richard P Wildes. Spatiotemporal multiplier networks for video action recognition. In *CVPR*, 2017.
69. Bowen Zhang, Limin Wang, Zhe Wang, Yu Qiao, and Hanli Wang. Real-time action recognition with enhanced motion vector CNNs. In *CVPR*, 2016.
70. Yi Zhu, Zhenzhong Lan, Shawn Newsam, and Alexander Hauptmann. Hidden two-stream convolutional networks for action recognition. In *ACCV*, 2019.
71. Zejia Weng, Zuxuan Wu, Hengduo Li, Jingjing Chen, and Yu-Gang Jiang. Hcms: Hierarchical and conditional modality selection for efficient video recognition. *TOMM*, 2021.
72. Zhaofan Qiu, Ting Yao, and Tao Mei. Learning spatio-temporal representation with pseudo-3d residual networks. In *ICCV*, 2017.
73. Saining Xie, Chen Sun, Jonathan Huang, Zhuowen Tu, and Kevin Murphy. Rethinking spatiotemporal feature learning: Speed-accuracy trade-offs in video classification. In *ECCV*, 2018.
74. Ali Diba, Vivek Sharma, and Luc Van Gool. Deep temporal linear encoding networks. In *CVPR*, 2017.
75. Bolei Zhou, Alex Andonian, Aude Oliva, and Antonio Torralba. Temporal relational reasoning in videos. In *ECCV*, 2018.
76. Ali Diba, Mohsen Fayyaz, Vivek Sharma, Amir Hossein Karami, Mohammad Mahdi Arzani, Rahman Yousefzadeh, and Luc Van Gool. Temporal 3d convnets: New architecture and transfer learning for video classification. *arXiv preprint arXiv:1711.08200*, 2017.
77. Pauline C Ng and Steven Henikoff. Sift: Predicting amino acid changes that affect protein function. *NAR*, 2003.
78. Josef Sivic and Andrew Zisserman. Efficient visual search of videos cast as text retrieval. *TPAMI*, 2008.
79. Zhongwen Xu, Yi Yang, and Alex G Hauptmann. A discriminative cnn video representation for event detection. In *CVPR*, 2015.
80. Antoine Miech, Ivan Laptev, and Josef Sivic. Learnable pooling with context gating for video classification. *arXiv preprint arXiv:1706.06905*, 2017.
81. Christopher M Bishop and Nasser M Nasrabadi. *Pattern recognition and machine learning*. 2006.
82. James Philbin, Ondrej Chum, Michael Isard, Josef Sivic, and Andrew Zisserman. Lost in quantization: Improving particular object retrieval in large scale image databases. In *CVPR*, 2008.
83. Hervé Jégou, Florent Perronnin, Matthijs Douze, Jorge Sánchez, Patrick Pérez, and Cordelia Schmid. Aggregating local image descriptors into compact codes. *TPAMI*, 2011.
84. Florent Perronnin, Jorge Sánchez, and Thomas Mensink. Improving the fisher kernel for large-scale image classification. In *ECCV*, 2010.
85. Tommi Jaakkola and David Haussler. Exploiting generative models in discriminative classifiers. In *NeurIPS*, 1998.
86. Paul Over. Trecvid 2013–an overview of the goals, tasks, data, evaluation mechanisms and metrics. 2013.
87. Shoou-I Yu, Lu Jiang, Zexi Mao, Xiaojun Chang, Xingzhong Du, Chuang Gan, Zhenzhong Lan, Zhongwen Xu, Xuanchong Li, Yang Cai, et al. Informedia@ trecvid 2014 med and mer. In *NTVREW*, 2014.
88. Relja Arandjelovic, Petr Gronat, Akihiko Torii, Tomas Pajdla, and Josef Sivic. Netvlad: Cnn architecture for weakly supervised place recognition. In *CVPR*, 2016.

89. Sami Abu-El-Haija, Nisarg Kothari, Joonseok Lee, Paul Natsev, George Toderici, Balakrishnan Varadarajan, and Sudheendra Vijayanarasimhan. Youtube-8m: A large-scale video classification benchmark. *arXiv preprint arXiv:1609.08675*, 2016.
90. Joe Yue-Hei Ng, Matthew Hausknecht, Sudheendra Vijayanarasimhan, Oriol Vinyals, Rajat Monga, and George Toderici. Beyond short snippets: Deep networks for video classification. In *CVPR*, 2015.
91. Z. Wu, Y.-G. Jiang, X. Wang, H. Ye, and X. Xue. Multi-stream multi-class fusion of deep networks for video classification. In *ACM MM*, 2016.
92. Sepp Hochreiter and Jürgen Schmidhuber. Long short-term memory. *Neural Computation*, 1997.
93. Junyoung Chung, Caglar Gulcehre, KyungHyun Cho, and Yoshua Bengio. Empirical evaluation of gated recurrent neural networks on sequence modeling. *arXiv preprint arXiv:1412.3555*, 2014.
94. J. Donahue, L. A. Hendricks, S. Guadarrama, M. Rohrbach, S. Venugopalan, K. Saenko, and T. Darrell. Long-term recurrent convolutional networks for visual recognition and description. In *CVPR*, 2015.
95. Qing Li, Zhaofan Qiu, Ting Yao, Tao Mei, Yong Rui, and Jiebo Luo. Action recognition by learning deep multi-granular spatio-temporal video representation. In *ICMR*, 2016.
96. Hugo Touvron, Matthieu Cord, Matthijs Douze, Francisco Massa, Alexandre Sablayrolles, and Hervé Jégou. Training data-efficient image transformers & distillation through attention. In *ICML*, 2021.
97. Daniel Neimark, Omri Bar, Maya Zohar, and Dotan Asselmann. Video transformer network. In *ICCV*, 2021.
98. Iz Beltagy, Matthew E Peters, and Arman Cohan. Longformer: The long-document transformer. *arXiv preprint arXiv:2004.05150*, 2020.
99. Jacob Devlin, Ming-Wei Chang, Kenton Lee, and Kristina Toutanova. Bert: Pre-training of deep bidirectional transformers for language understanding. *arXiv preprint arXiv:1810.04805*, 2018.
100. Anurag Arnab, Mostafa Dehghani, Georg Heigold, Chen Sun, Mario Lučić, and Cordelia Schmid. Vivit: A video vision transformer. In *ICCV*, 2021.
101. Yanghao Li, Chao-Yuan Wu, Haoqi Fan, Karttikeya Mangalam, Bo Xiong, Jitendra Malik, and Christoph Feichtenhofer. Mvitv2: Improved multiscale vision transformers for classification and detection. In *CVPR*, 2022.
102. Wangmeng Xiang, Chao Li, Biao Wang, Xihan Wei, Xian-Sheng Hua, and Lei Zhang. Spatiotemporal self-attention modeling with temporal patch shift for action recognition. In *ECCV*, 2022.
103. Yuxuan Liang, Pan Zhou, Roger Zimmermann, and Shuicheng Yan. Dualformer: Local-global stratified transformer for efficient video recognition. In *ECCV*, 2022.
104. Bolin Ni, Houwen Peng, Minghao Chen, Songyang Zhang, Gaofeng Meng, Jianlong Fu, Shiming Xiang, and Haibin Ling. Expanding language-image pretrained models for general video recognition. In *ECCV*, 2022.
105. Junting Pan, Ziyi Lin, Xiatian Zhu, Jing Shao, and Hongsheng Li. St-adapter: Parameter-efficient image-to-video transfer learning. In *NeurIPS*, 2022.
106. Ivan Laptev and Tony Lindeberg. Local descriptors for spatio-temporal recognition. In *SCVMA*, 2006.
107. Ivan Laptev, Marcin Marszalek, Cordelia Schmid, and Benjamin Rozenfeld. Learning realistic human actions from movies. In *CVPR*, 2008.
108. Yu-Gang Jiang, Zuxuan Wu, Jun Wang, Xiangyang Xue, and Shih-Fu Chang. Exploiting feature and class relationships in video categorization with regularized deep neural networks. *TPAMI*, 2017.
109. Gunnar A Sigurdsson, Gül Varol, Xiaolong Wang, Ali Farhadi, Ivan Laptev, and Abhinav Gupta. Hollywood in homes: Crowdsourcing data collection for activity understanding. In *ECCV*, 2016.

110. Raghav Goyal, Samira Ebrahimi Kahou, Vincent Michalski, Joanna Materzynska, Susanne Westphal, Heuna Kim, Valentin Haenel, Ingo Fruend, Peter Yianilos, Moritz Mueller-Freitag, et al. The" something something" video database for learning and evaluating visual common sense. In *ICCV*, 2017.
111. Joao Carreira, Eric Noland, Andras Banki-Horvath, Chloe Hillier, and Andrew Zisserman. A short note about kinetics-600. *arXiv preprint arXiv:1808.01340*, 2018.
112. Farzaneh Mahdisoltani, Guillaume Berger, Waseem Gharbieh, David Fleet, and Roland Memisevic. Fine-grained video classification and captioning. *arXiv preprint arXiv:1804.09235*, 2018.
113. Joao Carreira, Eric Noland, Chloe Hillier, and Andrew Zisserman. A short note on the kinetics-700 human action dataset. *arXiv preprint arXiv:1907.06987*, 2019.
114. Kishore K Reddy and Mubarak Shah. Recognizing 50 human action categories of web videos. *MVA*, 2013.
115. Chunhui Gu, Chen Sun, David A Ross, Carl Vondrick, Caroline Pantofaru, Yeqing Li, Sudheendra Vijayanarasimhan, George Toderici, Susanna Ricco, Rahul Sukthankar, et al. Ava: A video dataset of spatio-temporally localized atomic visual actions. In *CVPR*, 2018.
116. Dima Damen, Hazel Doughty, Giovanni Maria Farinella, Sanja Fidler, Antonino Furnari, Evangelos Kazakos, Davide Moltisanti, Jonathan Munro, Toby Perrett, Will Price, et al. Scaling egocentric vision: The epic-kitchens dataset. In *ECCV*, 2018.
117. Xing Zhang, Zuxuan Wu, Zejia Weng, Huazhu Fu, Jingjing Chen, Yu-Gang Jiang, and Larry S Davis. Videolt: large-scale long-tailed video recognition. In *ICCV*, 2021.
118. Noureldien Hussein, Efstratios Gavves, and Arnold WM Smeulders. Timeception for complex action recognition. In *CVPR*, 2019.
119. Zheng Shou, Dongang Wang, and Shih-Fu Chang. Temporal action localization in untrimmed videos via multi-stage cnns. In *CVPR*, 2016.
120. Zheng Shou, Jonathan Chan, Alireza Zareian, Kazuyuki Miyazawa, and Shih-Fu Chang. Cdc: Convolutional-de-convolutional networks for precise temporal action localization in untrimmed videos. In *CVPR*, 2017.
121. Ke Yang, Peng Qiao, Dongsheng Li, Shaohe Lv, and Yong Dou. Exploring temporal preservation networks for precise temporal action localization. In *AAAI*, 2018.
122. Yuanjun Xiong, Yue Zhao, Limin Wang, Dahua Lin, and Xiaoou Tang. A pursuit of temporal accuracy in general activity detection. *arXiv preprint arXiv:1703.02716*, 2017.
123. Tianwei Lin, Xu Zhao, Haisheng Su, Chongjing Wang, and Ming Yang. Bsn: Boundary sensitive network for temporal action proposal generation. In *ECCV*, 2018.
124. Tianwei Lin, Xu Zhao, and Zheng Shou. Single shot temporal action detection. In *ACM MM*, 2017.
125. Xiaolong Liu, Qimeng Wang, Yao Hu, Xu Tang, Shiwei Zhang, Song Bai, and Xiang Bai. End-to-end temporal action detection with transformer. *TIP*, 2022.
126. Chen-Lin Zhang, Jianxin Wu, and Yin Li. Actionformer: Localizing moments of actions with transformers. In *ECCV*, 2022.
127. Maheen Rashid, Hedvig Kjellstrom, and Yong Jae Lee. Action graphs: Weakly-supervised action localization with graph convolution networks. In *WACV*, 2020.
128. Limin Wang, Yuanjun Xiong, Dahua Lin, and Luc Van Gool. Untrimmednets for weakly supervised action recognition and detection. In *CVPR*, 2017.
129. Mihir Jain, Amir Ghodrati, and Cees GM Snoek. Actionbytes: Learning from trimmed videos to localize actions. In *CVPR*, 2020.
130. Sujoy Paul, Sourya Roy, and Amit K Roy-Chowdhury. W-talc: Weakly-supervised temporal activity localization and classification. In *ECCV*, 2018.
131. Linjiang Huang, Yan Huang, Wanli Ouyang, and Liang Wang. Relational prototypical network for weakly supervised temporal action localization. In *AAAI*, 2020.
132. Phuc Xuan Nguyen, Deva Ramanan, and Charless C Fowlkes. Weakly-supervised action localization with background modeling. In *ICCV*, 2019.
133. Zheng Shou, Hang Gao, Lei Zhang, Kazuyuki Miyazawa, and Shih-Fu Chang. Autoloc: Weakly-supervised temporal action localization in untrimmed videos. In *ECCV*, 2018.

134. Phuc Nguyen, Ting Liu, Gautam Prasad, and Bohyung Han. Weakly supervised action localization by sparse temporal pooling network. In *CVPR*, 2018.
135. Yuan Yuan, Yueming Lyu, Xi Shen, Ivor W Tsang, and Dit-Yan Yeung. Marginalized average attentional network for weakly-supervised learning. *arXiv preprint arXiv:1905.08586*, 2019.
136. Jiyang Gao, Zhenheng Yang, Kan Chen, Chen Sun, and Ram Nevatia. Turn tap: Temporal unit regression network for temporal action proposals. In *ICCV*, 2017.
137. Jiyang Gao, Zhenheng Yang, and Ram Nevatia. Cascaded boundary regression for temporal action detection. In *BMVC*, 2017.
138. Huijuan Xu, Abir Das, and Kate Saenko. R-c3d: Region convolutional 3d network for temporal activity detection. In *ICCV*, 2017.
139. Yu-Wei Chao, Sudheendra Vijayanarasimhan, Bryan Seybold, David A Ross, Jia Deng, and Rahul Sukthankar. Rethinking the faster r-cnn architecture for temporal action localization. In *CVPR*, 2018.
140. Yue Zhao, Yuanjun Xiong, Limin Wang, Zhirong Wu, Xiaoou Tang, and Dahua Lin. Temporal action detection with structured segment networks. In *ICCV*, 2017.
141. Tianwei Lin, Xiao Liu, Xin Li, Errui Ding, and Shilei Wen. Bmn: Boundary-matching network for temporal action proposal generation. In *ICCV*, 2019.
142. Haisheng Su, Weihao Gan, Wei Wu, Yu Qiao, and Junjie Yan. Bsn++: Complementary boundary regressor with scale-balanced relation modeling for temporal action proposal generation. In *AAAI*, 2021.
143. Junyi Shen, Ma Li, and Jikai Zhang. Temporal action detection methods based on deep learning. *International Journal of Pattern Recognition and Artificial Intelligence*, 36(03):2252005, 2022.
144. Shaoqing Ren, Kaiming He, Ross Girshick, and Jian Sun. Faster r-cnn: Towards real-time object detection with region proposal networks. *Advances in neural information processing systems*, 28, 2015.
145. Thomas N Kipf and Max Welling. Semi-supervised classification with graph convolutional networks. *arXiv preprint arXiv:1609.02907*, 2016.
146. Runhao Zeng, Wenbing Huang, Mingkui Tan, Yu Rong, Peilin Zhao, Junzhou Huang, and Chuang Gan. Graph convolutional networks for temporal action localization. In *Proceedings of the IEEE/CVF international conference on computer vision*, pages 7094–7103, 2019.
147. Mengmeng Xu, Chen Zhao, David S Rojas, Ali Thabet, and Bernard Ghanem. G-tad: Sub-graph localization for temporal action detection. In *Proceedings of the IEEE/CVF Conference on Computer Vision and Pattern Recognition*, pages 10156–10165, 2020.
148. Yueran Bai, Yingying Wang, Yunhai Tong, Yang Yang, Qiyue Liu, and Junhui Liu. Boundary content graph neural network for temporal action proposal generation. In *Computer Vision–ECCV 2020: 16th European Conference, Glasgow, UK, August 23–28, 2020, Proceedings, Part XXVIII 16*, pages 121–137. Springer, 2020.
149. Shyamal Buch, Victor Escorcia, Bernard Ghanem, Li Fei-Fei, and Juan Carlos Niebles. End-to-end, single-stream temporal action detection in untrimmed videos. In *BMVC*, 2019.
150. Fuchen Long, Ting Yao, Zhaofan Qiu, Xinmei Tian, Jiebo Luo, and Tao Mei. Gaussian temporal awareness networks for action localization. In *CVPR*, 2019.
151. Chuming Lin, Chengming Xu, Donghao Luo, Yabiao Wang, Ying Tai, Chengjie Wang, Jilin Li, Feiyue Huang, and Yanwei Fu. Learning salient boundary feature for anchor-free temporal action localization. In *CVPR*, 2021.
152. Joseph Redmon, Santosh Divvala, Ross Girshick, and Ali Farhadi. You only look once: Unified, real-time object detection. In *CVPR*, 2016.
153. Joseph Redmon and Ali Farhadi. Yolo9000: better, faster, stronger. In *CVPR*, 2017.
154. Wei Liu, Dragomir Anguelov, Dumitru Erhan, Christian Szegedy, Scott Reed, Cheng-Yang Fu, and Alexander C Berg. Ssd: Single shot multibox detector. In *ECCV*, 2016.
155. Chen Sun, Sanketh Shetty, Rahul Sukthankar, and Ram Nevatia. Temporal localization of fine-grained actions in videos by domain transfer from web images. In *ACM MM*, 2015.

156. Piotr Bojanowski, Rémi Lajugie, Francis Bach, Ivan Laptev, Jean Ponce, Cordelia Schmid, and Josef Sivic. Weakly supervised action labeling in videos under ordering constraints. In *ECCV*, 2014.
157. De-An Huang, Li Fei-Fei, and Juan Carlos Niebles. Connectionist temporal modeling for weakly supervised action labeling. In *ECCV*, 2016.
158. Alexander Richard, Hilde Kuehne, and Juergen Gall. Weakly supervised action learning with rnn based fine-to-coarse modeling. In *CVPR*, 2017.
159. Hilde Kuehne, Alexander Richard, and Juergen Gall. Weakly supervised learning of actions from transcripts. *CVIU*, 2017.
160. Sanath Narayan, Hisham Cholakkal, Fahad Shahbaz Khan, and Ling Shao. 3c-net: Category count and center loss for weakly-supervised action localization. In *ICCV*, 2019.
161. Julien Schroeter, Kirill Sidorov, and David Marshall. Weakly-supervised temporal localization via occurrence count learning. In *ICML*, 2019.
162. Ashraful Islam and Richard Radke. Weakly supervised temporal action localization using deep metric learning. In *WACV*, 2020.
163. Mingfei Gao, Yingbo Zhou, Ran Xu, Richard Socher, and Caiming Xiong. Woad: Weakly supervised online action detection in untrimmed videos. In *CVPR*, 2021.
164. Yandong Wen, Kaipeng Zhang, Zhifeng Li, and Yu Qiao. A discriminative feature learning approach for deep face recognition. In *ECCV*, 2016.
165. Fadime Sener and Angela Yao. Unsupervised learning and segmentation of complex activities from video. In *CVPR*, 2018.
166. Anna Kukleva, Hilde Kuehne, Fadime Sener, and Jurgen Gall. Unsupervised learning of action classes with continuous temporal embedding. In *CVPR*, 2019.
167. Guoqiang Gong, Xinghan Wang, Yadong Mu, and Qi Tian. Learning temporal co-attention models for unsupervised video action localization. In *CVPR*, 2020.
168. Vahdani E and Tian Y. Deep learning-based action detection in untrimmed videos: a survey. *TPAMI*, 2022.
169. M Jain, Van Gemert, J, H Jégou, P Bouthemy, and C Snoek. Action localization with tubelets from motion. In *CVPR*, 2014.
170. Jasper RR Uijlings, Koen EA Van De Sande, Theo Gevers, and Arnold WM Smeulders. Selective search for object recognition. *IJCV*, 2013.
171. Gkioxari G and Malik J. Finding action tubes. In *CVPR*, 2015.
172. Hou R, Chen C, and Shah M. Tube convolutional neural network (t-cnn) for action detection in videos. In *ICCV*, 2017.
173. Saha S, Singh G, Sapienza M, Torr P, H, and Cuzzolin F. Deep learning for detecting multiple space-time action tubes in videos. *arXiv preprint arXiv:1608.01529*, 2016.
174. Yang Z, Gao J, and Nevatia R. Spatio-temporal action detection with cascade proposal and location anticipation. *arXiv preprint arXiv:1708.00042*, 2017.
175. Ye Y, Yang X, and Tian Y. Discovering spatio-temporal action tubes. *VCIR*, 2019.
176. X. Peng and C. Schmid. Multi-region two-stream r-cnn for action detection. In *ECCV*, 2016.
177. V. Kalogeiton, P. Weinzaepfel, V. Ferrari, and C. Schmid. Action tubelet detector for spatio-temporal action localization. In *CVPR*, 2017.
178. X. Yang, X. Yang, M.-Y. Liu, F. Xiao, L. S. Davis, and J. Kautz. Step: Spatio-temporal progressive learning for video action detection. In *CVPR*, 2019.
179. C.-Y. Wu, C. Feichtenhofer, H. Fan, K. He, P. Krahenbuhl, and R. Girshick. Long-term feature banks for detailed video understanding. In *CVPR*, 2019.
180. C. Sun, A. Shrivastava, C. Vondrick, K. Murphy, R. Sukthankar, and C. Schmid. Actor-centric relation network. In *ECCV*, 2018.
181. Y. Zhang, P. Tokmakov, M. Hebert, and C. Schmid. A structured model for action detection. In *CVPR*, 2019.
182. M. Tomei, L. Baraldi, S. Calderara, S. Bronzin, and C. Cucchiara. Stage: Spatio-temporal attention on graph entities for video action detection. *CVIU*, 2019.
183. J. Ji, R. Krishna, F.-F. Li, and J. C. Niebles. Action genome: Actions as compositions of spatio-temporal scene graphs. In *CVPR*, 2020.

184. J. Pan, S. Chen, Z. Shou, J. Shao, and H. Li. Actor-context-actor relation network for spatio-temporal action localization. In *CVPR*, 2021.
185. R. Girdhar, J. Carreira, C. Doersch, and A. Zisserman. Video action transformer network. In *CVPR*, 2019.
186. Y. Jiang, J. Liu, R. Zamir, G. Toderici, I. Laptev, M. Shah, and R. Sukthankar. Thumos challenge: Action recognition with a large number of classes. 2014.
187. Hilde Kuehne, Ali Arslan, and Thomas Serre. The language of actions: Recovering the syntax and semantics of goal-directed human activities. In *CVPR*, 2014.
188. Yansong Tang, Dajun Ding, Yongming Rao, Yu Zheng, Danyang Zhang, Lili Zhao, Jiwen Lu, and Jie Zhou. Coin: A large-scale dataset for comprehensive instructional video analysis. In *CVPR*, 2019.
189. Antoine Miech, Dimitri Zhukov, Jean-Baptiste Alayrac, Makarand Tapaswi, Ivan Laptev, and Josef Sivic. Howto100m: Learning a text-video embedding by watching hundred million narrated video clips. In *ICCV*, 2019.
190. Jiyang Gao, Chen Sun, Zhenheng Yang, and Ram Nevatia. Tall: Temporal activity localization via language query. In *ICCV*, 2017.
191. Lisa Anne Hendricks, Oliver Wang, Eli Shechtman, Josef Sivic, Trevor Darrell, and Bryan Russell. Localizing moments in video with natural language. In *ICCV*, 2017.
192. Hao Zhang, Aixin Sun, Wei Jing, and Joey Tianyi Zhou. Temporal sentence grounding in videos: A survey and future directions. *IEEE Transactions on Pattern Analysis and Machine Intelligence*, 2023.
193. Runzhou Ge, Jiyang Gao, Kan Chen, and Ram Nevatia. Mac: Mining activity concepts for language-based temporal localization. In *2019 IEEE Winter Conference on Applications of Computer Vision (WACV)*, 2019.
194. Meng Liu, Xiang Wang, Liqiang Nie, Xiangnan He, Baoquan Chen, and Tat-Seng Chua. Attentive moment retrieval in videos. In *SIGIR*, 2018.
195. Aming Wu and Yahong Han. Multi-modal circulant fusion for video-to-language and backward. In *IJCAI*, 2018.
196. Shaoxiang Chen, Wenhao Jiang, Wei Liu, and Yu-Gang Jiang. Learning modality interaction for temporal sentence localization and event captioning in videos. In *ECCV*, 2020.
197. Yitian Yuan, Tao Mei, and Wenwu Zhu. To find where you talk: Temporal sentence localization in video with attention based location regression. In *AAAI*, 2019.
198. Yitian Yuan, Lin Ma, Jingwen Wang, Wei Liu, and Wenwu Zhu. Semantic conditioned dynamic modulation for temporal sentence grounding in videos. *TPAMI*, 2020.
199. Xuguang Duan, Wenbing Huang, Chuang Gan, Jingdong Wang, Wenwu Zhu, and Junzhou Huang. Weakly supervised dense event captioning in videos. In *NeurIPS*, 2018.
200. Zhijie Lin, Zhou Zhao, Zhu Zhang, Qi Wang, and Huasheng Liu. Weakly-supervised video moment retrieval via semantic completion network. In *AAAI*, 2020.
201. Yijun Song, Jingwen Wang, Lin Ma, Zhou Yu, and Jun Yu. Weakly-supervised multi-level attentional reconstruction network for grounding textual queries in videos. *arXiv preprint arXiv:2003.07048*, 2020.
202. Wenfei Yang, Tianzhu Zhang, Yongdong Zhang, and Feng Wu. Local correspondence network for weakly supervised temporal sentence grounding. *TIP*, 2021.
203. Minghang Zheng, Yanjie Huang, Qingchao Chen, and Yang Liu. Weakly supervised video moment localization with contrastive negative sample mining. In *AAAI*, 2022.
204. Niluthpol Chowdhury Mithun, Sujoy Paul, and Amit K Roy-Chowdhury. Weakly supervised video moment retrieval from text queries. In *CVPR*, 2019.
205. Minuk Ma, Sunjae Yoon, Junyeong Kim, Youngjoon Lee, Sunghun Kang, and Chang D Yoo. Vlanet: Video-language alignment network for weakly-supervised video moment retrieval. In *ECCV*, 2020.
206. Reuben Tan, Huijuan Xu, Kate Saenko, and Bryan A Plummer. Logan: Latent graph co-attention network for weakly-supervised video moment retrieval. In *WACV*, 2021.
207. Zhu Zhang, Zhou Zhao, Zhijie Lin, Xiuqiang He, et al. Counterfactual contrastive learning for weakly-supervised vision-language grounding. In *NeurIPS*, 2020.

208. Meng Liu, Xiang Wang, Liqiang Nie, Qi Tian, Baoquan Chen, and Tat-Seng Chua. Cross-modal moment localization in videos. In *ACM MM*, 2018.
209. Songyang Zhang, Jinsong Su, and Jiebo Luo. Exploiting temporal relationships in video moment localization with natural language. In *ACM MM*, 2019.
210. Ke Ning, Ming Cai, Di Xie, and Fei Wu. An attentive sequence to sequence translator for localizing video clips by natural language. *TMM*, 2019.
211. Bin Jiang, Xin Huang, Chao Yang, and Junsong Yuan. Cross-modal video moment retrieval with spatial and language-temporal attention. In *ICMR*, 2019.
212. Yawen Zeng, Da Cao, Xiaochi Wei, Meng Liu, Zhou Zhao, and Zheng Qin. Multi-modal relational graph for cross-modal video moment retrieval. In *CVPR*, 2021.
213. Huijuan Xu, Kun He, Bryan A Plummer, Leonid Sigal, Stan Sclaroff, and Kate Saenko. Multilevel language and vision integration for text-to-clip retrieval. In *AAAI*, 2019.
214. Shaoxiang Chen and Yu-Gang Jiang. Semantic proposal for activity localization in videos via sentence query. In *AAAI*, 2019.
215. Shaoning Xiao, Long Chen, Songyang Zhang, Wei Ji, Jian Shao, Lu Ye, and Jun Xiao. Boundary proposal network for two-stage natural language video localization. In *AAAI*, 2021.
216. Daizong Liu, Xiaoye Qu, Jianfeng Dong, and Pan Zhou. Adaptive proposal generation network for temporal sentence localization in videos. *arXiv preprint arXiv:2109.06398*, 2021.
217. Da Zhang, Xiyang Dai, Xin Wang, Yuan-Fang Wang, and Larry S Davis. Man: Moment alignment network for natural language moment retrieval via iterative graph adjustment. In *CVPR*, 2019.
218. Bingbin Liu, Serena Yeung, Edward Chou, De-An Huang, Li Fei-Fei, and Juan Carlos Niebles. Temporal modular networks for retrieving complex compositional activities in videos. In *ECCV*, 2018.
219. Songyang Zhang, Houwen Peng, Jianlong Fu, and Jiebo Luo. Learning 2d temporal adjacent networks for moment localization with natural language. In *AAAI*, 2020.
220. Jingyuan Chen, Xinpeng Chen, Lin Ma, Zequn Jie, and Tat-Seng Chua. Temporally grounding natural sentence in video. In *EMNLP*, 2018.
221. Hao Wang, Zheng-Jun Zha, Liang Li, Dong Liu, and Jiebo Luo. Structured multi-level interaction network for video moment localization via language query. In *CVPR*, 2021.
222. Soham Ghosh, Anuva Agarwal, Zarana Parekh, and Alexander Hauptmann. Excl: Extractive clip localization using natural language descriptions. *arXiv preprint arXiv:1904.02755*, 2019.
223. Chujie Lu, Long Chen, Chilie Tan, Xiaolin Li, and Jun Xiao. Debug: A dense bottom-up grounding approach for natural language video localization. In *EMNLP-IJCNLP*, 2019.
224. Long Chen, Chujie Lu, Siliang Tang, Jun Xiao, Dong Zhang, Chilie Tan, and Xiaolin Li. Rethinking the bottom-up framework for query-based video localization. In *AAAI*, 2020.
225. Runhao Zeng, Haoming Xu, Wenbing Huang, Peihao Chen, Mingkui Tan, and Chuang Gan. Dense regression network for video grounding. In *CVPR*, 2020.
226. Xiang Fang, Daizong Liu, Pan Zhou, Zichuan Xu, and Ruixuan Li. Hierarchical local-global transformer for temporal sentence grounding. *arXiv preprint arXiv:2208.14882*, 2022.
227. Shaoxiang Chen and Yu-Gang Jiang. Hierarchical visual-textual graph for temporal activity localization via language. In *ECCV*, 2020.
228. Yi-Wen Chen, Yi-Hsuan Tsai, and Ming-Hsuan Yang. End-to-end multi-modal video temporal grounding. In *NeurIPS*, 2021.
229. Jingyuan Chen, Lin Ma, Xinpeng Chen, Zequn Jie, and Jiebo Luo. Localizing natural language in videos. In *AAAI*, 2019.
230. Hao Zhang, Aixin Sun, Wei Jing, and Joey Tianyi Zhou. Span-based localizing network for natural language video localization. *arXiv preprint arXiv:2004.13931*, 2020.
231. Hao Zhang, Aixin Sun, Wei Jing, Liangli Zhen, Joey Tianyi Zhou, and Rick Siow Mong Goh. Parallel attention network with sequence matching for video grounding. *arXiv preprint arXiv:2105.08481*, 2021.
232. Xinli Yu, Mohsen Malmir, Xin He, Jiangning Chen, Tong Wang, Yue Wu, Yue Liu, and Yang Liu. Cross interaction network for natural language guided video moment retrieval. In *SIGIR*, 2021.

233. Yawen Zeng. Point prompt tuning for temporally language grounding. In *SIGIR*, 2022.
234. Mingfei Gao, Richard Socher, and Caiming Xiong. Weakly supervised natural language localization networks, 2020.
235. Jiabo Huang, Yang Liu, Shaogang Gong, and Hailin Jin. Cross-sentence temporal and semantic relations in video activity localisation. In *ICCV*, 2021.
236. Thomas G Dietterich, Richard H Lathrop, and Tomás Lozano-Pérez. Solving the multiple instance problem with axis-parallel rectangles. *AI*, 1997.
237. Zheng Wang, Jingjing Chen, and Yu-Gang Jiang. Visual co-occurrence alignment learning for weakly-supervised video moment retrieval. In *ACM MM*, 2021.
238. Daizong Liu, Xiaoye Qu, Yinzhen Wang, Xing Di, Kai Zou, Yu Cheng, Zichuan Xu, and Pan Zhou. Unsupervised temporal video grounding with deep semantic clustering. *arXiv preprint arXiv:2201.05307*, 2022.
239. Jinwoo Nam, Daechul Ahn, Dongyeop Kang, Seong Jong Ha, and Jonghyun Choi. Zero-shot natural language video localization. In *ICCV*, 2021.
240. Guolong Wang, Xun Wu, Zhaoyuan Liu, and Junchi Yan. Prompt-based zero-shot video moment retrieval. In *ACM MM*, 2022.
241. Alec Radford, Jong Wook Kim, Chris Hallacy, Aditya Ramesh, Gabriel Goh, Sandhini Agarwal, Girish Sastry, Amanda Askell, Pamela Mishkin, Jack Clark, et al. Learning transferable visual models from natural language supervision. In *ICML*, 2021.
242. Michaela Regneri, Marcus Rohrbach, Dominikus Wetzel, Stefan Thater, Bernt Schiele, and Manfred Pinkal. Grounding action descriptions in videos. *TACL*, 2013.
243. Marcus Rohrbach, Michaela Regneri, Mykhaylo Andriluka, Sikandar Amin, Manfred Pinkal, and Bernt Schiele. Script data for attribute-based recognition of composite activities. In *ECCV*, 2012.
244. Bart Thomee, David A Shamma, Gerald Friedland, Benjamin Elizalde, Karl Ni, Douglas Poland, Damian Borth, and Li-Jia Li. Yfcc100m: The new data in multimedia research. *Communications of the ACM*, 2016.
245. Ran Xu, Caiming Xiong, Wei Chen, and Jason J Corso. Jointly modeling deep video and compositional text to bridge vision and language in a unified framework. In *AAAI*, 2015.
246. Atsuhiro Kojima, Takeshi Tamura, and Kunio Fukunaga. Natural language description of human activities from video images based on concept hierarchy of actions. *IJCV*, 2002.
247. Chun-Chet Tan, Yu-Gang Jiang, and Chong-Wah Ngo. Towards textually describing complex video contents with audio-visual concept classifiers. In *ACM MM*, 2011.
248. Marcus Rohrbach, Wei Qiu, Ivan Titov, Stefan Thater, Manfred Pinkal, and Bernt Schiele. Translating video content to natural language descriptions. In *ICCV*, 2013.
249. Anna Rohrbach, Marcus Rohrbach, Wei Qiu, Annemarie Friedrich, Manfred Pinkal, and Bernt Schiele. Coherent multi-sentence video description with variable level of detail. In *GCPR*, 2014.
250. Sergio Guadarrama, Niveda Krishnamoorthy, Girish Malkarnenkar, Subhashini Venugopalan, Raymond Mooney, Trevor Darrell, and Kate Saenko. Youtube2text: Recognizing and describing arbitrary activities using semantic hierarchies and zero-shot recognition. In *ICCV*, 2013.
251. Jesse Thomason, Subhashini Venugopalan, Sergio Guadarrama, Kate Saenko, and Raymond Mooney. Integrating language and vision to generate natural language descriptions of videos in the wild. In *COLING*, 2014.
252. Subhashini Venugopalan, Huijuan Xu, Jeff Donahue, Marcus Rohrbach, Raymond Mooney, and Kate Saenko. Translating videos to natural language using deep recurrent neural networks. In *NAACL HLT*, 2015.
253. S. Venugopalan, M. Rohrbach, J. Donahue, R. Mooney, T. Darrell, and K. Saenko. Sequence to sequence - video to text. In *ICCV*, 2015.
254. Vasili Ramanishka, Abir Das, Dong Huk Park, Subhashini Venugopalan, Lisa Anne Hendricks, Marcus Rohrbach, and Kate Saenko. Multimodal video description. In *ACM MM*, 2016.
255. Qin Jin, Jia Chen, Shizhe Chen, Yifan Xiong, and Alexander Hauptmann. Describing videos using multi-modal fusion. In *ACM MM*, 2016.

256. Jun Xu, Ting Yao, Yongdong Zhang, and Tao Mei. Learning multimodal attention LSTM networks for video captioning. In *ACM MM*, 2017.
257. Chiori Hori, Takaaki Hori, Teng-Yok Lee, Ziming Zhang, Bret Harsham, John R Hershey, Tim K Marks, and Kazuhiko Sumi. Attention-based multimodal fusion for video description. In *ICCV*, 2017.
258. Xiang Long, Chuang Gan, and Gerard De Melo. Video captioning with multi-faceted attention. *TACL*, 2018.
259. Xin Wang, Yuan-Fang Wang, and William Yang Wang. Watch, listen, and describe: Globally and locally aligned cross-modal attentions for video captioning. *arXiv preprint arXiv:1804.05448*, 2018.
260. Junbo Wang, Wei Wang, Yan Huang, Liang Wang, and Tieniu Tan. M3: Multimodal memory modelling for video captioning. In *CVPR*, 2018.
261. Wenjie Pei, Jiyuan Zhang, Xiangrong Wang, Lei Ke, Xiaoyong Shen, and Yu-Wing Tai. Memory-attended recurrent network for video captioning. In *CVPR*, 2019.
262. Xishan Zhang, Ke Gao, Yongdong Zhang, Dongming Zhang, Jintao Li, and Qi Tian. Task-driven dynamic fusion: Reducing ambiguity in video description. In *CVPR*, 2017.
263. Sheng Liu, Zhou Ren, and Junsong Yuan. Sibnet: Sibling convolutional encoder for video captioning. In *ACM MM*, 2018.
264. Shaoxiang Chen and Yu-Gang Jiang. Motion guided region message passing for video captioning. In *ICCV*, 2021.
265. Li Yao, Atousa Torabi, Kyunghyun Cho, Nicolas Ballas, Christopher Pal, Hugo Larochelle, and Aaron Courville. Describing videos by exploiting temporal structure. In *ICCV*, 2015.
266. Chenggang Yan, Yunbin Tu, Xingzheng Wang, Yongbing Zhang, Xinhong Hao, Yongdong Zhang, and Qionghai Dai. Stat: Spatial-temporal attention mechanism for video captioning. *TMM*, 2019.
267. Shaoxiang Chen and Yu-Gang Jiang. Motion guided spatial attention for video captioning. In *AAAI*, 2019.
268. Ziqi Zhang, Yaya Shi, Chunfeng Yuan, Bing Li, Peijin Wang, Weiming Hu, and Zheng-Jun Zha. Object relational graph with teacher-recommended learning for video captioning. In *CVPR*, 2020.
269. Junchao Zhang and Yuxin Peng. Object-aware aggregation with bidirectional temporal graph for video captioning. In *CVPR*, 2019.
270. Boxiao Pan, Haoye Cai, De-An Huang, Kuan-Hui Lee, Adrien Gaidon, Ehsan Adeli, and Juan Carlos Niebles. Spatio-temporal graph for video captioning with knowledge distillation. In *CVPR*, 2020.
271. Yingwei Pan, Ting Yao, Houqiang Li, and Tao Mei. Video captioning with transferred semantic attributes. *arXiv preprint arXiv:1611.07675*, 2016.
272. Yingwei Pan, Tao Mei, Ting Yao, Houqiang Li, and Yong Rui. Jointly modeling embedding and translation to bridge video and language. In *CVPR*, 2016.
273. Liang Sun, Bing Li, Chunfeng Yuan, Zhengjun Zha, and Weiming Hu. Multimodal semantic attention network for video captioning. In *ICME*, 2019.
274. Hobin Ryu, Sunghun Kang, Haeyong Kang, and Chang D Yoo. Semantic grouping network for video captioning. In *AAAI*, 2021.
275. Jesus Perez-Martin, Benjamin Bustos, and Jorge Pérez. Attentive visual semantic specialized network for video captioning. In *ICPR*, 2021.
276. Hanhua Ye, Guorong Li, Yuankai Qi, Shuhui Wang, Qingming Huang, and Ming-Hsuan Yang. Hierarchical modular network for video captioning. In *CVPR*, 2022.
277. Bairui Wang, Lin Ma, Wei Zhang, Wenhao Jiang, Jingwen Wang, and Wei Liu. Controllable video captioning with pos sequence guidance based on gated fusion network. In *ICCV*, 2019.
278. Jingyi Hou, Xinxiao Wu, Wentian Zhao, Jiebo Luo, and Yunde Jia. Joint syntax representation learning and visual cue translation for video captioning. In *ICCV*, 2019.
279. Qi Zheng, Chaoyue Wang, and Dacheng Tao. Syntax-aware action targeting for video captioning. In *CVPR*, 2020.

280. Jincan Deng, Liang Li, Beichen Zhang, Shuhui Wang, Zhengjun Zha, and Qingming Huang. Syntax-guided hierarchical attention network for video captioning. *TCSVT*, 2021.
281. Jesus Perez-Martin, Benjamin Bustos, and Jorge Pérez. Improving video captioning with temporal composition of a visual-syntactic embedding. In *WACV*, 2021.
282. Xin Wang, Wenhu Chen, Jiawei Wu, Yuan-Fang Wang, and William Yang Wang. Video captioning via hierarchical reinforcement learning. In *CVPR*, 2018.
283. Ramakanth Pasunuru and Mohit Bansal. Reinforced video captioning with entailment rewards. *arXiv preprint arXiv:1708.02300*, 2017.
284. Yangyu Chen, Shuhui Wang, Weigang Zhang, and Qingming Huang. Less is more: Picking informative frames for video captioning. In *ECCV*, 2018.
285. Lijun Li and Boqing Gong. End-to-end video captioning with multitask reinforcement learning. In *WACV*, 2019.
286. Bang Yang, Yuexian Zou, Fenglin Liu, and Can Zhang. Non-autoregressive coarse-to-fine video captioning. In *AAAI*, 2021.
287. Fenglin Liu, Xuancheng Ren, Xian Wu, Bang Yang, Shen Ge, Yuexian Zou, and Xu Sun. O2na: An object-oriented non-autoregressive approach for controllable video captioning. *arXiv preprint arXiv:2108.02359*, 2021.
288. Xian Zhong, Yi Zhang, Shuqin Chen, Zhixin Sun, Huantao Zheng, and Kui Jiang. Dual-scale alignment-based transformer on linguistic skeleton tags for non-autoregressive video captioning. In *ICME*, 2022.
289. Qinghao Ye, Guohai Xu, Ming Yan, Haiyang Xu, Qi Qian, Ji Zhang, and Fei Huang. Hitea: Hierarchical temporal-aware video-language pre-training. *arXiv preprint arXiv:2212.14546*, 2022.
290. Mingkang Tang, Zhanyu Wang, Zhenhua Liu, Fengyun Rao, Dian Li, and Xiu Li. Clip4caption: Clip for video caption. In *ACM MM*, 2021.
291. Huaishao Luo, Lei Ji, Ming Zhong, Yang Chen, Wen Lei, Nan Duan, and Tianrui Li. Clip4clip: An empirical study of clip for end to end video clip retrieval. *arXiv preprint arXiv:2104.08860*, 2021.
292. Paul Hongsuck Seo, Arsha Nagrani, Anurag Arnab, and Cordelia Schmid. End-to-end generative pretraining for multimodal video captioning. In *CVPR*, 2022.
293. Luowei Zhou, Yingbo Zhou, Jason J Corso, Richard Socher, and Caiming Xiong. End-to-end dense video captioning with masked transformer. In *CVPR*, 2018.
294. Yehao Li, Ting Yao, Yingwei Pan, Hongyang Chao, and Tao Mei. Jointly localizing and describing events for dense video captioning. In *CVPR*, 2018.
295. Pradipto Das, Chenliang Xu, Richard F Doell, and Jason J Corso. A thousand frames in just a few words: Lingual description of videos through latent topics and sparse object stitching. In *CVPR*, 2013.
296. Marcus Rohrbach, Anna Rohrbach, Michaela Regneri, Sikandar Amin, Mykhaylo Andriluka, Manfred Pinkal, and Bernt Schiele. Recognizing fine-grained and composite activities using hand-centric features and script data. *IJCV*, 2015.
297. Atousa Torabi, Christopher Pal, Hugo Larochelle, and Aaron Courville. Using descriptive video services to create a large data source for video annotation research. *arXiv preprint arXiv:1503.01070*, 2015.
298. Anna Rohrbach, Marcus Rohrbach, Niket Tandon, and Bernt Schiele. A dataset for movie description. In *CVPR*, 2015.
299. Anna Rohrbach, Atousa Torabi, Marcus Rohrbach, Niket Tandon, Christopher Pal, Hugo Larochelle, Aaron Courville, and Bernt Schiele. Movie description. *IJCV*, 2017.
300. Yuncheng Li, Yale Song, Liangliang Cao, Joel Tetreault, Larry Goldberg, Alejandro Jaimes, and Jiebo Luo. TGIF: A new dataset and benchmark on animated GIF description. In *Proceedings of the IEEE Conference on Computer Vision and Pattern Recognition*, pages 4641–4650, 2016.
301. Xin Wang, Jiawei Wu, Junkun Chen, Lei Li, Yuan-Fang Wang, and William Yang Wang. Vatex: A large-scale, high-quality multilingual dataset for video-and-language research. In *ICCV*, 2019.

302. Kishore Papineni, Salim Roukos, Todd Ward, and Wei-Jing Zhu. Bleu: a method for automatic evaluation of machine translation. In *ACL*, 2002.
303. Satanjeev Banerjee and Alon Lavie. METEOR: An automatic metric for MT evaluation with improved correlation with human judgments. In *ACLW*, 2005.
304. Chin-Yew Lin. ROUGE: A package for automatic evaluation of summaries. In *ACL*, 2004.
305. Ramakrishna Vedantam, C Lawrence Zitnick, and Devi Parikh. Cider: Consensus-based image description evaluation. In *CVPR*, 2015.
306. Stephen Robertson. Understanding inverse document frequency: on theoretical arguments for IDF. *Journal of documentation*, 2004.
307. Peter Anderson, Basura Fernando, Mark Johnson, and Stephen Gould. Spice: Semantic propositional image caption evaluation. In *ECCV*, 2016.
308. Aapo Hyvärinen and Erkki Oja. Independent component analysis: algorithms and applications. *Neural networks*, 2000.
309. J Hans van Hateren and Dan L Ruderman. Independent component analysis of natural image sequences yields spatio-temporal filters similar to simple cells in primary visual cortex. *Proceedings of the Royal Society of London. Series B: Biological Sciences*, 1998.
310. Jarmo Hurri and Aapo Hyvärinen. Simple-cell-like receptive fields maximize temporal coherence in natural video. *Neural Computation*, 2003.
311. Aapo Hyvärinen, Jarmo Hurri, and Patrick O Hoyer. *Natural image statistics: A probabilistic approach to early computational vision.* Springer, 2009.
312. Quoc V Le, Will Y Zou, Serena Y Yeung, and Andrew Y Ng. Learning hierarchical invariant spatio-temporal features for action recognition with independent subspace analysis. In *CVPR*, 2011.
313. Laurenz Wiskott and Terrence J Sejnowski. Slow feature analysis: Unsupervised learning of invariances. *Neural Computation*, 2002.
314. Hossein Mobahi, Ronan Collobert, and Jason Weston. Deep learning from temporal coherence in video. In *ICML*, 2009.
315. Will Zou, Shenghuo Zhu, Kai Yu, and Andrew Ng. Deep learning of invariant features via simulated fixations in video. In *NeurIPS*, 2012.
316. Ilya Sutskever and Geoffrey Hinton. Learning multilevel distributed representations for high-dimensional sequences. In *AISTATS*, 2007.
317. Graham W Taylor, Rob Fergus, Yann LeCun, and Christoph Bregler. Convolutional learning of spatio-temporal features. In *ECCV*, 2010.
318. MarcAurelio Ranzato, Arthur Szlam, Joan Bruna, Michael Mathieu, Ronan Collobert, and Sumit Chopra. Video (language) modeling: a baseline for generative models of natural videos. *arXiv preprint arXiv:1412.6604*, 2014.
319. Tomas Mikolov, Martin Karafiát, Lukas Burget, Jan Cernockỳ, and Sanjeev Khudanpur. Recurrent neural network based language model. In *Interspeech*, 2010.
320. Tomas Mikolov, Kai Chen, Greg Corrado, and Jeffrey Dean. Efficient estimation of word representations in vector space. *arXiv preprint arXiv:1301.3781*, 2013.
321. Vignesh Ramanathan, Kevin Tang, Greg Mori, and Li Fei-Fei. Learning temporal embeddings for complex video analysis. In *ICCV*, 2015.
322. Xiaolong Wang and Abhinav Gupta. Unsupervised learning of visual representations using videos. In *CVPR*, 2015.
323. Deepak Pathak, Ross Girshick, Piotr Dollár, Trevor Darrell, and Bharath Hariharan. Learning features by watching objects move. In *CVPR*, 2017.
324. Xiaolong Wang, Kaiming He, and Abhinav Gupta. Transitive invariance for self-supervised visual representation learning. In *ICCV*, 2017.
325. Longlong Jing, Xiaodong Yang, Jingen Liu, and Yingli Tian. Self-supervised spatiotemporal feature learning via video rotation prediction. *arXiv preprint arXiv:1811.11387*, 2018.
326. Spyros Gidaris, Praveer Singh, and Nikos Komodakis. Unsupervised representation learning by predicting image rotations. *arXiv preprint arXiv:1803.07728*, 2018.
327. Ishan Misra, C Lawrence Zitnick, and Martial Hebert. Shuffle and learn: unsupervised learning using temporal order verification. In *ECCV*, 2016.

328. Basura Fernando, Hakan Bilen, Efstratios Gavves, and Stephen Gould. Self-supervised video representation learning with odd-one-out networks. In *CVPR*, 2017.
329. Hsin-Ying Lee, Jia-Bin Huang, Maneesh Singh, and Ming-Hsuan Yang. Unsupervised representation learning by sorting sequences. In *ICCV*, 2017.
330. Dejing Xu, Jun Xiao, Zhou Zhao, Jian Shao, Di Xie, and Yueting Zhuang. Self-supervised spatiotemporal learning via video clip order prediction. In *CVPR*, 2019.
331. Donglai Wei, Joseph J Lim, Andrew Zisserman, and William T Freeman. Learning and using the arrow of time. In *CVPR*, 2018.
332. Sagie Benaim, Ariel Ephrat, Oran Lang, Inbar Mosseri, William T Freeman, Michael Rubinstein, Michal Irani, and Tali Dekel. Speednet: Learning the speediness in videos. In *CVPR*, 2020.
333. Simon Jenni, Givi Meishvili, and Paolo Favaro. Video representation learning by recognizing temporal transformations. In *ECCV*, 2020.
334. Ross Goroshin, Joan Bruna, Jonathan Tompson, David Eigen, and Yann LeCun. Unsupervised learning of spatiotemporally coherent metrics. In *ICCV*, 2015.
335. Dinesh Jayaraman and Kristen Grauman. Slow and steady feature analysis: higher order temporal coherence in video. In *CVPR*, 2016.
336. Pulkit Agrawal, Joao Carreira, and Jitendra Malik. Learning to see by moving. In *ICCV*, 2015.
337. Uta Buchler, Biagio Brattoli, and Bjorn Ommer. Improving spatiotemporal self-supervision by deep reinforcement learning. In *ECCV*, 2018.
338. Dahun Kim, Donghyeon Cho, and In So Kweon. Self-supervised video representation learning with space-time cubic puzzles. *TPAMI*, 2019.
339. Unaiza Ahsan, Rishi Madhok, and Irfan Essa. Video jigsaw: Unsupervised learning of spatiotemporal context for video action recognition. In *2019 IEEE Winter Conference on Applications of Computer Vision (WACV)*, 2019.
340. Gustav Larsson, Michael Maire, and Gregory Shakhnarovich. Learning representations for automatic colorization. In *ECCV*, 2016.
341. Richard Zhang, Phillip Isola, and Alexei A Efros. Colorful image colorization. In *ECCV*, 2016.
342. Carl Vondrick, Abhinav Shrivastava, Alireza Fathi, Sergio Guadarrama, and Kevin Murphy. Tracking emerges by colorizing videos. In *ECCV*, 2018.
343. Nitish Srivastava, Elman Mansimov, and Ruslan Salakhudinov. Unsupervised learning of video representations using lstms. In *ICML*, 2015.
344. Carl Vondrick, Hamed Pirsiavash, and Antonio Torralba. Anticipating visual representations from unlabeled video. In *CVPR*, 2016.
345. Ali Diba, Vivek Sharma, Luc Van Gool, and Rainer Stiefelhagen. Dynamonet: Dynamic action and motion network. In *ICCV*, 2019.
346. Wilson Yan, Yunzhi Zhang, Pieter Abbeel, and Aravind Srinivas. Videogpt: Video generation using vq-vae and transformers. *arXiv preprint arXiv:2104.10157*, 2021.
347. Aaron Van Den Oord, Oriol Vinyals, et al. Neural discrete representation learning. *NeurIPS*, 2017.
348. Michael Mathieu, Camille Couprie, and Yann LeCun. Deep multi-scale video prediction beyond mean square error. *arXiv preprint arXiv:1511.05440*, 2015.
349. Carl Vondrick, Hamed Pirsiavash, and Antonio Torralba. Generating videos with scene dynamics. In *NeurIPS*, 2016.
350. Sergey Tulyakov, Ming-Yu Liu, Xiaodong Yang, and Jan Kautz. Mocogan: Decomposing motion and content for video generation. In *CVPR*, 2018.
351. Alexey Dosovitskiy, Philipp Fischer, Eddy Ilg, Philip Hausser, Caner Hazirbas, Vladimir Golkov, Patrick Van Der Smagt, Daniel Cremers, and Thomas Brox. Flownet: Learning optical flow with convolutional networks. In *ICCV*, 2015.
352. Eddy Ilg, Nikolaus Mayer, Tonmoy Saikia, Margret Keuper, Alexey Dosovitskiy, and Thomas Brox. Flownet 2.0: Evolution of optical flow estimation with deep networks. In *CVPR*, 2017.
353. Nawid Sayed, Biagio Brattoli, and Björn Ommer. Cross and learn: Cross-modal self-supervision. In *German Conference on Pattern Recognition*, 2018.

354. Senthil Purushwalkam and Abhinav Gupta. Pose from action: Unsupervised learning of pose features based on motion. *arXiv preprint arXiv:1609.05420*, 2016.
355. Relja Arandjelovic and Andrew Zisserman. Look, listen and learn. In *ICCV*, 2017.
356. Andrew Owens and Alexei A Efros. Audio-visual scene analysis with self-supervised multisensory features. In *ECCV*, 2018.
357. Jiangliu Wang, Jianbo Jiao, Linchao Bao, Shengfeng He, Yunhui Liu, and Wei Liu. Self-supervised spatio-temporal representation learning for videos by predicting motion and appearance statistics. In *CVPR*, 2019.
358. Ting Yao, Yiheng Zhang, Zhaofan Qiu, Yingwei Pan, and Tao Mei. Seco: Exploring sequence supervision for unsupervised representation learning. In *AAAI*, 2021.
359. Vidit Kumar. Unsupervised learning of spatio-temporal representation with multi-task learning for video retrieval. In *NCC*, 2022.
360. Simon Jenni, Alexander Black, and John Collomosse. Audio-visual contrastive learning with temporal self-supervision. In *AAAI*, 2023.
361. Michael Gutmann and Aapo Hyvärinen. Noise-contrastive estimation: A new estimation principle for unnormalized statistical models. In *AISTATS*, 2010.
362. Aaron van den Oord, Yazhe Li, and Oriol Vinyals. Representation learning with contrastive predictive coding. *arXiv preprint arXiv:1807.03748*, 2018.
363. Pierre Sermanet, Corey Lynch, Yevgen Chebotar, Jasmine Hsu, Eric Jang, Stefan Schaal, Sergey Levine, and Google Brain. Time-contrastive networks: Self-supervised learning from video. In *ICRA*, 2018.
364. Christoph Feichtenhofer, Haoqi Fan, Bo Xiong, Ross Girshick, and Kaiming He. A large-scale study on unsupervised spatiotemporal representation learning. In *CVPR*, 2021.
365. Daniel Gordon, Kiana Ehsani, Dieter Fox, and Ali Farhadi. Watching the world go by: Representation learning from unlabeled videos. *arXiv preprint arXiv:2003.07990*, 2020.
366. Jinpeng Wang, Yuting Gao, Ke Li, Yiqi Lin, Andy J Ma, Hao Cheng, Pai Peng, Feiyue Huang, Rongrong Ji, and Xing Sun. Removing the background by adding the background: Towards background robust self-supervised video representation learning. In *CVPR*, 2021.
367. Ceyuan Yang, Yinghao Xu, Bo Dai, and Bolei Zhou. Video representation learning with visual tempo consistency. *arXiv preprint arXiv:2006.15489*, 2020.
368. Aadarsh Sahoo, Rutav Shah, Rameswar Panda, Kate Saenko, and Abir Das. Contrast and mix: Temporal contrastive video domain adaptation with background mixing. In *NeurIPS*, 2021.
369. Jiangliu Wang, Jianbo Jiao, and Yun-Hui Liu. Self-supervised video representation learning by pace prediction. In *ECCV*, 2020.
370. Simon Jenni and Hailin Jin. Time-equivariant contrastive video representation learning. In *ICCV*, 2021.
371. Peihao Chen, Deng Huang, Dongliang He, Xiang Long, Runhao Zeng, Shilei Wen, Mingkui Tan, and Chuang Gan. Rspnet: Relative speed perception for unsupervised video representation learning. In *AAAI*, 2021.
372. Debidatta DwibediR, Jonathan Tompson, Corey LynchR, and Pierre Sermanet. Self-supervised representation learning for continuous control. 2018.
373. Kaiming He, Haoqi Fan, Yuxin Wu, Saining Xie, and Ross Girshick. Momentum contrast for unsupervised visual representation learning. In *CVPR*, 2020.
374. Tian Pan, Yibing Song, Tianyu Yang, Wenhao Jiang, and Wei Liu. Videomoco: Contrastive video representation learning with temporally adversarial examples. In *CVPR*, 2021.
375. Jue Wang, Gedas Bertasius, Du Tran, and Lorenzo Torresani. Long-short temporal contrastive learning of video transformers. In *CVPR*, 2022.
376. Tengda Han, Weidi Xie, and Andrew Zisserman. Video representation learning by dense predictive coding. In *ICCV Workshop*, 2019.
377. Ankesh Anand, Evan Racah, Sherjil Ozair, Yoshua Bengio, Marc-Alexandre Côté, and R Devon Hjelm. Unsupervised state representation learning in atari. In *NeurIPS*, 2019.
378. R Devon Hjelm, Alex Fedorov, Samuel Lavoie-Marchildon, Karan Grewal, Phil Bachman, Adam Trischler, and Yoshua Bengio. Learning deep representations by mutual information estimation and maximization. *arXiv preprint arXiv:1808.06670*, 2018.

379. Ishan Dave, Rohit Gupta, Mamshad Nayeem Rizve, and Mubarak Shah. Tclr: Temporal contrastive learning for video representation. *Computer Vision and Image Understanding*, 2022.
380. Liangzhe Yuan, Rui Qian, Yin Cui, Boqing Gong, Florian Schroff, Ming-Hsuan Yang, Hartwig Adam, and Ting Liu. Contextualized spatio-temporal contrastive learning with self-supervision. In *CVPR*, 2022.
381. Haofei Kuang, Yi Zhu, Zhi Zhang, Xinyu Li, Joseph Tighe, Sören Schwertfeger, Cyrill Stachniss, and Mu Li. Video contrastive learning with global context. In *ICCV*, 2021.
382. Bo Fang, Wenhao Wu, Chang Liu, Yu Zhou, Dongliang He, and Weipinng Wang. Mamico: Macro-to-micro semantic correspondence for self-supervised video representation learning. In *ACMMM*, 2022.
383. Rui Qian, Tianjian Meng, Boqing Gong, Ming-Hsuan Yang, Huisheng Wang, Serge Belongie, and Yin Cui. Spatiotemporal contrastive video representation learning. In *CVPR*, 2021.
384. Michael Dorkenwald, Fanyi Xiao, Biagio Brattoli, Joseph Tighe, and Davide Modolo. Scvrl: Shuffled contrastive video representation learning. In *CVPR*, 2022.
385. Chengxu Zhuang, Tianwei She, Alex Andonian, Max Sobol Mark, and Daniel Yamins. Unsupervised learning from video with deep neural embeddings. In *CVPR*, 2020.
386. Pavel Tokmakov, Martial Hebert, and Cordelia Schmid. Unsupervised learning of video representations via dense trajectory clustering. In *ECCV*, 2020.
387. Vivek Sharma, Makarand Tapaswi, M Saquib Sarfraz, and Rainer Stiefelhagen. Clustering based contrastive learning for improving face representations. In *IEEE International Conference on Automatic Face and Gesture Recognition*, 2020.
388. Saquib Sarfraz, Vivek Sharma, and Rainer Stiefelhagen. Efficient parameter-free clustering using first neighbor relations. In *CVPR*, 2019.
389. Salar Hosseini Khorasgani, Yuxuan Chen, and Florian Shkurti. Slic: Self-supervised learning with iterative clustering for human action videos. In *CVPR*, 2022.
390. Brian Chen, Andrew Rouditchenko, Kevin Duarte, Hilde Kuehne, Samuel Thomas, Angie Boggust, Rameswar Panda, Brian Kingsbury, Rogerio Feris, David Harwath, et al. Multimodal clustering networks for self-supervised learning from unlabeled videos. In *ICCV*, 2021.
391. Jie Xu, Huayi Tang, Yazhou Ren, Liang Peng, Xiaofeng Zhu, and Lifang He. Multi-level feature learning for contrastive multi-view clustering. In *CVPR*, 2022.
392. Bruno Korbar, Du Tran, and Lorenzo Torresani. Cooperative learning of audio and video models from self-supervised synchronization. In *NeurIPS*, 2018.
393. Pedro Morgado, Nuno Vasconcelos, and Ishan Misra. Audio-visual instance discrimination with cross-modal agreement. In *CVPR*, 2021.
394. Jiutong Wei, Guan Luo, Bing Li, and Weiming Hu. Inter-intra cross-modality self-supervised video representation learning by contrastive clustering. In *ICPR*, 2022.
395. Antoine Miech, Jean-Baptiste Alayrac, Lucas Smaira, Ivan Laptev, Josef Sivic, and Andrew Zisserman. End-to-end learning of visual representations from uncurated instructional videos. In *CVPR*, 2020.
396. Yonglong Tian, Dilip Krishnan, and Phillip Isola. Contrastive multiview coding. In *ECCV*, 2020.
397. Tengda Han, Weidi Xie, and Andrew Zisserman. Self-supervised co-training for video representation learning. In *NeurIPS*, 2020.
398. Jean-Baptiste Alayrac, Adria Recasens, Rosalia Schneider, Relja Arandjelović, Jason Ramapuram, Jeffrey De Fauw, Lucas Smaira, Sander Dieleman, and Andrew Zisserman. Self-supervised multimodal versatile networks. In *NeurIPS*, 2020.
399. Hassan Akbari, Liangzhe Yuan, Rui Qian, Wei-Hong Chuang, Shih-Fu Chang, Yin Cui, and Boqing Gong. Vatt: Transformers for multimodal self-supervised learning from raw video, audio and text. In *NeurIPS*, 2021.
400. Alec Radford, Karthik Narasimhan, Tim Salimans, Ilya Sutskever, et al. Improving language understanding by generative pre-training. *OpenAI*, 2018.
401. Hangbo Bao, Li Dong, Songhao Piao, and Furu Wei. BEit: BERT pre-training of image transformers. In *ICLR*, 2022.

402. Kaiming He, Xinlei Chen, Saining Xie, Yanghao Li, Piotr Dollár, and Ross Girshick. Masked autoencoders are scalable vision learners. In *CVPR*, 2022.
403. Mark Chen, Alec Radford, Rewon Child, Jeffrey Wu, Heewoo Jun, David Luan, and Ilya Sutskever. Generative pretraining from pixels. In *ICML*, 2020.
404. Pascal Vincent, Hugo Larochelle, Isabelle Lajoie, Yoshua Bengio, Pierre-Antoine Manzagol, and Léon Bottou. Stacked denoising autoencoders: Learning useful representations in a deep network with a local denoising criterion. *JMLR*, 2010.
405. Deepak Pathak, Philipp Krahenbuhl, Jeff Donahue, Trevor Darrell, and Alexei A Efros. Context encoders: Feature learning by inpainting. In *CVPR*, 2016.
406. Chen Wei, Haoqi Fan, Saining Xie, Chao-Yuan Wu, Alan Yuille, and Christoph Feichtenhofer. Masked feature prediction for self-supervised visual pre-training. In *CVPR*, 2022.
407. Rui Wang, Dongdong Chen, Zuxuan Wu, Yinpeng Chen, Xiyang Dai, Mengchen Liu, Yu-Gang Jiang, Luowei Zhou, and Lu Yuan. Bevt: Bert pretraining of video transformers. In *CVPR*, 2022.
408. Hao Tan, Jie Lei, Thomas Wolf, and Mohit Bansal. Vimpac: Video pre-training via masked token prediction and contrastive learning. *arXiv preprint arXiv:2106.11250*, 2021.
409. Agrim Gupta, Stephen Tian, Yunzhi Zhang, Jiajun Wu, Roberto Martín-Martín, and Li Fei-Fei. Maskvit: Masked visual pre-training for video prediction. In *ICLR*, 2023.
410. Xinyu Sun, Peihao Chen, Liangwei Chen, Thomas H Li, Mingkui Tan, and Chuang Gan. Masked motion encoding for self-supervised video representation learning. 2022.
411. Zhan Tong, Yibing Song, Jue Wang, and Limin Wang. Videomae: Masked autoencoders are data-efficient learners for self-supervised video pre-training. *arXiv preprint arXiv:2203.12602*, 2022.
412. Christoph Feichtenhofer, Haoqi Fan, Yanghao Li, and Kaiming He. Masked autoencoders as spatiotemporal learners. *arXiv preprint arXiv:2205.09113*, 2022.
413. Wele Gedara Chaminda Bandara, Naman Patel, Ali Gholami, Mehdi Nikkhah, Motilal Agrawal, and Vishal M Patel. Adamae: Adaptive masking for efficient spatiotemporal learning with masked autoencoders. In *CVPR*, 2023.
414. Rohit Girdhar, Alaaeldin El-Nouby, Mannat Singh, Kalyan Vasudev Alwala, Armand Joulin, and Ishan Misra. Omnimae: Single model masked pretraining on images and videos. *arXiv preprint arXiv:2206.08356*, 2022.
415. Sunil Hwang, Jaehong Yoon, Youngwan Lee, and Sung Ju Hwang. Efficient video representation learning via masked video modeling with motion-centric token selection. *arXiv preprint arXiv:2211.10636*, 2022.
416. Limin Wang, Bingkun Huang, Zhiyu Zhao, Zhan Tong, Yinan He, Yi Wang, Yali Wang, and Yu Qiao. Videomae v2: Scaling video masked autoencoders with dual masking. In *CVPR*, 2023.
417. Aditya Ramesh, Mikhail Pavlov, Gabriel Goh, Scott Gray, Chelsea Voss, Alec Radford, Mark Chen, and Ilya Sutskever. Zero-shot text-to-image generation. In *ICML*, 2021.
418. Patrick Esser, Robin Rombach, and Bjorn Ommer. Taming transformers for high-resolution image synthesis. In *CVPR*, 2021.
419. Navneet Dalal and Bill Triggs. Histograms of oriented gradients for human detection. In *CVPR*, 2005.
420. Heng Wang, Alexander Kläser, Cordelia Schmid, and Cheng-Lin Liu. Dense trajectories and motion boundary descriptors for action recognition. *IJCV*, 2013.
421. Richard S Sutton, David McAllester, Satinder Singh, and Yishay Mansour. Policy gradient methods for reinforcement learning with function approximation. 1999.
422. Zhiwu Qing, Shiwei Zhang, Ziyuan Huang, Xiang Wang, Yuehuan Wang, Yiliang Lv, Changxin Gao, and Nong Sang. Mar: Masked autoencoders for efficient action recognition. *arXiv preprint arXiv:2207.11660*, 2022.
423. Yi Liu, Limin Wang, Yali Wang, Xiao Ma, and Yu Qiao. Fineaction: A fine-grained video dataset for temporal action localization. *TIP*, 2022.

424. Kristen Grauman, Andrew Westbury, Eugene Byrne, Zachary Chavis, Antonino Furnari, Rohit Girdhar, Jackson Hamburger, Hao Jiang, Miao Liu, Xingyu Liu, et al. Ego4d: Around the world in 3,000 hours of egocentric video. In *CVPR*, 2022.
425. Frederik Ebert, Chelsea Finn, Alex X Lee, and Sergey Levine. Self-supervised visual planning with temporal skip connections. In *ICRL*, 2017.
426. Sudeep Dasari, Frederik Ebert, Stephen Tian, Suraj Nair, Bernadette Bucher, Karl Schmeckpeper, Siddharth Singh, Sergey Levine, and Chelsea Finn. Robonet: Large-scale multi-robot learning. *arXiv preprint arXiv:1910.11215*, 2019.
427. Andreas Geiger, Philip Lenz, Christoph Stiller, and Raquel Urtasun. Vision meets robotics: The kitti dataset. *IJRR*, 2013.
428. Sergey Ioffe and Christian Szegedy. Batch normalization: Accelerating deep network training by reducing internal covariate shift. In *ICML*, 2015.
429. Sijie Zhu, Taojiannan Yang, Matias Mendieta, and Chen Chen. A3d: Adaptive 3d networks for video action recognition. *arXiv preprint arXiv:2011.12384*, 2020.
430. Andrew G Howard, Menglong Zhu, Bo Chen, Dmitry Kalenichenko, Weijun Wang, Tobias Weyand, Marco Andreetto, and Hartwig Adam. Mobilenets: Efficient convolutional neural networks for mobile vision applications. *arXiv preprint arXiv:1704.04861*, 2017.
431. François Chollet. Xception: Deep learning with depthwise separable convolutions. In *CVPR*, 2017.
432. Saining Xie, Ross Girshick, Piotr Dollár, Zhuowen Tu, and Kaiming He. Aggregated residual transformations for deep neural networks. In *CVPR*, 2017.
433. Xiangyu Zhang, Xinyu Zhou, Mengxiao Lin, and Jian Sun. Shufflenet: An extremely efficient convolutional neural network for mobile devices. In *CVPR*, 2018.
434. Andrew Howard, Andrey Zhmoginov, Liang-Chieh Chen, Mark Sandler, and Menglong Zhu. Inverted residuals and linear bottlenecks: Mobile networks for classification, detection and segmentation. In *CVPR*, 2018.
435. Chenxu Luo and Alan Yuille. Grouped spatial-temporal aggregation for efficient action recognition. In *ICCV*, 2019.
436. Dongliang He, Zhichao Zhou, Chuang Gan, Fu Li, Xiao Liu, Yandong Li, Limin Wang, and Shilei Wen. Stnet: Local and global spatial-temporal modeling for action recognition. In *AAAI*, 2019.
437. Bichen Wu, Alvin Wan, Xiangyu Yue, Peter Jin, Sicheng Zhao, Noah Golmant, Amir Gholaminejad, Joseph Gonzalez, and Kurt Keutzer. Shift: A zero flop, zero parameter alternative to spatial convolutions. In *CVPR*, 2018.
438. Huasong Zhong, Xianggen Liu, Yihui He, and Yuchun Ma. Shift-based primitives for efficient convolutional neural networks. *arXiv preprint arXiv:1809.08458*, 2018.
439. Linxi Fan, Shyamal Buch, Guanzhi Wang, Ryan Cao, Yuke Zhu, Juan Carlos Niebles, and Li Fei-Fei. Rubiksnet: Learnable 3d-shift for efficient video action recognition. In *ECCV*, 2020.
440. Quanfu Fan, Chun-Fu Richard Chen, Hilde Kuehne, Marco Pistoia, and David Cox. More is less: Learning efficient video representations by big-little network and depthwise temporal aggregation. In *NeurIPS*, 2019.
441. Chao-Yuan Wu, Ross Girshick, Kaiming He, Christoph Feichtenhofer, and Philipp Krahenbuhl. A multigrid method for efficiently training video models. In *CVPR*, 2020.
442. Tengda Han, Weidi Xie, and Andrew Zisserman. Turbo training with token dropout. In *BMVC*, 2022.
443. Diederik P Kingma and Jimmy Ba. Adam: A method for stochastic optimization. In *ICLR*, 2015.
444. Lijian Lin, Xintao Wang, Zhongang Qi, and Ying Shan. Accelerating the training of video super-resolution. *arXiv preprint arXiv:2205.05069*, 2022.
445. Feng Cheng, Mingze Xu, Yuanjun Xiong, Hao Chen, Xinyu Li, Wei Li, and Wei Xia. Stochastic backpropagation: A memory efficient strategy for training video models. In *CVPR*, 2022.

446. Soravit Changpinyo, Piyush Sharma, Nan Ding, and Radu Soricut. Conceptual 12m: Pushing web-scale image-text pre-training to recognize long-tail visual concepts. In *CVPR*, 2021.
447. Xianhang Li, Huiyu Wang, Chen Wei, Jieru Mei, Alan Yuille, Yuyin Zhou, and Cihang Xie. In defense of image pre-training for spatiotemporal recognition. In *ECCV*, 2022.
448. Limin Wang, Zhan Tong, Bin Ji, and Gangshan Wu. Tdn: Temporal difference networks for efficient action recognition. In *CVPR*, 2021.
449. Max Bain, Arsha Nagrani, Gül Varol, and Andrew Zisserman. Frozen in time: A joint video and image encoder for end-to-end retrieval. In *ICCV*, 2021.
450. Dongxu Li, Junnan Li, Hongdong Li, Juan Carlos Niebles, and Steven CH Hoi. Align and prompt: Video-and-language pre-training with entity prompts. In *CVPR*, 2022.
451. Junke Wang, Dongdong Chen, Zuxuan Wu, Chong Luo, Luowei Zhou, Yucheng Zhao, Yujia Xie, Ce Liu, Yu-Gang Jiang, and Lu Yuan. Omnivl: One foundation model for image-language and video-language tasks. In *NeurIPS*, 2022.
452. William L Briggs, Van Emden Henson, and Steve F McCormick. *A multigrid tutorial*. 2000.
453. Armin Kappeler, Seunghwan Yoo, Qiqin Dai, and Aggelos K Katsaggelos. Video super-resolution with convolutional neural networks. *TCI*, 2016.
454. Xintao Wang, Kelvin CK Chan, Ke Yu, Chao Dong, and Chen Change Loy. Edvr: Video restoration with enhanced deformable convolutional networks. In *CVPRW*, 2019.
455. Kelvin CK Chan, Xintao Wang, Ke Yu, Chao Dong, and Chen Change Loy. Basicvsr: The search for essential components in video super-resolution and beyond. In *CVPR*, 2021.
456. Seungjun Nah, Sungyong Baik, Seokil Hong, Gyeongsik Moon, Sanghyun Son, Radu Timofte, and Kyoung Mu Lee. Ntire 2019 challenge on video deblurring and super-resolution: Dataset and study. In *CVPR*, 2019.
457. Chao Jia, Yinfei Yang, Ye Xia, Yi-Ting Chen, Zarana Parekh, Hieu Pham, Quoc Le, Yun-Hsuan Sung, Zhen Li, and Tom Duerig. Scaling up visual and vision-language representation learning with noisy text supervision. In *ICML*, 2021.
458. Chen Ju, Tengda Han, Kunhao Zheng, Ya Zhang, and Weidi Xie. Prompting visual-language models for efficient video understanding. In *ECCV*, 2022.
459. Ziyi Lin, Shijie Geng, Renrui Zhang, Peng Gao, Gerard de Melo, Xiaogang Wang, Jifeng Dai, Yu Qiao, and Hongsheng Li. Frozen clip models are efficient video learners. In *ECCV*, 2022.
460. Mengmeng Wang, Jiazheng Xing, and Yong Liu. Actionclip: A new paradigm for video action recognition. *arXiv preprint arXiv:2109.08472*, 2021.
461. Xin Liu, Silvia L Pintea, Fatemeh Karimi Nejadasl, Olaf Booij, and Jan C van Gemert. No frame left behind: Full video action recognition. In *CVPR*, 2021.
462. Zuxuan Wu, Caiming Xiong, Chih-Yao Ma, Richard Socher, and Larry S Davis. Adaframe: Adaptive frame selection for fast video recognition. In *CVPR*, 2019.
463. Yulin Wang, Zhaoxi Chen, Haojun Jiang, Shiji Song, Yizeng Han, and Gao Huang. Adaptive focus for efficient video recognition. In *ICCV*, 2021.
464. Bruno Korbar, Du Tran, and Lorenzo Torresani. Scsampler: Sampling salient clips from video for efficient action recognition. In *ICCV*, 2019.
465. Wenhao Wu, Dongliang He, Xiao Tan, Shifeng Chen, and Shilei Wen. Multi-agent reinforcement learning based frame sampling for effective untrimmed video recognition. In *ICCV*, 2019.
466. Ruohan Gao, Tae-Hyun Oh, Kristen Grauman, and Lorenzo Torresani. Listen to look: Action recognition by previewing audio. In *CVPR*, 2020.
467. Boyang Xia, Zhihao Wang, Wenhao Wu, Haoran Wang, and Jungong Han. Temporal saliency query network for efficient video recognition. In *ECCV*, 2022.
468. Amir Ghodrati, Babak Ehteshami Bejnordi, and Amirhossein Habibian. Frameexit: Conditional early exiting for efficient video recognition. In *CVPR*, 2021.
469. Hehe Fan, Zhongwen Xu, Linchao Zhu, Chenggang Yan, Jianjun Ge, and Yi Yang. Watching a small portion could be as good as watching all: Towards efficient video classification. In *IJCAI*, 2018.

470. Yulin Wang, Yang Yue, Yuanze Lin, Haojun Jiang, Zihang Lai, Victor Kulikov, Nikita Orlov, Humphrey Shi, and Gao Huang. Adafocus v2: End-to-end training of spatial dynamic networks for video recognition. In *CVPR*, 2022.
471. Yue Meng, Chung-Ching Lin, Rameswar Panda, Prasanna Sattigeri, Leonid Karlinsky, Aude Oliva, Kate Saenko, and Rogerio Feris. Ar-net: Adaptive frame resolution for efficient action recognition. In *ECCV*, 2020.
472. Chuofan Ma, Qiushan Guo, Yi Jiang, Ping Luo, Zehuan Yuan, and XIAOJUAN QI. Rethinking resolution in the context of efficient video recognition. In *NeurIPS*, 2022.
473. John Schulman, Filip Wolski, Prafulla Dhariwal, Alec Radford, and Oleg Klimov. Proximal policy optimization algorithms. *ArXiv*, abs/1707.06347, 2017.
474. Eric Jang, Shixiang Gu, and Ben Poole. Categorical reparameterization with gumbel-softmax. *arXiv preprint arXiv:1611.01144*, 2016.
475. Yulin Wang, Yang Yue, Xinhong Xu, Ali Hassani, Victor Kulikov, Nikita Orlov, Shiji Song, Humphrey Shi, and Gao Huang. Adafocusv3: On unified spatial-temporal dynamic video recognition. In *ECCV*, 2022.
476. Junke Wang, Xitong Yang, Hengduo Li, Li Liu, Zuxuan Wu, and Yu-Gang Jiang. Efficient video transformers with spatial-temporal token selection. In *ECCV*, 2022.
477. Evangelos Kazakos, Arsha Nagrani, Andrew Zisserman, and Dima Damen. Epic-fusion: Audio-visual temporal binding for egocentric action recognition. *ICCV*, 2019.
478. Rameswar Panda, Chun-Fu Chen, Quanfu Fan, Ximeng Sun, Kate Saenko, Aude Oliva, and Rogério Schmidt Feris. Adamml: Adaptive multi-modal learning for efficient video recognition. *ICCV*, 2021.
479. Zuxuan Wu, Caiming Xiong, Yu-Gang Jiang, and Larry S. Davis. Liteeval: A coarse-to-fine framework for resource efficient video recognition. In *NeurIPS*, 2019.
480. Hengduo Li, Zuxuan Wu, Abhinav Shrivastava, and Larry S. Davis. 2d or not 2d? adaptive 3d convolution selection for efficient video recognition. *CVPR*, 2020.
481. Ximeng Sun, Rameswar Panda, Chun-Fu Chen, Aude Oliva, Rogério Schmidt Feris, and Kate Saenko. Dynamic network quantization for efficient video inference. *ICCV*, 2021.
482. Quentin Berthet, Mathieu Blondel, Olivier Teboul, Marco Cuturi, Jean-Philippe Vert, and Francis Bach. Learning with differentiable pertubed optimizers. In *NeurIPS*, 2020.
483. Alec Radford, Jeffrey Wu, Rewon Child, David Luan, Dario Amodei, Ilya Sutskever, et al. Language models are unsupervised multitask learners. *OpenAI Blog*, 2019.
484. Hugo Touvron, Thibaut Lavril, Gautier Izacard, Xavier Martinet, Marie-Anne Lachaux, Timothée Lacroix, Baptiste Rozière, Naman Goyal, Eric Hambro, Faisal Azhar, et al. Llama: Open and efficient foundation language models. *arXiv preprint arXiv:2302.13971*, 2023.
485. Hugo Touvron, Louis Martin, Kevin Stone, Peter Albert, Amjad Almahairi, Yasmine Babaei, Nikolay Bashlykov, Soumya Batra, Prajjwal Bhargava, Shruti Bhosale, et al. Llama 2: Open foundation and fine-tuned chat models. *arXiv preprint arXiv:2307.09288*, 2023.

Index

Z. Wu, Y.-G. Jiang, *Deep Learning for Video Understanding*, Wireless Networks, https://doi.org/10.1007/978-3-031-57679-9

Earth and Environmental Sciences Library

Earth and Environmental Sciences Library (EESL) is a multidisciplinary book series focusing on innovative approaches and solid reviews to strengthen the role of the Earth and Environmental Sciences communities, while also providing sound guidance for stakeholders, decision-makers, policymakers, international organizations, and NGOs.

Topics of interest include oceanography, the marine environment, atmospheric sciences, hydrology and soil sciences, geophysics and geology, agriculture, environmental pollution, remote sensing, climate change, water resources, and natural resources management. In pursuit of these topics, the Earth Sciences and Environmental Sciences communities are invited to share their knowledge and expertise in the form of edited books, monographs, and conference proceedings.

More information about this series at https://link.springer.com/bookseries/16637

Abdelazim Negm · Liliana Zaharia ·
Gabriela Ioana-Toroimac
Editors

The Lower Danube River

Hydro-Environmental Issues and Sustainability

Springer

Editors
Abdelazim Negm
Faculty of Engineering
Zagazig University
Zagazig, Egypt

Liliana Zaharia
Faculty of Geography
University of Bucharest
Bucharest, Romania

Gabriela Ioana-Toroimac
Faculty of Geography
University of Bucharest
Bucharest, Romania

ISSN 2730-6674 ISSN 2730-6682 (electronic)
Earth and Environmental Sciences Library
ISBN 978-3-031-03864-8 ISBN 978-3-031-03865-5 (eBook)
https://doi.org/10.1007/978-3-031-03865-5

This Springer imprint is published by the registered company Springer Nature Switzerland AG
The registered company address is: Gewerbestrasse 11, 6330 Cham, Switzerland

Preface

The Danube River is one of the largest European watercourses, namely the second in terms of length and basin area. It crosses the continent from west to east over a length of almost 3,000 km, from the Black Forest (Schwarzwald) Mts. (in Germany), to the Black Sea, serving as important natural corridor for socio-economic and biogeographic connections. The Danube River flows through 11 countries and many larger or smaller cities (including 4 country capitals) and villages have developed along its course. They are closely dependent on the water resources and related services provided by the river. At the same time, they are exposed to the hydrological hazards, such as the water excess (i.e., floods) and scarcity (i.e., low-waters). Furthermore, the Danube River and the adjacent floodplain host a rich and valuable biodiversity that are dependent on the quantitative and qualitative characteristics of the Danube River.

The Danube River Basin (DRB) covers about 10% of Europe and exhibits a great diversity of natural, socio-economic and political conditions. It overlaps the territories of 19 countries, being considered the most international basin in the world. Over the last century, the Danube River and its catchment were increasingly affected by human pressures (e.g., water and land uses, engineering works, water pollution etc.). They have led to more or less severe alterations of the quantitative and qualitative features of waters, and changes in morphology of the river channel and related floodplain. These pressures also impaired the functionning of the aquatic and floodplain ecosystems.

Due to the great importance of the Danube River for society and the environment, knowing its characteristics is of high scientific and practical interest for the Danubian countries. Therefore, the Danube River and its catchment have been the subject of numerous scientific publications, both at the scale of the whole river or basin, and focusing on specific issues at smaller spatial scales. An overview of the main publications on the hydrological features in the Danube River Basin is presented in Chapter "Flow Variability of the Lower Danube River: An Up-to-Date Overview" in this book. In recent years, two comprehensive books were published by Springer publishing House, dedicated to the Danube River Basin. The first one is titled "Hydrological Processes of the Danube River Basin. Perspectives from the Danubian Countries" and was edited by Mitja Brilly in 2010. It address issues related mainly to the hydrological features within the Danube River Basin. The second one, titled "The Danube River Basin", was edited by Igor Liska in 2015. It provides information on the qualitative features (chemical, biological and hydromorphological) of waters in the Danube River Basin.

This book addresses a complex topic, different from the two publications mentioned above: it focuses on the hydro-environmental issues of the Lower Danube River (LDR). The limits of this last sector of the Danube River are debatable, as shown in the Chapter "Flow Variability of the Lower Danube River: An Up-to–Date Overview". In this volume, the Lower Danube River was considered from its entrance in Romania (at Baziaș) to the Danube delta (not included in this study), over a length of about 1,000 km (see Figure 1 in Chapter "Flow Variability of the Lower Danube River: An Up-to-Date Overview"). In this sector, the Danube River is the natural border between Romania (on the one hand) and Serbia, Bulgaria, Republic of Moldova and Ukraine (on the other hand).

The information on the characteristics of the aquatic environment in the downstream sector of the Danube River is relatively sparse, so the editors and the authors decided to collaborate to produce this book focused on the Lower Danube River. It provides current findings and new knowledge about this area to the global research community and all those interested.

The book gives an overview on some of the major issues faced by the lower sector of the Danube River that was severely impacted by human pressures, especially since the second part of the last century. The two large dams on the Lower Danube River (Iron Gates I and II) and the damming of its major tributaries in this sector altered the flow regime and diminished significantly the sediment load. The narrowing forced by anthropogenic levees as well as the decrease in sediment load led to changes in riverbed and fluvial islets morphology and morphometry. These adjustments translated by alterations of aquatic habitats and biocenoses. The Lower Danube is the most polluted sector of the river, because it collects pollutants emitted by both upstream countries and those bordering the lower Danube watercourse. According to the Water Framework Directive rules, this sector was categorized as *at risk* due to pollution with nutrients and hazardous substances along its entire length on the territory of riparian countries.

In addition to the anthropogenic impacts on the Lower Danube River, there are those induced by the climate change (e.g., modifications in flow regime and in magnitude and frequency of extreme phenomena such as floods and hydrological droughts).

In this complex context, the management of the Danube River Basin (including its lower part) aims to ensure a balance between anthropogenic pressures and natural processess, in order to mitigate their negative impacts on society and environment. This remains a constant challenge and the scientific researches can provide the information support for river and basin management plans at different spatial scales.

This volume brings together contributions of authors from countries sharing the Lower Danube River and related basin. It contains 19 chapters addressing topics related to hydro-environmental issues in this geographic area, as previously outlined.

They were grouped in four parts in sequence, as follows: (1) Hydrological and Hydromorphological Processess, including four chapters; (2) Physico-Chemical Features and Quality of the Hydro-Environment, including three chapters; (3) Climate and Water Related Hazards, including six chapters and (4) Sustainable Management and Governance of the Hydro-Environment, including six chapters. In the following, we will indicate why each of the 19 chapters is presented in this book by focusing on its unique achievements.

The chapter "Flow Variability of the Lower Danube River: An Up-to-Date Overview" highlights the spatio-temporal variations of the average, maximum and minimum annual and monthly discharges of the LDR, at several gauging stations located on the Romanian bank of the river. The studied sector extends between the stations Baziaş and Ceatal Izmail (named Ceatal Chilia in Romania), over a length of about 1,000 km. The analyzed periods range from 44 years (1976–2019) to more than 170 years (1840–2012). In addition to the rigorous analysis of the Danube's flow variability at different time-scales, the chapter also includes an overview of the present state of hydrological knowledge and water management in the Danube River Basin, with a focus on its lower sector. The chapter contains 77 references and 23 figures.

The chapter "Dynamics of Islands and Danube River Channel Along Vedea-Călăraşi Sector (1856–2019): Hydrogeomorphological Approach" is devoted to analyze the Vedea-Călăraşi reach (about 135 km long) of the LDR along the border between Romania and Bulgaria. The authors are trying to answer the question "what has been the dynamics of the islands in both countries in the last century and a half, since the end of the Little Ice Age, taking into account the numerous engineering works that the Danube had suffered?". The new information based on the conducted investigations updates and complements the previous ones on the recent hydrogeomorphology of the LDR. The chapters contains 59 references and 12 figures.

The chapter "Hydro-sedimentary Modeling and Fluvial Morphological Processes Along the Lower Danube River (Giurgiu-Olteniţa-Călăraşi Reach)" reports the morphological changes of the LDR for a better understanding of the hydromorphological behaviour of the Danube fluvial system along the Giurgiu—Călăraşi reach in Romania (corresponding to Ruse—Silistra reach in Bulgaria). The chapter contains analysis of the temporal and spatial morphological changes of the river channel. Water and sediment data for a period of 8 years (between 2008 and 2015) were used for the numerical model, setting up a 1D sediment transport model and calibrate it. Also, the chapter compares results with field observations and provides understanding of the complex physical processes of the hydrodynamics of sediment transport and analysis of the morphological changes (aggradation/degradation) and sediment loads in space and time, over the study period. The chapter contains 89 references and 32 figures.

The chapter "Hydro-Environmental Specifics of the Lower Danube Bulgarian Tributaries" gathers the existing knowledge with new investigated features of the Bulgarian Danube tributaries and provides useful scientific information for decision and policy makers. Three important issues have been addressed in this chapter,

(i) basic description of the Bulgarian Danube plain, considering the main geographical features (topography, climate, hydrology, etc.); (ii) legislative and management brief review, with regard to the implementation of Bulgarian and European directives (e.g. Water Framework Directive, Floods Directive), and related documents; and (iii) hydro-environmental specification concerning the Bulgarian tributaries of the LDR, their qualitative state and extreme events. The chapter contains 33 references and 16 figures.

The chapter "Water Temperature Variability in the Lower Danube River" presents a time series analysis of the water temperature recorded between 2001 and 2016. Additionally, a one-year ahead forecasting has been provided at three monitoring sections located on the Romanian side of the LDR i.e., Pristol (RO2), Chiciu (RO4) and Reni (RO5), based on data from Trans-National Monitoring Network (TNMN) of the Danube River database. The chapter contains 64 references and 7 figures.

The chapter "Variability of Nutrient Concentrations Along the Lower Danube River" provides information on the temporal and spatial variation of several nutrient concentrations (NH_4-N, NO_2-N, NO_3-N, PO_4-P and total phosphorous) between 1996 and 2017, based on data recorded at five monitoring stations located along the Lower Danube River (between km 1,071 and km 132), belonging to the Trans-National Monitoring Network (TNMN), namely: Baziaș, Pristol, Oltenița, Chiciu and Reni. The dependence of the selected nutrient contents on some hydrological and physio-chemical parameters of water (e.g. discharge, temperature, dissolved oxygen concentration) was also investigated. The chapter contains 61 references and 9 figures.

The chapter "Human Impacts on Water Resources in the Lower Danube River Basin in Serbia" presents the water resource uses including water supply, hydropower use, navigation, tourism, recreation and fishing, as well as water quality, pollution and protection of water resources. The authors assess the water quality in general and for different purposes. They used various indices and compared them to show which ones provide the added value to the water quality topic. The chapter contains 170 references and 10 figures.

The chapter "Using Köppen Climate Classification Like Diagnostic Tool to Quantify Climate Variation in Lower Danube Valley for the Period 1961–2017" provides an analysis of the Annual Climate Types (ACT) that identifies the long-term climate variability in Lower Danube Valley. The author uses temperature and precipitation monthly data for the period 1961–2017 (57 years) from 10 meteorological stations located in the Bulgarian part of Danube valley. The chapter contains 34 references and 6 figures.

The chapter "Observed Changes in the Temperature and Precipitation Regime Along the Lower Danube River" analyzes the monthly average temperatures and the monthly precipitation for five weather stations that are representative for the geographical location along the LDR: Drobeta Turnu Severin, Calafat, Zimnicea, Călărași and Galați. The authors highlight the changes occurred in the air temperature and precipitation regime along the LDR in Romania, for the period 1961–2019, based on both the meteorological data from surface measurements and MODIS satellite images. The chapter contains 37 references and 13 figures.

The chapter "A SPEI-Based Approach to Drought Hazard, Vulnerability and Risk Analysis in the Lower Danube River Region" investigates the climatic characteristics of drought for different Standardized Precipitation-Evapotranspiration Index (SPEI) timescales (a) to determine the intensity of drought hazard, (b) to estimate the degree of vulnerability to drought and (c) to determine the drought risk "hotspots" within the study region. Therefore, the authors have been analyzed the drought hazard and vulnerability to droughts in the LDR region to provide information about drought characteristics over the 1981–2019 period in the counties (Romania) and administrative districts (Bulgaria) located northward and southward along the LDR. The chapter contains 124 references and 4 figures.

The chapter "Synoptic Conditions Associated with Floods and Highest Discharges on Lower Danube River (1980–2010)" focuses on a thirty-year interval, between 1980 and 2010, when three major flood events occurred on the Lower Danube River, reaching their peak on: January 21, 1998, March 15, 2006 and June 21, 2010. Additionally, the authors clarify the large-scale atmospheric circulation conditions at continental scale. They bring new information on local and regional atmospheric circulation patterns preceding the major flood events and highest discharges pushing forward the knowledge on the regional weather patterns over the analyzed region. This aspect is very important for improving the linkages between weather and hydrological forecast. The chapter contains 35 references and 17 Figures.

The chapter "Assessment of Soil Erosion and Torrential Flood Susceptibility: Case Study—Timok River Basin, Serbia" investigates the contribution of the Timok River in total suspended sediment discharge of the Danube River, and assesses the susceptibility to soil erosion and torrential floods in the Timok River Basin, by using Erosion Potential Method (EPM) and Flash Flood Potential Index (FFPI), estimated in GIS environment. The used approaches represent a great potential for gross erosion prediction and sediment transport assessment at the river basin or regional scale. Additionally, the authors assessed the potential damage in urban areas, agricultural land and traffic communications (roads and railways) and the degree of torrential flood susceptibility in various watercourses in the Timok River Basin. The outcomes have significant interest for practical issues such as integrated water management projects, sustainable and land-use planning, spatial planning, forest ecosystems and environmental protection, sediment management, etc. The chapter contains 48 references and 6 figures.

The chapter "Hydrological Extremes Anomalies and Trends in Lower Danube Basin: Case Study—Romanian Drainage Area Between Siret and Prut Rivers" analyzes the extreme hydrological tendencies, registered between 1955 and 2018, in eastern Romania. Two data subsets were extracted, of 32 years in length, each: 1955–1986 and 1987–2018. Therefore, the analysis of extreme hydrological anomalies represents a continuation of research undergone in this field, within a region of the Lower Danube basin, in the context of ever-increasing effects of the extreme manifestations of regional climate change. Investigating extreme hydrological anomalies can aid to better understand the impact of climate change on rivers flow at regional scale. The chapter contains 41 references and 10 figures.

The chapter "A Transdisciplinary Approach Using Danube River Multi-connectivity in Wetland Management" coveres social and economic implications, considering the relationship between the natural heritage resources of the geographical subunits and the existence of the socio-economic system. The discussions include the principles and Lower Danube region examples to serve as a useful guide to ecological restoration. They serve as part of a new effort that goes beyond the current concept of natural resources conservation toward a deeper concept of restoring of "environment life"—an ecologically viable state where ecosystems are self-sustaining and improve the functioning and quality of services over time. Extensive knowledge of the risks and threats arising from the disruption of continuity and connectivity in the Danube area presents multiple opportunities to counter them. This requires an inter- and trans-disciplinary approach, as well as effective communication and cooperation between specific stakeholders in this field. The chapter contains 51 references and 13 figures.

The chapter "Anthropogenic Changes and Biodiversity Protection and Conservation Along the Lower Danube River Valley" presents an overview of the anthropogenic activities within Lower Danube River Valley, in particular Romanian part, since the end of the nineteenth century, which caused significant changes in terms of water flow, flooding regime, sediments load, morphology and biodiversity. The high degree of artificialization of the LDR coincided with the communist regime, although the damming of rivers, channelization, draining of the large floodplains has been widely practiced in many other countries. Therefore, the authors presented a review of human interventions from the last century that lead to alteration, degradation and irreversible losses of habitats along the LDR valley, restoration projects of former floodplain areas and biodiversity protection and conservation actions carried out over the area in the last decades. Additionally, the change of the political regime in all countries that overlap the LDR Basin gave the opportunity to both governments and non-governmental organizations to act for restoring of the degraded and damaged ecosystems and habitats in the LDR, as well as for the conservation of the biodiversity within. The chapter makes a review of all types of protected areas that were designed in the last three decades and nowadays they compose a very complex network. The chapter contains 129 references and 13 figures.

The chapter "Land Management Practices Favoring Environmental Conservation in the Danube Lower Valley (Romania)" demonstrates how the abandonment of the irrigation systems can generate wetlands supporting wildlife conservation within the Danube Lower Valley. For the case study in Romania, the authors established the following objectives: (i) identify the spatial distribution of protected areas (SCI and SPA) within the Danube Lower Valley, created in landscapes with abandoned irrigation systems, based on GIS techniques; (ii) model the dynamic of artificial and natural land cover classes within the Danube Lower Valley through landscape ecology metrics and (iii) explore the implications of abandoned irrigation systems on biodiversity, based on cross-referencing the available scientific biogeographical literature. The chapter contains 39 references and 7 figures.

The chapter "The Danube River: Between Conservation and Human Pressures in the Iron Gates Natural Park" analyzes the integration of competing social and economic objectives with the conservation process of the Iron Gates Natural Park (IGNP), with a special focus to the Danube River. The authors proposed for this a three-staged analysis: (i) assessing the current social and economic situation of the communities living within IGNP; (ii) classifying the conservation objectives and measures presented in the Management Plan of the IGNP, and (iii) identifying public perceptions on the conservation status of the area. The authors focused on describing in this chapter an integrated perspective, including both socio-economic and conservation realities specific to the IGNP, as prior studies have analyzed just one of these perspectives. The chapter contains 58 references and 10 figures.

The chapter "Citizen Science for the Danube River—Knowledge Transfer, Challenges and Perspectives" examines the activities with citizen involvement in the region and the best practices that can be implemented for environmental monitoring. The educational, social and economic barriers in citizen science projects success and impact are discussed. This chapter also explores the dynamics between the involved parties (scientists, citizens, water managers and policy-makers). The potential tools that can be used to optimize public participation programs were identified. The chapters contains 93 references and 4 figures.

The chapter "Stakeholders' Interests and Participation in the Sustainable Use of the Lakes Along the Danube Floodplain. A Romanian Sector as Case Study" proposes a comprehensive framework to establish and implement stakeholder's interests and participation in sustainable use of the Danube Floodplain lakes (Romania), from stakeholder identification to their level of involvement and provides new and original information on this topic, derived from own research. In this sense, assessing stakeholders' involvement can contribute to the sustainable use of lakes by paying attention to the increasing involvement of public institutions with decision making power in the sustainable use of floodplain lakes. The chapter contains 72 references and 10 figures.

The editors want to express their special thanks to all who contributed to make this high-quality volume a real source of knowledge by presenting the latest findings in the Hydro-Environmnetal aspects of the Lower Danube River. We would like to thank all the authors for their invaluable contributions. Without their patience and effort in writing and revising the different versions to satisfy the high-quality standards of Springer, it would not have been possible to produce this book and make it a reality. Much appreciation and great thanks are also owed to the reviewers of the chapters and the editors of the Earth and Environmental Sciences series at Springer, for the constructive comments, advice and the critical reviews. Acknowledgments are extended to include all members of the Springer team who have worked long and hard to produce this volume.

The volume editors would be happy to receive any comments to improve future editions. Comments, feedback, suggestions for improvement or new chapters for next editions are most welcome and should be sent directly to the volume editors.

Zagazig, Egypt — Abdelazim Negm
Bucharest, Romania — Liliana Zaharia
Bucharest, Romania — Gabriela Ioana-Toroimac
January 2022

Contents

Hydrological and Hydromorphological Processess

Flow Variability of the Lower Danube River: An Up-to-Date Overview

Liliana Zaharia, Gabriela Ioana-Toroimac, Gabriela-Adina Moroşanu, Elena Ţuchiu, Gabriela Osaci-Costache, and Abdelazim Negm

Abstract This chapter provides an up-to-date overview of the flow variability of the Lower Danube River, on a length of about 1,000 km, from the entrance in Romania (at Baziaş) to the beginning of the delta (at Ceatal Izmail or Ceatal Chilia, in Romania). It highlights the spatio-temporal variation of the average, maximum and minimum annual and monthly discharges of the Danube River, at several gauging stations located on the Romanian bank of the river. The analyzed periods range from 44 years (1976–2019) to more than 170 years (1840–2012). Between Baziaş and Ceatal Izmail, the multiannual discharge of the Danube River increased by almost 1,000 m^3/s (from 5,551 to 6,516 m^3/s, during the period 1840–2012), as a result of tributaries' contribution from the riparian countries (Romania, Serbia, Bulgaria, Republic of Moldova, Ukraine). The flow regime of the Lower Danube River shows the highest discharges in spring and early summer (April–June, with the peak in April) and the lowest discharges in late summer–autumn (August-November, with minimum

L. Zaharia (✉) · G. Ioana-Toroimac · G. Osaci-Costache
Faculty of Geography, University of Bucharest, Nicolae Bălcescu Blvd. 1, 010041 Bucharest, Romania
e-mail: liliana.zaharia@geo.unibuc.ro

G. Ioana-Toroimac
e-mail: gabriela.toroimac@geo.unibuc.ro

G. Osaci-Costache
e-mail: gabriela.osacicostache@geo.unibuc.ro

G.-A. Moroşanu
Institute of Geography of the Romanian Academy, Str. Dimitrie Racoviţă 12, 023993 Bucharest, Romania
e-mail: gabriela.morosanu@unibuc.ro

E. Ţuchiu
National Administration "Romanian Waters", Str. Ion Câmpineanu 11, 030167 Bucharest, Romania
e-mail: elena.tuchiu@rowater.ro

A. Negm
Water and Water Structures Engineering Department, Faculty of Engineering, Zagazig University, Zagazig 44519, Egypt
e-mail: amnegm@zu.edu.eg; amnegm85@yahoo.com

A. Negm et al. (eds.), *The Lower Danube River*, Earth and Environmental Sciences Library, https://doi.org/10.1007/978-3-031-03865-5_1

in September). During the period 1931–2019, the highest maximum discharges occurred during the historical flood in 2006. They reached 15,800 m^3/s at Baziaș and 15,900 m^3/s at Ceatal Izmail, but at some intermediate gauging sections the discharges exceeded 16,000 (e.g. 16,300 m^3/s, at Giurgiu, 16,200 m^3/s, at Oltenița). During the same period, the minimum discharges decreased up to 1,040 m^3/s at Baziaș (in 1949) and 1,790 m^3/s at Ceatal Izmail (in 1947). The two large dams and reservoirs built on the Lower Danube River within the hydroelectric and navigation systems Iron Gates I and II, did not significantly impair the water flow of the Danube River, but mostly the sediment flux.

Keywords Discharge · Spatio-temporal variability · Flood · Danube River · Romania

List of Acronyms and Symbols

DRB	Danube River Basin
DRBMP	Danube River Basin Management Plan
DRPC	Danube River Protection Convention
EC	European Commission
EEC	European Economic Community
EU	European Union
FRMD	Flood Risk Management Directive
FRMP	Flood Risk Management Plan
HENS	Hydroelectric and Navigation System
h.s.	Hydrometric station
ICDPR	International Commission for the Protection of the Danube River
IHP/UNESCO	UNESCO International Hydrological Program
LDR	Lower Danube River
NARW	National Administration "Romanian Waters"
NIHWM	National Institute of Hydrology and Water Management
WFD	Water Framework Directive

1 Introduction

Since ancient times, the large rivers have been the main water source for the various needs of the riparian settlements, supporting their socio-economic development. Furthermore, they fostered the connections between human communities along the rivers and between geographical regions. This is also the case of the Danube River, a major European fluvial and polarization axis [1], which crosses the continent from west to east over a length of 2,857 km [2], from the Black Forest Mts. (Schwarzwald), in Germany, to the Black Sea, in Romania and Ukraine. The Danube River Basin (DRB) extends over an area of 801,463 km^2 [2], i.e. about 10% of Europe's territory (in [3], the Danube's length is 2,826 km and the catchment area is 817,000 km^2). The Danube catchment overlaps the territories of 19 countries and hosts approx. 80.5 million people, being considered the most international basin in the world [2]. The largest share of the total area and the number of inhabitants belongs to Romania (about 30% and 22% respectively). Due to its large size, the basin exhibits a wide variety of natural, socio-economic and political conditions [4]. It has a rich ecological variety and holds the highest freshwater biodiversity in Europe [2, 5]. The Danube River serves as important West–East corridor for species migration, connecting different biogeographic zones [6].

Three main sectors of the watercourse (and of the basin) can be distinguished along the Danube River: the Upper, the Middle and the Lower Danube (Fig. 1). They differ substantially in their features and are separated by two gorges on the Danube River: the Davin Gate (between the Upper and Middle sector) and the Iron Gates (between the Middle and Lower sector). The Upper Danube has a length of about 620 km from the source to its confluence with the Morava River, at Bratislava [7]. The related basin has a predominantly mountainous relief, reaching 4,052 m a.s.l. in the south, in the Alps [7]. The Middle Danube drains the large depression of the Pannonian plain over a length of about 930 km, but the related basin overlaps, at its extremities, mountainous areas (i.e. the Alps, the Dinarides, the Carpathians) [3, 7]. Along the middle sector, Danube River receives three major tributaries (Drava, Tisza, and Sava rivers) that substantially increase its flow. The Lower Danube extends from

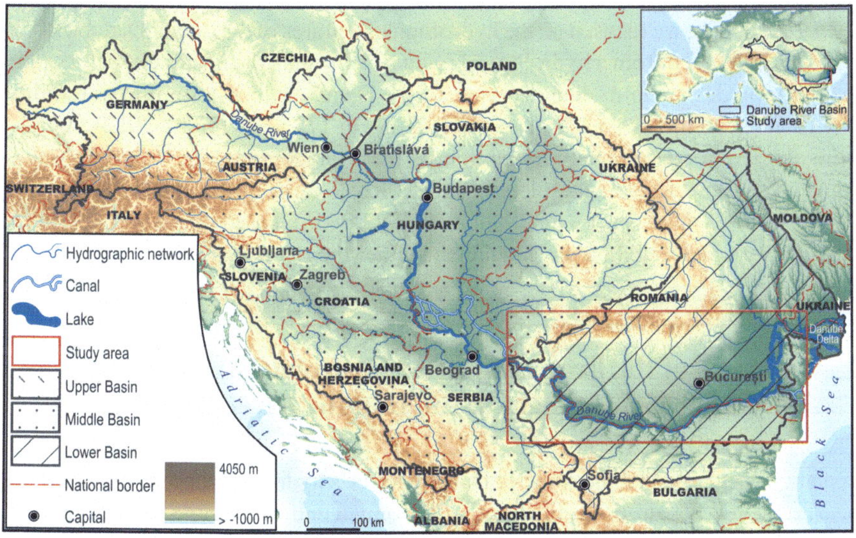

Fig. 1 The Danube River Basin and the location of the study area (the red frame marks the limits of the Lower Danube River sector, as considered in this chapter and the book). The digital terrain model was extracted from [9]

the Iron Gates to the Danube Delta and the Black Sea. Its exact limits are debatable, as discussed in Sect. 3.1. According to [7], the length of the Lower Danube is about 863 km, without the delta, that is considered a separate river section. The Lower Danube crosses the Romanian-Bulgarian lowlands, bordered by the Carpathians Mts. (in the north) and the Balkan (Stara Planina) Mts. (in the south). The Danube Delta cover a total area of about 6,750 km^2 in Romania and Ukraine [8].

The Danube River crosses 11 countries and 4 capitals (Wien, Bratislava, Budapest and Belgrade) serving as a major waterway that connects Central Europe and South-east Europe. Along the Danube River, several large cities and hundreds of small towns and villages are located. Their existence and development are closely dependent on the Danube River, which is used for various purposes (domestic and industrial water, irrigation, hydropower generation and navigation).

The DRB is one of the most human impacted large river catchments in the world [10]. The anthropogenic pressures altered the hydro-sedimentary flows and channel morphodynamics, as well as the water quality and ecological status [7]. In addition to the human pressures, the hydrological features of the Danube River are impacted by climate change-related processes [7, 11]. In this complex context, the investigation of the hydrological features of the Danube River is of high interest both scientifically and, especially, practically, to provide up-to-date information useful for adequate water and basin management.

Because of their great importance for society and the environment, the Danube River and its catchment were intensively studied, both at the scale of the whole river or basin, and focusing on specific issues and sectors/areas. General information on the wide-basin environmental features is found in several monographic works such as [12–15].

In this chapter, we focused on the hydrological features of the Lower Danube River (LDR), less studied from a hydrological point of view than the upstream sectors. The chapter provides an up-to-date overview of the Danube's long-term flow variability over periods ranging from 44 years (1976–2019) to 173 years (1840–2012). It highlights the temporal and spatial variation of the average, maximum and minimum discharges from the entrance of the Danube River in Romania (at Baziaș) to the beginning of its delta (over a length of about 1,000 km), based on data from several gauging stations located on the Romanian side of the river.

2 Overview of the Present State of Hydrological Knowledge and Water Management in the Danube River Basin

It can be considered that the study of Danube River's hydrology, based on systematic measurements, began in the first part of the nineteenth century. The international nature of the river has required the scientific transboundary cooperation, in the field of hydrology, between the countries sharing the Danube catchment. This cooperation began in 1961 and continues to this day. Since 1975, it has been held in the framework water line [42]. Another particular feature is the presence of a well-developed floodplain on the left (Romanian) side, with variable width (2–15 km) [29]. Before the extensive engineering works carried out in the second half of the twentieth century (embankments and draining), the floodplain included numerous lakes [1, 33].

Downstream of the Iron Gates I dam, a second Hydroelectric and Navigation System, named (in Romania) Iron Gates II, was built on the Danube River. It is also jointly managed by Romania and Serbia. This system includes two dams located on the branches of the river that surround the *Ostrovul Mare* island: a dam is on the main branch of the Danube River (at km 862.8) and the second one, on the secondary branch (named Gogoșu), at km 875 [39]. The construction works started in 1977 and the system became fully operational in 1986 [38, 39]. The total capacity of the power stations increased from 486 MW (the initial capacity), to about 600 MW (after upgrading the turbines) [38, 39, 41].

Between Drobeta-Turnu Severin and Călărași, the Danube River receives several tributaries, among which the most important are: Jiu, Olt and Argeș rivers (in Romania), Timok (at the Bulgarian-Serbian border) and Iskar and Yantra rivers (in Bulgaria), with multiannual discharges ranging from 31 m^3/s (Timok River) to 174 m^3/s (Olt River) (Table 1).

(3) The sector between Călărași and Brăila cities (about 200 km in length) has as particular feature the branching of the Danube River into several arms and

Table 1 Main tributaries of the Lower Danube River and their specific features

River	Side	Length (km)	Catchment area (km^2)	Average discharge (m^3/s)
Cerna	Left	87[a]	1,360[a]	23.9[a]
Timok	Right	180	4,630	31
Lom	Right			
Jiu	Left	339	10,080	86
Ogosta	Right	147[b]	3,157[b]	18[b]
Iskar	Right	368	8,684	54
Vit	Left	189[c]	3,252[c]	14.6[c]
Olt	Right	615	24,050	174
Osam	Left	314[c]	2824[c]	13.1[c]
Vedea	Right	224[d]	5,430[d]	13.8[e]
Yantra	Left	285	7,879	47
Argeș	Right	350	12,550	71
Ialomița	Right	417	10,350	45
Siret	Right	559	47,610	240
Prut	Right	950	27,540	110

Data Sources [8], excepting superscripts a–d, taken from other sources, as follows:
[a][37]; [b][43]; [c][44]; [d][45]; [e][46]

the presence of two large islands (named *bălți*, in Romanian), a reason why this stretch is still called the "*bălți*" sector. The two large islands are: *Balta Ialomiței* (to the south, with a length of about 100 km, maximum width of 15 km and area of 880 km^2) and *Balta Brăilei* (to the north, with a length of 60 km, maximum width of 20 km and area of 960 km^2 [29] (Fig. 2). These islands were mostly embanked and drained to use the land for agriculture.

Unlike the previous sectors where the Danube River has a general flow direction from west to east forming the border between Romania (on the left side) and Serbia and Bulgaria (on the right side), in this sector the river flows from south to north, entirely within the Romanian territory. Between Călărași and Brăila cities, the Danube River has very few tributaries, the most important being Ialomița River, with Qav = 45 m^3/s [8] (Table 1).

(4) The last sector extends from Brăila city to the division of the Danube River into two arms at Ceatal Izmail (or Ceatal Chilia, in Romania) within Pătlageanca village, right before the entrance in the Danube Delta. In this sector, the river has a single channel. Between the cities of Brăila and Galați the river flows to the north, and downstream Galați, the flow direction changes by 90°, the river heading towards east. At the end of the nineteenth century, the river channel in this sector was subjected of important engineering works to facilitate the navigation of heavy ships (maritime) between the Black Sea and the harbor of Brăila. Within this sector (almost 90 km long) also called Maritime Danube [1, 47], the river collects from the left side the waters of two of its largest tributaries in the lower course, Siret (Qav = 240 m^3/s) and Prut (Qav = 110 m^3/s) rivers [8] (Tabel 1).

Along the LDR, the climate is of temperate-continental type, but with different influences induced by the regional atmospheric circulation: Mediterranean and oceanic influences in the west (determined by Atlantic and Mediterranean cyclones), while in the eastern part there are Pontic (induced by the vicinity of the Black Sea and the cyclones formed in its area) and arid influences (generated by the Euro-Asian anticyclones). The large aquatic surfaces of the Danube River and related floodplain determine local peculiarities and the existence along the watercourse, of a specific "Danubian" topoclimate [29]. The morphology and orientation of the valley influence the direction and speed of the air mass circulation. More detailed information on the climate along the LDR are provided by the Chapters "Using Köppen Climate Classification Like Diagnostic Tool to Quantify Climate Variation in Lower Danube Valley for the Period 1961–2017" and "Observed Changes in the Temperature and Precipitation Regime Along the Lower Danube River" of this book.

The favorable living conditions along the LDR and the natural resources offered by the river and its floodplain favoured the intense population of the adjacent area and the development of numerous settlements on both banks of the Danube River. Some of these are regional centers and county residences with over 50,000 inhabitants (e.g. Drobeta-Turnu Severin, Giurgiu, Călărași (in Romania) and even over 100,000 inhabitants (e.g. Brăila and Galați, in Romania, and Ruse in Bulgaria). Danube River is a key natural resource for the existence and development of not only riverside settlements, but also of a larger area, due to the multipurpose uses of the river: water supply (for agriculture/irrigation, industry, domestic consumption), fishing, hydropower generation, fluvial navigation, tourism and recreation. The use of the watercourse for fluvial transport favoured the connexions between the countries from the different sectors of the Danube. All these uses are closely dependent on variations in river's level and flow. In extreme cases (e.g. floods and low waters), the flow fluctuations can cause serious damage to society. Furthermore, the LDR and the adjacent floodplain host a rich and valuable biodiversity, equally dependent on the fluctuation of Danube's water [15].

3.2 Overview of the Hydrological Knowledge on the Lower Danube River

Because of the crucial economic and environmental importance of the LDR, the riparian countries were concerned with conducting hydrological studies providing information on the flow variability, essential for water and related risks management. In Romania, comprehensive information on the hydrological features of the LDR are provided by two monographic publications (both in Romanian): *Dunărea între Baziaş şi Ceatal Izmail. Monografie hidrologică* (*The Danube between Baziaş and Ceatal Izmail. Hydrological monograph*) [48] and *Geografia Văii Dunării Româneşti* (*Geography of the Romanian Danube Valley*) [49]. Likewise, more or less synthetic information on the hydrological characteristics (including flow variability) of the LDR are included in dedicated chapters in several books such as [29, 50–53].

In addition to the volumes (such as those mentioned above) including hydrological information on the LDR, there are numerous papers published in scientific journals and conference proceedings, which address topics related to the flow variability in the Lower Danube River, of which we mention: [1, 28, 54–68]. An analysis of the variation of the average, maximum and minimum flows at several stations along the LDR, in the period 1931–2016 is performed in [69].

Information on the hydrological features of the Lower Danube River is also found in the publications carried out at the wide-basin scale, as already mentioned in the previous Sect. 2.

This chapter completes the existent information on the Danube's flow in its lower sector, providing an updated overview on the long-term variability of the average, maximum and minimum discharges at several gauging stations in this sector.

3.3 Data on the Analyzed Hydrometric Stations

The analysis carried out in this chapter is mainly based on processing flow data series recorded at hydrometric/gauging stations (h.s.) located along the LDR, on the left (Romanian) side, belonging to the Romanian hydrometrical network. As already mentioned, the studied sector extends between the hydrometric stations of Baziaș, in the western extremity, at the entrance of the Danube River in Romania (at km 1,072 away from its mouth) and Ceatal Izmail in the eastern extremity of the studied sector (at 80 km from the Danube's mouth). In Romania and in some publications (e.g. [58, 59, 67, 69]), Ceatal Izmail is named Ceatal Chilia. Between the two extreme stations, we also considered some intermediate hydrometric stations located in key positions, namely upstream of the main tributaries of the Danube River, as follows (Fig. 2): Gruia h.s., located downstream the HENS Iron Gates II, controlling the western part of the studied sector; Bechet h.s., situated downstream of the Jiu River mouth (the most important tributary in the western half of the Lower Danube River); Turnu Măgurele h.s., situated at 82 km away from Bechet, downstream of the confluences with the tributaries Olt (in Romania), and Osam in Bulgaria; Oltenița h.s., located downstream of the Argeș River mouth; Brăila and Grindu hydrometric stations, situated before and respectively after the embouchure of the Siret River, the largest tributary of the Lower Danube River (in terms of discharge and area of its catchment). The last h.s. (Ceatal Izmail), is located downstream of the Prut River mouth (the third largest tributary of the Lower Danube River) and just before the beginning of the delta. Data on the gauging stations are presented in Table 2, and their location is shown in Fig. 2.

The processed data include values of the average, maximum and minimum discharges (annual and monthly) covering different time-periods:

- 1840–2012 for the average multiannual discharges at all selected stations; the data were extracted from [28];
- 1931–2016 for the maximum and minimum multiannual discharges at Baziaș, Gruia, Giurgiu, Oltenița, Brăila and Ceatal Izmail; the data were extracted from [69];
- 1931–2019 for the annual average, maximum and minimum discharges at Baziaș, Oltenița and Ceatal Izmail;

Table 2 Data on the analyzed hydrometric stations

Hydrometric station	Altitude (m.a.s.l.)	Distance from the Danube R. mouth (km)	River basin area (km^2)
Baziaș	64	1,072.4	570,896
Gruia	29	851	580,100
Bechet	22	679	603,586
Turnu Măgurele	19	597	654,000
Oltenița	16	429.7	684,803
Brăila	1	169.4	726,000
Grindu	0.8	139.6	775,500
Ceatal Izmail	0.6	79.6	776,883

Source NARW database

- 1976–2019 for the monthly average, maximum and minimum discharges at Baziaş, Olteniţa and Ceatal Izmail.

For the last two periods, the data come from the National Administration Romanian Waters (NARW) database, which includes the National Institute of Hydrology and Water Management (NIHWM) and Water Basin Administrations (WBAs) databases.

Based on the streamflow data, classical statistical parameters (e.g. minimum, average, maximum, coefficient of variation) were computed for different time-scales and their spatio-temporal variability was analyzed. The linear trends in the variation of streamflow were investigated by using the statistical non-parametric test of Mann-Kendall coupled with the non-parametric Sen's method for the magnitude of the trend. The level of significance (α) of the identified trend was considered at 0.001, 0.01, 0.05 and 0.1 [70]. To identify possible influences of the Iron Gates I and II hydroelectric systems on the Danube's flow, comparisons of flow parameters between the periods pre- and post-commissioning of the systems were performed.

4 Spatial Variation of the Multiannual Flow

In this section we investigated the long-term variation of the average, maximum and minimum discharges along the LDR.

To highlight the features of the Danube's **average flow** variation, we analyzed the multiannual discharges recorded at gauging stations located at the beginning and end of the studied sector (Baziaş and respectively Ceatal Izmail), as well as data from several intermediate stations: Gruia, Bechet, Turnu Măgurele, Olteniţa, Brăila and Grindu. The analyzed period covers more than 170 years (1840–2012). The data were extracted from [28] and were indirectly estimated, by extending the length of the data series on daily levels and discharges measured at gauging stations located on the Lower Danube River (as shown in [57]). Even if these estimated data are not officially validated, due to the very long period, in the case of the multiannual average flow, the errors (resulted by comparing the estimated and recorded discharges) are reduced. Consequently, we considered that the results reflect well the real situation at the selected stations.

As expected, the average multiannual flow of the Danube increases from upstream to downstream, as the basin area and the input of tributaries grow. Thus, there is an increase between the two extreme stations of almost 1,000 m^3/s (namely 934 m^3/s), from 5,551 m^3/s at Baziaş, to 6,516 m^3/s, at Ceatal Izmail (Fig. 3). The most important difference between successive stations occurs between Brăila and Grindu hydrometric stations (about 260 m^3/s) due to the significant contribution of the Siret River. An important flow increase is also noticed between Bechet and Turnu Măgurele gauging stations (about 210 m^3/s). In this sector the Danube River receives its second largest tributary in the lower course (Olt River, on the left side) and a few smaller tributaries on the right side (e.g. Iskar, Vit, Osam). Upstream Gruia h.s., although

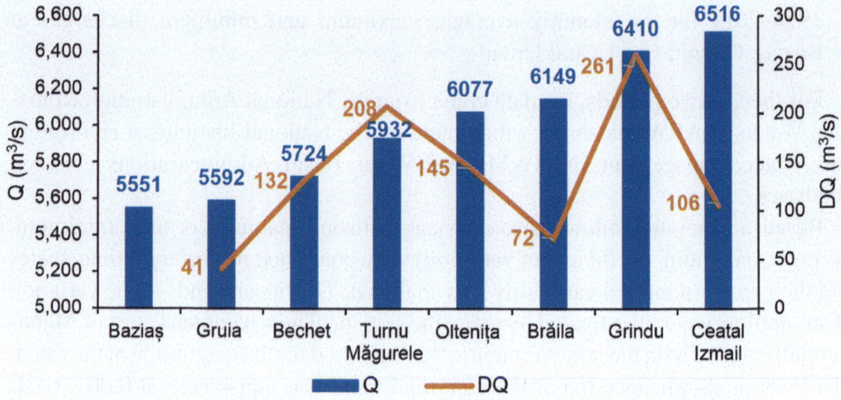

Fig. 3 Variation of the average multiannual discharges (Q) at hydrometric stations located on the Lower Danube River and of the differences (DQ) between successive stations (1840–2012)

the distance to Baziaș is over 200 km (about 1/5 of the length of the studied sector), the increase in flow is small (only 41 m^3/s), because of the relatively few tributaries, with low discharges.

In the case of extreme flows (maximum and minimum), we analyzed a shorter period (1931–2016), based on the data recorded at gauging stations (validated within the national hydrometric network), extracted from [69]. We no longer considered the period 1840–2012 because we found some significant differences between the recorded maximum and minimum discharges and those estimated by extension in [28]. The analyzed stations are: Baziaș, Gruia, Giurgiu, Oltenița, Brăila and Ceatal Izmail.

During the considered period, the **maximum multiannual discharges** were recorded at all analyzed stations during the historical flood in the spring of 2006 (Fig. 4a), when the Danube River reached 15,800 m^3/s at Baziaș, being the highest discharge recorded in the entire flow monitoring period at this station (since 1838). As result of the tributaries' inputs, the discharge increased further at Giurgiu, where it reached a maximum of 16,300 m^3/s, but below this station, because of flood-peak attenuation due to Danube's overflow, the maximum discharge decreased at Oltenița h.s. to 16,200 m^3/s, and at Brăila, to 15,800 m^3/s. Due to the high volume of water brought by the tributaries Siret and Prut, the flow increased downstream, reaching 15,900 m^3/s at Ceatal Izmail (Fig. 4a). It should be noted, therefore, that the maximum flows do not exhibit a progressive increase from upstream to downstream, as would be expected, since they are influenced by the morphology of the river channel and floodplain, as well as by the presence of the levees along the river and their management. Thus, in the low-lying and non-embanked sectors, the river overflows into the floodplain, leading to the flood-peak attenuation, as it happened during the historical flood of 2006, downstream of Giurgiu h.s. Flood attenuation because of the river overflowing generally occurs at flows of over 13,000 m^3/s [58].

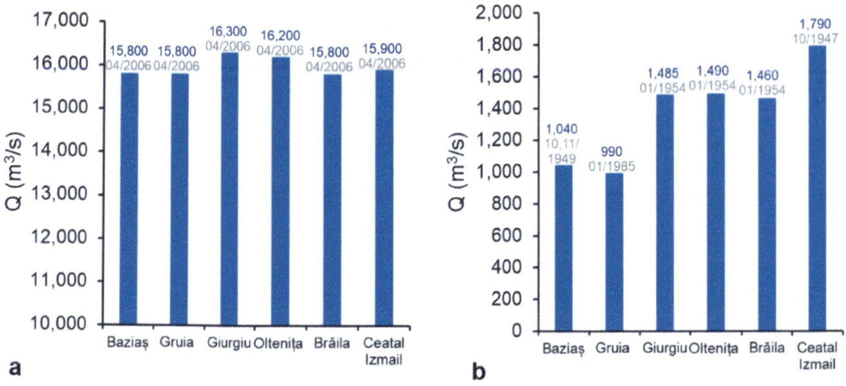

Fig. 4 Variation of the maximum (a) and minimum (b) multiannual discharges (Q) at hydrometric stations located on the Lower Danube River and data (month and year) of their occurrence (1931–2016)

If other sources and periods are considered, the values of the maximum multiannual discharges may differ. Thus, the data estimated in [28] for the period 1840–2012, indicated, in the case of stations from the eastern end of the sector, values higher than those recorded, namely: 17,525 m^3/s at Brăila, 21,347 m^3/s at Grindu and 21,867 m^3/s at Ceatal Izmail. At the last two stations the maximums were noted in July 1897, and at the others, in April 2006. In other sources quoted in [58], at Ceatal Izmail, different maximum discharge values were found. Thus, the highest flow seems that occurred in July 1897, but with quite different values: 35,000, 19,347, 19,233, 17,700 m^3/s. A very high discharge (28,300 m^3/s) is also mentioned in 1891 [58]. However, the values estimated for the period before 1930 have not been validated, and are only informative.

The **minimum multiannual discharges** generally show an increase from upstream to downstream, from approx. 1,000 m^3/s at stations located in the western part of the Lower Danube River, to almost 1,800 m^3/s at Ceatal Izmail (Fig. 4b). The lowest minimum discharge was noticed at Gruia h.s., in January 1985. Because this station is located only a few km downstream of the HENS Iron Gates II (fully put into operation in 1986), it is highly likely that this minimum value was caused by water retention behind the two dams, during the system construction. At half of the analyzed stations, the minimum multiannual discharges occurred in 1954, which indicates a very dry year in the lower Danube River basin. At Bazias, where the Danube's flow reflects the climatic conditions in the middle and upper basins, the year with the lowest flow was 1949. At Ceatal Izmail, the lowest discharge was recorded in October 1947. During the year 1947 the lower Danube basin (especially its eastern part) experienced a severe drought.

According to the data estimated by extending the series of daily flows in [28], the lowest annual minimum flows in the lower course of the Danube River reached: 1,018 m^3/s in 1901 at Baziaş, 1,015 m^3/s in 1858 at Gruia, 1,075 m^3/s in 1858 at Giurgiu, 1,051 m^3/s in 1920 at Olteniţa, 1,112 m^3/s in 1908 at Brăila and 1,553 m^3/s

in 1863 at Ceatal Izmail. These data obtained indirectly indicate lower discharges than those measured in the period 1931–2016. We again specify that due to errors that may affect data before 1931, they should be considered informative.

5 Variability of the Average Flow

The average flow synthetically expresses, through its specific parameters, the water resources of a river for a given period. Its variability along the river is dependent on all the natural and anthropogenic factors controlling the runoff, that act differently over space and time. In the absence of anthropogenic interventions, the climatic factor is determinant for the temporal flow variation.

In this section of the chapter, we investigated the inter- and intra-annual variation of the average flow, at three gauging stations located in relevant positions on the LDR: Baziaș (at the beginning of the studied sector), Oltenița (controlling about half of the Danube Lower Basin) and Ceatal Izmail, located at the entrance in the Danube delta. The analyzed period is 1931–2019 for the annual discharges, while for the monthly discharges, the period is 1976–2019. At Ceatal Izmail we also considered the period 1931–2019 to identify the impact of the Iron Gates I and II Hydroelectric and Navigation Systems on the monthly discharges.

5.1 Interannual Flow

The interannual variability of the average streamflow mainly reflects the climatic conditions of the years and the anthropogenic quantitative pressures on water through its various uses and related engineering works (e.g. dams, intakes). In Table 3 the basic statistic parameters of the average annual discharges data-series at the selected gauging stations are summarized, and in Figs. 5 and 6, their variation during the period 1931–2019 is shown.

A normal increase can be noticed for the considered statistical parameters, from upstream to downstream. The coefficient of interannual variation (Cv) is relatively

Table 3 Basic statistical parameters of the average annual discharges at hydrometric stations on the Lower Danube River (1931–2019)

Parameter/station	Baziaș		Oltenița		Ceatal Izmail	
	Value	Year	Value	Year	Value	Year
Average	5,536	–	6,056	–	6,509	–
Maximum	7,980	1941	8,830	1941	9,498	2010
Minimum	3,770	1990	4,130	1990	4,252	1990
Cv	0.17	–	0.17	–	0.18	–

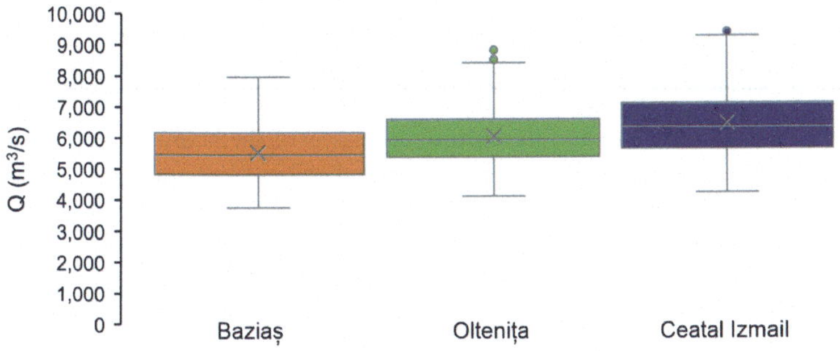

Fig. 5 Box-plots of the average annual discharges of the Danube River at Baziaș, Oltenița and Ceatal Izmail hydrometric stations (1931–2019) (X sign is for the average value)

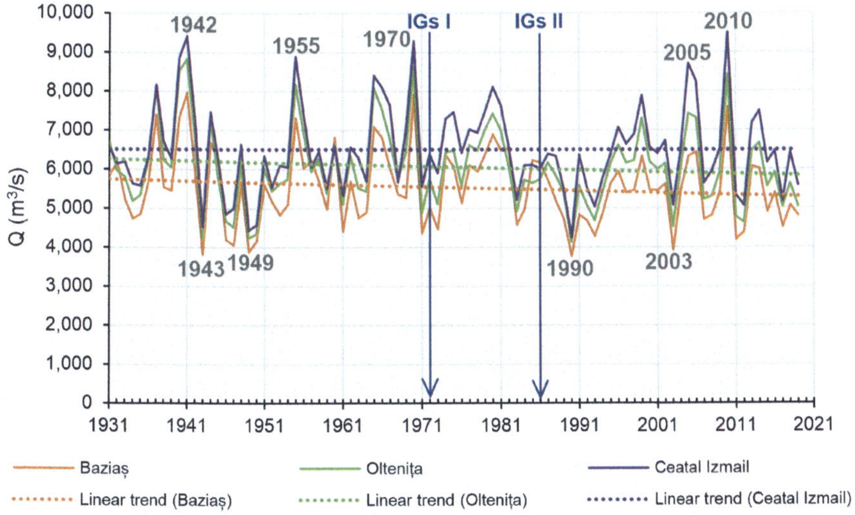

Fig. 6 Variation of the average annual discharges of the Danube River at Baziaș, Oltenița and Ceatal Izmail hydrometric stations (1931–2019). IGs = Iron Gates

constant (Cv = 0.17–0.18). The highest values of the average annual flows occurred in 1941 at Baziaș and Oltenița, and in 2010 at Ceatal Izmail, in the context of large magnitude floods in those years. Very high values of the average annual flows were also reached in 1955, 1970, and 2005 (Fig. 6), with significant floods occurring on the Lower Danube River basin tributaries. The lowest average annual discharges were noticed in 1990 at all analyzed stations, but very low values were also recorded in 1943, 1947, 1949, 2003 and 2011 (Fig. 6).

Figure 6 indicates slight linear downward trends at Baziaș and Oltenița stations during the analyzed period, while at Ceatal Izmail h.s. the data series is quasi-stationary. According to the Mann-Kendall test, the linear trends are statistically non-significant. The Sen's slope is low, of 40–50 m^3/s per decade at Baziaș and

Oltenița (negative) and only about 6 m^3/s at Ceatal Izmail (positive). Our results concerning long-term trends are similar to those found in previous studies performed at stations on the LDR (e.g. [21, 54, 59, 68, 71]). However, there are some differences that can be mainly attributed to the different lengths of the analyzed periods and to the employed method of trend analysis (graphical or based on statistical tests).

The comparative analysis of the average annual flows in the period prior to the entry into operation of the Iron Gates II HENS (i.e., 1931–1986) and in the subsequent interval (1987–2019), indicated a relatively small decrease of the average runoff, with only 5% at Oltenița and 2% at Ceatal Izmail. These low rates suggest that the impact of the HENS functioning on the annual average runoff is not significant.

To identify changes in the average flow of the LDR due to anthropogenic impact, in [62] three periods were separated and analyzed between 1931 and 2010: 1931–1964 (with reduced changes in the natural regime, so quasi-natural flow regime), 1965–1984 (with important changes in the river channel and floodplain—transition flow regime), and 1985–2010 (with reduced modifications—current flow regime). The results showed slight linear downward trend in the first two periods and upward trend in the third period. The increase found in the last period was also attributed to the more intense climatic variability during this period. At all analyzed stations, the highest average runoff was recorded during the transition regime (1965–1984) [62].

In [59] and [69], based on the analysis of the decadal average flows at hydrometric stations along the LDR in the periods 1931–2010 and 1931–2016 respectively, it was shown that the highest values were recorded in the decade 1961–1970 (excepting for the Baziaș station) as a result of floods that occurred in the LDB in 1970. High average decennial discharge was also found in the period 2001–2010, because of the large-scale floods of 2005, 2006 and 2010. Decades with low average flows were 1941–1950 and 1981–1990, as a result of severe droughts during these intervals (e.g. 1946–1947, 1990).

The analysis performed in [21], on the long term 30-year discharges (from the nineteenth century to the first decade of the 21th century) at Reni h.s. (located about 30 km upstream of Ceatal Izmail) showed a decrease of about 100 m^3/s in the period 1976–2005 compared to the period 1961–1990. The analysis of the long term 10-year discharge, indicated a significant decrease (of about 700 m^3/s) in the decade 1981–1990, compared to 1971–1980, followed by a gradual increase in the following decades, so that between 2001 and 2010, the average flow was similar to those of 1971–1980 decade.

If the influence of the impoundments and engineering works along the LDR on the liquid flow is not very obvious, this influence is significant in the case of sediment flux. Thus, the downward trend of suspended sediment flow is evident at all stations downstream of the Iron Gates I and II dams: at Ceatal Izmail, the suspended sediment load decreased drastically, by approx. 2/3, from almost 1700 kg/s (the multiannual average from 1931 to 1964), to an average of almost 600 kg/s between 1985 and 2010 [62].

5.2 *Intra-annual Flow*

During a year, the Danube's discharge is uneven due to the temporal fluctuation of climatic and anthropogenic factors. Thus, the streamflow is closely dependent especially on the variation of main climatic factors (precipitation, air temperature, evaporation and evapotranspiration, snow depth, atmospheric circulation) at basin-wide and regional scales, but the anthropogenic activities (especially water uses) can alter the natural flow regime. Relevant information on the flow regime of the Danube River and its main tributaries can be found in several previous studies such as [21, 47, 63, 72]. This chapter highlights Danube's annual flow regime at Baziaş, Olteniţa and Ceatal Izmail hydrometric stations between 1976 and 2019. To identify the impact of the HENSs Iron Gates I and II on the monthly discharges' variation, at Ceatal Izmail h.s. we analyzed a longer time period, between 1931 and 2019.

At the entrance in the lower sector (at Baziaş), the Danube River's flow regime reflects the action of flow control factors (mainly, the climatic conditions) from the upper and middle sectors of the basin. Downstream of the Iron Gates gorge, a determinant role in the variation of the Danube's flow during the year have the climate in the lower basin and the flow regime of main tributaries in this sector. To these, the major human pressures on the Danube River (e.g. the water uses and the HENSs Iron Gates I and II) are added.

Because of the location of the Danube River basin in the temperate climate region, the general pattern of the intra-annual flow reflects the features of this climate with four seasons. At the same time, the topographic characteristics of the basin also play an important role in defining the flow regime, more precisely the presence of mountainous areas with altitudes exceeding 2000 m a.s.l. favouring the snow accumulation during the cold season. At altitudes over 3000 m a.s.l. there are also glaciers. To these are added the climatic influences generated by the proximity of wide aquatic environments (e.g. Mediterranean and Black seas).

As shown in Fig. 7a, the most important water volumes (10–12% of the average annual volume) are carried by the Danube during spring months (March–May, with

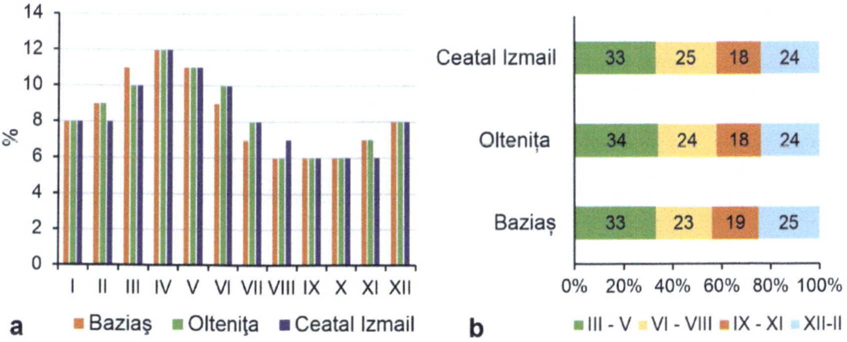

Fig. 7 Monthly (**a**) and seasonal (**b**) distribution of the average flow along the Lower Danube River, at Baziaş, Olteniţa and Ceatal Izmail (in % from the average annual volumes, 1976–2019)

maximum in April) and in June, due to the rich rainfall in these months, often associated with the snow melting. The spring is the season with the highest flow, of about 1/3 (33%) of the average annual volume of the Danube River (Fig. 7b). The lowest average volumes (6–7%) are noticed in August-November, autumn being the driest season from a hydrological point of view (18–19% of the average annual volume of water of the Danube River) (Fig. 7b). The monthly and seasonal distribution of the average flow is almost similar at all analyzed stations along the Lower Danube River, with no noticeable spatial differences.

Based on the monthly and seasonal distribution of the average volumes of water during the year, the Lower Danube flow regime can be described as continental nivo-pluvial, with a snow melting supply from March to June [21]. Because of the large size of the watershed, the flow regime is relatively stable along the Lower Danube River and the main tributaries in this last sector (mainly those from the left-side, draining the Carpathian Mountains), do not change the regime defined at Baziaş h.s. by all the runoff control factors in the upstream basin [21, 47].

In Fig. 8 the variations of the monthly average discharges of the Danube River at the selected stations are shown, based on the specific statistical parameters of box-plots (during the period 1976–2019). These reflect the specific flow regime, as

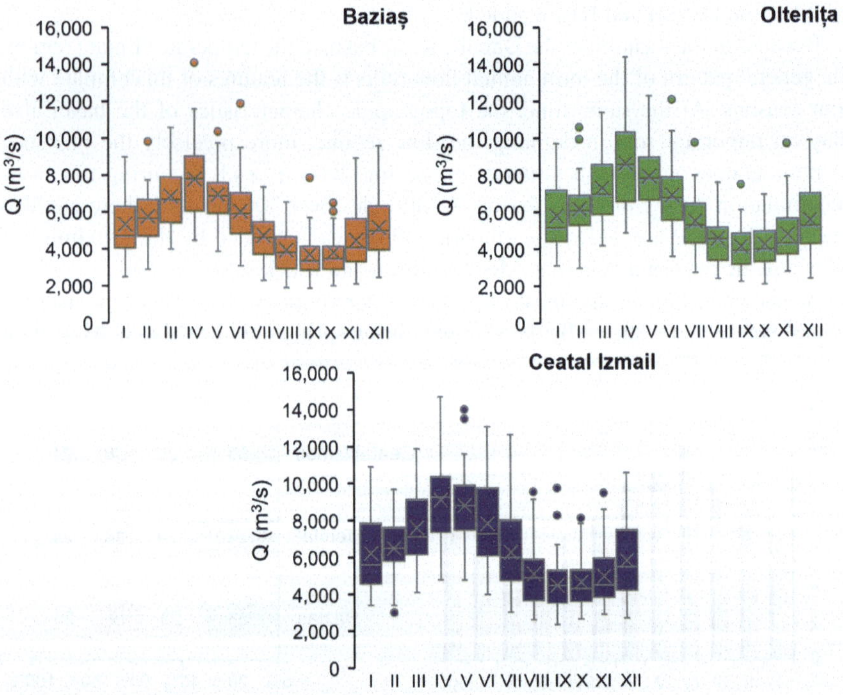

Fig. 8 Box-plots of the average monthly discharges of the Danube River at Baziaş, Olteniţa and Ceatal Izmail hydrometric stations (1976–2019) (X sign is for the average value)

mentioned above, with high waters in spring and early summer and low waters in late summer and autumn. In general, there is a greater variation of the monthly discharges at Ceatal Izmail h.s. This could be explained by the particular regime of the last two major tributaries (Siret and Prut rivers) draining the eastern extremity of the Danube basin, with a drier continental climate.

As shown above, the construction of the two hydroelectric and navigation systems (Iron Gates I and II) and their operation did not significantly impact the average annual flow. In the case of monthly discharges, according to [73], at Orșova h.s. (located few km upstream the Iron Gates I dam), changes in the period 1970–2005, compared to 1921–1960 were noticed, due to the functioning of the Iron Gates I HENS and to its flow regularizing role. These changes consist of higher discharges during the months with low waters (September–October and December–February) and lower discharges during the periods with high waters (e.g. May–August). Significant differences between the natural and controlled regimes were noticed during the flood events [73].

The comparative analysis of the multiannual monthly averages of the Danube River's flow at Ceatal Izmail in the period before (1931–1986) and after (1987–2019) the put into full operation of the HENS Iron Gates II, indicated a decrease of the monthly discharge by −4 ... −15% in the warm period (May–September) and slight increases, of up to 2–4% in January, April and October–November (Fig. 9). These modifications in the annual flow regime could be caused by the operation of the two large hydroelectric systems, but the changes in the intra-annual variability of climatic parameters could also be also responsible for this [62].

The comparison of the flow regime in the period 1981–2005 with that of the period 1956–1980, based on Pardé-coefficients at stations along the Danube River, indicated for its lower sector (at Ceatal Izmail) a decrease of the Pardé variability flow coefficient (by −20%), showing a loss in monthly flow variability [21]. It was also found that the maximum flow occurrence in spring was reduced from two months

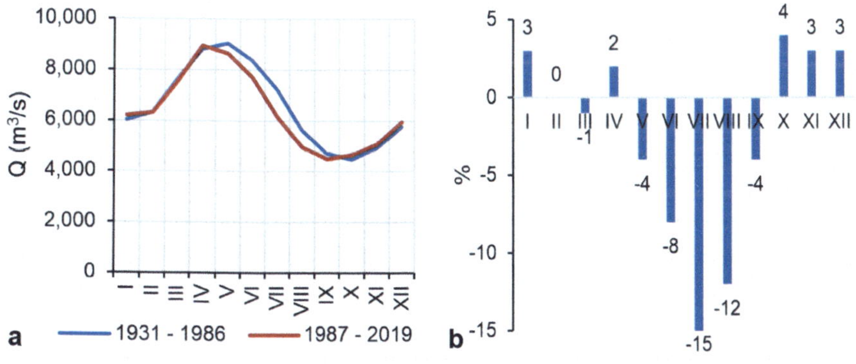

Fig. 9 Variation of the average monthly discharges of the Danube River at Ceatal Izmail before (1931–1986) and after (1987–2019) the commissioning of the hydroelectric and navigation system Iron Gates II (**a**) and the rate of changes in the second period compared to the first period (**b**)

to one month [21]. Both climate change and anthropogenic influences could be responsible for these changes.

6 Maximum Flow Variability

The main factor responsible for the maximum flow in the LDR is the climate at basin-wide and regional scale. The rich rainfall in the spring months, coupled with the snow melting (the snow can persist in the high mountain areas until early summer), leads to spring–summer floods and high waters during this period, as shown in Sect. 5.2.

Several previous studies have addressed the topic of the maximum flow and floods on the Danube River as a whole or only in its lower sector. Among the most recent and relevant we mention: [58, 60, 63, 65, 67, 69, 74]. A comprehensive analysis of the maximum flow and floods within the Danube River Basin basin was carried out in [21]. As mentionned in Sect. 2, a valuable publication on flood risk along the Danube River is the "Danube Atlas - Flood Hazard and Risk Maps" (2012) [27]. Several studies on the topic of floods are the outputs of the projects carried out through the cooperation of Danubian countries within the framework of IHP/UNESCO, such as: [21, 72, 74, 75], as quoted in [68].

In this chapter, we investigated the variation of the maximum flow of the LDR based on the analysis of the annual and monthly maximum discharges at Baziaș, Oltenița and Ceatal Izmail during the period 1931–2019 for the annual values, and between 1976 and 2019, for monthly discharges.

6.1 Maximum Annual Discharges

As mentioned in Sect. 4, the highest maximum discharges in the period 1931–2019 were registered at all analyzed stations during the historical flood in the spring of 2006 (Figs. 4a and 10) and Table 4. At Baziaș h.s. the flood-peak reached 15,800 m^3/s (the highest since the beginning of hydrological measurements in Romania, in 1838). At Oltenița h.s., the maximum discharge during the flood in 2006 reached 16,200 m^3/s, while at Ceatal Izmail the flood-peak was lower (15,900 m^3/s), due to the Danube's overflow and flood attenuation.

The variation of the basic statistical parameters of the time-series of maximum annual discharges between 1931 and 2019 at Baziaș, Oltenița and Ceatal Izmail gauging stations is shown by the box-plots in Fig. 11. The variation coefficient of the annual maximum discharges is constant and relatively low (Cv = 0.18) (Table 4).

The magnitude of the floods, as indicated by the ratio between the maximum annual discharge (Qmaxa) and the multiannual average discharge (Qavma), differed along the LDR. Thus, in 2006 this ratio reached the maximum values, of 2.9 at Baziaș, 2.7 at Oltenița and 2.4 at Ceatal Izmail. High rations (more than 2.3) were

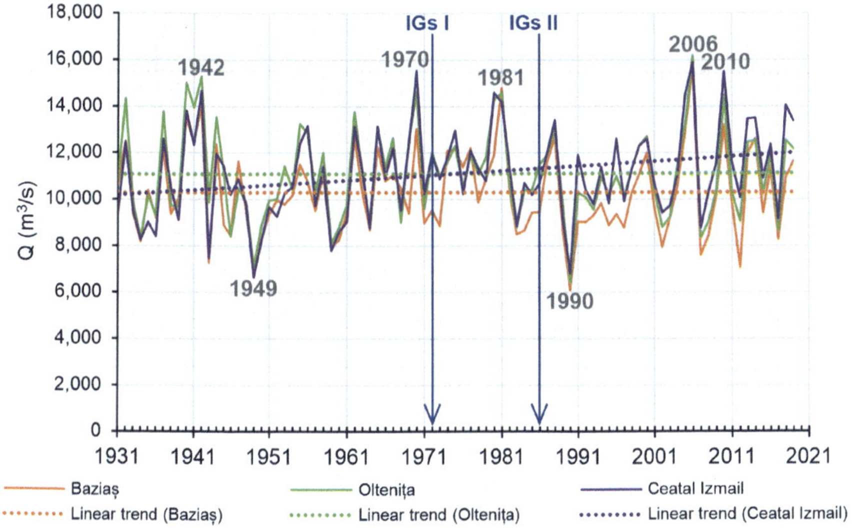

Fig. 10 Variation of the maximum annual discharges at hydrometric stations along the Lower Danube River (1931–2019) and their linear trends

Table 4 Basic statistical parameters of the maximum annual discharges at gauging stations on the Lower Danube River (1931–2019)

Parameter/Station	Baziaş		Olteniţa		Ceatal Izmail	
	Value	Year	Value	Year	Value	Year
Med	10,308		11,096		11,105	
Max	15,800	2006	16,200	2006	15,900	2006
Min	6,100	1990	6,420	1990	6,660	1949
Cv	0.18		0.18		0.18	

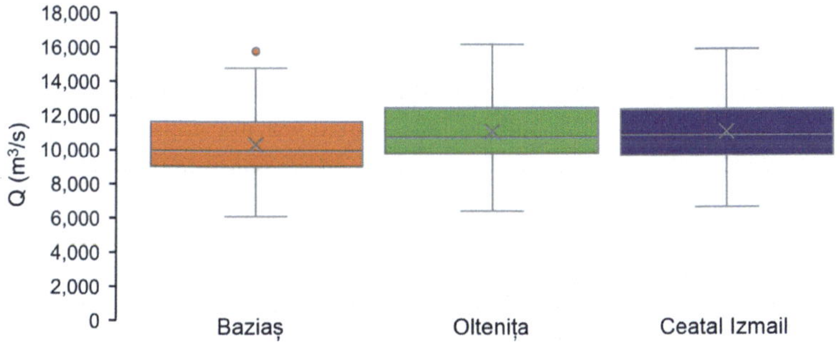

Fig. 11 Box-plots of the maximum annual discharges of the Danube River at Baziaş, Olteniţa and Ceatal Izmail hydrometric stations (1931–2019) (X sign is for the average value)

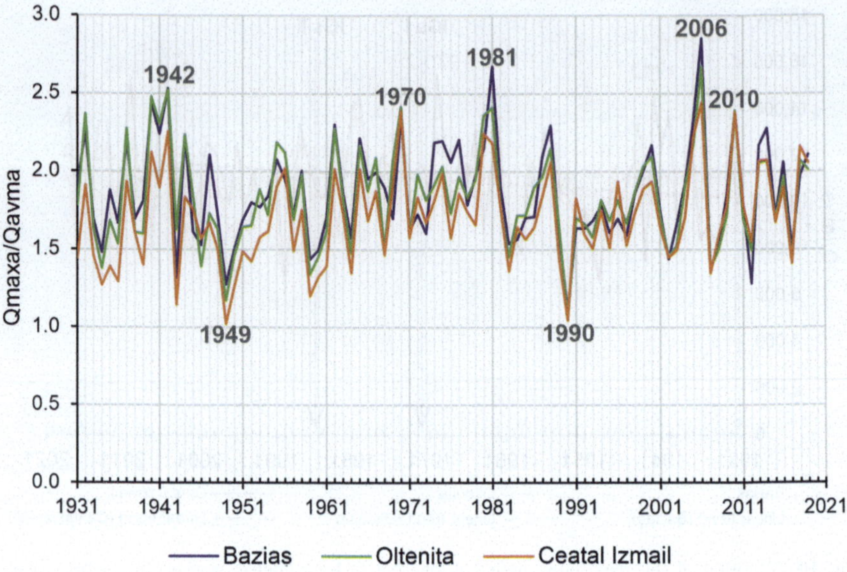

Fig. 12 Variation of the ratio between the maximum annual discharges (Qmaxa) and the multiannual average discharge (Qavma) at hydrometric stations along the Lower Danube River (1931–2019)

also noticed during the floods occurred in 1942, 1970, 1981 and 2010. In 1949 and 1990, the maximum annual discharges were low, similar to the average multiannual discharge, the value of the Qmaxa/Qavma ratio being almost 1 (Fig. 12).

For the period prior to 1931, scientific literature indicates maximum annual discharges (based generally on indirect estimations) at hydrometric stations along the LDR, of more than 20,000 m^3/s, as mentioned in Sect. 4.

Between 1931 and 2019 at Baziaș and Oltenița, no significant linear trends were detected in the variation of the maximum annual discharges. At Ceatal Izmail, an upward trend at $\alpha = 0.05$ level of significance was found, with an average Sen's slope of about 200 m^3/s per decade. In a previous study [21], during the period 1931–2005, the upward trend found at Ceatal Izmail was not statistically significant. The major floods occured after 2005 (in 2006, 2010) led to a significant increase in the maximum annual flows. According to [21], at Orșova (located in Iron Gates gorge area, about 120 km downstream of Baziaș h.s., and a few kilometres upstream the Iron Gates I dam) and Reni (located close to Ceatal Izmail h.s.), in the decade 2001–2020, the average of the maximum annual discharges of the Danube River was higher than in previous decades, namely 1981–1990 and 1991–2000.

The comparison of the average of the maximum annual discharges between 1931 and 1986 (before the commissionning of the Iron Gates II HENS) with the average from the period 1987–2019 indicated, at the Oltenița h.s., a small decrease of approx. 200 m^3/s (almost 2%) in the second period, while at Ceatal Izmail h.s. an increase of about 630 m^3/s (nearly 6%) was found. As a result, a slight reduction of the flood

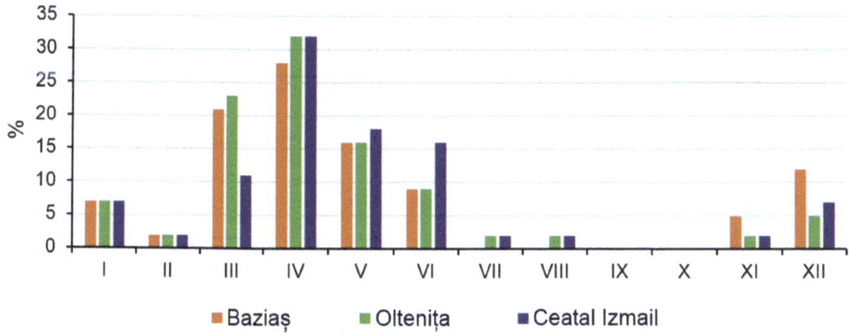

Fig. 13 The monthly distribution of the annual flood-peaks at hydrometric stations along the Lower Danube River (1931–2019, in % from the total number of years)

peaks could be deduced at Olteniţa, which is no longer found at Ceatal Izmail, where maximum flow is strongly influenced by the input of the last two large tributaries (Siret and Prut rivers).

Most frequently, the peaks of the annual floods occurred in April (28–32% of the analyzed years), March (up to 23%) and May (16–18%), while in September–October no floods occurred (Fig. 13). However, there are some notable spatial differences between the selected stations. Thus, at Baziaş h.s., the share of floods in December and November is almost double compared to the other two stations, which can be explained by the influence of the Mediterranean climate on the runoff in this section. At Ceatal Izmail there is a much lower frequency (about half) of annual floods in March compared to Baziaş and Olteniţa and much higher (almost double) in June. This situation could be related to the climatic influence induced by the proximity of the Black Sea (where retrograde cyclones are active in summer) and of the Carpathians (which cause a delay in the water supply from snow melting in late spring and early summer).

6.2 *Maximum Monthly Discharges*

The variation of the maximum monthly discharges of the LDR and of their multiannual averages at the analyzed stations (between 1976 and 2019), illustrates, once again, the occurrence of the highest floods between April and July (Fig. 14). This is due to the abundant precipitations during these months in conjunction with the snow and glaciers melting (in the upper basin of the Danube, overlapping the Alps).

During the mentionned period, maximum monthly discharges at all analyzed stations along the LDR were recorded in January 2010, March 1981, April 2006, July 2010, October 2014 and November 1998 (Table 5). In the other months, spatial differences were noticed, probably caused by the contribution of tributaries in the lower sector or by the phenomenon of overflow and attenuation of flood-peaks between

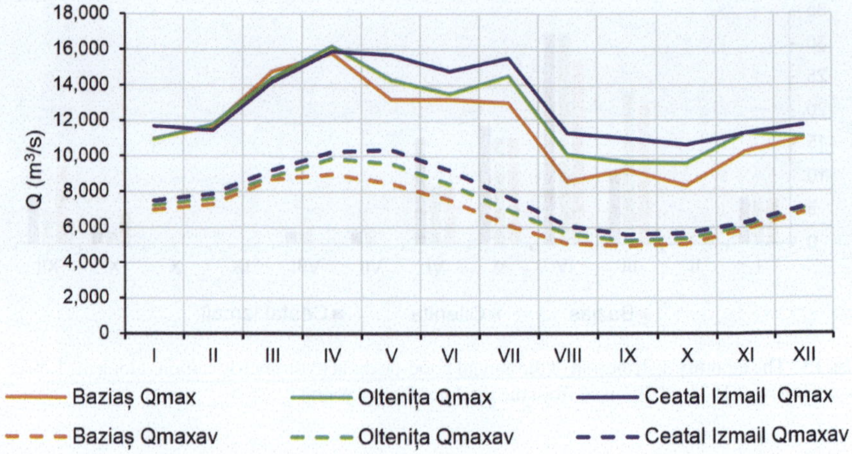

Fig. 14 Variation of maximum monthly discharges (Qmax) and of their averages (Qmaxav) at hydrometric stations along the Lower Danube River (1976–2019)

Table 5 Maximum monthly discharges and the years of their occurrence at hydrometric stations along the Lower Danube River (1976–2019)

Month /	**Baziaș**		**Oltenița**		**Ceatal Izmail**	
Station	Q (m^3/s)	Year	Q (m^3/s)	Year	Q (m^3/s)	Year
I	11,000	2010	10,960	2010	11,700	2010
II	11,700	1977	11,800	1979	11,460	1977
III	14,800	1981	14,400	1981	14,200	1981
IV	**15,800**	2006	**16,200**	2006	**15,900**	2006
V	13,210	2006	14,300	1980	15,700	2006
VI	13,200	2010	13,500	1980	14,770	2010
VII	13,000	2010	14,490	2010	15,500	2010
VIII	8,600	1991	10,100	1997	11,300	2005
IX	9,250	2014	9,660	2014	11,000	2005
X	8,360	2014	9,610	2014	10,640	2014
XI	10,280	1998	11,300	1998	11,300	1998
XII	11,000	1981	11,100	1981	11,740	2010

Red font—highest discharge. Colorful background—year that repeats. White background—year without repetition

stations. For example, in May, at Oltenița h.s., the maximum discharge was not registered during the historical flood of 2006, as it was the case at the other two stations, but it occured in 1980. A similar situation was noticed in June, when at Oltenița the maximum discharge was reached in 1980, while at Baziaș and Ceatal Izmail it was

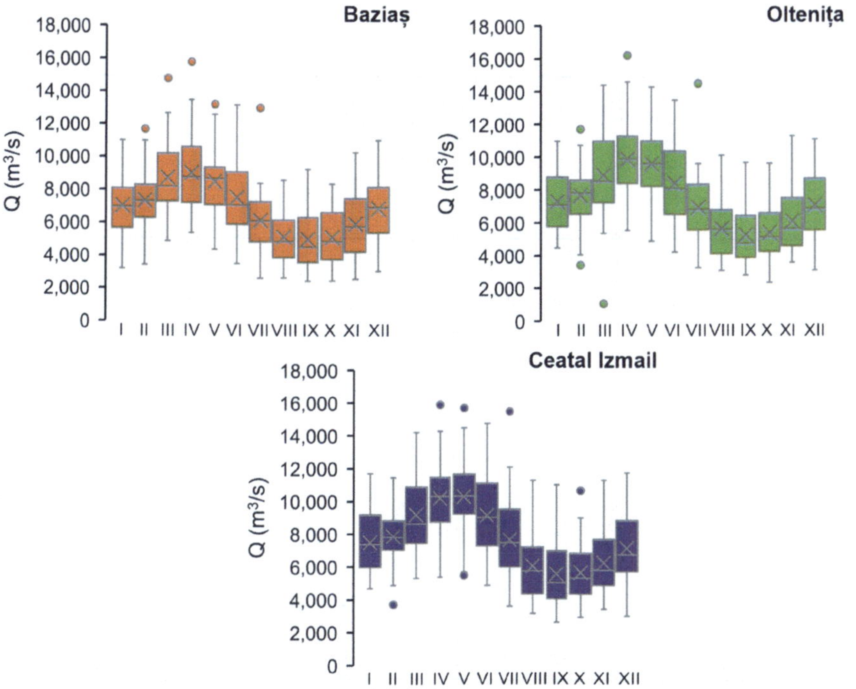

Fig. 15 Box-plots of the maximum monthly discharges of the Lower Danube River at Baziaş, Olteniţa and Ceatal Izmail hydrometric stations (1976–2019) (X for the average value)

recorded in 2010. In August the maximum discharges occured in different years at the three selected stations (Table 5).

The variation of the basic statistical parameters of the maximum monthly discharges time-series between 1976 and 2019 at Baziaş, Olteniţa and Ceatal Izmail gauging stations is shown in the form of box-plots in Fig. 15.

In order to highlight the magnitude of the maximum monthly discharges, we determined the ratio between them and the average monthly discharges, during the period 1976–2019. The values of these ratios are high (over 2.2, up to 2.5–2.7) in July–November (Fig. 16), when low waters characterize the Danube's flow regime.

7 Minimum Flow Variability

The low minimum flow is associated with the time periods with little or no rainfall within the Danube River Basin. As showed in Sect. 5.2, in the annual flow regime of the LDR, the low flow, and consequently the minimum discharges, are specific for the late summer-early autumn period, namely during the months of August–November (6–7% from the average annual volume).

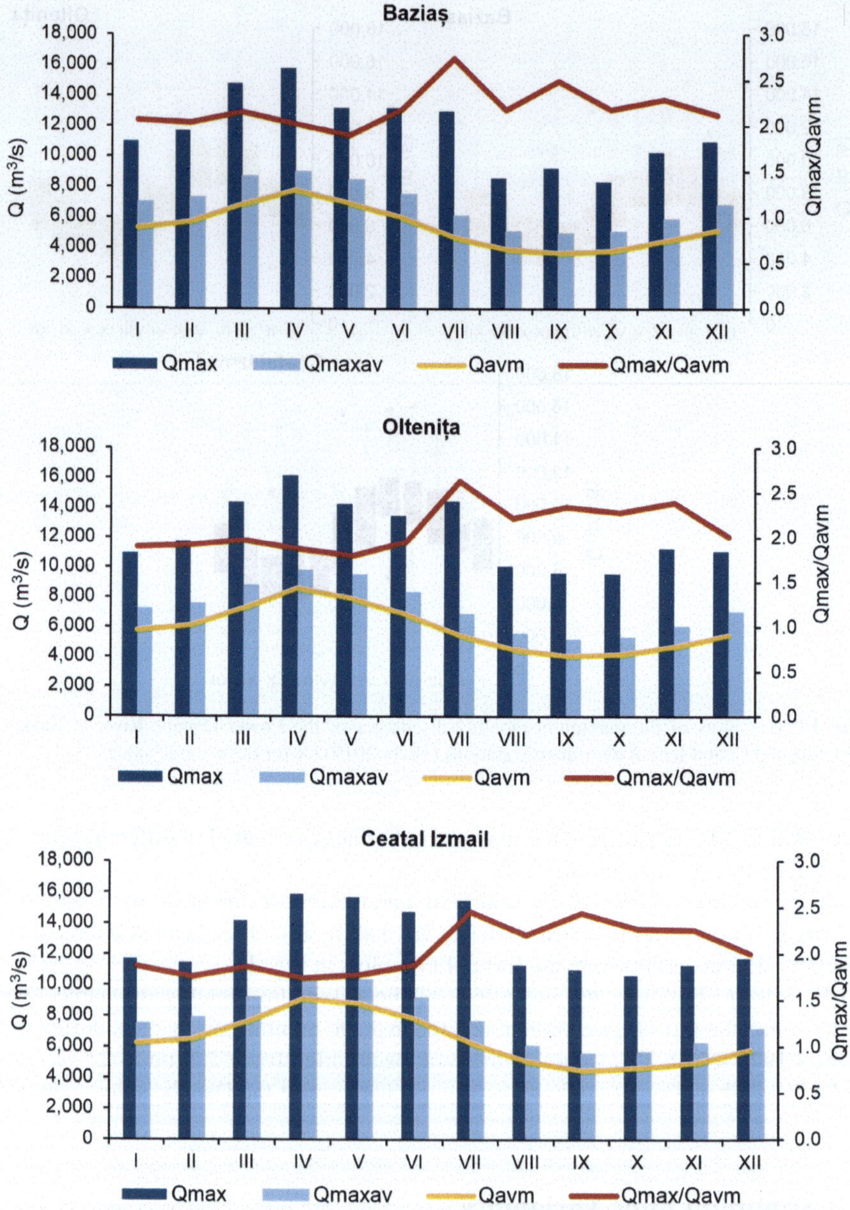

Fig. 16 Variation of the maximum monthly discharges (Qmax), of their average values (Qmaxav) and of the ratio between the maximum monthly discharges (Qmax) and the average monthly discharges (Qavm) along the Lower Danube River, at Baziaș, Oltenița and Ceatal Izmail (1976–2019)

In the last decade, relevant information on the minimum flow of the Lower Danube River from different time-periods is found in several publications such as: [58, 60, 63–65, 67, 69]. In this chapter, we presented up-to-dated data and results on the variations of minimum flow along the LDR, at Baziaş, Olteniţa and Ceatal Izmail gauging stations, based on the analysis of the annual and monthly minimum discharges during the period 1931–2019 for annual values and 1976–2019, for the monthly ones.

7.1 Minimum Annual Discharges

In Sect. 4 we showed that the multiannual minimum discharges recorded at gauging stations located along the LDR (during the period 1931–2016), decreased to approx. 1,000 m^3/s at the western part of the studied sector, and almost 1,800 m^3/s at Ceatal Izmail (Fig. 4b). Very low discharges were recorded in dry years at basin-wide scale and also in the lower sector, such as 1947, 1949, 1954, 1961 and 2003 (Fig. 17). Before 1931, dry years were noticed in 1858, 1863, 1901, 1908 and 1920, when very low minimum flows were estimated in [28] (see Sect. 4).

The variation of the basic statistical parameters of the time-series of minimum annual discharges between 1931 and 2019 at Baziaş, Olteniţa and Ceatal Izmail hydrometric stations is illustrated by the box-plots in Fig. 18. The increase of these parameters from upstream to downstream is obvious, as also shown in the Table 6.

The interannual variation of the minimum annual discharges is relatively small (Cv = 0.26) and similar at all three study gauging stations. During the period between

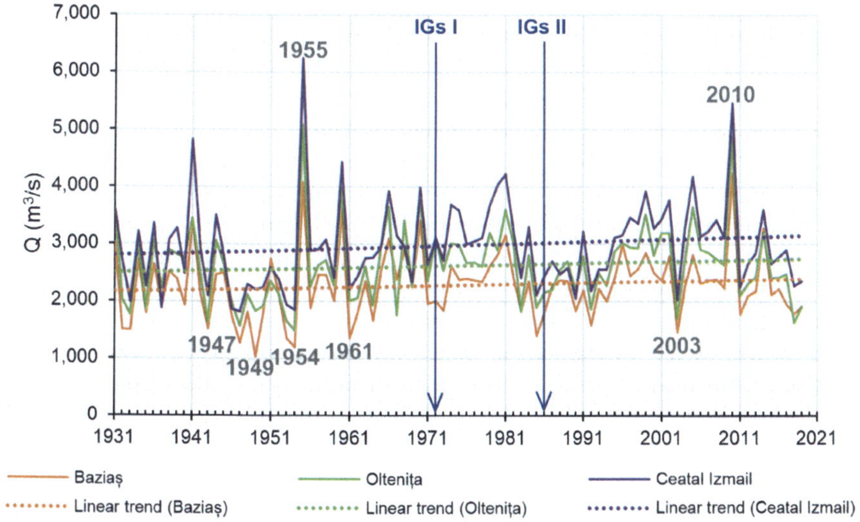

Fig. 17 Variation of the minimum annual discharges at hydrometric stations along the Lower Danube River (1931–2019) and their linear trends

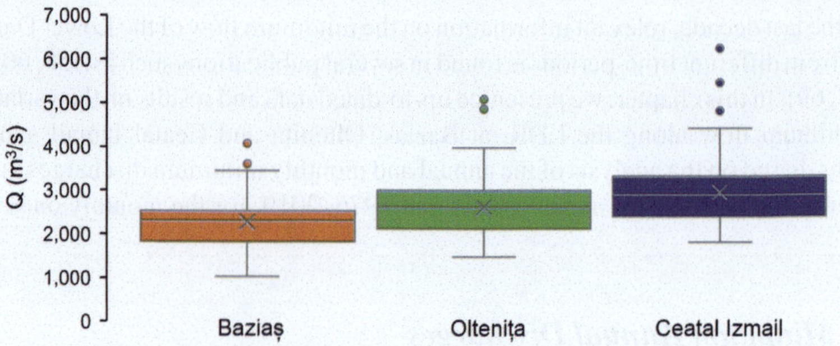

Fig. 18 Box-plots of the minimum annual discharges of the Danube River at Baziaș, Oltenița and Ceatal Izmail hydrometric stations (1931–2019) (X for the average value)

Table 6 Basic statistical parameters of the minimum annual discharges at hydrometric stations on the Lower Danube River (1931–2019)

Parameter/station	Baziaș		Oltenița		Ceatal Izmail	
	Value	Year	Value	Year	Value	Year
Med	2,282		2,625		2,978	
Max	4,250	2010	5,090	1955	6,250	1955
Min	1,040	1949	1,490	1954	1,820	1947
Cv	0.26		0.26		0.26	

1931 and 2019 the lowest minimum annual discharges occurred in 1947, 1949, 1954 and 1961. In these years and several other years (e.g., 1985, 2003), the ratio between the minimum annual discharges and the multiannual average discharges at the analyzed hydrometric stations ranged between 0.2 and 0.3 (Fig. 19).

During the period 1931–2019 at Baziaș and Oltenița stations, slight upward linear trends in the variation of the minimum annual discharges can be observed (Fig. 17), but they are not statistically significant. At Ceatal Izmail h.s., the Mann-Kendall test showed an upward trend at $\alpha = 0.1$ level of significance, with an average Sen's slope of about 55 m^3/s per decade. In the period 1961–2010, in [64] linear upward trends were found at most gauging stations along the LDR, with more pronounced slopes downstream of Oltenița h.s., but the statistical significance of these trends was not established. In [58] a trend of slight increase of minimum annual discharges of the Danube River was noticed at Baziaș and Ceatal Izmail between 1931 and 2010, but the trend was established only based on the analysis of the graphic. Similar results, also based on the analysis of the graphical linear trend, but for a longer period (1931–2016) are mentioned in more recent studies [67, 69].

Comparing the average maximum annual flows from 1931 to 1986 (before the commissioning of the HENS Iron Gates II) with those from the period 1987–2019, we found, both in the case of Oltenița and Ceatal Izmail stations, a very slight increase, of approx. 3–4% (80–90 m^3/s) in the second period. These results show that the functioning of the hydroelectric system does not significantly impact the

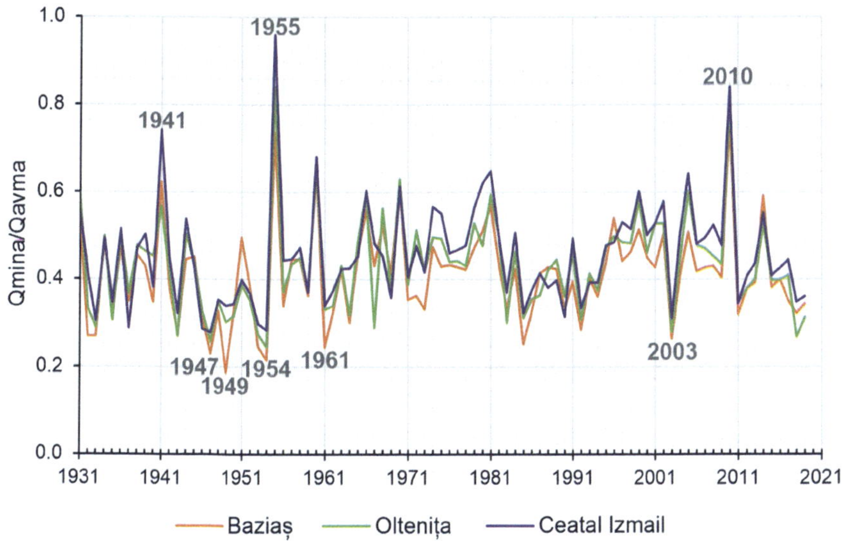

Fig. 19 Variation of the ratio between the minimum annual discharge (Qmina) and the average multiannual discharge (Qavma) at hydrometric stations along the Lower Danube River (1931–2019)

minimum annual flow of the LDR. In addition, the impact of climate changes could be responsible for the flow alteration.

The analysis of the monthly distribution of the minimum annual discharges during the period 1931–2019 indicates that these values most frequently occured in the late summer-autumn period, more precisely from August to November, with the maximum occurrence (23–26% of the total number of analyzed years) found in September and November (Fig. 20). This is due to the low precipitations in this period of the year. Frequencies of up to 9% were also found in December and January, in the climatic context characterized by water freezing phenomena and predominantly solid

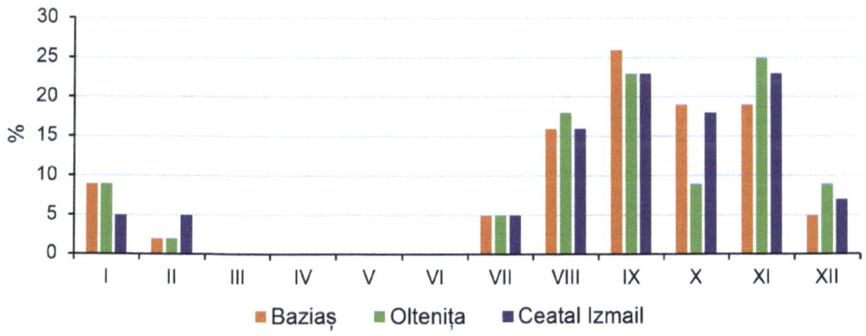

Fig. 20 The monthly distribution of the minimum annual discharges at hydrometric stations along the Lower Danube River (between 1931 and 2019, in % from the total number of the years)

precipitation. Between March and June, characterized by high waters, no minimum annual discharges occurred during the studied period. A similar monthly distribution was found in [63] at Ceatal Izmail between 1921 and 2010. At Baziaş, over a shorter period (1991–2008), no minimum annual discharges were recorded between December and May [63].

7.2 *Minimum Monthly Discharges*

The minimum monthly discharges were analyzed based on instantaneous and averaged values recorded at Baziaş, Olteniţa and Ceatal Izmail hydrometric stations, during the period 1976–2019. Both types of values have exhibited a variation consistent with that of the monthly average flow, namely, high discharges in spring and early summer, while in late summer, autumn and winter the lowest values were recorded (Fig. 21).

During the analysed period, in some months were registered minimum values at all stations along the LDR: January 1985, May 2007, June 1993, July and September 2003 (Table 7). Overall, in the LDR sector, the period between July and September 2003 stands out with the highest frequency of occurrence of minimum monthly flows. In months such as March, October, November, there is a heterogeneity regarding the years when minimum flows occured along the Danube River, which can be explained by the stronger influence of regional climatic conditions and of the tributaries during these periods.

The variation of the basic statistical parameters of the time-series of minimum monthly discharges at the three analyzed stations, in the period 1976–2019, is shown

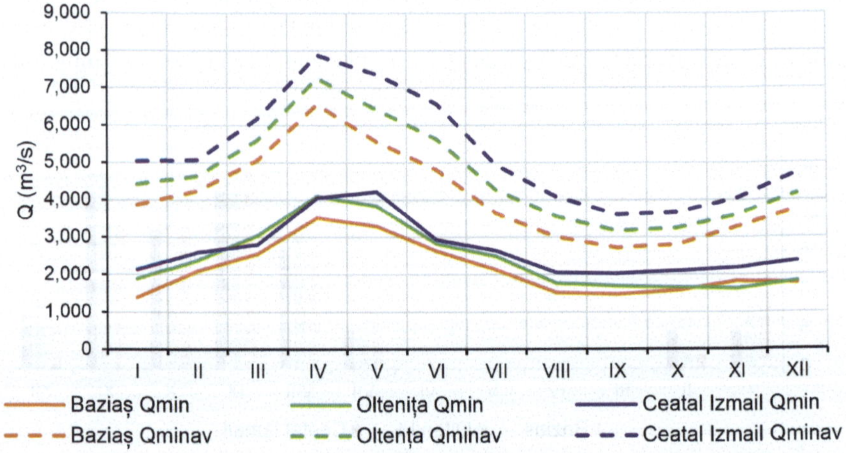

Fig. 21 Variation of minimum monthly discharges (Qmin) and of their averages (Qminav) at hydrometric stations along the Lower Danube River (1976–2019)

Table 7 Minimum monthly discharges and the years of their occurrence at hydrometric stations along the Lower Danube River (1976–2019)

Month / Station	Baziaș Q (m³/s)	Year	Oltenița Q (m³/s)	Year	Ceatal Izmail Q (m³/s)	Year
I	**1,400**	1985	1,890	1985	2,140	1985
II	2,100	2012	2,380	2012	2,590	1989
III	2,550	1993	3,040	1993	2,800	1989
IV	3,530	1991	4,100	1991	4,060	1990
V	3,300	2007	3,840	2007	4,220	2007
VI	2,630	1993	2,830	1993	2,930	1993
VII	2,132	2003	2,490	2003	2,640	2003
VIII	1,520	2003	1,780	2003	2,050	1990
IX	1,474	2003	1,700	2003	**2,030**	2003
X	1,580	1992	1,660	2018	2,100	2003
XI	1,830	1953	**1,630**	2018	2,180	1985
XII	1,780	2011	1,820	1983	2,360	2011

Blue font—lowest discharge. Colorful background—year that repeats.
White background—year without repetition

in the form of box-plots in Fig. 22. They generally indicate larger dispersions in spring values (especially in April) and lower dispersions in fall (September–October).

In Fig. 23, the variation of the minimum monthly discharges and their averaged values are shown, compared to the average monthly discharges recorded between 1976 and 2019 at the three selected stations. In order to highlight the severity of the low flow, we calculated the ratio between the minimum monthly discharges (Qmin) and the average monthly discharges (Qavm) during the period 1976–2019 at the three stations. The lowest values of the ratios (0.3–0.4) can be noticed in the winter and autumn months (Fig. 23).

8 Conclusion and Recommendation

This chapter provides an up-to-date overview of the spatial and temporal variability of the Lower Danube River flow, over a length of almost 1000 km, from the entry in Romania (at Baziaș) to the Danube Delta (at Ceatal Izmail, named Ceatal Chilia in Romania). The analyzes are based on the processing of the discharge time-series from several gauging stations located on the Romanian side of the Danube River (belonging to the national hydrometric network). The results highlight the temporal and spatial variations of the average, maximum and minimum discharges (annual

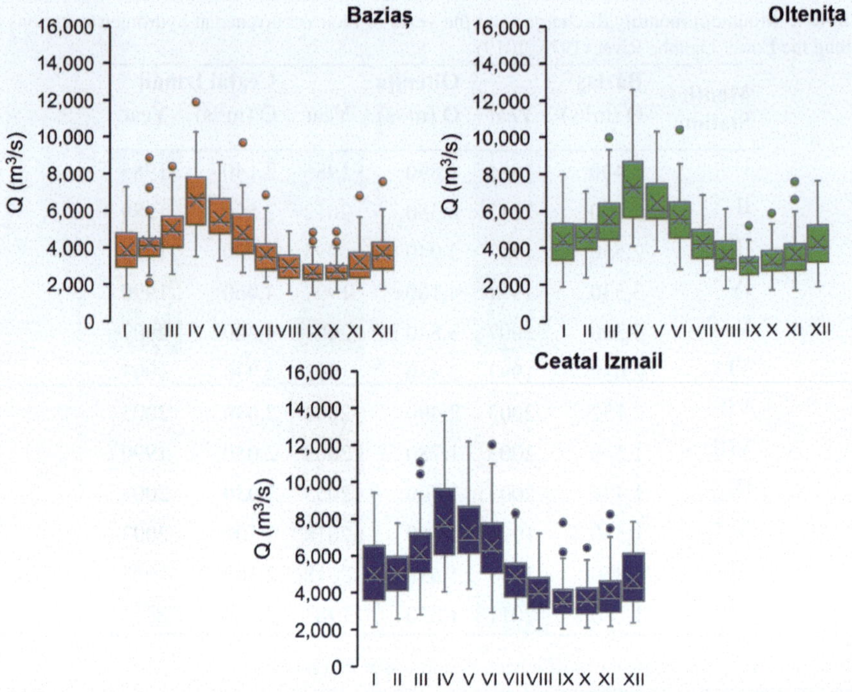

Fig. 22 Box plots of the minimum monthly discharges of the Danube River at Baziaş, Olteniţa and Ceatal Izmail hydrometric stations (1976–2019) (X sign is for the average value)

and monthly) for different time-periods, ranging from 44 years (1976–2019), to more than 170 years (1840–2012).

Along its lower sector, the Danube River average flow increases by about 1,000 m^3/s, from 5,551 m^3/s, at Baziaş to 6,516 m^3/s, at Ceatal Izmail (during the period 1840–2012). This is due to the contribution of tributaries draining the riparian countries (Romania, Serbia, Bulgaria, Republic of Moldova Ukraine), of which the most important (in term of discharges and catchment area) are the rivers Siret, Olt, Prut and Jiu. Between 1931 and 2019, the maximum discharges of the Danube River reached the highest values during the historical flood of spring 2006. They ranged between 15,800 m^3/s at Baziaş and 16,300 m^3/s at Giurgiu. At Ceatal Izmail, the flood-peak was lower than at the upstream stations, because of the flow attenuation by flooding the large floodplain. During the very dry years of the period 1931–2019, the Danube minimum discharges decreased below 2,000 m^3/s, namely up to 1,040 m^3/s at Baziaş (in 1949) and 1,790 m^3/s at Ceatal Izmail (in 1947).

The intra-annual flow, as shown by the variations of the monthly discharges, is mainly determined by the climate variability within the Danube basin and anthropogenic water uses (the most important in the lower sector are the hydroelectric and navigation systems Iron Gates I and II). In the specific climatic conditions of the Danube basin (especially in the middle and lower sector), the flow regime of the

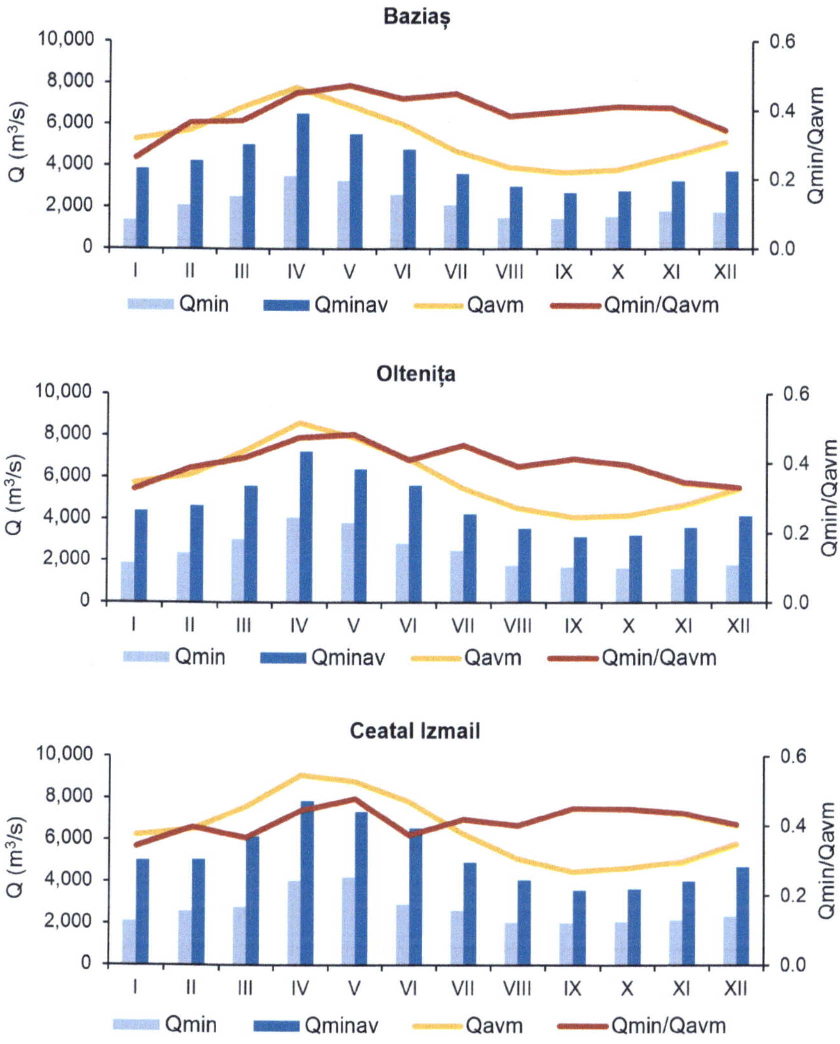

Fig. 23 Variation of the minimum monthly discharges (Qmin), of their average values (Qminav) and of the ratio between the minimum monthly discharges (Qmin) and the average monthly discharges (Qavm) along the Lower Danube River, at Baziaş, Olteniţa and Ceatal Izmail (1976–2019)

LDR at the analyzed gauging stations exhibits high waters (10–12% from the total annual average water volume, in each month) in spring and early summer (March–June, with the maximum in April). The low waters are specific for the late summer–autumn period (August–November) with monthly rates of 6–7% of the total annual average water volume, and minimum values (6%) in September–October. Therefore, the Danube River has the richest flow during the spring (33% from the total annual average water volume), while it decreases up to 18–19% in autumn. In summer and winter, the flow rates are quite similar (23–25%). The hydroelectric and navigation systems Iron Gates I and II do not appear to significantly influence the Danube River annual flow, but mostly the sediment flux. Some possible influences could be noticed in the intra-annual flow variation. Thus, the multiannual averages monthly discharges of the Danube River at Ceatal Izmail, in the period prior to commissioning the Iron Gates I and II hydropower systems, compared to those in the period after the systems became functional (i.e., 1931–1986 versus 1987–2019), showed in the second period average discharges lower by −5 … −15% in May–September and higher, up to 2–3%, in January, April and October–November. These alterations may also be caused by climate change, but it is difficult to assess separately the role of each anthropogenic and climatic factor.

In general, between 1931 and 2019, no significant linear trends in the annual flow variation (average, maximum and minimum) were identified at the analyzed gauging stations. However, at Ceatal Izmail h.s., an upward trend (at $\alpha = 0.05$ level of significance) was found in maximum annual discharges variation. A similar positive trend was also found at Ceatal Izmail h.s. in minimum annual discharges variation, at $\alpha = 0.1$ level of significance.

Due to the major societal and environmental importance of the Danube's flow variability, the studies on the hydrological features are of high interest both scientifically and practically. Such studies provide valuable information for the management of water resources and hydrological hazards. In the context of the environmental changes (especially climatic) affecting the Danube River Basin, rigorous investigations are recommended to detect their impacts on flow variation, at different spatial and temporal scales. Therefore, the analysis of updated long-time data series of recorded discharges allows the detection of the current trends in the flow variability, while the projections based on different climate change scenarios allow the assessment of the flow variability in the future. Information on observed and future possible changes in flow variability is essential for designing and implementing adaptation strategies in the water field. Because in the DRB, the climate changes are considered a major threat that is likely to cause significant impacts on water resources [24], the ICPDR adopted the Strategy on Adaptation to Climate Change in December 2012 (updated and revised in 2018). It provides an outline of the climate change scenarios for the DRB and the expected water-related major impacts, as well as the guiding principles on adaptation and integration into ICPDR activities [76, 77]. Because in the Middle and Lower Danube River basins the information on climate change impacts is relatively sparse, it is recommended to fill in these knowledge gaps through approaches covering all relevant hydrological parameters [76]. In this context, we intend to further investigate the Lower Danube River's flow variability to detect possible changes on a finer time scale (e.g. monthly) and in the frequency of extreme phenomena.

Acknowledgements This is to acknowledge the equal contribution of the first four authors (Liliana Zaharia, Gabriela Ioana-Toroimac, Gabriela-Adina Moroşanu and Elena Ţuchiu) in data processing, analysis, designing and writing of the chapter. The penultimate author (Gabriela Osaci-Costache) is acknowldged for her valuable contribution to the drawing of figures, and the last author (Abdelazim Negm) for his constructive comments and completions that improved the chapter's overall quality.

References

1. Gâștescu P (2017)The Danube-hydrographic polarisation European axis. State-of-the-art. Riscuri şi catastrofe XVI, 20(1):9–23
2. ICPDR (2009) Danube River Basin District management plan. Part A—basin-wide overview. Doc. No. IC/151, Final Version 14 December 2009, Vienna. https://www.icpdr.org/main/sites/default/files/DRBM_Plan_2009.pdf. Accessed 25 Jun 2021
3. Schiller H, Miklós D, Sass J (2010) The Danube River and its basin physical characteristics, water regime and water balance. In: Brilly M (ed) Hydrological processes of the Danube River Basin. Springer, Dordrecht, pp 25–78. https://doi.org/10.1007/978-90-481-3423-6_2
4. Liska I (2015) Managing an international river basin towards water quality protection: the Danube case. In: Liska I (ed) The Danube River Basin. Hdb Env Chem, vol 39. Springer-Verlag, Berlin, Heidelberg, pp 1–20. https://doi.org/10.1007/698_2015_388
5. Stagl JC, Hattermann FF (2016) Impacts of climate change on riverine ecosystems: alterations of ecologically relevant flow dynamics in the Danube River and its major tributaries. Water 8(566). https://doi.org/10.3390/w8120566
6. Feldbacher E, Paun M, Reckendorfer W, Sidoroff M, Stanica A, Strimbu B, Tusa I, Vulturescu V, Hein T (2016) Twenty years of research on water management issues in the Danube Macro-region—past developments and future directions. Sci Total Environ 572:1297–1306. https://doi.org/10.1016/j.scitotenv.2015.12.133
7. Habersack H, Hein T, Stanica A, Liska I, Mair R, Jager E, Hauer C, Bradley C (2016) Challenges of river basin management: current status of, and prospects for, the River Danube from a river engineering perspective. Sci Total Environ 543:828–845. https://doi.org/10.1016/j.scitotenv.2015.10.123
8. ICPDR (2011) The Danube River basin. Facts and figures. https://www.icpdr.org/main/danube-basin-facts-and-figures-brochure. Accessed 27 Jun 2021
9. USGS (2011) Global multi-resolution terrain elevation data. https://edcintl.cr.usgs.gov/downloads/sciweb1/shared/topo/downloads/GMTED/Global_tiles_GMTED/300darcsec/bln/E000/. Accessed 27 Nov 2021
10. Hein T, Schwarz U, Habersack H, Nichersu I, Preiner S, Willby N, Weigelhofer G (2016) Current status and restoration options for floodplains along the Danube River. Sci Total Environ 543:778–790. https://doi.org/10.1016/j.scitotenv.2015.09.073
11. Chapman DV, Bradley C, Gettel GM, Hatvani IG, Hein T, Kovács J, Liska I, Oliver DM, Tanos P, Trásy B, Várbíró G (2016) Developments in water quality monitoring and management in large river catchments using the Danube River as an example. Environ Sci Policy 64:141–154. https://doi.org/10.1016/j.envsci.2016.06.015
12. Liepolt R (ed) (1965–1967) Limnologie der Donau. E. Schweizerbart'sche Verlagsbuchhandlung, Stuttgart
13. Popp N (1985) Fluviul Dunărea. Editura Științifică și Enciclopedică, Bucharest
14. Gâştescu P (1998) Danube River: hydrology and geography. In: Herschy RW, Fairbridge RW (eds) Encyclopedia of hydrology and water resources. Kluwer Academic Publishers, Dordrecht, Boston, London, pp 159–161
15. Sommerwerk N, Hein T, Schneider-Jacoby M, Baumgartner C, Ostojic A, Siber R, Bloesch J, Paunovic M, Tockner K (2009) The Danube River basin. In: Tockner K, Robinson CT, Uehlinger U (eds) Rivers of Europe. Academic Press, Oxford, pp 59–112
16. Miklós D (2010) History and results of the hydrological co-operation of the countries sharing the Danube Catchment (1971–2008). In: Brilly M (ed) Hydrological processes of the Danube

River basin. Perspectives from the Danubian countries. Springer, Dordrecht, pp 1–23. https://doi.org/10.1007/978-90-481-3423-6_1
17. Prohaska S, Brilly M, Kryžanowski A (2020) Cooperation of hydrologists from the Danube River Basin. Hydrol Earth Syst Sci Discuss. https://doi.org/10.5194/hess-2020-66
18. RZD (Regionale Zusammenarbeit der Donauländer) (1986) Die Donau und ihr Einzugsgebiet – Eine hydrologische Monographie. Bayerische Landesanstalt für Wasserwirtcshaft, München
19. Brilly M (ed) (2010) Hydrological processes of the Danube River basin. Perspectives from the Danubian countries. Springer, Dordrecht
20. Liska I (ed) (2015) The Danube River basin. Hdb Env Chem, vol. 39. Springer-Verlag, Berlin Heidelberg
21. Pekárová P, Miklánek P (eds) (2019) Flood regime of rivers in the Danube River basin. Follow-up volume IX of the Regional Co-operation of the Danube Countries in IHP UNESCO. IH SAS, Bratislava. https://doi.org/10.31577/2019.9788089139460
22. ICPDR (1998) Convention on cooperation for the protection and sustainable use of the Danube River (Danube River Protection Convention). https://www.icpdr.org/flowpaper/app/services/view.php?doc=DRPC%20English%20ver.pdf&format=pdf&page={page}&subfolder=default/files/. Accessed 28 Jun 2021
23. ICPDR (2021) International commission for the protection of the Danube River. About us. https://www.icpdr.org/main/icpdr/about-us. Accessed 28 Jun 2021
24. ICPDR (2015) The Danube River Basin District Management Plan, Part A Basin-wide overview. http://www.icpdr.org/main/sites/default/files/nodes/documents/drbmp-update2015.pdf. Accessed 28 Jun 2021
25. ICPDR (2015) Flood risk management plan for the Danube River Basin District. https://www.icpdr.org/flowpaper/app/services/view.php?doc=1stdfrmp-final.pdf&format=pdf&page={page}&subfolder=default/files/nodes/documents/. Accessed 26 Jun 2021
26. ICPDR (2021) Danube FloodRisk Project. https://www.icpdr.org/main/activities-projects/danube-floodrisk-project. Accessed 26 Jun 2021
27. ICPDR (2012) Danube Atlas. Hazard and Risk maps. http://www.icpdr.org/icpdr/static/dfr-maps/map_embed.html. Accessed 27 Jun 2021
28. Bondar C, Iordache G (2017) Sediment transport on the Danube River in the Romanian border area—characteristics. Rev Roum Géogr/Rom Journ Geogr 61(1):3–17
29. Zăvoianu I, Bugă D, Pișota I (coords) (2005) Dunărea, lunca, delta și complexul lacustru Razim-Sinoie. In: Badea et al. (eds) Geografia României, vol. V., Câmpia Română, Dunărea, Podișul Dobrogei, Litoralul românesc al Mării Negre și Platforma Continentală. Editura Academiei Române, pp 488–643
30. Stagl J, Hattermann FF (2015) Impacts of climate change on the hydrological regime of the Danube River and its tributaries using an ensemble of climate scenarios. Water 7:6139–6172. https://doi.org/10.3390/w7116139
31. Miklós D, Neppel F (2010) Palaeogeography of the Danube and its catchment. In: Brilly M (ed) Hydrological processes of the Danube River Basin. Springer, Dordrecht, pp 79–124. https://doi.org/10.1007/978-90-481-3423-6_3
32. WWF International (2011) Lower Danube Green Corridor agreement. http://danube.panda.org/wwf/web/search/details.jsp?pid=41. Accessed 5 Jul 2021
33. Constantinescu Ş, Achim D, Rus I, Giosan L (2015) Embanking the lower Danube: from natural to engineered floodplains and back. In: Hudson P, Middelkoop H (eds) Geomorphic approaches to integrated floodplain management of lowland fluvial systems in North America and Europe. Springer, New York, pp 265–288. https://doi.org/10.1007/978-1-4939-2380-9_11
34. WWF (2012) The Lower Danube Green Corridor, Fact sheet 2012. https://wwfint.awsassets.panda.org/downloads/wwf_factsheet_ldgc.pdf. Accessed 5 Jul 2021
35. Guvernul României (2006) Hotarare nr. 1.208 din 6 septembrie 2006 pentru aprobarea Programului de redimensionare ecologica si economica in sectorul romanesc al Luncii Dunarii si a finantarii acestuia. Monitorul oficial 786 din 18 septembrie 2006. http://www.monitoruljuridic.ro/act/hotarare-nr-1-208-din-6-septembrie-2006-pentru-aprobarea-programului-de-redimensionare-ecologica-si-economica-in-sectorul-romanesc-al-luncii-dunarii-si-a-finantarii-acestuia-emitent-75173.html. Accessed 5 Jul 2021

36. Interreg (2021) Danube floodplain. http://www.interreg-danube.eu/approved-projects/danube-floodplain. Accessed 5 Jul 2021
37. Zaharia L, Diaconu D, Sârbu I (2003) Activități practice de hidrologie în arealul Porțile de Fier. Editura Credis, Bucharest
38. Hidroelectrica (2021) SHEN Portile de Fier I si II. https://www.hidroelectrica.ro/article/f5e9cc4f-2c02-3701-0316-76d5dc27d358. Accessed 10 Jul 2021
39. Comoglio C (2011) Scoping mission at Iron Gates I and II dams (Romania and Serbia). Preliminary assessment of the feasibility for providing free passage to migratory fish species. Report, FAO. http://awsassets.panda.org/downloads/34_2011_fao_mission_report_fish_migration_iron_gates_int.pdf. Accessed 10 Jul 2021
40. Zaharia L, Ioana-Toroimac G (2013) Romanian Danube River management: impacts and perspectives. In: Arnaud-Fassetta G, Masson E, Reynard E (eds) European continental hydrosystems under changing water policy. Verlag Dr, Friedrich Pfeil, München, pp 159–170
41. Popp G (1996) România. Geografie hidroenergetică. Editura Presa Universitară Clujeană, Cluj-Napoca
42. Goda L (2010) The Fords of the Danube. In Brilly M (ed) Hydrological processes of the Danube River basin. Springer, Dordrecht, pp 343–366. https://doi.org/10.1007/978-90-481-3423-6_11
43. Stoyanova T, Traykov I (2011) Assessment of the ecological status of Ogosta River, North-western Bulgaria, based on the Macrozoobenthos and the general physical and chemical quality elements. Acta zool bulg, Suppl 7:173–178
44. Seymenov K (2020) Spatial and temporal aspects of high streamflow periods within the Danube drainage basin in Bulgaria. Eur J Geogr 11(1):006–022. https://doi.org/10.48088/ejg.k.sey.11.1.6.22
45. AQUAPRPOIECT (1992) Atlasul cadastrului apelor din România, Partea I, Ministerul Mediului, Bucharest
46. Posea G, Iordan I, Zăvoianu I (coords) (2005) Câmpia Română. In: Badea et al. (eds) Geografia României, vol. V., Câmpia Română, Dunărea, Podișul Dobrogei, Litoralul românesc al Mării Negre și Platforma Continentală. Editura Academiei Române, pp 27–487
47. Kovács P (2010) Characterization of the runoff regime and its stability in the Danube catchment. In: Brilly M (ed) Hydrological processes of the Danube River Basin. Springer, Dordrecht, pp 143–174. https://doi.org/10.1007/978-90-481-3423-6_5
48. Comitetul de Stat al Apelor (1967) Dunărea între Baziaş şi Ceatal Izmail. Monografie hidrologică. ISCH, Bucharest
49. Institutul de Geologie și Geografie al Academiei R.S.R. (1969) Geografia Văii Dunării Româneşti. Editura Academiei, Bucharest
50. Ujvari I (1972) Geografia apelor României. Editura Ştiinţifică, Bucharest
51. Pișota I (1983) Dunărea. In: Badea et al. (eds) Geografia Romaniei, vol I—Geografia fizică. Editura Academiei, Bucharest, Romania, pp 346–351
52. Academia Română, Institutul de Geografie (2005) România. Spațiu, societate mediu. Editura Academiei Române, Bucharest
53. Pătru I, Zaharia L, Oprea R (2006) Geografia Fizică a României. Climă, ape, vegetație, soluri. Editura Universitară, Bucharest
54. Rîmbu N, Boroneanţ C, Buţă C, Dima M (2002) Decadal variability of the Danube River flow in the lower basin and its relation with the North Atlantic Oscillation. Int J Climatol 22:1169–1179. https://doi.org/10.1002/joc.788
55. Levashova EA, Mikhailov VN, Mikhailova MV, Morozov VN (2004) Natural and human-induced variations in water and sediment runoff in the Danube River mouth. Water Resour 31:235–246. https://doi.org/10.1023/B:WARE.0000028693.49051.d9
56. Bondar C (2008) Bilanţul morfologic pe Dunăre in sistemul lacurilor de retenţie Porţile de Fier 1 şi 2 pentru intervalele 1971–2005, respectiv 1985–2005. Geo-Eco-Marina 14(1):35–40
57. Bondar C, Iordache G (2012) Expanding in the past (up to 1840) of daily water levels and water discharge of the Danube at gauging stations and sections on the Romanian border sector. INHGA—International Scientific Conference, Bucharest, 8–10 October 2012

58. Gâştescu P, Ţuchiu E (2012) The Danube River in the lower sector in two hydrological hypostases—high and low waters. Riscuri şi catastrofe 10(1):1–19
59. Gâştescu P, Ţuchiu E (2012) The Danube River in the pontic sector—hydrological regime. In: Gâştescu P, Lewis W Jr, Breţcan P (eds) Water resources and wetlands. Conference Proceedings, 14–16 September 2012, Tulcea
60. Borcia C, Teodor S, Radulescu C, Petrea C, Blendea V (2012) Aspects of the manifestation of hazards on the Danube. INHGA—International Scientific Conference, Bucharest, 8–10 October 2012
61. Bondar C, Borcia C (2013) Morphological processes on the Romanian Danube riverbed. INHGA—Scientific Conference, Bucharest, 23–26 September 2013
62. Borcia C, Teodor S, Rădulescu C, Petrea C, Ciucă R, Blendea V (2014) Aspects of variability of hydrological parameters as a result of human impact and climate change in the lower Danube. INHGA—Scientific Conference, Bucharest, 10–11 November 2014
63. Hristova N (2014) Seasonal river flow variability of the middle and lower Danube and its tributaries. Forum geografic. Studii și cercetări de geografie și protecția mediului XIII(2):164–177. https://doi.org/10.5775/fg.2067-4635.2014.063.d
64. Borcia C, Teodor S, Rădulescu C, Ciucă R, Blendea V, Petrea C (2016) Issues regarding the manifestation of low water hydrological regime for the Danube. INHGA—International Scientific Conference, Bucharest, 11–12 November 2016
65. Bondar C, Borcia C, Rădulescu C, Blendea V, Ciucă R, Petrea C (2017) Extremely hydrological data on the Danube. INHGA—International Scientific Conference, Bucharest, 14–15 November 2017
66. Marin AF (2017) Hydrological and morphological changes of the Danube River between Giurgiu and Oltenita cities (Romania). Cinq Continents 7(15):99c111
67. Gâștescu P, Țuchiu E (2020) Hydrographycal, hydrologycal and qualitative characteristics water with references on the pontic sector. Riscuri și catastrofe XX, 27(2):9–29
68. Pekarova P, Meszaros J, Miklanek P, Pekar J, Prohaska S, Ilić A (2021) Long-term runoff variability analysis of rivers in the danube basin. Acta hortic regiotecturae, special issue Nitra, 37–44. https://doi.org/10.2478/ahr-2021-0008
69. Țuchiu E (2020) Starea corpurilor de apă de pe cursul inferior al Dunării între Baziaş şi Isaccea. Editura Transversal, Târgoviște
70. Salmi T, Määttä A, Anttila P, Ruoho-Airola T, Amnell T (2002) Detecting trends of annual values of atmospheric pollutants by the Mann-Kendall test and Sen's slope estimates—the Excel template application MAKESENS. Publications on Air Quality 31, Finnish Meteorological Institute, Helsinki, Finland
71. Dai A, Qian T, Trenberth KE, Milliman JD (2009) Changes in continental freshwater discharge from 1948 to 2004. J Climate 22(10):2773–2792. https://doi.org/10.1175/2008JCLI2592.1
72. Belz JU, Goda L, Buzás Z, Domokos M, Engel H, Weber J (2004) Flow regime of River Danube and its Catchment (Follow-up volume no. VIII/2 to the Danube Monograph). Koblenz, Baja
73. Zaharia L (2010) The Iron Gates reservoir—aspects concerning hydrological characteristics and water quality, Lakes, reservoirs and ponds. Romanian Journal of Limnology 4(1–2):52–69
74. Prohaska S, Ilić A (2010) Coincidence of flood flow of the Danube River and its tributaries. In: Brilly M (ed) Hydrological processes of the Danube River basin—perspectives from the Danubian countries. Springer, Dordrecht, pp 175–226. https://doi.org/10.1007/978-90-481-3423-6_6
75. Stănescu VA, Ungureanu V, Mătreaţă M (2004) Regional analysis of annual peak discharges in the Danube catchment (follow-up volume no. VII to the Danube Monograph). INMH, Bucharest
76. ICPDR (2013) Strategy on adaptation to climate change. http://www.icpdr.org/flowpaper/app/services/view.php?doc=icpdr_climate-adaptation-strategy.pdf&format=pdf&page={page}&subfolder=default/files/nodes/documents/. Accessed 20 Dec 2021
77. ICPDR (2019) Climate change adaptation strategy. https://www.icpdr.org/flowpaper/app/services/view.php?doc=icpdr_climatechangeadaptationstrategy_2.pdf&format=pdf&page={page}&subfolder=default/files/nodes/documents/. Accessed 20 Dec 2021

Dynamics of Islands and Danube River Channel Along Vedea-Călăraşi Sector (1856–2019): Hydrogeomorphological Approach

Florina Grecu, Gabriela Ioana-Toroimac, Gabriela Osaci-Costache, Liliana Zaharia, Octavian Cocoş, Kamel Hachemi, and Lidia Sălăjan

Abstract The objective of this chapter is to analyze the Vedea-Călărași reach (about 135 km long) of the Lower Danube River, along the border between Romania and Bulgaria, from a hydrogeomorphological perspective. The research question is: what has been the dynamics of the islands in the last century and a half, since the end of the Little Ice Age, after the numerous engineering works that the Danube River has suffered? For this purpose, we used a variety of historical cartographic documents and recent satellite imagery that were georeferenced and overlapped. Further, the contours of the islands were digitized. We found that the Lower Danube River preserved its islands on approximately one third of the total area of its channel both in 1856 and 2019. The total area of the islands covered by vegetation has increased by almost 29.4% (11.1 km^2), while the area devoid of vegetation, including the sandbars, has decreased by 51% (5.1 km^2) in comparison with the situation in 1856. This finding indicates either a less important morphogenic character of present-day floods or the

---

F. Grecu · G. Ioana-Toroimac (✉) · G. Osaci-Costache · L. Zaharia · O. Cocoș
Faculty of Geography, University of Bucharest, 1 Nicolae Bălcescu, 010041 Bucharest, Romania
e-mail: gabriela.toroimac@geo.unibuc.ro

F. Grecu
e-mail: grecu@geo.unibuc.ro

G. Osaci-Costache
e-mail: gabriela.osacicostache@geo.unibuc.ro

L. Zaharia
e-mail: liliana.zaharia@geo.unibuc.ro

O. Cocoș
e-mail: octavian.cocos@geo.unibuc.ro

K. Hachemi
Laboratory MEDIATIONS, Sorbonne University, Paris-Est University,
Maison de la Recherche, 28 rue Serpente, 75006 Paris, France
e-mail: kamel.hachemi@u-pec.fr

L. Sălăjan
National Institute of Hydrology and Water Management,
97E Șoseaua București-Ploiești, 01366 Bucharest, Romania
e-mail: lidia.salajan@yahoo.com

A. Negm et al. (eds.), *The Lower Danube River*, Earth and Environmental Sciences Library, https://doi.org/10.1007/978-3-031-03865-5_2

fact that floods have been less frequent than in 1856 and before. A more detailed analysis of a 10 km-long river reach (Albina-Paraschiva-Slaviavin-Garvan islands) showed that between 1856 and 1953 the Danube River channel and its islands suffered numerous alterations (i.e. the islands either joined together or merged with the left bank and consequently underwent variations in area). Later on, between 1953 and 2019, the hydrosystem became relatively stable. By surveying the Ciocănești-Devnya islands, we found that their areas remained relatively stable (covered by vegetation) pre- versus post- historic flood of April 2006. In perspective, the use of more indicators and recent satellite imagery at good temporal and spatial resolution could help us to better understand the evolution of the islands situated along the Lower Danube River.

Keywords Fluvial islands · Diachronic analysis · Historical maps · Satellite imagery · Floods

1 Introduction

The concept of hydrogeomorphology has been defined in the early 1970s [1]. It was later used especially by the French school of geomorphology to name a methodology for delimiting the flood plain [2–5]. The concept of hydrogeomorphology refers to the relation between the water loaded or not with sediment and the fluvial forms and processes, exacerbated during floods [6]. To point out the transfer of surface water and materials through landscapes, the concept of hydrogeomorphic connectivity emerged. It is successfully applied in the field of catchment hydrology and fluvial geomorphology, but also in other fields, such as social sciences [7].

Fluvial islands are emerged landforms of river channels which generally result from sediment accumulation and are continuously adjusted by the water flow through fluvial processes of erosion/accumulation; they may be covered by vegetation and may persist in the same position for a long time [6]. A fluvial island is defined as a land mass within a river channel that is separated from the floodplain by water on all sides, exhibits some stability [8], and remains exposed during bankfull flow (whereas a bar may be submerged) [9]. Islands are the result of the interplay between flows, sediments and woody vegetation [10]. Vegetation is generally a good indicator of stability [11]. Fluvial islands are important in both hydrologic and biotic capacities, and can therefore be indicators of the general health and energy of the hydrosystem [12]. Islands area may vary depending on the water level especially during floods as it was previously demonstrated [13, 14].

The large rivers of Europe lost their islands over the period of major human interference [10]. The Rhone in France, the Rhine in Germany, the Waal in Netherlands, the Danube in Austria, and so on, are examples of European rivers characterized before 1900 by multiple channel reaches with wooded islands [10]. Along the Danube River in Austria, during the last two centuries, besides other adjustments, the (gravel or sand) bars and vegetated islands decreased by 94 and 97%, respectively [15].

According to the same authors, embankment and channelization altered these rivers during the eighteenth and nineteenth centuries; additionally, lock-and-dam systems were introduced to improve the navigation along the large rivers. These changes of anthropic origin are set against the background of the Little Ice Age that corresponds to baseline period of cartographic and documentary records [10].

The Lower Danube River also suffered changes due to the large-scale embankments, mostly accomplished at the middle of the twentieth century, as well as to the construction of the dams Iron Gates I (1964–1972) and Iron Gates II (1977–1984) [16]. Additionally, the floodplain and the islands were afforested with Euro-American poplar trees [17]. A synthesis of the historical human pressures on the Lower Danube River and its floodplain is presented by Strat et al. (Chapter "Anthropogenic Changes and Biodiversity Protection and Conservation Along the Lower Danube River Valley" in this book). Concerning the dynamics of the islands, Constantinescu et al. [18] established that their number, density and area generally decreased during the last century. Other authors [19–24] showed that the major islands reacted differently (increased or decreased) to the changes of the last century and a half. Marin and Armaș [25] and later Marin [26] argued that the islands generally became more elongated and migrated downstream. Yet, the occurrence of fluvial islands, their morphometry, and their response to human impact vary considerably along the Danube, as the river crosses various physiographic and geomorphologic units. Therefore, the study of various reaches of the Lower Danube River can contribute to the better understanding of the river's behavior. Goda [27] concluded that systematic measurements should be carried out on selected island groups of the Lower Danube and their related fords in order to improve the knowledge about these special processes of river morphology.

The objective of this chapter is to analyze the Vedea-Călărași reach (about 135 km long) of the Lower Danube River, along the border between Romania and Bulgaria (Fig. 1), where the islands belong to one of the two countries. More precisely, we aimed to answer to the question: what has been the dynamics of the islands in the last century and a half, since the end of the Little Ice Age, taking into account the numerous engineering works that the Danube River has suffered? The results of our investigation provide new information that updates and complements the previous ones on the hydrogeomorphology of the Lower Danube River during the last century and a half.

To simplify things, in this study we use the term "island" both for the fluvial islands covered by vegetation and for the barren sandbars. The vegetation criterion [28–31] is mentioned on every occasion when it is important to understand thoroughly the river processes.

2 Study Area

The reach studied in this chapter extends over 135 km in length (12.5% of Lower Danube's length), between the Vedea commune and Călărași city located both on the Romanian side of the Danube River (Fig. 1).

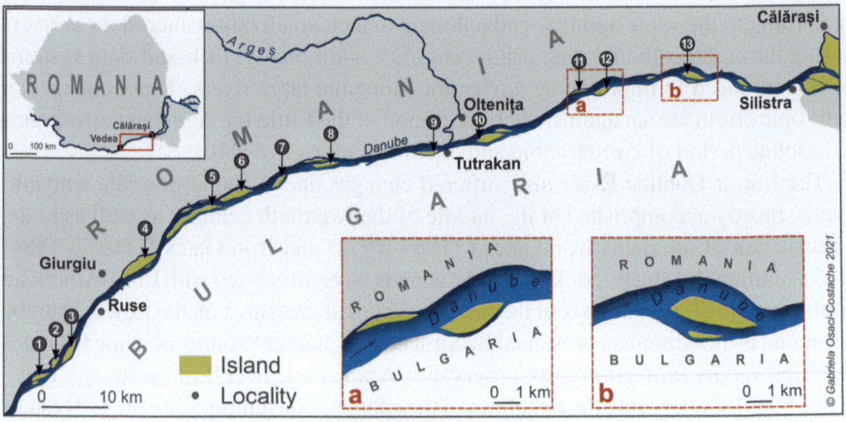

Fig. 1 Location of the study area—Vedea-Călărași reach along the Lower Danube River: **a** case study Albina-Paraschiva-Slaviavin-Garvan islands; **b** case study Ciocănești-Devnya islands (the number and names of the islands are listed in Table 3)

On this section the average multiannual discharge of the Danube River is around 6,000 – 6,100 m^3/s, but the extreme values of the streamflow varied widely between approximately 1,010 m^3/s (at Zimnicea located about 50 km upstream this section) and 17,303 m^3/s, at Oltenița and Chiciu-Călărași (Table 1) [32]. The highest discharges were reached during the historical flood of April 2006, while the lowest flows rates were recorded in January 1858 (at Zimnicea, Giurgiu and Oltenița) and in December 2020 at Chiciu-Călărași [32]. The hydrological regime in the study area is characterized by high waters in April–May and low waters in September–October [33]. The multiannual average of suspended sediment load increase between the extremities of the study reach from 1,102 kg/s (at Zimnicea) to 1,379 kg/s (at Chiciu-Călărași), mainly as result of the tributaries inputs, of which the most important are Argeș River (in Romania), and Yantra River (in Bulgaria), which annually

Table 1 Multiannual values of liquid and solid flow of the Danube River in the studied reach (1840–2012)

Gauging station	Qav (m^3/s)	Qmax/year (m^3/s)	Qmin/year (m^3/s)	Rav (kg/s)	Rmax (kg/s)	Rmin (kg/s)	Rgav (kg/s)
Zimnicea	5,991	16,919/2006	1,010/1858	1,102	2,631	150	11.7
Giurgiu	6,011	17,000/2006	1,030/1858	1,211	2,989	141	11.9
Oltenița	6,077	17,303/2006	1,060/1858	1,291	2,962	170	11.9
Chiciu-Călărași	6,107	17,303/2006	1,041/1920	1,379	3,624	172	11.0

Qav = multiannual average discharge; Qmax = multiannual maximum discharge; Qmin = multiannual minimum discharge; Rav = multiannual average of suspended sediment load; Rmax = multiannual maximum of suspended sediment load; Rmin = multiannual minimum of suspended sediment load; Rgav = multiannual average bedload (dislodged coarse sediment). Source of data: [32]

Table 3 Dynamics of the island areas and their land cover between 1856 and 2019

Number in Fig. 1	Year for the name of island[a]	Island	1856		2019	
			Area (km^2)	Land use (%)	Area (km^2)	Land use (%)
1	1856	Dinului	1.8	g—34.5 s—65.5	2.5	f—86.5 s—13.5
2	1856	Kamedin	2.5	g—48.1 s—51.9	1.0	f—95.8 s—4.2
3	1856	Gole	2.0	g—0.1 s—99.9	2.1	f—100
	1979	Liuleak				
4	1856	Mokan	1.4	f—79 s—21	6.5	f—100
	1979	Mocanu				
5	1856	Mik	2.4	f—87.5 s—12.5	5.3	f—100
	1979	Aleko				
6	1856	Lunga	3.0	f—90.1 s—9.9	2.9	f—100
	1979	Lung				
7	1979	Saceanlak	0.2	f—100	1.2	f—100
8	1856	Beker	3.3	f—88.3 s—11.7	1.5	f—100
	1979	Goliam Brașlen				
9	1979	Radețki	–	–	1.2	f—100
10	1979	Kosui	0.5	f—100	3.1	f—100
11	1979	Spanțov (rest of Albina)	–	–	0.4	f—100
12	1856	Garvan	1.1	f—100	1.3	f—100
	1979	Cibalaka Gargalaka				
13	1979	Ciocănești	0.3	f—100	2.4	f—100
Total (including other smaller islands)	–	–	37.7	f, g	48.8	f, g
			10.0	S	4.9	S
			47.7	All	53.7	All

f = forest, g = grassland, s = bare sediment
[a]According to documents in Table 2

that the most significant increases occurred especially around the oldest islands, which is probably a pattern of island evolution along the Vedea-Călărași reach.

At the same time, the land use suffered a number of changes. Bare sediment areas were larger in 1856 than in 2019 on several islands, as shown in Table 3. In 1856, some of them were covered by grasslands (e.g. Dinului, Kamedin, Gole). At present, the forest has become dominant on all the investigated islands.

Based on these facts, we can draw the conclusion that the Lower Danube River was a less dynamic hydrosystem in 2019 than in 1856. At the end of the Little Ice Age, in the nineteenth century, the higher than usual winter temperatures and the spring precipitations probably caused the high incidence of floods [50]. Floods twice the mean annual discharge occurred on the Lower Danube in June 1839, May 1845, and May 1853 [43]. The flood of 1845 is the historic flood at Călărași, on the Danube [32]. This finding indicates either a less important morphogenic character of present-day floods or the occurrence of less frequent floods when compared to the situation of 1856. This may also suggests that sediment load was higher a century ago. Yet, this evolution of the islands is not entirely natural—afforestation and other engineering works [51] may complicate the understanding of the driving factors in this situation.

Additionally these quantitative results may be influenced by the scale/resolution of the used documents. We certainly detected more emerged areas in 2019 than in 1856. For further studies, it is advisable to set a threshold of the minimum area that can be seen on the *Fligely Map* to be also used on Google satellite imagery.

3.2 Case Study of Albina-Paraschiva-Slaviavin-Garvan Islands

To better understand the dynamics of the island areas and find other patterns of evolution, we focused on the Danube reach between the communes of Chiselet (Romania) and Popina (Bulgaria). The reach (about 10 km long) was delimited based on the maximum extension of the islands (or their „print" in the case of the islands that disappeared or changed their position and morphometric features, as shown in [47]). Upstream, the limit passes west of Spanțov Island according to its form in 1953, while downstream it is situated east of Paraschiva Island according to its form in 1898 (including the surrounding channel, in order to compute the braiding index).

Along this reach, the left bank has lower altitudes (13–15.6 m) than the right bank (generally 15–20 m, but sometimes even up to 118.7 m). The entire left bank is protected by a levee, while the right one is embanked, but only where it is very low.

In 1856, there were four big islands (Albina, Paraschiva, Cibukluk and Garvana) and five small ones (located in the north-eastern part of Albina Island). Based on their diachronic dynamics, the islands can be grouped in three clusters: the Albina islands group (Albina and other small islands located in its surroundings), the Paraschiva islands group and the Slaviavin-Garvan islands group (Slaviavin, Garvan and other small islands) (Table 4 and Fig. 3). These islands mostly disappeared and only small parts of them can still be seen. The Danube River also created small sandbars. No island was formed by avulsion.

The total area of the islands decreased from 8 km^2 in 1856 to 2.4 km^2 in 2019, that is by 70.3% in 164 years (Fig. 4). All the three groups of islands lost a part of their area (Fig. 5). Albina and Paraschiva islands disappeared, but Garvan continued to exist. The decrease concerned both the vegetated area and the bars. Yet the proportion

Table 4 Spatial dynamics of Albina-Paraschiva-Slaviavin-Garvan islands (the toponyms were extracted from the cartographic documents listed in Table 2)

Group	Component islands of groups					
	1856	1898	1953	1979	1987	2019
Albina islands group	Albina	Albina	Disapeared	Disapeared	Disapeared	Disapeared
	Five small islands merging with Albina in 1898		No name—west (rest of Albina)	No name—west (rest of Albina)	Spanțov (rest of Albina)	Spanțov (rest of Albina)
			No name—est (rest of Albina)	No name—est (rest of Albina)		
	–	–	–	Small bank, no name	Small bank, no name, south of Spanțov	disapeared
Paraschiva islands group	Paraskiva	Paraschiva	Disapeared	Disapeared	Sand bar	Sand bar
Slaviavin-Garvan islands group	Cibukluk	Ciubucluc	Slaveavin	Cibaklaka-Gargalaka	Slaviavin	Garvan
	Garvana	Gargalâc	Garvan		Garvan	
	–	–	–	Small bar upstream of Garvan		
	–	Small bar upstream of Ciubucluc	Disapeared	Disapeared	Disapeared	Disapeared
		–	–	–	–	Small bar upstream of Garvan
Total number of islands	9	5	3	5	5	4

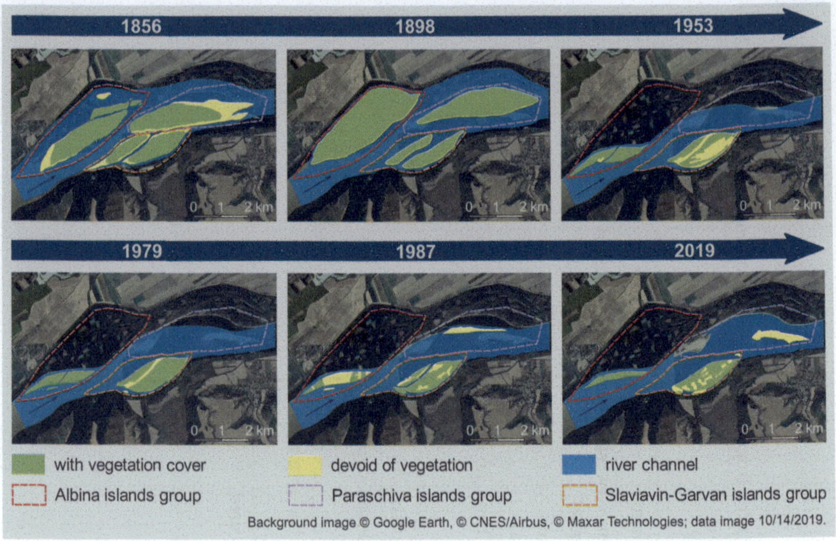

Fig. 3 Albina-Paraschiva-Slaviavin-Garvan case study—river channel and island dynamics (1856–2019)

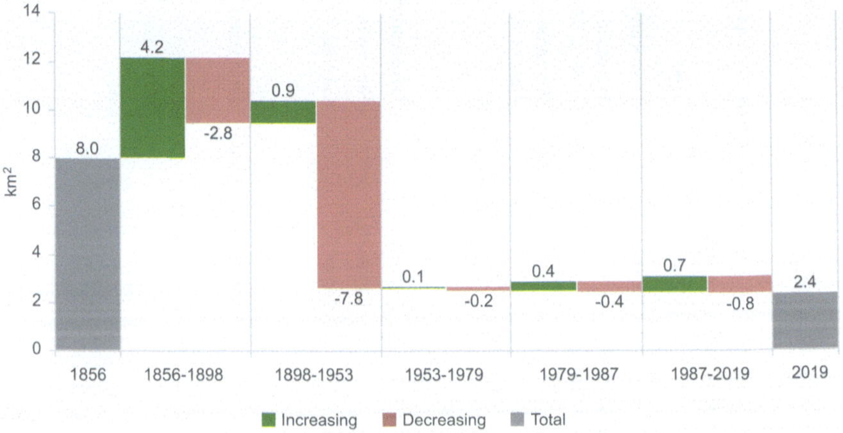

Fig. 4 Albina-Paraschiva-Slaviavin-Garvan case study—variations of the total area of the islands (1856–2019)

of vegetation versus bars generally remained the same (Fig. 6). Figures 4, 5 and 6 show the dynamics of the island area per period.

Between 1856 and 1898, the total area of the islands increased by 18.4%, from 8 to 9.5 km^2. The highest increase concerned the Albina Island, which extended towards the left bank and included the neighboring small islands, thus growing from

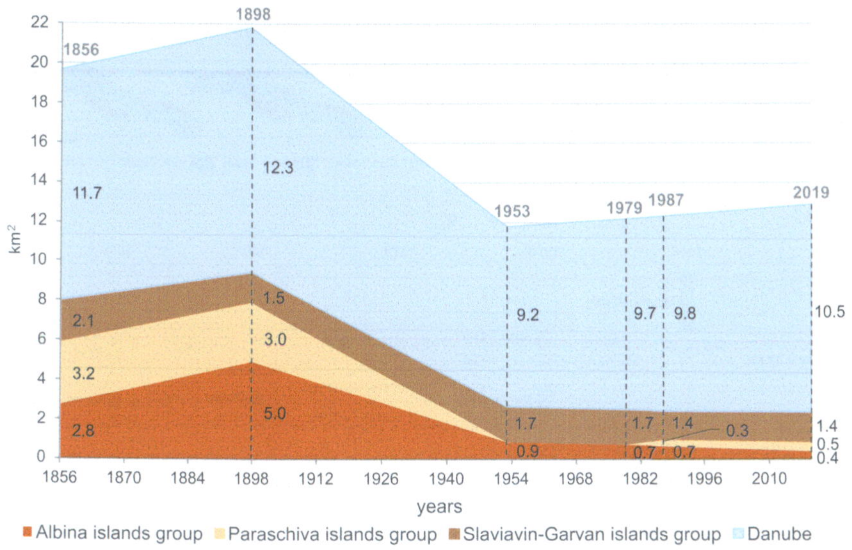

Fig. 5 Albina-Paraschiva-Slaviavin-Garvan case study—variations of islands area per group of islands (1856–2019)

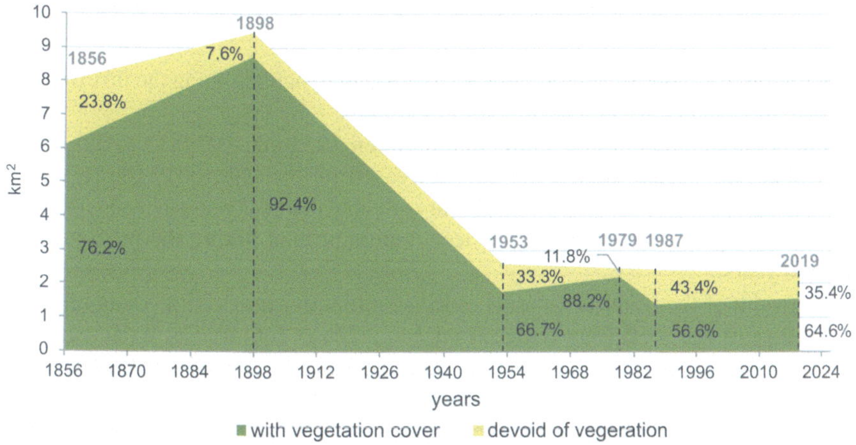

Fig. 6 Albina-Paraschiva-Slaviavin-Garvan case study—variations of the land cover of the islands: vegetation versus bare sediment (1856–2019)

2.2 km^2 in 1856 to almost 5 km^2 in 1898. During this time interval, the islands grew in size to the detriment of the water (Fig. 7).

During the next 55 years (1898–1953), the total area of the islands decreased by 72.4% (or 1.3% per year), reaching 2.6 km^2 in 1953. The decrease was mostly due to the merging of the Albina and Paraschiva islands group with the left bank. This kind

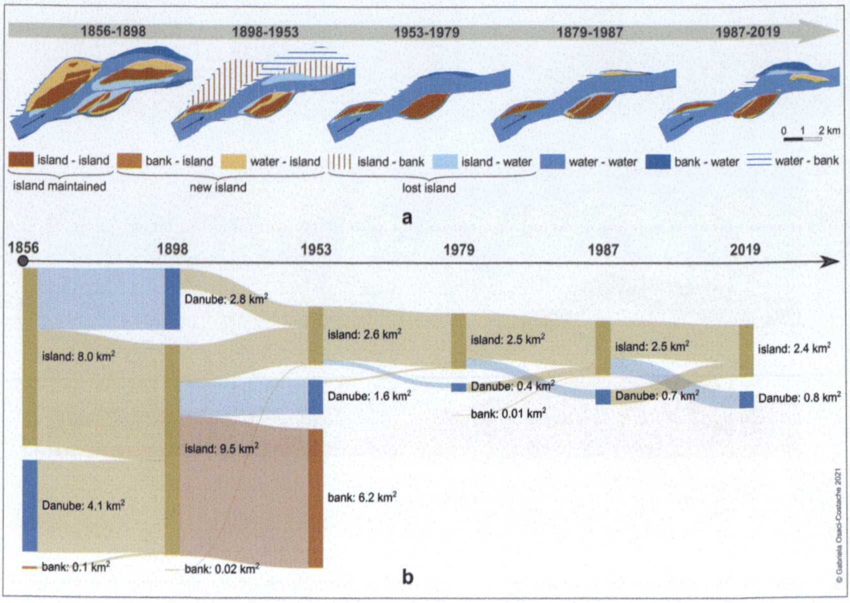

Fig. 7 Albina-Paraschiva-Slaviavin-Garvan case study—islands dynamics by conversion per phase (binary approach): **a** schemes of conversions (1856–2019); **b** areas in km^2

of evolution may represent another pattern of island dynamics along the investigated reach of the Danube River.

The decrease continued at a low rate between 1953 and 1979 (from 2.6 to 2.4 km^2), then kept steady until 1987 and decreased further until 2019 (down to 2.3 km^2). Since the mid of the twentieth century, all conversions (water-islands, islands-water, water-bank, bank-water) were less important than during the previous one hundred years.

The width of the Danube River recorded a similar dynamics (Fig. 8): an increase between 1856 and 1898 (on average from 2 to 2.2 km), followed by a decrease (on average from 1.2 to 1.3 km) due mostly to the merging of the islands with the left bank, a process that continued until 2019. As far as the area is concerned (Fig. 5), the period 1856–1898 was dominated by erosion, accumulation was active between 1898 and 1953, while between 1953 and 2019 the erosion intensified again. A demonstrative example is shown in Fig. 9. The left bank of the Danube is affected by erosion and it retreated about 0.7 km between 1953 and 2019; the eroded part corresponds to the ancient Paraschiva Island.

The braiding index decreased from 3.2 in 1856 to 2.2 in 2019 (Fig. 9). The decrease was quasi-continuous with the exception of a small increase in 1987. This evolution generally corresponds to the number of islands which were in existence during the study period: 9 in 1856 and 4 in 2019.

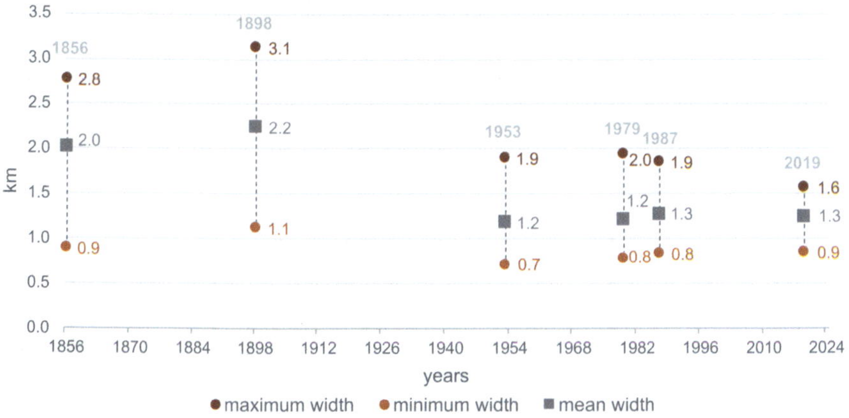

Fig. 8 Albina-Paraschiva-Slaviavin-Garvan case study—variations of the Danube main channel width (1856–2019) based on cross section profiles every 20 m

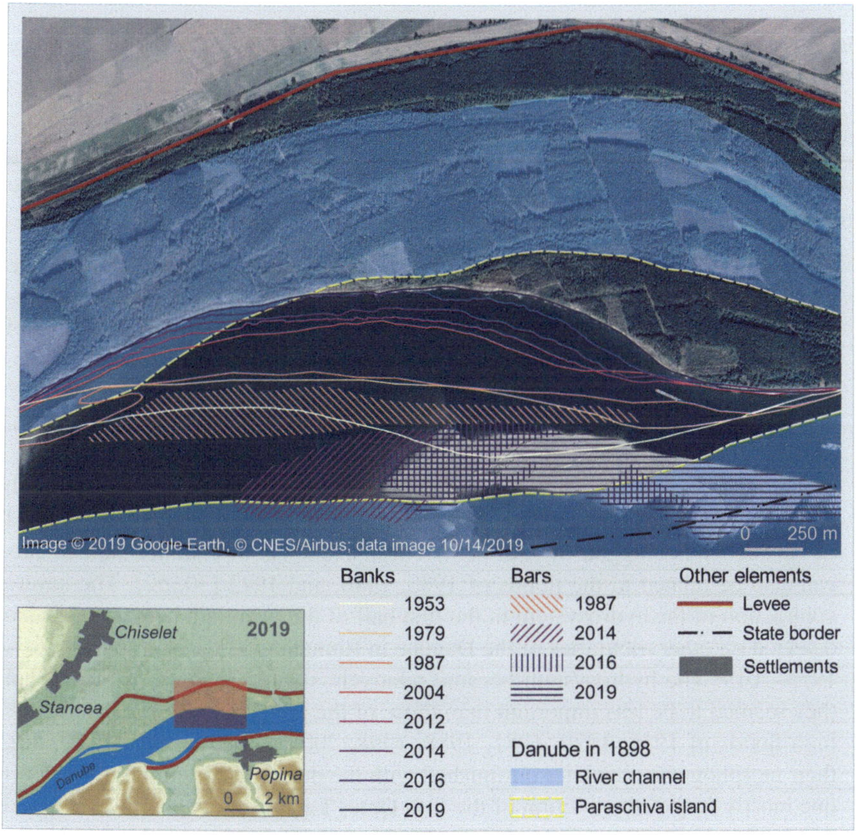

Fig. 9 Left bank erosion corresponding to the ancient Paraschiva Island (1953–2019)

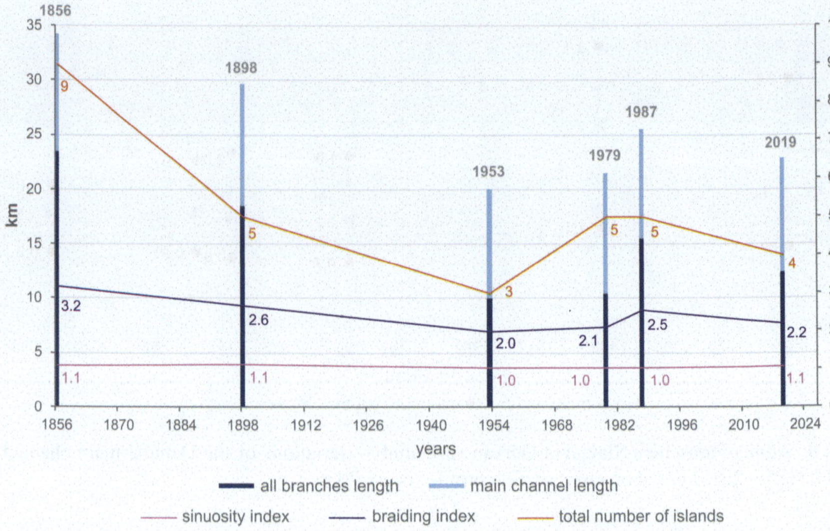

Fig. 10 Albina-Paraschiva-Slaviavin-Garvan case study—variations of the Danube River braiding index (1856–2019)

At this time scale, it is difficult to retrace the driving forces responsible for this evolution of the Danube River channel and its islands, yet a few hypotheses can be proposed:

- 1856–1898. The number of islands decreased, their area expanded while the channel width shrank. The islands probably merged due to the available sediment load from bank erosion. This could be due to the heavy floods of 1895 and 1897 [30], but also to the significant floods of March 1871, April and July 1876, January–February and May 1879, and April 1888 [43]. The flood of July 1897 was the historic flood downstream, recorded just before the river enters the delta [32].
- 1898–1953. Numerous alterations occurred in the first part of the twentieth century. The islands merged with the bank, which indicates that the small channel in between was clogged by sediment. These examples of channel adjustments can also be related to the floods of 1930, 1940, and 1942 [30, 52]. The severe contraction of the hydrosystem in the first half of the twentieth century was also detected on other tributaries of the Danube in Romania [53].
- 1953–2019. The hydrosystem became relatively stable. Changes occurred, but they seemed to be less important than those of the previous century. Despite the high floods of 1955, 1970, 1981, 1988, 1998, 2005, 2006, and 2010 [30, 52], their morphogenic character diminished with the decrease of sediment transport due mostly to the construction of the Iron Gates I and II dams and the disrupted connectivity with the floodplain [18, 30, 31, 54]. The effect of the Iron Gates dams can be seen in Fig. 6, at the time when bare sediment had a low percentage in the total area of the islands. Meanwhile, the river channel probably incised [37].

Other analyses and interpretations of the Danube River channel dynamics and driving factors along the Vedea-Călărași reach were conducted by Gogoașe-Nistoran et al. (Chapter "Dynamics of Islands and Danube River Channel Along Vedea-Călărași Sector (1856–2019): Hydrogeomorphological Approach" in this book).

4 Recent Dynamics of the Islands in Relation with Hydrological Variability

To better understand the driving forces behind the channel dynamics of the Danube, we analyzed the area of the islands against the hydrological variability occurring during the period 1995–2010.

In order to examine the hydrological variability of the Danube River, we used the mean daily discharges at Zimnicea station, which were downloaded from the Global Runoff Data Center.

4.1 The Features of the Islands Under Average Hydrological Conditions

To analyze the dynamics of the river channel, we digitized the islands with and without vegetation based on satellite imagery. We used Synthetic Aperture Radar (SAR) images (ERS1/2 and ENVISAT satellites) for the Vedea-Oltenița reach (source European Space Agency) and Landsat 5TM for Oltenița-Călărași reach (source Earth Observing System). We selected scenes from summer-early autumn of 1995 versus 2010, which corresponded to discharges around or above the mean multiannual value (Table 5).

Table 5 Satellite scenes used in sub-section 4.1.

Reach	Document	Resolution (m)	Land surveys (mm/dd/yyyy)	Data source	Danube discharge at Zimnicea (m^3/s)[a]
Vedea-Oltenița	SAR	12.5	July 6, 1995	European Space Agency	8,580
			July 22, 2010		6,290
Oltenița-Călărași	Landsat 5 TM	30	September 13, 1995	Earth Observing System	6,380
			August 21, 2010		6,550

[a]*Data source* Global Runoff Data Center

The methodology of using SAR imagery for analyzing the Danube islands is explained by Hachemi et al. [55–57]. These ascending images cover an area of 100 km (range) × 102.5 km (azimuth). The acquisition is nocturnal, from an azimuth direction upwards. The scene is illuminated to the right, in lateral view with an incidence angle of 23°, in the "C" wavelength band (5.65 cm), and a vertical polarization (VV). Hachemi et al. [55] produced: calibrated, filtered, geo-referenced, and ortho-rectified SAR amplitude images of the Vedea-Oltenița reach with a resolution of 12.5 m. Then, colored compositions were created in order to detect changes and to facilitate interpretation. Finally, the islands were extracted and digitized. The digitization was semi-automatic and then controlled by hand, pixel by pixel.

With Landsat imagery (spatial resolution 30 m, temporal resolution 15 days), the analysis consisted in manually delimiting the islands from water and digitizing the islands contour with or without vegetation. The newly created vectors were exported in QGIS 3.6.1, where their projection was transformed from WGS84 to Stereo 70.

We extracted the number and area of islands on all the analyzed documents. Then, we compared the two time series of islands area by the non-parametric test of Mann–Whitney to detect statistical differences at $p < 0.05$.

Compared to the mean multiannual discharge of approximately 6,000 m^3/s at Zimnicea (Table 1), we found high discharges in April 2006, July 2010, and April 2005 (Fig. 11). In 2006 and 2010, the Lower Danube experienced historic floods, while in 2005, major floods occurred on its tributaries [58]. In terms of mean daily value, the historical discharge reached at Zimnicea 16,400 m^3/s on 23–24 April 2006 (Fig. 11). However, 56% of all the mean daily values were below the mean annual flow rate of the studied period.

The largest 18 islands were all present during the studied period (1995–2010), while the number of small islands increased (from 9 in 1995 to 16 in 2010). The total area of the islands recorded a slight increase of 9% between 1995 (41.5 km^2 of which 3.8 km^2 small islands) and 2010 (45.2 km^2 of which 5.5 km^2 small islands). This process is more obvious for the Vedea-Oltenița reach (Fig. 11). Downstream,

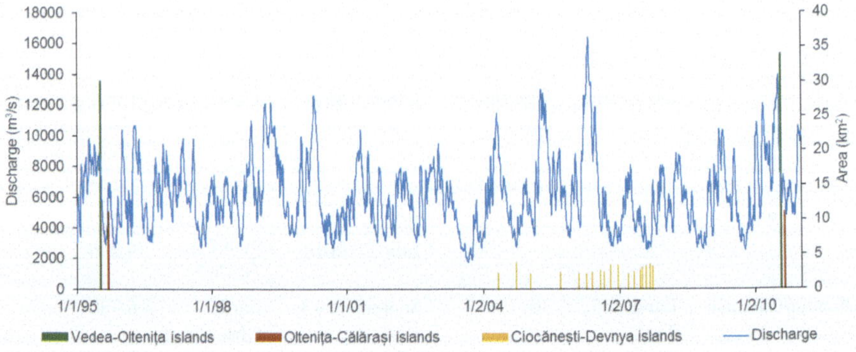

Fig. 11 Variations of the islands area in relation with hydrological variability (1995–2010, Danube daily mean discharge at Zimnicea)

between Oltenița and Călărași, the islands either slightly increased or decreased. Overall, according to the Mann-Whitney test, at $p < 0.05$, there is no statistical change between the islands area in 1995 and 2010.

The slight increase of all the islands along the Vedea-Oltenița reach could be related to the Danube River discharge at the moment of the satellite passage over the study area. The higher flow rate and probably water level in 1995 could explain the less emerged area of the islands, while lower discharge corresponds to a larger emerged area. The small bars probably formed and grew through the sediment supply provided by the Danube floods of April 2006, July 2010, but also of 2005 on its tributaries.

Overall, we conclude that the major floods of the Danube did not dramatically transformed the islands. The Danube islands along the Vedea-Călărași reach appeared to be in a relative equilibrium in terms of erosion and deposition processes in 2010 versus 1995, except for migrating downstream [55].

4.2 Case Study of Ciocănești-Devnya Islands Pre-versus Post-floods

In order to better understand the relationship between Danube islands dynamics and hydrological variability, we choose to focus on Ciocănești-Devnya islands during the interval pre- and post-2006 flood. We chose to work on Ciocănești Island because it was classified as being a natural island by [51]. At low-waters, it is connected with its neighbor—Devnya Island—and therefore we analyzed the two islands together.

In spring 2006, the Danube River recorded the highest discharges since 1840, reaching in the study sector 16,900–17,300 m^3/s (Table 1). As a result of the large floods occurred in 2006 but also in 2005 on Danube tributaries, the mean annual suspended sediment load reached more than 18 million t in 2006 and about 15 million t in 2005 at Zimnicea, higher than the multiannual average of about 13 million t/year for the period 1986–2014 [59].

In order to better understand the impact of the historic flood of 2006 on these islands, we selected all available Landsat 5TM scenes (source Earth Observing System) for the time interval 2004–2007 (overall 18 scenes). We delimited the contour of islands according to the methodology explained above.

The maximum area of the two islands was detected on 21 September 2004 when it reached 3.9 km^2 at a river discharge of 2,870 m^3/s (i.e. low-waters) (Figs. 11 and 12). The minimum area occurred during the flood peak—2.3 km^2 at 12,600 m^3/s on 4 April 2006 (i.e. high-waters). The area of the two islands was also 2.3 km^2 at mean value of the discharge—6,230 m^3/s on 6 March 2007. The bars devoid of vegetation can be seen only at low-waters while the area covered by forest is apparent during all phases of the hydrological regime. We noticed a kind of cyclic evolution of the islands area between high-waters and low-waters. Yet, the lowest area was detected

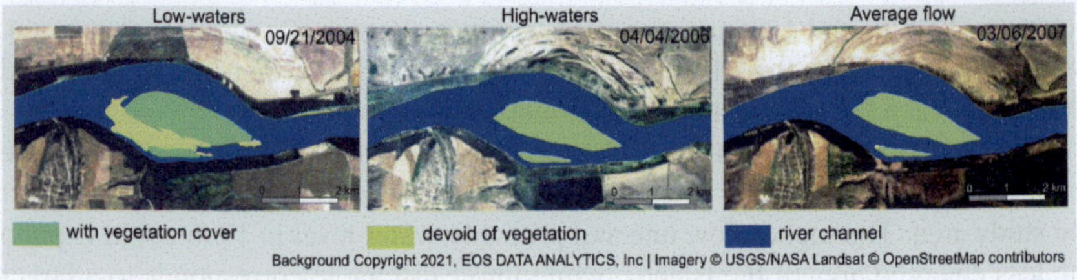

Fig. 12 Spatial dynamics of Ciocănești-Devnya islands in relation with hydrological variability

before the historic floods of 2006, therefore we can not draw any conclusions about its role in the formation of the islands.

This small difference between areas at mean and high discharges may be due to the canopy of the forest. Actually, it is the canopy that we see from satellite. Another pattern of island dynamics along the investigated reach could be the relative stability of the emerged areas, despite the floods occurring in the region.

5 Conclusion

The aim of this chapter was to investigate the dynamics of the Lower Danube islands and channel by focusing on the Vedea-Călărași sector. We used various cartographic documents dating since the year 1856, as well as recent satellite imagery (2019) to explain the islands and channel dynamics over the last century and a half.

The area of the islands in 2019 represents almost one third of the total area of the channel similar to the situation in 1856. This finding suggests a relative equilibrium despite numerous transformations in the river basin which may be due to the low-energy river type of the Lower Danube River section [40]. As the Lower Danube channel pattern depends mostly on suspended sediment load, islands are likely to continue to form through bank erosion. The Albina-Paraschiva-Slaviavin-Garvan islands were subjected to a detailed analysis, by taking into account several cartographic documents, in order to identify their spatial and temporal dynamics during the period 1856–2019. We also investigated the recent dynamics of the islands in relation with hydrological variability through the case study of the Ciocănești-Devnya islands whose dynamics was analyzed during the period 2004–2007, including the historical floods occurred on Lower Danube River (in 2006) and its tributaries (in 2005).

Overall, three patterns of historical evolution were found: (i) the expansion of the older islands mostly through lateral accretion, as in the case of those listed in Table 3; (ii) the merging of the islands with the bank (and also with each other) as in the case of Albina and Paraschiva islands; (iii) the increase of the islands area in the last decades, which was detected through the methods used in this study (overall analysis of the study area since 1856, case study of Albina-Paraschiva-Slaviavin-Garvan islands since 1856, interpretation of SAR imagery 1995–2010). Other three patterns characterize the present-day evolution of islands: (i) the increase of the islands by deposition upstream, as in the case of Ciocănești-Devnya islands; (ii) the persistence of vegetated islands under flood conditions, as in the case of Ciocănești-Devnya islands; (iii) important variations of small bars (some of them are forming while others are disappearing).

The number, density and morphometry/dimensions of islands depend on their genesis, sediment supply, water level, biopedoclimatic constraints and other local factors. We also noticed that under a quasi-natural regime (1856–1953), both the Danube River channel and its islands suffered numerous alterations. Later on, under a modified hydrological regime (1953–2019), the hydrosystem became relatively stable and the islands were colonized by vegetation.

6 Recommendation

To improve and complete the findings of this chapter, we have a few recommendations for further studies on the hydrogeomorphological dynamics of the Lower Danube River.

- Search for maps or even drawings in archives in order to better understand the trajectories of the Danube River channel and its islands more than a century ago.
- Use other indicators to characterize the Danube River channel and its islands, therefore better comprehend their dynamics.
- Use recent satellite imagery at good resolution for greater accuracy of river processes.
- Focus on the age of the forest in order to understand the extension of certain landforms, which is similar to the recommendation of Chelu et al. [24].
- Extend the analysis to other reaches of the Lower Danube River.
- Survey the effects of climate change on fluvial islands in the context of the increasing air temperatures along the Lower Danube Valley (according to Constantin et al.—Chapter "Observed Changes in the temperature and Precipitation Regime Along the Lower Danube River" in this book).

Acknowledgements We are grateful to Stefan Dorondel and Stelu Șerban for fieldwork and discussion on islands along this sector of the Danube River. The first three authors (Florina Grecu, Gabriela Ioana-Toroimac and Gabriela Osaci-Costache) contributed equally to the work in this chapter.

References

1. Scheidegger AE (1970) Theoretical geomorphology. Springer-Verlag, Berlin
2. Masson M, Garry G, Ballais JL (1996) Cartographie des zones inondables. Approche Hydrogéomorphologique. Editions Villes et territoires, Paris
3. Arnaud-Fassetta G, Fort M (2004) La part respective des facteurs hydroclimatiques et anthropiques dans l'évolution récente (1956–2000) de la bande active du Haut Guil, Queyras, Alpes françaises du Sud. Méditerranée 1–2:143–156
4. Malavoi JR, Bravard JP (2010) Elements d'hydromorphologie fluviale. http://www.onema.fr/hydromorphologie-fluviale. Accessed 15 Jan 2017
5. Ballais JL, Chave S, Dupont N, Masson E, Penven MJ (2011) La méthode hydrogéomorphologique de détermination des zones inondables. Physio-Géo 5. https://doi.org/10.4000/physio-geo.3307
6. Grecu F (2018) Geomorfologie dinamică pluvio-fluvială. Teorie şi aplicaţii. Editura Universitară, Bucharest
7. Parsons AJ, Bracken L, Poeppl RE, Wainwright J, Keesstra SD (2015) Introduction to special issue on connectivity in water and sediment dynamics. Earth Surf Process Landf 40:1275–1277. https://doi.org/10.1002/esp.3714
8. Osterkamp WR (1998) Processes of fluvial island formation, with examples from Plum Creek, Colorado and Snake River, Idaho. Wetlands 18(4):530–545. https://doi.org/10.1007/BF03161670
9. Wyrick JR, Klingeman PC (2011) Propose fluvial island classification scheme and its use for river restoration. River Res Appl 27:814–825. https://doi.org/10.1002/rra.1395
10. Gurnell AM, Petts GE (2002) Island-dominated landscapes of large floodplain rivers, a European perspective. Freshw Biol 47:581–600. https://doi.org/10.1046/j.1365-2427.2002.00923.x
11. Kollmann J, Vieli M, Edwards PJ, Tockner K, Ward JV (1999) Interactions between vegetation development and island formation in the Alpine River Tagliamento. Appl Veg Sci 2(1):25–36. https://doi.org/10.2307/1478878
12. Beechie TJ, Liermann M, Pollock MM, Baker S, Davies J (2006) Channel pattern and river-floodplain dynamics in forested mountain river systems. Geomorphology 78:124–141. https://doi.org/10.1016/j.geomorph.2006.01.030
13. Summer W, Klaghofer E, Hintersteiner K (1996) Trends in soil erosion and sediment yield in the alpine basin of the Austrian Danube. In: Walling DE, Webb BW (eds) Erosion and sediment yield: global and regional perspectives. IAHS Publication no. 236, pp 473–479.
14. Kiss T, Andrasi G (2014) Morphological classification and changes of islands on the Drava river, Hungary-Croatia. Carpathian J Earth and Environ Sci 9(3):33–46
15. Hohensinner S, Lager B, Sonnlechner C, Haidvogl G, Gierlinger S, Schmid M, Krausmann F, Winiwarter V (2003) Changes in water and land: the reconstructed Viennese riverscape from 1500 to the present. Water Hist 5:145–172. https://doi.org/10.1007/s12685-013-0074-2
16. Zaharia L, Ioana-Toroimac G (2013) Romanian Danube River management: impacts and perspectives. In: Arnaud-Fassetta G, Masson E, Reynard E (eds) European continental hydrosystems under changing water policy. Verlag Dr. Friedrich Pfeil, München, p 159–170
17. Hinkov G, Glogov P, Kachova V, Georgieva ML, Anghelus C, Ciuvat AL (2020) A methodological approach for monitoring of riparian forests in protected areas on Danube islands. Silva Balcanica 21(2):53–69. https://doi.org/10.3897/silvabalcanica.21.e54963
18. Constantinescu S, Achim D, Rus I, Giosan L (2015) Embanking the Lower Danube: from natural to engineered floodplains and back. In: Hudson PF, Middelkoop H (eds) Geomorphic approaches to integrate floodplain management of lowland fluvial systems in North America and Europe. Springer, New York, pp 265–288
19. Grigorie Ş (2013) Valea Dunării în sectorul Olteniţa-Călăraşi, studiu hidrogeomorfologic. PhD thesis, University of Bucharest, Bucharest

20. Grecu F, Ioana-Toroimac G, Grigorie (Omrani) Ş, Ghiţă C (2017) The dynamics of the Danube channel over the last 150 years. In: Proceedings of Romanian geomorphology symposium Universitatea Alexandru Ioan Cuza din Iaşi, Iaşi, 11–14 May 2017
21. Iordache G, Bondar C, Anghel S, Malageanu M (2017) Morphological changes of the fluvial islets from the Danube River (Turcescu Islet—Cernavodă sector) between 1908 and 2016. Geo-Eco-Marina 23:197–204
22. Iordache G, Bondar C, Anghel S, Malageanu M (2018) Evolution of danubian islets from Balta Ialomitei hydrographic system (Km 345–241) between 1908–2016. Sci Ann Danube Delta Inst 23:41–50. https://doi.org/10.7427/DDI.23.06
23. Iordache G, Bondar C, Anghel S, Malageanu M, Dudu A (2019) Using of GIS techniques to assess the spatial changes of the fluvial Islands in the Corabia Area—Bechet Harbor Sector (Km 626–679) between 1910–2017. Sci Ann Danube Delta Inst 24:251–260. https://doi.org/10.7427/DDI.24.27
24. Chelu A, Falkowski T, Brach M (2018) Morphological dynamics of the islands on the Lower Danube River in the Călărași-Cernavodă sector and GIS error assessment. Riscuri și Catastrofe XVII, 23(2):51–70
25. Marin AF, Armaș I (2016) Shape characteristics of fluvial islets based on GIS techniques. A case study: the Danube's islets between Giurgiu and Olteniţa. Forum Geographic XV(2): 133–139. https://doi.org/10.5775/fg.2067-4635.2016.047.d
26. Marin A (2019) Studiul morfodinamicii ostroavelor în relație cu factorii de control fluviali prin tehnici geospațiale. Aplicație în sectorul dunărean Giurgiu-Oltenița. Editura Universitară, Bucharest
27. Goda L (2010) The Fords of the Danube. In: Brilly M (ed) Hydrological processes of the Danube River basin. Springer, Dordrecht, pp 343–366. https://doi.org/10.1007/978-90-481-3423-6_11
28. Wyrch JR (2005) On the formation of fluvial islands. PhD thesis, Oregon State University
29. Picco L, Ravazzolo D, Reinato R, Lenzi MA (2014) Characteristics of fluvial islands along three gravel-bed rivers of north-eastern Italy. Cuadernos de Investigación Geográfica 40(1):53–66. https://doi.org/10.18172/cig.2505
30. Marin AF (2017) Hydrological and morphological changes of the Danube River between Giurgiu and Oltenița cities (Romania). Cinq Continents 7(15):99–111
31. Marin AF (2018) Inventarierea insulelor fluviale în sectorul Dunărean Giurgiu/Ruse – Oltenița/Tutrakan (România-Bulgaria). Geographia Napocensis XII(2):41–47
32. Bondar C, Iordache G (2017) Sediment transport on the Danube River in the Romanian border area—characteristics. Rev Roum Géogr/Rom. J Geogr 61(1):3–17
33. Pătru I, Zaharia L, Oprea R (2006) Geografia fizică a României. Climă, ape, vegetație, soluri. Editura Universitară, Bucharest
34. Bondar C (2008) Hydromorphological balance of the Danube River Channel on the Sector between Bazias (km 1072.2) and Danube Delta Inlet (km 80.5). International of expert conference on the safety of navigation and environmental security in transboundary context the Black Sea Basin, Odesa, Ukraine, 24–26 June 2008
35. Pekárová P, Gorbachova L, Mitkova VB, Pekár J, Miklanek P (2019) Statistical analysis of hydrological regime of the Danube River at Ceatal Izmail Station. IOP Conf Ser: Earth Environ Sci 22:012035. https://doi.org/10.1088/1755-1315/221/1/012035
36. Borcia C, Teodor S, Radulescu C, Petrea C, Ciuca R, Blendea V (2014) Aspects of variability of hydrological parameters as a result of human impact and climate change if the lower Danube. INHGA—Scientific Conference, Bucharest, 10–11 November 2014
37. Water Research Institute (2019) Long-term morphological development of the Danube in relation to the sediment balance. http://www.interreg-danube.eu/uploads/media/approved_project_output/0001/39/6299a3c18cf54d8ef7e02150844e27358da8038d.pdf. Accessed 30 Nov 2021
38. Habersack H, Hein T, Adrian Stanica A, Liska I, Mair R, Jäger E, Hauer C, Bradley C (2016) Challenges of river basin management: current status of, and prospects for, the River Danube from a river engineering perspective. Sci Total Environ 543(A):828–845. https://doi.org/10.1016/j.scitotenv.2015.10.123

39. Vladev D, Krastev T, Stankova S, Chenkova N, Penerliev M (2008) Natural-geographic potential of north-eastern Bulgaria and its meaning for the cross-border cooperation with Romania-Phorum. Geogr Stud Environ Prot Res 7:233–247
40. Schumm SA, Meyer DF (1979) Morphology of alluvial rivers of the Great Plains. Great Plains Agric Council 91:9–14
41. Castro JM, Thorne CR (2019) The stream evolution triangle: integrating geology, hydrology, and biology. River Res Appl 35(4):315–326. https://doi.org/10.1002/rra.3421
42. INHGA (2016) Planul de management al riscului la inundații. http://www.inhga.ro/documents/10184/121027/12+PMRI+Dunare.pdf/a85c0b75-dfb7-4197-a457-a03fd64b6556. Accessed 28 Jun 2021
43. Gâștescu P, Țuchiu E (2012) The Danube River in the Pontic sector—hydrological regime. In: Gâștescu P, Lewis W Jr, Brețcan P (eds) Water resources and wetlands. Conference Proceedings, 14–16 September 2012, Tulcea
44. Armaș I, Necea D, Miclăuș C (2019) Fluvial terrace formation and controls in the Lower River Danube, SE Romania. Quatern Int 504(10):5–23. https://doi.org/10.1016/j.quaint.2018.03.031
45. Gatta G (2011) Analisi metrica di cartografia antica in ambiente digitale. Bollettino dell'Associazione Italiana di Cartografia 141–142:41–53
46. Mastronunzio M (2011) Analisi cartometrica di due mappe di confine (1776–1886) del Trentino pre-unitario. Bollettino dell'Associazione Italiana di Cartografia 143:49–59
47. Armaş I, Osaci-Costache G, Braşoveanu L (2014) Forest landscape history using diachronic cartography and GIS. Case study: Subcarpathian Prahova Valley, Romania. In: Crăciun C, Boştenaru-Dan M (eds) Planning and designing sustainable and resilient landscapes. Springer Geography Springer Science+Business Media, Dordrecht, pp 73–86. https://doi.org/10.1007/978-94-017-8536-5_6
48. Osaci-Costache G, Armaș I (2016) Lost landscapes: in search of cartographic evidence. In Boștenaru Dan M, Crăciun C (eds) Space and time visualisation. Springer, Cham, pp 35–62. https://doi.org/10.1007/978-3-319-24942-1_3
49. Ioja IC, Osaci-Costache G, Breuste J, Hossu CA, Gradinaru S, Onose DA, Nita MR, Skokanová H (2018) Integrating urban blue and green areas based on historical evidence. Urban Forestry & Urban Greening 34:217–225. https://doi.org/10.1016/j.ufug.2018.07.001
50. Perșoiu I, Perșoiu A (2018) Flood events in Transylvania during the medieval warm period and the little ice age. Holocene 29(1):85–96. https://doi.org/10.1177/0959683618804632
51. Danubeparks (2021). WILDisland map. https://wildisland.danubeparks.org/#map. Accessed 30 Nov 2021
52. Pătruț SM (2010) Riscul inundațiilor în Bazinul Dunării. PhD thesis, University of Bucharest, Bucharest
53. Ioana-Toroimac G (2014) La dynamique hydrogéomorphologique de la rivière Prahova : fonctionnement actuel, évolution récente et conséquences géographiques. Editura Universitară, Bucharest
54. Ollero A, Ibisate A, Granado D, Real de Asua R (2015) Channel responses to global change and local impacts: perspectives and tools for floodplain management, Ebro River and Tributaries, NE Spain. In: Hudson PF, Middelkoop H (eds) Geomorphic approaches to integrate floodplain management of lowland fluvial systems in North America and Europe. Springer, New York, pp 27–52
55. Hachemi K, Grecu F, Ioana-Toroimac G, Grigorie (Omrani) Ş, Ozer A, Kuzucuoglu C (2020) Contribution of SAR radar imagery in the study of the dynamics of the Danube island system, Giurgiu-Călăraşi sector, Romania. Adv Model Anal B 63(1–4):1–6. https://doi.org/10.18280/ama_b.631-401
56. Hachemi K, Grecu F, Ioana-Toroimac G, Grigorie-Omrani S, Ozer A, Kuzucuoglu C (2021) The utility of morphometric parameters extracted from SAR radar images in the monitoring of the dynamics of the Danube Island System, Giurgiu-Călăraşi sector, Romania. Int J Des Nat Ecodynamics 16(1):13–19. https://doi.org/10.18280/ijdne.160103
57. Hachemi K, Grecu F, Ioana-Toroimac G, Constantin (Oprea) DM, Ozer A (2021) The diachronic analysis of island dynamics along the Vedea-Oltenița Danube river sector using SAR imagery. Med. Geosc Rev 3:159–173. https://doi.org/10.1007/s42990-020-00042-5

58. Grecu F, Zaharia L, Ioana-Toroimac G, Armaş I (2017) Floods and flash-floods related to river channel dynamics. In: Rădoane M, Vespremeanu-Stroe A (eds) Landform dynamics and evolution in Romania. Springer, Cham, pp 821–844. https://doi.org/10.1007/978-3-319-32589-7_33
59. Budapest University of Technology and Economics, University of Natural Resources and Life Sciences Vienna (eds) (2019) Danube sediment. Analysis of sediment data collected along the Danube. http://www.interreg-danube.eu/uploads/media/approved_project_output/0001/27/af1a5cf355224c606b709fd738ebbc96902beeda.pdf. Accessed 30 Nov 2021

Hydro-sedimentary Modeling and Fluvial Morphological Processes Along the Lower Danube River (Giurgiu-Olteniţa-Călăraşi Reach)

Daniela-Elena Gogoașe-Nistoran, Andreea Florentina Marin, Iuliana Armaș, Cristina Sorana Ionescu, Georgiana-Mihaela Tudor, and Andrei Cozma

Abstract The study aims to provide an advancement in understanding the complex hydro-sedimentary behaviour and morphological processes of the Danube fluvial system. A 120 km-in length reach is chosen for the study along the Romanian-Bulgarian border, subject to intense anthropogenic influences and multiple, interconnected controlling factors. The reach stretches between the Romanian towns Giurgiu and Călărași (corresponding to Bulgarian towns of Ruse and Silistra). To this end, hydro-sedimentary modeling and diachronic analysis of planform dynamics of the channel width on the maps, aerial photos and satellite images are used. To analyse the induced changes under both hydrodynamic and sedimentary regimes, a coupled numerical model was set up. Topobathymetric GIS data are used to build a digital elevation model (DEM) of the area and extract a 1D model's geometry in HEC-RAS (USACE) software. Observed flow/stage hydrographs and sediment discharge series at the gauging stations are used to calibrate the model. Sediment transport rate, annual sediment budget, spatial and temporal variation of thalweg elevation in

D.-E. Gogoașe-Nistoran (✉) · C. S. Ionescu
Faculty of Power Engineering, Department of Hydraulics, Hydraulic Machines and Environmental Engineering, University Politehnica of Bucharest, 313 Spl. Independentei, sector 6, 060042 Bucharest, Romania

C. S. Ionescu
e-mail: cristina.ionescu@onmicrosoft.upb.ro

A. F. Marin · I. Armaș
Faculty of Geography, Department of Geomorphology-Pedology-Geomatics, University of Bucharest, Bd. Regina Elisabeta 4-12, sector 3, 030018 Bucharest, Romania
e-mail: andreea.marin@unibuc.ro

I. Armaș
e-mail: iuliana.armas@unibuc.ro

G.-M. Tudor
National Institute of Hydrology and Water Management, Șos. București-Ploiești 97E, sector 1, 013686 Bucharest, Romania
e-mail: georgiana.tudor@hidro.ro

A. Cozma
JBA Consulting Romania, 34 Aviator Petre Crețu Str, sector 1, Bucharest, Romania
e-mail: andrei.cozma@gbaconsulting.ro

A. Negm et al. (eds.), *The Lower Danube River*, Earth and Environmental Sciences Library, https://doi.org/10.1007/978-3-031-03865-5_3

each cross-section of the model geometry are obtained from numerical simulations and, together with the information obtained from diachronic analysis and navigation maps, are used to reveal the spatio-temporal morphological dynamics of the Danube River in the studied area.

Keywords Control factors · Hydraulic model · Sediment transport model · Diachronic analysis · Erosion-deposition · Fluvial morphology

1 Introduction

Danube River is one of Europe's major fluvial system which has long been affected by various natural factors (tectonic, geological, climate, hydrological) and human activities. As a response to the complex interaction of all these governing conditions, the river channel's hydro-morphological characteristics suffered multiple changes. In the last decades there has been an increasing scientific interest to follow the hydrological and sediment transport trends, together with the morphological alterations of the Danube River, both internationally (e.g. [1–5]) and also, nationally (e.g. [6–11]).

Similarly to other large European rivers such as the Rhine, Volga, Arno, Loire, Po [12–14], the Lower Danube, is characterized by a decreasing sediment transport rate during the last decades. The Danube River Basin's main human impacts are flood control, hydropower generation, and navigation. River engineering works such as embankment dykes and dams interrupt the longitudinal and lateral connectivity and prevent the sediment supply and transport from the upstream tributaries and reaches to the downstream areas [15]. Observed morphological changes are complex, depending on location and hydrological regime: from lateral erosion and vertical incision reaches (downstream the reservoirs, in the narrow areas) [16] to deposition ones (in the reservoirs, on sand bars, banks, islets), creating problems to fluvial navigation during low flow periods.

According to Panin and Jipa [17] the natural mean multi-annual suspended sediment load at the Danube Delta apex (Ceatal Izmail) has decreased by more than 50% after 1960.

The degree of difficulty and time-consuming activities for measuring sediment transport rates and particle grading are well-known [15]. Therefore, properly calibrated numerical models help to approximate missing data over the simulation period, this way offering a better understanding of the coupled hydrodynamic-morphologic phenomena. Numerical models for sediment transport (1D to 3D) have evolved during the last decades. In order to realistically reproduce the real phenomena they have to be properly calibrated. Even though lack in spatial complexity, 1D numerical models are still successfully used for long river reaches and long period of times (years, decades), provided they are based on detailed topographic and bathymetric data. Their advantages are: simple computations and calibration, ease of parametrisation, few necessary hydrologic data for calibration and validation, shorter computation time. However, they cannot capture variations of hydrodynamic and morphological alterations across the river channel. Such a software is HEC-RAS (Hydrologic Engineering Centre, River Analysis System, version 5.07).

The chapter's overall objective is to compare the results obtained from a hydro-sedimentary model with the observed and reported morphological changes for a better understanding of the hydromorphological behaviour of the Danube fluvial system along the Giurgiu-Călărași reach in Romania (corresponding to Ruse-Silistra in Bulgaria). To this end, several specific objectives have been defined:

1. To analyse the temporal and spatial morphological changes of the river channel.
2. To process water and sediment data during the analysis period of 8 years, between 2008 and 2015, necessary for the numerical model.
3. To set up a 1D sediment transport model and calibrate it.
4. To compare results with field observations and understand the complex physical processes of the hydrodynamics of sediment transport.
5. To analyse morphological changes (aggradation/degradation) and sediment loads in space and time, over the study period.

2 Site Description

The Danube River is one of the most important rivers in Europe, being the second longest after Volga River. Its source is in the Black Forest (Germany) and flows towards the southeast for 2870 km to the Black Sea [10]. Based on its bed slope, the river course can be divided into three sub-regions: Upper Danube—from its headwaters to Bratislava City (Slovakia), Middle Danube—from Bratislava City (Slovakia) to the Iron Gate dam (Romania) and Lower Danube—from Drobeta-Turnu Severin, rkm 931 (chainage along Danube is referenced at Sulina river mouth into the Black Sea) to Ceatal Izmail (rkm 80.5).

Along the low energy, multi-thread anabranching Lower Danube stretch [18], two major Hydro Power Plants (HPP) were built in 1971 (Iron Gate I at rkm 943) and 1985 (Iron Gate II at rkm 863) downstream the Iron Gates Gorges of Carpathian Mountains. Downstream the HPPs, the Lower Danube flows across a large plain, where the river channel is wider, less deep, with multiple vegetated islets (mainly composed of sand and silt), before dividing into branches and flowing into a swampy delta.

The study reach has about 120 km in length, and is situated at the Romanian-Bulgarian border, between the towns of Giurgiu and Călărași-Romania (Ruse and Silistra in Bulgaria) (Fig. 1). The valley has an asymmetric morphologic development, with a low and wide, terraced, left Southern part of the Lower Danube Plain, and a higher and terraced (Bulgarian), right side [19, 20]. Terraces were cut due to the uplift in the Bulgarian sector of the Moesian platform delineated by two NW-SE-trending faults passing throughout Giurgiu-Kubrat-Vetrino fault [21]. The

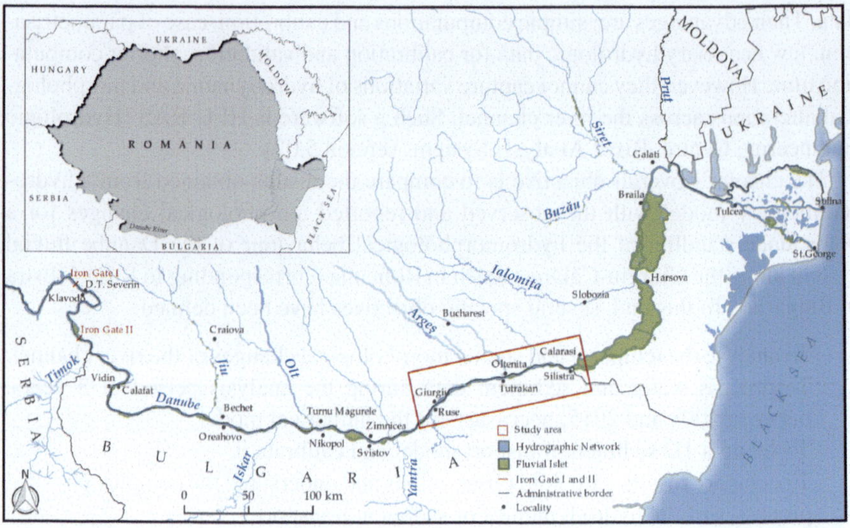

Fig. 1 The study reach along Danube River: Giurgiu-Călăraşi reach

anabranching index estimated for 10 km long sub-reaches over the study reach is between 1.5 and 2 [18].

The principal characteristics of the main tributaries of the study reach on lower Danube are presented in Table 1. The gravel-bed tributaries from the right bank, Timok, Jantra and Iskar, raise the natural coarsening in the Lower Danube's sand-bed channel, in areas downstream their mouths (local effect).

Table 1 Principal characteristics of main tributaries along the study reach on Lower Danube [22, 23]

Crt. No	Danube tributary	Distance of Danube confluence to Black Sea mouth at Sulina	Side of confluence	Mean multi-annual discharge at confluence	No. of HPP (>1 MW)/reservoirs	Catch'ment area	Total length
		rkm	–	m^3/s	Number	km^2	km
1	Timok	846	Right	31	1/1	4,630	180
2	Jiu	694	Left	86	5/61	10,080	339
3	Iskar	636	Right	54	10/10	8,684	368
4	Olt	604	Left	174	23/25	24,050	615
5	Iantra	537	Right	47	–	7,879	285
6	Argeş	428	Left	71.2	17/11 (1966–1997)	12,550	350

HPP—Hydro Power Plants

Giurgiu-Călăraşi is a typical lowland, river reach, where processes of erosion and deposition are alternating. It has numerous vegetated islets and channel bars, which divide the river channel into multiple active branches, characteristic to the anabranching (anastomosing) fluvial systems, with a prevalence of suspended solid load [18, 24].

The average depth along the study reach is about 7 m for the mean multi-annual flow, Q_m, varying between 5.1 and 9.4 m from Q = 4,000 m^3/s to Q = 14,000 m^3/s, respectively). The average bottom slope value is about 0.05 ‰ and the river channel width varies between 500 and 2500 m for the mean multi-annual discharge. Due to the area's high risk of flooding embankments were built—mostly along the lower, Romanian bank to protect agricultural areas [10]. However, at high flows (with their peak having a recurrence interval greater than 50–100 years), the enclosed floodplain areas (mainly agricultural) are flooded.

Argeş River is the only important tributary within the study reach. It drains the Carpathians' curvature region, with the highest suspended sediment yield in Romania [12, 25]. Along its lower, anastomosed section, it flows through erodible alluvial deposits. The river's flow and sediment regimes were significantly influenced by the 38 reservoirs built mainly for hydropower and flood protection purposes [12, 18, 26]. First dam from the confluence with Danube is at 85 km distance (Mihăileşti).

The Romanian gauging stations (GS) along the Giurgiu-Călăraşi reach, are shown in Fig. 2: Giurgiu GS (rkm 493.05), Olteniţa GS (rkm 430, which is 2 km downstream of the confluence with Argeş River), Chiciu-Călăraşi (rkm 379), very close to the corresponding Bulgarian one—GS Chiciu-Silistra (rkm 375). Last two are located just upstream the nodal bifurcation where Danube River splits in two: the old Danube and Borcea branch. Călăraşi GS (rkm 365) is located on the Borcea branch and

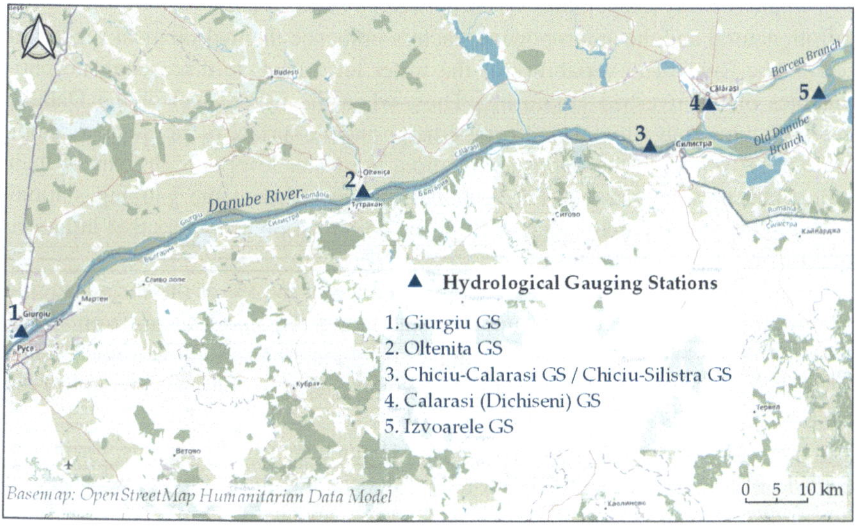

Fig. 2 Romanian gauging stations used for the analysis along the Giurgiu-Călăraşi reach

Izvoarele GS (rkm 348.6) is located on the old Danube. In the following analyses two subreaches were considered for the Giurgiu-Călărași reach: the upstream one, from Giurgiu to Oltenița and the downstream one, from Oltenița to Călărași.

3 Fluvial Morphology Factors Along the Danube Study Reach

Sediment regime of a river stretch is influenced by processes and actions at the level of the whole basin, as a result, being dependent mainly on land use pattern, hydraulic structures, river training works and water management strategies.

The fluvial morphology along the study reach is influenced by natural and anthropogenic factors, having synergic effects. The analysis of these factors was divided for the current study into 3 time periods: (I) 1920–1970, (II) 1971–1990 and (III) 1991–2015. Bondar [27] showed a series of changes in terms of flow and a drop in sediment discharge over the lower Danube during the aforementioned study periods, as a consequence of both natural and anthropogenic factors.

3.1 Natural Factors

The natural factors influencing Danube River morphology are: climate (mainly rainfall and temperature), flow variability, geology, lithology, vegetation etc. [28].

At the local level of the study reach, the annual rainfall slightly exceeds 500 mm [29], whereas water temperature varies between 0 and 29 °C.

Both natural and the anthropogenic factors influence the hydrological regime of the Danube River. The variability of the water level leads to a seasonal morphodynamics of the riverbed landforms. Thus, when the water level of the Danube decreases, numerous alluvial landforms emerge and the surface of the islets increases temporarily. While, in the opposite situation, the water level is high, part of their surface becomes submerged. The Danube flow recorded at the gauging stations located in the studied area is the result of the combination of factors that influence the flow from the upstream sector.

The mean multi-annual water discharge decreased by 3% at the three gauging stations along the study reach: Giurgiu, Oltenița and Chiciu-Călărași, from 1931–1970 to 1985–2005 [27]. As for the impact of climate change on extreme events, as for other European fluvial systems, they are expected to continue to amplify the droughts, followed by intensified floods [30].

In terms of lithology, deposits in the lower Danube are usually loesses, formed by weathering processes of mainly flysch and local rocks and transported by water and/or wind from upstream sources (main river and its tributary system), or even Black Sea beaches. Generally, the deposits include: typical loess (quartz coarse silt, very fine sand), sandy loess (fine and medium sand) and clay-loess (clay-size particles) [31, 32].

Vegetation plays an important role in defining the hydro-geomorphological character of a river, ensuring a series of climatic, hydrological and ecological functions [33–36], as well as in increasing bank stability. In the studied Danube reach, the riparian vegetation is composed predominantly of popular forests (*Populus nigra* and *Populus alba*) and willows (*Salix alba*). In some places *Fraxinus Pennsylvanica* and *Ulmus laevis* appear [37].

3.2 Anthropogenic Factors

The anthropogenic interventions to the Danube waterways' natural course introduced lateral and longitudinal disruptions to the fluvial system continuity, changing the hydrological regime, morphology of the riverbed and the floodplain ecosystem. The hydropower developments on Danube and its tributaries along all its Upper, Middle and Lower sections (on the Lower Danube being the Iron Gates 1—rkm 943/944, put into operation in 1971 and the Iron Gates 2—rkm 863, put into operation in 1985); and the hydropower cascade on Olt and Argeș, flood embankments, dredging works for navigation, sediment mining and shipyards maintenance, water management measures, intakes for irrigation systems and vegetation cutting (Fig. 3), all had a major impact on flow variability and downstream sediment transport over the analysed reach [9, 10, 12].

Measurements of inflow and outflow sediment discharge from Iron Gates reservoirs, performed between 1974 and 2015 showed the reservoirs retain about 80% of the entering loads [40]: an average of 16.2 Mt/year enter the reservoirs from upstream and 2.9 Mt/year are discharged downstream [40]. In Fig. 4 a decrease of the mean multi-annual suspended sediment rate may be seen at Giurgiu GS, especially in the first three years after the entry into operation of the two Iron Gates hydropower plants and hydropower works on tributaries. In the two linear fitting equations, x is

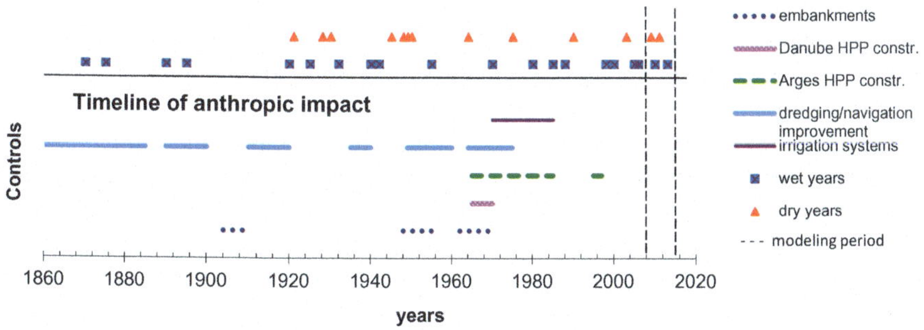

Fig. 3 Timeline of anthropic impact along Danube study reach and modeling period (HPP = hydropower plant) [9, 10, 38, 39]

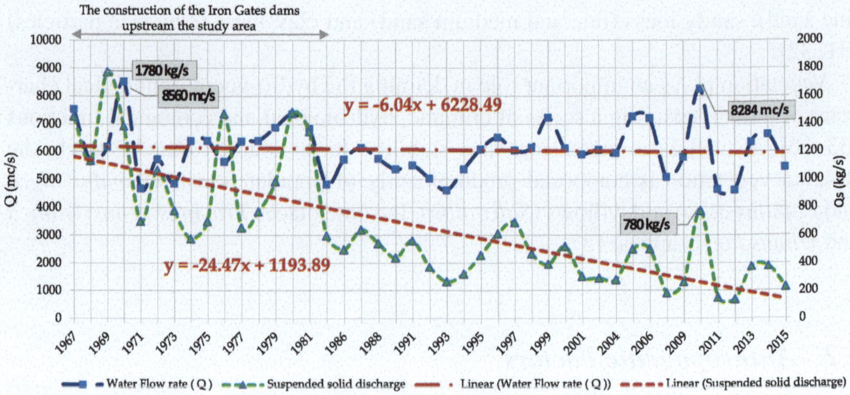

Fig. 4 Time evolution of water discharge (Q) and suspended solid discharge (Q_{SS}) at Giurgiu GS between 1967 and 2015

the year numbering in which the initial element of the sequence is 0 corresponding to first year, 1967. Most of the Argeș reservoirs are in cascade, and already silted in high percentages of their capacity (over 70%) [12]. Some authors [10, 12, 41] estimate an average of 67% reduction in sediment load after the commissioning of Argeș hydropower dams. However, these structural works have a major importance in flood attenuation and protection of important inhabited areas.

Prior to the building of *flood embankments* along the banks of the study reach, the Danube floodplain (containing ponds and marshes) was subjected to periodic flooding, which brought into the flooded areas new fertile material, good for agriculture [10]. After long debates between engineers, agronomists and biologists on choosing the best solution, an extensive embankment and drainage works along lower Danube started. More than 700 km of flood embankments were built between 1948 and 1985 along lower Danube [27]. Currently, 73% in length of the Iron Gates-Călărași reach is embanked [10, 42]. In the Danube floodplain between Giurgiu and Călărași, five agricultural enclosures (precincts) were created [43, 44] (Fig. 5) by longitudinal and transverse embankments Table 2. Some embankment dykes (levees), designed using a higher probability of flow exceedance, were built before 1940. Since the enclosures were still flooded, they had to be heightened afterwards based on a higher design flow. Therefore, in Table 2 the design flow and probability of exceedance are given for both cases.

By embanking a river, the floods are constrained to flow within the confinement at increased water levels and velocity. By confining the flow within the embankments, the stream consumes its high energy during floods through longitudinal profile incision and changes in planform morphology. This can lower the water table leading to groundwater depletion, thereby exacerbating droughty conditions [46] and leading to secondary soil salinization and ecological problems [47]. After the embankments, the drainage works of the Danube Floodplain followed, consisting of networks of channels and pumping stations [37]. The embankment systems are very costly to maintain

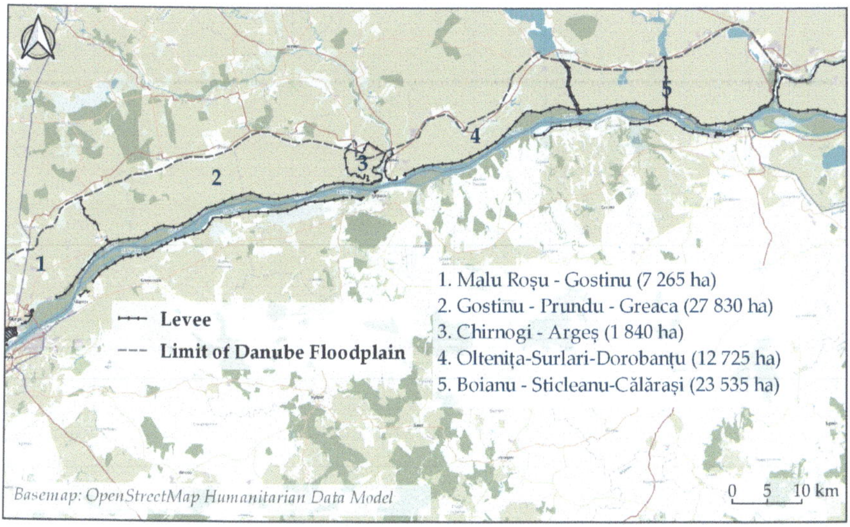

Fig. 5 Embankment dykes (levees) and enclosed areas between Giurgiu and Călăraşi [45]

and risky during prolonged flood periods. For example, in April–May 2006, several flood dykes breached, such as: Ghidici-Rast-Bistreţ (rkm 719–745), Bechet-Dăbuleni (rkm 667–679) and Olteniţa-Surlari-Dorobanţu (rkm 401–429) in the study area [39]. The last embankment dyke failed after long period of high-water stages, and severely flooded inhabited areas, farm animals, isolated buildings and large agricultural areas.

The riverbed dredging works carried out on the Danube are closely related to two activities: (i) maintaining the optimal depths for navigation and (ii) sand and gravel mining (rkm 494.5–400, rkm 520–522.5 and rkm 510.5–508) [48]. As reported by [18], between rkm 862—and rkm 300, about 2.5 mm^3 alluvial material was dredged between 1920 and 1970. In the period 1971 and 1990, were dredged about 24.34 mm^3 and 17.37 mm^3 in the period 1991–2017. Mean multi-annual volume of dredged sediments along all lower Danube, between 1961 and 2005 is 2.5 mm^3/year [49]. Despite dredging works have an impact on river morphology, sediment dredged volumes over the study period could not be considered in the model because there are multiple companies performing such works with no integrated public database [50].

Another negative effect on morphology along the study reach was given by riparian and island vegetation cutting. Vegetation has an important stabilising role against erosion of the channel islets and riverbanks during floods.

All of these impacts led to: (i) reduced suspended sediment discharge, which dropped more than 4 times (from 1630 to 1890 kg/s in the period 1840–1900 to 434–454 kg/s in the period 1985–2006) along the study reach [17, 27, 51]; (ii) unnatural variations in quantity and grain size of bed sediments; (iii) planform and vertical morphological changes of the river channel; (iv) a decreased number of morphological units (islets and bars) and their total emerged surface [37].

Table 2 Flood embankments built in the study reach [42]

Crt. No	Name of embanked enclosure	Length (km)	Average height (m)	Enclosed area (ha)	Year of putting into operation	Design flow	Flow exceedance probability
1	Malu Roșu-Gostinu	18.8	5	7,265	1928/1965	– /15,770	– /2%
2	Gostinu-Greaca-Argeș	21.5	5	27,830	1966	16,500	1%
3	Greaca-Chirnogi-Argeș	19.5	5	1,840	1966	16,800	1%
4	Oltenița-Dorobanțu (Manastirea, Spanțov)	31.15	4.5–5		1928/1964	12,450/16,900	– /1%
5	Tatina/Surlari (Spanțov)/Dorobantu (Mostiștea)	3.35/2.25/8	5.5/4.5/5.8	12,750	1964/1964/1976	–	1%/1%/3%
6	Boianu Sticleanu Călărași	34.2	4	23,535	1971	–	10%

4 Data Processing

4.1 Data Processing for the Hydro-sedimentary Model

Sediment transport modelling requires multiple data, whose processing and management are very time consuming. Moreover, missing data has to be approximated by correlation functions. All data series needed for simulations and model calibration/validation, and many of the plots were prepared using a visual utility engine for data storage and processing. This utility, namely HEC-DSS Vue [52], is compatible with the hydraulic and sediment transport software, HEC-RAS [53] (version 5.07).

4.1.1 Hydrologic Data

Hydrologic data used for the modelling was acquired at the gauging stations (GS) from Giurgiu, Oltenița, Chiciu-Călărași and Călărași. The last two being very close, they share stage measurements and even sediment particle properties. Mean multi-annual discharges and historical maxima and minima for Danube over the 1913–2010 period are summarized in Table 3 [9, 18, 42, 54]. For Argeș River, the mean multi-annual discharge was obtained by summing up the corresponding values of all its tributaries upstream the confluence with Danube: Arges (over 1950–2006), Dâmbovița (1988–2006), Colentina (2006–2012) and Sabar (1958–2004) [55, 56].

Data consist in daily mean discharges (Fig. 6), water surface elevations (Fig. 7) from the period 2008–2015 [57]. All ground and water surface elevations in the paper are measured in m and referenced to Black Sea Sulina datum (m BS). Low-flow navigable water levels (LNWL) required to determine the navigation floodway in Fig. 7 are defined by the Danube Commission and the Administration of the Lower Danube (in Galati, Romania), as the minimum water level for navigation (corresponding to a discharge, Q_{min}, with an exceedance probability of 94%, over a period of more than 40 years) [50]. Floods usually occur in May–June (and have variable duration, sometimes extending over several months), whereas low waters periods usually occur in September (Fig. 6). Last historic flood over Danube occurred in 2006 and lasted for five months (February–June).

Considering the mean multi-annual discharge of the Argeș River represents only 1.16% of the corresponding Danube discharge at Oltenița GS and also taking into account the tributary natural discharge variability is highly reduced by the upstream hydropower development, in the numerical simulations the lateral inflow was considered to be constant and equal to the mean-multi-annual value of 71.2 m^3/s from Table 3.

Table 3 Main flow parameters at the gauging stations [9, 18, 42, 54]

Crt. No	Gauging station name	Distance to Black Sea mouth (Sulina)	$Q_{\text{mean multi-annual}}$	Q_{max} (1913–2010)	Date	Q_{min} (1913–2010)	Date
		rkm	m^3/s	m^3/s	–	m^3/s	–
1	Giurgiu	493.1	6,044	16,300	16.04.2006	1,485	01.1954
2	Argeș tributary confluence	431.8	71.2	2,580	22.06.1979	10	–
3	Oltenița	429.8	6,115	16,200	05.1897	1,490	01.1954
4	Chiciu-Călărași	379.58/375	6,133	16,200	16.04.2006	1,530	1947
5	Călărași/on Borcea branch, at 5 km from split	96.4 (Borcea branch)	946	3,203	2.05.1845	132	6.10.1958
6	Izvoarele/on old Danube	348.6	5,148	15,751	16.04.2006	1,858	6.10.1858

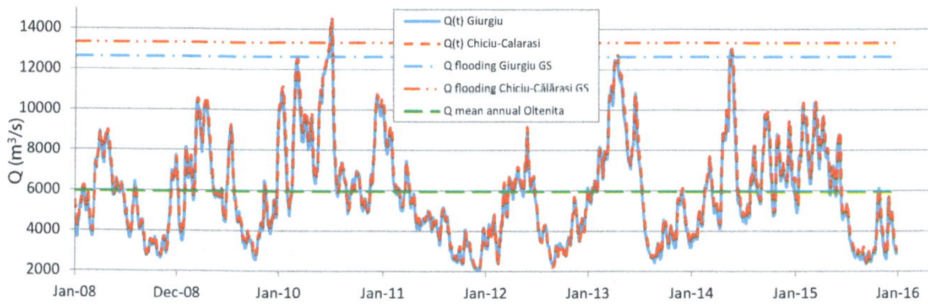

Fig. 6 Measured discharge values at the three GSs along the studied reach in the period 2008–2015. Dashed lines represent flooding flows and dotted lines the mean multi-annual flows [57]

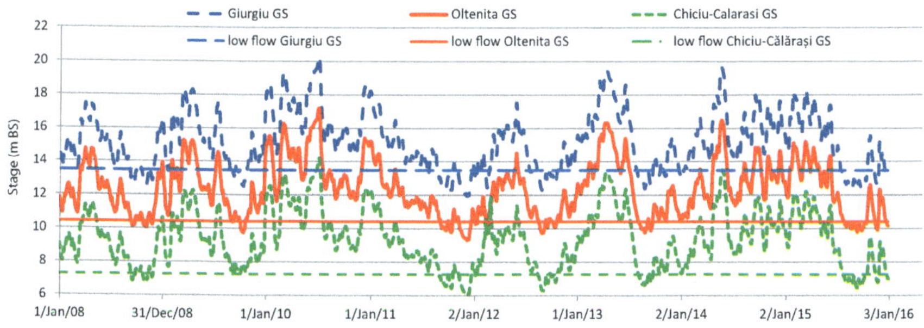

Fig. 7 Observed daily water surface elevations at the three GSs along the studied reach during the period 2008–2015. Dotted lines represent the low-flow stages [57, 58]

4.1.2 Sediment Data

It is notorious that sediment data are usually rare, being difficult to be measured [59, 60]. This is why modellers are often required to make assumptions and assess these assumptions' sensitivity and uncertainty throughout the modelling process [61].

The *nature* of sediment particles transported by the river is mainly silica, with a relative density of 1.65 kg/dm^3. The *main grain class* of Danube bed-material is sand with a median diameter d_{50} from 0.1 to 2 mm. However, smaller and larger classes are also encountered from clays and silts (with mean diameter $d_{50} < 0.032$ mm) to coarse gravels (up to 16 mm) [40, 50, 51]. Sediments with diameters over 0.08 mm are considered by Bondar [27] the most important grain class for the geomorphologic processes affecting the riverbed.

From the point of view of *transport mechanisms*, smaller particles (clay, silt, fine and medium sand) are transported exclusively by suspension (*wash load* and *suspended load*) under the influence of current turbulence, depending on the water discharge. The coarser fractions are carried at higher flows as *bed load* by dragging, rolling, saltation or partially lifting *in suspension* when the water current velocity and corresponding shear stress exceed their critical (entrainment) limits [51, 62]. At Chiciu-Călărași GS, for example, the highest fraction (75%) of total sediments are

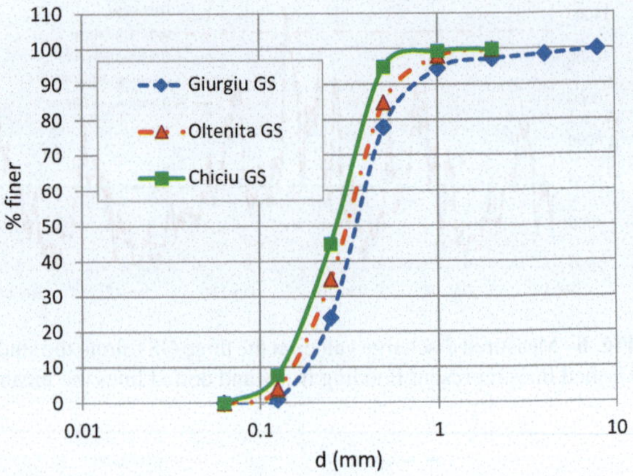

Fig. 8 Riverbed sediment grading curves at the gauging stations from the study reach used in the sediment transport model

clay and silt transported as wash load and about 25% is sand, transported either in suspension (21%) or as bed load (4%) [49].

Daily mean sediment suspended discharge (Q_{ss}) values are calculated at the GS depending on discharge and turbidity, and therefore include errors from measuring other variables (such as stage, as well as velocity and suspended sediment discharge periodically sampled in multiple verticals of a cross-section). Bed material is sampled from a surface layer of the riverbed with a specially developed equipment (by the National Administration "Apele Române"/National Institute of Hydrology and Water Management) [38].

Mean riverbed sediment grading curves used in the model at the three GSs are shown in Fig. 8. Characteristic diameters were calculated with the grain size distribution statistics calculator of Parker [63] and resulting d_{50} are: 0.023 mm for Giurgiu GS, 0.025 mm for Oltenița GS and 0.024 mm for Chiciu-Călărași GS. These grading curves were interpolated along the study reach to account for *spatial distribution* in the simplified modeling scenario.

In terms of channel *depth stratification*, the only available information is from Dumitrescu [64], who analysed at Bechet GS (rkm 689) sediment grading curves of core samples taken from riverbed, for 5 slices, each of 1 m depth. The authors did not find important differences in sediment grading of the riverbed for the top 4 m, but found an increase of the mean diameter for the deepest slice (4–5 m depth), where $d_{50} = 0.6$ mm. This erodible depth limit of the riverbed was considered the same for modelling the study reach. No data on sorting, nor on armouring or depth of an active layer was available.

Fluvial morphological alteration associated with sediment transport are usually related to a lack or to a surplus of *bed load discharge* (Q_b) and/or to anthropogenic impact upon the natural processes [15]. However, this component of the *total sediment transport rate* (Q_t) is notoriously difficult to be measured. Over the study period no series values were available. Rákóczi [65] estimated bed load share from the total sediment transport represent about 1% along low-land sections of Danube, whereas Bondar and Iordache [51] and Batuca and Buta [66] state a 2–5% share of suspended

de Tragere" (scale 1:20 000), which dates from 1889 to 1916 and are distributed by the Romanian community geo-spatial.org, (ii) Romania Topographical Map (scale 1:25 000) from 1974 to 1980, available for the teachers and researchers from the University of Bucharest, through the WMS service of the OpenGIS project, and (iii) satellite imagery of the Landsat 8 from 2013, downloaded from USGS EarthExplorer website. The authors are aware of the various discharge conditions depicted at the time of mapping or satellite image capture. The cartographic documents were chosen from the following considerations: the better accuracy of these maps compared with other historical cartographic materials and free access to all these materials. The satellite images were searched for the early summer months (June or July) of the years 2008–2015, the interval for which the hydro-sedimentary model was made. The image from 2013 was chosen for the reasons related to the availability of the products and the cloud cover.

5 Methods

5.1 Sediment Transport Numerical Modelling

Over more than the last century the scientific community has devoted extensive effort to predict the sediment transport rate by different equations. However, given the complexity of the two-phase, sediment-water flow field with variable distribution range of particles in terms of size and shape, no particular empirical formula has been universally accepted in the field for prediction of solid transport rate [72]. The challenge of sediment transport modelling is further complicated by the influence of model parameters on the computed sediment discharge.

Even though 2D and 3D sediment transport models are currently used, 1D models are still the best choice for long period of simulation times (over a decade or more) and long reaches (provided good accuracy topobathymetric data are available), when observed data for calibration are limited at the gauging stations. They have the advantage of smaller computation times, less computation resources and easier model setup. Taking into account that spatial sediment data are often rare (usually they are measured only at the gauging stations), building a more complex (2D, 3D) model with approximate data does not add value to the study. However, the simple 1D models have their drawbacks in terms of approximations and limitations; for example they are not able to capture hydro-morphological phenomena across the flow direction, in bifurcations, confluences, or around islands. Therefore, for such detail areas, where both horizontal components of the velocity vectors are comparable, a higher complexity model should be used, with boundary conditions resulted from the 1D model. One of the 1D/2D computational software is HEC-RAS (version 5.07), which has a good graphic interface in GIS environment with an experienced community of users.

5.1.1 1D Hydro-sedimentary Mathematical Model

HEC-RAS is a coupled hydraulic and sediment equations model that solves the Saint-Venant equations (for water, considered a Newtonian liquid at solid volumetric concentrations less than 5–10%), together with the Exner sediment continuity equation, the system being solved over a control volume:

$$\begin{cases} \frac{\partial h}{\partial t} + \frac{\partial V h}{\partial x} = 0 \\ \frac{\partial (V h)}{\partial t} + \frac{\partial (V^2 h)}{\partial x} = -\frac{1}{2} g h \frac{\partial h}{\partial x} - g h \frac{\partial \eta}{\partial x} - C_f V^2 \\ (1 - \lambda_p) \frac{\partial \eta}{\partial t} = -\frac{\partial q_t}{\partial x} + D_S - E_S \quad q_t = q_t(V) \end{cases}, \tag{1}$$

in which h is the hydraulic depth (in m), x—the coordinate along flow direction, V is the mean velocity (in m/s), λ_p is the active bed sediment porosity (nondimensional, necessary to translate mass change of water and sediment mixture into volume change), η is—the bed elevation, Q_t—the volume transport rate of total sediment (suspended and bedload, i.e. by rolling, dragging or saltation) q_t—the volume transport rate of total sediment per unit stream width = Q_t/B (in m^2/s), D_s is—the deposition rate (volume rate per unit time per unit bed area stored from the water column on the bed, in m/s), E_s is—the erosion rate (volume rate per unit time per unit bed area removed from the bed by suspension into the water column, in m/s), g is—the gravity and C_f is—the bed friction coefficient. The Exner equation translates the difference between inflowing load ($Q_{t\ in}$) and outflowing load ($Q_{t\text{-}out}$) from the control volume associated with each cross-section into bed change, eroding (in case of sediment deficit) or depositing (in case of sediment surplus).

While the hydrodynamic routing is performed from the downstream boundary of the modelled reach to the upstream boundary (backwater simulations), the sediment routing is computed with the flow (upstream to downstream). The non-linear equations are solved through the finite difference method. Computation time steps should be low enough so that bed changes should not produce computation instabilities.

To Eqs. (1), other mathematical formulations are added to the system, to turn the hydrodynamics into sediment transport capacity of the river. This depends on the excess stream energy over a critical value (in terms of shear stress or stream power), for which the flow begins to selectively move bed particles, depending on their diameter, shape and sorting. The modeler has to choose: (i) the most appropriate transport mathematical formula for the river hydro-sedimentary specific conditions [73]; (ii) the bed vertical sorting or armouring method and (iii) the equation expressing the fall velocity of particles. Since all these functions are empirical and obtained under certain conditions, the model has to be calibrated in order to reproduce the sediment flow as realistically as possible.

Different empirical transport functions are defined in the literature for the non-cohesive particles, used as simplifying assumptions for the study reach. In HEC-RAS library [74], one may choose from the Ackers and White, Engelund and Hansen, Laursen-Copeland, Meyer, Peter and Muller, Toffaleti, Yang models (Table 5). The choice of Laursen-Copeland equation [75, 76] was made based on river sediment

Table 5 Range of main input values for sediment transport functions

		Ackers and White (1973)	Engelund and Hansen (1976)	Laursen-Copeland (1958, 1989)	Meyer, Peter and Muller-Toffaleti (1948, 2003, 2007)	Toffaleti (1969)	Yang (1973, 1984)
1	Laboratory flume/field	Flume	Flume	Flume and field	Flume	Flume	Field
2	Particles diameter (mm)/d_{50}	Uniform non-cohesive sands and fine gravels (0.04–7)/N.A	Sand with substantial suspended load; N.A./$d_{50} \in 0.19 \div 0.93$	Sand and gravel 0.08–0.7 (field)/0.011 ÷ 29 (flume), variable	Sand and mostly gravel 0.4 ÷ 29/N.A	Sand, 0.062–4/$d_{50} \in 0.095 \div 0.76$	Sand 0.062 ÷ 1.7 and gravel 2.5 ÷ 7
3	Dependent variable	Q_{st}	Q_{st}	Q_{st}	Q_{sb}	Q_{sb}, Q_{ss}	Q_{st}
4	Flow characteristics	V, h, u^*, d	V, h, S, d	V, h, u^*, S, d, ω	V, u^*_B, d, h, S	V, d, S	V, u^*, d, S, ω

Where: u^* = shear velocity (m/s), V = mean velocity (m/s), h = hydraulic depth (m), S = energy slope (−), d = characteristic/median diameter (m), ω = fall velocity (m/s), Q_{st} = total sediment load (mass) transport rate (kg/s), Q_{sb} = bedload sediment transport rate (kg/s), Q_{ss} = suspended load sediment transport rate (kg/s) characteristics and previous modelling results obtained along the study reach [29, 72, 77] This function computes the total sediment discharge based on mean velocity, flow depth, energy slope, sediment gradation, shear and fall velocities.

The other two functions relate to physical processes limiting the sediment continuity equation. To compute critical fall velocity, the Rubey function [78] was chosen from Toffaleti, Van Rijn, Report 12 and Dietrich models [74], based on grain size distribution and bed characteristics. The last mathematical formulation accounting for sorting, mixing and armouring mechanisms from the mobile bed surface layer was chosen to be Thomas (from the Copeland and active layer's formulations) [74].

5.1.2 Model Geometry

347 cross-sections were extracted from the DEM in Ras Mapper (the GIS environment under the 2D version of HEC-RAS 5.07), at a mean distance of about 350 m, following the crossing bathymetric trajectories (in order to reduce the interpolation errors). An example of the 1D geometry mesh of the model is shown in Fig. 15.

The red points in Fig. 15 representing main channel bank limits (where roughness coefficient changes from the main channel values to the floodplain ones), were initially placed from orthophotos and satellite images and then adjusted in each cross-section, according to bank inflection points and, after initial simulations, according to discharge values [79].

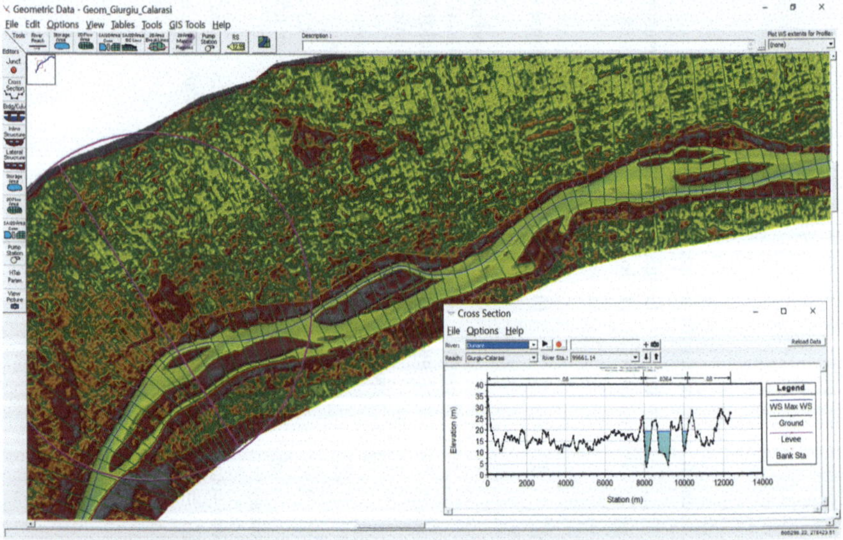

Fig. 15 Extracting the 1D model geometry in HEC-RAS by cutting the DEM with cross-sections (detail over the study reach)

Table 6 Boundary and initial conditions for the hydraulic and sediment transport model

Model	Condition type	Location	Used/necessary data
Hydraulic model	Boundary conditions	Upstream cross-section (Giurgiu GS)	Flow hydrograph (daily records)
		Danube confluence with Argeș River	Tributary inflow
		Downstream cross-section (Chiciu-Călărași GS)	Stage hydrograph (daily records)
	Initial condition		Water surface profile computed for initial value of the discharge
	Calibration coefficient	All cross-sections	Manning roughness coefficients for the channel and floodplain
Sediment transport model	Boundary conditions	Upstream cross-section (Giurgiu GS)	Inflow sediment rating curve ($Q_{st} = f(Q)$) Sediment load-gradation curves
		Danube confluence with Argeș River	Lateral sediment rating curve ($Q_{st} = f(Q)$) Sediment load-gradation ranges
	Initial conditions and transport parameters	All cross-sections	Bed sediment grading curves Maximum erodible depth/width Water temperature (daily records) Transport function—Laursen-Copeland Sorting method—Exner 5 Fall velocity method—Rubey

5.1.3 Boundary and Initial Conditions

The boundary and initial conditions used in the hydraulic and sediment transport model are summarized in Table 6.

5.2 GIS Analysis of Cartographic and Imagery Sources

Historical maps and satellite imagery were used to extract a series of geospatial data, necessary for the diachronic analysis and various thematic maps. To improve the

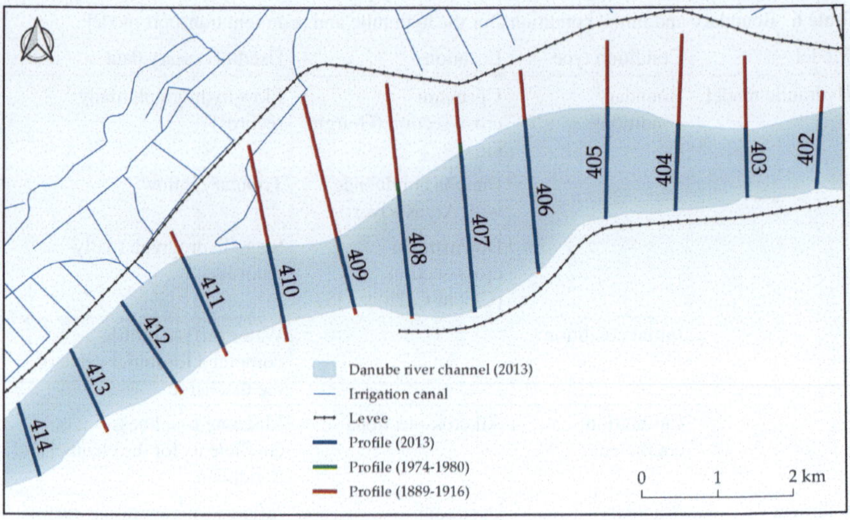

Fig. 16 Cross-section profiles for estimating river-channel width

spatial resolution of the satellite imagery, from 30 to 15 m, a pan-sharpening was performed, using the panchromatic band.

All geospatial data were processed, digitised and analysed in the open-source application QGIS. To analyse spatial-temporal planar changes of the riverbed relevant morphometric parameters were calculated, such as: the channel width and the number of fluvial islets and their emerging area. The width of the Danube River main channel in the study area was estimated using a GIS analysis of the polygons created for the riverbed from the historical maps and satellite images, aforementioned in 4.2. The widths were measured using a set of cross-sectional profiles drawn every river kilometre. An example of how the Danube channel widths were calculated is shown in Fig. 16.

The area and number of the geomorphological landforms of the river channel (islets with vegetation) were calculated from the cartographic and satellite image materials using the function *$area* from QGIS software.

6 Results

6.1 Model Calibration

A sediment transport model relying on hydraulic parameters (flow, depths, velocities, shear stresses etc.) has to start with a well calibrated hydraulic model, which should accurately reproduce the flow over the whole range. Initial estimates of the global Manning roughness coefficients of the main channel and floodplain along the two

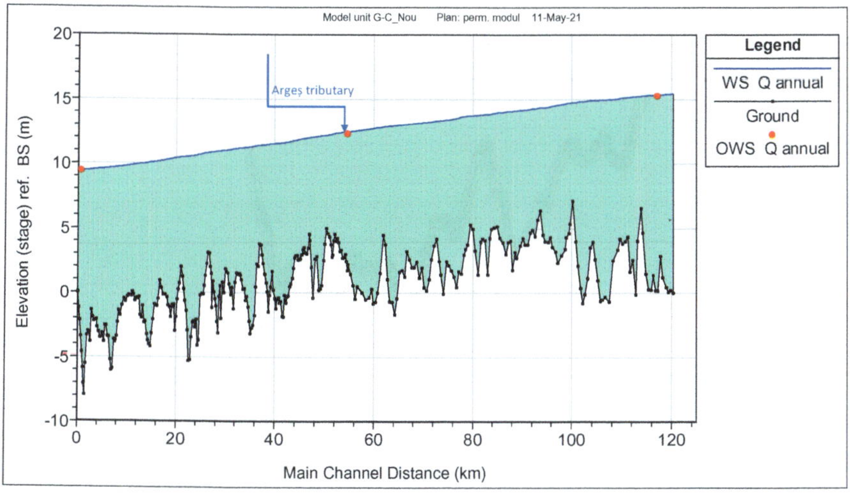

Fig. 17 Water surface profile for the mean annual flow (distance is measured from downstream Chiciu-Călăraşi GS); WS = water surface, OWS = observed water surface

sub-reaches were obtained under steady flow regime simulations for the multi-annual discharges, by calibrating the model at Olteniţa and Chiciu-Călăraşi on observed water stages (Fig. 17).

Hydraulic model calibration under unsteady flow conditions was performed by using a flow-varying adaptive roughness factors procedure of [80] for the year 2008, and validated on various hydrologic events over the entire study period 2009–2015. The model accuracy was assessed on deterministic metrics such as the Mean error (ME) and Root Mean Square Error (RMSE) in the form of following objective functions [80]

$$ME = \frac{1}{n}\sum_{i=1}^{n}(z_{icomp} - z_{iobs}). \tag{2}$$

$$RMSE = f = \sqrt{\frac{1}{n}\sum_{i=1}^{n}(z_{icomp} - z_{iobs})}. \tag{3}$$

In Eqs. (2) and (3) n = the number of data, $z_{i\text{comp}}$ = computed and $z_{i\text{obs}}$ = observed water levels both at the same GS and at the same time.

Calibration was performed by adjusting the flow-roughness factor until these functions were minimised at Olteniţa and Giurgiu GS. Global roughness values vary roughly between 0.027 and 0.028 for Giurgiu-Olteniţa reach and 0.029–0.03 for the Olteniţa-Călăraşi reach within the main channel and from 0.078 to 0.086 for the areas in the floodplain. In Fig. 18 are shown some model calibration performances in terms of computed vs observed stages and discharges from 2008, whereas in Fig. 19 are

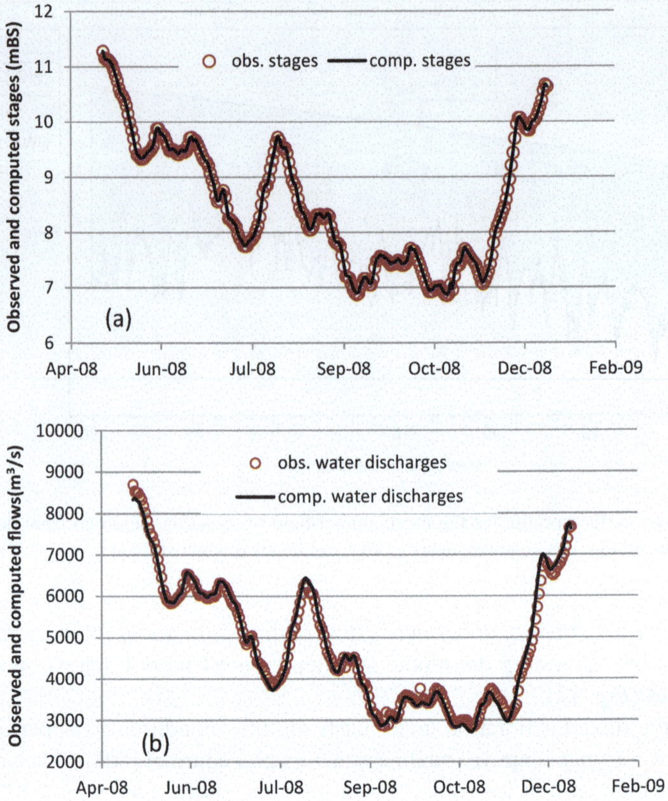

Fig. 18 Calibration results: (**a**) computed and measured stages at Giurgiu GS. Simulation results: (**b**) computed and measured flows at Chiciu-Călărași GS

shown the model validation performances in terms of difference between computed and observed stages with discharge and time at Giurgiu GS and Oltenița GS.

Most absolute stage errors lie between ±0.2 m, which means very good relative errors in terms of depth (of under 2–5%). Therefore, the hydraulic model can be considered to adequately reproduce real flows along the study reach over the whole range of possible discharges.

No second bathymetry was available during the study period, therefore the model calibration was performed by comparing computed with observed sediment discharge values at Oltenița and Chiciu-Călărași GSs. Since sediment transport is strongly non-linear and most of the load and bed changes are concentrated in relatively brief periods of high flows, models require small computational time steps. This is why a sensitivity analysis was performed at decreasing values the computational time step (from 1 h to 5 min). The highest value was retained for further simulations, from which, no differences in bed elevation were noticed by further decreasing it. Best results obtained with the Laursen-Copeland function are shown in Figs. 20 and 21

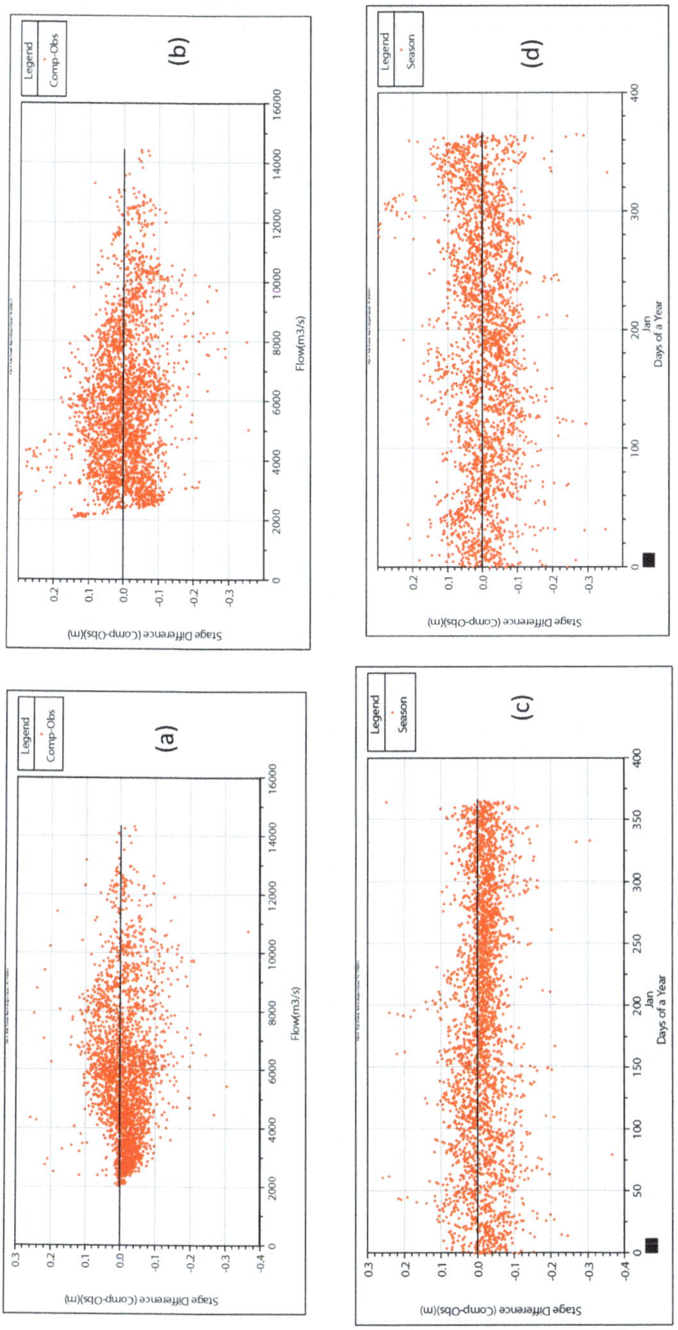

Fig. 19 Validation results: stage difference (computed—observed) with discharge (**a**) and (**b**) and with time (**c**) and (**d**) at Giurgiu GS and at Oltenița GS

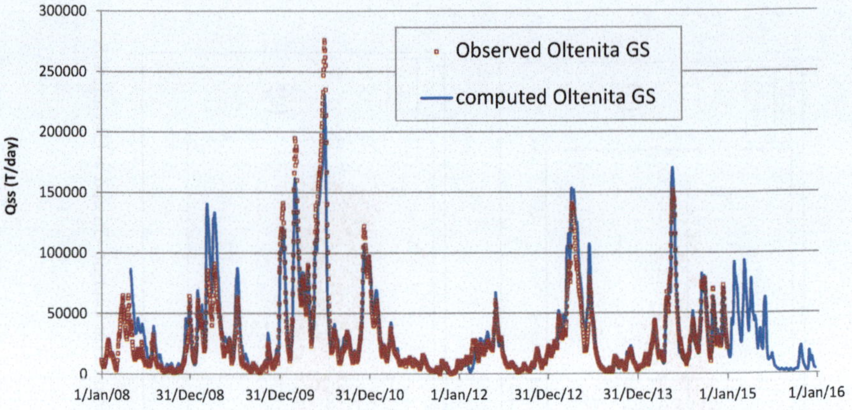

Fig. 20 Comparison between observed and computed total sediment discharge at Oltenița GS

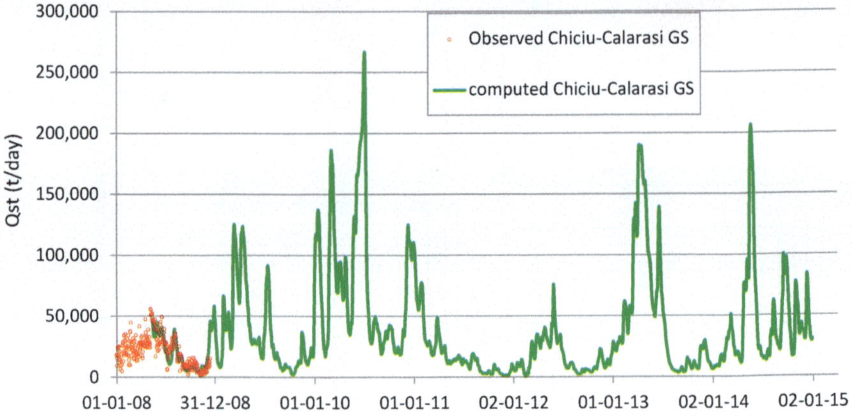

Fig. 21 Observed versus computed total sediment discharges at Chiciu-Călărași GS

at Oltenița and Chiciu-Călărași GS, respectively. For the downstream GS observed solid discharge values were available only for the first year.

Given the lack of data on bedload discharge and the simplifying assumptions on tributary inflow, only a qualitative comparison between observed and computed sediment discharges could be performed for the year 2008 at Chiciu-Călărași GSs and for the years 2008–2014 at Oltenița GS. One may observe the highest differences are obtained during flood. This is expected in case of a strongly non-linear model, for which sediment discharge is a power function of water flow.

6.2 *Planform Morphological Changes*

The results of channel width measurements performed on maps and images (Fig. 22) show the study reach is part of the same evolution trend as other rivers in Europe, by narrowing or widening to maintain planform stability [45, 81, 82]. One can identify reaches where the river channel widens, due to erosion in the concave bank, such as near Gostinu village (rkm 471–480) (Fig. 23). Also, one may observe reaches where the river narrows down, such as the one near Giurgiu town (Fig. 22), due to structural works built since the XVIIIth century [44], and near Stancea-Chiselet-Manastirea village (Fig. 23) (rkm 403–414), due to the Mostiștea River alluvial fan. Fluvial processes were altered because of structural works that have been built in this area, especially between 1963 and 1978.

(a)

(b)

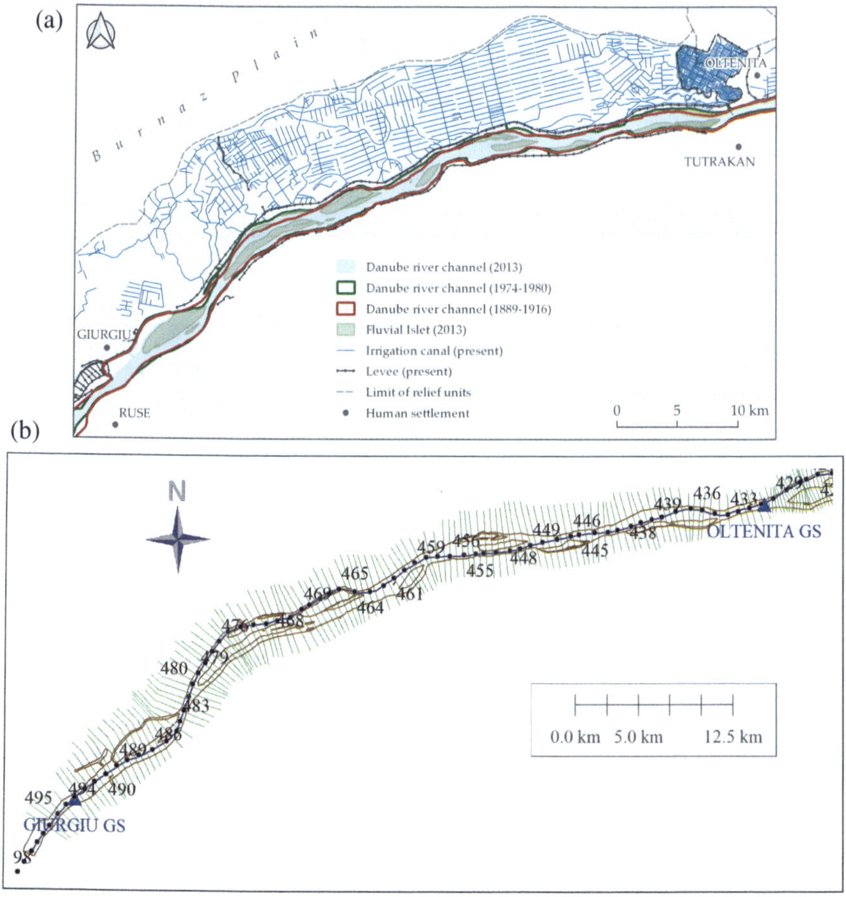

Figure 22 **a** Planform changes of the Danube River channel between Giurgiu-Oltenița. **b** Chainage in rkm and islets

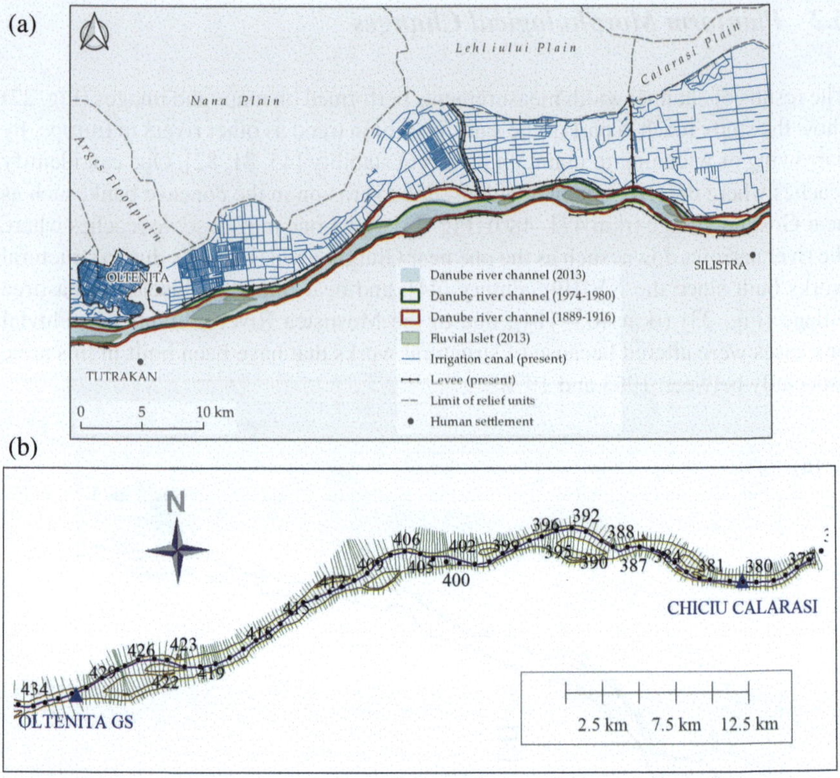

Figure 23 **a** Planform changes of the Danube River channel between Oltenița and Călărași. **b** Chainage in rkm and islets

Posner-Nenciu and Armas [44] have identified the following design mistakes: (i) the wrong location choice of the first flood embankment of 1928; (ii) during the crest raising works of 1963, the embankment did not follow the convex and concave meanders of the channel banks, which favoured erosion in the convex area; (iii) the distance to which an embankment should be built from the river bank was shortened by about 100 m; (iv) in order to build the embankment, riparian vegetation has been cut [37].

Channel width is one of the important morphological features of a stream that can be related to lateral migration of the fluvial banks or to the presence/absence of islets. In Figs. 24 and 25 is shown the measured river channel width displaying the alternating widening and narrowing sub-reaches, from rkm 499 down to rkm 375, during different time periods.

In Fig. 24 one may observe frome the river channel width measurements performed on maps and satellite images, that between 1889–1916 and 1974–1980 periods there are sub-reaches where the Danube channel narrowed down to 1.8 km

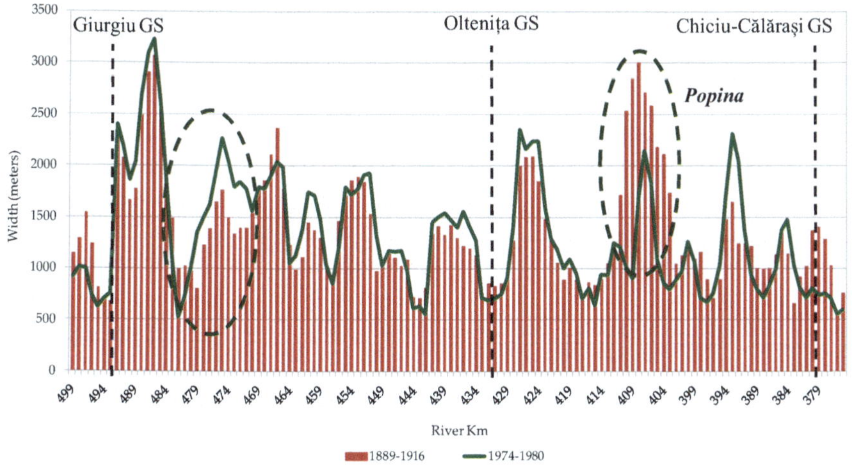

Fig. 24 Changes in the width of the active channel on historical maps called "Planurile Directoare de Tragere" (1889–1916) and Romania Topographical Map (1974–1980)

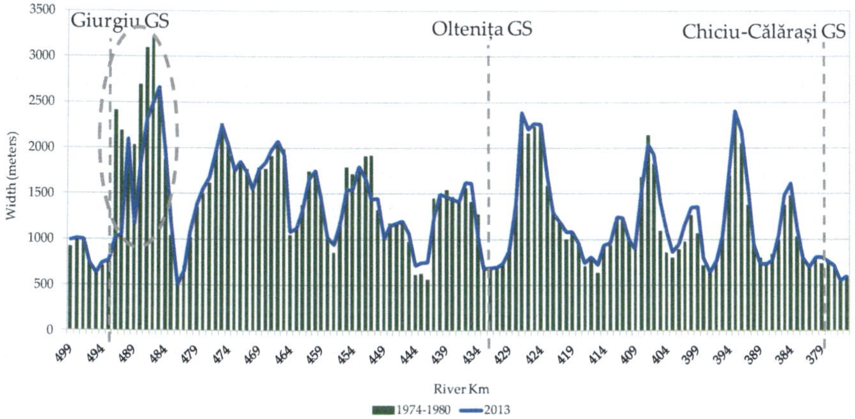

Fig. 25 Changes in the width of the active channel on Romania Topographical Map (1974–1980) and Landsat 8 Image (2013)

(for example rkm 409) and reaches where the Danube channel widened by over 0.5 km (for example rkm 475–473 or rkm 393–392).

Figure 25 shows a comparison between measured widths from the 1974–1980 period and 2017 year. Significant changes can be observed between rkm 492 and rkm 483 downstream Giurgiu port), where the river channel narrows down to 1 km (for example rkm 487).

These morphological changes should be analysed in a hydraulic context. Therefore, with this purpose, the computed ratio width/depth has been represented in Fig. 26 as a function of velocity for the mean multi-annual discharge. One may observe from

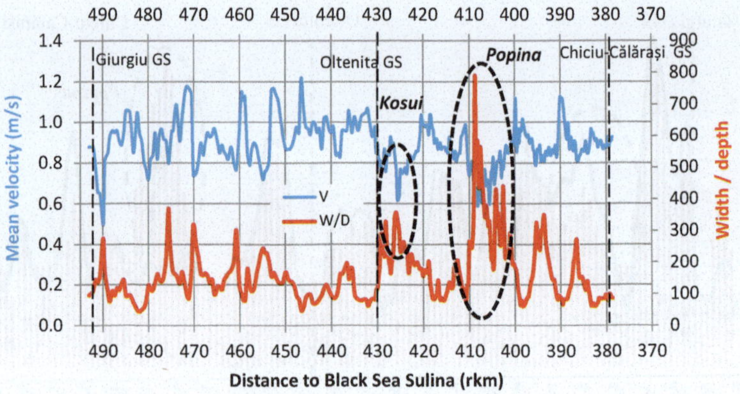

Fig. 26 Computed velocities and width/depth ratio for the mean multi-annual flow. Along the study reach (highlighted navigation critical spots and observed morphological alterations)

Figs. 24, 25 and 26 the aforementioned critical sub-reaches with a deposition trend correspond to areas where channel width has high width/depth ratios and where the velocities are very low and vice-versa.

Over the last century the number of fluvial islets from Giurgiu and Călăraşi decreased, varying from 43 in 1889–1916 to 37 in 1974–1980 and 2013. However, one may notice from Fig. 27 that the upstream Giurgiu-Olteniţa reach has a higher fluvial islet density than the downstream Olteniţa-Călăraşi one. The total area of these landforms varied in time from 34.09 km^2 (1889–1916) to 35.46 km^2 (for the interval 1974–1980—after the commissioning of the dams) and to 34.91 km^2 for year 2013. An important role in islets morphodynamics is given by flood frequency

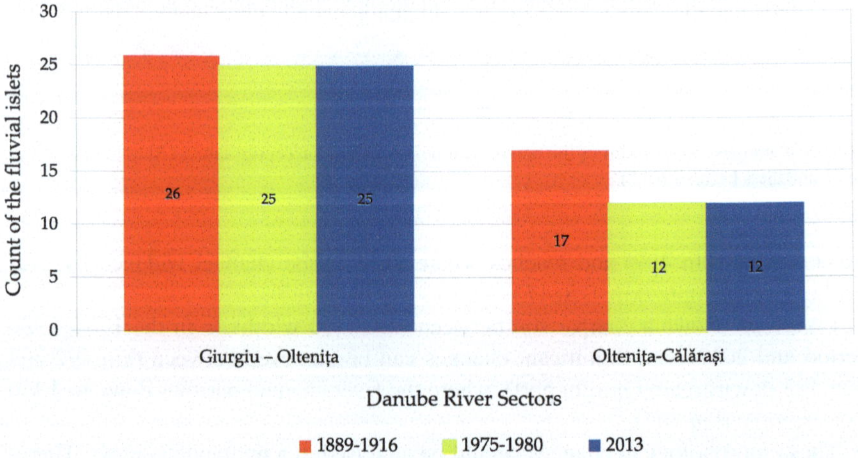

Fig. 27 Time evolution of the number of islets along Giurgiu-Olteniţa and Olteniţa-Chiciu Călăraşi

and intensity [11], as the morphological units act as sediment sources and location of deposition during these periods, particularly along sediment deficit river reaches, such as the study one.

6.3 *Vertical Morphological Changes*

In Fig. 28 one may see that the mean riverbed slope calculated from the bed elevation profile (thalweg line) is lower along the upstream reach than along the downstream reach. This difference in stream gradient is expected to lead to increased hydromorphological alteration and corresponding erosion-deposition dynamics along the downstream reach. A thalweg elevation raise of more than 5 m can be observed downstream the confluence with Argeș River. This is a well-known navigation bottleneck and shallow area, next to Kosui islet, where processes of aggradation are produced mainly due to the tributary sediment inflow, entailing downstream enlargement of the river channel (see Figs. 24 and 25). Another critical spot for navigation is between rkm 401 and 408 (next to Popina settlement/Bulgaria), emphasized in Figs. 24, 26 and 28, where a submerged sand bar needs periodic dredging. Both critical areas have been documented and investigated through the Fast Danube European project [83]. Overall degradation of up to few meters was documented by the DanubeSediment project along Giurgiu-Călărași study reach between 1962 and 2017 [50], with local aggradation reaches, mostly downstream the Argeș confluence.

The downward movements in the Tutrakan depression enforced the confluence location of Argeș with the Danube (Fig. 29). The large sediment loads transported by the tributary in the past have built a submerged alluvial fan, which increased the bed slope along the Oltenița-Călărași reach. Intense erosions down to bare rock (4 m in height) upstream Argeș confluence were recorded as a fluvial response to the cascade of dams, embankments and complex channel works from the late '80 s and also to the intensive sediment mining along the downstream stretch after 1990. All these led to dramatical alteration of the Argeș River morphological equilibrium [84] with possible consequences in the Argeș-Danube morphodynamical feedback response.

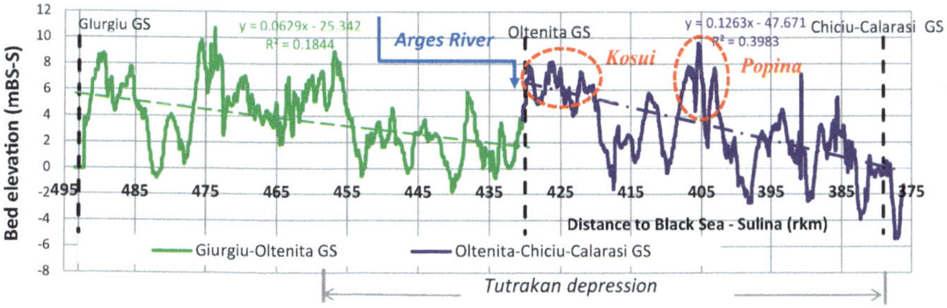

Fig. 28 Thalweg line and bed slope over the Giurgiu-Călărași reach

Fig. 29 Simplified tectonic map of the Lower Danube study area (adapted after [85]). IMF—Intramoesian Fault (Tinosu-Fierbinți-Călărași), COF—Capidava-Ovidiu Fault, BDF—Balchik-Dulovo Fault, KVF—Kubrat-Vetrino Fault, LF—Lit Fault, OF—Osam Fault

On the other hand, the tectonic uplift near Chiciu-Călărași (at rkm 370) most probably triggered Danube River to split and create the left Borcea branch, thus reducing the bed slope and velocity and further decreasing the total width downstream rkm 370 (from about 1000 to 650 m along the Old Danube and to 170 m along the Borcea branch). These led to intensified aggradation processes along the Old Danube (right branch) [18]. Because of that, the reach downstream of rkm 370 is very critical for fluvial navigation.

In Fig. 30 computed annual sediment erosion deposition patterns are shown for two typical years: one wet (2010) and one dry (2011). One may see these processes are much more intense over the downstream reach and locally reversed between the dry and wet years, this mechanism being part of the equilibrium morphological adaptation of the fluvial system [86–88]. The observed pattern is a very small time-segment of a slow aggradational cycle, due to boundary conditions (e.g., sea level rise). Therefore, in a cumulative plot like the one from Fig. 31, over a period of many years (2008–2015 for the computation case) results depend very much on the predominance of dry/wet years. Therefore, a long-time scale of analysis is needed to assess the impact of climate change on river morphology.

Cumulatively, in space and over the 8-year study period, the upstream reach proves to be much more stable morphologically on a timescale of the order of decade, whereas the downstream reach has a slight degradational trend. This trend has also been notified by other studies [40]. Batuca and Buta [66] reported for the two-decade period between 1980 to 2000 an overall bed degradation of 0.65 m at rkm 504 (in cross-section over 2.5 m erosion of the left bank and a 2 m deposition on the right bank), a local erosion more than 2 m at rkm 476 and an overall aggradation of 0.5–0.75 m at Oltenița GS. For this Arges-Danube confluence region near Oltenita GS a 2D model should be set up, since the riverbed displays a large deposition area of 5 m in height on the left bank and a 2–3 m erosion in depth on the right bank.

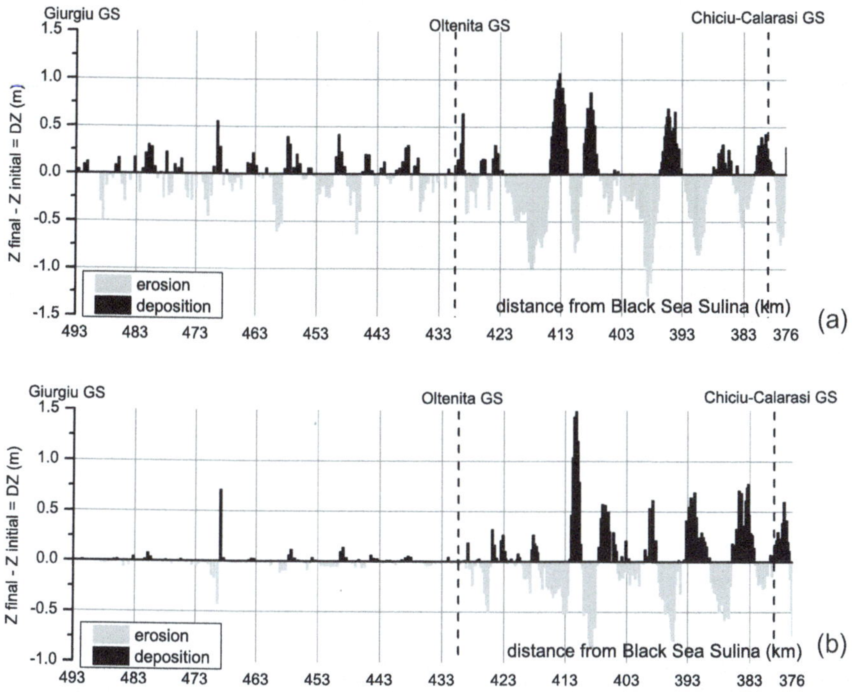

Fig. 30 Annual erosion–deposition over the study reach for a a) wet year (2010) and b) a dry year (2011)

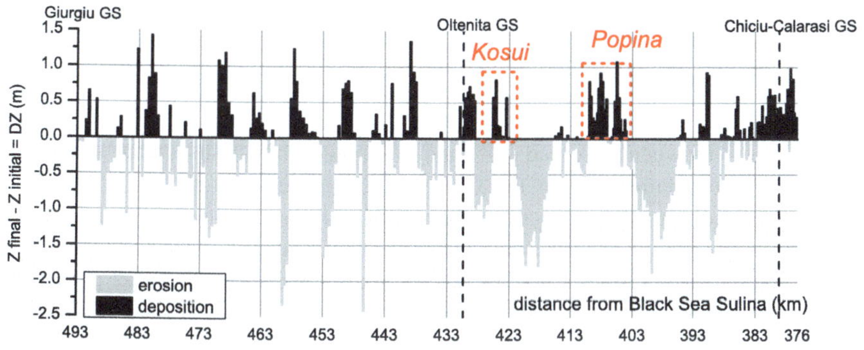

Fig. 31 Cumulative vertical changes along Oltenița Călărași reach observed during the study period (2008–2015)

Intensive local erosions have also been observed at rkm 389–391 between 2008 and 2017, when another bathymetry was performed [40]. Navigable maps and reports on improving navigation conditions along Danube [70, 89] show corresponding fairway bottlenecks or shallow areas (critical spots with low water depths) at the following locations: rkm 472–476 (near Gostinu islet), rkm 460–463 (near Mishka islet), rkm

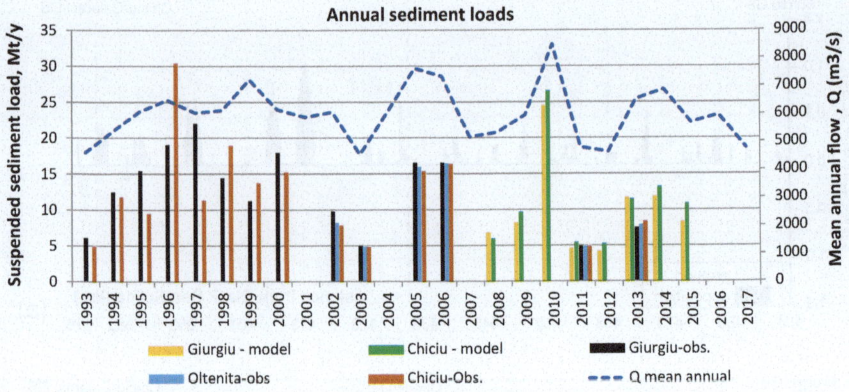

Fig. 32 Sediment budgets at the gauging stations. Comparison between observed and computed annual sediment loads (budgets)

455–458 between Giurgiu and Oltenita GSs and rkm 425–428 (downstream Oltenita, near Kosuy ilsets), rkm 420–422, rkm 410–412 (near islet Albina) rkm 404–407, rkm 401–399 and rkm 382–385 between Olteniţa and Chiciu-Călăraşi GSs.

However, given the chaotic nature of the sediment transport processes and of the simplifying assumptions of the 1D numerical model, erosion/deposition plots in Figs. 30 and 31 should be interpreted with caution. This is mainly because degradation/aggradation processes do not occur in fact uniformly over the wetted perimeter of a cross section.

6.4 Sediment Budgets

A comparison between observed [18] and computed annual suspended sediment loads at gauging stations along the Giurgiu-Călăraşi reach is shown in Fig. 32. Unfortunately, observed data were available for only two years (2011 and 2013) during the study period. However, the mean multi-annual suspended sediment load over the study period (2008–2015) and corresponding value from another period of observations (1993–2006) are of the same order of magnitude: at Giurgiu GS 10.2 and 12.8 Mt/year respectively, while at Chiciu-Călăraşi GS 11.2 and 12.5 Mt/year, respectively.

7 Conclusion

As most of the European large rivers, the long-term sediment balance over the Lower Danube has strongly been disturbed mainly due to the combined impacts of flood protection measures, dredging works and hydropower dams. Consequently, there has been a reported solid discharge decrease o nearly 1/4 of the values prior to 1900 [17, 27, 51].

By combining a longer time scale morphological analysis with a detailed spatial investigation over a shorter time from a sediment transport model, present study offers an advancement in understanding river morphological processes over the Giurgiu-Călărași reach under hydrodynamical context, considering also the natural and anthropogenic drivers.

The hydro-sedimentary model presented in the paper proves how complex and difficult a task is to predict sediment discharges in natural rivers, even when all the necessary data are available and the complexity of the model is the lowest (1D). This type of modelling requires careful data management, knowledge of the algorithms, skilful model application and calibration and contentious interpretation of results.

Spatial and temporal variation of hydrodynamic parameters and bed elevation in each cross-section of the 1D model geometry, together with annual cumulative sediment loads were obtained from numerical simulations. The results highlight the areas with dominant erosion / deposition processes due to the solid transport occurring during floods.

Results show the upstream Giurgiu-Oltenița sub-reach is more sediment-balanced than the downstream Oltenița-Călărași one, the entire reach having a slight overall deficit in sediments over the modeling period 2008–2015. Areas with intense local erosion confirm the general incising trend of a disrupted continuity fluvial system exposed to human disturbances. Riverbed degradation also lowered the groundwater table and led to changes in floodplain ecology reported by other studies.

Critical areas of local aggradation have also been observed from both the diachronic image analyses and from the hydro-sedimentary model results. The most important one is downstream the confluence with Argeș River, where the tributary has created over the time a submerged alluvial fan. Before the construction of the cascade of hydropower dams, Argeș brought into the fluvial system large quantities of sediments. This contributed to an increased slope along the downstream Oltenița-Călărași sub-reach and to increased channel morphodynamic behaviour. Most of the deposition areas found into the study are also reported by the online navigation maps as shallow waters or bottlenecks causing navigation problems, especially during low-flow periods. The diachronic image analysis proved additional long-term morphological planform alterations, such as: channel width modifications, islets migration and changes of their emerged area.

The interdisciplinary study is a step forward in searching for different points of view on the hydrodynamics of sediment transport and morphological response of the Danube fluvial system. An improvement of the sediment model could be obtained if the computed morphology is to be compared with a subsequent bathymetry (such as the one from 2017). However, one should not forget about the limitations of a simplified, one-dimensional model.

8 Recommendation

The following recommendations could contribute to a better understanding of the complex hydro-sedimentary behaviour of river systems subjected to multiple anthropogenic stressors:

- to design a national database, integrating activities such as mining of riverbed aggregates for construction works or local governments' needs, and dredging for navigation purposes. This database should use information provided in environmental and water management consents and/or permits. Such a database should be publicly available to the scientific community.
- to set-up a GIS database for locating the aforementioned activities and to monitor them over time.
- to develop 2D and 3D models, that could more accurately simulate and track hydro-morphological processes of rivers. These models should be based on the improved data sources, that provide information on anthropogenic influences and their interconnected controlling factors.

Bibliography

1. Pišút P (2002) Channel evolution of the pre-channelized Danube River in Bratislava, Slovakia (1712–1886). Earth Surf Proc Land 27(4):369–390. https://doi.org/10.1002/esp.333
2. Hohensinner SJ (2011) Spatio-temporal habitat dynamics in a changing Danube River landscape 1812–2006. River Res Appl 27(8):939–955. https://doi.org/10.1002/rra.1407
3. Ricaurte LB (2012) The distribution and environmental state of vegetated island within human-impacted European rivers. Freshw Biol 57(12):2539–2549. https://doi.org/10.1111/fwb.12026
4. Mladenovic MK (2013) Sediment regime of the Danube River in Serbia. Int J Sedim Res 28(4):470–485. https://doi.org/10.1016/S1001-6279(14)60006-8
5. Nagy JK (2016) Hydrological and morphological changes of the Lower Danube near Mohács, Hungary. J Environ Geogr 5(1–4):1–6. https://doi.org/10.1515/jengeo-2016-0001
6. Bondar C, Buta C (1995) Trends of the water, sediment and salt discharges on the Danube over the Romanian reach. Romanian J Hydrol Water Resour 2(1–2):61–66
7. Bondar C (2000) Hydromorphological processes on the lower Danube riverbed. Proceedings of international geophysical conference and exposition. Romanian Soc Geophys 7(suppl):428–431
8. Vișinescu I (2014) Incintele indiguite la Dunare– probleme actuale si de perspectiva. Paper presented at the S.C.D.A. Symposium, Brăila, 23 May 2014
9. Gâștescu P, Țuchiu E (2012) The Danube River in the lower sector in two hydrological hypostases—high and low waters. In: International conference water resources and wetlands 2012. 14–16 September 2012. Tulcea, Romania. https://www.limnology.ro/water2012/abstract/A/A1.html

10. Constantinescu Ş, Achim D, Rus I, Giosan L (2015) Embanking the Lower Danube: from natural to engineered floodplains and back. In: Hudson P, Middelkoop H (eds) Geomorphic approaches to integrated floodplain management of lowland fluvial systems in North America and Europe. Springer, New York, pp 265–288. https://doi.org/10.1007/978-1-4939-2380-9_11
11. Grecu F, Zaharia L, Ioana-Toroimac G, Armas I (2017) Floods and flash-floods related to river channel dynamics. In: Radoane M, Vespremeanu-Stroe A (eds) Landform dynamics and evolution in Romania. Springer, Cham, pp 821–844. https://doi.org/10.1007/978-3-319-32589-7_33
12. Rădoane MR (2005) Dams, sediment sources and reservoir silting in Romania. Geomorphology 71(1–2):112–125. https://doi.org/10.1016/j.geomorph.2004.04.010
13. Rinaldi MS (1998) Bed-level adjustments in the Arno River, central Italy. Geomorphology 22(1):57–71. https://doi.org/10.1016/S0169-555X(97)00054-8
14. Latapie AC (2014) Assessing channel response of a long river influenced by human disturbance. CATENA 121:1–12. https://doi.org/10.1016/j.catena.2014.04.017
15. Habersack HJ (2013) The status of the Danube River sediment regime and morphology as a basis for future basin management. Int J River Basin Manag 11(2):153–166. https://doi.org/10.1080/15715124.2013.8151
16. Surian NR (2003) Morphological response to river engineering and management in alluvial channels in Italy. Geomorphology 50(4):307–326. https://doi.org/10.1016/S0169-555X(02)00219-2
17. Panin N, Jipa D (1998) Danube river sediment input and its interaction with the North-western Black Sea. Estuar Coast Shelf Sci 54(3):551–562. https://doi.org/10.1006/ecss.2000.0664
18. Interreg (2019a) Long-term morphological development of the Danube in relation to the sediment balance. Danube Transnational Programme Danube Sediment, Water Research Institute, Bratislava. http://www.interreg-danube.eu/uploads/media/approved_project_output/0001/39/6299a3c18cf54d8ef7e02150844e27358da8038d.pdf
19. Possea GB (2005) Geografia României Vol. 5. Câmpia Română, Dunărea, Podişul Dobrogei, Litoralul Românesc al Mării Negre şi Platforma Continentală. Editura Academiei Române, București
20. Nikolov V (2015) Geomorfological Map of Danube Valley 1:200 000. The stretch between Russe and Tutrakan. Geological Institute at Bulgarian Academy of Sciences, Sofia
21. Enciu P (2010) Danube valley geological structure, neotectonic activity and evolution during the Pliocene—Pleistocene time. Scientific Annals, School of Geology. Proceedings of the XIX CBGA Congress, Aristotle University of Thessaloniki, Thessaloniki, Greece
22. ICPDR (International Commission for the Protection of the Danube River) (2014a) Danube River Basin District: Hydropower Plants (HPP). Available at https://www.icpdr.org/main/sites/default/files/nodes/documents/map_hydropower_june_2014.pdf . Accessed 10 May 2021
23. ICPDR (International Commission for the Protection of the Danube River) (2009b) Danube basin: facts and figures. https://www.icpdr.org/flowpaper/viewer/default/files/nodes/documents/icpdr_facts_figures.pdf. Accessed 12 May 2021
24. Rădoane M, Persoiu I, Cristea I, Chiriloaiei F (2013) River channel planform changes based on successive cartographic data. A methodological approach. Revista de Geomorfologie 15:69–88. https://www.geomorfologie.ro/wp-content/uploads/2015/08/revista-de-geomorfologie-nr-15-2013-radoane.pdf
25. Zaharia L, Grecu F, Ioana-Toroimac G, Neculau G (2011) Sediment transport and river channel dynamics in Romania—variability and control factors. In: Manning AJ (eds) Sediment transport in aquatic environments, IntechOpen, Rijeka, Croatia, pp 293–316. https://www.intechopen.com/books/sediment-transport-in-aquatic-environments/sediment-transport-and-river-channel-dynamics-in-romania-variability-and-control-factors
26. Gogoaşe Nistoran DE (2017) A simple method to assess theoretical hydropower potential of a river. In: 10th international symposium on Advanced Topics in Electrical Engineering (ATEE), Bucharest, March 2017, p 764. https://doi.org/10.1109/ATEE.2017.7905068

27. Bondar C (2008) Hydromorphological balance of the Danube River Channel on the Sector, between Bazias (km 1072.2) and Danube Delta Inlet (km 80.5). In: International expert conference on "The safety of navigation and environmental security in a transboundary context in the Black Sea basin", Odessa (Ukraine), pp 1–13
28. Mitof I (2016) Resursele de apa din nord-estul Campiei Romane (intre raurile Călmățui si Buzău). Editura Universitara, Bucharest
29. Marin AF (2017) Studiul morfodinamicii ostroavelor în relație cu factorii de control fluviali prin tehnici geospațiale. Aplicație în sectorul dunărean Giurgiu—Oltenița. PhD Dissertation, University of Bucharest
30. Bondar C, Borcia C, Radulescu C, Blendea B, Ciuca R, Petrea C (2017) Date hidrologice extreme pe Dunare. In: Conferinţa Ştinţifică Anuală a INHGA, Bucureşti, pp 53–66
31. Conea A, Balley R, Canarache A (1972) Guidebook to excursions of the INQUA Loess Symposium in Romania. Geological Institute Bucharest
32. Smalley I, Leach J (1978) The origin and distribution of the loess in the Danube Basin and associated regions of East-Central Europe—a review. Sed Geol 21(1):1–26. https://doi.org/10.1016/0037-0738(78)90031-3
33. Hickin E (1984) Vegetation and river channel dynamics. Can Geogr 28(2):111–126. https://doi.org/10.1111/j.1541-0064.1984.tb00779.x
34. Hupp C, Osterkamp W (1996) Riparian vegetation and fluvial geomorphic processes. Geomorphology 14(4):277–295. https://doi.org/10.1016/0169-555X(95)00042-4
35. Erskine L (2003) The relation between riparian vegetation, bank erosion and channel pattern. Magela Creek, Northern Territory. Internal Report 419, Australian Government—Department of the Environment and Heritage https://www.environment.gov.au/science/supervising-scientist/publications/internal-reports/relationship-between-riparian-vegetation-bank-erosion-and-channel-pattern-magela-creek
36. Gurnell A, Petts G, Hannah D, Smith B, Edwards P, Kollmann JWJ, Tockner K (2001) Riparian vegetation and island formation along the gravel-bed Fiume Tagliamento, Italy. Earth Surf Processes Landforms 26(1):31–62. https://doi.org/10.1002/1096-9837(200101)26:1%3c31::AID-ESP155%3e3.0.CO;2-Y
37. Marin AF (2017) Hydrological and morphological changes of the Danube River between Giurgiu and Oltenita cities (Romania). Cinq Continents. Revue Roumaine de Geographie 7(15):99–111. http://cinqcontinents.geo.unibuc.ro/7/7_15_Marin.pdf
38. Interreg (2018) Sediment monitoring in the Danube River. Danube transnational programme Danube Sediment. Budapest University of Technology and Economics, Budapest. http://www.interreg-danube.eu/uploads/media/approved_project_output/0001/27/659489792a6c2b58c4e322ac8c609943565c3095.pdf
39. ICPDR (International Commission for the Protection of the Danube River) (2008). The analysis of the Danube floods 2006. https://www.icpdr.org/main/sites/default/files/The%20Analysis%20of%20the%20Danube%20Floods%202006%20FINAL.pdf. Accessed 12 May 2021
40. Interreg (2018c) Analysis of sediment data collected along the Danube. Danube Transnational Programme Danube Sediment, Budapest University of Technology and Economics, Budapest. http://www.interreg-danube.eu/uploads/media/approved_project_output/0001/27/af1a5cf355224c606b709fd738ebbc96902beeda.pdf. Accessed 12 May 2021
41. Prăvălie R (2012) Amenajările de pe râul Argeș: între necesitate energetică și impact asupra reliefului. In: Lucrari si rapoarte de cercetare, vol III, Orșova, Octombrie 2011. Workshop: Hydro-Geomorphological Systems, Editura Universitatii din Bucuresti, Bucuresti, p 107. http://geo.unibuc.ro/2015/201304/lucrari_rapoarte_cerc_vol_III_2012.pdf
42. NWARW (National Waters Authority "Romanian Waters") (2016) Flood risk management plan. Danube Fluvial System (in Romanian), Bucharest
43. Botzan M, Haret C, Stanciu I, Vișinescu I, Buhociu L (1991) Valorificarea hidroameliorativă a Luncii Dunării Românești și a Deltei, Editura Redacția de propagandă tehnică agricolă, Bucuresti
44. Posner-Nenciu C, Armas I, (2015) Historical perspective on structural methods for flood protection in Lower Danube. Georeview Sci Ann Stefan cel Mare Univ Suceava Geogr Ser 25(1):1–12(12)

45. Marin AF (2019) Studiul morfodinamicii ostroavelor în relație cu factorii de control fluviali prin tehnici geospațiale. Aplicație în sectorul dunărean Giurgiu-Oltenița. Editura Universitară, București. Ed. Universitara, București
46. Edwards PJ (2019) Toward a better understanding of recurrence intervals, bankfull, and their importance. J Contemp Water Res Educ 166(1):35–45. https://doi.org/10.1111/j.1936-704x.2019.03300.x
47. Vlad LI (2013) The effects of river regularization, embankment and draining on the ecology of major water meadows. Present Environ Sustain Dev 7(1):235–242. http://www.pesd.ro/articole/nr.7/PESDVOL7NR12013/21TEORREADOTEOMWM27052013235242.pdf
48. Modev S, Kirilova S (2006) Sediment related environmental problems at the Lower Danube. Energie und Umwelt im Donauraum: Umwelttagung des Vereins für Ökologie und Umweltforschung, Passau, September 2005. In: Umwelt - Schriftenreihe für Ökologie und Ethologie, 32, Wien: Facultas-Verl. 2006, 116 S. p 73–89
49. Egis Eau—CNR Ingénierie (2017) BTF 57356S, Studiu de fezabilitate pentru soluții tehnice alternative/complementare privind lucrările ce se vor executa în punctul critic 01 Bala din cadrul proiectului "Îmbunatățirea condițiilor de navigație pe Dunăre între Călărași și Brăila rkm 375–175" https://www.afdj.ro/sites/default/files/bala_raport_solutii_tehnice_v8.5_ro.pdf. Accessed 12 May 2021
50. Interreg (2019d) Data analyses for the sediment balance and morphological development of the Danube. Danube Transnational Programme Danube Sediment, Water Research Institute, Bratislava. http://www.interreg-danube.eu/uploads/media/approved_project_output/0001/37/96aa5f0389e23a912c019a07ffa96a21d63348de.pdf
51. Bondar C, Iordache G (2016) Sediment transport on the Romanian section of the Danube River. Geo-Eco-Marina 22:29–49. https://doi.org/10.5281/zenodo.889809
52. US Army Corps of Engineers (2021) HEC-DSS Vue documentation. Accessed 12 May 2021. https://www.hec.usace.army.mil/confluence/dssvuedocs/latest
53. US Army Corps of Engineers (2020) HEC-RAS river analysis system, hydraulic reference manual. https://www.hec.usace.army.mil/software/hec-ras/documentation/HEC-RAS_6.0_ReferenceManual.pdf. Accessed 12 May 2021
54. Mailat E (2010) Water resources in the upper hydrographic basin of the River Argeș. In: Gâștescu P, Brețcan P (eds) Water resources from Romania. Vulnerability to the pressure of man's activities. 11–13 June 2010, Târgoviște (Romania). Editura Transversal, Târgoviște, pp 110–117
55. Zaharia L, Ioana-Toroimac G, Cocoș O, Ghița F, Mailat E (2016) Urbanization effects on the river systems in the Bucharest City region (Romania). Ecosyst Health Sustain 2(11). https://doi.org/10.1002/ehs2.1247
56. Cocoș O (2006) Managementul apei in municipiul Bucuresti. Ars Docendi, Bucuresti
57. ICPDR-TNMN (International Commission for the Protection of the Danube River-Trans National Monitoring Network) (1996). https://www.icpdr.org/main/activities-projects/tnmn-transnational-monitoring-network
58. Feodorov CO (2020) Delta Dunării. https://edelta.ro/. Accessed 12 May 2021.
59. Barkdoll BD (2008) Sediment modeling: issues and future directions. J Hydraul Eng 134(3):285–285. https://doi.org/10.1061/(asce)0733-9429(2008)134:3(285)
60. Claude NR (2012) Estimating bedload transport in a large sand-gravel bed river from direct sampling, dune tracking and empirical formulas. Geomorphology 179:40–45. https://doi.org/10.1016/j.geomorph.2012.07.030
61. Gibson SC, Comport B, Corum Z (2017) Calibrating a sediment transport model through a gravel-sand transition: avoiding equifinality errors in HEC-RAS models of the Puyallup and White Rivers. In: World Environmental and Water Resources Congress 2017. 21–25 May 2017, Sacramento, California. https://doi.org/10.1061/9780784480625.017
62. Chanson H. (2004) The hydraulics of open channel flow. An introduction. Oxford, Elsevier.
63. Parker G (2006) 1D sediment transport morphodynamics with applications to rivers and turbidity currents (e-book). http://hydrolab.illinois.edu/people/parkerg/morphodynamics_e-book.htm. Accessed 12 May 2021

64. Dumitrescu VB (2012) Research on the simulation of the alluvial transport on the numerical model and the estimation of the morphological modifications, with application on the Danube riverbed for a proposed sector, in the period of 2005–2010. In: International conference water resources and wetlands 2012. 14–16 September 2012. Tulcea, Romania. https://www.limnology.ro/water2012/Proceedings/014.pdf
65. Rákóczi L (2010) Sediment regime of the River Danube (1956-1985). In: Brilly M (ed) Hydrological processes of the Danube River Basin. Perspectives from the Danubian countries. Springer, pp 298–304
66. Batuca D, Buta C (2002) Morphodynamic aspects of the Danube River from Corabia to Oltenita in the period of the time from 1980 to 2000. In: XXIst conference of the Danubian countries. Bucharest, Romania. http://www.ksh.fgg.uni-lj.si/Proceedings/XXI-Conf_Bucharest_2002_Proceedings/stinga.htm
67. Interreg (2019b) Sediment balance assessment for the Danube. Danube Transnational Programme Danube Sediment, Water Research Institute, Bratislava. http://www.interreg-danube.eu/uploads/media/approved_project_output/0001/39/d7f9e88e194b7dcea22b51235d653c50d358b7ae.pdf
68. Popescu CL (2019) Mathematical modelling and interpretation of results for siltation phenomena of rivers in the vicinity of reservoirs. E3S web of conferences. https://doi.org/10.1051/e3sconf/20198507004
69. Nachtnebel HP (2008) Sediment transport: a key issue for transboundary water management in the Danube basin. In: 5th international SedNet conference urban sediment management and port redevelopment & sediment in river basin management plans. 27–29 May 2008, Norwegian Geotechnical Institute, Oslo, Norway. https://sednet.org/download/2%20Hans-Peter%20Nachtnebel.pdf
70. Ministry of Environment and Waters (2008) Technical assistance for improvement of navigation conditions on the Romanian—Bulgarian common sector of the Danube and accompanying studies. https://www.moew.government.bg/en/technical-assistance-for-improving-the-navigation-conditions-of-the-romanian-bulgarian-common-sector-of-the-danube-an-the-accompanying-studies/
71. Pătru G (2014) Influenta amenajarilor hidroenergetice din sectorul aval al Dunarii asupra regimului sedimentelor. Modelare hidro-sedimentara pe sectorul Oltenita-Călărași. Dissertation, University Politehnica of Bucharest
72. Gogoașe-Nistoran DE (2017) One dimensional sediment transport model to assess channel changes along Olteniţa-Călăraşi reach of Danube River, Romania. Energy Procedia 112:67–74.
73. Nakato T (1990) Tests of selected sediment-transport formulas. J Hydraul Eng 116(3):362–379. https://doi.org/10.1061/(asce)0733-9429(1990)116:3(362)
74. USACE (2020) Hydrologic engineering center, HEC-RAS sediment transport. https://www.hec.usace.army.mil/software/hec-ras/documentation/HEC-RAS_6.0_1DSedimentUserManual_(BetaRelease).pdf Accessed 12 May 2021
75. Laursen E (1958) Total sediment load of streams. J Hydraul Div 84(1):1–36. https://doi.org/10.1061/JYCEAJ.0000158
76. Copeland RT, Thomas WA (1989) Corte Madera Creed Sedimentation Study. Numerical model investigation. TR-HL-89-6, US Army Engineer Waterways Experiment Station. https://apps.dtic.mil/dtic/tr/fulltext/u2/a207175.pdf
77. Grigorie-Omrani Șt (2014) Valea Dunării, între Oltenița și Călărași, PhD Thesis, University of Bucharest, Bucharest, Romania
78. Rubey W (1933) Settling velocities of gravel, sand, and silt particles. Am J Sci s5–25(148):325–338. https://doi.org/10.2475/ajs.s5-25.148.325
79. Navratil O, Albert MB (2006) Determination of bankfull discharge magnitude and frequency: comparison of methods on 16 gravel-bed river reaches. Earth Surf Process Landf Wiley Interscience 31(11):1345–1363. https://doi.org/10.1002/esp.1337
80. Bessar MA (2020) Uncertainty analysis of a 1D river hydraulic model with adaptive calibration. Water 12(2):561. https://doi.org/10.3390/w12020561

81. Winterbottom S (2000) Medium and short-term channel planform changes on the Rivers Tay and Tummel, Scotland. Geomorphology 34(3–4):195–208. https://doi.org/10.1016/S0169-555X(00)00007-6
82. Feier I (2010) Reconstituirea evoluţiei geomorfologice a văii Someşul Mic în Holocen, PhD Dissetration, Facultatea de Geografie - Geologie „Al. I. Cuza", Iaşi
83. Jacobs (2019) Imbunatatirea conditiilor de navigatie pe sectorul comun romano-bulgar al Dunarii. Raport HRO/050/R/20190620. http://www.anpm.ro/documents/12220/40416825/FAST_Danube_Memoriu+prezentare.pdf/834112e5-01e9-47d0-8378-2413627636af
84. Ciortan R (2015) Aspecte privind infrastructura pentru navigatie din Romania. Bull AGIR 3:143–146
85. Enciu P (2010) Danube valley geological structure, neotectonic activity and evolution during the Pliocene–Pleistocene time. In: Scientific annals, school of geology. Aristotle University of Thessaloniki, Greece. Proceedings of the XIX CBGA Congress. Special volume 100, pp 183–190
86. Pittaluga B, Luchi R, Seminara G (2014) On the equilibrium profile of river beds. J Geophys Res Earth Surf 119(2):317–332. https://doi.org/10.1002/2013JF002806
87. Armaş I, Necea D, Miclăuş C (2018) Fluvial terrace formation and controls in the Lower River Danube, SE Romania. Quatern Int 504:5–23. https://doi.org/10.1016/j.quaint.2018.03.031
88. Leopold LB (2006) A view of the river. Harvard University Press
89. Navionics (2020). Navionics and Garmin plotter charts. Retrieved May 10, 2021, from https://webapp.navionics.com/

Hydro-Environmental Specifics of the Lower Danube Bulgarian Tributaries

Mila Chilikova-Lubomirova

Abstract Significant part of the Danube River function relates to the crucial connections with its tributaries. This is linked to the water resources, organic and inorganic substances, and biological elements provision. Such interaction is absolutely natural and impact on the current state of the Danube River. Thus, of great importance is to have good knowledge and understanding about the connected processes. For this purpose, many factors such as climate, physico-geographical, land and hydrological specifics, anthropogenic features, etc. must be considered. But of especial importance is to observe also additional aspects as extreme or rare events, such as droughts or floods occurrences, which in many cases impact harmfully on the rivers ecosystems state and society. Thus many additional measures were proposed to mitigate these effects. Connected initiatives cover both legislative and practical aspects, to guarantee best results obtaining. For this purpose constantly are developed various scientific investigations and practical solutions, management approaches, programmes and strategies.

The present chapter is focused on the specific issues in the area of the Bulgarian tributaries of the Lower Danube River. The following topics have been addressed:

- basic description of the Bulgarian Danube plain, considering the main geographical features (topography, climate, hydrology, etc.);
- legislative and management brief review, with regard to the implementation of Bulgarian and European directives (e.g. Water Framework Directive, Floods Directive), and related documents;
- hydro-environmental specification concerning the Bulgarian Tributaries of the Lower Danube River, their qualitative state and extreme events.

For their better clarification, in the chapter briefly are presented some of the most applicable methods that are used worldwide, in the European Union and Bulgarian practice. Described aspects are examined with regard to the needs and functions of the ecosystem. Presented issues are illustrated by the results of original assessments and practical examples from both real flood and drought investigations.

M. Chilikova-Lubomirova (✉)
Fluid Mechanics Department, Institute of Mechanics—Bulgarian Academy of Sciences, Acad. G. Bontchev Str., Bl. 4, 1113 Sofia, Bulgaria
e-mail: milasemail@yahoo.com

A. Negm et al. (eds.), *The Lower Danube River*, Earth and Environmental Sciences Library, https://doi.org/10.1007/978-3-031-03865-5_4

Keywords Bulgarian Danube tributaries · Lower Danube River · Water management · Floods · Droughts · Water quality

1 Introduction

The Bulgarian part of Lower Danube covers 5.90% of the Danubian basin. In the Danube River inflow seven main and several small Bulgarian rivers. They contribute to the Danube quantitive and qualitative status. For this reason, it is important to know their general specifics, and how they are changed in time. This is connected to constant monitoring and investigation of main hydrological and environmental issues, to the analysis of current state and trends, to obtaining of reliable understanding on their behavior. Because of their importance is well to consider also the local specifics of climate, topography, land cover, urbanization, anthropogenic features, etc.

In this context, the present chapter aims to characterize the Bulgarian Danube plain and the main rivers draining this area, in order to highlight their general features and the main challenges related to their behavior. For the purpose there briefly are outlined the main hydrological specifics, with focus on the extreme events like floods and droughts. As a result of such hydrological hazards, significant effects on the environment and ecosystems often occur. But additional impacts are also observed. For example, floods have both negative and positive impact, depending on the extent of overflow. Observed impacts are connected to the inundation of large areas with loss of crop production and farms damages, disruptions of river banks, infrastructure, buildings and settlements, and/or social stress. Droughts are also part of the hydrological cycle. They occur after prolonged precipitation deficit and lead to serious water stress or shortages of water with effects on the agricultural and farms production, industry, and households.

To mitigate such events different measures and actions were designed and implemented in practice. They cover both management and legislative solutions. To clarify such implementation, in this chapter briefly are considered main horizontal and vertical actions that are developed. The analysis covers Bulgarian regulation and legislative acts. They transposed the approved European legislative measures, as Bulgaria is a European Union member.

Based on such regulation, different practical solutions are also implemented. They are briefly presented, by characterizing the most applicable approaches in the country. They are connected also to provision and maintenance of good hydro-physical quality of the river systems. Such aspects are analyzed for the Bulgarian Danube tributaries.

All of presented issues are provided based on the ecosystem approach. They correspond to the purposes and instruments connected to the Danube River Protection Convention and existing legal and research documents. In many cases existing knowledge is insufficient [1–10, etc.] This study gather the existing knowledge with new investigated features of the Bulgarian Danube tributaries and provide useful scientific information for decision and policy makers.

2 Bugarian Danube Plain: General Features

The Danube River is a northern borderline of Republic of Bulgaria with Romania (Fig. 1). In the western Bulgarian part it starts from border between Bulgaria and Serbia. It flows 470 km to the east, before reaching Dobrudja Region. On the North is situated Romania and on the South is the Danube plain that extends to the Balkan Mountains. Numerous islands are situated along the Danube Rriver.

The Danubian floodplain is part of the Danubian Plain, developed on both sides of the Danube River (in Bulgaria and Romania), part of the morphostructure of the Moesian plate. The relief is characterized by lowlands and plateau forms, asymmetric and canyon-like river valleys, dry valleys, landslides, and karst forms. The hilly and plateau-like character of the relief is divided by the valleys of the rivers crossing it. The diverse relief allows division of the Bulgarian plain into three sectors.

The western sector starts from the border between Bulgaria and Serbia, and extends to the Vit River. It is the lowest one, with an altitude of about 130 m amsl To the west of the Lom River, the valleys have a canyon-like character, and to the east, they are asymmetrical. The slope of this part of the Danube valley is in a north-easterly direction. In this direction are oriented the valleys of the larger tributaries of the Danube River as Lom, Ogosta, Iskar, Vit, etc. Following the slope, the rivers shifted their beds to the east, eroding the right slopes of their valleys. As a result, asymmetrical valleys resulted—with steep right and sloping left banks. The river valleys divide the loess plateaus.

The middle sector extends between the rivers Vit and Jantra. It is transitional part with mean altitude of about 138 m a.s.l. In its western part is located the asymmetrical valley of the Vit River. To the east are the wide valleys of Osam and Jantra Rivers. The

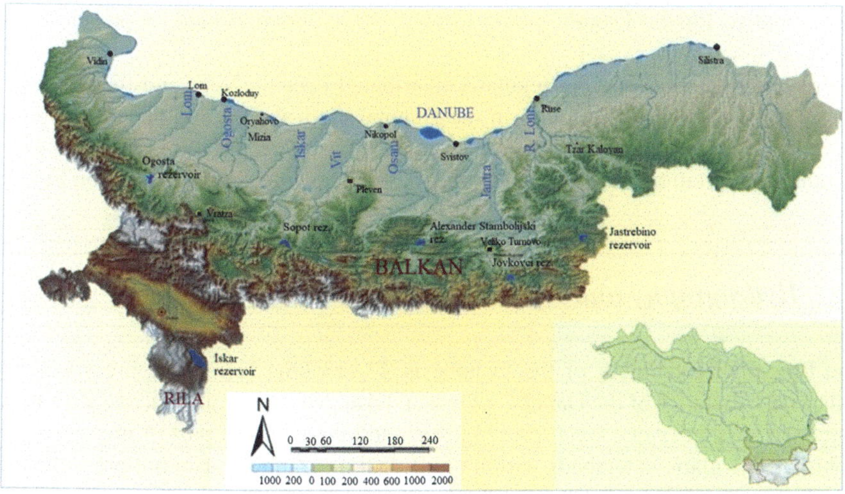

Fig. 1 Bulgarian Danube basin

relief there is hilly. Along the Danube River, there are the most extensive alluvial lowlands. In this part are observed areas with active landslides along the Danube bank.

The eastern sector is characterized by a typical plateau-hilly relief and a higher altitude. Rusenski Lom River is the last main Bulgarian tributary of the Danube. In the eastern part of this sector, there are five more rivers, but their water periodically reach the Danube River. Karst landforms of the relief are common for this part of the Danube Plain. Active landslides are also observed on the Danube bank there. The mean altitude of the Danube plain is 178 m amsl.

2.1 Climate of the Danubian Plain

Despite the small territory (111,000 km^2) Bulgaria has a variable and complex climate that is influenced by the strongly contrasted Continental and Mediterranean climates, as well as by the topography and land coverage specifics. Mountains in most cases act as barriers of the air masses, creating a sharp contrast in weather over a relatively short period. Such effect is caused by the Balkan Mountains. It hampers the Mediterranean influences and as a result the continental climate predominates in the Danube plain. The flat relief that is open to the north favors the continental air masses easily to flow into the territory of the Danube plain. As a result during springs, summers and autumns moist air masses easily occur, as winters fall under the Eastern European anticyclone, that brings cold Arctic air mass and abundant snowfalls [11, 12].

In the Danube Plain summers are typically warm and winters very cold [13] with January the coldest month while July is the warmest. The average annual temperature is 11.4 °C. Details about the average air monthly temperatures for a weather station representative for the Danube plain, located in Oryahovo village, is presented in Fig. 2.

Precipitations are almost irregularly distributed over the year and the territory. The average annual precipitation in the Danube Plain is 520 mm. The maximum is registered in June and the minimum—in February. The variation of the average monthly precipitation measured at Oryahovo weather station are presented on Fig. 2.

2.2 Hydrography and Hydrology

The Bulgarian catchment of the Danube is 47,413 km^2 [1]. There are six main Bulgarian tributaries of the Danube—Ogosta, Iskar, Vit, Osam, Jantra, and Rusenski Lom, and smaller rivers, west from Ogosta, and in the eastern part of the sector, in Dobruja. The main watersheds within the Bulgarian Danube basin are presented in Fig. 3.

As a result of the topography, the mentioned rivers are relatively short—with a maximum lengths of about 370 km. Their catchments are relatively small, with

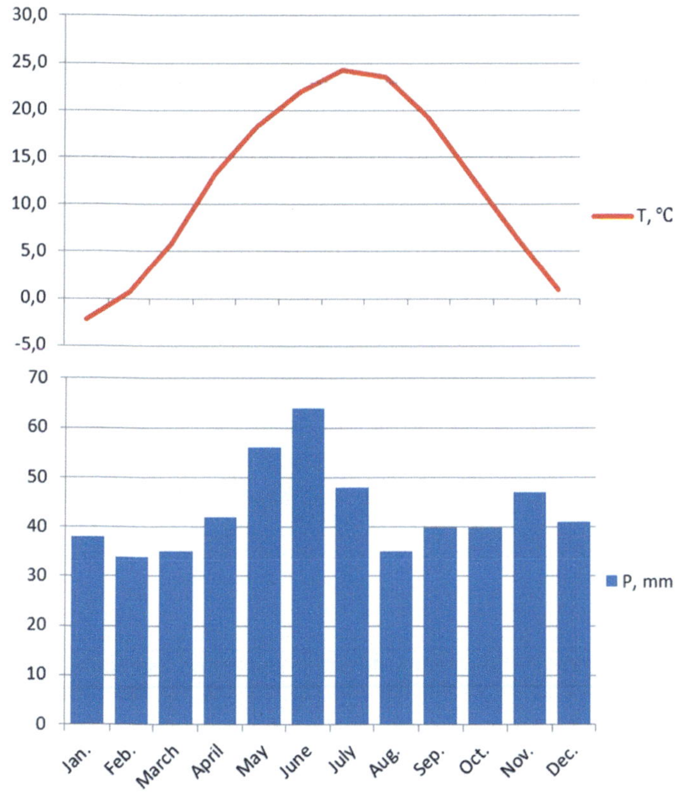

Fig. 2 Average air monthly temperatures (T) and average monthly precipitation (P) in Oryahovo weather station, 1931–1980, data source [13]

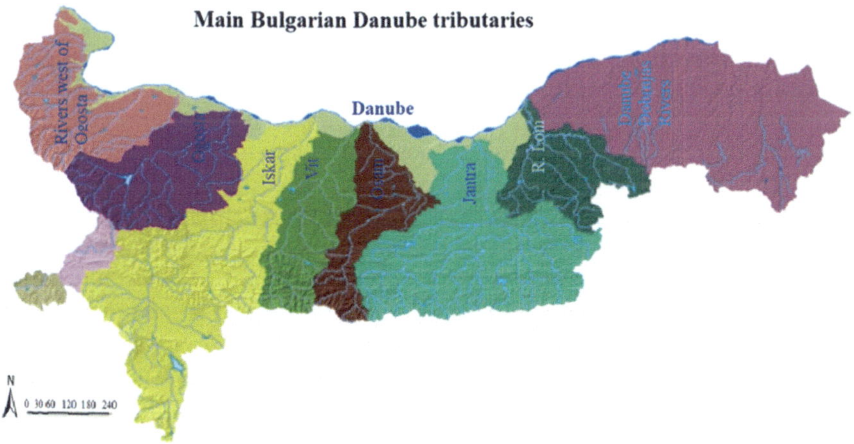

Fig. 3 Bulgarian Danube basin and the main Bulgarian watersheds [1]

Table 1 The main Bulgarian Danube tributaries (based on [1])

Name	Length (km)	Source	Catchment area (km^2)
Lom	92.5	Midzur peak, Balkan Mountain	1140
Ogosta	141.1	Vraza glava peak, Balkan Mountain	3157.1
Iskar	368	The confluence of Cherni and Beli Iskar rivers, Rila Mountain	8646
Vit	189	The confluence of Cherni and Beli Vit rivers, Balkan Mountain	3225
Osam	314	The confluence of Beli and Cherni Osam rivers, Balkan Mountain	2824
Jantra	285.5	Atovo padalo peak, Balkan Mountain	7862
Rusenski Lom	50 (197 km with the main tributaries)	The confluence of Cherni Lom and Beli Lom rivers	2947

areas of about 3,000 km^2, excepting the Iskar and Jantra catchments extended on about 8000—9000 km^2 (Table 1). The sources of some of the rivers are smaller watercourses that after confluence formed the main rivers. Most of them accumulate their discharges from the northern slopes of the Balkan Mountains. Close to the Danube River, the topography is flat, covered by arable land, permanent grassland, or permanent crops. A particular case is the Iskar River that is the only river that sources south of the Balkans—from the Rila Mountain. In Table 1 are presented the main characteristics of mentioned rivers, based on the information published in the River Basin Management Plan [1]. They are showed also on Fig. 1.

The flow regime of the Bulgarian Danube Tributaries is closely connected to many factors like local climate and topography, land cover, soil specifics, and existing engineering works. Climatic factors are of especial importance. Major role play the precipitations (in liquid and solid form), and most of the rivers have rainfall-snow supply. The richest precipitation in the area is registered in the highest altitudes of the Balkan Mountains—in the west and the central part of the mountains (over 1000–1200 mm yearly). As a result, the runoff is high in this area. Northerly of the Balkan, due to the decrease in altitude, the annual precipitation diminishes up to 650–700 mm, and consequently the runoff also decrease. Precipitations and snowmelt are determinative for the river maximums flow. As a result are observed high waters in spring and low waters in summer and autumn. Typically maximums discharges are registered in March–April, as a result of the snowmelt. Low waters are observed from June–July to October–November, and the minimum discharges often occur in August-October. There are some rivers that flow in karstic basins (Vit River and Danube Dobrujas Rivers). Practically the Danube Dobrujas Rivers reach the Danube

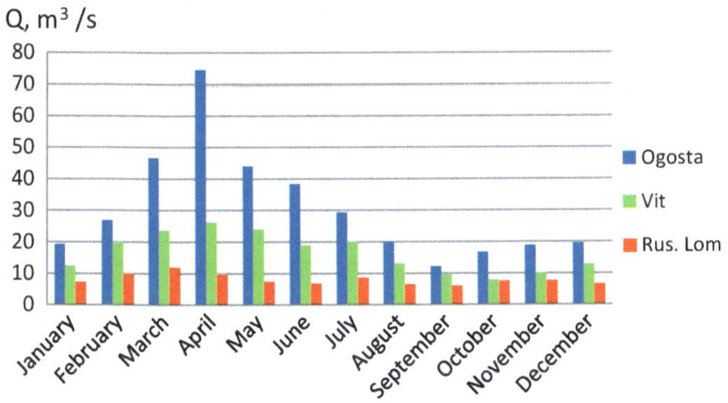

Fig. 4 Hydrograph of the rivers Ogosta, Vit and Russenski (Rus.) Lom (2001–2010)

by their underground flow. Hydrographs for the period 2001–2010 for the rivers Ogosta, Vit and Russenski Lom are presented in Fig. 4.

The quantitative status of the surface water is determined by regular monitoring of the rivers runoff in the Danube plain. In 2018 the annual volume of the Bulgarian Danube basin was 9.752×10^6 m^3 [11]. This volume is changed in time. It is closely connected to the rivers flow regime and extreme events occurrences.

2.3 *Biodiversity of the Danubian Plain*

Biodiversity in the area is very rich. Several important natural protected areas are located the Danubian plain.

The most famous is the Srebarna Natural Reserve, which includes the freshwater lake Srebarna and its surroundings, located on the Via Pontica—a bird migration route between Europe and Africa. It is a home of 139 plant species, 39 mammal, 21 reptile and amphibian and 10 fish species. Almost 179 bird species nest on its territory [14]. Of importance is also the Persina Natural Park, situated along the Bulgarian valley of the Danube River, including also the Persin island. They are engaged in conservation and restoration of the Danube wetlands and biodiversity protection. There are many small islands planned also for conservation both inside and outside the parks borders.

Of interest is also the Rusenski Lom Nature Park, located on the tributary Rusenski Lom. It protects a scenic canyon that is a habitat for many cliff breeding birds and bats. It is in charge with the protected site Kalimok-Brushlen marches. It is a home of about 242 birds, more of them rare and endangered species. Within the protected area can be found about 14 species of reptiles, also several amphibians [15]. The Natura 2000 sites located within the Bulgarian Danube floodplain are showed in Fig. 5.

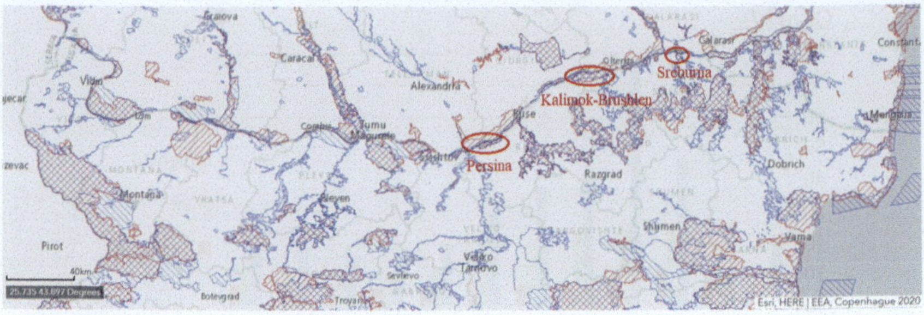

Fig. 5 Natura 2000 sites in Bulgarian Danube floodplain, covering Bird Directive Sites (SPA) and Habitat Directive sites (based on [16])

2.4 Socio-Economic Aspects and Water Uses

The Danube plain cover territories situated in the districts Vidin, Montana, Vratsa, Pleven, Veliko Tarnovo, Ruse, Silistra, Lovech, Gabrovo, Shumen, and Sofia province. Main population centers are the towns Ruse, Vidin, Svistov, Silistra, Nikopol, Kozloduy, Lom, etc., with main centers Pleven, Vratza, Lovech, Gabrovo, Shumen and Veliko Turnovo. There is no high rate of urbanization. In the past many of the lakes and swamps situated along the Danube River were dried up. They were converted into agricultural land. In the middle of XX century the whole territory was planned for agricultural use. In order to increase the agricultural yield in the past were designed huge branches of irrigation systems. To provide water for different socio-economical needs in the territory many reservoirs were constructed on the Danube tributaries—generally for irrigation purposes. Reservoirs are situated mainly in the mountain areas. Nowadays within the Danube plain operate following irrigation systems branches: Vidin, Mizia, Middle Danube and Lower Danube. They manage, operate and ensure the integrated use of water. Today the agricultural works are smaller than in the period of XX century, but are still very active. Significant water transfer between irrigation branches and river watersheds is not longer active. Details about main Danube plain territory reservoirs are presented in Table 2 and illustrated on Fig. 1.

Table 2 List of the main reservoirs on the Bulgarian Danube tributaries [1]

Rezervoir	Municipality	Volume, in m^3
Iskar	Sofia City	673,000,000
Ogosta	Montana	506,000,000
Alexander Stambolijski	Gabrovo, Veliko Turnovo	205,500,000
Jovkovci	Veliko Turnovo	92,179,000
Jastrebino	Turgoviste	63,000,000
Sopot	Lovech	61,800,000

Existing dams within the Danube plain serve for water regulation. But downstream the dams, the streamflow is not much affected by this regulation works. As a result, in some cases floods and inundations occurred in the floodplain, result of significant precipitations. To protect and mitigate the vulnerable areas from this hazards, some of rivers were corrected, or embanked.

3 Legislation and Management Issues

Leading principle in Bulgaria regarding the environment is the accepted ecosystem approach concerning management and legislation. As a member state of the EU, Bulgaria transpose in the legislation the EU policies. The main legislative documents are:

- Water Framework Directive (WFD 2000), Directive 60/2000/EC [17];
- Directive 2007/60/EC (Floods Directive, 2007) [18];
- Directive 92/43/ECC (The Habitats Directive, 1992) [19];
- Convention on Biological Diversity (CBD, 1992) [20];
- Common implementation strategy for the WFD, River Basin Management in a changing climate (CIS TR-2009–040, 2009) [21], etc.

They are implemented in practice along with the National legislation. The following legislative documents are also in force:

- Environment protection Law (2014) [22];
- Water Law (2015) [23];
- Regulation No 1/11.04.2011 for water monitoring [24];
- Regulation No H-4 from 14.09.2012 for surface water characterization [25];
- Regulation of terms and conditions for technical and safe exploitation of dams and related facilities, and their technical state control (2016) [26];
- Biodiversity Law (2017) [27];
- Protected Areas Law (2013) [28];
- National strategies and documents (National programmes in the field of protection and sustainable development of waters, such us: Marine Strategy; National Strategy for Management and Development of the Water Sector; etc.).

The main institutions in charge with the environmental issues at National Level are the Ministry of Environment and Water (MOEW) and connected to it four river Basin Directorates, the Executive Environment Agency—Bulgaria (ExEA-Bulgaria), administration with the Minister of Environment and Water, and sixteen Regional Inspectorates of Environment and Water. The river basin Directorates are separated on a river basin principle. The Danubian plain is under the authority of the Danube River Basin Directorate, having its headquarters in Pleven city. Its main activities are related to:

- development and implementation of the Plans for River Basin Management [1] and Flood Risk Management Plan [2] and their update;
- provision of sufficient quality and quality of water for the needs of the population, economy, and ecosystems;
- control of water resources and discharges of wastewater and development of monitoring systems;
- mitigation of adverse climate impacts, associated with floods, droughts, and water scarcity;
- inland and transboundary waters protection and sustainable management.

Their work is connected also to coordination with the existing strategies and plans for the International Basin of the Danube River, national concepts, strategies, plans and programs, regional strategies, plans and programs, municipalities plans and programs (as the Municipalities Development Plans, Municipalities Programmes for Waste activities management, etc.), plans for protected areas management, and water protected zones management.

4 Hydro-Environmental Issues of the Bulgarian Tributaries of the Lower Danube

Considering the holistic approach for the river systems study of importance is to gather together all factors that are determinative for the rivers environmental function. For this purpose of importance is to observe the qualitative and quantitative status of the rivers, but also to consider the occurring of extreme events. The knowledge about such phenomena is crucial as in many cases they lead to serious modifications in the rivers state and harmful effects on the surroundings. This requires broad and general understanding of such occurrences, centering on their main characteristics and study approaches.

4.1 Qualitative State of the Rivers Water Bodies

The hydro-environmental specifics of the Danube plain are detected by the proper monitoring of main hydro-environmental elements. In Bulgaria, the responsibility of this activity on the national level belongs to the Executive Environmental Agency (ExEA-Bulgaria). It is part of the administration with the Ministry of Environment and Water. It is dealing with the management, coordination, and information flow as regards the control and environmental protection and regularly monitors the environmental components and factors on the whole country. Regarding waters, regularly physico-chemical and hydro-biological monitoring is provided. Emission control and hydromorphological monitoring are also provided. Results are publically available and periodically published in Water status reports.

To determine the physico-chemical status of surface waters by regular monitoring the following indicators are measured: dissolved oxygen, biochemical oxygen demand, ammonium and nitrate nitrogen, phosphorous. Detected physico-chemical quality elements show that most of the results determine high and good quality. In Fig. 6 are presented the average annual concentrations of the indicators mentioned above, for main Bulgarian Danube tributaries, in 2019, based on ExEA-Bulgaria data. The results show that the measured elements for the rivers Vit, Osam, Iskar and Jantra vary from high, to good, and moderate status [8].

In the Bulgarian Danube River Basin for the period 1996–2020, the concentrations of DO, NH_4-N, NO_3-N, BOD_5, and PO_4-P decreased comparing with the previous years. Recently results of the monitoring showed a long-term upward trend in the surface water quality [8]. Figure 7 represents the variability of the annual quality parameters, estimated with regard to 1996, adopted as a base.

Some of results are used in the preparation of the Danube River Basin Management Plan [1]. There are accounted all the provided investigations considering the ecological status of the water bodies, and the indicators impact. As a result, the ecological quality status of the water bodies was determined, as illustrated in Fig. 8. In the period 1999–2020 progressive improvement of the subsurface chemical waters quality was also observed.

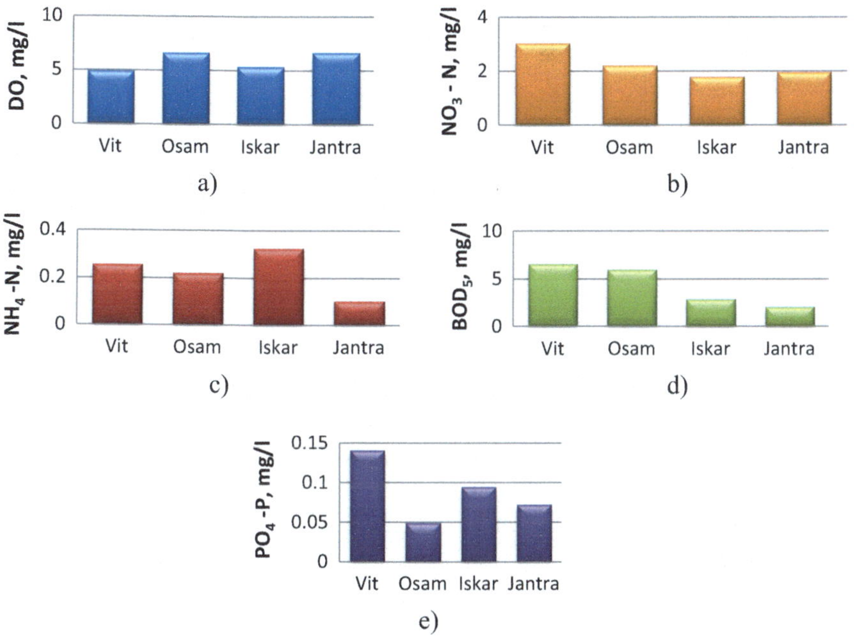

Fig. 6 Average annual concentrations of (a) dissolved oxygen (DO), (b) nitrate nitrogen (NO_3-N), (c) ammonium nitrogen (NH_4-N), (d) biochemical oxygen demand (BOD_5), and (e) phosphorous (PO_4-P) in 2019 for main Bulgarian Danubian tributaries (based on data from ExEA-Bulgaria, 2020)

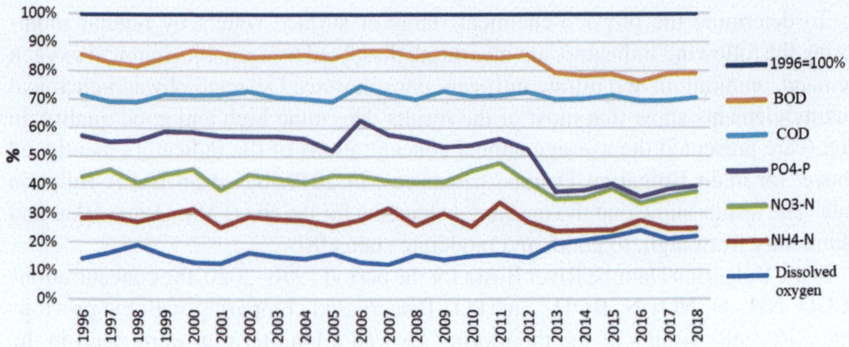

Fig. 7 Variability in concentration of main indicators of the chemical status of surface water in Bulgarian Danube River Basin, estimated with regard to 1996, adopted as a base [1]

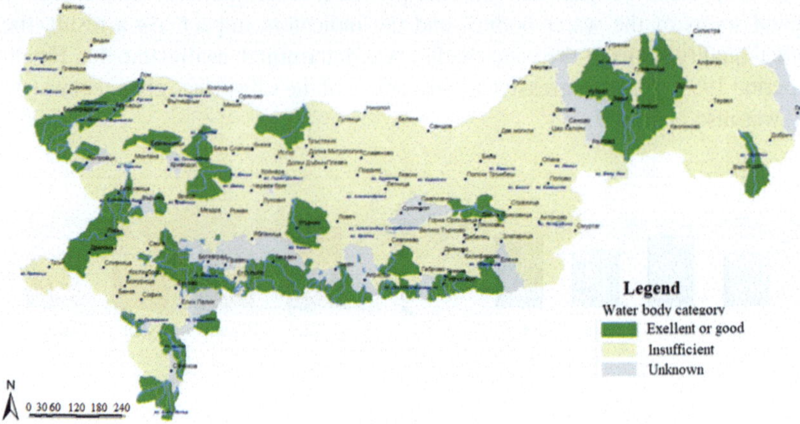

Fig. 8 Surface ecological water bodies status in the Bulgarian Danube River Basin [1]

The hydro-biological state of rivers is determined by direct monitoring in selected monitoring points in the summer-autumn season. Evaluation is presented in five classes—high, good, moderate, poor, and bed status. For 2018 the Benthic Invertebrate status of rivers in the monitoring points of the Bulgarian Danube River Basin in percentage was as follow: 18%—in high status, 11%—in good status, 11%—in moderate status, 19%—in poor status, and 41%—in bad status [8]. Samples were taken by one-off samplings.

4.2 Extreme and Risk Events

As a part of the hydrological cycle, in the river systems behavior are observed also unusual events, classified as extremes or risk events. Such events are floods and droughts. In many cases, they can be accompanied to harmful consequences on nature, ecosystems, human activities, artificial structures, economy, social life, etc. Their occurrence is a natural phenomenon but can be also a result of human activities, or of both natural and human factors. Thus, in the process of their investigation it is important to properly consider all related factors, as well long term effects connected to the climate influences like climate change, in order to assess the associated risks. For this purpose are accepted two main approaches—investigations on past events and development of models for future scenarios. In Bulgaria such approaches are performed generally for flood events. Droughts are less analyzed—generally concerning climate factors and meteorological drought. Connected to this topic, below are briefly presented main issues related to the Bulgarian Danube tributaries. They are based on the existing legal documents and investigations in the area presenting also the potential for future studies.

4.2.1 Flood Events

Floods are processes characterized by temporary covering by water of land not normally covered by water, excluding sewerage systems [17]. Generally occur in coastal and inland territories. Coastal areas are exposed to coastal floods and storm surge, thus inland areas are exposed to river floods, inland flooding and flash floods.

River floods occur in rivers and stream channels. They happen slowly or suddenly as a result of excessive rainfall, thunderstorm or snowmelt. The water level rises and overflows the riverbanks covering with discharges connected floodplain. Inland floodings happen over several days of steady rainfall, after a short and intense period of rainfall, or after significant snowmelt. They are observed also when water ways get blocked by debris, ice or dams. Often are referred to urban floodings. Flash floods mostly happen as a result of extremely intensive rainfall over a short period of time, but can occur as technogenic damage—in case of dams or levee break. They are characterized also by violently torrents that happen with little or no warning. Often rip through river beds, urban streets, canyons. As a result of all presented floods connected territories such as floodplain and riparian areas, urbanized areas, and agricultural land are highly endangered. For this reason in many cases they are accepted as hazardous or risk events.

In the Bulgarian Danube plain in the last decades, several serious floods occurred. One of the most damaging of them hit Mizia town and surroundings in August 2014. It was caused by Skat River, a tributary of Ogosta River (Fig. 9).

This region is vulnerable to floods and flooding because of the climate and flat terrain specifics. In recent years many such events occurred:

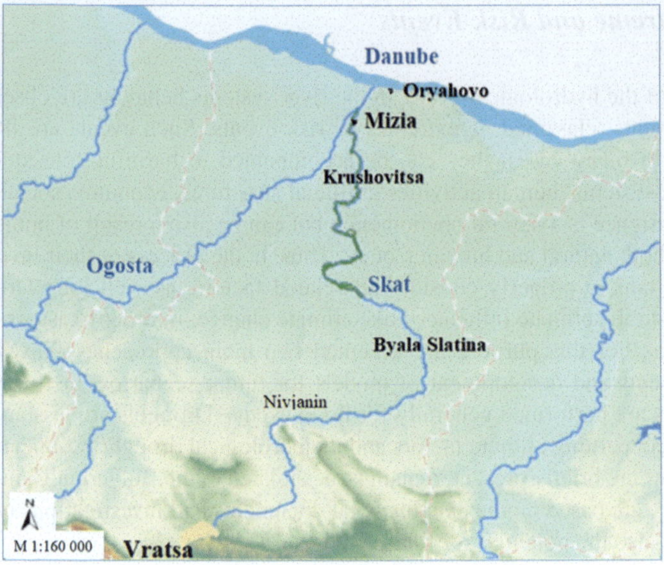

Fig. 9 The Skat River and the settlements affected by floods in August 2014

- on 15.04.2003: flood outside the Mizia town, caused by significant precipitations and snowmelt; consequences: flooded agricultural land;
- in August 2005: flood outside of Mizia town generated by important precipitation; consequences: flooded agricultural land, damaged pedestrian footpaths, flooded pumping station, water supply and sewage OOD Vratsa, damaged roof of an Early Childhood Centre "Zdravetz";
- on 16.05.2010: flood in the town area, caused by significant precipitation and snowmelt;
- 16.06.2010: flood in the town area, caused by significant precipitation and snowmelt; consequences: destroyed asphalt road;
- on 1–2 August 2014: extreme flood caused by significant long lasting precipitation, with serious consequences.

After July 22, 2014 the Northern Bulgaria territory was under transiting low pressure area. Almost daily it was raining in the Vratsa region, till July 31. In the first two days of August heavy rainfall was registered in the area. The Agency Hail Suppression presented that between 16 o'clock on July 31 and 9 o'clock on August 1, significant precipitation was recorded with a center situated west from the Bjala Slatina town, with maximums in the villages Malorad and Devene. As a result, the whole Skat river catchment was saturated and a significant surface runoff supplied the Skat river and caused floods and floodings in provinces Vratsa, Borovan, Bjala Slatina, and Mizia. The most affected were the Mizia town and Krushovitsa village, located on the river downstream (Fig. 9).

As a result of the flood many properties in Mizia and the surrounded areas were underwater—agricultural land and farmyards, streets, gardens, cellars, etc. (Fig. 10) The river levee was broken in two places. More than 1200 houses were affected, more than 700 houses needed renovation, 103 buildings and other structures collapsed. The water supply and sewage, electricity, and transport infrastructure were damaged. Crop and animal production was lost. More than 800 people were evacuated, and two persons were found dead [29].

Another serious flood affected the Tzar Kaloyan town on August 6, 2007. As a result of significant precipitations and the flood occurred on Hlebarovska River the town was underwater (Fig. 11). Part of the nearest situated micro reservoir Ezerche

Fig. 10 Mizia town flooded in August, 2014 (Author Chilikova-Lubomirova)

Fig. 11 Tzar Kaloyan town after the flood 2007 (Author Chilikova-Lubomirova)

Fig. 12 Areas with significant potential flood risk in Bulgarian Danube River Basin [2]

was broken. A flood wave hit the town and consequently 15 houses were destroyed, other properties were underwater and seven victims occurred [29].

On a national scale measures for floods prevention and risk mitigation are connected generally with the preparation and implementation of the Flood Risk Management Plan—Danubian Basin [2], and with related documents in the context of the Preliminary evaluation of the flood risk and flood hazard maps. In this context, were determined and published main flood hazard areas with significant potential flood risk for the main Danube tributaries based on significant floods happened in the past with potential to happen again (Fig. 12).

Flood Risk Management Plans present also models for the areas vulnerable to floods with probability of occurrence 1% (Fig. 13).

Additionally, the municipalities designed various measures for mitigation and prevention purposes connected to human health protection, increased protection of the environment, higher protection of the critical infrastructure and business, for enhancing the preparedness and reactions of the population face to floods, etc.

4.2.2 Hydrological Droughts

Drought is a natural phenomenon, result of significant precipitation deficiency. Its occurrences are irregularly distributed in time and space. Practically this is a multi-aspect event, associated with all components of the water cycle, characterized by meteorological, soil moisture, and hydrological impacts. In practice it is estimated with regard to past events or future scenarios.

Most often it is associated to the forecasts connected to the expected future climate. For this purpose, usually are developed climate projections for main parameters as temperature, precipitations and extreme events. Such projections were developed also

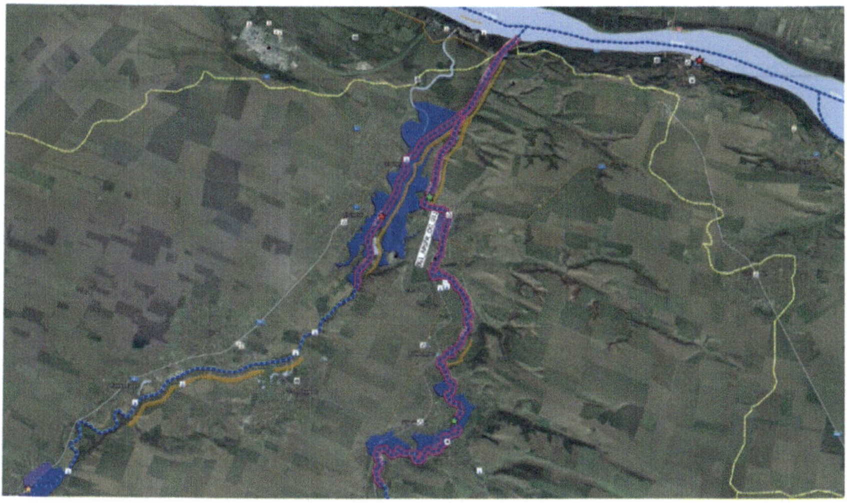

Fig. 13 Modeled Mizia flood territories, for floods with probability of occurrence of 1% [2]

for Bulgaria using the new Representative Concentration Pathways (RCPs) scenarios of the Intergovernmental Panel on Climate Change (IPCC). They were performed for RCP 8.5 (pessimistic scenario) and RCP 4.5 (intermediate scenario) compared to a reference period (1976–2005). For the Danube plain, the results show rise of mean temperature for summer months by 1.7 °C and decreasing of precipitation by 12% (RCPs 4.5 for the horizon 2021–2050) and rise of temperature by 2.1 °C and decreasing of precipitation by 10.6% (RCPs 8.5 for horizon 2021–2050). For the winter months the climate projections show: rise by 0.7 °C of mean temperature and rise of precipitation with 15.5% (RCPs 4.5 for horizon 2021–2050), and rise of temperature by 0.8 °C and rise of precipitation with 2.5% (RCPs 8.5 for horizon 2021–2050) [1]. There is no available information about such systematic investigations on streamflow, but following are presented results of scientific investigation.

For description of the ultimate low hydrological state is accepted the term "hydrological drought". It is applied to less than normal amounts of water represented by low water levels in streams, reservoirs, lakes, etc. Practically surface water drought is accepted as a natural water deficiency over-sufficiency long period in relation to the average value, resulting from precipitation. In this context, for its characterization it was accepted a similar index to those used for to the atmospheric drought investigation, namely the Standardized Runoff Index—SRI [30, 31]. As mentioned above, it is similar to the Standardized Precipitation Index—SPI [32, 33] and has common theoretical base. Evaluations are conducted after selection of a representative time scale for the investigated drought. With regard to this scale, the abnormal water state is evaluated transforming the streamflow data into a standardized evaluation. Usually drought is evaluated on monthly (SRI 1), seasonal (SRI 3) or yearly (SRI 12) base. For the evaluation, the following scale proposed by the Expert Group on Water Scarcity and Drought, EC [7, 29] was used:

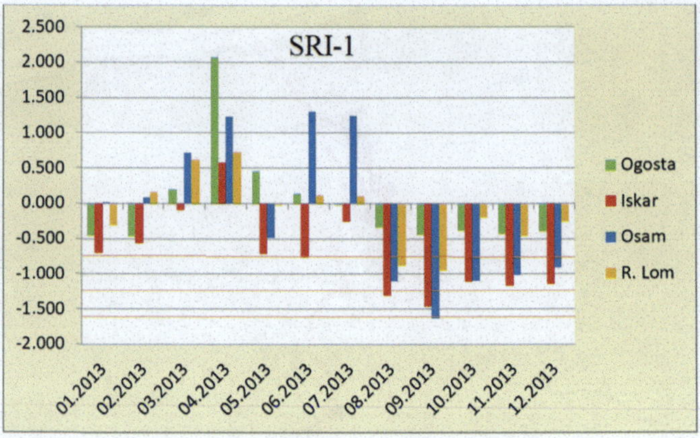

Fig. 14 Monthly SRI values for the rivers Ogosta, Iskar, Osam and Rusenski (R.) Lom in 2013

$-0.84 \geq SRI \geq -1.28$ (a) corresponds to moderate drought;
$-1.28 \geq SRI \geq -1.65$ (b) corresponds to severe drought;
$SRI < -1.65$ (c) corresponds to extreme drought.

On this base drought evaluation was performed by computing the SRI 1 values in 2013, at monthly scale for some main tributaries of the Danube River, by using the discharges data recorded at the monitoring stations located in the lower part of the selected watercourses: Ogosta River—HMS 16850, Iskar River—HMS 18850, Osam River—HMS 2200, and Rusenski Lom River—HMS 31830. The results are presented in Fig. 14 and show that during the year 2013, severe and moderate hydrological droughts occured.

Severe drought was observed in the case of the Iskar River during the months of August and September, and in the case of the Osam River, during September. Moderate drought was observed in the period October–December for the rivers Iskar and Osam, as well as August for the Osam River, and August and September for Russenski Lom River. But drought was not detected for Ogosta River. This shows once again that the drought phenomenon is irregularly distributed in time and space, and it is closely connected to the local conditions. Even neighborhood situated rivers as Ogosta and Iskar represent different behavior with regard to the connected topography and climate conditions.

To clarify the drought process some additional investigations focused on the Rusenski Lom River were also performed. They aimed to investigate the river‘s behavior in time. For this purpose the monthly mean discharge at the station HMS 31830 for the period 1982–2012 was used. The hydrograph (Fig. 15) show high variability of the average monthly discharges.

Hydrological drought investigations were also performed on Rusenski Lom River by the implementation of the Standardized Runoff Index SRI-1. During the period 1982–2012 there were observed cases with moderate, severe, and extreme droughts. The most affected by drought periods were 1993–1994, 1990–1991, and 1985–1986.

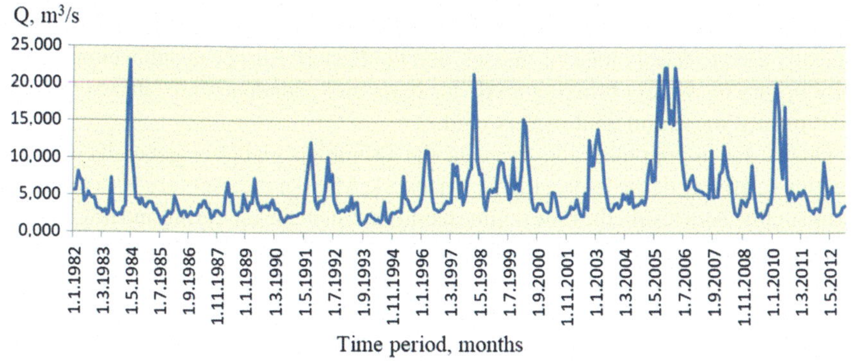

Fig. 15 Hydrograph of the average monthly discharges of Rusenski Lom River (HMS 31830) for the period 1982–2012

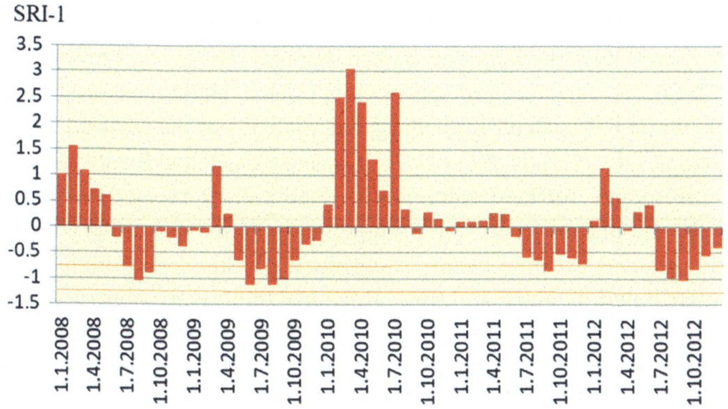

Fig. 16 Monthly SRI values for the Rusenski Lom River (HMS 31830) (2008–2012)

Droughts were observed also in 2008–2009, 2011, and 2012 (Fig. 16). The most frequent drought periods are during the summer. Extreme droughts occur during August and September, and severe drought can be observed often in June, July, September, and October.

5 Conclusion and Recommendation

The Lower Danube state is influenced significantly by the tributaries and their behaviors. To clarify the problem, in the chapter briefly are presented main Bulgarian tributaries of the Lower Danube River and their major specifics. The analyzed rivers are: Ogosta, Iskar, Vit, Osam, Jantra, and Rusenski Lom. Their watersheds are located mainly in the Danubian plain, extended in the northern Balkan mountains, under the influence of the continental climate. As a result, in springs, summers, and autumns typically are observed moist air masses, and the Eastern European anticyclone typically occurs in winters bringing arctic air mass and abundant snowfalls. Summers are typically warm and winters very cold.

Due to the topography, the rivers in the Danubian plain are generally shorts, creating relatively dense hydrographical network. Most of the rivers are with rainfall-snow supply. As a result, the flow regime has high waters during spring and low waters during summer. Some of the analyzed rivers are strongly influenced by anthropogenic works (some of them are corrected or modified with levees and dams).

The Danube River and the adjacent floodplain are home to many species (plants, mammals, reptiles, amphibians, and fishes). The biodiversity is very rich, thus in the area several important natural protected areas were designed.

The hydro-environmental features of the Danube plain are determined by proper monitoring of the main hydro-environmental elements. Regarding waters, regularly is provided physico-chemical, hydro-biological, and hydromorphological monitoring. Emission control is also provided. Main chemical indicators that are controlled are dissolved oxygen (DO), biochemical oxygen demand (BOD_5), ammonium (NH_4-N) and nitrate (NO_3-N) nitrogen, phosphorous (PO_4-P). Their measured values show that the rivers water bodies have high and good quality. For the period 1996–2020 the concentrations of DO, NH_4-N, NO_3-N, BOD_5, and PO_4-P had lower values compare with the previous years. Surface water quality results show a long-term upward trend.

The Danube tributaries hydro-environmental specifics are closely related to the river flow regime and extreme events occurrence (floods and droughts), which are irregularly distributed in space and time. Recently there were observed large floods with severe damages (affected agricultural land, households, infrastructure, as well as economical, and life loses). Observed flood occurrences are not connected to the typical high flow conditions. In the studied area drought events were also observed with a grade from extreme to moderate.

To mitigate the harmful hydro-climatic consequences, different measures were designed. They are connected both to the adopted legislation and management practices. As a European Community Member State, Bulgaria accepts and strictly implements the EC legislation. In this regard, main tools designed and applied in practice are the River Basin Management Plans (RBMPs) and Flood Risk Management Plans (FRMPs) and connected documents. They account both the local specifics and national regulations. All adopted measures are adapted to the existing ecosystems, and for this purpose, the ecosystem approach is widely applied in the designed measures.

Presented material gives valuable information on the current Bulgarian Danube tributaries state. It properly complements the ongoing monitoring with river basin management measures. But long-term measures require good understanding of the on-going processes, and investigations on flood risk are timely and appropriate. Regarding the flood risk management, the extension of preventive measures is recommended. This means, among others, to consider the early warning systems implementation, to develop all connected research and investigations, to create a decision support system (DSS), etc. This will ensure better reaction helping to harmful effects mitigation, protection of the population and its goods, as well as the ecosystems preserving. Floods reduction measures also need to be improved. This means to observe the options for water reservoirs usage for floods retention, and also to implement additional nature based retention measures when needed. This will help the ecosystems protection and their natural state ensuring.

Development of scenarios for future climate and its hydrological impacts is another useful measure for negative weather behavior mitigation. They can be considered as a first step for future hydrological state detection. To be effective, they need a systematical analysis and river basin specifics accounting. For this purpose, of importance is to have a good knowledge about the past events behavior. On this basis, some potentially vulnerable territories in the future can be detected.

This study presented also information on the past hydrological droughts occurred on some Bulgarian Danube tributaries. They are scientifically grounded and can be used in future for the Danube basin hydrological drought analysis. To be effective there is a need of more detailed drought study that can be in the center of the future Drought Risk Management Plans Development.

The present chapter provides valuable information on the main Bulgarian Danube tributaries and management practices that can be useful for decision and policy makers in the area for future more effective river basin and hydrological risks management.

Acknowledgements This work has been carried out in the framework of the National Science Program "Environmental Protection and Reduction of Risks of Adverse Events and Natural Disasters", approved by the Resolution of the Council of Ministers № 577/17.08.2018 and supported by the Ministry of Education and Science (MES) of Bulgaria (Agreement № D01-230/06.12.2018).

References

1. RBMP-DRBD (River Basin Management Plan—Danube River Basin Directorate) (2016) River Basin Management Plan (2016–2021), Danube River Basin. http://www.bd-dunav.org/content/upravlenie-na-vodite/plan-za-upravlenie-na-rechniia-baseyn/. Accessed 22 Feb 2021
2. FRMP-DRBD (Flood Risk Management Plan—Danube River Basin Directorate) (2016) Flood Risk Management Plan (2016–2021), Danube River Basin. http://www.bd-dunav.org/content/upravlenie-na-vodite/upravlenie-na-riska-ot-navodneniia/. Accessed 22 Feb 2021
3. Chilikova-Lubomirova M (2016) Drought: recent definitions, impacts and mitigation. J Balk Ecol 19(3):229–238
4. Vitanov NK, Dimitrova ZI (2015) Risk analysis. Technological, political and other risks. Qualitative analysis of risk I. "Vanio Nedkov", Sofia
5. Kolcheva K (2019) Water allocation and climate character and adaptation. Vono delo 1/2. STUVA, Sofia
6. Ivanova-Radovanova P (2020) Assessment of the natural landscapes as a prerequisite for environmental risk assessment and preparation of management plans. In: Proceedings of 1st International conference on environmental problems and disasters RISKs. Part II. Sofia, pp 512–519
7. Belev G (2021) Ecological problems in south watersheds on Struma River. SocioBrains J 78:155–162. SMART IDEAS—WISE DECISIONS Ltd., Sofia

8. ExEA-Bulgaria (2020) National Report on the State of Environment in Bulgaria, Ministry of Environment and Water and Executive Environmental Agency
9. European waters—current status and future challenges—a synthesis (2012) EEA Report No 9/2012, EEA, Copenhagen
10. COM 673 Final (2012) A Blueprint to Safeguard Europe's Water Resources [SWD(2012) 382 Final], Brussels, 14 Nov 2012, EC
11. Alexandrov V (2005) On the soil drought in Bulgaria. Thematic reports on Project 0004350 Capacity building for sustainable land management in Bulgaria. Sofia
12. Donchev D, Karakashev H (2004). Topics on physical and social economic geography of Bulgaria. Ciela, Sofia (in Bulgarian)
13. Velev St (1990) Climate in Bulgaria. Narodna Prosveta, Sofia
14. Sreburna National Park, https://srebarnalake.com/. Accessed 25 Feb 2021
15. Kalimok-Brushlen marches, http://natura2000.moew.government.bg/Home/ProtectedSite?code=BG0000377&siteType=HabitatDirective. Accessed 25 Feb 2021
16. Natura 2000, https://natura2000.eea.europa.eu/. Accessed 25 Feb 2021
17. WFD (2000), Directive 60/2000/EC of the European Parliament and of the Council of 23 October establishing a framework for Community action in the field of water policy (Water Framework Directive). Official Journal of the European Communities. EC. 22.12.2000
18. Floods Directive (2007) Directive 2007/60/EC of the European Parliament and of the Council of 23 October 2007 on the assessment and management of flood risk (Text with EEA relevance). Official Journal of the European Communities. EC. 6.11.2007
19. The Habitats Directive (1992) Directive 92/43/ECC. Council Directive 92/43/EEC of 21 May 1992 on the conservation of natural habitats and of wild fauna and flora. OJEC No 1, 206/7, 22.07.92
20. CBD (1992) Convention on Biological Divercity. United Nations 1992. UN
21. CIS TR-2009-040 (2009) Common implementation strategy for the WFD, River Basin Management in a changing climate. Technical Report—2009-040. EC
22. Environment Protection Law (2020) Sate Gazette No 91/ 25 Sep 2002, last amend. SG. No 102/1 Dec 2020
23. Water Law (2020) State Gazette No 67/27 July 1999 Effective from 28 Jan 2000, last amend. SG. No 102/ 1 Dec 2020
24. Regulation No 1 (2011) for water monitoring from 11.04.2011 last amend. SG. No 20/15 March 2016
25. Regulation No H-4 (2012) for surface water characterization from 14.09.2012, last amend. SG 79/23 Sep 2014
26. Regulation of terms and conditions for technical and safe exploitation of dams and associated facilities, and their technical state control (2016)
27. Biodiversity Law (2018) SG 77/9 Aug 2002, last amend. SG. 98/27 Nov 2018
28. Protected Areas Law (2019) SG. No 133/11 Nov 1998, last amend. S.G No 1/3 Jan 2019
29. EM-DAT, https://www.emdat.be/. Accessed 25 Feb 2021
30. Shukla S, Wood AW (2008) Use of a standardized runoff index for characterizing hydrologic drought. Geophys Ressearch Lett 35:L02405. https://doi.org/10.1029/2007GLO32487
31. Merida AA, Ureta JM, Trevino JG, Chilikova-Lubomirova M (2014) Water resources and society with respect to water scarcity and drought. In: Proceedings Dooge Nash International Symposium, Ireland, pp 223–232
32. McKee TB, Doesken NJ, Kleist J (1993) The relationship of drought frequency and duration to time scales. Preprints, Eight Conf Appl Climatol: 170–184. Anaheim, CA, Am Meteor Soc
33. Guttman N (1999) Accepting the Standardized Precipitation Index: A calculation algorithm. J Am Water Resour Assoc, JAWRA 35(2):311–322

Physico-Chemical Features and Quality of the Hydro-Environment

Water Temperature Variability in the Lower Danube River

Daniel Dunea, Petre Brețcan, Gheorghe Șerban, Dănuț Tanislav, Elena Țuchiu, and Ștefania Iordache

Abstract The chapter presents the analysis of the water temperature variability in the Lower Danube River. Temperature of water is one of the most important quality indicators for river ecosystems, which controls many physical and biogeochemical processes within the water body. All the aquatic species have the specific water temperature ranges for growth and development, thus, significant variations of water temperature may cause harmful consequences to the aquatic ecosystems. Surface waters present high variations of temperature depending on spatio-temporal variability and environmental conditions. Gradual rising of the surface waters temperature has a favorable influence on the water properties because this facilitates the natural water purification. An important influencing factor is the discharge of heated wastewaters directly in the streams, which can cause the reduction of dissolved

D. Dunea (✉)
Faculty of Environmental Engineering and Food Science, Department of Environmental Engineering, Valahia University of Târgoviște, Aleea Sinaia No. 13, 130004 Târgoviște, Romania
e-mail: dan.dunea@valahia.ro

P. Brețcan (✉)
Faculty of Humanities, Department of Geography, Valahia University of Târgoviște, Aleea Sinaia No. 13, 130004 Târgoviște, Romania
e-mail: petre.bretcan@valahia.ro

G. Șerban
Faculty of Geography, Babeș-Bolyai University, Cluj-Napoca, Romania
e-mail: gheorghe.serban@ubbcluj.ro

D. Tanislav
Faculty of Humanities, Department of Geography, Valahia University of Târgoviște, Târgoviște, Romania
e-mail: dan.tanislav@valahia.ro

E. Țuchiu
National Administration "Romanian Waters", Bucharest, Romania
e-mail: elena.tuchiu@rowater.ro

Ș. Iordache
Faculty of Environmental Engineering and Food Science, Department of Food Engineering, Valahia University of Târgoviște, Târgoviște, Romania
e-mail: stefania.iordache@valahia.ro

A. Negm et al. (eds.), *The Lower Danube River*, Earth and Environmental Sciences Library, https://doi.org/10.1007/978-3-031-03865-5_5

oxygen content. In this regard, we present a time series statistical analysis of the water temperature recorded between 2001 and 2016 in three monitoring sections located on the Romanian side of the Lower Danube i.e., Pristol (RO2), Chiciu (RO4), and Reni (RO5) using monitoring data from the Transnational Monitoring Network of the Danube River (TNMN) database. Despite some differences between the monitoring sections determined by the local hydrological, climatic, and topographical conditions, a relative constancy of the water temperature was observed on the entire analyzed period. However, the obtained trendlines show that the water temperature increased from 2001 to 2016, this pattern being more evident in the southernmost control section (Chiciu-RO4). The SARIMA model provided a comprehensive description of the spatiotemporal variations of the water temperature but more complex approaches for improving water monitoring and modeling in the Lower Danube are required to integrate them in process-based analysis.

Keywords Environmental monitoring · Water temperature · Lower Danube River · Time series analysis · Seasonal trend

1 Introduction

Water temperature is a key indicator for river ecosystems, which controls many physical and biogeochemical processes within the water body, such as the reaeration processes [1, 2], decomposition of organic matter [3], nitrification dynamics [4] etc. All the aquatic species (e.g., zooplankton, phytoplankton, fish, and insects) have the specific water temperature ranges for growth and development, thus, significant variations of water temperature may cause harmful consequences to the water ecosystems [5]. Biological activity and growth of aquatic organisms are clearly influenced by the temperature [6]. The number of individuals is affected when the temperatures reach too far below or above this optimal range particularly on longer periods.

Water temperature is an important control of water density, oxygen solubility (physical characteristics), nutrient mineralization (biochemical), organism growth and biological behavior of fluvial hydrosystems [4, 7, 8]. Higher water temperatures usually increase the rate of chemical and metabolic reactions. The aquatic life in a stream generally suffers because of warm stream water, which holds less dissolved oxygen compared to cool water that may affect various species living in that stream. At higher temperatures, some compounds may become more toxic to existing aquatic life. Persistently warmer temperatures favor the release of excess nutrients into the water [9]. Aquatic plants and algae undergo modified rates of photosynthesis according to the temperature variations. The rate of photosynthesis increases with the rising of temperature in the presence of an optimal quantity of nutrients. The upper limit range for fish species is from 38 °C and from 50 °C for aquatic insects to 73 °C for blue-green algae [10]. The temperatures should not exceed 32 °C in warm water streams, and 20 °C in cold streams, respectively. Excessive heat in summer is favoring fish death due to low oxygen levels in water that are occurring concomitantly with elevated temperatures [11].

Moreover, the thermic condition is one important supporting physical parameter of the biological quality elements defined for evaluation ecological status of surface water bodies. In the Romanian ecological status classification system, thresholds are established in relation to thermic polluted water discharges (less than 21.5 °C in salmonid waters and less than 28 °C in cyprinid waters downstream to discharge point) [12].

Several key factors such as air temperature fluctuations, solar radiation characteristics, meteorological episodes and particular stream and watershed conditions determine the variations of a stream temperature. There are some intrinsic conditions of the stream namely origin, morphology, velocity, land cover/land use, type of vegetation, and the coverage of impervious surfaces [13]. It is obvious that a shaded tight shoreline will diminish the solar radiation impact on the water heating, while a denuded shoreline combined with a wide shallow riverbed will undergo the opposite [14].

Overall, the temperature of a stream is strongly dependent on the watershed's climate. Usually, the temperature is colder in the upper part of the stream near its origin and warmer near its outlet. This natural trend may be disturbed in a particular section due to other sources of water from snow melting or precipitations, effluents discharged in the main stream, river discharge, and shading vegetation (riparian patches on shorelines). Indeed, riparian vegetation is able to reduce the contribution of solar radiation by shadowing effect regulating the diurnal temperature fluctuations of streams' water [15, 16]. Previous studies pointed out that the removal of neighboring riparian buffers could clearly lead to an increase in streams' temperature [17]. The composition and structure of the vegetation surrounding the water body establish the intensity of the temperature lowering effect [18]. In two wetlands of small river basins located in the south of Romania, the air temperature at ground level was lower under the canopy than outside of it showing differences of 3.2 °C in the upstream region and of 5.1 °C in the downstream section. Somehow, the dense herbaceous closed canopy was more efficient in lowering ground level temperatures than the trees scattered canopy [19]. Similar reports originate from Switzerland where a decrease of 1.2 °C under the canopy has been observed at ground-level [20], and from New Zeeland where the average water temperature at clear-cut locations was 3.2 °C higher than in areas where the stream was shaded by the pine forests [21].

The shadowing effect of riparian vegetation is essential as high water temperatures affect the integrity of the aquatic ecosystem putting at risk the existing biota [22]. An increase of 2 °C in atmospheric temperature translates in water temperature rising with 1.3 °C in barren location and of only 0.8 °C in areas with riparian vegetation [23].

The vegetation from riverbanks can reduce the stream temperatures up to 10 °C depending on the climatic zone [24]. It can also diminish the daily and seasonal variations in the temperature during the summer periods of low flows especially for unshaded streams [10]. Many species of fish and invertebrates encumber stressful conditions between these periods of warm water in unshaded small streams. The shading effect of riparian buffers is less important in wider rivers, in which water temperature is more frequently dependent on the runoff characteristics, the exposure to solar radiation and the heat of groundwater entering the river from aquifers [16]. Usually, higher temperatures occur from the removal of the trees belonging to the riparian zone. A potential rehabilitation can be obtained when structured replanting is performed. Nonetheless, the runoff from various surfaces can contribute to the increasing of the stream temperature [11].

Because phreatic aquifers are under the direct influence of atmospheric precipitations, the groundwater level will reflect their dynamics with a certain delay depending on the precipitations amount falling on the surface of the hydrogeological basin of the aquifer, as well as the thickness of the unsaturated area [25]. Subfluvial aquifers ensure the natural discharge of rivers together with the saturated area on the slopes during the periods without precipitations. The situation changes during periods of high flows when the water level in the river is higher than the water level in the aquifer and the exchanges take place in the opposite direction, feeding the unsaturated zone of the major riverbed. In arid areas, if the groundwater level is very low, rivers can lose water completely when the riverbed is made of sand and gravel [26]. Industrial uncontrolled discharges, insufficient treatment of wastewaters, local modifications of hydrology due to dams or channels, and changes in land cover/land use in the watershed are other exogenous factors that modify consistently the water temperature of streams. It was found that water temperature rises more quickly compared to air temperature in agricultural areas in the absence of major dams, and more slowly in areas with forests and dams [10]. The occurrence of deviations in the thermal regime is related to the discharge of wastewaters and/or effluents combined to solar irradiance and groundwater inputs [11]. In [27] it was pointed out that hypolemnetic release dams can lower stream water temperature by releasing colder water from the bottom of a reservoir, or increase it by releasing warmer water from the epilimnion on both local and regional scales. Aquatic organisms have evolved to cope with such regular fluctuations through the regulation of their metabolism and adaptation of reproductive cycles.

Finally, it can be stated that water temperature is one of key characteristics of the water streams because it regulates most of the processes related to the water quality from controlling physical properties such as water viscosity and infiltration rates, physical reactions, chemical equilibria and reaction rates to biological characteristics like habitat suitability and growth rate. The water temperature has an impact on human activities like recreation, aquaculture, and navigation patterns [28]. Withal, temperature is a variable in water systems modeling. Both air and water temperatures are useful for the description of advective processes (e.g. evaporation, evapotranspiration) in water quality models (e.g. SWAT model) because the water cycle involves the exchange of energy leading to temperature changes [29].

Some technical solutions for cooling the water could be the temporarily water storage in off-stream reservoirs before release back to the river [30]. Furthermore, multiple off-take structures exist, which can take cool water from deep or from its warmer surface in a reservoir. The impact of cool water released from high dams can be diminished by such structures.

The study of Earth's water balance and energy processes and the heat transfers occurring at the surface waters–atmosphere–biosphere interface require the detailed assessment of ground surface temperature [30].

In this context, we present a time series analysis of the water temperature recorded between 2001 and 2016 and one-year ahead forecasting at three monitoring sections located on the Romanian side of the Lower Danube River i.e., Pristol (RO2), Chiciu (RO4), and Reni (RO5), based on data from Transnational Monitoring Network of the Danube River (TNMN) database. The studied region has a significant aquatic biodiversity and a historic cultural heritage.

2 The Study Area: General Features

Together with the Danube Delta, the Lower Danube River, which flows beyond 1000 km through several countries including Romania, is an important biodiversity region in the world sheltering numerous rare and endangered habitats and species. The freshwater ecosystems of the Lower Danube River provide needful environmental services and numerous opportunities for the sustainable development of local communities [31].

The study area is located between Gura Văii and Pătlăgeanca (Fig. 1), with a total length of 931 km. It represents the lower sector of the Danube River, which can be finely divided into two major sub-sectors: Gura Văii-Călărași and Călărași - Pătlăgeanca. From a morphological point of view, the two sectors are characterized by the asymmetry of the banks and reduced slopes (0.02–0.07%), the Bulgarian bank being steep and with 200 m higher compared to the Romanian one [32, 33].

The general altitude of the floodplain decreases from upstream to downstream from 43–45 m upstream to 2–3 m upstream of the entrance to the Danube Delta and the width of the floodplain varies from 2–3 km to 24 km in Balta Brăilei, respectively 30 km near the Călmățui channel. In the Gura Văii-Călărași sector, the low slopes cause a decrease in the meandering of the minor riverbed alternating with the appearance of islands ("*ostroave*") [34, 35]. It has widths of 0.5–1.5 km, water depths of 4–10 m and water speed <2 m/s. The Călărași - Pătlăgeanca sector is defined by very low slopes (0.02–0.04%), which determines a decrease of the water velocity (0.8–0.6 m/s) depending on the discharge and the unraveling in two main channels that close the Ialomița and Brăila wetlands.

Because of its geographical position in the south (between parallels of 44° and 45° N lat.), a long duration of sunlight (2250–2500 h of sunshine) occurs with increased values of global solar radiation (>207 W/cm^2), which determines the appearance of a specific "*Danubian*" topoclimate [36]. The influence of the large water mass attenuates the thermal contrasts in winter / summer, in the hot season the average monthly temperatures reaching 23 °C, while in the cold season the values are negative between −1 and −2.8 °C. Based on the precipitation variations in the Lower Danube Basin recorded at ten weather stations, determine variations of the tributary rivers

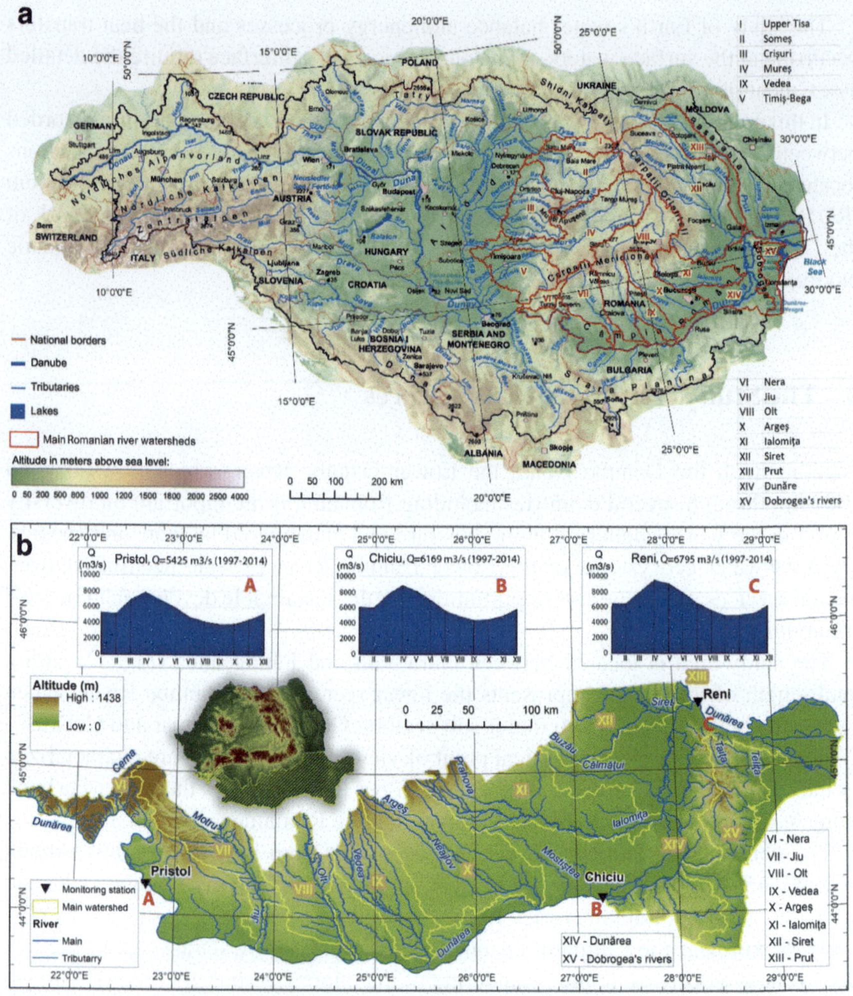

Fig. 1 The study area: (**a**) the Danube River basin and the main Romanian watersheds (processed after ICPDR relief and topography map—https://www.icpdr.org/main/sites/default/files/Map02_Relief_and_Topography.pdf); (**b**) The Lower Danube (Dunărea) River between Gura Văii and Pătlăgeanca and its main tributaries in Romanian side (monitoring sections considered in this study: A—Pristol; B—Chiciu; C—Reni)

discharge and differences of discharge between the Orșova upstream station and the Ceatal downstream station [36].

Therefore, the longitudinal arrangement of the Danube watercourse in most parts of the analyzed sector combined with the large volume of water, make the Danube River a "*climate barrier*" that moderates the excesses of climatic influences generated by the general circulation. This is because the general circulation is influenced in the west by Atlantic and Mediterranean cyclones, while the Eurasian anticyclones determine an accentuated aridity in the east [37, 38].

In the analyzed sector, the hydrological regime of the Danube River is influenced by the inflow of tributaries, determining an increase of the discharge from 5425 m^3/s (at Gruia) to 6795 m^3/s at Isaccea (Fig. 1b) [39, 40]. The Romanian bank has the most important contribution to the increase of discharges with its rivers' influxes. Based on the size of the discharge, the main tributaries are Siret (225 m^3/s), Olt (180 m^3/s), Jiu (93 m^3/s), Arges (70 m^3/s), and Prut (65 m^3/s), while from the Bulgarian bank it receives the following tributaries: Iskar (57 m^3/s), Timok (39 m^3/s), Iantra (40 m^3/s), Ogosta (18 m^3/s), and Vit (13 m^3/s) [41, 42].

3 Methodology

Time series analysis (TSA) is an important tool of statistics providing suitable indicators for time-dependent data including the possibility of time series forecasting [43]. Some of the commonly used techniques are random moving averages, random walks, trend models, seasonal exponential smoothing, Boltzmann composite lattice, and autoregressive parametric models and its variants [43]. The moving average, exponential smoothing, and *Auto-Regressive Integrated Moving Average* (ARIMA) are linear models that are used in the forecasting of water parameters. Considering that the time series is stationary, the predictions of values are linear functions of past observations. Box and Jenkins (1976) [44] formulated the ARIMA model based on the autoregressive and moving average parameters using three types of parameters: the autoregressive parameter, the moving average parameters and the differentiation passes. A parametric model relating the most recent data value to the previous data values and previous noise gives the best forecast for future data. The major ARIMA model weakness relies upon the assumption that the examined time series is linear and stationary with no structural changes [43, 44].

3.1 Description of the SARIMA Model

SARIMA is an extension to ARIMA, which supports the direct modeling of the seasonal component of the univariate data series. ARMA and ARIMA features are briefly described below to better understand the particularities of SARIMA.

ARMA (*Auto—Regressive—Moving-Average*) models are a combination of autoregressive models (AR) with moving average models (MA). These were introduced by Box and Jenkins (1976). The ARMA model of order (p, q) is given by the following equation [44]:

$$X_t = \alpha_1 X_{t-1} + \cdots + \alpha_p X_{t-p} + e_t + \beta_1 e_{t-1} + \cdots + \beta_q e_{t-q}$$

where $\{\alpha_i\}$ are AR parameters, $\{\beta_i\}$ are MA parameters, p is the order of the AR process, q is the order of the MA process, X_t are the terms of the time series, and e_t are the terms that represent the error. The e_t error terms are assumed to be white noise i.e. a succession of independent and identically distributed random variables, with zero mean. The abovementioned relation can be written as:

$$\left(1-\sum_{i=1}^{p}\alpha_i L^i\right)X_t=\left(1-\sum_{j=1}^{q}\beta_j L^j\right)e_t$$

where L^i is the delay operator $L^i\,X_t = X_{t-i}$.

The calibration of an ARMA model requires that the time series is stationary, which indicates that there are no systematic changes in average and variance. One of the advantages of ARMA is that it involves fewer parameters than MA or AR taken alone [45]. ARMA (p, q) processes are also known as Box-Jenkins non-seasonal stationary processes [46].

ARMA models are very useful for processing stationary time series, but most environmental processes, including water temperature dynamics, vary from season to season, and these seasonal fluctuations can cause problems in its use. It is necessary to eliminate the non-stationary variation sources in order to be able to use ARMA type models [47]. A relatively simple method is to subtract seasonal and annual averages from each corresponding month to avoid seasonal and trend influences. ARMA model was used in the forecasting of the monthly inflow in a dam reservoir [48].

A more complex method was developed to capture the seasonal fluctuations and order to overcome the shortcomings of ARMA (p, q). This is the ARIMA (*Auto-Regressive Integrated Moving Average*) model [49].

An *non-seasonal* ARIMA model is actually a classic "ARIMA (p, d, q)" model, where:

- p is the number of autoregressive terms,
- d is the number of non-seasonal differences,
- q is the number of delayed prediction errors in the prediction equation.

It is important to identify the differentiation order needed to make the time series stationary to identify the most appropriate configuration of ARIMA model for a specific time series, after which it is determined whether AR and MA terms are required to correct the autocorrelation that remains in the differentiated series [48].

The *Seasonal ARIMA*—SARIMA is based on an ARIMA model, to which additional seasonal terms are added. The SARIMA model is formally represented as follows [49]:

$$\text{ARIMA}(p,d,q)\,\times\,(P,D,Q)$$

where:

- P is the number of seasonal autoregressive terms (**S**easonal **A**uto **R**egresive),
- D is the number of seasonal differences,
- Q is the number of seasonal moving averages (**S**easonal **M**oving **A**verage).

The equation has the following form:

$$\Phi_P\left(L^s\right)\varphi(L)\nabla_s^D\nabla^d X_t = \Theta_Q\left(L^s\right)\varphi(L)e_t$$

where $\{X_t\}$ denotes the nonstationary time series, $\{e_t\}$ the Gaussian white noise, s the time series interval, polynomials $\varphi(L)$ and $\theta(L)$ of orders p and q are the ordinary autoregressive and moving average components, $\Phi_P(L^s)$ and $\Theta_Q(L^s)$ the seasonal autoregressive and moving average components with P and Q orders, ∇_d and $\nabla^D{}_s$ represent the ordinary and seasonal difference components, while L indicates the backshift operator. [47].

In Box and Jenkins' (1976) [44] approach to time series modeling, there are three steps:

- *Identification.* In this step, it must be determined whether the model is stationary or not. A transformation can be applied to the model to remove the influence of the trend. At this stage, too large fluctuations can be smoothed by using logarithmic values instead of real values.
- *Estimation.* At this stage, the SARIMA model is calibrated using empirical autocorrelation functions and partial autocorrelation functions. The importance of the parameters is checked.
- *Verification.* The last step is the analysis of the residues.

In this chapter, we used the abovementioned steps to evaluate the forecasting capabilities of SARIMA for one-year ahead forecasting (12 values) using long time series of averaged monthly measurements (2001–2016) and the forecasted series was compared to the real values recorded in 2017. Regarding the utilization of SARIMA for river water parameters forecasting, there are several works in literature that report its forecasting performances [e.g. 50–54]. However, the utilization of SARIMA for temperature forecasting of water streams, especially for large rivers such as the Danube River, is seldom approached in reported research.

3.2 *Description of the Datasets*

The data monitored on Danube River in the selected control section have been collected within the *Transnational Monitoring Network of the Danube River* (TNMN) under the auspices of the International Commission for the Protection of the Danube River (ICPDR). Since 2000, TNMN evaluates the status of the water bodies based on the Water Framework Directive 2000/60/EC. It is a long-term goal that envisages the recovering of the water quality by reducing the pollutant loads in the

Danube River, as well as in its main tributaries. The data are recorded by the laboratories belonging to the National Administration "Romanian Waters" (NARW), which send the data to the Hydrological Institute from Bratislava (Slovakia). The quality control is ensured through laboratory-specific procedures and inter-comparison trials within the ICPDR scheme [29].

Within the TNMN monitoring program, the relevant hydrological and physico-chemical parameters are recorded. Water temperature is measured usually with bimonthly frequency (Hg—Thermometer analytical method) according to the applicable water quality standards.

The data on water temperature of the Danube River used in this study were recorded in three monitoring sections, on the Romanian side of the Danube River, considered relevant for the study area: Pristol (RO2), located at 44.21418 N, 22.67613 E, km 834; Chiciu (RO4), located at 44.12757 N; 27.26771 E, km 375 and Reni (RO5), located at 45.46324 N; 28.23190 E, km 132. Data was normalized for SARIMA model by computing the average of the monthly values or keeping the single value (if one value was reported in the TNMN database) to obtain just one representative value for seasonal setup and forecasting of a monthly output. Consequently, the time series contained 192 values in each monitoring point for data processing. Three time series for each monitoring section ($n = 192$) from January 2001 to December 2016 were used in the Time Series Analysis (TSA) procedure. STATISTICA software [55] was applied to analyze the recorded time series data, in order to detect and evaluate temporal patterns in water temperature. Least Significant Differences (LSD) test was used to perform the multiple comparisons for identifying the statistically significant differences between the water temperature time series. The Kruskal–Wallis test was considered to check the statistical differences between the medians of diffrerent time series. The data from all the columns was first combined and ranked from smallest to largest. The average rank was then computed for the data in each column [56].

4 Water Temperature Variability

4.1 Variation of Danube's Temperatures Between 2001 and 2016

Temperature variability may provide a sound basis for characterizing the surface streams. Figure 2 presents the variation of the raw data series (2001–2016) and the corresponding trendlines of monthly averages for all three sections, RO2, RO4 and RO5. A higher amplitude of water temperature is visible for the RO4 section compared to the other points.

A comparison of the main statistics of the time series recorded between 2001 and 2016 in the monitoring sections of the Danube River (Romanian bank) is showed in Fig. 3.

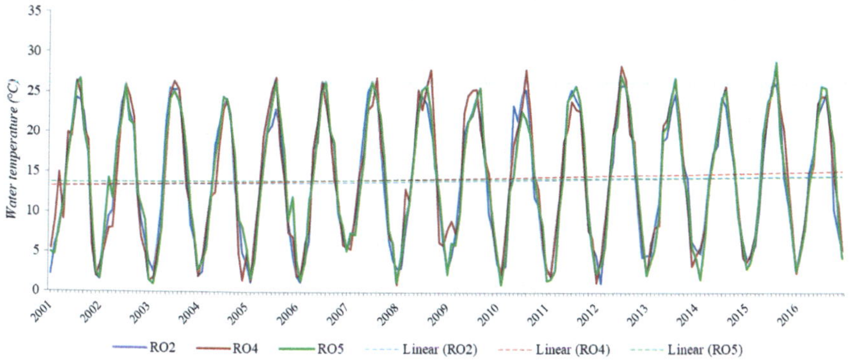

Fig. 2 Water temperature recorded between 2001 and 2016 in the monitoring sections on the Lower Danube River (monthly averages) and the corresponding linear trendlines. *Data source* TNMN

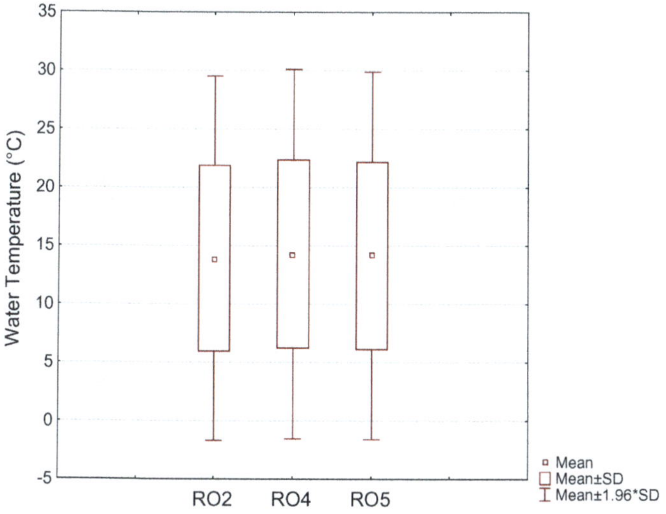

Fig. 3 Descriptive statistics of the time series of water temperature (°C) recorded between 2001 and 2016 in the monitoring sections on the Lower Danube River (Romanian bank)

As expected, the multiannual average from the RO2 section (13.9 °C) is lower than the one from the other two sections (14.3 and 14.2 °C). In the upper part of the lower Danube (RO2—km 834), the multiannual average temperature was with 0.4 and 0.3 °C lower compared to the subsequent control sections located at km 375 and km 132, respectively closer to the river mouth. There were no statistically significant differences between any pair of means at the 95.0% confidence level (LSD – DL5% = ± 1.6 °C). The Kruskal–Wallis test showed that there is not a statistically significant difference amongst the medians at the 95.0% confidence level ($p = 0.87$).

Table 1 Central tendency, dispersion and distribution parameters for the time series of monthly averages of water temperature (°C) recorded between 2001 and 2016 in the control sections on the Lower Danube River ($n = 192$)

Danube River section	RO2	RO4	RO5
Arithmetic average	13.9	14.3	14.2
Geometric average	11.0	11.4	11.1
Median	13.7	14.0	14.2
Minimum	1.2	1.0	1.0
Maximum	26.3	28.5	29.0
Amplitude	25.1	27.5	28.0
Std. Dev.	8.0	8.1	8.0
Confidence SD −95%	7.2	7.3	7.3
Confidence SD +95%	8.8	9.0	8.9
Coef. of Var. (%)	57.1	56.4	56.6
Skewness	0.0	0.1	0.0
Kurtosis	−1.4	−1.4	−1.3

Table 1 shows the central tendency, dispersion, and distribution statistical parameters for the time series of monthly averages of water temperature (2001–2016). The maximum value (29 °C) and the highest amplitude (28.0 °C) were found in RO5. The lowest maximum value (26.3 °C) was recorded in RO2, which showed an amplitude of 25.1 °C, with 2.9 °C lower than the one from RO5.

The coefficients of variation for 2001–2016 time series were 57.1% (RO2), 56.4% (RO4) and 56.6% (RO5). Skewness and kurtosis describe the data distribution for both shape and symmetry. The indicators revealed a normal distribution for all the time series.

Tables 2, 3, and 4 present the annual statistical indicators for the entire period (from January 2001 to December 2016) in each monitoring section. When looking at the synthetic indicators obtained from the dataset, more fluctuations were present between years and between control sections. The highest averages for water temperature were recorded in RO2 with 14.7 °C in 2007, 16.1 °C in 2009 for RO4 and 14.9 °C in 2012 and 2015 in RO5. The smallest values occurred in 2012 for RO2 (1.2 °C), and 1 °C in 2008 (RO4), in 2003 and 2010 (RO5). The absolute maximum monthly average was 29 °C reached in RO5 in 2015, 28.5 °C in RO4 in 2012, and 26.3 °C in RO2 in 2006 and 2007.

However, the multiple comparison procedure applied to determine which averages are significantly different from which others showed that there were no statistically significant differences between any pair of averages at the 95.0% confidence level for each monitoring section (Difference limits for statistical significance: LSD − DL5% = ± 6.6 °C for RO2; ± 6.7 °C for RO4 and RO5).

Consequently, despite some differences between the monitoring sections determined by the local hydrological, climatic, and topographical conditions, a relative constancy of the water temperature was observed in the entire analyzed period. However, the trendlines presented in Fig. 2 show that the water temperature increased

Table 2 Annual statistical parameters of the water temperature (°C) (2001–2016) for RO2 monitoring section (Pristol, km 834—Romanian bank); $n = 12$

Year	2001	2002	2003	2004	2005	2006	2007	2008	2009	2010	2011	2012	2013	2014	2015	2016
Arithmetic average	13.4	13.9	14.4	13.3	12.4	13.2	14.7	14.1	14.4	13.8	14.2	14.4	14.4	14.1	14.4	13.7
Geometric average	10.2	11.2	11.3	10.2	9.3	9.3	12.3	11.4	12.0	10.5	10.9	10.7	12.4	12.4	12.0	11.1
Median	14.2	12.0	13.3	11.8	13.1	14.0	13.7	14.6	15.6	12.1	13.6	15.0	15.1	15.1	12.8	12.8
Minimum	2.1	1.6	2.6	1.9	1.3	1.3	4.3	3.0	3.0	2.8	2.4	1.2	4.7	5.0	4.3	2.5
Maximum	24.4	25.4	25.6	24.3	23.0	26.3	26.3	24.7	24.2	25.5	25.5	26.1	25.0	24.5	26.2	25.0
Std. Dev.	8.4	8.0	9.0	8.2	7.9	9.1	8.3	8.0	7.7	8.9	8.7	9.1	7.2	6.8	8.4	8.2
Coef. Var.	62.5	57.2	62.5	61.9	63.1	68.7	56.8	56.7	53.2	64.4	61.7	62.9	50.1	48.4	58.3	59.5
Skewness	0.0	0.1	0.1	0.0	−0.1	0.1	0.2	−0.1	−0.2	0.1	0.0	0.0	−0.1	0.1	0.2	0.1
Kurtosis	−1.7	−1.4	−1.9	−1.6	−1.7	−1.5	−1.6	−1.6	−1.7	−1.8	−1.6	−1.6	−1.5	−1.3	−1.6	−1.7

Table 3 Annual statistical indicators of the water temperature (°C) (2001–2016) for RO4 section (Chiciu, km 375—Romanian bank)—$n = 12$

	2001	2002	2003	2004	2005	2006	2007	2008	2009	2010	2011	2012	2013	2014	2015	2016
Arithmetic average	14.7	13.3	13.4	12.3	14.0	13.3	14.6	15.0	16.1	14.6	13.7	15.4	14.3	14.3	15.8	14.3
Geometric average	11.9	10.8	9.5	9.0	10.9	10.0	12.6	11.1	14.4	11.8	10.7	11.8	11.1	12.2	13.3	11.8
Median	17.0	10.1	11.8	12.3	13.3	12.8	15.4	15.3	15.9	14.4	13.8	16.5	13.0	14.8	16.2	14.1
Minimum	2.0	3.5	1.5	1.5	1.5	2.0	5.5	1.0	7.5	2.3	2.0	1.3	2.3	4.5	4.0	2.5
Maximum	26.5	26.0	26.5	24.0	27.0	26.5	27.0	28.0	25.5	28.0	24.0	28.5	26.5	26.0	28.2	24.7
Std. Dev.	8.1	8.4	9.2	8.1	8.5	8.5	7.6	9.3	7.4	8.3	8.1	9.0	8.7	7.5	8.5	7.8
Coef. Var.	55.2	63.4	68.9	66.1	61.1	64.3	51.9	62.1	46.1	56.5	59.6	58.3	61.3	52.5	53.5	54.6
Skewness	−0.1	0.5	0.2	0.1	0.2	0.1	0.2	0.0	0.1	0.0	−0.1	−0.1	0.0	0.1	0.0	0.0
Kurtosis	−1.4	−1.6	−1.5	−1.4	−1.4	−1.5	−1.3	−1.5	−1.9	−1.1	−1.7	−1.3	−1.7	−1.3	−1.6	−1.3

Table 4 Annual statistical indicators of the water temperature (°C) (2001–2016) for RO5 section (Reni, km 132—Romanian bank)—$n = 12$

Years	2001	2002	2003	2004	2005	2006	2007	2008	2009	2010	2011	2012	2013	2014	2015	2016
Arithmetic average	14.1	14.8	13.7	13.8	14.1	13.7	14.6	14.7	14.6	13.0	13.7	14.9	13.9	13.8	14.9	14.5
Geometric average	11.3	12.4	9.1	11.2	11.1	10.4	12.6	11.5	12.0	10.1	9.2	11.7	10.9	10.9	12.1	11.7
Median	14.0	13.6	14.1	14.1	13.8	12.9	14.3	14.2	15.0	13.6	12.7	15.2	13.4	15.3	14.0	15.2
Minimum	2.5	1.8	1.0	2.8	1.8	1.5	5.3	1.3	2.3	1.0	1.5	2.4	2.2	1.7	3.0	2.8
Maximum	26.8	26.0	25.3	24.5	26.5	26.5	26.5	26.0	25.8	22.8	26.0	27.3	27.0	25.5	29.0	26.0
Std. Dev	8.4	7.1	9.4	7.9	8.2	8.5	8.0	8.3	7.9	7.2	9.7	8.8	8.6	8.0	8.7	8.5
Coef. Var	59.3	48.2	68.6	57.2	58.0	61.8	54.4	56.7	54.3	55.6	70.9	59.1	61.5	57.9	58.5	58.3
Skewness	0.2	−0.2	−0.1	0.0	0.0	0.1	0.3	0.0	0.0	−0.2	0.0	−0.1	0.1	−0.1	0.1	0.0
Kurtosis	−1.3	−0.7	−1.8	−1.6	−1.3	−1.3	−1.7	−1.2	−1.4	−1.2	−1.8	−1.5	−1.5	−1.4	−1.3	−1.6

from 2001 to 2016, this pattern being more evident in the southernmost control section (Chiciu—RO4).

4.2 *Application of SARIMA Model to Forecast Water Temperature*

Latest studies have warned that water temperatures have increased in the major European rivers by 1–3 °C in the last 100 years [57]. Based on the long-term projections, it is believed that the increase in air temperature will clearly influence the river and lake surface water temperatures [58]. An increase of the average river water temperatures by 1.6–2.1 °C is expected for all major European rivers including the Danube River during 2071–2100 compared to the 1971–2000 period [57]. Furthermore, the stress to river fauna and flora will increase because the number of days with water temperatures above 25 °C will expand from 2–15 days (2001–2010 reference period) to 32–75 days (2071–2100) [59].

Furthermore, the expected impact of climate change is the increasing of water temperature, following the future increase in air temperature. In the Danube River Basin, depending on local or regional context, an increase of 1–2 °C of water temperature is estimated during summer. With higher temperatures, the surface water bodies' ecological status is expected to be worse and more frequent. In addition, intense algal blooms may appear [60].

To assess such potential scenarios and the associated impact, it is important to test and apply forecasting tools for different time windows. For this reason, various configurations of SARIMA model have been tested to find the most suitable results in terms of forecasting using the time series from RO2, RO4, and RO5 ($n = 192$).

Plots of the original data portray an overview of how the time series are commonly performing and whether seasonal differencing is needed. The suitability of the SARIMA model was evaluated based on the degree of accuracy required from the forecast, the amount of time and resources available, the amount and type of available data, and how far ahead is required to forecast.

Figure 4 presents the best fitting model i.e. SARIMA (1,0,1)(2,0,2) for one year ahead forecasting for the time series recorded in RO2 section. The model performs quite well when compared to the real observations from 2017 especially in the first 8 months of the year (Fig. 7).

SARIMA (1,0,1)(5,0,1) was considered the most suitable configuration for the time series recorded in RO4 and RO5 (Figs. 5 and 6). When comparing with the real observations, the model tended to underestimate the water temperature in the first part of the year especially in RO 4 section (Fig. 7).

Following the procedures presented in [45], the best fitted model was selected based on how the forecasts used past data to determine the variation of the forecast errors (smallest root mean squared error) and to calculate limits within which a future value of the series will stand with a given probability.

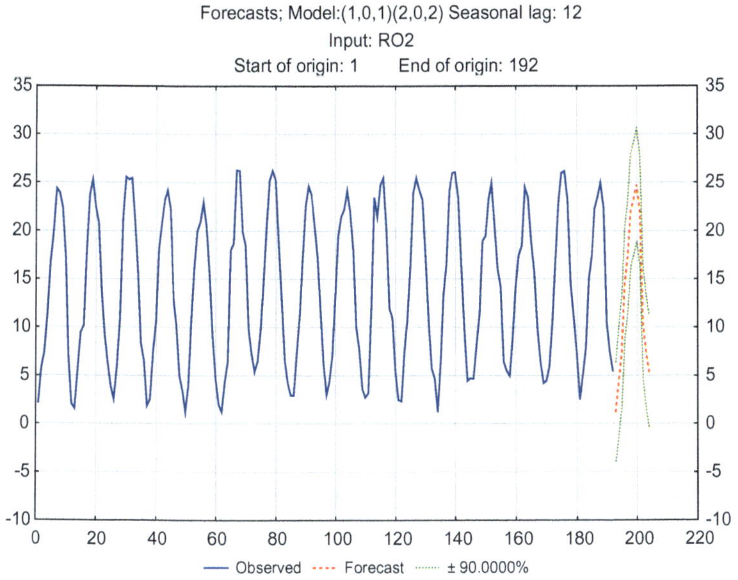

Fig. 4 One year ahead forecasting for the time series recorded in RO2 (2001–2016) using SARIMA (1,0,1)(2,0,2) (x-axis: month; y-axis: water temperature, in °C)

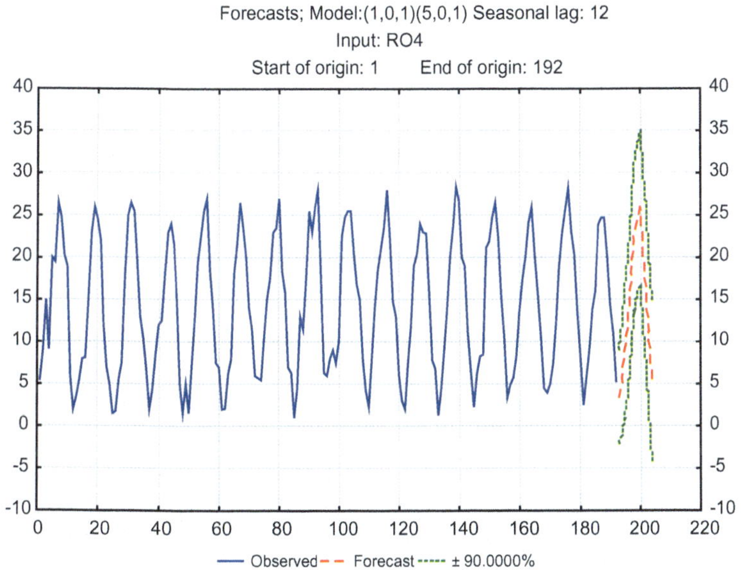

Fig. 5 One year ahead forecasting for the time series recorded in RO4 (2001–2016) using SARIMA (1,0,1)(5,0,1) (x-axis: month; y-axis: water temperature, in °C)

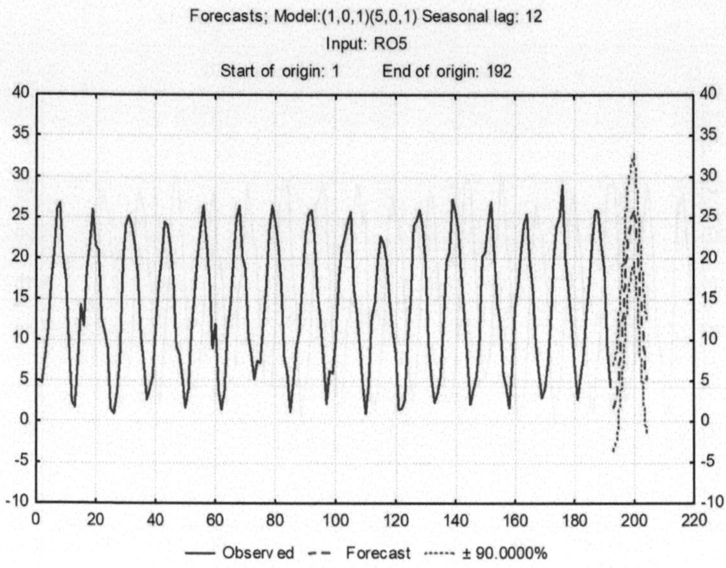

Fig. 6 One year ahead forecasting for the time series recorded in RO5 (2001–2016) using SARIMA (1,0,1)(5,0,1) (x-axis: month; y-axis: water temperature, in °C)

Overall, finding a suitable configuration of SARIMA guarantees a proper forecasting of water temperature and the statistical model can provide satisfactory results based on solely past observations of temperature. A multivariate model or hybrid neural network involving other exogenous factors could increase the accuracy of the forecasting for non-stationary and/or incomplete time series.

5 Conclusion and Recommendation

This study provides information on the water temperature variability in the Lower Danube River from 2001 to 2016 using observations from three monitoring sections (Pristol, Chciu and Reni) on the Romanian bank. The general trend shows a slight increase in water temperature during the 15 years period, which is in agreement with previous reports [57].

SARIMA model can summarize the forecasting results and present them in such a way that a specialist can more easily understand the patterns within the data describing time series that exhibit non-stationary evolutions across and within seasons. Without such tools, it is difficult to evaluate the trends of various indicators such as water temperature. One challenge is to capture all relevant seasonality trends and repeating patterns for proper forecasting i.e. the longer the forecast, the harder it might prove for SARIMA model to predict accurately. It is required that the seasonal time series are stationary and have no missing data. The disadvantage

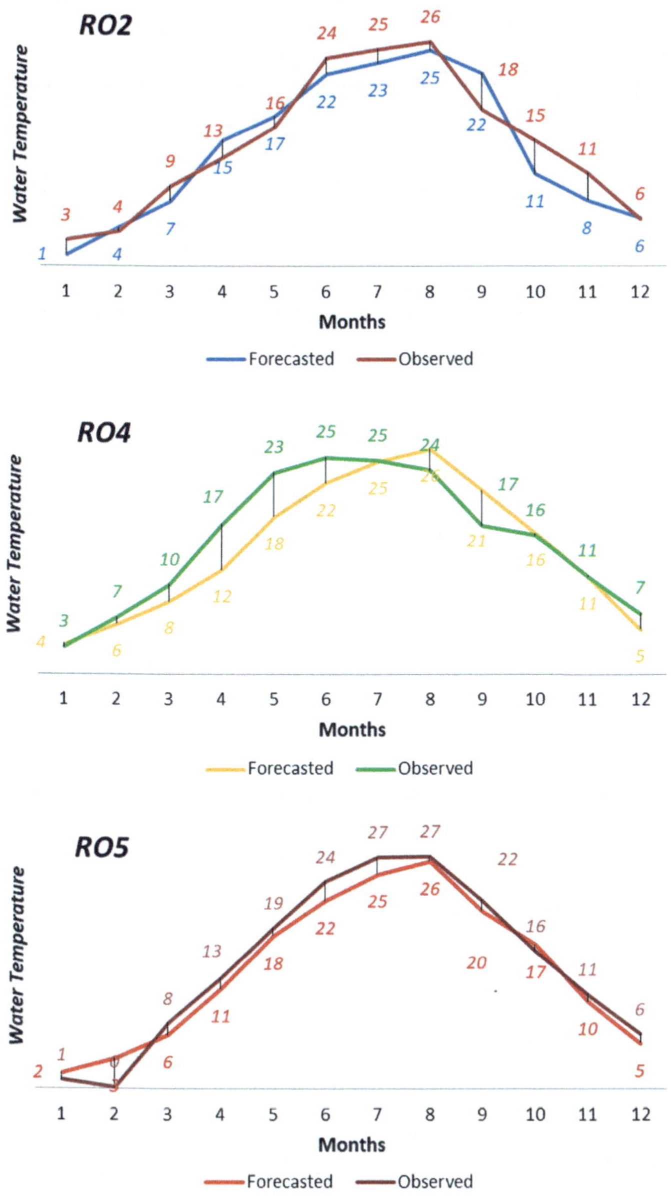

Fig. 7 Plots of observed vs. forecasted values of water temperature (°C) using the modeled time series and the real values recorded in 2017 in each control section selected on the Lower Danube River

of the SARIMA model is that it can only extract linear relationships within the time series data.

Most of the scenarios regarding global warming indicate increases in average stream water temperatures. These predicted increases are corroborated with significant daily and seasonal variations. An increase of water temperature has a wide-ranging impact on stream ecology and can be considered an important environmental issue linked to global warming [61]. Therefore, finding proper modeling tools able to forecast the water temperature of large rivers is important and should be a priority.

The present study has some limitations because the available dataset relied on a limited set of observations (2–3 measurements/month) due to the monitoring plan and logistics. A potential improvement of this aspect could be the data fusion with satellite observations and derived data of water temperature. A potential source of data that will be used in our future approaches is the Climate Data Store from Copernicus [62] to overcome the usability gap.

Based on the TSA and the evaluation of SARIMA model, a comprehensive description of the spatiotemporal variations of the water temperature was achieved but more complex approaches for improving water monitoring and modeling in the Lower Danube River are required to integrate them in process-based analysis. This will support the assessment of the interaction between different factors that influence the thermal dynamics including the anthropogenic activities and the effect of temperature variability on ecological processes from biochemistry to primary production, food web interactions, and community assemblages [63].

The implementing of an Early Warning System (EWS) [64] coupled with a process-based model on the Lower Danube hydrological basin would significantly improve the surveillance of the ecological status and a better linking between water temperature dynamics and river ecological functions.

Acknowledgements This work was supported by a grant of the Romanian National Authority for Scientific Research, CNDI—UEFISCDI, project number PN-III-1.2-PCCDI-2017-0721 (https://inter-aspa.ro/).

References

1. Gualtieri C, Gualtieri P, Doria GP (2002) Dimensional analysis of reaeration rate in streams. J Environ Eng 128(1):12–18
2. Bondar-Kunze E, Kasper V, Hein T (2021) Responses of periphyton communities to abrupt changes in water temperature and velocity, and the relevance of morphology: a mesocosm approach. Sci Total Environ 768:145200
3. Hill BH, Elonen CM, Herlihy AT, Jicha TM, Mitchell RM (2017) A synoptic survey of microbial respiration, organic matter decomposition, and carbon efflux in U.S. streams and rivers. Limnol Oceanogr 62(1):S147–S159. https://doi.org/10.1002/lno.10583
4. Fritz KM, Schofield KA, Alexander LC et al (2018) Physical and chemical connectivity of streams and riparian wetlands to downstream waters: a synthesis. J Am Water Resour Assoc 54(2):323–345. https://doi.org/10.1111/1752-1688.12632

5. Dai Y, Hein T, Preiner S, Reitsema RE, Schoelynck J (2020) Influence of water temperature and water depth on macrophyte–bacterioplankton interaction in a groundwater-fed river. Environ Sci Pollut Res 27(12):13166–13179
6. Dang AT, Kumar L, Reid M (2020) Modelling the potential impacts of climate change on rice cultivation in Mekong Delta, Vietnam. Sustainability 12(22):9608
7. Skowron R (2017) Water temperature in investigations of Polish lakes. Limnol Rev 17(1):31–46
8. Wolf KL, Noe GB, Ahn C (2013) Hydrologic connectivity to streams increases nitrogen and phosphorus inputs and cycling in soils of created and natural floodplain wetlands. J Environ Qual 42(4):1245–1255. https://doi.org/10.2134/jeq2012.0466
9. Whitehead PG, Wilby RL, Battarbee RW, Kernan M, Wade AJ (2009) A review of the potential impacts of climate change on surface water quality. Hydrol Sci J 54(1):101–123. https://doi.org/10.1623/hysj.54.1.101
10. Dodds WK (2002) Freshwater ecology concepts and environmental applications. Academic Press, San Diego
11. Caissie D (2006) The thermal regime of rivers—a review. Freshw Biol 51:1389–1406
12. OECD (2021) Data sheets for surface water quality standards—Annex 1. https://www.oecd.org/env/outreach/38205662.pdf. Accessed 20 July 2021
13. Cai Y, Zhang H, Zheng P et al (2016) Quantifying the impact of land use/land cover changes on the urban heat island: a case study of the natural wetlands distribution area of Fuzhou City, China. Wetlands 36:285–298. https://doi.org/10.1007/s13157-016-0738-7
14. Rutherford JC, Blackett S, Blackett C, Saito L, Davies-Colley RJ (1997) Predicting the effects of shade on water temperature in small streams. NZJ Mar Freshw Res 31(5):707–721. https://doi.org/10.1080/00288330.1997.9516801
15. Osborne LL, Kovacic DA (1993) Riparian vegetated buffer strips in water-quality restoration and stream management. Freshw Biol 29(2):243–258
16. Bowler DE, Mant R, Orr H et al (2012) What are the effects of wooded riparian zones on stream temperature? Environ Evid 1:3. https://doi.org/10.1186/2047-2382-1-3
17. Feller MC (1981) Effects of clearcutting and slashburning on stream temperature in southwestern British Columbia 1. J Am Water Resour Assoc 17(5):863–867
18. Garner G, Malcolm IA, Sadler JP, Hannah DM (2017) The role of riparian vegetation density, channel orientation and water velocity in determining river temperature dynamics. J Hydrol 553:471–485
19. Dunea D, Iordache Ș, Iordache V, Purcoi L, Predescu L (2019) Monitoring of the evapotranspiration processes in riparian grasslands. Sci Papers Ser A Agron 62(1):278–285
20. Roth TR, Westhoff MC, Huwald H, Huff JA, Rubin JF, Barrenetxea G, Vetterli M, Parriaux A, Selker JS, Parlange MB (2010) Stream temperature response to three riparian vegetation scenarios by use of a distributed temperature validated model. Environ Sci Technol 44(6):2072–2078
21. Boothroyd IK, Quinn JM, Langer EL, Costley KJ, Steward G (2004) Riparian buffers mitigate effects of pine plantation logging on New Zealand streams: 1. Riparian vegetation structure, stream geomorphology and periphyton. For Ecol Manag 194(1–3):199–213
22. Purcoi EL (2020) Assessing the health and role of riparian vegetation under anthropogenic impacts. A comparative study of two degraded river basins from Romania. Dissertation, King's College London, University of London
23. Trimmel H, Weihs P, Leidinger D, Formayer H, Kalny G, Melcher A (2018) Can riparian vegetation shade mitigate the expected rise in stream temperatures due to climate change during heat waves in a human-impacted pre-alpine river? Hydrol Earth Syst Sci 22(1):437–461
24. Allan JD (1995) Stream ecology: structure and function of running waters. Chapman & Hall, London
25. Wu WY, Lo MH, Wada Y et al (2020) Divergent effects of climate change on future groundwater availability in key mid-latitude aquifers. Nat Commun 11:3710. https://doi.org/10.1038/s41467-020-17581-y
26. Dobrot R (2020) Lecții de Hidrologie și Hidrogeologie. Ed. Didactică și Pedagogică S.A., București

27. Murdock JN (2008) Stream management. In: Jørgensen SE, Fath BD (eds) Encyclopedia of ecology. Elsevier
28. Loperfido JV (2014) Surface water quality. In: Ahuja S (ed) Streams and rivers: scaling and climate change, comprehensive water quality and purification, 1st edn, vol 4. Elsevier, pp 87–105
29. Dunea D, Brețcan P, Tanislav D, Șerban G, Teodorescu R, Iordache S, Petrescu N, Țuchiu E (2020) Evaluation of water quality in Ialomița river basin in relationship with land cover patterns. Water 12:735
30. Garcia M, Ridolfi E, Di Baldassarre G (2020) The interplay between reservoir storage and operating rules under evolving conditions. J Hydrol 590:125270
31. Csagoly P, Magnin G, Hulea O (2016) Lower Danube green corridor. In: Finlayson C, Milton G, Prentice R, Davidson N (eds) The wetland book. Springer, Dordrecht. https://doi.org/10.1007/978-94-007-6173-5_251-1
32. Mihăilescu V (coord) (1969) Geografia Văii Dunării Românești, Edit. Academiei Române, București
33. Zaharia L, Ioana-Toroimac G (2013) Romanian Danube River management: impacts and perspectives. European Continental Hydrosystems under Changing Water Policy. München, Verlag Dr. Friedrich Pfeil, pp 159–170
34. Hachemi K, Grecu F, Ioana-Toroimac G, Constantin DM, Ozer A (2020) The diachronic analysis of island dynamics along the Vedea-Oltenița Danube river sector using SAR imagery. Med Geosc Rev 1–15
35. Hachemi K, Grecu F, Ioana-Toroimac G, Grigorie (Omrani) Ş, Ozer A, Kuzucuoglu C (2020) Contribution of SAR radar imagery in the study of the dynamics of the Danube island system, Giurgiu-Călăraşi sector, Romania. Adv Model Anal B 63(1–4):1–6. https://doi.org/10.18280/ama_b.631-401
36. Dobrică V, Demetrescu C, Mareș I et al (2018) Long-term evolution of the Lower Danube discharge and corresponding climate variations: solar signature imprint. Theor Appl Climatol 133:985–996. https://doi.org/10.1007/s00704-017-2234-2
37. Bogdan O (1978) Fenomene climatice de iarna si de vara, Edit. Stiintifica si Enciclopedica, Bucuresti
38. Bogdan O (1989) Inversiunile de temperatură cu privire specială asupra celor care se produc pe suprafețele de apă. SCGGG-geografie, XXXVI
39. Gâştescu P, Ţuchiu E (2012) The Danube river in the pontic sector-hidrologycal regime. In: Water resources and wetlands. Conference proceedings, pp 14–16
40. Tuchiu E (2020) Starea corpurilor de apă de pe cursul inferior al Dunării între Baziaş şi Isaccea. Edit. Transversal, Targoviste
41. Zaharia L (1993) Câteva observaţii asupra scurgerii medii a unor râuri tributare Dunării româneşti. An Univ Bucur ser. Geogr 42:73–80
42. Gâștescu P, Zăvoianu I, Pisota I, (2005) Apele în Geografia României, vol. V. Edit Academiei Române, București
43. Dunea D, Iordache S (2016) Analyzing the impact of airborne particulate matter on urban contamination with the help of hybrid neural networks. In: Garcia Rosa JL (ed) Artificial neural networks. IntechOpen. https://doi.org/10.5772/63109
44. Box G, Jenkins G (1976) Time series analysis: forecasting and control. Holden-Day, San Francisco
45. Dunea D, Iordache S (2011) Time series analysis of the heavy metals loaded wastewaters resulted from chromium electroplating process. Environ Eng Manage J 10(3):421–434
46. Chatfield C (1996) The analysis of time series, an introduction, 5th edn. Chapman and Hall, London
47. Pohoață A, Dunea D (2014) Statistical methods and mathematical algorithms for estimating children's exposure to particulate matter pollution. In: Iordache S, Dunea D (eds) Methods for the assessment of air pollution with particulate matter to children's health. Ed. Matrix Rom, Bucharest

48. Valipour M, Banihabib ME, Mahmood S, Behbahani R (2013) Comparison of the ARMA, ARIMA, and the autoregressive artificial neural network models in forecasting the monthly inflow of Dez dam reservoir. J Hydrol 476:433–441
49. Nau FR (2005) Introduction to ARIMA: nonseasonal models. https://people.duke.edu/~rnau/411arim.htm. Accessed 20 July 2021
50. Schoch AL, Schilling KE, Chan KS (2009) Time-series modeling of reservoir effects on river nitrate concentrations. Adv Water Resour 32(8):1197–1205
51. Moeeni H, Bonakdari H, Ebtehaj I (2017) Integrated SARIMA with neuro-fuzzy systems and neural networks for monthly inflow prediction. Water Resour Manage 31(7):2141–2156
52. Fathian F, Mehdizadeh S, Sales AK, Safari MJS (2019) Hybrid models to improve the monthly river flow prediction: integrating artificial intelligence and non-linear time series models. J Hydrol 575:1200–1213
53. Han P, Wang PX, Zhang SY, Zhu DH (2010) Drought forecasting based on the remote sensing data using ARIMA models. Math Comput Model 51(11–12):1398–1403
54. Anda A, Soos G, da Silva JAT, Kozma-Bognar V (2015) Regional evapotranspiration from a wetland in Central Europe, in a 16-year period without human intervention. Agric For Meteorol 205:60–72
55. STATISTICA software Statsoft. Inc., Tulsa, OK, USA (2007) https://www.statistica.com/en/
56. Murphy BL, Morrison RD (eds) (2015) Introduction to environmental forensics, 3rd edn. Academic Press. https://doi.org/10.1016/C2012-0-01202-1
57. EEA (2021) European Environment Agency, Indicator assessment: water temperature. https://www.eea.europa.eu/data-and-maps/indicators/water-temperature-2/assessment. Accessed 20 July 2021
58. Marszelewski W, Pius B (2019) Effect of climate change on thermal-ice regime of shallow lakes compared to deep lakes: Case study of lakes in the temperate zone (Northern Poland). J Limnol 78(1). https://doi.org/10.4081/jlimnol.2018.1763
59. ICPR (2014) Estimation of the effects of climate change scenarios on future Rhine water temperature development. Extensive Version. ICPR Report. http://www.iksr.org/fileadmin/user_upload/Dokumente_de/Taetigkeitsberichte/214_en.pdf. Accessed 20 July 2021
60. van Vliet MTH et al (2013) Global river discharge and water temperature under climate change. Glob Environ Change 23(2):450–464. https://doi.org/10.1016/j.gloenvcha.2012.11.002
61. Water Enciclopedia (2021) http://www.waterencyclopedia.com/Re-St/Stream-Ecology-Temperature-Impacts-on.html#ixzz6jA7P8PmB. Accessed 20 July 2021
62. Climate Data Store—Copernicus (2021) https://cds.climate.copernicus.eu/. Accessed 20 July 2021
63. Ouellet V, St-Hilaire A, Dugdale SJ, Hannah DM et al (2020) River temperature research and practice: recent challenges and emerging opportunities for managing thermal habitat conditions in stream ecosystems. Sci Total Environ 736:139679
64. Crețescu I, Kovács Z, Cîmpeanu SM (2016). Monitoring of surface water status in the Lower Danube Basin, In: Bucur D (ed) River basin management. IntechOpen. https://doi.org/10.5772/64399

Variability of Nutrient Concentrations Along the Lower Danube River

Liliana Zaharia, Elena Ţuchiu, Gabriela Ioana-Toroimac, Gabriela-Adina Moroşanu, Abdelazim Negm, and Iuliana Pintilie

Abstract This chapter presents information on the temporal and spatial variation of several nutrient concentrations (NH_4-N, NO_2-N, NO_3-N, PO_4-P and total phosphorous) over the course of 22 years (1996–2017), based on concentrations recorded at five monitoring stations located along the Lower Danube River (between km 1,071 and km 132), belonging to the TransNational Monitoring Network (TNMN), namely: Baziaş, Pristol, Olteniţa, Chiciu and Reni. The dependence of the selected nutrient contents on some hydrological and physico-chemical parameters of water (e.g. discharge, temperature, dissolved oxygen concentration) was also investigated.

The results show that, among the analyzed nutrients, the highest multiannual average concentrations are found for NO_3-N (1.579 mg/l) and NH_4-N (0.238 mg/l) at Chiciu monitoring station, with generally increasing values from upstream toward downstream, while NO_2-N concentration is low (less than 0.05 mg/l) and relatively constant along the Lower Danube River. Phosphorus species have multiannual

L. Zaharia (✉) · G. Ioana-Toroimac
Faculty of Geography, University of Bucharest, Nicolae Bălcescu Blvd. 1, 010041 Bucharest, Romania
e-mail: liliana.zaharia@geo.unibuc.ro; zaharialili@hotmail.com

G. Ioana-Toroimac
e-mail: gabriela.toroimac@geo.unibuc.ro

E. Ţuchiu · I. Pintilie
National Administration "Romanian Waters", Str. Ion Câmpineanu 11, 030167 Bucharest, Romania
e-mail: elena.tuchiu@rowater.ro

I. Pintilie
e-mail: iuliana.pintilie@rowater.ro

G.-A. Moroşanu
Institute of Geography of the Romanian Academy, Str. Dimitrie Racoviţă 12, 023993 Bucharest, Romania
e-mail: gabriela.morosanu@unibuc.ro

A. Negm
Water and Water Structures Engineering Department, Faculty of Engineering, Zagazig University, Zagazig 44519, Egypt
e-mail: amnegm@zu.edu.eg; amnegm85@yahoo.com

A. Negm et al. (eds.), *The Lower Danube River*, Earth and Environmental Sciences Library, https://doi.org/10.1007/978-3-031-03865-5_6

average concentrations ranging from less than 0.15 mg/l, up to less than 0.05 mg/l in the lower part of the studied sector. The mean annual concentrations of nutrients have experienced a general downward linear trend during the period 1996–2017, as a result of a combination of factors. An important role in nutrient load decline was played by the measures to reduce pollution, implemented within the European Union and Danube River Basin. A direct/positive dependance (statistically significant) of the nutrient content on Danube discharge was generally found, and an inverse dependance on water temperature.

Although significant decreases in nutrient concentration were noticed, further implementation of measures for reducing nutrient emissions in the Danube River Basin is still required.

Keywords Nutrients · Trends · Spatio-temporal variability · Lower Danube River

List of Acronyms

DRB	Danube River Basin
DRBMP	Danube River Basin Management Plan
DRPC	Danube River Protection Convention
DWQM	Danube Water Quality Model
EC	European Commission
EU	European Union
ICDPR	International Commission for the Protection of the Danube River
JDS	Joint Danube Survey
LDR	Lower Danube River
MONERIS	MOdelling Nutrient Emissions into RIver Systems
NARW	National Administration "Romanian Waters"
RBMP	River Basin Management Plan
SWAT	Soil and Water Assessment Tool
TNMN	TransNational Monitoring Network
WFD	Water Framework Directive
WQI	Water Quality Indices

1 Introduction

Nutrients (mainly nitrogen and phosphorus) are essential elements for living organisms, but in high concentrations they can alter the quality of the surface and ground water bodies, impairing aquatic ecosystems, human health and socio-economic activities [1–3]. Nutrient enrichment of surface waters can lead to their eutrophication with negative effects on aquatic biodiversity and the functioning of the hydro-environments, thus disturbing the ecosystem services they provide. Likewise, high nutrient loads in waters can cause serious diseases to humans, affecting both adults and children (even infants) and animals [4]. Pollution of waters with nutrients limits or even hinders their use for different socio-economical needs (e.g. drinking water supply, fisheries, recreation etc.) [5].

In recent decades, the contamination of waters with nutrients has become a world-wide issue [6]. They are closely monitored and considered key elements for water quality management due to their major role in eutrophication [7]. Water quality is especially altered in large and intensely populated watersheds, with well-developed and diverse socio-economic activities, as is the case of the Danube River Basin (DRB). Eutrophication of large rivers, due to the increase in nutrient concentration, is one of the major issues in water quality facing the European countries [8] and one of the main challenges in water management. According to the Water Framework Directive—WFD (2000/60/EC), the European Union (EU) member states must achieve the "good" status of the water bodies by 2027, at the latest, by implementing measures and actions to reduce pollutant emissions. Nutrients are a key category of parameters listed by WFD used in water status assessment, belonging to the group of physico-chemical elements that support the biological components. Nutrients are considered substances that contribute to eutrophication [9].

The tool for applying the provisions of WFD within EU Member States, with references to nutrients, is the River Basin Management Plan, prepared at the river basin district scale and covering a six year cycle. At the scale of the Danube River Basin, the International Commission for the Protection of the Danube River (ICPDR) developed the first "Danube River Basin Management Plan" (1st DRBMP) in 2009, which was updated in 2015 and 2021 (2nd DRBMP and 3rd DRBMP, respectively). The plan includes detailed information on the anthropogenic pressures impacting the water bodies within the Danube Basin, on the ecological and chemical status of the water bodies, as well as on measures and actions required to be undertaken by the Danube countries in order to achieve the environmental objectives [5, 10]. With regard to the nutrient pollution issue, the ICPDR's basin-wide vision, as defined in the 2nd DRBMP, is "the achieving of a balanced management of nutrient emissions from point and diffuse sources in the Danube River Basin so that the Danube and Black Sea waters will not be threatened or impacted by eutrophication" [5].

In the above-mentioned context, this chapter aims to provide new and up-to-date information on the variation in nutrient concentrations and their dependency on key hydrological and physico-chemical parameters, in the Lower Danube River. This information contributes to the enhancement of the knowledge on this issue of major interest for water quality protection and management in the study area. The chapter focuses on the analyses of the spatial and temporal variability of measured concentrations of five forms of nutrients along the Lower Danube River over a period of 22 years (1996–2017), based on data processing from 5 monitoring stations which are part of the TransNational Monitoring Network (TNMN).

The analyzed species of nutrients are: ammonium-nitrogen (NH_4-N), nitrite-nitrogen (NO_2-N), nitrate-nitrogen (NO_3-N), orthophosphate-phosphorus (PO_4-P) and total phosphorous (TP). For the selected monitoring stations, the dependence between the nutrient contents on the one hand, and corresponding daily discharges, water temperature and dissolved oxygen on the other hand, was also investigated. These findings contribute to a better understanding of the role played by streamflow and water physico-chemical parameters in the variation of the nutrient concentrations.

2 Overview on the Nutrients Pollution Issue in the Danube River Basin

Pollution by nutrients has been recognized as one of the four significant water management issues and among the main pressures affecting surface water status within the DRB [5]. About 65% of the Danube River length was categorized as being *at risk* due to nutrient pollution [11]. The nutrient load in the Danube River peaked around the late 1980s, causing severe eutrophication that impaired the Danube Delta and the adjacent Black Sea coastal area [12–14]. Since then, a decline in emissions and nutrient concentrations has been noticed. It can be the result of the measures implemented in the DRB, to protect freshwater and marine ecosytems [15], as well as of the reduction of the polluting industrial and agricultural activities, in the former communist countries from the Danube basin (e.g. Romania). By signing the Danube River Protection Convention (DRPC) in 1994 (entering into force in 1998), the contracting parties agreed to contribute to the reduction of the pollutants loads entering the Black Sea from sources in the Danube River Basin [16]. According to the Memorandum of Understanding adopted by the International Commission for the Protection of the Black Sea (ICPBS) and the ICPDR in 2001, the Danube countries agreed to take measures to decrease the nutrient emmissions. Hence, as the Black Sea ecosystems return to a status similar to that of the 1960s [17].

The nutrients entering the Danube River come both from natural and anthropogenic sources. The most important, quantitatively, are the anthropogenic ones, including point sources (e.g. municipal, industrial and agricultural wastewater emissions which are untreated or partially treated), and diffuse sources spread throughout the catchment area (mainly originating from agriculture, due to the use of fertilizers) [5, 7, 18, 19]. The basin-wide nutrient emissions entering the surface water bodies in the period 2009–2012 in the DRB were estimated in [5] at 605,000 tons per year total nitrogen (TN) and 38,500 tons per year total phosphorous (TP). The greatest contribution to the total emissions come from diffuse pathways: by 84% for TN and 67% for TP. The highest rates of nutrients emissions are attributable to agriculture (42% for TN and 28% for TP) and urban water management (25% for TN and 51% for TP) [5].

In terms of pollution, the most affected is the lower stretch of the Danube River, which collects pollutants, including nutrients, emitted by both upstream countries and those bordering the lower Danube watercourse. Therefore, the Lower Danube River was categorized as *at risk* due to pollution with nutrients and hazardous substances along its entire length on the territory of riparian countries (Serbia, Bulgaria and Romania), according to the WFD provisions [11].

Due to the major importance of the Danube River in Europe, in recent decades, many scientific studies and projects were dedicated to its water quality. In terms of nutrients (considered individually or together with other water quality parameters) the researches were conducted both at the scale of the whole Danube River/basin and at regional level, in different stretches of the river or sectors of its catchment. Among the recent publications providing information on nutrients at large scale (entire Danube River or basin-wide), we mention: [13, 17, 20, 21, etc.].

Detailed information on nutrients in waters (sources, loads, pathways, variability etc.), as well as on Danube's water status are provided by the Danube River District Management Plan (adopted in 2009) and its updates (adopted in 2015 and 2021). Large volume of data on water quality parameters (including nutrients) of the Danube River and its major tributaries were collected and analyzed during the Joint Danube Surveys (JDS), four research expeditions organized and conducted by the ICPDR in 2001, 2007, 2013, and 2019.

In order to enhance the knowledge on nutrient emissions, their pathways and in-stream loads within the DRB, several modeling tools were developed and implemented, such as the Danube Water Quality Model (DWQM) which quantifies Danube's in-stream loads of nitrogen and phosphorous and the MONERIS Model (MOdelling Nutrient Emissions into RIver Systems), for assessing the nutrient emissions within the basin [5, 22]. Such models were used for the assessment of nutrient emissions in studies (e.g. [23]), research projects (e.g. daNUbs—DAnube NUtrients Black Sea) as well as within the DRBMP [24] and its updates. In [25], water and nutrient fluxes in the DRB were simulated with SWAT model and the results were validated based on the observed data. The SWAT model could be an usefull tool for providing support to the implementation of the European Environmental Directives in the water quality field [25].

In recent years, many studies were dedicated to the nutrients in the Lower Danube River in different monitoring stations along the river (e.g. in Bulgaria and Romania), both individually and together with other chemical and biological parameters: in [8] the analysis focused on spatio-temporal analysis of nutrient pollution in the Lower Danube River, in Călărași-Brăila sector (375–175 river km), during one year (September 2012–September 2013); in [12] the long-term (during 55 years) variability of nutrient contents (dissolved inorganic species of nitrogen and PO_4-P loads) and other parameters in the Lower Danube River was analyzed (a Bulgarian-Romanian stretch, between 376 and 554 river km); in [14] the distribution of the main macrozoobenthic groups along the Bulgarian stretch of the Danube River (in 16 sampling sections between 382 and 844 river km) was investigated in relation to the nutrient load, during the period of low water in August–September 2013; the nutrient content (NH_4-N, NO_2-N, NO_3-N, PO_4P-and TP) in the Lower Danube River between 347 and 182 river km was studied in [19] in the period January 2013—December 2014, in order to assess the ecological status of the river, in terms of nutrients; in [26], the distribution and variability of chlorophyll-*a* were analyzed in relation to nutrient concentrations (total nitrogen and total phosphorus), based on samples taken from the Danube River between the kilometers 347 and 182, during September 2012–August 2014. Complex and detailed analyzes of the nutrient content variation (considered separately or together with other water quality parameters) in the Lower Danube River (between Baziaș and Isaccea) were made in [7, 27, 28]. In [7], the analysis was focused only on the nutrient loads, at for 6 monitoring stations (located between 132 and 1,071 river km), belonging to the TNMN for a period of 10 years (2006–2015). In [27], the research concerned the nutrient species (nitrates and orthophosphates) together with other general indicators and oxygenation conditions used for physico-chemical quality assessment, during the period 1996–2015, for the same stations analyzed in [7]. In [28], the variation of several physico-chemical parameters of water quality was investigated (including nitrates, ammonium, nitrites and orthophosphates) at Reni (132 river km) correlated with the Danube flow measured at Issacea gauging station (100.2 river km), during the period 1996–2014, with a focus on 2003 (with low waters) and 2006 (with high waters). Based on data provided by the ICPDR, the nutrient contents and their trends were analized in [29] at Reni monitoring station. In [30], information on the multiannual average (1996–2015) concentrations of several nutrients (NH_4-N, NO_3-N and total phosphorus) is presented for 8 monitoring stations along the Lower Danube River (Romanian side), between the kilometers 0 (Sf. Gheorghe) and 1,071 (Baziaș). A detailed analysis of the spatial and temporal variation of seven forms of nutrients at the monitoring stations located along the Lower Danube River, is performed in [31].

In [32], 14 physical–chemical parameters (including NH_4-N, NO_2-N, NO_3-N, total nitrogen, PO_4-P-and total phosphorus) were analyzed, based on data from TNMN, at 9 monitoring stations along the Lower Danube River, between Baziaș (at 1,071 river km) and the confluence with the Black Sea, through the three arms of the Danube River (Chilia, Sulina and Sf. Gheorghe), by using different statistical methods, during the period 1996–2017. This period is similar with the one used in our study, but the approach is quite different. In [33] the seasonality and correlations between Water Quality Parameters (including nutrient species) were investigated at Chiciu station (375 river km), based on water samples collected during January 2010–December 2012.

In several papers on the Lower Danube River, the nutrients were integrated into different water quality indices and/or multivariate statistical analysis, such as [34–37], based on data from monitoring stations located on the Danube River in Serbia and [38–47], using data from stations located in Romania. An overview on the Danube River water quality and its trends over the past decades are provided in a review paper based on the open access Web of Science database (see [48]).

The present study completes and updates the existing information and assessments on nutrients concentrations and their variability in the lower Danube River through a complex analysis for over 20 years, ending by 2017.

3 Study Area and Location of the Monitoring Stations

The investigations carried out in this chapter are focused on the Lower Danube River (LDR), with a length of about 940 km. The studied sector extends between 1,071 river km (at Baziaș, located at the entry of the Danube River in Romania) and 132 river km (at Reni, situated at 50 km upstream of the delta entrance; the river length is considered from its mouth). Along this sector, the Danube River is bordered by Romania, Republic of Moldova and Ukraine, on the left side, and by Serbia and Bulgaria, on the right side (Fig. 1).

The average multiannual discharge of the Danube River is about 5,350 m^3/s at Baziaș and it increases to 6,450 m^3/s before the delta entrance [49]. In the upper part of the sector there are two large dams and adjacent reservoirs, impacting the Danube's hydro-sedimentary fluxes and channel dynamics: Iron Gates I and Iron Gates II (reservoir known as *Ostrovul Mare*). These dams and the adjacent reservoirs are parts of important hydropower and navigation systems managed in collaboration between Romania and Serbia.

Along the LDR (without the delta), the Romanian water authorities have been delineated 4 water bodies (WB): the two reservoirs, Iron Gates I—(WB 1) and Iron Gates II—(WB 2) and two river water bodies: Iron Gates II—Chiciu streach (WB 3) and Chiciu—Isaccea streach (WB 4) [50], including two large islands (named *bălți,* in Romanian), namely, *Balta Ialomiței* and *Balta Brăilei* (Fig. 1).

This study is based on water quality data from 5 monitoring stations, located along the Lower Danube River, on the Romanian (left) bank, namely: Baziaș, Pristol, Oltenița, Chiciu and Reni (Fig. 1 and Table 1). According to the TNMN, the full names of the stations Pristol, Oltenița and Chciu are: Pristol/Novo Selo, upstream

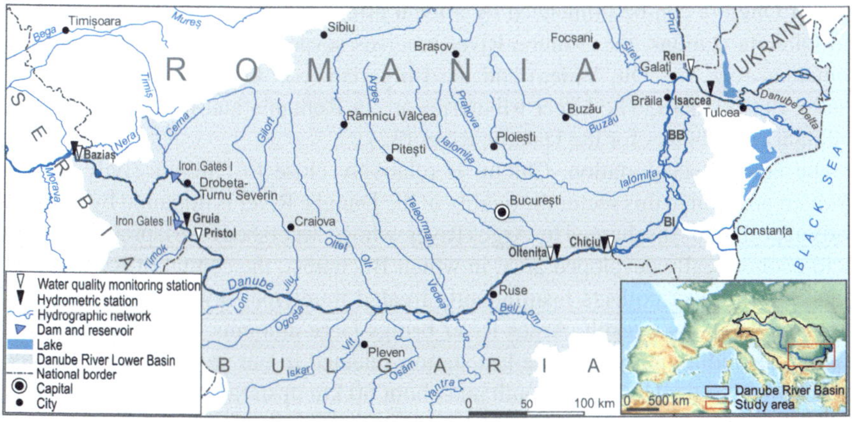

Fig. 1 The study area and the location of the analyzed monitoring stations (BI—*Balta Ialomiței*; BB—*Balta Brăilei*)

Table 1 Data on the analyzed water quality monitoring stations (according to [52])

Monitoring station	Water body	Altitude (m.a.s.l.)	Station code[a]	Distance from the Danube R. mouth (km)	River basin area (km^2)	Monitoring frequency for nutrients
Baziaș	WB 1	70	RO 1	1,071	570,896	Monthly
Pristol	WB 3	31	RO 2	834	580,100	Twice a month
Oltenița	WB 3	16	RO 3	432	676,150	Monthly
Chiciu	WB 4	13	RO 4	375	698,600	Twice a month
Reni	WB 4	4	RO 5	132	805,700	Twice a month

[a]According to TNMN

Argeș (Oltenița) and Chiciu/Silistra, respectively. In this study we used the simplified station names, as mentioned in Table 1. The monitoring profiles include samples collected on the left and right banks and in the middle of the river.

Baziaș station is located at the entrance of the Danube River in Romania (at km 1,071 away from its mouth). Pristol station is situated about 90 km downstream of the Iron Gates I dam and 10 km downstream of the Iron Gates II (Ostrovul Mare) dam. Oltenița monitoring station is located 400 km downstream of Pristol, about 100 m upstream of the mouth of the Argeș River, a Romanian tributary of the Danube affected by intense pollution. Due to the small distance between the monitoring station and the Argeș River mouth, the pollutants' concentrations (including nutrients) at Oltenița can be influenced by a local effect of backwater. Between Pristol and Oltenița stations, the Danube River receives several tributaries from both the Romanian (e.g. Jiu, Olt, Vedea) and the Bulgarian (e.g. Lom, Ogosta, Iskar, Vit, Osam, Yantra) sides. The larger tributaries are in Romania and they are important pollutant contributors for the Danube River [32].

The next selected station, Chiciu, is relatively close to Oltenița (only 57 km downstream), but in this sector the quality of the Danube River is impaired by the high loads of pollutants discharged by Argeș River, whose watershed overlaps an intensely socio-economically developed area, in which Bucharest, the capital of Romania, is situated. This river collects insufficiently treated municipal sewerage waters from Bucharest [47]. As a result, Argeș River brings large amounts of pollutants to the Danube, including nutrients. The last station selected in our study is Reni, located at 132 km from the Danube's mouth and about 50 km upstream from the beginning of the delta. Between the stations of Chiciu and Reni, the Danube River receives three important tributaries: Ialomița, Siret and Prut. The last two are large, cross-border rivers, draining territories in Romania, Ukraine and the Republic of Moldova. These rivers bring to the Danube important discharges (240 m^3/s and respectively

110 m^3/s multiannual average discharges, according to [51]) and significant nutrient loads [47].

4 Data and Methods

As mentioned above, the analysis in this chapter is based on processing data recorded in five monitoring stations located along the LDR, belonging to Romania, namely: Baziaș, Pristol, Oltenița, Chiciu and Reni (Fig. 1). Data on these stations are presented in Table 1. The monitoring profile at each station includes 3 sample collection sites: left bank, right bank and middle.

The selected stations are part of the Danube TransNational Monitoring Network (TNMN) of the ICPDR. This network was launched in 1996 to provide information on water quality and pollutant loads in the major rivers within the Danube watershed. The legal base for establishing such a monitoring network is the Convention for the protection and sustainable use of the Danube River. Since 2000, once the Water Framework Directive came into force, the purpose of the TNMN has been extended to evaluate the ecological and chemical status of water bodies [11].

The laboratories belonging to water management, environmental and research institutions from the Danube River Basin countries participate in the process of the sampling and analysing the relevant water parameters and pollutants. In Romania, the laboratories of the National Administration Romanian Waters from Craiova (Jiu Water Basin Administration) and Constanța (Dobrogea—Litoral Water Basin Administration) are involved in the Danube River monitoring program and contribute to the Danube water quality database. The quality of the TNMN data is guaranteed both by internal specific procedures of the laboratories and through their participation in the inter-comparison schema of the ICPDR [7].

The main data processed in this chapter are the concentrations (in mg/l) of five forms of nutrients (dissolved and total): ammonium-nitrogen (NH_4-N), nitrite-nitrogen (NO_2-N), nitrate-nitrogen (NO_3-N), orthophosphate-phosphorus (PO_4-P) and total phosphorous (TP). The frequency of the samples is monthly (at Baziaș and Oltenița) and bi-monthly (at Pristol, Chiciu and Reni) (Table 1). In addition to the nutrient concentrations, our study also considered data on corresponding daily discharge (in m^3/s), water temperature (in °C) and dissolved oxygen content (in mg/l), recorded on the same days when the nutrient sampling took place (accessed from the TNMN and the Danube River Basin Water Quality Database [52]). The general analysis period extends from 1996 to 2017 (excepting the year 2016 for the TP at Baziaș, Pristol, Oltenița and Chiciu).

Based on concentrations of nutrients recorded in different sites of the monitoring profile (left bank, right bank and middle), statistical classical parameters (minimum, average, maximum) were computed for different time periods (multiannual, annual, monthly) and their spatio-temporal variability was analysed. The trends in the variation of average monthly and annual concentrations of nutrients were investigated by using the Mann–Kendall statistical non-parametric test. The level of significance (α) of the identified trend was considered at 0.001, 0.01, 0.05 and 0.1 [53].

To identify the dependence of nutrient load on streamflow, we correlated during the analyzed period, the nutrient concentrations recorded in the middle of the monitoring profile, with the discharges recorded in the same days with the nutrient sampling, at the nearest hydrometric stations (Baziaș, Gruia, Oltenița, Chiciu-Călărași and Issaccea), located 2–32 km away from the water quality monitoring stations (Fig. 1). In addition, the relationships between the daily nutrient concentrations (in the middle profile of the monitoring station) on the one hand, and water temperatures and dissolved oxygen concentrations on the other hand, in the selected quality monitoring stations were investigated. The intensity and statistical significance of the dependence between the considered variables were tested by using the Spearman correlation coefficient, a non-parametric measure of monotonic correlation (linear or not) between two variables [54]. This coefficient is usually adopted for datasets having non-normal distribution (with high skewness), because it is not sensitive to outliers [13, 33]. This is the case with our data series, most of them being highly positively skewed, with skewness higher than 2. The high skewness found for the analyzed datasets is consistent with [55]. In the quoted paper, it is showed that the water quality data are commonly skewed and considered that nonparametric procedures are more efficient than parametric ones in cases where the skewness is high and the data series is large. Positive values of the Spearman coefficient show a direct correlation between the associated variables (both move in the same direction), while negative coefficients indicate an opposite situation (inverse correlation). The Spearman correlation coefficient was considered statistically significant if p-value (probability that test is significant) is lower than 0.05 ($p < 0.05$).

5 Results

The results highlight two main issues: (i) the temporal and spatial variation (at different time scales) of the nutrient concentrations at the five selected stations along the Lower Danube River and (ii) the relationships between the nutrient concentrations on the one side and hydrological and physico-chemical characteristics of the Danube River, on the other side. The results will be interpreted in the section dedicated to discussions.

5.1 Multiannual Nutrient Concentrations and Their Spatial Variation

Based on the values of the multiannual concentrations (average, maximum and minimum) of the selected nutrients recorded at the five studied monitoring stations between 1996 and 2017, we investigated their spatial variation along the lower sector of the Danube River.

The highest average concentrations were found for NO_3-N (1.579 mg/l) and NH_4-N (0.238 mg/l) at Chiciu station (Fig. 2a and Table 2). Generally, the nitrogen species present an increase of their concentrations from upstream to downstream. The lowest average concentrations were found for NO_2-N (less than 0.05 mg/l), with a spatial variation relatively stable along the Danube River. The highest value (0.043 mg/l) identified at Oltenița station is debatable, because in October–December 2011 very high concentrations of NO_2-N were recorded, but it is difficult to explain them, as we shall see in Sect. 5.2. If we do not consider the values recorded in October–December 2011, the multiannual average concentrations of NO_2-N at Oltenița is 0.023 mg/l.

Unlike nitrogen, the annual average concentrations of the phosphorus species decrease from upstream to downstream, with the highest value noticed at Pristol (0.083 mg l for PO_4-P and 0.13 mg/l for total phosphorous) (Fig. 2a). The spatial variation of the basic statistical parameters (mean, maximum and minimum) of the series of annual average concentrations of nutrients recorded at the selected stations during the studied period (1996–2017) is shown in Fig. 3. At Oltenița station, for NO_2-N we considered both the case with the extremely high values measured in 2011, and without these data.

The maximum multiannual absolute values of nutrient contents have experienced a great variability along the LDR (Fig. 2b). The highest concentrations were recorded for NO_3-N (6.00 mg/l) at Oltenița (on April 4, 2000), for TP (4.07 mg/l) at Chiciu (on November 25, 2004) and for NH_4-N (3.986 mg/l) at Baziaș (on April 19, 2005) (Table 2).

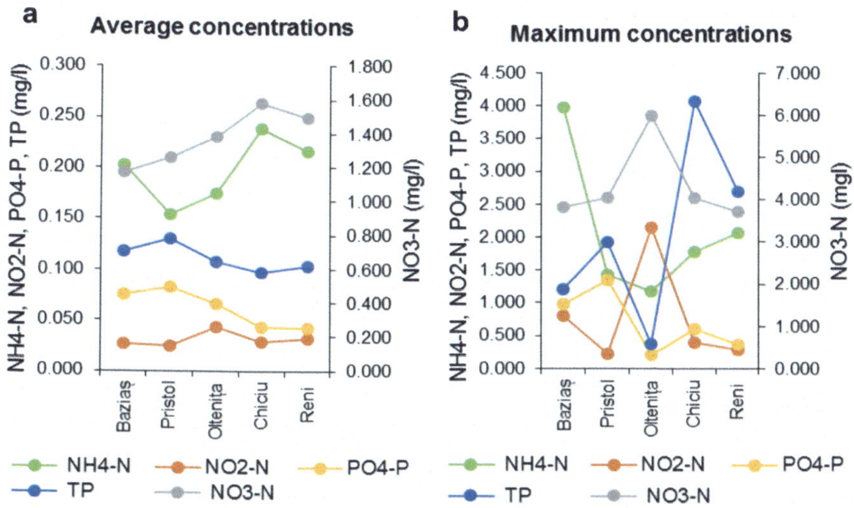

Fig. 2 Average (a) and maximum (b) multiannual nutrient concentrations at the selected monitoring stations along the Lower Danube River (1996–2017; data unavailable for TP in 2016 at Baziaș, Pristol, Oltenița, Chiciu)

Table 2 Descriptive main statistics of the nutrient concentrations (mg/l) recorded at the studied monitoring stations along the Lower Danube River (1996–2017)

Parameter	Statistics	Baziaş	Pristol	Olteniţa	Chiciu	Reni
NH_4-N	Average	0.203	0.154	0.175	0.238	0.216
	Maximum	3.986	1.440	1.183	1.780	2.070
	Minimum	0.000	0.000	0.000	0.000	0.000
NO_3-N	Average	1.174	1.258	1.384	1.579	1.493
	Maximum	3.810	4.036	6.000	4.041	3.720
	Minimum	0.075	0.059	0.000	0.000	0.000
NO_2-N	Average	0.027	0.025	0.043* (0.023**)	0.029	0.032
	Maximum	0.813	0.225	2.152* (0.151**)	0.412	0.300
	Minimum	0.000	0.000	0.000	0.000	0.000
PO_4-P	Average	0.076	0.083	0.066	0.043	0.042
	Maximum	0.980	1.360	0.210	0.611	0.368
	Minimum	0.000	0.000	0.000	0.000	0.000
TP***	Average	0.118	0.130	0.108	0.097	0.103
	Maximum	1.200	1.930	0.380	4.070	2.700
	Minimum	0.000	0.010	0.011	0.006	0.000

*With data of 2011; **Without data of 2011; ***Data unavailable in 2016 at Baziaş, Pristol, Olteniţa and Chiciu

Regarding the minimum multiannual absolute values, in the database are recorded null (zero) values for each nutrient species in most of the monitoring stations, except for NO_3-N at Baziaş and Pristol, and TP at Pristol, Olteniţa and Chiciu (Table 2).

5.2 *Interannual Variation of Nutrient Average Concentrations*

The analysis of the nutrient concentrations variation (annual averages between 1996 and 2017) along the LDR highlights some aspects common to all the selected stations and some particularities. Below we present the situations for each selected nutrient form.

Ammonium-nitrogen (NH_4-N)

As a common feature in the variation of the average annual concentrations of NH_4-N, an obvious decline is found after the year 2005, at all analyzed stations (Fig. 4). High peaks are noticeable at all stations in 1998 and at almost all of them in 2004–2006. In the upper part of the analyzed sector of the Danube River (at Baziş, Pristol and Olteniţa), we identified the highest peak in 2000. At Olteniţa and the downstream stations (Chiciu and Reni) significant ammonium concentrations were recorded in

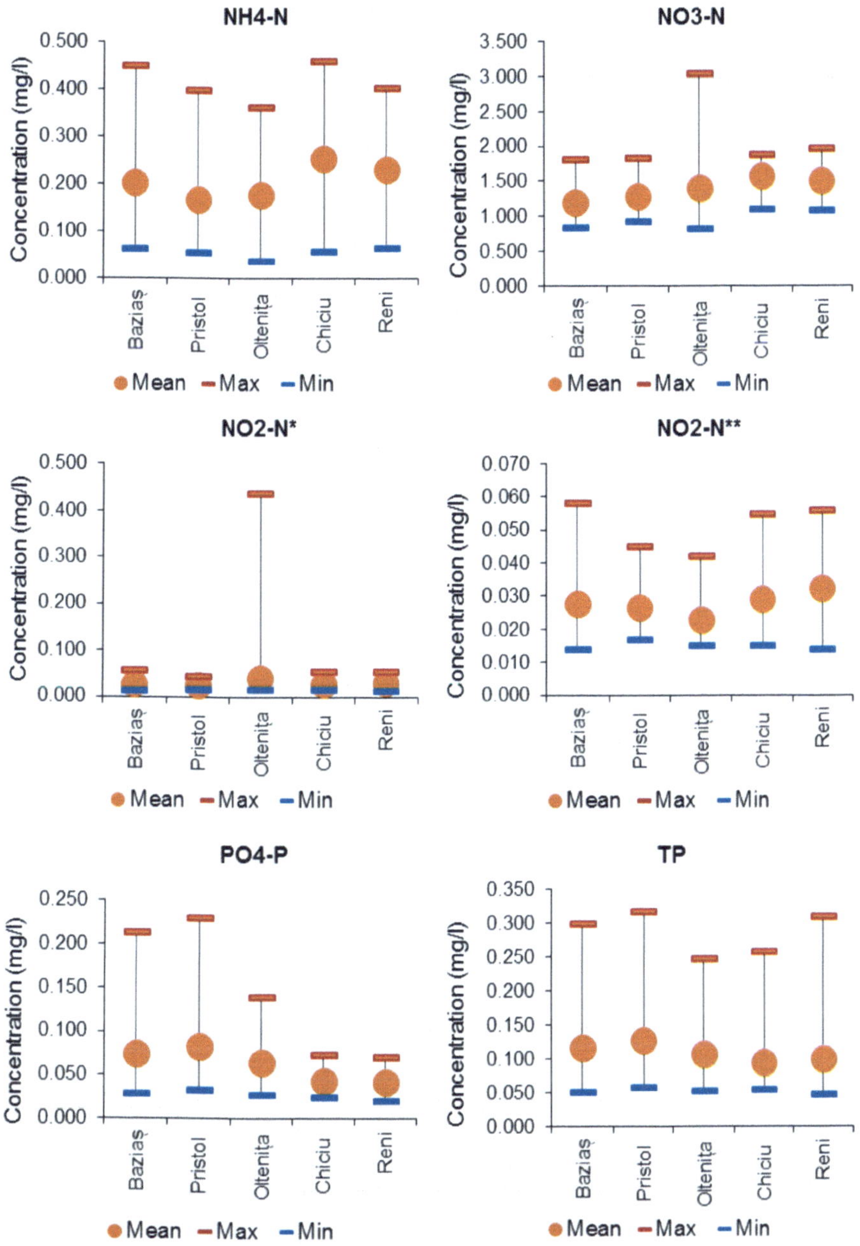

Fig. 3 Variability of annual average concentrations of nutrients at the selected monitoring stations along the Lower Danube River (1996–2017). Data unavailable for TP in 2016 at Baziaş, Pristol, Olteniţa and Chiciu. *With data of 2011. **Without data of 2011

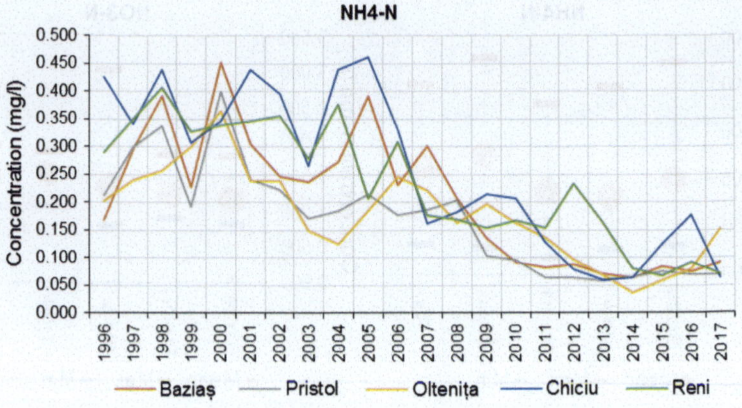

Fig. 4 Annual average concentrations of ammonium-nitrogen (NH_4-N) at the selected monitoring stations along the Lower Danube River (1996–2017)

2001–2002. After 2007, the most important peaks were noticed at the stations located in the lower sector of the study area (Chiciu and Reni), in 2009–2010 and 2016 at Chiciu, and in 2012 at Reni. The variability of the annual concentrations of ammonium in Danube River at Reni is generally different compared to the upstream stations, probably because of the significant input of water and the flow regime of the two last major tributaries of the Danube River, Siret and Prut.

Throughout the analyzed period, the highest average annual concentrations of NH_4-N were recorded in most years at Chiciu, with maximum values of over 0.450 mg/l (Fig. 4), and the lowest, generally at Oltenița and Pristol.

Nitrite nitrogen (NO_2-N)

In the analysis of the variation of nitrites, two situations were considered that were imposed by the identification of very high values of concentrations at Oltenița during October–December 2011 (over 1.5–2 mg/l). These led to an annual average of 0.437 mg/l, much higher than that of the other stations (Fig. 5a). The verifications carried out, based on the analysis of the concentrations at Chiciu station (located less than 50 km downstream of Oltenița), and on the concentrations of the Danube tributaries in the vicinity of the Oltenița station (Argeș in Romania and Yantra and Losenski Lom, in Bulgaria), did not show peculiar concentrations that could explain the very high values from Oltenița. Accordingly, in the analysis of the variation of the average annual concentrations at Oltenița, we considered both the situation with the year 2011 and without it (Fig. 5b). In the latter case, the highest values can be found at Chiciu and Oltenița, illustrating the slight increase from upstream to downstream. Regarding the peaks, we noticed the heterogeneity of the values, as it was impossible to identify periods or years with high peaks at all stations. However, the period 2001–2004 should be noted, with high values of concentrations, especially at the stations in the lower sector (Reni and Chiciu). Baziaș station stands out in the

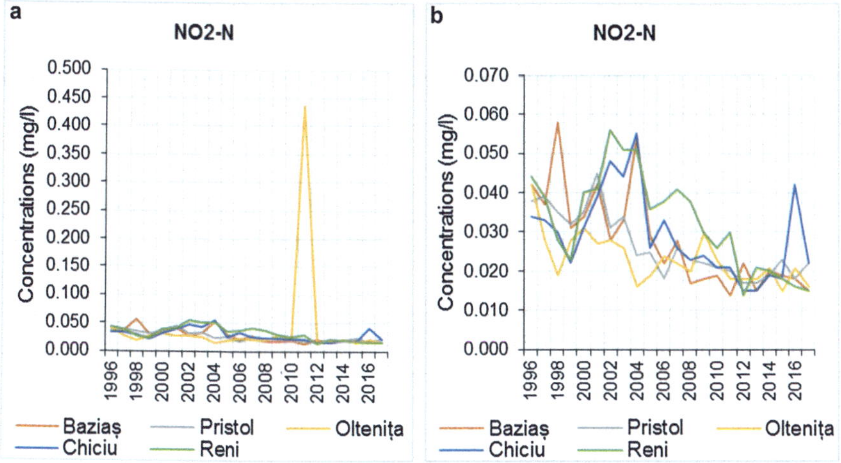

Fig. 5 Annual average concentrations of nitrite nitrogen (NO_2-N) at the selected monitoring stations along the Lower Danube River (1996–2017). (a) Considering the records in October–December 2011 at Olteniţa; (b) Without the records in October–December 2011 at Olteniţa

year 1998, when no high values were found at the other stations, which could be caused by a local NO_2-N input.

Nitrate nitrogen (NO_3-N)

The analysis of the variation of the annual average concentrations of NO_3-N along the LDR indicates their relative uniformity and stationarity in the studied period (Fig. 6). Contents generally higher than at the other stations were noticeable at Chiciu and Reni, which illustrates, also in the case of this nutrient, the slight increase from upstream to downstream. It is noted the year 2000 at Olteniţa, with very high average concentration, that could be explained by the NO_3-N contribution of the Argeş River, in the context of the lowflow of the Danube River. No significant decline in NO_3-N concentrations was observed after 2005, as found in the case of the other nitrogen forms analyzed above.

Orthophosphate-phosphorus (PO_4-P)

Unlike the nitrogen forms, whose average annual concentrations generally increase from upstream to downstream (the highest values being identified in Chiciu and Reni), in the case of PO_4-P, the lowest values were recorded, during most years, at the downstream stations. Regarding the highest average annual concentrations, they were found in the upstream sector of the lower Danube (at Baziaş, Pristol and Olteniţa) (Fig. 7). The spatio-temporal variation of the PO_4-P content showed significant peaks in 2008–2009 at stations from the upper part of the analyzed sector (Baziaş and Pristol) and in 2002 at Olteniţa. Over the period 1996–2017, there is no obvious trend in the variation of the average annual concentrations of PO_4-P, but since 2010 there is a certain uniformity, with similar values at all stations, of around 0.05 mg/l (Fig. 7).

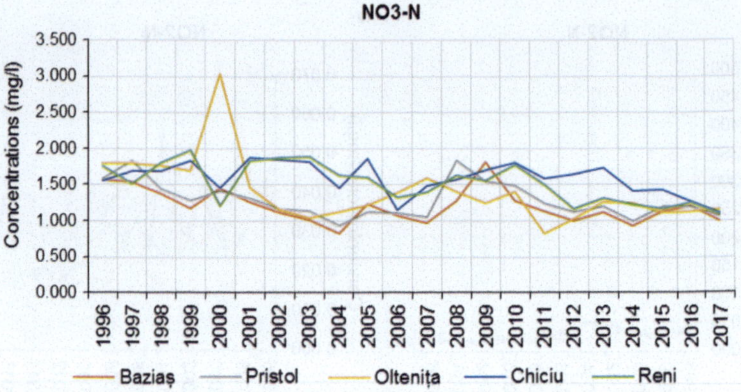

Fig. 6 Annual average concentrations of nitrate nitrogen (NO_3-N) at the selected monitoring stations along the Lower Danube River (1996–2017)

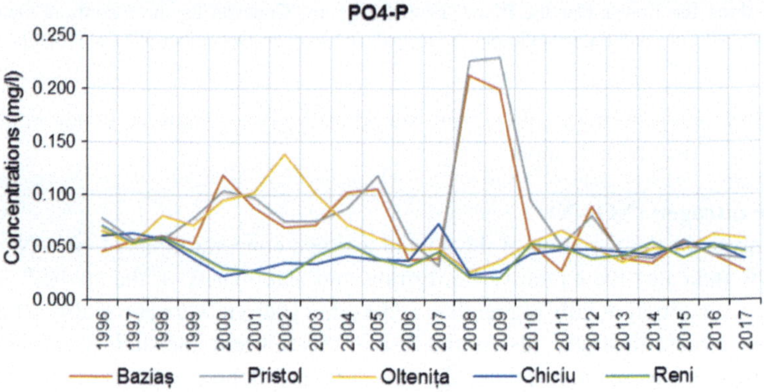

Fig. 7 Annual average concentrations of orthophosphate-phosphorus (PO_4-P) at the selected monitoring stations along the Lower Danube River (1996–2017)

Total phosphorus (TP)

The average annual TP concentrations present, over the analyzed period, a very heterogeneous variation, with sometimes quite large differences from one year to another. There are some high peaks in the 2008–2009 (over 0.25 mg/l) at the upstream stations (Baziaș and Pristol) and 2005, with high concentrations at all stations, but especially at the downstream ones (Chiciu and Reni), while in the middle sector (at Oltenița) the highest peak was in 2002 (Fig. 8). After 2010, a reduction of the average annual concentrations of TP and their relative homogenization was noticed, with a variability similar to that of the period before 2000.

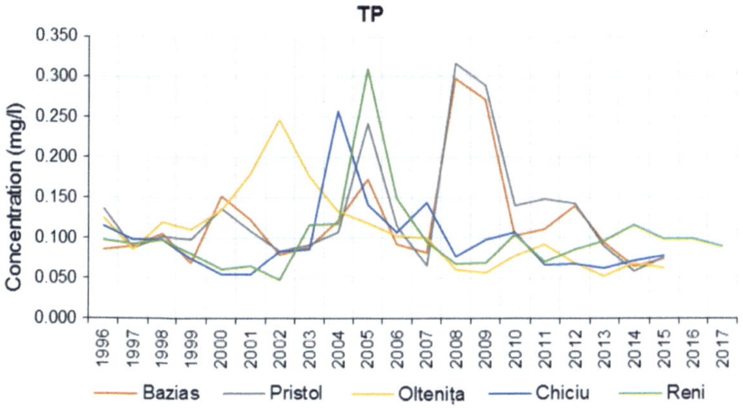

Fig. 8 Annual average concentrations of total phosphorus (TP) at the studied monitoring stations along the Lower Danube River (1996–2017). Data unavailable in 2016 at Baziaş, Pristol, Oltenita and Chiciu

5.3 *Trends in Average Annual Concentrations of Nutrients*

In order to identify possible trends in the variation of the average annual concentrations of nutrients between 1996 and 2017, the non-parametric Mann–Kendall test was applied. It indicated significant decreasing trends (at 0.001–0.05 level of significance), for most of the stations and parameters analyzed, especially for nitrogen compounds (Table 3). For phosphorus forms, the trends were also found to be downward, but mostly statistically insignificant, except for Oltenița station. The only situations in which upward trends have been identified are at Chiciu for PO_4-P and at Reni, for TP, but these were also statistically insignificant (Table 3).

Table 3 Linear trends in average annual concentrations of nutrients in the Lower Danube River and their statistical significance, as identified by using the Mann–Kendall test (1996–2017)

Monitoring station	NH_4-N	NO_2-N	NO_3-N	PO_4-P	TP
Baziaş	↓ (0.001)	↓ (0.001)	↓ (0.05)	↓	↓
Pristol	↓ (0.001)	↓ (0.001)	↓ (0.05)	↓	↓
Oltenița	↓ (0.001)	↓ (0.01)* ↓ (0.01)**	↓ (0.01)	↓ (0.05)	↓ (0.01)
Chiciu	↓ (0.001)	↓ (0.01)	↓ (0.1)	↑	↓
Reni	↓ (0.001)	↓ (0.001)	↓ (0.001)	↓	↑

*With data of 2011; **Without data of 2011

Note The values in parentheses show the level of significance (α) of the linear trend. ↓—Downward trend; ↑—Upward trend. Data unavailable in 2016 for TP at Baziaş, Pristol, Oltenița and Chiciu

5.4 Monthly Variation of Average Nutrient Loads

During the year, nutrient concentrations are variable, depending on several natural (e.g. river flow regime, water temperature) and anthropogenic (nutrient inputs from different sources) factors. The analysis of the average monthly concentrations of the nutrients (determined by averaging all the measurements from each month at the selected stations in the period 1996–2017), shows that, in general, the nutrient concentrations are high in the cold period of the year (between November–April). This could be explained by the richer precipitation and more intense soils washing. In the warm season (between May–October), concentrations are lower due to primary producers' nutrient consumption through the development of phytoplankton and algal growth. The nitrites-nitrogen (NO_2-N), however, have an inverse variation, with high concentrations during the warm period (Fig. 9). The highest monthly average concentrations of nitrogen species are noticed in the downstream sector of the Lower Danube (at Chiciu and Reni), while in the case of phosphorus compounds, the upstream stations (Baziaș and Pristol) have the highest monthly average concentrations (Fig. 9).

5.5 Trends in Average Monthly Nutrient Concentrations

By applying the nonparametric Mann Kendall statistical test, linear trends in the average monthly nutrient concentrations variation at selected monitoring stations on the Lower Danube River were identified. The test results are presented in Table 5 (in Annex) for each considered nutrient and indicate the following aspects:

- NH_4-N concentrations exhibit statistically significant decreasing trends in all months of the year and in all monitoring stations;
- significant downward trends are also noticed in the case of NO_2-N for all/almost all months, especially in the upper half of the studied sector (at Baziaș and Pristol) and at the last downstream station (at Reni); at Oltenița and Chiciu the decreasing trends are statistically significant only in a few months (generally during July–December);
- statistically significant decreasing trends are also noticeable in most cases for the NO_3-N concentrations, but during a smaller number of months, except for the Reni station, where the Mann–Kendall test indicated significant decreases in 10 months of the year;
- in the case of phosphorus species, the identified linear trends are, in their great majority, decreasing, but they are statistically insignificant, except for the Oltenița station, where they are significant, especially for total phosphorus; in the downstream stations (Chiciu and Reni) we also identified some upward trends in phosphorus concentrations (especially in the cold months of the year), but without statistical significance.

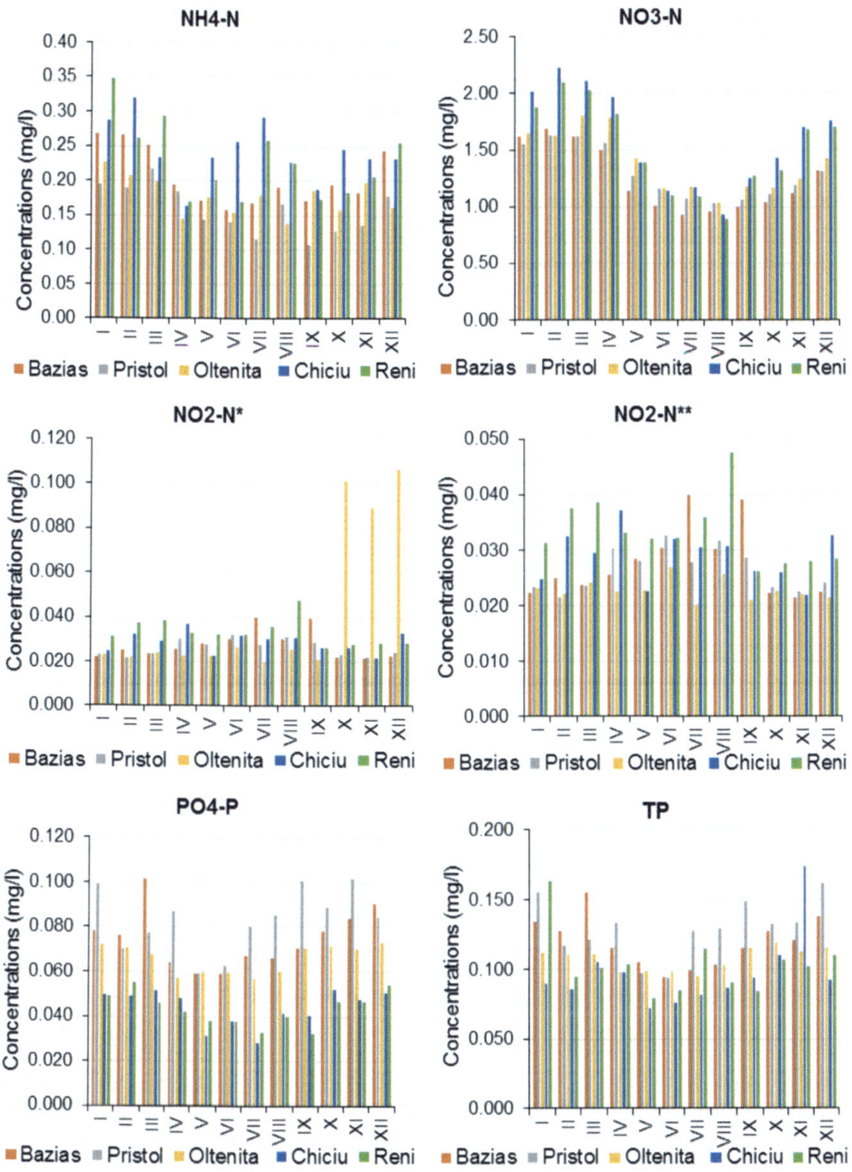

Fig. 9 Monthly average concentrations of nutrients at selected monitoring stations along the Lower Danube River (1996–2017). Data non available for TP in 2016 at Baziaş, Pristol, Olteniţa and Chiciu. *With data of October–December 2011; **Without data of October–December 2011

5.6 Correlations Between Nutrient Concentrations and Characteristics of the Danube Water

Variation of nutrient concentrations in the Danube River is the result, on the one hand, of the inputs from natural and anthropogenic sources, and on the other hand, of certain hydrological and physico-chemical characteristics of the watercourse which action as control variables. The connection between these and water quality parameters are relatively less studied. There are several studies on this topic in large European river catchments, including the DRB [13, 33, 56], that we will talk about in discussions section.

In this study, we investigated the links between the selected forms of nutrients on the one hand, and hydrological and physico-chemical parameters of water on the other, by using the Spearman correlation coefficient. More precisely, we correlated the daily nutrient concentrations recorded at the selected stations (in the middle point of the monitoring station) with the daily discharges, water temperatures and dissolved oxygen concentrations recorded in the same days with the nutrients records at the quality monitoring station and at the closest hydrometric station: Baziaș, Gruia, Oltenița, Chiciu-Călărași and respectively Isaccea (see the section Data and Methods). The Spearman correlation coefficient was considered statistically significant if the p-value is lower than 0.05.

The correlations' results are shown in Table 4 and the conclusions are summarized below.

Correlations between nutrients and daily discharges:

- most of the Spearman coefficients are positive, showing a direct dependence between the variables (larger discharges lead to higher nutrient concentrations);
- statistically significant correlations were identified at all stations, between the streamflow and NO_3-N concentrations, while for other nutrients (excepting NH_4-N), significant correlations are mostly found at the last two stations (Chiciu and Reni).

Correlations between nutrients and water temperature:

- generally negative correlations, statistically significant were detected for most of stations and parameters (Table 4). This can be explained by the fact that high temperatures lead to lower nutrient concentrations, by favoring the development of phytoplankton/algal growth with high nutrient consumption, while in periods of low temperatures, nutrient consumption is low, in the absence of a developed phytoplankton;
- in the case of NO_2-N, significant correlations were found only at Pristol (positive) and Chiciu (negative), while for NO_3-N, significant negative relationships were identified at all monitoring stations.

Correlations between nutrients and dissolved oxygen concentrations:

- both positive and negative relationships were found, but most of them were positive (Table 4);

Table 4 Correlations between nutrient concentrations and characteristics of the Danube's water (discharges, temperatures and dissolved oxygen concentrations)

Relation	Baziaş-Baziaş		Pristol-Gruia		Olteniţa-Olteniţa		Chichiu-Chiciu-Călăraşi		Reni-Isaccea	
	Spearman rho	p-value	Spearman rho	p-value	Spearman rho	p-value	Spearman rho	p-value	Spearman rho	p-value
Discharge										
NH_4-N	−0.006	0.911	0.23	0	0.106	0.262	0.088	0.081	0.061	0.195
NO_2-N	0.026	0.646	0.044	0.373	0.049	0.605	0.2	0	0.137	0.004
NO_3-N	0.305	0	0.303	0	0.261	0.005	0.355	0	0.266	0
PO_4-P	−0.025	0.66	−0.019	0.701	−0.055	0.561	0.141	0.005	0.039	0.415
TP	−0.022	0.699	−0.017	0.732	0.023	0.81	0.171	0.001	0.147	0.002
Water temperature										
NH_4-N	−0.218	0	−0.331	0	−0.074	0.264	−0.095	0.048	−0.07	0.147
NO_2-N	0.09	0.093	0.201	0	−0.019	0.779	−0.212	0	−0.061	0.203
NO_3-N	−0.528	0	−0.425	0	−0.369	0	−0.687	0	−0.651	0
PO_4-P	−0.155	0.004	0.003	0.954	−0.143	0.031	−0.264	0	−0.251	0
TP	−0.209	0	−0.052	0.283	−0.103	0.121	−0.162	0.001	−0.201	0
Dissolved oxygen concentration										
NH_4-N	0.082	0.155	0.166	0.002	0.049	0.465	−0.117	0.017	−0.035	0.469
NO_2-N	−0.164	0.004	−0.301	0	−0.027	0.682	0.036	0.464	−0.033	0.504
NO_3-N	0.507	0	0.426	0	0.26	0	0.475	0	0.489	0
PO_4-P	0.046	0.43	−0.076	0.166	0.073	0.276	0.154	0.001	0.236	0
TP	0.126	0.029	0.034	0.539	0.058	0.384	0.025	0.605	0.231	0

Statistical significant at $p < 0.05$ in grey

- statistically significant correlations were detected for almost half of cases (13 correlations of the total 25; see the Table 4), of which 10 are positive;
- significant indirectly (negative) corellations were identified at Baziaş, for NO_2-N and at Chiciu, for NH_4-N;
- the link between NO_3-N and dissolved oxygen exhibits a significant positive correlation coefficient at all stations;
- PO_4-P and TP are statistically positively correlated with the content of dissolved oxygen at the stations located on the downstream sector of the Lower Danube, Chiciu and Reni (only for TP);

These results show that in the LDR aquatic system, the higher dissolved oxygen concentrations leads to higher concentrations of nutrient oxidized forms (NO_3-N and PO_4-P), these being also stable and bioavailable nutrient species.

6 Discussions

In this chapter, we investigated the temporal and spatial variation of the concentrations of several species of nutrients at five monitoring stations located along the Lower Danube River, over a distance of about 940 km, between Baziaş and Reni.

Next, we will summarize and discuss the results with a focus on three main issues: (i) general features of the multiannual nutrient concentrations and their spatial variation; (ii) temporal dynamics and trends in nutrient loads, and (iii) dependence of nutrient concentrations on hydrological and physico-chemical characteristics of the Danube River.

6.1 General Features of Multiannual Nutrient Concentrations and Their Spatial Variation

Our results have shown that, among the analyzed nutrients, the highest multiannual average concentrations have been found for NO_3-N (1.579 mg/l) and NH_4-N (0.238 mg/l) at Chiciu station. The nutrient concentrations generally increase upstream toward downstream, which could be explained by the increase in Danube's flow, corroborated with the contribution of pollution sources, and mainly of some highly polluted tributaries, such as Argeș and Ialomița (in Romania), and Timok (in Bulgaria). As showed in [32], the pollution levels in Ialomița and Argeș rivers are much higher than in the Danube.The low quality of Argeș River is mainly altered by ammonium and phosphorus, because of the pollution with insufficiently treated municipal wastewater, originating from the Bucharest wastewater treatment plant. This polluted water is discharged into Dambovița River (the main tributary of Argeș River), downstream Bucharest city [57, 58]. During the period between 1996 and 2015, Argeș River contributed to the nitrate load of the Danube River with a multiannual average NO_3-N concentration of 2.04 mg/l [27].

The highest concentrations of NH_4-N and NO_3-N were identified at the penultimate analyzed station (Chiciu) at not at the last one downstream (Reni). This could be explained through the polluted inflow of the Argeș River approximately 50 km upstream Chiciu station.

The NO_2-N content is low (less than 0.05 mg/l) and relatively constant along LDR.

In the case of phosphorus species, their contents are quite low (less than 0.15 mg/l) and at the downstream station they are slightly reduced (at about 0.100 mg/l for TP and less than 0.05 mg/l for PO_4-P). The decrease becomes noticeable downstream of Pristol, and is more obvious in the case of PO_4-P (Fig. 2a). This could be explained by the sedimentation process for TP, as well as by the algal bloom (especially for PO_4-P) in the lower part of the studied sector of the Danube River. Given that the Pristol monitoring station is located downstream from the Iron Gates I and Iron Gates II dams, it is possible that the sedimentation in reservoir of phosphorus (adsorbed in suspended solids) and the eutrophication negatively influence the concentrations of PO_4-P and TP [7]. A study which investigated the nutrient and sediment retention capacity of the Iron Gate I Reservoir, based on weekly measurements (during

9 months in 2001, from February to October) of nitrogen and phosphorus species and of total suspended solids (TSS) showed that the reservoir does not play a major role in the retention of TN and TP, but only 1% for TN and 12% for TP of the incoming load were retained during the analyzed period [59]. The quoted study considers that the nutrient storage in large reservoirs cannot be taken for granted, and the reduction of nutrient concentrations in the LDR is caused by the cumulative effect of a large number of dams along the Danube and its triburaries.

6.2 *Temporal Variation and Trends in Nutrient Content of the Lower Danube River*

The temporal variation of nutrient concentration is determined, on the one hand, by natural factors, and on the other hand, by anthropogenic ones. Among the natural factors, the determinants are the hydro-climatic ones that induce variations of liquid and sediment flow and the physico-chemical properties of water (e.g. temperature, dissolved oxygen concentration, suspended sediments, etc.).

The nutrient content dependence on hydro-climatic conditions can be noticed in the years and extreme periods in terms of rainfall and runoff. Thus, considering the average annual concentrations of nutrients, high values were found in years with rich rainfall in the Danube basin and significant floods on the Danube River and its direct tributaries, such as 2005, 2006 and 2010. In the opposite direction, in the years 2011–2012, characterized by poor rainfall and low discharges of the Danube River, nutrient concentrations were also reduced [7]. It was observed that discharge is a major factor controlling the temporal variation of nutrient content [44]. However, the variation of the streamflow can only partially explain that of the nutrient content, since the latter can be significantly influenced by anthropogenic pressures, more precisely, the input of pollutants.

Over the whole period analyzed (1996–2017) we detected general linear decreasing trends in the variation of the average annual nutrient concentrations. They were statistically significant at all stations analyzed for nitrogen species, while in the case of phosphorus, a statistically significant decrease was identified only at Oltenița. Obvious decreases in nutrient concentrations were noticed at all stations, for most nutrients, after 2005. The exception is NO_3-N, for which the average annual concentration shows slight variations throughout the analyzed period.

For the stations in the upstream part of the analyzed sector (Baziaș and Pristol), in the case of phosphorus species, in the context of the period with their low concentrations, the years 2008 and 2009 stand out, with very high values, which could be due to some local sources of pollution (from Romania, Serbia or Bulgaria).

Over the course of a year, the nutrient content is variable, being influenced by hydro-climatic regime, water properties (e.g. temperature, dissolved oxygen content, suspended sediments), as well as anthropogenic pollution inputs.

Our analysis generally identified higher concentrations of nutrients in the cold period of the year (November–April) and lower ones in the warm season (May–October), due to lower and respectively higher consumption by phytoplankton. An exception is NO_2-N, which experiences inverse variations, with higher concentrations in the warmer months, most probably due to the reduction of nitrates to nitrites in less oxygenated conditions.

This variation reflects the inverse relationship between nutrient concentration and water temperature and the direct dependence on dissolved oxygen content, which is higher during the cold period of the year.

For the low nutrient content values in the summer-early autumn period, the low flow specific to the Danube and the high consumption of nutrients by the primary producers (well-developed phytoplankton during this period) may be responsible.

The trends we have identified are, in general, similar with those found in some previous studies. For instance, at Reni it was found a decreased content of NH_4-N between 1990 and 2012 and oscillating N and P transports since the 2000s [29]. The spatio-temporal variation of Water Quality Indices (WQI) in the Lower Danube River and its major tributaries in Romania (Jiu, Olt, Argeș, Ialomița, Siret, Prut), during the same period with our study (1996–2017), was performed in [47], based on annual means of 10 parameters, including ammonium nitrogen-ammonium (NH_4-N), nitrogen-nitrates (NO_3-N) and total phosphorus (TP). The cited study showed that the parameters with the highest contribution to WQI are ammonium and total phosphorus and the trend analysis indicated that water quality has improved significantly at the stations located along the Danube River (including those analyzed in our study) between 1996 and 2017. The tributaries Argeș, Ialomița, Siret and Prut are more polluted than the Danube River. The most altered are Argeș and Ialomița rivers whose water quality is significantly lower than the quality of the Danube River.

In the case of tributaries, water quality has not significantly improved, Argeș and Ialomița rivers still being sources of nutrients for the Danube River. However, due to the relatively low flow of these tributaries compared to the Danube River, their impact on the quality of the Danube water is not very high and it decreases downstream [47, 60].

The Joint Danube Survey 3, organized by ICPDR in 2013, also highlighted that, with respect to nutrients, Danube water quality has improved compared to previous surveys made in 2001 and 2007: in the Lower Danube River, the total nitrogen concentrations have decreased significantly, while total phosphorus had a slight decrease [60].

The analysis of pollution by nutrients along the Danube River (including its lower stretch) over the period 2001–2009, showed a general decreasing trend for all nutrient forms (except for few locations) [13].

For the period 2006–2015, in [7] it was also found a downward trend in the concentrations of the nutrient forms analyzed in our study, except for the case of PO_4-P, which experienced a slight increase. Based on the review of relevant papers on the Danube water quality, included in the Web of Science database, the reduction in nutrient loads was noticed since 1990, due to political, economic and water quality management changes in European countries [48].

The general decreasing trends identified in our study for annual average values, mainly for nitrogen species in all/most monitoring stations analyzed, reflect monthly trends. As shown in Sect. 5.5, statistically significant decreases were generally identified for average ammonium and nitrite concentrations in all/almost all months of the year. In the case of PO_4-P and TP, the linear trends identified are mostly decreasing, but statistically insignificant, except for the Oltenița station, where they are significant, especially for TP.

The decline of nutrient contents in the Danube's water during the last years can be explained by a reduction of pollution in the DRB following the implementation of a series of measures and actions aimed at protecting water quality and reducing polluting emissions, such as: the enhancing municipal and industrial wastewater treatment, reduction of nutrient emissions in agriculture throughout the Danube watershed, application of advanced technologies, environmentally friendly in the industry and agricultural practices, the use of phosphates free detergents etc. [5, 60]. Such measures have been applied in the specific legislative context of the European Union, in accordance with the regulations and goals set by the WFD and other EU Directives, of which the most relevant are: the urban wastewater treatment Directive (91/271/EEC), the Directive concerning the protection of waters against pollution caused by nitrates from agricultural sources (91/676/EEC), the Directive on integrated pollution, prevention and control (96/61/EC), the Directive on industrial emissions (2010/75/UE). As a result of the measures implemented at basin-wide scale, the 2nd DRBM (2015) reported that nutrient emissions were significantly lower in comparison to those of the 1st DRBM Plan (in 2009): total N emissions declined by 12%, and total P emissions decreased by 34% [5].

In Romania, after the fall of communism, in December 1989, there was a severe decline in industrial and agricultural activities which was reflected in the reduction of emissions of pollutants (including nutrients) generated by these two major economic areas and, therefore, in the better quality of waters. With its entry into the EU (in 2007), Romania has aligned itself with the policies of the EU, including in terms of water quality protection and aquatic ecosystem conservation. Furthermore, as a country located in the DRB, Romania must implement the measures established at the DRB level, such as those provided in the Danube River Basin District Management Plan and its updates. A major objective of this plan is to reduce pollution in general and, in particular, nutrient pollution.

6.3 Dependence of Nutrient Concentrations on Hydrological and Physico-Chemical Characteristics of the Danube River

As previosuly mentioned, the variation of nutrients in the Danube water is influenced by several natural and anthropogenic driver factors. It is difficult to accurately quantify the role of each factor, but some estimates of the intensity of the links between nutrient concentration and different water or basin characteristics can be made. As shown in Sect. 5.6., there are several studies on this topic in large European river catchments, including the Danube basin, e.g. [13, 36, 56].

The influence of driver factors varies spatially. In [56], the relationships between several water quality parameters was investigated (including total nitrogen, orthophosphates and total phosphorus) and four main natural and anthropogenic drivers (air temperature, streamflow, agriculture and population) in three large European river basins (i.e., Adige, Ebro, Sava) during the period between 1990 and 2015. The study highlighted the complex relationships between the considered drivers and the analyzed water quality parameters, with spatial differences of the identified relationships.

In the present chapter, we investigated, by using the Spearman rank correlation, the links between the daily values of the selected nutrients, on the one hand and those of the daily flow, water temperature and dissolved oxygen concentrations, on the other hand. The results generally indicated a direct dependence between nutrient concentrations and streamflow, as in many cases the correlations were found to be statistically significant (especially for NO_3-N). Predominantly positive and statistically significant correlations (for almost half of cases) were also identified between nutrient contents and dissolved oxygen concentrations (the NO_3-N content and dissolved oxygen has a significant positive correlation coefficient at all stations).

Water temperature is generally inversely related to nutrient concentration. Thus, high temperatures favor the development of phytoplankton/algal with high nutrient consumption, which reduces their concentrations, while in the cold period of the year, against a background of low phytoplankton development/activity, the nutrient content is higher. The higher dissolved oxygen content determines the presence of oxidized nutrient forms in higher concentrations (nitrates and orthophosphates).

Several previous studies have also approached the relationship between nutrients and hydrological/physico-chemical characteristics of the Danube's waters (including in its lower sector). Thus, in [13], using the Spearman correlation coefficient, the authors investigated the dependence between forms of nutrient concentrations and discharges for 22 monitoring stations located along the Danube River, of which 8 in the lower sector, based on TNMN data from 2001 to 2009. Our results are largely similar to those in the cited study, indicating, in general, the direct dependence between flow and most nutrients. However, some differences were also found. The most notable is in the case of NO_2-N: in [13] negative (but statistically insignificant) correlations were obtained at 3 of the 5 stations also analyzed by us (Baziaș, Oltenița and Chiciu), while our results indicated only positive correlations between discharge and NO_2-N load, of which two are statistically significant (at Chiciu and Reni). Some differences were also found in terms of the statistical significance of the correlations. The differences between the results can be attributed to different periods of analysis and to the factors that influenced/determined (sometimes locally) the nutrient contents in those periods.

7 Conclusion

In this study we investigated the spatial and temporal variation of nutrient concentrations, listed by WFD as key physico-chemical indicators in water ecological status assessment. The analysis focused on the Lower Danube River, categorized as being *at risk* due to nutrient pollution [11].

We investigated the variation of several species of nitrogen (NH_4-N, NO_2-N, NO_3-N) and phosphorus (PO_4-P and TP), at five monitoring stations located along the Lower Danube River, on a length of about 940 km bewteen (Baziaș, 1071 river km) and (Reni, 132 river km), based on data collected over 20 years (1996–2017) within the framework of the TNMN. In order to understand the role of different drivers of the nutrient variation, we also investigated the dependence of the selected nutrient contents on some hydrological and physico-chemical parameters of water.

Due to the large size of the Danube River and its basin, there is a significant diversity of natural and anthropogenic drivers both at the level of the catchment and the watercourse. Therefore, the spatial and temporal variation of nutrients in the Danube's waters is highly complex and does not always respect predictable patterns and laws.

Upon entering the lower sector (at Baziaș) the concentration of nutrients in the Danube is determined by emissions from upstream countries. Downstream, nutrients are added from diffuse and point sources of pollution along the LDR, as well as from tributaries in this sector, which can cause significant fluctuations in Danube's nutrient contents over time.

For nitrogen species (except for NO_2-N), an increase in average multiannual concentrations from upstream to downstream has generally been identified. This could be explained by the increase in Danube flow and the input of pollutants brought by major tributaries of the Danube, such as Argeș, Ialomița, Siret and Prut (in Romania), and Russenski Lom, Iskar and Yantra (in Bulgaria), as well as due to local pollution sources.

Among the nutrients analyzed, the highest multiannual average concentrations have been found for NH_4-N and NO_3-N, while NO_2-N content is low and relatively constant along the Lower Danube River. The phosphorus forms also exhibited low concentrations, with a very slight decrease downstream.

In the variation of the annual average concentrations, general linear trends of statistically significant decrease for nitrogen species were identified in most of the analyzed monitoring stations, which reflect the trends at the monthly level. In the case of PO_4-P and TP, the identified linear trends are mostly decreasing, but statistically insignificant.

The decline of nutrient contents in the Danube water in recent years (especially evident after 2005) can be the result of the combination of several factors and measures for water quality protection, adopted and implemented in the European Union and within the Danube River Basin, such us: improvement of municipal/industrial wastewater treatment, reduction of fertilizers in agriculture, retention of nutrients in reservoirs, economic decline of former communist countries in the DRB etc. [61].

The study also investigated the links between nutrient variation and some characteristics of the Danube River, such as flow, water temperature and dissolved oxygen concentration. The results indicated the complex relationships between nutrient loads and considered factors, with spatial differences of the detected relationships due to the influence of local factors.

A direct link between nutrient content and streamflow was generally noticed, reflected by statistically significant correlations (especially for NO_3-N). Predominantly positive and statistically significant correlations (for almost half of cases) were also identified between nutrient load and dissolved oxygen. Regarding water temperature, in most cases, an inverse relationship with nutrient concentration has been found.

8 Recommendation

Although a substantial reduction of nutrient concentrations in the Lower Danube River has been noticed, further efforts are still needed to decrease the nutrient contents, in order to meet the water quality objectives set by the WFD and the DRPC, namely to achieve the good status of water bodies. Furthermore, the ICPDR's basin-wide vision is that the Danube and Black Sea should no longer be impaired by eutrophication.

Therefore, further implementation of measures aimed at reducing nutrient inputs into surface waters and groundwater in the DRB is necessary. Because diffuse pathways are dominant in the total nutrient emissions (84% for TN and 67% for TP, according to [5]), measures addressing land management are crucial. Particulary attention should be paid to the reduction of the emissions of nutrients generated by agriculture. Of high importance are also the mesures aiming at reducing nutrient point source emissions, particularly improving the urban and industrial wastewater quality, through the use of appropriate nutrient removal technology.

Since, compared to the Danube River, in the case of its main tributaries in the lower sector water quality has not significantly improved, it is imperative to act to reduce pollution of these water courses. This is especially necessary for Argeș River, highly polluted, because of the collection of partially treated wastewater originating from the Bucharest wastewater treatment plant.

Monitoring the water quality parameters of the Danube River and its tributaries must be continued rigorously, as knowledge of their variation in time and space is crucial for assessing the water quality and for adopting protection measures against pollution. These measures should be in line with the qualitative characteristics of the different sectors of the Danube River. The analysis of the temporal dynamics of the pollutant concentrations provides to the authorities responsible with water quality management valuable information regarding the efficiency of the pollution reduction measures [31].

Scientific studies on nutrient content in aquatic systems, based on data from available databases and measurement campaigns conducted by researchers, are of great interest to provide and improve the scientific basis for integrated river basin management. Due to the variability of the water quality parameters and their determinants, addressing this topic remains a permanent challenge in the Danube River Basin.

Acknowledgements This is to acknowledge the equal contribution of the first four authors (Liliana Zaharia, Elena Țuchiu, Gabriela Ioana-Toroimac and Gabriela-Adina Moroșanu) in data processing, drawing figures, designing and writing of the chapter. The penultimate author (Abdelazim Negm) is acknowldged for his constructive comments and completions that improved the chapter's overall quality, and the last author (Iuliana Pintilie) for her valuable contribution to the data processing.

Annex

See Table 5.

Table 5 Trends in average monthly nutrient concentrations in Lower Danube River at studied monitoring stations (1996–2017)

Month	Baziaş		Pristol		Oltenita		Chiciu		Reni	
	Trend	Sign.	Trend	Sign.	Trend	Sign.	Trend	Sign.	Trend	Sign.
NH4-N										
I	↓	**	↓	*	↓	*	↓	**	↓	**
II	↓	**	↓	**	↓	*	↓	***	↓	**
III	↓	**	↓	***	↓	*	↓	***	↓	**
IV	↓	***	↓	***	↓	***	↓	*	↓	**
V	↓	***	↓	***	↓	***	↓	***	↓	***
VI	↓	***	↓	***	↓	**	↓		↓	*
VII	↓	***	↓	**	↓	***	↓	***	↓	**
VIII	↓	**	↓	***	↓	***	↓	*	↓	
IX	↓	*	↓	***	↓	***	↓	***	↓	+
X	↓	**	↓	***	↓	**	↓	*	↓	**
XI	↓	*	↓	***	↓	**	↓	***	↓	***
XII	↓	***	↓	***	↓	***	↓	**	↓	**
NO2-N										
I	↓	+	↓		↑		↓		↓	+
II	↓	**	↓	*	↓		↓		↓	*
III	↓	***	↓	*	↑		↓		↓	
IV	↓	**	↓	***	↓		↓	*	↓	**
V	↓	*	↓		↓		↑		↓	+
VI	↓	*	↓	**	↓		↓		↓	*
VII	↓	*	↓	**	↓	**	↓		↓	*
VIII	↓	*	↓	**	↓	***	↓	**	↓	**
IX	↓	***	↓		↓		↓		↓	**
X[a]	↓	***	↓	**	↓	*	↓	**	↓	**
XI[a]	↓	*	↓	***	↓	(+)	↓		↓	
XII[a]	↓	*	↓	**	↓		↓	*	↓	+
NO3-N										
I	–		↓		↑		↓		↓	*
II	↓		↑		↓		↓	*	↓	**
III	↓	+	↓	*	↓		↓		↓	
IV	↓	**	↓	**	↓		↓	*	↓	*
V	↓		↓	*	↓		↓		↓	**
VI	↓		↓		↑		↑		↓	
VII	↓	**	↓		↓	**	↓		↓	+
VIII	↓		↓		↓		↓		↓	+
IX	↓		↓		↓	+	↓	*	↓	***
X	↓		↑		↓	*	↓		↓	*
XI	↓		↓		↓	**	↓	*	↓	***

(continued)

Table 5 (continued)

XII	↓		↓	*	↓		↓	**	↓	**
PO_4-P										
I	↓		↓		↓		↑		↑	
II	↓		↓	+	↓		↑		↑	
III	↓		↓		↓	+	↑		↓	
IV	↓	*	↓		↓	*	↑		↓	
V	↓		↓	*	↓		↑		↓	
VI	↓		↓		↓	*	↓		↓	
VII	↓		↓		↓	***	↓		↓	
VIII	↓		↓		↓	*	↓	*	↓	
IX	↓		–		↓		↓		↑	
X	↑		↓		↓		↑		↑	
XI	↓		↓	+	↓		↑		↑	
XII	↓	*	↓		↓	+	↑		↑	
TP										
I	↓		↓		↓	+	↓		↑	
II	↓		↓		↓		↑		↑	
III	↓		↓		↓	*	↓		↑	
IV	↓		↓		↓	**	↓		↓	
V	↓		↓		↓	*	↓		↓	
VI	↓		↓		↓	**	↓		↓	
VII	↓		↓		↓	***	↓		↑	*
VIII	↓		↓		↓	**	↓	*	↓	
IX	↑		–		↓	+	–		↑	
X	↑		↑		↓		↓	+	↑	
XI	↓		↓		↓	+	↓		↓	
XII	↓		↓		↓	**	↓		↑	

↓ Downward trend; ↑Upward trend; – No trend (stationarity). Level of significance (α) of the linear trend: *** = 0.001; ** = 0.01; * = 0.05; + = 0.1; empty cell = level of significance less then 0.1 (no statistical significant trend). Data non available in 2016 for P total at Baziaș, Pristol, Oltenița and Chiciu

[a]At Oltenița for NO_2-N the trends are similar for October and Decembre whether or not we consider the records in 2011, but in November it is a downward trend at $\alpha = 0.1$ level of significance (marked in brakets) if data from 2011 are not considered

References

1. Sprague LA, Lorenz DL (2009) Regional nutrient trends in streams and rivers of the United States, 1993–2003. Environ Sci Technol 43(10):3430–3435. https://doi.org/10.1021/es803664x
2. Li S, Li J, Zhang Q (2011) Water quality assessment in the rivers along the water conveyance system of the Middle Route of the South to North Water Transfer Project (China) using multivariate statistical techniques and receptor modeling. J Hazard Mater 195:306–317
3. Duan W, Takara K (2020) Estimation of nutrient and suspended sediment loads in the Ishikari River. In: Impacts of climate and human activities on water resources and quality. Springer, Singapore, pp 127–159. https://doi.org/10.1007/978-981-13-9394-5_7
4. Rantanen PL, Mellin I, Keinänen-Toivola MM, Ahonen M, Vahala R (2018) The seasonality of nitrite concentrations in a chloraminated drinking water distribution system. Int J Environ

Res Public Health 15:1756. https://doi.org/10.3390/ijerph15081756

5. ICPDR (2015) The Danube River Basin district management plan, Part A Basin-wide overview. http://www.icpdr.org/main/sites/default/files/nodes/documents/drbmp-update2015.pdf. Accessed 30 Apr 2021
6. Jolánkai Z, Krisztián Kardos MK, Clement A (2020) Modification of the MONERIS nutrient emission model for a lowland country (Hungary) to support river basin management planning in the Danube River Basin. Water 12(859):1–28. https://doi.org/10.3390/w12030859
7. Ţuchiu E (2019) Impact of human activities on nutrient loads in the Lower Danube water bodies. Riscuri și Catastrofe XVIII 24(1):7–30
8. Radu VM, Diacu E, Ionescu P, György D (2015) Spatio-temporal characterization of nutrient pollution in Lower Danube area. Rev Chim 66(5):601–606
9. European Council (2000) Directive 2000/60/EC of the European Parliament and of the Council establishing a framework for the Community action in the field of water policy. https://eur-lex.europa.eu/resource.html?uri=cellar:5c835afb-2ec6-4577-bdf8-756d3d694eeb.0004.02/DOC_1&format=PDF. Accessed 30 Apr 2021
10. Ministry of Waters and Forests, National Administration "Romanian Waters" (2015) Updated National management plan related to section from the international Danube River Basin which is included in the Romanian territory 2016–2021
11. ICPDR (2009) The Danube River Basin district management plan, Part A, Basin-wide overview. http://www.icpdr.org/flowpaper/app/services/view.php?doc=DRBM_Plan_2009.pdf&format=pdf&page={page}&subfolder=default/files/. Accessed 30 Apr 2021
12. Kalchev R, Ionica D, Beshkova M, Botev I, Sandu C (2008) Long-term and seasonal changes of nutrients, seston and phytoplankton concentrations in the Lower Danube (Bulgarian-Romanian stretch). Large Rivers 18(1–2):25–43. https://doi.org/10.1127/lr/18/2008/25
13. Hamchevici C, Udrea I (2015) Pollution by nutrients in the Danube basin. In: Liska I (ed) The Danube River Basin. Hdb Env Chem 39. Springer, Heidelberg, pp 39–59. https://doi.org/10.1007/698_2014_311
14. Ihtimanska M, Varadinova E, Kazakov S, Hristova R, Naumova S, Pehlivanov L (2014) Preliminary results about the distribution of macrozoobenthos along the Bulgarian Stretch of the Danube River with respect to loading of nutrients, heavy metals and arsenic. Acta Zool Bulg Suppl 7:165–171
15. Kovacs A, Zavadsky I (2021) Success and sustainability of nutrient pollution reduction in the Danube River Basin: recovery and future protection of the Black Sea Northwest shelf. Water Int 46(2):176–194. https://doi.org/10.1080/02508060.2021.1891703
16. ICPDR (1994) Danube River protection convention. https://www.icpdr.org/main/sites/default/files/DRPC%20English%20ver.pdf. Accessed 25 July 2021
17. Popovici M (2015) Nutrient Management in the Danube River Basin. In Liska I (ed) The Danube River Basin. Hdb Env Chem 39. Springer, Heidelberg, pp 23–38. https://doi.org/10.1007/698_2014_311
18. Gasparotti C (2014) The main factors of water pollution in Danube River Basin. EuroEconomica 33:1–15
19. Anghel AM, Marinescu F, Ilie M, Ghita G, Ionescu P, Marcus I, Tociu C, Dumitru FD, Mitiu M, Popescu I, Szep R, Laslo L, Holban E (2017) Advanced processing of environmental data for establishing the ecological status of the Lower Danube water in terms of nutrients. J Environ Prot Ecol 18(3):853–861
20. Sommerwerk N, Hein T, Schneider-Jacoby M, Baumgartner C, Ostojic A, Siber R, Bloesch J, Paunovic M, Tockner K (2009) The Danube River Basin. In: Tockner K, Robinson CT, Uehlinger U (eds) Rivers of Europe. Academic Press, Elsevier, Oxford, pp 59–112
21. Jaruskova D, Liska I (2011) Statistical analysis of trends in organic pollution and pollution by nutrients at selected Danube River stations. J Environ Monit 13:1435–1445
22. van Gils J (2015) The Danube water quality model and its application in the Danube River Basin. In Liska I (ed) The Danube River Basin. Hdb Env Chem 39. Springer, Heidelberg, pp 61–84. https://doi.org/10.1007/698_2015_335

23. Schreiber H, Behrendt H, Constantinescu LT, Cvitanic I, Drumea D, Jabucar D, Juran S, Pataki B, Snishko S, Zessner M (2005) Nutrient emissions from diffuse and point sources into the River Danube and its main tributaries for the period of 1998–2000—results and problems. Water Sci Technol 51(3–4):283–290. https://doi.org/10.2166/wst.2005.0602
24. ICPDR (2021) DAnube NUtrients Black Sea. https://www.icpdr.org/main/activities-projects/danubs-danube-nutrients-black-sea. Accessed 22 Nov 2021
25. Malagó A, Bouraoui F, Vigiak O, Grizzetti B, Pastori M (2017) Modelling water and nutrient fluxes in the Danube River Basin with SWAT. Sci Total Environ 603–604:196–218. https://doi.org/10.1016/j.scitotenv.2017.05.242
26. Ilie M, Marinescu F, Ghita G, Anghel AM, Deák G, Raischi M (2017) Assessment of nutrients-chlorophyll-a relationship in the Lower Danube River. Int J Adv Chem Engg, Biol Sci (IJACEBS) 4(1):15–19. https://doi.org/10.15242/IJACEBS.C0117029
27. Ţuchiu E (2018) Danube—qualitative characteristics of the water in the pontic sector. Riscuri și Catastrofe XVII 23(2):7–25
28. Ţuchiu E (2019) Impact of climate change on the water quality of the Danube River lower sector. Riscuri și Catastrofe XIX 25(2):115–130
29. Cozzi S, Ibáñez C, Lazar L, Raimbault P, Giani M (2019) Flow regime and nutrient-loading trends from the largest South European watersheds: implications for the productivity of mediterranean and Black Sea's coastal areas. Water 11(1):1–27. https://doi.org/10.3390/w11010001
30. Gâștescu P, Ţuchiu E (2020) Hydrographycal, hydrologycal and qualitative characteristics water with references on the pontic sector. Riscuri și catastrofe XX 27(2):9–29
31. Ţuchiu E (2020) Starea corpurilor de apă de pe cursul inferior al Dunării între Baziaş şi Isaccea. Editura Transversal, Târgoviște
32. Frîncu RM, Iulian O (2020) Long-term water quality trends in the Lower Danube (1996–2017). UPB Sci Bull Series B 82(4):147–158
33. Frîncu RM, Omocea C, Eni CI, Ungureanu EM, Iulian O (2020) Seasonality and correlations between water quality parameters in the Lower Danube at Chiciu for the Period 2010–2012. Rev Chim 71:449–455
34. Jakovljevic D (2012) Serbian and Canadian water quality index of Danube River in Serbia in 2010. J Geogr Inst JC 62:1–18. https://doi.org/10.2298/IJGI1203001J
35. Mladenović-Ranisavljevića I, Takića L, Vukovićb M, Nikolićb D, Živkovićc N, Milosavljevićd P (2012) Multi-criteria ranking of the Danube water quality on its course through Serbia. Serbian J Manag 7(2):299–307
36. Takić L, Mladenović-Ranisavljević I, Vasović D, Đorđević L (2017) The assessment of the Danube River water pollution in Serbia. Water Air Soil Pollut 228:380. https://doi.org/10.1007/s11270-017-3551-x
37. Mladenović-Ranisavljević I, Žerajić SA (2018) Comparison of diferent models of water quality index in the assessment of surface water quality. Int J Environ Sci Technol 15:665–674. https://doi.org/10.1007/s13762-017-1426-8
38. Iticescu C, Murariu G, Georgescu L, Burada A, Topa MC (2016) Seasonal variation of the physico-chemical parameters and water quality index (WQI) of Danube water in the transborder Lower Danube area. Rev Chim 67(9):1843–1849
39. Radu VM, Ivanov AA, Ionescu P, Deák G, Tudor M (2016) Development of a multiparametric quality index for water quality monitoring. Environ Eng Manag J 15(5):1069–1074
40. Ismail A, Robescu D (2017) Chemical water quality assessment of the Danube River in the loer course using water quality indices. UPB Sci Bull 12:51–62
41. Paun I, Chiriac FC, Marin NM, Cruceru LV, Pascu LF, Lehr CB, Ene C (2017) Water quality index, a useful tool for evaluation of Danube River raw water. Rev Chem 6(8):1732–1739
42. Popa P, Murariu G, Timofti M, Georgescu LP (2018) Multivariate statistical analyses of Danube River water quality at Galați, Romania. Environ Eng Manag J 17:1249–1266
43. Iticescu C, Georgescu LP, Murariu G, Topa C, Timofti M, Pintilie V, Arseni M, Timofti M (2019) Lower Danube water quality quantified through WQI and multivariate analysis. Water 11(6):1305. https://doi.org/10.3390/w11061305

44. Ismail AH, Robescu D (2019) Application of multivariate statistical techniques in water quality assessment of Danube River, Romania. Environ Eng Manag J 18:719–726
45. Ismail AH, Robescu D (2019) Assessment of water quality of the Danube River using water quality indices technique. Environ Eng Manag J 18:1727–1737
46. Calmuc M, Calmuc V, Arseni M, Topa C, Timofti M, Georgescu LP, Iticescu CA (2020) Comparative approach to a series of physico-chemical quality indices used in assessing water quality in the Lower Danube. Water 12(11):3239. https://doi.org/10.3390/w12113239
47. Frîncu RM (2021) Long-term trends in water quality indices in the Lower Danube and tributaries in Romania (1996–2017). Int J Environ Res Public Health 18:1665. https://doi.org/10.3390/ijerph18041665
48. Mănoiu VM, Crăciun AI (2021) Danube River water quality trends: a qualitative review based on the open access web of science database. Ecohydrol Hydrobiol 21(4):613–628. https://doi.org/10.1016/j.ecohyd.2021.08.002
49. Zaharia L, Ioana-Toroimac G (2013) Romanian Danube River management: Impacts and perspectives. In: Arnaud-Fassetta G, Masson E, Reynard E (eds) European continental hydrosystems under changing water policy. Verlag Friedrich Pfeil, München, pp 159–170
50. National Administration "Romanian Waters" (NARW) (2016) Planul Național de Management actualizat aferent porțiunii naționale a bazinului hidrografic internațional al fluviului Dunărea. http://www.mmediu.ro/app/webroot/uploads/files/2016-04-12_PNMBH_sinteza_planuri%20bazinale.pdf. Accessed 27 July 2021
51. ICPDR (2005) The Danube River Basin district, Part A, Basin-wide overview. https://www.icpdr.org/main/sites/default/files/nodes/documents/danube_basin_analysis_2004.pdf. Accessed 28 July 2021
52. ICPDR (2021) Danube River Basin water quality database (1996–2017). https://www.icpdr.org/wq-db/. Accessed 22 Nov 2021
53. Salmi T, Määttä A, Anttila P, Ruoho-Airola T, Amnell T (2002) Detecting trends of annual values of atmospheric pollutants by the Mann-Kendall test and Sen's slope estimates—the Excel template application MAKESENS. Publications on Air Quality 31, Finnish Meteorological Institute, Helsinki
54. Artusi R, Verderio P, Marubini E (2002) Bravais-Pearson and Spearman correlation coefficients: meaning, test of hypothesis and confidence interval. Int J Biol Markers 17(2):148–151
55. Hirsch RM, Alexander RB, Smith RA (1991) Selection of methods for the detection and estimation of trends in water quality. Water Resour Res 27(5):803–813. https://doi.org/10.1029/91WR00259
56. Diamantini E, Lutz SR, Mallucci S, Majone B, Merz R, Bellina A (2018) Driver detection of water quality trends in three large European river basins. Sci Total Environ 612:49–62. https://doi.org/10.1016/j.scitotenv.2017.08.172
57. Zaharia L, Ioana-Toroimac G, Cocoş O, Ghiţă FA, Mailat E (2016) Urbanization effects on the river systems in the Bucharest City region (Romania). Ecosyst Health Sustain 2(11):e01247. https://doi.org/10.1002/ehs2.1247
58. Ionescu P, Ivanov AA, Radu VM, Deak G, Diacu E, Marcu E, Anghel AM (2019) Quality assessment of some freshwater resources located in Bucharest and surrounding areas II. Water quality assessment of Arges and Dambovita Rivers. Rev Chim 70:3638–3643
59. Teodoru C, Wehrli B (2005) Retention of sediments and nutrients in the Iron Gate I Reservoir on the Danube River. Biogeochemistry 76:539–565
60. ICPDR (2015) Joint Danube survey 3: what the river told us. http://danubesurvey.org/jds3/jds3-files/nodes/documents/icpdr_jds3-public-report.pdf. Accessed 21 Dec 2020
61. Chapman DV, Bradley C, Gettel GM, Hatvani IG, Hein T, Kovács J, Liska I, Oliver DM, Tanos P, Trásy B, Várbíró G (2016) Developments in water quality monitoring and management in large river catchments using the Danube River as an example. Environ Sci Policy 64:141–154. https://doi.org/10.1016/j.envsci.2016.06.015

Human Impacts on Water Resources in the Lower Danube River Basin in Serbia

Dejana Jakovljević, Ana Milanović Pešić, and Dragana Miljanović

Abstract Due to favourable living conditions, the Lower Danube River Basin in Serbia has constantly been populated since prehistory, which has caused different impacts on the environment. This chapter aims to address human impacts on water resources in this area. These impacts involve the use of aquatic resources for water supply, hydropower, navigation, fishing, tourism and recreation. The multiple purposes of the Đerdap Hydropower and Navigation System (also known as Iron Gate) are also presented. Human activities cause changes in the hydrochemistry and living conditions for aquatic organisms in the Danube River and its tributaries. Water quality and pollution was assessed using the water quality indices, including the Serbian Water Quality Index (SWQI), Canadian Water Quality Index (CWQI), Agri-food Water Quality Index (AFWQI), and the Water Pollution Index (WPI). Measurements are performed on three hydrological stations: Tekija, Brza Palanka and Radujevac. Results show that water quality depends on parameters used in different indices. However, general conclusion is that the lowest water quality is recorded at Radujevac, which is the farthest downstream. Anthropogenic pollution sources include Copper Mine in Majdanpek, industrial zone in Mosna, the production of phosphoric and mineral fertilizers in Elixir Prahovo, untreated wastewater and landfills, emissions from road traffic and navigation, pesticides and other chemicals from agriculture. Besides significant multifunctional role of the Đerdap Hydropower and Navigation System, the construction of Đerdap reservoir had negative impact on migratory fish species. The chapter also addressed issues on protection and restoration of water resources. In this context, protective measures and international projects jointly implemented by Serbia and Romania are presented and the role of the Đerdap National Park for the conservation of water resources was emphasized.

D. Jakovljević · A. Milanović Pešić (✉) · D. Miljanović
Geographical Institute "Jovan Cvijić" of the Serbian Academy of Sciences and Arts, Belgrade, Republic of Serbia
e-mail: a.milanovic@gi.sanu.ac.rs

D. Jakovljević
e-mail: d.jakovljevic@gi.sanu.ac.rs

D. Miljanović
e-mail: d.miljanovic@gi.sanu.ac.rs

A. Negm et al. (eds.), *The Lower Danube River*, Earth and Environmental Sciences Library, https://doi.org/10.1007/978-3-031-03865-5_7

Keywords Human impacts · Water Quality Indices · Lower Danube River Basin · Đerdap Gorge · Serbia

1 Introduction

Water resources and aquatic ecosystems are affected by population increase, economic development and urbanization. All these stressors affect the integrity of aquatic ecosystems and sustainable use of water resources. Land use changes, such as increasing urbanization and deforestation, decrease groundwater recharge and increase flood hazards. Agricultural practices, such as irrigation, strongly impact the availability of freshwater for humans and ecosystems [1]. Agricultural land use degrades water resources by increased diffuse inputs of sediments, pesticides and nutrients [2]. Effluents from wastewater treatment plants, untreated sewage and industry severely affect water quality [3]. Increased urban land use can change the amount and variety of pollutant runoff, increase water temperature, leading to the loss of riparian vegetation and degradation of aquatic habitats [2]. Rivers are among the most affected ecosystems [4]. Anthropogenic influences have considerable effects on large river systems, resulting in multiple and severe hydromorphological alterations changes in sediment and nutrient flux [5, 6]. Rivers are threatened by pollution, water abstraction, river damming and channelization [4]. River channelization measures alter the fluvial morphology in the most direct form. When a meandering river is transformed into a straight channel, the whole ecosystem in the main river arm and the alluvial reaches beside it is affected by the hydraulic effects of channelization. The construction of dams is a local severe intervention with remote upstream and downstream impacts on the river system. Dams capture sediment moving down the river, causing severe downstream consequences, such as the erosion of fine sediment and degradation of habitats for aquatic species [7, 8]. Transforming rivers to reservoirs has far-reaching impacts, including production of organic matter, biodiversity, and changes in function and services provided by aquatic ecosystems [9]. Dammed reservoirs serve as a sink for contaminants [10]. The hydrological alteration, including river regulation, impoundments, and channelization, causing floodplains to be disconnection from the main river, may also significantly impact nutrient cycling [11]. Rivers are also used for water supply, fisheries [12] and hydropower [1]. Hydropower generation causes a major pressure on river ecosystems. Through damming, water abstractions, hydropeaking, hydropower plants affect aquatic habitats by altering discharge regimes and fragmenting river channels [1, 13]. Hydropeaking (discontinuous release of turbined water due to the peaks of energy demand) causes artificial discharge fluctuations downstream of reservoirs with a harmful impact on aquatic ecology, e.g. the relocation of organisms to a potentially less suitable habitats, as well as physiological, mechanical, or predatory stress [14]. Human activities can negatively impact fish habitats; the same applies to commercial or recreational fisheries, which can use stocks in an unsustainable way [15]. Please check and confirm the author names and initials are correct. Also, kindly confirm the details in the metadata are correct. We confirmed.

Significant pressures on the Danube River include organic pollution and pollution with nutrients, and hazardous substances, as well as hydromorphological alterations, and other issues, such as changes of quality and quantity of sediments, and appearance of alien invasive species. Organic and nutrient pollution is generated from urban wastewater, industry and agricultural sources. Hydromorphological alteration involves interruptions of river continuity and morphological alterations, wetland or floodplain disconnection and hydrological alterations [16]. The alterations of Danube's morphology involve engineering approaches to create a single straightened channel accompanied with changing the depth or width of the river. In order to improve inland navigation, flood protection and hydropower generation, the Danube River has been narrowed, channelized, disconnected from floodplains and morphologically degraded [5]. The impact of these works has been boosted by the effects of land use changes, such as agricultural intensification and forestry development. Engineering works also cause river bed erosion and lateral erosion (e.g. downstream of Đerdap I and II) [6]. The construction of hydropower impoundments changes river systems by disrupting the connection between the river and backwater, changing the shoreline, and stabilizing previously dynamic water levels [17]. Impoundments are marked by deposition and excessive sedimentation and remobilization of fine sediments during severe floods [6]. This ecological situation is reflected in the alteration of riverine habitats, leading to the decline of the biodiversity of species and affect certain faunal associations, especially fish assemblages [17], such as sturgeon [5].

The main issue addressed in the chapter is the human impact on water resources in Lower Danube River Basin. The chapter aims to present water resource uses including water supply, hydropower use, navigation, tourism and recreation and fishing, as well as water quality, pollution and protection of water resources. In order to assess water quality (general and for different purposes), various indices were used and compared, which provide the added value of this study.

2 Study Area

Out of the total length of the Danube River (2,857 km), 588 km (20.58%) are located in Serbia. Its basin area covers 801,463 km^2, of which 81,506 km^2 (10.17%) belong to Serbia [18]. The Danube River enters Serbia near the settlement of Batina (km 1,433, at a height of 81 m above sea level), and exits at the mouth of the Timok River (km 845.5, at a height of 28 m above sea level). It is the border river with Croatia (138 km, km) and Romania (230 km) [19]. In terms of natural features, the watercourse of the Danube River in Serbia can be divided into the Pannonian, Đerdap and Western Pontic sectors. The lengths of these sectors are determined based on the Navigation Chart of the Danube River in the Republic of Serbia (km 1433.1–km 845.5) [20]. In the Pannonian sector, which stretches from the Hungarian–Serbian border (km 1,433) to the settlement of Golubac (km 1,042), reaching the length of about 391 km, the

Danube River is a plain water course. The Đerdap sector encompasses the course of the Danube River through the Đerdap Gorge (also known as Iron Gate) from Golubac (km 1,042), to the settlement of Sip (km, 939) and it is 103 km long [20]. In this part, the Danube River used to be a mountain river. Now, the river velocities are significantly reduced. Downstream from Sip (km 939) to the confluence of the Timok River into the Danube River (km 845), on the Serbian–Bulgarian border, there is a 94 km long Western Pontic sector. In this part, the Danube River has the characteristics of a plain river. This chapter deals with the Lower Danube River in Serbia, which covers the Đerdap and Western Pontic sectors, and has a total length of 197 km (Fig. 1). In administrative terms, this area in Serbia covers 3,018 km^2 in the territories of four municipalities (Golubac, Majdanpek, Kladovo and Negotin) and has a population of 84,708 inhabitants [21].

According to the 1948 Danube Convention, the Đerdap sector of the Danube River is 117 km long: from the settlement of Vinci (upstream from Golubac) to the settlement of Kostol (downstream from Kladovo). However, based on geomorphological criteria, the Danube River enters the Đerdap Gorge near the Golubac fortress (km 1,042) and exits near the settlement of Sip (km, 939) [22]. In this 103 km long sector, the river flows through a composite valley, including four gorges (Golubac, Gospođin Vir, Kazan and the Sip) and three alternating valleys (Ljupkovska Valley, Donjomilanovačka Valley and Oršavska Valley). Before the construction of the Đerdap I Hydropower and Navigation System, the Danube River was a mountain river with large falls in this sector, 180–2,200 m wide, with a river flow speed of 18 km/h [23]. With vortex erosion between rocks, it cut giant pot holes, causing

Fig. 1 Map of study area

significant difficulties in navigation. In one of them, near the rock of Pjatra Lunga (Long Stone) in the gorge of Gospođin Vir, the greatest depth of the Danube River and the greatest river depth in Europe (82 m) was measured. After the construction of the Đerdap reservoir, the water velocity was reduced to 1.08 km/h, and many rocks in the riverbed were submerged [23]. The largest bays of the Danube River in the Đerdap sector are located along the Romanian coast, at the mouth of the river Cerna near Orșova (4.5 km long and up to 2.1 km wide) and on the Serbian coast at the mouth of the Porečka River (4.25 km long, up to 0.62 km wide) [22]. They were formed after the construction of the Đerdap reservoir in places where water submerged the lower parts of river valleys.

The Đerdap reservoir, formed by partitioning the Danube River in the Đerdap Gorge is the largest lake in Serbia. It was constructed at km, 943 in 1964–1972. The dam has two side parts and a central space with 14 overflow fields, which drain excess water from the reservoir. It is 61 m wide and 1,278 m long [24]. Hydropower Plant Đerdap I (HPP Đerdap I) is located on it. The formation of the reservoir improved conditions for upstream and downstream navigation because the waters submerged underwater rocks. When the water level on the Danube River changes, its shoreline shifts, and all morphometric indicators change. At high water levels, Đerdap reservoir covers an area of 253 km^2 (163 km^2 on the Serbian side and 90 km^2 on the Romanian side) [25] and the maximum volume of the reservoir reaches 2.8×10^9 m^3 of water [26]. The width is different in its parts: it is the smallest in the gorge Mali Kazan (about 180 m) and the largest in the Donjomilanovačka Valley (about 2,200 m). There are also different data regarding the maximum depth, ranging from 82 to 92 m. The maximum water transparency is 3–4 m. The largest island in the lake is Moldova, located 10–35 m above the lake surface opposite Golubac and it belongs to Romania [22]. The submergence of the coastal zone caused changes in the territorial distribution of settlements. The reservoir completely or partially submerged the settlements of Donji Milanovac, Mosna, Malo Golubinje, Veliko Golubinje, Tekija, Sip, Dobra, Brnjica, Golubac and Usje. The population and more valuable buildings from these settlements were relocated. Cultural and historical monuments (Lepenski Vir, Trajan's Way, the fortress on the island of Ada Kale) were submerged, while Trajan's Plaque was raised above the level of the lake [22].

Downstream from the Đerdap Gorge at km 862.8 of the Danube River it was built a second dam and reservoir in 1977–1985. The second dam was built 80 km downstream from the first one for additional power production and more flexibility of the joint operation of the two power plants [27].

The major tributaries of the Lower Danube River in Serbia include: Brnjica River (25.6 km), Porečka River (19.1 km), Boljetinska River (16 km) and Dobranjska River (12.8 km) [22] in Đerdap sector, and Timok River (202 km), Jesenička River (40 km), Zamna River (35 km), Slatinska River (23 km) [23] and Podvrška River (20.4 km) [22] in the Pontic sector.

Based on data from the nearest hydrological station where discharge is measured, Veliko Gradište (located upstream of the Đerdap sector), the average annual discharge of the Danube River in this sector is 5,460 m^3/s. The highest waters occur in April (7,793 m^3/s), and the lowest in October (3,637 m^3/s) [22]. It is estimated that the mean multiannual discharge of the Danube at the exit from Serbia is approximately 5,500 m^3/s [28]. Based on the above-mentioned data, it can be concluded that the Danube River is abundant in water in April–May, and the poorest in water in September–October. The Danube's tributaries in this area, and especially the Timok River, have the highest discharges in March and April, and the lowest in August and September.

In Serbia, the Đerdap Gorge, together with the Danube River and Đerdap reservoir, was granted the status of a national park in 1974. The Serbian part of the Đerdap National Park stretches along the right bank of the Danube River over the territory of three municipalities (Golubac, Majdanpek and Kladovo), covering an area of 637.68 km^2 [29] and it is the largest national park in Serbia. It is referred to as the "river national park" [30], keeping in mind that a significant part (7.8% of the total area of the NP) is the Danube River [22]. The Đerdap Gorge is the most striking natural phenomenon in the National Park. It is the longest incising composite gorge in Europe [22, 29, 31, 32].

The Đerdap National Park is internationally recognized as an Important Plant Area (IPA): 57 highly complex and diverse forest communities and 1080 plant species have been identified in the national park. The Park was also declared an internationally Important Bird Area (IBA), since about 255 species of birds are present, as well as a Prime Butterfly Area (PBA), thanks to the presence of 104 species of daily butterflies [22]. Finally, the Đerdap area was declared in 2020 the eleventh and largest area in Serbia that has been included in the list of Wetlands of International Importance (Ramsar sites). The Đerdap Ramsar area covers a total area of 665.25 km^2. It includes the Đerdap National Park and the Internationally Important Bird Area Mala Vrbica, which is outside the boundaries of the National Park. It has gained the status of an Internationally Important Area thanks to the presence of habitats of wetland birds [33–35]. The Đerdap National Park is also a member of the Danube River Network of Protected Areas [36].

The Đerdap Geopark is the first area in Serbia to be inscribed on the UNESCO Global Geoparks Network in 2020. It covers an area of 1,330 km^2 and includes the territory of the Đerdap National Park and Danube hinterland (parts of the Kučaj and Miroč mountain massifs) by 692 km^2. It includes four municipalities: Golubac, Majdanpek, Kladovo and Negotin [32].

3 Water Management Framework in Serbia

3.1 Legal Framework

Establishing a national water legislation framework compliant to the requirements of the European water legislation (known as the *Acquis*) is one of the priorities in the European Union accession process.

The water sector in Serbia is regulated by numerous laws, of which the main one is the Water Law [37]. Its implementation is supported by relevant bylaws related to water (Regulations and Rulebooks), such as: the Regulation on emission limit values of pollutants in water and deadlines for their achievement [38]; Rulebook on parameters of the ecological and chemical status of surface waters, and parameters of the chemical and quantitative status of ground waters [39]; Regulation on limit values of pollutants in surface and ground waters and sediments and deadlines for their achievement [40]; Regulation on limit values of priority and priority hazardous substances that pollute the surface waters and deadlines for their achievement [41]; Rulebook on the method and conditions for measuring the quantity and testing the quality of wastewater and the content of the report on the performed measurements [42]; Decision on determining the boundaries of water areas [43]; Rulebook on defining the methodology for designing the vulnerability map and flood risk map [44], and etc. (more information about all bylaws is available on the internet portal of the Water Directorate [45]).

The transposition of the *Acquis* (European legislation) into national legislation is presented in the National Programme for the Adoption of the *Acquis*—Third Revision [46]. Apart from the mentioned document, the transposition of the EU environmental legislation (environmental *Acquis*) into national legislation and the required institutional framework to implement that legislation, as well as the estimated the total cost of environmental approximation, are presented in the National Environmental Approximation Strategy for the Republic of Serbia [NEAS] [47]. In the Water Sector Approximation Strategy [48], the framework for approximation of EU water legislation, regulations that are relevant for the Republic of Serbia, are defined in an accompanying document to the NEAS. The aforementioned National Programme for the Adoption of the *Acquis*—Third Revision [46] presents the current situation regarding the transposition and implementation of the EU environmental legislation (Section 3.27), including water management directives (Subsection 3.27.4). The directives that have not been fully transposed into the national legislation (Water Law and accompanying bylaws) have been singled out. As highlighted in the National Environmental Approximation Strategy for the Republic of Serbia [47], based on the experience from previous EU enlargement processes, the transitional period involves only Heavy Investment Directives, which require significant financial resources (about € 4.1 billion), such as the Urban Waste Water Treatment Directive and the Nitrates Directive.

According to [49] difficulties related to EU Urban Waste Water Treatment Directive implementation in Serbia are associated with poor sewage systems and unsatisfactory wastewater treatment (the lack or wastewater treatment or inadequate procedures), which is reflected in the quality of wastewater from households and industry. It is loaded with organic matter and nutrients, as well as with hazardous substances. As the authors conclude, untreated wastewaters are one of the most important threats to surface water in Serbia. Namely, in 2017, about 62.2% of urban waste-water was collected by public sewage systems, 13.9% of which was collected by public systems with treatment (1.3% of the population is connected to primary treatment, 9.2% to secondary treatment, and only 3.4% to the most advanced, tertiary treatment), and 48.3% to the systems without purification [50]. Given the level of sanitation of urban settlements, it is evident that Serbia is lagging significantly behind European countries, as indicated by the data presented in [49].

In Serbia, 50 wastewater treatment plants were built in settlements with more than 2,000 inhabitants and 32 plants are active. Only a few wastewater treatment plants operate according to their design criteria [24]. According to the 2011 Census, settlements with up to 2,000 inhabitants account for 90.5% of the total number of settlements, with about 25% of the total population, while the population of settlements with more than 2,000 inhabitants (9.5% of the total number of settlements) accounts for about 75% of the total population [51]. The national settlements network is dominated by the capital city, owning 16.23% of the total population and 27.45% of the population of urban settlements [52]. These data reveal the spatial distribution of the population, i.e. an unbalanced population development and regional inequality [51–54].

Results of analysis of nitrate concentrations trends in Serbian watercourses for two decades (1998–2007 and 2008–2017) according to [55], show the importance of Serbia's obligations (related to reduce water pollution caused by nitrates from agricultural sources) in the implementation of the provisions of the Nitrates Directive in the EU accession process.

Environmental legislation, primarily in relation to water protection from pollution, is of particular importance for achieving the good status of water. The following laws are particularly important: Law on Environmental Protection [56], Law on Environmental Impact Assessment [57], Law on Strategic Environmental Impact Assessment [58] and Law on the Integrated Prevention and Control of Environmental Pollution [59]. The aforementioned National Programme for the Adoption of the *Acquis*—Third Revision [46] outlines further steps in the process of harmonizing with the EU's environmental legislation (Section 3.27), sections Horizontal legislation (Subsection 3.27.2) and Industrial pollution (Subsection 3.27.6), where the implementation of Directive 2010/75/EU of the European Parliament and the Council of 24 November 2010 on industrial emissions (integrated pollution prevention and control) is of special importance in terms of water protection (to reduce pollution from industrial facilities). It has been partially transposed into the Law on Integrated Prevention and Control of Environmental Pollution [59] and bylaws.

Spatial development policy is closely related to environmental protection policy [60] and, accordingly, to water management policy. As already mentioned, the Law on Planning and construction [61], which determines the conditions and the land use mode towards preventing conflicts of the physical space, is also important for water protection, as well as for the protection of organisms living in aquatic ecosystems, and accommodation of water infrastructure in physical space. In addition, land use planning is an important instrument for reducing flood risk. In contrast, according to [62], an uncontrolled urban development in Serbia increases the vulnerability of urban areas to natural disasters.

There are other laws which regulate particular aspects of water policy (i.e. water quality protection and the protection of natural/hydrological heritage, water use, protection against the harmful effects of water, monitoring, etc.). These are the Law on Nature Conservation [63], Law on Protection and Sustainable Use of Fish Resources

[64], Law on Public Health [65], Law on Navigation and Ports on Inland Waterways [66], Energy Law [67], Law on Waste Management [68], Law on Meteorological and Hydrological Activities [69], Law on Communal Activities [70], Law on Local Government [71], Law on Disaster Risk Reduction and Emergency Situations Management [72], and another laws.

3.2 *Institutional Framework*

The institutional framework for the water sector, consisting of institutions and national, regional and local government bodies, and public water-related companies, ensures successful transposition and implementation of water-related EU directives, i.e. the implementation of adopted laws. The competence of relevant institutions is defined by the Water Law [37].

The Ministry of Agriculture, Forestry and Water Management, in accordance with the Law on Ministries [73], is responsible for the water sector, together with the Water Directorate (an administrative unit in the Ministry). The Serbian Environmental Protection Agency (SEPA), responsible for air and water quality monitoring, reporting on the status of the environment, and other activities in the domain of environmental protection according to the Law on Environmental Protection [56] and Law on Ministries [73], is part of the Ministry of Environmental Protection.

Reporting on water quality in Serbia is defined by national legislation. In accordance with the 2011 Law on Ministries [74], the implementation of the annual surface and groundwater quality monitoring programme has been transferred from the competence of the Republic Hydrometeorological Service (RHMS) of Serbia to the competence of the Serbian Environmental Protection Agency (SEPA) under the Ministry of Environment Mining and Spatial Planning. The Agency publishes water quality data. Annual water quality reports have been published since 1965 [75]. According to the authors, in 2012, the SEPA started the implementation of a surface and groundwater status monitoring programme in accordance with several bylaws harmonized with the Water Framework Directive. Quantitative monitoring including data collection about water levels, discharges and temperatures of surface waters, as well as water tables and temperatures of groundwater is performed by the RHMS of Serbia.

Two monitoring programmes (one for the Đerdap I reservoir in 1978, and the other for the Đerdap II reservoir, in 1985) were established to ensure the monitoring of the environmental impacts of the Đerdap Hydropower and Navigation System on the Danube River and effects of the protection measures (Mladenović and Radosavljević, 2013, cited in [13]).

In addition to the Ministry of Agriculture, Forestry and Water Management and the Ministry of Environmental Protection, other ministries are also involved in the activities in the water sector, in accordance with the Law on Ministries [73], e.g. the Ministry of Health, the Ministry of Construction, Transport and Infrastructure, and the Plovput Waterways Directorate (responsible for the maintenance and development of inland waterways in the Republic of Serbia with international and interstate navigation regimes—the Danube, Sava and Tisa Rivers), the Ministry of Mining and Energy—Energy Agency, Ministry of Interior—Sector for Emergency Situations, etc.

At the local level, local self-government units, such as secretariats (for the territory of the City of Belgrade), or directorates, departments and other units in other local self-governments (cities and municipalities) are responsible for activities in the water sector [24].

Three public water management companies (WMC) are responsible for water management activities on the territory of Serbia: Srbijavode WMC, Vode Vojvodine WMC and Beogradvode WMC [24]. Public Water Management Company Srbijavode has three water management centres [76], of which Sava–Danube center (for the Sava and Danube water areas outside the territory of the Autonomous Province of Vojvodina), includes Low Danube Basin in Serbia. The Jaroslav Černi Institute for the Development of Water Resources is a leading research organization in Serbia in the water sector [77].

3.3 Planning Framework

The Water Law [37] stipulates the development of the Water Management Strategy for the territory of the Republic of Serbia, a planning document which indicates the long-term directions of water management in this country. As highlighted in the Water Management Strategy for the territory of the Republic of Serbia until 2034 [24], the adoption of this planning document ensures continuity in the long-term planning of the water sector. Until the adoption of the Strategy, issues related to water management were regulated by the document Water Management Plan of the territory of the Republic of Serbia [78]. The current Water Management Strategy defines the long-term directions of water regulation and water use, water protection from pollution and waterway regulation, as well as protection against the harmful effects of water, in accordance with the EU water legislation.

The Spatial Plan of the Republic of Serbia from 2021 to 2035 [79] as the basic document of spatial planning and development in this country, defines, inter alia: the concept of the long-term water infrastructure development so as to ensure the rational use, regulation and protection of waters; the concept of protection and improvement of environmental status based on the conservation of major compartments of the environment (e.g. air, water and soil); and the development concepts for other thematic fields that regulate some of the issues related to water, as well as development concepts for other sectors. According to [61], regional plans as well as plans of local government units (municipalities) also deal with water infrastructure development.

The National Strategy for Sustainable Use of Natural Resources and Goods [80] stands out as a strategic document relevant for the water sector. The goal of the Strategy is to improve the economic development through an efficient use of natural resources, which involves less intensive use, while reducing the environmental impact. In the context of water protection refers, it implies an economic growth that does not put pressure on water resources (reduced water use) and does not cause water pollution.

The importance of strategic development direction's for individual segments in the water sector, from water use to water protection, is also highlighted in other national documents/sectoral strategies (for example: development strategies on industry, tourism, agriculture, waste management, water transport, etc.). Namely, the intertwining of public policies reflects the functional interdependencies within the natural system, as well as those between social systems (institutions) and the fields of public policies—sectoral integration [60]. Thus, for example, providing a sustainable solution to the issue of waste management through the construction of modern infrastructure would prevent the contamination of surface and groundwater from landfills as a serious form of pollution in Serbia. It is necessary to emphasize the role of education as one of the most important strategies to increase awareness of environmental problems [e.g. 81, 82], i.e. on water pollution, as well as the importance of education in protection against natural disasters, i.e. for natural disaster preparedness [e.g. 83, 84].

4 Human Use of Water Resources

The importance of the Danube River as a water resource for various uses has been recognized since ancient days. This is evidenced by numerous archaeological sites at various locations; several of them can be found in the Đerdap Gorge. The most famous is the archeological site Lepenski Vir (9500 BC), which shows that humans inhabited this area in an early age due to favorable living conditions [22]. In the past, the Danube River was mostly used for navigation and fishing. Danube River in the Đerdap Gorge was a natural spawning ground for sea fish in fresh waters: beluga sturgeon, trout, sturgeon, sterlet. The river species caught by fishermen include catfish, perch, carp, bream, barbel, chub etc. Caviar was obtained from the caught sea fish, which had a significant share in the fish catch, especially after World War I. With the construction of the HPPs Đerdap I and Đerdap II, fish migration routes were cut off, due to which river fish species now prevail in the fish catch in the Danube River.

Nowadays, the Danube River as a water resource is used for different purposes: water supply of settlements, hydropower production, industry, agriculture, tourism and recreation.

4.1 Water Supply

In Serbia, the water supply of the population ranges between acceptable and good [28] and this situation is also reflected in the municipalities of the Lower Danube Basin. In the municipal centres of the Lower Danube River, water supply is organized from regional or local water supply systems. The inhabitants of rural areas receive drinking water from public water supply systems operated by municipalities, local water supply systems built and maintained by the communities, or from their own wells.

In the territory of the four municipalities in the Lower Danube River discussed in this chapter, groundwater from local sources (often accumulated in karst massifs) is prevailingly used for water supply. Alluvial aquifers, the Danube River and reservoirs are also used. According to the data of the Statistical Office of the Republic of Serbia for the 2009–2019 period [85] the average amount of extracted water in the municipalities of the Lower Danube River in Serbia was 8.37×10^6 m^3/year, out of which 4.27×10^6 m^3/year was delivered for drinking water supply, which is about 51% of the total extracted water. Based on data from 2019, it may be calculated that 25,674 households are connected to various water supply systems, accounting for about 81.46% of the total households. This is slightly lower than Serbia's average, showing that 86.71% of the total households are connected to drinking water supply systems. Based on the amount of delivered drinking water and the number of households connected to water supply systems in the analyzed period, the average specific water consumption in the Lower Danube River Basin in Serbia was calculated to be 170.1 l/inhabitant/day. This value is above Serbia's average, which was 148 l/ inhabitant/day in the previous period [28]. However, the consumption is lower than the average consumption in European countries, which is 200–300 l/ inhabitant/day [86]. For a more detailed analysis of water supply in individual municipalities, values were calculated based on the data provided by the Statistical Office of the Republic of Serbia for 2009–2019 [85]. The amount of water consumed as drinking water in the municipalities of the Lower Danube River in Serbia is shown in Fig. 2. A significant decrease in the supplied drinking water can be observed in the municipalities of Kladovo and Majdanpek after 2011. According to [85], the number of households connected to water supply systems in Kladovo

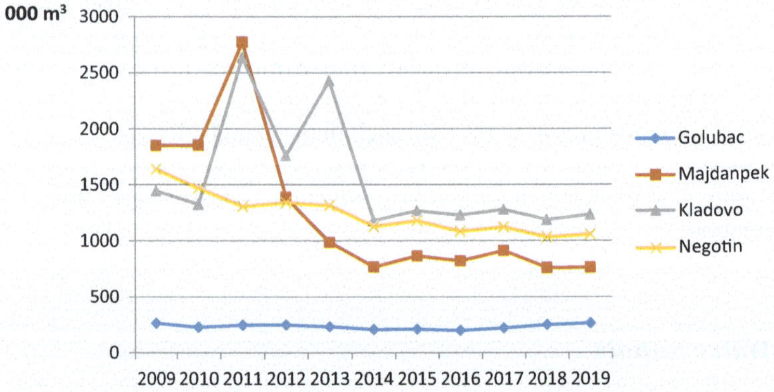

Fig. 2 Supplied drinking water by municipalities in the Lower Danube River Basin in Serbia (*Source* of data [85])

Municipality was around 10,500 in the period 2009–2011, while since 2012 there are about 7,100 households. The decrease in the number of households by about 3,400 contributed to lower consumption of drinking water. In Majdanpek Municipality, no trend was noticed in the change of number of households in the analyzed period. Therefore a clear reason for the decrease in drinking water consumption can not be stated. One of the reasons could be the decrease in the number of inhabitants in this municipality. However, the latest data in the number are from the 2011 Census, so now it is difficult to determine the exact number of inhabitants.

In the Golubac Municipality, the groundwater springs is used for water supply. There is a central water supply system for the municipal centre at Golubac and five suburban settlements along the Danube River. Other 18 settlements are supplied from rural water supply systems relying on local water intakes and springs [87]. According to the data provided by [85], it was calculated that the average amount of extracted water was 0.84×10^6 m^3/year. Out of that amount, 0.23×10^6 m^3/year on average was delivered as drinking water (about 27.4%). Based on data for 2019, it is calculated that 70.7% of households in this municipality are connected to various water supply systems. The specific water consumption in this municipality was 107.03 l/inhabitant/day in the analyzed period.

The Majdanpek Municipality is supplied with water from reservoirs on the rivers Veliki Pek and Mali Pek, Danube River and local springs in the mountainous area. It is noteworthy that the second-largest settlement in this municipality, Donji Milanovac, is supplied from a drinking water treatment plant with a capacity of about 40 l/s. This plant uses the Danube water as raw water [88]. In 2009–2019 period, the average amount of extracted water in the Majdanpek Municipality was 2.9×10^6 m^3/year, out of which 1.25×10^6 m^3/year was delivered as drinking water (about 43.1%). Based on data for 2019, it is calculated that 82.8% of households in this municipality are connected to water supply systems. Specific water consumption was 221.22 l/inhabitant/day in the 2009–2019 period.

In the Kladovo Municipality, 18 settlements have water supply systems, two of them have combined water supply, while three settlements do not have water supply systems [22, 89]. The supply with drinking water to the municipal center of Kladovo and two surrounding settlements is based on groundwater. The average amount of extracted water for 2009–2019 in the Kladovo Municipality was 2.51×10^6 m^3/year, out of which 1.54×10^6 m^3/year was delivered as drinking water (about 61.3%) of the total extracted water. Based on data for 2019, it is calculated that 92% of households in this municipality are connected to water supply systems. Also, it has been calculated that in the Kladovo Municipality, the specific water consumption was 222.77 l/inhabitant/day in the analyzed period. A higher water consumption and the fact that the water supply system in Kladovo does not meet current needs are a consequence of the high loss in the network, which amounts to 40% [90].

The population of the Negotin Municipality is supplied with drinking water from local sources through groundwater exploitation, as well as from the Timok Regional Water Supply System, Bor–Zaječar Subsystem [91]. This subsystem for water supply consists of two reservoirs in Timok River Basin. In Negotin Municipality, 27 settlements are supplied with drinking water from different water supply systems, while

12 settlements do not have water supply systems as their construction is merely planned [92]. In the 2009–2019 period, the average amount of extracted water in the Negotin Municipality was 2.12×10^6 m^3/year, out of which 1.25×10^6 m^3/year was delivered for as drinking water (about 59%). According to the data for 2019, it is calculated that 76.9% of households in this municipality are connected to the water supply systems. The specific water consumption in the Negotin Municipality amounted to 120.3 l/inhabitant/day in 2009–2019 period. The main problem is the frequent lack of water during summer.

Generally, in Serbia, problems related to technological and economic development, as well as underdeveloped awareness on water protection reflect in impaired water quality in water supply systems [93]. Inadequate water quality is more present in small settlements, and this also applies to the municipalities of Lower Danube River. Another problem in Serbia are illegal wells, mostly in rural and suburban areas, built for individual household water supply [93]. This problem is observed in rural settlements of the Lower Danube River Basin as well.

4.2 Hydropower Use

In Serbia, the Danube River has a great hydropower potential, which is estimated at 10,000 GWh/year in total, while the specific potential amounts to 38.46 GWh/km per year [91]. Particularly favourable conditions for the hydropower potential use can be found in the Lower Danube River, in the area of the Đerdap Gorge. Thanks to its morphological characteristics and the flow greater than 5,000 m^3/s, attention has so far been mostly focused on exploiting hydropower potential in this sector. The former Socialist Federal Republic (SFR) of Yugoslavia and the Socialist Republic (SR) of Romania jointly undertook to construct a dam and a reservoir in the Đerdap sector of the Danube River in the 1960s. The Đerdap Hydropower and Navigation System is the largest hydro-technical structure on the Danube River in Serbia and Europe. It consists of two flow hydropower plants HPP Đerdap I and HPP Đerdap II.

The HPP Đerdap I (Iron Gate I; Fig. 3) is the largest hydropower plant in Serbia and the largest hydro-engineering facility on the Danube River. It is located at km 943 km of the Danube River, 10 km upstream from Kladovo. In mid-1964, preparation works for its construction were undertaken and the first hydro generating units were put into operation in 1971. According to the Agreement on Construction and Exploitation between the SFR of Yugoslavia and the SR of Romania, the HPP Đerdap I was designed and built so that each side has one power plant, one ship lock and seven spillways [26]. The two power plants are connected so that power generators on the Serbian side can deliver electric power to the network on the Romanian side, and vice versa, if such a need arises. The total length of the structure is 1,278 m, whereas its height is 34.8 m [24]. The two-level ship lock is 310 m long and 34 m high [27]. Its depth at the threshold is 5.5 m; it has an under-keel clearance of 13.5 m and is suitable for river and sea ships with a carrying capacity of 5,000 t. On the Serbian side, there

Fig. 3 Hydropower Plant Đerdap I (Photo taken from the archaeological site of Diana. *Source* Geographical Institute "Jovan Cvijić" SASA)

are six hydropower generating units: two with a capacity of 171 MW and four revitalizing hydro units of 190 MW each [26]. The revitalization of the hydropower generating units and adaptation of the ship lock began in 2009, and was completed in 2021. The nominal active power of the power plant is now 1,102 MW and the total controlled flow amounts to 5,040 m^3/s. The average electricity production is about 5,500 GWh/year [94].

The HPP Đerdap II (Iron Gate II) is located 80 km downstream from the HPP Đerdap I, at a distance of 862.8 km from the mouth of the Danube River, at the profile Kusjak–Ostrovul Mare. This is another power plant on the Danube River that was jointly constructed by the SFR of Yugoslavia and the SR of Romania. It was completed in 1985 as a multipurpose hydro-engineering facility, just like the HPP Đerdap I. HPP Đerdap II consists of the main power plant, two additional power plants, two ship locks, two overflow dams and two power distribution facilities. The dam is located on the main course of the Danube River and it is 1,009 m long and 330 m wide [27]. The hydropower generating units were put into operation between 1985 and 2001. On the Serbian side, there are ten hydropower generating units with a total installed capacity of 270 MW (10 × 27 MW). The installed flow rate amounts to 4,200 m^3/s and the average electricity production is around 1,500 GWh/year [26]. The adaptation of the ship lock is also underway, and the overhaul of all 10 hydro units is planned to be completed over the next 10 years. This will increase the installed production capacity of the HPP Đerdap II by 50 MW. Before revitalization the average electricity production in the HPPs Đerdap I and II was about 6,989 GWh. After the revitalization of hydropower generating units in the HPP Đerdap I, the annual electricity production amounted to 7,072 GWh (18% of electricity production in Serbia) [2694]. The remaining hydropower potential of the Danube River in Serbia can be used only in accordance with the criteria related to the multipurpose use of water and environmental protection, taking into account the international character of the river.

The formation of Đerdap reservoir has caused some geographical changes (in the microclimate, vegetation, reduction of water velocities, content of chemical compounds in the water, groundwater regime, river regime, ice regime, water quality etc.). After the formation of this reservoir, the hydrological regime of the Danube River in the slow zone (up to Veliko Gradište at high waters) has been significantly changed, while some river mouths and settlements were submerged [90, 95, 96]. In the slow zone, water velocities are reduced and depths are increased depending on the natural regime. As a result of these changes, the sediment transport power is significantly reduced, due to which sediments are deposited in the reservoir [97]. Deposited sediments reduce the useful volume of the reservoir, which affects flood protection, electricity production and water quality. After the construction of HPP Đerdap I, the ice regime on the Danube River changed. Under the natural regime, due to hydraulic and morphological conditions, the Đerdap sector was in danger of congestion (ice jam). Later, this critical part moved upstream, to the slow zone, between Veliko Gradište and Novi Sad.

4.3 Navigation

The Danube River, as the important European waterway E80 (also known as the Danube Corridor, Cprridor VII), about 196 km long, increases the accessibility of this area. The Danube Corridor is a link between West, Central and East Europe. It allows navigation of the largest river cruisers, and it is also recognized as one of the nine multimodal TransEuropean transport corridors (TEN–T) networks. The Danube River enables the spatial and functional integration of this area into a transnational context (Danube Strategy); however, this great potential for development is still insufficiently exploited in terms of navigation and tourism. After the construction of the Đerdap Hydropower and Navigation system, conditions for navigation have been improved by building appropriate ship locks. However, the intensity of the national river traffic is still low. In this sector, the Danube River belongs to the waterways of the highest category (class VII), i.e. it serves as a waterway for cargo ships reaching up to 285 m in length, ranging between 33 m and 34.2 m in width, having a deep draft from 2.5 m to 4.5 m [22]. River–sea ships with a carrying capacity of 5,000 t can sail, provided that the preconditions for navigation of these vessels are met downstream from HPP Đerdap II. At the dam of HPP Đerdap I, the minimum depth at the threshold is 5 m, which enables the simultaneous transfer of a convoy composed of tugs (pushers) and nine pushers with a total carrying capacity of 14,500–27,000 t, or two ships with a carrying capacity of 5,000 t [19]. These conditions are met by

Table 1 Cargo handling and passenger transport on the Danube River in Serbia

Year	Cargo handling (t)	Passenger transport	
		Number of port calls	Embarked/disembarked passengers
2015	5,006,191.80	903	114,932
2016	9,936,455.21	955	119,125
2017	11,071,071.84	996	131,780
2018	12,324,912.06	1,150	157,901
2019	14,168,142.05	1,542	208,797

Source of data [98]

building ship locks, due to which navigation is now determined by the ice regime. Information about cargo handling and passenger transport by river cruises on the Danube River, towards the Black Sea are shown in Table 1. Based on the data, it can be concluded that there was a constant significant increase in cargo handling and passenger transport over the five-year period. Compared to 2015, cargo handling was almost three times higher, the number of port calls increased by almost 70%, and the number of passengers was by about 80% greater.

The accompanying facilities at the Danube waterway in this area include harbours for tourism (Tekija, Donji Milanovac, Kladovo and Brza Palanka) and transportation purposes: Jelenske Stene, shipyard in Kladovo and Prahovo and moorings in Novi Mihajlovac (Fig. 4). Also, some unregulated parts of the riverside are used for boat anchorage.

Comparative analyses of large river navigation systems in Europe (Danube, Volga, Rhine and Elbe) have shown that the Danube River is potentially the most cost-effective traffic route and, at the same time, a traffic route with a great potential to increase traffic intensity [90]. In 2019, the Chamber of Commerce and Industry of Serbia, in cooperation with the Port Governance Agency, adopted a strategic document defining a network of marinas on international, interstate and national waterways in the Lower Danube Basin. The Port Governance Agency [98] prepared the studies on the port area of marinas in Kladovo and Golubac, which are the basis for the further development of these sites towards making marinas an unavoidable and equally significant segment of waterway passenger transport.

4.4 *Tourism and Recreation*

Tourism development in this area begins in the period when the HPP Đerdap I was built (1964–1970). At that time, ports were modernized and the infrastructure for workers' accommodation was built. A major influence on tourism development between Serbia and Romania was the construction of a road across the dam and the establishment of a border crossing [22]. Tourist values are numerous and could be classified into two main categories: natural assets and cultural heritage. Natural

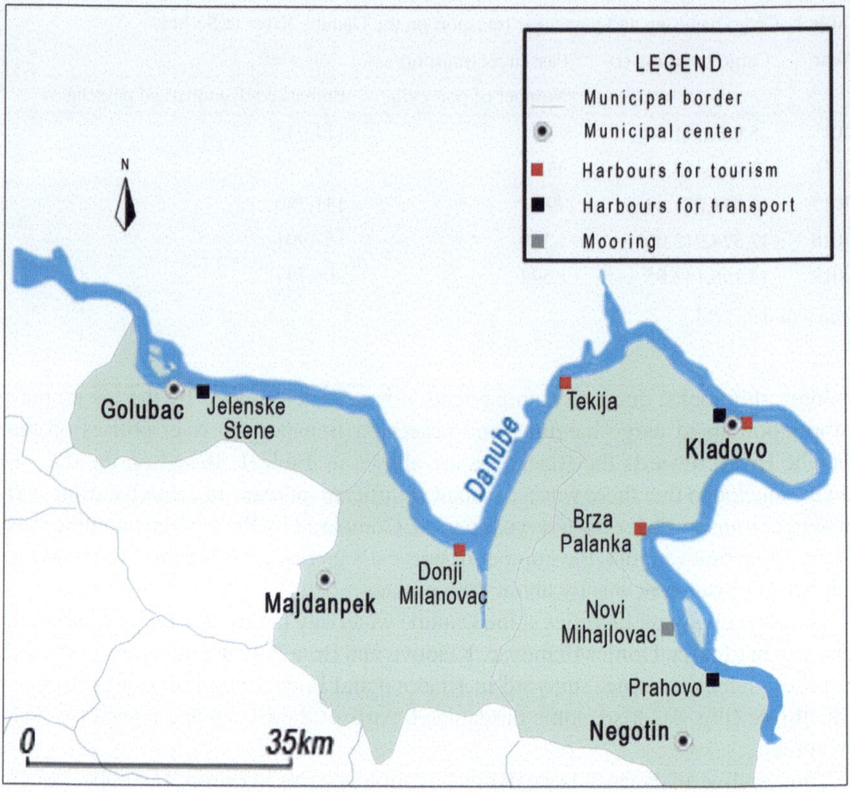

Fig. 4 Map of harbours in the Lower Danube River in Serbia

wealth includes: geological profiles (the geological column in Pesača, the geological profile in the canyon of Boljetinska Reka, and the geological profile Greben [99]); geomorphological features (e.g. Đerdap Gorge, in Fig. 5a, the most visited part of the Danube valley; the cave Velika Pećina in Duboka; the canyon of the Vratna River with its three natural stone bridges—Vratnjanske prerasti; the canyon of the river Zamna; the natural stone bridge Šuplja Stena); the geomorphological–hydrological natural monument Beli Izvorac (tufa deposits with waterfalls) [22, 100, 101]; hydrological features (the Danube River with the Đerdap reservoir and tributaries, as well as waterfalls) [22, 96, 102]; biogeographical elements (forest ecosystems, endemic and relict species) and protected areas (Đerdap National Park, as well as other protected natural assets, such as the confluence of the Timok River into the Danube) [103]. Cultural heritage includes archaeological sites, religious and cultural monuments, artistic, ethnographic and festive values. The most important cultural heritage includes the archaeological sites (e.g. Lepenski Vir, in Fig. 5b; Trajan's Bridge; Trajan's Plaque etc.). There are also fortresses at Golubački Grad and Kladovo (Fetislam), which are the most important medieval fortress in this sector of the Danube valley [99, 101, 104–106]. These natural and anthropogenic values

Fig. 5 **a** Đerdap Gorge **b** Museum of Lepenski Vir (*Source* Geographical Institute "Jovan Cvijić" SASA)

enable the development of different forms of tourism. The most pronounced are recreational and cultural tourism. The diversity of flora and fauna is favourable for the development of hunting and fishery tourism. Various water activities (such as swimming, water sports, sailing) as forms of river and nautical tourism are also present. Ecotourism is part of different programmes, such as bird-watching, hiking trails to viewpoints. Anthropogenic values have contributed to the development of city tourism (especially congress tourism), while festive tourism is developed both in rural and urban settlements. Over the past years, some households have specialized in rural tourism, based on authentic ethnographic motifs. Touristic activities are managed by touristic organizations in Golubac, Majdanpek (with the branch in Donji Milanovac), Kladovo and Negotin. In a regional context, the largest part of tourist traffic takes place in the municipalities of Majdanpek and Kladovo, which are the most developed in terms of tourism in the Lower Danube River area in Serbia, while tourist traffic is the least intensive in the municipalities of Negotin and Golubac [90].

Among the areas that have a potential for tourism development in the Lower Danube region in Serbia, the Đerdap National Park (NP) is especially distinguished. Keeping in mind that admission is free and that most tourists stay there for a day, it is difficult to determine the intensity of the tourist traffic. The data on the tourist traffic in the Đerdap NP for the 2010–2017 period presented in [99] have shown that local tourists have the greatest share in the total tourist traffic, while the number of international tourists is significantly lower (10.4–19.6% in total). Compared to the total tourist traffic in Serbia, the share of tourists visiting this NP is insignificant (1.89–3.09%). Tourist traffic measured by the number of overnight stays amounts to only 1.1–2.9 nights, suggesting that tourism potential in this area is insufficiently exploited. Similar conclusions about the insufficiently used potential of the Danube River for the geotourism development are made by [107]. Projects such as the *Fortresses on the Danube* and *Awake the Danube* have laid a good foundation for the formation of a unique nautical tourist product on the rivers, combining cultural and nature tours [98].

On the other hand, it is important to highlight that tourism development has also caused some pressures and negative consequences on water resources. Water quality was also affected by wastewater from the Golubac Municipality including hotels, restaurants, accommodation and other catering facilities. Due to the lack of a wastewater treatment plant, wastewater directly discharges into the Danube River [102].

4.5 Fishing

Fishing is practised in the following locations: from Golubac to Brnjica, Dobra, Donji Milanovac, Tekija, Karataš, and from Brza Palanka to the confluence of the Timok River into the Danube River [90]. According to [22], there are about 40 species of fish (seven under strict protection and 21 protected species) in the Đerdap sector of the Danube River. These include: sturgeon, pike, Prussian carp, carp, barbel, brook trout, mallard, chub, European bitterling, nase, carp bream, catfish, perch, etc. Fishing destinations in this area are favourable for the growth of perch, carp and catfish [90]. Two types of fishing are presented: sport (recreational) and commercial fishing. Commercial fishing is forbidden 500 m upstream of the Porečka River mouth, and 500 m downstream the HPP Đerdap I, while recreational fishing is forbidden near the HPP Đerdap I [22].

As a consequence of the closure of the Đerdap dams, endangered medium-distance migratory fish, such as sterlet (*Acipenser ruthenus*) and huchen, Danube salmon (*Hucho hucho*) and large-distance migratory species, such as starred sturgeon (*Acipenser stellatus*) and beluga or great sturgeon (*Huso huso*) became extinct in the Upper Danube River [108]. Along with these fish species belonging to sturgeon, many other native species, such as common carp (*Cyprinus carpio*), common barbell (*Barbus barbus*), zander (*Stizostedion lucioperca*) and wels catfish (*Silurus glanis*), have been declining after the construction of the dam [109]. Apart from the negative impacts that the Đerdap dams have on river flow regulation, these fish species are impacted by unsustainable and illegal fishery and pollution [97, 110]. The anthropogenic habitat modification made this area suitable for invasive species [111].

5 Water Quality

Water quality plays a crucial role in all aspects of human life and activities and it is highly relevant for ecosystem sustainability.

5.1 Pollution Sources and Pollutants

In the studied area, water quality is affected by the HPPs Đerdap I and Đerdap II, an industrial zone in Mosna (electrical industry, wood industry and metal haberdashery industry), a copper mine in Majdanpek (Fig. 6), belonging to Copper Mining and Smelting Complex Bor (RTB), and Elixir Prahovo (Industry of Chemical Products) as well as wastewater from settlements, agriculture, road traffic, navigation and minor local sources of pollutants. The major pollution sources from mining and industry are shown in Table 2.

Pollution is generated mainly by the activities of these companies. The main activity of the RTB Bor is opencast mining and quarrying. Furthermore, the facility Sumporna produces basic organic chemicals, such as acids, including chromic acid, hydrofluoric acid, phosphoric acid, nitric acid, hydrochloric acid, sulphuric acid, sulphurous acid and oleum, on an industrial scale. The activities of Elixir Prahovo are focused on the production of phosphorus, nitrogen or potassium-based fertilizers on an industrial scale [112]. According to the investigations about environmental impact assessment of Elixir Prahovo from 2008 to 2013, wastewater from the factory's drain network (total wastewater from the factory grounds—wastewater collector) had a direct impact on the quality of Danube River [113].

Along with these companies, there are other potential anthropogenic pollution sources but the data about their emission into water are not available. These companies include: the Pig Farm Ramski Rit in Veliko Gradište and the Pig Farm Mustapić

Fig. 6 Copper mine Majdanpek (*Source* Geographical Institute "Jovan Cvijić" SASA)

Table 2 Mining and industry pollution in the Lower Danube River Basin in Serbia

Company	Facility	Pollutant
RTB	Copper mine Majdanpek open pit	Cu and compounds Pb and compounds
RTB	Copper mine Bor, open pit Cerovo	Zn and compounds
RTB	Copper mine Bor, open pit Veliki Krivelj	Zn and compounds Ni and compounds Cu and compounds Cd and compounds Pb and compounds
RTB	Copper mine Bor, open pit Jama	Cd and compounds Cu and compounds Ni and compounds Zn and compounds Pb and compounds
RTB	Sumporna	Cu and compounds
Elixir Prahovo	Elixir Prahovo	Fluorides Cd and compounds Total Phosphorus Zn and compounds

Source of data [112]

in Kučevo, which have facilities for the intensive poultry and pig growing [112]. Also, a potential source of pollution is the shipyard in Kladovo/Rhein–Donau Yard shipyard (in bankruptcy), the main activities of which include building new ships; repairs, overhauls and conversions of existing vessels; as well as the production of equipment and facilities for shipping and offshore industries. It is located on the Danube bank near Kladovo. In addition, it is noteworthy that the large industrial zone of Drobeta–Turnu Severin on the Romanian side towards Kladovo contributes to the pollution of the Danube.

Along the Lower Danube River in Serbia, collection and disposal of municipal wastewater through the sewage network is only partially organized. The greatest part of the infrastructure construction, which includes sewerage networks, took place in the Kladovo Municipality. The coverage of the municipal center of Kladovo with the sewerage network is about 80%, while the remaining 20% is solved by septic tanks [89]. In addition to the center, a complete sewerage network was built in five other settlements in this municipality. The municipal center of Golubac is covered by the sewerage network. However, these facilities are not available in other settlements within the municipality. In the town of Negotin, the drainage of municipal wastewater through the sewerage network is ensured only in one part of the town, whereas in other parts wastewater collection is managed through individual water-permeable septic tanks. The sewerage network is partially built in the municipal center of Majdanpek and in two smaller settlements. A fecal collector has not been built and fecal water is drained into Mali Pek River through a sedimentation tank. The settlement Boljetin, which is located in a strictly protected zone in the Đerdap National Park, has a big problem with sewage wastewater because over 90% of households do not have a septic tank and wastewater flows directly into the Boljetinska River, a tributary of the Danube River [90].

Wastewater treatment plants in the Lower Danube River area in Serbia do not exist or are not operational. In Golubac and Majdanpek, no facilities have been built; construction has begun in Kladovo, while in Negotin, the existing plant is not operational. The wastewater treatment plant in Kladovo was designed in the 1980s, when the construction of the main facilities was undertaken, but it has not been put into operation yet. Currently, fecal sewage from the settlement flows directly into the Danube River through the emergency outlet in the incomplete facility. In Brza Palanka, there is a sewage network which brings wastewater to the fecal pumping station, which pumps it into the treatment plant. Through the overflow system, the deposited water is discharged into the Grobljanski Potok, which flows into the Danube River [89]. A wastewater treatment plant with a capacity of 5,000 m^3 was built in Negotin. However it should be completely renovated and upgraded, as it is currently not operational. The construction of a sewerage network and a wastewater factory in Prahovo is planned, but currently there are no financial resources for its maintenance and operation. After rough treatment, wastewater is discharged into the melioration canal, into the Timok River, and the Danube River [92].

The amount of total discharged wastewater and the wastewater discharged into sewerage systems between 2013 and 2019 by municipalities located along the Lower Danube River in Serbia are presented in Fig. 7a, b, c, d. It is noteworthy that the total discharged wastewater is the sum of the wastewater discharged into wastewater collection systems and the estimated amount of wastewater discharged

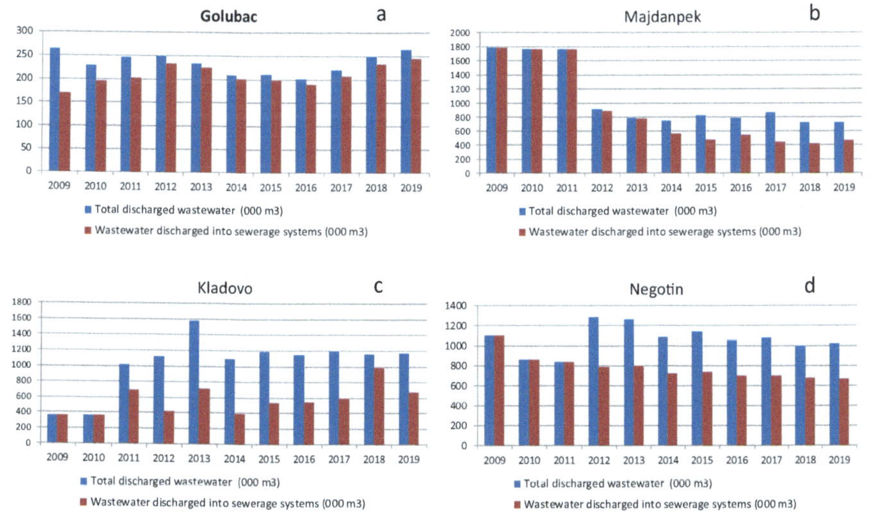

Fig. 7 Total discharged wastewater and the wastewater discharged into sewerage systems by municipalities located along the Lower Danube River in Serbia (*Source* of data [85])

into other recipients. Figures 7a–d show that since 2013 more detailed estimates of wastewater discharged into other recipients are made. A significant decrease in the amount of discharged wastewater is observed only in the Majdanpek Municipality. Since 2013, in the municipalities of Majdanpek and Negotin, more than half of the discharged wastewater is conducted through wastewater drainage systems, while the worst situation is in Golubac, where this percentage varied significantly and usually amounted to about 40%, indicating the lack of a sewerage network.

Besides communal wastewater, it is important to highlight that the water quality of the Danube River is significantly affected by the Timok River, which is its largest tributary in the Lower Danube area in Serbia. Namely, the Timok River is from time to time severely polluted (III or IV water quality class) and it is endangered by organic and inorganic pollution [114]. This is a consequence of the untreated communal wastewater and the wastewater from the Bor mining industrial complex [114, 115]. Also, its tributary the Borska River is an out-of-class watercourse in terms of quality, due to wastewater from the mining process and wastewater from metallurgical and chemical processes in the industrial plants in Bor, which is directly discharged into this river [116, 117]. In addition, the water quality of the Danube tributaries in the Đerdap sector (Boljetinska, Brnjička, Porečka rivers) is especially endangered by nutrients, organic and inorganic pollution (due to the discharge of untreated municipal wastewater and drainage water from agriculture) and heavy metals (due to the breach of the dam at the flotation landfill Valja Fundata near Majdanpek). Some of the local wells are endangered by illegal landfills (in Sip), sewage discharges (in Kladovo, Tekija) and poor sanitation of the settlements [118]. All these facts have a significant influence on the water quality of the Lower Danube River in this area and downstream, in Romania and Bulgaria.

Combustion products from road traffic and navigation also impair the Danube's water quality. Illegal and industrial landfills belonging to the HPP Đerdap I and industrial zone in Mosna directly contaminate water resources. Furthermore, untreated municipal and industrial wastewater [10, 118] as well as drainage water from agriculture are released into watercourses [22]. Agriculture, especially in the Municipality of Golubac, leads to eutrophication and contamination of water with heavy metals, nitrogen and phosphorus due to agrochemical use [118].

5.2 *Assessment of the Water Quality*

The assessment of water quality is an important task and prerequisite for the protection and sustainable use of water resources. In order to examine water quality, various mathematical and statistical methods based on the use of water quality indices have been applied in many studies. The application of different water quality indices helps in the assessment of water quality and the ecological status of water resources, as well as in the identification of possible factors/sources that affect water bodies.

Table 3 SWQI categories

SWQI	Descriptor
90–100	Excellent
84–89	Very Good
83–72	Good
39–71	Bad
0–38	Very Bad

Source [148]

5.2.1 Data and Methodology

Data on water quality were obtained from the *Yearbooks III Water Quality* [119] published by the RHMS of Serbia and the SEPA for the 2009–2018 period, collected at the Tekija, Brza Palanka and Radujevac hydrological stations. The parameters were measured once monthly using relevant standard methods applied by the RHMS of Serbia. Frequency of recorded values was $N = 12$ per year.

Data were processed using the following types of the Water Quality Index (WQI): Serbian Water Quality Index (SWQI), Canadian Water Quality Index (CWQI), Agri-food Water Quality Index (AFWQI), and the Water Pollution Index (WPI). SWQI was calculated monthly. Based on monthly SWQI values, calculations of averaged yearly SWQI values were performed. Averaged yearly SWQI values were used for calculation of averaged SWQI value for ten-year (2009–2018) period. CWQI, AFWQI and WPI values were calculated yearly. Calculation of averaged ten-year period CWQI and AFWQI values were performed based on yearly values of these indices.

The Water Quality Index (WQI) methodology was developed by the Scottish Development Department in 1976, and it is often used for water quality assessment [120–126]. Types of the WQI, applied in this study were also used for water quality assessment in previous investigations: Serbian Water Quality Index (SWQI) in [127–133]; Canadian Water Quality Index (CWQI) in [126, 127, 129, 134–136]; Agri-food Water Quality Index (AFWQI) in [129, 137] and Water Pollution Index (WPI) in [114, 116, 138–147].

The mentioned WQI types rely on a different set of water quality parameters and they are used to calculate and compare obtained results in order to provide a better insight into the water quality status in study area.

Serbian Water Quality Index (SWQI)

Serbian Water Quality Index (SWQI) is an officially accepted methodology for water quality assessment in Serbia, developed by SEPA. The SWQI value is dimensionless—a single number, ranging from 0 to 100 (best quality) within the five categories, presented in Table 3.

The SWQI methodology uses ten quality parameters: oxygen saturation, biochemical oxygen demand (BOD), ammonium, pH, total nitrogen oxides, orthophosphate, suspended solids, temperature, conductivity and the most probable number of coliform bacteria (E.Coli/MPN). Each of these parameters has the value q_i and the weight unit w_i [148, 149]. The SWQI is calculated as a sum of the values of each parameter:

$$\text{SWQI} = \sum q_i \times w_i \tag{1}$$

This methodology is simple for application and can be used as indicator of urban sustainability. SWQI presents information about spatial distribution of surface water quality downstream of municipal wastewater discharge [150]. The main limitation is the small number of parameters and the possibility to calculate the index when some parameters are missing and even when only one parameter is available. SWQI is adequate for the evaluation of organic pollution, but it does not provide information about inorganic pollution, because parameters of heavy metal concentrations are not included [127].

Canadian Water Quality Index (CWQI)

The Canadian Water Quality Index (CWQI) was developed by the Canadian Council of Ministers of the Environment, based on the British Columbia Ministry of Environment formulation, in 1995 [151]. The CWQI is calculated based on the parameters presented in the Annex 1. Most of these parameters have their objectives (limit values) defined. This methodology makes possible to calculate the index even if some parameters are missing.

The Canadian Water Quality Index 1.0 Calculator (EXCEL application) is used to perform calculations using this methodology [151]. CWQI is based on three factors of water quality that relate to water quality objectives:

Scope (F_1):—the percentage of water quality variables that do not meet the objectives in at least one sample ("failed variables").

$$F_1 = \left(\frac{\text{Number of failed variables}}{\text{Total Number of variables}}\right) \times 100 \tag{2}$$

Frequency (F_2):—the percentage of individual tests that do not meet the objectives ("failed tests").

$$F_2 = \left(\frac{\text{Number of failed tests}}{\text{Total number of tests}}\right) \times 100 \tag{3}$$

Amplitude (F_3): The number of failed test that do not meet the objectives. F_3 is calculated in three steps [151], as follows:

1. The number of times by the value of the variable is greater than (or less than, when the objective is a minimum) the objective is termed as "excursion". When the test value must not exceed the objective:

$$\text{excursion}_i = \left(\frac{\text{Failed Test Value}_i}{\text{Objective}_1}\right) - 1 \quad (4)$$

For the cases in which the test value must not be less than the objective:

$$\text{excursion}_i = \left(\frac{\text{Objective}_j}{\text{Failed Test Value}_i}\right) - 1 \quad (5)$$

2. The collective amount is calculated by summing the excursions of individual tests from their objectives and dividing by the total number of tests. This ratio is referred to as the normalized sum of excursions, or *nse*.

$$nse = \frac{\sum_{i=1}^{n} \text{excursion}_i}{\neq \text{of tests}} \quad (6)$$

3. F_3 ranges between 0 and 100 and is calculated as follows:

$$F_3 = \left(\frac{nse}{0.01\,nse + 0.01}\right) \quad (7)$$

When all factors are obtained, CWQI is calculated by summing up the three factors. In this model, the index changes are in direct proportion to changes in all three factors:

$$\text{CWQI} = 100 - \frac{\sqrt{\text{F}_1^2 + \text{F}_2^2 + \text{F}_3^2}}{1.732} \quad (8)$$

For each CWQI range a descriptive quality indicator has been defined [151], with the following ranges: excellent (95–100), good (80–94), fair (65–79), marginal (45–64) and poor (0–44):

- Excellent—there is no threat to the water quality; conditions are very close to natural or pristine level;
- Good—there is a minor threat or impairment; conditions rarely depart from natural or desirable levels;
- Fair—water quality is usually protected but occasionally threatened; conditions sometimes depart from natural or desirable levels;
- Marginal—water quality is frequently threatened; conditions often depart from natural or desirable levels;
- Poor—water quality is almost always threatened; conditions usually depart from natural or desirable levels [151].

Agri-food Water Quality Index (AFWQI)

Agri-food Water Quality Index (AFWQI) has also been developed by the Canadian Council of Ministers of the Environment, based on the British Columbia Ministry of Environment formulation, in 1995 [152]. The AFWQI methodology is based on the parameters, which upper limits are presented in Annex 2.

For each AFWQI range, a descriptive quality indicator has been defined ranging from poor (0–44), marginal (45–64), fair (65–79), good (80–89), very good (90–94), excellent (95–100). The AFWQI index provides information on the suitability of water for agricultural purposes, such as irrigation and livestock watering, and it is important for the assessment of pesticide pollution. The program used for AFWQI calculation is Agri-food Water Quality Index 1.0 Calculator. Calculation is performed in the same way as for the Canadian Water Quality Index [153]. In both cases, it is possible to calculate the indices even if some parameters are missing. Both indices share the same limitation: it is impossible to calculate the index in a single measurement (it is necessary to perform at least four measurements).

Water Pollution Index (WPI)

The Water Pollution Index (WPI) is an arithmetical method for integrating parameters to assess the chemical and ecological status of inland waters [154, 155]. Its advantage is the possibility to combine different parameters (physical, chemical, biological); also, there is no limitation as to the number or types of the used parameters. Therefore, the WPI is widely applied as an indicator in the evaluation of the water quality status in different water bodies, which allows for a simple and objective interpretation of results. It allows a simple and objective interpretation of results.

According to [154] the WPI is the sum of the ratios of the measured annual average value of parameters (Ci) and the prescribed maximum values for water quality class I (SFQS) for each parameter, divided by the number of used parameters (n):

$$\mathrm{WPI} = \sum_{i=1}^{n} \frac{\mathrm{Ci}}{\mathrm{SFQS}} x \frac{1}{n} \tag{9}$$

The calculated WPI values for watercourses can be classified into six different classes (Table 4).

The WPI involves a comparative analysis between the average annual concentration values of the observed parameters and the limit concentration values determined in the ecological classification for class I according to the national legislation. The standard threshold values for all parameters of the ecological status are defined for each country. In Serbia, they are established at the national level by several abovementioned regulations [39–41].

To calculate the WPI in this study, data relating to 18 physical, chemical and biological parameters, collected between 2009 and 2018 were used. Based on [39],

Table 4 Water quality classification according to WPI

WPI	Class	Characteristics
≤ 0.3	I	Very pure
0.31–1.0	II	Pure
1.01–2.0	III	Moderately polluted
2.01–4.0	IV	Polluted
4.01–6.0	V	Impure
>6.01	VI	Heavily impure

Source [154]

rivers in Serbia are divided into six types. The Danube, which is the subject of this study, belongs to Type 1; threshold values of the some parameters analyzed for this river type are shown in Annex 3.

The threshold values for other analyzed parameters in this study are presented in Annexes 4 and 5. For these parameters there are unique threshold values for classes, i.e. the rivers are not divided by different types.

A combination of indices was used in order to mitigate the limitations of individual indices and to obtain as precise results as possible. This approach provides more information about water quality and could be applied in various purposes.

5.2.2 Results and Discussion

Judging by the monthly parameters values for a ten-year period (2009–2018), averaged SWQI values were very good at all stations (84 at Tekija, 87 at Radujevac and 88 at Brza Palanka). However, an analysis of SWQI values for individual years reveals oscillations at all stations ranging from good to excellent (83–91 at Radujevac, 85–92 at Brza Palanka, and 74–91 at Tekija). The highest SWQI values at all stations were recorded during the same year (2012), while the lowest SWQI values for individual stations were recorded during different years (2013 at Tekija, 2016 at Radujevac and 2017 at Brza Palanka). Exceptions were noticed for the Tekija station during summer, when bad SWQI values were calculated (69 in August 2010 and 70 in June 2014). High water temperatures caused a decline in oxygen saturation in both cases. Furthermore, increased concentrations of suspended solids in June 2014 and an increased number of coliform bacteria in August 2010 also contributed to the decline of SWQI values.

The main sources of suspended solids are flotation tailings, metallurgical and heating plants [22]. The Đerdap reservoir is also an important source of suspended solids, as well as organic contaminants [3].

During the same period, at same stations, the CWQI values for overall water quality were significantly lower than SWQI values (Fig. 8a, b, c), ranging from marginal (50 at Radujevac) to fair (66 at Tekija and 68 at Brza Palanka). Oscillations were also greater for individual years, compared to SWQI, and they ranged from

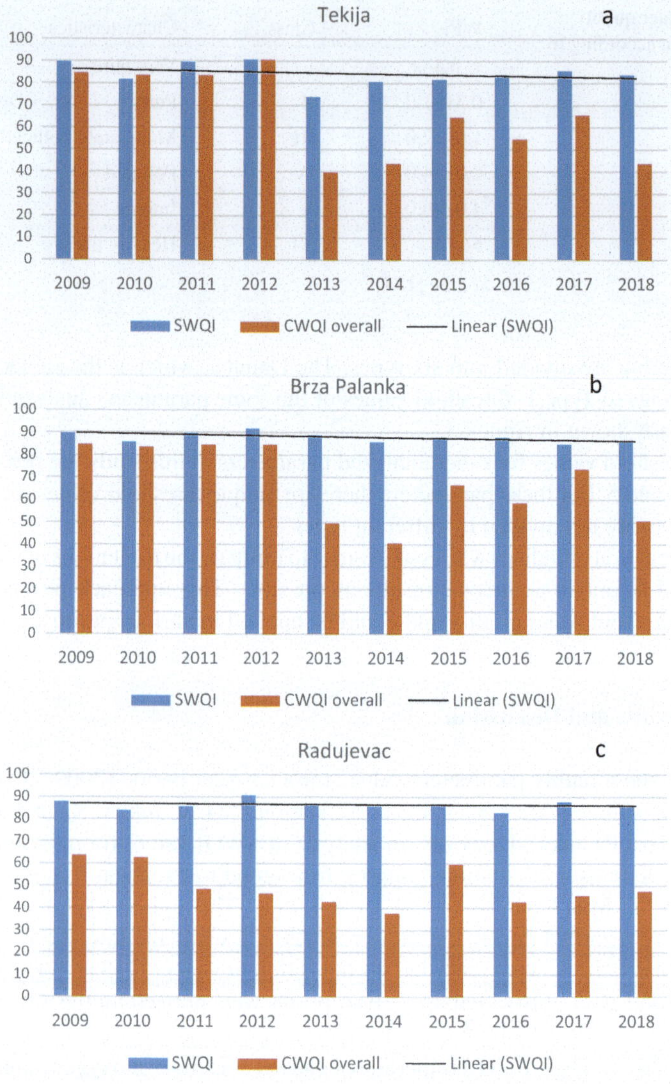

Fig. 8 SWQI and CWQI at Tekija (a), Brza Palanka (b) and Radujevac (c) stations on the Lower Danube River in Serbia

poor to good (41–85 at Brza Palanka and 40–91 at Tekija) and from poor to marginal (38–64 at Radujevac). The highest CWQI value was recorded at Tekija 2012, while the lowest value for the same station was recorded in 2013, which was in the line with SWQI. The highest CWQI value at Brza Palanka was recorded in 2012, as well as in 2009 and 2011, while the highest CWQI value at Radujevac was recorded in 2009. The lowest CWQI values at Radujevac and Brza Palanka were recorded in 2014.

The differences between the two indices could be explained by increased metal concentrations, which are calculated by CWQI. Between 2009 and 2012, good CWQI value was due to the missing parameters of metal concentration for the stations at Brza Palanka and Tekija. For both stations, the variable with the most failed tests was dissolved oxygen. The variables with the highest *nse* were turbidity (2010 for both stations and 2009 for Brza Palanka), and dissolved oxygen (2011 and 2012 for both stations, and 2009 for Tekija). The increased metal concentration had the greatest impact on the CWQI decline from 2013 to 2018 for Brza Palanka and Tekija, and throughout the whole study period (2009–2018) for Radujevac. In 2013–2018, the variable with the highest *nse* was aluminium for the Brza Palanka and Tekija stations, and for Radujevac, in 2011–2018. The variables with the most failed tests were mainly aluminium and copper for the Tekija (2013–2018) and Radujevac stations (2009–2018), and, in some cases, chromium. The variables with the most failed tests were dissolved oxygen, for Brza Palanka, and aluminium and copper in three cases.

Increased copper concentrations, which caused the decline of water quality, were mainly a result of the flotation process at the copper mine in Majdanpek [156] during which copper was released; it further reached water via soil and air. It also generated from untreated wastewater of the copper mine in Bor [117]. Organic pollution was caused by untreated wastewater, illegal landfills and inadequate sanitation in settlements [118].

The CWQI values for drinking water were good (Fig. 9) throughout the ten–year period for all stations (86 at Radujevac, 88 at Brza Palanka and 89 at Tekija). Some oscillations were recorded during individual years and they ranged from fair to excellent (79–96 at Brza Palanka and 72–100 at Tekija) and from fair to good (72–92 at Radujevac). The years with the highest CWQI for drinking water were 2012 (at Tekija and Radujevac) and 2017 (at Tekija and Brza Palanka). The lowest values were recorded in 2014 (at Brza Palanka and Radujevac) and 2013 (at Tekija). The variables that affected the CWQI for drinking water the most were iron and turbidity. Iron was the variable with the most failed tests and with the highest *nse* for all years with two exceptions: during 2010, turbidity was the variable with the most failed tests and with the highest *nse*, while manganese had the highest *nse* in 2017. In 2009 and 2010, the variable with the greatest number of failed tests and the highest *nse* was turbidity, while the pH was the variable with the most failed tests and the highest *nse* in 2011 for Brza Palanka and Tekija, as well as for Brza Palanka in 2012. Turbidity was the variable with the most failed tests and the highest *nse* in 2016 and 2017 for Brza Palanka and with the most failed tests in 2016 for Tekija and 2018 for Brza Palanka. In 2013, turbidity and iron had the most failed tests for Brza Palanka. In all other years, iron was the variable with the most failed tests for Brza Palanka and Tekija, with two exceptions: 2012 and 2017, when there were no failed tests for Tekija.

The copper mine in Majdanpek and the industrial zone in Mosna also produce other heavy metals, such as iron and chromium, as well as suspended solids, which cause increased turbidity [22]. Iron also is generated from open pits Veliki Krivelj and Jama of copper mine in Bor [157]. Besides these pollution sources, the Đerdap reservoir acts as a sink for fine sediments and pollutants [158].

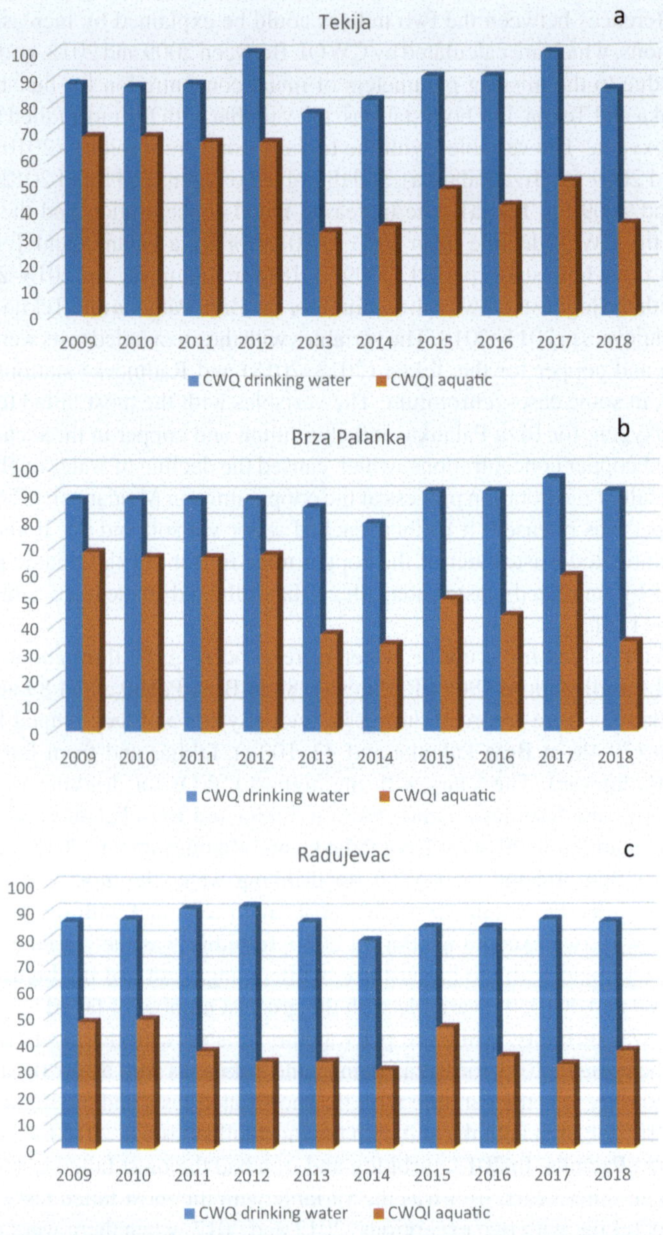

Fig. 9 CWQI for drinking water and aquatic life at Tekija (a), Brza Palanka (b) and Radujevac (c) stations on the Lower Danube River in Serbia

Over the ten-year period, the CWQI values for aquatic life (Fig. 9a, b, c) were marginal (51 at Tekija and 52 at Brza Palanka) and poor (38 at Radujevac). Oscillations during individual years ranged from poor to fair (33–68 at Brza Palanka and 32–68 at Tekija), and from poor to marginal (30–49 at Radujevac). The years with the highest CWQI for aquatic life were 2009 (at the Brza Palanka and Tekija) and 2010 (at Tekija and Radujevac), while the years with the lowest CWQI were the same as those when the CWQI the overall water quality was the lowest. The variables with the most failed tests and the those with the highest *nse* were in line with the CWQI for the overall water quality (for all stations and all years) with the following exceptions: dissolved oxygen was the variable with the highest *nse* in 2009 and 2010, at Brza Palanka, and 2010, at Tekija, whereas turbidity was the variable with the highest *nse* for the overall water quality.

Over the ten-year period, the CWQI for irrigation (Fig. 10 a, b, c) showed high values: good (94 at Radujevac) and excellent (98 at Tekija and 99 at Brza Palanka). Minor oscillations were observed in individual years, ranging from good to excellent (94–100 at Brza Palanka, 88–100 at Tekija, and 92–95 at Radujevac). Lower values were recorded in 2013 (Tekija and Radujevac) and 2014 (Brza Palanka). Chromium caused a slight decline in the CWQI for irrigation in all cases when CWQI had a value lower than 100. The only exception was Radujevac in 2017, when manganese was the variable with the most failed tests and the highest value of *nse*. In all other years, chromium had the most failed test and the highest *nse* for Radujevac, Brza Palanka (2014 and 2018), and Tekija (2013 and 2018). The CWQI for livestock was excellent (100) at all stations and in all years. CWQI did not reach the maximum value only in Radujevac in 2010 due to a slightly increased arsenic value in one measurement, causing a minor decline in CWQI (95). Increased chromium concentrations could be explained by the presence of chromic acid, which is the product of the facility Sumporna within the Copper Mining and Smelting Complex Bor (RTB) [112].

A comparison of the CWQI for irrigation and the AFWQI for irrigation reveals some similarities, but also some differences (Fig. 10a, b, c). Over the ten-year period, AFWQI values were mainly in line with CWQI for irrigation and were very good (89 at Tekija and 93 at Radujevac) and excellent (97 at Brza Palanka). Oscillations during individual years for Brza Palanka were not high and they ranged from good to excellent (88–100). However, significant oscillations were recorded for Tekija and Radujevac and they ranged from fair to excellent (68–100 at Tekija and 70–100 at Radujevac). The highest AFWQI value was recorded in 2010 and 2011 for all stations, as well as in 2009 for Brza Palanka and Radujevac, and in 2013, 2015 and 2016 for Brza Palanka. The lowest AFWQI values were recorded in 2014 at Tekija, 2015 at Radujevac, and in 2017 at Brza Palanka. Lower AFWQI values were mainly caused by the increased number of Coliforms total and Coliforms fecal. These variables had the most failed tests and the highest *nse*. Exceptions are registered in two single cases at Tekija: in 2009, linuron exceeded the limit in one measurement, and in 2013, cadmium was also above the limit in one measurement. Apart from these cases, manganese and Coliform total were the variables with the most failed tests in 2017 for Radujevac. Linuron is a herbicide that can also be found in the sediment of

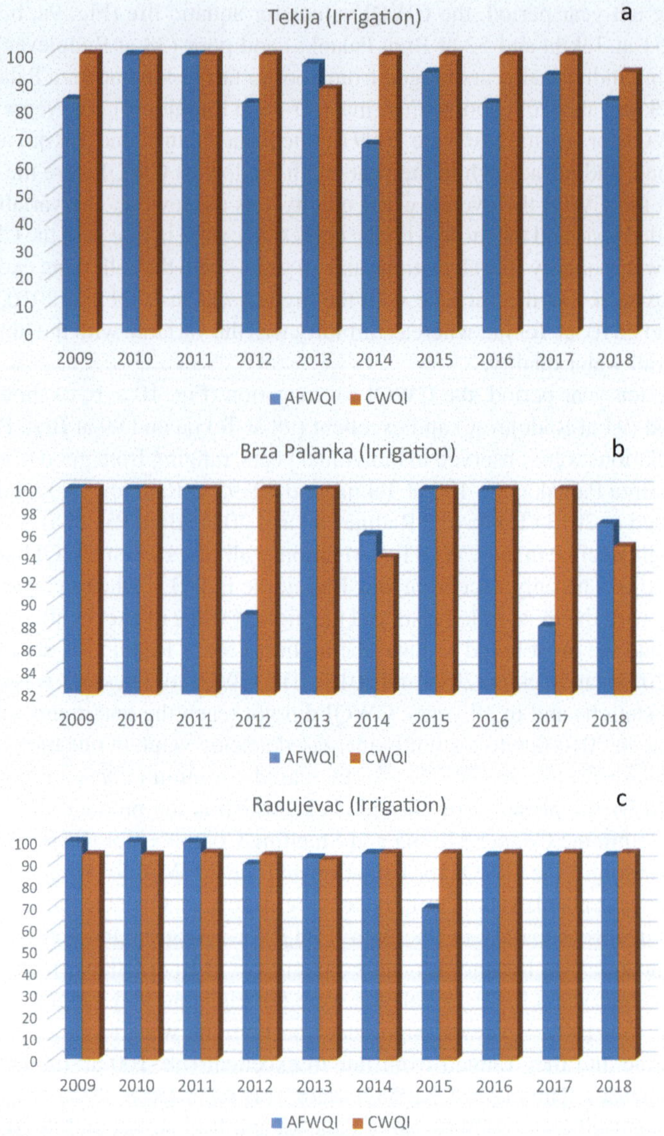

Fig. 10 AFWQI and CWQI for irrigation at Tekija, Brza Palanka and Radujevac stations on the Lower Danube River in Serbia

the Đerdap reservoir. The primary source of this environmental pollutant is its use in agriculture [3].

The AFWQI for livestock was in line with the CWQI for livestock and it usually had the maximum value (100). In 2012 at Tekija and Radujevac, as well as in 2011 and 2014 at Radujevac, increased concentrations of phenols (in one measurement each year) caused a slight decline and AFWQI did not reach the maximum value: it was 96 at Tekija and 97 at Radujevac. Slightly increased arsenic concentration in one measurement in 2010 at Radujevac caused a slight decline in the AFWQI value (97), which was also in the line with the CWQI in the same case.

Some WPI analyses for the Serbian part of the Danube River have been presented in several previous studies. Based on data from ten hydrological stations the WPI was calculated for the entire Serbian part of the Danube River for 2014 [145]. The WPI was calculated for the Hydrosystem Danube–Tisa–Danube based on data from nine hydrological stations between 2004 and 2009 [138] and for the Đerdap sector of the Danube for a ten-year period (2007–2016) based on data from three hydrological stations [102]. As far as the water quality and pollution of the Danube's tributaries in Serbia are concerned, the WPI index was calculated for the Timok River at four hydrological stations [114], and for its tributary Borska Reka at one hydrological station [116] for two periods: 1993–1996 and 2006–2009.

In order to assess the ecological status of water quality of the Lower Danube River in Serbia in the present study, WPI was calculated based on the yearly parameters values from the aforementioned three hydrological stations. The results of the study conducted in 2009–2018 are presented in Table 5 and they were used to determine water pollution levels in this area. The results indicate that class III (moderately polluted water) was the most frequent in the Lower Danube River for a ten-year period. These results are similar to the literature data, which indicates moderate pollution of the Danube River in Serbia [145, 159]. Based on the analysis of WPI according to the locations of hydrological stations, the worst quality is at the Tekija, which is located in the Đerdap National Park. It is important to highlight that pollution of the Lower Danube River (especially in the Đerdap region) is different from the pollution of other Danube River sectors because of the relief, a small number of tributaries and small discharge. However, the Danube River receives pollutants from the upper sector of the river basin, where large industrial centres are located [160].

At the Tekija hydrological station, the WPI values ranged from 1.02 to 2.30, corresponding to class III (moderately polluted water) and class IV (polluted water). In addition to the aforementioned facts, one of the reasons for this situation was the insufficiently developed and poorly regulated utility infrastructure in the Đerdap National Park [102]. None of the accommodation and catering capacities of within the Đerdap National Park had their own wastewater treatment systems and they discharged wastewater into the city sewage. Also, as already pointed out, the municipalities have not fully resolved the issue of wastewater disposal.

At the Brza Palanka hydrological station, located downstream from the Đerdap I dam, the WPI values were in the 0.72–1.80 range, i.e. from class II (pure water) to class III (moderately polluted water). It is noticeable that class IV was not recorded in this period. One of the reasons for this could be the dam, which became an artificial barrier depositing sediment, accumulating pollutants and creating significant environmental problems [160]. The Đerdap dams and reservoirs significantly influence the sediment transport in two ways: on the one hand, they are a trap for suspended sediments [5, 27], and on the other, they are an important nutrient sink and deposition area for hazardous toxic matters causing pollution [161]. During the 1972–1994

Table 5 WPI values and water classes at Tekija, Brza Palanka and Radujevac stations on the Lower Danube River in Serbia

	2009	2010	2011	2012	2013	2014	2015	2016	2017	2018
Tekija										
WPI	1.96	2.21	1.52	1.26	1.25	2.30	1.18	1.51	1.02	1.63
Class	III	IV	III	III	III	IV	III	III	III	III
Characteristics	Mod. polluted	Polluted	Mod. polluted	Mod. polluted	Mod. polluted	Polluted	Mod. polluted	Mod. polluted	Mod. polluted	Mod. polluted
Brza Palanka										
WPI	1.51	1.56	0.77	0.72	0.91	1.03	1.29	0.83	1.80	0.95
Class	III	III	II	II	II	III	III	II	III	II
Characteristics	Mod. polluted	Mod. Polluted	Pure	Pure	Pure	Mod. polluted	Mod. polluted	Pure	Mod. polluted	Pure
Radujevac										
WPI	1.33	1.25	1.11	1.05	2.90	1.00	1.17	1.06	0.97	1.09
Class	III	III	III	III	IV	II	III	III	II	III
Characteristics	Mod. polluted	Mod. polluted	Mod. polluted	Mod. polluted	Polluted	Pure	Mod. polluted	Mod. polluted	Pure	Mod. polluted

period, about 325 million tonnes of sediments (10% of the entire reservoir) were retained by the Đerdap dams, leading to a strong decline in suspended sediment transport along the Lower Danube River [162]. However, the process of the sediment deposition does not exhibit constant features: it takes place within the river channel only at low discharges, while at high discharges of the Danube River (when the hydraulic regime is close to natural), fluvial erosion takes place and removes previously–deposited sediment from the riverbed [27].

At the Radujevac hydrological station, located close to the place where the Danube River exits Serbia, the WPI values ranged from 0.97 to 2.90, i.e. from class II (pure water) to class IV (polluted water). It is noteworthy that at this station, the WPI values most frequently indicated class III (moderately polluted water), while class IV was recorded once and class II twice in a ten-year period. A higher degree of pollution was recorded at Radujevac than at Brza Palanka, as an upstream station. This is partly a consequence of the production in Elixir Prahovo (former IHP Prahovo), a factory for the production of phosphoric acid and mineral fertilizers.

The analysis of the WPI values over years for the same stations (Table 5) did not reveal any distinct trend on the Lower Danube River. Higher pollution levels were recorded at Tekija in 2010 and 2014, and at Radujevac, only in 2013. It is important to emphasize that at Brza Palanka, a lower degree of water pollution was noticed than at the other two stations. Also, a trend of water quality improvement was recorded in 2011–2013, when, based on the WPI value, water belonged to class II (clean water).Kindly check and confirm Table 10 citation is provided. But relevant table is missing. Kindly provide relevant table or delete this citationWe have corrected it. It is Table 5.

Taking into consideration the parameters used to calculate the WPI at all three hydrological stations, it can be concluded that out of the 18 analyzed parameters, six (pH, Mn, Ni, Cu, Hg and SO_4^{2-}) always belonged to class I. The values of four analyzed parameters (suspended solids, S, COD, Cd) mainly belonged to class I, with minor occasional deviations. In the 2009–2018 period, two analyzed parameters (oxygen saturation and BOD) belonged to class I and class II by half of the period, and two analyzed parameters (Fe and coliform bacteria) were classified as class I to class III. The other four analyzed parameters (dissolved oxygen, NO_2^{2-}, NH_4-N, PO_4^{3-}) deviated more or less from the permissible values for class I, thereby affecting the WPI values. Therefore, it can be concluded that the elements that served as the indicators of organic pollution had a significant impact on water pollution. Our results are reflective of other results reported in the literature. Calculation of nutrient loads (dissolved inorganic nitrogen—DIN), orthophosphates (PO_4^{3-}) and total phosphorus (TP) for Serbian part of the Danube River and its main tributaries in 2001–2010 and comparison to stations in other Danube countries showed that values mainly overlap with the general increasing trend in the downstream direction [163]. Concentrations of PO_4^{3-} and TP were higher at Radujevac compared with the upstream stations and belonged to the III class in terms of values. According to the WPI results for the Danube River in 2014, concluded that PO_4^{3-} and TP, the indicators of organic pollution, were the major nutrients that influenced the primary production in aquatic ecosystems of the Danube River [145]. Also, a chemical water quality assessment of the Danube River in the sector between the area 2 km downstream from HPP

Đerdap I to Drobeta–Turnu Severin showed that the most important variables that affected water quality were NH_4-N , TP, temperature and total suspended solids[164], indicating organic pollution. It is noteworthy that $NO_2^{2,-}$ NH_4-N and PO_4^{3-} have an important role in the formation of nutrition loads in rivers, causing eutrophication, oxygen depletion and water quality deterioration.

Other studies found anthropogenic sources of pollution in the area of the Đerdap reservoir [3, 10, 165]. Hagemann et al. [10] found organic pollution in the sediments. Matić Bujagić et al. [3] reported the presence of four classes of emerging pollutants in the sediment core (pharmaceuticals, pesticides, steroids and perfluorinated compounds). Pollution indices revealed that the concentration of minor elements (inorganic pollutants) in sediments had anthropogenic sources in some periods [165].

Calmuc et al. [126] also found the existence of pollution sources from agricultural and industrial activities. Their results also show differences between WPI, CCME WQI and WQI. According to WPI, water quality was ranked as class II (pure quality), according to CCME WQI, 98% of the sampling stations were ranked as class II, while according to WQI, 53% of the sampling stations were ranked as class III and 47% as class II. Based on this study, WQI should be applied in the evaluation of potentially toxic pollutants, CCME WQI in areas with permanent sources of pollution, while WPI should be used in the general characterisation of watercourses. It was emphasized that the most suitable index for water quality assessment is WQI because it takes into account various types of pollution sources [126].

6 Protective Measures by International Projects

The main aim of international projects in this area is to establish fish migration routes. Sturgeon fish species are considered endangered and nearly extinct. The International Commission for the Protection of the Danube River (ICPDR) has adopted the Sturgeon Strategy with the aim of ensuring the recovery and survival of sturgeon in the Danube River Basin. Three critically endangered species of the sturgeon include beluga or great sturgeon (*Huso huso*), Danube or Russian sturgeon (*Acipenser gueldenstaedti*), and starred sturgeon (*Acipenser stellatus*). Both short-distance migration populations (spawning downstream of Đerdap Gorge) and long-distance migrating populations (spawning upstream the Đerdap Gorge) are blocked in the spawning migration by HPPs Đerdap I and II. The priority is to take measures to save the genetic fingerprint of sturgeon in the Danube River Basin, specifically those sturgeon species that are nearly extinct and the few individuals born upstream of Đerdap Gorge. When they return to Đerdap Gorge, they are trapped because their migration upstream is blocked. The ICPDR Sturgeon Strategy seeks to involve stakeholders in the discussion and facilitate dialogue with the operators of Đerdap hydropower and navigation industry and infrastructure sector players in order to solve the issues related to interruptions/barriers and other hydromorphological alterations [166].

One of the projects is *Restoring fish migration routes in the Danube River Basin*, based on the grant agreement between the ICPDR and the Commission's Directorate–General for Regional and Urban Policy (DG REGIO). It aims to preserve fish stocks at the Romanian–Serbian border. This action is an important part of the central objective of the EU Strategy for the Danube Region: saving the Danube sturgeon from extinction. The Danube sturgeon (*Acipenser gueldenstaedti*) is endangered due to the construction of the Đerdap dams: the disruption of the river continuity constitutes an obstacle for migratory fish, including the sturgeon species, Danube salmon (*Hucho hucho*) and European eel (*Anguilla anguilla*).

According to the EU Strategy for the Danube Region, the EU Water Framework Directive, the EU Habitats Directive and the Bern Convention, stakeholders and international experts provided the framework for the development of specific conservation measures. The first phase, from 2011 to 2016, facilitated dialogue between the ICPDR, stakeholders and EU Commission—represented by DG REGIO and The Directorate–General for Environment (DG ENV). The second phase is the feasibility study. The third and fourth phases regard technical design (2021–2023) and implementation (2024 and onwards). In line with the feasibility study, the ICPDR will coordinate and implement activities jointly with the Danube Delta National Institute for Research and Development (DDNI) from Romania and the Jaroslav Černi Institute for Water Resources Development (JCI) from Serbia [167].

A similar project is an initiative *We Pass*, which aims to facilitate fish migration in the Danube River Basin with the focus on preservation and reestablishment of the migration routes of endangered fish species in the Danube River and its tributaries, specifically at the Đerdap Gorge. This initiative is set up by the ICPDR, JCI, DDNI, CDM Smith OAK consultants, and the Norwegian Institute for Nature Research (NINA). One of the projects in the scope of this initiative is *Study on environmental and ecological thematics in the framework of macro–regional strategies (MRS) and policy coordination with the Directorate–General for Neighbourhood and Enlargement Negotiations (DG NEAR) and DG ENV: Support for the implementation of the Feasibility Study analyzing options for characteristic Danube fish migration at Đerdap I & II*. The Serbian side has proposed the construction of a third, smaller hydropower plant (Đerdap III) with the assurance of no adverse impacts on the ecosystem area. The specific objective of this project is to restore and preserve water and soil quality, and biodiversity in the Danube Region. Addressing the blockage of a fish migration route around the Đerdap dams is the central concern in preserving biodiversity in the entire Danube, before and beyond the Serbian–Romanian border area. In order to achieve these objectives, the following tasks are proposed: project management, analysis of the current situation and data gathering, monitoring fish behavior at Đerdap Gorge, communication activities, data quality assurance and quality checks, 3D basis model [168].

The Đerdap National Park in Serbia and the Iron Gate Nature Park in Romania were the subject of the transnational cooperation project BioREGIO Carpathians (2011–2013). The objectives of the project were conservation, restoration and valorization of the Carpathian ecological continuum in order to enable wild animals to live in coexistence with modern society. The projects applied a multi–disciplinary approach (physical, legal and socio–economic) to identify the most influential barriers regarding connectivity throughout the Carpathians. A part of this project was the case study *The Lynx in the Pilot Area Đerdap National Park (Serbia)/Iron Gate Nature Park (Romania).* The aim of the study was to find the Least Coast Paths (LCP) for Lynx in this area and potential barriers. The final recommendation which related to this case study concentrated on animal–vehicle collision (highlightening the absence of mitigation structures and the driving behaviour—special focus on the road 25-1 in Đerdap National Park). This study could help in identifying the most important corridors for both wildlife movements and the human–wildlife coexistence [169]. Crossborder cooperation of the Đerdap National Park is also realized through IPA crossborder programmes between Serbia, Romania and Bulgaria and it is related to the formation of the European Green Belt (i.e. the Lower Danube Green Corridor along the borders with Romania), the European Road of Culture (Danube Road), the Euroregion "Danube 21", etc. [118].

7 Conclusion and Recommendation

Water resources around the world are overexploited due to population growth and an increased demand for water. Large rivers, such as the Danube River, are especially under pressure, because of multipurpose use. This chapter deals with the human impact and pressures on the Lower Danube River Basin in Serbia, which is presented through an overview of the water resources use in this area. The water resources of the Lower Danube River are used for hydropower production, navigation, water supply of settlements, agriculture, industry, fishing, tourism and recreation. Đerdap Hydropower and Navigation System has especially significant role for multipurpose Danube's water use, with very favourable conditions for hydropower production and navigation. However, these potentials are not completely valorized. Groundwater from local sources (often accumulated in karst massifs) is prevailingly used for water supply. Tourist values are numerous including natural assets and cultural heritage. All these uses affect water quality and aquatic ecosystems. Fishing is also impacted by Đerdap dams. In this study, assessment of water quality in Lower Danube River in Serbia was made at three stations (Tekija, Brza Palanka and Radujevac), based on the following indices: Serbian Water Quality Index (SWQI), Canadian Water Quality Index (CWQI), Agri-food Water Quality Index, and Water Pollution Index (WPI). Depending on the parameters used, the results show some oscillations and variations. However, general conclusion is that water quality at the Radujevac station is somewhat lower because the station is the farthest downstream. The low values of the CWQI for aquatic life suggest that living organisms are especially endangered. The major anthropogenic pollution sources are the Copper Mine in Majdanpek, the industrial zone in Mosna, the production of phosphoric acid and mineral fertilizers in Elixir Prahovo, untreated wastewater and landfills, emissions from road traffic and navigation, pesticides and other chemicals from agriculture. Along with pollution, aquatic ecosystems are also affected by hydromorphological alterations, such as the construction of the Đerdap reservoir, which had a negative impact on migratory fish species. Several international projects seek to address the decreased fish population by establishing fish migration routes.

Considering the above-mentioned water supply of municipalities in the Lower Danube Region in Serbia, the recommendations would be: the reconstruction of the existed water distribution network (to reduce losses in the water supply network); revitalization and protection of reservoirs; the reconstruction of the plants for drinking water production; establish group water supply systems in the rural settlements of this municipality (especially in the hinterland of the Danube River); the protection of the water sources in order to prevent water pollution from mines; the supply of industrial facilities with technological water provided from local watercourses with mandatory recirculation of technological water.

In the context of water protection, these recommendations among others, include the following: construction of wastewater treatment plants, both for wastewaters from households in municipal centers and in industrial plants for industrial wastewater; increasing the level of coverage of the population by sewage systems; control of the use of mineral and organic fertilizers and pesticides in order to reduce the pollution from agriculture; introduction of good agricultural practice; reduce water pollution by sediments and associated pollutants from agricultural land (phosphorus, nitrogen, pesticides, herbicide, etc.) through soil erosion control; improving municipal waste management through the construction of sanitary landfills and the closure of uncontrolled (illegal) dumping sites; implementation of the Guiding principles on sustainable hydropower development in the Danube River Basin [170], which, among other things, includes a technical upgrade of existing hydropower plants, strategic planning of new hydropower development, as well as ecological restoration measures to ensure sustainable use of hydropower; the improvement of navigation conditions; construction and maintenance of marinas, harbours and anchorages [22]; the reactivation of a national river fleet; increasing awareness among the population on environmental problems through environmental education and by organizing various events, such as the Danube Day in order to raise awareness of the social impact on water pollution.

Acknowledgements This study was financially supported by the Ministry of Education, Science and Technological Development of the Republic of Serbia (Contract No. 451-03-68/2022-14/200172).

The authors address special thanks to colleague Dr. Milovan Milivojević from Geographical Institute "Jovan Cvijić" of the Serbian Academy of Sciences and Arts for technical support in the preparation of the map.

Author Contributions All authors contributed equally to this work.

Conflict of Interest Authors declare that no conflict of interest.

Annex 1: CWQI Parameters with Limit Values

		Overall		Drinking		Aquatic		Recreation		Irrigation		Livestock	
Variables	Units	Lower	Upper	Lower	Upper	Lower	Upper	Lower	Upper	Lower	Upper	Lower	Upper
Color	TCU		15		15								15
Turbidity (Turb)	NTU		1		1								
Dissolved Oxygen (DO)	mg/l	9.5				9.5							
pH		6.5	8.5	6.5	8.5	6.5	9	5	9				
Calcium (Ca)	mg/l		1000										1000
Sodium (Na)	mg/l		200		200								
Sulphate (SO_4^{2-})	mg/l		500		500								1000
Chloride (Cl^-)	mg/l		110		250						110		
Fluoride (F^-)	mg/l		1		1.5		1.2				1		1
Nitrate, Nitrite (NO_3^- NO_2^-)	mg/l		100										100
Aluminum (Al)	mg/l		0.005				0.005				5		5
Arsenic (As)	mg/l		0.005		0.025		0.005				0.1		0.025
Barium (Ba)	mg/l		1		1								
Cadmium (Cd)	mg/l		0.005		0.005						0.0051		0.08
Cromium (Cr)	mg/l		0.001		0.05		0.001				0.0049		0.05
Copper (Cu)	mg/l		0.002		1		0.002				0.2		0.5
Iron (Fe)	mg/l		0.3		0.3		0.3				5		
Mercury (Hg)	μg/l		0.003		1		0.1						0.003
Manganese (Mn)	mg/l		0.05		0.05						0.2		
Molybdenum (Mo)	mg/l		0.073				0.073						
Nickel (Ni)	mg/l		0.025				0.025				0.2		1

(continued)

(continued)

		Overall		Drinking		Aquatic		Recreation		Irrigation		Livestock	
Variables	Units	Lower	Upper	Lower	Upper	Lower	Upper	Lower	Upper	Lower	Upper	Lower	Upper
Lead (Pb)	mg/l		0.001		0.01		0.001				0.02		0.05
Zinc (Zn)	mg/l		0.03		5		0.03				1		50

Source Adapted from [151]

Annex 2: AFWQI Parameter Limits

Parameter	Unit	Irrigation	Livestock
Aldicarb	μg/l	54.9	11
Aluminum (Al)	μg/l	5,000	5,000
Arsenic (As)	μg/l	100	25
Atrazine	μg/l	10	5
Beryllium (Be)	μg/l	100	100
Boron (B)	μg/l	6,000	5,000
Bromacil	μg/l	0.2	1,100
Bromoxynil	μg/l	0.33	11
Cadmium (Cd)	μg/l	5.1	80
Calcium (Ca)	μg/l		1,000,000
Captan	μg/l		13
Carbaryl	μg/l		1,100
Carbofuran	μg/l		45
Chloride (Cl^-)	μg/l	700,000	
1,2-Dichloroethane	μg/l		5
1,1,2-Trichloroethene (TCE)	μg/l		50
Chlorothalonil	μg/l	5.8	170
Chlorpyrifos	μg/l		24
Trivalent Chromium (Cr(III))	μg/l	4.9	50
Hexavalent Chromium (Cr(VI))	μg/l	8	50
Cobalt (Co)	μg/l	50	1,000
Coliforms, fecal (E. Coli)	in 100 ml	100	
Coliforms Total	in 100 ml	1,000	
Copper (Cu)	μg/l	1,000	5,000
Cyanazine	μg/l	0.5	10
Deltamethrin	μg/l		2.5
Dibromochloromethane	μg/l		100
Dicamba	μg/l	0.006	122
Dichlorobromomethane	μg/l		100
Dichloromethane	μg/l		50
Diclofop-methyl	μg/l	0.18	9
Diisopropanolamine (DIPA)	μg/l	2,000	
Dimethoate	μg/l		3
Dinoseb	μg/l	16	150
Ethylbenzene	μg/l		2.4

(continued)

(continued)

Parameter	Unit	Irrigation	Livestock
Fluoride (F^-)	μg/l	1,000	2,000
Glyphosate	μg/l		280
Hexachlorobenzene	μg/l		0.52
Iron (Fe)	μg/l	5000	
Lindane	μg/l		4
Linuron	μg/l	0.071	
Lithium (Li)	μg/l	2,500	
Manganese (Mn)	μg/l	200	
4-chloro-2-methyl-phenoxy (MCPA)	μg/l	0.025	25
Mercury (Hg)	μg/l		3
Metolachlor	μg/l	28	50
Metribuzin	μg/l	0.5	80
Molybdenum (Mo)	μg/l	50	500
Nickel (Ni)	μg/l	200	1,000
Nitrate + Nitrite ($NO_3^- + NO_2^-$)	μg/l		100,000
Nitrite (NO_2^-)	μg/l		10,000
Phenols (C_6H_6O)	μg/l		2
Phenoxy herbicides	μg/l		100
Picloram	μg/l		190
Selenium (Se)	μg/l	50	50
Simazine	μg/l	0.5	10
Sulfolane	μg/l	500	
Sulphate	μg/l		1,000,000
Tebuthiuron	μg/l	0.27	130
Tetrachloromethane	μg/l		5
Toluene	μg/l		24
Total Dissolved Solids (TDS)	μg/l	3,500,000	3,000,000
Triallate	μg/l		230
Tribromomethane	μg/l		100
Tributyltin	μg/l		250
Trichloromethane	μg/l		100
Tricyclohexiltyn	μg/l		250
Trifluralin	μg/l		45
Triphenyltin	μg/l		820
Uranium (U)	μg/l	10	200
Vanadium (V)	μg/l	100	100
Zinc (Zn)	μg/l	5,000	50,000

Source: Adapted from [152]

Annex 3: Threshold Values for the Parameters of the Ecological Status for Rivers Belonging to Type 1 in Serbia (for the Danube River)

Parameter	Unit	Prescribed maximum values between ecological status classes			
		I–II	II–III	III–IV	IV–V
Type 1					
Chemlical and physico-chemical parmeters of ecological status assessment					
pH value (pH)		6.5–8.5	6.5–8.5	6.5–8.5	<6.5 or >8.5
Dissolved oxygen (DO)	mg/l	8.5 2	7.0	5.0	4.0
Biochemical oxygen demand (BOD)	mg/l	2	5	8	20
Ammonium (NH_4-N)	mg/l	0.1	0,3	0,8	1.0
Orthophosphate (PO_4^{3-})	mg/l	0.02	0.1	0.2	0,.5
Biological parameters of ecological status assessment					
Saprobic index (S)		2.10	2.65	2.90	3.20
Microbioogical parameters of ecological status assessment					
Coliform bacteria (CB)	nb/100 ml	500	10,000	100,000	1,000,000

Source [39]

Annex 4: Limit Values of Pollutants in Surface Waters in Serbia

Parameter	Unit	Prescribed maximum values between classes				
		I	II	III	IV	V
Suspended solids (SS)	mg/l	25	25	–	–	–
Oxygen saturation (OS)	%	90–110	90–110	90–110	90–110	90–110
Chemical oxygen demand (COD)	mg O_2/l	5	10	20	50	>50
Nitrite (NO_2^{2-})	mg N/l	0.01	0.03	0.12	0.3	>0.3
Sulfate (SO_4^{2-})	mg/l	50	100	200	300	>300
Metals						
Iron (Fe)	μg/l	200	500	1000	2000	>2000
Manganese(Mn)	μg/l	50	100	300	1000	>1000

(continued)

(continued)

Parameter	Unit	Prescribed maximum values between classes				
		I	II	III	IV	V
Copper (Cu)	μg/l	5 ($T = 10$) 22 ($T = 50$) 40 ($T = 100$) 112 ($T = 300$)	5 ($T = 10$) 22 ($T = 50$) 40 ($T = 100$) 112 ($T = 300$)	500	1000	>1000

Source [40]

Annex 5: Limit Values for Priority and Priority Hazard Substances in Surface Waters in Serbia

Priority hazard substance	Average annual concentration (AAC) (μg/l)	Maximum allowed concentration (MAC) (μg/l)
Mercury (Hg)	–	0.07
Cadmium (Cd)	<0,08 (Class I) 0,08 (Class II) 0,09 (Class III) 0,15 (Class IV) 0,25 (Class V)	<0,45 (Class I) 0,45 (Class II) 0,6 (Class III) 0,9 (Class IV) 1,5 (Class V)
Nickel (Ni)	4	34

Source [41]

References

1. Pletterbauer F, Melcher A, Graf W (2018) Climate change impacts in riverine ecosystems. In: Schmutz S, Sendzimir J (eds) Riverine ecosystem management: science for governing towards a sustainable future. Springer Open, Cham, pp 203–223
2. Trautwein C, Pletterbauer F (2018) Land use. In: Schmutz S, Sendzimir J (eds) Riverine ecosystem management: science for governing towards a sustainable future. Springer Open, Cham, pp 241–252
3. Matić Bujagić I, Grujić S, Laušević M, Hofmann T, Micić V (2019) Emerging contaminants in sediment core from the Iron Gate I reservoir on the Danube River. Sci Total Environ 662:77–87. https://doi.org/10.1016/j.scitotenv.2019.01.205
4. Sendzimir J, Schmutz S (2018) Challenges in riverine ecosystem management. In: Schmutz S, Sendzimir J (eds) Riverine ecosystem management: science for governing towards a sustainable future. Springer Open, Cham, pp 1–16

5. Habersack H, Jäger E, Hauer C (2013) The status of the Danube River sediment regime and morphology as a basis for future basin management. Int J River Basin Manag 11(2):153–166. https://doi.org/10.1080/15715124.2013.815191
6. Habersack H, Hein T, Stanica A, Liska I, Mair R, Jäger E, Hauer C, Bradley C (2016) Challenges of river basin management: Current status of, and prospects for, the River Danube from a river engineering perspective. Sci Total Environ 543:828–845. https://doi.org/10.1016/j.scitotenv.2015.10.123
7. Vukovic D, Vukovic Z, Stankovic S (2014) The impact of the Danube Iron Gate Dam on heavy metal storage and sediment flux within the reservoir. CATENA 113:18–23. https://doi.org/10.1016/j.catena.2013.07.012
8. Hohensinner S, Hauer C, Muhar S (2018) River morphology, channelization, and habitat restoration. In: Schmutz S, Sendzimir J (eds) Riverine ecosystem management: science for governing towards a sustainable future. Springer Open, Cham, pp 41–65
9. Schmutz S, Moog O (2018) Dams: ecological impacts and management. In: Schmutz S, Sendzimir J (eds) Riverine ecosystem management: science for governing towards a sustainable future. Springer Open, Cham, pp 111–127
10. Hagemann L, Kašanin-Grubin M, Gajica G, Štrbac S, Šajnović A, Jovančićević B, Vasić N, Schwarzbauer J (2019) Four decades of organic anthropogenic pollution: a compilation for Djerdap Lake sediments. Serbia. Water Air Soil Pollut 230:246. https://doi.org/10.1007/s11270-019-4277-8
11. Weigelhofer G, Hein T, Bondar-Kunze E (2018) Phosphorus and nitrogen dynamics in riverine systems: human impacts and management options. In: Schmutz S, Sendzimir J (eds) Riverine ecosystem management: science for governing towards a sustainable future. Springer Open, Cham, pp 187–202
12. Haidvogl G (2018) Historic milestones of human river uses and ecological impacts. In: Schmutz S, Sendzimir J (eds) Riverine ecosystem management: science for governing towards a sustainable future. Springer Open, Cham, pp 19–39
13. Panić M, Urošev M, Milanović Pešić A, Miljanović D (2014) Hydropower development in Serbia: current state, perspectives and challenges. In: Burke L (ed) Hydroelectricity and power electronics: environmental impact, emerging technologies, and challenges. Nova Science Publishers, New York, pp 1–54
14. Greimel F, Schülting L, Graf W, Bondar-Kunze E, Auer S, Zeiringer B, Hauer C (2018) Hydropeaking impacts and mitigation. In: Schmutz S, Sendzimir J (eds) Riverine ecosystem management: science for governing towards a sustainable future. Springer Open, Cham, pp 91–110
15. Unfer G, Pinter K (2018) Recreational fisheries: the need for sustainability in fisheries management of alpine rivers. In: Schmutz S, Sendzimir J (eds) Riverine ecosystem management: science for governing towards a sustainable future. Springer Open, Cham, pp 253–270
16. International Commission for the Protection of the Danube River (ICPDR) (2015) The Danube River Basin Management Plan. https://www.icpdr.org/flowpaper/app/#page=1. Accessed 22 Apr 2022
17. Waidbacher H, Drexler S-S, Meulenbroek P (2018) Danube under pressure: hydropower rules the fish. In: Schmutz S, Sendzimir J (eds) Riverine ecosystem management: science for governing towards a sustainable future. Springer Open, Cham, pp. 473–489
18. International Commission for the Protection of the Danube River (ICPDR) (2020) Countries of the Danube Basin. https://www.icpdr.org/main/danube-basin/countries-danube-river-basin. Accessed 20 Nov 2020
19. Uredba o utvrđivanju Prostornog plana područja posebne namene međunarodnog vodnog puta E 80—Dunav (Panevropski koridor VII) (Decree on Determing the Spatial Plan of the Special Purpose Area of International Waterway E 80—the Danube (Pan-European Corridor VII), Official Gazette of the Republic of Serbia 14/2015
20. Directorate for Inland Waterways Plovput (2020) Navigation Chart of the Danube River in the Republic of Serbia (km 1433.1–km 845.5). Directorate for Inland Waterways Plovput, Belgrade. https://danubeportal.com/application/cache/uploads/charts/paper/RS/118/1560771184.pdf. Accessed 20 Nov 2020

21. Statistical Office of the Republic of Serbia (2011) Popis stanovništva, domaćinstava i stanova 2011. u Republici Srbiji — prvi rezultati: Knjiga 540 (2011 Census of Population, Households and Dwellings in the Republic of Serbia—First Results Book 540). http://media.popis2011.stat.rs/2011/prvi_rezultati.pdf. Accessed 20 Oct 2020
22. Milanović Pešić A (ed) (2015) Leksikon Nacionalnog parka Đerdap (Lexicon of the Đerdap National Park). Public Enterprise Official Gazette, Public Enterprise "Đerdap National Park", Geographical Institute "Jovan Cvijić" SASA, Belgrade
23. Gavrilović Lj, Dukić D (2014) Reke Srbije (Rivers of Serbia). Zavod za udžbenike i nastavna sredstva, Belgrade
24. Strategija upravljanja vodama na teritoriji Republike Srbije do 2034. godine (Water Management Strategy for the Territory of the Republic of Serbia until 2034), Official Gazette of the Republic of Serbia 3/2017
25. Stanković S (2005) Jezera Srbije (Lakes of Serbia). Zavod za udžbenike i nastavna sredstva, Belgrade
26. Javno preduzeće "Elektroprivreda Srbije" (2020) Public Enterprise "Elektroprivreda" Serbia. Branches. http://www.eps.rs/cir/Pages/Ogranci.aspx Accessed Dec 10 2020
27. Babic Mladenovic M, Kolarov V, Damjanovic V (2013) Sediment regime of the Danube River in Serbia. Int J Sediment Res 28(4):470–485. https://doi.org/10.1016/S1001-6279(14)60006-8
28. Urošev M, Kovačević-Majkić J, Štrbac D, Milanović Pešić A, Milijašević D, Jakovljević D, Petrović A (2017) Vode Srbije (Water of Serbia). In: Radovanović M (ed) Geografija Srbije (Geography of Serbia) (Special issues 91). Geographical Institute "Jovan Cvijić" SASA, Belgrade, pp 160–235
29. Zakon o nacionalnim parkovima (Law on National Parks), Official Gazette of the Republic of Serbia 84/2015, 95/2018
30. Public Enterprise Đerdap National Park (2020) About the Park. https://npdjerdap.rs/o-parku/ Accessed 20 Oct 2020
31. Amidžić L, Krasulja S, Belij S (eds) (2009) Protected natural areas in Serbia. Ministry of Environmental Protection, Institute for Nature Conservation of Serbia, Belgrade
32. Maran Stevanović A (2017) Activities on the establishment of Đerdap Geopark (Serbia) and candidature of the area to the UNESCO Global Geoparks Network. Bulletin Natl Hist Museum 10:7–28. https://doi.org/10.5937/bnhmb1710007M
33. Public Enterprise Đerdap National Park (2020) Područje Đerdap proglašeno za Ramsarsko područje (The "Đerdap" area designated as Ramsar site) https://npdjerdap.rs/podrucje-djerdap-proglaseno-za-ramsarsko-podrucje/. Accessed 20 Dec 2020
34. Institute for Nature Conservation of Serbia (2020) Đerdap designated as a the wetland of international importance. https://www.zzps.rs/wp/djerdap-proglasen-za-medjunarodno-znacajno-vlazno-podrucje-ramsarsko-podrucje/. Accessed 20 Dec 2020
35. Ministry of Environmental Protection (2020) Đerdap jedanaesto Ramsarsko područje u Republici Srbiji (Đerdap eleventh Ramsar site in Serbia) https://www.ekologija.gov.rs/saopstenja/vesti/djerdap-jedanaesto-ramsarsko-podrucje-u-republici-srbiji. Accessed 20 Dec 2020
36. DANUBEPARKS (2020) Associations https://www.danubeparks.org/association Accessed 20 Dec 2020
37. Zakon o vodama (Water Law), Official Gazette of the Republic of Serbia 30/2010, 93/2012, 101/2016, 95/2018, 95/2018
38. Uredba o graničnim vrednostima emisije zagađujućih materija u vode i rokovima za njihovo dostizanje (Regulation on Emission Limit Values of Pollutants in Water and Deadlines for Their Achievement), Official Gazette of the Republic of Serbia 67/2011, 48/2012, 1/2016
39. Pravilnik o parametrima ekološkog i hemijskog statusa površinskih voda i parametrima hemijskog i kvantitativnog statusa podzemnih voda (Rulebook on Parameters of the Ecological and Chemical Status of Surface Waters, and Parameters of the Chemical and Quantitative Status of Ground Waters), Official Gazette of the Republic of Serbia 74/2011
40. Uredba o graničnim vrednostima zagađujućih materija u površinskim i podzemnim vodama i sedimentu i rokovima za njihovo dostizanje (Regulation on Limit Values of Pollutants in

Surface and Ground Waters and Sediments and Deadlines for Their Achievement), Official Gazette of the Republic of Serbia 50/2012
41. Uredba o graničnim vrednostima prioritetnih i prioritenih hazardnih supstanci koje zagađuju površinske vode i rokovima za njihovo dostizanje (Regulation on Limit Values of Priority and Priority Hazardous Substances that Pollute the Surface Waters and Deadlines for Their Achievement), Official Gazette of the Republic of Serbia 24/2014
42. Pravilnik o načinu i uslovima za merenje količine i ispitivanje kvaliteta otpadnih voda i sadržini izveštaja o izvršenim merenjima (Rulebook on the Method and Conditions for Measuring the Quantity and Testing the Quality of Wastewater and the Content of the Report on the Performed Measurements), Official Gazette of the Republic of Serbia 33/2016
43. Odluka o određivanju granica vodnih područja (Decision on Determining the Boundaries of Water Areas), Official Gazette of the Republic of Serbia 92/2017
44. Pravilnik o utvrđivanju metodologije za izradu karte ugroženosti i karte rizika od poplava (Rulebook on Defining the Methodology for Designing of Vulnerability Map and Flood Risk Map), Official Gazette of the Republic of Serbia 13/2017
45. Ministry of Agriculture, Forestry and Water Management, Republic Water Directorate (2020) Podzakonska akta (Bylaws). http://www.rdvode.gov.rs/podzakonska-akta.php
46. Ministry of European Integration of the Republic of Serbia (2018) Nacionalni program za usvajanje pravnih tekovina Evropske Unije – treća revizija (National Programme for the Adoption of the *Acquis*—Third revision). https://www.mei.gov.rs/upload/documents/nacionalna_dokumenta/npaa/npaa_2018_2021.pdf. Accessed 20 Nov 2020
47. Nacionalna strategija za aproksimaciju u oblasti životne sredine za Republiku Srbiju (National Environmental Approximation Strategy for the Republic of Serbia), Official Gazette of the Republic of Serbia 80/2011
48. Ministry of Environment, Mining and Spatial Planning of the Republic of Serbia (2012) Strategija aproksimacije za sektor voda (Water Sector Approximation Strategy). Ministry of Environment, Mining and Spatial Planning of the Republic of Serbia, Belgrade
49. Veljković N, Petrović Z, Šotić A, Cibulić V (2018) Perspective on implementation of the EU Urban Waste Water Treatment Directive in Serbia). Voda i sanitarna tehnika 47(1):5–16. https://www.sepa.gov.rs/download/radovi/2018/UrbanWasteWaterTreatmentDirectiveInSerbia.pdf. Accessed 10 Oct 2020
50. Lekić D, Perunović Ćulić T (eds) (2019) Izveštaj od stanju životne sredine u Republici Srbiji za 2018 godinu (The Report on the Environmental State in the Republic of Serbia in 2018). http://www.sepa.gov.rs/download/Izvestaj2018.pdf. Accessed 10 Oct 2020
51. Drobnjaković M, Spalević A (2017) Naselja Srbije (The settlements in Serbia). In: Radovanović M (ed) Geografija Srbije (Geography of Serbia) (Special issues 91). Geographical Institute “Jovan Cvijić” SASA, Belgrade, pp 566–613
52. Živanović Z, Tošić B, Nikolić T, Gatarić D (2019) Urban system in Serbia—the factor in the planning of balanced regional development. Sustainability 11(15):4168. https://doi.org/10.3390/su11154168
53. Miljanović D, Miletić R, Đorđević J (2010) Regional inequality in Serbia as development problem. Acta Geogr Slov 50(2):253–275. https://doi.org/10.3986/AGS50204
54. Drobnjaković M, Panić M, Đorđević J (2015) Traditional undeveloped municipalities in Serbia as result of regional inequality. Eur Plan Stud 24(5):926–949. https://doi.org/10.1080/09654313.2015.1129396
55. Veljković N, Denić Lj, Jovičić M, Dopuđa Glišić T, Domanović M (2020) Komparativna analiza trendova koncentracija nitrata u vodotokovima Srbije (Comparative trend analysis nitrate concentrations in Serbian watercourses). Voda i sanitarna tehnika (Water Sanitary Technol) 60(1–2):25–29. http://www.sepa.gov.rs/download/radovi/2020/KOMPARATIVNA_ANALIZA_TRENDOVA_NITRATA-Nebojsa_Veljkovic.pdf. Accessed 10 Nov 2020
56. Zakon o zaštiti životne sredine (Law on Environmental Protection), Official Gazette of the Republic of Serbia 135/2004, 36/2009, 72/2009, 43/2011, 14/2016
57. Zakon o proceni uticaja na životnu sredinu (Law on Environmental Impact Assessment), Official Gazette of the Republic of Serbia 135/2004, 36/2009

58. Zakon o strateškoj proceni uticaja na životnu sredinu (Law on Strategic Environmental Impact Assessment), Official Gazette of the Republic of Serbia 135/2004, 88/2010
59. Zakon o integrisanom sprečavanju i kontroli zagađivanja životne sredine (Law on the Integrated Prevention and Control of Environmental Pollution), Official Gazette of the Republic of Serbia 135/2004, 25/2015
60. Miljanović D (2018) Geosistemske osnove upravljanja životnom sredinom (Geosystem foundations of environmental governance) (Unpublished Doctoral Dissertation). University of Belgrade, Belgrade, Faculty of Geography
61. Zakon o planiranju i izgradnji (Law on Planning and Construction), Official Gazette of the Republic of Serbia 72/2009, 81/2009, 64/2010, 24/2011, 121/2012, 42/2013, 50/2013, 98/2013, 132/2014, 145/2014, 83/2018, 31/2019, 37/2019, 9/2020
62. Kovačević-Majkić J, Panić M, Miljanović D, Miletić R (2014). Vulnerability to natural disasters in Serbia: spatial and temporal comparison. Nat Hazards 72(2):945–968. https://doi.org/10.1007/s11069-014-1045-3
63. Zakon o zaštiti prirode (Law on Nature Conservation), Official Gazette of the Republic of Serbia 36/2009, 88/2010, 91/2010, 14/2016, 95/2018
64. Zakon o zaštiti i održivom korišćenju ribljeg fonda (Law on Protection and Sustainable Use of Fish Resources), Official Gazette of the Republic of Serbia 118/2014, 95/2018
65. Zakon o javnom zdravlju (Law on Public Health), Official Gazette of the Republic of Serbia 15/2016
66. Zakon o plovidbi i lukama na unutrašnjim vodama (Law on Navigation and Ports on Inland Waterways), Official Gazette of the Republic of Serbia 73/2010, 121/2012, 18/2015, 96/2015, 96/2016, 104/2016, 113/2017, 41/2018, 95/2018, 37/2019, 9/2020
67. Zakon o energetici (Energy Law), Official Gazette of the Republic of Serbia 145/2014, 95/2018
68. Zakon o upravljanju otpadom (Law on Waste Management), Official Gazette of the Republic of Serbia 36/2009, 88/201 14/2016, 95/2018
69. Zakon o meteorološkoj i hidrološkoj delatnosti (Law on Meteorological and Hydrological Activities), Official Gazette of the Republic of Serbia 88/2010
70. Zakon o komunalnim delatnostima (Law on Communal Activities), Official Gazette of the Republic of Serbia 88/2011, 104/2016, 95/2018
71. Zakon o lokalnoj samoupravi (Law on Local Government), Official Gazette of the Republic of Serbia 129/2007, 83/2014, 101/2016, 47/2018
72. Zakon o smanjenju rizika od katastrofa i upravljanju vanrednim situacijama (Law on Disaster Risk Reduction and Emergency Situations Management), Official Gazette of the Republic of Serbia 87/2018
73. Zakon o ministarstvima (Law on Ministries), Official Gazette of the Republic of Serbia 128/2020
74. Zakon o ministarstvima (Law on Ministries), Official Gazette of the Republic of Serbia 16/2011
75. Veljković N, Denić Lj, Dopuđa Glišić T, Jovičić M, Milovanović J, Domanović M (2015) Reporting on surface water quality in Serbia based on national legislation and cooperation with European Union. Voda i sanitarna tehnika 45(3–4):23–32. http://www.sepa.gov.rs/download/radovi/2015/KvalitetPovrsinskihVodaSrbije_112015.pdf.
76. Javno vodoprivredno preduzeće (JVP) "Srbijavode" (Public Water Management Company (PWMC) "Srbijavode") (2017) Statut JVP "Srbijavode" Beograd (Statute PWMC "Srbijavode" Belgrade). http://www.srbijavode.rs/web/images/dokumenti/Statut-2017-godina.pdf. Accessed 1 Oct 2020
77. Institut za vodoprivredu "Jaroslav Černi" (2020). The Jaroslav Černi Institute for the Developmant of Water Resources (JCI). https://www.jcerni.rs/ Accessed 20 Oct 2020
78. Uredba o utvrđivanju vodoprivredne osnove Republike Srbije (Decree on Determining Water Management Plan of the Republic of Serbia), Official Gazette of the Republic of Serbia 11/2002
79. Ministry of Construction, Transport and Infrastructure of the Republic of Serbia (2021) Prostorni plan Republike Srbije od 2021 do 2035. godine - Nacrt (Spatial Plan of the Republic

of Serbia from 2021 to 2035—Draft). https://www.mgsi.gov.rs/sites/default/files/PPRS%20Nacrt.pdf
80. Nacionalna strategija održivog korišćenja prirodnih resursa i dobara (National Strategy of Sustainable Use of Natural Resources and Goods), Official Gazette of the Republic of Serbia 33/2012
81. Stanišić J, Maksić S (2014) Environmental education in Serbian primary schools: challenges and changes in curriculum, pedagogy, and teacher training. J Environ Educ 45(2):118–131. https://doi.org/10.1080/00958964.2013.829019
82. Stern MJ, Powell RB, Dawn H (2014) Environmental education program evaluation in the new millennium: what do we measure and what have we learned? Environ Educ Res 20(5):581–611. https://doi.org/10.1080/13504622.2013.838749
83. Kovačević-Majkić J, Milošević MV, Panić M, Miljanović D, Ćalić J (2014) Risk education in Serbia. Acta Geogr Slov 54(1):163–178. https://doi.org.https://doi.org/10.3986/AGS54305
84. Bosschaart A, van der Scheec J, Kuipere W, Schoonenboom J (2016) Evaluating a flood-risk education program in the Netherlands. Stud Educ Eval 50:53–61. https://doi.org/10.1016/j.stueduc.2016.07.0020
85. Statistical Office of the Republic of Serbia (2010–2020) Municipalities and regions in the Republic of Serbia, 2009–2019 https://www.stat.gov.rs/en-US/publikacije/?d=13&r=. Accessed 20 Oct 2020
86. United Nations development Programme (UNDP) (2006) Human Development Report 2006. Beyond Scarcity: Power, Poverty and the Global Water Crisis. New York: United Nations Development Programme. ISBN 0–230–50058–7. https://www.undp.org/content/undp/en/home/librarypage/hdr/human-development-report-2006.html. Accessed 20 Nov 2020
87. Prostorni plan opštine Golubac (Spatial Plan of the Golubac Municipality) (2011) Arhiplan d.o.o, Aranđelovac
88. Prostorni plan opštine Majdanpek (Spatial Plan of the Majdanpek Municipality) (2012) Direkcija za izgradnju, urbanizam i građevinsko zemljište Smederevo JP, Smederevo
89. Prostorni plan opštine Kladovo (Spatial Plan of the Kladovo Municipality) (2006) Centar za planiranje urbanog razvoja- CEP d.o.o, Beograd
90. Scientific Research Center (2007) Master plan turističke destinacije "Donje Podunavlje" (Master Plan for Tourist Destination Donje Podunavlje). https://mtt.gov.rs/download/sektor-za-turizam/master-planovi/MasterplanDonjePodunavlje.pdf. Accessed 10 Dec 2020
91. Jaroslav Černi Institute for the Development of Water Resources (JCI) (2001) Vodoprivredna osnova Republike Srbije (Water Management Plan of the Republic of Serbia). Jaroslav Černi Institute for the Development of Water Resources, Belgrade
92. Prostorni plan opštine Negotin (Spatial Plan of the Negotin Municipality) (2011) Faculty of Geography. University of Belgrade, Belgrade
93. Milanović Pešić A, Jakovljević D, Radovanović M (2017) Water supply issues in Serbia. In: Materials of the international scientific-practical conference "Water resources, hydraulic facilities and environment" (part I). "AZERSU" OJSC, Baku, pp 450–454
94. Public Enterprise Electric Power Industry of Serbia (2019) http://www.eps.rs/cir/Pages/Ogranci.aspx. Accessed 20 Nov 2020
95. Smederevac Lalić MM, Kalauzi AK, Regner SB, Lenhardt MB, Naunovic ZZ, Hegediš AE (2017) Prediction of fish catch in the Danube River based on long-term variability in environmental parameters and catch statistics. Sci Total Environ 609:664–671. https://doi.org/10.1016/j.scitotenv.2017.07.177
96. Brankov J (2020) Water resource management in national parks in Serbia—towards an integrated protection and sustainable tourism use. In: Milanović Pešić A, Jakovljević D (eds) Water resources management: methods, applications and challenges. Nova Science Publishers, New York, pp 195–230
97. Jovičić K, Višnjić Jeftić Ž, Jarić I, Subotić S, Hegediš A, Lenhardt M (2018) Literature survey on fish tissues contamination by heavy metals and elements in the Danube River from 1433–845 rkm. In: Poleksić V, Rašković B, Marković Z (eds) Conference Proceedings of the VIII International Conference Water & Sea. Faculty of Agriculture, University of Belgrade,

International Aquatic Veterinary Biosecurity Consortium Ludwig-Maximilians-University Munich, Belgrade, pp 386–393
98. Port Governance Agency (2019) Annual Bulletin. Nb.4, Marinas new development opportunity
99. Milanović Pešić A (2019) Identification and valorisation of Danube as a hydrological touristic attractiveness in Serbia—Example of National park Djerdap. In: Proceedings of the VII International scientific-practical conference "Innovative aspects of the development service and tourism". Faculty of Social and Cultural Service and Tourism, Stavropol State Agrarian University, Stavropol, pp 84–90
100. Belij M, Đurđić S, Stojković S (2018) The evaluation of geoheritage for geotourism development: case study on the potential Geopark Djerdap. Collection of papers—Faculty of Geography, University of Belgrade 66(2):121–132. https://doi.org/10.5937/zrgfub1802121B
101. Lukić D, Petrović MD, Vuković DB (2018) Potential of geotourism for regional development: the case of "Iron Gates" park in Serbia. R-ECONOMY 4(4):158–166. https://doi.org/10.15826/recon.2018.4.4.021
102. Milanović Pešić A, Brankov J, Milijašević Joksimović D (2020) Water quality assessment and populations' perceptions in the National park Djerdap (Serbia): key factors affecting the environment. Environ Dev Sustain 22:2365–2383. https://doi.org/10.1007/s10668-018-0295-8
103. Institute for Nature Conservation of Serbia (2020) Pregled zaštićenih područja u Republici Srbiji (Overview of protected areas in Republic of Serbia). https://www.zzps.rs/wp/pdf/centralniregistar/2020%203%20Pregled%20zasticenih%20podrucja%20Srbije.pdf. Accessed 20 Nov 2020
104. Terzić A (2014) Perspektive razvoja kulturne rute "Tvrđave na Dunavu" u funkciji obogaćivanja turističke ponude Srbije (Development perspectives of the cultural route "Fortress along the Danube" for enriching the tourist offer in Serbia) (Special Issues 88). Geographical Institute "Jovan Cvijić" SASA, Belgrade
105. Belij M (2017) Cultural tourist motives in national parks in Serbia. In: Filipović D, Gosar A, Koderman M, Đurđić S (eds) Tourism in protected areas of nature in Serbia and Slovenia. University of Belgrade, Belgrade, Faculty of Geography, pp 113–128
106. Lukić D, Petrović M (2020) Uloga objekata geonasleđa u turizmu Podunavlja Srbije (The role of geoheritage sites in tourism of Serbian Podunavlje Region) (Special Issues 97). Geographical Institute "Jovan Cvijić" SASA, Belgrade
107. Tomić N, Marković SB, Antić A, Tešić D (2020) Exploring the potential for geotourism development in the Danube region of Serbia. Int J Geoheritage Parks 8:123–139. https://doi.org/10.1016/j.ijgeop.2020.05.001
108. Seliger C, Zeiringer B (2018) River connectivity, habitat fragmentation and related restoration measures. In: Schmutz S, Sendzimir J (eds) Riverine ecosystem management: science for governing towards a sustainable future. Springer Open, Cham, pp 171–186
109. Macura B, Bojović P, Petrić I, Ćosić N, Tadić M, Jarić I, Knežević J, Šripić J, Jarić M (2013) Local communities and management of the Djerdap protected area in Serbia. In: Healy H, Martínez-Alier J, Temper L, Walter M, Gerber JF (eds) Ecological economics from the ground up. Routledge, Abingdon, pp 366–389
110. Lenhardt M, Višnjić-Jeftić Ž, Navodaru I, Jarić I, Vassilev M, Gačić Z, Nikčević M (2012) Fish stock management cooperation in the lower Danube Region: a case study of sturgeons and pontic shad. In: Lagutov V (ed) Environmental security in watersheds: the Sea of Azov. NATO science for peace and security series c: environmental security. Springer, Dordrecht, pp 127–140. https://doi.org/10.1007/978-94-007-2460-0_7
111. Smederevac Lalić M, Regner S, Lenhardt M, Nikolić D, Cvijanović G, Jaćimović M, Hegediš A (2019) Review of allochthonous fish species with the marine origine in Serbian freshwater system. Studia Marina 32(1):33–46. https://doi.org/10.5281/zenodo.3274548
112. Serbian Environmental Protection Agency (SEPA) (2014) PRTR registar (PRTR register) http://www.sepa.gov.rs/PRTR/search.html?lang=de. Accessed 10 Jan 2021

113. Faculty of Mining and Geology, University of Belgrade (2015) Elixir Prahovo, Industrija Hemijskih Proizvoda, DOO Prahovo—Non-Technical Summary of the Environmental Impact Assessment Study on the Phosphogypsum Storage. http://www.mmediu.ro/app/webroot/uploads/files/2015-04-08_Sumar.pdf. Accessed 10 Jan 2021
114. Brankov J, Milijašević D, Milanović A (2012) The assessment of the surface water quality using the Water Pollution Index: a case study of the Timok River (The Danube River Basin), Serbia. Arch of Environ Prot 38:49–61. https://doi.org/10.2478/v10265-012-0004-x
115. Šerbula S, Stanković V, Živković D, Kamberović Ž, Gorgievski M, Kalinović T (2016) Characteristics of wastewater streams within the Bor copper mine and their influence on pollution of the Timok River, Serbia. Mine Water Environ 35:480–485. https://doi.org/10.1007/s10230-016-0392-6
116. Milijašević D, Milanović A, Brankov J, Radovanović M (2011) Water quality assessment of the Borska river using the WPI (Water Pollution Index) method. Arch Biol Sci 63(3):819–824. https://doi.org/10.2298/ABS1103819M
117. Gardić VR, Petrović JV, Đurđevac-Ignjatović LV, Kolaković SR, Vujović SR (2015) Procena uticaja rudničkih drenažnih i komunalnih otpadnih voda na kvalitet površinskih voda u Boru i okolini (Impact assessment of mine drainage water and municipal wastewater on the surface water in the vicinity of Bor). Hem Ind 69(2):165–174. https://doi.org/10.2298/HEMIND140128031G
118. Prostorni plan područja posebne namene Nacionalnog parka Đerdap (Spatial Plan for the Special Purpose Area of the Đerdap National Park) (2013) Resource document. https://www.rapp.gov.rs/sr-Latn-CS/zasticena-i-turisticka-podrucja/cid294-83243/prostorni-plan-podrucja-posebne-namene-nacionalnog-parka-Đerdap. Accessed 10 Dec 2020
119. Republic Hydrometeorological Service of Serbia (2009–2018). Yearbook III-Water Quality (2009–2018). Republic Hydrometeorological Service of Serbia, Belgrade
120. Lumb A, Sharma TC, Bibeault JF (2011) A review of genesis and evolution of Water Quality Index (WQI) and some future directions. Water Qual Expo Health 3:11–24. https://doi.org/10.1007/s12403-011-0040-0
121. Karbassi AR, Mir Mohammad Hosseini F, Baghvand A, Nazariha M (2011) Development of Water Quality Index (WQI) for Gorganrood River. Int J Environ Res 5(4):1041–1046. https://doi.org/10.22059/ijer.2011.461
122. Rubio-Arias H, Contreras-Caraveo M, Quintana RM, Saucedo-Teran RA, Pinales-Munguia A (2012) An overall Water Quality Index (WQI) for a man-made aquatic reservoir in Mexico. Int J Environ Res Public Health 9:1687–1698. https://doi.org/10.3390/ijerph9051687
123. Iticescu C, Georgescu LP, Topa CM (2013) Assessing the Danube Water Quality Index in the City of Galati, Romania. Carpath J Earth Env 8(4):155–164. Accessed 10 Dec 2020
124. Venkatramanan S, Chung SY, Lee SY, Park N (2014) Assessment of river water quality via environmentric multivariate statistical tools and Water Quality Index: a case study of Nakdong River Basin, Korea. Carpath J Earth Env 9(2):125–132. http://www.ubm.ro/sites/CJEES/viewTopic.php?topicId=426. Accessed 10 Dec 2020
125. Şener Ş, Şener E, Davraz A (2017) Evaluation of water quality using Water Quality Index (WQI) method and GIS in Aksu River (SW Turkey). Sci Total Environ 584–585:131–144. https://doi.org/10.1016/j.scitotenv.2017.01.102
126. Calmuc M, Calmuc V, Arseni M, Topa C, Timofti M, Georgescu LP, Iticescu C (2020) A comparative approach to a series of physico-chemical quality indices used in assessing water quality in the Lower Danube. Water 12(11):3239. https://doi.org/10.3390/w12113239
127. Jakovljević D (2012) Serbian and Canadian Water Quality Index of Danube River in Serbia in 2010. J Geogr Inst Cvijic 62(3):1–18. https://doi.org/10.2298/IJGI1203001J
128. Veljković N (2013) Sustainable development indicators: case study for South Morava River Basin. Hem Ind 67(2):357–364. https://doi.org/10.2298/HEMIND111226059V
129. Jakovljević D (2015) Geoekološke determinante zaštite i revitalizacije tekućih voda u funkciji održivog razvoja AP Vojvodine (Geoecological determinants of the protection and revitalization of water courses in the terms of the sustainable development of AP Vojvodina) (Unpublished Doctoral Dissertation). Faculty of Geography, University of Belgrade, Belgrade

130. Walker D, Jakovljević D, Savić D, Radovanović M (2015) Multi-criterion water quality analysis of the Danube River in Serbia: A visualisation approach. Water Res 79:158–172. https://doi.org/10.1016/j.watres.2015.03.020
131. Pantelić M, Dolinaj D, Savić S, Leščešen I, Stojanović V (2016) Water quality and population standpoints as factors influencing the utilization for agricultural purposes of the Great Backa Canal, Serbia. J Environ Sci Manage 19(2):8–14. https://ovcre.uplb.edu.ph/journals-uplb/index.php/JESAM/article/view/141/123
132. Milijašević Joksimović D, Gavrilović B, Lović Obradović S (2018) Application of the Water Quality Index in the Timok River Basin (Serbia). J Geogr Inst Cvijic 68(3):333–344. https://doi.org/10.2298/IJGI180610007M
133. Milanović Pešić A, Jakovljević D, Milijašević Joksimović D (2020) Water resources management in Serbia. In: Milanović Pešić A, Jakovljević D (eds) Water resources management: methods, applications and challenges. Nova Science Publishers, New York, pp 231–299
134. Hurley T, Sadiq R, Mazumder A (2012) Adaptation and evaluation of the Canadian Council of Ministers of the Environment Water Quality Index (CCME WQI) for use as an effective tool to characterize drinking source water quality. Water Res 46:3544–3552. https://doi.org/10.1016/j.watres.2012.03.061
135. Jakovljević D, Lozanov-Crvenković Z (2015) Water quality changes after Kraljevo earthquake in 2010. Nat Hazards 79(3):2033–2053. https://doi.org/10.1007/s11069-015-1943-z
136. Jakovljević D (2020) Assessment of water quality during the floods in May, 2014. Serbia. J Geogr Inst Cvijic 70(3):215–226. https://doi.org/10.2298/IJGI2003215J
137. Blessing NRW, Benedict S (2017) Computing principles in formulating water quality informatics and indexing methods: an ample review. J Comput Theor Nanosci 14(4):1671–1681. https://doi.org/10.1166/jctn.2017.6531
138. Milanović A, Milijašević D, Brankov J (2011) Assessment of polluting effects and surface water quality using water pollution index: a case study of Hydro-system Danube-Tisa-Danube. Serbia. Carpath J Earth Env 6(2):269–277
139. Nachiyunde K, Ikeda H, Tanaka K, Kozaki D (2013) Evaluation of portable water in five provinces of Zambia using a Water Pollution Index. Afr J Environl Sci Technol 7(1):14–29. https://www.ajol.info/index.php/ajest/article/download/88667/78259
140. Al Obaidy AHMJ, Talib AH, Zaki SR (2015) Application of Water Pollution Index for assessment of Tigris River ecosystem. Int J Adv Res 3(2):219–223
141. Proskurnin OA, Komaristaja BN, Smirnova SA (2015) Calculation of the surface Water Pollution Index in the evaluation of environmental component of product life cycle. ScienceRise 5/2(10):32–35. https://doi.org/10.15587/2313-8416.2015.42379
142. Wang Q, Wu X, Zhao B, Qin J, Peng T (2015) Combined multivariate statistical techniques, Water Pollution Index (WPI), and Daniel trend test methods to evaluate temporal and spatial variations and trends of water quality at Schanchong River in the Northwest Basin of Lake Fuxian, China. PLoS ONE 10(4):e0118590. https://doi.org/10.1371/journal.pone.0118590
143. Popović N, Đuknić J, Čanak Atlagić J, Raković M, Marinković N, Tubić B, Paunović M (2016) Assessment of the ecological status of rivers: a case study of the Sava River. Serbia. Acta Zool Bulgar 68(1):97–102
144. Li R, Zou Z, An Y (2016) Water quality assessment in Qu River based on fuzzy Water Pollution Index method. J Environ Sci 50:87–92. https://doi.org/10.1016/j.jes.2016.03.030
145. Takić Lj, Mladenović-Ranisavljević I, Vasović D, Đorđević Lj (2017) The assessment of the Danube River water pollution in Serbia. Water Air Soil Pollut 228:380. https://doi.org/10.1007/s11270-017-3551-x
146. Hassan AA, Dawood AS, Al-Mansori NJ (2018) Assessment of water quality of Shatt Al-Basrah Canal using Water Pollution Index. Int J Eng Technol 7(4.19):757–762. https://doi.org/10.14419/ijet.v7i4.19.27994
147. Hossain M, Patra PK (2020) Water Pollution Index—a new integrated approach to rank water quality. Ecol Indic 117:106668. https://doi.org/10.1016/j.ecolind.2020.106668
148. Veljković N, Lekić D, Jovičić M, Pajčin N (2010) Internet application Serbian Water Quality Index. Voda i sanitarna tehnika 40(3):15–18

149. Stojanović Z, Deršek Timotić I, Novaković B, Veljković N (2018) Surface water quality assessment in Serbia—Water Quality Index and ecological status comparison. Int J Mod Eng Res 8(10):7–14
150. Veljković N, Perunović T, Jovičić M, Denić Lj, Dopuđa Glišić T, Redžić N, Radić S (2020) Indikator urbane održivosti na nacionalnom nivou – studija slučaja za Srbiju (Urban sustainable indicator at the national level—case study for Serbia). Voda i sanitarna tehnika 50(3–4):47–52. http://www.sepa.gov.rs/download/radovi/2020/IndikatorUrbaneOdrzivostiNV.pdf. Accessed 20 Nov 2020
151. Canadian Council of Ministers of the Environment (CCME) (2005) Canadian Water Quality Index 1.0 calculator. http://www.env.gov.nl.ca/env/waterres/quality/background/cwqi.html.Accessed 20 Oct 2020
152. Canadian Council of Ministers of the Environment (CCME) (2008) Agri-food Water Quality Index 1.0. calculator. https://www.gov.nl.ca/eccm/waterres/quality/background/afwqi/. Accessed 20 Oct 2020
153. Water Resources Management Division (WRMD) (2008) Agri-food Water Quality Index (AFWQI) Help manual. Water Resources Management Division, Department of Environment and Conservation, Newfoundland, Labrador https://www.gov.nl.ca/eccm/files/waterres-quality-background-afwqi.pdf. Accessed 20 Oct 2020
154. Lyulko I, Ambalova T, Vasiljeva T (2001) To integrated water quality assessment in Latvia. International workshop on information for sustainable water management, MTM (Monitoring Tailor-Made) III, Netherlands, 449–452
155. Filatov N, Pozdnyakov D, Johannessen O, Pettersson L, Bobylev L (2005) White Sea: its marine environment and ecosystem dynamics influnced by global change. Springer & Praxis Publishing, Chichester
156. Šerbula S, Ristić A, Manasijević S, Doli N (2015) Teški metali u otpadnim vodama Rudnika bakra Majdanpek (Heavy metal ions in the wastewater of the Majdanpek Copper Mine). Zaštita materijala 56(1):52–58. https://scindeks-clanci.ceon.rs/data/pdf/0351-9465/2015/0351-94651501052S.pdf
157. Bogdanović GB, Trumić MŽ, Stanković V, Antić DV, Trumić MS, Milanović Z (2013) Rudničke vode iz rudnika RTB Bor — resurs za dobijanje bakra ili zagađivač životne sredine (Mine waters from Mining & Smelting Basin Bor—a resource of the recovery of copper or polluter of the environment). Reciklaža i održivi razvoj 6:41–50. www.ror.tf.bor.ac.rs
158. Vuković Ž, Vuković D, Radenković M, Stanković S (2012) A new approach to the analysis of the accumulation and enrichment of heavy metals in the Danube River sediment along the Iron Gate reservoir in Serbia. J Serb Chem Soc 77(3):381–392. https://doi.org/10.2298/JSC110217169V
159. Voza D, Vukovic M, Takić Lj, Nikolić Dj, Mladenovic-Ranisavljevic I (2015) Application of multivariate statistical techniques in the water quality assessment of Danube river. Serbia Arch Environ Prot 41(4):96–103
160. Milanović A, Kovačević-Majkić J, Milivojević M (2010) Water quality analysis of Danube River in Serbia—pollution and protection problems. Bull Serb Geogr Soc 90(2):47–68. https://doi.org/10.2298/GSGD1002047M
161. Dogterom J (2001) Assessment practices and environmental status of 10 transboundary rivers in Europe. International Water Assessment Centre under the UNECE Convention on Protection and Use of Transboundary Watercourses and International Lakes. Helsinki, 1992. International Water Assessment Centre/riza (Pub.).
162. International Commission for the Protection of the Danube River (ICPDR) (2006) Management problems of sediment quality and quantity in the Danube River Basin, Draft Issue Paper. International Commission for the Protection of the Danube River, Vienna
163. Urošev M, Milanović Pešić A, Kovačević Majkić J (2013) Биогенная нагрузка Дуная и ее основных притоков в Сербии (Assessment of nutrient load on Danube River and its main tributaries in Serbia). In: Sobol SV (ed) Proceedings from the 15th International scientific and industrial forum "Great Rivers" 2013 (Volume 1). NNGASU, N. Novgorod, pp 48–52

164. Ismail A, Robescu D (2017) Chemical water quality assessment of the Danube River in the lower course using water quality indices. U.P.B. Sci Bull Ser B 79(4):51–62. https://www.scientificbulletin.upb.ro/rev_docs_arhiva/rez90c_587393.pdf Accessed 20 Oct 2020
165. Kašanin-Grubin M, Hagemann L, Gajica G, Štrbac S, Jovančićević B, Vasić N, Šajnović A, Djogo Mračević S, Schwarzbauer J (2020) Geochronological investigation of the Danube Djerdap Lake sediments (Serbia): sedimentology and inorganic composition. Environ Geochem Health 42:693–707. https://doi.org/10.1007/s10653-019-00403-6
166. International Commission for the Protection of the Danube River (ICPDR) (2018) ICPDR Sturgeon Strategy. https://www.icpdr.org/main/sites/default/files/nodes/documents/om-20_-_3.4_icpdr_sturgeon_strategy_endorsed_version_final.pdf. Accessed 20 Oct 2020.
167. International Commission for the Protection of the Danube River (ICPDR) (2020) Restoring fish migration routes in the Danube River Basin—feasibility study on Iron Gate dams commissioned to explore joint solutions. https://www.icpdr.org/main/restoring-fish-migration-routes-danube-river-basin-feasibility-study-iron-gate-dams-commissioned. Accessed 20 Nov 2020
168. WePass (2020) Facilitating Fish Migration and Conservation at the Iron Gates. https://www.we-pass.org. Accessed 20 Nov 2020
169. Favilli F, Hoffmann C, Elmi M, Ravazzoli E, Streifeneder T (2015) The BioREGIO Carpathians project: aims, methodology and results from the "Continuity and Connectivity" analysis. Nat Conserv 11:95–111. https://doi.org/10.3897/natureconservation.11.4424
170. International Commission for the Protection of the Danube River (ICPDR) (2013) Guiding Principles for Sustainable Development of Hydropower in the Danube Basin. https://www.icpdr.org/flowpaper/viewer/default/files/nodes/documents/icpdr_hydropower_final.pdf. Accessed 20 Nov 2020

Climate and Water Related Hazards

Using Köppen Climate Classification Like Diagnostic Tool to Quantify Climate Variation in Lower Danube Valley for the Period 1961–2017

Hristo Popov

Abstract This chapter present climate condition in Lower Danube Valley in North Bulgaria. It used average monthly and average annual data on air temperature and precipitation from 10 meteorological stations located evenly on the study territory. In order to reveal the dynamics of climate change, are data calculated 30-year, 20-year, 10-year moving averages, which determined the affiliation to climate types according to the Köppen classification. The climatic types for each individual year are also calculated. The results reveals of dry steppe climate (BSk) in the eastern and central parts of the study area for some of the years. The presence of Mediterranean influence has been registered, which is defined by the Csa and Csb indices. At the same time, the Mediterranean influence is not revealed when use the 30- and 20- years moving average. When use 10- year moving average the presence of Mediterranean influence is established in only 2 stations for the period in second half of 80's and first half of 90's of the twentieth century. According to the 10-year moving averages in the Knezha region, we register a continental climate with a cold winter (Dfa) for the period of the 60s and early 70s. According to Annual Climate Type (ACT) study area is under Potential Aridity Condition (PAC) for half of study period.

Keywords Annual Climate Types (ACT) · Potential Aridity Condition (PAC) · Moving averages · Climate changes · Köppen classification · Lower Danube Valley

1 Introduction

Classification of climates has been the subject of numerous studies since the end of nineteenth century. Scientists usually have been using two types of classifications: statistical classifications [1–4] and genetic classifications [5–7]. Classifications of Köppen [1, 2], De Martonne [3] and Viers [4] are generally based on average data of temperature and precipitation for periods over 30 years and define thresholds corresponding to large biogeographic and bioclimatic domains; for this reason, they

H. Popov (✉)
Faculty of Geology and Geography, Department of Climatology, Hydrology and Geomorphology, Sofia University St. Kliment Ohridski, Sofia, Bulgaria
e-mail: hpopov@gea.uni-sofia.bg

A. Negm et al. (eds.), *The Lower Danube River*, Earth and Environmental Sciences Library, https://doi.org/10.1007/978-3-031-03865-5_8

are also often referred to as physiognomic. The second group—classifications of Alissow [5], Pédelaborde [6] and Flohn [7] defines climates by the succession of the usual types of weather, that is to say the dynamic characteristics of the climate resulting from the combinations between the different meteorological parameters. The first group is criticized for a too reductive view of the climate, the average being, by definition "the value that never occurs", and the second group for a somewhat abstract or complex presentation of the climate as so many instantaneous states of the atmosphere in a given place [8–11].

Among the classifications of climates, the one of W.P. Köppen is the most famous and one of the most used in the world. It was developed in 1900 from the global vegetation maps available at the time, the author seeking to match climatic data with the boundaries of major plant domains.

It is often regarded as the first attempt at a quantitative climatic classification of climate together with of Thornthwaite [12] and Trewartha [13]. A simple combination of monthly average temperature and precipitation data is quickly applicable and generalizable. Hantel [14], Essenwanger [15] and Peel [16] have proposed adaptations of the thresholds and limits proposed by W. Köppen and, more recently, many studies have taken up this way of approaching the climate to illustrate its current and future evolution. Kottek et al. [17] made a updated World Map of the Köppen-Geiger climate classification. Rubel et al. [18] observed and projected climate shifts for the period 1901–2100. Beck et al. [19] present current and future Köppen-Geiger climate classification maps at 1-km resolution. Allam et al. [20] and Gallardo et al. [21] used Köppen—Trewartha classification in different scenarious of regional climate models. Lohmann et al. [22] used Köppen climate classification as a diagnostic tool for represent results from general circulation models. Taking up the thresholds and limits proposed by Köppen and his successors, several authors have also tried to correct the static aspect of this classification by applying it not to climatic 30 years averages, but for decadal averages or for each year. Chen and Chen [23] used Köppen classification to compare spatially stability of main climate types on three times scales, interannual, interdecadal and 30 years, for the period 1901–2010. The expression "Annual Climate Type" (ACT) has thus been used to define the climatic atmosphere of a given year [24]. The classification of W. Köppen has thus already been applied by Planchon and Rosier [25] to define these ACT in the Argentine Northwest, Quénol et al. [26] and Eveno et al. [27] in France and Dubreuil et al. in Brazil [11].

In Bulgaria there are many studies about precipitation, air temperatures, aridity and hydro-meteorological drought [28–31] which object is Danube Valley. Chenkova and Nikolova [28] observed air temperature and precipitation variability in North-eastern Bulgaria. Vlăduţ et al. [29] research aridity in southern Romania and northern Bulgaria and thermal continentality in same areas [30]. Few authors used Köppen classification in their study. Kirov [32] used Köppen classification together with other classifications to describe the climate in Bulgaria. Topliyski [33] used Köppen classification calculated monthly averages for the period 1931–1970. Popov [34] studies climate changes in Struma and Mesta valleys using Köppen climate types and calculate 30 years moving averages for the period 1931–2012.

The aim of the research is to analyze the ACT, which composite longterm climate variability in Lower Danube Valley.

2 Data and Methods

This research uses temperature and precipitation monthly data for the period 1961–2017 (57 years) from 10 meteorological stations located in the Bulgarian part of Danube valley (Table 1).

Five stations are located on the river floodplain—Vidin, Lom, Oryahovo, Ruse and Silistra. Another 5 stations are in the catchment area—Vratza, Kneja, Pleven, Razgrad and Shumen. There are 4 stations with data gaps, but there are data for period of at least 43 years (shortest period of observation is in Kneja 1961–2003). Silistra is the other station with temperature data gaps. There are missing temperature data for the period 1987–1993 and missing precipitation data after 2013.

Monthly data are used to calculate 30 years, 20 years and 10 years moving average of the temperature and precipitation. Annual Climate Type (ACT) using Köppen climate classification is used to calculate frequency of every climate type.

Using the results we define Potential Aridity Condition (PAC) index to describe climate conditions in study area.

The Köppen climate classification consists of five major groups. In study area are present only 3 of them. There are a number of sub-types under each major group, as listed in Table 2. While all the major groups except B are determined by temperature only. All the sub-types in current chapter are decided based on the combined criteria relating to seasonal temperature and precipitation. Therefore, the

Table 1 Coordinates and period of observation of meteorological stations used in study

Station	Latitude N (°)	Longtide E (°)	Altitude (m)	Period of observation
Vidin	43.59	22.51	35	1961–2017
Lom	43.49	23.14	35	1961–2013
Vratza	42.12	23.32	358	1961–2017
Oryahovo	43.44	23.58	124	1961–2017
Kneja	43.30	24.05	120	1961–2003
Pleven	43.20	24.35	163	1961–2017
Razgrad	43.31	26.14	206	1961–2017
Ruse	43.84	25.95	44	1961–2017
Shumen	43.16	26.56	216	1961–2010
Silistra	44.11	27.27	16	1961–2013 for precipitation 1961–1986 & 1994–2013 temperature

Table 2 Köppen climate classification scheme symbols description table. 1st—main climate type, 2nd—Subtype for B (arid) climate or precipitation regime type for C (temperate) climate and D (continental) climate, 3rd—subtype based on the temperature regime

1st	2nd	3rd	Type
B (Arid)	S (Steppe)	h (Hot)	BSk—Cold steppe climate
		k (Cold)	
C (Temperate)	w (Dry winter)		Cwa—Dry-winter humid subtropical climate
	f (No dry season)		Csa—Mediterranean with hot summer
	s (Dry summer)		Cfa—Subtropical humid
		a (Hot summer)	Cwb—Dry-winter subtropical highland climate
		b (Warm summer)	Csb—Mediterranean with warm summer
		c (Cold summer)	Cfb—Temperate Oceanic climate
D (Continental)	w (Dry winter)		Dwa—Hot summer continental climates
	f (No dry season)		Dsa—Hot and dry summer continental climates
	s (Dry summer)		Dfa—Hot-summer humid continental climate
		a (Hot summer)	Dwb—Warm summer continental or hemiboreal climates
		b (Warm summer)	Dsb—Warm and dry summer continental climates
		c (Cold summer)	Dfb—Warm humid continental climates

classification scheme as a whole represents different climate regimes of various temperature and precipitation combinations.

The dry climate B is determined by the annual mean precipitation and temperature, as well as the annual cycle of precipitation. Different sub-types distinguish between arid (desert) and semi-arid areas and further seasonal difference in precipitation conditions. The mild temperate C represents the climate with the lowest monthly mean temperature between −3 °C and +18 °C, while the different seasonal precipitations give rise to the four sub-types. The snow climate D has the lowest monthly mean temperature equal or lower than −3 °C, where as the sub types are decided based on the seasonal precipitation. Finally the polar climate E has the highest monthly mean temperature equal or lower than +10 °C, and the two sub-types further divide the major group into two temperature conditions. The last major type is tipical for subpolar and polar latitude and high mountains which are not part of study area.

3 Results and Discussion

When are used 30-years moving average, the study area obviously could be separated by two subareas. Territories in the river's floodplain from Vidin to Silistra absolutely (100%) belong to the "Cfa" climate type. Central part of the Danube valley, which is between the Osam River and the Skat River, is described as a humid subtropical climate.

The second subarea its more variable according to the Köppen climate classification. It includes territories which are not so close to the river. In the western part the humid subtropical climate (Cfa) still have great influence (67.9%). But a temperate oceanic climate type (32.1%), which is typical for Western Europe, have been found here. The eastern part of the Danube valley, which includes Ludogorie and Shumen plateaus, mainly belongs to the temperate oceanic climate type (Cfb) (90–96%). The tendency to increase summer temperatures leads to a slow shift of the climate type to a subtropical humid (Cfa) for the last 20–30 years (Tables 2 and 3).

Using 20-years moving average the data are showing some changes. The area which fully (100%) belongs to a humid subtropical climate (Cfa) decreased its territory in the western part of the valley (around the Skat River—Kneja station) and increased its territory in the eastern part (Cfa 8–29%). The area around Ludogorie and Shumen plateaus is more influenced by subtropical humid climate (Cfa).

Some of the areas, which according to the 30-years moving average are entirely in the subtropical humid climate zone, are transiting (due to rising summer temperatures) from Temperate Oceanic Cfb to the hot summer variant of Cfa (subtropical humid). The eastern part of the Danube valley could also be noticed warming and transition to the subtropical humid climate. For the region at the foot of the Shumen Plateau this transition is established after 1983, and at the foot of the Ludogorie Plateau after 1996. On the border with the Fore-Balkan zone, in the western part of the valley, the transition from temperate oceanic to subtropical humid climate shifted from 1970 to 1975 (Tables 2 and 4).

We can get even more detailed picture when analyzing the data from the 10-years moving averages (Tables 2 and 5). The Danube lowlands from Oryahovo/Lom to Silistra remain for the whole period in the zone of the subtropical humid climates (Cfa).

In the western part of the Danube lowlands (Vidin region) we register periods with temperate oceanic (Cfb) (1967–1979) and mediterranean climate (Csa) (1984–1993).

In the border area with the Fore-Balkans in the western part of the Danube plain, we register a transition from Cfb to Cfa after 1979. The Knezha region has calculated a transition from Temperate continental Dfa to Temperate oceanic Cfb, and after 1979 to Subtropical humid Cfa climate. The territories at the foot of the Ludogorie plateau went to a Subtropical humid climate Cfa after 1998, and at the foot of the Shumen plateau this change is registered after 1983 (Table 5).

The classification of the data for ACT gives us the greatest detail. For each station the number of climatic subtypes varies between 7 and 12. For the Danube lowlands

Table 3 30-years moving average of Köppen Climatic index in Lower Danube Valley for the period 1961–2017

	Vidin	Oryahovo	Lom	Ruse	Silistra	Vratza	Kneza	Pleven	Razgrad	Shumen
1961–1990	Cfa	Cfa	Cfa	Cfa	Cfa	Cfb	Cfa	Cfa	Cfb	Cfb
1962–1991	Cfa	Cfa	Cfa	Cfa	Cfa	Cfb	Cfa	Cfa	Cfb	Cfb
1963–1992	Cfa	Cfa	Cfa	Cfa	Cfa	Cfb	Cfa	Cfa	Cfb	Cfb
1964–1993	Cfa	Cfa	Cfa	Cfa	Cfa	Cfb	Cfa	Cfa	Cfb	Cfb
1965–1994	Cfa	Cfa	Cfa	Cfa	Cfa	Cfb	Cfa	Cfa	Cfb	Cfb
1966–1995	Cfa	Cfa	Cfa	Cfa	Cfa	Cfb	Cfa	Cfa	Cfb	Cfb
1967–1996	Cfa	Cfa	Cfa	Cfa	Cfa	Cfb	Cfa	Cfa	Cfb	Cfb
1968–1997	Cfa	Cfa	Cfa	Cfa	Cfa	Cfb	Cfa	Cfa	Cfb	Cfb
1969–1998	Cfa	Cfa	Cfa	Cfa	Cfa	Cfb	Cfa	Cfa	Cfb	Cfb
1970–1999	Cfa	Cfa	Cfa	Cfa	Cfa	Cfa	Cfa	Cfa	Cfb	Cfb
1971–2000	Cfa	Cfa	Cfa	Cfa	Cfa	Cfa	Cfa	Cfa	Cfb	Cfb
1972–2001	Cfa	Cfa	Cfa	Cfa	Cfa	Cfa	Cfa	Cfa	Cfb	Cfb
1973–2002	Cfa	Cfa	Cfa	Cfa	Cfa	Cfa	Cfa	Cfa	Cfb	Cfb
1974–2003	Cfa	Cfa	Cfa	Cfa	Cfa	Cfa	Cfa	Cfa	Cfb	Cfb
1975–2004	Cfa	Cfa	Cfa	Cfa	Cfa	Cfa	No data	Cfa	Cfb	Cfb
1976–2005	Cfa	Cfa	Cfa	Cfa	Cfa	Cfa	No data	Cfa	Cfb	Cfb
1977–2006	Cfa	Cfa	Cfa	Cfa	Cfa	Cfa	No data	Cfa	Cfb	Cfb
1978–2007	Cfa	Cfa	Cfa	Cfa	Cfa	Cfa	No data	Cfa	Cfb	Cfb
1979–2008	Cfa	Cfa	Cfa	Cfa	Cfa	Cfa	No data	Cfa	Cfb	Cfb
1980–2009	Cfa	Cfa	Cfa	Cfa	Cfa	Cfa	No data	Cfa	Cfb	Cfa
1981–2010	Cfa	Cfa	Cfa	Cfa	Cfa	Cfa	No data	Cfa	Cfb	Cfa
1982–2011	Cfa	Cfa	Cfa	Cfa	Cfa	Cfa	No data	Cfa	Cfb	No data
1983–2012	Cfa	Cfa	Cfa	Cfa	Cfa	Cfa	No data	Cfa	Cfb	No data
1984–2013	Cfa	Cfa	Cfa	Cfa	Cfa	Cfa	No data	Cfa	Cfb	No data
1985–2014	Cfa	No data	No data	Cfa	Cfa	Cfa	No data	Cfa	Cfb	No data
1986–2015	Cfa	No data	No data	Cfa	Cfa	Cfa	No data	Cfa	Cfb	No data
1987–2016	Cfa	No data	No data	Cfa	Cfa	Cfa	No data	Cfa	Cfa	No data
1988–2017	Cfa	No data	No data	Cfa	Cfa	Cfa	No data	Cfa	Cfb	No data

the variety of climatic subtypes is less—between 7 and 10. For the plain areas far from the river the number of subtypes for ACT is between 10 and 12.

The presence of dry years with type "B" is more typical for the territories of the Danube lowlands and decreases out of the river floodplain. The largest number of dry years is in the eastern part. For the foothills of the Shumen Plateau there are no dry years according to the climate indices of Köppen (Figs. 1 and 2).

For the Danube lowlands, the distribution of ACT is dominated by Cfa & Csa. As for the Kozloduy lowland Csa is dominant. For the whole study area, the Mediterranean climate Csa remains between 12.3 and 38.5% of the study period. The climate of the cold steppes ACT is present everywhere except at the foot of the Shumen plateau (Figs. 3 and 4).

The analysis of the main types of climate reveals the presence of cold "D" climates for the whole territory. For the Danube lowlands the percentage of years with cold winters varies between 8 and 24%, and for the plain part far from the river is between 16 and 47%. Cold or snow climates "D" is present mainly by two sub-types—Hot-summer humid continental climate (Dfa) and Hot and dry summer continental climates (Dsa). If compare subtypes climate continental climate with dry and hot summer (Dsa) prevail (16% of ACT) in Knezha region.

Table 4 20-years moving average of Köppen Climatic index in Lower Danube Valley for the period 1961–2017

	Vidin	Oryahovo	Lom	Ruse	Silistra	Vratza	Kneza	Pleven	Razgrad	Shumen
1961–1980	Cfa	Cfa	Cfa	Cfa	Cfa	Cfb	Cfb	Cfa	Cfb	Cfb
1962–1981	Cfa	Cfa	Cfa	Cfa	Cfa	Cfb	Cfb	Cfa	Cfb	Cfb
1963–1982	Cfa	Cfa	Cfa	Cfa	Cfa	Cfb	Cfb	Cfa	Cfb	Cfb
1964–1983	Cfa	Cfa	Cfa	Cfa	Cfa	Cfb	Cfb	Cfa	Cfb	Cfb
1965–1984	Cfa	Cfa	Cfa	Cfa	Cfa	Cfb	Cfb	Cfa	Cfb	Cfb
1966–1985	Cfa	Cfa	Cfa	Cfa	Cfa	Cfb	Cfb	Cfa	Cfb	Cfb
1967–1986	Cfa	Cfa	Cfa	Cfa	Cfa	Cfb	Cfb	Cfa	Cfb	Cfb
1968–1987	Cfa	Cfa	Cfa	Cfa	Cfa	Cfb	Cfb	Cfa	Cfb	Cfb
1969–1988	Cfa	Cfa	Cfa	Cfa	Cfa	Cfb	Cfb	Cfa	Cfb	Cfb
1970–1989	Cfa	Cfa	Cfa	Cfa	Cfa	Cfb	Cfa	Cfa	Cfb	Cfb
1971–1990	Cfa	Cfa	Cfa	Cfa	Cfa	Cfb	Cfa	Cfa	Cfb	Cfb
1972–1991	Cfa	Cfa	Cfa	Cfa	Cfa	Cfb	Cfa	Cfa	Cfb	Cfb
1973–1992	Cfa	Cfa	Cfa	Cfa	Cfa	Cfb	Cfa	Cfa	Cfb	Cfb
1974–1993	Cfa	Cfa	Cfa	Cfa	Cfa	Cfb	Cfa	Cfa	Cfb	Cfb
1975–1994	Cfa	Cfa	Cfa	Cfa	Cfa	Cfa	Cfa	Cfa	Cfb	Cfb
1976–1995	Cfa	Cfa	Cfa	Cfa	Cfa	Cfa	Cfa	Cfa	Cfb	Cfb
1977–1996	Cfa	Cfa	Cfa	Cfa	Cfa	Cfa	Cfa	Cfa	Cfb	Cfb
1978–1997	Cfa	Cfa	Cfa	Cfa	Cfa	Cfa	Cfa	Cfa	Cfb	Cfb
1979–1998	Cfa	Cfa	Cfa	Cfa	Cfa	Cfa	Cfa	Cfa	Cfb	Cfb
1980–1999	Cfa	Cfa	Cfa	Cfa	Cfa	Cfa	Cfa	Cfa	Cfb	Cfb
1981–2000	Cfa	Cfa	Cfa	Cfa	Cfa	Cfa	Cfa	Cfa	Cfb	Cfb
1982–2001	Cfa	Cfa	Cfa	Cfa	Cfa	Cfa	Cfa	Cfa	Cfb	Cfb
1983–2002	Cfa	Cfa	Cfa	Cfa	Cfa	Cfa	Cfa	Cfa	Cfb	Cfa
1984–2003	Cfa	Cfa	Cfa	Cfa	Cfa	Cfa	Cfa	Cfa	Cfb	Cfa
1985–2004	Cfa	Cfa	Cfa	Cfa	Cfa	Cfa	No data	Cfa	Cfb	Cfa
1986–2005	Cfa	Cfa	Cfa	Cfa	Cfa	Cfa	No data	Cfa	Cfb	Cfa
1987–2006	Cfa	Cfa	Cfa	Cfa	Cfa	Cfa	No data	Cfa	Cfb	Cfa
1988–2007	Cfa	Cfa	Cfa	Cfa	Cfa	Cfa	No data	Cfa	Cfb	Cfa
1989–2008	Cfa	Cfa	Cfa	Cfa	Cfa	Cfa	No data	Cfa	Cfb	Cfa
1990–2009	Cfa	Cfa	Cfa	Cfa	Cfa	Cfa	No data	Cfa	Cfb	Cfa
1991–2010	Cfa	Cfa	Cfa	Cfa	Cfa	Cfa	No data	Cfa	Cfb	Cfa
1992–2011	Cfa	Cfa	Cfa	Cfa	Cfa	Cfa	No data	Cfa	Cfb	No data
1993–2012	Cfa	Cfa	Cfa	Cfa	Cfa	Cfa	No data	Cfa	Cfb	No data
1994–2013	Cfa	Cfa	Cfa	Cfa	Cfa	Cfa	No data	Cfa	Cfb	No data
1995–2014	Cfa	No data	No data	Cfa	Cfa	Cfa	No data	Cfa	Cfb	No data
1996–2015	Cfa	No data	No data	Cfa	Cfa	Cfa	No data	Cfa	Cfa	No data
1997–2016	Cfa	No data	No data	Cfa	Cfa	Cfa	No data	Cfa	Cfa	No data
1998–2017	Cfa	No data	No data	Cfa	Cfa	Cfa	No data	Cfa	Cfa	No data

Second letter in index of Köppen climate classification is define by precipitation regime. In study areas between 40 and 51% of the years are without dry season. In west part of Danube lowlands and plains prevail humid years "f" ACT without dry season. In Danube lowlands humid years or without dry season are between 41 and 48% of the period. In Kozloduy lowland (between Oryahovo and Lom) and foothills of Shumen plateu prevails years with dry summer (48%). Eastern than Kozloduy lowland area till Ruse lowland dry summer years become equal with humid years (43,9%). Equal number of dry summer years ("s" 40%) and humid years ("f" 40%) observed in Knezha region. East from Ruse lowland years become dryer—only 1/3 of years were with dry summer, but 20% of years belong to dry steppe (Bsk) ACT. With averaged data not registered, but using ACT we register years with dry winter "w". Between of 4 and 18% of years in south part of Lower Danube valley are with

Table 5 10-years moving average of Köppen Climatic index in Lower Danube Valley for the period 1961–2017

	Vidin	Oryahovo	Lom	Ruse	Silistra	Vratza	Kneza	Pleven	Razgrad	Shumen
1961–1970	Cfa	Cfa	Cfa	Cfa	Cfa	Cfb	Dfa	Cfa	Cfb	Cfb
1962–1971	Cfa	Cfa	Cfa	Cfa	Cfa	Cfa	Dfa	Cfa	Cfb	Cfb
1963–1972	Cfa	Cfa	Cfa	Cfa	Cfa	Cfb	Dfa	Cfa	Cfb	Cfb
1964–1973	Cfa	Cfa	Cfa	Cfa	Cfa	Cfb	Dfa	Cfa	Cfb	Cfb
1965–1974	Cfa	Cfa	Cfa	Cfa	Cfa	Cfb	Cfb	Cfa	Cfb	Cfb
1966–1975	Cfa	Cfa	Cfa	Cfa	Cfa	Cfb	Cfb	Cfa	Cfb	Cfb
1967–1976	Cfb	Cfa	Cfa	Cfa	Cfa	Cfb	Cfb	Cfa	Cfb	Cfb
1968–1977	Cfb	Cfa	Cfa	Cfa	Cfa	Cfb	Cfb	Cfa	Cfb	Cfb
1969–1978	Cfb	Cfa	Cfa	Cfa	Cfa	Cfb	Cfb	Cfa	Cfb	Cfb
1970–1979	Cfb	Cfa	Cfa	Cfa	Cfa	Cfb	Cfb	Cfa	Cfb	Cfb
1971–1980	Cfa	Cfa	Cfa	Cfa	Cfa	Cfb	Cfb	Cfa	Cfb	Cfb
1972–1981	Cfa	Cfa	Cfa	Cfa	Cfa	Cfb	Cfb	Cfa	Cfb	Cfb
1973–1982	Cfa	Cfa	Cfa	Cfa	Cfa	Cfb	Cfb	Cfa	Cfb	Cfb
1974–1983	Cfa	Cfa	Cfa	Cfa	Cfa	Cfb	Cfb	Cfa	Cfb	Cfb
1975–1984	Cfa	Cfa	Cfa	Cfa	Cfa	Cfb	Cfb	Cfa	Cfb	Cfb
1976–1985	Cfa	Cfa	Cfa	Cfa	Cfa	Cfb	Cfb	Cfa	Cfb	Cfb
1977–1986	Cfa	Cfa	Cfa	Cfa	Cfa	Cfb	Cfb	Cfa	Cfb	Cfb
1978–1987	Cfa	Cfa	Cfa	Cfa	Cfa	Cfb	Cfb	Cfa	Cfb	Cfb
1979–1988	Cfa	Cfa	Cfa	Cfa	Cfa	Cfa	Cfa	Cfa	Cfb	Cfb
1980–1989	Cfa	Cfa	Cfa	Cfa	Cfa	Cfa	Cfa	Cfa	Cfb	Cfb
1981–1990	Cfa	Cfa	Cfa	Cfa	Cfa	Cfa	Cfa	Cfa	Cfb	Cfb
1982–1991	Cfa	Cfa	Cfa	Cfa	Cfa	Cfa	Cfa	Cfa	Cfb	Cfb
1983–1992	Cfa	Cfa	Cfa	Cfa	Cfa	Cfa	Cfa	Cfa	Cfb	Cfb
1984–1993	Csa	Cfa	Cfa	Cfa	Cfa	Cfa	Cfa	Cfa	Cfb	Cfb
1985–1994	Cfa	Cfa	Cfa	Cfa	Cfa	Cfa	Cfa	Csa	Cfb	Cfb
1986–1995	Cfa	Cfa	Cfa	Cfa	Cfa	Cfa	Cfa	Cfa	Cfb	Cfb
1987–1996	Cfa	Cfa	Cfa	Cfa	Cfa	Cfa	Cfa	Cfa	Cfb	Cfb
1988–1997	Cfa	Cfa	Cfa	Cfa	Cfa	Cfa	Cfa	Cfa	Cfb	Cfb
1989–1998	Cfa	Cfa	Cfa	Cfa	Cfa	Cfa	Cfa	Cfa	Cfb	Cfb
1990–1999	Cfa	Cfa	Cfa	Cfa	Cfa	Cfa	Cfa	Cfa	Cfb	Cfb
1991–2000	Cfa	Cfa	Cfa	Cfa	Cfa	Cfa	Cfa	Cfa	Cfb	Cfb
1992–2001	Cfa	Cfa	Cfa	Cfa	Cfa	Cfa	Cfa	Cfa	Cfb	Cfa
1993–2002	Cfa	Cfa	Cfa	Cfa	Cfa	Cfa	Cfa	Cfa	Cfb	Cfa
1994–2003	Cfa	Cfa	Cfa	Cfa	Cfa	Cfa	Cfa	Cfa	Cfb	Cfa
1995–2004	Cfa	Cfa	Cfa	Cfa	Cfa	Cfa	No data	Cfa	Cfb	Cfa
1996–2005	Cfa	Cfa	Cfa	Cfa	Cfa	Cfa	No data	Cfa	Cfb	Cfa
1997–2006	Cfa	Cfa	Cfa	Cfa	Cfa	Cfa	No data	Cfa	Cfb	Cfa
1998–2007	Cfa	Cfa	Cfa	Cfa	Cfa	Cfa	No data	Cfa	Cfa	Cfa
1999–2008	Cfa	Cfa	Cfa	Cfa	Cfa	Cfa	No data	Cfa	Cfa	Cfa
2000–2009	Cfa	Cfa	Cfa	Cfa	Cfa	Cfa	No data	Cfa	Cfa	Cfa
2001–2010	Cfa	Cfa	Cfa	Cfa	Cfa	Cfa	No data	Cfa	Cfa	Cfa
2002–2011	Cfa	Cfa	Cfa	Cfa	Cfa	Cfa	No data	Cfa	Cfa	No data
2003–2012	Cfa	Cfa	Cfa	Cfa	Cfa	Cfa	No data	Cfa	Cfa	No data
2004–2013	Cfa	Cfa	Cfa	Cfa	Cfa	Cfa	No data	Cfa	Cfa	No data
2005–2014	Cfa	No data	No data	Cfa	Cfa	Cfa	No data	Cfa	Cfa	No data
2006–2015	Cfa	No data	No data	Cfa	Cfa	Cfa	No data	Cfa	Cfa	No data
2007–2016	Cfa	No data	No data	Cfa	Cfa	Cfa	No data	Cfa	Cfa	No data
2008–2017	Cfa	No data	No data	Cfa	Cfa	Cfa	No data	Cfa	Cfa	No data

dry winter. Most common years, between 12 and 18%, with dry winter registered in west part of plains of study area. In Danube lowlands and plateus in eastern part of plains years with dry winter are between 4 and 9%.

If we summarise years with ACT of dry preriods ("s" + "w") with dry years "Bsk", or exlude humid years with second "f" ACT, we will receive a number (or %) of years

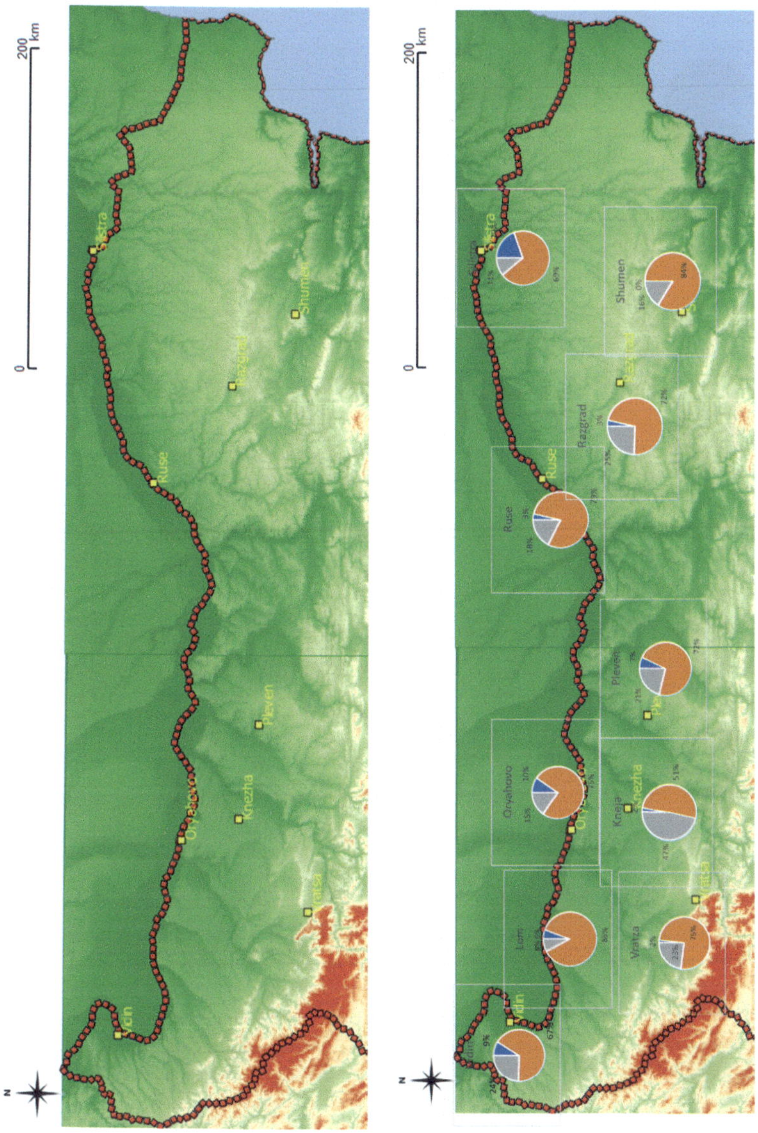

Fig. 1 Spatial distribution of main Annual Climate Type (ACT) in Lower Danube valley (North Bulgaria) (zoom of charts in Fig. 2)

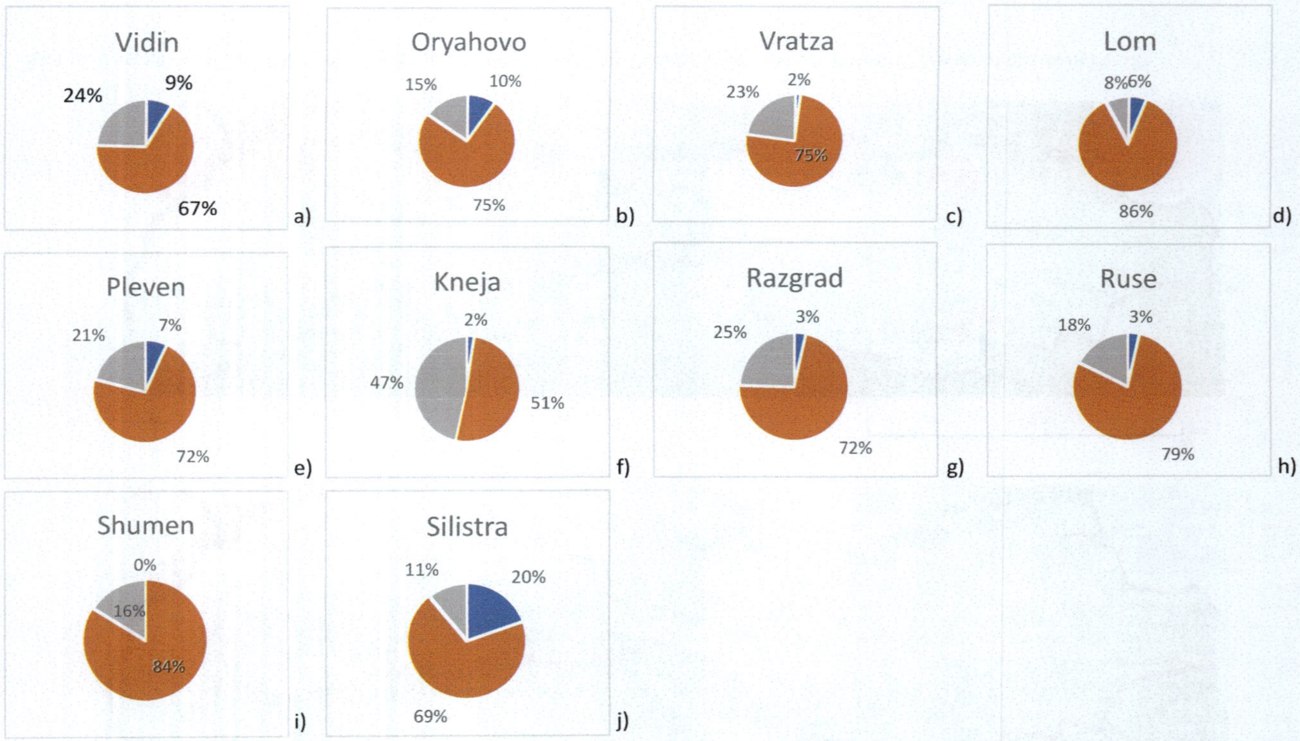

Fig. 2 Distribution of main climate types by Köppen classification (B [dry climate]—blue, C [temperate climate]—orange, D [continental climate]—gray) using Annual Climate Type (ACT) in Lower Danube Valley (North Bulgaria) for the period 1961–2017

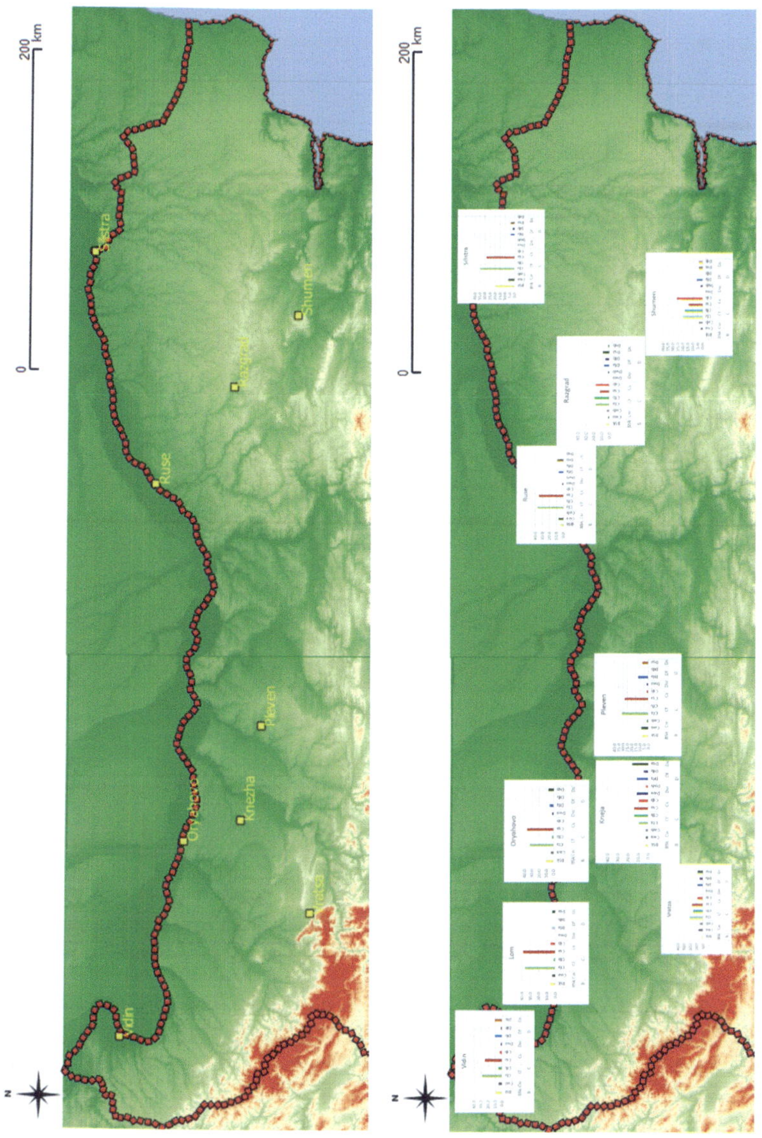

Fig. 3 Spatial distribution of detailed Annual Climate Type (ACT) in Lower Danube Valley (North Bulgaria) (zoom of charts in Fig. 4)

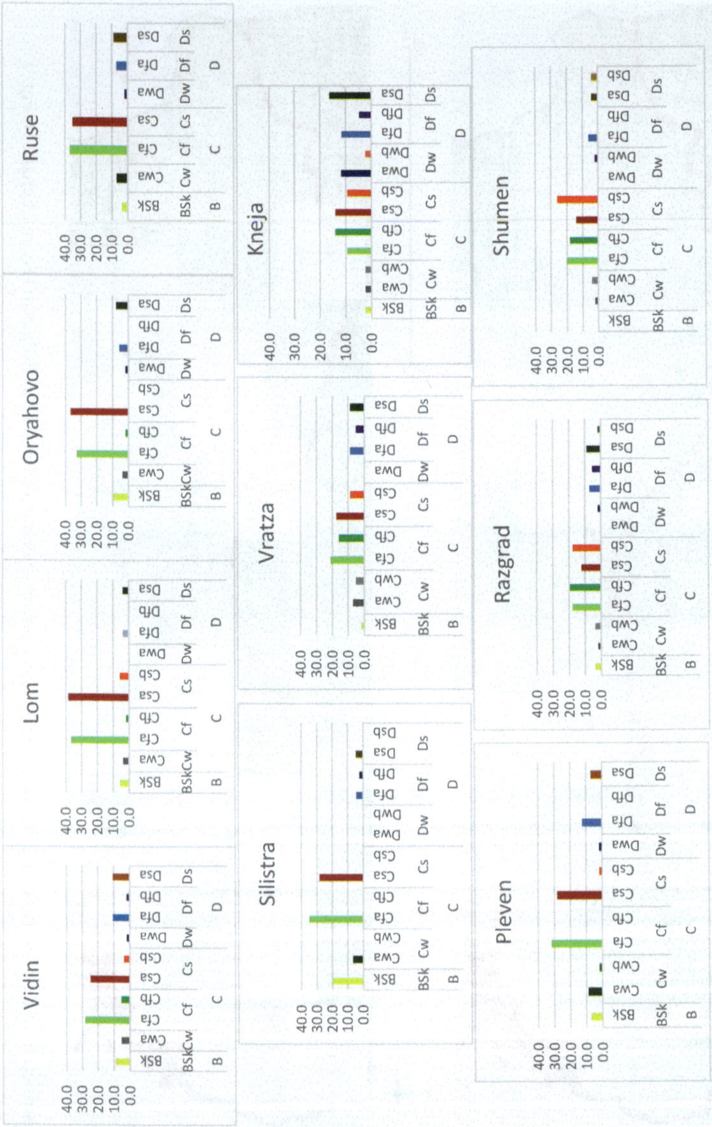

Fig. 4 Quantitive distribution of detailed Annual Climate Type (ACT) in Lower Danube Valley (North Bulgaria). Ox = climate types and Oy = %

with Potential Aridity Condition (PAC). In south part of Lower Danube valley years with PAC is between 49 and 60%. In west part of study area and Ludogorie plateu PAC is around 50% (between 49% for Vratza and 51.4% for Vidin, Razgrad 51%). For other part of Danube lowlands inner plains and foothills of Shumen plateu years with PAC are between 56 and 60% (Figs. 5 and 6).

4 Conclusion and Recommendation

The lowlands of the Danube plain are in the area with Subtropical Wet Climate (Cfa) for most of the time of the selected period. A detailed study, using 10-years moving averages and for ACT, shows that this is an area in which combinations of individual climatic factors lead to a wide variety of climate types according to the Köppen climate classification. Due to this greater detail, the significant Mediterranean influence (Csa + Csb) is registered, which is masked when the monthly data are averaged. This Subtropical Mediterranean type (Csa + Csb) is dominant for the floodplain of the Danube in the Kozloduy lowland. In combination with the temperate continental steppe climate (BSk) they form about 50% of the time in ACT. This may explain the natural steppe landscape in these areas. In the east and in the west along the Danube valley, the Mediterranean influence is not dominant and covers between 25 and 35% of the years in the studied period.

Averaged data masked years with dry winter "w" but using ACT we registered it. Between of 4 and 18% of years in south part of Lower Danube valley are with dry winter. Between 12 and 18% of years with dry winter registered in west part of plains of study area. In Danube lowlands and plateus in eastern part of plains years with dry winter are between 4 and 9%.

Using summarise years with ACT of dry preriods we can calculate a number (or %) of years with Potential Aridity Condition (PAC). In west part of study area and Ludogorie plateu PAC is around 50%. For teritory east from Vidin lowland of Danube lowlands, plains and foothills of Shumen plateu years with PAC are between 56 and 60%.

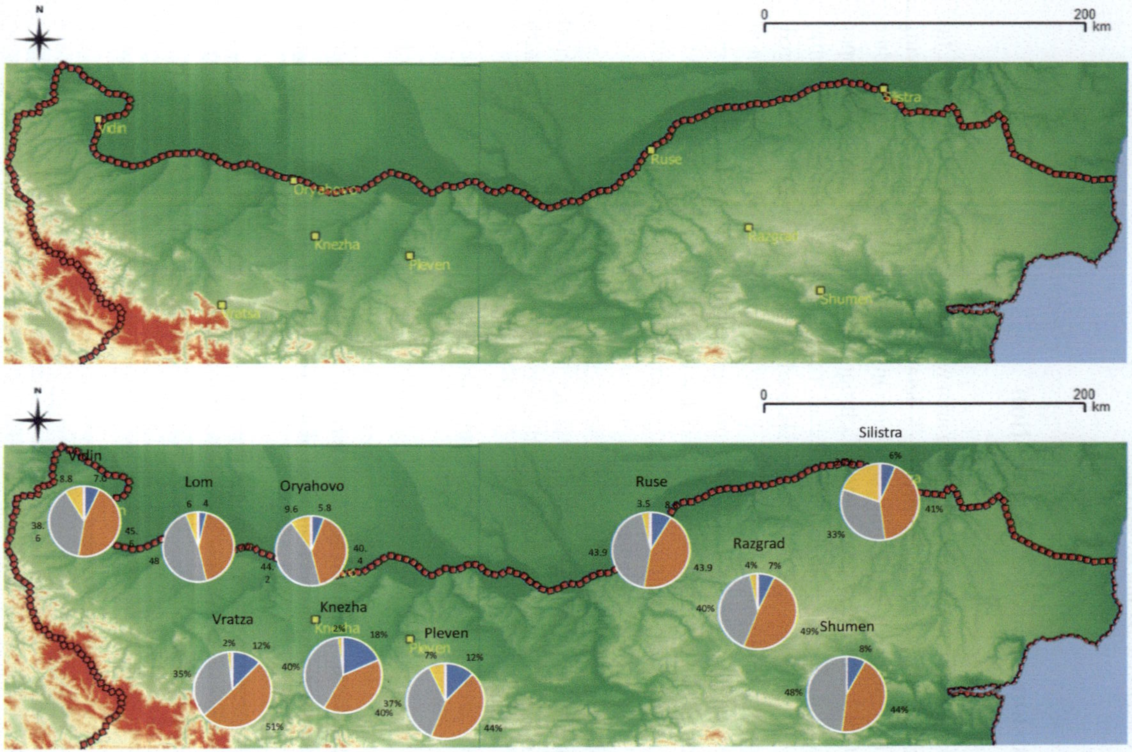

Fig. 5 Spatial distribution of Köppen climate classification sub-types (%) define by precipitation regime (Bsk (dry climate)—yellow, f (humid or without dry season)—orange, s (dry summer)—gray, w (dry winter)—blue) using Annual Climate Type (ACT) in Lower Danube valley (North Bulgaria) for the period 1961–2017

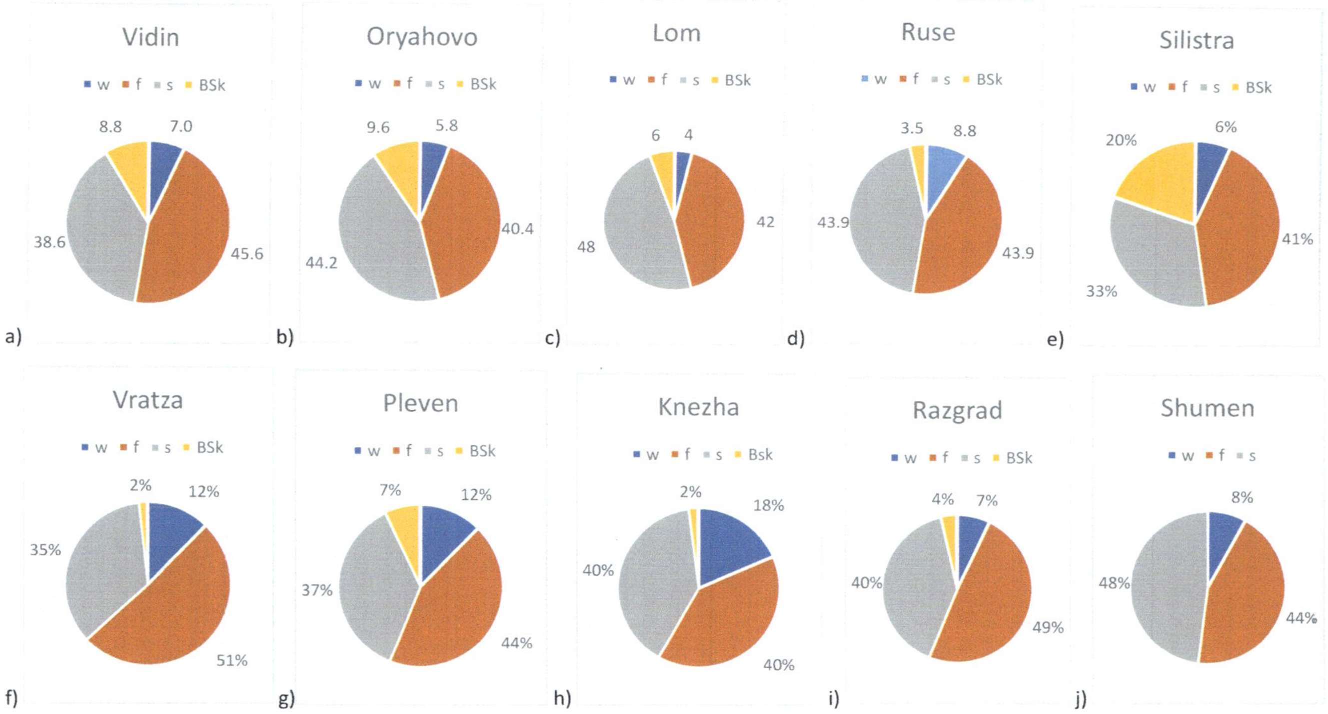

Fig. 6 Distribution of Köppen climate classification sub-types (%) define by precipitation regime (Bsk [dry climate]—yellow, f [humid or without dry season]—orange, s [dry summer]—gray, w [dry winter]—blue) using Annual Climate Type (ACT) in Lower Danube valley (North Bulgaria) for the period 1961–2017. (a–e station in Danube lowlands; f–j plains outside Danube lowland and near plateus

Acknowledgements This work has been carried out in the framework of the National Science Program "Environmental Protection and Reduction of Risks of Adverse Events and Natural Disasters", approved by the Resolution of the Council of Ministers No 577/17.08.2018 and supported by the Ministry of Education and Science (MES) of Bulgaria (Agreement No Д01-363/17.12.2020).

We thank to prof. Liliana Zaharia from University of Bucharest and prof. Abdelazim Negm for invitation and opportunity to publish in this prestige book. We thank to prof. Vincent Dubreuil from University of Rennes for nice discusions about Köppen climate classification. We also thank to editor prof. Abdelazim Negm and the anonymous reviewers for their constructive comments.

References

1. Köppen W (1900) Versuch einer Klassifikation der Klimate, vorzugweise nach ihren Beziehungen zur Pflanzenwelt. Geogr Z 6:657–679
2. Köppen W (1901) Versuch einer Klassifikation der Klimate, vorzugweise nach ihren Beziehungen zur Pflanzenwelt. Meteorol Z 18:106–120
3. De Martonne E (1942) Nouvelle carte mondiale de l'indice d'aridité. Ann Géogr 51(288):241–250
4. Viers G (1968) Éléments de climatologie. Nathan, Paris
5. Alissow BP (1954) Die Klimate der Erde. Deutch, Ubers, Berlin
6. Pédelaborde P (1957) Le climat du bassin Parisien – Essai d'une méthode rationnelle de climatologie physique. Génin, Paris
7. Flohn H (1969) Climate and weather. World Univ, Library, McGraw-Hill, New York
8. Péguy CP (1970) Précis de climatologie. Masson, Paris
9. Lamb HH (1977) Climate: present, past, and future, vol 2. Climatic history and the future. Methuen & Co Ltd., London
10. Sanderson M (1999) The classification of climates from Pythagoras to Koeppen. Bull Amer Meteor Soc 80:669–673
11. Dubreuil V, Fante KP, Planchon O, Neto JLS (2017) Les types de climats annuels au Brésil : une application de la classification de Köppen de 1961 à 2015. EchoGéo 41:1–27
12. Thornthwaite CW (1933) The climates of the earth. Geogr Rev 23:433–440
13. Trewartha GT (1937) An introduction to weather and climate. McGraw-Hill Book Company, New York
14. Hantel M (1989) Climatology, series Landolt-Bцrnstein, numerical data and functional relationships in science and technology. new series, vol 4C2, The present global surface climate, Springer, Berlin, pp 117–474
15. Essenwanger OM (2001) Classification of climates, world survey of climatology 1C, general climatology. Elsevier, Amsterdam
16. Peel MC, Finlayson BL, McMahon TA (2007) Updated world map of the Köppen-Geiger climate classification. Hydrol Earth Syst Sci 11:1633–1644
17. Kottek M, Grieser J, Beck C, Rudolf B, Rubel F (2006) World map of the Köppen-Geiger climate classification updated. Meteorol Z 15:259–263
18. Rubel F, Kottek M (2010) Observed and projected climate shifts 1901–2100 depicted by world maps of the Köppen-Geiger climate classification. Meteorol Z 19:135–141
19. Beck HE, Zimmermann NE, McVicar TR, Vergopolan N, Berg A, Wood EF (2018) Present and future Köppen-Geiger climate classification maps at 1-km resolution. SCIENTIFIC DATA, 5:180214, https://doi.org/10.1038/sdata.2018.214 https://www.nature.com/articles/sdata2018214.pdf
20. Allam A, Moussa R, Najem W, Bocquillon C (2019). Mediterranean specific climate classification and future evolution under RCP scenarios. https://doi.org/10.5194/hess-2019-381

21. Gallardo C, Gil V, Hagel E, Tejeda C, de Castro M (2013) Assessment of climate change in Europe from an ensemble of regional climate models by the use of Köppen – Trewartha classification. Int J Climatol 33:2157–2166
22. Lohmann U, Sausen R, Bengtsson L, Cubasch U, Perlwitz J, Roeckner E (1993) The Koppen climate classification as a diagnostic tool for general circulation models. Clim Res 3:177–193
23. Chen D, Chen HW (2013) Using the Köppen classification to quantify climate variation and change: an example for 1901–2010. Environ Dev 6:69–79. https://doi.org/10.1016/j.envdev.2013.03.007
24. Brisse H, Grandjouan R, De Ruffray P (1982) Les types de climats annuels, un mode d'expression des gradients climatiques intйgrant les variations interannuelles. Météorol VIe série 31:39–81
25. Planchon O, Rosier K (2005) Climat et variabilité climatique dans le Nord-Ouest de l'Argentine: problèmes posés et analyse durant la deuxième moitié du vingtième siècle. Ann Assoc Int Climatol 2:55–76
26. Quénol H, Planchon O, Wahl L (2008) Méthodes d'identification des climats viticoles. Bull Soc Géogr 51:127–137
27. Eveno M, Planchon O, Oszwald J, Dubreuil V, Quenol H (2016) Variabilitй et changement climatique en France : analyses au moyen de la classification de Köppen et des «types de climats annuels». Climatol 13:47–70
28. Chenkova N, Nikolova N (2015) Air temperature and precipitation variability in Northeastern Bulgaria on the background of climate change. Therm Sci. https://doi.org/10.2298/TSCI150430104C
29. Vlăduţ A, Nikolova N, Licurici M (2017) Aridity assessment within southern Romania and northern Bulgaria. Hrvatski Geogr Glas 79(2):5–26. https://doi.org/10.21861/HGG.2017.79.02.01
30. Vlăduţ A, Nikolova N, Licurici M (2018) Evaluation of thermal continentality within southern Romania and northern Bulgaria (1961–2015). Geofiz 35. https://doi.org/10.15233/gfz.2018.35.1
31. Radeva K, Nikolova N, Gera M (2018) Assessment of hydro-meteorological drought in the Danube Plain, Bulgaria. Hrvatski Geogr Glas 80(1):7–25. https://doi.org/10.21861/HGG.2018.80.01.01
32. Kirov K (1929) Climatic scheme of Bulgaria. A compilation of the Bulgarian academy of science. Book XXV. Nat Sci 11:57–60 [in Bulgarian]
33. Topliyski (2006) Climate of Bulgaria. Amstels, Sofia [in Bulgarian]
34. Popov H (2018) Local climates of Vardar, Struma and Mesta valleys (Balkan Peninsula) according to the modified Köppen climate classification. Bull Serb Geogr Soc https://doi.org/10.2298/GSGD180428005P

Observed Changes in the Temperature and Precipitation Regime Along the Lower Danube River

Dana Maria Constantin, Irina Onţel, Adrian Amadeus Tişcovschi, Anişoara Irimescu, Elena Grigore, Raul Gabriel Ilea, and Gabriela Dîrloman

Abstract Knowing the space–time variability of the temperature and precipitation provides the ability to objectively assess its effects on the water resources. The Lower Danube River represents a significant part of the surface water resources of Romania and it is of crucial importance for the socio-economic development of the riparian localities. In this chapter, we will analyze the variability and changes occurred in the time series of the average air temperatures and precipitation along the Lower Danube River, for the period 1961–2013. The meteorological data are extracted

D. M. Constantin (✉)
Faculty of Geography, University of Bucharest, 1 Nicolae Bălcescu Avenue, 010041 Bucharest, Romania
e-mail: oprea.constantin@geo.unibuc.ro

I. Onţel · A. Irimescu
Remote Sensing and Satellite
Meteorology, National Meteorological Administration, 97 Bucureşti-Ploieşti Street, 013686 Bucharest, Romania
e-mail: irina.ontel@meteoromania.ro

A. Irimescu
e-mail: anisoara.irimescu@meteoromania.ro

A. A. Tişcovschi · E. Grigore
Faculty of Geography, University of Bucharest, 1 Nicolae Bălcescu Avenue, 010041 Bucharest, Romania
e-mail: adrian.tiscovschi@unibuc.ro

E. Grigore
e-mail: grigore.elena@geo.unibuc.ro

R. G. Ilea
Faculty of Geography, Doctoral School and
Center of Operational Forecasting, University of Bucharest and National Meteorological Administration, 1 Nicolae Bălcescu Avenue, 010041 Bucharest, Romania
e-mail: raul-gabriel.ilea@drd.unibuc.ro

G. Dîrloman
"Nicolae Kretzulescu" Commercial High School, 17 Hristo Botev Avenue, 030233 Bucharest, Romania
e-mail: gabriela.dirloman.69@kretzulescu.ro

A. Negm et al. (eds.), *The Lower Danube River*, Earth and Environmental Sciences Library, https://doi.org/10.1007/978-3-031-03865-5_9

from the Romanian daily gridded climatic dataset (ROCADA) and the archive of the National Meteorological Administration. These data come from five weather stations which are representative for the Lower Danube River. Also, based on MODIS satellite images, the land surface temperature parameter (LST) was extracted. Most of the Lower Danube River is characterized by an annual average temperature of 11°C and by annual precipitation between 484.9 and 647.4 mm. Both the average annual air temperature and the amounts of the annual precipitation decrease from the West to the East, as a consequence of the increase in the same direction of the degree of the continental influences and the weakening of the thermic convection above the water. There were observed tendencies to increase the air temperature at all the weather stations both at the annual, as well as in spring, summer, June, July and August, being in line with those at the continental level. The trends of precipitation amounts are not as significant as those of air temperature, so the precipitation regime in the study area can be considered as stationary.

Keywords Temperature · Precipitation · Changes and Trends · Lower Danube River · Romania · Contents

1 Introduction

The current global warming has led to a dynamics of the hydrological cycle, so that the effects of the climate change on the water resources are reflected in: the water supply, the water quality, the water requirements and in the extreme events (floods and droughts) [1].

The climate change of the last decades in the hydro-climatic system in Romania has been highlighted in numerous studies [2–5]. These changes are manifested especially by the appearance of the drought phenomenon [6–9]. There has been reported a significant warming in most regions of Romania, at an annual average level of about 0.5°C, since 1901 to present, and an intense inter-decade variability in terms of precipitation, with successions of dry and rainy periods [10, 11].

In recent years, more attention has been paid to identify and analyze the climate change based on the satellite imagery. Products obtained by remote sensing methods such as land surface temperature (LST) aid to understand the local climate characteristics and trends [12]. The annual and seasonal LST product highlight the variability and climate change manifests itself mainly through high-temperature and drought [13]. The space–time variation of LST provides useful information regarding the impact of temperature on the agricultural production. Also, it helps disaster monitoring and early warning and ecological protection [14]. Moreover, LST datasets are important elements for improving global hydrological and climate prediction models [15]. The remote sensing improves the limitations of the meteorological observations [16]. Therefore, various organizations need to pay attention to these satellite products [17].

The South-Eastern Danube region is expected to become much warmer than in the last decades (higher than 1 °C [18]) and with less precipitation (possibly variations in the order of 70 mm/year [19]). According to The International Commission for the Protection of the Danube River (ICPDR) Strategy on Adaptation to Climate Change [20], in the Danube River Basin "*climate change is likely to cause significant impacts on water resources and can develop into a significant threat*".

Knowing the space–time variability of the temperature and precipitation system provides the ability to objectively assess the effects on the water resources. The high temperatures can affect the quality of the river water (the decreased dissolved oxygen, the eutrophication etc.) and the less precipitation can cause problems in ensuring the safe flow.

The main purpose of this chapter is to investigate the changes occurred in the air and surface land temperature, as well as in the precipitation regime along the Lower Danube River in Romania, for the period 1961–2019, based on both the meteorological data from surface measurements and those based on the MODIS satellite images.

2 Data and Methods

This study is based on the processing of the monthly average temperatures and the monthly precipitation for five weather stations (WS) representative for geographical location along the Lower Danube River: Drobeta (Dr.) Turnu (Tr.) Severin, Calafat, Zimnicea, Călărași and Galați (Fig. 1).

The meteorological data used are extracted from ROCADA: a gridded daily climatic dataset for the interval 1961–2013 [21, 22] which belongs to the National Meteorological Administration (NMA). Also, there were used meteorological data records from the five weather stations, covering the interval 2014–2019, the data being extracted from the archive of the NMA.

The remote sensing products used in this study have been MOD11A1.006 [23] and MYD11A1.006 Daily Global – 1 km for Land Surface Temperature (LST) [24] assessment along the Lower Danube River, for the period 2000–2019. The LST average for each day has been calculated combining the four images recorded by the MODIS sensor, installed on the Terra and Aqua satellites. Then, the Kelvin degrees have been converted to Celsius degrees and the monthly average LST has been calculated. The data set for January 2000 is missing from the analysis due to the lack of data collection.

A summary of the climatological and remote sensing data is presented in Table 1.

Also, in the analysis of the climatic variability of the study area, there have been calculated the standardized anomalies for the months of January and July, considered typically months from the climatic point of view, and at the year level.

Standardized anomalies for air temperature, precipitation and LST have been calculated as it follows (1):

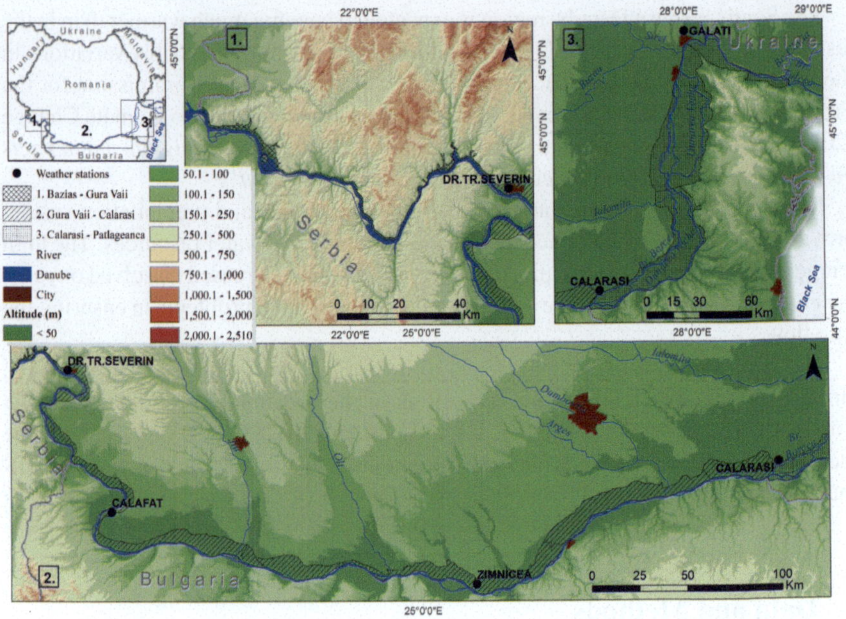

Fig. 1 Location of the whole study area in Romania (top left) and of the three analyzed sectors: (1) Baziaș-Gura Văii or the Danube Gorge; (2) Gura Văii-Călărași and (3) Călărași-Pătlăgeanca

Table 1 Summary of the data used in this chapter

Product Name	Type/Bands	Time period	Resolution	Units	Dataset Provider
ROCADA	Average air temperature	1961-2013	11 km	°C	ROCADA [22]
	Precipitation			mm	
Meteorological data	Average air temperature	2014-2019	Weather Stations	°C	National Meteorological Administration [25]
	Precipitation			mm	
MOD11A1.006	LST_Day_1km	2000-02-01 to 2019-12-31	1 km	Kelvin	NASA LP DAAC at the USGS EROS Center [23]
	LST_Night_1km				
MYD11A1.006	LST_Day_1km	2002-07-04 to 2019-12-31	1 km	Kelvin	NASA LP DAAC at the USGS EROS Center [24]
	LST_Night_1km				

$$x_{anomaly} = \frac{x - \mu}{\sigma} \tag{1}$$

where:

$x_{anomaly}$ – is the standardized anomaly;

x – is the value for the month;

μ – is the average value over the 1961–2019 for the meteorological data and 2000–2019 time period for LST;

σ – is the standard deviation value over the 1961–2019 for the meteorological data and 2000–2019 time period for LST.

The analysis of the changes in the average regime of temperature and precipitation as well as of the LST parameter is completed by the calculation of the linear trends and their statistical significance at annual, monthly and seasonal time scale, by using the Mann-Kendal statistic test and Sen's slope, by means of the MAKESENS – application [26].

3 Study Area: General Geographical Features and Climatic Peculiarities

This section presents, in the first part, a brief physical-geographical characterization of the study area, and in the second part, we investigate and highlight the specific features of the air and land surface temperatures, as well as of precipitation regime, based on meteorological data and satellite images, for the period 1961–2019.

3.1 Geographical Location and Geomorphological Aspects

The Lower Danube River extends downstream from Baziaș (settlement located at the entry of the river in Romania) to its mouth in the Black Sea and is also called the Pontic sector of the Danube River [27]. This lower course has a length of 1075 km, which represents 37% of the river's total length. The valley in the lower course of the river are usually asymmetrical, the right side being about 200 m higher than the left side [28]. The geomorphological characteristics of the valley, the floodplain and riverbed have imposed its division into four sectors: the Iron Gates Gorge, Gura Văii-Călărași, Călărași-Pătlăgeanca and the Danube Delta [28].

In this chapter, there will be analyzed the specific climatic particularities and identified the changes in the thermic and pluviometric regime for the first three sectors of the lower course of the Danube River (Fig. 1).

The Iron Gates Gorge sector is the longest and most spectacular gorge in the entire course of the river that extends between Baziaș and Gura Văii. It is characterized by a succession of narrow and wide sectors and by the construction of the Iron Gates I reservoir, part of the hydro-power and navigation system with the same name [27].

The Gura Văii-Călărași sector is characterized by a floodplain without a continuous development with a maximum width of 10–15 km, on which there were, in the past, numerous lakes, ponds and parallel watercourses, while in the river channel, there are numerous islands [28].

The Călărași-Pătlăgeanca sector is highlighted by the largest width of the floodplain in the country, with an average of 15–20 km, reaching a maximum of 25–30 km.

Also, the river branches into two main arms that include two large islands called "bălți" (puddles) [27, 28].

In this chapter, we considered the Romanian side of the Lower Danube River, including the river channel, the riparian floodplain, as well as the immediate edge of the shore or slope and those areas that are under the varied and variable influence of the river, where the alignment of the localities dependent on the Danube is found (Fig. 1).

3.2 *Climatic Peculiarities*

Due to its geographical location, the lower course of the Danube River is characterized by a temperate continental climate with Mediterranean influences in the Western half of the course and aridity in the Eastern half. These climatic influences together with the influence of the large water surfaces materialize in some specific features of the main meteorological parameters, which individualize the Danube corridor compared to the surrounding areas. In order to identify and highlight the features of the temperature and precipitation regime along the Lower Danube River, the monthly averages values at the five representative weather stations (Fig. 1) were analyzed.

3.2.1 Air Temperature and Precipitation

Along the Lower Danube River, there can be observed some differences in terms of the climatic regime of the analyzed parameters. Thus, both the average annual air temperature and the annual amount of precipitation decrease from the West to the East by 0.9°C in the case of temperature (11.7°C at Dr. Tr. Severin, and 10.8°C at Galați), respectively 162.5 mm in the case of precipitation (647.4 mm at Dr. Tr. Severin, and 484.9 mm at Galați) for the period 1961–2019 (Fig. 2).

This decrease from the West to the East along the Lower Danube is explained by the transition from the Mediterranean influences, specific to the areas of Dr. Tr. Severin, Calafat and Zimnicea weather stations towards those of aridity, specific to Călărași and Galați areas.

The monthly air temperature regime outlines the classic evolution regime for most weather stations in Romania, in which the lowest value of the monthly average is recorded in January, and the highest value is reached in July. The highest value of January is registered at Dr. Tr. Severin WS, of -0.3°C, and the lowest value is registered at Galați WS of -1.9°C. This thermic difference between the two weather stations is the direct result of the climatic influences, as mentioned above. Between the two weather stations, the monthly average temperature for January oscillates between -0.7°C at Calafat, -1.8°C at Zimnicea and -1.3°C at Călărași. The higher value from Călărași weather station, compared to the one from Zimnicea, is the result of the influence of the water moderator role which is felt more strongly at Călărași WS as a result of the increase of the river water surface near the station. July, the

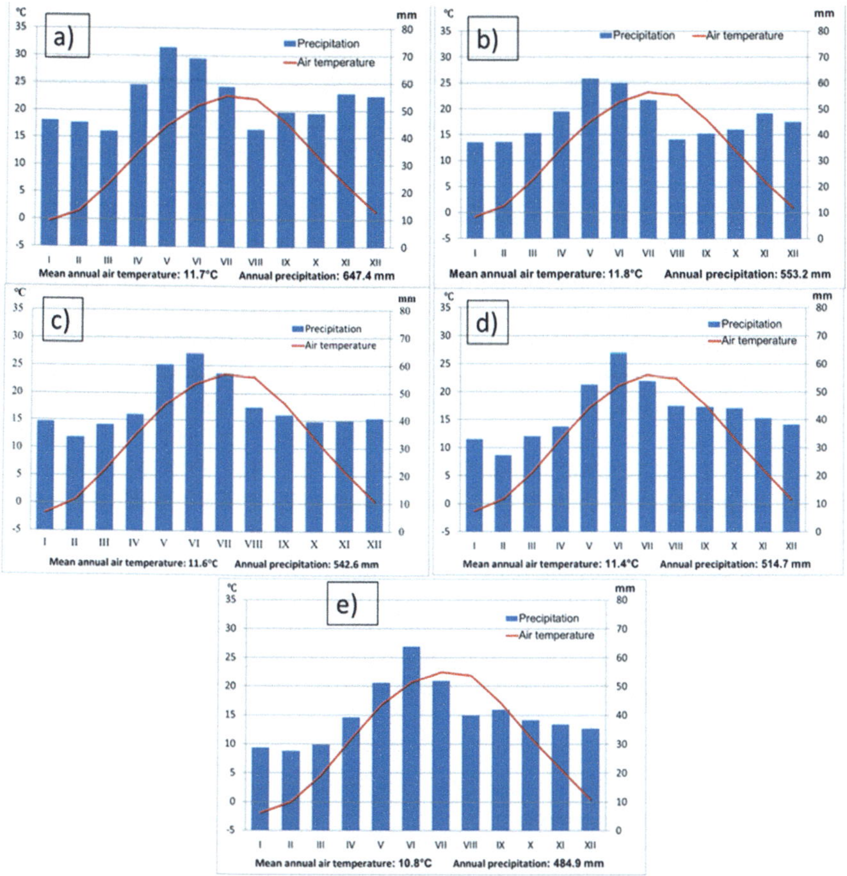

Fig. 2 The annual regime of the monthly average air temperature and monthly average precipitation at the weather stations along the Lower Danube River for the period 1961–2019: (a) Dr. Tr. Severin; (b) Calafat; (c) Zimnicea; (d) Călărași; (e) Galați

warmest month of the year, is characterized by air temperatures lower than 23.0°C only at the extremities, at the Dr. Tr. Severin, of 22.8°C, and at Galați, of 22.5°C, while the values for the other stations exceed 23.0°C (23.3°C at Calafat, 23.5°C at Zimnicea and 23.0°C at Călărași) (Fig. 2). Temperatures higher than 23.0°C are the highest monthly average values of July in the country, due to the latitude, Zimnicea being the most Southern locality in the country, and to the influence of the tropical hot air advections.

The annual precipitation regime is characterized by a complex distribution along the Lower Danube valley as a result of the higher share, in their genesis, of the atmospheric circulation. Dr. Tr. Severin is characterized by monthly precipitation amounts between 42.2 mm in March and 73.1 mm in May, for the period 1961–2019. At Calafat, the minimum monthly amount of precipitation is recorded in January of

37.1 mm, and the highest average monthly value is more than 61.7 mm to mention the month. For the other weather stations, the lowest average amount of precipitation per month is recorded in February, being 33.8 mm at Zimnicea, 27.3 mm at Călărași and 27.7 mm at Galați (Fig. 2). The highest monthly amount of precipitation for the weather stations Zimnicea, Călărași and Galați is registered in June being: 64.3 mm, 63.9 mm and 63.8 mm respectively (Fig. 2). There is a significant decrease in precipitation from the West to the East, due to the increase in continentalism. The highest monthly amount is recorded in June, sometimes in May, often in the form of showers, due to the development of thermic convection [29]. Also, the maximum monthly values of the precipitation quantities are the result of the frequent penetration of the humid air masses, of oceanic origin, on the Romanian territory [28].

Differences between the weather stations along the Lower Danube can also be seen in terms of Quantile values. The third quantile (Q_3) shows a value about 12.3°C for the first weather stations from West (Dr. Tr. Severin, Calafat) and it gradually decreases to the east, reaching at Galați WS about 11.5°C (Fig. 3). Also, the median and the first quantile (Q_1) have values much lower at Galați WS compared to other stations. Differences between the average annual temperatures registered at the five stations can be observed also in terms of the shape of the boxplots. The distribution of the values registered at the Galați WS, mainly as a result of the extreme values, gives it a longer shape compared to the other stations (Fig. 3).

In terms of annual precipitation, the values of the third quantile (Q3) as well as of median and the first quantile (Q_1), decrease from West to East. Therefore, the lowest values can be observed at Galați WS (Fig. 4).

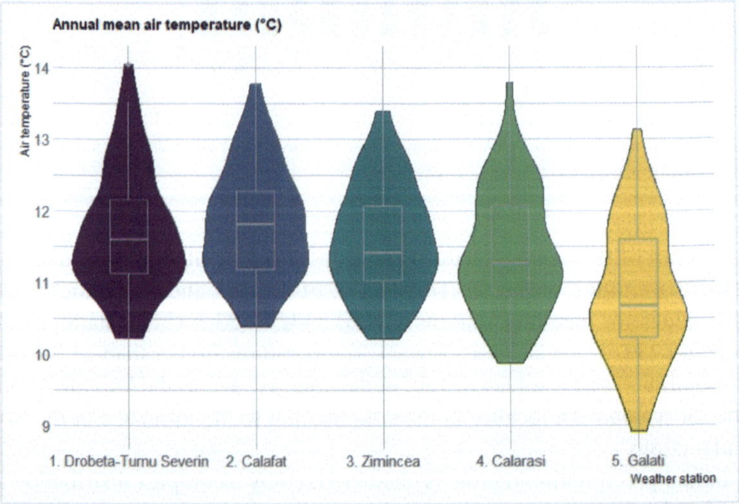

Fig. 3 The annual average air temperature boxplots at the weather stations along the Lower Danube River for the period 1961–2019

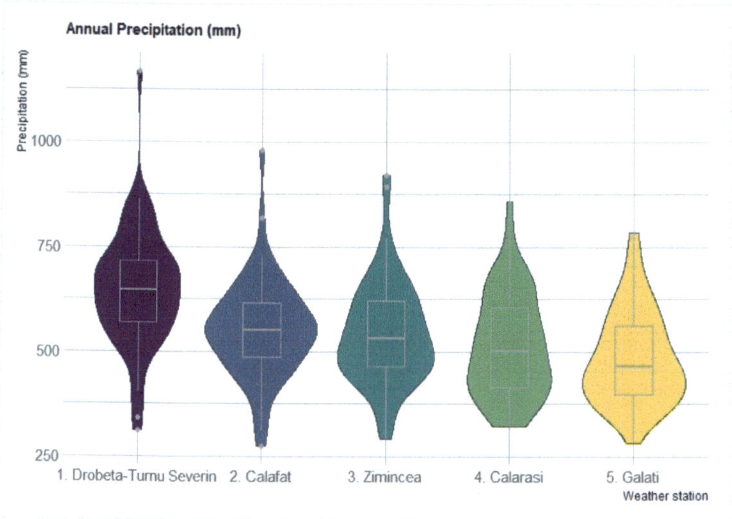

Fig. 4 The annual precipitation boxplots at the weather stations along the Lower Danube River for the period 1961–2019

3.2.2 Péguy Climograms

The Péguy climograms (Fig. 5) show the correlations between the monthly average values of temperature and precipitation, thus, highlighting the monthly precipitation-temperature characteristics in the study area, from the five weather stations analyzed, for the period 1961–2019. The shape of the polygons obtained, the longer it is, the more it indicates the presence of a climate with higher thermic and precipitation contrasts. December, January and February (winter) are cold and humid at all weather stations, due to the influences of continental cold air and to the water surfaces (Fig. 5). August is an arid month for the Calafat and Călărași weather stations and at the limit for the other weather stations (Fig. 5).

3.2.3 Land Surface Temperature (LST)

The complexity of the active structure (terrestrial) surface along the Lower Danube Valley is also highlighted by the space–time analysis of the climatic parameter, the land surface temperature (LST) from satellite images.

The values of this parameter decrease in January along the Lower Danube Valley from the West to the East. Thus, in the Western sector, Baziaș-Gura Văii and the first half of the Gura Văii-Călărași sector, the temperature of the active surface has values between -2°C and 0°C, and in the second half of the Gura Văii-Călărași sector and the Călărași-Pătlăgeanca sector, the temperature is between −4°C and −2°C (Fig. 6a). In July, the temperature at the land surface is distributed relatively evenly along the

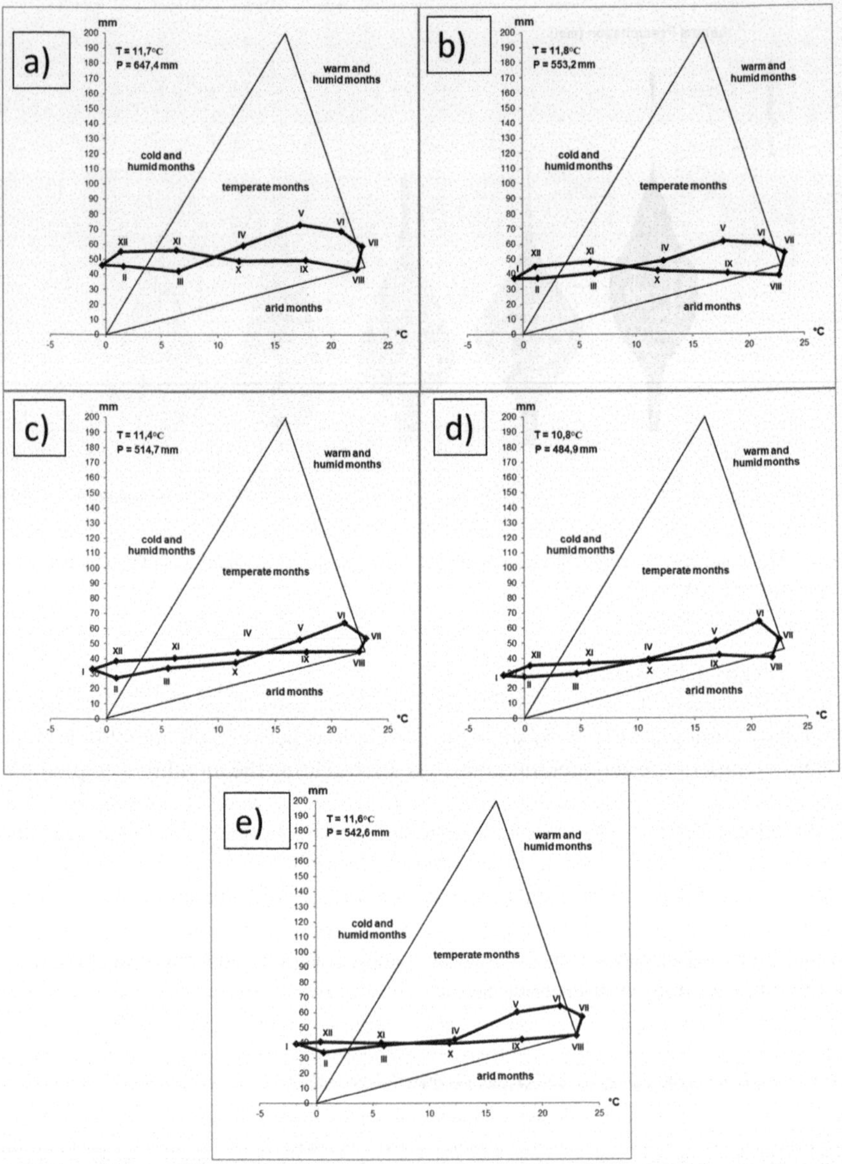

Fig. 5 The Péguy climograms at the weather stations along the Lower Danube River for the period 1961–2019: (a) Dr. Tr. Severin; (b) Calafat; (c) Zimnicea; (d) Călărași; (e) Galați. T—the annual mean air temperature (°C); P—the annual precipitation amount (mm)

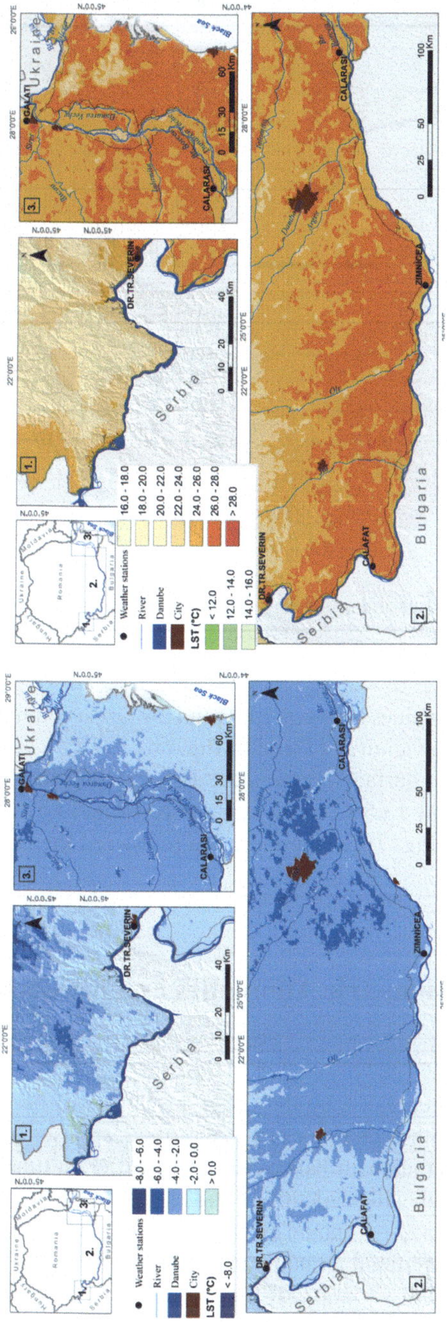

a) LST mean for January 2000-2019

b) LST mean for July 2000-2019

Fig. 6 The spatial distribution of LST values for a) January and b) July along the Lower Danube River, for the period 2000–2019

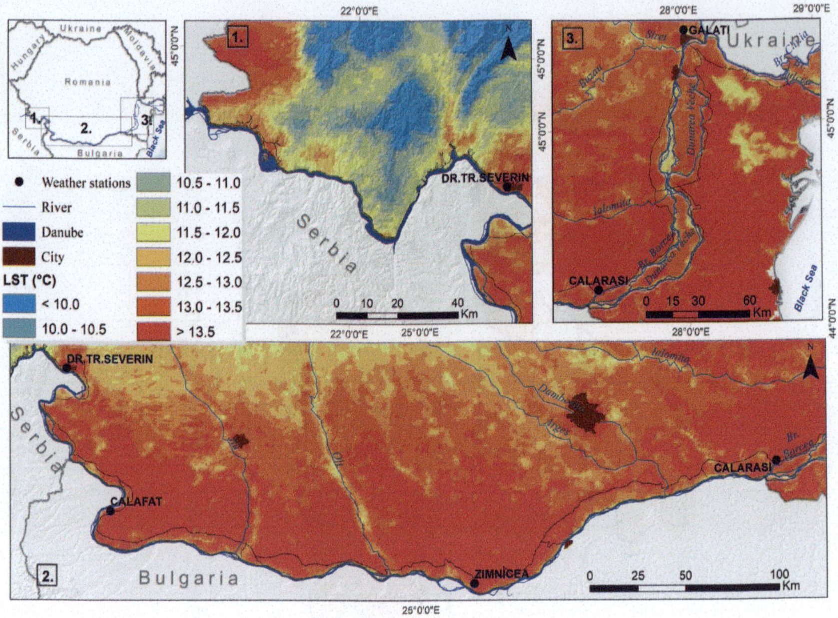

Fig. 7 The spatial distribution of the annual average values of LST along the Lower Danube River, for the period 2000–2019

Lower Danube Valley, with values between 24°C and 26°C, exceeding 26°C in some areas, and even 28°C, in the vicinity of Calafat (Fig. 6b).

The annual average land surface temperature (LST) along the Lower Danube Valley is between 13°C and 13.5°C, sometimes exceeding 13.5°C, especially in Gura Văii-Călărași sector (Fig. 7), where there are registered the highest values in the South of the country.

4 Thermal and Pluviometric Anomalies and Decennial Variability

The climate variability is generated by the different nature of the climate system components that interact with each other [10]. And, the responses of these components to internal and/or external disturbances have very different time intervals [30]. In this chapter sequence, there will be identified and analyzed the anomalies observed in the variability of air temperature (annual averages and monthly averages in January and July) and precipitation (annual amounts and monthly averages in January and July) as well as the decennial variation of the two climatic parameters along the Lower Danube Valley, during the period 1961–2019, in order to detect changes in their variability.

4.1 *Annual and Monthly (January and July) Anomalies*

The analysis of the anomalies of the main climatic elements is very important, being a tool for the climate diagnosis [31]. When the temperature anomaly is positive, it indicates that the observed temperature was higher than the reference value, thus, the weather being warmer. On the contrary, when the anomaly is negative, it indicates that the observed temperature was lower than the reference value, thus, making the weather colder.

Over the last 20 years, the frequency of positive annual thermic anomalies has increased at all the weather stations along the Lower Danube Valley (Fig. 8). Starting with the year 2000, the positive annual anomalies were registered in 17–19 years,

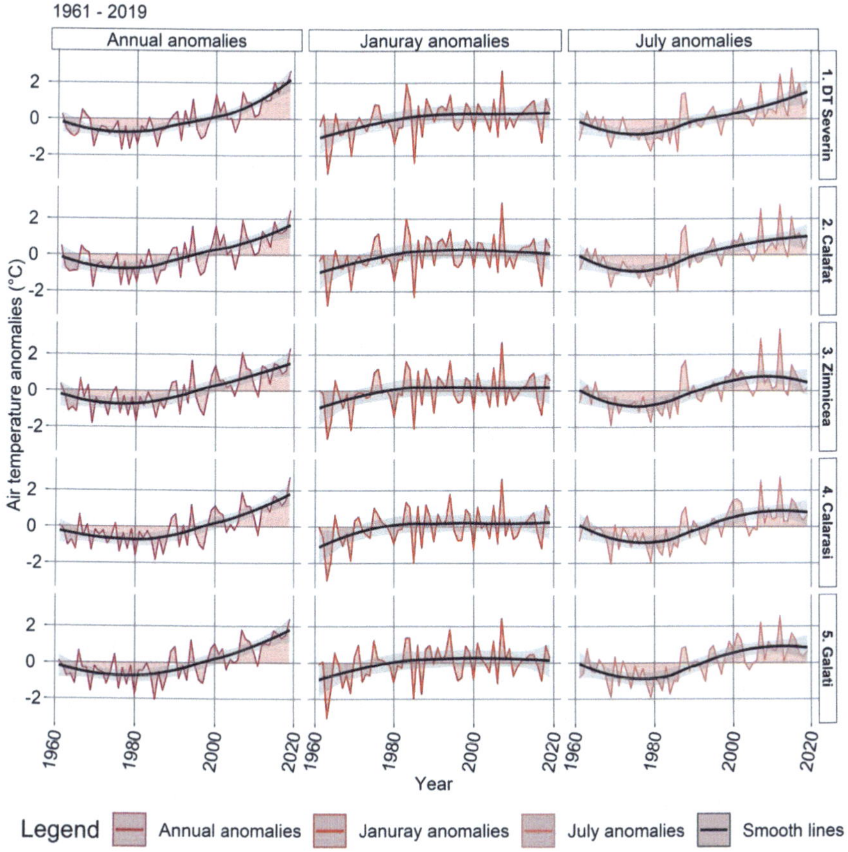

Fig. 8 The annual and monthly (January and July) air temperature anomalies (°C) at the weather stations along the Lower Danube River for the period 1961–2019: (1) Dr. Tr. Severin; (2) Calafat; (3) Zimnicea; (4) Călărași; (5) Galați

while before this year; the positive annual anomalies were registered in only 9–12 years. Therefore, the number of years with high temperatures is significantly higher in the last two decades compared to the previous decades of the period 1961–2019.

In January, for the analyzed period, the positive anomalies are not as frequent as in the case of July, along the Lower Danube Valley, these alternating with the negative ones (Fig. 8). In July, there was an evolution of anomalies similar to those calculated annually, the positive anomalies being more frequent in the last two decades of the analyzed period, compared to the decades before 2000 (Fig. 8).

The precipitation anomalies highlight no change in their frequency in the last twenty years, unlike the temperature anomalies for the analyzed period. The positive anomalies alternate with the negative ones, both annually and in the two months (January and July), respectively (Fig. 9).

The land surface temperature anomaly (LST), calculated for LST average within the three sectors of the Lower Danube, has highlighted an increase in positive values in the last ten years (Fig. 10). Except for the year 2000, when the annual LST anomaly exceeded 2°C, the year 2019 registered the highest value of the LST anomaly in the three sectors of the Lower Danube (Fig. 11). The values from the last two sectors (Gura Văii-Călărași and Călărași-Pătlăgeanca) were higher than in the first sector (Baziaș-Gura Văii) (Fig. 11). 2019 was the warmest year, in Romania, for the period 1961–2019, according to the National Meteorological Administration [32].

At the level of monthly values, it is found that positive values have been frequently recorded in the recent years, especially in the autumn and winter months (Fig. 10).

For January and July, the biggest positive anomalies of LST were registered in 2007. The year 2007, according to the National Meteorological Administration is the fourth warmest year in Romania, for the period 1961–2019 [32]. The influence of water on the land temperature variations can be observed on the products obtained from satellite images, as a result of the role of thermic regulator that the aquatic surface has as a thermic regulator. Thus, in January 2007, the highest values of the anomalies were recorded near the Danube, in the form of a narrow band (Fig. 12a), while in July 2007, lower values of the anomaly were recorded in the wetlands, such as the Bistreț Lake area and the Danube floodplain in the Corabia area (Fig. 12b). However, in the Călărași-Pătlăgeanca sector, this effect of water is not so obvious on the satellite images.

4.2 The Thermo-Pluviometric Variation from one Decade to Another

The changes in the air temperature and precipitation variation can also be identified by the analysis of the decennial averages. These averages for successive ten-years reflect well the long-term variability from one decade to another [33]. The decennial

along the Lower Danube Valley. In the last two decades of the analyzed period, the decadal average values of January are in an increasing trend, being higher than the multiannual average of the period 1961–2019 (Table 3). In July, the ten-year variation shows no significant changes. These changes are less spatially consistent compared to air temperature (Table 3).

The changes detected in the thermo-pluviometric parameters in the last decades have also been captured by the MODIS satellite images, based on the LST products, for the period 2000–2019. Thus, the annual average of the decade 2011–2019 is 1–2°C higher compared to the decade 2001–2010 on almost the entire length of the Lower Danube Valley. The biggest differences can be seen in the Calafat-Călărași sector (Fig. 13top).

The average of January in the decade 2011–2019 registered a decrease in the Eastern sector of the valley (Călărași-Pătlăgeanca) and in some areas of the sector Calafat-Călărași compared to the average of the decade 2001–2010 (Fig. 13 middle). The same situation can be observed in July, the average values in the second decade are slightly lower than in the first decade (Fig. 13 bottom).

5 Trends of the Thermal and Pluviometric Regime

The purpose of this part is to identify the general trends, in the time series of air temperature, atmospheric precipitation and land surface temperature (LST) along the Lower Danube Valley. This identification will be based on the non-parametric Mann–Kendall test [35, 36], which shows the statistical significance and the risk of error of the linear trend in the analyzed data series, contrary to the null hypothesis, namely not to have the trend [37]. The level of significance represents the probability of rejecting the null hypothesis when it is true. A significance level (α) is acceptable the closer it is to the value 0 [10]. In the application of the Mann–Kendall test, four significances levels were considered: $\alpha = 0.001$, $\alpha = 0.01$, $\alpha = 0.05$ and $\alpha = 0.1$ [26].

Considering the air temperature, according to Table 4, a statistically significant increase at a significance level $\alpha = 0.001$ was found, for all weather stations along the Lower Danube Valley, both in June, July and August, as well as at the annual, spring and summer level, for the period 1961–2019. These rising air temperature trends are in line with those at the continental level [37].

The trends of precipitation amounts compared to those of air temperature are much less clear and not significant. Thus, the amounts of precipitation falling in the study area can be considered generally as stationary in time (Table 5).

The non-parametric Mann–Kendall test indicates few statistically significant upward trend for $\alpha = 0.01$ in October at Zimnicea and Galați weather stations (Table 5). Also, the positive trend was registered for the autumn season, but statistically significant for $\alpha = 0.05$ at Zimnicea WS and $\alpha = 0.1$ at Călărași WS (Table 5).

Decreasing trends, statistically significant for $\alpha = 0.05$ at Zimnicea WS and $\alpha = 0.1$ at Galați WS were identified in August (Table 5). Also, it was found that the

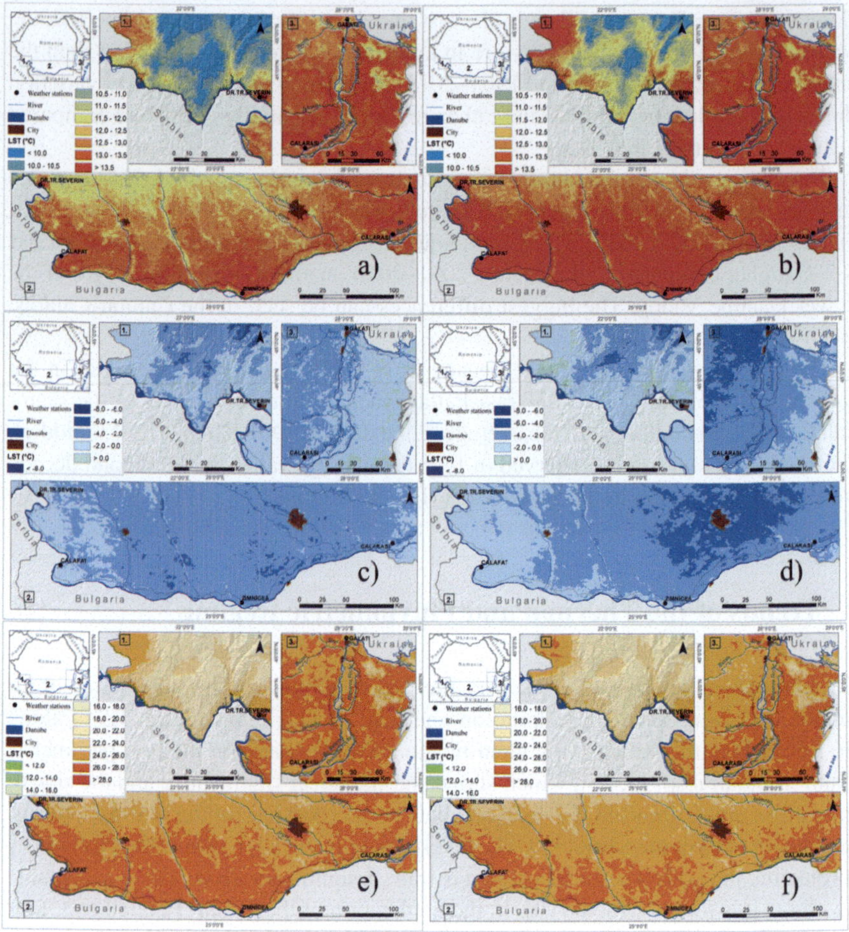

Fig. 13 The average decadal variation of LST at the year level (top), January (middle) and July (bottom) along the Lower Danube River, for the decades 2001–2010 (a, c and e) and 2011–2019 (b, d and f): (1) Baziaș-Gura Văii sector, (2) Gura Văii-Călărași sector; (3) Călărași-Pătlăgeanca sector

decreasing tendencies of the monthly average precipitation amounts predominate, but without being statistically significant according to the Mann–Kendall test (Table 5).

In the last 20 years, a tendency of temperature increase was detected at the land surface temperature (LST) in all the three sectors of the Lower Danube Valley. However, this increase was statistically significant for $\alpha = 0.05$ in September and in the autumn season, according to the Mann–Kendall test (Table 6). A statistically significant increase was registered for April in Baziaș-Gura Văii and Gura Văii-Călărași sectors. At annual level, the statistically significant positive trend was registered only for the Baziaș-Gura Văii sector (Table 6).

Table 4 The trends of the average monthly, annual and seasonal air temperature (°C) according to the non-parametric Mann–Kendall test for weather stations along the Lower Danube River (1961–2019)

No years	Time series	Dr. Tr. Severin		Calafat		Zimnicea		Călărași		Galați	
		Z	SS	Z	SS	Z	SS	Z	SS	Z	SS
59	I	2.35	*	1.63		1.57		1.75	+	1.65	+
59	II	1.83	+	1.40		1.67	+	1.88	+	2.21	*
59	III	2.45	*	2.58	**	2.58	**	2.59	**	2.75	**
59	IV	2.41	*	1.77	+	1.32		1.56		1.90	+
59	V	1.94	+	1.67	+	1.70	+	2.38	*	2.84	**
59	VI	4.26	***	4.17	***	3.87	***	4.29	***	4.09	***
59	VII	5.05	***	4.70	***	4.22	***	4.36	***	4.53	***
59	VIII	4.62	***	4.17	***	4.50	***	4.45	***	5.05	***
59	IX	1.22		0.99		1.98	*	2.69	**	2.45	*
59	X	1.48		0.51		1.05		1.35		1.70	+
59	XI	0.50		0.20		0.49		0.63		0.47	
59	XII	1.86	+	1.49		1.22		1.07		1.39	
59	Annual	5.21	***	4.94	***	4.81	***	4.72	***	4.97	***
59	Spring	3.90	***	3.41	***	2.98	**	3.41	***	4.22	***
59	Summer	5.39	***	5.28	***	5.17	***	5.65	***	5.87	***
59	Autumn	1.61		0.85		1.52		2.12	*	2.46	*
58	Winter	2.41	*	1.89	+	1.64		1.97	*	2.11	*

***$\alpha = 0.001$ level of statistical significance; **$\alpha = 0.01$ level of statistical significance; *$\alpha = 0.05$ level of statistical significance; + $\alpha = 0.1$ level of statistical significance; Z—Test Z; SS—Statistical Significance

Downward trends in the LST climate parameter were found in May and July in all the sectors of the Lower Danube Valley, but these were not statistically significant, according to the Mann–Kendall test (Table 6).

6 Conclusion and Recommendation

The influence of the river waters, the complexity of the terrestrial surface and the diversity of the climatic influences materialize in some climatic particularities specific to the Lower Danube Valley, which individualizes it within the country.

In the present study, the termo-pluviometric features and changes observed in the variation of air and land surface temperatures and of precipitation amounts for the period 1961–2019 were highlighted. Following this analysis, it was found that both the average annual air temperature and the annual amount of precipitation decrease from the West to the East along the Lower Danube Valley. Although the Danube River is a permanent source of air humidity, it was found, based on the Péguy climograms and boxplots, the presence of a climate with high thermic and precipitation contrasts

Table 5 The trends of average monthly, annual and seasonal precipitation amounts (mm) according to the Mann–Kendall non-parametric test for the weather stations along the Lower Danube River (1961–2019)

No years	Time series	Dr. Tr. Severin		Calafat		Zimnicea		Călărași		Galați	
		Z	SS	Z	SS	Z	SS	Z	SS	Z	SS
59	I	0.42		0.76		1.06		2.39	*	1.60	
59	II	0.08		-0.34		-0.78		-0.69		-0.73	
59	III	-0.37		0.27		1.09		1.22		1.07	
59	IV	0.47		-1.12		-0.48		1.05		0.16	
59	V	-0.20		0.19		0.69		0.03		-0.86	
59	VI	-0.35		-0.05		0.48		0.33		0.01	
59	VII	-0.09		1.22		-1.14		-0.60		0.12	
59	VIII	-0.10		0.98		-2.31	*	-0.67		-1.83	+
59	IX	0.68		0.94		1.24		1.15		0.58	
59	X	0.90		1.63		2.66	**	1.49		2.58	**
59	XI	-0.08		-0.03		0.10		0.89		0.37	
59	XII	-0.89		-0.80		-0.26		0.64		0.48	
59	Annual	-0.07		1.09		0.46		1.62		0.69	
59	Spring	-0.77		-0.25		0.44		0.75		0.27	
59	Summer	-0.35		0.75		-1.14		-0.26		-0.92	
59	Autumn	0.93		1.31		2.05	*	1.67	+	1.60	
58	Winter	-0.44		-0.38		0.06		1.14		1.23	

***$\alpha = 0.001$ level of statistical significance; **$\alpha = 0.01$ level of statistical significance; *$\alpha = 0.05$ level of statistical significance; + $\alpha = 0.1$ level of statistical significance; Z—Test Z; SS—Statistical Significance

that intensifies in the West–East direction, as a result of the transition from specific Mediterranean influences at the Baziaș-Gura Văii sector and a significant part of the Gura Văii-Călărași sector to those of aridity that characterize the rest of the Gura Văii-Călărași sector (the area of the Călărași weather station) and the Călărași-Pătlăgeanca sector (in the eastern part of the Danube valley). Considering the average annual land surface temperature (LST), the Gura Văii-Călărași sector registers the highest values in the South of the country, and implicitly in the study area.

The ten-years average variation in the air temperature showed progressive increases towards the end of the analyzed period, the last decade being the warmest, similar to the global level. This progressive increase was also recorded by the land surface temperature. In terms of precipitation amounts, the variation of the decadal average values indicates increases and decreases from a decade to another, but without significant changes. The general trends in the variability of the temperature and precipitation, identified with the non-parametric Mann–Kendall test, indicates for the air and land surface temperatures much clearer changes than those of the precipitation amounts and consistent with those of the continental and global level [37]. For the analyzed period, it was found for all weather stations in the study area, a statistically significant increase at the level of $\alpha = 0.001$ of the air temperature for

Table 6 The trends of the average monthly, annual and seasonal land surface temperature (LST) (°C) according to the Mann–Kendall non-parametric test along the Lower Danube River (2000–2019)

No years	Time series	Baziaş-Gura Văii		Gura Văii-Călăraşi		Călăraşi-Pătlăgeanca	
		Z	SS	Z	SS	Z	SS
19	I	0.3		0.1		-0.7	
20	II	0.9		0.0		0.0	
20	III	0.6		0.2		0.1	
20	IV	2.0	*	1.7	+	0.4	
20	V	-0.8		-0.6		-0.2	
20	VI	0.2		0.0		0.2	
20	VII	-0.7		-1.5		-0.6	
20	VIII	0.9		0.0		0.0	
20	IX	2.0	*	2.4	*	2.6	*
20	X	0.6		0.8		0.4	
20	XI	1.2		1.1		1.2	
20	XII	1.4		1.5		1.5	
20	Annual	2.0	*	1.5		0.7	
20	Spring	1.1		0.5		-0.2	
20	Summer	0.6		-0.5		-0.9	
20	Autumn	2.2	*	2.3	*	2.1	*
19	Winter	1.2		0.8		0.6	

***$\alpha = 0.001$ level of statistical significance; **$\alpha = 0.01$ level of statistical significance; *$\alpha = 0.05$ level of statistical significance; + $\alpha = 0.1$ level of statistical significance; Z—Test Z; SS—Statistical Significance

June, July, August, spring, summer and at the annual level. The trend of land surface temperature was registered at the statistical level of $\alpha = 0.05$ for all the sectors of the Lower Danube Valley, in September and autumn. The trends of the average monthly precipitation are predominantly decreasing, but not statistically significant.

The analysis of the climate variability and the identification of changes in the variability of the main climate parameters are of great scientific and practical interest. They are valuable tools for assessing the impact of climate changes on both the environment and society. These analyzes, at different spatial scales, can be the basis of strategic management decisions to reduce and combat the impact of the current global warming. Also, these decisions can be taken in connection with the specifics of the local communities and the seventeen Sustainable Development Goals (SDGs) of the 2030 Agenda for Sustainable Development. The integration of climate change measures into national policies for the development of specific adaptation methods should be based on scientific studies. In order to achieve such national strategies depending on the predominant political, historical, cultural and ecological circumstances, a better collaboration between academic institutions (researchers) and stakeholders is necessary.

References

1. Hoegh-Guldberg, O, Jacob D, Taylor M, Bindi M, Brown S, Camilloni I, Diedhiou A, Djalante R, Ebi KL, Engelbrecht F, Guiot J, Hijioka Y, Mehrotra S, Payne A, Seneviratne SI, Thomas A, Warren R, and Zhou G (2018) Impacts of 1.5°C of Global Warming on Natural and Human Systems. In: Global Warming of 1.5°C. An IPCC Special Report on the impacts of global warming of 1.5°C above pre-industrial levels and related global greenhouse gas emission pathways, in the context of strengthening the global response to the threat of climate change, sustainable development, and efforts to eradicate poverty p 138
2. Dumitrescu A, Bojariu R, Birsan M-V, Marin L, Manea A (2015) Recent climatic changes in Romania from observational data (1961–2013). Theoret Appl Climatol 122:111–119. https://doi.org/10.1007/s00704-014-1290-0
3. Vlăduţ A, Onţel I (2013) Summer air temperature variability and trends within Oltenia Plain. In: Journal of the Geographical Institute "Jovan Cvijic"
4. Zaharia L, Ioana-Toroimac G, Perju E-R (2020) Hydrological impacts of climate changes in Romania. In: Negm AM, Romanescu G, Zeleňáková M (eds) Water resources management in Romania. Springer, Cham, pp 309–351
5. Birsan M-V, Zaharia L, Chendes V, Branescu E (2012) Recent trends in stream flow in Romania (1976–2005). Romanian Rep Phys 64:275–280
6. Cheval S, Busuioc A, Dumitrescu A, Birsan M-V (2013) Spatiotemporal variability of the meteorological drought in Romania using the standardized precipitation index. In: EGU General Assembly Conference Abstracts. pp EGU2013–7085
7. Páscoa P, Gouveia CM, Russo AC, Bojariu R, Vicente-Serrano SM, Trigo RM (2020) Drought impacts on vegetation in southeastern Europe. Remote Sensing 12:1–21. https://doi.org/10.3390/rs12132156
8. Constantin (Oprea) MD, Ionac N, Marinică I, Vătămanu VV, Marinică AF, Grigore E, Bogan E, Onțel I (2019) The agro-climatic resources of the year 2015 and their impact on the agricultural crops in the South–West of Romania 19:81–92
9. Constantin (Oprea) MD, Marinică AF, Marinica I, Zaharia L, Grigore E, Ontel I, Ioana-Toroimac G, Tișcovschi AA (2019) The thermal and pluvial features of the agricultural year 2018 in the South-Western Romania and the effects on the crops. pp 543–548
10. Busuioc A, Caian M, Cheval S, Bojariu R, Boroneant C, Baciu M, Dumitrescu A (2010) Variabilitatea şi schimbarea climei în România. Pro Universitaria, ISBN: 978–973–129–549–7
11. Mateescu E (2018) 100 de recorduri climatice în România, Conferința Națională Schimbări climatice globale, in Romanian, Bucharest
12. Dang T, Yue P, Bachofer F, Wang M, Zhang M (2020) Monitoring land surface temperature change with Landsat images during dry seasons in Bac Binh. Vietnam. Remote Sensing 12:1–21. https://doi.org/10.3390/rs12244067
13. Luintel N, Ma W, Ma Y, Wang B, Subba S (2019) Spatial and temporal variation of daytime and nighttime MODIS land surface temperature across Nepal. Atmospheric and Oceanic Science Letters 12:305–312. https://doi.org/10.1080/16742834.2019.1625701
14. Yan Y, Mao K, Shi J, Piao S, Shen X, Dozier J, Liu Y, Ren H, Bao Q (2020) Driving forces of land surface temperature anomalous changes in North America in 2002–2018. Sci Rep 10:1–13. https://doi.org/10.1038/s41598-020-63701-5
15. Zhao B, Mao K, Cai Y, Shi J, Li Z, Qin Z, Meng X, Shen X, Guo Z (2020) A combined Terra and Aqua MODIS land surface temperature and meteorological station data product for China from 2003 to 2017. Earth System Science Data 12:2555–2577. https://doi.org/10.5194/essd-12-2555-2020
16. Jaber SM, Abu-Allaban MM (2020) MODIS-based land surface temperature for climate variability and change research: the tale of a typical semi-arid to arid environment. European Journal of Remote Sensing 53:81–90. https://doi.org/10.1080/22797254.2020.1735264
17. NourEldeen N, Mao K, Yuan Z, Shen X, Xu T, Qin Z (2020) Analysis of the spatiotemporal change in land surface temperature for a long-term sequence in Africa (2003–2017). Remote Sensing 12. https://doi.org/10.3390/rs12030488

18. European Commission. Joint Research Centre (2015) Water scenarios for the Danube river basin: elements for the assessment of the Danube agriculture energy water nexus. Publications Office, LU
19. European Commission. Joint Research Centre (2020) Water scenarios for the Danube River Basin: future challenges and preparedness: a foresight study to inform water management in the Danube River Basin. Publications Office, LU
20. ICPDR River Basin Management Expert Group (2013) ICPDR strategy on adaptation to climate change. https://www.icpdr.org/main/publications/climate-change-adaptation-strategy-danube-basin-adopted. Accessed on 15 July 2021
21. Dumitrescu A, Birsan M-V (2015) ROCADA: a gridded daily climatic dataset over Romania (1961–2013) for nine meteorological variables. Nat Hazards 78:1045–1063. https://doi.org/10.1007/s11069-015-1757-z
22. Birsan M-V, Dumitrescu A (2014) ROCADA: Romanian daily gridded climatic dataset (1961–2013) V1.0. Administratia Nationala de Meteorologie, Bucuresti, Romania
23. Wan Z, Hook S, Hulley G (2015) MOD11A1 MODIS/Terra land surface temperature/emissivity daily L3 global 1km SIN grid V006. NASA EOSDIS Land Processes DAAC. Accessed on 15 Jan 2021
24. Wan Z, Hook S, Hulley G (2015) MYD11A1 MODIS/Aqua land surface temperature/emissivity daily L3 global 1km SIN grid V006. Accessed on 15 Jan 2021
25. National Meteorological Administration (2020) Meteorological data from the archive of the National Meteorological Administration
26. Salmi T, Määttä A, Anttila P, Ruoho-Airola T, Amnell T (2002) MAKESENS-application for trend calculation. Finnish Meteorological Institute, Helsinki
27. Patru IG, Zaharia L, Oprea R (2006) Geografia fizica a Romaniei – Clima, apa, vegetatie, soluri. Editura Universitară, Bucharest
28. Posea G, Zavoianu I (2005) Geografia României, V - Câmpia Română, Dunărea, Podişul Dobrogei, litoralul românesc al Mării Negre şi Platforma Continentală, V. Romanian Academy, Bucharest
29. Dragotă C-S (2006) Precipitațiile excedentare în România. Editura Academiei Române, București
30. Bojariu R, Chitu Z, Dascălu SI, Gothard M, Velea LF, Burcea R, Dumitrescu A, Burcea S, Apostol L, Amihaesei V, Marin L, Crăciunescu VS, Irimescu A, Mătreață M, Niță A, Bîrsan MV (2021) Schimbările climatice – de la bazele fizice la riscuri și adaptare. Ediție revăzută și adăugită. Printech, Bucharest, ISBN: 978–606–23–1275–6
31. NOAA (2021) Global surface temperature anomalie. https://www.ncdc.noaa.gov/monitoring-references/faq/anomalies.php. Accessed on 10 July 2021
32. National Meteorological Administration (2021) Raport anual 2020. https://www.meteoromania.ro/despre-noi/raport-anual/raport-anual-2020/. Accessed on 15 July 2021
33. Constantin (Oprea) MD (2013) Relația climă-poluarea mediului înconjurător în arealul municipiului Slatina. Universitară, Bucharest
34. World Meteorological Organization (2021) 2020 was one of three warmest years on record. https://public.wmo.int/en/media/press-release/2020-was-one-of-three-warmest-years-record. Accessed on 10 July 2021
35. Mann HB (1945) Non-parametric test against trend. Econometrica 13:245–259. https://doi.org/10.2307/1907187
36. Kendall MG (1975) Rank correlation methods. Griffin, London
37. Ontel I (2018) Monitorizarea și evaluarea fenomenelor de uscăciune și secetă în Câmpia Olteniei. Thesis, University of Bucharest

A SPEI-Based Approach to Drought Hazard, Vulnerability and Risk Analysis in the Lower Danube River Region

Nina Nikolova, Dana Magdalena Micu, Alexandru Dumitrescu, Kalina Radeva, Monica Paraschiv, Sorin Cheval, and Leonid Todorov

Abstract Natural hazards, including droughts, are processes and phenomena that can trigger a negative impact on the environment, society and various economic sectors. The present chapter aims to identify spatial peculiarities of drought characteristics (frequency, duration, affected area) and to analyse drought hazard, vulnerability and risk in the Lower Danube region. The study area includes administrative regions from Romania (counties) and Bulgaria (districts) located along the Danube River, which is the common administrative border between the two countries. The northward and southward Danube territories are part of the most important agricultural areas of both countries, where natural landscapes have been significantly transformed by anthropogenic activities which contributed to the removal of the natural vegetation and its replacement with cultivated plants and urban areas. Drought characteristics and associated hazards were analysed using the Standardized Precipitation-Evapotranspiration Index (SPEI-3, 6, 12) for the period 1981–2019. Population density and land cover/land use data were taken into account in the drought vulnerability assessment. Drought hazard and vulnerability were considered in the drought risk evaluation which allowed the identification of the regional drought "hotspots". Results show a very high level of drought risk associated to short-term drought (SPEI-3) in the central and eastern parts of the study region. In the case of long-term drought (SPEI-12), a reduction in areas showing a very high drought risk level is observed. The administrative regions located in the western part of the study area have very low and low levels of drought risk.

N. Nikolova (✉) · K. Radeva · L. Todorov
Faculty of Geology and Geography, Sofia University St. Kliment Ohridski, 15 Tsar Osvoboditel Blvd, 1504 Sofia Center, Bulgaria
e-mail: nina@gea.uni-sofia.bg

K. Radeva
e-mail: kradeva@gea.uni-sofia.bg

L. Todorov
e-mail: leo_todorov@gea.uni-sofia.bg

D. M. Micu · A. Dumitrescu · M. Paraschiv · S. Cheval
Meteo Romania (National Meteorological Administration), București-Ploiești no. 97, Sector 1, 013686 Bucharest, Romania
e-mail: gabriela.paraschiv@meteoromania.ro

A. Negm et al. (eds.), *The Lower Danube River*, Earth and Environmental Sciences Library, https://doi.org/10.1007/978-3-031-03865-5_10

Keywords Drought duration · Hazard · Vulnerability · Risk · Romania · Bulgaria

1 Introduction

Drought is a complex hydroclimatic hazard associated to water deficit that can last for several days, months or even years. It has a low onset and can cause a wide variety of short-term to long-term impacts in many sectors such as agriculture and livestock farming [1], ecosystems [2, 3], energy production and price [4], which can persist after the event end [5]. Drought can occur in any region or climate, and can produce impacts at multiple spatial scales ranging from local to regional, national or even continental [6].

Due to its complex nature and wide range of possible impacts, in many sectors drought has always been given constant attention throughout the scientific community, in an effort to understand its meaning (definitions), triggered effects, but also to develop methodologies for quantifying its characteristics (frequency, onset, end, duration and severity), tracing its periodicity, and estimating impacts over given historical observational periods. Drought can be defined in several different ways depending on the purpose, criteria for identification of dry events and foreseen impacts, e.g. a long-term decrease in water availability at a given time in a given area [e.g. 7, 8]; a permanent and extensive reduction in area below the average natural water availability, which may affect all elements of the water cycle [e.g. 9]. Drought differs from aridity, low flow, water scarcity and desertification [10]. Both drought and aridity phenomena depict dryness-prone environments, but the two concepts are clearly delimited [11]. While aridity is a feature of average climate within environments with long-term state of dryness, drought refers to abnormal temporal deficiencies in moisture in a given environment and is a cause of climate variability (short time scale) [12]. On the other hand, water scarcity (or desertification) is perceived as a temporary (or long-term) water imbalance with anthropogenic causes, due to an unsustainable use of water resources influenced by water managers [13].Today's understanding of drought is oriented towards its multi-dimensional nature, resulting in different types of manifestations grouped into meteorological, hydrological, and pedological characteristics defined by the physical aspects of the propagating hazard (e.g. meteorological or soil moisture drought), but also by the potentially impacted sector (e.g. agricultural drought, socio-economic drought) [12, 14].

Climate change was found to influence the characteristics and distribution of seasonal drought events throughout Europe, with an increase in frequency and severity especially in Southern Europe (summer and autumn), and a decrease in Northern Europe (winter, spring) [15, 16]. The increases observed in air temperature and the fluctuations in the amount and distribution patterns of seasonal rainfalls produced visible changes in the hydrological cycle, affecting water storage, groundwater recharge, soil moisture and water resource availability for the needs of society, the environment and business activities [17, 18]. In recent years, various regions across Europe were affected by 'dry' hazards, which emerged through the joint action of prolonged periods with below normal precipitation and record-breaking summer temperatures [19–21]. These hazards occurred in cascade, as drought events coupled with heat waves and wildfires, and resulted in substantial negative impacts. Relevant examples include the most recent drought events of the twenty-first century (e.g. droughts of summers 2003, 2010, 2012, 2015, 2018), already comprehensively documented, which affected vast territories and determined yield losses and productivity decreases for certain crops [22–26], soil dryness and land degradation processes [27–29], increased demand for cropping irrigations [30, 31], widespread wildfires [32–34] and numerous heat-related deaths [35–37]. The dimension of the economic impact of drought across Europe can be impressive. Previous estimations have shown that the events recorded over the 1976–2006 period determined an average economic impact of about 100 billion €, twice as much over the 1991–2006 period, and an exceptional cost of 8.7 billion € in 2003 alone [e.g. 38–40]. The MunichRe's NatCatSERVICE1 database indicates annual losses associated to drought events of around 1.3 €billion over the last 35 years [41]. In other estimations, the average annual economic consequences of droughts recorded between 1990 and 2010 in Europe have drastically risen to 6.2 €billion per year [42]. The impacts of dry hazards proved to be cumulative and show an apparent intensification trend amid climate change, in relation to an increasing occurrence probability due to projected future warming [43–46].

While drought development in time and space depends on multiple complex factors (e.g. atmospheric processes, surface energy budget, land use), its propagation is mainly climate-dependent [47, 48]. Additionally, anthropogenic drivers (e.g. intensive water use, deficient water management) can further exacerbate the pre-existing dry conditions and increase social vulnerability [e.g. 49–51]. The complexity of drought (characteristics, causes, impacts) is well captured in a vast body of literature available worldwide, which grew significantly over recent decades in relation to the implications of global warming on precipitation and drought. However, these effects may vary substantially by region [52], in connection to the enhanced moisture holding capacity of the atmosphere resulting in precipitation increase in some regions, to the offset role of evapotranspiration especially in dry regions, and to the balance between atmospheric radiative cooling and the latent heating of the atmosphere [53, 54].

Numerous studies already provide good evidence on the link between climate change and drought. At global scale, long-term drought tendency is still being debated in literature [55, 56]. In Europe, while some regions, such as the Mediterranean, act as emergent "hot spots", where both frequency and severity of drought show a visible increase since the 1950s [57–60], others, such as Eastern Europe, show no clear tendencies of drought evolution [58]. Conversely, Northern Europe exhibits clear wetting patterns and less severe dry episodes [61–63].

Drought trends in the catchment of one of Europe's most important water ways (the Danube River) are inconclusive, showing both decreases and increases [59], in relation to the oscillatory behavior of precipitation extremes and changes in the persistence of atmospheric circulation patterns over the North Atlantic [64]. Nevertheless, in recent decades, the extreme rainfall deficits recorded in 2003, 2007, 2012 and 2015 produced a wide variety of impacts throughout large areas across the

Danube Catchment such as low flows, disruptions in water-borne transportation, reduced hydropower production, crop failures, and decreased vegetation activity [65–70]. Based on an extensive analysis of impacts produced by the historical droughts recorded between 1981–2016 in Danube Catchment countries, [71] the Romanian-Bulgarian Danube Floodplain area was classified as a region of high drought impact, in relation to agricultural and forestry activities, although only a slight increase in the number of severe droughts has been observed. Many studies focused on the characteristics, causes and impacts of meteorological and hydrological droughts both in Romania and Bulgaria. These countries are prone to drought occurrence, but they are only assigned a medium level of water scarcity hazard (20% chance for drought to occur in the coming 10 years) [72].

In Bulgaria, many drought studies focused on various methodological approaches for drought research [73, 74], drought climatology and drought types [75–80], and drought impact [81]. In recent years, drought risk was the main topic in several studies conducted at regional scale [e.g. 82, 83]. Observational data indicate that drought frequency at national scale has increased over the past two decades [74, 82, 84]. A significant water deficit has also been observed in a large part of Bulgaria for many years. This phenomenon is related to the shortening of the snow cover period, as well as to prolonged periods without rainfall and high temperatures in the summer period, which caused rapid evaporation of water from plants and soil. The intensification of agriculture and the unification of habitats, construction of drainage systems, as well as dense, impermeable development of urbanized areas, resulted in the acceleration of water circulation in river catchments, contributing to an increase in the frequency of droughts countrywide. As a result of these activities, the natural retention capacity in river catchments was reduced. In addition, global warming also causes an increase in temperature in Bulgaria, especially in the cold half-year [85, 86]. This, in turn, entails an increase in field evaporation in winter and spring, and a decrease in infiltration of groundwater alimentation in the cool half-year. As a result, the water resources available in the warm half-year are lower, which causes problems in water supply to various users. Across half of Bulgaria's territory, water deficits appear periodically and, most often and to the greatest extent, affect the Danube plain [87]. In this region, the risks associated with the lack of access to water of adequate quantity and quality are currently at a peak, and, taking into account the observed direction of climate change, this condition may worsen even more [88, 89]. Thus, an active drought risk management policy is necessary in Bulgaria to ensure the safety of water resources in the face of current climate threats and further expected changes.

In Romania, drought is acknowledged as a specific country risk [90]. Numerous studies documented various aspects related to drought and climate aridity, focusing on different past observational periods and different spatial scales such as: climatology, variability and trends in relation to climate change [e.g. 91–95], analysis methodologies [e.g. 96–100], multidecadal variability in connection to large-scale circulation patterns [e.g. 101], variability of Danube river flow in the lower basin [e.g. 102] or climate change effects on water deficit and associated impacts in agriculture [e.g. 103–105]. Other recent studies also aimed to evaluate the socio-economic effects of drought and associated vulnerability in the southern lowlands of Romania, prone to drought occurrence, including some sectors of the Romanian Danube Valley [e.g. 103, 106–108]. In most previous research studies, the Lower Danube Floodplain area was so far given limited attention in spite of its great exposure to this phenomenon amid the current climate, but also from the perspective of the changing climate projected over the next decades, when an increasing frequency of both drought and low flows is expected throughout the entire Danube River Basin [109].

The present study aims to analyse the drought hazard and vulnerability to droughts in the Lower Danube River region and to provide information about drought characteristics over the 1981–2019 period in the counties (Romania) and administrative districts (Bulgaria) located northward and southward along the Danube River, using a Standardized Precipitation Evapotranspiration Index (SPEI) -based approach. The research objectives of this study are the following: (1) to investigate the climatic characteristics of drought for different SPEI timescales; (2) to determine the level of drought hazard associated to different SPEI timescales at NUTS3 level; (3) to estimate the level of vulnerability to drought and 4) to determine the drought risk "hotspots" by drought risk within the study region.

2 Study Region

The Romanian-Bulgarian Danube Floodplain region is located in the Lower Danube Basin, along the common administrative border of the two countries, covering parts of the Romanian Plain (the Lower Danube Plain) and the Pre-Balcanic Tableland. The study region stretches between 45.7–45.6° N latitude and 21.3–28.0° E longitude, and covers a 2 to 2278 m elevation range. The Danube crosses the region along 748 km and plays the role of a vital waterway for both wildlife and local communities. Over the past decades, the region was subject to major environmental and landscape transformations including the conversion of wetlands to agriculture, removal of riparian forests, building of flood defence infrastructure (e.g. river dykes) and river pollution. Most of these transformations triggered subsequent impacts on the flooding regimes and eutrophication, which ultimately affected the ability of these delicate riparian forest ecosystems to regenerate [110]. The region is part of the Lower Danube Green Corridor (established in 2000) along which actions to protect and restore wetland biodiversity and to reconnect the river to its natural flooding areas are foreseen. Said actions aim to reduce the risk of major flooding in populated areas and to provide benefits to local economies (e.g. through fisheries, tourism) and to ecosystems along the river [111].

The region has a typical continental climate with pronounced aridity especially in summer [112, 113], frequent frosty winters and little snow, and is sensitive to climate change and associated hydro-meteorological extremes such as drought, floods and heat waves, which encompass a broad range of impacts on agriculture, irrigation, forestry, biodiversity and ecosystems, water related energy production and navigation. The southern and eastern parts of the Danube River Basin (overlapping the study region) show the largest exposure to drought compared to other sections of

Fig. 1 The study region and location of selected cities

the catchment [114], especially in the lowland areas with elevation below 200–300 m. According to the International Commission for the Protection of the Danube River (ICPDR), the most important drought events which affected agriculture and the hydrological regime in the Lower Danube River Basin were recorded in 1992/1993, 1996, 2003 and 2015.

In this study, drought hazard, vulnerability and risk were aggregated in relation to the administrative units corresponding to NUTS3 (counties in Romania and administrative districts in Bulgaria). The selected counties and districts from the Romanian-Bulgarian Danube Floodplain region are shown in Fig. 1. The climatic characteristics of the drought phenomenon were discussed for a number of 18 cities located on both sides of the Danube River, in drought-prone areas.

3 Data and Methods

3.1 Data Sets

This study is based on monthly air temperature and precipitation data recorded over the 1981–2019 period, extracted from ERA5-Land global land-surface database available from the Climate Change Service (C3) of the EU Copernicus Programme (https://cds.climate.copernicus.eu/cdsapp#!/dataset/reanalysis-era5-land-monthly-means?tab=overview). This database was recently released (2019) and made publicly available in the Climate Data Store of C3S with a horizontal resolution of 11 km.

The assessment of vulnerability to drought was made at NUTS 3 level, aggregating population data extracted from the latest available censuses in Romania and in Bulgaria (2011), as well as land cover/land use data sourced from CORINE Land Cover (CLC2018) records available through the Land Monitoring Service (LMS) of the COPERNICUS Programme.

3.2 Methods

The results are based on the computation of the Standardized Precipitation-Evapotranspiration Index (SPEI) with the R SPEI library (https://cran.r-project.org/web/packages/SPEI/index.html). This library gridded data on air temperature and precipitation were used to compute the reference evapotranspiration estimated based on the Thornthwaite method. SPEI was selected for the drought analysis in this study due to the advantages it provides, i.e. it considers air temperature effects (PET) on drought severity and its sensitivity to global warming [115]. This study focuses on the moderate (MD, SPEI between −1.00 and −1.49), severe (SD, −1.50 < SPEI >−1.99) and extreme (ED, SPEI <−2.00) drought events, and considers that a drought event begins in the first month when SPEI values become negative (or go below the corresponding threshold for each drought type) and end in the last month, when SPEI exceeds 0 (or the corresponding threshold for each drought type). The threshold for identifying dry events is 0 instead of −1, to allow the inclusion of dry events of low duration [116]. The three timescales used for SPEI computation were retained for analysis in relation to their relevance to meteorological drought (SPEI 3 months), hydrological drought (SPEI 12 months) and agriculture (SPEI from 3 to 6 months) [58].

The gridded series of SPEI were used in the diagnostic analysis of regional drought characteristics (frequency, duration and coverage of affected areas) at seasonal and annual time scales. The SPEI was calculated at each grid cell, which were further averaged at the county level. Hereafter, the occurrence of drought events is given by the number of consecutive months with negative SPEI values. The gridded SPEI series were used to analyze drought characteristics such as the *frequency* (the number of months with negative SPEI), *duration* (the number of consecutive months with negative SPEI) and *affected area* (pixels with negative SPEI values, expressed in % relative to the total area of NUTS3 territory).

Trends and their statistical significance in drought SPEI-based characteristics (frequency, duration and affected area) were evaluated using the non-parametric Mann–Kendall test, relative to the significance level of at least 10% (p-value < 0.1).

Drought Hazard Index (DHI) quantifies the probability of occurrence of negative SPEI grid points corresponding to each timescale (3, 6 and 12 months) annually, which was averaged for the entire study period and further classified into five classes of drought hazard, using the quantile method, from very low to very high.

Drought vulnerability. The impact magnitude of a drought depends on the vulnerability of the exposed assets and sectors (yield, people, high-water consumption industries, natural ecosystems). Factors determining the vulnerability to drought change in space and time, which leads to various consequences of this phenomenon. Droughts result from the complex interaction of biophysical (e.g. meteorological) and human (social and economic) factors. In this study, the selection of indicators for quantifying drought vulnerability was conducted from the perspective of two main aspects: i) a significant part of the study region is covered by agricultural crops or natural vegetation, ii) and agriculture is considered the most vulnerable sector to adverse meteorological and climatic phenomena, including drought in both countries.

Population data refer to population density (expressed as number of inhabitants/km^2), as a relevant component of element at risks, reflecting the size of population affected by drought. Land cover data used in the quantification of drought vulnerability refer to the share (%) of four land use/land cover (LULC) classes in the total area of the selected NUTS3 regions, which are relevant to agricultural activities across the region (share of arable lands/permanent crops, pastures, forests and shrubs and herbaceous associations). Based on the expert knowledge and peculiarities of the territory these classes were grouped, weighted and assigned to five drought vulnerability classes, ranging from very low to very high, as for DHI (Table 1).

The weights assigned to the selected CLC2018 types followed the Analytic Hierarchy Process according to other previous approaches [e.g. 117, 118]. In Table 1 the values in the row show the degree of vulnerability to drought of a given type of land use/land cover compared to the types in the column. For example, arable land is 3 times more vulnerable to drought than forests. Based on the points obtained for each LULC type, its percentage of the total number of points is calculated. According to AHP, the most vulnerable type of LULC to drought is the group of Arable land, Permanent crops, Heterogeneous agricultural areas and Inland wetlands, and the least vulnerable is the group of Scrub and/or herbaceous vegetation associations.

The relative share of the area of each LULC class for each administrative region was multiplied by the assigned weights and then summed up. Thus, a LULC-based composite indicator of vulnerability to drought of each administrative district was obtained.

In order to compare the values given in different scales and units, and to aggregate them and calculate the drought vulnerability index (DVI), the population density data and integrated LULC indicator data for districts was normalized using the following formula:

$$X = \frac{X_i - X_{min}}{X_{max} - X_{min}},$$

where X is the normalized value of the indicator; X_i is the value of the given indicator (population density or land use/land cover indicator) for district "i", X_{max} is the

Table 1 Drought vulnerability (DV) indicators and weights

LULC types	Arable land; Permanent crops; Complex cultivation patterns; Inland wetland	Pastures; Land principally occupied by agriculture, with significant areas of natural vegetation	Forests	Scrub and/or herbaceous vegetation associations	Total	Weight %
Arable land; Permanent crops; Complex cultivation patterns; Inland wetland	1.00	2.00	3.00	4.00	10.0	45
Pastures; Land principally occupied by agriculture, with significant areas of natural vegetation	0.50	1.00	2.00	3.00	6.5	29
Forests	0.33	0.50	1.00	2.00	3.8	17
Scrub and/or herbaceous vegetation associations	0.25	0.33	0.50	1.00	2.1	9
Total					22.4	100

maximum value of indicator X_i, and X_{min} is the minimum value of indicator X_i according to [107].

The Drought Vulnerability Index (DVI) was calculated as the mean value of normalized population density (PD) data and the integrated land use/land cover indicator (LULC) using the following formula:

$$DVI = \frac{PD + LULC}{2}$$

The values of DHI and DVI are grouped into five classes according to 1-st – 5-th quantiles (1-st very low – 5-th very high), and regional distribution maps for drought hazard and vulnerability were developed. Several publications [119–121] consider drought hazard and vulnerability as components of drought risk assessment.

Drought risk was calculated through the multiplication of drought hazard and drought vulnerability for each timescale of SPEI (3, 6 and 12 months). In the present study, the assessment of drought risk was conducted at the level of administrative districts, to identify critical drought-prone areas (hereafter also referred to as "hotspots") in the study area.

4 Results and Discussions

4.1 Drought Characteristics

4.1.1 Drought Frequency

Drought is a recurrent hydro-metrological extreme event across the region in relation to the precipitation deficits recorded throughout the 1981–2019 period. The total number of drought events decreases with the size of drought timescale, ranging from 36 events (12 months) to 761 events (6 months), and to 1056 events (3 months). Annually, the southern part of the region (e.g. Russe, Razgrad, Silistra) exhibits a higher number of drought events lasting in general 3 to 6 months, relevant to meteorological drought and agriculture (Table 2). Moderate droughts are best represented across the study region, with maximum frequencies of more than 50 events/year (3 months) at Giurgiu and more than 20 events/year (6 months) in locations such as Pleven, Veliko Tarnovo, Silistra, Craiova, Slatina, and Calarasi.

As expected, the lowest annual frequency is specific to drought events over 12 months (Montana 12, of moderate type). Severe droughts (SD) reach maximum frequencies of over 30–40 events/year for both SPEI3 and SPEI6 only at Russe, Zimnicea, Razgrad, Oltenita and Alexandria.

Seasonally, there are notable differences in the number of drought events as a function of the considered drought timescale. In the southern part of the region, spring and autumn, as well as spring and summer are the seasons showing the highest number of drought events, accumulated over 3- and 6-month periods, respectively (with a total number of events of over 141–156 and 149–209 events/period, respectively), whereas drought events associated to the 12-month timescale appear to be more frequent in summer and fall, with total number of events of over 153–160/per period (Table 3). In the region's northern areas, the seasonal drought frequency is slightly lower, with spring, summer and fall having the highest frequency of drought events on the 3-month timescale (192–208 events/period), spring and summer for drought events on the 6-month timescale (194–355 events/period), and summer and fall for drought events over 12 months (174–182 events/period).

Similarly to the annual scale, the number of moderate drought (MD) events is best represented in both northern and southern parts of the Lower Danube region, especially in spring and summer (3 and 6 months) or summer and fall (12 months). The number of severe drought (SD) events per season over 1981–2019 does not exceed 200 in none of the seasons, showing the highest values in spring, summer and fall for 3 and 12 months, and winter and fall for 6 months. Although rarer, with a total number of extreme drought events per period varying from 104 (3 months) to 249 (6 months), extreme drought (ED) events occurred most frequently in summer (3 and 6 months) and winter (12 months).

Throughout the Lower Danube study region, the highest (total) number of drought events per period was found in spring for the 3-month timescale (Bechet 26), spring and summer for the 6-month timescale (Slatina 24 and Giurgiu 24, respectively) and

Table 2 Annual drought frequency in the Romanian-Bulgarian Lower Danube region for different SPEI timescales (SPEI3, SPEI6 and SPEI12 months) over the 1981–2019 period (data was derived from the gridded dataset)

Danube floodplain areas	3 months				6 months				12 months			
	MD	SD	ED	Total frequency	MD	SD	ED	Total frequency	MD	SD	ED	Total frequency
Vidin	28	24	0	52	12	12	0	24	0	0	0	0
Montana	27	16	12	55	1	12	12	25	12	0	0	12
Vratca	12	19	0	31	0	13	12	25	0	0	0	0
Pleven	19	12	12	43	25	12	12	49	0	0	0	0
Veliko Tarnovo	41	28	0	69	24	25	0	49	0	0	0	0
Russe	43	48	0	91	13	36	12	61	0	0	0	0
Razgrad	31	36	0	67	13	36	12	61	0	0	0	0
Silistra	39	36	0	75	27	24	12	63	0	0	0	0
Drobeta Turnu Severin	24	0	0	24	12	0	0	12	0	0	0	0
Calafat	28	24	0	52	12	12	0	24	0	0	0	0
Bechet	19	24	0	43	1	24	0	25	0	0	0	0
Zimnicea	31	36	0	67	1	48	0	49	0	0	0	0
Craiova	40	0	0	40	24	0	0	24	0	0	0	0
Slatina	31	0	0	31	24	12	0	36	0	0	0	0
Alexandria	31	36	0	67	13	48	0	61	0	0	0	0
Giurgiu	55	36	0	91	13	24	12	49	0	0	0	0
Oltenita	43	36	0	79	13	36	12	61	0	0	0	0
Calarasi	43	36	0	79	27	24	12	63	0	0	0	0
North (RO)	**240**	**219**	**24**	**483**	**115**	**70**	**72**	**357**	**0**	**0**	**0**	**0**
South (BG)	**345**	**228**	**0**	**573**	**140**	**228**	**36**	**404**	**12**	**0**	**0**	**12**

Table 3 Seasonal drought frequency in the Romanian-Bulgarian Lower Danube region for different SPEI timescales (3, 6 and 12 months) over the 1981–2019 period

Danube floodplain areas	3-months				6-months				12-months			
	DJF	MAM	JJA	SON	DJF	MAM	JJA	SON	DJF	MAM	JJA	SON
Vidin	13	24	19	24	16	21	11	16	14	10	15	16
Montana	16	25	18	24	16	19	12	15	16	15	24	21
Vratca	17	20	15	18	17	21	20	14	15	11	17	18
Pleven	12	18	17	19	13	22	20	16	17	9	24	20
Veliko Tarnovo	13	15	20	18	12	15	17	16	13	11	22	21
Russe	14	20	15	19	14	19	17	17	13	16	21	19
Razgrad	11	18	17	19	13	16	20	17	21	13	21	21
Silistra	13	16	20	19	17	15	17	19	21	17	16	17
Drobeta Turnu Severin	14	20	19	20	22	21	20	17	17	14	17	23
Calafat	13	24	19	24	16	21	20	16	14	10	15	16
Bechet	14	26	19	20	15	22	16	17	15	16	20	21
Zimnicea	11	20	19	18	15	18	21	17	15	16	23	20
Craiova	16	23	24	19	15	20	16	17	17	18	19	18
Slatina	13	22	19	17	16	24	12	17	16	18	17	15
Alexandria	10	19	18	18	14	19	21	15	14	19	17	16
Giurgiu	13	19	18	19	14	16	24	17	16	18	16	17
Oltenita	14	18	20	18	16	18	18	18	19	18	14	18
Calarasi	13	17	19	19	15	15	20	21	18	17	16	18
North (RO)	**131**	**208**	**194**	**192**	**158**	**194**	**188**	**172**	**161**	**164**	**174**	**182**
South (BG)	**109**	**156**	**141**	**160**	**118**	**148**	**134**	**130**	**130**	**102**	**160**	**153**

summer and fall for 12 months (Montana and Pleven 24, and Drobeta Turnu Severin 23, respectively).

In general, irrespective of drought timescales, the number of drought events per season did not exceed 3. Peak frequencies of drought events across the region were recorded generally during the mid- and late-1990s, and the early to mid-2000s.

4.1.2 Drought Duration

The total duration of drought events (expressed for all drought events of negative SPEI or for different drought categories, for the whole study period) lasting up to

3 months are the most frequent across the region, with a share of about 35% in the total number of events over the 1981–2010 period (Table 4). At local scale (selected cities inside and outside the Danube Floodplain), the total annual drought duration accumulated over the entire study period decreases with the increase of the timescale and does not show any notable differences between the southern and northern parts of the Lower Danube region. The total drought duration varies from 71 to 85 months for the 3-month timescale (e.g. Montana 85; Vidin, Bechet, Calafat 81), 67 to 81 months for 6 months (e.g. Calarasi, Drobeta Turnu Severin 81; Silistra 80), and between 58 and 82 months for the 12-month timescale (e.g. Montana 82, Craiova 75).

The maximum duration of drought spells is generally below 4 months for the 3-month timescale, 5 months for the 6-month timescale, and 7 months for the 12-month scale. The drought spells lasted the most, up to 6 consecutive months, for the 3- and 6-month timescales (e.g. Craiova over December 2001-May 2002 interval/severe drought; Pleven over July-December 2008/moderate drought; Alexandria, Giurgiu and Oltenita over Apr-Sep.2007/extreme drought). One of the most significant drought events associated with the most persistent SPEI 12-month episode in the 1981–2019 period (SD type) was identified at Calarasi and lasted from Nov.2008 to Jan.2010 (15 consecutive months). Other highly persistent drought events of 7–8 consecutive months, which were classified as severe or extreme droughts based on SPEI, showing relevance for the hydrological regime (associated to the 12-month timescale) were recorded in June 1990-January 1991 at Russe and Zimnicea, October 2000-April 2001 at Vidin and Calafat, and June 2009-January 2010 at Silistra. Although less persistent, with a maximum duration of only 3 consecutive months (3-month SPEI timescale), the drought event of July–October 2003 is referenced in literature as an extreme hydrological drought event in the Danube River Basin [122]).

This drought event contributed to the lowest water levels of the Danube River over the past 160 years and triggered major economic losses, e.g. the shutdown of Unit 1 of the Cernavoda nuclear power plant, which caused a 10% loss in the station's electricity output [123]; significant disruptions of inland navigation causing losses of 2.5–3.0 mil. $ [124]. In general, the occurrence of lengthy drought spells (3- and 6-months) was prevalent during the summer-fall months (about 23% of the long-lasting events), followed by those starting in spring and ending in summer (about 11%) posing a great threat to field crops and their yields, especially to rain-fed crops such as maize, which cover large agricultural areas in both southern and northern parts of the study region. The occurrence of long-lasting drought spells is linked to the significant precipitation deficits recorded during the early 1990s (1992–1993), in 2000–2003, 2007–2008, 2011–2012 and 2019, generally overlapping periods of intensified warming (e.g. 2000, 2007, 2012, 2019) that affected the entire study region.

Noticeably, severe and extreme droughts (SD and ED, respectively) on the 3- and 6-month timescales, tend to occur more frequently especially in the northern part of the region, within the Danube Floodplain, but also in its surrounding plain areas (e.g. Drobeta Turnu Severin, Craiova, Slatina, Giurgiu, Calarasi), compared to the southern part, where such types of drought events have been recorded more sparsely (e.g. Vidin, Russe, Silistra).

Table 4 Duration of drought events in the Romanian-Bulgarian Lower Danube region for different SPEI timescales (3, 6 and 12 months) over the 1981–2019 period. Drought types are: moderate (MD), severe (SD) and extreme (ED)

	3 months				6 months				12 months			
	Max (months)	Date	Drought type	Total duration (months)	Max (months)	Date	Drought type	Total duration (months)	Max (months)	Date	Drought type	Total duration (months)
Vidin	5	December 2001–April 2002	SD	81	4	February–May 1992 May–August 2008 May–August 2007 September–December 2011 July–October 2000	MD SD ED	68	7	October 2000–April 2001	ED	58
Montana	4	November 1992–February 1993	MD	85	5	June–October 2000	ED	69	8	September 2000–April 2001	ED	82
Vratca	3	December 1992–February 1993	MD	74	5	June–October 2000	ED	74	8	September 2000–April 2001	ED	68
Pleven	3	Aug.-Oct.1993 July–October 2007, October–December 2019	MD	73	6	July–December 2008	MD	75	6	November 2000–April 2001	ED	72
Veliko Tarnovo	3	December 1992–Feburary 1993	MD	74	4	April–July1985	MD	70	6	April–September 2009	MD	71
Russe	3	March–May1983	SD	76	6	April–September 2007	ED	72	8	June 1990–January 1991	MD	71
Razgrad	4	November 1992–Feburary 1993	MD	71	4	July–October 2012	MD	70	7	Apr–October 1985	MD	74
Silistra	3	March–May 1994 February-April 2007, August–October 2008, July–October 2009 October–December 2019	MD SD	77	5	March–Juy 1990	MD	80	7	July 2009–January 2010	SD	71

(continued)

Table 4 (continued)

	3 months				6 months				12 months			
	Max (months)	Date	Drought type	Total duration (months)	Max (months)	Date	Drought type	Total duration (months)	Max (months)	Date	Drought type	Total duration (months)
Drobeta Turnu Severin	4	Feburary–May 2002	ED	72	5	December 1998–April 1999	MD	81	5	Dec.2011-Apr.2012	SD	70
Calafat	5	September1992–Jan.1993 December 2001–April 2002	MD SD	81	4	July–October 2000	ED	68	7	October 2000–April 2001	ED	58
Bechet	4	October 2001–January 2002	MD	81	4	April–July 2007	ED	72	6	July–December 2019	MD	77
Zimnicea	5	January–May 2007	SD	75	4	April–July 2007	ED	76	8	Jun.1990–January 1991	MD	77
Craiova	6	December 2001–May 2002	SD	82	5	Jun.-Oct.2000	SD	67	8	August 2012–March 2013	SD	75
Slatina	4	February–May 1994	SD	73	4	June–September 2003 September–December.2012	SD ED	71	6	July–December 2012	SD	67
Alexandria	4	February–May 2007	SD	71	6	Aprli–September 2007	ED	70	4	August–November 2012	SD	67
Giurgiu	3	June–August 2003 April–June 2007	SD	77	6	April–September 2007	ED	70	5	September 1990–January 1991	MD	69
Oltenita	4	September–Decmber 2019	MD	78	6	April–September 2007	ED	76	8	April 2009–November2009	SD	68
Calarasi	3	June–August 2003 March–May 1994	SD	78	5	July–November 2009	SD	81	15	November 2008–January 2010	SD	69

The *area affected by drought* allowed the identification of the most critically drought-prone areas throughout the region. Examining the average coverage of dry SPEI at regional scale (NUTS3), we found a higher level of similarity between the northern and southern parts of the study region. On average, drought affected in general 30 to 50% of the territory of each county or district located along the Danube, on all SPEI timescales (Table 5).

The frequency of months with total drought coverage (negative SPEI in all grid points of selected counties and districts of the region) generally falls between 30 and 42% in all SPEI timescales. The maximum number of consecutive months with total drought coverage at NUTS3 level was found to increase proportionally with the SPEI timescale as follows:

- 3-month timescale: up to 11–13 consecutive months during the drought events recorded from March 2000 to March 2001 (e.g. Vidin, Montana, Mehedinti, Dolj, Razgrad, Vratca, Olt), August 2001 to June 2002 (e.g. Vidin, Girugiu, Vratca), November 2006 to September 2007 (e.g. Dolj, Silistra, Calarasi, Russe, Teleorman, Giurgiu, Vratca, Razgrad) and Marsh 2008 to January 2009 (e.g. Silistra, Calarasi, Giurgiu);

Table 5 Drought affected areas (% of the NUTS3 regions) in the Romanian-Bulgarian Lower Danube region for different SPEI timescales (3, 6 and 12 months) over the 1981–2019 period

Counties/districts	3 months		6 months		12 months	
	Average (%)	Total area affected cases (%)	Average (%)	Total area affected cases (%)	Average (%)	Total area affected cases (%)
Vidin	38.9	38.9	48.4	36.5	47.9	35.0
Montana	48.5	37.4	48.7	34.4	46.4	30.8
Razgrad	49.1	41.5	50.9	40.6	50.6	38.0
Silistra	50.3	42.3	48.8	40.4	50.9	40.8
Pleven	50.2	36.5	50.6	34.0	48.0	30.6
Veliko Tarnovo	50.3	33.8	51.1	33.1	49.5	33.3
Vratsa	49.9	37.2	49.0	33.8	45.8	29.1
Russe	49.0	37.4	51.5	37.0	49.9	36.3
Caras Severin	50.2	30.8	49.8	33.1	48.1	30.8
Mehedinti	48.2	35.9	49.1	36.1	47.5	32.3
Dolj	49.7	34.6	48.7	35.5	46.4	33.5
Olt	49.6	32.9	50.2	32.7	48.3	34.8
Teleorman	49.6	36.3	50.6	34.0	49.1	35.9
Giurgiu	49.8	39.5	50.9	37.6	50.2	39.7
Calarasi	48.6	38.9	48.4	37.0	49.7	39.3
North (RO)	**49.4**	**35.6**	**49.7**	**35.1**	**48.5**	**35.2**
South (BG)	**48.3**	**38.1**	**49.9**	**36.2**	**48.6**	**34.2**

- 6-month timescale: up to 16–21 consecutive months during the drought events recorded from December 1992-Marsh 1994 (Montana, Vratca), May 2008 to January 2010 (Silistra, Calarasi, Giurgiu) and August 2018 to December 2019 (Silistra and Calarasi);
- 12-month timescale: 22–27 consecutive months e.g. from June 1992 to July 1995 (Vidin, Montana, Mehedinti, Dolj, Olt) and exceptionally up to a peak of 42 consecutive months during the drought interval November 2006-April 2010 (Silistra and Calarasi).

4.1.3 Observed Changes in Drought Climatic Characteristics Over the 1981–2019 Period

The changes observed in air temperature and precipitation are well reflected in the changing characteristics of drought throughout the region.

Drought events tend to become more frequent (especially moderate and severe types) at the 3-month timescale, especially in the northern part of the study region. In terms of mean regional values, the observed increase in drought frequency is about 0.87 events decade^{-1} (p-value < 0.05). This positive trend is suggestive for a growing exposure to meteorological droughts.

The duration of droughts across the region is on a visible upward trend on all selected SPEI timescales, although not statistically significant, as follows:

- increase both in the northern (0.54 months decade^{-1}) and especially in the southern (0.65 months decade^{-1}) parts of the region (3-month timescale);
- increase in both parts of the region, higher in the north (0.83 months decade^{-1}) than in the south (0.40 months decade^{-1}) (6-month timescale);
- increase only in the southern part of the region (0.54 months decade^{-1}) and no trend in the northern part.

The area affected by drought shows the most important changes over the 1981–2019 period. The region exhibits significant positive trends in both parts of the region, growing in magnitude with the SPEI timescale. Estimated trend slopes in the share of drought-affected area in relation to the total Bulgarian NUTS3 territory increase from 3.8% decade^{-1} (3 months) to 4.8% decade^{-1} (6 months), and to 6.4% decade^{-1} (12 months). In comparison, the share of drought affected area in relation to the total Romanian NUTS3 territory increases from 1.9% decade^{-1} (3 months) to 3.5% decade^{-1} (6 months) and to 7.3% decade^{-1} (12 months). Noticeably, the observed trends are statistically significant only for SPEI12 in both parts of the region (p-value < 0.1).

4.2 Aggregated Drought Hazard Index

The Drought Hazard Index (DHI) was generated through the aggregation within the boundaries of NUTS3 units (counties or districts) of gridded SPEI data (11 km spatial resolution) on 3-, 6- and 12-month timescales over the 1981–2019 period. DHI is directly related to the cumulative probability of drought occurrence (negative SPEI) in each administrative unit located in the northern and southern parts of the selected Lower Danube region. DHI distribution maps show a greater sensitivity to drought in the eastern half of the study region on all SPEI timescales, including both Romanian and Bulgarian administrative units (Fig. 2).

Overall, drought hazard stays at the highest levels in the eastern half of the Lower Danube region, both northward and southward, on all SPEI timescales. Drought hazard associated to the 3-month timescale is particularly high in five administrative districts located southward such as Pleven, Razgrad and Silistra, which were assigned a "very high" drought hazard level, and Veliko Tarnovo, and Russe ("high"). Comparatively, in the northward Danube sector, there is only one county where the drought hazard level is "high" (Teleorman). In the rest of the investigated territory, most regions have a "low" or "very low" level of drought hazard (Calarasi, Montana, Vratca, Caras Severin, Vidin, Giurgiu), and only a few are assigned a "moderate" level of hazard (Mehedinti, Dolj, Olt).

The "very high" and "high" drought hazard levels are assigned to seven regions (Veliko Tarnovo, Russe, Pleven and Razgrad in the south, and Teleorman and Giurgiu in the north) for the 6-month timescale and six regions (Veliko Tarnovo, Russe, Razgrad and Silistra in the south, and Calarasi and Girugiu in the north) for the 12-month timescale. For the longest SPEI timescale (12 months), with relevance for the Danube's hydrological regime, drought hazard remains "moderate" at Pleven, Vidin and Mehedinti, and "low" or "very low" at Dolj, Montana and Vratsa.

4.3 Drought Vulnerability

The analysis of the territorial distribution of DVI shows that in 24.2% of the study area the vulnerability to drought is "very high", whereas about 22% has a "low" level of vulnerability. The central and eastern districts northward Danube show a "very high" or a "high" level of vulnerability (Fig. 3). The "very high" vulnerability to drought in these districts is a result of the high share (between 73 and 74%) of agricultural land, including arable land (non-irrigated areas and rice fields), permanent crops (vineyards and fruit trees) and complex arable land, which have an integrated LULC indicator corresponding to a high level of vulnerability (Table 6). These areas also include the majority of wetlands (inland marshes) along the Danube, which are particularly dependent on water availability and drought occurrences. This part of the Lower Danube region has a high population density which determines "high" and "very high" vulnerability levels.

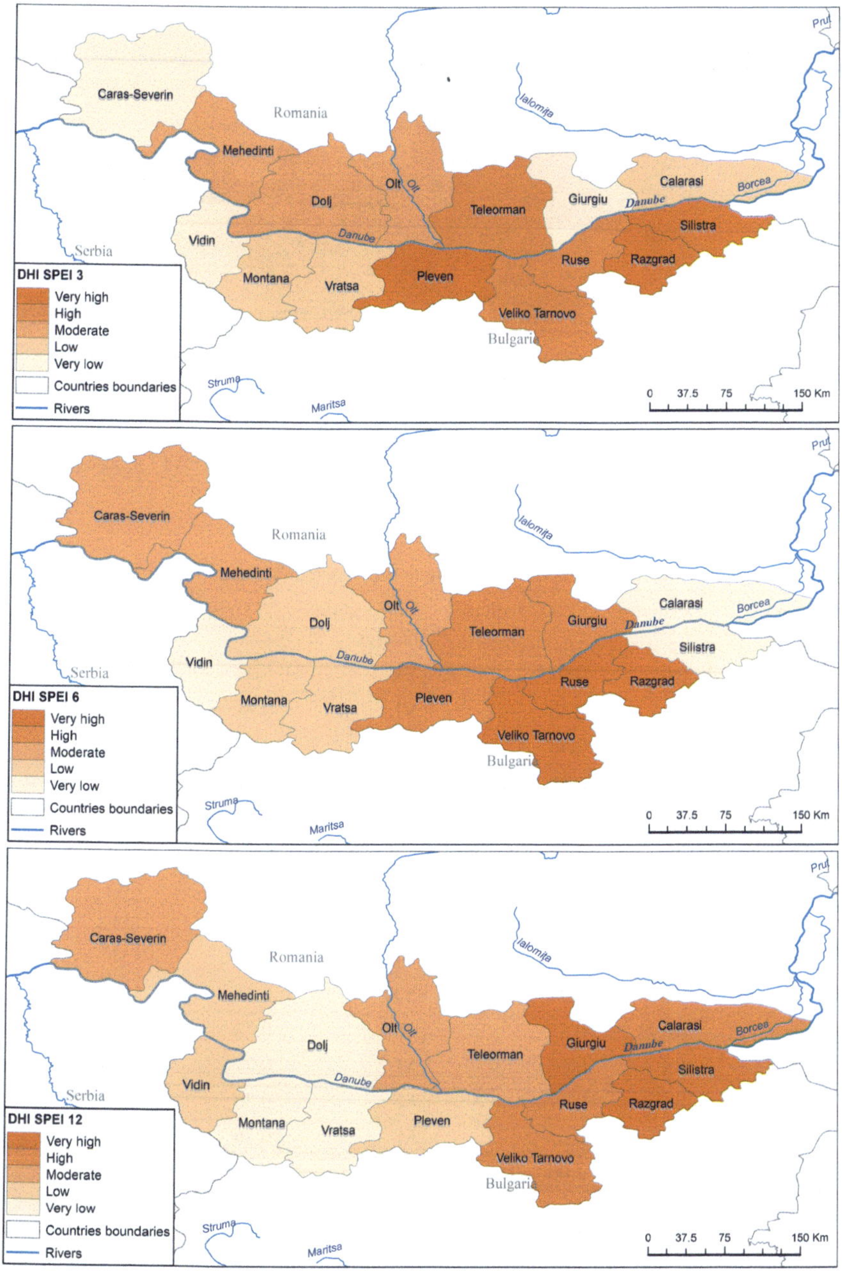

Fig. 2 Aggregated drought hazard index (DHI) in counties (RO) and administrative districts (BG) based on 3-, 6- and 12-month SPEI timescales (1981–2019), in the Lower Danube River region

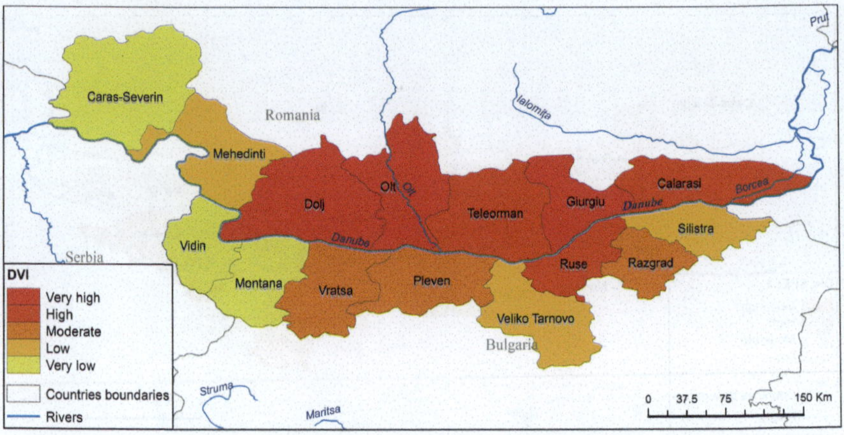

Fig. 3 Drought vulnerability in the Lower Danube River region based on DVI for administrative units (counties - RO and districts - BG)

Table 6 Drought vulnerability classes according to the selected indicators

NUTS 3	Country	LULC indicator	Vulnerability classes	PD indicator	Vulnerability classes
Caras-Severin	Ro	0.01	VL	0.08	VL
Mehedinti	Ro	0.38	VL	0.38	M
Vidin	Bg	0.48	L	0.01	VL
Montana	Bg	0.45	L	0.13	VL
Dolj	Ro	0.81	H	1.00	VH
Vratsa	Bg	0.58	L	0.27	L
Olt	Ro	0.83	H	0.73	H
Pleven	Bg	0.81	H	0.38	M
Veliko Tarnovo	Bg	0.44	VL	0.37	M
Teleorman	Ro	0.94	VH	0.54	H
Giurgiu	Ro	0.81	H	0.73	H
Ruse	Bg	0.65	M	0.77	VH
Razgrad	Bg	0.67	M	0.30	L
Silistra	Bg	0.62	M	0.16	L
Calarasi	Ro	1.00	VH	0.48	H

A "very high" level of vulnerability according to the population density is also observed in Ruse district, although here the most vulnerable agricultural areas have a small share of the entire district territory compared to the three abovementioned districts. This determines the inclusion of Ruse district in the fourth vulnerability class ("high"). The "high" vulnerability to drought is also observed in the easternmost administrative district (Calarasi), as well as in Teleorman. In these administrative units, the share of territories highly vulnerable to drought is very high (85 and 80% of the territory, respectively). Consequently, the vulnerability level based on LULC is very high (Table 6).

A "low" vulnerability level was found in three districts located in the western part of the Lower Danube region, which cover about 22% of the entire investigated region (Fig. 3). This level of drought vulnerability is a result of the low vulnerability associated to LULC and of the very low vulnerability determined by population density (Table 6). In other NUTS3 regions, arable land decreases and the area occupied by forests increases (e.g. in Caras-Severin, about 59% of the territory is covered by forests).

The concentration of areas with higher vulnerability to drought in the northward Danube sector reflects the impact of the underlying surface and of the relief. In the northern part of the study region, arable land prevails, whereas in the southern part, with a hilly relief, the share of less drought vulnerable territories increases in relation to more extensive areas covered by forests and natural vegetation.

4.4 Drought Risk

Drought risk assessment and mapping are two of the most important elements of any action plans or strategies for the mitigation of drought effects. In this paper, the drought risk index (DRI) is developed based on drought hazard (DHI) and drought vulnerability (DVI). Based on the assessment of DHI and DVI, drought risk maps are generated for each SPEI timescale (3, 6 and 12 months) (Fig. 4). Each administrative district is categorized into five groups similarly to DHI and DVI (i.e. very low, low, moderate, high and very high).

On the SPEI3 timescale, the districts showing the highest levels of drought risk are concentrated in the middle part of the investigated region. These districts are Dolj, Olt, Teleorman, Pleven, Razgrad and Ruse, which account for about 42% of the total investigated Lower Danube region. Only one administrative district was assigned a "high" level of drought risk (Silistra). Two administrative regions have a "moderate" risk level (Veliko Tarnovo and Calarasi), whereas the rest of the investigated region has "low" or "very low" drought risk.

Similarly to SPEI3, drought risk on the SPEI6 timescale shows that the "high" and "very high" drought risk regions are generally located in the central part of the investigated region. These administrative regions are Teleorman, Girgiu, Ruse, Olt, Pleven and Razgrad. They represent 36.6% of the total region. Regions with a "moderate" drought risk level are more numerous when compared to SPEI3 (Mehedinti, Dolj, Vratsa and Veliko Tarnovo). The districts that were assigned "low" and "very low" levels of drought risk are Caras Severin, Vidin, Montana, Calarasi and Silistra.

Drought risk associated to SPEI12 correlates with the DHI (SPEI12) and the DVI (SPEI12) distribution maps, which show that the "high" and "very high" drought risk is found in the administrative regions located in the northern and eastern half

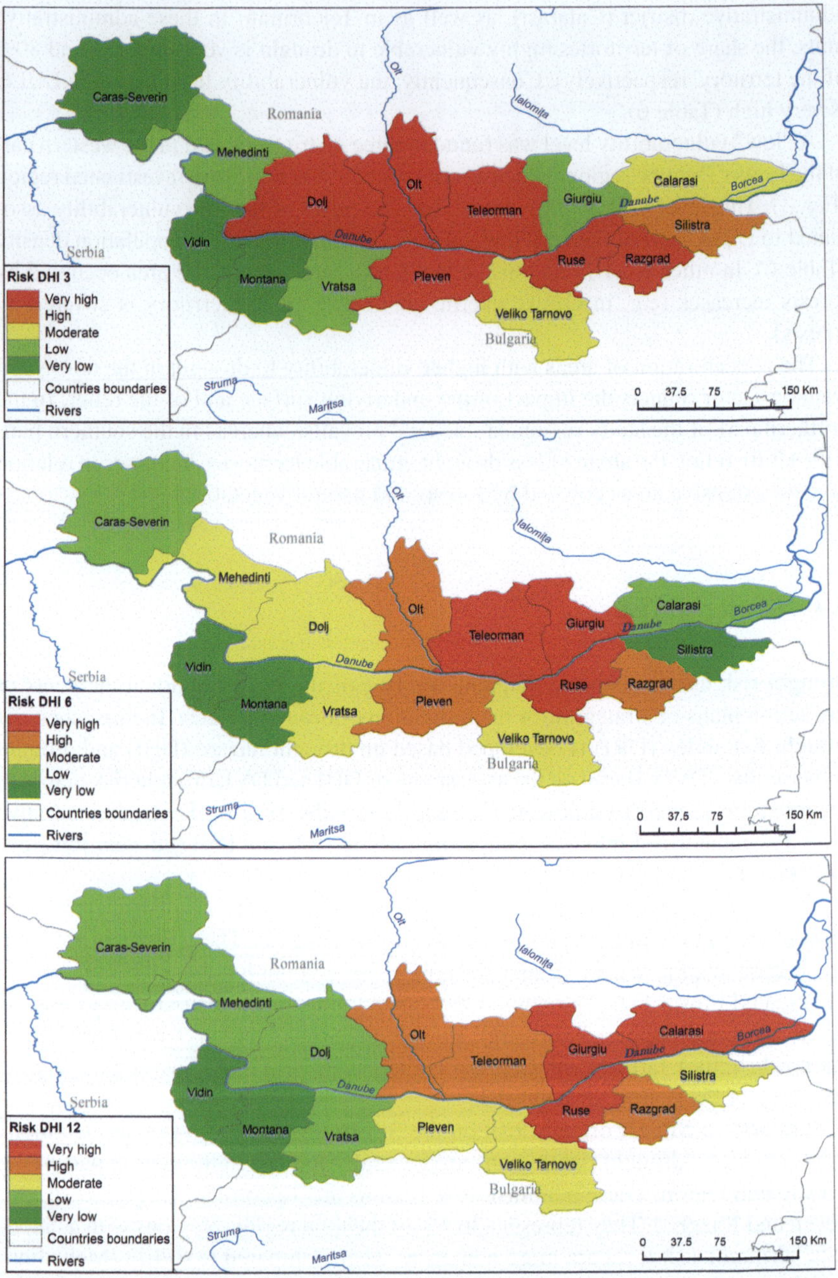

Fig. 4 Drought risk in the Lower Danube River region for administrative units (counties - RO and districts - BG) based on drought hazard maps on 3-, 6- and 12-month SPEI timescales (1981–2019) and drought vulnerability map

of the investigated region (Olt, Teleorman, Girugiu, Calarasi, Ruse and Razgrad), accounting for about 37% of its territory.

The "moderate" drought risk spreads across the southern part of the Lower Danube region in the districts of Pleven, Veliko Tayrnovo and Silistra. The sectors with "low" and "very low" levels of drought risk are generally concentrated in the western part of the region (Caras Severin, Mehedinti, Dolj, Vidin, Montana and Vratsa). This correlates with the results of the drought hazard and drought vulnerability assessment for this part of the study region.

Overall, on all timescales, DRI values show a particularly high risk of drought in Olt and Teleorman (Romania), as well as in Ruse and Razgrad (Bulgaria). This situation emerges from the greater frequency of droughts coupled with the high population and high agricultural activity in this regions.

5 Conclusion

This paper analysed drought hazard, drought vulnerability and drought risk over the 1981–2019 period, using a SPEI-based approach, for the cities and counties/administrative regions located northward and southward Danube in its lower basin.

Drought in the Lower Danube region occurs all throughout the year (especially during the summer-fall and spring–summer months) and produces a wide range of effects on the environment as well as in various water-dependent economic sectors (e.g. agriculture, navigation, hydro- and nuclear electricity production), as shown by some of the most recent events recorded in the 1990s, 2000–2003, 2007–2008, 2011–2012 and 2019. Drought has been found to affect the region 30 to 50% on all SPEI timescales (Table 5). The frequency of months with total drought coverage generally falls between 30 and 42% on all SPEI timescales. Drought increased in frequency and duration throughout the region over the 1981–2019 period, although trends are mostly not statistically significant. Furthermore, the area affected by drought also shows positive trends across the region.

The DHI hazard distribution reveals a greater sensitivity to drought in the eastern half of the study region on all SPEI timescales. Drought hazard decreases with the SPEI timescale. Razgrad, Russe, Veliko Tarnovo, Teleorman, Giurgiu and Calarasi are in general the counties/administrative regions showing the highest level of drought hazard ("high" or "very high").

Due to the larger area of arable land and the higher population density in the northern part of the study area (on the territory of Romania), vulnerability to drought is higher compared to most southern administrative units (on Bulgarian territory). "High" and "very high" vulnerability based on LULC indicator of the eastern districts determines high values of the integrated drought vulnerability (DVI), while "low" and "very low" vulnerability is related to low population density of western districts, which determines the overall low vulnerability to drought of the western part of the study area.

The drought risk distribution generally correlates with the distribution of drought hazard and drought vulnerability in the investigated area, showing a "high" level of drought risk in eastern and northern-central administrative units.

6 Recommendation

The results from the present study can be used by the researchers working on issues related to the natural, economic and social dimensions of drought. Beneficiaries of this research would also be decision-makers in formulating and choosing policies and measures for adaptation to climate change as well as in risk management and land use planning. The accurately identifying drought hazard and vulnerability at a regional scale can lead to the development and implementation of a wide range of measures and programs to reduce the negative impact of future droughts. Future work will be directed to expanding the scope of research and assessment of socio-economic dimension of drought.

Acknowledgements This work has been carried out in the framework of the National Science Program "Environmental Protection and Reduction of Risks of Adverse Events and Natural Disasters", approved by the Resolution of the Council of Ministers № 577/17.08.2018 and supported by the Ministry of Education and Science (MES) of Bulgaria (Agreements № Д01-322/18.12.2019 and № Д01-363/17.12.2020).

References

1. Ceglar A, Medved-Cvikl B, Morán-Tejeda E, Vicente Serrano SM, Kajfež-Bogataj L (2012) Assessment of multi-scale drought datasets to quantify drought severity and impacts in agriculture: a case study for Slovenia. Int J Spat Data Infrastruct Res 7:464–487
2. Buras A, Rammig A, Zang C (2020) Quantifying impacts of the drought 2018 on European ecosystems in comparison to 2003. Biogeosciences 17:1655–1672
3. Schuldt B, Buras A, Arend M, Vitasse Y, Beierkuhnlein C, Damm A, Gharun M, Grams TEE, Hauck M, Hajek P, Hartmann H, Hiltbrunner E, Hoch G, Holloway-Phillips M, Körner C, Larysch E, Lübbe T, Nelson DB, Rammig A, Rigling A, Rose L, Ruehr NK, Schumann K, Weiser F, Werner C, Wohlgemuth T, Zang CS, Kahmne A (2020) A first assessment of the impact of the extreme 2018 summer drought on Central European forests. Basic Appl Ecol 45:86–103
4. Erfurt M, Glaser R, Blauhut V (2019) Changing impacts and societal responses to drought in southwestern Germany since 1800. Reg Environ Change 19:2311–2323
5. Wilhite DA, Svoboda MD, Hayes MJ (2007) Understanding the complex impacts of drought: a key to enhancing drought mitigation and preparedness. Water Resour Manag 21(5):763–774
6. Dai A (2011) Drought under global warming: A review. Clim Change 3:45–65
7. Beran M, RodierJA, (1985) Hydrological aspects of drought, Studies and reports in hydrology 39. UNESCO-WMO, Paris
8. Hisdal H, Tallaksen LM (2000) Drought event definition. Technical Report. No. 6. Assessment of the Regional Impact of Droughts in Europe. University of Oslo, Department of Geophysics

9. Tallaksen L M, Van Lanen HAJ (2004) Hydrological drought. Processes and estimation methods forstreamflow and groundwater. Developments in Water Science. Vol. 48. Elsevier Science B.V., Amsterdam
10. Van Loon AF (2015) Hydrological drought explained. WIREsWater 2:359–392. https://doi.org/10.1002/wat2.1085
11. Sohoulande Djebou DC (2017) Bridging drought and climate aridity. J Arid Environ 144:170–180
12. Mishra AK, Singh VP (2010) A review of drought concepts. J Hydrol 391:202–216
13. Van Loon AF, Van Lanen HAJ (2013) Making the distinction between water scarcity and drought using an observation-modeling framework. Water Resour Res 49:1483–1502
14. UNDRR 2019. Global Assessment Report on Disaster Risk Reduction (GAR). Available via https://www.droughtmanagement.info/literature/UNIVERSITYofOSLO_Drought_Event_Definition_2000.pdf. Accessed 27 Sep 2021
15. Spinoni J, Gustavo N, Vogt JV (2017) Pan-European seasonal trends and recent changes of drought frequency and severity. Global Planet Change 148:113–130
16. Vicente-Serrano SM, Lopez-Moreno JI, Beguería S, Lorenzo-Lacruz J, Sanchez-Lorenzo A, García-Ruiz JM et al (2014) Evidence of increasing drought severity caused by temperature rise in southern Europe. Environ Res Lett 9(4):044001
17. Riedel T, Weber TKD (2020) Review: The influence of global change on Europe's water cycle and groundwater recharge. Hydrogeol J 28:1939–1959
18. Stagl J, Mayr E, Koch H, Hattermann F F, Huang S (2014) Effects of climate change on the hydrological cycle in Central and Eastern Europe. In: Rannow S., Neubert M. (eds) Managing protected areas in Central and Eastern Europe under climate change. Advances in global change research, vol 58. Springer, Dordrecht
19. Sutanto C, Vitolo C, Di Napoli M, D'Andrea HAJ, Lanen V (2020) Heatwaves, droughts, and fires: exploring compound and cascading dry hazards at the pan-European scale. Environ Int 134:105276
20. Russo A, Gouveia CM, Dutra E, Soares PMM, Trigo RM (2019) The synergy between drought and extremely hot summers in the Mediterranean. Environmental Research Letters 14(1)
21. Barriopedro D, FischerEM LJ, Trigo R, Garcia-HerreraR, (2011) The hot summer of 2010: redrawing the temperature record map of Europe. Science 332:220–224
22. EEA (2019) Climate change adaptation in the agriculture sector in Europe, EEA Report 04. Available via https://www.eea.europa.eu/publications/cc-adaptation-agriculture. Accessed on 9 Jan 2021
23. Kim W, Iiyumi T, Nishimori M (2019) Global patterns of crop production losses associated with drought from 1983 to 2009. J Appl Meteorol Climatol 58(6):1233–1244
24. Páscoa P, Gouveia CM, Russo A, Trigo RM (2017) The role of drought on wheat yield interannual variability in the Iberian Peninsula from 1929 to 2012. Int J Biometeorol 61:439–451
25. Potopová V, Boroneanţ C, Boincean B, Soukup J (2016) Impact of agricultural drought on main crop yields in the Republic of Moldova. Int J Climatol 36(4):2063–2082
26. Bastos A, Gouveia CM, Trigo RM, Running SW (2014) Analysing the spatio-temporal impacts of the 2003 and 2010 extreme heatwaves on plant productivity. Eur Biogeosci 11:3421–3435
27. Peters W, Bastos A, Ciais P, Vermeulen A (2020) A historical, geographical and ecological perspective on the 2018 European summer drought. Phil Trans R Soc B 375:20190505
28. Samaniego L, Thober S, Kumar R, Wanders N, Rakovec O, Pan M, Zink M, Sheffield J, Wood EF, Marx A (2018) Anthropogentic warming exacerbates European soil moisture droughts. Nat Clim Chang 8(5):421–426
29. EEA (2012) Climate Change, Impacts and Vulnerability in Europe 2012, an Indicator-based Report Office for Official Publications of the European Union, Luxembourg
30. Rey D, Holman IP, Knox JW (2017) Developing drought resilience in irrigated agriculture in the face of increasing water scarcity. Reg Environ Change 17:1527–1540
31. Weatherhead EK, Knox JW, Hess TM, Daccache A (2015) Exploring irrigation futures—developments in demand forecasting. Outlook Agr 44(2):119–126

32. von Hardenberg TMJ, AghaKouchak, A, Carmen Llasat M, Provenzale A, Trigo RM (2017) On the key role of droughts in the dynamics of summer fires in mediterranean Europe. Scient Rep 7(81):1–11
33. Grumm RH (2011) The central European and Russian heat event of July-August 2010. BAMS:1285–1296
34. Witte JC, Douglass AR, da Silva A, Torres O, Duncan RLBN (2011) Nasa a-train and terra observations of the 2010 Russian wildfires. Atmos Chem Phys 11:9287–9301
35. Di Napoli C, Pappenberger F, Cloke HL (2019) Verification of heat stress thresholds for a health-based heatwave definition. J Appl Meteorol Climatol 58:1177–1194
36. Ionita M, Tallaksen LM, Kingston DG, Stagge JH, Laaha G, Van Lanen HAJ, Scholz P, Chelcea SM, Haslinger K (2017) The european 2015 drought from a climatological perspective. Hydrol Earth Syst Sci 21:1397–1419
37. Robine JM, Cheung SLK, Roy SL, Van Oyen H, Griffiths C, Michel JP, Herrmann FR (2008) Death toll exceed 70,000 in Europe during the summer of 2003. CR Biol 331:171–178
38. Tsakiris G, Nalbantis I, Vangelis H, Verbeiren B, Huysmans M, Tychon B, Jacquemin I, Canters F, Vandehaegen S, Engelen G, Poelmans L, De Backer P, Batalaan O (2013) A system-based paradigm of drought analysis for operational management. Water Resour Manag 27(15):5281–5297
39. Maxwell JT, Soulé P T (2011) Drought and other driving forces behind population change in six rural counties in the United States. Southeastern Geographer V(51): 133–148
40. European Commission (2007b) Water scarcity and droughts, second Interim Report, June 2007. Available via http://ec.europa.eu/environment/water/quantity/pdf/comm_droughts/2nd_int_report.pdf. Accessed 09 Mar 2015
41. Cammalleri C, Arias-Muñoz C, Barbosa P, de Jager A, Magni D, Masante D, Mazzeschi M, McCormick N, Naumann G, Spinoni J, Vogt J (2020) A revision of the Combined Drought Indicator (CDI) as part of the European Drought Observatory (EDO). Nat Hazards Earth Syst Sci Discuss [preprint]
42. EEA (2010c) Mapping the impacts of natural Hazards and technological accidents in Europe: an overview of the last decade. Technical report No 13/2010, European Environment Agency (EEA), Publications Office of the European Union, Luxembourg
43. Leng G, Hall J (2019) Crop yield sensitivity to global major agricultural countries to droughts and the projected changes in the future. Sci Total Environ 654:811–821
44. Zscheischler J, Seneviratne SI (2017) Dependence of drivers affects risks associated with compound events. Sci Adv 3:e1700263
45. Forzieri G, Feyen L, Russo S, Vousdoukas M, Alfieri L, Outten S, Migliavacca M, Bianchi A, Rojas R (2016) Cid Multi-hazard assessment in Europe under climate change. Clim Change 137:105–119
46. Dai A (2013) Increasing drought under global warming in observations and models. Nat Climate Change 3:52–58
47. Stahl K, Hisdal H (2004) Hydroclimatology. In: Tallaksen LM, Lanen HAJ van (eds) Hydrological Drought—processes and estimation methods for streamflow and groundwater. Developments in water science, 48, Elsevier Science, Amsterdam
48. Sheffield J, Wood E (2011) Drought: past problems and future scenarios. Earthscan, London, Washington
49. WadaY, van Beek LPH, Wanders N, Bierkens MFP, (2013) Human water consumption intensifies hydrological drought worldwide. Environ Res Lett 8:1–14
50. Döll P, Fiedler K, Zhang J (2009) Global-scale analysis of river flow alterations due to water withdrawals and reservoirs. Hydrol Earth Syst Sci 13:2413–2432
51. Vörösmarty CJ, Green P, Salisbury J, Lammers RB (2000) Global water resources: vulnerability from climate change and population growth. Science 289:284–288
52. Donat MG, Alexander LV, Herold N, Dittu AJ (2016) Temperature and precipitation extremes in century-long gridded observations, reanalyses, and atmospheric model simulations. J Geophys Res 121:11174–11189

53. Funk C, Harrison L, Alexander L, Peterso P, BehrangiA HG (2019) Exploring trends in wet-season precipitation and drought indices in wet, humid and dry regions. Environ Res Lett 14(11):115002
54. Pendergrass AG, Hartmann DL (2014) The atmospheric energy constraint on global-mean precipitation change. J Clim 27:757–768
55. Trenberth KE, Dai A, van der Schrier G, Jones PD, Barichivich J, Briffa KR, Sheffield J (2014) Global warming and changes in drought. Nat Clim Change 4(1):17–22
56. Sheffield J, Wood EF, Roderick ML (2012) Little change in global drought over the past 60 years. Nature 491(7424):435–438
57. Páscoa P, Gouveia CM, Russo A, Trigo RM (2017) The role of drought on wheat yield interannual variability in the Iberian Peninsula from 1929 to 2012. Int J Biometeorol 6:439–451
58. Spinoni J, Naumann G, Vogt JV, Barbosa P (2015) The biggest drought events in Europe from 1950 to 2012. Journal of Hydrology: Regional Studies 3:509–524
59. Spinoni J et al. (2016) Meteorological droughts in Europe: Events and impacts—past trends and future projections. JRC Technical Report (Luxembourg: Publications Office of the EU). Available via http://publications.jrc.ec.europa.eu/repository/bitstream/JRC100394/lb-na-27748-en-n.pdf. Accessed 21 Jan 2021
60. Hoerling M, Eischeid J, PerlwitzJ QX, Zhang T, Pegion P (2012) On the increased frequency of Mediterranean drought. J Clim 25(6):2146–2161
61. Kingston DG, Stagge JH, Tallaksen LM, Hannah DM (2015) European-scale drought: understanding connections between atmospheric circulation and meteorological drought indices. J Clim 28(2):505–516
62. Seneviratne SI (2012) Climate science: historical drought trends revisited. Nature 491(7424):338–339
63. Bordi I, Fraedrich K, Sutera A (2009) Observed drought and wetness trends in Europe: an update. Hydrol Earth Syst Sci 13(8):1519–1530
64. Willems P (2013) Multidecadal oscillatory behaviour of rainfall extremes in Europe. Clim Change 120(4):931–944
65. Páscoa P, Gouveia C, Ana R, Bojariu R, Vicente-Serrano S, Trigo R (2020) Drought impacts on vegetation in Southeastern Europe. Remote Sensing 12:2156
66. Hänsel S, Ustrul Z, Łupikasza E, Skalak P (2019) Assessing seasonal drought variations and trends over Central Europe. Water Resour 127:53–57
67. Ionita M et al (2017) The European 2015 drought from a climatological perspective. Hydrol Earth Syst Sci 21:1397–1419
68. Van Lanen HAJ, Laaha KDG, Gauster T, Ionita M, Vidal J-P, Vlnas R, Tallaksen LM, Stahl K, Hannaford J, Delus C, Fendekova M, Mediero L, Prudhomme C, Rets E, Romanowicz RJ, Gailliez S, Wong WK, Adler J, Blauhut V, Caillouet L, Chelcea S, Frolova N, Gudmundsson L, Hanel M, Haslinger K, Kireeva M, Osuch M, Sauquet E, Stagge JH, Van Loon AF (2016) Hydrology needed to manage droughts: the 2015 European case. Hydrol Process 30:3097–3104
69. Colesca SE, Ciocoiu CN (2013) An overview of the Romanian renewable energy sector. Renew Sustain Energy Rev 24:149–158
70. Sepulcre-Canto G, Horion S, Singleton A, Carrao H, Vogt J (2012) Development of a combined drought indicator to detect agricultural drought in Europe. Nat Hazards Earth Syst Sci 12:3519–3531
71. Jakubínský J, Bláhov M, Bartosova L, Steinerová K, Balek J, Dížková P, Semeradova D, Alexandru D, Bardarska G, Bokal S, Borojević G, Bucur A, Cindric K, Barbu A, Debre B, Đorđević M, Đurić I, Mircea B, Gatarić S, Trnka M (2019) Repository of drought event impacts across the Danube Catchment Countries between 1981 and 2016 using publicly available sources. Acta Universitatis Agriculturae et Silviculturae Mendelianae Brunensis 67:925–938
72. ThinkHazard platform. https://thinkhazard.org/en/. Accessed Apr 2021

73. Alexandrov V (ed) (2011) Methods for monitoring, assessment and impact of drought in Bulgaria. Sofia
74. Koleva E, Alexandrov V (2008) Drought in the Bulgarian low regions during the 20th century. Theoret Appl Climatol 92:113–120
75. Petkova B, Kuzmova K, Berova, M (2019) The main abiotic stress factors limiting crop cultiva-tion and production in Bulgaria. Climate changes, drought, water deficit and heat stress. Agricultural Sciences/Agrarni Nauki 11(26):13–22
76. Radeva K, Nikolova N, Gera M (2018) Assessment of hydro-meteorological drought in the Danube Plain. Bulgaria. Hrvatski geografski glasnik 80(1):7–25
77. Popova Z, Ivanova M, Martins D, Pereira LS, Doneva K, Alexandrov V, Kercheva M (2014) Vulnerability of Bulgarian agriculture to drought and climate variability with focus on rainfed maize systems. Nat Hazards 74(2):865–886
78. Popova Z, Ivanova M, Pereira L, Alexandrov V, Kercheva M, Doneva K, Martins D (2015) Droughts and climate change in Bulgaria: assess-ing maize crop risk and irrigation requirements in relation to soil and climate region. Bulgarian J Agr Sci 21(1):35–53
79. Nikolova N, Alieva G, Voislavova I (2012) Drought periods in non-mountainous part of South Bulgaria on the background of climate change. Ge-ographica Pannonica 16(1):18–25
80. Alexandrov V (2006) Soil drought monitoring (review). In UNDPGEF, Building capacity for sustainable land management in Bulgaria, 20052008, C.42
81. Knight CG, Raev I, Staneva M P (2004) Drought in Bulgaria: A contemporary analog of climate change. Ashgate Studies in Environmental Policy and Practice, Aldershot
82. Radeva K, Nikolova N (2020) Hydrometeorological Drought hazard and vulnerability assessment for Northern Bulgaria. Geographica Pannonica 24(2):112–123
83. Bunea A (2019) Drought risk assessment by means of Drought Hazard and vulnerability indices in Muntenia region. Riscuri şi catastrophe XIX(25,2):73–84
84. Nikolova N (2013) The impact of the drought on the main crops cultivated in Northeastern Bulgaria. Forum Geografic 12(1):16–24. https://doi.org/10.5775/fg.2067-4635.2013.016.i
85. Alexandrov V, Schneider M, Koleva E, Moisselin JM (2004) Climate variability and change in Bulgaria during the 20th century. Theor Appl Climato 79:133–149. https://doi.org/10.1007/s00704-004-0073-4
86. Rachev N, Dimitrova D (2016) Changes in average temperatures and precipitation in Bulgaria for the period 1995–2012. Ann. of Sofia University "St. Kliment Ohridski." Faculty of Physics 109:1–25
87. Hristova H, Radeva K (2011) Regional analysis of water stress in Bulgria. In: Zlatunova D et al (eds) Seven International conference „Global changes". Sofia University Press, Sofia
88. DMCSEE (2012) Summary of the Result of the DMCSEE Project, Co-Financed by the South East Europe TransnationalCooperation Programme; Drought Management Centre for South-East Europe. DMCSEE, Ljubljana
89. Bisselink B, RooAd B, J, Gelati E, (2018) Future projections of water scarcity in the Danube River Basin due to land use, water demand and climate change. Journal of Environmental Geography 11:25–36
90. Country Report Romania (2016) EUROPEAN COMMISSION Brussels, 26.2.2016 SWD (2016) 91 final
91. Cheval S, Dumitrescu A, Birsan M-V (2017) Variability of the aridity in the South-Eastern Europe over 1961–2050. CATENA 151:74–86
92. Cheval S, Busuioc A, Dumitrescu A, Birsan M-V (2014) Spatiotemporal variability of meteorological drought in Romania using the standardized precipitation index (SPI). Climate Res 60:235–248
93. Croitoru AE, Piticar A, Imbroane AM, Burada DC (2013) Spatiotemporal distribution of aridity indices based on temperature and precipitation in the extra-Carpathian regions of Romania. Theor Appl Climatol 112:597–607
94. Stefan S, Ghioca M, Rimbu N, Boroneant C (2004) Study of meteorological and hydrological drought in southern Romania from observational data. Int J Climatol 24:871–881

95. Barbu I, Popa I (2003) Drought monitoring in Romania. Forestry Technical Publishing House, Câmpulung Moldovenesc
96. Ionita M, Scholz P, Chelcea S (2016) Assessment of droughts in Romania using the Standardized Precipitation Index. Nat Hazards 81(3):1483–1498
97. Paltineanu C, Mihailescu I, Seceleanu I et al (2007) Using aridity indices to describe some climate and soil features in Eastern Europe: a Romanian case study. Theor Appl Climatol 90:263–274
98. Paltineanu C, Mihailescu IF, Prefac Z, Dragota C, Vasenciuc F, Claudia N (2009) Combining the standardized precipitation index and climatic water deficit in characterizing droughts: a case study in Romania. Theor Appl Climatol 97:219–233
99. Ghioca M (2009) Drought monitoring using self-calibrating Palmer's indices in the Southwest of Romania. Rom Rep Phys 61(1):151–164
100. Cheval S, Popa I, Baciu M, Breza T (2003) Spatial and temporal variability of the standardized precipitation index in Romania. Scientific Papers, Institutul National de Meteorologie si Hidrologie, Bucharest
101. Ionita M, Rimbu N, Chelcea S, Patrut S (2013) Multidecadal variability of summer temperature over Romania and its relation with Atlantic Multidecadal Oscillation. Theor Appl Climatol 113:305–315
102. Rimbu N, Boroneant C, Buta C, Dima M (2002) Decadal variability of the Danube river flow in the lower basin and its relation with the North Atlantic Oscillation. Int J Climatol 22:1169–1179
103. Pravalie R, Patriche CV, Sirodoev I, Bandoc G, Dumitrascu M, Peptenatu D (2016) Water deficit and corn productivity during post-socialist period. Case study: southern Oltenia drylands. Romania. Arid Land Res Manag 30:239–257
104. Sandu I, Mateescu E, Vătămanu V V (2010) Schimbări climatice în România şi efectele asupra agriculturii. Editura Sitech, Craiova
105. Lupu AB, Ionescu FC, Borza I (2010) The phenomenon of drought and it's effects within Romania. Research Journal of Agricultural Science 42(4):102–109
106. Lupu L, Boroneanţ C, Dogaru D (2018) Evaluation of the socioeconomic effects of drought in the Turnu-Măgurele - Giurgiu sector of the Romanian Danube Valley. Rev Roum Géogr/Rom Journ Geogr 62(1):49–70
107. Dumitraşcu M, Mocanu I, Mitrică B, Dragotă C, Grigorescu I, Dumitrică C (2017) The assessment of socio-economic vulnerability to drought in Southern Romania (Oltenia Plain). International Journal of Disaster Risk Reduction 27:142–154
108. Bălteanu D, Popovici D-S, A, Dumitraşcu M, Kucsicsa G, Grigorescu I, (2013) Land use and crop dynamics related to climate change signals during the post-communist period in the south Olteania. Romania Proc Rom Acad Series B 15(3):265–278
109. ICDPR (2013) ICPDR Strategy on Adaptation to Climate Change. ICPDR—International Commission for the Protection of the Danube River
110. Schneider E, Dister E, Döpke M (eds) (2009) An Atlas for the Lower Danube Green Corridor. Ettlingen: WWF Germany, 47 pages, http://www.wwf.de/fileadmin/fm-wwf/Publikationen-PDF/Donauatlas_ Gesamt.pdf. Accessed 22 Feb 2021
111. Mansourian S, Doncheva N, Valchev K (2019) Lessons learnt from 20 years of floodplain forest restoration: the Lower Danube landscape. WWF (available at https://wwfeu.awsassets.panda.org/downloads/lessons_learnt_from_20years_of_floodplain_forest_restoration_the_lower_danube_landscap_1.pdf. Accessed on 22 Jan 2021
112. Topliisky D (2006) Climate of Bulgaria. Amstels Foundation, Sofia
113. Mitkov S, Topliisky D (2018) Climate change in Bulgaria, represented by complex indices. Annual of Sofia University "St. Kliment Ohridski", Faculty of Geology and Geography, Book 2- Geography:25–38
114. ICPDR (2018) Revision and Update of the Danube Study. Final Report. https://www.icpdr.org/main/resources/climate-changeadaptation-update-danube-study. Accessed Apr 2021
115. Vicente-Serrano SM, Begueria S, Lopez-Moreno JI (2010) A multi-scalar drought index sensitive to global warming: The Standardized Precipitation Evapotranspiration Index – SPEI. J Climate 23:1696–1718

116. Domínguez-Castro F, Vicente-Serrano SM, Tomás-Burguera M, Peña-Gallardo M, Beguería S, El Kenawy A, Luna Y, Morata A (2019) High spatial resolution climatology of drought events for Spain: 1961–2014. Int J Climatol 39(13):5046–5062
117. SaatyR (1987) The analytic hierarchy process—what it is and how it is used. Mathematical Modelling 9(3–5:161–176
118. Nikolova V, Zlateva P (2018) Geoinformation approach for complex analysis of natural Hazard. ISPRS—International Archives of the Photogrammetry, Remote Sensing and Spatial Information Sciences XLII-3:375–381
119. Downing TE, Bakker K (2000) Drought discourse and vulnerability. Chapter 45. In: Wilhite DA (ed) Drought: A global assessment, Natural Hazards and Disasters Series. Routledge Publishers, Abingdon
120. Rajsekhar D, Singh VP, Mishra AK (2015) Integrated drought causality, hazard, and vulnerability assessment 20 for future socioeconomic scenarios: An information theory perspective. J Geophys Res Atmos 120:6346–6378. https://doi.org/10.1002/2014JD022670
121. World Bank (2019) Assessing drought hazard and risk: principles and implementationGuidance. World Bank, Washington
122. Mikhailova MV, Mikhailov VN, Morozov VN (2012) Extreme hydrological events in the Danube River Basin of the last decades. Water Resour 39(2):161–179
123. Nuclearelectrica Annual Report (2003) Avalaible via https://www.nuclearelectrica.ro/wp-content/uploads/2014/08/f49fraport_anual_2003_-_snn_sam2.pdf. Accessed on 24 Jan 2021
124. Bălteanu D, Sima M, Chendeş V, Micu D, Boroneanţ C (2020) Climate change and floods: European and national context. in Bălteanu D, Micu D, Chendeş V (eds) Hazard, vulnerability and resilience to floods in Romania. Approaches at different spatial scales, Publishing House of the Romanian Academy, Bucharest

Synoptic Conditions Associated with Floods and Highest Discharges on Lower Danube River (1980–2010)

Lucian Sfîcă, Andreea-Diana Damian, Adrian Grozavu, Andrei-Ion Niță, and Marius-Victor Bîrsan

Abstract In this chapter, we explore in detail the synoptic conditions associated to the floods occurred on the Lower Danube River (from the entrance of the river in Romania through the Iron Gates gorge to the Danube Delta), as well as to the highest discharges recorded at Ceatal Izmail hydrometric station, before the entrance of the river into the deltaic region. The floods along this sector represent a response to the atmospheric circulation conditions over the entire Danube River basin, and therefore they can picture the synoptic conditions leading to high amounts of precipitation over the central and south-eastern part of Europe. The analysis investigated three flood events recorded along the Romanian side of the Danube River during the period 1980–2010, which generally corresponds to the current climate conditions.

In order to understand the triggering role of the atmospheric conditions for the floods occurrence, we have analyzed each flood in association with the phases of the most important teleconnections manifesting at continental scale—the Arctic Oscillation (AO) and North Atlantic Oscillation (NAO)—but also with regional atmospheric circulation conditions assessed using Gross Wetter Typen (GWT) method derived from COST733 catalogue.

The diachronic analysis takes in consideration the atmospheric circulation from the occurrence day of the flood peak back to three months prior to the hydrological event. Generally, the events are preceded by more positive phases of NAO and more negative values for the AO index especially within the three months' period before the hydrological event. These conditions indicate on the long term the role of anticyclonic blocking conditions at continental level inducing a prolonged interval with atmospheric instability over the Danube catchment area, while on the short-term, zonal conditions can lead to cyclonic activity enhancing the increase of the river discharge. The results are reinforced by the GWT analysis which brings other valuable information depending on the season. In this way, we can see that during winter and early spring the south-westerly circulation can lead to warm advection and the

L. Sfîcă (✉) · A.-D. Damian · A.-I. Niță
Faculty of Geography and Geology, Alexandru Ioan Cuza University of Iași, Carol I Blvd. 20A, Iași, Romania
e-mail: lucian.sfica@uaic.ro

A. Grozavu · A.-I. Niță · M.-V. Bîrsan
Meteo Romania (National Meteorological Administration), Bucharest, Romania

A. Negm et al. (eds.), *The Lower Danube River*, Earth and Environmental Sciences Library, https://doi.org/10.1007/978-3-031-03865-5_11

rapid melting of the snowpack especially in the mountain area, while in summer the atmospheric circulation types inducing large scale convection represent the main trigger for the hydrological events.

The chapter presents detailed information structured on the following sub-sections: (1) overview of major flood events and historical discharges in the Lower Danube River; (2) weather associated with hydrological events for 1980–2010; (3) the methods used to assess the atmospheric circulation (teleconnection and GWTs) and (4) the classification of atmospheric circulation leading to major floods and highest discharges.

Keywords Floods · Historical Discharges · Atmospheric Circulation · Lower Danube River · Romania

1 Introduction

Floods represent, by far, the most important hydro-climatic risk phenomenon that generates most casualties and significant material damage over the temperate zone [1]. There are three main factors that may cause this kind of events at global scale: weather factors (such as excess rainfall, snow and ice melting), partially weather factors (river runoff, tides and marine storms), and other factors like earthquakes, landslides and damaged dams [2]. It is therefore obvious that the weather conditions play the most important role in floods occurrence and the better we understand the weather conditions associated to these events, the better we are able to mitigate their effects and minimize their impacts.

It was assessed that floods have affected more than two billion people worldwide between 1998 and 2017 and the most vulnerable ones were those living on the riversides, in old houses and buildings, and especially the citizens of states poorly prepared for this type of hazard. Another important fact is that 80 $\div$ 90% of the natural disasters on Earth were caused by floods, followed by drought, heat waves and severe storms and all of these kind of events continue to increase in frequency and intensity because of the various weather extremes manifesting more frequently as an effect of climate changes [1].

In many European countries extreme events (excess rainfall, floods) are frequent, but it is considered that they tend to develop more on the eastern part of the continent, comprising the Lower Danube Basin, which is considered a hotspot for floods [3].

In the Danube Basin, the topic of floods is of very high importance due to the high density of human population and the numerous cities along its main course leading to severe impact of flood events on local or regional economy. The Danube is the second longest river in Europe, after the Volga, measuring 2860 km from its sources (from the Black Forest Mountains of Germany) to its mouth (at the Black Sea) [4]. It crosses the European continent from west to east through nine countries (Germany, Austria, Slovakia, Hungary, Serbia, Bulgaria, Romania, the Republic of Moldova and

Ukraine) receiving numerous tributaries through which drains a catchment area of 805,300km^2 between 8 and 30° east longitude and between 42 and 50° north latitude (Fig. 1).

Morpho-hydrographic differentiations require the division of the Danube course into three distinct sectors (Fig. 1): (1) the Upper Danube, with a length of 1060 km, from the springs to the Devin Gate (confluence with Morava, in Slovakia), has an average slope of 0.6–0.9 m/km, and most of the tributaries it collects have springs on the northern flank of the Alps; (2) the Middle Danube, also called the Pannonian sector, stretches between Devin Gate and Baziaș (entrance to Romania), has a length of 725 km, an average slope of less than 0.1 m/km and collects the most important tributaries throughout this basin (Drava, Sava and Tisa) which also define the type of water regime for the lower sector; 3) the Lower Danube, also called the Pontic sector, stretches over a distance of 1075 km from Baziaș to the Black Sea, has a very low riverbed slope (on average between 0.04 and 0.07 m/km), most tributaries which it collects having their sources in the Romanian Carpathians and Balkan Mountains [4].

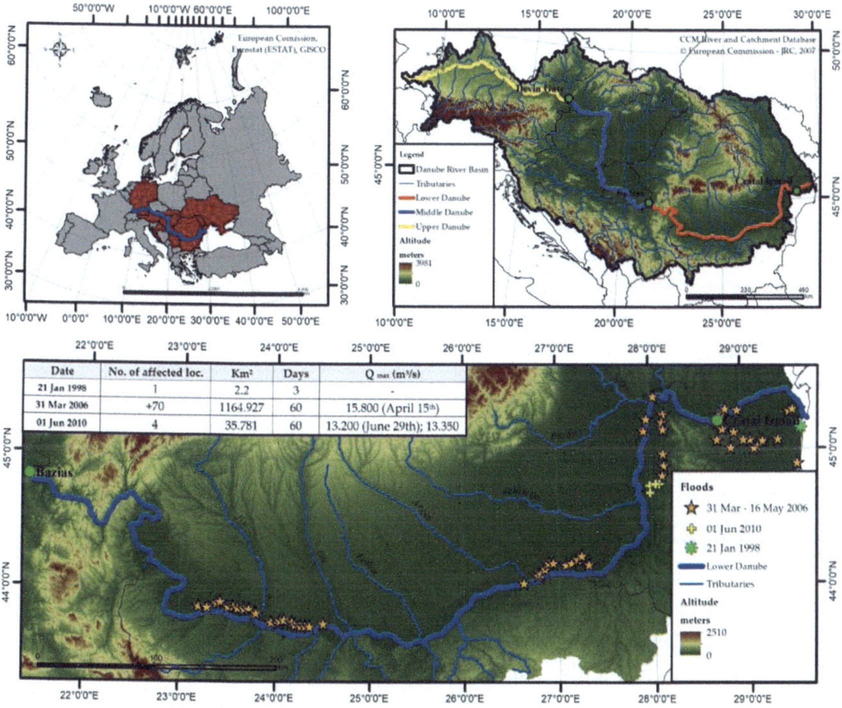

Date	No. of affected loc.	Km²	Days	Q max (m³/s)
21 Jan 1998	1	2.2	3	-
31 Mar 2006	+70	1164.927	60	15.800 (April 15th)
01 Jun 2010	4	35.781	60	13.200 (June 29th); 13.350

Fig. 1 Study area and the flooded areas during the main flood events from 1980–2014 (*Data source* [16, 17])

The Lower Danube sector, has been affected over time by various important flood episodes, which were responsible for many deaths and extraordinary material loss worth over €25 million [5].

The aim of this study is to carry out a comprehensive research over the general weather conditions associated to the major flood events and the highest discharges in the Lower Danube River, with details on the synoptic conditions that preceded them. To achieve this goal a large amount of data was analysed such as climatic teleconnections—e.g. the Arctic Oscillation (AO) and the North-Atlantic Oscillation (NAO)—reanalysis data (e.g. NCEP/NCAR, ERA-Interim) and a classification of atmospheric circulation types—Gross Wetter Typen (GWT)—was derived from COST 733 software, in order to depict at a daily level, the atmospheric circulation conditions at continental scale.

There are many research studies dealing with the synoptic context of the major flood events and with extreme discharge values. The main approach in this regard was based in assessing the correlation between the climatic teleconnections, such as the Arctic and the North-Atlantic Oscillations, and many others, on one side, and the precipitation and streamflow variability on other side, as well as was in studying how these elements could help to predict future meteorological and hydrological events [6, 7].

Regional and country-wide studies show that the Romanian territory is affected by the action of these two indices through the fluctuation in discharge levels of the major rivers on the continent. NAO is more intense in this part of Europe [8], between the variability of this index and the multi-annual variation of the pluviometrical and hydrological regimes, as well as the flood events on the Danube's Lower Sector, being a very strong connection [5, 9–12]. Besides clarifying these large-scale atmospheric circulation conditions at continental scale, we bring new information on local and regional atmospheric circulation patterns preceding the major flood events and highest discharges pushing forward the knowledge on the regional weather patterns over the analysed region. This aspect is very important for improving the linkages between weather and hydrological forecast.

2 Overview of Major Flood Events and Highest Discharges in Lower Danube River (1980–2010)

Two of the most considerable flood events ever recorded in the Lower Danube sector were those from April 2006 and June-July 2010.

The first one reached a historical discharge of 15,800 m^3/s at Baziaș hydrometric station and the overall induced damage was exceptional on the entire course of the river. The total flooded area was about 88,900 ha, starting with Ghidici–Rast–Bistreț sector and ending with Ostrov—Pecineaga sector, and two areas were flooded on purpose to reduce the water level on the Fetești—Cernavodă sector [13, 14]. The 2006 flood event was due to the continental hydro-meteorological context from the previous months, especially February and March [13], when the snow melting from the Upper and Middle sectors of the Danube River superimposed with the rainfall intensity from April had a large impact on the flow of the Lower sector in this last month. The discharge values were very high at Baziaș gauging station and they exceeded the defence levels. This event was one of the greatest in period with recorded data and had a probability of occurrence of 1% [15]. The high discharges had some major consequences. There were approximately 17,600 ha of flooded terrain and four precincts in distress as a side effect of the collapse and rupture of the dam, and more than that, for the security of other localities and to diminish the flood wave, it was extremely necessary to inundate other 51.765 ha of land. The total damage associated with the 2006 flood was estimated at 0.3 billion EUR [3], which means that there were over 3000 houses and 16,000 household annexes damaged, 15,000 evacuated residents, 8.4 km/597 km of national/local roads flooded, and 255 footbridges broken-down. A total number of 12 counties in Romania were afflicted by the floods [16].

Even though the flood event from 2010 had lower discharge than in 2006, it surpassed more important flood levels in the downstream sector of Călărași, but this was an effect of the high inflows of the Danube's large tributaries, such as Siret and Prut. The 2010 event consisted of two flood waves, the first one being recorded in mid-June (13.200 m^3/s) and the second one in the beginning of July (13.350 m^3/s).

The third major flood event occurred between 1980 and 2010 was in January 1998. It affected mainly the Danube delta, representing a typical hydro-meteorological winter event.

As for the highest discharge values recorded between 1980 and 2010, we selected the monthly maximum discharges recorded at Ceatal Izmail hydrometric station, according to the database of the National Institute of Hydrology and Water Management (NIHWM) from Bucharest. As we can observe in Table 1, the highest values were registered in April 2006, when, as discussed, one of the most important flood episode occurred in Romania in last century. After that, the next in line are those corresponding to the 2010 flood event, occurred in July. However, we should underline that according to the database of the National Institute of Hydrology and Water Management (NIHWM) from Bucharest, the highest of all discharge values was recorded at the Călărași Chiciu station (16,200m^3/s), in April 2006.

According to the "Romanian Waters" National Administration (RWNA) [17], from 1965 to 2012, at Baziaș gauging station, there were some extremely high discharges on the Lower Danube River as follows: 20 years that had maximum flow between 10,000 and 13,000 m^3/s, 3 years with maximum values between 13,000 and 15,000 m^3/s, and 1 year that had over 15,000 m^3/s.

A characteristic of these high discharges is that the majority of them happened in April and in other spring–summer months, April and May being the months with the most frequent occurrence of flood peaks in the Lower Danube River. In most years, at the end of the spring season, rainfall is much more intense and along with the snow melting it can cause serious issues. Also, not only the spring rainfall can be dangerous, but the beginning of the summer can create significant damage too, through its frequent high amounts of precipitation [13, 16].

Table 1 The highest discharges recorded at Ceatal Izmail gauging station (1980–2010) and the weather conditions prior to their occurrence over the Danube river basin

Date	Monthly Q_{max} (m^3/s)	Weather background	Key-region within the Danube Basin
06 June 1987	12000	Large scale convection	South Part of Pannonian Basin
21 April 1988	13400	Snow melting and high amount of precipitation	High amount of snow in Carpathians and Alps during March
27 November 1998	11300	High amount of precipitation during autumn	Pannonian Basin in September and Bulgaria in November
08 May 1999	12300	High amount of precipitation in early spring with high soil moisture during winter	Pannonian Basin and Balkan Mountains
21 April 2000	12600	Snow melting in March and high amount of precipitation	Pannonian Basin and Northern Carpathians
29 April 2004	11100	High amount of precipitation after rapid snow melting in March	Dinaric Mountains and Southern Pannonian Basin
02 May 2005	14500	High amount of precipitation during April	Eastern Carpathians
25 April 2006	15900	Snow melting and high amount of precipitation	Eastern Alps and Pannonian Basin
21 April 2009	11550	Snow melting	High amount of snow in late March over the Alps, Dinaric and Balkan Mountains
06 July 2010	15500	Large scale convection	Northern Carpathians

Source Data [16, 19]

3 Data and Methods

3.1 Floods and Highest Discharges Data

The current analysis is firstly focused on a thirty-year interval, between 1980 and 2010, during which three major flood events occurred on the Lower Danube River, reaching their peak on: January 21, 1998, March 15, 2006 and June 21, 2010 (according to the public data from the RWNA's website [17]).

Based on the maps showing the affected areas in the flood plain of the river and the Danube Delta in 2006, created by the Remote Sensing and GIS Laboratory of the National Meteorological Administration in Romania [18], and the floods database from the RWNA, we managed to locate every flooded area from these three events and the results are shown in Fig. 1.

Besides these three events we have also used the maximum monthly discharges recorded at Ceatal Izmail gauging station. We have chosen only this station due to its location as the last station before the entering of the Danube River into its deltaic region. According to the RWNA [17] the highest discharges are considered all the values >10,000 m^3/s which is almost twice the multiannual mean of the Danube at this station. Between 1980 and 2010 a number of ten events of this kind were recorded (Table 1). The meteorological/synoptic context associated to these events have been described by using the archived products of Global Forecast System from wetter3.de

[19] and other products derived using tools of NCEP/NCAR database [20]. Based on these data, we have been able to assess the weather background of these events (Table 1), and to indicate the key-regions of the events, referring to those regions where the high amount of precipitation/snow melting or both have been recorded.

The flood events occurred in 2006 and 2010 were also considered in the analysis of the highest discharges, starting from the day of the flood peak and not the first day of the flood events. The flood event from January 1998 was not considered in the analysis of the highest discharges, because it was generated an ice jam phenomenon due to a prolonged episode of positive air temperature anomalies over the Danube River Basin following cold winter conditions.

3.2 Climatic and Teleconnection Data

In order to describe the weather associated with the three flood events and the highest discharges between 1980 and 2010, we used the reanalysis data sets from NCEP/NCAR [20]. As well, a series of cartographic products have been made using the E-OBS dataseries [21] using R software and ArcGis packages. We assumed that previous weather conditions over the Danube Basin River represented a trigger for the flood event. For this purpose, composite maps for key weather elements for floods occurrence have been selected (precipitation amount, air temperature anomalies at ground level and in altitude or sea level pressure and geopotential height) depending on the period of the year. The maps were produced for representative time intervals before the floods's occurrence and manifestation in the Lower Danube Basin.

The next step was to analyze the two climatic indices—the AO and NAO. The daily database was downloaded from the Climate Prediction Center's website [22]. We calculated the average values according to five intervals (3, 10, 14, 30 and 90 days) for the three days during which flood events reached their peaks to see which time period had the most significant contribution before the flood episodes occurred.

We applied the same methodology, as used for the days with flood events, for the days which recorded the highest discharges from Ceatal Izmail, where the Danube Delta starts to form. The database from this hydrometric station was obtained from the NIHWM, Bucharest.

3.3 Atmospheric Circulation Classification

In order to understand more precisely the synoptic patterns that lead to the largest floods and highest discharges occurred on the Lower Danube River, in this study we have used a classification of atmospheric circulations derived from the COST733 software [23]. We have considered that Gross Wetter Typen (GWT) classification, which is an objective method based on threshold criteria, fits well with the task being known as a classification that explains best the precipitation amount over a specific region [24, 25].

The daily mean-sea-level-pressure at 12 UTC (MSLP), data from ERA-Interim reanalysis [26], was used as input data for this classification. The main idea for this classification is to characterize the circulation patterns in terms of zonality, meridionality, and also vorticity of the large scale mean sea level pressure field [25].

In this manner, an objective classification of the daily synoptic pattern was conceived for the continental region covering the entire Danube Basin River. Figure 2 displays the 18 atmospheric circulation types derived from this approach. They include two main circulation types considered for Center Low (type 17), and Center High (type 18) over the study domain and also a couple of cyclonic circulation types (from T1 to T8) and anticyclonic circulation types (from T9 to T16). The classification was carried for the period 1979–2016, from which only 1980–2010 data were selected.

Generally, we must underline that the anticyclonic conditions are prevalent over the Danube River Basin at annual level with all the corresponding types (9–16 and 18), summing up 65.7% of the days along the year. It is known that these atmospheric conditions inhibit the precipitation occurrence. In the meantime, we can observe that some cyclonic circulation types are also connected with anticyclonic conditions centered on the north of the Danube River Basin (type 6), or east and north-east of the continent (types 7 and 8) that indicate the blocking of cyclonic conditions over the Danube Basin. They sum up 12% annually and can induce high amounts of precipitation in the central and eastern part of Europe. Pure cyclonic conditions instead are assessed at 4.2% annually (circulation type 17).

4 Results

4.1 Weather Conditions Associated to Major Hydrological Events in the Lower Danube

The three events of major floods in the Lower Danube River from 1980 to 2010 have been caused by very high amounts of precipitation during the warm season (June 2010), rapid melting of snowpack during winter (January 1998) and high amounts of precipitation occurred synchronously with the intensive melting of snowpack in the mountainous region (April 2006). Besides these events, the highest discharges of the Danube River at Ceatal Izmail can occur during all seasons, but they are specific for the spring, with April recording 5 of the 10 analyzed highest discharges.

Further, we give some factual details of the weather associated to these major flood events in order to better understand the meteorological background of their occurrence.

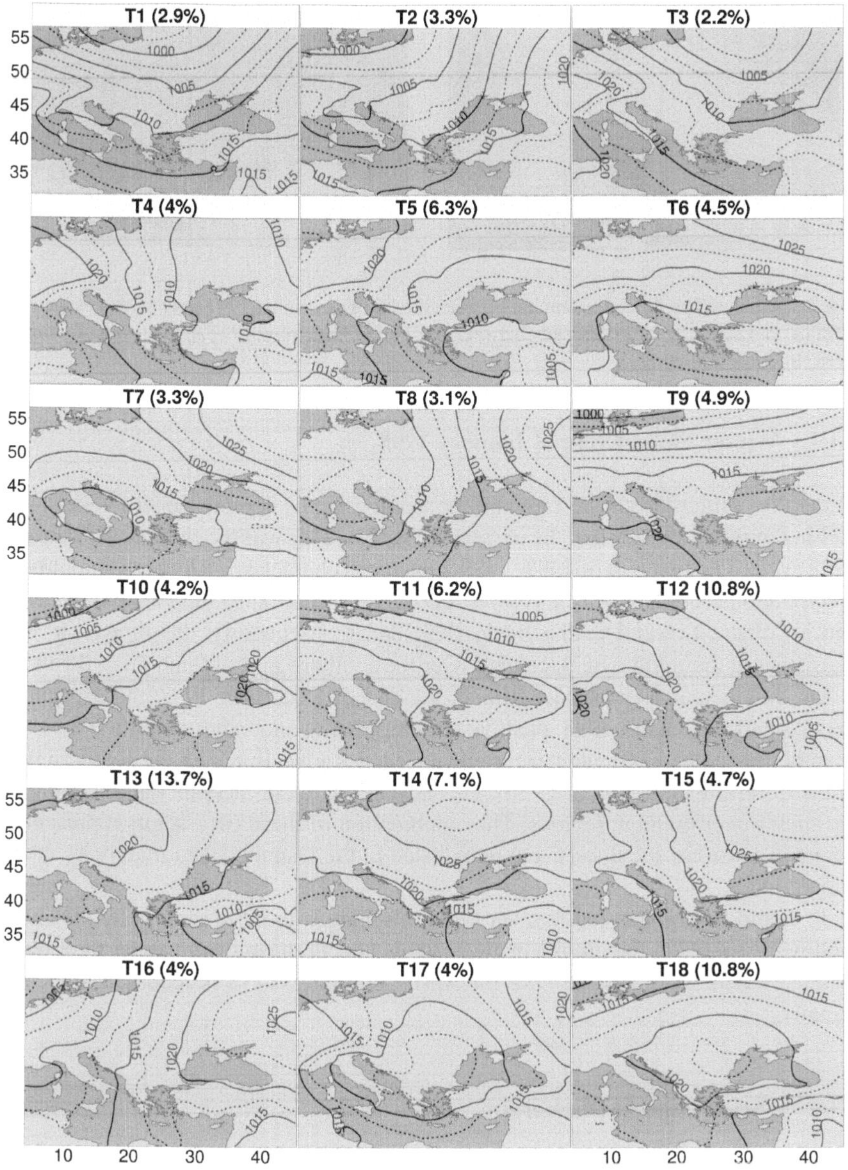

Fig. 2 Types (T) of atmospheric circulation according to GWT classification focused on Lower Danube river using based on Sea Level Pressure (hPa); cyclonic (T1-8) and anticyclonic (T9-16) subtypes are separated based on the cardinal position of the pressure centers towards the center of the domain (N-north, NE-north-east, E-east, SE-south-east, S-south, SW-south-west, W-west, NW-north-west) next to cyclonic (T17) and anticyclonic (T18) types when these pressure centers are located over the territory of Romania (*Data source* [26])

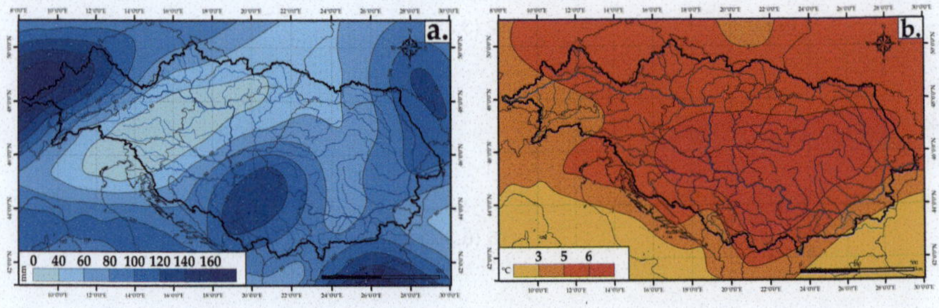

Fig. 3 Cumulated precipitation amount between the 1st of December 1997 and the 20th of January 1998 (**a**) and air temperature anomaly between the 25th of December 1997 and the 20th of January 1998 (**b**) (*Data source* [20])

4.1.1 Winter Flood Event of January 1998

Generally, winter floods are rare and the leading factor is represented by the rapid melting of the snowpack and the ice jam formation. This was the case for the January 1998 event. The first part of 1997–1998 winter was characterized by high amounts of precipitation recorded especially over the Balkan Mountains, the Pannonian Basin and all of the Lower Danube Basin (Fig. 3a). Consequently, the snowpack was consistent in the middle of December over the southern part of Pannonian Basin.

On this meteorological background, the end of December 1997 and especially the first half of January 1998, positive air temperature anomalies amplified over the Danube Basin due to an intensive tropical air mass advection over the southern part of the European continent (Fig. 3b). This advection determined the rapid melting of the snow cover on large regions. The combination of these two factors represented the main cause for the flood event manifested in the Danube Delta mainly as an ice jam flooding.

It should be emphasized that very high discharges induced mainly by the snow melting restricted to the mountain area can also occur during spring over the Lower Danube Basin, as it was the case for April 2009 (11,550 m^3/s recorded at Ceatal Izmail).

4.1.2 Spring Highest Discharges

April 2006. The amount of precipitation in the Danube Basin River, cumulated throughout the entire 2005–2006 winter, was slightly above the normal, varying from less than 100 mm in East Carpathian to > 250 mm in the middle of the Pannonian Basin (Fig. 4a). Even if not impressive, this amount of precipitation was associated with strong negative anomalies in the field of air temperature above all the European continent (Fig. 4b). Therefore, most of the precipitation was cumulated as a consistent snowpack in the mountain area from the Alps to the Carpathians.

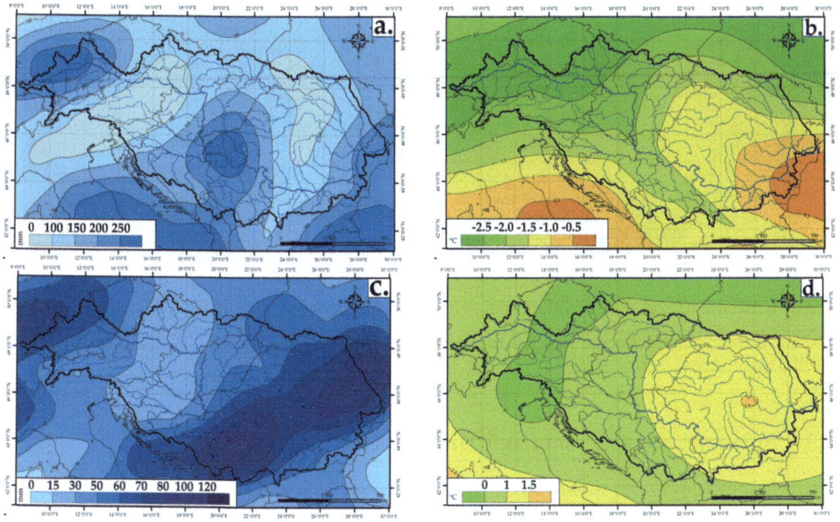

Fig. 4 Cumulated precipitation amount (a) and air temperature anomaly (b) between the 1st of December 2005 and the 15th of March 2006; cumulated precipitation amount between the 16th of March and the 15th of April 2006 (c) and air temperature anomaly at 850 hPa geopotential height between the 15th and the 31st of March 2006 (d) over the Danube River Basin (*Data source* [20])

In these conditions, the second half of March brought a sudden warming over the entire Danube River Basin leading to a rapid and generalized melting of the snow cover. Moreover, the warming was felt especially in altitude, in the middle mountain ranges leading to the increase in river discharges. Additionally, to the snow melting, the interval from the 15th of March to the 15th of April the South and East Carpathians, next to the Pannonian Basin, recorded high precipitation amounts. These factors, combined, represented the cause of the 2006 flood event which can be considered as a typical spring flood event.

In this regard, a key element for this kind of early and middle spring floods or historical discharges is given by a consistent snowpack during March, especially in the mountain area over the entire Danube River Basin. Actually, analyzing all the months of March before the April hydrological events we can observe that their composite maps indicate a positive anomaly in the field of atmospheric precipitation for the entire Danube Basin (Fig. 5a), presenting a maximum over its upper Basin. In the same time, those months were below normal in term of temperature anomaly, contributing to the accumulation of snow in the mountain areas (Fig. 5b).

The same weather pattern as that of April 2006 was associated to other very high discharges recorded at Ceatal Izmail such as April 1988, April 2000, April 2004 and April 2009 (Table 1).

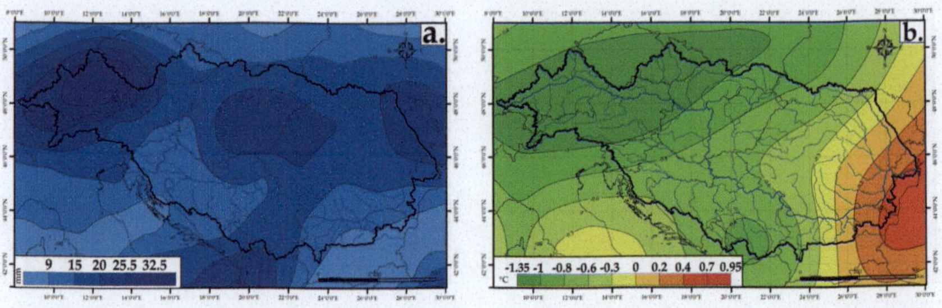

Fig. 5 Mean precipitation amount anomaly (a) and air temperature anomaly (b) during the month of March prior to April hydrological events from 1988, 2000, 2004, 2006 and 2009 within the Danube Basin (*Data source* [20])

4.1.3 Summer Flood Events and High Discharges

June 2010. May 2010 was extremely rich in precipitation all over the European continent and especially over its central-eastern part. The monthly amount of precipitation reached more than 150 mm over the Northern Carpathians in Slovakia and over all the Pannonian Basin. Consequently, the soil moisture was very high at the end of this month over these regions (Fig. 6a), so that the precipitation recorded in June over the same region lead to very high discharges of the tributaries, such as Prut and Siret rivers. In this period these rivers recorded impressive discharges such as the historical maximum of 2850 m^3/s reached by Siret River [27], a value representing more than 10 times the mean annual discharge of this river.

The high amounts of precipitation recorded during the warm season are determined mainly by the atmospheric instability which can induce convection on large spatial scale. Additionally, orographic convection on the outer slopes of the mountains can enhance very high amounts of precipitation. This was the case of the Eastern Carpathians during the 2010 summer event, recording a maximum over the area of Beskids Mountains, over the Upper Siret and Prut Basins.

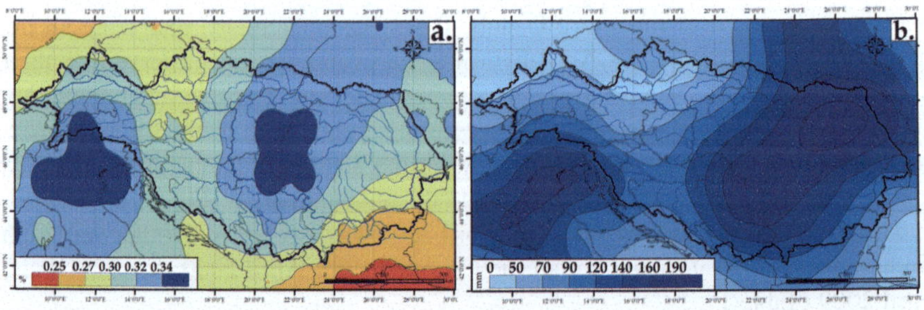

Fig. 6 The soil moisture anomaly during May 2010 (a) and precipitation amount during June 2010 (b) over the Danube River Basin (*Data source* [20])

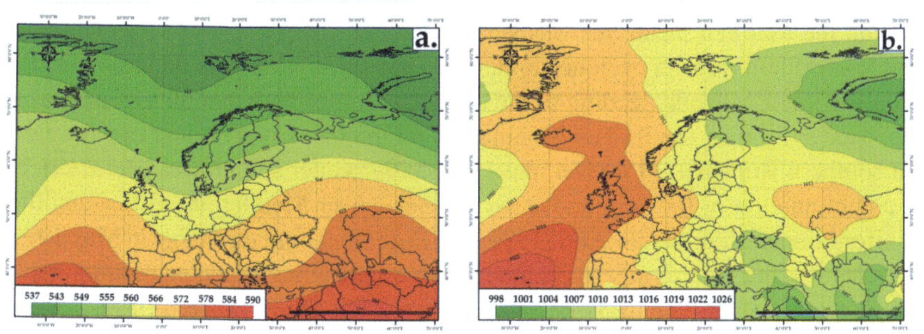

Fig. 7 The 500 hPa geopotential height for May–June 2010 (a) and mean sea level pressure for June 2010 (b) at continental scale (*Data source* [20])

In this situation a very important role is played by the atmospheric circulation in the upper layers of the troposphere, the instability being triggered mostly by cut-off lows or long-wave troughs [28], which led to a negative anomaly in the field of 500 hPa geopotential height over the central part of the continent. This is shown by the composite map of 500 hPa geopotential height, which displays a long wave trough structure, centered over Western Europe with the Danube Basin subject to a south-westerly flow (Fig. 7a). These upper structures—imposing intensive convergence at the surface—lead progressively to the decrease of the atmospheric pressure at sea level (Fig. 7b). Through this process, May–June 2010 was characterized by a very extended low-pressure area over the central and eastern part of the continent. In the same time, a ridge extended over the North Atlantic indicate the manifestation of an anticyclonic blocking at continental scale [29]. Moreover, this is a dynamic pattern related clearly with the negative phase of the so-called East Atlantic-West Russia teleconnection index manifesting over the Euro-Atlantic domain [30].

Thus, it is clear that the 2010 event is typical from a synoptic point of view for warm season floods and also for the historical discharges on the Lower Danube River. In fact, other very high discharges occurred at Ceatal Izmail in quasi-similar weather conditions at continental scale, such as June 1987 (12,000 m^3/s).

4.1.4 Autumn High Discharge of November 1998

Besides the summer and spring floods presented above, a very original case of high discharge on the Lower Danube is represented by November 1998 (11,300 m^3/s, recorded at Ceatal Izmail). Even if the summer of 1998 was mostly dry over the entire Danube River Basin, the early autumn brought an impressive amount of precipitation in September on the Pannonian Basin and in November over the Balkan Mountains and Lower Danube plain.

Firstly, in September the precipitation amounts were determined by a very high and unusual atmospheric instability caused by long wave troughs developed over Central Europe, propagating towards the east of the continent. During November instead, the Mediterranean Sea was subject for intense cyclogenesis that caused a couple of Mediterranean cyclones to develop and to move across the northern part of the Balkan Peninsula, leading to the accumulation of high amounts of precipitation, especially over the Balkan Mountains and the low lands of northern Bulgaria and southern Romania. Moreover, the 1998–1999 winter season was characterized by a continuation of the very active cyclogenesis over Mediterranean Sea and this manifestation during March–April 1999 lead to another high discharge in May 1999.

Generally, the Mediterranean cyclones induce high amounts of precipitation, especially on the Balkan Peninsula on the southern flank of the Danube Basin and the month of November is a typical one for their development and manifestation in the Lower Danube [31]. In this regard we underline that the Danube Basin receives important precipitation amounts from two cyclogenesis regions: the Gulf of Genoa and the Aegean Sea [32] that are active during the cold season from October to April. While the cyclones from the first region can affect the entire Danube Basin, but especially the Alps and Dinaric mountains, those from the second region can bring important precipitation amounts mainly over the Balkan Mountains and the southern part of the Danube Basin.

4.2 Large Scale Circulation Patterns Associated to Major Flood Events and High Discharges in the Lower Danube Basin (1980–2010)

Generally, it is well known that positive anomalies in the field of atmospheric precipitation over different regions within the Danube River Basin are associated with blocking conditions prevailing over large parts of the Basin [33].

For this reason, it is not surprising that for the flood events analyzed here (Table 2), the NAO was mostly in its negative phase especially during the last two flood events. It is to be noticed that the negative AO and NAO conditions prevailed not only during the event, but also over the 90 days before the occurrence, especially for AO. This aspect underlines that the flood events represent a cumulated effect of long persistency of some synoptic patterns over the continent, such as blocking activity.

Positive phases of NAO characterized the previous period only for January 1998 flood event, and this is normal since the positive phase of NAO is associated with positive anomaly in the field of air temperature over the continent [34] and that lead to intensive snow melting, as presented earlier.

The same general aspects can be observed on the results from the analysis performed on all the historical discharges events (Fig. 8). We can see even more clearly in this case that AO tends toward negative phases, while NAO is found more closely to neutral conditions.

These blocking conditions can lead especially to the disruption of the cold polar vortex in middle and high troposphere that determines the isolation of cut-off lows

Table 2 Mean values of AO and NAO indices during the 3, 10, 14, 30 and 90 days' intervals before the major flood events on the Lower Danube River at Ceatal Izmail (1980–2010)

	NAO / AO					
Date		3 days	10 days	14 days	30 days	90 days
21.01.1998	0.41/−3.01	0.39/−2.42	0.31/−2.80	0.26/−2.94	0.27/−0.94	−0.34/−0.75
15.03.2006	−0.37/−2.60	−0.09/−1.37	−0.10/0.11	−0.39/−0.55	−0.24/0.07	0.13/-0.61
01.06.2010	−0.31/−0.28	−0.45/−0.66	−0.81/−0.96	−0.59/−0.88	−0.59/−0.93	−0.46/−0.50

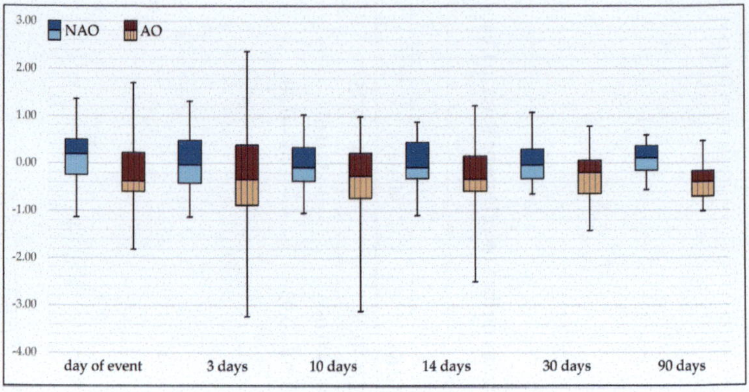

Fig. 8 The distribution of the mean values of the AO and NAO indices during the 3, 10, 14, 30 and 90 days' intervals before the high discharges at Ceatal Izmail (1980–2010) (*Data source* [22])

over central, southern or south-eastern Europe which are capable to induce high atmospheric instability over diverse regions [28] from the Danube Basin.

Actually, these cut-off lows can determine high amounts of precipitation all over the Danube Basin, and depending on their central position, they can induce high amounts of precipitation especially on the slopes of some mountain ranges under the effect of the orographic convection, as it was the case for June 2010 for the eastern flank of Eastern Carpathians [27].

Also, some types of atmospheric blocking can induce Mediterranean cyclogenesis. Depending on their tracks, the Mediterranean cyclones can determine high amounts of precipitation especially on the mountain range, from the southern side of the Danube River Basin (Alps, Dinaric, Balkan Mountains). For instance, the so-called Vb cyclones are known to produce intensive precipitation especially over the Alps, the upper Danube Basin or the northern parts of the Dinaric Mountains [35], while the so-called IIb cyclones are causing similar precipitation over the northern part of the Balkan Peninsula.

The blocking conditions associated to the negative phases of AO and NAO are known to be related to specific types of atmospheric circulation on continental and regional scale. Due to this reason, in order to bring a clearer image on the mechanism of weather pattern associated with/induced by the flood events and historical discharges we will further explore these aspects using the results of GWT objective classification of atmospheric circulation.

4.3 GWT Atmospheric Circulation Types Associated to Major Flood Events on Lower Danube River

In the 10-day interval (Fig. 9) before the analyzed hydrological events, the most frequent types of GWT atmospheric circulation were the T6 and the T17 (Fig. 2). The T6 circulation type is representative for the situation in which the Mediterranean cyclones affect the southern part of the Danube Basin, while the T17 is characteristic for a low-pressure system centered over the Lower Danube Basin. Moreover, the two circulation types can represent either the phases of the Mediterranean cyclones evolution over the Balkan Peninsula or the sea level pressure response of the cut-off low conditions prevailing over the south-eastern part of Europe.

As well, it can be observed that some south-westerly circulations over southern Europe (T10 and T11) are overrepresented for the 10 days' interval. These atmospheric circulations determine an important amount of precipitation over the Dinaric Mountains causing high discharges on Sava and Drava rivers, which are, as already mentioned, two of the main tributaries of the Danube in its middle sector. As well, these kind of south-westerly flows are associated with massive warm air advections over the same region leading to rapid snow melting in Balkan Peninsula mountain.

Also, it is not surprising that north-westerly or anticyclonic conditions expressed by T12 and T16 are not common in the 10-day interval before the flood event, being clearly underrepresented before the events comparing with their long time frequency during the same period.

The 90-day interval before the flood event (Fig. 9) was also characterized by cyclonic activity above normal, the T17 being the most overrepresented among all circulation types. The T7-8 circulation types are also overrepresented indicating the

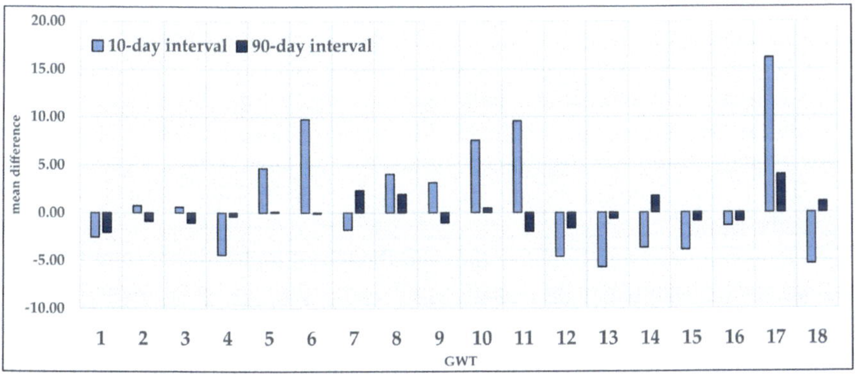

Fig. 9 The frequency of the 18 types of atmospheric circulation (GWT) during the 10-day and 90-day interval before the three flood events from January 1998, April 2006 and July 2010 over Danube River basin compared with the long-term frequency in the analyzed period (positive/negative values indicate overrepresentation/underrepresentation of circulation types)

same manifestation of Mediterranean cyclones over southern Europe. The anticyclone related types didn't occur as much as the other ones, but that doesn't mean they were inexistent. We observed a slight increase in frequency of the T18, which characterizes the opposed situation of the T17, in this case an anticyclone being located over the Lower Danube. The slight overrepresentation of these anticyclonic circulation types, as shown by T18 or T14 as well, indicate in fact the role of the blocking activity over the continent in the period before the flood event on the Lower Danube. In brief, the flood event can occur after sudden shifts of weather pattern from anticyclonic to cyclonic circulation types reigning over the Danube Basin.

4.4 GWT Atmospheric Circulation Types Associated to the Highest Discharges on Lower Danube River

Analyzing the 10 cases of the highest discharges recorded between 1980 and 2010 at Ceatal Izmail, and additionally the ice jam flooding of January 1998, we observed simply that the meteorological mechanism of flooding differs substantially between seasons.

Therefore, we have identified some features of the meteorological background which are presented in Table 1:

- during *winter* events high discharges are associated with the warm air advection over large parts of the continent, inducing the generalized melting of the snow cover;
- during the *early and middle part of the spring* occur the most highest discharges, mainly as a combination between snow melting following a positively abnormal snowpack during March and a couple of epsiodes of heavy precipitation determined especially by the action of the Mediterranean cyclones;
- *the late spring events* are mainly caused by long wave trough propagation towards the eastern part of the continent inducing atmospheric instability over central and south-eastern Europe, but also the Mediterranean cyclones can be involved;
- *the summer events* are determined exclusively by large scale convection over central Europe induced directly especially by cut-off lows isolated over the Danube Basin;
- *the late autumn events* cumulate the contribution of high soil moisture in some years and high amount of precipitation produced especially by the Mediterranean cyclones.

The GWT analysis allows us to understand in some details the weather mechanism that support these hydrological events (Fig. 10).

(a) For the *winter events,* the snow melting—which is the key-element of a hydrological event in this period of the year—was determined by a mixture of circulation types (Fig. 10a), such as T8 and T10, causing a very strong southerly advection over the entire Danube Basin. In this regard, we can observe that the

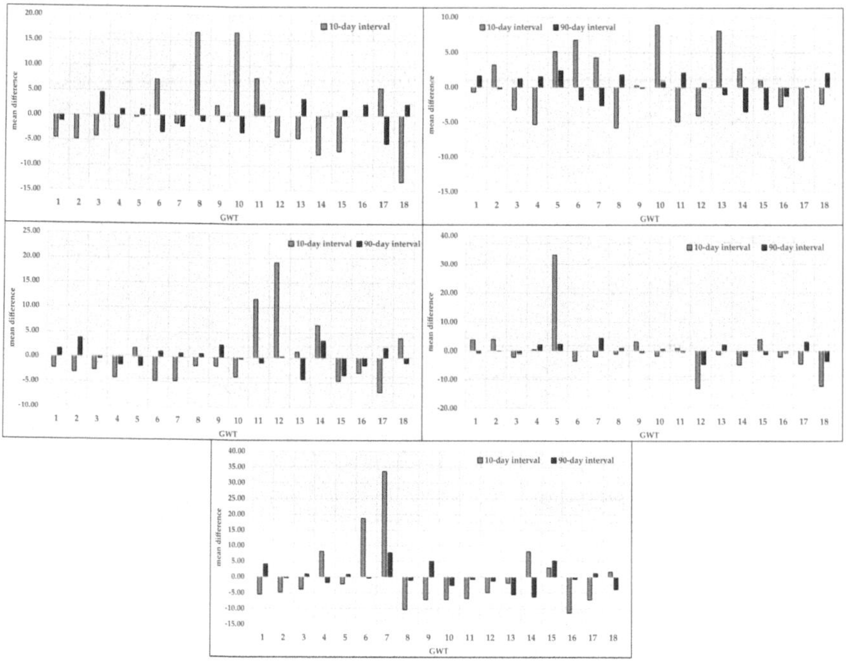

Fig. 10 The frequency of the 18 types of atmospheric circulation during the 10-day and 90-day intervals before the recording of the highest 10 discharges registered at from Ceatal Izmail hydrologic station (1980–2010) compared with the long-term frequency in the analyzed period (positive/negative values indicate overrepresentation/underrepresentation of circulation types) for winter (a), first half of spring (b), second half of spring (c), summer (d) and autumn (e)

region with the highest positive anomaly of air temperature (Fig. 11) is located over the region with the highest amount of precipitation during December and January (Fig. 3a), leading to a rapid snow melting in that area.

In the same time the synoptic conditions in the 90 days before the event is not showing any abnormal conditions, thus the event is mostly related to rapid snow melting in the 10 days before the event.

(b) For *April* we can identify two groups of atmospheric circulations that are overrepresented during the 10 days before the hydrological event. Firstly, we observed a high frequency of atmospheric circulations types 6 and 7 that display the main sea level pressure features of the manifestation of the Mediterranean cyclones. Secondly, circulation types 13, 14 and 15 are more frequent as usual in these intervals and indicate mostly anticyclonic conditions that favor the persistency of warm air masses advected from previous southerly advections over the Danube Basin. While the first group of atmospheric circulations cause high amounts of precipitation, the second group determines warm air advection and have the capacity to induce rapid snow melting especially over the mountain area of the southern limits of Danube Basin.

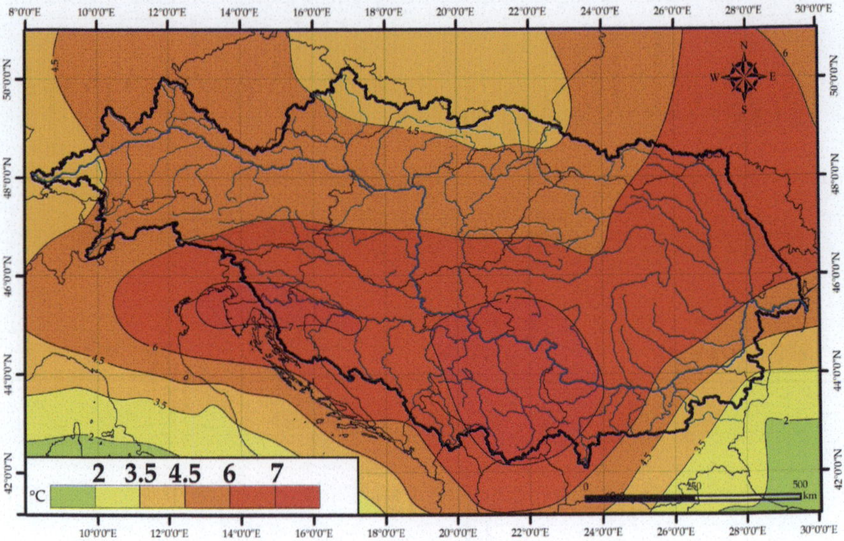

Fig. 11 Mean air temperature anomalies for 14–16 and 18–19 January 1998, days with various southerly circulation types (8, 10) over the Danube Basin; Data source: [20]

As discussed before, the rapid warming due to warm air advections in the first half of April, following a month of March which is rich in snow and colder than usual, represents the key element for the spring hydrological events. Therefore, rapid snow melting occurs especially in the mountain area of the Danube Basin being caused by a combination of southerly circulation types (2, 7 and 10) that are overrepresented in the 10-day interval before the April events (Fig. 12).

For the 90 days before the event we spot a high frequency of the circulation type 5 (common for Mediterranean cyclones) or 8 (low pressure over central Europe) that are capable of bringing important quantities of precipitation over large regions inside the Danube Basin. The high frequency of these weather patterns increases especially the snow amount in the mountains during the winter months, but also during March (Fig. 13).

(iii) *Late spring events* are characterized during the 10 days before the occurrence by the overrepresentation of T5 related mostly to Mediterranean cyclones and the high frequency of T12 showing the presence of a ridge structure over central Europe. Within the days with circulation type 5 the highest precipitation amount was recorded in the northern part of the Pannonian Basin and north-west Romania (Fig. 14a). However, the amounts were rather small and do not represent the key element of the event.

From a meteorological perspective, the key feature of these events, is given by the high frequency of circulation types 1 and 2 during the 90 days before the event. These conditions are related to long wave trough and low pressure area over the central part of Europe that determine high amounts of precipitation

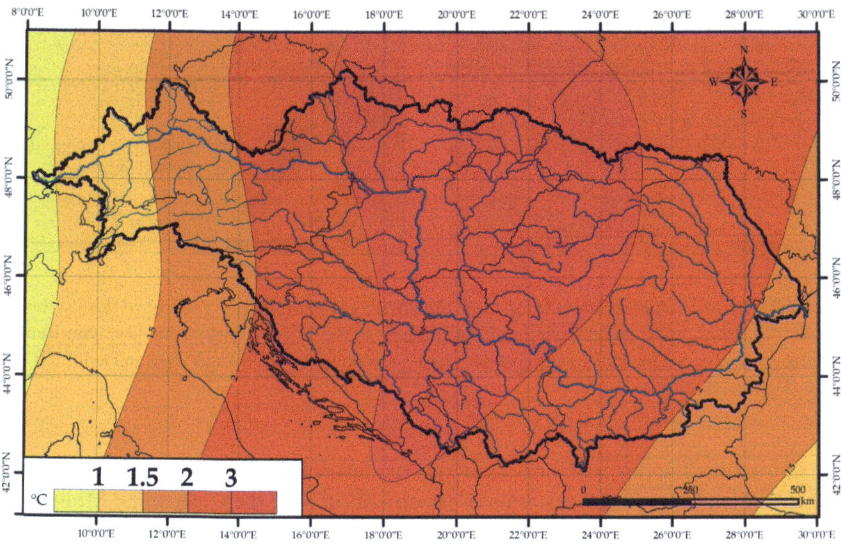

Fig. 12 Mean air temperature anomalies at 850 hPa associated with 10 days with southerly circulation types 2, 7 and 10 during the 10 days before the April hydrological events in Danube River Basin (*Data source* [20])

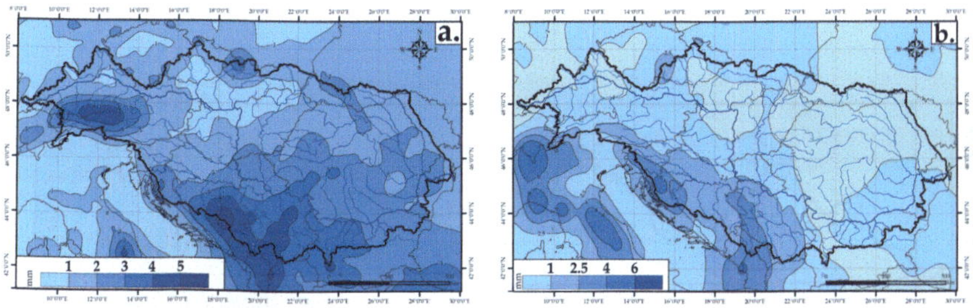

Fig. 13 Daily precipitation amount for the 32 days with circulation type 5 recorded during the 90 days before the spring hydrological events (a) and daily precipitation amount for the 10 days with circulation type 6 and 7 recorded during the 10 days before the April hydrological events between 1980 and 2010 in Danube River Basin (*Data source* [20])

especially over the Dinaric Mountains and the upper part of the Danube Basin (Fig. 14b). A large part of these quantities were recorded as snow during the late winter.

iv During the *summer event*, the 10 days before the hydrological event seem to play the most important role, being characterized by the persistency of the 5th type of GWT atmospheric circulations (Fig. 10d). This type of weather pattern indicates a low-pressure area centered over the Eastern Mediterranean and southern Anatolia and high-pressure conditions over central and western Europe. This is

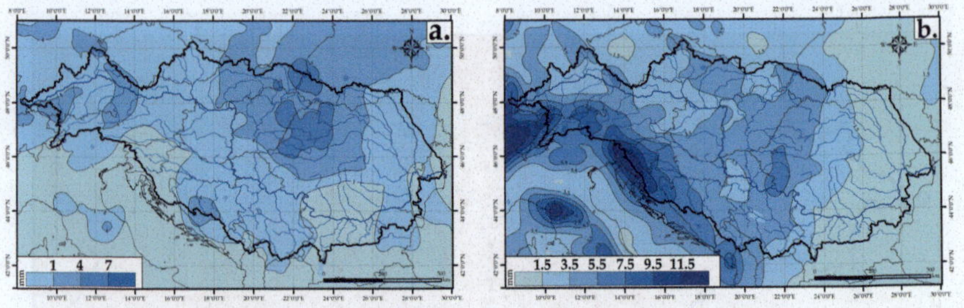

Fig. 14 Daily precipitation amount for the 2 days with circulation type 5 recorded during the 10 days before the event (a) and the precipitation amount for the 17 days with circulation type 1 and 2 for 90 days (b) before the late spring hydrological events between 1980 and 2010 in Danube River Basin (*Data source* [20])

commonly the result of the persistency of cut-off low structures over the Danube Basin leading to high amounts of precipitation over the Carpathian Mountains by the mechanism previously explained for the 2010 event (Fig. 15).

This is easily seen on the composite maps at continental scale that show a long wave trough with its axis prolonged from the Baltic Sea towards the eastern part of the Mediterranean Sea, and a cut-off low structure located in the region of Romania. This long wave trough induces an extensive negative anomaly of the geopotential height over the Mediterranean Sea (Fig. 16a), where it contributes

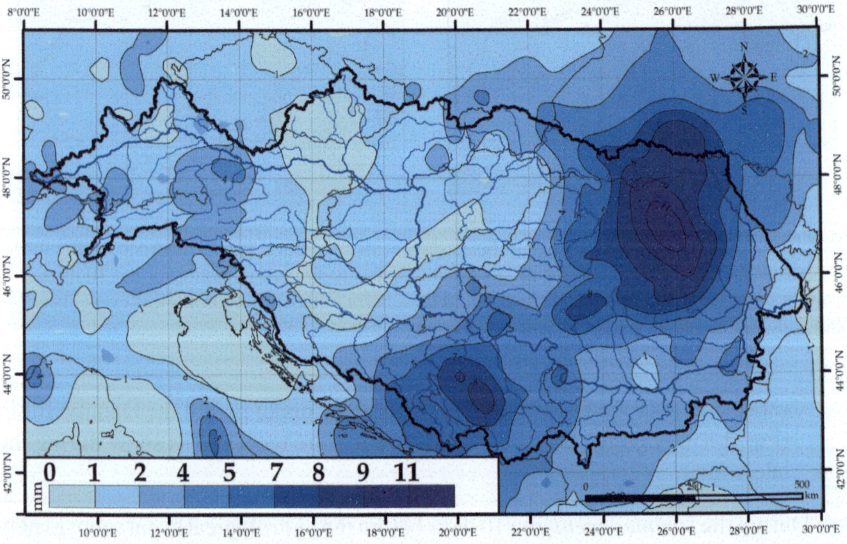

Fig. 15 Daily precipitation amount for the 9 days with circulation type 5 recorded during the 10 days before the summer hydrological events from June 2006 and June-July 2010 in Danube River Basin (*Data source* [20])

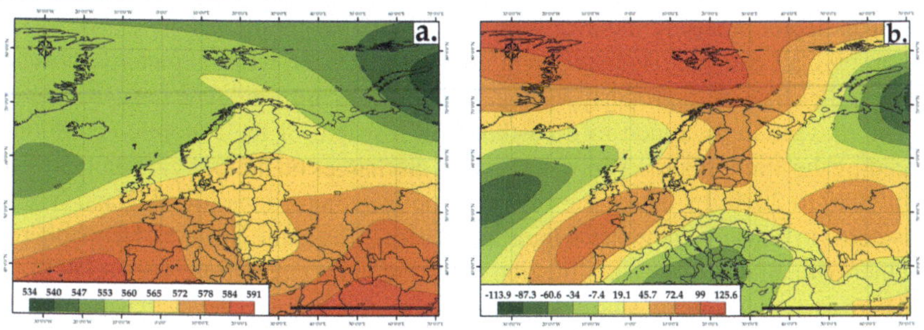

Fig. 16 The geopotential height at 500 hPa (a) and the anomaly of geopotential height at 500 hPa for the days with the circulation type 5 within the 10 days before the event (b) between 1980 and 2010 at European scale (*Data source* [20])

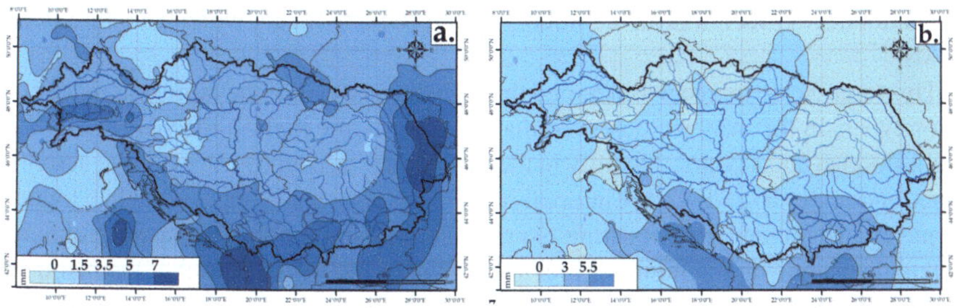

Fig. 17 Daily precipitation amount for the 13 days with circulation type 6 and 7 recorded during the 90 days before the event (a) and the precipitation amount for the 4 days with circulation type 7 for 10 days before the autumn hydrological event (b) between 1980 and 2010 in Danube River Basin (*Data source* [20])

to the development of a vast depression inclusing also the region of Romania (Fig. 16b).

(e) For *autumn*, the event from November 1998 underlines the role of the long term weather conditions for the occurrence of the historical discharge.

During the 90 days before the event (Fig. 10 e) T6 and T7 were persistent and determined extensive high amounts of precipitation over the Danube Basin (Fig. 17a, b). As discussed for spring, these circulation types indicate the action of Mediterranean cyclones.

5 Conclusion

It is obvious that the excessive precipitation amount represents the main cause of the major flood events occurred in the Danube Basin between 1980–2000. Therefore, if we can depict precisely the evolution of the synoptic conditions on a large scale, and observe all connections between flood occurrence and weather patterns, we can increase the chances to properly manage and mitigate this kind of events.

The present study aimed to do a thorough research of the synoptic conditions preceding the major floods events and highest discharges on the Lower Danube River from 1980 to 2010, by analyzing two climatic teleconnections (the AO and NAO) considered to be a key-element for the climate variability in the Northern Hemisphere, together with an objective classification of atmospheric circulation types (GWT).

There is a plethora of studies at continental and national level that prove the significant correlation between the increase of streamflow and floods occurrence in the southern part of Europe, and the negative phase of the two indices. Our analysis showed that also the highest discharges and flood events on the Lower Danube River occurred mostly during the negative phases of the AO and NAO. This underlines the role of anticyclonic blocking conditions reigning at continental scale for the high amounts of precipitation recorded over the Danube Basin, which generate large floods.

This large-scale image was detailed through the analysis of the atmospheric circulation types. In this way we observed that, in general, the highest discharges and the major flood events occurred mainly in the days characterized by the cyclonic activity types, and in the 10 and 90-day intervals these situations also happened with a higher frequency. Deep troughs persisting over the central and eastern Europe are able as well to determine high amount of precipitation over the Upper Danube Basin and an intensive south-westerly flow over the southern part of the Danube Basin leading to high amounts of precipitation from the Dinaric and Balkan Mountains.

It is somehow surprising that anticyclonic conditions persistent over the Danube Basin are present before the hydrological events, meaning that some high discharges and flood events were preceded by dry periods after which the transition from extremely dry to extremely wet weather was very rapid.

Generally, the most common events specific for the month of April are triggered by a combination of snow melting and high amounts of precipitation and we have shown both the weather patterns leading to intensive warm air advection, but also to high amounts of precipitation. The most impressive hydrological events instead are those occurring during the summer and are determined by intense cut-off low conditions prevailing over the Danube Basin.

6 Recommendation

In brief, identifying precisely the GWT circulation types associated with the occurrence of extreme hydrological events enhances the overall capacity to predict such events in advance, being able to sustain a long-term hydrological forecast on the Lower Danube. Further studies could apply the same analysis to other sectors along the Danube for an integrated hydrological forecast system at catchment level.

Acknowledgements Andreea-Diana Damian acknowledges the doctoral school of Geosciences for the support of her PhD study.

Author Contributions Lucian Sfîcă and Andreea-Diana Damian are equally the main authors of the current study.

Conflict of Interest Authors declare no conflict of interest.

Copyrights issues and permission to republish the copyrighted published materials. All the figures with no citations are the authors work and the figures with citations has no copyrights issues (it is the public domain) and the authors have obtained the permissions of the copyrighted figures/tables from the first publisher(s).

References

1. World Health Organization (2018) Floods. https://www.who.int/health-topics/. Accesed on Feb 1, 2021
2. Romanescu G (2018) Floods: Calamity or normality? case studies: Prut and Siret hydrographical basins, Romania (in Romanian), Târgoviște, Romania
3. Chendeș V, Micu D, Sima M, Ion MB (2015) A database design of major past flood events in Romania from national and international inventories. Conference: Air and Water Components of the Environment, Cluj-Napoca, Romania
4. Pisota I, 1983, Dunarea (The Danube), In: Badea et al. (Eds.) Geografia Romaniei, vol I - Geografia fizica (Geography of Romania, vol I – Physical Geography), 346–351, Edit. Academiei, Bucharest, Romania.
5. Pătruț S (2010) Exterme events (floods) on the Lower Danube in conjunction with the North Atlantic Oscillation (NAO) and the pressure at sea level (SLP). Vulnerability to the pressure of man's activities, Water Resources from Romania
6. Pekar J, Pekarova P, Miklanel P (2020) Development of relations between values of NAO index and discharge series in the Danube Basin. Int J of Climatol 22(10):1169–1179. https://doi.org/10.1002/joc.788
7. Mareș I, Mareș C, Stanciu P (2006) Variability of the discharge level in the Danube Lower Basin and Teleconnection with NAO
8. Mareș C, Mihăilescu M, Mareș I (2012) NAO influence on the Danube Lower Basin. Conference: the V-th International Scientific Conference in Water, Climate and Environment, Balwois
9. Mareș I, Mareș C (2002) Mihăilescu M (2002) NAO impact on summer moisture variability across Europe. Physics and Chemestry of the Earth, Parts A/B/C 27(23–24):1013–1017. https://doi.org/10.1016/S1474-7065(02)00135-3
10. Armaș I, Mendeș D, Omrani ȘG, Posner C (2015) The North Atlantic Oscillation influence on the climate and flow variability of the Lower Danube Valley, between the towns of Oltenița and Călărași, Romania. Forum Geografic 14:134–147. https://doi.org/10.5775/fg.2067-4635.2015.111.d
11. Rîmbu N, Boroneanț C, Buță C, Dima M (2002) Decadal variability of the Danube river flow in the Lower Basin and its relation with the North Atlantic Oscillation. International J of Climtol 22(10):1169–1179. https://doi.org/10.1002/joc.788
12. Rîmbu N, Dima M, Lohmann G, Ștefan S (2004) Impacts of the North Atlantic Oscillation and the El Niño/Southern Oscillation on Danube river flow variability. Geophys Res Lett, 31(23), https://doi.org/10.1029/2004GL020559
13. Romanian Waters National Administration, Flood Risk Management Plan—The Danube River (http://www.inhga.ro/documents/). Accessed Feb 1, 2021.

14. Bădăluță MC (2010) Water management basics, Technical sciencies, Vol. 229, Ed. Orizonturi Universitare, Romania, ISBN: 9736384446, 9789736384448
15. Spahiu MM, Ștefan N, Sălăjan L (2010) Effects if extreme hydrological phenomena on the Danube River—Călărași-Hârșova sector between 2003 and 2006. Conference: Water Resources from Romania. Vulnerability to the pressure of man's activities, Conference Proceedings, ISBN: 978–606–8042–65–2, 11–13 June 2010;
16. Romanian Waters National Administration, Report—Flood Risk Preliminary Assessment. The Danube River, within the National Institute of Hydrology and Water Management (http://arhiva.rowater.ro/). Accessed Feb 1, 2021
17. Romanian Waters National Administration's website. https://www.rowater.ro. Accessed on July 4, 2020
18. Remote Sensing and GIS Laboratory of the National Meteorological Administration in Romania website. https://www.inmh.ro. Accessed on 2 July, 2020
19. Global Forecast System' products. http://www1.wetter3.de/. Accessed on February, 2021
20. Physical Sciencies Laboratory, NCEP/NCAR reanalysis' website. https://psl.noaa.gov/data/gridded/data.ncep.reanalysis.html. Accessed on Sep 29, 2020
21. European Cliamte Assessment & Dataset, E-OBS gridded dataset. https://www.ecad.eu/download/ensembles/download.php. Accessed on October 4, 2020
22. Climate Prediction Center. https://www.cpc.ncep.noaa.gov. Accessed on July 15, 2020
23. Philipp A, Bartholy J, Beck C, Erpicum M, Esteban P (2010) Fettweis X (2010) Cost733cat—A database of weather and circulation type classifications. Phys Chem Earth Parts A/B/C 35:360–373
24. Beck C (2000) Circulation dynamic variability in the North Atlantic—Europe region since 1780 (in german). Institute for Geography, Würzburg, Würzburg, Germany.
25. Beck C, Jacobeit J, Jones PD (2007) Frequency and within-type variations of large-scale circulation types and their effects on low-frequency climate variability in Central Europe since 1780.Int. J Clim 27:473–491
26. Dee DP, Uppala SM, Simmons AJ, Berrisford P, Poli P, Kobayashi S, Andrae U, Balmaseda MA, Balsamo G, Bauer P et al (2011) The ERA-Interim reanalysis: Configuration and performance of the data assimilation system. Quart J Roy Meteor Soc 137:553–597
27. Romanescu G, Mihu-Pintilie A, Stoleriu CC, Carboni D, Paveluc LE, Cimpianu CI (2018) A Comparative Analysis of Exceptional Flood Events in the Context of Heavy Rains in the Summer of 2010: Siret Basin (NE Romania) Case Study. Water 10:216
28. Dobri RV, Sfîcă L, Ichim P, Harpa GV (2017) The distribution of the monthly 24-hour maximum amount of precipitation in Romania according to their synoptic causes. Geogr Tech 12(2):67–72. https://doi.org/10.21163/GT_2017.122.06
29. Masato G, Hoskins BJ, Woollings T (2013) Winter and summer Northern Hemisphere blocking in CMIP5 models. J of Clim 26(18):7044–7059
30. Lim Y (2015) The East Atlantic/West Russia (EA/WR) teleconnection in the North Atlantic: climate impact and relation to Rossby wave propagation. Clim Dyn 44:3211–3222. https://doi.org/10.1007/s00382-014-2381-4
31. Garaba L, Sfîcă L (2015) Climatic features of the romanian territory generated by the action of mediterranean cyclones. Conference: International Geographical Conference "Dimitrie Cantemir" 39(1): 11–24, ISSN: 1222–989X. Iași, Romania. https://doi.org/10.15551%2Flsgdc.v39i1.1012
32. Campins J, Genoves A, Picornell MA, Jansa A (2011) Climatology of Mediterranean cyclones using the ERA-40 dataset. Int J of Climat 31(11):1596–1614. https://doi.org/10.1002/joc.2183
33. Prăvălie R, Piticar A, Roșca B, Sfîcă L, Bandoc G, Tiscovschi A, Patriche CV (2019) Spatio-temporal changes of the climatic water balance in Romania as a response to precipitation and reference evapotranspiration trends during 1961–2013. CATENA 172:295–312. https://doi.org/10.1016/j.catena.2018.08.028

34. Sfîcă L, Voiculescu M (2014) Possible effects of atmospheric teleconnections and solar variability on tropospheric and stratospheric temperatures in the Northern Hemisphere. J of Atmos and Sol-Terr Phys 109:7–14. https://doi.org/10.1016/j.jastp.2013.12.021
35. Messmer M, Raible CC, Gómez-Navarro JJ (2020) Impact of climate change on the climatology of Vb cyclones. Tellus A: Dyn Meteorol and Ocean 72(1):1–18. https://doi.org/10.1080/16000870.2020.1724021

Assessment of Soil Erosion and Torrential Flood Susceptibility: Case Study—Timok River Basin, Serbia

Slavoljub Dragićević, Stanimir Kostadinov, Ivan Novković, Natalija Momirović, Marko Langović, Tomislav Stefanović, Milan Radović, and Radislav Tošić

Abstract The territory of Serbia is vulnerable to natural hazards. The most frequent and destructive hazard regarding huge material damage, loss of human lives, and environmental problems are the torrential floods. The main objective of this study is susceptibility assessment to soil erosion and torrential floods in the Timok River Basin using Erosion Potential Method (EPM) and Flash Flood Potential Index (FFPI). The erosive processes in the Timok River Basin belong to the medium erosion (Z_{av} = 0.42), which represents the third category of devastation. More than half of the basin is in the category of very weak and weak erosion, but also the category of medium erosion is geospatial dominant (45.71%). The Timok River Basin is endangered by soil erosion, with 812 registered torrents, and it stands out in comparison with other rivers of Eastern Serbia. In the inventory of torrential flood events, which was made for the territory of Serbia for the period 1915–2013, 40 torrential floods and 21 casualties were recorded within the Timok River Basin. According to the number of casualties, the Timok River Basin is ranked on second place in the Serbian territory.

S. Dragićević (✉) · I. Novković · M. Langović · M. Radović
Faculty of Geography, University of Belgrade, Studentski trg 3/3, 11000 Belgrade, Serbia
e-mail: slavoljub.dragicevic@gef.bg.ac.rs

I. Novković
e-mail: ivan.novkovic@gef.bg.ac.rs

M. Langović
e-mail: marko.langovic@gef.bg.ac.rs

M. Radović
e-mail: milan.radovic@gef.bg.ac.rs

S. Kostadinov
Faculty of Forestry, University of Belgrade, Kneza Višeslava 1, 11030 Belgrade, Serbia
e-mail: stanimir.kostadinov@sfb.bg.ac.rs

N. Momirović · T. Stefanović
Institute of Forestry, Kneza Višeslava 3, 11030 Belgrade, Serbia

R. Tošić
Faculty of Natural Sciences and Mathematics, University of Banja Luka, Bosnia and Herzegovina, Mladena Stojanovića 2, 78000 Banja Luka, Bosnia and Herzegovina
e-mail: rtosic@blic.net

A. Negm et al. (eds.), *The Lower Danube River*, Earth and Environmental Sciences Library, https://doi.org/10.1007/978-3-031-03865-5_12

Analysis of the FFPI values revealed that the class of very high susceptibility to torrential floods is registered on 2.24%, and a class of high susceptibility on 43.05% of the total basin area. This shows that 45.29% of the Timok River Basin is highly susceptible to torrential floods which indicates that this data should be seriously taken into consideration. The medium susceptibility class occupies 49.47% while 5.24% of the total river basin area belong to the low class. Therefore, only 5% of the basin are not significantly threatened by torrential floods. It was determined that the average annual rate of sediment transport in the Timok River Basin is about 833,682.83 t yr^{-1}. In addition to the damage caused to agricultural land and infrastructure, erosion sediments and torrential floods contribute to significant siltation of the Danube River.

Keywords Soil erosion · Torrential floods · Erosion Potential Method (EPM) · Flash Flood Potential Index (FFPI) · Susceptibility · Timok River Basin · Serbia

1 Introduction

The territory of Serbia is vulnerable to various types of natural hazards. Floods and torrential floods are considered the most frequent and the most significant natural hazards within the territory of Serbia [1, 2]. Based on the available information obtained from various reliable sources in Serbia, it was registered 848 flood events and over 133 casualties in the period 1915–2013 [3].

Torrential floods occur in the hilly-mountainous regions in Serbia, endangered by intensive soil erosion processes, which are favored by the forest degradation and destruction. In such areas, special hydrological conditions are established, which change runoff regime, and are characterized by large flood waves and long periods of drought after that. Forest ecosystems are a powerful agent in reducing flood peak frequency, soil erosion and sediment transport intensity, in a certain drainage basin [4, 5]. The annual sediment yield for the Republic of Serbia amounts to 45,607,559.00 t yr^{-1} which is four times more than the value of geological erosion [6].

According to the sediment management concept, the interdependence between erosion and sedimentation phenomena is important and needs to be investigated. Most of the sediment produced within the territory of Serbia, entrained along with the river network, reaches the main watercourses that flow into two large rivers—Sava and Danube. Sediment transport in rivers is related to the erosion processes within the corresponding river basins. The sediment transport rate, in a given river, can be expressed in terms of unit-area sediment yield. Serbia's river sediment budget was approximated based on existing sediment database, taking into account the erosion map and hydrological characteristics of significant watercourses. Downstream from Belgrade, after the confluence of the Velika Morava River, the Danube's yearly average suspended sediment load is of 18.5 million tons, most of which remains within the Iron Gates I reservoir [7]. It was estimated that this reservoir retains up to 77% of the suspended sediment volume supplied by the Danube River [8]. Sediment transport over the dam is directly dependent on the water flow, and it varied from 0.95

(in the dry year 1990) to 6.62 million tons per year (in 2006) [9]. The annual average amount of suspended sediment supplied by the Danube River in its delta decreased by more than 50% after the Iron Gate I dam construction [10, 11]. Unfortunately, apart from short term sporadic measurements of suspended sediment load on individual tributaries in the Timok River Basin, there is a lack of sediment transport monitoring.

The Timok River Basin is endangered by soil erosion, with 812 registered torrents [12] and it stands out in comparison with other rivers in Eastern Serbia. Considering the number of registered torrential floods, the Timok River Basin is in sixth place in Serbia, with 40 recorded events and more than 21 casualties. Regarding to the number of registered casualties, the Timok River Basin is on the second place, right after the Južna Morava River Basin. Previous research indicated that the frequency of torrential floods occurrence within a period 1915–2013 has a significant increment: the number of registered events is more than doubled comparing the latest period (1991–2010) with the first one (1929–1960) [13–15]. The most catastrophic torrential flood in the Timok River Basin occurred in May 1915 and caused the loss of 21 human lives. This was one of the most catastrophic floods in the twentieth century in Serbia, which affected the river basins of Beli, Crni and Veliki Timok. The emphasized phenomenon of torrential floods in the Timok River Basin is explained by intensive erosion processes and high density of torrents. In addition to the damage caused to agricultural land and infrastructure, erosion sediments and torrential floods contribute to significant siltation of the Danube River.

In this context, the present chapter aims to investigate the contribution of the Timok River in total suspended sediment discharge of the Danube River, and to assess the susceptibility to soil erosion and torrential floods in the Timok River Basin, by using Erosion Potential Method (EPM) and Flash Flood Potential Index (FFPI), estimated in GIS environment. The approaches for using Erosion Potential Model (EPM) [16] in combination with GIS techniques represent a great potential for gross erosion prediction and sediment transport calculations at the river basin scale or regional scale.

In Serbian territory, several papers indicate that soil erosion is the major problem associated with land use and climate changes [6, 17–19], but measured data on the sediment transport in torrential watercourses are insufficient (except in some scientific-research studies). The Republic Hydrometeorological Service of Serbia (RHMSS) performs measurements of suspended sediment concentration and sediment transport only on large alluvial watercourses. In the Timok River Basin the intensity of soil erosion and torrential flood susceptibility have not been explored much. In this context, knowing the contribution of the Timok River in total suspended sediment discharge to the Danube River, we identified the areas endangered by accelerated erosion and torrential floods, as well as the areas with different degrees of susceptibility to the occurrence of these natural hazards.

Beside the agricultural land, in the Timok River Basin, the road network is considerably endangered by torrents since dense network of torrents intersects with roads on many locations. To prevent the potential damage in urban areas, agricultural land and traffic communications (roads and railways), the degree of torrential flood susceptibility in various watercourses in the Timok River Basin was assessed. By implementing appropriate measures to mitigate erosion, the amount of sediment that will reach the Danube River will be reduced. The data presented in this paper has significant interest for practical issues such as integrated water management projects, sustainable and land-use planning, spatial planning, forest ecosystems and environmental protection, sediment management, etc.

2 Study Area

The Timok River Basin is located in the eastern part of Serbia (Fig. 1) and covers 5.13% of the country's area. The Timok River is classified as a middle-sized river in the territory of Serbia according to its length (218.2 km) and the basin area (4,529.51 km²). In longitude, the watershed extends between 21°39′06,8″ E and, 22°40′32,3″ E, and in latitude, between 43°17′54,5″ N and, 44°12′57,9″ N. The highest point in the basin is located on the Stara Planina Mountain, at 2,077 m a.s.l., while the lowest altitude is found at the confluence of rivers Veliki Timok and Danube (the lowest point in the Republic of Serbia), at 28 m a.s.l. (Table 1). The height difference between these two extreme elevation points is 2,049 m.

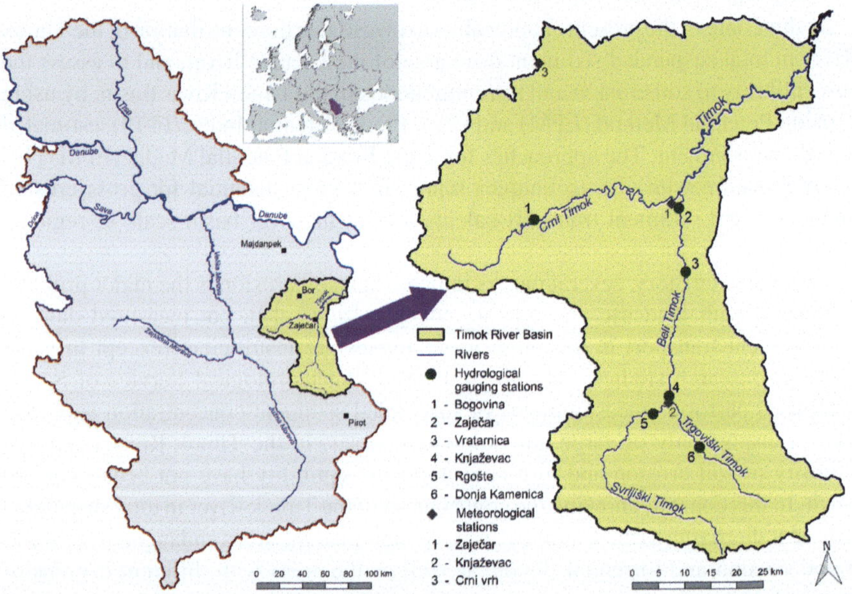

Fig. 1 Location of the Timok River Basin in Serbia

Table 1 The basic morphometric parameters of the Timok River Basin

Parameter	Code	Unit	Timok River
Area	A	km^2	4,529.5
Perimeter	P	km	480.5
Highest point	Hp	m.a.s.l	2,077.0
Confluence point	Cp	m.a.s.l	28.0
Mean altitude	Am	m.a.s.l	487.6
Length of the main river	L	km	218.2
Mean channel slope	Sm	%	0.94
Average drainage basin differences	D	km	0.46
Drainage density	Rd	$km\ km^{-2}$	0.61

2.1 *Geological and Geomorphological Characteristics*

The Timok River Basin is built of the most diverse metamorphic, igneous and sedimentary rocks. This area had a complex and long geological evolution that can be traced from the Paleozoic, through the Mesozoic and Cenozoic to the formations that were shaped in the recent past. In this long period, there was an irregular change of depositional and terrestrial environments, which resulted in a very diverse geological structure and lithology.

The analysis of hypsometric characteristics of the Timok River Basin (based on DEM 100 m resolution) shows that only 8.94% (404.84 km^2) of its territory is located at the altitude of less than 200 m a.s.l., while altitudes from 200 to 500 m a.s.l. cover 49.39% (2,237.15 km^2) of the total basin area. Altitudes up to 500 m a.s.l. cover 58.33% of the Timok River Basin (2,642 km^2), altitudes between 500 and 1,000 m a.s.l. accounts for 38.35% (1,737.23 km^2), those between 1,000 and 2,000 m a.s.l. occupy 3.32% (150.24 km^2) and altitudes over 2,000 m a.s.l. cover only 0.001% of the Timok River Basin (0.04 km^2). Based on the presented data, it was calculated that the average elevation of the Timok River Basin is 487.6 m a.s.l.

The hypsometric map (Fig. 2) shows that the altitudes lower than 500 m a.s.l. (respectively hilly relief) are the most dominant in the Timok River Basin. This is followed by low-mountainous relief and middle-mountainous relief from 1,000 to 2,000 m a.s.l. These three ranges account 96.58% of the basin's territory. Presence of altitudes higher than 500 m a.s.l., in the Timok River Basin, receiving the largest amounts of precipitation, is especially important for the intensity of surface runoff and the occurrence of torrential floods.

By analyzing the slope gradient map of the Timok River Basin (Fig. 2), it was determined that the values up to 10° are spread over 49.12% of the total territory, while slope gradients of 10–20° cover 30% of the basin area. Areas characterized by slope gradients greater than 20°, cover 20.82% of the Timok River Basin. These slopes are very important for the formation of torrential flood waves and increased intensity of soil erosion. The average slope angle of the study area is 12.3°.

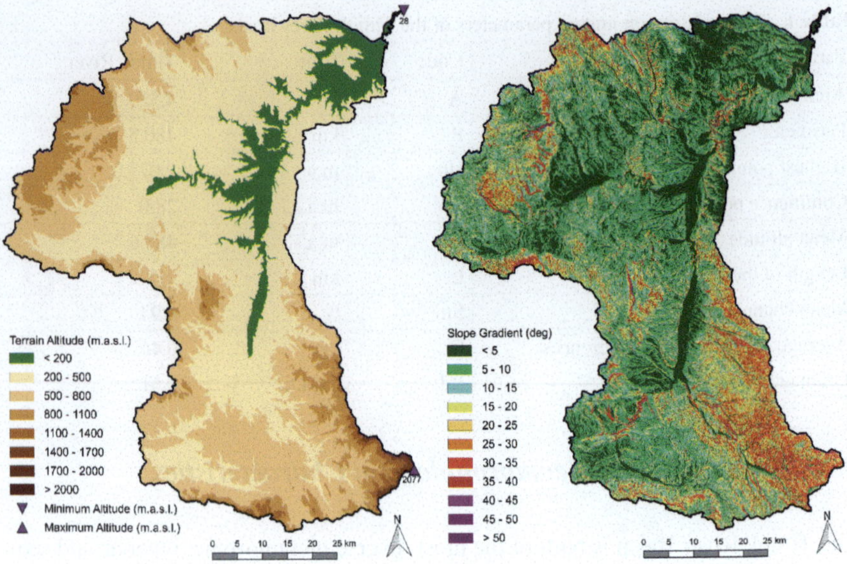

Fig. 2 Hypsometric (at left) and slope gradient map (at right) of the Timok River Basin

2.2 Climatological Characteristics

Climatological characteristics are one of the most important factors controlling the intensity and type of erosion and torrential processes. Three meteorological stations are located in the Timok River Basin: Zaječar, Knjaževac, and Crni vrh (Fig. 1). Based on the available data, for the period (1968–2017), the most significant meteorological parameters for the development of erosion processes, the air temperature and rainfall, were analyzed.

In the Timok River Basin, during the analyzed period, the highest average annual air temperature was recorded at the Zaječar meteorological station (10.9 °C), while the lowest was registered at the Crni vrh (6.5 °C). The coldest month at all stations is January, while July is the warmest month at stations on lower altitudes (Zaječar and Knjaževac), while the warmest month at the Crni vrh station is August (Table 2). The annual temperature amplitude ranges from 20.3 °C (Crni vrh) to 22.5 °C (Zaječar). The average annual air temperature for the whole Timok River Basin is 9.1 °C. At the beginning of the April, a sudden increase of air temperature in the mountainous parts of the Timok River Basin is observed. This increase intensifies the snow melting process, which further leads to the increase of river discharges and to the occurrence of torrential floods.

The amount, frequency, and type of precipitation have a major impact on river discharge, flood occurrence, intensity of the erosion processes. Based on the data from the meteorological stations Zaječar, Knjaževac, and Crni vrh, the pluviometric regime for the period 1968–2017 was analyzed. The annual amount of precipitation

Table 2 Average monthly and annual temperatures (°C) in the Timok River Basin (1968–2017) [20]

Meteorological stations	Altitude (m)	I	II	III	IV	V	VI	VII	VIII	IX	X	XI	XII	Year
Zaječar	144	− 0.6	1.3	5.8	11.4	16.5	20.1	21.9	21.2	16.5	10.5	5.1	0.8	10.9
Knjaževac	250	− 0.6	1.4	5.7	11.1	16.1	19.6	21.3	20.7	16	10.3	5.2	1.05	10.7
Crni vrh	1,027	− 3.7	− 2.7	0.9	6.1	11.0	14.2	16.5	16.6	12.2	7.3	1.7	− 2.3	6.5

Table 3 Average monthly and annual precipitation (mm) in the Timok River Basin (1968–2017) [20]

Meteorological stations	Altitude (m)	I	II	III	IV	V	VI	VII	VIII	IX	X	XI	XII	Year
Zaječar	144	42	41	42	49	66	66	56	46	44	52	54	51	583
Knjaževac	250	40	42	45	51	65	63	59	43	46	50	51	50	605
Crni vrh	1,027	44	43	50	69	92	108	81	62	71	61	57	54	792

is higher with the increase of the altitude. Thus, the highest amounts of annual precipitation were recorded at the Crni vrh (792 mm), while the lowest at the Zaječar station (583 mm) (Table 3). The highest amount of monthly precipitation is registered in June in the northern part of the Timok River Basin (Zaječar and Crni vrh), and in May in the southern part (Knjaževac).

In the Timok River Basin, the annual precipitation varies from 600 to 800 mm. The area of lower precipitation values (below 600 mm) mainly corresponds to the lower part of the Timok River Basin and its tributaries. The average annual precipitation for the whole watershed in the period 1968–2017 is 673 mm.

2.3 *Hydrological Features*

Timok River is also known as the Veliki Timok River, due to its origin from the Beli Timok (right side tributary) and the Crni Timok (left side tributary) that confluence near the city of Zaječar (Fig. 1). The total length of the Timok River (with components Beli Timok and Trgoviški Timok) is 218.2 km, while its basin covers an area of 4,529 km^2. The last 15.5 km of its course, before the confluence into the Danube River, the Timok River represents a border river between Serbia and Bulgaria [21]. In the Timok River Basin, there are 10 active hydrological gauging stations, where measurements of the most important hydrological parameters are performed. However, only six station (Fig. 1) continuously register river discharges in the period of 50 years (1968–2017). The basic characteristics of these stations are presented in the Table 4.

The Mann–Kendall test was used to identify the existence of a trend in the variability of the mean annual discharges. The test results indicated that annual discharges have a dominant decreasing trend in the period 1968–2017, which is in accordance with the majority of rivers in the Republic of Serbia [22, 23]. Out of 6 investigated hydrological gauging stations in the Timok River Basin, an increasing trend of mean annual discharges was not recorded at any station, while a decreasing trend was observed in all six hydrological profiles, mainly without a statistical significance. A moderate statistically significant decreasing trend was registered at the Donja Kamenica station (Trgoviški Timok River) with an average decrease rate of 0.029 m^3s^{-1} $year^{-1}$.

Table 4 The basic data for hydrological gauging stations in the Timok River Basin [20]

	Hydrological gauging stations	Rivers	Altitude (m)	Basin area (km^2)	Annual mean discharge (m^3s^{-1}) for period 1968–2017
1	Bogovina	Crni Timok	221.57	467	5.5
2	Zaječar	Beli Timok	124.41	2,150	11.1
3	Vratarnica		149.76	1,771	9.43
4	Knjaževac		208.71	1,242	7.65
5	Rgošte	Svrljišk Timok	225.96	618	2.98
6	D. Kamenica	Trgoviški Timok	270.17	360	3.13

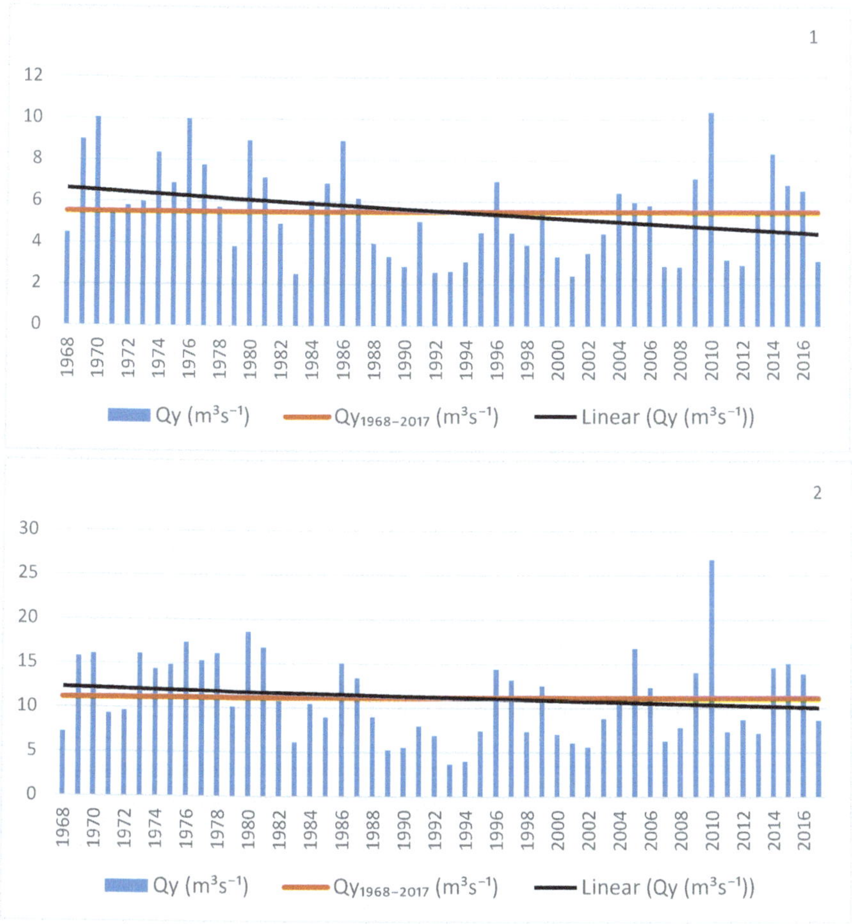

Fig. 3 Hydrographs of mean annual discharges (Qy in m^3s^{-1}) and their linear trends for station Bogovina on Crni Timok (1), and Zaječar on Beli Timok (2)

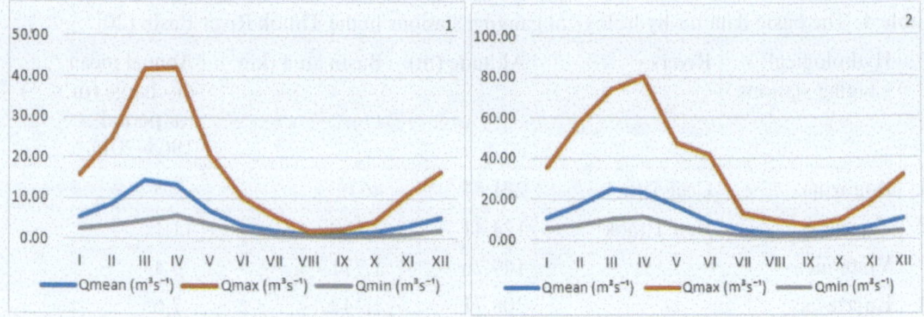

Fig. 4 Hydrographs of mean, maximum, and minimum monthly discharges for two selected stations Bogovina on Crni Timok (1), and Zaječar on Beli Timok (2) in the period 1968–2017

In Fig. 3 variations of mean annual discharge values for the two main stations on the Crni Timok (at Bogovona) and Beli Timok (at Zaječar), in the period 1968–2017, were shown. The highest values of mean annual discharge, at the station Bogovina (Crni Timok River), were reached in 2010 (10.36 m^3s^{-1}), 1970 (10.1 m^3s^{-1}), and 1976 (10 m^3s^{-1}) (two times higher than average for the entire observed period, 5.5 m^3s^{-1}). The lowest annual discharges were registered in 2001 (2.44 m^3s^{-1} which is about 55% less than average). Similar data were measured at the Zaječar station with highest mean annual discharges in 2010 (27.5 m^3s^{-1} which is two times higher comparing to the average of 11.1 m^3s^{-1}), while the lowest discharge values were measured in 1994 (only 2.65 m^3s^{-1} or about three times less than average).

The monthly and seasonal runoff distribution over a year is rather uneven. The amplitude of the discharge shows that the maximum average monthly discharge is around 7 times higher than the minimum average monthly discharge (stations Bogovina and Zaječar), indicating an uneven flow regime (Fig. 4). The overland flow and groundwater are the lowest in summer and autumn—in August, September, and October (as a consequence of summer—automn droughts) while in spring months (April and March) have the highest values (as a consequence of the snow melting in the highest parts of the basin and significant precipitations in spring). Therefore, all watercourses in the Timok River Basin belong to the rainy-snowy water regime. In some parts of the basin, the relationship between surface and underground runoff is different. It has been established that with the increase of altitude underground runoff increases at the expense of surface runoff.

One of the most specific year in the Timok River Basin, in terms of high wateriness and occurrence of floods, was 2010, when the maximum daily discharge for the entire researched period, was recorded, on February 21 at Zaječar station (297 m^3s^{-1}). It occurred as a consequence of heavy few-hour precipitation and snow melting process, especially in the subbasin of the Beli Timok River. The city of Zaječar was the most affected by the 2010 flood, when more than 600 households were endangered by this natural disaster. The minimum daily discharge, at the same station, was recorded on November 10, 1975 (only 0.15 m^3s^{-1}). This fact shows that the discharge of the Timok River is unbalanced and that the river has the torrential flow.

2.4 Land Use

The analysis of the land cover (based on Corine Land Cover—CLC database in 2012) shows that out of the total number of classes that characterize the land cover in Serbia, 21 CLC classes are presented in the Timok River Basin. The CLC class 311 (deciduous forests) dominates, covering 39.36% of the total area, followed by class 242 (complex of agricultural areas) with 18.82%, class 243 (agricultural areas with a significant share of natural vegetation) with 15.17% and class 324 (woody-shrub vegetation) with 13.95% of the total area of the Timok River Basin. Agricultural areas (CLC classes 242, 243 and 211) cover slightly less than 40% of the total basin, which is a significant area from the aspect of erosion protection.

The forested and agricultural areas cover almost the same percentage of the Timok River Basin (about 40%). Bearing in mind that a large part of the basin is covered by coppice (forests) with a low density, it results that there are significant predispositions in the basin for the development of intensive water erosion processes.

2.5 Socio-Economical Features

The total population in the Timok River Basin is 205,444 (according to the 2011 Census), and the average population density is around 35 inhabitants/km^2. Demographic resources in this area are weakened by the negative population increase rate and migration process. Population redistribution is characterized to a lesser extent by migrations towards regional city centers—Bor, Majdanpek, Zaječar, Pirot, and to a much greater extent by migration to the other more developed centers in the Republic of Serbia, as well as to the more developed countries in Europe. As a consequence, a large number of smaller settlements is facing demographic extinction.

The economic development of this area is based primarily on the existence of a great diversity of natural conditions and resources. Favorable climatological and hydrographical conditions, geological characteristics and fertile land have contributed to the development of agriculture. Mineral resources, copper and gold ores, and deposits of gravel, sand, fire brick and other building stones have been exploited for centuries in the Timok River Basin.

One of the greatest development potential of the settlements in the Timok River Basin represent their advantageous location in the terms of transport infrastructure, especially road and railway network.

3 Methods

Research of the erosion intensity and susceptibility to torrential floods in the Timok River Basin include the following activities: the study of physical characteristics of the river basin, analysis of vegetation cover and creation of land use map; analysis of the soil erosion intensity and creation of soil erosion map with Erosion Potential Model (EPM); calculation of average annual gross erosion and sediment transport with Erosion Potential Model (EPM); assessment of susceptibility of basins to torrential floods by using the Flash Flood Potential Index (FFPI).

In Serbia, as well as in the other countries in the Balkan region [24–27] the estimation of soil erosion potential is generally achieved with the Erosion Potential Model (EPM) [16]. The analytical equation for calculation of the annual volume of detached soil (gross erosion) due to water erosion [16] is:

$$W_{year} = T \cdot H_{year} \cdot \pi \sqrt{Z^3} \cdot F \tag{1}$$

where: W_{year} is average annual gross erosion (m^3yr^{-1}), T is temperature coefficient in the form: $T = (0.1 \cdot t + 0.1)^{0.5}$, where t is the mean annual air temperature (°C), H_{year} is the average yearly precipitation (mm), F is the river basin area (km^2), and Z is the soil erosion coefficient, which can be estimated using corresponding tables or can be calculated with the following equation [16]:

$$Z = Y\,X\left(\phi + \sqrt{I}\right) \tag{2}$$

in which, Y is the soil erodibility coefficient, X is soil protection coefficient, ϕ is erosion and stream network developed coefficient and I is average slope steepness of the basins, in degree.

The quantitative values of the erosion coefficient (Z) have been used to separate erosion intensity to classes or categories. The calculation of Z enables observations of changes in the soil erosion intensity since the formula includes the parameters which have been changed by the performed erosion control work.

After the total annual gross erosion rates are calculated, it is necessary to determine the sediment delivery ratio (R_u) for actual sediment transport calculation. Gavrilovic [16] has suggested the following equation for assessing of the sediment delivery ratio:

$$R_u = \frac{\sqrt{O\,D}}{0.2\,(L + 10)} \tag{3}$$

where, O is the perimeter of the basin (km), D is the average drainage basin difference (km) and L is the length of the watershed (km).

Finally, the average annual sediment transport (G, in m^3yr^{-1}) was calculated as [16]:

$$G_{year} = W_{year}\,R_u \tag{4}$$

Total sediment transport includes suspended sediment and bedload. The percentage of bedload is determined by the equation [16]:

$$\tau = \frac{Z(Yv - 1)}{\pi Ys} \quad (5)$$

where, τ is the percentage of bedload sediment transport in the torrent basin, Z is the erosion coefficient, Y_v is the volumetric weight of bedload sediments (t m^{-3}), Y_s is the volumetric weight of suspended sediments (t m^{-3}).

Using EPM model in combination with GIS techniques represent a great potential for gross erosion prediction and sediment transport assessment at the river basin or regional scale [17–19]. In this study, GIS was successfully integrated with EPM model aiming to determine surfaces with the same quantitative erosion category, but also to define the soil erosion coefficient (Z) for each erosion polygon.

Susceptibility to the occurrence of torrential flood was estimated by using the Flash Flood Potential Index (FFPI) [28]. This method is based on the fact that there is an unbreakable bond between this hazard and certain physical-geographical characteristics of the territory on which it occurs. The value of the proposed index is in the spatial representation of the areas with a flash flood risk, therefore, giving the possibility to prevent the negative effects [29–37]. Calculation of FFPI is performed according to the formula [28]:

$$\text{FFPI} = \frac{a_1\,M + a_2\,S + a_3\,L + a_4\,V}{\sum_{n=1}^{n=1} a_n} \quad (6)$$

where: M = slope index; S = soil type index; L = land use index; V = vegetation density index; a_n = sum of weightings. Index values are within the range 1 to 10 (from least to most susceptible). In this case all weightings had value of 1, which means that the next formula is used:

$$\text{FFPI} = \frac{M + S + L + V}{4} \quad (7)$$

The slope index is calculated according to the following formula:

$$M = 10^{n/30} \quad (8)$$

where n = slope in %. If n is greater or equal to 30%, then M value is 10.

Considering the fact that there are no soil structure and texture data available for investigated river basin, to calculate soil type index, values from 1 to 10 were given to certain soil types, based on their characteristics which are significant for the development of torrential floods. Land use index is calculated on the basis of CORINE Land Cover 2012 database, where values from 1 to 10 were given to certain types of land cover, depending on the characteristics important for torrential processes. Vegetation density index is obtained by analysis of multispectral images from the LANDSAT 8 satellite, TOA-radiance corrected, and calculating BSI (Bare Soil Index). Due to the fact that the vegetation density index is ranging from 1 to 10, the correlation with BSI values is performed, and the resulting formula is [36]:

$$V = 6.597\ \text{In}(\text{BSI} + 1) + 7.118 \tag{9}$$

After FFPI calculation, the classification of results on the four classes of susceptibility to torrential floods was carried out (very high, high, medium and low susceptibility). Based on the spatial distribution of FFPI values in the basin, all watercourses are classified in 4 classes, representing the potential for flash floods occurrence under appropriate conditions.

4 Results

4.1 Soil Erosion Intensity

Torrential floods are known to be related to the intensity and spatial distribution of erosion processes in the Timok River Basin. It is of major importance to point out the recent state of soil erosion intensity, because it represents a significant factor of sediment production and transport through torrents, and also a condition for floods occurrence, backfilling culverts and damage to traffic infrastructure.

The Timok River Basin is exposed to strong erosion and therefore stands out compared to other rivers in eastern Serbia. It is possible to distinguish different erosional forms caused by the action of water erosion: uneroded soil, surface erosion, rill erosion, as well as gully erosion. These processes are spread over the whole surface of the Timok River Basin, but their intensity is different depending on the dominance of the controlling factors.

The map of soil erosion intensity (Fig. 5) indicates the distribution of erosion processes, i.e., the spatial vulnerability and degradation of the investigated area by erosion. In the evolution of the soil erosion process, the laminar transfer of fine material (loose silt particles, gravel and fines) occurs first and lasts as long as the water has a low velocity. With the increase of kinetic energy, linear forms of denudation are formed, the intensity and removal of the soil is intensified and accelerated. By reaching watercourses as the basic elements of the lower erosive base, the eroded material modifies the mechanical and accumulative fluvial process, by forming various forms of fluvial relief. This is especially pronounced in the alluvial plains of active meandering rivers, which are characterized as one of the most dynamic and sensitive elements of the fluvial landscape. As a result of sediment transport and accumulation on the convex side of river courses, new forms of fluvial relief (e.g. river bars) are formed. Changes in soil erosion intensity in the upper parts of river basins have a direct impact on accumulative fluvial form modifications in the downstream sectors [38]. Due to the different size of the area that covers and the diversity of relief forms in the Timok River basin, the process of soil erosion can justifiably be included as the dominant geomorphological processes in the investigated area.

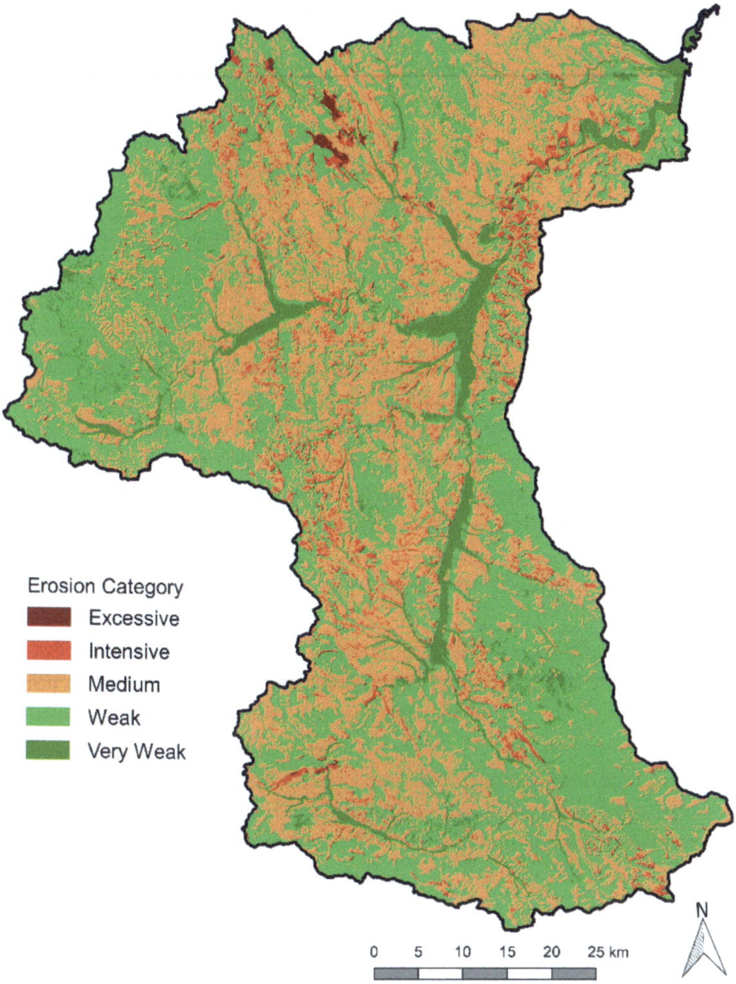

Fig. 5 Map of soil erosion intensity in the Timok River Basin

The prevalence of excessive (I) and intensive (II) erosion categories is small, and the most common areas are affected by the processes of medium (III), weak (IV) and very weak (V) categories of erosion (Table 5). According to the data of erosion coefficient and categories of erosion, these processes in the Timok River Basin belong to medium erosion ($Z_{av} = 0.42$), i.e., to the III category. More than half of the basin area is in the category of very weak (V) and weak (IV) erosion, but the category of medium erosion is also geospatially very represented (45.71%). The presence of the category of medium erosion is related to the parts of the basin with a more pronounced vertical fragmentation of the relief, with a significant slope gradient, without a quality forest cover. This arrangement of the category of medium erosion

Table 5 Erosion categories in the Timok River Basin

Erosion category	Area [km^2]	Percent of the total area [%]
Excessive erosion (I)	11.70	0.26
Intensive erosion (II)	146.78	3.24
Medium erosion (III)	2,070.60	45.71
Weak erosion (IV)	2,011.55	44.41
Very weak erosion (V)	288.89	6.38
Total	4,529.51	100.00

gives opportunities for generation, i.e., production of sediments that will enhance the torrent characteristics of the existing watercourses.

The most endangered part of the investigated area is the subbasin of the Trgoviški Timok River (525 km^2, which drains the western slopes of the Stara Planina Mt.) and all its numerous tributaries. In the mentioned basin 216 torrents are registered, which are characterized by high relief energy. Unappropriated use (devastation) of forests caused intensive water processes erosion. This subbasin is followed by other parts of the Beli Timok River Basin (without the Trgoviški Timok River Basin) with 151 registered torrents [12].

Based on the erosion map of the Timok River Basin (Fig. 5), it can be noticed that the distribution of the erosion process intensity is related the lithological units in the analyzed basin. Volcanolastics, ultramafics, flysch, and Neogene sediments dominate in areas with medium, strong and excessive erosion. Lands that are prone to erosion appear on the mentioned lithological units. Forests were intensively deforested in the past, while today most forests are not adequate, from the point of view of erosion control and management.

4.2 *Sediment Transport Assessment*

Apart from short term sporadic measurements of suspended sediment load on individual tributaries in the Timok River Basin, there is a lack of sediment transport monitoring. To determine the production and sediment transport in the investigated basin, the method of erosion potential was used [16, 39].

The Republic Hydrometeorological Service of Serbia has been measuring the transport of suspended sediment load on hydrological station Zaječar (Beli Timok River), in the period 1958–2002. These data were processed by the Institute of Water Management "Jaroslav Černi" and as a result, the average annual sediment transport of 326,000.0 t yr^{-1} was obtained. The area of the Beli Timok River basin is around 50% of the total area of the Timok River Basin [40].

By using the method of erosion potential, for the entire Timok River Basin, the following results were obtained:

- the average annual gross erosion, $W_{year} = 3{,}206{,}472.41\ t\ yr^{-1}$;
- the sediment delivery ratio, $R_u = 0.26$;
- the average annual sediment transport, $G_{yr} = 833{,}682.83\ t\ yr^{-1}$.

In the value of total annual sediment transport, bedload sediments are represented with 173,101.17 t yr^{-1} (20.76%) and suspended sediments with 660,581.66 t yr^{-1} (79.24%). It is noticed that the sediment delivery ratio in the basin is relatively small, which means that large amounts of produced, eroded sediments are retained in different parts of the basin. The main reasons for this are certainly the elongation of the basin and the low mean slope of the river channel Sm = 0.94%. It can be assumed that every year an average of 833,682.83 t yr^{-1} of total sediment will reach the Danube River.

The analysis and comparison of the results of the suspended sediment transport for the entire Timok River Basin and the measured suspended sediment discharge for the Beli Timok River Basin is done in order to check the reliability of the assessment method. The measured suspended sediment transport for the Beli Timok River Basin is 326,000.00 t yr^{-1}. If this sum is multiplied by two, since the area of the Beli Timok River Basin is about 2 times smaller than the area of the Veliki Timok River Basin, we will get the total of 652,000.00 t yr^{-1}, which is a very close amount of suspended sediment transport calculated by EPM method for the whole Timok River Basin (660,581.66 t yr^{-1}). It can be concluded that the estimated amount of total and suspended sediment transport for the Timok River Basin is realistic, but it can't be concluded that the calculation method and measured data are always exact.

4.3 Torrential Floods Susceptibility

Torrential floods, as the most common cause of floodings in the studied basin, occur as a consequence of intense precipitation or sudden melting of the snow cover. The main characteristic of the torrential floods is rapid flow with high concentrations of sediments, short duration and great damage. As a result of high erosion within the basin, sediments are displaced from the area of the watershed to the hydrographic network and further transported in river channels to the Danube River.

Based on the *Flash Flood Potential Index* (FFPI), we assessed and mapped the susceptibility of the Timok River Basin to the occurrence of torrential floods (Fig. 6).

According the classification of the obtained FFPI values, we found that the class of very high susceptibility is represented on 101.36 km^2 (i.e., on 2.24% of the Timok River Basin), and the class of high susceptibility on 1,950.05 km^2 (43.05% of its total area). This shows that 45.29% of the Timok River Basin is very susceptible to torrential floods and this data should be taken seriously. The class of medium susceptibility occupies 49.47%, and the class of low susceptibility 5.24% of the total basin area. Thus, only 5% of the basin is not significantly threatened by torrential floods. Bearing in mind the occurrence of increasingly common rainfalls in recent decades in the form of heavy rains, natural characteristics of the Timok River Basin

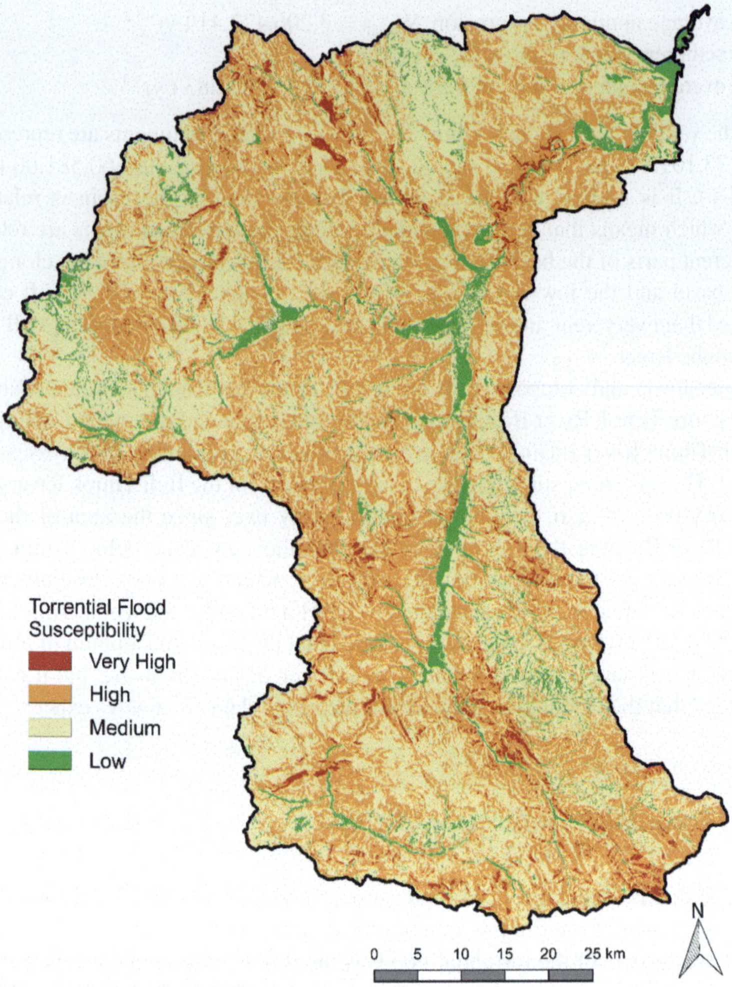

Fig. 6 Torrential flood susceptibility map of the Timok River Basin

and analysis of the FFPI index, it can be concluded that there is a real threat in the Timok River Basin by the torrential floods.

5 Discussion

The analysis of physical conditions in the Timok River Basin has unequivocally shown that this area, due to its geomorphological and hydrological characteristics, but also to the land use, is prone to the occurrence of torrential floods. The Timok

River Basin is a good example of acquiring almost all conditions for the occurrence of frequent and large floods on the biggest watercourses—Timok, Beli Timok, Crni Timok, Svrljiški Timok and Trgoviški Timok. Except the floods on the Timok River, perhaps an even bigger problem presents the floods on numerous tributaries, because they all have characteristics of typical torrents. On these smaller watercourses defense against torrential floods is significantly different and more difficult than the flood defense in the case of larger watercourses.

Compared to the Kolubara River Basin in Serbia, which is characterized by similar morphometric and hydrological characteristics, and known for catastrophic floods in 2010 and 2014 [41, 42], the results show that the Timok River Basin is more significantly threatened by erosion and torrential floods. According to the erosion coefficient and intensity erosion category, the Kolubara River Basin belongs to the category of weak erosion ($Z_{av} = 0.35$), i.e. the fourth category of intensity/destruction, while the Timok River Basin belongs to the medium erosion ($Z_{av} = 0.42$), i.e., the third category. Also, 25% of the Kolubara River Basin is very susceptible to torrential floods [43], while 45.29% of the Timok River Basin is very susceptible and this data should be taken seriously. It is necessary to emphasize that within the territory of the Timok River Basin there is a network of 924.7 km of roads of first and second category which are threatened by torrential floods. This is confirmed by the fact that there are 539 registered intersections between roads and torrents which are characterized as a potential torrential flood risk locations, of which 431 (80%) are determined as very endangered intersection locations [44]. As concerns the Kolubara River Basin there are 523 intersections between roads and torrents, with potential risk locations of which 58% are identified as very endangered [43].

Compared to the results of similar research, the Timok River Basin is determined as one of the most threatened by torrential floods. In Ukrina River Basin (Bosnia and Herzegovina), 40.86% of its area belong to the high and very high susceptibility category [36], in Bâsca River Basin (Romania) 29% [45], in Jošanica River Basin (Serbia) 21% [30], in the Sărățel River Basin (Romania) about 20% [32], etc.

The soil erosion leads to permanent landscape degradation. The eroded sediments cause turbidity of water (mechanical pollution) which generate ecological problems in aquatic ecosystems with harmful consequences for humans as well. Together with the sediments from the slopes, various substances (nutrients, mineral and organic fertilizers, pesticides, etc.) get into rivers (and later into lakes and reservoirs), causing chemical water pollution and the alteration of the water quality index [46–48]. Such water is not safe for water supply and often cannot be used for industrial purposes, nor for irrigation.

All these problems occur in the Danube River, as a collector of the Timok River. The huge amounts of sediment that reach the Danube River cause the silting of the riverbed which can influence the navigation. The amount of suspended load can cause environmental problems and endanger the survival of river flora and fauna in the downstream sector. On the other hand, torrential floods endanger settlements, transport infrastructure, industrial facilities, agricultural land, and the progress of society as a whole.

In order to diminish the torrential flood occurrence and mitigate their negative consequences, the only successful manner is the prevention. This consists of constant control of soil erosion processes in a river basin applying an integrated system for erosion and torrential floods monitoring. The integrated system consists of optimum land use, appropriate erosion control works (biological, bio-technical, and technical), and erosion control measures. Having in mind on-going climate changes, the increase of extreme precipitation, intensifying of erosion and torrential processes, and as consequences, increasing of torrential floods frequency can be expected [43].

In order to diminish the torrential erosion, also it is necessary to apply appropriate administrative measures such as: prohibition of deforestation, as well as of destruction of meadows and pastures, cultivation on areas with a slope greater than 7%, etc.

6 Conclusion

The results showed that 49.21% of the Timok River Basin is endangered by strong categories of erosion. This is in line with the values of the FFPI index which indicated that 45.29% of the Timok River Basin is very susceptible to torrential floods and this data should be taken seriously. The class of medium susceptibility to torrential floods occupies 49.47%, and the class of low susceptibility, 5.24% of the total basin area. It can be concluded that there is an actual endangerment from torrential floods in the Timok River Basin, and only 5% of its area is not significantly threatened by torrential floods. In Serbia, the measured sediment transport data in torrential watercourses is insufficient (except in some scientific-research studies). As a result of the conducted analyzes and calculations, we can estimate that an average sediment amount of 833,682.83 t yr^{-1} reach the Danube River. The findings and conclusions of this research would provide base for future water resources management and design of flood mitigation measures in context of changing environment in the Timok River Basin.

Finally, an integrated Timok River Basin management plan, with the proper selection of land use and organization of erosion control works, would bring a balanced runoff regime of the river. Furthermore, the flood peaks and flood occurrence would be diminished, as well their destructive effects. Such activities would lead to the enhancement of the basin's hydrological behavior and, consequently to the mitigation of the risks induced by the torrential floods and soil erosion within the Timok River Basin.

7 Recommendation (Future Tasks)

Although the torrential floods are natural induced disaster, the human activities also significantly contribute to their occurrence and intensity. The following measures of soil erosion and torrential flood control and risk mitigation seem appropriate:

1. The implementation of a *Decision Support System* (DSS) for the optimal coordination of torrential flood prevention or mitigation activities.
2. Designing web-platforms for professional data users with an interactive access to hydro-meteorological and hazard information.
3. The creation of a new soil erosion map of Serbia based on the Erosion Potential method.
4. Continuous erosion and torrent control measures in watersheds.
5. The preparation of Plans of Identifying Erosion Regions for each municipality in Serbia.
6. The preparation of Plans of Torrential Flood Control.
7. The preparation of an inventory of torrents for each watershed in Serbia.
8. The documentation of performed erosion and torrential flood control works.
9. Real-time monitoring of rainfall and river discharge (automatic station) for the improvement of forecast and early warning systems.

Acknowledgements This study was supported by The Ministry of Education, Science and Technological Development of the Republic of Serbia within Project 43007 ('Studying climate change and its influence on the environment: impacts, adaptation and mitigation—subproject: Frequency of torrential floods and degradation of soil and water as a consequence of global changes').

Author Contributions All authors contributed equally to this work.

Conflict of Interest Authors declare that no conflict of interest.
Copyrights issues and permission to republish the copyrighted published materials. All the figures with no citations are the authors work.

References

1. Dragićević S, Filipović D, Kostadinov S, Ristić R, Novković I, Živković N, Andjelković G, Abolmasov B, Šećerov V, Djurdjić S (2011) Natural Hazard Assessment for land-use planning in Serbia. International Journal of Environmental Research 5(2):371–380
2. Dragićević S, Novković I, Carević I, Živković N, Tošić R (2011) Geohazard assessment in the Eastern Serbia. Forum geografic 10(1):10–19
3. Petrović A, Kostadinov S, Dragićević S (2014) The inventory and characterisation of torrential flood phenomenon in Serbia. Pol J Environ Stud 23(3):823–830
4. Kostadinov S, Mitrović S (1994) Effect of forest cover on the stream flow from small watersheds. J Soil Water Conserv 49(4):382–386

5. Kostadinov S (2010) Forests in Serbia as the factor of Soil and water protection against degradation in the conditions of global climate change. In: Alexandrov V, Gajdusek MF, Knight CG, Yotova A (eds) Global environmental change: challenges to science and society in Southeastern Europe. Springer Science+Business Media, pp 77–190
6. Kostadinov S, Košanin O, Petrović A, Dragićević S (2017) Extreme climate events and erosion control in headwater catchments of Serbia. In: Krecek J, et al. (ed), Ecosystem services of headwater catchments. Capital Publishing Company, New Delhi, India, Copublished by Springer International Publishing, Cham, Switzerland, pp 217–224. https://doi.org/10.1007/978-3-319-57946-7_19
7. Petković S, Dragović N, Marković S (1999) Erosion and sedimentation problems in Serbia. Hydrol Sci J 44:63–78. https://doi.org/10.1080/02626669909492203
8. Mladenović MB, Kolarov V, Damjanović V, Radosavljević P (2017) Comparison of the "natural" and the present sediment regime of the Danube River. Water Res Manag 7:1–11
9. Mladenović MB, Kolarov V, Vanja D (2013) Sediment regime of the Danube River in Serbia. Int J Sedim Res 28(4):470–485
10. Zaharia L, Grecu F, Ioana-Toroimac G, Neculau G (2011) Sediment transport and river channel dynamics in Romania. Variability and control factors. In: Manning O (Ed) Sediment transport in aquatic environments. InTech, pp 293–316. https://doi.org/10.5772/21416
11. Zaharia L, Ioana-Toroimac G (2013) Romanian Danube River management: impact and perspectives. In: Arnaud-Fassetta, G, Masson, E, Reynard, E (ed) European continental hydrosistems under changing water policy. Munchen, pp 159–170
12. Gavrilović Lj, Dukić D (2002) Reke Srbije. Zavod za udžbenike i nastavna sredstva. Beograd. p 218
13. Ristić R, Radić B, Nikić Z, Trivan G, Vasiljević N, Dragićević S, Živković N, Radosavljević Z (2011) Erosion control and protection from torrential floods in Serbia – spatial aspects. Spatium 25:1–6. https://doi.org/10.2298/SPAT1125001R
14. Ristić R, Kostadinov S, Abolmasov B, Dragićević S, Trivan G, Radić B, Trifunović M, Radosavljević Z (2012) Torrential floods and town and country planning in Serbia. Nat Hazard 1(12):23–35
15. Petrović A, Dragićević S, Radić B, Milanović A (2015) Historical torrential flood events in the Kolubara river basin. Nat Hazards 79(1):537–547
16. Gavrilović S (1972) Engineering of torrents flows and erosion. J "Izgradnja", Spec. Edition, p 292 (in Serbian)
17. Tošić R, Dragićević S, Lovrić N (2012) Assessment of soil erosion and sediment yield changes using erosion potential model – case study: Republic of Srpska (BiH). Carpathian J Earth Environ Sci 7(4):147–154
18. Kostadinov S, Zlatić M, Dragićević S, Novković I, Košanin O, Borisavljević A, Lakićević M, Mlađan D (2014) Anthropogenic influence on erosion intensity changes in the Rasina river watershed - Central Serbia. Fresenius Environ Bull 23(1a):254–263
19. Kostadinov S, Braunović S, Dragićević S, Zlatić M, Dragović N, Rakonjac N (2018) Effects of erosion control works: case study - Grdelica Gorge, the South Morava River (Serbia). Water 10(8):1094. https://doi.org/10.3390/w10081094
20. Republic Hydrometeorological Service of Serbia (RHSS). Meteorological and hydrological yearbooks (1968–2017); RHSS, Belgrade, Serbia
21. Blagojević B, Langović M, Novković I, Dragićević S, Živković N (2020) Water resources of Serbia and its utilization. In: Negm A, Romanescu G, Zelenakova M (eds) Water resources management in Balkan Countries. Springer Water. Springer, Cham. pp 213–247. https://doi.org/10.1007/978-3-030-22468-4_9
22. Kovačević-Majkić J, Urošev M (2014) Trends of mean annual and seasonal discharges of rivers in Serbia. J Geogr Inst "Jovan Cvijić", SASA 64(2):143–160.
23. Langović M, Manojlović S, Čvorović Z (2017) Trends of mean annual river discharges in the Zapadna Morava River Basin. Bull Serbian Geogr Soc 97(2):19–45. https://doi.org/10.2298/GSGD1702019L

24. Dragičević N, Karleuša B, Ožanić N (2016) A review of the Gavrilović method (erosion potential method) application. Građevinar 68:715–725
25. Tošić R, Lovrić N, Dragićević S (2019) Assessment of the impact of depopulation on soil erosion: Case study-Republika Srpska (Bosnia & Herzegovina). Carpathian J Earth Environ Sci 14:505–518
26. Dragićević S, Martinović M, Sibinović M, Novković I, Tošić R, Babović S (2014) Recent changes of the erosion intensity caused by socio-demographic and land use changes (case study: Knjaževac municipality, Serbia). In: Zlatic M, Kostadinov S (eds) Advances in GeoEcology, n°43, Challenges: Sustainable Land Management – Climate Change. CATENA VERLAG GMBH, Germany, pp 271–286
27. Zorn M, Komac B (2009) Response of soil erosion to land use change with particular reference to the last 200 years (Julian Alps, Western Slovenia). Revista de Geomorfologie 11:39–47
28. Smith G (2003) Flash flood potential: determining the hydrologic response of FFMP
29. Basins to Heavy Rain by Analyzing Their Physiographic Characteristics. http://www.cbrfc.noaa.gov/papers/ffp_wpap.pdf. Accessed on 21 Sept 2020
30. Novković I, Dragićević S, Živković N, Tošić R, Čvorović Z (2018) Vulnerability assessment of the Jošanička river basin to torrential floods and forest fires. In: Zlatic M, Kostadinov S (eds) Advances in GeoEcology, n°45, Soil and water resources protection in the changing environment. CATENA, Soil Sciences. Schweizerbart Science Publishers, Germany, pp 65–78
31. Prăvălie R, Costache R (2014) The analysis of the susceptibility of the flash floods genesis in the area of the hydrographical basin Bâsca Chiojdului River. Forum Geografic 8(1):39–49. https://doi.org/10.5775/fg.2067-4635.2014.089.i
32. Costache R, Pravalie R, Mitof I, Popescu C (2015) Flood vulnerability assessment in the low sector of Saratel Catchment. Case study: Joseni Village. Carpathian J. Earth Environ Sci 10(1):161–169
33. Zaharia L, Costache R, Prăvalie R, Minea G (2015) Assessment and mapping of flood potential in the Slănic catchment in Romania. J Earth Syst Sci 124(6):1311–1324 http://link.springer.com/article/10.1007/s12040-015-0608-3
34. Zaharia L, Costache R, Prăvălie R, Ioana-Toroimac G, (2017) Mapping flood and flooding potential indices: a methodological approach to identifying areas susceptible to flood and flooding risk. Case study: the Prahova catchment (Romania), Front Earth Sci: 1–19. https://link.springer.com/article/10.1007/s11707-017-0636-1
35. Tošić R, Lovrić N, Dragićević S, Manojlović S (2018) Assessment of torrential flood susceptibility using GIS Matrix method: case study-Vrbas river basin (B&H). Carpathian J Earth Environ Sci 13(2):369–382
36. Lovrić N, Tošić R, Dragićević S, Novković I (2019) Assessment of torrential flood susceptibility: case study - Ukrina river basin (B&H). Bull Serbian Geogr Soc 99(2):1–16. https://doi.org/10.2298/GSGD1902001L
37. Dragomir A, Tudorache AV, Costache R (2020) Assesement of flash-flood susceptibility in small river basins. Present environment and Sustainable Development 14(1):119–131
38. Langović M (2020) Investigations of the lateral channel migration: a case study of the South Morava River. Bull Serbian Geogr Soc 100(1):1–21. https://doi.org/10.2298/GSGD2001001L
39. Kostadinov S (2008) Torrential flows and erosion. Faculty of Forestry, University of Belgrade, Belgrade (in Serbian)
40. The Jaroslav Černi Institute for the Development of Water Resources (2003) Study of erosion production and sediment transport on the territory of Serbia (In Serbian)
41. Dragićević S, Ristić R, Živković N, Kostadinov S, Tošić R, Novković I, Borisavljević A, Radić B (2013) Floods in Serbia in 2010 – case study: The Kolubara and Pcinja River Basins. In: Loczy D (ed), Geomorphological impacts of extreme weather: case studies from Central and Eastern Europe. Springer Geography, pp 155–170. https://doi.org/10.1007/978-94-007-6301-2_10
42. Dragićević S, Živković N, Novković I, Petrović A, Tošić R, Milevski I (2016) Hydrological and suspended sediment regime in the Kolubara River during the extreme year of 2014. Revista de Geomorfologie 18:26–38. https://doi.org/10.21094/rg.2016.054

43. Kostadinov S, Dragicevic S, Stefanovic T, Novkovic I, Petrovic A (2017) Torrential flood prevention in the Kolubara river basin. J Mt Sci 14(11):2230–2245
44. Kostadinov S, Dragicevic S, Stefanovic T, Novkovic I, Momirovic N, Langovic M, Radovic M (2020) A study of the vulnerability of I and II category state roads from the occurrence of floods and torrential flows in the Timok river basin. Contracting authority: PE Roads of Serbia. Project holder: Institute of Forestry & Faculty of Geography, Belgrade.
45. Minea G (2013) Assessment of the flash flood potential of Bâsca River Catchment (Romania) based on physiographic factors. Cent Eur J Geosci 5(3):344–353. https://doi.org/10.2478/s13533-012-0137-4
46. Dragićević S, Nenadović S, Jovanović B, Milanović M, Novković I, Pavić D, Lješević M (2010) Degradation of Topciderska river water quality (Belgrade). Carpathian J Earth Environ Sci 5(2):177–184
47. Brankov J, Milijašević D, Milanović A (2012) The assessment of the surface water quality using the water pollution index: a case study of the Timok River (the Danube River Basin) Serbia. Archives of Environ Prot 38(1):49–61
48. Milijašević Joksimović D, Gavrilović B, Lović Obradović S (2018) Application of the water quality index in the Timok River basin (Serbia). J Geogr Inst Cvijic 68(3):333–344

Hydrological Extremes Anomalies and Trends in Lower Danube Basin: Case Study—Romanian Drainage Area Between Siret and Prut Rivers

Ionuţ Minea, Marina Iosub, Andrei Enea, Daniel Boicu, and Oana-Elena Chelariu

Abstract Hydrological extremes, as a manifestation of the natural flow regime, have negative impacts on the environment through the effects they produce. Floods cause significant fatalities and economic losses and leave their long-term psychological imprint on local communities. Hydrological droughts also represent complex hydrological phenomena, that have direct negative effects on agricultural productivity. In this study, analysis of the hydrological extremes' including anomalies and trends was carried out across 25,000 km^2 in the eastern part of Romania, between the last two major tributary of the Danube River, namely Siret and Prut. This study investigates the anomalies and trends associated to extreme hydrological events, for both high and low flow events, using data from hydrometric stations in the Siret-Prut area, by applying quantile perturbation method (QPM) and innovative trend analysis (ITA) method, respectively. Also, a magnitude corresponding to each type of hydrological extremes, associated with ITA, was calculated. Data from 11 hydrometric stations was processed, spanning over 64 years of recordings (between 1955 and 2018). Results reveal more trends than expected, related to random occurrence for most of the measures of extreme flow characteristics. Annual and spring maximum flows show a decreasing trend in flow magnitude (for 90% of the hydrometric stations analyzed) with magnitudes ranging between −0.089 m^3/s and −3,070 $m^{3/}$s on annual level. Low flow magnitudes exhibit both increasing trends (in winter, for 55% of hydrometric stations), and decreasing trends during the spring season (for 64% of hydrometric stations). The results reveal that important investments in the associated infrastructure are required to reduce the impact of hydrological extremes, for both high and low flows.

I. Minea (✉) · M. Iosub · A. Enea · D. Boicu · O.-E. Chelariu
Faculty of Geography and Geology, Department of Geography, "Alexandru Ioan Cuza" University of Iaşi, 700506 Iaşi, Romania
e-mail: ionutminea1979@yahoo.com

M. Iosub
Integrated Center of Environmental Science Studies in the North Eastern Region—CERNESIM, "Alexandru Ioan Cuza" University of Iaşi, Iaşi, Romania

O.-E. Chelariu
Institute of Interdisciplinary Research—ICI-UAIC, "Alexandru Ioan Cuza" University of Iaşi, Iaşi, Romania

A. Negm et al. (eds.), *The Lower Danube River*, Earth and Environmental Sciences Library, https://doi.org/10.1007/978-3-031-03865-5_13

Keywords Hydrological extremes · Floods · Hydrological droughts · Anomalies · Trends · Magnitudes · Lower Danube Basin

1 Introduction

Climatic change can generate significant alterations of hydrological dynamics, both at a local and regional level, emphasizing a general tendency of amplification of extreme hydrological events [1]. Due to this, it is of high importance to identify the anomalies and trends of extreme hydrological events, such as droughts and floods. In addition, this type of analysis aids in increasing the protection level of inhabited areas. This can be achieved at different scales, from the lowest level, associated with a hydrographical basin, up to the scale of a geographical region, and also different environments, from a mountainous region to regional climate areas.

Extreme flow values of rivers, and the corresponding floods and droughts, induce severe hydrological phenomena. These are caused mainly by regional meteorological anomalies, and modified by natural processes, occurring at a drainage basin level, and also by human intervention. Considering this, the effects of global climate change, along with local fluctuations, can generate an unequally distributed response in the manifestation of hydrological extreme events. Therefore, identifying anomalies and tendencies for maximum and minimum flows is highly important [2].

Hydrological research on identifying anomalies and tendencies for hydro-climatic parameters have been mainly driven by the analysis of extreme values of atmospheric precipitation [3], as well as maximum flow rate of rivers [4]. Several results are associated with the impact of climate change upon atmospheric precipitation in climatic conditions specific to the warm [5] and temperate [6] regions. Others have addressed identifying the response of river runoff, to climatic change [7, 8], and also identifying the causes that induce the temporal oscillations of river flows [9].

Determining the tendencies in the aforementioned studies has been based on data sets of medium lengths of seasonal and annual extremes of climatic and hydrologic parameters, by using statistical methods, such as Mann–Kendall test, Sen's slope and linear regression. Identification of anomalies for extreme climatic and hydrologic parameters has been performed by applying the quantile perturbation method [10, 11]. More novel statistical approaches involve identifying the graphical tendencies of climatic and hydrologic parameters, in relation to previous events (average and extreme), by using the Innovative Trend Analysis (ITA) method [12]. This method can be easily applied for extreme meteorological and hydrological events [13], due to the fact that it does not rely on the restrictive statistical assumption [14].

In relation to the current study area, located in the Lower Danube River basin, in Romania, between Siret and Prut rivers, several studies estimating the tendencies of the different hydro-climatic and hydro-geological parameters have been accomplished, based on the Mann–Kendall test and Sen's slope linear regression [15, 16]. The entire region has been analyzed from a climate change perspective, and their effects on the flows of rivers [17], of atmospheric precipitations and extreme temperatures [18, 19], as well as the spatial and temporal evolution of the climatic water balance [20], and the tendencies of the groundwater level [21]. Recent scientific approaches addressed the analysis of water resource vulnerability in the region to the impact of climatic change and anthropogenic activities [22], and also identified the anomalies and tendencies of maximum flows for a number of rivers in the region [23].

Therefore, the analysis of extreme hydrological anomalies represents a continuation of research undergone in this field, within a region of the Lower Danube basin, in the context of ever-increasing effects of the extreme manifestations of regional climate change. Investigating extreme hydrological anomalies can aid in a better understanding of key factors which influence regional climatic processes, with direct effects upon the local, hydrological ones. Furthermore, it can act as a starting ground in creating an efficient water resource management system, in different scenarios of climatic trends.

2 Study Area

The Lower Danube basin encompasses a series of areas that overlap the north-western region of the Black Sea. Each area is clearly separated from a geographic point of view, by the tributary network of the Danube River. The study area chosen for identifying the extreme hydrological anomalies extends between Siret and Prut rivers, overlapping the Moldavian Plateau (Fig. 1). These are the last two major tributaries flowing on the left-hand side of the Danube, right before the entry in the deltaic region. They cover a surface of over 20,000 km^2, which amounts for 2.5% of the Danube River basin. The hydrographic network includes 392 rivers, with a cumulated length of 7696 km, and an average drainage density of 0.38 km/km^2 [24]. The temperate continental climate that characterizes this area, has multi-annual average temperatures varying between 7 °C and 8 °C in the northern region and between 9 °C and 10 °C in the southern part [25]. Consequently, this reveals a nonuniform spatial and temporal distribution of the total amount of precipitation, which has a direct impact on the river flows. Despite the fact that the annual amount of precipitations exceeds 650 mm in the northern area and decreases below 480 mm in the southern part, the seasonal variations indicate the fact that the precipitations occurring during the summer season, are 2–3 times more abundant, compared to the ones registered during the winter season [26].

The flow of inland rivers with lengths of less than 50 km registers significantly low values, with flow rates lower than 1 m^3/s [27]. The hydrological regime manifests through maximum flow rates during the spring and minimum values during the end of the summer season/beginning of autumn [28]. The maximum flow in the spring season is generated by abundant precipitation and snow melt. However, during

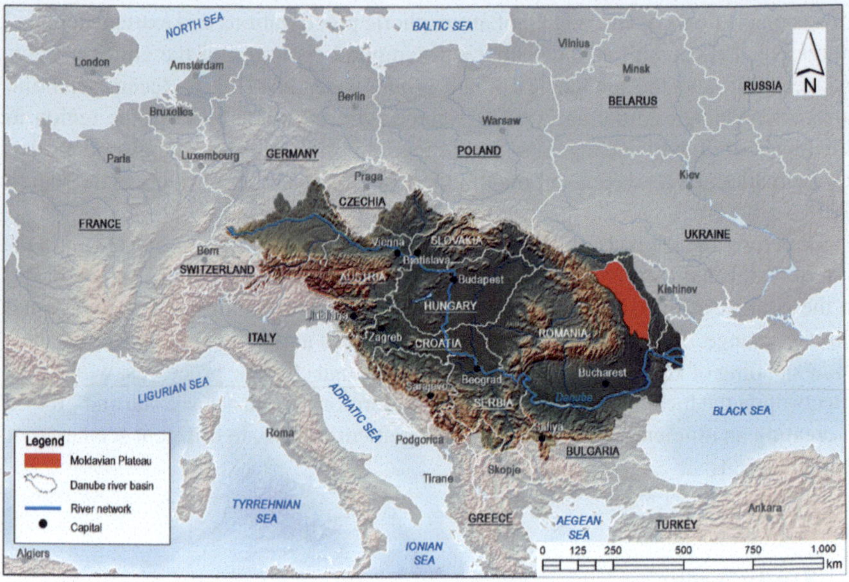

Fig. 1 Location of the study area in Romania and in the Danube River basin

an entire year, the hydrological regime suffers from a series of anomalies. These are generated by the effects of regional climate conditions (high flow values in the summer, or prolonged hydrological droughts in the autumn) [29, 30].

3 Data Base and Methodology

In order to analyze anomalies and extreme hydrological tendencies, 5 sets of data corresponding to maximum and minimum seasonal and annual flow rates were used. The data has been obtained from 11 hydrometric stations. These were chosen, according to the geographical conditions from the upstream drainage basins, where large scale hydrotechnical constructions (such as large man-made reservoirs, embankments, river rerouting, irrigation systems etc.) have not been conducted (Fig. 2). All hydrometric stations are under the management of Prut-Bârlad Water Basin Administration. The time span, for which data was available ranges between 1955 and 2018. Furthermore, all datasets are complete and do not have any data gaps across the entire period.

There were situations where data related to the natural flow of rivers could not be used. Considering the anthropic influences, the maximum and minimum seasonal and annual natural flow rates have been reconstructed. They were based on the minimum and maximum monthly and annual flow rates, registered at the hydrometric stations, by using the following formula:

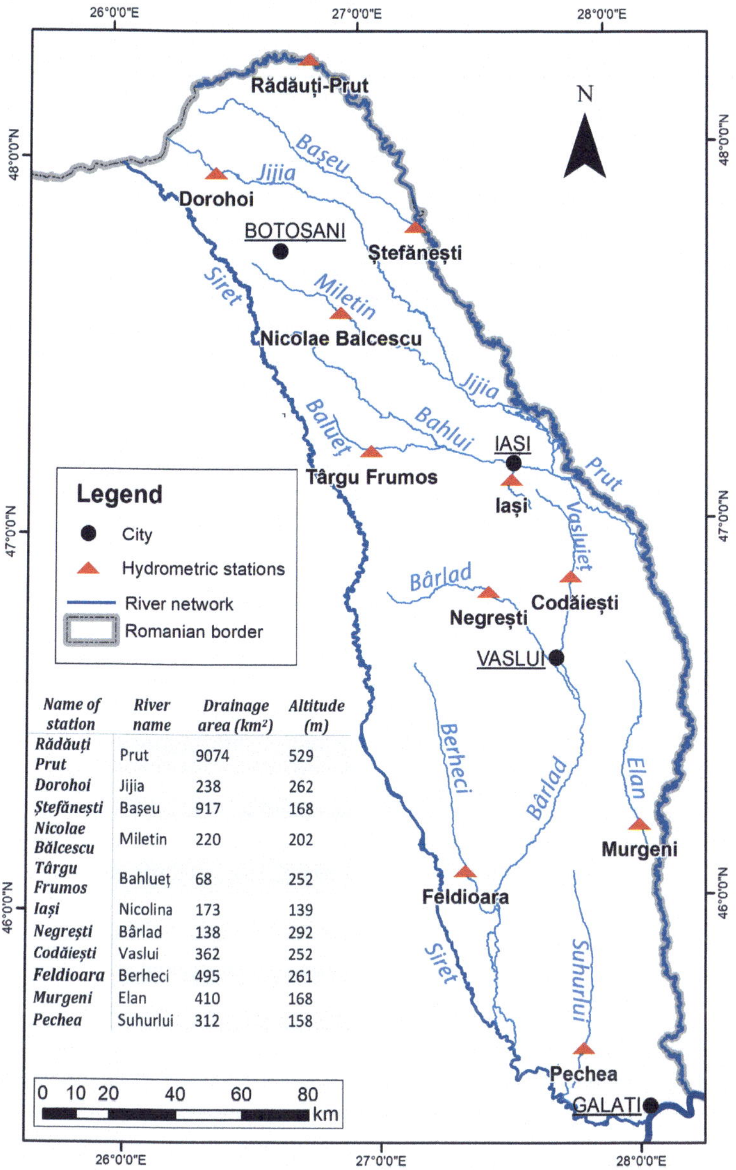

Name of station	River name	Drainage area (km²)	Altitude (m)
Rădăuți Prut	Prut	9074	529
Dorohoi	Jijia	238	262
Ștefănești	Bașeu	917	168
Nicolae Bălcescu	Miletin	220	202
Târgu Frumos	Bahlueț	68	252
Iași	Nicolina	173	139
Negrești	Bârlad	138	292
Codăiești	Vaslui	362	252
Feldioara	Berheci	495	261
Murgeni	Elan	410	168
Pechea	Suhurlui	312	158

Fig. 2 Location of the hydrometric stations in the study area

$$Qnat = Qm + \sum Qc - \sum Qr$$

where:

Qnat—natural flow rate (m^3/s);
Qm—measured flow rate (m^3/s);
$\sum$ Qc—the sum of flow rates, used for several purposes, upstream of the hydrometric station;
$\sum$ Qr—the sum of reconstructed flow rates, used for several purposes, upstream of the hydrometric station (including the addition of flow rates, originated from rerouted courses from neighboring hydrographic basins, or water sourced from infiltration). This formula is applied in cases where, upstream of the hydrometric station there are only significant water consuming use-cases.

In the case of association with several lakes, the aforementioned formula also includes the sum of accumulation/flow rates in/from the accumulation ponds located upstream of the hydrometric station, or for flow rates originated from precipitation, snow melt, or lost through evaporation:

$$Qnat = Qm + \sum Qc - \sum Qr + \sum Wa/Wd + \sum Q_p - \sum Q_E + \sum Q_{GH}$$

where:

Qnat—natural flow rate at a given hydrometric station (m^3/s);
Qm—measured flow rate, measured at the hydrometric station (m^3/s);
$\sum$ Wa/Wd—sum of accumulation/discharge flow rates in/from the accumulation ponds located upstream of the hydrometric station;
$\sum Q_p$—sum of flow rates originating from atmospheric precipitation, registered on the surface of accumulation ponds, located upstream of the hydrometric station;
$\sum Q_E$—sum of flow rates derived from evaporation occurring in the reservoirs located upstream of the hydrometric station;
$\sum Q_{GH}$—sum of flow rates originating from ice formed on the surface of the reservoirs located upstream of the hydrometric station.

In order to identify the anomalies, the quantile perturbation method (QPM) was applied. This method emphasizes the changes occurring at a quantile level, for different time periods, in correspondence to the entire period of analyzed data. The method was applied for maximum and minimum monthly values, registered at the aforementioned hydrometric stations. Therefore, the highest three values of minimum and maximum flow rates for each season and also on an annual basis, have been taken into account. These values were considered as the reference threshold for identifying anomalies. An important step in the statistical analysis stage was the selection of a number of shorter time periods, from the entire time span. This is relevant for emphasizing anomalies. Considering that the entire data set spans for over 64 years, the length of the smaller time periods analyzed was chosen to span over 10 years, for comparison purposes. This was considered justified and appropriate, due to previous studies, performed on the same type of hydrological parameters [31]. Furthermore, the anomalies are not generated by the occurrence of a random value, or a suite of random values. This was taken into account due to the fact that they could induce significant modifications on the entire data base [32]. Furthermore, the fact that the statistical analysis for the anomalies is based on shorter time intervals required applying the following principles: (i) the time interval between consecutive anomalies that have been identified (both minimum and maximum) is longer than the length of the shorter adjacent time period analyzed; (ii) the presence of larger or more reduced anomalies, compared to a given threshold must be validated; (iii) if, in the range of the same analyzed time period, for more neighboring hydrometric stations, the same minimum and maximum variation of anomalies is detected, they can be considered correct, from a statistical point of view.

In order to identify the tendencies, the Innovative trend analysis (ITA) graphical method was used. This method can be applied on different extreme values of hydro-climatic parameters, and is suitable for data sets that do not depend on restrictive statistical assumptions, during the analysis, such as the common parametric and nonparametric trend tests [33]. Applying this method involves overcoming 2 stages:

1. During the first stage, the entire data set is split into two equal subsets, sorted in an ascending manner;
2. During the second stage, the two subsets are graphically represented into a bidimensional system, of cartesian type, in the form of scatter points. These points are compared to a median line. After the graphical representation, there are points which are placed above the median line, they emphasize the increasing trends, while the points located below the median line reveal the decreasing trends. If the points concentrate along the median line, they do not indicate tendencies. Corresponding to this method, Şen [34] suggests a formula used to calculate the slope of the tendency:

$$S = \frac{2(\overline{y_2} - \overline{y_1})}{n}$$

where: s is slope of trend, $\overline{y}_1$, $\overline{y}_2$ are the averages of the first and second series, and n is the total number of the data. Further applications of the ITA method have allowed Wu and Qian [3] to develop an index, for estimating the intensity of the tendency:

$$\mathrm{PI} = \frac{1}{n}\sum_{i=1}^{n} \mathrm{PL}_i = \frac{1}{n}\sum_{i=1}^{n} \frac{10(y_i - x_i)}{\overline{x}}$$

where: PI expresses the intensity of the tendency, x_i is the i_{th} value of first ordered sub-series and $y_{i\,1}$ is the i_{th} value of the second ordered sub-series and $\overline{x}$ is the average of x_i.

In this study to analyze the extreme hydrological tendencies, registered between 1955 and 2018, two data subsets were extracted, of 32 years in length, each: 1955–1986 and 1987–2018.

4 Extreme Hydrological Anomalies

The dynamic of the extreme hydrological anomalies, based on the annual data series, manages to reveal several decline or growth periods, which can be correlated to the periods characterized by the abundance or lack of atmospheric precipitation. Therefore, regarding the annual maximum flow rates, a constant growth period can be observed, related to all hydrometric stations, reaching maximum values in the mid-1970s (Fig. 3).

This period of positive anomalies regarding annual maximum flow rates, is associated with an intervalof pluviometric maximum. The values of the anomaly factor range between 1.2 (at Rădăuți Prut hydrometric station), and 2.7 (at Nicolina hydrometric station). This characterizes the extent of the entire country, between 1965 and 1975 [25]. Therefore, at all 11 stations that were analyzed during this period, maximum values of the anomaly factor were recorded, for maximum annual flow rates. On one side, these values are lower at Rădăuți Prut hydrometric station, due

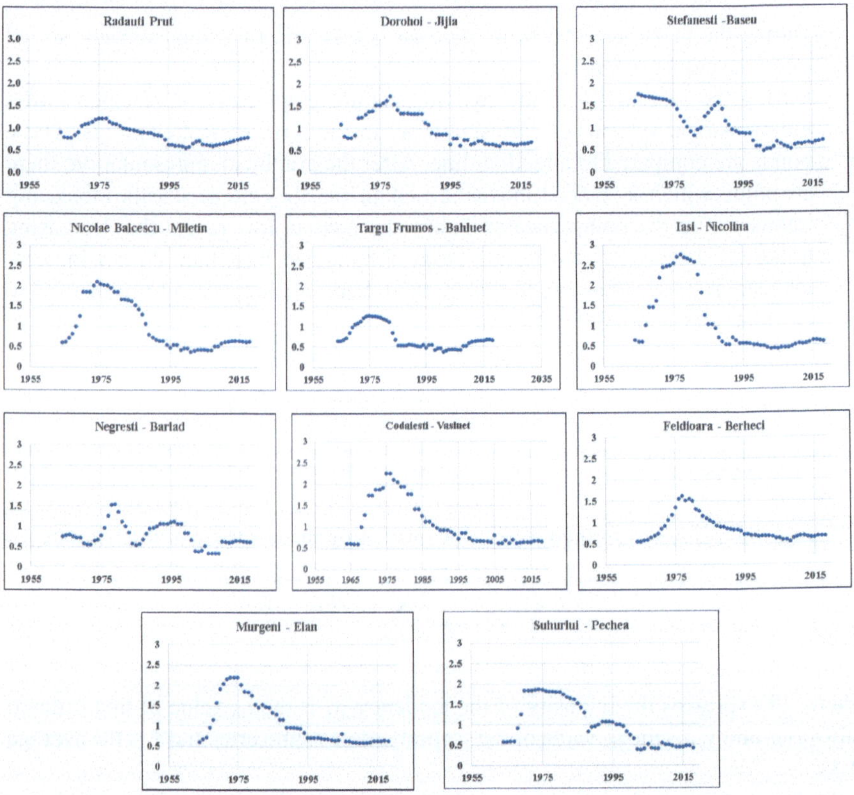

Fig. 3 Anomaly factor variation for annual maximum flow rates for the analyzed hydrometric station

to the large surface area of the upstream drainage basin. On the other side, there are higher values at the hydrometric stations with significantly smaller drainage basins (Iași, Codăiești, Murgeni, Pechea). After this time period, for 8 out of the total 11 analyzed stations, a constant decrease in the values of the anomaly factor could be observed for maximum flow rates, which is more pronounced at the hydrographic stations located in the central area (Nicolae Bălcescu, Iași, Codăiești), and less at the following stations: Rădăuți Prut, Dorohoi, Târgu Frumos and Feldioara. Regarding the central-northern part of the study area, a significant tendency of precipitation decreasing can be observed in the 1970–1995 period [29]. Furthermore, the constant tendency in the decrease of values, after the middle of the 1970s, can also be correlated with a series of changes that occurred in the catchment that have generated significant reductions of the maximum flow rates (reservoirs, water supply system, irrigation etc.).

For three out of the 11 hydrometric stations analyzed (Ștefănești, Negrești, Pechea), a period of significant increases of the positive anomaly factor could still be observed between 1985 and 1997, which were generated by climatic oscillations, which are specific to the temperate continental climate area, located in the lower Danube basin. During the 2005–2015 period, the anomaly factor tendencies for maximum annual flow rates reveal a slight increase (for 8, out of the 11 stations), based on a modest increase in the precipitation quantity, especially in the warm season [15].

The largest value of the positive anomaly factor in the case of maximum annual flow rates, has been registered at Iași hydrometric station (2.75), in 1977, while the lowest value is 0.35, at Negrești station, between 2006–2009. This value has emerged, as a result of the occurrence of a period with severe hydrological drought [35].

According to the seasonal data series dynamics of the anomalies, they reveal the same oscillations as the annual ones, for 8 of the total analyzed hydrometric stations. The seasonal values of the anomaly factor are significantly higher than the annual ones, reaching 4.94, at Nicolae Bălcescu station, during the summer of 1977, and 4.84 at Codăești station, in the summer of 1978. The lowest seasonal values of the anomaly factor (0.15) have been calculated for Nicolae Bălcescu and Codăiești, in the winter of 1995.

Generally speaking, during the winter season, the anomalies of the maximum flow rates have revealed lower values, before 1975 (Fig. 4). Following this particular year, there is a time interval with significantly higher values. After this period, the 1990 and 2018 period would register extremely low values of the anomaly factor for the maximum flow rates, at the following hydrometric stations: Nicolae Bălcescu, Târgu Frumos, Iași, Codăiești, Murgeni, and Pechea.

During the spring season, when the river regime in this region is highly dependent on snowmelt phenomenon which is frequently associated with significant events of rainfall, an increase of the values for the anomaly factor could be observed, for all hydrometric stations, during the 1970–1980 decade. For the stations located in the northern part of the study area (Rădăuți Prut, Dorohoi, and Ștefănești), this time interval of positive anomalies extends up to the beginning of the 1990s. During the last decades, there is a slight tendency for decrease of the anomaly factor corresponding

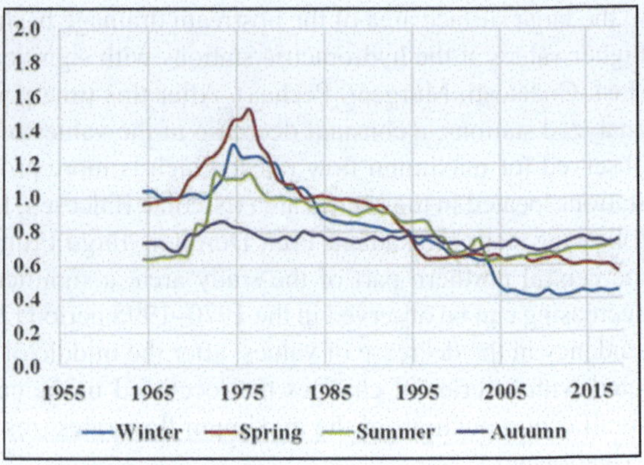

Fig. 4 Seasonal variation of the maximum flow rate anomaly factor

to maximum flow rate values for all the stations in the region, valid for the spring season.

During the summer season, for 8, out of the total 11 stations, there are also tendencies for increases in the anomaly factor values, in the 1970–1980 decade. In addition, at all of the 8 stations, maximum positive values for the anomaly factor are registered (and they vary from 2.29, at Dorohoi hydrometric station in 1975, to 4.94, at Nicolae Bălcescu in 1977). A second growth period for the anomaly factor for the maximum flow rates of Dorohoi and Iași, is identified between 1985 and 1990. In the last two decades, for 9 out of the 11 stations that were analyzed, a general decrease tendency of the maximum flow rates anomaly factor is visible, with a slight increase tendency in the last 5 years.

For the autumn season, a similar model of evolution for the maximum flow rate anomaly factor can be observed, as was the case of the winter season, with decreased values before 1975, followed by slightly higher values. However, there is no evidence of a generalized tendency for growth or decrease during the last few decades.

As far as the minimum annual flow rates are concerned, the analysis of the variation of the anomaly factor reveals a general characteristic, specific for the entire study area. This means that there is no visible specific decrease or increase period of the anomaly factor, for the entire region (Fig. 5). Several general conclusions can be written down from the aforementioned, which cannot be explained through correlation with the dynamic of the climatic elements, or geographic features of the hydrographic basins, inside which the analyzed hydrometric stations are located.

Therefore, at 3 out of the total 11 hydrometric stations (Rădăuți Prut, Ștefănești and Negrești), there is a visible, relatively constant evolution of the anomalies, associated with minimum annual flow rates, with no significant oscillations. As far as the other 8 stations are concerned, between 1964–1980, there are no significant oscillations of the anomaly factor, for minimum annual flow rates. Following 1980, significant variations start to appear, with an obvious tendency of increase in the values of the anomaly factor for the minimum annual flow rates at Dorohoi, Nicolae Bălcescu, Târgu Frumos, Iași, Codăiești, Murgeni and Pechea hydrometric stations.

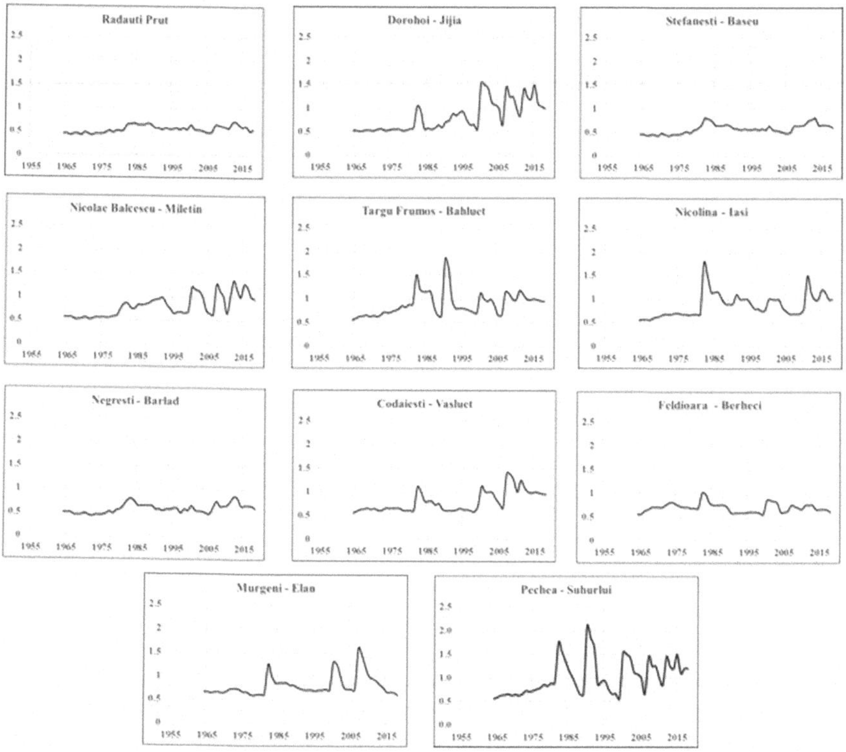

Fig. 5 Variation of minimum annual flow rates anomalies factor for analyzed hydrometric stations

This increasing tendency is generated by the succession of atmospheric drought periods, which have affected the entire region, starting from 1982, and continuing with the 1990–1992, 1996–1998, 2000, 2007, 2012 and 2015 [36].

It is obvious that the years characterized by droughts, are also reflected in the variations of the anomaly factor for the minimum annual flow rates, which reveal the largest values at Târgu Frumos (1.85, in 1990), and Iași (1.78, in 1982). The lowest values of the anomaly factor for the minimum annual flow rates were recorded between 1969–1972, at Rădăuți Prut, Ștefănești and Negrești hydrometric stations (0.42).

Furthermore, from a seasonal perspective, a significant tendency for increase or decrease of the anomaly factor for minimum annual flow rates could not be observed, for the winter and autumn seasons (Fig. 6). Only in the spring season, the significant variations of this factor can be observed, at 8 of the 11 analyzed hydrometric stations, with a slight increase tendency in the last decades. As was the case of the maximum annual flow rates, as far as the minimum annual flow rates are concerned, the anomaly factor values are significantly higher, reaching 5.06 at Târgu Frumos station, in the summer of 1990, or 5.01 at Iași hydrometric station, in the summer of 1982.

Fig. 6 The seasonal variation of the anomaly factor for the minimum flow rates

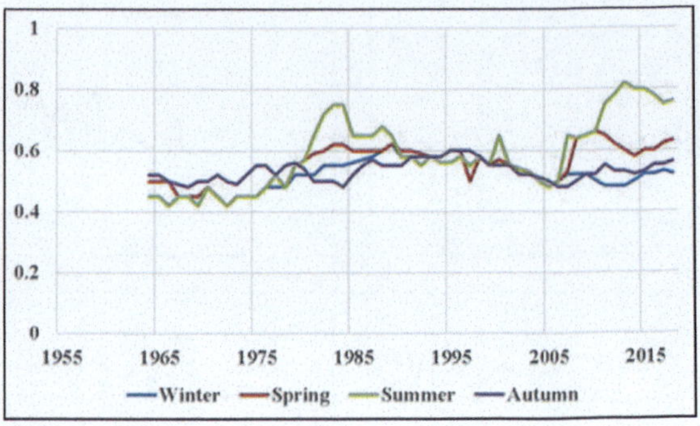

5 Hydrological Extremes Trends

The detailed analysis of the maximum and minimum flow rates, by using the ITM analysis, reveals a series of specific conclusions.

In the case of maximum annual flow rates, it is graphically emphasized that at 8 of the total of 11 analyzed stations, there is a decrease tendency of the maximum flow rate values in the last decades (Fig. 7). The calculus for the tendency slope reveals that this situation is applicable for 10 hydrometric stations. Therefore, the only exception is Rădăuți Prut station, where both by graphical methods, and means of slope calculations, there is an obvious increase tendency (the slope has a value of 5.328 m^3/s) (Table 1). This station is located in the vicinity of the Carpathian mountainous region, which is the main source of input for Prut river, and reflects the wet climate conditions, that are more specific to Baltic influences. The other stations reflect the conditions specific to a hydrological regime corresponding to the temperate-continental climate. This regime characterizes the lower sector of the Danube drainage basin, with a significant input of rainfall during the warm season. This is caused by the dynamic of the air masses, above the Black Sea area, which can generate extreme floods [37], but also has a very oscillating behavior.

The slope of the tendency is decreasing, for the maximum annual flow rates, and varies between −1.576 m^3/s, at Negrești station, and −0.009 m^3/s, at Pechea station, with higher values, at the stations located in the central area (Negrești), and north-eastern area (Ștefănești).

From a seasonal perpective, there is a slight tendency for increased values, as far as maximum flow rates are concerned. This is especially valid during the last decades, in the winter season, at Rădăuți Prut, Nicolae Bălcescu and Codăiești hydrometric stations (but with slopes of low values (0.144 m^3/s, 0.320 m^3/s, and 0.230 m^3/s respectively). There is also a decrease tendency for values in the southern part of the region (Feldioara, Murgeni and Pechea) (Fig. 8).

The most significant decreases can be observed during the spring and summer seasons, when all the stations in the central and southern parts of the region register slopes between −0.913 m^3/s at Negrești (for the spring season), and −0.643 m^3/s at

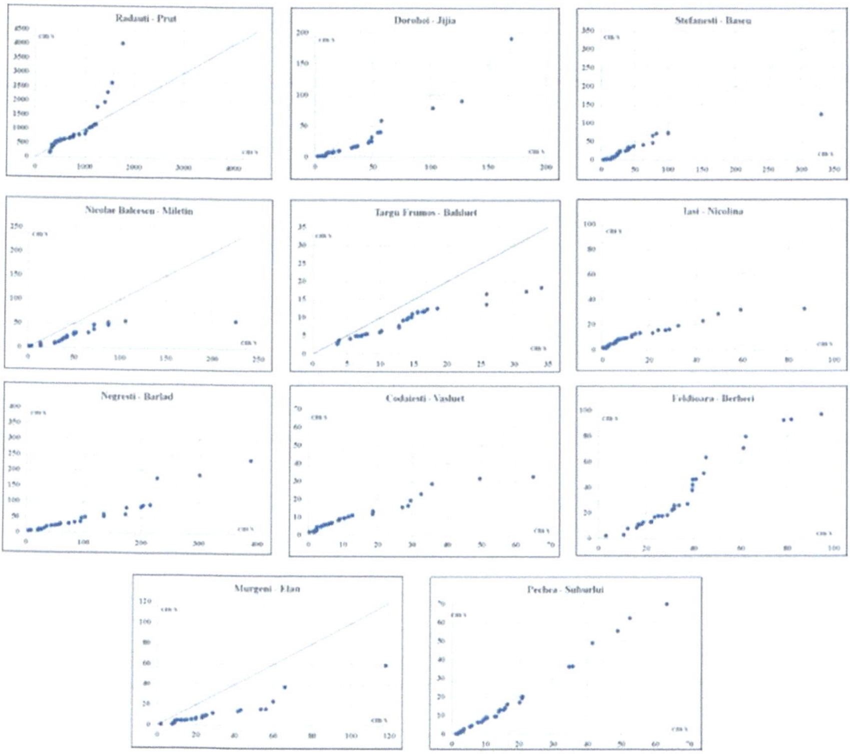

Fig. 7 Trends of annual maximum flows for analyzed hydrometric stations

Nicolae Bălcescu (for the summer season). The only station that registers a tendency increase in the maximum flow rates, for all seasons is Rădăuți Prut (Table 1).

The intensity of these increases and decreases, which is reflected through the PI index, reveals that the tendency increase of the maximum flow rates, on a yearly level, are visible only at Rădăuți Prut station (+1.074 m^3/s of the PI index value) (Table 2). From a seasonal perspective, the increase in maximum flow rates corresponds to PI index values of +6.445 m^3/s at Codăești hydrometric station in spring, of + 1.310 m^3/s at Nicolae Bălcescu, in winter, and +2.722 m^3/s at Negrești, during the autumn (Table 2). The PI values for the decrease tendencies range from −3.535 m^3/s at Murgeni, during the winter season, −3.016 m^3/s at Nicolae Bălcescu, in spring, and −1.403 m^3/s, at Murgeni, in autumn.

The tendencies for the minimum annual flow rates are constant in the last decades, at all hydrometric stations, except for Rădăuți Prut. Here, a slight tendency in increase of the values for the minimum annual flow rate values can be observed (Fig. 9), while the tendency slope reaches a value of +0.063 m^3/s.

On a seasonal level, the tendencies of the minimum flow rates register a decrease, which is insignificant from a statistical point of view, at Negrești and Murgeni, during the winter and spring seasons. Therefore, except for Rădăuți Prut station, the

Table 1 Slope for seasonal and annual maximum and minimum rivers flow in the study area

		Maximum flow slope (m^3/s)				
	Hydrometric stations	Winter	Spring	Summer	Autumn	Annual
1	Radauti Prut	0.144	0.018	2.139	3.348	5.328
2	Dorohoi	−0.085	−0.376	−0.088	−0.016	−0.332
3	Stefanesti	0.001	−0.315	−0.346	0.036	−0.545
4	Nicolae Balcescu	0.032	−0.712	−0.643	−0.068	−0.799
5	Targu Frumos	−0.058	−0.139	−0.094	−0.001	−0.147
6	Iasi	−0.006	−0.065	−0.118	−0.011	−0.165
7	Negresti	−0.200	−0.913	−0.986	0.153	−1.576
8	Codaiesti	0.023	0.025	−0.074	−0.019	−0.106
9	Feldioara	−0.175	−0.219	−0.068	−0.024	−0.054
10	Murgeni	−0.187	−0.312	−0.318	−0.026	−0.484
11	Pechea	−0.063	−0.077	−0.081	−0.003	−0.009
		Minimum flow slope (m^3/s)				
	Hydrometric stations	**Winter**	**Spring**	**Summer**	**Autumn**	**Annual**
1	Radauti Prut	0.079	0.120	−0.055	0.015	0.063
2	Dorohoi	0.000	−0.001	0.000	0.000	0.000
3	Stefanesti	0.001	0.000	0.001	0.001	0.001
4	Nicolae Balcescu	0.000	−0.001	0.000	0.000	0.000
5	Targu Frumos	0.001	0.000	0.001	0.002	0.001
6	Iasi	0.002	0.001	0.001	0.001	0.001
7	Negresti	−0.001	−0.003	−0.001	0.000	0.000
8	Codaiesti	0.001	0.001	0.001	0.001	0.001
9	Feldioara	0.001	−0.001	0.000	−0.001	0.000
10	Murgeni	−0.001	−0.001	0.000	0.000	0.000
11	Pechea	0.000	0.000	0.000	0.000	0.000

increases and decreases of minimum seasonal flow rates are insignificant, from a statistical perspective (Fig. 10).

However, while analyzing the values of the PI index, there is an increase of the minimum annual flow rates, between +4.819 m^3/s, at Iași, and +0.356 m^3/s at Negrești. From a seasonal perspective, the PI index reveals a growth intensity of the minimum flow rates, which reaches the highest values at Iasi (+5.430 m^3/s) during winter season, and at Targu Frumos (+6.649 m^3/s) in autumn. Significant negative values of the PI index are registered at Murgeni (−2.161 m^3/s, in the winter season), and at Nicolae Bălcescu hydrometric station (−1.036 m^3/s in the summer season) (Table 2).

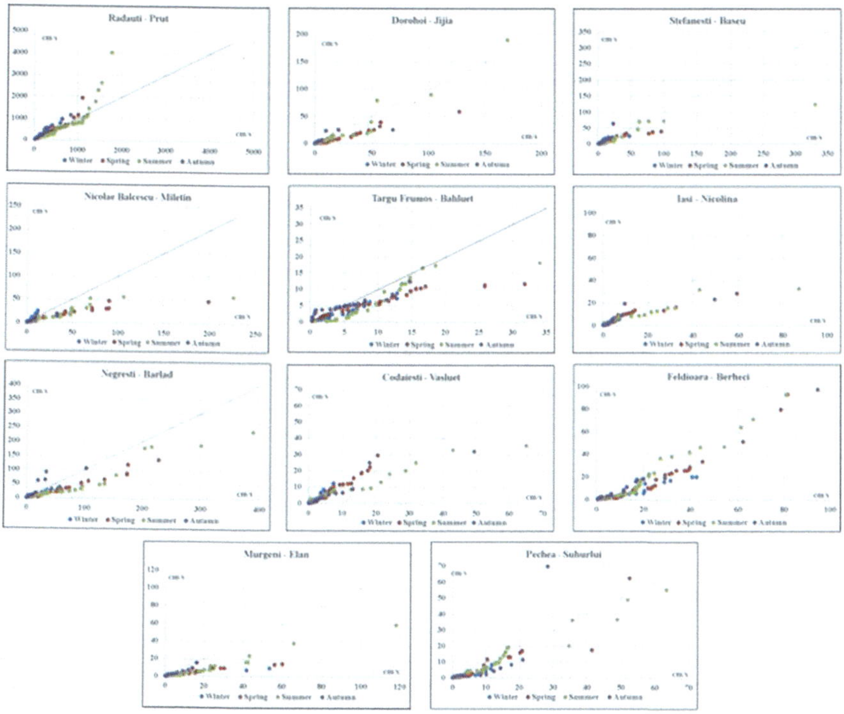

Fig. 8 Trends in seasonal maximum flow for the analyzed hydrometric stations

6 Discussions

The area between Siret and Prut rivers In Romania is known in the European Union, as one of the least developed, from an economical stand point, and was affected by the frequent series of floods and flash flood events, during the last decades (1996, 2005, 2008 and 2010), or by the more intense episodes of hydrological drought (2000, 2007 and 2012).

Based on the aforementioned, there is also the ever-increasing impact of current climate change, to be mentioned. The changes in climatic regime, in Romania, fit in the global context, with several regional particularities, regarding the air temperature, and also precipitation regime [19].

By means of a series of natural mechanisms, there is an obvious aridization tendency, both gradually, and cumulatively, which can be observed through the increase of the occurrence frequency of extreme air temperature and precipitation values. In addition to this, there is also the tendency of concentration of precipitation in shorter time intervals, with a direct effect on the increase of maximum flow, and the occurrence of more frequent floods [38]. Previous studies have revealed that more than half of the rivers that drain the eastern part of Romania have an increase tendency of the annual average and seasonal runoff [15].

Table 2 PI values for seasonal and annual maximum and minimum river flow

		PI Maximum flow (m^3/s)				
	Hydrometric stations	Winter	Spring	Summer	Autumn	Annual
1	Radauti Prut	0.153	0.006	0.458	2.660	1.074
2	Dorohoi	−2.761	−2.649	−0.689	−0.631	−1.529
3	Stefanesti	0.002	−2.235	−2.033	0.941	−1.976
4	Nicolae Balcescu	1.310	−3.016	−2.661	−1.972	−2.605
5	Targu Frumos	−1.807	−2.041	−1.795	−0.023	−1.706
6	Iasi	−0.451	−1.270	−1.797	−0.436	−1.637
7	Negresti	−2.021	−2.478	−2.196	2.722	−2.495
8	Codaiesti	1.793	6.445	−1.203	−0.665	−1.219
9	Feldioara	−2.677	−1.400	−0.473	−0.463	−0.251
10	Murgeni	−3.535	−3.584	−2.869	−1.403	−3.070
11	Pechea	−2.911	−1.465	−0.917	−0.144	−0.089
		PI Minimum flow (m^3/s)				
	Hydrometric stations	**Winter**	**Spring**	**Summer**	**Autumn**	**Annual**
1	Radauti Prut	0.960	0.609	−0.271	0.095	0.786
2	Dorohoi	−0.013	−0.582	−0.614	−0.367	−1.108
3	Stefanesti	1.314	−0.032	1.725	1.939	2.462
4	Nicolae Balcescu	−0.308	−1.726	−1.036	1.306	1.848
5	Targu Frumos	3.446	−0.207	1.423	6.649	4.631
6	Iasi	5.430	1.559	1.329	2.811	4.819
7	Negresti	−1.158	−1.369	−0.752	−0.457	0.356
8	Codaiesti	3.069	2.069	1.396	2.643	3.718
9	Feldioara	0.671	−0.722	−0.169	−0.707	0.479
10	Murgeni	−2.161	−1.001	0.467	−0.877	−0.557
11	Pechea	−0.175	0.008	−0.448	−0.070	0.636

During the last decades, according to the official reports, there have been periods of severe hydrological and meteorological phenomena (floods, flash floods, droughts), mainly during the summer season, from June, to August. At the same time, across the rivers which drain the Siret-Prut region, a hydrological regime which has been predominantly characterized by a constant deficit, starting from 2011, up to 2016, with a severe registered deficit, during 2015. Their effects have been seen in the decreased agricultural production caused by the reduction of irrigation or the decrease in the amount of drinking water for water supply.

This has prompted the Romanian government to approach several quantitative and qualitative aspects concerning the effects of climate change, in order to emphasize the synergy between the watershed management plans, and the management plans which address the flood risk. This measure consisted in updating the sections of the National Management Plan, which address the international sector of the Danube

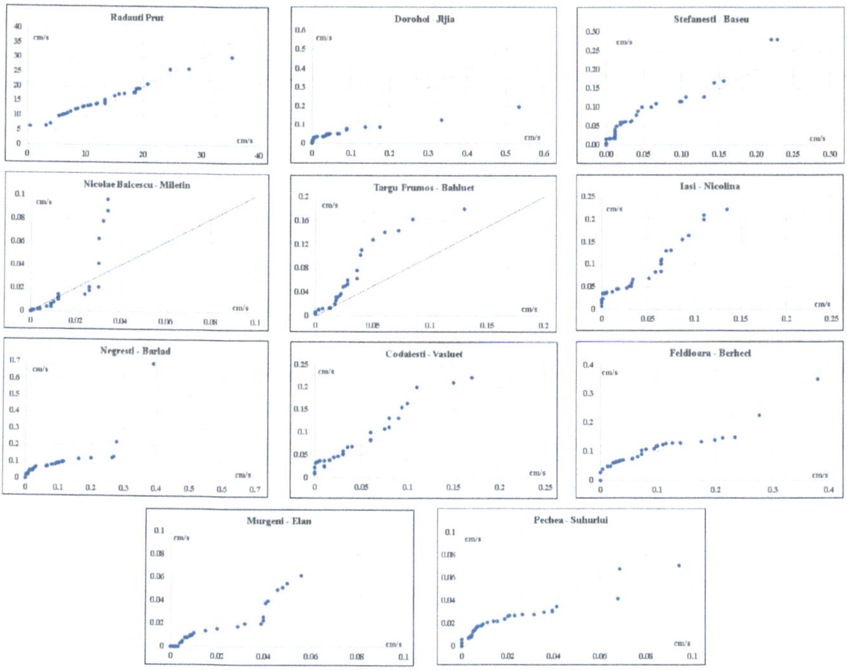

Fig. 9 Trends of annual minimum flow for the analyzed hydrometric stations

River, located in Romania, during 2016–2021, which was performed by the Romanian National Water Administration (2018) [24]. These measures are applied, in conjunction with the analysis and evaluation effects of climatic change on the water resources, which are regulated through the Water Framework Directive 2000/60/EC [39], and continued in the Water Framework Directive 2007/60/EC [40], and also in the IPCC [41].

7 Conclusion

The assessment of the anomalies and extreme hydrological tendencies has emphasized that the previous studies addressing this area have only roughly estimated a series of anomalies and tendencies, for different hydrological and climatic parameters. However, they did not take into account the changes that have occurred over extended periods of time.

This chapter analyzes the detection of anomalies and tendencies which are associated to extreme hydrological events, both for high flow rate values, and for minimum flow rates events, by applying the quantile perturbation method (QPM) and the innovative trend application method (ITA), with a corresponding PI index for evaluating

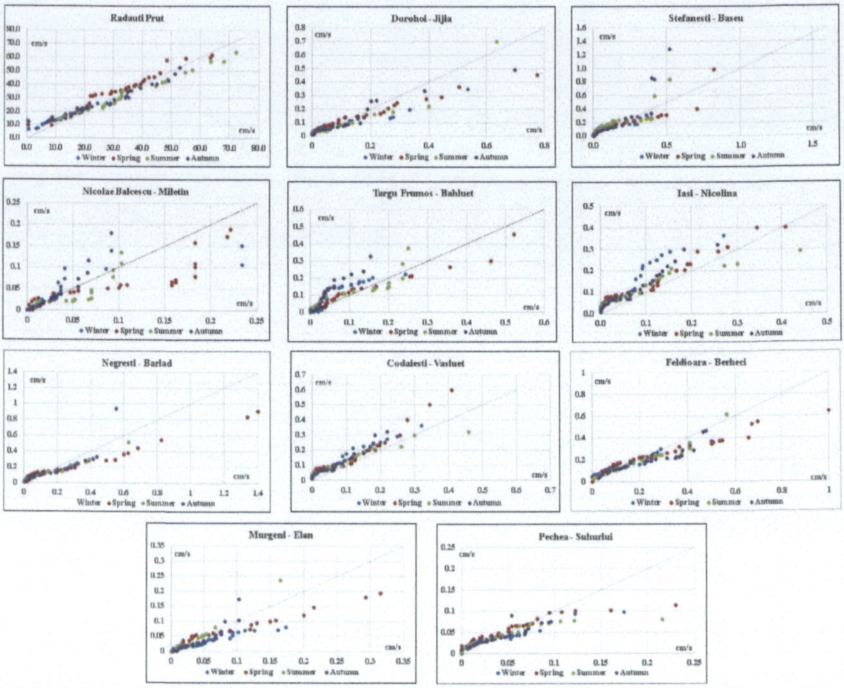

Fig. 10 Trends of seasonal maximum flow for the analyzed hydrometric stations

the tendency intensity. Based on the obtained results, several relevant conclusions can be summarized. Firstly, the anomalies of the maximum annual and seasonal flow rates are significantly more pronounced than those of the minimum annual and seasonal flow rates. Secondly, the most significant positive anomalies, regarding the maximum annual flow rates occured during the 1968–1975 period, which has been an important interval from a pluviometric perspective, for the entire country. After this period, there is a constant decrease in the values of the anomaly factor for maximum flow rates, which is more pronounced at the hydrometrical stations located in the central-southern region, and more attenuated, for the hydrometric stations located in the northern part.

Regarding the overall tendencies, it has been emphasized that the maximum annual flow rates, as well as the ones in spring reveal a decreasing tendency (for 90% of the analyzed stations), with intensities varying from −0.089 m^3/s and −3.070 m^3/s. The values of the minimum monthly and seasonal flow rates show increasing tendencies (in the winter, for 55% of the hydrometrical stations), as well as decreasing tendencies, during the spring season (for 64% of the hydrometrical stations in the region).

The current results of the assessment of the anomalies and tendencies of the minimum and maximum flow rates, stand as a starting ground for developing efficient

Sustainable Management and Governance of the Hydro-Environment

A Transdisciplinary Approach Using Danube River Multi-connectivity in Wetland Management

Iulian Nichersu, Iuliana Nichersu, Adrian Constantinescu, and Alexandru Nichersu

Abstract Policies and management plans in the Danube River basin were first developed and applied at its entire scale in the second part of the twentieth century and were exclusively based on the principles of neoclassical economy. These principles were translated in a high number of economic and social objectives, some of which that have been identified as driving forces for the wetlands of the Lower Danube System behind structural and functional changes. The transformations to which the Lower Danube System has been subjected and the absence of multi-connectivity in planning were marked by an insufficiency of support tools at decision-making level. Wetlands play a dual role in terms of planning, acting both as risk-inhibitor and services enabler when inclusive multi-dimensional planning is performed. Given the relevant spatial and temporal scale of the current context, capitalizing on scientific knowledge, experiences and information of local communities in future policymaking is crucial. The productivity and stability of ecosystems depend directly on their ability to ensure energy transfer both intra-, but especially inter-system. The lateral and longitudinal analysis of ecosystems in the Lower Danube System, regarded as both dynamic and nonlinear systems, as well as production units, is based on a conceptual framework aimed at their multi-connectivity. This ensures long-term processes whose variability and diversity are essential for the stability and productivity of these units. The study includes social and economic implications, considering the relationship between the natural heritage resources of the geographical sub-units and the existence of the socio-economic system.

I. Nichersu (✉) · I. Nichersu · A. Constantinescu
Institutul Naţional de Cercetare Dezvoltare Delta Dunării (INCDDD), 165 Babadag Street, 820112 Tulcea, Romania
e-mail: iulian.nichersu@ddni.ro

I. Nichersu
e-mail: iuliana.nichersu@ddni.ro

A. Constantinescu
e-mail: adrian.constantinescu@ddni.ro

A. Nichersu
Institut für Technologie, Institute for Automation and Applied Informatics (IAI), Eggenstein-Leopoldshafen, 76344 Karlsruhe, Germany
e-mail: alexandru.nichersu@kit.edu

A. Negm et al. (eds.), *The Lower Danube River*, Earth and Environmental Sciences Library, https://doi.org/10.1007/978-3-031-03865-5_14

Keywords Wetland management · Multi-connectivity · Human intervention and impacts · Lower Danube River

1 Introduction

Watercourses on the surface of the earth are each unique. Their dimensions differ from each other in terms of lengths, depths, water flow, widths of the riverbed, floodplains extent, by sediments or alluvium quantities and types, vegetation and plant associations in their basin, by the nature of the riverbed and river mouth, and the geological and climatological conditions. Even the same river can be subject to large fluctuations during a year (in terms of depths, flows, widths of riverbed, etc.).

During rainy periods or during the sudden melting of snow in temperate climates, their waters can become large, fast and murky (full of sediments). In many places, riparian floodplains are not able to store the sudden increase in water quantities, flowing over and covering large areas with water. Today, in part due to the globalization of mass media, we often hear of extreme meteorological events that involve the destruction of crops, flooding of settlements situated in floodplains, or even breaking of dikes and levees and the destruction of transport infrastructure. Water related disasters especially in the paradigm of climate change increase the vulnerability of anthropic systems and require adaptation and mitigation.

As is stated in [1] the benefits of ecosystem services are purification of both air and water, crop pollination, nutrient cycle, waste decomposition and soil generation and renewal. In addition to this easily attributable direct contributions, ecosystems moderate environmental and climatic conditions. These invisible yet paramount contributions stabilize the climate, reduce the risk of extreme weather events, mitigate droughts and floods, and protect soils from erosion. These impacts and services manifests themselves locally, regionally and globally, and are visible through conflicts between stakeholders at these different levels.

Prof. Einstein H. A. states that change in wetlands is a topic that should not be dealt with lightly, as its impact can be wide and in-depth. He is quoted on the topic in [2] as saying: "*If we change a river, we usually do some good somewhere and "good" in quotation marks. That means we achieve some kind of a result that we are aiming at but sometimes forget that the same change, which we are introducing, may have widespread influences somewhere else. I think if, out of today's emphasis of the environment, anything results for us it is that it emphasizes the fact that we must look at a river or a drainage basin or whatever we are talking about as a big unit with many facets. We should not concentrate only on a little piece of that river unless we have some good reason to decide that we can do that.*"

Continuity and/or connectivity of ecosystems as variables of external manifestation play a significant role on the services that beneficiaries receive. Quantifying the associated "intrinsic value" of these services is best performed from a multidimensional perspective of causality, functionality, hierarchy and system development process. For this process to be performed the theory of scale dependence, for the relationship between the model and the multiplication process, along with the metapopulation theory, presented in [3], are used. This explain the complexity of self-organization in ecosystems and use the principle of aggregation and decentralization of management to the most appropriate value.

To be able to proceed with this manuscript the following taxonomy was defined. Continuity is fulfilled when an uninterrupted connection exists, while connectivity refers to the degree of connection or condition thereof, of elements of the same or similar type, including both structural and functional components. The presentation of the two definitions helps define a set of principles used in the specifics of the approach presented in this paper that looks at the relationship within the system or between systems. This helps define connectivity as a norm with both strategic and operational value. In addition, it can be deduced that the value of continuity and connectivity in the ecosystem context, is that such an approach is axiomatic. Accepting and promoting continuity as a principle highlights the self-confirming character of the respective dimension. The role and importance of connectivity in achieving ecosystems continuity is ensured by the very nature of the theory and practice of the flow of matter and energy.

It must be said that the multitude of spatial, physical and thematic meanings behind connectivity, combined with the complexity of the Lower Danube System have, during the evolution of the theory and practice presented in this manuscript, helped configuring various fundamental concepts in ecological restoration. The main disciplines involved are regional geography and landscape ecology and include principles that constitute methodological landmarks for the presented method. One of the central ideas concerning connectivity is that it acts as a support axis in ecological restoration.

Connectivity must persist both in the system and in the processes of external manifestation, both at the level of operational functioning and at the relational level. At the level of designing system development directions in conservation and environmental protection policies, the role of connectivity is indisputable. Respecting this principle favors the shaping of a clear vision on the perspective of sustainable and social development by means of ecosystem and adaptive management. This also determines the quantities of various resources and means required as a means to an end. The presence of connectivity generates interaction relationships between life and the environment.

As exhibited in [4] and again in [5], the theme of continuity and connectivity in landscape ecology as a science of studying and influencing the spatial model of landscapes and its ecological consequences, has become, at an accelerated and exponential pace, a pressing concern both in theory and in applications. It has now been established both as a field of study as well as a new ecological paradigm.

In essence, connectivity is a field of study in landscape ecology [6], which is both inter- and trans-disciplinary that focuses on the spatial modeling of landscape elements and its relationships with ecological processes at different scales in space and time. At the same time, it keeps a strong relationship with System Theory and Geography. No matter what aspect of the field one focuses on, be they biophysical, socio-economic or both, the connectivity paradigm helps the functioning of ecosystems and brings phenomena into perspective by integrating model, process, scale and hierarchy. Concerning ecological restoration, connectivity is a key issue as well as a relevant research topic.

In [2] the six fundamental concepts of eco-geographic analysis were defined. They are imperative when dealing with modeling of river basins and large rivers. These notions are:

- the watercourse is only one part of the system;
- the system is dynamic;
- the system is functioning complexly;
- there are geomorphological thresholds and when they are exceeded an abrupt change may result;
- time scale (retrospective versus perspective);
- spatial scale (hierarchy, holarchy).

The Lower Danube (Fig. 1)—in layman words, or the Pontic Danube System—in scientific literature [7], is somewhat a paradoxical entity. Although it crosses different relief units in terms of age and tectonic mobility (starting with the Carpathian orogeny unit and ending with the craton unit), the local peculiarities of the floodplain evolution have caused interruptions or changes in the transverse profile. This happens without reaching the total change of the arrangement of the component elements.

The Danube bends and wetlands have changing dimensional and quantitative relationships between them [9]. This materialize in their varied distribution, from one place to another, and depend on both general factors that hatch them (frequency, intensity and duration of overflows) and the other features of the local factors. The general and particular characteristics of the floodplain relief are a direct result of the floods' peaks and intensity, considering frequency, magnitude and duration.

The Lower Danube System is the most dynamic hydrogeographical unit in Romania [10], in terms of systemic methodology. The interactions occur between natural components, flows of matter and energy (hydrological and climatic), on the one hand and human intervention (with an ever higher intensity in recent decades), on the other hand. Today, about 80% of the Romanian sector of the Danube floodplain is embanked and transformed by desiccation and drainage processes [9]. From a

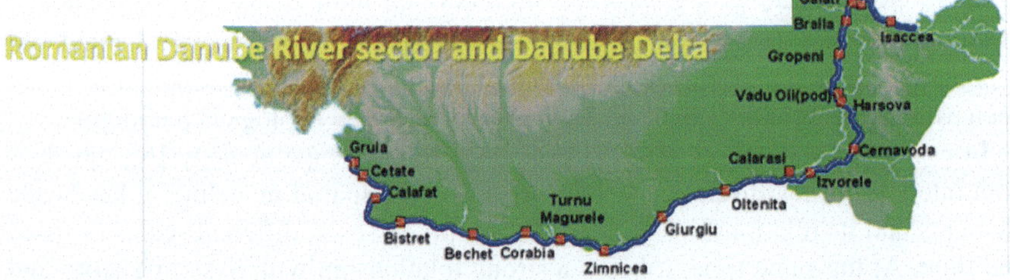

Fig. 1 Relief units and hydrological monitoring points in the Lower Danube System [8]

geographical point of view, the Lower Danube System is a complex system. Spatial–temporal geographical evolution stages in its history are dependent on correlative interference. Its current configuration is conclusively reflected by this.

The Lower Danube section stretches over 885 river kilometers from the point where the Danube breaks through the Carpathian Mountains and passes through the Walachian Lowlands. Its major tributaries in that section include the Siret and the Prut rivers. The last section is represented by the arms of the Danube Delta [11].

The Lower Danube Wetland System [12], which is also referred to as the Lower Danube Valley [13], Danube River Corridor [9], Pontic Danube Valley [14] or Pontic Danube System [15] appears as a lateral water collection axis between the two large adjacent areas to the north and south that have a slight slope. All major tributaries that it cuts transversely or slightly diagonally determine the maintenance of a low transport power and the deposition at the outflow, in the Danube riverbed of fine alluvial materials.

As shown in [16], a river system model is dependent on the following global variables: weather, relief, geology (lithology, structure), climate, vegetation, hydrology, and morphology of the drainage network. Morphology is what gives it the dynamic character. Concurrently, given that the river system is unitary, the change of a component can result in complex changes of the entire system, composed from hydrographic areas, river corridors and watercourses, complex ecosystems changes that manifest themselves at different spatial scales [17].

Recent examples of biodiversity loss caused by climate and land use change, over-exploitation of natural resources, the introduction and spread of invasive alien species, have a significant socio-economic impact. At the same time, the upward trends in frequency/magnitude of the natural hazards and a more intense anthropogenic impact caused a degradation of worldwide ecosystem services. This is especially true in a region like the Lower Danube. Despite the positive effects of historical events such as the fall of the Iron Curtain in Europe (that slowed the pace and even stopped large hydroelectric and agricultural projects in the region), actions to halt biodiversity loss are needed. This strategic requirement to strengthen and expand adequate protection measures, using a modern legislative framework, needs to be synchronized with the theoretical basis. This integrated cooperation will provide for best practices and training in the dynamics and the ecological effects of spatial heterogeneity as well as the relationship between the landscape model and ecological and socio-economic processes at different scales in space and time.

The principles and Lower Danube region examples discussed in the next sections serve as a useful guide to ecological restoration. They serve as part of a new effort that goes beyond the current concept of natural resources conservation towards a deeper concept of restoring of "environment life"—an ecologically viable state where ecosystems are self-sustaining and improve the functioning and quality of services over time. Extensive knowledge of the risks and threats arising from the disruption of continuity and connectivity in the Danube area present multiple opportunities to counter them. This requires an inter- and trans-disciplinary approach, as well as effective communication and cooperation between specific stakeholders in this field.

2 Theoretical Aspects: The Multi-Dimensional Character of River Connectivity

In recent decades, spatial planning has become a major concern for scientific research, but also more inclusive towards all stakeholders, be they central or local authorities, companies or the general public. Space—as a support for human activities, but also natural capital—represents the objective and universal form of the existence of matter, which has the aspect of a continuous whole and express the order of coexistence of the real world [18]. Within space, the movement of matter takes place in time. In the human context of existence, space is defined by a double meaning: first, a condition for biological survival—a finite resource that explains its role in human history, and second a psychological necessity, space being perceived as a prime condition for individual freedoms, or a release from constraints and risks. Further, humans instinctive spatial abilities of spatial awareness, give it value in well-defined hierarchies and holarchy. From an anthropocentric perspective, the analysis of space is performed according to each personal "cultural" background—experience based and relationship based. This, in turn, allows people to organize space so that it corresponds to their requirements and social relations.

Additionally, the spatial dimension is vital for the support of the Socio-Economical System (SES), as space provides humans with appearance, mobility and experience. The hierarchy of spatial structures, through relational criteria is established in physical geography. Space creates the quality of the geographical landscape and provides SES with the elements of anthropogenic topography, integrated in the geographical space, which are in a direct relation.

Landscape mosaics, as defined in [18], are described by the landscape components of the patches, corridors, and the surrounding matrix. Further, patches, corridors and matrix directly influence spatial modeling and flows in a landscape while spatial scale also greatly affects landscape structure, heterogeneity and connectivity [19, 20]. This definition indicates a direct correlation between the organization of the space system and the inter- and trans-functional relations. For example, Fahrig and Merriam in [15] found that the frequency and persistence of local extinctions of *Peromyscus leucopus* (white-footed mouse) populations depend on the degree to which individual patches are isolated from each other. Models that make predictions about the dynamics of populations living in patches should take into account the spatial arrangement of these patches and should do so on the same time scale as local extinctions.

2.1 River Systems Description and Definitions

A river system is described in [16] as a model characterized by at least the following global variables: time, relief, geology (lithology, structure), climate, vegetation, hydrology, morphology of the drainage network, slope morphology.

In addition, the most significant initiating influences of changes at the level of a riverbed are, according to [21], natural or artificial changes in the: volume or flow of water passing through the river; volume or flow of solid or sediment characteristics available to the river; properties of the riverbed boundaries that affect its hydraulics and its vulnerability to erosion.

Landscapes, river basins, river corridors and watercourses are ecosystems encountered at different spatial and temporal scales. The structure of the area on which the variables and processes are manifested also requires a thorough description. The Federal Interagency Stream Restoration Working Group (FISRWG) defines the river corridor as an "ecosystem generally consisting of three major elements: the riverbed itself, the floodplain and the upper transition border to the river terraces, which works dynamically and valuable crossings of the landscape" [17].

FISRWG also classifies processes in river systems in 2001 [22], and defines the issues that need to be addressed through a systemic analysis. The classification and the corresponding research questions include five topics: hydrologic and hydraulic processes; geomorphic processes; chemical processes; biological processes; stream corridor functions and dynamic equilibrium.

According to [23], ecological studies deepen both horizontally and vertically perspectives of the notion of biome, the concept is now applied to ecosystem complexes with increasing territorial expansion. The essential criterion becomes more and more, the degree of complexity and not the dimensional attribute. In fact, it is the position or level from which we analyze a system: it can only be a component of another system, hierarchically superior or on the contrary, it is a complex of other subsystems, considered at a time and a certain level of organization, as elementary. Equally important is the direct inclusion of the three-dimensionality and therefore of the possible framework for the organization and hierarchy of the components of a geographical area.

In order to operationalize the analysis of this geographical space in the context of connectivity, the whole Lower Danube River can be considered a complex spatial system, biotic and abiotic, with a spatial organization that responds to specific functional laws. This system is an open system, characterized by a dynamic balance, in which there are permanent exchanges with the outside (inputs/outputs). Ludwig von Bertalanffy based the theory of open systems on this methodological basis. According to Bertalanffy's conception, first published in 1932 in Teoretische Biologie [24], open systems have constant exchanges. These transactions are of multiple types, for examples energy or matter with the outside. Based on the theory of open systems, Ludwig von Bertalanffy developed the general theory of systems, in which the main problem is to discover the laws that explain behavior, operations and development of systems by understanding their complex interconnected components and transfers.

2.2 River Connectivity

Already discussed and defined in the first section, the spatial connectivity of a watercourse takes into account lateral, longitudinal and vertical connectivity. A river system can also be described longitudinally as composed of three parts [25]. First is the upper part of the system, the river or drainage basin—this part of the system functions as a sediment supplier. The middle sector of the system, where the watercourse itself is a part of the system that functions as a sediment transfer zone. Lastly, the third section, is the lower part of the system, wither a delta, estuary, swamp, natural or artificial lake. The last part of the system also functions as a sediment deposition.

The three river sections described by Perks are usually not ideally separated, as under river current conditions, sediments can be accumulated, eroded and transported to lower sections. In addition, the three areas are linked by feedback loops. The focus of inter- and trans-disciplinary studies and analyzes lies with these loops. The three areas are organically connected and constitute a functional unitary system.

As shown in [15], the concept of fractals and scale of the system can be found in all of the three river areas in the geo-hydro-morphological system. In the previously mentioned publication, fractals proved ideal to determine change processes with historical river data and enhance the approach proposed in this manuscript.

To explain the relationships in between the three areas we state the following: if an average daily rainfall change occurs in the upper part (zone 1) of the river basin, this causes an increase in the rate of soil erosion with consequences in increasing the amount of sediment production of the basin. In the middle part (zone 2) sediment transfer will develop. This means that the area will consequently undergo a change in its state variables, so that the river will be able to transfer increased inputs with minimal energy expenditure. Similarly, the lower part (zone 3) will also be adjusted, as the characteristic shapes of sediment accumulations build up. As it can be seen, any change in a component of the ecosystem triggers, due to spatial connectivity, changes on the other two components and implicitly on the functionality of the system.

For a coherent understanding of the Lower Danube System, due to the spatial–temporal dynamics of the complexity of inter and trans system interactions, it is necessary to use an approach that works in a framework of integrative theoretical model of analysis, which allows changes, transformations, trends and identification of adjustments with in-depth understanding of the ecosystem.

Connectivity as an analysis model is based on the River Continuum Concept (RCC) first published in the Canadian Journal of Fisheries and Aquatic Sciences 40 years ago [26]. This concept integrates the most important theoretical foundations, such as the theory of habitat types [27], and the theory of entropy first described in [28]. At the same time, it revolutionized the research approach in streamflow [29] and postulated the theory so as to include mechanistic relationships between environmental parameters, energy inputs and biota composition. This allowed the field of streamflow to change from a descriptive to a predictive one. Another reason for its role in revolutionizing river ecology is that it marked a first interdisciplinary approach that encompassed several aspects of river systems, ranging from geomorphology, biology and biogeochemistry. As a result, over the past four decades, the RCC has become a key conceptual framework among river ecologists for testing basic and applied assumptions about species distribution and community structure, as well as mater and energy flow in river habitats, which have furthered new ideas in the field.

Using the RCC consists first in the evaluation of the subsystems within the Lower Danube complex system. Riverbeds and floodplains are constantly adjusting their areas due to environmental and anthropogenic factors. These factors are: water and sediments provided by the river basin upstream of the Lower Danube, climate and degradation of functional soil characteristics, pollution and land cover/land use change. In addition, another variable is the riverbed's response itself. This is unique to changes in the hydrological and sediment characteristics of water that can occur at different times and locations, requiring varying levels of matter and energy input.

Longitudinal connectivity. In a natural system, longitudinal connectivity implies that a watercourse (Danube River in this case) should respect the continuity principle within the variables described by Schumm and Lichty in [16, 30]. This occurs throughout the span of the river, from spring to discharge and is based on the RCC concept, which refers precisely to the balance of the physical gradient from source to discharge, chemical systems and biological communities, habitat continuity, inflow of organic materials and energy dissipation.

Longitudinal continuity is often affected by natural and artificial causes. The main human related ones are dams built for different purposes (water storage, energy production, etc.). Natural causes rarely occur and are usually accidental—storms, floods that carry huge amounts of matter, snow, dams created by beavers, etc.

Lateral connectivity. As defined in [31], lateral connectivity refers to the periodic flooding of the floodplain and the resulting exchange of water, sediment, organic matter, nutrients, and organisms. Lateral connectivity becomes especially important in the case of large rivers with wide floodplains. In order to discuss lateral connectivity research questions are required at the beginning of any ecological restoration project. These seek answers as a way of furthering the concept, for example:

Is the watercourse in contact with its floodplain (during floods, etc.*)?*

In a natural state, the river is connected to the floodplain, especially during floods. The river invades its floodplain with water, new sediments and a whole range of influences. The dried lakes become active and are filled with fresh water while further territories are covered with sediments.

Is there a connection between the aquatic and the terrestrial environment?

In most cases, there is a connection between aquatic and terrestrial environments, by the simple fact that they are next to each other. This respects the first law of geography: "*Everything is related to everything else, but near things are more related than distant thing*" [32]. As an example, water flowing through the capillarity of the soil provides a certain degree of humidity that influences the presence of specific vegetation and animals.

Is there a healthy waterfront area?

The riparian zone is the interface between the terrestrial environment and a river and in certain periods of time it becomes another type of water body. The ecosystems in this area are functional insofar as the structure and functions allow the maintenance of biodiversity, biotic integrity and ecological processes over time. Lateral connectivity is a premise of a healthy riparian biome.

Vertical connectivity. Vertical connectivity has a dual representation, both inside and outside the river system. Inside it stands from topographic evolutions of uplift and erosion of the riverbed, while externally by the connection between the atmosphere and groundwater.

The first connection type is visible in the course of the river in its floodplain and is in part due to the longitudinal connectivity. With the second case, the ability of water to circulate through soil, river, and air as liquid, vapor, or ice is important in storing and refreshing water. This exchange is usually seen as unidirectional—precipitation falling on the ground and then slowly being absorbed from the surface into the soil or percolating through the soil back into the river. An equally important transfer of water takes place in the opposite direction, i.e. from the river to the aquifer and from here by capillarity to the soil surface, where it can reach the atmosphere.

In this way, groundwater can contribute to river supply at certain times of the year and in certain places on the same river. Rivers can gain or lose water, in and around the surrounding aquifer, depending on their relative growth. Lowering the groundwater level by groundwater withdrawals can change this dynamic exchange in unforeseen ways [17].

Temporal connectivity. The river system also has temporal connectivity. This is of continuous physical, chemical and biological interactions over time, according to a predictable model [33]. This connectivity model is of paramount significance for the functioning of ecosystems. Over time, liquid and solid flows change, thus forming meanders. Further, canals and branches break of from the main channel. The channels change their trajectory and often intertwine ensuring balance in energy dissipation.

Danube levels and flows increase and decrease according to seasonal patterns, depending on precipitation and snowmelt (Figs. 2 and 3). The Lower Danube has a highwater hydrology in spring, the flows that decrease in summer, and become extremely low in autumn and in winter when we witness a first wave of flood lower in value. The water basin has adapted to these normal fluctuations and many organisms have evolved to depend on them.

Thermal connectivity. FISRWG showed in 2001 [22] that water temperature is a crucial factor in the health and development of water ecosystems. This is all the more important in the Lower Danube Complex System. First, dissolved oxygen decreases with increasing water temperature, and therefore the stress imposed by the process of decomposition of organic matter. Second, temperatures govern many of the physiological and biochemical processes, metabolism and reproduction rates, vital in stock stability. Third, many aquatic species can only tolerate a limited range

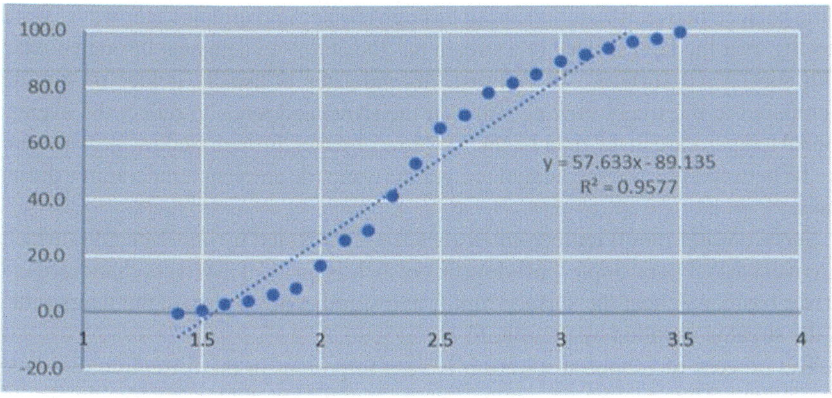

Fig. 2 Annual average frequencies of Danube water levels in Tulcea, 1932–2020

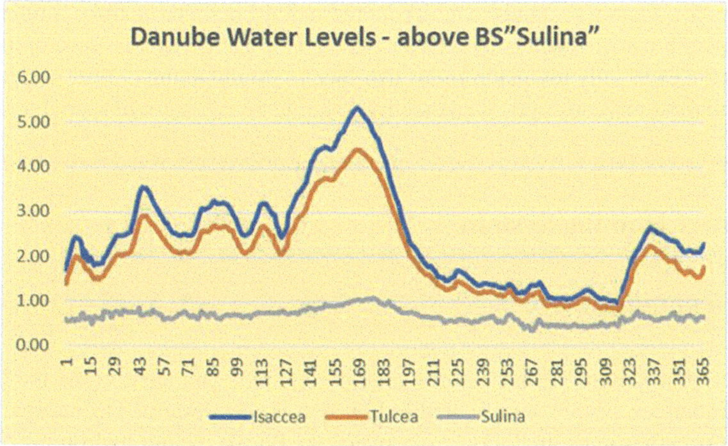

Fig. 3 Danube water levels in three key navigational sections of the maritime sector—Sulina, Tulcea and Isaccea

of temperatures and shifting maximum and minimum temperatures in system can have profound effects on species composition. Finally, temperatures affect many abiotic chemical processes, such as the rate of re-aeration, the absorption of organic matter by plants, the dissolution of chemicals present in sediment particles. High temperature can lead to increased stress due to toxic compounds, for which the dissolution fraction is higher than the bioactive fraction.

Water temperature in the river systems depends on the water temperature upstream, but also by the processes inside the water flow and the influential water temperature. The effects of water temperature are also addressed in lateral and vertical connectivity. The most important factor of water temperature connectivity in a river is the balance between water reaching through surface and groundwater ways. Water flowing over the land surface to a stream has the ability to gain heat through contact with sun-heated surfaces. In contrast, groundwater is usually colder in summer (compared to the freely flowing water in the river) and tends to reflect the average annual temperature in the river basin. The flow of water through shallow groundwater can be between the average annual temperature and the ambient temperatures during leaks.

Surface water runoff temperatures are strongly affected by impermeable surfaces in a river basin. For example, paved surfaces, such as many of the rivers channeled into a river basin, can heat the surface runoff and significantly increase the temperature of the streams supplied by the runoff.

The process of evapotranspiration is very important in the formation of advection clouds. The creation of embankments in the Lower Danube Floodplain has severely affected this process and altered the thermal balance. This is amplified by the conditions of climate change.

In addition, water is subjected to thermal charge by the direct effects of sunlight on water bodies. Therefore, land cover can increase or decrease basic flows, influencing the temperature at different levels. This is very important, both in the development of habitats but especially for anadromous fish species. Maintaining or restoring normal temperature by ensuring optical connectivity should be a concern of spatial planners.

3 Lower Danube System

3.1 Generalities

To proceed, a definition of the study area is required. The territory of the Lower Danube complex system is presented in Fig. 4. It is best described as the space that the river occupies in its final 931 km, between the Portile de Fier hydropower plants and the Black Sea. The river stands north from the Balkan space and separates it from Romania's core, as well as Moldavia's South and Ukraine's SSW region of Ismail. In the chosen perimeter—the natural capital has a productive capacity that is determined through its functional components. This is done so that regional planning avoids impacts such as degradation and destructuring under anthropogenic impact and to favor the sustainable use of its support capacity. Guaranteeing sustainable socio-economic development in the Lower Danube complex system area requires fore-knowledge of ecological sustainability, ecosystem integrity, environmental support capacity, regional, and local ecological ecosystems balancing.

The biological diversity, functionality and naturalness of the ecosystems in the Lower Danube complex system is a result of geological and topographical evolution over time. With a more limited but severe impact, is the succession of different "civilizations" that have often upset the balance of the components created by the natural evolution the initial environments. The desire to understand the current "crisis

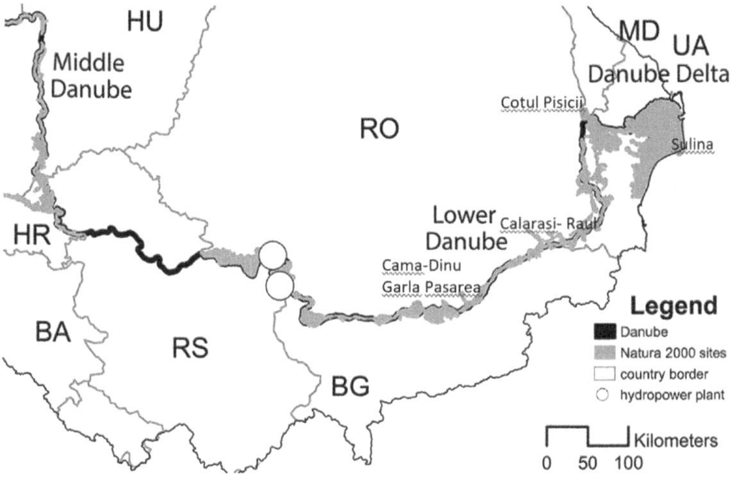

Fig. 4 Map of the study area including middle, and lower sections of the Danube; Natura 2000 sites; and hydropower plants that separate the two sections. Adapted/modified work from source [34]

of nature" also appears in the Danube floodplain context, and it is usually provided with different explanations for each geographical area with their distinct delimited natural units. The Lower Danube complex system, through its geographical position and evolution in historical time has been a space with diversity of landscapes that concentrates multiple categories of ecosystems. These can be classified per type, heterogeneity, spatial and temporal dynamics or stages of anthropization.

Space and habitat management, especially at the level of the Lower Danube Complex System, require the identification of methods and means of protection, conservation and social-economical management of ecosystems and landscapes [35]. The recent evolution of the landscapes under anthropogenic pressure (exploitation of natural resources, intensification of land use conversion to the detriment of natural ecosystems, profound transformation of grassland or aquatic ecosystems) leads to obvious decreases in productivity, but also to disruption of their functionality and productivity.

As is well known, the productivity and stability of ecosystems depend directly on the viability of their functioning, on providing physical support for the use of natural resources and on providing services to the system [36]. Ecosystems analysis uses dynamic, nonlinear systems to depict them as units of production undergoing long-term processes with variability and diversity. This modeling is essential for the proper planning of future stability and productivity of the System.

3.2 *Heterogeneity, Nonlinearity and Contingency of the Complex Space System of the Lower Danube*

Transversally, according to its characteristics and especially according to their local and regional differentiation, the Lower Danube is divided, according to [37], into three large sectors: the Iron Gates gorge, the terrace valley and the maritime Danube valley, each with specific characteristics. A different classification found in [7] was performed on the longitudinal axis, and sees the area divided in 3 large units:

- the minor riverbed, floodplain—with three characteristic strips:

    the high levee strip (shore levee),
    the transition strip (as part of the floodplain with average height) between the longitudinal levee (along the river) and
    the lower part of the floodplain, which is the third strip,

- the depressions, lake basins and terraces (often in this part are formed, right at the base of the terraces),
- the channels that cross the floodplain longitudinally for 50 km or more (e.g. Pasărea, Comasca or Somova Channels).

Usually, the lowest part of the floodplain is at their edge, just below the slope of the terrace or of the field. In some places, however, between the strip of depressions and the steep that delimits the floodplain, a higher strip (fourth) appears. This has the shape of a ridge with a very slight slope and a very smooth surface, which was formed due not to an increased intake of alluvium but to sedimentation processes. They start at the base of the terrace in the form of small organic overgrowth, small landslides and diluvia-colluvial accumulations, sorted and leveled by the lacustrine action during the floods. The dimensions of these strips vary from a few tens of meters to a few hundred meters. Their development is dependent on the local topography, such as peculiarities of the steepness that delimits the floodplain (height, fragmentation, slope processes), lake shoreline abrasion characteristics, sorting and leveling of crumbled materials and those brought in another way.

The Lower Danube floodplain appears uniform as it approaches its outflow, due to the continuous decrease of the relative altitudes from upstream to downstream and in relation to the amplitude of the floods. In addition, the continuous decrease of the levees' altitude is reflected by the gradual decrease of the level differences between the positive and negative forms within the floodplain. While the highest unevenness registered in the floodplain sector between Drobeta Turnu-Severin and the Jiu River mouth reaches 8–12 m, between Turnu-Măgurele and Giurgiu they are reduced to 5–6 m, and within Balta Borcei and Brăilei they do not exceed 2–3 m. These values are provided as results of the research project REELD [38]. If in the positive forms there is an inverse relationship between altitude and surface extension, in the negative forms (depressions, ponds), although it is less obvious, we can speak of a direct relationship. This is because in parallel with their surface increase, there is also a slight increase in depth.

The overall characteristics of the Lower Danube floodplain come from the general characteristic of the alluvial bed construction process (by direct alluvium and clogging). The different aspects and dimensions of the main forms (strips) are due to the variation of the general alluvial process intensity and clogging, as a direct reflection of the intensity of the overflow.

3.3 Regional and Local Characteristics of the Lower Danube Floodplain

In the general aspect of the different parts of the Lower Danube complex system, certain regional and local influences are noticed, which give specific characteristics to it, particularizing it as territorial individuality.

The floodplain and terraces of the Lower Danube River have been and are subject to continuous changes due to the destructive and constructive action of the river and the contribution of the main tributaries, as well as slope processes, suffocation, subsidence and wind processes and human influence too.

The complexity of the geological substrate, its mobility and the particularities of the geological evolution of the region, explain the very complicated course of the Danube with numerous changes of direction, large loops and meanders. The process of general uplift of the Holocene riverbed, through a massive accumulation of alluvium create favorable conditions for the appearance of meanders and riverbed braiding.

Transport, erosion and accumulation take place simultaneously [39], but on the entire Danube riverbed, the processes of transport and accumulation are dominant, through the continuous tendency of formation and uplift of the alluvial bed, due to the rise of the Black Sea level in Holocene. The intensity of river processes varies greatly during a year, depending on the hydrological regime, from the lowest levels, when the processes take place only within the river channel, to the maximum levels, when the processes extend over a larger area, including the floodplain.

Erosion is especially pronounced in the concavity of meanders, at upstream and in one of the sides of most of the islets, but it is observed in the sectors of rectilinear course. Erosion acts through swirls that cause shoreline changes, followed by bank landslides.

The usual forms of accumulation in the river channel are the islets, and the accumulations within the secondary arms are very frequent, with a tendency to anastomose and clog them, processes accentuated with the reduction of the slope.

The continuous tendency of raising the alluvium bed implies the development of general processes of alluvium and clogging, first of all of the low parts of the floodplain. The alluvium in the floodplain has an intermittent seasonal character, because the possibility of its manifestation appears only when reaching a certain level of the Danube waters, when between the river and the lake depressions inside the floodplain, a direct connection is established. Alluvium occurs either through channels, or small channels (at moderate elevations), or directly by crossing the shore levee at very high and exceptional water levels. The direct alluvium, over the shore levee, at the highest water levels is made in the surface, due to the reduction of the water speed when passing from the minor riverbed to the major one, in which there are many obstacles. This process directly results in the widening of the levees and the development of the transition floodplain strip to the detriment of the depression surfaces.

Another form of alluvium appears because of ruptures in the banks with accidental and short-term occurrences only at exceptional increases (in winter when Danube river frozen completely), but causes rapid local changes in the morphology of the floodplain. Stopping the ice causes a rapid rise in level and a sudden outflow of water to the depressions in the floodplain. Channeling primarily on existing channels and canals, with unusual speed, the waters break the levee in these places to open an increasingly wide path for drainage. Depending on the intensity and duration of the ice, as well as the structure of the levee shore, huge amounts of alluvium are introduced that cover with variable thicknesses the extended surfaces, sometimes of several kilometers. Although the frequency of this phenomenon is low, the contribution to the evolution of the floodplain is considerable due to the rapid change of the ratio between the high and the depressions within the floodplain. A single such alluvium is sufficient for clogging and disappearance of large areas of depression [7].

The action of the wind is manifested by the scattering of sands left to overflow by the Danube River and some of its tributaries, which it deposits in the form of dunes. The deflation process produces continuous changes of the dunes until they begin to set by covering with vegetation and solification. There are frequent cases of reactivation of the dunes by destroying the vegetal carpet and upsetting the thin horizon of incipient solification, caused by grazing, animal circulation, cultivation, deforestation.

3.4 Danube Floodplain Biome

Grafted on a relatively uniform hypsometry (except for the depth of the Danube arms), the Danube floodplain has a great diversity and abundance of plant and animal species, chained in a complex of aquatic and terrestrial biocenoses. Perhaps that is why the tendency to define the Danube floodplain as a biome has accentuated, thus emphasizing the biocenotic component of this territory. The definition of the biome in the Dictionary of the Romanian Language [40], considers it as an "ecological complex that is formed in relation to a certain environment". However, starting from the definition of the ecosystem, first made by Tansley in 1935 [41], as a mixed system consisting of a biotope occupied by a certain biocenoses. Odum, already in 1953 [42], furthers this concept and considers the biome an association of ecosystems with a complex functional aspect, corresponding to macroclimatic areas (e.g. tundra, steppe, savannah, tropical forests, etc.). In the same sense, starting with Odum's conception, Pierre quoted by [43], considers the biome as a superior biogeographic unit, on a zonal scale, forming a relatively homogeneous mass of plants and animals in accordance with the climate (e.g. tropical forests, tundra, etc.). In 1970, Whittaker quoted by [44] used the notion of a biome or major ecosystem for aquatic habitats (coastal, abyssal, etc.). In hindsight to these definitions, it can be stated that the floodplain of the Danube River is a typical example of a biome, a complex of aquatic ecosystems, both wetland and terrestrial.

4 Major Works and Impacts in the Lower Danube Complex System

The Danube River, together with its floodplain, is a very complex system of ecosystems, which provides habitat for a particularly rich flora and fauna and support for socio-economic activities. The map presented in Fig. 5 classifies the engineering works performed in the area during the last 150 years.

As Cristescu shows [45], the modification of a component of the system, exercised naturally or artificially, spontaneously triggers, according to the law of connection of objects in nature, the whole mechanism of modification of the other components. This is especially true in the context of changes in the Lower Danube and its floodplain.

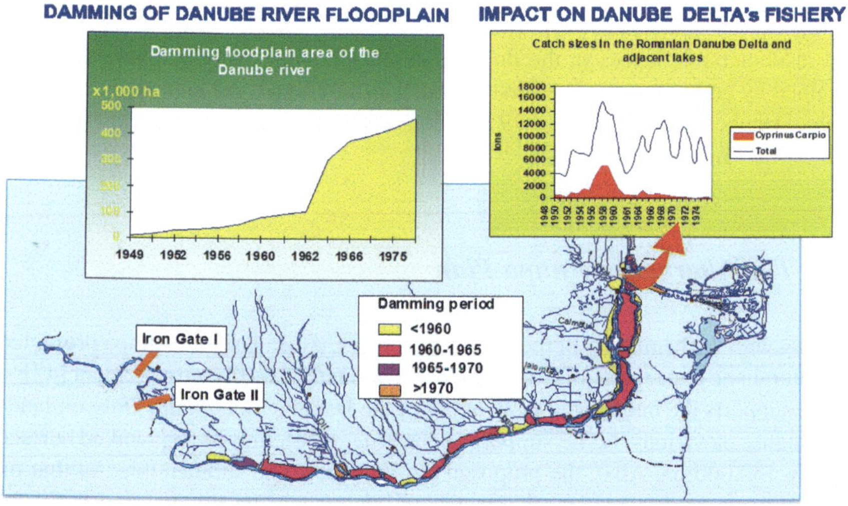

Fig. 5 Anthropogenic intervention (damming) in the Danube floodplain and impacts on the Danube Delta's fishery [48]

4.1 First Engineering Works

The first embankment works within the floodplain of the Danube were carried out during 1895 in the area of Mahmudia, on the Saint George branch, part of the Danube Delta, by Eng. Langeveldt (from the Netherlands) and Eng. Dithmer (from Denmark) [37]. This works failed because of the difficulties met in the desiccation of lands, reed clearing, and in the agricultural cropping. The work was abandoned after 250.000 Romanian lei (47.672 US $ in 1904 or 1,4 million US $ in 2021) have been spent. Historical records [46], mentioned by Saligni in 1904, show that the engineer Dithmer achieved success in embanking 1,700 ha in the Chirnogi Unit. During the next years (1905–1906) the works extended at the Mânăstirea Unit on 334 ha and at the Luciu-Giurgeni on an area of 3150 ha.

Large yields obtained on the arranged lands, lead to the issuance of a law in 1906, which authorized the Ministry of Agriculture and Lands to embank its domains in the floodplain of Danube. Because of the application of this law, between 1906 and 1908 an area of 1708 ha in the unit Spanțov (near Oltenița) was embanked. In 1910, 7100 ha were embanked, which, generally, were not dimensioned enough. Due to this fact, the inner desiccation was only achieved in the Unit Spanțov.

The exceptional yields obtained on this lands, facilitated the promulgation the "Law of improvement of the area liable to floods of the Danube" that would stipulate the almost complete embankment of the Danube floodplain with insubmersible and continuous dykes. The same law included the creation of the Special Service for Land Improvements, coordinated by the Eng. Anghel Saligny within the Ministry of Agriculture and Lands [46]. The objectives of this Service were to design safe and insubmersible levees in the floodplains, and to intensively cultivate the lands obtained by embankment. The First World War interrupted the embankment works in the Danube Floodplain, until 1925 when work was resumed at the Spanțov and Mânăstirea Units.

4.2 The Alternative Antipa Plan

Starting with the beginning of the twentieth century, Acad. Grigore Antipa promoted the naturalistic conception of extensive, predominantly fish-like arrangement [47]—which supports the interrupted embanking of the Danube Floodplain. Only on lands with higher elevations did he support embanking, restricting the embanked surface to only 130,000 ha. Also, the proposed levees were to be submersible, so that on average once every 10 years, the protected lands could be flooded and their use designed for complex fishing and forestry. There are certain arguments in favor of this thesis, which are still actual today:

- the inability of the unsinkable levees to ensure a certain defense of the lands and the inevitable catastrophes that would occur—exemplified by the ruptures of the levees observed over time;

- the alteration of the hydrological balance of the river which, without the floodplain, would raise its level to floods to such an extent that the ports, human settlements would be flooded, the dykes would be overtaken, and the river channel would lose its stability;
- the rapid decrease of soil fertility due to the cessation of the supply of fertilizers that the periodic floods ensured;
- current decreased fish production, due not only to the removal of permanent ponds, but also to the reducing of the flooded area, which is an area of natural filling of fish.

In opposition with the original technical concept of Saligny's department, which uses flood lands from the Danube floodplain for agriculture, the Board of State Fisheries, coordinated by the famous biologist Grigore Antipa [47], promoted the naturalist planning conception of extensive fisheries—which set partial embankments in the Danube floodplain. This was only to be implemented on lands with higher altitudes, using submersible proposed levees. The height of the terrain and the corresponding levees would ensure that every 10 years lands would be flooded. In addition, this would also disallow any other use outside of fisheries and forestry.

After the protests raised by the Board of State Fisheries, a new institution was set up, the Administration of Fisheries and Improvement of Floodplain Area (PARID). This institution also focused on land improvement. The new team included some amendments to the original plans, such as embankments in the Danube Delta, Balta Brăilei, Brateș Lake, and other low areas, that were left only to fishery uses. Within these areas, the method allowed Acad. Grigore Antipa to gain recognition for his ideas regarding the use of submersible levees [47]. One major concessions was given to Saligny and his department. This was the increase of the embanked area used for agricultural purposes to 200,000 ha. Also, the levees' levels were raised from 8–8.5 hydrograds to 8.5–9. Large shallow lakes and spawning areas were excepted from embankments.

In 1932, the scientist Grigore Antipa published the General Plan of improvement of the Lower Danube in the bulletin of the Division of Sciences of the Romanian Academy, where he exposed the conception of submersible embankment with dykes of maximum 8.5 hydrograds for the level of running waters on about 130,000 ha.

4.3 Hydrotechnical Interventions

During the last century, the floodplain of the Lower Danube was mostly embanked—almost 84%, namely 430,000 ha from a total of 513,900 ha [35]. Consequently, the ecosystems of this wetland were altered and mostly abolished. If the river dynamics can no longer create new ecosystems, the embanked sectors of the floodplain tend towards terrestrialization, understood as a phenomenon of changing the wetland ecosystem into a terrestrial ecosystem under anthropogenic impact.

In this sense, the following general aspects can be appreciated:

- the general characteristics of the floodplain relief are a direct result of the magnitude frequency and duration of floods and floodings;
- the almost complete embankment of the Danube floodplain with nonsubmersible levees has affected both the hydrogeomorphological system and the local and regional topoclimates, a phenomenon accentuated in the conditions of global climate change;
- as a result of the elimination of the floodplain, the nutrient retention capacity has been exceeded and since the 1970s, the Danube waters are affected by a strong eutrophication, which led to the reduction or loss of submerged macrophytes, change of specific algal spectrum and proliferation of competitive species, in conditions of excess nutrients (e.g. green–blue algae) [35]. These phenomena affect the trophic cycles, leading to a decrease in biological diversity through the disappearance of some species, many of which have high landscape and even economic value. Also, the hydromorphological processes with implications in the dimensioning of the river channel on which the drainage depends in good conditions, were affected, observing displacements of the Danube River currents;
- another action whose effects were not taken into account was the construction of dams and the formation of reservoirs necessary for power generation [34]. Their appearance led to the modification of the flood regime and to the decrease of the amount of alluvium transported by the Danube River, due to the decantation of the waters, having as effect major changes in the dynamics of the Romanian coast at the Black Sea. Another effect of the dams is the interruption of migration routes for the reproduction of sturgeon species with high economic value;
- the transformation of Danube floodplain ecosystems into terrestrial ecosystems has reduced their functions (ecological, economic, recreational, aesthetic and educational) to a single one—economic.

Today, approximately 430,000 ha of the Danube Floodplain area, in Romania (513,900 ha) are embanked (on a length of approx. 800 km) being arranged with drainage works and drainage premises (418,000 ha). In all cases, canals were constructed to intercept the runoff from the slopes, canals with pumping stations to evacuate the flood flows of the valleys with direct flow into the premises, as well as some ponds/lakes, such as Bistreț, Suhaia, Călărași, Bugeac, Oltina. The Danube, Vederoasa, Jijila, were sized for fish farming.

The effects of this action appeared much later [35], and were manifested by:

- eutrophication of the waters of the Danube Delta and partly of the northwest of the Black Sea, due to the elimination of the filtering effect of the floodplain and therefore the increase of nutrients, coming from intensive agriculture and untreated discharges of riparian cities, transported by Danube waters;
- the change of the specific fishing spectrum and the dramatic decrease of the fish populations with high economic value (especially of the sturgeons and even carp stocks) due to the disappearance of the areas with shallow water, propitious to the laying of caviar-eggs and the feeding of the young-fishes.

The natural areas were limited to only 83,900 ha distributed at the mouths of the tributaries of the Danube and within Insula Mică a Brăilei.

4.4 Major Works Prior to 1989. Construction of the Iron Gates (Porţile De Fier) I and Iron Gates II Dams

Prior to the change of regimes in Romania in 1989, which coincided with significant geopolitical change around the world, the major engineering works related to the Danube River followed communist ideology that impacted scientific perspectives and major decision making. The evolution of wildlife diversity in the lower Danube floodplain suffered extensively from two major technical interventions.

These hydrotechnical interventions resulted in a superior migration route of anadromous fish. The most affected were the sturgeon species, but also the Danube mackerel. Literature records [35], show that, in the past, these species migrated annually to the middle sector of the Danube, in the Pannonian Basin. Today the sturgeons cannot cross the Iron Gates I dam.

The construction of the Iron Gates I and Iron Gates II hydropower dams also had an effect on the taxonomic structures of the biocenoses upstream of the respective dams. The increase of the water depth, the transformation of the river sector into lakes, all favored the installation of stagnant species, which slowly eliminated the rheophilic forms. The most obvious changes occurred in the qualitative structure of plankton and ichthyofauna, which evolved into typical lake associations.

The temporal dimension of Lower Danube System embankment, emphasizes the correlation between economic/technological development and the intensity of the impact on the physical environment of terrestrial ecosystems. As it can be seen in Fig. 5, anthropogenic degradation of aquatic ecosystems is a ubiquitous reality with major implications.

The spatial dimension of the works performed in the Danube floodplain impacts the manipulation of the physical-geographical environment by affecting/inhibiting the natural processes—fluvial, wind and micro-climate or hydrological processes. Proper functioning of ecosystems leads to self-regulatory processes that maintain sustainable flows of matter and energy, which through anthropogenic changes in the floodplain cause degradation processes/morphological, chemical, hydrological or even biological changes, all creating pressure on the structure and functions of ecosystems.

The human impact on ecosystems is the central theme of numerous studies on the degree of anthropogenic degradation [35, 49]. Numerous evaluations and monitoring indicators have been developed to diagnose the state of ecosystems (naturalness index, productivity, land segregation, hierarchy of spatial distribution of ecological equipotential areas). Most of them indicate shifts in the land use as major contributor to changes in landscapes, ecosystems and the environment. Urban areas and related infrastructure are the fastest growing, with productive agricultural land coming in second. The recent evolution of the landscapes of the Lower Danube Floodplain under anthropogenic pressure (intensive exploitation of natural resources, intensification of built land—agricultural, fisheries to the detriment of natural ecosystems, profound transformation of riparian ecosystems) leads to obvious decrease in productivity, but and to disrupt their functionality. Space and habitat management, especially at the level of the Lower Danube floodplain requires the identification of ways and means of restoration, conservation, protection and social management of ecosystems and landscapes. Changes in land use have qualitatively changed the environment of some initially equipotential units, and topographically artificialized surfaces have influenced river hydraulics and increased risk of floods. The transformation of the Danube floodplain ecosystems into terrestrial ecosystems has reduced their functions (ecological, economic, recreational, aesthetic and educational) to a single one—economic.

5 Lessons and Perspectives Related to Connectivity in the Complex System of the Lower Danube

The projects developed in the complex system of the Lower Danube worked around the inter-governmental agreement of the Green Corridor of the Lower Danube established in 2000 by the governments of Bulgaria, Moldova, Romania, and Ukraine. This agreement was based on the theory of connectivity and aimed at creating a network of interconnected protected areas that could maintain and conserve biodiversity in the spatial context. This framework facilitated the projects in this paragraph.

The project results presented in the next sections were obtained by the implementation of projects within the Danube Delta National Institute for Research and Development.

5.1 The Green Corridor of the Lower Danube—Romanian Sector

A first project, developed in 2000—Elaboration of the Documentation for the "Green Corridor of the Lower Danube - Romanian Sector"—noted the importance of the role of the aquatic environment, and focused on the study of watercourse ecosystems. The objectives were to restore and conserve the biological diversity specific to the Lower Danube floodplain area. The project also promoted benefits for the population in the area by increasing its occupancy and developing traditional activities such as fishing and processing of fish products, reeds, wicker, wood, etc., increasing the tourist interest of the area and developing the necessary infrastructure. Figure 6 presents the proposed actions of the project in the Romanian sector.

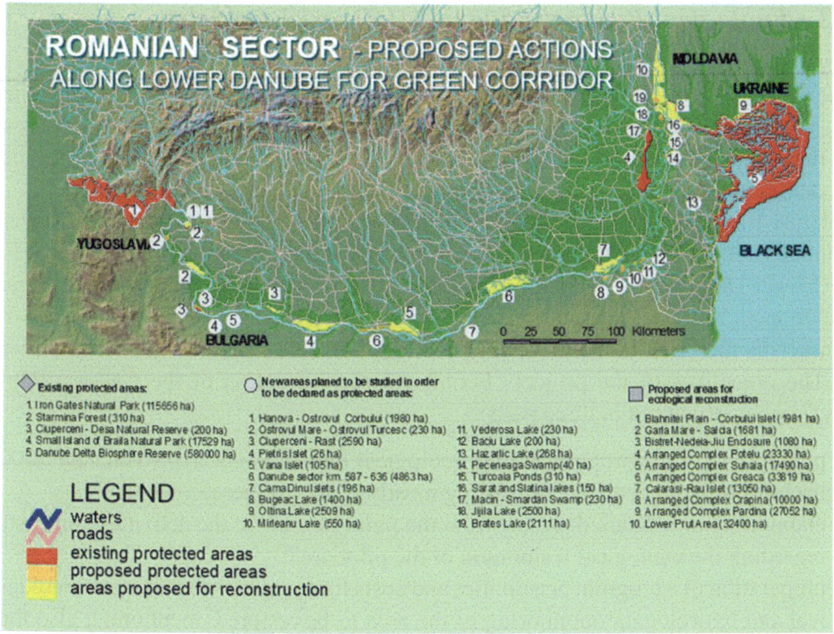

Fig. 6 Green Corridor of The Lower Danube—Romanian Sector [50]

The general objective of the above-mentioned project was to develop a regional ecological network in the Danube floodplain, integrated into the National Network of protected areas, by applying the concept of "coherence of spatial structures" throughout Danube Floodplain. Strictly protected areas (core areas), ecological corridors, buffer zones and ecological reconstruction areas were considered essential for the Green Corridor. The design of this complete ecological network included each of these components, on different levels of protection, adapted to local needs in order to implement integrated management plans.

A successful rehabilitation program in these regions should not be limited to technical, legal or economic measures, without conducting social/cultural anthropology studies, as it brings changes in fundamental cultural values and conditions, changes that protect the integrity of the ecosystem and the quality of human life.

The concept of harmonizing the requirements of environmental protection with the objectives of social development presupposes human awareness of the ecosystemic relations between biotic and abiotic resources. This harmonization requires the involvement of users/beneficiaries in the management of their own ecosystem—this cannot be achieved without knowing the historical conditions and social relations.

5.2 Ecological Reconstruction of Călăraşi-Răul Islet

The project "Ecological Reconstruction of Călărași-Răul Islet" was developed within the World Bank International Program "Pollution Control in Agriculture". It focused on the initial ecological status and documentation for the development of a hydraulic model for ecological reconstruction of the pilot area of ecological reconstruction—Călărași-Răul Islet (Fig. 7). The area had been previously evaluated in the Lower Danube Green Corridor Documentation, as an area with potential for rehabilitation in the Lower Danube floodplain—the Romanian sector, precisely due to its ecological importance, nutrient retention and recycling capacity, and the role of the islet in conserving the biodiversity.

The objectives of the project "Ecological reconstruction of the Călăraşi-Răul Islet", were structured in three parts:

- preparation of the proposal for the ecological restoration of approximately 3000 hectares of degraded land, consisting mostly of the former rice field;
- elaboration of a detailed project from the point of view of the activities and costs regarding the ecological restoration of the pilot area;
- preparation of a program of activities and costs for four years regarding the biological and hydrological monitoring of the area to be restored, mentioning also the essential indicators for monitoring and evaluation.

Unfortunately, this pilot project was only implemented in a part of the Călărași-Răul Islet—Reis Land Field of 3000 ha, which also made necessary the redesign of

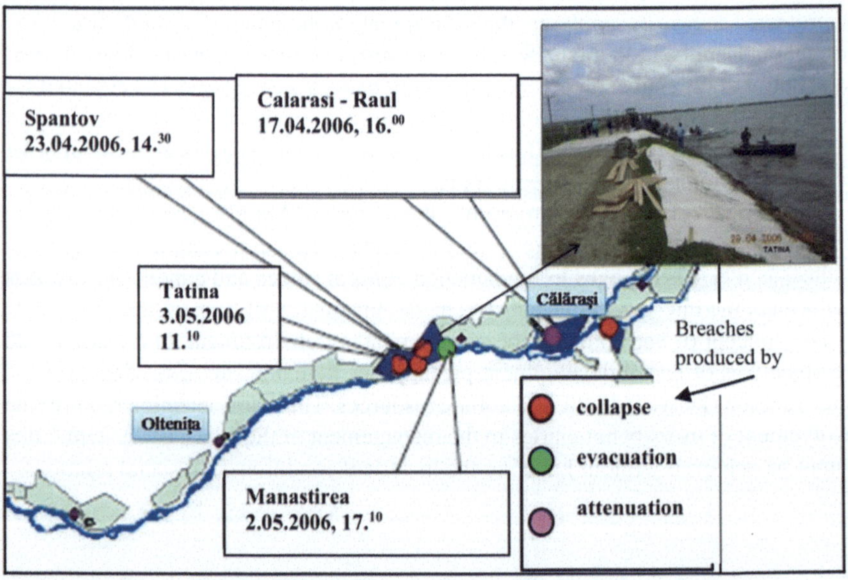

Fig. 7 Călărași-Răul Islet location [51]

the levees for the defense of the Agricultural Development in the entire islet. This led to raising costs, found to be unacceptable by the World Bank and the cancellation of future implementations. It is a conclusive example in which local connectivity is dependent on trans-systemic connections. The correct assessment of the impact of hydrotechnical works performed on watercourses requires a global and long-term ecosystemic approach, which involves the correlation of physical and geographical changes with their ecological consequences.

5.3 Protection of Wetlands of the Danube river—A Pilot Project for Cama Dinu Islets Area

The pilot project for Cama Dinu islets area (see Fig. 6), financed by the Phare CBC program, focused on ecological protection and ecological restoration options for the Romanian-Bulgarian section of the Lower Danube floodplain and the implementation of the EU Directives: Water and Birds and Habitats Framework Directives in the two countries. It used the Cama Dinu islets area as a pilot study.

The few areas of the Lower Danube floodplain that have no hydrotechnical works on them are considered areas with high biodiversity, resulting precisely from the interaction between the river and its floodplain. This includes the islets, which denotes the characteristics of connectivity. Also, the complexity of the relationship between different environmental parameters creates conditions for high biodiversity. The governing factor for the floodplain is to balance these parameters as the ecosystems are subject to permanent and cyclical changes in water levels, ranging from floods to extreme droughts.

This project laid the foundations for the implementation of the Natura 2000 Directive in the Complex System of the Lower Danube, but also created the premises for the study of connectivity between tributaries and river.

5.4 Restoration of the Water Course from the Danube Floodplain—Gârla Pasărea (Bird Channel)

The experience in the Cama Dinu islets was useful in developing the Feasibility Study entitled "Restoration of the water course from the Danube Floodplain—Gârla Pasărea (Bird Channel)—by reconnecting it between the Vedea River and the Danube" [19], that was elaborated in 2005 by a team of experts from Danube Delta National Institute for Research and Development Tulcea, Romania. The process of reconnecting the Vedea River—Pasărea River to the Danube, by ensuring historical lateral connectivity, raises the issue of protection of localities against floods for maximum water levels in the Danube, as well as the restoration of the hydraulic scheme of the entire unit. This solution can be considered only if there is a proposal to rehabilitate the entire Unit, located south of Gârla Pasarea (Pietroşani-Arsache Agricultural Precinct, in fact the former Arsache Meadow in the Danube floodplain) and also if the dykes that protect the localities will be resized (those localities in the north of Gârla Pasărea). This channel of over 40 km ensures a drain with a slope of 3 cm per km.

Hydraulic modeling and the Digital Terrain Model are the essential components in connectivity analysis. Based on these models, reconnection scenarios can be created—in the case of Gârla Pasărea, 5 scenarios were created with various flow rates at reconnection.

Reconnection to the Danube of the Gârla Pasărea (with or without water storage by side discharges) raises special problems both related to flood protection of localities for maximum values of the Danube level, and due to hydrotechnical schemes—Pumping stations, irrigation canals, dykes for compartmentalization of agricultural precincts, etc. An uninterrupted scenario can be considered only if this variant proposes a restoration of the entire unit and especially of the area located south of Gârla Pasărea (old Danube floodplain) and a resizing of the protection dykes of the localities located north of this one.

5.5 *More Room for the River in the Cat's Bend (Cotul Pisicii), Romania*

The project "More room for the river in the Cat's Bend (Cotul Pisicii), Romania" [18, 20], was developed within the Water Partnership Program of the Dutch Government in the context of the European Directives—Water Framework and Flood Risk Management. It provided supporting adaptive ecosystem management in floodable area management in the Lower Danube Valley. It sought to meet environmental issues in the formulation and implementation of these directives. The project developed integrated spatial planning sketches for the region of Cat's Bend (Fig. 8) in an interactive and participatory process with national, regional and local stakeholders, using the "Sketch Match" method developed by the Dutch Government Service for Land and Water Management (DLG).

Because the Danube River was "channelized" and closed laterally, through dykes/levees, there is no space left for polders that can help reduce the peak flow during periods of melting snow or torrential rains and for the functioning and sustainable services of ecosystems along the river. The Cat's Bend sees the Danube meet with two major tributaries, the Prut and the Siret rivers. Due to climate change and widespread deforestation and clearing of wetland vegetation, these peak flows occur more frequently, and also carry a larger volume of water in a shorter time. This happened during the 2005, 2006 and especially 2010 floods (Fig. 9), when much of the Hârşova region, located upstream of the bend, was affected by floods and even required evacuations.

The spatial planning solutions developed within the project are not limited only to this region of the project, but also target potential upstream and downstream

Fig. 8 Cat's Bend area targeted by the project—cartographic support Topo-military map 1985

effects at the qualitative level. Recurrent floods in recent years impose a new and different flood management policy. In the short term, anti-flood measures must be taken, paying attention to different land use functions: for example, flood protection near residential areas, nature, agriculture, maritime transport and other economic activities, such as tourism.

Traditional measures, such as the raising and strengthening of dykes relatively expensive, offer little chance of being effective. Water bodies need to be managed differently, in a way that gives back more space to watercourses and nature, through adaptive and ecosystem management to fight floods in the Danube Delta region. It must also pay sufficient attention to environmental and socio-economic issues and the mutual correlation between these functions and interests.

The initial objective of the project was to develop scenarios for flood protection in the Cat's Bend area. During the project, however, a multi-disciplinary approach was preferred, including all aspects such as regional socio-economic development and the improvement of the micro-climate (given the drought and high temperatures manifested during the summer).

An important result of this project is precisely the relevance of these components of connectivity for the identification of water management solutions, being perhaps even more relevant, compared to the initial objective, only for flood protection. This conclusion was the result of applying the interactive method of "Sketch Match", the representatives of various stakeholders analyzing together the current problems and developing possible solutions.

a,b – Galaţi - Port, July 2010

a, b –Smârdan locality, July 2010

Fig. 9 Impact of floods in 2010 in the Cat Bend region

The "Sketch Match" method is a working method developed by DLG (Rural Area Department of the Ministry of Agriculture, Netherlands) which works on the principle of working in a creative environment but under the pressure of conflicts of interest. In an interval from one day (minimum) to maximum three days, a series of participants who showed interest during the interview campaign, conducted during the preparation of the Sketch Match session (for example: citizens, policy makers, farmers) meet to analyze, define and find common solutions in the spatial planning process. The strong point of this method is group work.

The Sketch Match session took place in June 2009, with 45 persons which represent different regional and national stakeholders, to develop solutions to water management issues and to stimulate regional economic development, culminating in the integration of concepts into an integrated outline that was presented to the Ministry of Environment.

Thus, 3 solutions were integrated:

- reactivation of the former Channel (Gârla) Ciulineţ (with number 1 in the next figure) by reconnecting it on the existing segments—channel which existed on the edge of the Danube terrace and took over a flow from the river, during the floods. This concept is based on the principle of connecting a network of old arms and former floodplain territories (exterior delta) of the Danube, through a new connection channel with west–east orientation along the Macin Mountains. This new waterway may even encourage economic development and the microclimate in the villages at the base of the Macin Mountains, connecting them with the river and reactivating former lakes—Jijila and Crapina. In case of large volumes of water on the Danube, it will serve as a spillway contributing to lower water levels.

- drainage-irrigation system—this idea is based on the current irrigation system and drainage channels; aims to improve this system and connect it with the old network of dead channels (japșe) and arms. The infrastructure and plantings will be made following a parallel line with these canals and contribute to an improvement of the ecological and visual quality of the landscape. This concept particularly strengthens the potential for agricultural production and improves the microclimate.
- building a discharge canal (with number 3 in the next figure), south of Grindu. These measures contribute to flood protection, lowering the Danube water level. It is also possible to combine these measures with the system of drainage and irrigation canals (2) or even the Ciulinețul Channel (1), seen in Fig. 10.

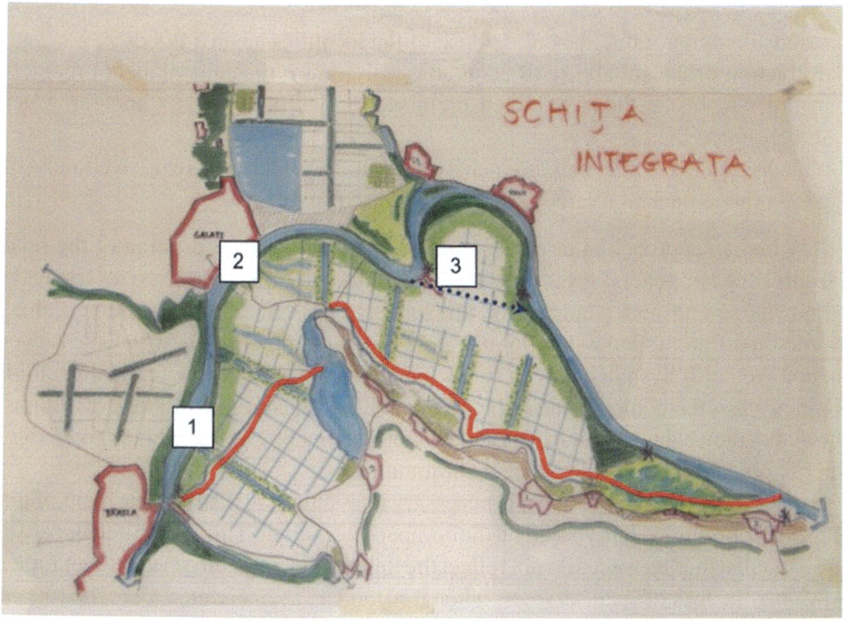

Fig. 10 Integrated solutions proposed in the Cat's bend

5.6 *The Ecological and Economic Resizing Program in the Romanian Danube Floodplain*

After the floods that occurred in the Danube Floodplain in 2006, the Ecological and Economic Resizing Program in the Romanian Sector of the Danube Floodplain was approved by Governmental Decision (GD) no. 1208/6 sept.2006. It represents the strategy for sustainable development and the reconsideration of the lines of defense against floods of the localities from the Danube floodplain. This strategy is based on the evaluation of the various flood suitability scenarios and on the public option. It created the synergy between the Lower Danube Green Corridor and flood risk management. The ecological and economic resizing program of the floodplain and Danube Delta precincts was designed and launched to assist the Romanian Government in the long-term strategic planning process. This was to achieve the objectives of the Water Framework Directive, as well as in the effective implementation of prevention, protection and mitigation tasks of flood effects, stipulated by the National Flood Risk Management Strategy.

The program, established as a decision tool, is structured on three levels—identification, evaluation and suitability—as follows:

- reconsideration of the lines of defense of the localities against floods;
- evaluation of the suitability of economic activities in the embankment precincts in order to resize them as mixed premises (agricultural/polders and for water storage);
- renaturation of some embankment precincts in order to create wetlands of conservative interest (Natura 2000—SCI and SPA areas).

Any human approach in terms of practice is a form of manifestation of the relationship between general group and individual interests that characterizes the level of knowledge and understanding of the surrounding reality, the system of values assumed at different spatial–temporal scales. The transformations to which the Lower Danube Wetlands System has been subjected, given the spatial scale of manifestation are relevant to the context stated above, marked by the absence of a real participatory democracy, which capitalizes on decision-making power, scientific knowledge, experiences and information of local communities.

From a scientific perspective it is necessary to assume a realistic assessment of the limits of economic systems based on the competitive market. This is required because, pure capitalist market forces do not reflect the impact of ecosystem services, but rather show only the loss of such services, when it is too late to intervene. Also, the role of natural capital as a vital factor of production is missing in market calculations. This indicates to the fact that private profit as a performance indicator must be abandoned, in order to make way for a multicriteria approach, in a holistic, integrative context.

Economic and ecological criteria must be complemented by socio-political criteria. We emphasize, in this context, that through the multifunctional arrangement of the Lower Danube floodplain are partially or totally replaced advantages appropriate to the individual time scale (generated by agro-systems) with advantages appropriate to the social time scale (generated by wetlands); therefore, the socio-political criteria can significantly correct the result of the analysis of the economic-ecological efficiency.

The Ecological and Economic Resizing Program in the Romanian Danube Floodplain sector approved by GD 1208/2006 is not only intended for protection against floods but has several other purposes, namely:

- sustainable development and introduction of integrated adaptive management;
- preserving the traditional forms of economic activity that do not contribute to the production of ecological imbalances;
- informing and educating the population regarding the value of the Lower Danube Complex System landscape and the need to conserve and protect the plant and animal species, of the respective landscapes; conservation of ecosystems and balanced use of renewable natural resources.

The objectives of the Program would be achieved by applying the concept of connectivity, respectively coherence of spatial structures—physical-geographical units, at the level of the entire Danube floodplain taking into account the following principles:

- protection and restoration of all key ecosystems and all-important species that make up the landscape and biological diversity of the Lower Danube floodplain, Romanian sector;
- promoting the principle of sustainable development in the Lower Danube floodplain through an integrated and adaptive management applied by professionals in ecology and economic activities, like agriculture, forestry and fishing;
- conservation of biological diversity and ecological reconstruction of damaged systems in the Lower Danube floodplain;
- highlighting the heritage of cultural and historical values in the Lower Danube Valley through the development of ecological tourism;
- compliance with the provisions of international conventions and programs on environmental protection;
- consolidation of the logistics of public education for the conservation of nature and landscape values in the Lower Danube Valley;
- cooperation of all European Danube nations.

The hydrological scenarios made with the hydraulic model of the Danube River considered the functions of the areas in the floodplain (Fig. 11) and focused on quantifying the reduction of the Danube level to maximum levels in the following 3 cases:

- flooding of agricultural embankment precincts in natural regime (there are no levees, the floodplain of the Danube is being restored);
- water storage in agricultural embankment precincts (water stocks);

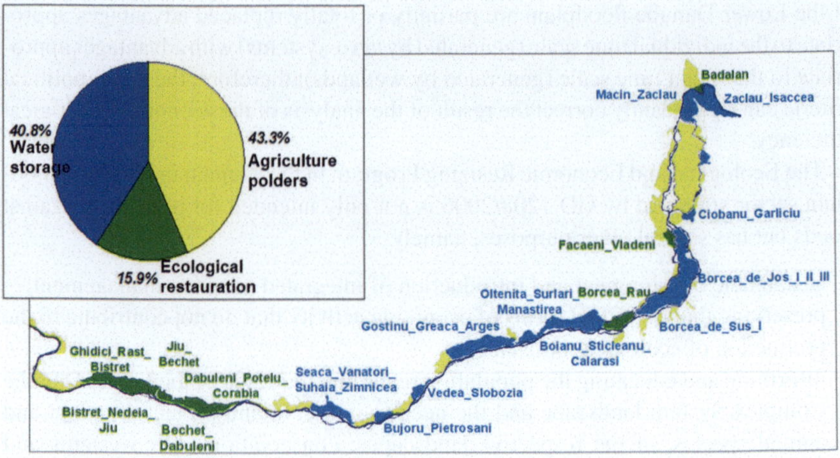

Fig. 11 Functions of agricultural/fishing enclosures considered for hydrological scenarios [38]

- mixed solution by storing water in some embankment precincts and flooding others in natural regime (renaturation).

The restoration of only 15.9% of the embankment surface of the Danube floodplain would mark a significant stage in the implementation strategy of the Program. Also, by promoting the concept of mixed enclosures on the 40.8%, polders with economic activity, agricultural/fishing and in at the same time, embankment precincts for water storage in case of floods, the implementation of the Flood Risk and Water Framework Directives in the Danube Floodplain are ensured.

5.7 *Danube Floodrisk Project—Stakeholder Oriented Flood Risk Assessment for the Danube Floodplain*

Another project aimed at extending the entire course of the Lower Danube River, the concept of synergy between the Green Corridor, Flood Risk Management and stakeholder involvement was the Danube Floodrisk project—Stakeholder oriented flood risk assessment for the Danube floodplain.

The overall objective of the Danube Floodrisk project was to develop and produce high-quality, stakeholder-oriented hazard and flood risk maps for the entire Danube River System to provide adequate information on spatial planning risks and economic demands. Risk information is the basis for sustainable development in regions along the Danube River. The key objective will only be achieved through intense transnational cooperation and stakeholder integration. The aim is to link scientific progress in harmonizing approaches and data with the practically targeted involvement of stakeholders and the end user. Vertical and horizontal cooperation are the two pillars of the project. The objectives of the project covered a number of aspects:

- develop a common method of flood risk mapping and harmonize data sources;
- production and provision of risk maps and risk information;
- integrate relevant stakeholders and users at different levels into the definition and implementation processes;
- involvement of various economic aspects in spatial planning such as land cover/use in the river basin, agriculture, tourism, energy supply or health services;
- relating flood risk mapping as a basis for spatial planning;
- development of model procedures for Flood Risk Management in the Danube countries and implementation of three pilot studies targeting the cities of Giurgiu, Cernavodă and Galați (see Fig. 12);
- feedback based on experiences of cooperation between partners on the implementation of EU directives, e.g. WFD, Flood Directive, using the platform of the ICPDR Expert Group on Flood Protection.

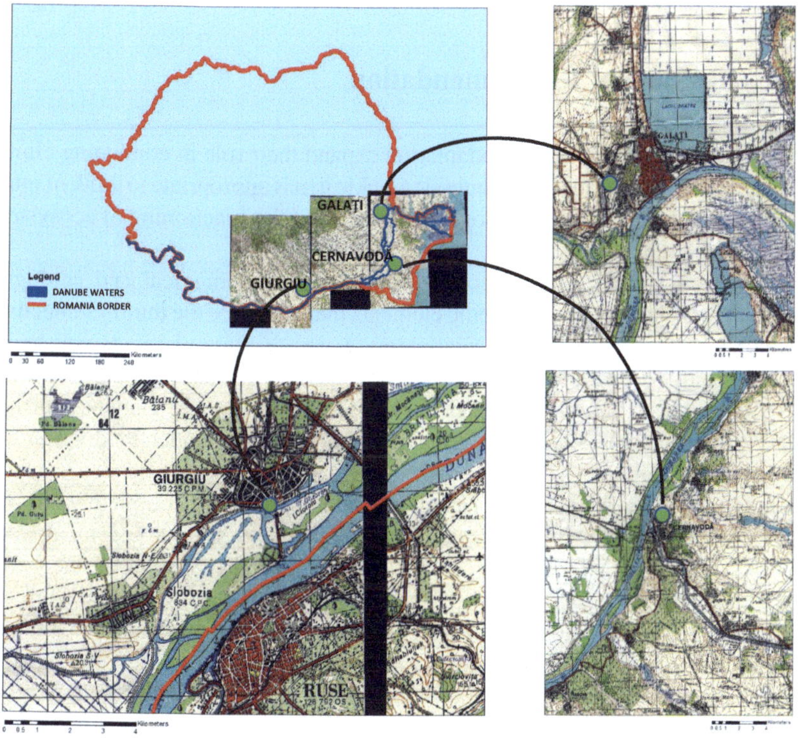

Fig. 12 Historical topographic maps used in the Danube Floodrisk project

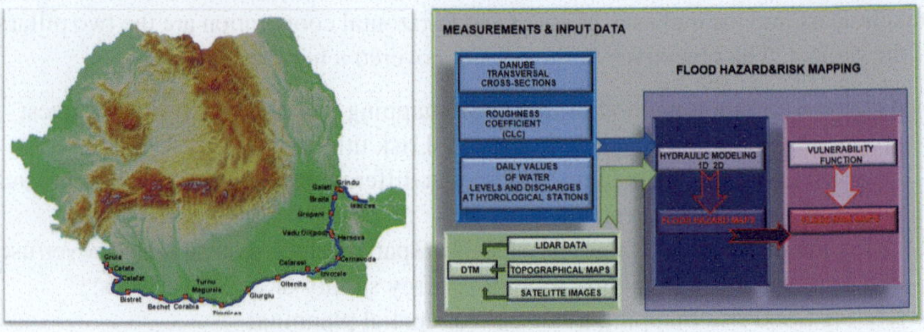

Fig. 13 Romanian Sector of the Danube between Gruia (km 851) and Isaccea (km 50)—left and the workflow structure and databases for the hydraulic model and hazard maps—right

In the project, the 1D_2D hydraulic model on the Romanian sector of the Danube between Gruia (km 851) and Isaccea (km 101) was performed. It required the spatial processing of several data layers in GIS format whose structure is presented in Fig. 13.

6 Conclusion and Recommendation

As environmental protection and adaptation expand their role in combating climate change and the energy transition required to tackle it, it is appropriate to think of multi-connectivity as a physical and biological way to meet the functioning of ecosystems and other current needs.

The Lower Danube forms a complex hydrogeomorphological and ecological system and presents a series of specific elements that influence the multi-connectivity and implicitly the functioning of ecosystems. Some of these elements are: hydrological cyclicity, determined by seasonal dependence, instability due to climatic and hydrological variations and land use conflict in the Lower Danube Corridor. However, these elements change regularly, and so does our understanding on how they are reflected in the system.

Due to human intervention, the Danube and its wetlands are completely different today from what they were 120 years ago, prior to the first hydrotechnical works performed. These transformations, to which the Complex System of the Lower Danube has been subjected, given the spatial scale of manifestation are relevant for the context presented above. The study area suffers from imbalances that human intervention has caused. In this situation, the multi-connectivity models the impact of proposed changed and helps with comprehensive planning. The proposed measures ensure a good circulation of energy and matter. This inter- and trans-connection can be a precise descriptor for all aspects related to environmental protection and quality.

The connectivity of the Danube system was affected at all levels: spatial, temporal, thermal/optical, ecological. The analysis performed in the presented studies aimed to identify future developments, trends, but also gaps in mentality, facing stakeholders in the Lower Danube area, their perception of the main coordinates of their activity in relation to ecosystems, precisely to form a basis for formulating proposals and solutions.

The combination of technical aspects aimed at creating optimal ecological conditions for sustainable development, and included economic aspects presenting the resources of natural capital with maximum economic efficiency that can be done within the Danube system, taking into account multiconnectivity.

The forecasting within the Danube system of the optimal ecological scenarios requires the substantiation of each component element of the entire system, approach at different scales—local and overall, based on rigorous calculations. This requires the use of modeling of multiconnectivity factors: spatial, temporal, thermal / optical and ecological, and the determination of scenarios, based on criteria related to multiconnectivity.

For a coherent interpretation and understanding, in terms of the continuity of spatio-temporal and hydro-climatic dynamics and complex of interactions, biological and anthropogenic transformations need to be approached in a transdisciplinary integration framework. This allows the assessment of trends and the identification of mitigators in the improvement of intra-and inter-system connectivity. This multidimensional audit consists of the following axis of connectivity: lateral, longitudinal, vertical, temporal thermal and socio-ecological.

The main recommendations in the use of multiconectivity as study approach in the Danube System analysis can be summarized as:

- two structural concepts must be taken into account: multi-connectivity, as a form of supporting continuity and resilience as a form of system development;
- the general parameters of morpho-hydrographic and climatic dynamics register significant changes and will have major effects both at global level and within the complex system of the Lower Danube;
- paradigm shifts are required in order to meet requirements such as circular economy, increased quality of life, physical and biological reintegration into nature;
- the intensity of interruptions in the systems connections differs between units but also within them;. their effect is global, as barriers, such as hydrotechnical dams (Iron Gates I and II) have effects both upstream and downstream;
- using participatory democracy in decision-making processes that capitalize on scientific knowledge, experiences and information of local communities;
- the main directions of action and change will create adaptative sustainable and inclusive management by using new technologies such as Artificial Neural Network modeling.

The presented research projects and engineering works provide us with more scientific and factual evidence on the limited capacity of natural capital to provide goods and services (energy and raw material crises, global warming and pollution). What is still required is political critical mass in decision-making. This can be triggered, as the past has shown, by further catastrophic floods in the plains or by external factors, such as climate change and international obligations that the Romanian government will take on due to EU and/or UN agreements.

Acknowledgements The authors address special thanks to Danube Delta National Institute and its researchers, for the opportunity to work and for scientific and technical support in Danube-related projects. The information presented in this chapter is the result of the teams involved in projects and was selected from the deliverables, reports, and results of the projects.

Author Contributions All authors contributed equally to this work.

References

1. Newcome J, Provins A, Johns H, Ozdemiroglu E, Ghazoul J, Burgess D, Turner K (2005) The economic, social and ecological value of ecosystem services: a literature review. Department for Environment, Food and Rural Affairs, London
2. Watson C, Biedenharn D, Scott S (1999) Channel rehabilitation: processes, design, and implementation. Engineer Research and Development Center, Vikinsburg
3. Opdam P (1991) Metapopulation theory and habitat fragmentation: a review of holarctic breeding bird studies. Landsc Ecol 5:93–106. https://doi.org/10.1007/BF00124663
4. Wu JG, Loucks OL (1995) From balance-of-nature to hierarchical patch dynamics: a paradigm. Q Rev Biol 7(4):439–466
5. Wu JG (2000) Landscape ecology: pattern, process, Scale and Hierarchy. Higher Education Press, Beijing
6. Bélisle M (2005) Measuring landscape connectivity: the challenge of behavioral landscape ecology. Ecology 86:1988–1995. https://doi.org/10.1890/04-0923
7. Badea L (1969) Geografia văii Dunării româneşti. Editura Academiei României. Bucuresti
8. Covăsnianu A, Tudose O-G, Cazacu M-M, Nichersu I, Memier M, Balin I (2010) R.E.E.L.D. (Economical and Ecological Reconstruction of the Danube Flood Plain) Campaign: airborne LIDAR data and GIS technique outputs. In: EGU General Assembly Conference Abstracts, Vienna, 2010, p 3982
9. Sârbu A, Janauer G, Schmidt-Mumm U, Filzmoser P, Smarandache D, Pascale G (2011) Characterisation of the potamal Danube River and the Delta: connectivity determines indicative macrophyte assemblages. Hydrobiologia 671:75–93. https://doi.org/10.1007/s10750-011-0705-5
10. Nichersu I, Staras M, Nichersu I (2015) Lower Danube 4D Reconnection - Strategic Framework for LU/Climate Change Adaptation. David Publ J Environ Sci Eng B 4:434–441. https://doi.org/10.17265/2162-5263
11. Vogel B, Pall K (2002) Nine geo-morphological Danube reaches. International Commission of River Danube Protection, Vienna
12. Cristofor S, Vadineanu A, Sarbu A, Postolache C, Dobre R, Adamescu M (2003) Long-term changes of submerged macrophytes in the Lower Danube Wetland System. Hydrobiologia 506:625–634. https://doi.org/10.1023/B:HYDR.0000008601.16757.35
13. Carozza J-M, Micu C, Mihail F, Carozza L (2012) Landscape change and archaeological settlements in the lower Danube valley and delta from early Neolithic to Chalcolithic time: a review. Quat Int 261:21–31. https://doi.org/10.1016/j.quaint.2010.07.017
14. Andrei M-T, Vartolomei F, Nedea P-S, Gherasim C, Pop I (2012) The Romanian Danube Valley – geopolitical aspects. Geopolit Hist Int Relat :398–409
15. Nichersu I, Nichersu I, Mierla M (2016) Applications of fractal geometry in morpho-hydrographic network of the Danube Delta potamology. RevCAD J Geod Cadastre :123–130

16. Schumm SA, Lichty RW (1965) Time, space and causality in geomorphology. Am J Sci 263:110–119
17. FISRWG (1998) Stream corridor restoration: principles, processes, and practices. Federal Interagency Stream Restoration Working Group (FISRWG)
18. Tetelea C (2017) Lower Danube river corridor - FLOODPLAIN RESTORATION OPPORTUNITY ANALYSIS. Invisible nature, WWF, Bucharest
19. Schneider E (2002) The ecological functions of the Danubian floodplains and their restoration with special regard to the Lower Danube. Large Rivers 141:129–149
20. Nichersu I, Marin E, Florentina S, Trifanov C (2010) Room for the river in Cat's Bend: application of the SketchMatch method in Romania. Dresden, Germany
21. Tilleard JW (2001) River channel adjustment to hydrologic change. PhD Thesis, Department of Civil Engineering, University of Melbourne
22. FISRWG (2001) Stream corridor restoration: principles, processes, and practices. Federal Interagency Stream Restoration Working Group (FISRWG)
23. Driga B-V, Gastescu P (2004) Delta Dunarii - sistemul circulatiei apei. Casa Cartii de Stiinta, Cluj-Napoca
24. Drack M, Apfalter W, Pouvreau D (2007) On the Making of a System Theory of life: Paul A Weiss and Ludwig von Bertalanffy's conceptual connection. Q Rev Biol 82:349–373. https://doi.org/10.1086/522810
25. Perks M (2013) Spatial and temporal dynamics of fine fluvial sediment transfer: implications for monitoring and management of upland river systems. PhD Thesis, Durham University
26. Vannote RL, Minshall GW, Cummins KW, Sedell JR, Cushing CE (1980) The river continuum concept. Can J Fish Aquat Sci 37:130–137. https://doi.org/10.1139/f80-017
27. Southwood TRE (1977) Habitat, the templet for ecological strategies? J Anim Ecol 46:337–365
28. Leopold LB, Langbein WB (1962) The concept of entropy in landscape evolution. Geological Survey Professional Paper 500-A, Washington D.C.
29. Minshall GW (1988) Stream ecosystem theory: a global perspective. J North Am Benthol Soc 7:263–288
30. Kennedy BA (1997) Schumm, S.A. and Lichty, R.W. 1965: time, space and causality in geomorphology. American Journal of Science 263, 110–19. Prog Phys Geogr Earth Environ 21:419–423. https://doi.org/10.1177/030913339702100305
31. Benke AC, Wallace JB, Harrison JW, Koebel JW (2001) Food web quantification using secondary production analysis: predaceous invertebrates of the snag habitat in a subtropical river. Freshw Biol 46:329–346. https://doi.org/10.1046/j.1365-2427.2001.00680.x
32. Tobler W (2005) A computer movie simulating urban growth in the Detroit region. Econ Geogr 46:234–240
33. Thorp JH, Thoms MC, Delong MD (2006) The riverine ecosystem synthesis: biocomplexity in river networks across space and time. River Res Appl 22:123–147. https://doi.org/10.1002/rra.901
34. Hein T, Schwarz U, Habersack H, Nichersu I, Preiner S, Willby N, Weigelhofer G (2015) Current status and restoration options for floodplains along the Danube River. Sci Total Environ 543:778–790. https://doi.org/10.1016/j.scitotenv.2015.09.073
35. Bănăduc D, Angela C-B, Pedrotti F, Cianfaglione K, Akeroyd JR (2020) Human impact on Danube Watershed biodiversity in the XXI century. Springer, Cham
36. Mace GM, Norris K, Fitter AH (2012) Biodiversity and ecosystem services: a multilayered relationship. Trends Ecol Evol 27:19–26. https://doi.org/10.1016/j.tree.2011.08.006
37. Petrescu N (1971) Dezvoltarea agriculturii în zonele îndiguite şi desecate. Academia de Ştiințe Agricole și Silvice, Bucuresti
38. Nichersu I, Constantinescu A, Trifanov C (2009) The use of models in flood risk management of the Lower Danube. Danube News. Bulletin of the International Association for Danube Research (IAD) / Informationsblatt der Internationalen Arbeitsgemeinschaft Donauforschung (IAD). Donau Aktuell 11(9):13–15
39. Wolman MG (1967) A cycle of sedimentation and erosion in urban River channels. Geogr Ann Ser Phys Geogr 49:385–395. https://doi.org/10.1080/04353676.1967.11879766

40. Sala M (2008) The dictionary of the Romanian language. Roman Philol 62:185–189
41. Bayne BL (2017) Chapter 10 - Oysters and the ecosystem. In: Bayne B (ed) Developments in aquaculture and fisheries science. Elsevier, pp 703–834
42. Odụm EP (2001) Ecosystem, concept of. In: Levin SA (ed) Encyclopedia of biodiversity, 2nd edn. Academic Press, Waltham, pp 59–63
43. Nicole J-M (1970) George, Pierre. Dictionnaire de la géographie, Paris, Presses universitaires de France, 1970, 448 pages. Cah Géographie Qué 14:287–288. https://doi.org/10.7202/020926ar
44. Mucina L (2019) Biome: evolution of a crucial ecological and biogeographical concept. New Phytol 222:97–114. https://doi.org/10.1111/nph.15609
45. Cristescu C (2015) Morfologia albiilor şi conceptul de restaurare a râurilor. Thesis chapter, Facultatea de Silvicultură și Exploatări Forestiere, Brașov
46. Nenciu Posner C, Armas I (2015) Historical perspective on structural methods for flood protection in Lower Danube. Georeview 25:1–13. https://doi.org/10.4316/GEOREVIEW.2015.25.1.171
47. Antipa G (1895) Studii asupra pescariilor din Romania. Imprimeria Statului, București
48. Constantinescu A, Nichersu I, Trifanov C, Nichersu I, Mierla M (2012) Morphometric analyze for flood hazard map using DTM built with LIDAR and Echo-sounder data in Danube Delta. In: EGU General Assembly Conference Abstracts, Vienna, pp 8875
49. Hein T, Funk A, Pletterbauer F, Graf W, Zsuffa I, Haidvogl G, Schinegger R, Weigelhofer G (2019) Management challenges related to long-term ecological impacts, complex stressor interactions, and different assessment approaches in the Danube River Basin. River Res Appl 35:500–509. https://doi.org/10.1002/rra.3243
50. Dumitrescu V, Carsmariu A (2014) The Lower Danube Green Corridor. Contributions to the analysis of the restoration opportunities for certain surveyed wetlands in the Danube floodplain, diversified by types of potential. In: Gâștescu P, Włodzimierz M, Brețcan P. (eds.), 2nd International Conference - Water resources and wetlands. 11–13 September, 2014 Tulcea (Romania), pp 166–175
51. Omrani Ş, Grecu F (2016) Fluvial landforms seen as causes of the Danube floods case study: Oltenita - Calarasi Sector. J Eng Stud Res 21. https://doi.org/10.29081/jesr.v21i1.39

Anthropogenic Changes and Biodiversity Protection and Conservation Along the Lower Danube River Valley

Daniela Strat, Simona Mihăilescu, and Iuliana Florentina Gheorghe

Abstract The rivers and their floodplains are integrated systems. The biodiversity of the Lower Danube River (LDR), in terms of species and habitats, is strongly linked with its hydro-geomorphic-diversity and the natural regions it passes. Human activities, directly and indirectly, are the primary cause which has induced changes in hydrologic regime, longitudinal and lateral connectivity, floodplain geomorphology and function, biodiversity of the river waters and riparian zone. During the twentieth century, particularly after World War II, the LDR has undergone alteration of physical habitat, significant landscape changes, and ecological loss as a result of hydropower damming works and their associated water reservoirs, floodplain embankment, wetlands drainage, chemical pollution, eutrophication, and invasion of exotic species. The extensive embankments and drainage work along LDR in Romania converted about 80% of the annual flooded zone of the floodplain area primarily into agricultural region, obviating its essential connection with the river. Few areas, including reed marshes, meadows, floodplain forests, large shallow lakes, fluvial islands, and the braided section of the river named "the Small Island of Brăila", have been preserved in natural regime in order to preserve valuable samples of biodiversity, hydro-morpho dynamic processes, and particular fluvial landforms. Most of them are ecotonal areas that have an increased and extremely dynamic biodiversity. This increased turnover of species is exacerbated by anthropogenic factors, which sometimes they can negatively influence certain species of fauna, such as sturgeons, modifying their habitats for reproduction, feeding and resting. After the 1990s, due to the change of the political system in Romania and following integrated programs of the Danube Riparian States, some areas of the engineered floodplain are subject to ecological restoration and integrated management in order to provide convenient

D. Strat (✉)
Faculty of Geography, University of Bucharest, 1 Nicolae Bălcescu Blvd, 010041 Bucharest, Romania
e-mail: daniela.strat@geo.unibuc.ro

S. Mihăilescu
Institute of Biology Bucharest, Romanian Academy, 296 Splaiul Independenţei, 060031 Bucharest, Romania

I. F. Gheorghe
Ecological University of Bucharest, 1G General Vasile Milea Blvd, 061341 Bucharest, Romania

A. Negm et al. (eds.), *The Lower Danube River*, Earth and Environmental Sciences Library, https://doi.org/10.1007/978-3-031-03865-5_15

ways of reconciliation between nature and human society for a sustainable development. The currently Ramsar and Natura 2000 sites network designed along the LDR provides the national and international legal framework of protection and conservation of wildlife and its habitats. The objectives of this chapter are to present a review of: (1) human interventions from the last century that lead to alteration, degradation, and irreversible losses of habitats along the LDR valley, (2) restoration projects of former floodplain areas, and (3) biodiversity protection and conservation actions carried out over the area in the last decades.

Keywords Lowe Danube floodplain · Anthropogenic changes · Biodiversity conservation · Protected areas · River restoration

1 Introduction

The Danube River on his way, from its headwater sources to the Black Sea, crosses the Europe along the 2857 km being a link between Central and South-Eastern Europe, the river that defines and integrates Europe [1]. The drainage basin of Danube River covers an area of 801,463 km^2, which encompasses approximately 10% of mainland Europe, with specific and valuable eco-systems. With 19 countries within its watershed which have different cultural, political, and environmental legacies [1] and being natural border for eight of these, the Danube River is the most international river basin in the world [2]. The Danube is the most important fluvial system within European Union in terms of length, watershed (11 EU Member States are Danubian countries), water and sediment discharge [2], and also it is the largest tributary into the Black Sea. It connects five biogeographical regions, as can be seen on the map of the biogeographical regions of Europe [3]. Around 83 million people inhabit in Danube River Basin [1], which represents over 10% of total Europe population.

Geographically, and not only, the Danube River Basin is divided into three main sections and it's the delta (Fig. 1). From its source, in the Black Forest Mountains, until its confluence with the Morava River, the place named Devin Gate or Porta Hungarica, is the Upper Danube River section. From Devin Gate, near Bratislava, to the downstream of the Iron Gates Gorge extends the Middle Danube. The Lower Danube is formed by the Romanian-Bulgarian lowlands, followed by the large Romanian floodplains that extend from downstream of Călărași-Silistra localities (border between Romania and Bulgaria) until Ceatal Pătlăgeanca, where the Danube delta begins.

Along its last 1075 km, before to meet the Black Sea, the Danube River passes and borders the south and southeast Romanian territory, starting from Baziaș. Therefore, this course section of the river with a catchment area of 218,387 km^2 [4], which mainly overlaps with Romanian territory, is named the "Romanian" Danube.

Between downstream of Iron Gates Gorge (Figs. 2 and 3), the point where the Lower Danube River sector begins, and Ceatal Pătlăgeanca, the first bifurcation of the Danube River, which is the place that marks the apex of its delta, the river is

Fig. 1 The Danube River Basin and the Lower Danube Valley section between Gruia and Pătlăgeanca (Map courtesy Florian Bodescu)

accompanied by a morphological floodplain that progressively becomes wider [5], and which, prior to the major human interventions produced in the second half of twentieth century, it was very complex and divers from hydro-geo-morphological and ecological point of view, which included numerous shallow lakes rich in fish and other forms of aquatic wildlife, riparian forests, meadows and unflooded terrains used locally for agriculture [6, 7].

The first attempt to establish an international association for Danube research was in 1935, with the famous Romanian zoologist and hydrobiologist Grigore Antipa, the Austrian ichthyologist Adolf Cerny, and the Hungarian Danube researchers Rezsö Maucha and Emil Unger as initiators [8, 9]. Their great project was abandoned because of the economic and geopolitical circumstaces of those times, but it was not forgotten. Two decades later, in 1956, at the 13th International Congress of Limnology held in Helsinki, taking into account the economic importance and internationality of the Danube River, it was proposed an international program of limnological reaserch of the river with participation of all riveran countries. Also, in same year 1956, it was established the International Association for Danube Research (IAD). Based on the field surveys of the Danube River from its source until it meets the Black Sea, carried out by IAD in 1960 and 1961, it was published in 1967 the first limnological monograph of the whole Danube River [10]. The study, which was performed by a multinational team, was imperative in scientific terms, because it was

Fig. 2 The Danube River in the Iron Gates Gorge section. In the background can be seen one of the narrowest points of the Danube Gorge that is named *clisură* by locals (Photo: Daniela Strat)

Fig. 3 The Lower Danube River with its floodplain and delta. Along the floodplain section with the maximum width, named *Bălților zone* or islands section, from Silistra to Giurgeni-Vadu Oii, is the first island, the area named *Balta Ialomiței*. Between Giurgeni-Vadu Oii and Brăila is the *Balta Brăilei*. It consists of the Big Island of Brăila, which was turned into polder with extended agricultural lands and several small rural communities, and the Small Island of Brăila that has remained under free flood regime. Nowadays, the Small Island of Brăila is a protected area with national and international designation: natural park, Natura 2000 site, and Ramsar site, Wetland of International Importance. The map was redrawn after [18, 19]

the only opportunity for scientists to collect data and capture the Lower Danube in his natural state, before being transformed and modified by embankment works, draining and damming that were carried out in the following decades [8]. Similar, the Romanian Academy published a limnological monograph of the the *Romanian Danube* [11], which is comparable to that made by IAD in terms of its comprehensiveness and scientific details, and also a geographycal monograph [5]. All these monographs are extremely important from scientific point of view because they had captured the natural state of the Lower Danube Valley. In addition, must be mentioned the first integrated study of the floodplain, meaning mainly the strip of land adjoining the river channel that annually is alternately the subject of a flood phase and dry phase, named "Regiunea inundabilă a Dunării și mijloacele de a o pune în valoare" (The floodable plain of the Danube River and the ways to capitalize on it), based on about 20 years of researches performed by Antipa on the Lower Danube, aiming mainly to find the best ways for the economic use of each hidrogeomorphic units of the Danube floodplain [6].

Despite of many and diverse anthropogenic interventions carried out for over a century and which have caused floodplain degradation, changes in river flow and sediment load, habitat degradation, loss of habitats and decline of biodiversity, along the Lower Danube Valley still exist areas with high degree of naturality, some of them being nearly undisturbed, and highly ecologically sections, such as fluvial islands, natural banks, and floodplain remnants [12–14].

Comparing to other rivers from other parts of Europe and North America, whose floodplains are used extensively for agricultural purpose or have become highly urbanized, and therefore they have become entirely functionally extinct [4], the floodplain of the Lower Danube has a great potential to conserve and preserve many habitats and to restore and reconstruct ecologically significant areas. All these "islands of naturality" remembers us what it was irreversibly lost and why it is urgently needed to preserve and conserve them.

After decades of isolation behind of the Iron Curtain during the Cold War, it was possible that the Lower Danube Valley to be part of integrated European and transboundary projects on economic, social and environment issues. Taking into account, since 2007, the most part of the Lower Danube Valley overlaps on the territories of two member states of European Union, it became part of the EU Strategy for the Danube Region, following adoption of the EU 2020 Biodiversity Strategy [15]. The Natura 2000 sites established along the Lower Danube Valley, together with the various other categories of protected areas (national parks, biosphere reserves and other national, regional and international protected areas), from both UE and non-UE nations, forms an ecological network that defines a hotspot of Biodiversity for Europe, and fulfill the aim of the Pan-European Biological and Landscape Diversity Strategy for establishing corridors to link core conservation areas and to permit exchange of species between sites [16].

The aim of this chapter was to present an overview concerning the anthropogenic activities within Lower Danube River Valley, in particular in Romanian section, since the end of the nineteenth century which caused significant changes in terms of water flow, flooding regime sediments load, morphology, and biodiversity. The high degree of artificialization of the Lower Danube River coincided with the communist regime, although the damming of rivers, channelization, draining of the large floodplains were has been widely practiced in many other countries.

The change of the political regime in all riverine countries that overlap on the Lower Danube River Basin gave the opportunity to both governments and non-governmental organizations to act for restoring of the degraded and damaged ecosystems and habitats in the Lower Danube River as well as for the conservation of the biodiversity within. The chapter summarizes these actions and make a review concerning all types of protected areas that were designed in the last three decades and nowadays they compose a very complex network. For this reason, the paper gives valuable information to all who are interested in nature conservation or their work is related with it.

2 The Study Area

The first topographic map that reveals the floodplain extension of the Lower Danube River and some of its hydro-geo-morphological units is the Lesser and Greater Wallachia Map from 1790 [17]. Indubitably, the first detailed topographic map which accurately shows The Lower Danube River Valley is Wallachia Map at a scale of 1: 57,600 as a result of Austrian military survey during the Crimean War. The Austrian Map, with its Romanian version named Szatmári Map, depicts accurately the hydro-geo-morphological diversity of the Danube River channel and its floodplain, and often is mentioned as a milestone of its natural state in comparative studies on evolution both under natural and anthropogenic factors.

On the Romanian territory, geographically, the Lower Danube Valley is divided in subsections, according to hydrological complexity of the river channel and geomorphological complexity of its floodplain [5, 6, 18, 19]. Both, the river and floodplain make a big ecological corridor.

The floodplain, as a geomorphological unit, is asymmetrically developed, being prevalent on the left side of the river and progressively it becomes wider from west to east (Fig. 3). Overall, the main area of the Lower Danube floodplain (92%) belongs to Romanian territory, covering an area of 540,000 ha [18]. Its diversity in terms of features, micro-topography, ecosystems, and bio-hydro-geo-morphological processes was precisely and comprehensively described for the first time at the beginning of twentieth century by Antipa [6] who frequently used the Romanian word *baltă* with two meanings, referring both to the floodable area of the Danube River floodplain and the large shallow lakes extant in it, and later by [5, 11, 18–20].

Downstream of Drobeta-Turnu Severin, the floodplain breadth ranges from 4 km, at Ciuperceni, to over 10 km the confluence of the Danube with the Jiu River, respectively [5, 18]. This breadth of the floodplain remains almost constant until Zimnicea and this former natural floodplain section contained many lakes and large shallow lakes which in Romanian are named *bălți*. Then, on a length of 60 km, between Zimnicea and Giurgiu, the valley is narrowing, the floodplain reaching only 2 km wide near to Giurgiu. After this narrow section, the valley gradually widens again, until Călărași, where reaches the maximum width—13 km [5, 18]. Also, this floodplain section had several large lakes and shallow lakes, of which have been partially preserved lake Suhaia.

Along the Danube River channel between Drobeta Turnu Severin and Călărași, the quite common features of the riverine landscape are fluvial islands of various size, named in Romanian *ostrove*. They are the equivalent of the *aits*, the fluvial islands that were described and named on the River Thames [21]. Being the creation of fluvial processes, these islands are formed by the deposit of fluvial sediments in the water, which accumulate over time until they emerge and are stabilized by vegetation. Apart from the islands formed as a result of the natural processes in the active river channel, there are islands that have formed as a result of the river regulation, dredging, and the hydrological alterations that followed the different engineering works made to provide the optimal navigation conditions [22].

The number of fluvial islands is quite numerous. Based on the recent islands inventory provided by the Danube River Network of Protected Areas [22], along the Danube River section between Rkm 845 and Rkm 80, there are 179 fluvial islands, which cover a significant area. Thus, on the river section that is border between Romania and Bulgaria, at the beginning of the twenty-first century, the estimated area of the fluvial islands belonging to Romanian was about 11,063 ha, and islands on the Bulgarian Danube side, covered an area of 10,713 ha [23].

Some of these islands of the Danube River, the oldest and largest ones, have been transformed in agricultural lands and forest plantations, but many others, such as Albina, Cianu Nou, Ciocăneşti, Fermecatu, Halalambie, Păcuiu lui Soare, Şoimu, Trămşani, Turcescu from Romania, which represent around 51% of the area of all islands area [23] are covered by natural vegetation according to their age and stages of ecological succession. For this reason, these islands are not only biodiversity hotspots and wildlife refuges requiring integrate conservation actions, but also, based on the presence of a certain plant species on an island, it can be determined the flow condition of the river in the area. Also, fluvial islands are sensitive indicators which could be used in assessment of the ecological state of the river [24] and least but not last "because islands and rivers are so fundamentally linked, any river restoration strategies must incorporate islands as well" [25, p. 815].

However, from all 147 islands clasified as completely natural, representing true river wilderness, that were inventoried along the entire Danube River [22], 68 islands are along the Lower Danube River and 36 in the Danube delta. Many of these islands were designed protected areas and all of them are elements of the green infrastructure in the Danube River basin, and because they are among the most important habitats for intact river dynamics and the home of characteristic species, the cross-sectoral Danube WILDisland corridor programme has been established to protect them [22, 26].

From Chiciu/Silistra, where Danube changes its watercourse from west–east to northward, until downstream of the Brăila city, the Danube is an anabranching river, with main stem side arms and river islands (Fig. 3). In this section, before damming and regulating works, the former floodplain of Danube has reached the maximum width, over 22 km [5, 18]. This large section of the Danube Valley was named in the past the *Bălților zone*, a vernacular name which was used to describe along the Danube Valley an extended area predominantly covered by shallow lakes, reed marshes, dead waters, oxbows, and river bars, and that is largely inundated during flooding events and then, since second half of twentieth century, after it was drained and largely transformed into agricultural lands, the area is known as the islands zone. In fact, the whole area is a large internal delta fluvial system that has evolved between the Romanian Plain and the Dobrogea Plateau, similar to the area developed downstream to the Devin Gate as well as with those that formed by Chilia arm within Danube Delta.

However, the *Bălților zone* or islands zone, geomorphologically, consists of two large islands resulted as avulsion processes. They are scroll dominated floodplain areas formed by the meandering of branches, with shallow lakes, puddles, oxbows, meandering branch channels, wetlands areas occurred as a result of hintered drainaje, and several other smaller islands that are fluvial islands developed from bars.

The first big island of the *Bălților zone*, named *Balta Ialomiței*, is delineated by two main anabranches, Borcea Arm, on the west side, and Ostrov or Dunărea Veche (Old Danube), on the east side (Fig. 3). The two arms rejoin for a short distance downstream of Hârșova town, at Giurgeni-Vadu Oii. From Dunărea Veche branch diverts the navigable Danube-Black Sea canal, a shortcut of 64.4 km to the Black Sea that crosses the south of Dobrogea Plateau from Cernavodă to Constanța harbour.

Then, after approximately four kilometers, the main stem of the river splits again, into three branches—Cremenea or New Danube, at west, the Vâlciu arm, in the middle, and the Măcin/Old Danube, on the right side. These arms rejoin at Smârdan, a small village downstream of Brăila. Each major branch of Danube has developed anabranching behavior on longer or shorter channel sections. Between the Vâlciu and Old Danube arms there is the second large island, named *Balta Brăilei* in the past (Fig. 3) and currently Insula Mare a Brăilei (the Big Island of Brăila), after it was entirely drained and transformed into agricultural land. On the western side, between the Vâlciu arm and main Danube branch Cremenea, it has been formed the Small Island of Brăila (*Insula Mică a Brăilei* or *Balta Mică a Brăilei*). It is remnant of the former large natural floodplain from this section of the river, seasonally flooded, with lotic side arms and dead arms, channels and oxbow lakes, stagnant water bodies (Fig. 4), and reed marshes.

Downstream of Brăila until Pătlăgeanca (Rkm 90), the Danube River is flowing through a single deep channel, accompanied by a wide floodplain, more or less symmetrically developed on the both sides. Its course has also a large valley, with riverine forests, shallow lakes and extensive reed marshes. Because of that, this section is named Balta Isaccea. In this section, typical for the left side of valley are the two *limans*—Cahul and Ialpug—which are particular lakes, developed along the lower courses of its tributaries that are not able to flow into Danube River because their mouths have been closed by fluvial sediment bars. Overall, on the left side of the Lower Danube, on the Romanian territory, between Gruia (Rkm 851), downstream of Iron Gates II dam, and Isaccea (Rkm 108), the geomorphological floodplain covers an area of 530,500 ha [27].

Fig. 4 The Small Island of Brăila: a swampy area with willows (left) and a very shallow stagnant water (right), called *japșă* by locals, following spring floods, with typical vegetation composed by *Cicuta virosa*, *Nymphoides peltata*, *Carex* sp., and *Schoenoplectus lacustris* (Photo: Iuliana Florentina Gheorghe)

On the right side of the Danube River, in Serbia and Bulgaria, its floodplain consists in a narrow and discontinuous strip. The entire morphological floodplain of the Lower Danube from the four riverine countries (Bulgaria, Romania, Moldova, and Ukraine) covers an area of 803,300 ha, of which the remaining active floodplain, including the main channel with fluvial islands, represents 214,300 ha [28].

The current active floodplain of the lower Danube River is limited to the areas between the river banks and flood protection dikes behind which are the polders. It is a narrow strip with flood channels, small lakes and oxbows, sand banks and marshes.

However, at the European level, based on a combination of abiotic factors—ecoregion, mean water slope, substratum composition, geomorphology, and water temperature—ten Danube section types were identified along the river. Of these, to Lower Danube River it coresponds two section types: Western Pontic Danube (Rkm 943 Drobeta-Turnu Severin—Rkm 375.5 Chiciu/Silistra) and Eastern Vallachian Danube (Rkm 375.5 Chiciu/Silistra—Rkm 100 Isaccea) [29].

3 Biodiversity of the Lower Danube Valley

Because of the Lower Danube River Valley has a considerable west-southeast extension, its life communities and vegetation composition are not only defined by the ecologically determining factors such as the hydrological and morphological dynamics, but also by the continentally gradients of the climate [30]. This becomes obvious from the comparison of the floodplain vegetation occurring along the varying sections of the Danube River, consists of many pontic-pannonian and continental but also submediterranean, thermophilic species. The vegetation along the river in the remnant active floodplain areas, and in particular in many fluvial islands, is in natural succession. Its zonation spans from pioneer vegetation to alluvial woodland.

The morphological structure of the valley provides a mosaic of biotopes for animals and plants.

From the first map and comprehensive description of the vegetation, we find out that along the Lower Danube floodplain there were forests with softwood forest species (willows and poplars), lakes, reeds, levees, floating reed carpets "plaur" and forest enclaves composed of ash meadow with the rare liana *Periploca graeca* [31].

The majority of the natural forested areas from Romanian Danube floodplain, including the restricted areas of hardwood forests, was deforested and replaced with agricultural lands during 1960–1980 period. Aerial images (Google Earth) show that a belt of riverine forests is still stretch along the orographic left bank of Danube on the strip of land between the river bank and dam, although most of these forests are hybrid poplar plantations, which is a common practice in other floodplains from European countries as well, especially since these plantations can be subsidized through European Union funds [32]. Poplar plantations are extended on many fluvial islands, too. However, large areas of remarkable remnant areas of old-growth riverine forests, rich in woody species and lianas, are preserved in the Lower Danube Valley on several fluvial islands and downstream from Călărași, in particular in the area of the Balta Mică a Brăilei Natural Park.

Under natural conditions, the three species distribution is according to their adaptive performances to periodically floods, grain size of substratum, excess of moisture in soil. The flood level is expressed in hydro grades, one hydro grades, corresponds to the tenth part of the difference between the lowest and highest water level. For the Lower Danube floodplain, the shrub vegetation develops above hydro-grade 3, white willow settlements have been recorded between 3 and 6 hydro grades, swampy alder areas occur between 7–8 hydro grades, and between 9–10 hydro grades it is a mixed shrub vegetation composed of oak, elm, and poplar [33]. Then, between 6 and 9 hydro grades there are willow coppices, which occur in more elevated areas, and pure ash woods as well as settlements of alder [30].

Along the river banks and on the shore islands, the forests are dominated by willow species, the white willow (*Salix alba*) being the most common species (Figs. 4 and 5). Depending on the altitude and the level of the flood, besides the white willow, the stands are formed poplar (*Populus nigra, Populus alba*), elm (*Ulmus laevis*), and ash meadow (*Fraxinus angustifolia*). The herbaceous layer consists of species that are tolerate fluctuating water levels such as great yellowcress (*Rorippa amphibia*), water pepper (*Polygonum hydropiper*), and fen ragwort (*Senecio paludosus*). In the very high areas, unexposed to flooding and with deep water ground table, there are hardwood floodplain forests composed of stands of common oak (*Quercus robur*), Balkan oak (*Quercus pedunculiflora*), elm (*Ulmus carpifolia*), hairy ash (*Fraxinus pallisae*), and scattered individuals of cherry plum (*Prunus cerasifera*), forest apple (*Malus sylvestris*), and European wild pear (*Pyrus pyraster*).

Because the embankment works avoid overflows into floodplain during spring floods, the fluvial islands are more exposed to flooding in terms of area surface and duration, changing the soil proprieties and water ground table, thus largely diminishing suitable habitats for oaks. Therefore, on islands with unmanaged natural vegetation and no direct anthropogenic interference there is a serious decline of oak,

Fig. 5 Willow forest with trees covered by woody climbing plants species: Wild grapevine, common hop, and old man's beard (Photo: Daniela Strat)

without new established stands. Also, oak stands existent before the embankments have degenerated and remained only older, less sensitive specimens.

The layer of shrubs consists of blackberry (*Rubus caesius*), which in some area is dominant and can reaches one meter in height. There is also gray willow (*Salix cinerea*), osier (*S. purpurea*), almond willow (*S. triandra*), bloody dogwood (*Cornus sanguinea*), and more rarely oneseed hawthorn (*Crataegus monogyna*), black elderly (*Sambucus nigra)*, and dogrose (*Rosa canina*). Unfortunately, the bare surfaces and not they are colonized by false indigo-bush (*Amorpha fruticosa*). In fact, this highly competitive species that has an exceptional success of invasiveness forms usually mono-dominant communities by replacing native species, as in case of *Salix triandra,* and altering the structure of native plant communities.

Among the woody climbing plant that hangs from trees (Fig. 5) are common the wild grapevine (*Vitis sylvestris*), ivy (*Hedera helix*), common hop (*Humulus lupulus*), old man's beard (*Clematis vitalba*), and the silk vine (*Periploca graeca*), which is a rare plant species in Romania, with scattered occurrence in Dobrogea and Lower Danube Valley.

The diversity of grass layer is according to substrate type and soil water balance. The mesophytic species *Glecoma hederacea, Tanacetum vulgare, Althaea officinalis, Asparagus officinalis, Artemisia vulgaris, Arctium lappa, Calystegia sepium, Solanum dulcamara, Sonchus asper, Inula British, Rumex pulcher, Potentilla reptans,*

Taraxacum officinale, Torilis japonica colonize dried terrains, and the plant species *Lycopus europaeus, L. exaltatus, Stachys palustris, Galium palustre, Lysimachia nummularia, Polygonum mite, P. hydropiper, Lythrum Salicaria, Euphorbia palustris, Oenanthe aquatica, Scutellaria galericulata, Iris pseudacorus, Sium latifolium* occur in areas with excessive moisture and temporary stagnant waters.

Apart of typical floodplain forests and poplar plantation there are the black locust (*Robinia pseudoacacia*) plantations from Desa-Ciuperceni region. The black locust was used, since 1852, for afforestation of the mobile sand dunes and protection of agricultural lands against mobile sands [34].

The shallow lakes and ponds host a great diversity of macrophytes (Fig. 6), typical for European temperate lowlands (*Nymphaea alba, Nuphar lutea, Trapa natans, Nymphoides peltata, Sagittaria sagittifolia Stratiotes aloides, Butomus umbellatus, Sparganium emersum Potamogeton* sp., *Typha angustifolia, Typha latifolia, Schoenoplectus lacustris, Cyperus sp., Carex sp.*), including the Global endangered carnivorous plant species *Aldrovanda vesiculosa* [35], and the European protected aquatic fern water-clover (*Marsilea quadrifolia*) (Fig. 7).

The reed *(Phragmites australis)* forms dense and extensive stands and floating fen due to its propagation either by rhizome or from seeds. This helophyte acts as a climax plant species because it has adaptive strategies that inhibit the growth and survive of

Fig. 6 A pond dominated by water soldiers (*Stratiotes aloides*), water lily (*Nymphaea alba*), and water caltrop (*Trapa natans*). Most of the litoral zone of the pond is covered by reed stands and floating fen, bordered by a forest composed by willow, poplar, and alder (Photo: Daniela Strat)

Fig. 7 *Marsilea quadrifolia*. Once widespread in wetlands of all big rivers from Europe, this aquatic fern with its four leaf clover configuration also became a rare and threatened plant species along the Lower Danube Valley during the second half of the 20th century because of its habitat loss and destruction as a result of anthropogenic impacts (Photo: Daniela Strat)

other species that compose wetland vegetation. Along the secondary tributaries and narrow channels there is a vegetation belt that consists of parallel strips of rush and reed, the latter being a successional species which replaces *Typha angustifolia* that has previously colonized deep waters [36]. Their associated species are *Butomus umbellatus*, *Carex sp.*, *Schonoplectus littoralis*, *Cicuta virosa*, *Calystegia sepium*, *Solanum dulcamara*, *Iris pseudacorus*, *Sparganium erectum*, *Epilobium hirsutum*, *Lythrum salicaria*, *Alisma plantago aquatica*, *Symphytum officinale*. However, the reed is an autogenic ecosystem engineer species in floodplain because it alters irreversibly the environment through its own physical structures and its living and dead tissues, is a real habitat-modifying organism [37] and also plays a key role as bio-geo-morphologic agent in evolution of shallow waters and back waters thought its particular bioconstructions (the *plaur*) formed in shallow lakes [38]. The highest degree of naturalness is reached on the several fluvial islands, which ranges from the first herbaceous colonizers to white willow and black poplar forests [39–41].

Fauna consists in an impressive number of both terrestrial and aquatic species, some of them being predominantly restricted to the Danube River, as in case of the snail *Theodoxus danubialis*, which is a Pontic fauna element. Also, the ecosystems from Lower Danube River provides habitats for many other species protected under Habitats Directive such as: the large copper butterfly (*Lycaena dispar*), fish (*Gobio albipinnatus, Gobio kessleri-Romanogobio kessleri, Rhodeus sericeus amarus*), amphibians and reptiles (*Bombina bombina, Triturus dobrogicus, Emys orbicularis*), and mammals (*Spermophilus citellus, Lutra lutra*).

Among the fish species, anadromous fish stand out: 4 sturgeon species (*Acipenser gueldenstaedtii, A. nudiventris, A. stellatus*, and *Huso huso*), the Pontic shad (*Alosa immaculata*), and the Black Sea shad (*A. tanaica*). Sturgeons are key element of life within the river basin, and an important part of the natural heritage of the entire Danube region. The over-fishing, the numerous dams and hydropower dams are the main cause of the dramatic decline of these migratory fish population in the past decades, which means that they are currently being assessed as critically endangered species at the regional [42] and international level [43]. More than that, the only viable populations of wild sturgeons in the European Union remained in the Lower Danube. For two native sturgeon species, which are vulnerable and globally critically endangered according to IUCN Red List [44, 45] there are manually restocking actions. It is about the sterlet *(Acipenser ruthenus)*, in Hungary, and the Russian sturgeon (*Acipenser gueldenstaedtii*), in Romania [46].

The Danube is a major flyway for many bird species and the existing wetlands along the Lower Danube Valley and the Danube Delta represents resting spots of migratory birds. Among the bird species that are listed in Annex 1 of the Birds Directive for which special protection areas have been designed are notable *Ardea purpurea, Botaurus stellaris Cygnus cygnus, Ciconia ciconia, Ciconia nigra, Egretta alba, Nycticorax nycticorax, Himantopus himantopus, Recurvirostra avosetta, Pelecanus crispus*, and *Platalea leucorodia.*

A recent study [47] along the Lower Danube Valley, including fluvial islands (the *ostroave*), revealed the occurrence of 46 mammal species. Among them are: *Erinaceus concolor, Talpa europaea, Sorex araneus, Rhinolophus hipposideros Myotis sp., Plecotus auritus, Vespertilio murinus Nyctalus noctula, Pipistrellus pipistrellus Miniopterus schreibersii, Lepus europaeus, Sciurus vulgaris, Spermophilus citellus, Muscardinus avellanarius, Arvicola terrestris, Microtus arvalis, Ondatra zibethicus, Mus musculus, Apodemus agrarius, Apodemus sylvaticus, Micromys minutus, Spalax leucodon, Myocastor coypus, Vulpes vulpes, Meles meles, Mustela nivalis, M. putorius, Felis silvestris, Sus scrofa,* and *Capreolus capreolus.* Their role in the local ecological equilibrium justifies their consideration as bioindicators of the habitat quality.

The golden jackal (*Canis aureus*) is a special case. It has become widespread along the Lower Danube Valley in last decades. This canid species, which had typically been distributed in the Balkan region until the end of nineteenth century, has crossed the Danube River [48, 49] and extends its areal range to north and western Europe in the second half of twentieth century [50, 51]. The massive colonization of Danube floodplain with golden jackal has started in 1990s [52], which may be associated with change in lands use because of socio-economic and political transformations occurred in Romania. It was reported that this extremely adaptive species is human-dependent in environmental preferences, with agricultural lands and mixed landscape between open areas and scrublands as suitable habitats [51] and correlated with the absence of the wolf [53]. Consequently, the golden jackal spread along the whole Danube corridor and its population has significantly increased in last decades [54]. According to recent data [53], the numerical density evolved from 0 individuals/10 km^2, in the Danube valley section between Iron Gates II and the Danube—Jiu River confluence, and 2–6 individuals/10 km^2 downstream of this section, in 2007, to 8–22 individuals/10 km^2 in 2018, with the highest densities in the south-west Danube valley section and Danube Delta.

4 Anthropogenic Changes in the Lower Danube Valley

4.1 An Historical Overview of the Human Impacts Along the Lower Danube Valley

After the first mention and geographical description of the *Istros*, the antique name of the Danube River, which belong to the Greek poet Hesiod (around 700 BC) and the Greek historian Herodotus (around 450 BC), probably, the first "scientific" field study of the Danube River Valley was performed during the European Scythian campaign of Darius I, when the river was bridged just before its deltaic area [55]. Six centuries later, the architect and military engineer Apollodorus of Damascus chose to build, downstream of the Iron Gates, the 1 km wide wooden famous bridge over Danube in 105 A.D, at the order of the Roman Emperor Trajan.

By the middle of ninetieth century, the Lower Danube Valley evolved under natural regime. Then, after the European Commission of the Danube was created in 1856, which was a post-Crimean war entity that had authority over the three mouths of the river and was responsible for facilitating navigation in the mouth and delta of the Danube, the direct anthropogenic alteration of the Lower Danube River has started and transformed it into "internationalized river" [7] due to both geopolitical context and economic reasons.

The channelization for navigation of the Sulina arm, the middle distributary of the Danube River in its delta, was the first large-scale work with significant morphological and hydrological effects. All engineering works that were designed by the English river and harbors engineer C.A. Hartley, "the father of Danube" [56], gave the opportunity to accurate and extensively scientific surveys of its arms and mouths, but also the river course section upstream of the delta was surveyed. Apart of the detailed mapping, there were performed researches to understand the behavior of this river concerning its water and sediment discharges, the cyclicality of flooding events and how they work.

Towards the end of nineteenth century and the beginning of twentieth century, the scientific investigations were extended on the Danube floodplain in terms of hydrology, geomorphology, biology and ecology, as the necessary basis for substantiating the plans for the capitalization and economic exploitation of its valuable resources, based on the multi-functionality of river-floodplain system [6, 7]. Based on the analysis of recoded water levels along lower Danube between 1895 and 1908, and field observations, Antipa [6] explained the importance of lateral exchange of water, nutrients and organisms between the river channel and its floodplain, and how flood events followed by low water levels control the river-floodplain system, flooding being the rule factor that support the high level of productivity.

It is necessary to point out that the Romanian polymath Grigore Antipa (1867–1945), a brilliant zoologist and hydrobiologist, as well as a pioneer of promoting the ecological economics principles, had pleaded with perseverance for an ecological approach of the use the Lower Danube floodplain, because "the river and its floodplain make a whole, a multi-functionality system, and not separate units" [7, p. 39]. His ideas anticipated the flood pulse concept that was described later by Junk et al. (1989) as a complementary concept of the river continuum concept [57], both applied to large river systems.

Antipa proposed a multipurpose exploitation of the floodplain—for fishing, agriculture, and forestry, according to its hydro-geo-morphological units and the yearly flooding regime of Danube River. It is about the rotating polder system, with minor changes to fluvial ecosystems, which in modern times corresponds to sustainable development. During dry years floodplain could be used as pasture, and in wet years, with big floods, it becomes an extensive natural fishery [6, 7]. Therefore, the sustainable technical solution proposed was the interrupted embankment of the river and submersible dams, which during the high floods to allow that all those defended areas to be flooded and thus their usage to be transmuted from agriculture in fishing, insisting how important is to maintain the connectivity of the river with its floodplain because this acts as a safety valve during high waters [7]. Antipa's ideas were advocated by engineer Vidrașcu, the one who mapped the Lower Danube and delta according to degree of inundability [58].

In opposition to these naturalistic principles, which, from the contemporary paradigms perspective, could describe Antipa's project as being designed in balance with the nature, the government agreed and ratified by law in 1910 the embankment works project with high unsubmersible dams and drainage system, designed by the engineer Anghel Saligny (1854–1925), based on his previous experience during building of the bridge over Danube River and Ialomița Island. The plan was to drain the floodplain for agricultural development. Because of financial reasons, the project was abandoned.

Major anthropic transformations along the Lower Danube took place between 1960 and 1989. After the World War II, in the context of the communist political regime which was installed in Romania, the idea of embankment of the Danube floodplain was taken up by the government and implemented. In reality, there was nothing original in this "grandiose" project as it was presented by the propaganda of the Romanian communist government, a tangible evidence that the New Man in the fight with the nature is victorious, is able to subdue it. The communist government just followed a paradigm that was already implemented in Western Europe and North America by years [59, 60].

More or less, the agricultural model and the technical solutions designed by engineer Saligny in 1910 were implemented and not the ecologically orientated ones proposed by Grigore Antipa. Therefore, up to the end of 1989, along to the left Danube River bank it was build a levee of 1158 km to prevent the cyclical annual floods over the floodplain, and the shallow lakes, ponds, and reed marshes were drained [61]. The main goal, driven by the communist government, was the transformation of the large Danube River floodplain into one of the highest productive agricultural area in Romania. If in the early 1950s more than 90% of the Lower Danube floodplain was an integrated complex of natural and seminatural ecosystems, four decades later the ratio was reversed: 430,000 ha, meaning 84% of whole area, were transformed in enclosures and used as agricultural land, mainly arable land [61]. Only 10% of it remained unpaired by agricultural encroachment, flood protection and regulation engineering works [27].

Consequently, fundamental changes in river flow and rivel channel morphology as well as in floodplain morphology and its hydrologic regime and lateral connectivity were made, valuable lentic and lotic ecosystems had disappeared [62–65] as in case of the "Balta Brăilei", where its area of 72,000 ha was dammed in 1964 and converted at a rate of 96% in agricultural land [61]. In the entire *Bălților zone*, compared to 1880, when the area occupied by lakes and other stagnant waters was 595 km^2, in 2005 it decreased by almost 85% [66].

Under natural conditions, the floodplain and lentic body waters exchanged water, nutrients and fauna with the river via natural channel network, the yearly floods maintained high biological productivity and biodiversity of the entire floodplain. Conversion of flooding areas for farming and others and cut-off of them from the river by dykes has led to decrease the storage capacity of Danube discharge and exacerbation of the floods peak as it happened during the 2005 and 2006 floods [67].

The flow regime of the Lower Danube has been dramatically modified after the two hydropower reservoirs were built—Iron Gates I and Iron Gates II—and a complex system of impoundments was created along its main tributaries—Olt, Argeș, Siret, and Prut rivers. Changes in water flow are accompanied by dramatical alteration of sediment load, all reservoirs being traps for sediments and nutrients. Thus, the sediment load decreased from about 40 million tons/year at the beginning of last century to 7.3 million tons/year today [1]. It was estimated that the Iron Gates reservoirs I and II reduce by more than 50% suspended solids in the Lower Danube [68, 69]. Therefore, the lateral erosion of the river channel has intensified, and river bed incision has increased [66].

Although some of the remaining patches of the nearly natural floodplain have been designed as protected areas, which are samples of the naturalness of fluvial processes and biodiversity hot spots, their integrity is threatened by hydrological disconnection because of the engineering works that have been performed to provide flood control navigation and hydropower generation [70].

The extensive embankment works along the Lower Danube River were preceded by extensive interdisciplinary studies [5, 18] which summarized the state of the Lower Danube River that was prevalent until the middle of the twentieth century. Now, these studies serve as hydro-morphological and hydrobiological reference conditions for the subsequent researches and assessments of human impacts, including work projects for ecological reconstruction of some areas which are suitable for a sustainable management.

4.2 *The Stages of the Anthropogenic Transformation of the Natural Danube Floodplain into Agricultural Land*

The main impacts on the Lower Danube Valley were caused by the embankment works that converted the floodplain into predominantly agricultural region. But if in Bulgaria, the floodplain already was embanked to a large extent (about 88,000 ha) previous to the World War II [27], in Romania the strongly anthropization process on extended areas took place in the second half of the twentieth century.

The embankment works started at the beginning of the twentieth century as an experiment. Initially, it was a surface area of 3172 ha, downstream of Oltenița town, and 3150 ha at Luciu Giurgeni, on the west side of the Bălților section [27]. Then, the area embanked at Spanțov (1780 ha) between 1906–1908 had become the field work of the first agricultural research station on embanked floodplains.

The technical approach of the built embankments, which was implemented by the Administration of State Fisheries, was based on the natural flood pulse promoted by Antipa [6], being mainly focused toward developing fisheries. By 1947, about 55,000 ha of natural floodplain had become a dammed area, partially protected from annual floods because the technical approach was that of submersible dams, and dominantly used as agricultural land [27]. Following delineation of the floodable region (530,500 ha) of the Lower Danube Valley, the section between Gruia (Rkm 851) and Isaccea (Rkm 100), its land use categories it was assessed in 1962 (Fig. 8).

Based on these assessments, the project plan for damming and draining of the largest part of the floodable region in the following decades was drawn up [27]. Accordingly, the intensive embankment works were performed during 1963–1971 period, when an area of 289,000 ha of floodplain was disconnected from the river. In the next two decades, other enclosures were constructed with a total surface of 41,800 ha, so that in 1990 the total embanked surface of the Lower Danube floodplain in Romania reached 430,800 ha, with 55% located on the left bank, 12% on the right bank, and 33% on the two big islands from the Bălților section, Island Ialomița and the Big Island of Brăila [27]. The remained active floodplain covered 111,796 ha, the largest area being maintained in the Bălților section, about 65% [71]. Consequently, the lands use within Lower Danube floodplain has changed significantly compared to the period before 1962. A study based on the assessment of the functionality of the floodplain in order to identify equipotential areas for flood-free future and sustainable development of the region, shows that in 2008, 73% of the Lower Danube floodplain was agricultural land and 17% were lands with forests and semi-natural areas [72].

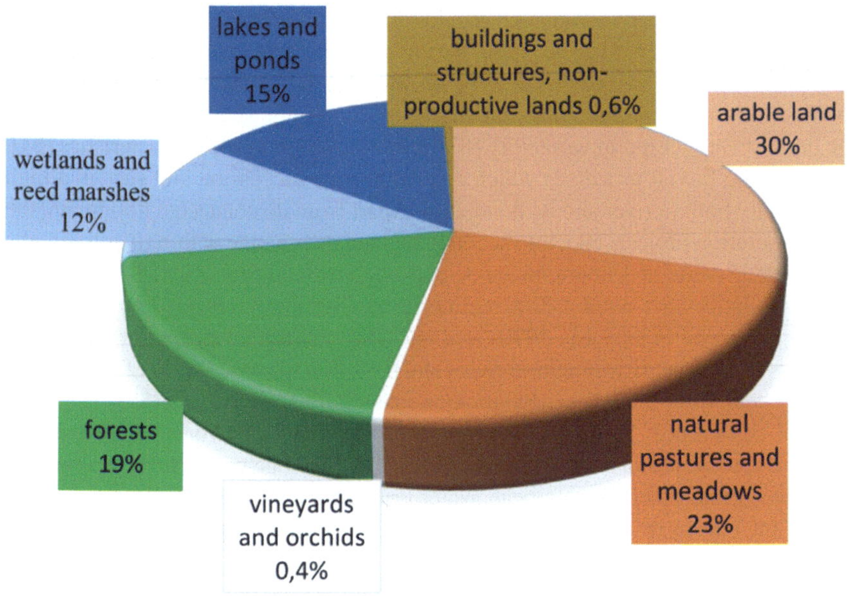

Fig. 8 Percentage of land use in the Lower Danube floodplain before 1962 in Romania. Data compilated from [27]

5 Lower Danube River Restoration

5.1 The Worldwide and European Context of the Lower Danube River Restoration

The restoration of rivers and wetlands is a global issue [4, 73–75]. In time it was demonstrated that in many cases the river regulations have not been a real success, both economically and ecologically, which led to a new paradigm toward the end of twentieth century: the development of "ecological" rivers [76, 77]. This means that where possible, ecosystems should be returned to their natural or original condition [76]. The main idea is that the regularized rivers could be restored by moving the embankments and turning them back into a meandering course although the restauration actions are far from simple engineering works [78].

According to Westra [79], the ecological restoration should be directed especially on restoring the abilities of ecosystems in order to continue their constant change and development unconstrained by past or present human interruptions. Therefore, restoring river ecology or restoring damaged river ecosystems to a state that is structurally and functionally comparable to the state they were in before the disturbance [80, 81] requires an interdisciplinary approach in line with the latest scientific advances that should promote healthy functioning of the river system. In this respect, a typical restoration project implies planning and design activities associated with the preconstruction phase, the construction phase, and the long term-post construction phase [82]. However, to be sustainable, a river restoration project needs to effectively recreate the functional characteristics of the river taking into account the flood-driven temporal variability [83], dynamic geomorphic characteristics [84], and dynamic fluvial processes which control habitat succession and sustain a high biodiversity both in river and its floodplain. Apart from disconnected floodplain, the river restoration projects take into account the land use change within river basin, the existence of dams and impoundments, and any artificial inter-/intra-basin transfer that profoundly alter natural flow regimes [85]. The first river restoration projects started in Europe in the early 1990s as a reaction of rising awareness of the permanent loss of the integrity of natural rivers and their floodplains [28]. For this purpose, the European Centre for River Restoration was established in 1995 [86]. The international conference "River Restoration 2000 - Practical Approaches" focused on practical approaches in river restoration throughout Europe within the framework of the implementation of the European Water Framework Directive [87].

The interest in restoring rivers and their related ecosystems has increased and tends to become an element of national and international policy strategy after 2000. In European Union, the conserving and restoring river and floodplain systems' flood capacity, biodiversity, and ecological status of adjacent water bodies represents a priority of the environmental and water policy as is stipulated by EU Water Framework Directive [88] and EU Flood Risk Directive [89]. The Habitats Directive [90], Birds Directive [91], and EU Biodiversity Strategy to 2020 [15], together with the Ramsar Convention on Wetlands [92] supports the conservation and restoration of floodplains. Therefore, well-planned, -designed and -managed projects of restoring river and wetlands systems can mitigate the negative consequences of changing land uses and of climate extremes, such as recurrent floods and droughts, and can also help to sustain all types of biodiversity [74].

A data analysis for 1989–2016 period, shows that in Europe were finalized 119 river restoration projects in 19 countries, the majority (51%) being designed and implemented by regional authorities or by other entities working on a regional scale, and then followed by national-scale entities (19%) and other consortia (18%), in the rest of the cases [73]. Among the successful projects, should be mentioned the restoration of two abandoned agricultural polders (3680 ha) from Danube Delta that were pilot project areas for ecological restoration, which was an important goal within the first management master plan of the Danube Delta Biosphere Reserve after it was designed in 1991 [93].

5.2 *Applied River Restoration and Ecological Reconstruction Along the Lower Danube Valley*

Toward the end of twentieth century, the Danube River was assed as one of the strongly affected rivers in Europe in terms of fragmentation by dams in the main channel and in its tributaries and of flow regulation [94]. Also, the Danube River was assessed as one of the most threatened ten major rivers in the world; the heaviest impacts are expected to be caused by the navigation infrastructure [95], following the European Union's plan to develop the Trans-European Networks for Transport (TEN-T) "Corridor VII" along the Danube [96]. In these circumstances, the necessity of Danube River restoration came naturally. The first restoration projects in Lower Danube River floodplain took place towards the end of the twentieth century [97].

In 2000, the Bulgaria, Romania, Moldova, and Ukraine, have banded together in an ambitious project in order to protect and restore one of Europe's most important wetland ecosystems—the Lower Danube River [98]. Therefore, these nations have committed to the Lower Danube Green Corridor (LDGC) project, at that time this trans-boundary nature conservation project being one of the Europe's largest international wetland preservation effort [99].

Later, for Romania and Bulgaria, which in the meantime have joined the European Union, the LDGC was the perfect framework for the implementation of the relevant EU legislation, in particular the Water Framework Directive, the Flood Directive, the Habitat Directive, the Birds Directive and the Renewable Resources Directive.

The basses for the LDGC agreement were the first assessment of the potential for restoration of the Lower Danube Floodplain [100]. Taking into consideration that based on an extended scientific assessment of biodiversity on behalf of World Wildlife Fund, the Lower Danube River and its delta was identified "as one of the world's most important ecoregions with a representative selection of the world's most outstanding and distinctive biological resources" [101–103] the creation of the LDGC was acknowledged by the Ramsar Convention as a "gift to the Earth" [104].

The designed green corridor, which stretch from downstream of Iron Gates Gorge, in the west, to the Black Sea, in the east, covers a total area of 18,344 km^2, from which the Lower Danube River and its floodplain, both in Romania and Bulgaria, covers 9,080 km^2 [105]. Within the LDGC, the conservation initiative aimed to protect an area of 11,574 km^2 from which 7,731.66 km^2 represents the already existing protected areas at the time of the establishment of the LDGC, including the Danube Delta Biosphere Reserve. In addition, 1,606.26 km^2 were allocated for new protected areas and other 2,236.08 km^2 include areas proposed to be restored to natural floodplain [105]. Later, to the existing network of protected areas were added new categories, such as Natura 2000 sites and Ramsar sites. However, in the first decade following the LDGC agreement, 11,740 ha of former wetlands were restored in four specific sites [28, 106]: two in Bulgaria (Rryahovo Ost and Belene Island), one in Romania (Călărași-Răul Island East), and one in Ukraine (Kugurluiskiy). Also, 1.4 million ha have been brought under some form of protection [105], and the

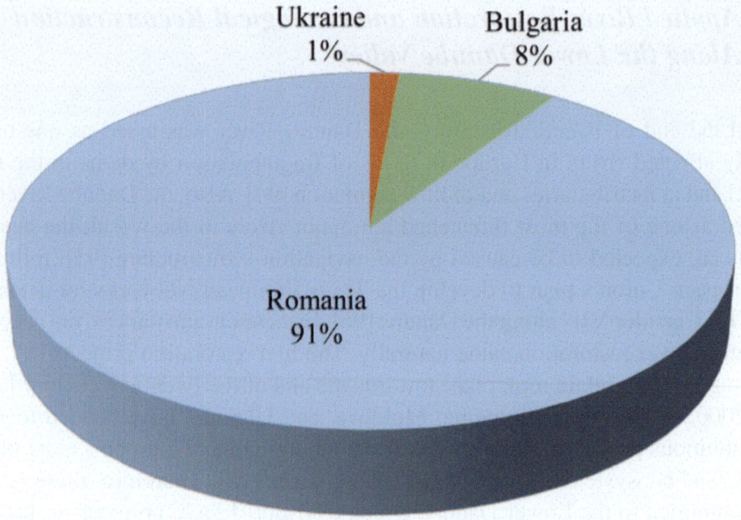

Fig. 9 Percentage of area planned for restoration in each Danube riverine country within Lower Danube Green Corridor. Total area planned for restoration: 497,950 ha. Data compilated from [28]

planned restored area increased as a result of the following official proposal of the Danube River Basin Management Plan [107].

The assessing of the restoration potential of Lower Danube floodplain, have revealed that potential areas for restoration cover 503,790 ha [13]. Thus, it was proposed that 15.9% of former flooded area to be restored, including two areas from the Danube Delta [72]. Therefore, a total area of approximately a half million ha (Fig. 9) within the Danube floodplain, mostly covering the agricultural polders from Romania, divided in 79 proposed and planed restoration sites, was planned for restoration [28].

A WWF report assessed that due to the rather "young" disconnection, compared with the upper and middle section of Danube, and the rather good hydro-morphological conditions, except the sediment budget altered by the Iron Gates dams that increases incision and instability of banks, as well as the hydrological regime, the restoration potential within Lower Danube Valley is still high, and there is no strong floodplain aggradation so far [28]. Typical projects implemented are side-channel reconnections, channel widening, bank revetment removal, and enlargement of floodplains by reconnecting former floodplains. In the second decade of this century several river restoration projects were implemented or are undergoing.

An eloquent example is the floodplain restoration in the Dunăreni-Bistreț area, as part of the Danube Floodplain Project that aimed to improve the transnational water management and flood risk prevention while maximizing benefits for biodiversity conservation [108]. Then, in December 2020 it was launched a reconstruction ecological project of an area of Danube floodplain on the Romanian territory, in the southeast part, at Gârla Mare-Vrata section, Mehedinți county [109]. By reconnecting the shallow lakes from former floodplain with the river, during the big floods, it is estimated that the area will retain over 5 mil m^3 water, and the habitats for spawning fish will be restored. Another restored site was Balta Geraiului, located at the Olt-Danube confluence, as a result of the cross-border project Green Borders, leaded by WWF, which created habitats for the two key waterbird species: ferruginous duck and pygmy cormorant [110].

However, the removal of dams and transformation of the entire embanked area of former morphological floodplain into a floodable area is difficult to achieve, controversial and debatable, considering the contemporary socio-economic circumstances, but also the hydrological changes produced along the entire Danube River. During April 2006, when along the Lower Danube were recorded historical flows and water levels, the largest in the 100 years [111], a quarter of the enclosed and drained area from former Lower Danube floodplain was inundated. Because of the huge pressure, dikes failed and seven enclosures, totaling about 88,000 ha, were naturally flooded. In addition, another two floodplain basins (15,607 ha) in the Bălților section, were flooded in a controlled manner through dike destruction.

Concerning to fish fauna, in the framework of the Bern Convention on the Conservation of European Wildlife and Natural Habitats [112], in 2005 a Sturgeon Action Plan was adopted in the Danube River basin in order to secure viable populations of all Danube sturgeon species by sustainable management and by restoration of their habitats and migratory movements, through national action and international action [113]. In this respect, the MEASURES program, runed by International Commission for the Protection of the Danube River and WWF Romania,aims to create ecological corridors on the entire Danube and its tributaries by identifying key habitats and initiating protection measures for sturgeons and other native migratory fish species, which migrate along the river as an essential part of their reproductive life-cycles [114].

Then, in 2018, a Pan-European Action Plan for Sturgeons was signed, for a comprehensive and integrated approach of conservation of eight endangered sturgeon species, of which four occur in Danube River, during the 2019–2029, both by in situ and ex situ measures [115]. However, the cross-border conservation is essential, and transnational cooperation is the tool which may be used to restore and preserve the connectivity of habitats along the Danube River.

6 Biodiversity Conservation

The protected areas network consists in areas of national, European and Global importance. It was built gradually, starting with the sixth decade of the twentieth century. Apart from national legislation, the EU directive and conventions, the international legislation and initiatives that are relevant to sites conservation and their associated species along the lower Danube River consists in: Biodiversity Convention, Ramsar Convention, Bonn Convention, World Heritage Convention, UNESCO's Man and Biosphere (MAB) Programme.

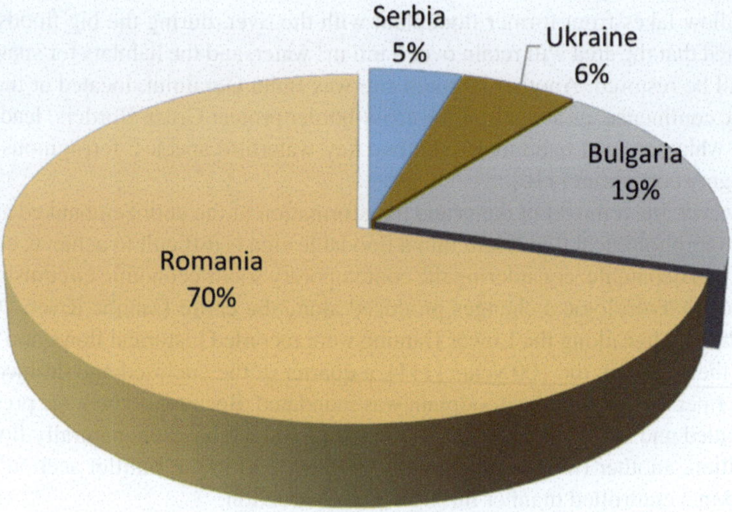

Fig. 10 Percentage of area of designed Important Bird and Biodiversity Areas from Lower Danube River Valley in each riverine country. Total area of the IBAs: 492,391 ha. Data compilated from [119]

Both, the Lower Danube Valley and Danube Delta are important for breeding, passage and wintering for large number of waterbirds and non-water birds. Therefore, following particular criteria [116–118] it has been designed 44 Important Bird and Biodiversity Areas (IBAs), which are places of greatest significance for conservation of world's birds [119]. The majority of IBAs are in Bulgaria (16) and Romania (25), two in Ukraine and one in Serbia. They cover an area of 492,391 ha (Fig. 10), which provide suitable habitats for over 160 birds species, including for nine wetland-dependent threatened species: black-winged pratincole (*Glareola nordmanni*), corncrake (*Crex crex*), Dalmatian pelican (*Pelecanus crispus*), ferruginous duck (*Aythya nyroca*), lesser white-fronted goose (*Anser erythropus*), pygmy cormorant (*Phalacrocorax pygmeus*), red-breasted goose (*Branta ruficollis*), white-headed duck (*Oxyura leucocephala*).

The key breeding species include the great white pelican (*Pelecanus onocrotalus*), Dalmatian pelican, glosy ibis (*Plegadis falcinellus*), black-crowned night heron (*Nycticorax nycticorax*), squacco heron (*Ardeola ralloides*), purple heron (*Ardea purpurea*), ferruginous duck [117]. Some of IBAs are extended beyond the geomorphological limit of the Danube Valley and include the confluence areas of the Danube River with its tributaries and parts of their floodplains (Fig. 11).

On the Dobrogea Plateau side of the Danube River, but also in Bulgaria and Ukraine, the designed IBAs include the fluvial lakes that have formed as a result of natural damming of mouth of small rivers that flow to Danube by a bar of fluvial sediments, named *liman*, a term which is used also for the fluvial lakes. Most of the IBAs served as basis for designation of the Special Protection Areas (SPAs) as part of the future Natura 2000 network in Romania and Bulgaria.

Fig. 11 The Jiu River at the confluence with Danube River, in November 2019. The Jiu-Danube confluence is designed Special Protection Area (ROSPA0023), according to Habitat Directive, and Ramsar site since 2013 (Photo: Simona Mihăilescu)

Ramsar sites or the Wetlands of International Importance or Ramsar sites, designed under "Convention on Wetlands of International Importance Especially as Waterfowl Habitat" [92] covers an area of 248,398 ha, of which almost 90% is in Romania (Fig. 12). The first designed Ramsar site in the Lower Danube Valley dates back to 1975 and was Srébarna Lake, in Bulgaria, with an area of 1464 ha. It was followed in 1995 by the two new designed sites in Ukraine—the Kartal and Kugurlui lakes. Currently, the Ramsar network along the Lower Danube Valley consists of 17 sites. All Ramsar sites that are located on the Romanian side of the Danube valley were established in the second decade of the twentieth century [120] and overlap with Natura 2000 sites, which are Special Protected Areas for rare and vulnerable birds and for regularly migratory species.

As a member states of the European Union, both Bulgaria and Romania designated special protected areas under two main directives that regulate nature conservation in EU. The Sites of Community Importance (SCIs), designed under Habitats Directive [90] and Special Protected Areas (SPAs), designed under Birds Directive [91] compose the Natura 2000 sites network. In this way each member state is obliged to guarantee favorable conservation status of the species and habitats for the protection of which those sites have been designated. The Natura 2000 network along the Lower Danube Valley comprises 64 sites [121], both SCIs and SPAs. All sites, 20 in Romania and 44 in Bulgaria, cover a total area of nearly 585,701 ha (Fig. 13). For the majority of them, their borders are extended beyond of the morphological floodplain.

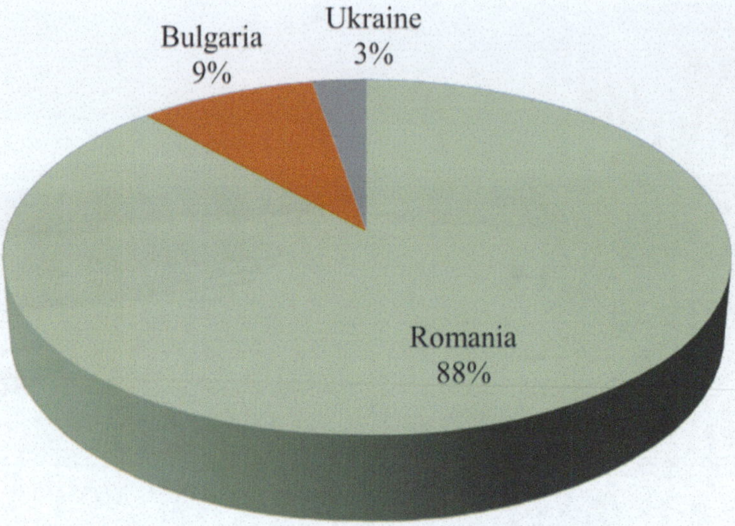

Fig. 12 Percentage of area of designed Ramsar sites in Lower Danube River Valley in each riverine country. Total area of the Ramsar sites: 248,398 ha. Data compiled from [120]

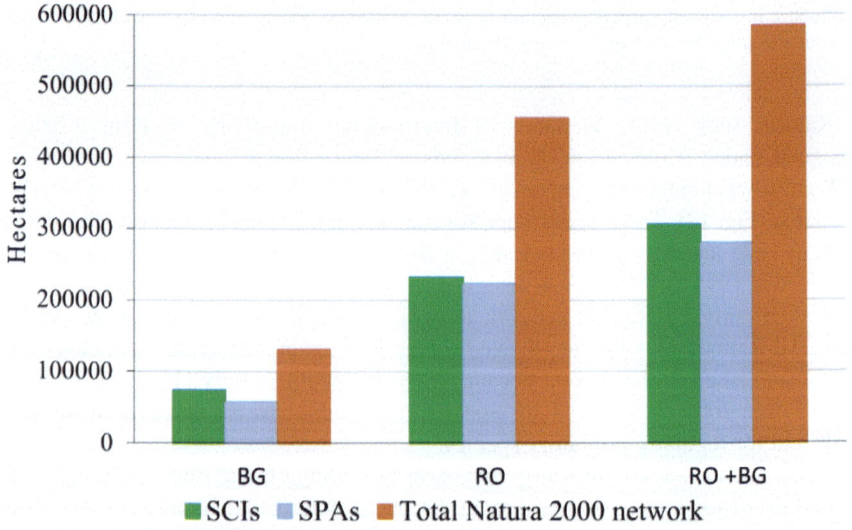

Fig. 13 Area of Natura 2000 sites in the Lower Danube River Valley. Data compiled from Natura 2000 standard data forms available at [122]

The area of SCIs is 306,010.6 ha, while the SPAs totalize 280,691.2 ha, but most of them are overlapped with each other. The designed SCIs support 22 habitat types of those listed on the Annex I of Habitats Directive, some of them being priority habitats (Table 1).

Table 1 Habitat types that have been identified along the Lower Danube Valley according to Annex I of Habitats Directive

Nr Crt	Habitat code	Habitat name
1	2160	Dunes with *Hippophae rhamnoides*
2	2190	Humid dune-slacks
3	2340[a]	Pannonic inland dunnes
4	3130	Oligotrophic to mesotrophic standing waters with vegetation of the *Littorelletea uniflorae* and/or of the *Isoëto-Nanojuncetea*
5	3140	Hard oligo-mesotrophic waters with benthic vegetation of *Chara* sp.
6	3150	Natural eutrophic lakes with Magnopotamion or Hydrocharion-type vegetation
7	3260	Watercourses of plain to mountain levels with *Ranunculion fluitantis* and *Calitricho-Batrachion* vegetation
8	3270	Rivers with muddy banks with *Chenopodion rubri* p.p. and *Bidention* p.p. vegetation
9	40A0[a]	Subcontinental peri-Pannonic scrub (*Syringo-Cotinion*)
10	40C0[a]	Ponto-Sarmatic deciduous thickets
11	62C0[a]	Ponto-Sarmatic steppes
12	6120[a]	Xeric sand calcareous grasslands (in association with non-coastal dune complexes)
13	6210	Smi-natural dry grasslands and scrubland facies on calcareous substrates (*Festuco-Brometalia*)
14	6430	Hydrophilous tall herb fringe communities of plains and of the montane to alpine levels
15	6440	Alluvial meadows of river valleys of the *Cnidion dubii*
16	6510	Lowland hay meadows (*Alopecurus pratensis, Sanguisorba officinalis*)
17	91AA[a]	Eastern white oak woods
18	91F0	Riparian mixed forests of *Quercus robur, Ulmus laevis* and *Ulmus minor*, *Fraxinus excelsior* or *Fraxinus angustifolia*, along the great rivers (*Ulmenion minoris*)
19	91I0[a]	Euro-Siberian steppic woods with *Quercus* spp.
20	91M0	Pannonian-Balkanic turkey oak–sessile oak forests
21	92A0	*Salix alba* and *Populus alba* galleries
22	92D0	Southern riparian galleries and thickets (*Nerio-Tamaricetea* and *Securinegion tinctoriae*)

Data compilated after [40, 42] and Natura 2000 standard data forms available at [122]
[a]European priority habitats according to Habitat Directive 92/43/EEC [90]

However, the total coverage of protected areas is over 70% for the active floodplain, including the Small Island of Brăila. There is only one biosphere reserve along the Lower Danube Valley. It is the Srébarna Biosphere Reserve, since 1977, in Bulgaria. As well as the Danube Delta, the Srébarna Lake is a World Heritage Site.

7 The Main Threats of the Abiotic and Biotic Environment

7.1 Navigation and Hydraulic Works

All countries that are crossed by the Danube use this river for inland navigation. In Romania, being the only waterway, this river plays an important role in the national transport system, especially, taking into account that the inland navigation has a share of more than 20% of all cargo in 2011 [123]. The water flow of the Lower Danube River shows seasonal variation, with the highest values in spring (April–May) and minimum amount in September. In the last two decades compared to the multiannual average, the water flow decreases in the main Lower Danube branch during the summer and autumn, which result in low water levels and low fairway depths, making navigation difficult in certain critical sections. For example, from the Iron Gates II (Rkm 863) until Călărași-Silistra point (Rkm 575), there are 38 sections with significant navigational constraints [39]. Yearly, the number of days with low waters, and thus low fairways, induced by climate changes along the several sectors of the Danube, downstream of Rkm 863, ranges from 7 to 70 [123].

There are many studies and projects concerning the navigation on the Danube, including some related to identification and mitigation the navigation difficulties on the Romanian-Bulgarian common sector [124]. In order to ensure a minimum navigable depth in the navigation channel of the critical sections (bottlenecks, low fairways) during the low-water period of summer/autumn, when the Danube water flow decreases substantially, five type hydraulic works were chosen as technical solution to be carried out during 2012–2022: bank protection, groins, chevrons, bottom sills, and dredging [39, 124]. The critical sections that are subject of hydraulic works overlap with the majority of the designed Natura 2000 sites along the Lower Danube River Valley, on both the Romanian as well as the Bulgarian side. However, of all 66 Natura 2000 sites designed along the Lower Danube Valley, within the borders of the 29 sites, of which 18 sites in Bulgaria and other 11 sites in Romania, were planned hydraulic works.

7.2 Invasive Alien Species

Invasive alien species is one of the major threats to native biodiversity in the Danube River basin. Its vulnerability to invasive species has increased since it was constructed the Rhine-Main-Danube canal. Thus, the Danube River is the Southern migration corridor of Ponto-Caspian species in Europe [125] and part of the European inland invasion network, with its mouths as "invasion gateways" [126]. The biological contamination of the Danube River is a serious issue given that after the three research expeditions performed in the first decade of twenty-first century along the river it has been identified six species of neophytes, 19 alien macroinvertebrates, and 15 non-native fish species [127].

There are reported two invasive aquatic macrophytes in the Lower Danube, *Vallisneria spiralis* and Canadian waterweed (*Elodeea canadensis*), but in the last years western waterweed (*Elodea nuttalii*) has progressively invaded the Danube Corridor from west to east, replacing Canadian waterweed [127]. Monospecific stands of *common* cocklebur (*Xanthium strumarium*) and indigo bush (*Amorpha fruticosa*) cover large areas along the river banks, both in Romania and Bulgaria [128], but these noxious species invaded meadows, shore of fluvial islands, and man-made habitats, also. On the Bulgarian side of the Danube valley it is mentioned that *Amorpha fruticosa* has been already broadly distributed in tree communities dominated by *Salix alba*, *Populus alba*, *P. nigra*, *Quercus robur*, and *Ulmus effuses* since the middle of twentieth century [128]. Nowadays, in Romania, the Danube valley, including delta, is the most affected area by *Amorpha fruticosa* [129].

Apart of the *Amorpha fruticosa*, the riparian forests, which correspond to habitats 92A0 *Salix alba* and *Populus alba* galleries and 91F0 Riparian mixed forests of *Quercus robur*, *Ulmus laevis* and *Ulmus minor*, *Fraxinus excelsior* or *Fraxinus angustifolia* along the great rivers, listed on Habitat Directive 92/43/EEC, are invaded by poplar hybrids from poplar plantations as well as individuals of tree of heaven (*Ailanthus altissima*), box elder (*Acer negundo*), white ash (*Fraxinus americana*), green ash (*Fraxinus pennsylvanica*), and black locust (*Robinia pseudoacacia*) which were also planted in the past because of their ecological versatility or melliferous flowers, as in case of the black locust. The devil's beggarticks (*Bidens frondosa*), daisy fleabane (*Erigeron annus*), rib-seeded sandmat (*Chamaesyce glyptosperma*), duckweed (*Portulaca oleracea*), the annual vines star cucumber (*Sicyos angulatus*), and bur cucumber (*Echinocystis lobate*) are present also within Danube corridor, both in natural and man-made habitats.

Within terrestrial ecosystems, the alien and invasive species penetrate mainly from old forest plantations located in vicinity of them. Riparian forests are invaded by poplar hybrids as well as individuals of *Amorpha fruticosa*, *Ailanthus altissima*, *Acer negundo*, *Fraxinus americana*, *Fraxinus pennsylvanica*, which were also planted in the past because of their ecological versatility, but now they have become noxious invaders, as in the case of *Amorpha fruticosa* and *Ailanthus altissima*.

Concerning the Neozoa species added to the original Danubian biota, the invasive aquatic macroinvertebrates that have been recorded in the Lower Danube River section are: the flat worm *Dugesia tigrina*, the aquatic bivalve mollusks *Corbicula fluminalis*, *Corbicula fluminea* (Asian mussel), and *Sinanodonta woodiana* (Chinese Pond Mussel), the tubificid worm *Branchiura sowerbyi*, and the spiny cheek crayfish *Orconectes limosus*. Nonindigenous fish recoded are *Ameiurus punctatus*, *Hypophthalmichthys molitrix*, *Hypophthalmichthys nobilis*, *Lepomis gibbosus*, *Perccottus glenii*, *Polyodon spathula*, and *Pseudorasbora parva*. It is expected that the number of Neozoa arrived from the Western European region, especially from the Rhine River, to increase in future [29]. Overall, in the Danube River, the benthic fauna assemblages are dominantly composed by nonindigenous, invasive or cosmopolitan elements that had been stabilized/naturalized for a long time [127].

8 Conclusion

Converting the Lower Danube floodplain with its high diverse ecosystem units into predominantly agricultural regions is the most devasting and abrupt anthropogenic transformation of the Lower Danube Valley although dramatic irreversible changes were made along the Iron Gates Gorge after construction of the two dams, the Iron Gates I, in 1972 and the Iron Gates II, in 1984. In addition, the navigation, chanel regularization and dredging have amplified and modified the dynamics of the area.

The remnant active floodplains areas, naturally, change in relation to the amount of precipitation coming from the entire Danube River basin and in relation to the Danube River discharge variation. Pedoclimatic conditions and hydro-geomorphological factors have a major influence on the dynamics of solid and liquid flows in strong correlation with sedimentation and erosion processes. Particular refugees and hotspots of biodiversity are fluvial islands.

Starting with 1990, a moment that coincides with the change of political regime in all riverine countries of the Lower Danube River, many actions were initiated for the protection and conservation of the wetlands. In order to minimize the effects of human activity and to conserve and preserve the special components of biodiversity, national, regional and, international measures were undertaken. As a result, nowadays, a network of various national and international protected areas and nature reserves overlaps the Lower Danube Valley.

Human activities, such as navigation, shore regularization, dredging, dams, hydropower plants have amplified and modified the dynamics of the area. In order to minimize the effects of human activity and to conserve the special components of biodiversity, in the Lower Danube Valley were declared a number of 64 Natura 2000 sites, both in Romania and Bulgaria, with a total area of 585,701 ha.

The Natura 2000 sites network materrialises the environmental policy of European Union, in order to ensure the long-term survival of Europe's most valuable and threatened species and habitats. The management of these protected areas having as main tool the management plan, together with the adequate evaluation study, tried to find a balance between the economic development of the area and the conservation of biodiversity.

These instruments are decisive for the financing of projects aiming at improving the navigation conditions on the Danube River, electricity production and the exploitation of resources. Also, significant areas from former active floodplain are planned to be restored and integrated in sustainable development projects.

The revitalization of agriculture in a conventional system will update the problem of diffuse pollution with nutrients and pesticides of surface waters, implicitly the Danube.

Despite all current actions, there is still a strong anthropogenic influence which, together with long-term climate change, will lead to major changes in the Lower Danube Valley. Converting the floodplain wetlands and lakes into predominantly agricultural regions is the most devastating and abrupt anthropogenic transformation of the Lower Danube Valley during the twentieth century.

9 Recomandation

The Lower Danube Valley is a very dynamic area. The terrestrial, semi-aquatic, and aquatic surfaces of the Lower Danube Valley naturally change in relation to the amount of precipitation coming from the entire Danube basin. Also, this sector of the Danube River represents an important biodiversity reservoir conserved based on the designation of several types of protected areas.

The management of all categories protected areas having as main tool the management plan together with the adequate evaluation study tried to find a balance between the development of the area and the conservation of biodiversity.

These instruments constrain the financing of projects for the improvement of the navigation conditions on the Danube, electricity production and the exploitation of resources.

Although there are concerns about harmonizing economic activities and conserving biodiversity, economic interests take precedence. That is why we recommend finding a balance between economic interests and biodiversity conservation by involving all the factors—authorities, stakeholders.

An important threat to biodiversity is the presence of invasive species; in most of the site management plans there are provided control measures. The appearance of invasive species is a dynamic process, so we recommend periodic inventory, monitoring and finding appropriate control measures. Periodic inventory, monitoring and control are recommended to be introduced in the process of reviewing and updating management plans.

In the last decade, the decrease in the use of fertilizers in agriculture has reduced the level of diffuse pollution. The revitalization of agriculture in a conventional system and development of industry will update the problem of diffuse pollution with nutrients, pesticides, and other pollutants both ground and surface waters, implicitly the Danube River. Consequently, we recommend the improvement of the legislation on the use of chemical fertilizers and the control on the treatment of industrial and domestic wastewater.

The development of new programs and implementation of already existing natural restoration projects will have benefic effects both in conserving biodiversity and mitigating the negative effects of the impredictibile devastating floods.

References

1. Sommerwerk N, Baumgartner C, Bloesch J, Hein T, Ostojic A (2009) The Danube River Basin. In: Robinson CT, Tockner K, Uehlinger U (eds) Rivers of Europe. Elsevier/Academic Press, Amsterdam, pp 59–112
2. ICPDR (2009) Danube River Basin District Management Plan (Vol. IC/151). International Comiision for the protection of the Danube River, Vienna
3. Roekaerts M (2002) The Biogeographical Regions Map of Europe. Basic principles of its creation and overview of its development. European Environment Agency
4. Tockner K, Stanford JA (2002) Riverine flood plains: present state and future trends. Environ Conserv 29(3):308–330. https://doi.org/10.1017/S037689290200022X
5. Mihăilescu V (1969) Geografia Văii Dunării Românești. Editura Academiei Române, București
6. Antipa G (1910) Regiunea inundabilă a Dunării și mijloacele de a o pune în valoare. Institul de Arte Grafice Carol Carol Göbl, București
7. Antipa G (1921) Dunărea și problemele ei științifice, economice și politice. Librăriile Cartea Românească și Pavel Suru, București
8. Bușniță T (1961) Studiul limnologic al Dunării, problemă actuală de interes internațional. Dezbaterile pe acesată temă în consfătuirile ținute la Viena. Hidrobiologia 2:7–28. Editura Academiei Republicii Socialiste România, București
9. Haidvogi G, Janauer G (2019) Six decades of scientific cooperation in the Danube River Basin. Danube News 21(40):5–9. https://www.danube-iad.eu/docs/DN_articles/2019-Six_decades_scientific_cooperation-DanubeNews-Haidvogl_Janauer.pdf. Accesed 19 Sept 2020
10. Liepold R (1967) Limnologie der Donau. Eine monographische Darstellung. Schweizerbart'sche Verlagsbuchhandlung, Stuttgart
11. Ardelean I, Arion-Prunescu E, Banu AC, Bănărescu P, Boisnard J, Brezeanu Gh, Bușniță Th (1967) Limnologia sectorului românesc al Dunării. Studiu monografic. Editura Academiei Republicii Socialiste România, București
12. Schneider E (2002) The ecological functions of the Danubian floodplains and their restoration with special regard to the Lower Danube. Arch Hydrobiol Large Rivers 13(1–2):129–149
13. Hein T, Schwartz U, Habersack H et al (2016) Current status and restoration options for floodplains along the Danube River. Sci Total Environ 543(Part A):778–790. https://doi.org/10.1016/j.scitotenv.2015.09.073
14. Mölder A, Schneider E (2011) On the beautiful diverse Danube? Danubian Floodplain forest vegetation and flora under the influence of river eutrophication. River Res Appl 27:881–894. https://doi.org/10.1002/rra.1403
15. COM (2011) Our life insurance, our natural capital: an EU biodiversity strategy to 2020. COM(244). http://eur-lex.europa.eu/legal-content/EN/TXT/PDF/?uri=CELEX:52011DC0244&from=EN. Accesed 7 June 2020
16. Council of Europe (1996) Pan-European biological and landscape diversity strategy. Strategic action plan 1996–2000. Nat Environ 74:1–66
17. https://mapire.eu. MAPIRE—Historical Maps Online. Lesser and Greater Wallachia (Oltenia and Muntenia) (1790). https://mapire.eu/en/map/romania-1790/. Accessed 5 June 2020
18. Banu AC (1967) Date geografice aupra bazinului și cursului inferior al Dunării. In: Banu AC (ed) Limnologia sectorului Românesc al Dunării. Studiu Monografic. Editura Academiei Republicii Socialiste România, București, pp 17–68
19. Posea, G (2005) Lunca. In: Posea G, Bogdan O, Zăvoianu I (eds) Geografia României: Câmpia Română, Dunărea, Podişul Dobrogei, Litoralul Românesc al Mării Negre şi Platforma Continentală, vol 5. Editura Academiei Române, București, pp 497–497
20. Enaceanu, V, Brezeanu, G, Prunescu-Arion, E (1967) Lunca inundabilă a Dunării. In: Banu AC (ed) Limnologia sectorului Românesc al Dunării. Studiu Monografic București: Editura Academiei Republicii Socialiste România, București, pp 372–426
21. Douglas I (1977) An introduction to systematic geomorphology. Humid landforms, vol 1. Australian National University Press, Canberra

22. http://wildisland.danubeparks.org. DANUBEPARKS (Danube River Network of Protected Areas). WILDisland—The Danube Wild Island Habitat Corridor. http://wildisland.danubeparks.org/#wildisland. Accessed 5 Dec 2020
23. LIFE06 NAT/RO/000177 (n.d.) GREENDANUBE—Conservation and integrated management of Danube. https://ec.europa.eu/environment/life/project/Projects/index.cfm?fuseaction=search.dspPage&n_proj_id=3111&docType=pdf. Accessed 2 Aug 2020
24. Sommerwerk N, Bloesch J, Paunović M, Baumgartner C, Venohr M, Schneider-Jacoby M, Hein T, Tockner K (2010) Managing the world's most international river: the Danube River Basin. Mar Freshw Res 61(7):736–748. https://doi.org/10.1071/MF09229
25. Wyrick JR, Klingeman PC (2010) Proposed fluvial island classification scheme and its use for river restoration. River Res Appl 27(7):814–825. https://doi.org/10.1002/rra.1395
26. DANUBEPARKS (2019) Ecological conectivity in the Danube River Basin. Future perspectives and guiding principles. Orth an der Donau. http://www.interreg-danube.eu/uploads/media/approved_project_output/0001/38/f13b76e9c045bbd128b19ab988cff27d5bb4428b.pdf. Accessed 10 Oct 2020
27. https://wwfeu.awsassets.panda.org (2010) Assessment of the restoration potential along the Danube and main tributaries, Vienna. https://wwfeu.awsassets.panda.org/downloads/wwf_restoration_potential_danube.pdf. Accessed 10 Oct 2020
28. Ioaniţoaia H, Dobre V, Moraru N (2007) Un secol (1906–2006) de lucrări de îndiguiri şi amenajări hidroameliorative în lunca Dunării. Hidrotehnica 52(1–2):41–45. https://old.anif.ro/aifcr/hioanitoaia/hioanitoaia1.htm. Accessed 2 Aug 2020
29. ICPDR (2005) The Danube River Basin District. International Commission for the Protection of the Danube River, Vienna. https://www.icpdr.org/main/sites/default/files/nodes/documents/danube_basin_analysis_2004.pdf. Accessed 29 Nov 2020
30. Schneider-Binder E (2009) Floodplain forests along the Lower Danube. TRSER 8:113–136. http://stiinte.ulbsibiu.ro/trser/trser8/113-136.pdf. Accessed 4 Oct 2020
31. Enculescu P (1924) Zonele de vegetaţie lemnoasă din România in raport cu condiţiile orohidrografice, climatice, de sol şi de subsol. Memoriile Institutului Geologic, Bucureşti
32. Freer-Smith P, Muys P, Bozzano B, Drössler L, Farrelly N, Jactel H, Korhonen J, Minotta G, Nijnik M, Orazio, C (2019) Plantation forests in Europe: challenges and opportunities. From Science to Policy 9. European Forest Institute. https://doi.org/10.36333/fs09
33. Stoiculescu, CD (2008) Reconstrucția ecologică a zonei inundabile a Dunării Românești. Green Steps SRL, București
34. Ciuvăţ AL, Abrudan IV, Blujdea V, Marcu C, Dinu C, Enescu M, Nuţă IS (2013) Distribution and peculiarities of black locust in Romania. Revista de silvicultură și cinegetică 18(32):76–85. https://www.cabi.org/ISC/FullTextPDF/2014/20143263765.pdf. Accessed 12 Aug 2020
35. Cross A, Adamec L (2020) Aldrovanda vesiculosa. The IUCN Red List of Threatened Species 2020, e.T162346A83998419. https://doi.org/10.2305/IUCN.UK.2020-1.RLTS.T162346A83998419.en. Accessed 7 June 2020
36. Botnariuc N, Boldor O, Şerbănescu G, Beldescu S (1964) Relații între stuf (Phragmites communis) și papură (Typha angustifolia) în zona inundabilă a Dunării. Hidrobiologia 5:165–176. Editura Academiei Republicii Socialiste România, București
37. Rooth J, Stevenson JC, Cornwell JC (2003) Increased sediment accretion rates following invasion by Phragmites australis: the role of litter. Estuaries Coast 26(2):475–483. https://doi.org/10.1007/BF02823724
38. Strat D, Vespremeanu E (2018) The evolution of Modern Sfantu Gheorghe sub-delta (Danube Delta, North-Western Black Sea) under biological factors control. In: Halicki W (ed) Wetlands: functions, services, importances and threats. Nova Science Publishers, New York, pp 155–188
39. Munteanu C, Gheorghe IF, Plăiașu R, Stan M, Tatole V, Murariu D (2009) Flora și fauna ostroavelor de pe Dunăre. Green Steps, Brașov
40. WWF (2009) An Atlas for the Lower Danube Green Corridor. Kraft-Druck Etlingen. https://www.grafillu.de/download/Atlas_der_unteren_Donau_WWF_Germany_2009.pdf. Accessed 14 Sept 2020

41. Munteanu CL, Enache V (2010) Conservation and integrated management on the Danube Islands, Romania: LIFE06NAT/RO/000177. Green Steps, Brașov
42. Mihăilescu S, Strat D, Cristea I, Honciuc V (2015) Raportul sintetic privind starea de conservare a speciilor și habitatelor de interes comunitar din România. Dobrogea, Constanța
43. IUCN Red List. Sturgeons. https://www.iucnredlist.org/search?query=Sturgeons&searchType=species. Accessed 3 Mar 2021
44. Gessner J, Freyhof J, Kottelat M (2010a) Acipenser gueldenstaedtii.The IUCN Red List of Threatened Species 2010: e.T232A13042340. https://doi.org/10.2305/IUCN.UK.2010-1.RLTS.T232A13042340.en. Accessed 3 Jan 2021
45. Gessner J, Freyhof J, Kottelat M (2010b) Acipenser ruthenus. The IUCN Red List of Threatened Species 2010: e.T227A13039007. https://doi.org/10.2305/IUCN.UK.2010-1.RLTS.T227A13039007.en. Accessed 3 Jan 2021
46. ICPDR (International Commission for the Protection of the Danube River) (2018) ICPDR Sturgeon Strategy. https://www.icpdr.org/main/activities-projects/sturgeons-danube-bas2in. Accessed 20 Oct 2020
47. Murariu D (2005) The state of the Mammals (Mamalia) along the Danube between Gârla Mare and Călărași (Romania). Trav Mus Natl Hist Nat Grigore Antipa 48:427–445
48. Călinescu R (1930) Şacali în România. Arh Olten 9(49–50):223–225
49. Călinescu R (1935) Mamifere găsite în Oltenia. Arh Olten 14(77–78):139–140
50. Arnold J, Hummer A, Heltai M, Murariu D, Spassov N, Hackländer K (2012) Current status and distribution of golden jackals Canis aureus in Europe. Mammal Rev 42(1):1–11. https://doi.org/10.1111/j.1365-2907.2011.00185.x
51. Spassov N, Acosta-Pankov I (2019) Dispersal history of the golden jackal (Canis aureus moreoticus Geoffroy, 1835) in Europe and possible causes of its recent population explosion. BDJ 7:e34825. https://doi.org/10.3897/BDJ.7.e34825
52. Plass J (2007) Dokumentation einer zweiten Einwanderungswelle des Goldschakals Canis aureus Linnaeus 1758 in Österreich aus den Jahren 2003–2006. Stutt Beitr Naturkd 17:55–68. https://www.zobodat.at/pdf/BNO_0017_0055-0068.pdf. Accessed 27 Aug 2020
53. Marfiuc L (2019) The position of the jackal among canine predators in Romania according to craniometry and dentition. Abstract PhD thesis. Universitatea Transilvania, Brașov. https://www.unitbv.ro/documente/cercetare/doctorat-postdoctorat/sustinere-teza/2019/marfiuc-liviu/Rezumat_Marfiuc.pdf. Accessed 14 Oct 2020
54. Şelaru N (2015) Istoria, răspândirea și înmulțirea șacalului auriu în România. Asociaţia Generală a Vânătorilor şi Pescarilor Sportivi din România, București. http://agvps.ro/docs/sacalul%20auriu.pdf. Accessed 20 Oct 2020
55. Shahbazi AS (1996) Darius iii. Darius I the Great. In: Enciclopedia Iranica, vol VII/1, pp 41–50. http://www.iranicaonline.org/articles/darius-iii. Accessed 8 June 2020
56. Hartley CW (1989) A biography of Sir Charles Hartley, Civil Engineer (1825–1915): the father of the Danube. The Edwin Mellen Press, Lewiston
57. Vannote RL, Minshall GM, Cummins KW, Sedell JR, Cushing E (1980) The river continuum concept. Can J Fish Aquat Sci 37:130–137
58. Vidrașcu I D (1921) Valorificarea regiunii inundabile a Dunării. Tipografia Urbană, București
59. Petts GE (1984) Impounded rivers: perspectives for ecological management. Wiley, Chichester
60. Thomson J (2002) Wetlands drainage, river modification, and sectoral conflict in the lower Illinois Valley, 1890–1930. Southern Illinois University Press, Carbondale
61. Bularda M, Vișinescu I (2014) Incintele indiguite la Dunăre - probleme actuale și de perspectivă. Brăila: Academia de Ştiințe Agricole și Silvice «Gheorghe Ionescu Sisești», Stațiunea de Cercetare-Dezvoltare Agricolă Brăila. http://www.asas.ro/wcmqs/noutati/interes-public/Incintele%20indiguite%20la%20Dunare%20-%20probleme%20actuale%20si%20de%20perspectiva.pdf. Accessed 2 June 2020
62. Bondar C, State I, Cernea D, Harabagiu E (1991) Waterflow and sediment transport of the Danube at its outlet into the Black Sea. Russ Meteorol Hydro 21(1):21–25
63. Panin N, Jipa D (1998) Danube river sediment input and its interaction with the north-western Black Sea. GEO-ECO-Marina 3:23–35. https://www.geoecomar.ro/website/publicatii/Nr.3-1998/4.NicolaePanin.pdf. Accessed 3 Oct 2020

64. Vădineanu A (2001) Lower Danube System. Observatorio Medioambiental 4:373–402. Universidad Complutense de Madrid
65. Panin N, Jipa D (2002) Danube River sediment input and its interaction with the North-Western Black Sea. Estuar Coast Shelf Sci 54(2):551–562
66. Constantinescu Ș, Achim D, Rus I, Giosan L (2015) Embanking the Lower Danube: from natural to engineered floodplains and back. In: Hudson P, Middelkoop H (eds) Geomorphic approaches to integrated floodplain management of lowland fluvial systems in North America and Europe. Springer, New York, pp 259–281. https://doi.org/10.1007/978-1-4939-2380-9_11
67. Ebert, S, Hulea, O, Strobel, D (2009) Floodplain restoration along the lower Danube: a climate change adaptation case study. Clim Devel 1(3:Lessons for climate change adaptation from better management of rivers):212–219. https://doi.org/10.3763/cdev.2009.0022
68. Teodoru C, Wehrli B (2005) Retention of sediments and nutrients in the Iron Gate I Reservoir on the Danube River. Biogeochemistry 76(3):539–565. https://doi.org/10.1007/s10533-005-0230-6
69. Kalchev R., Ionica D, Beshkova M, Botev I, Sandu C (2008) Long-term and seasonal changes of nutrients, seston and phytoplankton concentration in the Lower Danube (Bulgarian-Romanian stretch). Fund Appl Limnol 18. Large Rivers Suppl 166(1–2):25–43
70. Hein T, Funk PF, Graf W, Zsuffa I, Haidvogl G, Schinegger R, Weigelhofer G (2018) Management challenges related to long-term ecological impacts, complex stressor interactions, and different assessment approaches in the Danube River Basin. River Res Appl 35:500–509. https://doi.org/10.1002/rra.3243
71. Gâștescu P (2018, Sept 5–9) The Danube floodplain in the pontic sector: ecosystem services, anthropic modifications and ecological reconstruction. In: Gâștescu P, Brețcan P (eds) Water resources and wetlands, 4th International Conference Water Resources and Wetlands, Tulcea, pp 1–9
72. Trifanov C, Mihu-Pintilie A, Tudor M, Mierlă M, Doroftei M, Covaliov S (2020) Romanian Danube River floodplain functionality assessment. In: Negm AM, Romanescu G, Zeleňáková M (eds) Water resources management in Romania. Springer, Cham, pp 251–277. https://doi.org/10.1007/978-3-030-22320-5_8
73. Palmer M, Hondula KL, Koch BJ (2014) Ecological restoration of streams and rivers: shifting strategies and shifting goals. Annu Rev Ecol Evol Syst 45(1):247–269. https://doi.org/10.1146/annurev-ecolsys-120213-091935
74. Grygoruk M, Acreman M (2015) Restoration and management of riparian and riverine ecosystems: ecohydrological tools, experiences and perspectives. Ecohydrol Hydrobiol 15(3):109–110. https://doi.org/10.1016/j.ecohyd.2015.07.002
75. Szałkiewicz E, Jusik S, Grygoruk M (2018) Status of and perspectives on river restoration in Europe: 310,000 Euros per hectare of restored river. Sustainability 10(1)(129):1–15. https://doi.org/10.3390/su10010129
76. Rolston H (1988) Environmental ethics. Duties to and values in the natural world. Temple University Press, Philadelphia
77. Arthington AH, Naiman RJ, McClain ME, Nilsson C (2009) Preserving the biodiversity and ecological services ofrivers: new challenges and research opportunities. Freshw Biol 55(1):1–16. https://doi.org/10.1111/j.1365-2427.2009.02340.x
78. Olesen KW, Refsgaard A, Havno K (2000) Restoring river ecology—a complex challenge. In: Maione U, Majone-Lehto B, Monti R (eds) New trends in water and environmental engineering for safety and life. CRC Press, Rotterdam, pp 39–48
79. Westra L (1995) Ecosystem integrity and sustainability: the foundational value of the wild. In: Westra L, Lemons J (eds) Perspectives on ecological integrity. Environmental science and technology library, vol 5. Springer, Dordrecht, pp 12–33. https://doi.org/10.1007/978-94-011-0451-7_2
80. Cairns JJ, Heckman JR (1996) Restoration ecology: the state of Emerging Field. Annu Rev Energ Environ 21:167–189
81. Cairns J (2001) Sustainability, exceptionalism, and exemptionalism. Ecosyst Health 7(3):147–154. https://doi.org/10.1046/j.1526-0992.2001.01025.x

82. Darby S, Sear D (2008) Uncertainity in river restoration. Wiley, Chikester, West Sussex, UK
83. Tockner K, Malard F, Ward JV (2000) An extension of the flood pulse concept. Hydrol Process 14(16–17):2861–2883. https://doi.org/10.1002/1099-1085(200011/12)14:16/17%3c2861::AID-HYP124%3e3.0.CO;2-F
84. Lemons J, Victor R (2008) Uncertainity in river restoration. In: Darby S, Sear D (eds) River Restoration: managing the uncertainty in restoring physical habitat. Wiley, Chikester, West Sussex, UK, pp 3–14
85. Leroy Poff N, Zimmerman JK (2010) Ecological responses to altered flow regimes: a literature review to inform the science and management of environmental flows. Freshw Biol 55(1):194–2005. https://doi.org/10.1111/j.1365-2427.2009.02272.x
86. ECRR (European Centre for River Restoration). ECRR start and development. https://www.ecrr.org/About/ECRR. Accessed 5 Dec 2020
87. Fokkens B (2001) Preface. In: Nuland HJ, Cals MJR Conference of the river resoration. Proceedings, Wageningen, The Netherlands, 2000. Institute for Inland Water Management and Waste Water Treatment, Lelystad
88. WDF (2000, Dec 22) EU Water Framework Directive—Directive 2000/60/EC of the European Parliament and of the Council establishing a framework for the Community action in the field of water policy. Official Journal 327. https://eur-lex.europa.eu/legal-content/EN/TXT/?uri=CELEX:32000L0060. Accessed 12 Sept 2020
89. COM (2007) Directive 2007/60/EC of the European Parliament and of the Council of 23 October 2007 on the assessement and management of flood risks. Official Journal of the European Union L288/27. https://eur-lex.europa.eu/legal-content/EN/TXT/?uri=celex:32007L0060. Accessed 7 June 2020
90. Council Directive 92/43/EEC (1992, July 22) Council Directive 92/43/EEC of 21 May 1992 on the conservation of natural habitats and of wild fauna and flora. Official Journal L(206):0007–0050. https://eur-lex.europa.eu/legal-content/EN/TXT/?uri=celex%3A31992L0043. Accessed 29 Nov 2020
91. Directive 2009/147/E (2010, Jan 21) Directive 2009/147/EC of the European Parliament and of the Council of 30 November 2009 on the conservation of wild birds. Oficial Journal L(20):7–25. https://eur-lex.europa.eu/legal-content/EN/TXT/?uri=CELEX%3A32009L0147. Accessed 29 Nov 2020
92. www.ramsar.org. Ramsar (1971) Convention on Wetlands of International Importance especially as Waterfowl Habitat. https://www.ramsar.org/sites/default/files/documents/library/current_convention_text_e.pdf. Accessed 7 June 2020
93. Schneider E, Tudor M, Staraș M (2008) Ecological restoration in the Danube Delta Biosphere Reserve/Romania. Evolution of Babina polder after restoration works. WWF Germany/Danube Delta National Institute for Research and Development Tulcea, Kraft-Druck Ettlingen
94. Dynesius M, Nilsson C (1994, Nov 4) Fragmentation and Flow Regulation of River Systems in the Northern Third the World. Science, New Series, 266(5186):753–762. https://doi.org/10.1126/science.266.5186.753
95. Wong CM, Williams CE, Pittock J, Collier U, Schelle P (2007) World's top 10 rivers at risk. WWF International, Gland, Switzerland. https://wwfeu.awsassets.panda.org/downloads/worldstop10riversatriskfinalmarch13_1.pdf. Accessed 12 Oct 2020
96. ICPDR (2006, April 10) The Danube River Basin. International Commission for the Protection of the Danube River, Vienna, Austria. http://www.icpdr.org/main/danube-basin/river-basin. Accessed 20 Aug 2020
97. Mansourian S, Doncheva N, Valchev K, Vallauri D (2019) Experiences in Forest Landscape Restoration. Lessons learnt from 20 years of floodplain forest: the Lower Danube landscape. WWF-France
98. Declaration on the Cooperation for the Creation of a Lower Danube Green Corridor (2000, June 5) p. 1–3. http://awsassets.panda.org/downloads/ldgcdeclaration.pdf. Accessed 20 July 2020

99. wwf.panda.org (2010) Lower Danube Green Corridor Agreement. http://danube.panda.org/wwf/web/search/details.jsp?pid=4. Accessed 12 Sept 2020
100. DPRP (Danube Pollution Reduction Programm) UNPD/GEF (1999) Evaluation of wetlands an floodplain areas in the Danube River basin. WWF Danube Carpathian Programme and WWF Floodplain Institute (WWF-Germany). https://www.icpdr.org/main/sites/default/files/EVALUATIONWETLANDSFLOODPLAINAREAS.pdf. Accessed 12 July 2020
101. Olson DM, Dinerstein E (1998) The Global 200: a representation approach to conserving the Earth's most biologically valuable ecoregions. Conserv Biol 12(3):502–515
102. Olson DM, Dinerstein E (2002) The Global 200: priority ecoregions for global conservation. Ann Missouri Bot Gard 89:199–224
103. wwfeu.awsassets.panda.org. Declaration on the cooperation for the creation of a Lower Danube Green Corridor. Signed 5 June 2000, Bucharest, Romania. https://wwfeu.awsassets.panda.org/downloads/ldgcdeclaration.pdf. Accessed 8 June 2020
104. www.ramsar.org (2000) Creation of Lower Danube Corridor, and gift to the Earth. https://www.ramsar.org/news/creation-of-a-lower-danube-green-corridor-and-gift-to-the-earth. Accessed 8 June 2020
105. Csagoly P, Magnin G, Hulea O (2016) Lower Danube Green Corridor. In: Finlayson C, Milton G, Prentice R, Davidson N (eds) The Wetland Book. Springer, Dordrecht, pp 1–6. https://doi.org/10.1007/978-94-007-6173-5_251-1
106. Ştiucă R, Staraş M, Nichersu I (2010) Preparation of the Danube River's Revitalization of the finalized proposed projects for the assessment with the selection at least two projects per every standard criterion. Danube Delta National Institute for Research and Development, Tulcea
107. ICPDR (2009) Danube River Basin District Management Plan (Vol. IC/151). International ICPDR Comission for the protection of the Danube River, Vienna, p 105. https://www.icpdr.org/flowpaper/app/#page=1. Accessed 20 Aug 2020
108. Interreg. Danube Transnational Programme (2018) Danube Floodplain. Reducing the flood risk through floodplain restoration along the Danube River and tributaries. http://www.interreg-danube.eu/approved-projects/danube-floodplain. Accesed 8 June 2020
109. WWFRomania (2013) Organizaţia WWF a „rupt" graniţa bulgaro-română, pentru protejarea naturii. https://wwf.ro/noutati/comunicate-de-presa/organizatia-wwf-a-rupt-granita-bulgaro-romana-pentru-protejarea-naturii/. Accessed 20 Dec 2020
110. WWF Romania (2020). 50 de specii acvatice protejate vor beneficia de pe urma lucrărilor de reconstructie ecologică de la Gârla Mare. https://wwf.ro/dunare-delta-dunarii/50-de-specii-acvatice-protejate-vor-beneficia-de-pe-urma-lucrarilor-de-reconstructie-ecologica-de-la-garla-mare/. Accessed 20 Dec 2020
111. Wachter K, Toth S, Liska I (2007) The analysis of the Danube Floods 2006. An in depth analysis of the floods on the Danube and its main tributaries in 2006. International Commmision for the Protection of the Danube River, Vienna. https://www.icpdr.org/flowpaper/app/#page=1. Accessed 12 Sept 2020
112. Council of Europe (1979, Sept 19) Convention on the Conservation of European Wildlife and Natural Habitats. Bern, 19.IX.1979. European Treaty Series, pp 1–10. https://rm.coe.int/1680078aff. Accessed 29 Nov 2020
113. Bloesch J, Jones T, Reinartz R, Striebel B (2006) An action plan for the conservation of sturgeons (Acipenseridae) in the Danube River Basin. Oestere Wasser—Abfallwirtsch 58:81–88. https://doi.org/10.1007/BF03165708
114. Interreg. Danube Transnational Programme (2018) MEASURES. Managing and restoring aquatic EcologicAl corridors for migratory fiSh species in the danUbe RivEr baSin. http://www.interreg-danube.eu/approved-projects/measures. Accessed 20 Dec 2020
115. Council of Europe (2018) Pan-European action plan for Sturgeons. Strasbourg: T-PVS/Inf(2018)6. https://rm.coe.int/pan-european-action-plan-for-sturgeons/16808e84f3. Accessed 29 Nov 2020
116. Heat MF, Evans M, Hoccom DG, Payne AJ, Peet NB (2000) Important Birds Areas in Europe: priority sites for conservation, vol 1. BirdLife International, Cambridge, UK

117. Munteanu D (2000) Romania. In: Heath ME (ed) Important Bird Areas in Europe: priority sites for conservation, vol 2. Southern Europe. BirdLife International, Cambridge, UK, pp 421–502
118. BirdLife International (2001) Important Bird Areas and potential Ramsar Sites in Europe. The Netherlands: BirdLife International, Wageningen, p 136. https://www.birdlife.org/sites/default/files/attachments/ibas_ramsar_europe_0_0.pdf. Accessed 29 Nov 2020
119. www.birdlife.org. Birdlife International. https://www.birdlife.org/worldwide/programmes/sites-habitats-ibas-and-kbas. Accessed 7 Nov 2020
120. www.ramsar.org. The List of Wetlands of International Importance. https://www.ramsar.org/sites-countries. Accesed 28 Feb 2021
121. European Environment Agency (2019) Natura 2000 Network Viewer. https://natura2000.eea.europa.eu/. Accessed 7 Dec 2020
122. European Environment Agency (2019) Natura 2000 standard data forms. https://natura2000.eea.europa.eu/Natura2000/SDF.aspx. Accessed 5 Nov 2020
123. Scholten A, Rothstein B (2016) Navigation on the Danube—limitations by low water levels and their impacts. EUR 28374 EN, Luxemburg. https://doi.org/10.2788/236234
124. Europeaid/122137/D/SV/RO (2002) Technical assistance for improvement of navigation conditions on the Romanian—Bulgarian common sector of the Danube and accompanying studies. Technical memo required for issuing the environmental agreement (Ord. 860 / 2002). https://www.moew.government.bg/static/media/ups/tiny/file/Industry/EIA/Technical%20memoir-Dunav.pdf. Accesed 8 Dec 2020
125. Bij de Vaate A, Jazdzewski K, Ketelaars HA, Gollasch S, Van der Velde G (2002) Geographical patterns in range extension of Ponto-Caspian macroinvertebrate species in Europe. Can J Fish Aquat Sci 59:1159–1174. https://doi.org/10.1139/F02-098
126. Panov VE, Alexandrov B, Arbaciauskas K, Binimelis R, Copp GH, Grabowski M, Frances L, Leuven RSEW, Nehring, S, Paunović M, Semenchenko V, Son, M O (2010) Risk assessment of aquatic invasive species introductions via European Inland Waterways. In: Settele J, Penev L, Georgiev T, Grabaum R, Grobelnik V, Hammen V, Klotz S, Kotarac M, Kuhn I (eds) Atlas of biodiversity risk. Pensoft, pp 140–143
127. Paunović M, Csányi B, Simonović P, Zorić K (2015) Invasive alien species in the Danube. In: Liska I (ed) The Danube River Basin: the handbook of environmental chemistry, vol 39. Springer, Berlin, Heidelberg, pp 389–409. https://doi.org/10.1007/698_2015_376
128. Pedashenko HP, Apostolova II, Vassilev KV (2012) Amorpha fruticosa invasibility of different habitats in lower Danube. Phytol Balc 18(3):285–291
129. Ciuvăț AL, Vasile D, Dinu C, Apostol E, Apostol B, Petrițan A-M (2016) Valorisation possibilities of invasive indigo bush (Amorpha fruticosa L.) in Romania. Revista de silvicultură și cinegetică 21(39):96–99. https://www.researchgate.net/publication/318649291_Valorisation_possibilities_of_invasive_indigobush_Amorpha_fruticosa_L_in_Romania. Accessed 5 Dec 2020

Land Management Practices Favoring Environmental Conservation in the Danube Lower Valley (Romania)

Mioara Clius, Mihai Mustățea, Viorel Chendeș, and Mihai-Răzvan Niță

Abstract Deficient land management is common in rural landscapes of central and Eastern Europe, especially in the post-socialist period, and it represents a major source of economic losses. On the counterpart, increased areas of abandoned land favor the expansion of wetlands, suitable wildlife habitats, as in the case of the Danube Lower Valley in Romania. In our study, we hypothesize that weak land management can be a potential favoring factor for the renaturation of artificial landscapes, and future wildlife conservation. We aim to demonstrate how the abandonment of the irrigation systems can generate wetlands supporting wildlife conservation within the Danube Lower Valley. For the case study in Romania, we: (i) identified the spatial distribution of protected areas (SCI and SPA) within the Danube Lower Valley created in landscapes with abandoned irrigation systems; (ii) modeled the dynamic of artificial and natural land cover classes within the selected protected areas; (iii) explored the implications of abandoned irrigation systems on biodiversity and identified the factors that could lead to a potential return of wildlife habitats. Results indicate that after 2007, the Danube Lower Valley is dominated by the presence of SCI and SPA located in landscapes with abandoned irrigation systems. We have identified three main stages in terms of human influence: semi-natural state (1864–1950), intense modifications (1950–1990) and return to wilderness (after 1990). The main factors that could lead to a potential return of wildlife habitats are the presence of extreme hydrological or meteorological conditions, namely floods or climate change which could favor the extend of wetlands, abandonment of intensive agricultural practices

M. Clius (✉) · M. Mustățea · M.-R. Niță
Faculty of Geography, University of Bucharest, 1 Nicolae Bălcescu, 010041 Bucharest, Romania
e-mail: mioara.clius@geo.unibuc.ro

M. Mustățea
e-mail: mihai.mustatea@drd.unibuc.ro

M.-R. Niță
e-mail: mihairazvan.nita@g.unibuc.ro

V. Chendeș
National Institute of Hydrology and Water Management, 97E Bucharest-Ploiești, 013686 Bucharest, Romania
e-mail: viorel.chendes@hidro.ro

A. Negm et al. (eds.), *The Lower Danube River*, Earth and Environmental Sciences Library, https://doi.org/10.1007/978-3-031-03865-5_16

proper to generate the expansion of spontaneous secondary vegetation, and lack of general human activity. Our findings are valuable in identifying other areas with a similar pattern of human influence which are suitable candidates for future conservation and in the durable spatial planning of rural areas from the Danube Lower Valley.

Keywords Land management · Irrigation systems · Wilderness · Danube Lower Valley

1 Introduction

The analysis of land management practices, reflected in the land use and land cover of a certain area is a very good indicator for the human pressure on the environment, but in the same time a very challenging task when the subject of transformation has a synergic character [1]. Land transformations are processes which are often differently defined, measured, studied or compared across regions or across land use types and human activities [2]. The challenges of finding reliable data, selecting the adequate method of research or analysis, can all influence the results and the conclusions on the drivers and influencing factors of land management practices.

Literature and administrative trends are shifting away from traditional sectoral analysis of management practice for a single land use or similar land-uses, in favor of a strategic spatial planning, a participative collaboration tool for achieving sustainability targets by promoting coherent land management practices [3]. Such approaches have a high degree of novelty, but in historical times land management practices were determined mainly by net economic gain and promotion of a type of human use of natural resources, with examples such as creation of agricultural surfaces replacing wetlands, use of Danube water for irrigation systems or the development of hydrotechnical structures (such as Iron Gates reservoir dam) for using the hydropower and navigation potential of the river [4].

Traditional land management practices have a higher importance in areas where a mixture of landscapes and natural features are already existent. This is the case of the Danube River and its valley, which has traditionally been a space connecting cultures and creating networks between land, river, wetlands and the sea across a significant part of Europe [5]. The Romanian section, represented by the Lower Danube Valley is an interesting case, as it represents an area where the natural dynamic of the river has created a rich diversity of floodplain ecosystems. A large proportion of these wetland ecosystems have been historically lost (estimated 70% over the past two centuries), but are now re-emerging following conservation and restoration efforts overlapping the abandonment of traditional land management practices [6].

Land management transformations imposed from the state power but also expression of the internal resilience of the area faced with accelerated dynamic [7]. The morphological and hydrodynamic vulnerability of the Lower Danube Valley determined challenges in the establishment and management of conservation efforts [8].

Sometimes, the best conservation values are found presently in areas which were intensively transformed and abandoned, the process creating hotspots of biodiversity, especially for waterfowl.

The main factors that could lead to a potential return of wildlife habitats in areas with abandoned irrigation systems are the presence of extreme hydrological or meteorological conditions, namely floods or climate change which could favor the extend of wetlands, lack of intensive agricultural practices proper to generate the expansion of spontaneous secondary vegetation, and lack of general human activity which allows the increasing populations of mainly opportunistic and generalist wild life species.

At the European level wetlands are recognized as providing essential ecosystem services such as food production, water, controlling flooding and nutrients or providing spiritual values and opportunities for recreation [9]. At the same time, they are environmentally sensitive areas in which the slightest disturbances can affect the balance. Therefore, the approach is to connect them with the protected areas network, creating the instruments and mechanism of delivering effective protection to conservation values.

The current distribution of wetlands in the Lower Danube Valley is just a remnant of the initial wetlands systems, but the remaining ones are of extremely high conservation values [6]. Numerous programs and initiatives have been promoted for conservation in the Lower Danube Valley, starting with initial efforts such as the Environmental Program for the Protection of the Danube River Basin or the Danube River Convention from Sofia [10].

Currently, the Lower Danube Valley benefits from the attention of the European Union which initiated the Strategy of Danube Region [11], with the objectives of managing water resources and preventing flood risk, sustainable use of natural and cultural resources, but also the restoration and management of wetlands acting as ecological corridors for biodiversity conservation and delivery of ecosystem services. A series of projects (especially from the LIFE European program) were implemented between 2000 and 2010 in the Lower Danube Valley, focusing on habitat restoration and management, species conservation and ecological reconstruction.

Conservation through sites from the European protected areas network Natura 2000 represents the latest trend in the area, but while on paper they provide the needed protection level, in reality, they face numerous challenges in Romania in dealing with the uncontrolled exploitation of natural resources [12] or the reduced administrative capacity of institutions with competences in nature conservation [13] or lack of participation from other actors (such as non-governmental organizations or general public) [14].

There is a growing body of evidence that the Danube Lower Valley has both a social, environmental and economic importance for both local population or at regional and national level [15]. There is therefore a need of such studies which will explore the implications of abandoned irrigation systems on biodiversity and identify the factors that could lead to a potential return of wildlife habitats, based on cross-referencing the available scientific biogeographical literature, on field analysis or spatial analysis of data.

Precarious land management is common in rural landscapes of central and Eastern Europe, especially in the post socialist period, and it represents a major source of economic losses. The increasing areas of abandoned land favors the expansion of secondary land cover classes, such as wetlands, which could represent suitable habitats for numerous wildlife species, as in the case of the Danube Lower Valley, in southern Romania. In our study, we hypothesize that weak land management can be a potential favoring factor for the returning to the wilderness of artificial landscapes which once were natural areas with rich biodiversity, and by consequence, are suitable candidates for wildlife conservation in the future.

We aim to demonstrate how the abandonment of the irrigation systems can generate wetlands supporting wildlife conservation within the Danube Lower Valley. For the case study in Romania, we established the following *objectives*: (i) identify the spatial distribution of protected areas (SCI and SPA) within the Danube Lower Valley created in landscapes with abandoned irrigation systems, based on GIS techniques; (ii) model the dynamic of artificial and natural land cover classes within the Danube Lower Valley through landscape ecology metrics and (iii) explore the implications of abandoned irrigation systems on biodiversity, based on cross-referencing the available scientific biogeographical literature.

The originality of our study and its novelty compared to previous papers with similar issues is justified by: (a) a topical issue represented by the study of the expansion of wetlands and their potential to favor the establishment of protected areas for the conservation of aquatic habitats; (b) an original spatial and temporal landscape analysis based on a wide range of land cover data, such as historical military maps and satellite imagery; and (c) a set of useful recommendations, intended to promote the sustainable development of rural areas in the Lower Danube Valley, through the scientific and tourist capitalization of wetlands.

2 Drivers of Land Management Practices in the Danube Lower Valley

The Danube Lower Valley is located in the southern part of Romania and covers an area of 1,080,745 ha (Fig. 1) [16]. The valley is bordered by the Romanian Plain and Moldavian Plateau in the north, the Bulgarian Plateau in the south, the South-western Carpathian Mountains in the west, and the Dobrogea Plateau and the Black Sea in the east. Romania holds approximately 38% of the total length of the Danube (1075 km) and is geographically divided into several major subdivisions, such as the Danube Delta and the wetlands of Brăila and Ialomița (in Romanian, Balta Brăilei and Balta Ialomiței) in the east, respectively the Danube floodplain in the central and western areas [16]. Our analysis was focused on the Danube floodplain, and in addition, on the delta and neighboring lagoons, whereas the Danube terraces were not taken into account in the study.

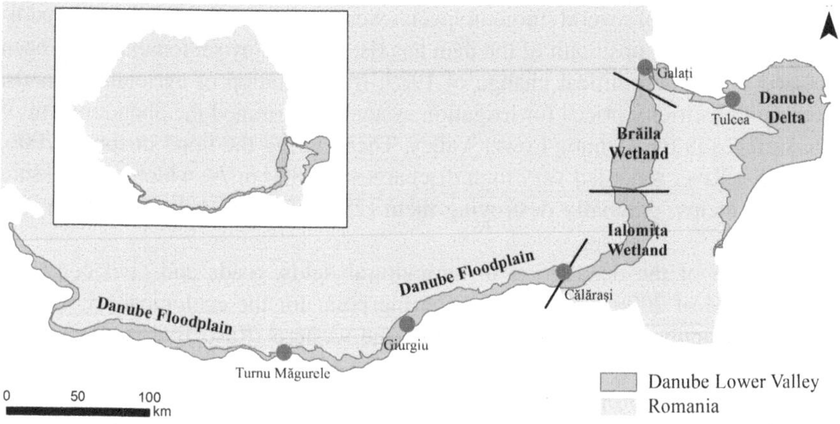

Fig. 1 Location of the Danube Delta, neighboring wetlands and floodplain in Romania. *Data source* Geografia României (2005) [16]

The Danube Lower Valley represents an important geographical region of Romania. Situated in the southern and eastern parts of the country, it ensures navigation with the center of Europe and also water resources for irrigation or other uses (especially for the eastern part of the Romanian plain and Dobrogea which have humidity deficit during summer). Historically, the area was gradually populated due to the high productivity of soils and fish resources. The natural landscape in the area was represented by extended wetlands, lakes or channels, economic and technological development determined changes, as the area becomes of interest for the Romanian agriculture. At the beginning of the twentieth century the economic value was limited to fishery and use of meadows for livestock, but following extensive works of embankments and damming much land was reclaimed from the wetlands and transformed in agricultural surfaces.

Beginning with the nineteenth century, extensive studies were realized for the creation of irrigation systems based on the Danube water, firstly through a water intake realized in the area of Drobeta-Turnu Severin and a channel flowing to east and providing water for parts of the Romanian plain [17]. This project failed for economic reasons, and recommendations of naturalists such as George Vâlsan (1915) and Grigore Antipa (1921) were directed to a moderate use of Danube waters and a reduced intervention. The increase of fisheries from the floodplain lakes, snagging of channels to ensure regulated overflow of the Danube, vineyard or pastoral use of areas, and the abandonment of embankment levees projects were among the recommendations made by the naturalists [17].

After 1950 the engineering approach of the area prevailed, and it suffered continuous and significant transformations, with the sole purpose of increasing the economic (and especially agricultural) potential of the area. The construction of the Iron Gates I and II reservoir dams had a major impact on the floodplain functioning. The construction of the dams also had a major impact on the environment, for example the breeding routes of several sturgeon species were permanently interrupted. In addition, the water level upstream of the dam has risen and many settlements have been relocated [18]. The political change of 1989, fragmentation of agricultural lands, increase of electricity prices for irrigation systems determined the abandonment of large surfaces in the Danube Lower Valley. Then, during the flood in spring 2006, the Danube River recorded very high discharges (15,800 m^3/s) which put pressure on embankments, gradually destroying them (28% of the Danube floodplain was affected) [18].

Regardless of the destruction of agricultural fields, roads and civil construction, the flood of 2006 represented a starting point for the ecological reconstruction of the Danube on large surfaces from the western (Rast, Bistreț, Bechet) or eastern (Mânăstirea, Oltina, Ostrov) sectors. The next year, following the accession of Romania to the European Union, in the Danube Lower Valley were established Natura 2000 sites with the role of maintaining an equilibrium between the exploitation of the area and nature conservation.

According to the geomorphological features of the valley, floodplain and river bed, the Danube Lower Valley in Romania can be characterized by four main sectors: (i) Porțile de Fier sector (Baziaș–Gura Văii) in which the Danube crosses the Carpathians and forms impressive gorges; (ii) Gura Văii–Călărași sector, with variable width of the floodplain, of 5–6 km until Calafat, followed by a larger section (10–15 km) developed especially in Romania and with numerous landforms and holms; (iii) Călărași–Pătlăgeanca sector, with a very large floodplain in the first part (over 20 km), where the Danube splits into large branches and (iv) Danube Delta sector in which the deltaic plain of the Danube is sustained by three main branches [16].

In terms of climate, the Lower Danube Valley has slightly differentiated characteristics from west to east, induced by the climatic influences (oceanic, Mediterranean or Pontic). Multiannual average annual temperatures are 11 °C, the duration of sun between 2250 and 2500 h, and annual precipitations decrease from west (700 mm) to east (445 mm) [19].

The valley spreads over three biogeographical regions: continental in the western and central part, steppe in the east, and pontic on a narrow portion located alongside the Black Sea coast [20]. The areas host the largest wetlands from Romania, protected within the Danube Delta Biosphere Reserve (starting from 1998) and the Wetland of Brăila National Park (established in 2001). Besides, there are numerous smaller marshland areas, many of them conserved under sites of community interest (SCI) and special areas of protection (SPA), being established in 2007 and 2011.

The most significant wetland in terms of biodiversity conservation from the study area is the Danube Delta, which shelters the largest populations of pelicans (*Pelecanus onocrotalus, Pelecanus crispus*) and white-tailed vultures (*Haliaeetus albicilla*) of Europe. Other rare and protected water birds are the black stork (*Ciconia nigra*) and the winter swan (*Cygnus cygnus*). Some of the rarest water mammals from Europe, such as the otter (*Lutra lutra*) and beaver (*Castor fiber*) are also protected in the Danube valley. These wetlands shelter habitats for various freshwater reptiles and amphibians, representative being the water turtle (*Emys orbicularis*) and the red-belly frog (*Bombina bombina*) [21].

In the natural regime, the floodplain was covered in intrazonal vegetation with forests, pastures and aquatic vegetation. Following the embankment and draining works, alongside the deforestation of forests with soft-wood (over 20,000 ha in counties Giurgiu, Călărași, Ialomița, Brăila), they were areas forested which introduced alien species, such as the American poplar (*Populus canadiensis*), abandoned in the last decades due to the reduced stability to flooding. Even after all the transformations, numerous surfaces of natural pastures remained, especially on the lower surfaces, with humidity excess and on sandy soils. The proximity of lakes and channels still holds large surfaces of hygrophyte and hydrophile vegetation (*Phragmites communns, Typha angustiffolia, Typha latifolia*) [16].

Despite its significance for biodiversity conservation, especially in the eastern section, due to the Danube Delta, the Danube Lower Valley is home to a diverse range of human activities. The historical pressure over the natural environment due to the development of mass agriculture led to the loss and degradation of wetlands. Nowadays, in the central and western section, the valley encompasses some of Romania's most productive arable lands, mainly in the wetlands of Brăila and Ialomița, where modern technologies are used in order to develop high productivity crops of corn, wheat and alfalfa. The rice crops which are located at the mouth of the Ialomița River, near the city of Țăndărei, form the most extended artificially permanent irrigated land from Romania. Also, the Danube Delta is the country's most important area for the production of reed and rush. Besides agriculture, another large-scale economic activity is represented by fishing, the Danube being the most important habitat in the European Union for sturgeons, valued for production of roe.

The Danube holds numerous small islands, many of them planted with poplar and willow forests and used for wood production [21]. The Danube Delta is one of the most important touristic regions of Romania due to the presence of wild areas, gastronomy and local culture.

In the Danube Lower valley, there are located a few major industrial centers, namely the largest steel plant from Romania, in the proximity of Galați city, and the only nuclear power plant in the country, at the outskirts of Cernavodă city [22]. If until the seventeenth century the area included only small rural settlements, the number increased gradually and occupations diversified (agriculture, crafting of reed, vineyards), although the living environment was still rather unattractive. The number of inhabitants per territorial administrative unit varies between less than 2,500 in the Danube Delta and Island of Brăila and over 50,000 in the cities of Tulcea, Galați, Călărași Giurgiu and Turnu Măgurele (Fig. 2) [23].

Since it is characterized by the presence of vast irrigation systems for agricultural cultures, high biodiversity wetlands and a dense concentration of protected areas, the Danube Lower Valley is a suitable candidate for analyzing the potential implications of land management practices on wild areas conservation. The changes into land management practices were directed to the reduction of humidity excess, and on the other hand for the protection of newly transformed agricultural lands to the high-water levels of the Danube, especially in the periods of April–June.

Before 1950, the land transformations in the floodplain of the Danube covered only a few thousand hectares and were represented mainly by earth embankments

Fig. 2 Number of inhabitants per territorial administrative unit in the Danube Lower Valley. *Data source* Tempo Online (2020) [23]

and regularization of watercourses. After 1950 land changes intensified, by the year 1960 no less that 60,000 ha were reclaimed for agriculture, between 1960 and 1965 another 227,000 ha and between the year 1965 and the year 1975 another 300,000 ha. In 1981 the total surface transformed in the area was over 860,000 ha represented by over 1100 km of embankments, irrigation channels and pumping stations [16].

Another major intervention was represented by the use of Danube water for irrigation of agricultural fields from the south and eastern parts of the Romanian plain. After 1989 agricultural lands were restituted in nature [24], generating a fragmentation of agricultural surfaces. This, together with economic challenges, abandonment and degradation of irrigation equipment, high energy costs and the increased frequency of drought periods represented premises for the abandonment of lands. In 2010 a mere 85,000 ha of agricultural lands remained irrigated (especially between Călărași and Pătlăgeanca).

3 Evaluating Land-Use Changes in the Danube Lower Valley

We modeled the spatial and temporal dynamics of land cover classes within the Danube Lower Valley, and identified the areas where wetlands have been replaced or have expanded, by using a landscape change analysis.

The methodological approach started with extracting spatial data regarding: (1) the wetland areas located within the Danube Lower Valley between 1864 and until 1950 through the Charta României Meridionale from 1864 [25] and the Military Survey Maps from 1915–1959 [26], (2) the distribution of irrigation systems and wetlands between 1950 and 1990, based on the Topographical Military Map of Romania from

1975 [27]. The three datasets were superimposed in order to identify the wetlands extend between 1864 and 1950, the areas where wetlands were restrained due to the development of irrigation systems, and remaining patches of large intact wetlands between 1950 and 1990.

For analyzing the landscape spatial and temporal changes from the Danube Lower Valley in the post-socialist period (after 1990), we have applied the Markov chains sequence model [28]. The Markov model is assessed in our study for identifying land cover transitions between two different time sequences. The model is a proper approach for our assessment since it can be used for quantifying landscape spatial changes on discrete data, as in case of the land cover classes' types. Also, this is a common method applied in a wide pallet of natural sciences, namely geography, biology and ecology.

The data used for applying the Markov model was Corine Land Cover—years 1990, 2000, 2006, 2012 and 2018 [29]. Based on the act that the model quantifies land changes from different time sequences, we have selected four time periods for our assessment: 1990–2000, 2000–2006, 2006–2012, and 2012–2018. The land cover classes within the Danube Lower Valley were reclassified into five major categories, according to the Corine Land Cover classification system: artificial surfaces, agricultural areas, forest and semi-natural areas, wetlands and water bodies. The Corine Land Cover data has potential mistakes concerning the identification of land cover classes, which are corrected from one year to another and influenced by the quality of the available satellite imagery, and these aspects could hamper the efficiency of our models [29].

A matrix was developed, which included all the potential conversions between the five categories of land cover classes for all of the four-time periods targeted. For all of the conversions registered, their areas expressed in hectares were quantified. Lastly, we have computed the total areas registered by the reclassified land cover classes according to each year, and compared them, in order to highlight their overall dynamics.

Our results reveal that, between 1864 and 1950, the wetlands covered an area of roughly 762,424 ha (70.5% of the entire Danube Lower Valley). The largest were the Danube Delta (315,520 ha), followed by the wetlands of Brăila (105,729 ha), Ialomița (87,485 ha), Iezeru Călărași (32,383 ha), Greaca (72,678 ha), Suhaia (37,561 ha), Potelu (28,879) and Bistreț (57,920 ha). Between 1950 and 1990, the areas covered by wetlands were restrained to just 243,344 ha (representing only 22.5% of the Danube Lower Valley), mainly due to the expansion of irrigation systems. Only three large continuous wetlands remained in 1990, such as the Danube Delta (210,676 ha, representing just 40.8% from the initial area until 1950), the Brăila Wetland (3756 ha—3.5% of the previous extend), and the Bistreț Wetland (5520 ha—totaling only 9.5% of the initial area) (Fig. 3).

After 1990, wetlands were characterized by expansion, yet only between 1990 and 2006 (from 243,536 ha to 257,675 ha). Afterward, a constant decrease was registered, from 249,664 ha in 2012 to 240,280 ha in 2018. From 2000 to 2006, wetlands have replaced especially semi-natural areas (13,623 ha) and water bodies (9751 ha).

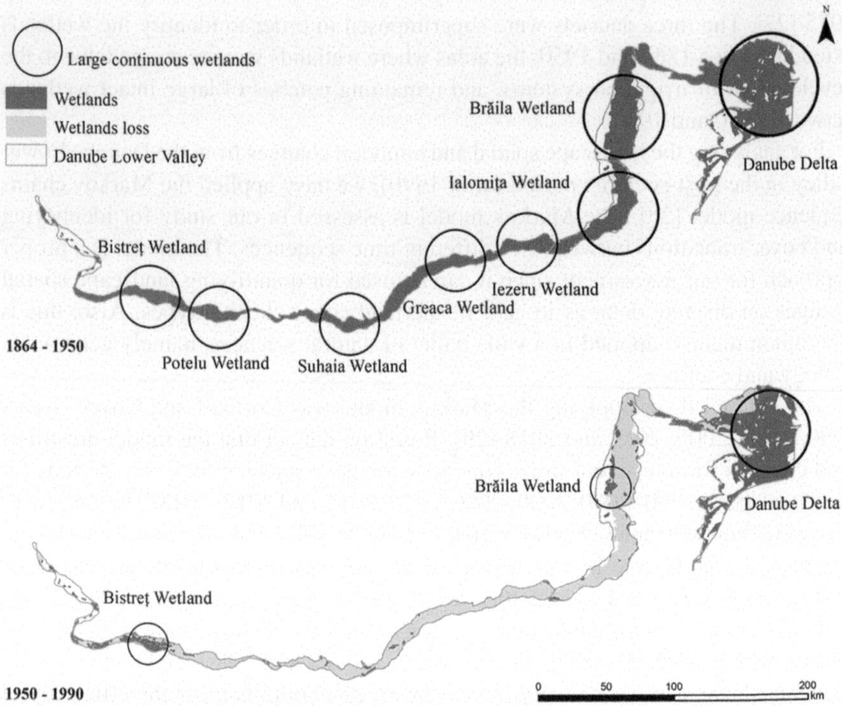

Fig. 3 Distribution of wetlands in the Danube Lower Valley between 1864–1950 and 1950–1990. *Data source* Charta României Meridionale (1864) [25]; Military Survey Maps (1915–1959) [26]; Topographical Military Map of Romania (1975) [27]

Another major expansion took place between 2006 and 2012, when 7205 ha of agricultural land converted into wetlands.

There were no expansions registered between 2012 and 2018. Between 1990 and 2018, the dominant land cover class, represented by agricultural areas, registered a continuous expansion (from 488,018 ha in 1990 to 511,897 ha in 2018). Arable land has replaced considerable areas of wetland primarily in 2000–2006 (7150 ha), 2012–2018 (9035 ha) and 2006–2012 (12,428 ha) (Table 1).

The areas covered by water bodies dropped from 178,047 ha in 1990 to 171,126 ha in 2006. In 2012, water bodies increased to 171,830 ha, and slightly decreased after to 171,757 ha in 2018. Water bodies have also replaced wetlands, yet on much smaller areas, namely 5589 ha between 2006 and 2012, followed by 4434 ha between 1990 and 2000.

Following a similar pattern, forests and semi-natural areas reduced from 144,082 ha in 1990 to 127,768 ha in 2006. A minor increase followed the next years, by reaching 131,623 ha in 2012, respectively 131,276 ha in 2018. Small surfaces of wetlands were converted into other semi-natural areas in 2000–2006 (4573 ha) and 2006–2012 (3705 ha). Artificial areas constantly decreased between 2000 and 2018 (from 27,217 ha to 25,524 ha). Between 2000 and 2006, they have registered their

Table 1 Land-cover conversions between 1990 and 2018

Land cover classes transitions	Area (hectares)			
	1990–2000	2000–2006	2006–2012	2012–2018
Unchanged artificial	*26,908*	*24,012*	*24,709*	*25,400*
Agricultural to artificial	251	1333	556	124
Forests to artificial	10	216	39	0
Wetlands to artificial	20	167	73	0
Water to artificial	28	142	23	0
Unchanged agricultural	*487,134*	*478,461*	*486,494*	*502,058*
Artificial to agricultural	77	2053	938	0
Forests to agricultural	237	10,299	1250	278
Wetlands to agricultural	113	7150	12,428	9035
Water to agricultural	264	332	1098	526
Unchanged forests	*143,399*	*118,176*	*123,135*	*131,250*
Artificial to forests	12	539	19	0
Agricultural to forests	350	3381	3713	26
Wetlands to forests	113	4573	3705	0
Water to forests	350	1099	1060	0
Unchanged wetlands	*242,856*	*229,874*	*235,880*	*240,280*
Artificial to wetlands	14	256	151	0
Agricultural to wetlands	199	4171	7205	0
Forests to wetlands	123	13,623	2852	0
Water to wetlands	847	9751	3576	0
Unchanged water	*176,558*	*166,105*	*165,369*	*171,304*
Artificial to water	40	357	53	0
Agricultural to water	84	479	327	0
Forests to water	313	1910	492	104
Wetlands to water	4434	2275	5589	349

Data source Environmental European Agency (1990, 2000, 2006, 2012, 2018) [29]

peak expansion over wetlands, by replacing 167 ha. Secondly, between 2006 and 2012, 73 ha of wetland were lost to artificial areas (Fig. 4).

Wetlands have expanded to the detriment of other semi-natural areas, water bodies and artificial areas, especially in the eastern section of the study area, within the Danube Delta and Brăila Wetland (both 1990–2000, 2000–2006 and 2006–2012). In opposition, expansions over agricultural areas were usually located in the western, in the proximity of the Bistreț Wetland (1990–2000 and 2000–2006) and central section, east of Suhaia Wetland (2000–2006 and 2006–2012) (Fig. 5).

A discussion of our results indicates the presence of three stages of human influence over Danube Lower Valley landscape: (1) semi-natural state (1864–1950),

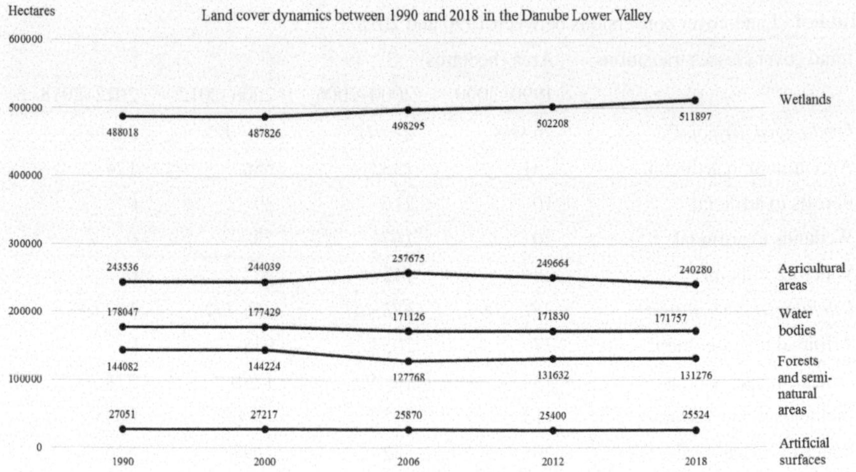

Fig. 4 Dynamics of the major land cover classes from the Danube Lower Valley across four time periods: 1990–2000, 2000–2006, 2006–2012, and 2012–2018. *Data source* Environmental European Agency (1990, 2000, 2006, 2012, 2018) [29]

(2) intense modifications (1950–1990) and (3) return to wilderness (after 1990). Until 1950, large intact wetlands, lakes and flooded forests dominated the landscape, forming a continuous wild land. These swamps were highly inaccessible and were considered insecure by people, since they sheltered bands of outlaws and packs of large carnivores, such as wolves [30]. Furthermore, the swamps were avoided since they were perceived as an unhealthy land, haunted by outbreaks of malaria. At that time, the valley was an aquatic area, where water stagnated for a long time and hampered economic exploitation. Yet, despite their apparently inaccessible character, various types of human activities with low impact on the environment were performed in the wetlands, such as hunting, fishing, honey collecting and even cattle grazing on peripheral grasslands [30].

In 1904, the first activities that generated major changes in the natural environment took place in the Danube floodplain in order to transform it into agricultural land, and the first levees were constructed near the town of Oltenița by several Dutch construction companies such as Schram, Dithmer and Langeweld. Previous, the embankment attempts were not successful due to the lack of hydrological data and the low adaptability of the constructions to the floodplain conditions [31].

The levees in the Danube floodplain were used to remove the lands from the influence of the river waters. Following these embankment works, the problem of formulating unitary norms for their design was raised, so that in 1910 the law for the development of lands in the Danube floodplain appeared, based on the engineering conception of Anghel Saligny. Following the conservative controversies initiated by Grigore Antipa [17] between 1912 and 1921, the embankment works were completely interrupted. Yet, the construction of levees restarted after the end of the First World War [31].

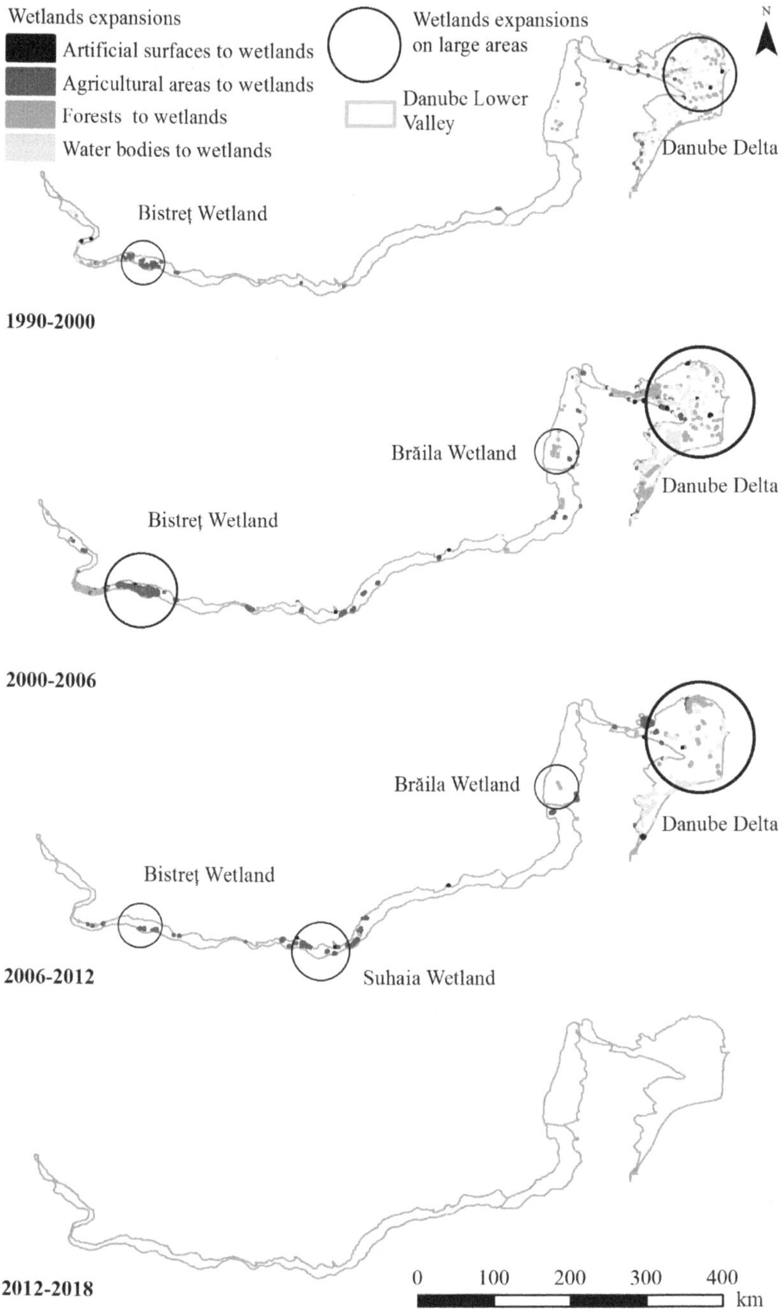

Fig. 5 Areas with wetlands expansion over other land cover classes between 1990 and 2018. *Data source* Environmental European Agency (1990, 2000, 2006, 2012, 2018) [29]

After 1950, in the communist period, the swamps were drained and embanked in order to be replaced by arable land. In the context of agricultural land expansion and agricultural development, hydro-amelioration works in the Danube floodplain continue and intensify since 1963 with drainage works, followed by irrigation in areas with moisture deficit, producing irreversible changes to environmental factors [31]. Overall, the works for agricultural exploitation of the lands were made in the following stages: (a) overall drainage which involved the construction of main canals for the collection and evacuation of excess water; (b) retail drainage and completion of the network of main canals with secondary drainage channels to intensify the collection and discharge of excess water; (c) irrigation arrangements, consisting of underground pipes and open canals, and (d) introduction of pipe drainage, associated with works of land leveling.

Drainage actions began after embankment. They consisted mainly in the execution of drainage channels that led to maintaining the groundwater level at reasonable levels [32]. Socialist decision-makers thought exclusively of engineering and did not consult hydrologists, biologists and ichthyologists. Therefore, Romania did not sign the Ramsar Convention (in 1975) dedicated on protecting wetlands, until after the revolution, on the year 1991. Until then, the Danube Delta Biosphere Reserve, established in the year 1979, was the only natural protected area with international nomination [33]. Overall, the socialist period was characterized by the highest degree of artificiality and maintenance of pressures on the natural ecosystems in the Danube floodplain.

After 1990, a massive wave of social and economic changes affected the agricultural activity in the valley. The irrigation systems were gradually abandoned. The first stage of mass conservation from the Danube valley took place in 1991, through the establishment of RAMSAR sites: Danube Delta, Small Wetland of Brăila, Bistreț, Călărași Lake, Suhaia, Blahnița, Borcea Arm, Calafat-Ciuperceni-Danube, Canarals of Hârșova, Bugeac-Iortmac Danube Islands, Jiu-Danube Confluence and Old Danube-Măcin Arm. Also, in 1991, the Danube Delta Natural World Heritage Site was established [33]. In addition, the Danube Lower Valley witnessed a wave of massive floods in 2006, which lead to the destruction of levees and flooding of arable land [18]. The floods represented the starting point for the expansion of wetlands.

Therefore, in 2007, after joining the European Union, Romania developed a network of environmentally/ecological protected areas, in order to develop a favorable conservation status for some of the rarest habitats and wildlife species from Europe. The Natura 2000 network encompassed Sites of Community Interest—SCI (established through the Habitats directive 92/43 of 1992 on the protection of Habitats, Fauna and Flora) and Special Protection Areas—SPA (established through the Birds Directive 79/409 of 1979 regarding the Conservation of Wild Birds). The Natura 2000 network was extended in 2011, respectively 2016. The new wetlands were included in protected areas, and wetland conservation and recovery reached its peak [21]. Although they benefit from the advantages of the community protection regime, the protected areas raise serious doubts concerning the efficiency of their administration, since many of them lack a proper management plan or internal zoning system [34].

4 Implications of Land Management for Conservation in the Danube Lower Valley

We also aimed to analyze if there is a spatial and temporal connection between the distributions of abandoned irrigation systems, wetlands expansion and protected areas (SCIs and SPAs) location within the Danube Lower Valley, after 1990, through the use of spatial super-imposing GIS techniques.

The first methodological step was to identify the categories of protected areas developed in the study area in the post-socialist period (after 1990) through the use of spatial data consisting in the location of SCIs and SPAs, from 2007 and 2011 [35]. Secondly, we have used the spatial linear data consisting in the location of irrigation systems from 1975 and converted it into polygons, with a buffer width of 10 m. Then, through the use of spatial function Erase from ArcGIS [36] we have performed a spatial intersection of SCIs/ SPAs with the irrigation systems, in order to identify the protected areas developed in 2007 and 2011, on irrigation systems that were abandoned after 1990.

The SCIs, SPAs and irrigation systems were superimposed over the spatial location of the areas where wetlands have expanded between 2000 and 2006, for identifying if there is a spatial connection between abandoned irrigation systems after 1990, wetlands expansion between 2000 and 2007, and protected areas (from 2007 and 2011) location within the Danube Lower Valley. We have selected the 2000–2006 time period because: (1) it precedes the years when the SCIs and SPAs have been created (2007 and 2011), and (2) it is characterized by the most significant expansions of wetlands. Lastly, for every type of wetland expansion identified between 2000 and 2006 (over agricultural areas, artificial surfaces, forests and semi-natural areas, and water bodies), we have computed the areas located within protected land, SCIs, SPAs, respectively SCIs and SPAs with abandoned irrigation systems.

Our results indicate the presence of only one SCI (Ciuperceni-Desa) in the Danube Lower Valley located in an area without abandoned irrigation systems. The site lies in the western extend of the valley, over an area of 16,731 ha. In opposition, we identified 17 SCIs with irrigation systems, totaling an area of 526,749 ha (48.7% of the entire Danube Lower Valley), the largest being: Danube Delta (445,318 ha), Jiu Valley (26,660 ha), Danube Stone Cliffs (24,098 ha), Brăila Wetland (20,650 ha), Măcin Arm (10,023 ha). The largest and most numerous sites are primarily concentrated in the eastern section of the valley (in the Danube Delta, and on the sections bordering the areas occupied once by the wetlands of Brăila and Ialomița).

Only 6 SPAs do not have abandoned irrigation systems. These SPAs cover an area of only 26,858 ha (2.5% of the Danube Lower Valley), and lie especially in the western part of the valley. The largest is Calafat-Ciupercei-Desa, covering 15,047 ha. Conversely, 27 SPAs are characterized by the presence of irrigation systems, and total 641,498 ha (59% of the valley). The largest are: Danube Delta and Razim-Sinoie (468,741 ha), Brăila Wetland (25,787 ha), Vedea-Danube (19,272 ha), Old Danube-Măcin Arm (15,139 ha), Danube-Ostroave (14,359 ha), Borcea Arm (12,913 ha), Oltenița-Ulmeni (12,180 ha), Sands of Dăbuleni (10,980 ha). As in case of the SCIs, these are distributed mainly in the proximity of the Danube Delta and adjacent river sections (Fig. 6).

Approximately 99.8% of the regions were forests and other semi-natural areas have been converted into wetlands are protected. Similarly, 98.4% of the areas where wetlands have replaced water bodies are concentrated in SCIs or SPAs. Furthermore, 78.8% of the areas where wetlands have replaced agricultural areas are included in protected land, whereas in case of conversions from artificial surfaces to wetlands, only 67.5% of them lie on protected areas. The results suggest that the expansions of wetlands over other semi-natural land cover classes (such as forest and water bodies) are primarily located in SCIs and SPAs (over 90% from the total converted areas, in both cases).

The conversions of artificial and agricultural areas into wetlands are also concentrated in protected areas, yet to a much lower extend (less than 80% in all of the two cases). By analyzing the overall areas were wetlands expanded at their peak period (2000–2006), and their distribution over protected land and abandoned irrigation systems, the results reveal that, 95.9% of them are located on protected areas (89% in SCIs, 90.7% in SPAs). Similarly, the largest percentages are concentrated on

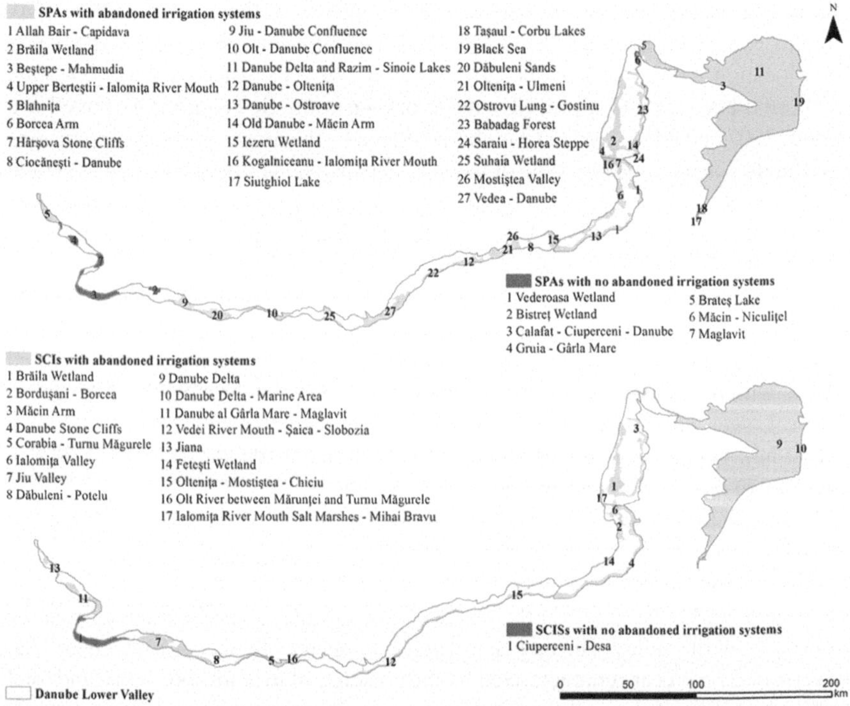

Fig. 6 Distribution of SCIs and SPAs with abandoned irrigation systems in the Danube Lower Valley after 1990. *Data source* Ministerul Mediului (2015) [35]

Table 2 Areas were wetlands expanded at their peak period (2000–2006), and the distribution over protected land and over areas with abandoned irrigation systems

Wetland expansion at their peak (2000–2006)	Area (hectares)					
	Overall Danube Lower Plain	Overall protected land	SCIs	SPAs	SCIs with abandoned irrigation systems	SPAs with abandoned irrigation systems
Artificial to wetlands	256	173	166	173	164	171
Agricultural to wetlands	4171	3290	2645	1872	2440	1633
Forests to wetlands	13,623	13,600	13,327	13,588	11,834	12,085
Water to wetlands	9751	9599	8625	9602	8624	8670
Overall wetland expansion	27,801	26,662	24,763	25,235	23,062	22,559

Data source Environmental European Agency (1990, 2000, 2006, 2012, 2018) [29]; Ministerul Mediului (2015) [35]

protected areas with abandoned irrigation systems (82.9% in SCIs, 81.1% in SPAs) (Table 2).

By superimposing, the areas were wetlands expanded at their peak period (2000–2006), the location of SCIs and SPAs, and the distribution of abandoned irrigation systems, the maps indicate that the three elements are spatially connected. The wetlands expansions are usually included in SPAs/ SPAs with abandoned irrigation systems, the only exceptions being a large area west of Bistreț Wetland where wetlands have replaced arable land. Similarly, the SCIs/ SCIs with abandoned irrigation systems also include the majority of new wetlands. Yet, there are several small areas where new wetlands are located outside protected land, such as east of Ciuperceni-Desa where forest and other semi-natural areas were converted into wetlands, and north of Măcin Niculițel where water bodies were converted into wetlands (Fig. 7).

Discussion of our results reveals the presence of a clear spatial connection between the distributions of abandoned irrigation systems, wetlands expansion and protected areas (SCIs and SPAs). The results indicate that weak land management could by a potential triggering factor for the process of returning to wilderness, and therefore, efficient in promoting wildlife conservation. Changes in land use from wetlands to agricultural land, development of irrigation systems and river embankment have been the main factors with a negative influence on biodiversity in the Danube floodplain, mainly due to the reduction and fragmentation of habitats available to different species, in the socialist period [37].

On the other hand, there are several factors that have favored the re-wilding of once natural ecosystems, such as the manifestation of extreme hydrological events,

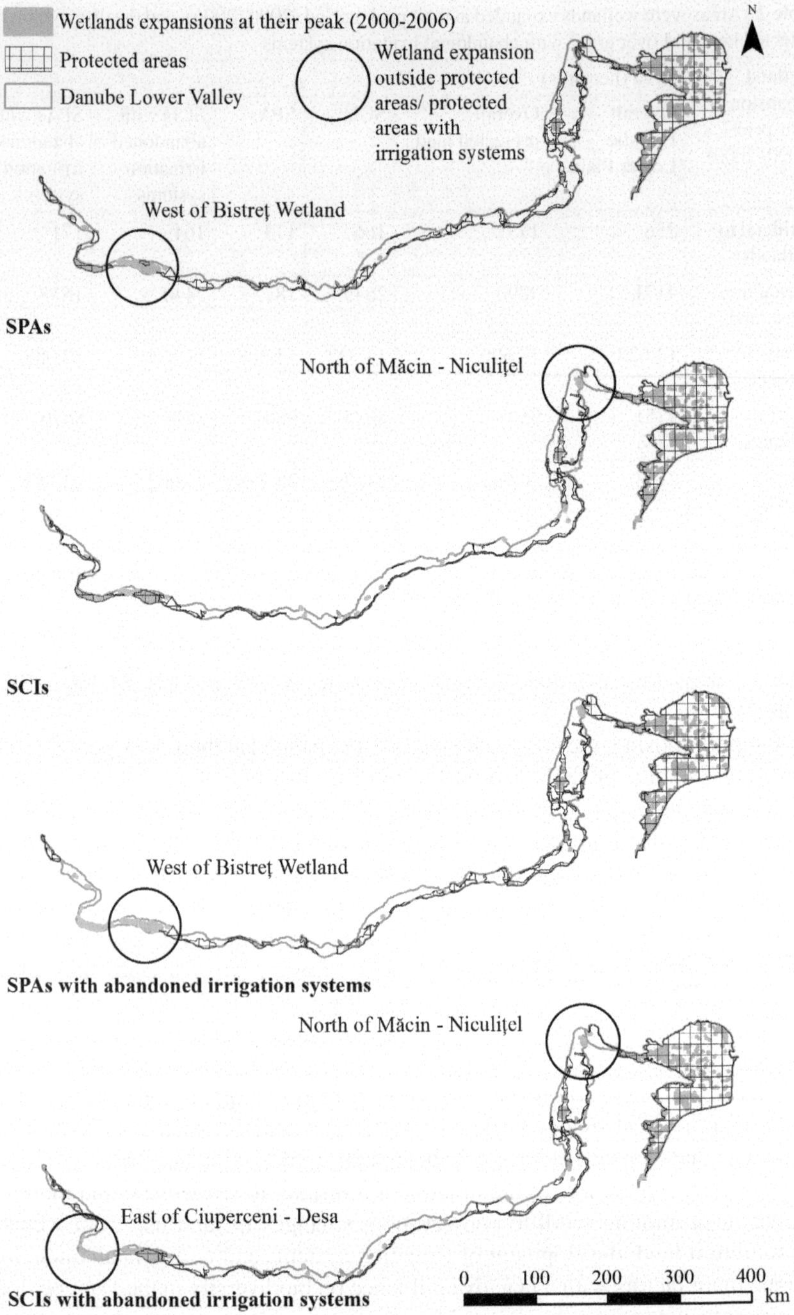

Fig. 7 The distribution of the areas were wetlands expanded at their peak period (2000–2006), the location of protected areas (SCIs and SPAs) and abandoned irrigation systems. *Data source* Environmental European Agency (1990, 2000, 2006, 2012, 2018) [29]; Ministerul Mediului (2015) [35]

namely floods, which have led to the extent of wetlands. The explanation for these changes consists in the fact that the natural hydrological regime in the Danube basin has been strongly modified by numerous hydrotechnical works (levees, drainage and irrigation systems), of which many also have a role of protection against floods. The river sectors in which the runoff regime is undisturbed by anthropogenic influence are restricted and are located on small tributaries. Despite these aspects, weak land management practices after 1990, such as the abandonment of irrigation systems have led to a decrease of human pressure over the landscapes ecological functions. Furthermore, within the Danube River, after 1990, floods occurred, being favored by a very high degree of embankment, long duration of high river levels, large surface of the river basin and reduced slope. The impact on the flow exerted by the levees is significant in the case of high waters, when the flood peak increases considerably compared to the flow that occurs in normal times. In 2006, high levels of the Danube fell into breaches in levees and flooded many areas of the floodplain. The waters remained captive for a long time. Prolonged stagnation of water in the depressed areas behind the levees later led to the development of wetlands [38].

Besides their importance from an economic point of view, being used for water supply and fishing, wetlands have an important role in regulating runoff and in conserving biodiversity, being habitats for a large number of species, especially water birds. Therefore, aquatic surfaces are constituted in important spaces where biological diversity can be conserved—*ecological supermarkets* [39]. The adaptation of wetland habitats to the threats generated by anthropogenic pressures involved taking complex measures in order to include them into a national system of protected areas, such as SCIs and SPAs. These measures aimed at creating protected areas as large as possible, establishing buffer outdoor areas to better filter out anthropogenic external influences, identifying ecological corridors to connect protected areas and ensuring sustainable land management to help maintain biodiversity [37].

5 Conclusion and Recommendation

The main factors that could lead to a potential return of wildlife habitats in areas with abandoned irrigation systems are the presence of extreme hydrological or meteorological conditions, namely floods or climate change which could favor the extend of wetlands, abandonment of intensive agricultural practices proper to generate the expansion of spontaneous secondary vegetation, and lack of general human activity.

Our findings could help valuing the protected areas natural potential, by increasing biodiversity management, wildlife conservation and connected touristic attractiveness.

In order to ensure the continuity of renaturalized ecosystems, we recommend limiting agricultural activities, especially intensive practices. Agricultural activities are also allowed in Natura 2000 sites, and in the event of a revival of irrigation systems (unlikely soon) most of the protected areas will not undergo major changes.

Also, for the sustainable development of the Danube Lower Valley, we suggest planning protected areas so that they have a spatial structure as compact as possible, respectively to seek their expansion in order to increase connectivity, because some sites are extremely isolated. In addition, we consider useful to stop the expansion of transport infrastructure in order to decrease landscape fragmentation, respectively to encourage residential expansion in compact structures, in order to maintain the areas occupied by agricultural land and open spaces at their present expansion.

References

1. Mihai B, Nistor C, Toma L, Săvulescu I (2016) High resolution landscape change analysis with CORONA KH-4B imagery: a case study from Iron Gates Reservoir Area. Procedia Environ Sci 32:200–210
2. Grădinaru SR, Iojă IC, Vânău GO, Onose DA (2020) Multi-dimensionality of land transformations: from definition to perspectives on land abandonment. Carpath J Earth Env Sci 15:167–177
3. Hersperger AM, Burgi M, Wende W, Bacau S, Gradinaru SR (2020) Does landscape play a role in strategic spatial planning of European urban regions? Landsc Urban Plan 194:103702
4. Năstase G, Şerban A, Năstase AF et al (2017) Hydropower development in Romania: a review from its beginnings to the present. Renew Sust Energ Rev 80:297–312
5. Teampău P (2020) Trouble in paradise: competing discourses and complex governance in the Romanian Danube delta. Mar Policy 112:103522
6. Csagoly P, Magnin G, Hulea O (2016) Danube River Basin. In: Finlayson C, Milton G, Prentice R, Davidson N (eds) The Wetland Book. Springer, Dordrecht, p 112
7. Tănăsescu M, Constantinescu S (2020) The human ecology of the Danube Delta: a historical and cartographic perspective. J Environ Manage 262:110324
8. Popescu I, Cioacă E, Pan Q, Jonoski A, Hanganu J (2015) Use of hydrodynamic models for the management of the Danube Delta wetlands: the case study of Sontea-Fortuna ecosystem. Environ Sci Policy 46:48–56
9. Gómez-Baggethun E, Tudor M, Doroftei M, Covaliov S et al (2019) Changes in ecosystem services from wetland loss and restoration: an ecosystem assessment of the Danube Delta (1960–2010). Ecosyst Serv 39:100965
10. Turnock D (2001) Cross-border conservation in East Central Europe: the Danube-Carpathian complex and the contribution of the WorldWide Fund for Nature. GeoJournal 55:655–681
11. Danube Region Strategy (2011) EU Strategy for the Danube Region. https://danube-region.eu/. Accessed 20 Jan 2021
12. Soran V, Biro J, Moldovan O, Ardelean A (2000) Conservation of biodiversity in Romania. Biodivers Conserv 9:1187–1198
13. Hossu CA, Ioja IC, Susskind LE, Badiu DL, Hersperger AM (2018) Factors driving collaboration in natural resource conflict management: evidence from Romania. Ambio 47:816–830
14. Nita A, Ciocanea CM, Manolache S, Rozylowicz L (2018) A network approach for understanding opportunities and barriers to effective public participation in the management of protected areas. Soc Netw Anal Min 8:31
15. Funk A, Martínez-López J, Borgwardt F, Trauner D, Bagstad KJ, Balbi S, Magrach A, Villa A, Hein T (2019) Identification of conservation and restoration priority areas in the Danube River based on the multi-functionality of river-floodplain systems. Sci Total Environ 654:763–777
16. Posea G, Bălteanu D, Cucu V, Gâştescu P, Ielenicz M (2005) Geografia României (Vol. 5). Romanian Academy Press, Bucharest

17. Antipa G (1921) Dunărea şi problemele ei ştiinţifice, economice şi politice. Lucrările Cartea Românească, Bucharest
18. Bălteanu D, Micu D, Chendeş V (2020) Hazard, vulnerabilitate şi rezilienţă la inundaţii în România. Abordări la diferite scări spaţiale. Romanian Academy Press, Bucharest
19. Administraţia Naţională de Meteorologie (2008) Clima României. Romanian Academy Press, Bucharest
20. European Environmental Agency (2016) Biogeographical regions. http://www.eea.europa.eu/data-and-maps/data/biogeographical-regions-europe-3. Accessed 16 Jan 2021
21. Brînzan T (2013) Catalogul habitatelor, speciilor şi siturilor Natura 2000 în România. Exclus Prod Press, Bucharest
22. Săgeată R (2014) Euroregiunile de cooperare transfrontalieră din bazinul inferior al Dunării. Studiu geografic. Romanian Academy Press, Bucharest
23. Tempo Online (2020). Baze de date statistice. http://statistici.insse.ro:8077/tempo-online/#/pages/tables/insse-table. Accessed 18 Jan 2021
24. Portal Legislativ (1991) Legea fondului funciar. http://legislatie.just.ro/Public/DetaliiDocumentAfis/1459. Accessed 20 Jan 2021
25. Charta României Meridionale—1864. http://charta1864.ro/charta.html. Accessed 25 Dec 2020
26. Geo-Spatial.org. Planuri Directoare de Tragere. http://geo-spatial.org/vechi/download/planurile-directoare-de-tragere. Accessed 18 Jan 2021
27. Agenţia de Informaţii Geospaţiale a Apărării Constantin Barozzi. Harta Topografică Militară 1975 scara 1: 250.000. www.geomil.ro. Accessed 25 Dec 2020
28. Pătru-Stupariu I (2011) Peisaj şi gestiunea durabilă a teritoriului. Aplicaţii la Culoarul transcarpatic Bran–Rucăr–Dragoslavele. University of Bucharest Press, Bucharest.
29. European Environmental Agency. Corine Land Cover 1990, 2000, 2006, 2012, 2018. https://land.copernicus.eu/pan-european/corine-land-cover/clc2018. Accessed 25 Dec 2020
30. Vărzaru S (1984) Prin Ţările Romane, Călători străini din secolul al XIX-lea. Sport-Turism Press, Bucharest
31. Grigoresu I (2010) Modificările mediului în Aria Metropolitană a Municipiului Bucureşti. Romanian Academy Press, Bucharest
32. Bărăscu F (1998) Modificările factorilor de mediu din lunca Dunării (sectorul Vedea-Argeş), în urma lucrărilor de îmbunătăţiri funciare. PhD Thesys. University of Bucharest
33. Dumitraşcu M, Grigorescu I (2016) Invasive terrestrial plant species in the Romanian protected areas: a geographical approach. Romanian Academy Press, Bucharest
34. Iojă C (2000) Metode şi tehnici de evaluare a calităţii mediului în Aria Metropolitană a Muncicipiului Bucureşti. University of Bucharest Press, Bucharest
35. Ministerul Mediului (2015) Baze de date GIS. http://www.mmediu.ro/articol/date-gis/434. Accessed 25 Dec 2020
36. ESRI. ArcGIS Pro. www.esri.com. Accessed 25 Dec 2020
37. Bălteanu D, Şerban M (2005) Modificările globale ale mediului. O evaluare interdisciplinară a incertitudinilor. Coresi Press, Bucharest
38. Bălteanu D, Sima M (2013) Evaluarea şi prevenirea hazardelor în Lunca Dunării: (Sectorul Calafat-Vidin-Turnu Măgurele-Nikopol). Universitaria Press, Craiova
39. Pătroescu M, Toma S, Sasaki L, Apostol G (2000) Priorities in re-habilitation pf rural landscape of the Romanian plain, southern Romania. Annals of University of Bucharest, XLIX

The Danube River: Between Conservation and Human Pressures in the Iron Gates Natural Park

Mihai-Răzvan Niţă, Athanasios Alexandru Gavrilidis, Andreea Niţă, Iulian Mihăiţă Niculae, and Maria Pătroescu

Abstract Designing natural protected areas is an effective approach for pausing biodiversity decline and habitat losses. Along its course, the Danube River has crafted unvaluable natural sanctuaries, most of them being targeted by nature conservation measures. At the same time, the Danube River has always created development opportunities for human settlement emerged along the river. The Iron Gates Natural Park (IGNP) is one of the largest natural protected areas along the Danube River and its objectives are not focussed only on biodiversity, but also at promoting sustainable development in the area. Despite being one of richest biodiversity hotspots in Europe, the human communities sheltered here struggle with every-day living, the region being socio-economic unrested. In this context, the fundamental objective of conservation strategies is to harmonize the human needs and requirements with the nature conservation focus. We have analysed the updated version of the Management Plan of IGNP, as our main focus was to identify how the socio-economic and environmental aspects are integrated for providing a sustainable development of the region while keeping its conservation status. We have also desired to assess whether the natural assets provided by the presence of the Danube River are used for the socio-economic progress by local communities. Furthermore, a perception analysis was conducted in the region, assessing the popularity of the conservation policies among local dwellers. The data were gathered from a survey undertook in 2020. Our results have indicated that imposed conservation policies can trigger discontent among the impoverished local communities. This current chapter offers a relevant image on how conservation relates with human pressure in the Iron Gates Natural Park. Our results may be considered relevant for identifying and promoting best or

M.-R. Niţă · A. A. Gavrilidis (✉) · A. Niţă · I. M. Niculae · M. Pătroescu
Faculty of Geography, University of Bucharest, 1 Nicolae Bălcescu, 010041 Bucharest, Romania
e-mail: athanasiosalexandru.gavrilidis@g.unibuc.ro

M.-R. Niţă
e-mail: mihairazvan.nita@g.unibuc.ro

A. Niţă
e-mail: andreea.nita@cc.unibuc.ro

I. M. Niculae
e-mail: mihaitaiulian.niculae@g.unibuc.ro

A. Negm et al. (eds.), *The Lower Danube River*, Earth and Environmental Sciences Library, https://doi.org/10.1007/978-3-031-03865-5_17

better practices related with natural conservation along the Danube and other riparian protected areas.

Keywords Danube · Biodiversity conservation · Socio-economic indicators · Conservation policies · Romania

1 Introduction

The Danube River is the environmental component which provided a multitude of services since ancient period to a significant number of European inhabitants. Whether it had a defensive or border role, providing food and a fertile meadow or it was used for transportation or leisure activities, the river fed the central and eastern Europeans' demands throughout the history. Therefore, the role of the Danube within the European society is undeniably of high importance [1].

Besides the societal impact, the river is a natural and efficient sanctuary for biodiversity as it carved a variety of landscapes through complex landforms. However, the historical evolution of Europe from social, cultural and economic perspectives, and the diversification of human needs had also implied changes within the natural balance of the Danube River, expressed through changes regarding the waterflow [2], riverbanks, landscapes [3] and biological diversity [4].

As the pressure on the ecosystems of the Danube River has become more complex and more aggressive, national and local governing bodies have established natural protected areas along the river in order to maintain the permanent flux of ecosystem services and preserve the riparian environments as wild as possible. This was the case for the region in which the Danube River separates the Carpathians from the Balkans (Fig. 1). This is one of the most spectacular landforms the river passes, situated at the junction between two of Danube's sectors—the Middle Basin and the Lower Basin [1]. The name "Iron Gates" relates to the gorge dug by the Danube River here as it passes through the Southern Carpathians [5], and it has been also attributed to the protected areas situated both on the northern (Romania) and southern (Serbia) banks of the river [6].

The region has passed throughout the history under several administrations and ways of governance. The time frames in which the most salient changes at a landscape level have emerged occurred during the Austrian-Hungarian administration, the Communist regime of Romania (1947–1989) and the contemporary period [3, 7–9]. One of the most evident change suffered by the river's meadows is undoubtedly the construction of the Iron Gates Hydropower and Navigation System, an ambitious project agreed between the Socialist Republic of Romania and the Socialist Federal Republic of Yugoslavia which started in 1964 and finished in 1972, consisting of a 60 m high and about 1 km long dam, being one of the largest hydro constructions in Europe [10].

The construction of the dam substantially altered the hydrological regime [2], with the reservoir behind it, expanding towards Belgrade's outskirts. The reservoir

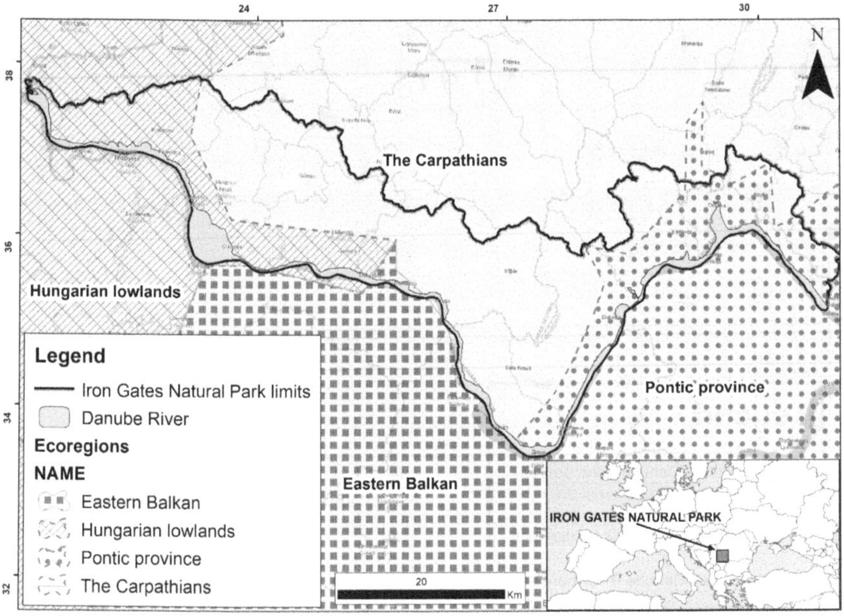

Fig. 1 Location of Iron Gates Natural Park within the European continent and the Ecoregions (according to European Environmental Agency)

has flooded the natural floodplain area and wetlands adjacent to the Danube River and some of the old human settlement along the way (Fig. 2) and extensive surfaces of arable land and pastures with a total flooded surface of about 100 km^2 [7].

Besides the construction of the Iron Gates hydropower plant, during the Romanian communist rule, several industrial activities took place in the region, ore extraction (coper and coal mainly) being the main focus [11]. However, despite these activities taking place, the area was restrictive for outsiders due to the controlled circulation of Romanians abroad and due to the numerous illegal attempts of border crossings through this region until 1990. Thus, despite the touristic potential of the region, highlighted by Patroescu and Vintila [12] in the late 90's, it was not until the late 2000's that the region has become popular for tourists and touristic infrastructures have emerged in the area.

The development of tourism relatively coincides with the conservation status of the region enforced in 2000. The Iron Gates Natural Park (IGNP) covers a surface of approximately 1156 km^2 and it includes territories from Caras-Severin and Mehedinti Counties [13]. Considering Romania's recent political and economic history and the negative perceptions of conservation efforts, the efficiency of IGNP management is still rather contentious [14]. The conservation status of the region was superimposed on the local structures in a classical central control process, with weak public participation [15], fact that still creates turmoil in the region. Is it notable that in

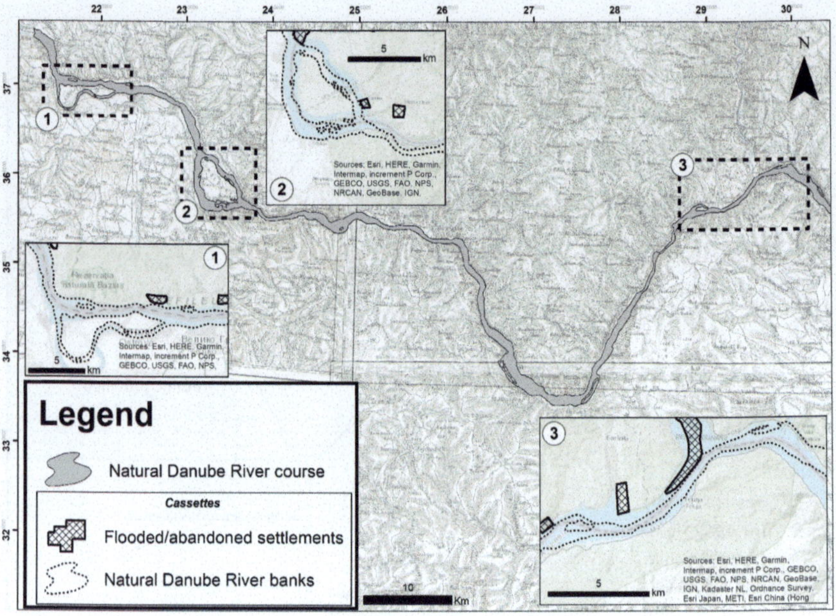

Fig. 2 The natural flow of the Danube until 1964 (main map) in contrast with the current morphology (cassettes). *Source of main base map* The Military mapping survey of Austria-Hungary (1:200,000) drawn between 1889–1915. http://lazarus.elte.hu/

Romania most of the protected areas are still established and managed by central authorities [16].

Some challenges regarding the conservation aims and human demands within IGNP come from external factors. One example consists of the cross-border cooperation with Djerdap National Park from Serbia, the open borders enhancing the contact between humans and prior isolated natural sanctuaries [17]. Besides that, the role of external bodies promoting conservation measures, such as the EU Commission through the Natura 2000 network or the International Commission for the Protection of the Danube River, implementing the Water Framework Directive [1] are adding more complexity on the issue of keeping a sustainable balance between human activities and natural conservation. Nevertheless, legislation and management directions imposed from Bucharest are also increasing the confusion and uneasiness among local communities, policy makers and stakeholders [18].

There are also internal factors in the IGNP, fostering the debate between conservation and human pressure, such as the level of ecological education and awareness among the wide public [19], the competition for resources and the collaboration related-processes favouring adaptive co-management [20] or the depopulation, ageing and social challenges characterizing human settlements in the area [3].

The social configuration of the area is widening the gap between the conservation aims and public acceptance. As Grodzinska-Jurczak and Cent [21] highlighted, the experience of the countries which were the first to implement European Union's directives on nature conservation show that they cannot be effectively applied without widespread public participation. Since deprived communities are considering the conservation status of their lands more as a drawback rather than an opportunity, conservation policies and their implementation will not be fully effective.

The main aim of this chapter is to analyse the integration of competing social and economic objectives with the conservation process of the Iron Gates Natural Park, giving a special focus to the Danube River. We proposed for this a three-staged analysis focusing on: (i) assessing the current social and economic situation of the communities living within IGNP; (ii) classifying the conservation objectives and measures presented in the Management Plan of the IGNP and (iii) identifying public perceptions regarding the conservation status of the area. We focused on describing in our study an integrated perspective, including both socio-economic and conservation realities specific to the Iron Gates Natural Park, as prior studies have analysed just one of these perspectives. This approach may provide an effective feedback for practitioners and policy makers concerned in enhancing the potential provided by the Danube River for harmonious development of local communities in relation with natural features.

2 Description of the Iron Gates Natural Park

The study area, characterized by wetlands, unique geological features and a sub-Mediterranean climate, represented a vast field of research due to the existing biodiversity, but also due to the environmental issues generated by the usage of natural resources and conflicts in governance [22, 23]. The status of the region has been officially established by Law no. 5 of 2000 regarding the plan for organizing the national territory, Sect. 3—protected areas [24, 25]. The protected area's official limits and total area are described by the Government Decision no. 230/2003 [26].

2.1 Overview of the Natural Features

Included in the "*natural park*" category [27], IGNP favours the sustainable use of resources and traditional activities, a large variety of the conservation objectives resulting from the historical interaction between the rich biodiversity and various forms of social and economic exploitation. The geodiversity of the region contributed in time to an increased variety of habitats, ecosystems, and landscapes.

The lithology of the region is dominated by crystallin and sedimentary rocks, with seldom magmatic intrusions of different ages. The limestones found here (e.g. Ciucaru Mare and Ciucaru Mic mountainous massifs) had a significant contribution in shaping the landscape, with numerous exokarst and endokarst complexes along with providing suitable habitats for calciphile and thermophilic species. The Iron Gates is characterized by a complexity of lithological and paleontological sites, considered to be unique along the Danube [28].

Most of the Danube's tributaries within the analysed area have their sources in the mountainous regions and are characterized by torrential flows. After the construction of the Iron Gates dam the confluence of the tributaries with the Danube River were flooded, forming small gulfs with deltaic patterns [3].

The biological diversity of the region has brought the attention of naturalists who in the late 19th and the early twentieth century signalled a new plant taxa - *Stipa danubialis* [29, 30], 17 new plant taxa unknown to be existing within Romanian territory at the time (e.g. *Minuartia hamata, Minuartia verna ssp. collina, Iris pallida* etc.) [31] and 25 rare taxa (e.g. *Dianthus pinifolius, Paeonia daurica, Verbena supina* etc.) vegetating exclusively in this region [32].

The biogeographical origin of species revealed by Mataca [32] highlights the presence of Eurasian (32%), Mediterranean (12%), European (12%), Pontic (10%) and Balkan (5%) features. The existence of some thermophilic communities was highlighted since 1957 [33]. Among the flagship species of fauna is the Hermann's tortoise (*Testudo hermanni* Boetgeri), an endangered species with a population rehabilitated in the area through a LIFE Nature project (LIFE Nature RO 7171) implemented by the University of Bucharest, which also analysed the distribution area of other species of reptiles [34, 35].

2.2 Conservation Policies—Timeline and Current Management

The biological richness of the Iron Gates region determined scientists and political figures in trying to establish here a protected area since the early twentieth century. In 1918 professor Victor Stanciu from Cluj-Napoca University made amends to the Agrarian Law of the time in which he aimed that all areas having special biological features, including parts of the Iron Gates region, to be expropriated and dedicated entirely for science [28]. However, this proposal failed to be enforced. In 1924 the geobotanist Alexandru Borza mentions the botanical reservations in the Danube's gorges, while later, in 1936 he urges the need for a special protection status towards the plant species *Prangos carinata* at the sunny slopes from Gura Vaii, area which is today included in IGNP.

The natural park includes other three Natura 2000 sites (ROSPA0026, ROSPA0080 and ROSCI0206) and a Ramsar site focusing on wetland conservation. In addition, the Iron Gates Natural Park includes 18 protected areas of national interest (Table 1), but also with different categories in the International Union for the Conservation of Nature classification. The conservation importance of the region is therefore sustained by the high number of protected areas overlapping in the Iron Gates. The internal zonation of the Iron Gates Natural Park included most of the above protected areas in zone with integral protection, limiting, at least on paper,

Table 1 List of natural reservations of national interest within Iron Gates Natural Park [28]

Name	Type	IUCN Cat.	Area (ha)	Habitats	Important species
Nera-Danube wetland	mixt	IV	10.0	Oligotrophic waters, eutrophic lakes, hydrofoil vegetation	*Iris pseudacorus; Butomus umbellatus; Populus alba; Salix alba; Egretta garzetta; Egretta alba; Alcedo atthis; Canis lupus; Lutra lutra; Emys orbicularis*
Baziaş	mixt	IV	170.9	Regional and local importance habitats	*Quercus cerris, Q. pubescens; Fraxinus ornus; Tilia tomentosa; Paeonia officinalis.*
Calinovat holm	birds	IV	24.0	Aquatic and wetlands	*Egretta alba; Egretta garzetta; Gavia stellata; Emys orbicularis;Bombina bombina; Cottus gobio; Aspius aspius etc.*
Martin steep from Divici valey	mixt	IV	5.0	Xerophile pastures, loess deposits	*Paeonia officinalis; Gladiolus imbricatus; Riparia riparia; Hirundo rupestris etc.*
Divici-Pojejena	birds	IV	498.0	Terrestrial and aquatic	*Ciconia nigra; Milvus migrans; Phalacrocorax pygmeus; Alcedo atthis; Hyla arborea; Zingel zingel; Cobitis elongata;*
Valea Mare	botanic	IV	1179.0	Potholes, Caves, Sinkholes, Aquatic habitats	*Fagus sylvatica; Carpinus orientalis; Tilia tomentosa; Genista ovata; Orchis pallens; Rhinolophus blasii; Hirundo rupestris; Vipera ammodytes; Testudo hermanni boettgeri*
Water cave from Polevii Valey	mixt	IV	3.2	Reef limestones	*Species of bats*
Moldova Veche holm	birds	IV	1627,0	Aquatic, Sandy wetlands	*Salix alba; Populus alba; Iris pseudacorus; Elodea canadensis; Cygnus cygnus; Buteo rufinus; Emys orbicularis*
Fossil site Sviniţa	paleontological	III	95.0	Limestone, Marl	*Amoniţi, brachiopode jurasice etc.*
Cazanele Mari and Cazanele Mici	mixt	IV	215.1	Limestone steeps, Potholes, Sinkholes	*Fagus sylvatica; Carpinus orientalis; Taxus baccata; Tulipa hungarica; Stipa aristella; Syringa vulgaris; Vipera ammodytes;*
Fossil site Bahna	paleontological	III	10.0	Limestone, Marls	*Fossil fauna of molluscs, foraminifera, coral or gastropods*
Duhovna hill	forestry	IV	50.0	Regional and local importance habitats	*Quercus petraea;*
Gura Văii-Vârciorova	mixt	IV	305.0	Terrestrial habits, Xerophile	*Quercus frainetto; Quercus dalechampii; Crataegus nigra; Cotinus coggygria; Verbascum varciorovae; Testudo hermanni boettgeri;*
Faţa Virului	botanic	IV	6.0	Lithophile	*Quercus virgiliana; Rubus severinensis; Celtis australis;*
Cracul Crucii	botanic	IV	2.0	Lithophile	*Minuartia capillacea; Cachrys ferulacea; Dianthus varciorovensis;*
Vărănic	mixt	IV	350.0	Xerophile, Limestone	*Carpinus orientalis; Fraxinus ornus; Syringa vulgaris; Vipera ammodytes; Testudo hermanni boettgeri*
Ogalnicului valey	botanic	IV	150.0	Xerophile	*Stipa danubialis; Pulsatilla grandis; Paeonia daurica; Testudo hermanni boettgeri*
Cracul Găioara	botanic	IV	5.0	Xerophile	*Ephedra distachya; Ceterach officinarum; Centaurea atropurpurea; Stipa danubialis*

human pressures to other type of areas (buffer or sustainable development areas) [36].

Recently there was an attempt that the Iron Gates Natural Park to be declared a UNESCO Biosphere Reservation. But while the legal and administrative documentation of this demarche was completed and included in Man and Biosphere meeting agenda that took place in Lima in 2016, the attempt failed.

3 Social and Economic Context within Iron Gates Natural Park

The Iron Gates Region is rich in natural resources, especially mining resources such as coal or copper, but it also provides breath-taking landscapes, attracting thousands of visitors every year. Both Romania and the Republic of Serbia have declared the Danube River banks and their surroundings as natural protected areas. Before being declared natural protected areas, there were intense mining activities in the region on both sides of the Danube—large copper extraction facilities at Moldova Nouă (RO) and Majdanpek (SRB) along with smaller coal mines or limestone quarries scattered on both banks.

It is a well-known phenomenon that human settlements located inside or in the proximity of protected areas are usually less developed and with reduced opportunities and access to infrastructure [37]. During the deindustrialization period and the abandonment of the mining activities (since 1990), the IGNP inherited a community of ex-miners, the region being characterized by high unemployment rates and social imbalances. For a long period, conservation efforts and socio-economic development have been considered incompatible due to a limited understanding of the relation that might exist between them [38].

The analysed region was populated since early ages as both banks of the Danube River are rich in archaeological sites, dating since the Mesolithic period [39]. On the Romanian side there are 10 archaeological sites and various undiscovered sites that archaeologists are still looking for (Fig. 3). These sites are the proof that this section of the Danube helped human community to flourish since the dawn of history. However, none of these historical sites from IGNP are exploited from a touristic perspective, being neglected by authorities, and hidden from visitors. This type of attitude diminishes the opportunity for connected activities to be developed in the region.

Using input data provided by the National Statistics Institute we could shape an overall situation of the socio-economic and demographic aspects describing the conditions of the communities located within IGNP. Then, using GIS techniques we mapped the results to be more comprehensive and to better highlight how the presence of the Danube River is exploited or underexploited by these communities. If we look at the demographic evolution within the area, it is obvious that the communities are aging, based on the last 13 years data (Fig. 4). We desired to capture the

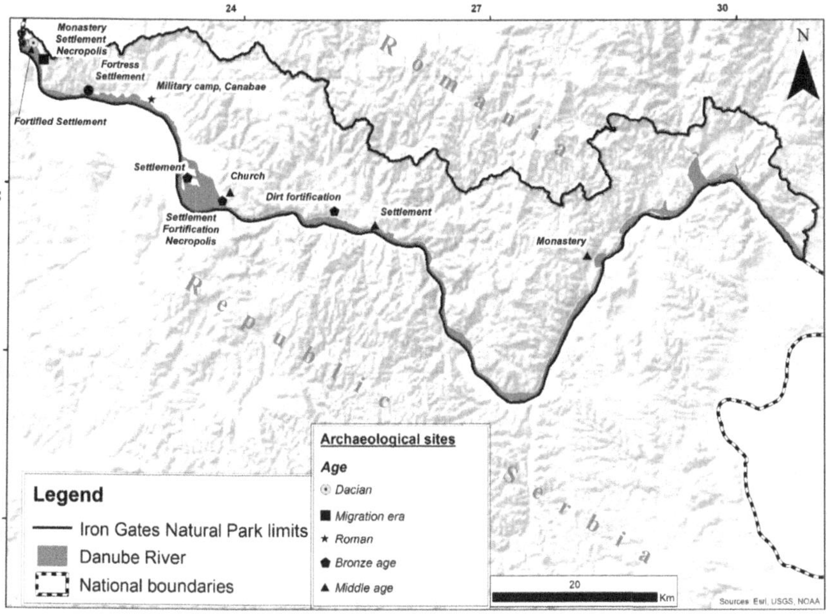

Fig. 3 Dispersion of archaeological sites discovered in the analysed region. *Data source* National Archaeological Repertoire – http://ran.cimec.ro/

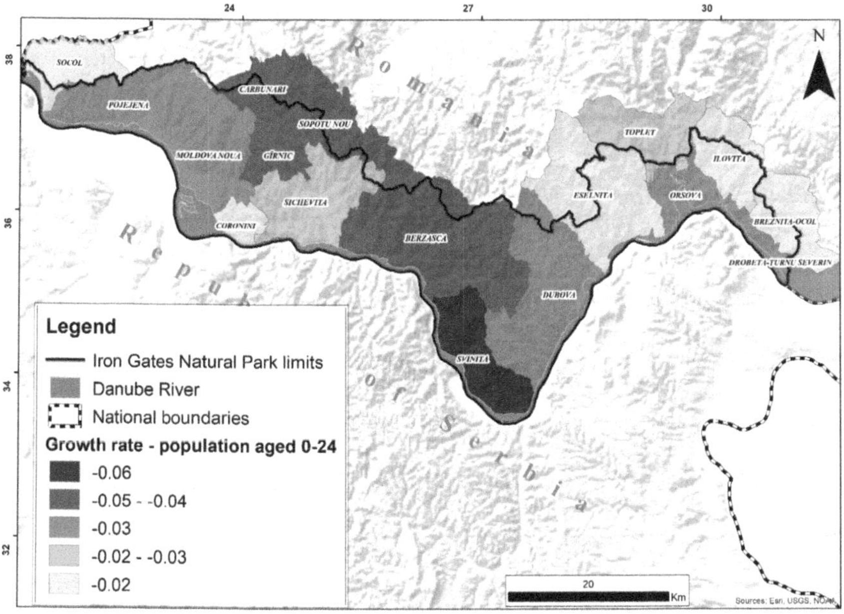

Fig. 4 Population growth rates on age clustered (Statistical data source from National Statistical Institute—time frame 2007–2019) [41]

demographic evolution in the region since 2007 when Romania officially became part of the European Union. It is a documented fact that from 2007, massive waves of Romanians left for Western European countries as they were seeking better wages and better living conditions [40]. The population growth rates based on age clusters shows how the trends regarding the youth and active population are negative.

This situation may be considered a ticking clock from a social perspective as in mid and long term the rural and small urban areas from the region could become deserted. Bijak, Kupiszewska [42] argue that without a combination policy aimed at increasing fertility and maintaining labour force within the aging communities, it will be almost impossible to face the socioeconomic challenges laying ahead. On the other hand, León-Ledesma and Piracha [43] argue that remittances sent by immigrants to their departed communities have a positive impact on productivity and employment, both directly and indirectly through its effect on investment. Thus, we cannot clearly establish whether the demographic trends within the Iron Gates communities is encouraging or not.

As the Danube River is the main feature of the IGNP, we assessed whether activities that could be endorsed by the presence of the river are represented here. Foremost, we looked at the aquaculture related businesses in the area and observed that, according to the National Commerce Register, in 2017 only two such businesses were officially registered in the area, while other agricultural related businesses were also poorly represented (Fig. 5). As this is a mountainous area, there are no proper

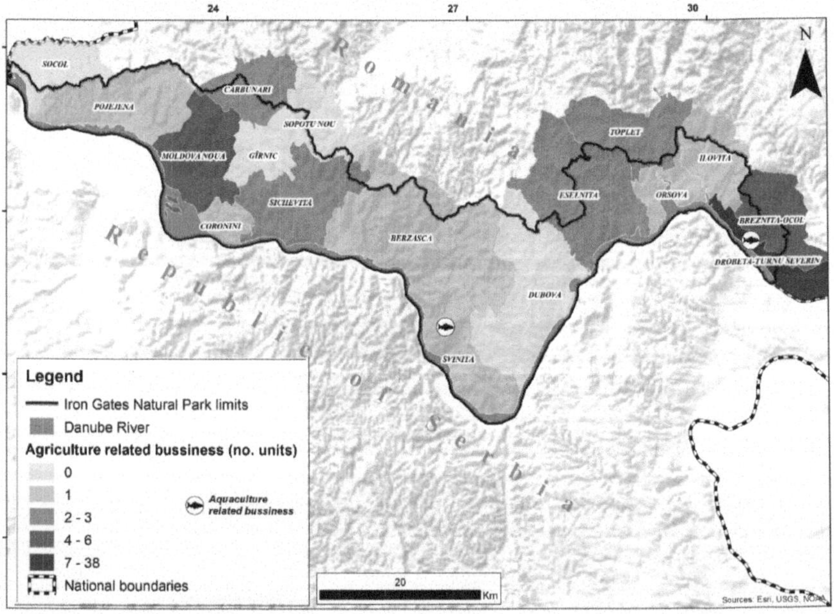

Fig. 5 Agricultural related business in the region with emphasis of aquaculture and fishing activities (referenced year: 2017). *Source* National Register of Commerce (www.onrc.ro)

soils for intensive agriculture. Most of the fertile lands were placed in the Danube floodplain which was covered with water after the construction of the Iron Gates dam.

According to EURSOTAT database, the amount of products derived from fishing and aquaculture increased in the last 10 years in Romania along with the incomes generated by this type of activity. An interesting aspect though is that the overall price per ton has decreased in the same period, which may explain the increasing production from the last five years.

The aquaculture production statistics from EUROSTAT and the National Statistics Institute recorded in 2010, shows that the production registered in the Iron Gates Natural Park, was around 45% at national level. During the next years, when the national production has increased, the percentage dropped, and while the area is under natural conservation legislation, the production could not be increased. Adding that the prices have also fallen, fishing and aquaculture businesses were no longer considered profitable, and these practices remained more as family traditions.

As we have mentioned before, during the communist period copper and coal extraction industries were the main job providing activities in the Iron Gates region, thus we could say that the region's economic profile was mainly industrial. However, since the shift towards a market economy in 1990, the mining industry has decayed, leaving behind several brownfields and huge tailing dumps. The former industrial activities induced important landscape changes in the region, affecting the ecosystems but at the same time provided numerous jobs which have also attracted outsiders to settle here.

The mining activity from Moldova Nouă harmed terrestrial habitats while the generated industrial waste has contaminated the Danube's waters as well as the atmosphere [44] causing transborder tensions with the neighbouring country, Serbia [45]. The gabbro extraction from Iuți triggered geomorphological processes, while marble exploitation from Văranic fragmented the habitats, especially Herman's tortoise habitat. These activities left a perceptive heritage within the IGNP, scattered throughout the region, raising challenges for conservationist or local policy makers.

According to the official data provided by the National Commerce Register, in 2017 most of the businesses taking place in the IGNP were related to the commercial and construction sectors and in few communities, businesses related to manufacturing industry were still taking place, while the extraction industry is poorly represented as a business sector, only in Drobeta Turnu Severin and Molodva Nouă. These businesses unfolded in the context of a negative growth rate of employees for most communities (Fig. 6). Thus, besides the demographic aging, the local human settlements are facing high unemployment rates, meaning that even the remaining active population is having a hard time finding a job in the region.

As the Danube is a navigable river, we assessed whether there are any businesses related to water transportation activities. The results showed that the main urban areas (Drobeta-Turnu Severin, Orșova, and Moldova Nouă) and three other rural settlement are home of such businesses. However, the non-water transportation businesses are better represented within the region, projecting the impression that the presence of

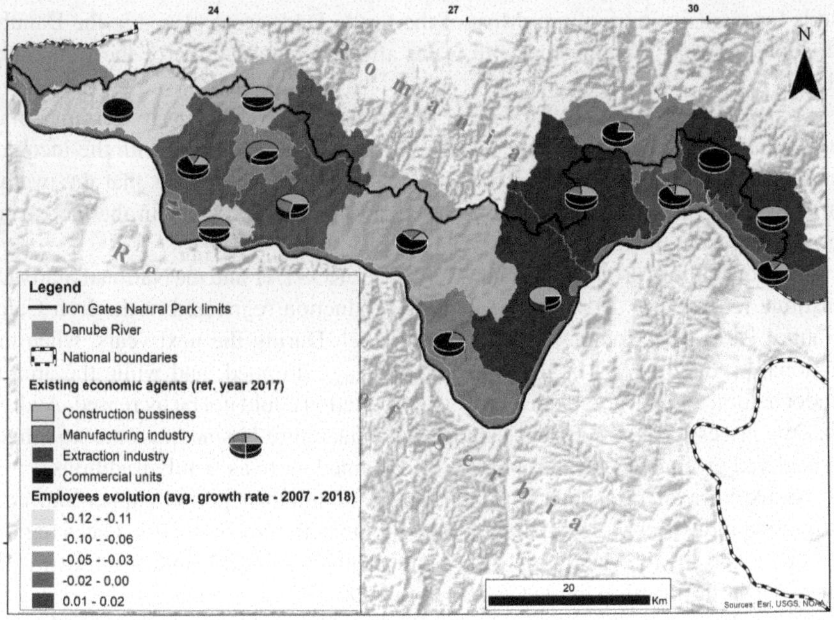

Fig. 6 Situation of the economic agents and employee's evolution. *Source of the employment data* National Institute of Statistics—2007–2019; economic data: National Register of Commerce—2017

the Danube in the proximity is again underappreciated. The statistics provided by the National Institute of Statistics regarding the charging and discharging of goods through internal water ports have shown that since 2008 a slight increase in both has been recorded at national level.

Since the Danube River is the only internal navigable waterway of Romania, these goods were charged and discharged in the harbours placed along the river. This increase of goods transportation could represent an opportunity for the Iron Gates communities to further develop. The official statistics made available by the Iron Gates Dam water locks service indicated that 15,422 boats passed through the Romanian locks in 2010 and 16,097 in 2011.

The 2011 harbour logs from Drobeta-Turnu Severin, the largest harbour in the Iron Gates region, indicate that ships from 20 countries, excluding Romania, have passed through the Iron Gates region (Fig. 7). The fact that the Danube is a vector for international trading is the major asset of this region. Even so, the communities from Iron Gates region cannot benefit from this context as no major route passes through this area and on-road heavy transportation is hardly supported by the existing roads. At the same time, no major economic activities producing enough goods to be exported through the nearby ports are taking place in the area. Thus, from this point of view, the presence of the Danube as a mean of international trading is hardly to make a positive difference for local communities.

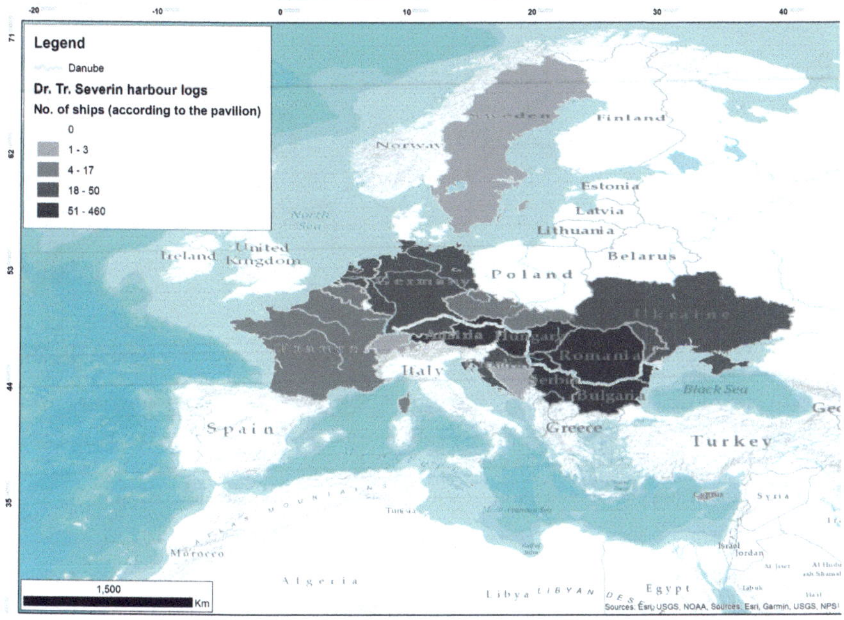

Fig. 7 The traffic intensity in the Iron Gates region based on the harbour logs from Drobeta–Turnu—Severin port—reference year 2011. *Source* Dr. Tr. Severin Harbour logs

There is however a business sector that seems to be better represented in the analysed area and that is the hospitality industry. The local communities having direct access to the Danube River have recorded a high number of tourist arrivals between 2001 and 2019 and they also recorded an increase of accommodation units according to the official economic statistics from 2017 provided by the National Institute of Statistics (Fig. 8). However, a field assessment made in 2013 aiming in counting the actual number of accommodation units in the region revealed a much higher number than the ones recorded in the official books, meaning that some of these units are practicing fraud through tax evasion. Unfortunately, tax evasion practices keep money from returning in local economies and goes hand in hand with black labour and in such a context the healthy socio-economic development of the Iron Gates communities is stagnating or even jeopardized.

The touristic activities in the Iron Gates Natural Park are not subject to common strategies or planning, thus we cannot talk about the practice of mass tourism. Also, the arriving tourist are focusing more on fishing activities, spending most of their time at their accommodations, neglecting the other natural landmarks which are placed within the mountainous region and further away from the Danube's water. The lack of international touristic ports in the region are preventing the influx of foreign tourists, thus this is a shortcoming when it comes to economic development. Local policy makers and stakeholders must focus in creating opportunities for the development of coherent touristic practices, thus focusing on developing the public

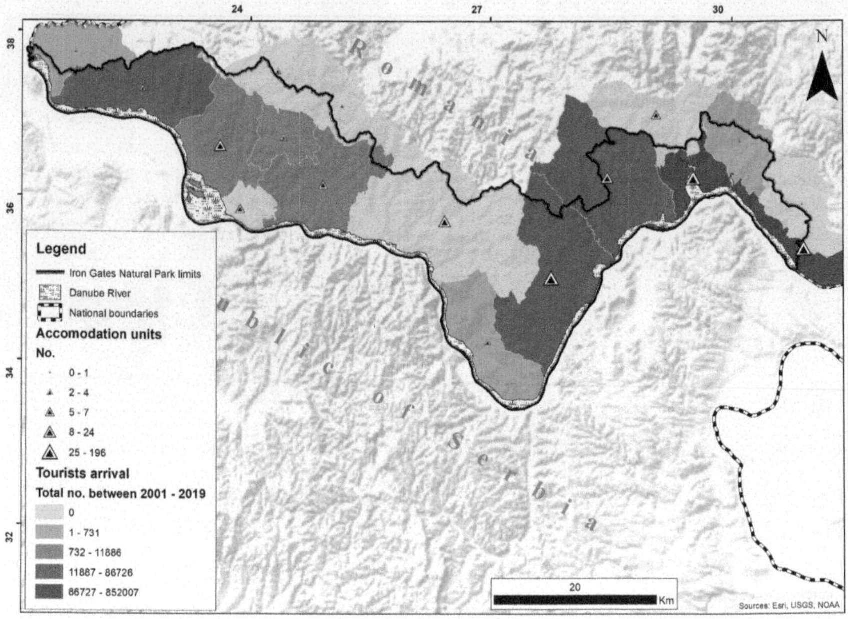

Fig. 8 Situation of touristic activities in the region. *Tourist arrival data source* National Institute of Statistics—2001–2019; Hospitality industry; National Register of Commerce—2017

and private services as well as the transportation infrastructure if they try to switch their economic profile towards profitable sustainable tourism as private investment are cashing out the profits in different areas than the ones they invested [46].

Since the official establishment of IGNP, the expansion of the built-up areas has started, rising the pressure over the riparian ecosystem as most of the new building were constructed adjacent to the Danube River or even in the watercourse as the shores were filled with soil and rocks to create proper grounds for constructions. This building practices led to an increase of the eutrophication process in the Danube's waters [47, 48] as few settlements have a centralized sewage system. Nevertheless, this type of pressure is jeopardizing the wetlands conservations goals as the nutrient pollution affects the population of amphibians [49, 50].

4 Conservation Measures in The Iron Gates Natural Park

Within this section, we focused our attention on the content included in two official documents based on which the Iron Gates Natural Park is managed: Management Plan and Park Regulation. We identified and analysed the general specific objectives, classifying them into three main categories according to their relationship with the

Danube River: (1) *directly addressing the Danube*, (2) *indirect relation with the Danube* and (3) *without any relation with the Danube*.

We have also realized a cross-check analysis for three additional classes regarding the measures in the documents, delineating: (i) *measures for conservation*, (ii) *measures for social aspects* and (iii) *measures for economy*. In the first stage we analysed the Integrated management plan of the Iron Gates Natural Park [28], as the official document establishing the general framework for activities in the park. The plan established 6 general objectives and 31 specific objectives (Table 2).

The management plan includes 133 measures and activities associated with the general and specific objectives presented above. A high number of these measures are considered active conservation measures for habitats and species within the protected area, while others address the conservation of cultural and natural patrimony, traditions and social-economic development of local communities, tourism, management etc. Only 4 measures are related directly to the Danube River (3%) and 18 measures are indirectly related to the Danube River (13.5%). The reduced number of measures directly involving the Danube can also be explained by the general character of most measures, requiring in-depth analysis of the description for identifying the relevant ones. Such measures include mapping and monitoring Danube habitats or monitoring the physical and chemical parameters of the Danube's water, regulating human activities with environmental impact on fish and amphibians.

From the measures found in the management plan, 28 targeted conservation measures, 79 are focused towards social, administrative and management activities, while only 26 are economic measures. The high number of measures in the social and administrative category can be explained by the typology of the protected area favouring the protection of cultural landscapes in relation to the social-economic development of local communities.

Analysing the provisions of the Regulation of the Iron Gates Natural Park [51] we identified and evaluated regulations enforced on human activities from the protected area. A series of activities along the Danube River are clearly nominated, such as the restrictions imposed on fisheries according to the zonation of the protected area. Since fishing is a traditional activity, it is encouraged in certain areas but only if it is practiced using traditional methods and techniques.

There are also regulations for species of wildlife flora and fauna status, especially focused on the bird species and their nesting areas within the wetlands situated along the Danube River. Complementary, the Regulation established a series of measures regarding other forms of human exploitation of natural resources (such as forestry, use of pastures, constructions, research, tourism) in all cases imposing traditional uses and promoting a reduced impact on the environment (including the Danube River). We have observed that the Regulation lacks measures of strengthening the cooperation between the administration and public authorities, general public, economic agents, NGO or other stakeholders, which would have increased the conservation potential in the region [52].

A special look should be given to human activities classified as threats to conservation activities in the Iron Gates Natural Park. Previous studies, such as the one conducted by Cucu, Niculae [13] have identified activities representing a very high

Table 2 General and specific objectives in the Iron Gates Natural Park

No.	General objectives	Specific objectives
1.	Conservation of natural patrimony	a) Improving the knowledge of species and habitats of community interest by realizing inventories, mapping, and evaluation studies b) Monitoring biodiversity in the area c) Conservation of integral protection areas in the natural park d) Maintaining and promoting a favourable conservation status for species and habitats of conservation interest e) Regulating activities and projects which can have a negative impact upon the protected area and elements of conservation interest
2.	Sustainable use of natural and cultural resources	a) Maintaining traditional uses of lands and natural resources b) Maintaining and promoting traditions and local activities
3.	Development of sustainable tourism	a) Development of sustainable tourism b) Development and rehabilitation of visiting infrastructures c) Diversification of touristic services d) Management of visitors and impact evaluation of tourism
4.	Increasing the awareness and education of interested parties	a) Increasing the information level of the public for the values of the natural protected area and the activity of administration b) Increasing the awareness of the public for the need for conservation c) Education of public for the values of protected area and an adequate environmental behaviour
5.	Efficient management of natural protected areas for achieving the targets	a) Ensuring financial, material, and human resources for an efficient management of natural protected area b) Involving the Scientific Council in the decision-making process c) Taking management decision based on consultations with local communities and other stakeholders d) Involving local communities and other stakeholders in the management process e) Involving volunteers in the management process f) Ensuring financial resources for the efficient management process g) Supplementing own financial resources for the implementation of objectives h) Elaborating strategic and planning documents for the implementation process i) Monitoring the implementation of the management plan and evaluating the efficiency of implemented measures j) Ensuring an adaptative management of the natural protected area k) Revision of the management plan l) Strengthening the institutional capacity of the park administration m) Improving the performance of park administration
6.	Improving the image of the Park Administration	a) Improving the communication level of the administration b) Increasing the acceptance of the protected area and management measures c) Increasing the promotion of the natural protected area d) Improving the image of the Park Administration and RNP-Romsilva (Romanian National Forestry Authority)

threat potential to conservation (human intrusion and disturbances, industrial activities, pollution), activities with high threat potential (agriculture, aquaculture, the use of biological resources, invasive species), medium threat potential (residential and commercial development, transport and services corridors, changes to natural systems) and activities with a low threat potential (geological events, climate changes or severe weather).

In addition to the institutions that are directly involved in the management of a protected area, the public represents an important actor in disseminating and understanding conservation measures, as well as for trying to minimize anthropogenic pressures. On the other hand, the assessment of the level of knowledge and perception of the local communities regarding the benefits of the existence of a protected area is of a special significance in the correct application of the conservation measures. From this perspective, there are various studies assessing public perception on water bodies and their benefits [53], but in our analysis we focused more on the conservation activities than the Danube River itself. Hence, in 2020, we carried out a sociological survey for a project financed through structural funds that envisages concrete conservation measures of biodiversity and landscape in Iron Gates Natural Park. We applied questionnaires both online and, in the field, and obtained 1040 answers. The survey aim was to determine the level of awareness of local communities and visitors on the need to protect the habitats of species in the area of the Iron Gates Natural Park (Fig. 9), and in this chapter, we further illustrate the public perception regarding the role of a protected area from the perspective of respondents.

The respondents of our survey were mainly inhabitants of the communities from the Iron Gates Natural Park, as follows: Orșova (31.34%), Eșelnita and Șvinița (16%) or Drobeta–Turnu Severin (6.25%). Even citizens of the country's capital, Bucharest were included in our survey, either as tourist in the area or tourist that have prior visited the area answering the on-line version of the questionnaire (5%). We have also received answers from people residing outside the Iron Gates Natural Park (3%). The survey was applied to 617 men and 423 women, aged as follows: 14–18 years (5%), 19–35 years (12%), 36–45 years (24%), 46–65 years (40%), over 65 years (19%). Regarding their level of education, it is noted that over 36% of respondents

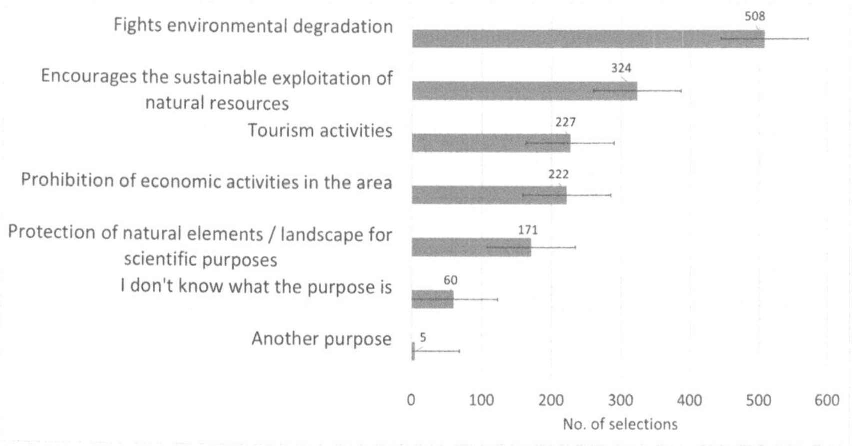

Fig. 9 The role of a protected area as perceived by the respondents

completed university or postgraduate studies (380 people) and over 35% high school studies, respectively almost 14% vocational studies.

The professional profile of the respondents covers a wide variety of occupations, which includes the following: students, teachers, administrators or commercial workers, employees involved in the hospitality industry, engineers, civil servants, accountants and economists, cashiers, mechanics, foresters, fishermen, police officers, entrepreneurs, self-employed or other categories of workers, as well as unemployed or retired people. This high diversity of institutions and occupations is well suited for the purpose of the study, as the management of protected areas requires the involvement of a high number of stakeholders while non-participative protected area management can often lead to conflicts [14].

The assessment regarding the knowledge level of the term "*natural protected natural area*" revealed that a majority of over 91% answered affirmatively. The main roles of a protected area perceived by the interviewees were the following: *protection of natural elements and landscape to combat environmental degradation* (approximately 49%), *encouraging sustainable exploitation of natural resources* (31%) followed by *tourism activities* (22%) and the *prohibition of economic activities in the wetlands area* (21%) (Fig. 9).

Some of the most well-known areas from Iron Gates Natural Park by the public, and which are also perceived as needing protection due to their high value are: Cazanele Mari and Cazanele Mici (81.44%), Danube River (49.42%), Moldova Nouă Holm (45%), the historical and cultural landscapes (32.56%), Calinovat Holm (27.11%), Almăjului-Locvei Mountains (26.15%), the Danube's wetlands (22.11%) and meadows (18.46%).

The main human activities in the area of the Iron Gates Natural Park indicated by the respondents to have a significant impact on biodiversity degradation are: the uncontrolled waste, grasslands burning, the wood and subsoil exploration, the construction of pensions and hotels, hunting, the road traffic and respectively, the Danube traffic. Other activities such as: agricultural activities, small boat rides, cruise ships, tourism, cultural artistic festivals and fairs or sport fishing are perceived as adding some pressure on the natural features but not as much as the ones mentioned before. The results of our survey are drawing attention to the need to further promote protected areas in general, and conservation measures for wetland habitats and specific protected biodiversity species in particular.

5 Conclusion

From the above analysis one can clearly observe that the establishment of the protected area changed the conditions in the Iron Gates region. In consistence with other parts where protected areas appeared, it sets up new regulations and restrictions in the region [54]. This can be perceived as both beneficial due to increased tourism and money from conservation activities but also negative due to restrictions imposed on economic agents [14].

Throughout this chapter, we aimed in providing a full scan of the socio-economic situation of the communities within the Iron Gates Natural Park and to determine if the Danube River, or in this case the Iron Gates reservoir can be considered an asset, an inconvenience or a forbidden fruit to be used for the future development of these communities. Given the geographic and historic context of the analysed area we considered that the vicinity of the Danube River could enhance the emergence of local businesses related to water transportation, resources' exploitation, leisure activities or tourism, while the natural park would act like a shock absorber of the potential challenges the ecosystems could be facing (Fig. 10).

The wide variety of problems in the region, from the reduced funding of conservation activities characterizing most Romanian protected areas [55], reduced capacity of the administration of protected areas [56], the decline of local population and loss of natural and cultural resources [5] or the poor collaboration between local stakeholders and public authorities [20], all raise serious challenges to the conservation of the area. The situation of the region as an EU-border with Serbia could also reduce the potential of collaboration, while they are not obliged to abide EU legislation [1]. Also, the lack of financial support from the Government to land owners within natural protected area ads more fuel on the conflict between conservationists and local communities [57, 58].

Through this study we showed that despite the potential provided by the Danube River in endorsing wealth and prosperity to its nearby human communities, the socio-economic situation and policy making at local and national level have not raised towards achieving this potential. As we desired through our study objectives,

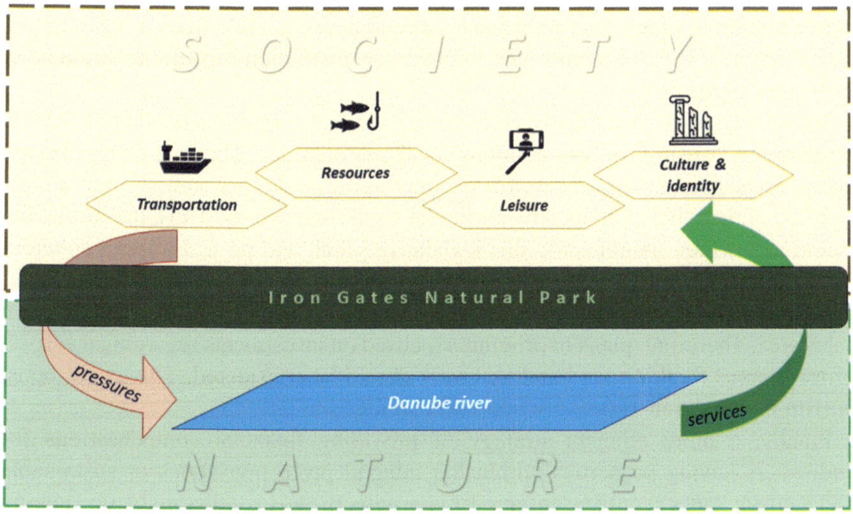

Fig. 10 The relation between the Danube and human activities in the context of nature conservation

we highlighted that it is quite challenging to keep the balance between natural conservation and economic development in a way that provides a win–win scenario for the local communities and conservationists along the Danube River.

The concluding remarks cannot really state that in the case of the Iron Gates region, the poor communities have remained poor, if not even became poorer due to the natural conservation policies enforced here but it is quite clear that the legal status of the area as well as the presence of the Danube River is highly underused and underappreciated by local policy makers. Even if the administrative documents ruling the natural protected area are tackling the socio-economic issues, aiming in endorsing, and boosting the local economies, the applicability of those goals is tough without the support of local policy makers and without a clear support, both financially and logistically from the central authorities.

6 Recommendation

Future concerns of the relevant stakeholders and policy makers for the Iron Gates regions should aim in providing more transparent means of communication towards the locals as the local public is having hazy thoughts about the conservation status of the region in regard with the permissions and restrictions. Our results have highlighted that the local inhabitants are emphasizing the restrictions induced by conservation status as important aspects of the precarious socio-economic situation in the area. Thus, Iron Gates Natural Park Administration should put more effort in spotlighting the opportunities raised by the conservation status, providing the proper support and information which may help local entrepreneurs to shift from a business-as-usual approach, which is confronted with various restrictions, to a more sustainable economic approach.

Nevertheless, the complexity of the legal status the Iron Gates Natural Park is troublesome, causing confusion among local communities. The Iron Gates Natural Park is overlapped by three Natura 2000 sites and several other natural reservations, each one of them being defined by particular legal frameworks. Thus, the Romanian central authorities should settle this legislative glitch and project a more coherent framework. Also, regional or national strategies could be oriented on using the presence of Danube River as an asset for socio-economic development within the Iron Gates area. Therefore, plans or programs focused on investments regarding transport infrastructures (both on land and by river) should be considered. This could boost tourism in the region, especially international tourism.

Finally, a more efficient strategy of providing financial compensations for landowners having properties within the integral protection zones or sustainable management zones should be enforced. Otherwise, the level of dissatisfaction among local communities towards the conservation status of the region will continue rising, making the management activities more demanding and the protected species and habitats more vulnerable to degradation.

Acknowledgment This work was supported by a grant of the Ministry of Research, Innovation and Digitization, CNCS - UEFISCDI, project number PN-III-P1-1.1-TE-2021-1067, within PNCDI III.

References

1. Csagoly P, Magnin G, Hulea O (2016) Danube River Basin. In: Finlayson C, Milton G, Prentice R, Davidson N (eds) The Wetland book. Springer, Dordrecht. https://doi.org/10.1007/978-94-007-6173-5_87-1
2. Gan TY, Zunic F, Kuo CC, Strobl T (2012) Flood mapping of Danube River at Romania using single and multi-date ERS2-SAR images. Int J Appl Earth Obs Geoinf 18:69–81
3. Niculae MI, Niţă MR, Vanău G, Ciocanea C, Gavrilidis A (2014) Spatial and temporal dynamic of Rural and Urban landscapes identified in the "Iron Gates" Natural Park. Transylvanian Rev Systematical Ecol Res 16(3):211–224
4. Buzea E (2011) Flooded areas and their importance in maintaining biodiverity. Meadow Lower Danube. J Wetlands Biodiversity 1:23–46
5. Lukić D, Petrović MD, Vuković DB (2018) Potential of geotourism for regional development: The case of "Iron Gates" park in Serbia. R-economy 4(3):158–166
6. Dragićević S, Mészáros M, Djurdjić S, Pavić D, Novković I, Tošić R (2013) Vulnerability of national parks to natural hazards in the Serbian Danube region. Pol J Environ Stud 22(4):75
7. Mihai B, Nistor C, Toma L, Săvulescu I (2016) High resolution landscape change analysis with CORONA KH-4B imagery: A case study from Iron Gates Reservoir Area. Procedia Environ Sci 32:200–210
8. Gatejel L (2016) Overcoming the Iron Gates: Austrian transport and river regulation on the Lower Danube, 1830s–1840s. Central Eur Hist 49(2):162–180
9. Olteanu F, Şerban I (2008) The systematization of the navigation and communication ways at Iron Gates until the end of the 19th century. Ann Univ Craiova-Ser Geogr 11:101–104
10. Năstase G, Şerban A, Năstase AF, Dragomir G, Brezeanu AI, Iordan NF (2017) Hydropower development in Romania: A review from its beginnings to the present. Renew Sustain Energy Rev 80:297–312
11. Popa ME, Predeanu G (2018) Coals of Romania: Geology, petrology and use. Int J Coal Geol 200:103–122
12. Patroescu M, Vintila G (1997) Natural, cultural and historical potential for tourism of the 'Iron Gates' National Park in Romania. Geographica Panonica 1
13. Cucu LA, Niculae MI, Pătroescu M (2013) Hierarchical analysis of the threats for species of community interest in the Iron Gates Natural Park. Romania. Forum geografic XI I(1):52–58
14. Nita A, Ciocanea CM, Manolache S, Rozylowicz L (2018) A network approach for understanding opportunities and barriers to effective public participation in the management of protected areas. Soc Netw Anal Min 8(1):1–12
15. Mikulcak F, Newig J, Milcu AI, Hartel T, Fischer J (2013) Integrating rural development and biodiversity conservation in Central Romania. Environ Conserv 40(2):129–137
16. Manolache S, Ciocănea CM, Rozylowicz L, Nita A (2017) Natura 2000 in Romania—A decade of governance challenges. Eur J Geogr 8(2):24–34
17. Turnock D (2001) Cross-border conservation in East Central Europe: The Danube-Carpathian complex and the contribution of the WorldWide Fund for Nature. GeoJournal 54:655–681
18. Manolache S, Nita A, Ciocanea CM, Popescu VD, Rozylowicz L (2018) Power, influence and structure in Natura 2000 governance networks. A comparative analysis of two protected areas in Romania. J Environ Manage 212:54–64
19. Soran V, Biro J, Moldovan O, Ardelean A (2000) Conservation of biodiversity in Romania. Biodivers Conserv 9:1187–1198

20. Hossu CA, Ioja IC, Susskind LE, Badiu DL, Hersperger AM (2018) Factors driving collaboration in natural resource conflict management: Evidence from Romania. Ambio 47(7):816–830
21. Grodzinska-Jurczak M, Cent J (2011) Expansion of nature conservation areas: Problems with Natura 2000 implementation in Poland? Environ Manage 47(1):11–27
22. Manolache S, Nita A, Hartel T, Miu IV, Ciocanea CM, Rozylowicz L (2020) Governance networks around grasslands with contrasting management history. J Environ Manage 273:111152
23. Nita A, Manolache S, Ciocanea CM, Rozylowicz L (2019) Real-World Application of Ego-Network Analysis to Evaluate Environmental Management Structures. In: Karampelas P, Kawash J, Özyer T (eds) From security to community detection in social networking platforms. ASONAM 2017. Lecture Notes in Social Networks. Springer, Cham. https://doi.org/10.1007/978-3-030-11286-8_1
24. Parlamentul României (2000) Lege nr 5 din 6 martie 2000 privind aprobarea Planului de amenajare a teritoriului naţional - Secţiunea a III-a - zone protejate. In: Monitorul Oficial al României NR. 152 din 12 aprilie 2000.
25. Ioja CI, Patroescu RL, Popescu VD, Verghelet M, Zotta ML, Felciuc M (2010) The efficacy of Romania's protected areas network in conserving biodiversity. Biol Conserv 143(11):2468–2476
26. Guvernul României (2003) HOTARARE DE GUVERN Nr. 230 din 4 martie 2003 privind delimitarea rezervatiilor biosferei, parcurilor nationale si parcurilor naturale si constituirea administratiilor acestora. In MONITORUL OFICIAL NR. 190 din 26 martie 2003: Bucharest.
27. Phillips A, Union WC (2002) Management guidelines for IUCN category V protected areas: Protected landscapes/seascapes, vol 9. IUCN Publications Services Unit, Cambridge
28. Iron Gates Natural Park Administration (2020) Management plan of the Iron Gates Natural Park and of Natura 2000 sites ROSCI 0206 Porţile de Fier, ROSPA 0026 Cursul Dunării Baziaş-Porţile de Fier and ROSPA 0080 Munții Almăjului-Locvei. Iron Gates Natural Park Administration, Orşova
29. Dihoru G, Roman N (1969) Une nouvelle espece du genre Stipa. Revue Roumaine de Biologie, Série Botanique 14(1):21–27
30. Martinovský J (1977) Clavis analytica nec non descriptiones breves taxorum generis Stipa in Europa centrali provenientium. Preslia 49(2):97–113
31. Beldie A (1979) Flora României: determinator ilustrat al plantelor vasculare, vol 2. Editura Academiei Republicii Socialiste România, Bucureşti
32. Mataca SS (2005) Parcul Natural Porţile de Fier. Flora, vegetaţia şi protecţia naturii. Editura Universitaria, Craiova
33. Călinescu R (1969) Biogeografia României. Editura Ştiinţifică, Bucureşti
34. Rozylowicz L, Dobre M (2010) Assessing the threatened status of Testudo hermanni boettgeri Mojsisovics, 1889 (Reptilia: Testudines: Testudinidae) population from Romania. North-Western J Zool 6(2):190–202
35. Spinelli O, Eriksson M, Houston J, Karling M, Küper B, Gazenbeek A, Guimaraes A, Fourli M, Fotiou S, Olmeda C, Patrimonio O, Raeymaekers G, Picchi S, Sundseth K, Thauront M, Zaghi D (2005) Integrated management of Natura 2000 sites. The contribution of LIFE-nature projects. In Goss S (ed) LIFE Focus, European Commission, European Commission Environment Directorate-General
36. Ministry of Agriculture Forests Water and Environment (2003) Order 552/2003 regarding the aproval of internal zoning of national parks and natural parks, from the perspective of the need to conserve biological diversity. Official Journal of Romania, Bucharest
37. Cojocariu L, Copăcean L, Popescu C (2019) Conservation of grassland habitats biodiversity in the context of sustainable development of mountain area of Romania. Appl Ecol Env Res 17(4):8877–8894
38. Ioras F (2003) Trends in Romanian biodiversity conservation policy. Biodivers Conserv 12:9–23

39. Mărgărit M, Radu V, Boroneanţ A, Bonsall C (2018) Experimental studies of personal ornaments from the Iron Gates Mesolithic. Archaeol Anthrop Sci 10(8):2095–2122
40. Roman M, Voicu C (2010) Some socio-economic effects of labour migration on the sending country. Evidence from Romania. Theor Appl Econ XVII(7):61–76
41. National Institute of Statistics (2019) Tempo. http://statistici.insse.ro:8077/tempo-online/#/pages/tables/insse-table. Accessed 1 March 2021
42. Bijak J, Kupiszewska D, Kupiszewski M (2008) Replacement migration revisited: Simulations of the effects of selected population and labor market strategies for the aging Europe, 2002–2052. Popul Res Policy Rev 27(3):321–342
43. León-Ledesma M, Piracha M (2004) International migration and the role of remittances in Eastern Europe. Int Migr 42(4):65–83
44. Cântar I, Chisăliţă I, Cadar N, Merce O, Turcu D (2015) The impact of mining activities from Moldova Noua on forests. J Hortic Forestry Biotechnol 19(1):207–215
45. Raischi S, Pană E, Holban E, Zamfir Ş, Raischi M, Deák G (2019) Analysis of air and soil quality regarding the cross-border impact risk of the mining tailings ponds in Moldova Nouă area. Int J Environ Sci Develop 10(6)
46. Ashley C, Boyd C, Goodwin H (2000) Pro-poor tourism: Putting poverty at the heart of the tourism agenda. Natural Resource perspectives 51:1–6
47. Gasparotti C, Rusu E, Dragomir S (2013) The impact of anthropogenic activities on the water quality in the Danube River Basin. International Multidisciplinary Scientific GeoConference SGEM 1:987–994
48. Teodoru C, McGinnis F, Wüest A, Wehrli B (2006) Nutrient retention in the Danube's Iron Gate reservoir. EOS Trans Am Geophys Union 87(38):385–392
49. Johnson PT, Chase JM, Dosch KL, Hartson RB, Gross JA, Larson DJ, Sutherland DR, Carpenter SR (2007) Aquatic eutrophication promotes pathogenic infection in amphibians. Proc Natl Acad Sci 104(40):15781–15786
50. Blaustein AR, Han BA, Relyea RA, Johnson PT, Buck JC, Gervasi SS, Kats LB (2011) The complexity of amphibian population declines: understanding the role of cofactors in driving amphibian losses. Ann N Y Acad Sci 1223(1):108–119
51. Iron Gates Natural Park Administration (2020) Regulation of the Iron Gates Natural Park, Natura 2000 sites ROSCI0206 Porţile de Fier, ROSPA0026 Cursul Dunării-Baziaş-Porţile de Fier, ROSPA0080 Munţii Almăjului-Locvei and other natural protection areas in the areas. Iron Gates Natural Park Administration, Orşova
52. Rozylowicz L, Nita A, Manolache S, Ciocanea CM, Popescu VD (2017) Recipe for success: A network perspective of partnership in nature conservation. J Nat Conserv 38:21–29
53. Ioana-Toroimac G, Zaharia L, Neculau G, Constantin DM, Stan FI (2020) Translating a river's ecological quality in ecosystem services: An example of public perception in Romania. Ecohydrol Hydrobiol 20(1):31–37
54. Teampău P (2020) Trouble in paradise: Competing discourses and complex governance in the Romanian Danube Delta. Mar Policy 112:103522
55. Dinculescu CG (2019) Biodiversity protection in Romania: Financial resources. In: Agrarian economy and rural development—Realities and perspectives for Romania. International Symposium. 10th edn. The Research Institute for Agricultural Economy and Rural Development (ICEADR), Bucharest
56. Ciocănea CM, Sorescu C, Ianoşi M, Bagrinovschi V (2016) Assessing public perception on protected areas in Iron Gates Natural Park. Procedia Environ Sci 32:70–79
57. Tudor CA, Iojă IC, Pătru-Stupariu I, Nită MR, Hersperger AM (2014) How successful is the resolution of land-use conflicts? A comparison of cases from Switzerland and Romania. Appl Geogr 47:125–136
58. Hossu CA, Ioja IC, Nita MR, Hartel T, B DL, Hersperger AM (2017) Need for a cross-sector approach in protected area management. Land Use Policy 69:586–597

Citizen Science for the Danube River—Knowledge Transfer, Challenges and Perspectives

Elfrida M. Cârstea, Cristina L. Popa, and Simona I. Donțu

Abstract Human changes on the Danube River have transformed it into a heavily engineered river, severely threatening its ecological status. The status and management strategies vary significantly between regions of the Danube River Basin. Citizen science approach can be the bridge in harmonizing water management practices across the Danube Basin and in recording large water quality datasets. This chapter reviews citizen science actions in the Lower Danube Basin and the available tools for citizens in this region. The study has shown that the activities with public involvement in this region have been supported by non-governmental organizations (NGOs) or independent citizen science platforms. Citizen science activities in this region are scarce and infrequent, and most of them without supervision from professional researchers or involvement from local authorities. Limited access to funds, lack of trust between participating groups and the restricted power of communities to voice concerns have been found as factors influencing citizen science activities. The scientists may be the missing link between policy makers, water managers and citizens, while providing the optimal tools and knowledge to all sectors. Professional scientists can collaborate with NGOs and build upon their extensive expertise and success in engaging with citizens and authorities.

Keywords Citizen science · Public involvement · River restoration · Pollution · Lower Danube

---

E. M. Cârstea (✉) · C. L. Popa · S. I. Donțu
National Institute of R&D for Optoelectronics, Atomistilor 409, 077125 Măgurele, Ilfov, Romania
e-mail: elfrida.carstea@inoe.ro

C. L. Popa
e-mail: cristina.popa@inoe.ro

S. I. Donțu
e-mail: simona.dontu@inoe.ro

A. Negm et al. (eds.), *The Lower Danube River*, Earth and Environmental Sciences Library, https://doi.org/10.1007/978-3-031-03865-5_18

1 Introduction

Before science became a paid profession, scientific research was generally undertaken by citizens, outside their main profession [1, 2]. Many countries have a long history of citizens doing science. For example, in Japan the timing of cherry blossom has been recorded for 1,200 years and phenology data for other plants and animal species have been registered for centuries [3]. In the United Kingdom, in 1789, Robert Marsham, a landowner, published observations over decades of plants and animals on his estate. In the USA, in 1851, Henry David Thoreau began his observations of animal life cycles, and 20 years later, the US National Weather Service started making forecasts using both professional and amateur generated data [4]. During the past 150 years, the role of non-professionals in science has been marginalized, which limited their activity around naturalist clubs. However, in recent years, scientists recognized the value of citizens in undertaking studies at scales that cannot be achieved by professionals alone [1]. Thus, citizens science activities have been increasingly included in environmental programs relating to biodiversity [5], freshwater quality [6–8], marine science [9, 10], conservation [11, 12], land-cover/land-use [13], climate adaptation [14], air pollution [15, 16], ecosystem services [17, 18] or flood control [19]. Nevertheless, limited citizen science programs have focused on the environmental quality of international rivers, such as the Danube River, even though they are an important resource for water supply, biodiversity, agriculture, industry, fishing, navigation, tourism and power generation [20].

The Danube River Basin has suffered significant changes in the chemical, physical and biological status due to point and diffuse pollution and to changes in land use and catchment water cycle [21]. Major sources of pollution have been identified as untreated wastewater, agricultural emissions and industrial releases. Eutrophication is evident at several sites along the Danube River, but it is more pronounced in the Lower basin, especially in the tributaries [21]. Monitoring the Danube River quality is critical because its ecological status is linked to human wellbeing and sustainable development. Organizational and legal frameworks, such as the International Commission for the Protection of the Danube River (ICPDR), and the European Union Water Framework and Flood Directives encourage and support public participation in decision making, but little is known about the level of public awareness and education regarding the Danube River Basin challenges. Effective water quality management of the Danube River relies on long-term, high quality data, at both regional and international scales. Nevertheless, databases and monitoring initiatives undertaken by professional researchers or environmental agencies are often incomplete [22]. Citizens can fill this information void and get involved in actions such as collecting water quality data, observing the environmental diversity or providing a perspective over the socio-economic pulse of the area. Citizen scientists may greatly contribute with large volumes of temporal and spatial data. They can analyse data, identify patterns and can provide support in detecting environmental changes or in monitoring the effectiveness of environmental management practices [12, 17, 23]. Despite these advantages, citizens have had minor involvement in environmental monitoring and management within the Danube River Basin, and the potential of this valuable resource has been largely ignored in the Lower Danube Basin. Moreover, according to Njue et al. [18], no citizen science projects on water quality and hydrology monitoring have been undertaken in East Europe. This leaves a large research gap that must be exploited in order to increase public education, create long-term, high frequency databases and allow citizens to voice their concerns and engage in decision making. In addition, there are no systematic reviews of citizen science programs on water quality in the Lower Danube River region.

This chapter examines the activities with citizen involvement in the region and the best practices that can be implemented for environmental monitoring. The educational, social and economic barriers in citizen science projects success and impact are discussed. This chapter also explores the dynamics between the involved parties (scientists, citizens, water managers and policy-makers). We also identify the potential tools that can be used to optimize public participation programs.

2 Citizen Science—Theory and Classifications

Citizen science is defined as "the engagement of non-professionals in scientific investigations – asking questions, collecting data, or interpreting results" [1]. Several terms occur in literature depending on the meaning of citizen science at individual and institutional levels: public participation in research projects, participatory science, participatory action research, crowd-sourcing data, mobile crowd sensing, crowd-based activities, volunteer-based monitoring, community science, community research, community-based monitoring, community participation, citizen surveillance, civic scientists, citizen observatories, etc.

The core aim of these programs is to "bring science to citizens and citizens to science" [24]. Citizen science programs lead to the democratization of the environment and science in general, by sharing information and expertise between professional researchers and the public. Thus, these programs increase scientific literacy, public involvement in local issues and social capital [22]. The connection and interaction between citizens and policy makers may help them explore their different world views and perceptions [25]. Improved citizen education, open access to scientific data and methods, and collaboration between citizens and policy makers also help communities to better understand environmental concepts, report pollution and push for greater environmental justice [26].

Citizen science actions may be initiated by professional scientists in collaboration with non-professionals, or by non-professionals alone. The citizen science programs may extract data from both structured and unstructured programs. Structured programs include specific monitoring and reporting tools used on defined parameters, locations and times, while unstructured programs involve information collected from social media, supplied unwillingly by participants [27]. The level of citizen participation varies depending on the nature of the project, and may be classified as, contractual, contributory, collaborative, co-created and autonomous

[1, 18, 24]. The contributory model is the most common form of participation [18]. Programs involving citizens can be further classified depending on the level of public participation, based on the typology provided by Pretty [28] (Table 1). The first three categories can be seen as types of non-participation and are the least likely to produce long lasting positive impact [28]. However, in the Lower Danube Basin, participation by consultation is the highest level of citizen involvement reached by most environmental research projects and agencies. According to Teodosiu et al. [29], there is institutional support to advance beyond this level and a general willingness to improve participation, but barriers such as lack of public education in water issues, limited public representation and poor communication between parties block citizen engagement.

3 Citizen Involvement in Environmental Programs on the Danube River

In general, the improvement of ecosystem quality through citizen science campaigns has mostly been inferred by protective actions, such as litter clean-up, or by monitoring studies (in-situ measurements and sample collection) [17, 22]. Citizen science actions on the Danube River Basin make no exception to this rule; however, they are scarce and largely infrequent (Table 2). Most of the actions were undertaken in the Upper Danube Basin and the least in the Middle and Lower parts. The programs from the Lower Danube Basin focused only on clean-up activities (Table 2). Two programs included environmental data from the Lower Danube Basin, as part of the worldwide network of ecosystems (Freshwater Watch and EarthEcho, Table 2).

3.1 Short-Term Programs

3.1.1 Litter Clean-Up

Two major litter clean-up initiatives have been launched in the Lower Danube River (Table 2). The campaigns "Let's do it, Danube" and "Clean Danube" were developed by non-governmental organizations (NGOs) and funded through private companies' donations. "Let's do it, Danube!" campaign was part of "Let's do it, Romania!" program and aimed to provide ecological education in schools and to involve volunteers in cleaning-up the river banks [31]. In 2014, over 9,000 schoolchildren and parents participated in the "Let's do it, Danube" campaign, reaching over 20,000 volunteers by 2015. During the COVID-19 pandemic, the NGO has encouraged citizens to report illegal waste dumping using a smartphone application, TrashOut.

"Clean Danube" program aimed to educate and involve volunteers to clean the Danube River of plastics [32], from the rural areas in Mehedinti county, Romania. In

Table 1 Forms of public participation to science (adapted from [1, 18, 24, 28])

Form of participation	Information distribution	Level of involvement	Decision making	Feedback	Nature of the CS* project
Manipulative participation	Partial (people's representative)	None	No	No	–
Passive participation	Partial (information belongs only to external professionals; decisions are communicated)	Low	No	No	
Participation by consultation	Yes	Low-average (people are consulted; external agents define both problems and solutions)	No	No	–
Participation for material benefits	Yes	Average (people participate by providing resources such as labour, in return for material incentives)	Partial (people often have no stake in deciding about the processes and in extending activities when incentives end)	N/A	Contractual
Functional participation	Yes	Average-high (people participate by forming groups to meet predetermined objectives related to the project)	Partial (often major decisions have already been made by external agents)	Yes	Contributory

(continued)

Table 1 (continued)

Form of participation	Information distribution	Level of involvement	Decision making	Feedback	Nature of the CS* project
Interactive participation	Yes	High (people participate in joint analysis, which leads to action plans and the formation of new local institutions or the strengthening of existing ones)	Yes (these groups take control over local decisions, and so people have a stake in maintaining structures or practices)	Yes	Collaborative Co-created
Self-mobilization	Yes (independent initiation)	High	Yes (actions can be supported by governments and NGOs; may challenge existing distributions of wealth and power)	Yes	Co-created Autonomous

* CS—citizen science

Table 2 Programs involving citizens in the Danube Basin (the programs that include the Lower Danube Basin are marked with italics)

Program	Aim	Country	Duration	Sampling points	Parameters	Training	Reference
Plastic Pirates	Monitoring	Germany	9 months	Multiple rivers, including Danube	River description, flow, litter on the banks and floating, microplastics	Booklet with instructions	[30]
Let's do it, Danube!	Clean-up	Romania	Ongoing from 2014	Danube	Litter quantity	Educational videos	[31]
Clean Danube	Clean-up	Romania	Ongoing from 2019	Danube	Litter quantity	Workshops, educational videos	[32]
Freshwater Watch	Monitoring	Worldwide	Ongoing	Major rivers across Europe including the Danube	Surroundings, vegetation, water uses, aquatic life, water flow and level estimation, nitrate, phosphate, turbidity, colour	Training booklet, sampling kit	[33]
EarthEcho	Monitoring	Worldwide	Ongoing since 2007	Multiple streams including the Danube River basin	pH, turbidity, oxygen	Booklet with instructions	[34]

(continued)

Table 2 (continued)

Program	Aim	Country	Duration	Sampling points	Parameters	Training	Reference
HydroCrowd	Spatial monitoring	Germany	1 day	280 streams, including the Danube basin	Sample collection GPS coordinates, stream width, colour, turbidity, precipitation prior to sampling, air temperature	Instruction sheet, training lectures	[35]
Drinkable Rivers	Monitoring	Hungary	-	Multiple rivers including the Danube	pH, temperature, chloride-, nitrate-, nitrite- and phosphate content, E. coli, electric conductivity and clarity		[36]
Scent	Spatial and temporal monitoring	Romania	2 years	The Danube Delta	Land-cover/land-use, images, water level, flow, soil moisture, air temperature	Training sessions, toolbox	[37]

addition, the organizers developed a pollution map where photographs were uploaded by citizens to report plastic pollution on the Danube banks. According to the platform, plastic pollution was reported in at least 15 locations, over a 50 km stretch. However, the "Clean Danube" program has been launched at the beginning of 2020 and has yet to report its impact considering the COVID-19 lockdown period.

So far, professional scientists have initiated no citizen science programs to study litter types and concentrations along the Lower Danube River banks. Research organizations could join the programs initiated by NGOs on litter clean-up and engage a group of volunteers, during the campaign, in collecting data. Building on the extensive experience of NGOs and their success in reaching citizens, researchers can obtain yearly observations on litter types, concentration and distribution, land use, river banks conditions etc. Researchers can also build on the knowledge of past public participation programs. In a citizen science project in Germany, Kiessling et al. [30] found that the Danube catchment had some of the most polluted riverbanks from the surveyed rivers. In most cases, they found a link between population density and the quantity of litter. On the Lower Danube River, researchers could target areas near cities with population over 25,000 inhabitants, which would allow access to a larger pool of volunteers and a comparison between cities regarding litter disposal on the river banks.

3.1.2 Data and Sample Collection

No citizen science programs were developed specifically for the Lower Danube River. The campaigns undertaken by citizens in the Lower Danube Basin were supported by independent platforms, as part of a worldwide network of aquatic ecosystems (Table 2). However, the involvement of regional professional scientists in such actions is not clear.

Independent citizen science platforms, such as the FreshWater Watch or EarthEco, provide significant support as a repository of data. They also supply ready-made measurement kits and instructions. FreshWater Watch [33] database contains over 25,500 records on water quality, including nitrate, phosphate, turbidity and visual observations. Despite its impact, this citizen science platform contains only one project on water quality measurements on the Danube River. The measurements were made by two volunteers kayaking the continent from London to the Black Sea, in 2018 [38]. The volunteers recorded observations in 35 locations along the Danube River, almost every 100 km. In the Lower part of the Danube River, the volunteers measured the water quality in seven locations until Giurgiu, during September 2018. They recorded that the river was surrounded by forest, it had evidence of aquatic life (fish, plants and birds, with no particular species being mentioned), the water level was low, nitrate values between 0.2–1 mg/L, phosphate values between 0.02–0.05 mg/L, turbidity below the threshold value (14 NTU), water colour green. They also found floating algae at the entrance of the Danube River in Romania, and foam on the water surface near Vidin. The volunteers also included photographs to support their observations. The platform provided automatic feedback on the measurements, with rough interpretation of the data and the ecological status of the river.

The EarthEcho Foundation [34] aims to recruit and engage citizens in protecting water resources. They provide basic and student testing kits for measuring pH, turbidity and oxygen and display the color-coded data on a map. Since 2007, nearly 1,700,000 participants provided data on over 79,000 water bodies. The EarthEcho platform shows that measurements were also made in the Lower Danube River Basin. However, the measurements are scarce and infrequent between years, not only in the lower part of the basin but over its full extent. The monitoring program may grow over the years with sufficient promotion through schools and local authorities.

The only citizen science project developed and coordinated by professional scientists was undertaken in the Danube Delta, the Scent Project [37]. Within this project, five campaigns were organized, between August 2018 and May 2019, to collect environmental data in flood prone regions, in the Danube Delta (Table 2). A number of 193 volunteers recorded over 18,000 data (for example, water level, surface flow velocity, soil moisture, air temperature) and submitted images of land cover/land use via smartphone applications (Scent Explore and Scent Measure). The success of the initiation was due to the organizing research structures dedicated to the study of the Danube Delta. However, no organization committed to develop citizen science activities in the Lower Danube Basin. Partnerships between research organizations could be developed to involve citizens in monitoring further regions of the Danube Basin. The citizens can use the same Scent smartphone applications or they can be provided with Freshwater Watch and EarthEcho kits to measure environmental parameters. In particular, monitoring campaigns can be included into school projects or assigned to undergraduate students in order to raise awareness and education in their local communities.

3.2 Long-Term Programs

Only one long-term citizen science program was undertaken in the Lower Danube Basin. Participants counted water birds at several sites along major rivers, including the Danube River [39]. However, within a period of over 30 years (1967–2009), most of the Danube River sites were visited for less than 5 years, and the data are inconsistent between years. Poor temporal resolution is a significant challenge in long-term programs [40], but it can be overcome if national environmental agencies and local organizations provide consistent resources through-out the program. Worldwide, long-term monitoring programs (>15 years), involving citizen scientists proved to be successful [41, 42]. However, these programs were supported and mandated by national and regional governments and thus the citizens had a clear purpose of their contribution towards decision making. A common effort from countries within the Lower Danube Basin could provide substantial funding for a long-term citizen science program, where each can maintain several monitoring stations across the basin and employ researchers to supervise local citizen science activities.

In addition, long-term monitoring programs may be developed with the help of local schools, as shown by Abbott et al. [42]. However, such programs require long-term commitment of the organizing institution and lasting collaboration with local organizations, such as schools or authorities. National and local official entities may be attracted to collaborate, later, in already established citizen science programs, which would reduce logistic costs and maintain the participants' motivation. These programs can expand over time to include more parameters that require training and additional equipment [41].

3.3 *Proposed Tools for Optimizing Program Implementation*

The rise of technology in the past decades has greatly enhanced the development of citizen science programs. Online media (smartphone applications, web platforms, social media) represent an effective means of engaging with volunteers, supervising activities, communicating results, and providing feedback and support [43]. Good communication leads to increased public understanding, participant objectivity, accountability, awareness and ownership of scientific projects and to stronger social ties. In addition, the right tools may attract particular groups, for example the interest of school children in citizen science activities may increase when smartphone applications are used [43]. Citizen science programs can be easily advertised and disseminated through social media, reaching more volunteers than ever before [44–47].

Online media are optimal for the citizens living in the Lower Danube Basin. Around 10 million persons, in Romania, use at least one social media platform [48], and most of them access social media from a smartphone [49]. The advantage of these approaches for advertising and collecting data is that they do not require any funds and help gather large pools of participants.

In addition to the above-mentioned approaches, smartphone applications can be used for data collection. These applications allow immediate access to data, provide a daily routine and can significantly increase the pool of participants. They can also facilitate feedback in real-time and connectivity to social media [50]. Figure 1 includes a list of smartphone applications (most of them are free) and tools for environmental monitoring, which can also be implemented in the Lower Danube Basin. However, these applications require basic knowledge of the English language, which may be a barrier for some volunteers. In addition, senior citizens may require more training in submitting data through smartphone applications or online platforms [51]. Despite potential language barriers, some smartphone applications (such as, TrashOut or iNaturalist) were frequently used by citizens from the Lower Danube Basin to report data.

FreshWater Watch and EarthEcho developed their own testing kits and host online platforms, where data can be submitted and visualised for free (Fig. 1). These ready-made testing kits could be the starting point for citizen science programs from the

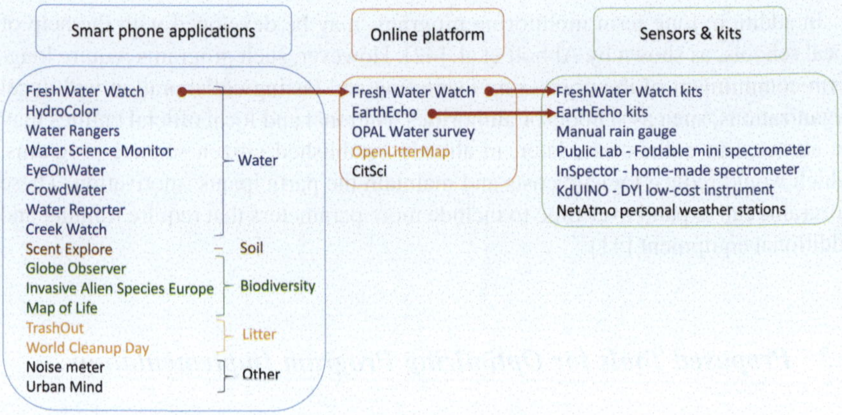

Fig. 1 Applications and tools suitable for implementation in the Lower Danube Basin

Lower Danube Basin, and they would not only simplify the protocol, but also facilitate comparisons with other databases. Past studies proved successful in implementing these testing kits [52–54]. Custom measurement kits could also be developed depending on the requirements of the study, which may reduce the costs of ready-made kits and the measurement time.

Sophisticated analyses are beyond the scope of a citizen science program. Starting with simple and easy parameters may be a good strategy to encourage citizens and increase their confidence in contributing to science, instead of starting with parameters that require complex training, which may overwhelm the volunteers. A simple strategy may also help volunteers to gradually connect with nature. Hence, the programs must be built over time and carefully strategized based on social and cultural prerequisites. The most common parameters measured by citizen scientists are nitrate, phosphate, water clarity and turbidity [36, 41, 55–57]. Also, citizens can be trained to make observations on shore vegetation and litter [55, 58]. In addition, citizens can survey river geomorphic features and dynamics (for ex. river bank profiles, features, terrestrial and aquatic vegetation, invasive species etc.) [59]. Apart from measuring parameters, citizens may report and monitor small scale pollution spills (oil, paint, wax deposits) [10]. In particular cases, citizens can collect samples and send them to professional researchers for laboratory measurements [44].

Low cost or DIY sensors can also be used by citizens to collect data (Fig. 1). These instruments can be constructed or installed even by school children, further developing their scientific skills [60, 61]. Apart from water quality monitoring, citizen scientists can provide data on hydrology. Authorities and researchers from the Lower Danube Basin may employ the support of locals to install manual rain gauges for precipitation records [51]. Also, researchers can install water level gauges in locations with easy access, such as public bridges, and also signboards with instructions, inviting visitors to send data [62]. The locals can also provide data on water level of small water bodies, while fishermen and boat tourist guides can measure water level on the Danube River. A virtual staff gauge, as a feature of the CrowdWater app [63] could be given to trained volunteers to record water level.

4 Factors Influencing Citizen Science Programs

Attempting to understand the underlying causes of the low number of citizen science programs in the Lower Danube Basin is intricate and requires an analysis of multiple factors. Firstly, the decisive factors influencing successful citizen science programs must be defined to offer a head start to future environmental programs. Secondly, it is important to discuss the means of increasing citizens' motivation to participate in science projects. Thirdly, policy support, and the educational and economic factors in the Lower Danube region may explain some of the issues regarding program implementation. Finally, the sometimes-problematic relationship between citizens, professional scientists, managers and policy-makers may act as a barrier in developing such programs.

4.1 Key Elements of Program Success

An important aspect that must be highlighted when analysing the projects based on citizen input in the Lower Danube Basin is the common practice of taking into consideration only the successful results to the detriment of unsuccessful projects. Citizen science groups may not publish failed approaches [25], although, even failed measurements can have an overall contribution, because it familiarizes citizens with science concepts, raises awareness and facilitates discussion with experts and policy makers [24]. To ensure success, a citizen science program must have a clear definition of goal and a clear sampling protocol with short sampling duration. The project should involve policy makers, an interdisciplinary team, and communication and data experts. It should also use digital tools for recording and submitting data, ensure quality control, acknowledge volunteer help and have realistic expectations [64].

Drawing on the design principles for citizen science projects outlined by San Llorente Capdevila et al. [65], two sets of success factors have been identified. The first set belongs to citizens, while the second set corresponds to institutions (Fig. 2). According to these principles, participants with previous experience and training are of great value to a project, not only because they are more likely to be highly committed to the mission and goal, and require less additional training, but they can also help new volunteers to gain knowledge and confidence in performing the tasks. However, citizens from the Lower Danube Basin have no previous experience or training. Knowing how to tap into citizen motivation will further guarantee project success. A survey can be launched prior to a program, to understand what would motivate citizens from the Lower Danube Basin to participate in citizen science activities. Generally, the public may enrol in environmental projects for various

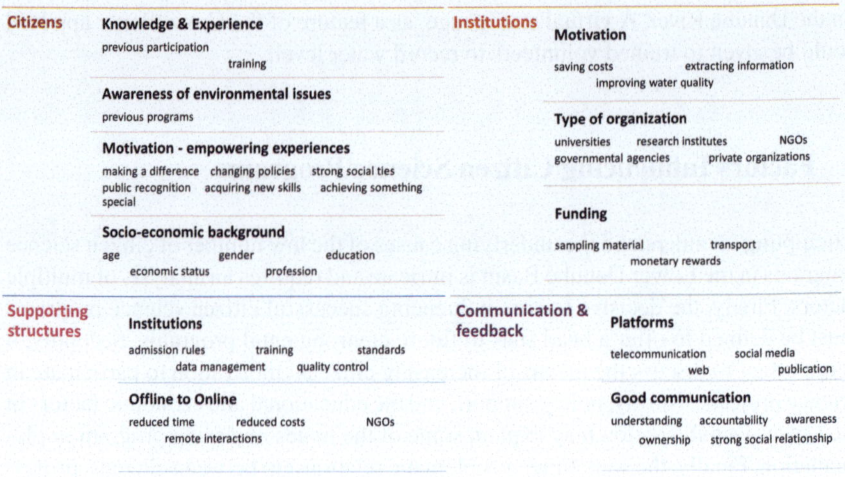

Fig. 2 Key elements of success and relational aspects in citizen science programs [17, 22, 24, 64, 65]

reasons including, the intrinsic desire to make a difference, the need to change policies or to create strong social networks, gaining public recognition, acquire a new skill or achieve something special [65]. "People tend to care for what they know and use" [66]. Based on this principle, citizen science programs with a high and clear social impact may have greater success in attracting participants compared to ones with abstract focus. Participation in the clean-up activities, from the Lower Danube basin programs was potentially driven by the same intrinsic motive (see Sect. 3). Volunteers were clearly aware of the environmental problems in the area and wanted to restore and preserve its beauty. The enthusiasm and experience of citizens that participated to these programs may be harnessed for future citizen science activities. Compensation may also increase citizen motivation. Forms of compensation include social acknowledgement and payment [67, 68]. FreshWater Watch, for example, acknowledged online the volunteers contributing with data from the Lower Danube Basin. Money rewards were also proposed as a solution to increase the number of participants [47]. Rewards, monetary or non-monetary require additional funds, which may not be available for the authorities and research institutions from the Lower Danube region. However, Freihardt [47] argued that monetary incentives may be necessary in low income, developing regions to obtain sufficient participation.

Institutions may be motivated to support a successful citizen science project for cutting costs with data collection, while obtaining large volumes of data (Fig. 2). Although savings can be made with a citizen science project, a consistent and adequate supply of resources is still needed to ensure long-term achievement and impact. Allocation of funds is determined by the type of institution that initiates and runs the citizen science program. However, issues may arise with supplying sufficient funds for citizen science projects in the Lower Danube Basin due to educational, economic and political pressures, as will be discussed in Sects. 4.2 and 4.3 (Fig. 2).

In addition to the factors outlined above, supporting structures and protocols of communication and feedback have been found as relational aspects to a successful citizen science project (Fig. 2). Supporting structures are important for the quantity and quality of data. The level of training, the type of observations requested from citizens, the quality control and management of data are critical to the success of the project. Another challenge relates to the limited involvement of participants due to the lack of incentives, excessive technical content, scarce sharing or repetitiveness of tasks [43]. Also, the data supplied by citizens may be fragmented and inaccurate if no data quality protocols are set. Moreover, if the process is not carefully monitored, the volunteer's biases or lack of objectivity may interfere with the information [17, 22]. Over time, this may lead to a lack of trust between involved parties [22]. Some of these challenges may be overcome if data collection shifts from offline to online using the tools described in Sect. 3.3, which further reduces costs, allows remote interaction and publication of data.

4.2 Educational and Economic Levels

A complex analysis of research, educational and economic circumstances in the Lower Danube Basin is beyond the scope of this study. However, some aspects must be mentioned as it has been shown that the level of participation in citizen science projects is highly influenced by the educational and economic status of the citizens. According to Sauermann et al. [69], citizen science projects, generally, tended to attract participants with a high educational level and with above average income. This means that citizen science activities are unlikely to reach marginalized communities, limiting their access to environmental education and science in general, and further widening the level of social exclusion.

There are high differences between regions along the Danube River in terms of economy and the spending levels on education, research and innovation. Although poverty has declined in the Lower Danube Basin in the past decades, Romania and Bulgaria still remain among the poorest in Europe [70]. For example, in 2017, almost 36% of the Romanian population were at risk of poverty or social exclusion [71]. In addition, Romania has the second highest urban–rural income gap in the EU, due to insufficient access to funding, ageing population and lack of technology and skills [71]. Furthermore, higher unemployment rates were reported in the second half of the Lower Danube River compared to first section, reflecting the low economic situation of the region. Lower unemployment rates were found in Bulgaria compared to Romania [72]. Considering the economic situation in the Lower Danube Basin, monetary rewards may be considered as an option to stimulate public participation in science programs.

Moreover, research and education sectors have been underfunded for decades in Romania, in comparison to other countries. In 2017, Romania invested in education 2.8% of GDP, which was the lowest rate in EU. As a result, in terms of higher education attainment, Romania is significantly below the EU average [73]. In 2012,

Romania had the lowest percentage of citizens with high education degree (15.4%) from the countries of the Danube River Basin, being significantly exceeded by Bulgaria with a rate of 24.0% [72]. Also, the percentage of early leavers from education exceeded the EU average, while the percentage of underachievers in science was double compared to the EU average. Moreover, the participation of adults in learning was below 1%, in 2018, ten times lower than the EU average [73]. Involvement of citizens in research projects would help them learn new skills and create strong social networks. It would also familiarize citizens with science and environmental concepts, and would increase confidence in their abilities, ultimately leading to improved participation to education and greater sustainable development in the Lower Danube Basin.

In the research & innovation sectors, Romania invested, in 2017, the least amount of funds (0.5% of GDP), from the European countries, almost 6 times less than Austria or Germany [74, 75]. This leads to limited funds that can be allocated to citizen science programs. Developing countries have other priorities in allocating funds, for example on building capacities or on increasing the level of innovation, until they have enough funds to invest in novel research [76]. Romania also suffers from a severe shortage of researchers in the R&D sector, with ten times less researchers per million people compared to Denmark, the highest-ranking country, and with three times less than Bulgaria [75]. Consequently, there are less researchers available for complex monitoring activities or for accessing remote locations within the Lower Danube Basin. Citizens can, temporarily at least, fill this void by collecting data and samples from distinct regions. In addition, the environment of political and cultural intolerance towards research represents a barrier in advancing science [76], and this can change only if successful citizen science programs are undertaken in the Lower Danube Basin. Insufficient funding towards education and science reflects not only on how citizens are formed, but also in the number of qualified professional researchers. In an analysis of the educational system in the Danube River Basin, Irvine et al. [77] found a lack of coherence between skill development strategy and environmental, economic, social and sustainable development policies. Thus, current educational programs do not provide the necessary pool of water professionals with both technical and stakeholders relational skills (including citizens) to ensure smooth transition of the Danube River Basin to sustainability [77].

The level of digital economy in the Lower Danube Basin is also important for the development of citizen science programs. Limited access to technology may prevent public participation [69] and low-income areas generally have limited access to internet or smartphones [47]. In particular, Romanian rural areas have lower internet coverage compared to urban areas [75]. Although the internet coverage is fragmented in the Lower Danube Basin, there are regions with full coverage, which can be exploited using the latest reporting tools (Fig. 1). In areas with poor internet coverage, printed leaflets and questionnaires may be distributed, through community structures, to counteract these issues [47]. Also, in areas with poor phone reception, applications should be developed to allow participants to store the data offline until reception improves [50].

4.3 Policy Support

A three-fold path of policy support has been identified for implementing citizen science and public involvement in Danube River monitoring and conservation (Fig. 3). The first set of policies addresses ecosystem health in general. The second set of policies refers strictly to the Danube River monitoring and protection, whilst the third set of policies allows public consultation before being launched. The organizations that developed and implemented these policies are: the United Nations (UN), the European Union (EU) and the International Commission for the Protection of the Danube River (ICPDR). They have recognized the strength of citizen science programs as crucial components in reaching UN Sustainability Goals, in achieving good ecological status of water bodies and in preserving and increasing ecosystem biodiversity.

Citizen science could support the UN Sustainable Development Goal 6 (Clean Water and Sanitation), indicator 6.3.2 (proportion of bodies of water with good ambient water quality) [54, 78]. In addition, the Convention on Biological Diversity, under the UN Environment Program (UNEP), aimed to "raise public awareness of the values of biodiversity and the steps they can take to conserve and use it sustainably"

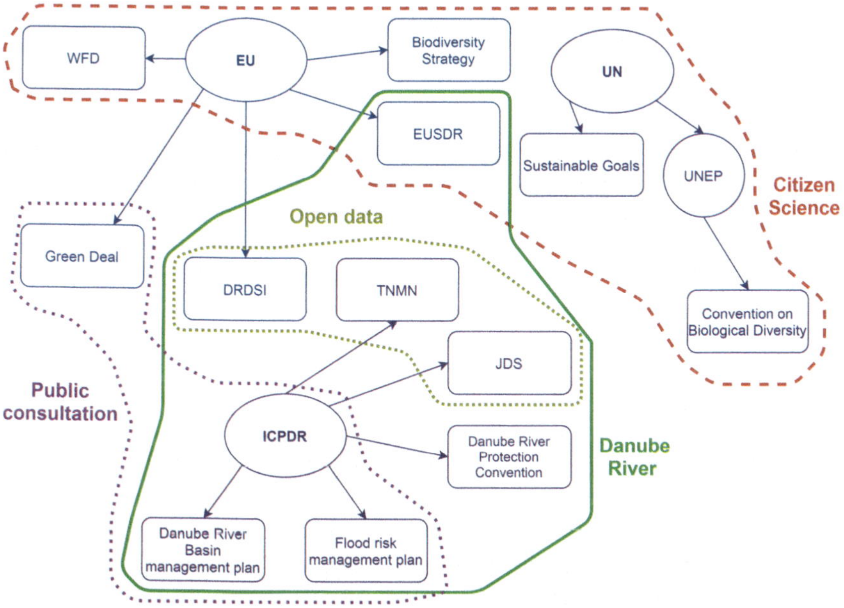

Fig. 3 Current policies regarding ecosystem health, monitoring and conservation, and their field of impact. UN—United Nations; UNEP—UN Environment Program; EU—European Union; WFD—Water Framework Directive; EUSDR—EU Strategy for the Danube Region; DRDSI—Danube Reference Data and Services Infrastructure; ICPDR—International Commission for the Protection of the Danube River; TNMN—Transnational Monitoring Network; JDS—Joint Danube Survey

(Target 1). However, UNEP could play a more active role in supporting countries to develop such programs, which may help in raising awareness and building databases [78].

The EU mentioned that, in the Water Framework Directive (WFD), public participation was "a key component for ensuring the WFD's success", and that the EU Biodiversity Strategy for 2020 encouraged "the active involvement of civil society at all levels of implementation" and recognized "the value of citizen science initiatives for gathering high-quality data while mobilizing citizens to get involved in biodiversity conservation activities" [79]. The EU developed the Danube Reference Data and Services Infrastructure (DRDSI) project in order to offer policy makers a repository of high-quality data from professional research studies [85]. Within a period of three years, DRDSI project identified over 10,000 datasets, including from Romania. However, the platform has been archived in 2019. A common platform, even informal, to collect data from the Danube Basin would provide citizens and researchers a tool for data storage and open access to information. Long-term data support could be provided by strong organizations, such as the EU Strategy for the Danube Region (EUSDR) and the ICPDR that are directly involved in preserving and improving the Danube River quality.

In 2010, the EU established the EUSDR with the roles of addressing common challenges within this macro-region, aligning funding, providing improved coordination and of finding new ideas [80, 81]. Within the Priority Area 10, of the EUSDR, citizen science projects contribute to Action 7 (To strengthen the involvement of civil society and local actors in the Danube Region), Target 7.1 (Supporting the empowerment of young people for participation in the development of the Danube Region through strategic guidance and the implementation of macro-regional networks) and Target 7.2 (Building capacities on participatory governance and involvement of civil society and local actors in cooperation with the Danube Local Actors Platform, the Danube Civil Society Forum, and/or further experts and stakeholders). Recently, in 2020, the EUSDR initiated the process of embedding priority areas into EU programs for 2021–2027. This step not only emphasizes the importance of involving and educating citizens in Danube River environmental issues, but also facilitates access to funding [81].

The ICPDR was established over 25 years ago in order to improve the state of Danube River Basin water bodies. It builds upon decades of agreements between countries, starting with the Declaration of the Danube Countries to Cooperate on Questions Concerning the Water Management of the Danube in 1985, also known as the Bucharest Declaration. Later, the Danube River Basin countries agreed to implement a shared water quality monitoring system, to address cross-border pollution, and to protect and preserve habitats [82]. Hence, in 1996, ICPDR launched the Transnational Monitoring Network (TNMN) with over 100 stations across the Danube and its main tributaries [83]. Also, ICPDR organizes regular water quality monitoring campaigns along the Danube River, within the WFD and under the Joint Danube Survey program. The reports and data of the surveys are publicly available [83, 84]. The ICPDR have organized public consultation sessions, over the last decade and a half, with civil society and stakeholders regarding the development of strategies for water and environmental management. The most important work programs are the Danube River Basin and the Flood Risk Management Plans. Also, public consultations for policies within the EU Green Deal framework are underway. Nevertheless, it is not clear if the ICPDR consultation announcements reach the citizens from the Lower Danube River region. To ensure a larger dissemination of their messages and roles towards the civil society and stakeholders, the ICPDR has established a group of Observers from across the Danube Basin, but with limited responsibility and decision power. Thus, mechanisms have been developed in order to help citizens to voice their concerns. However, there is no local group to establish a connection between citizens and policy makers.

The response of the policy makers may be slow, despite public pressure or expectations, depending on the cultural and political landscapes [85]. The EU has reinforced the need to continue funding for citizen science projects, in the Horizon Europe program [86]. With the support of EUSDR and ICPDR, the EU could direct research funds to citizen science actions, either coordinated by professional researchers or by autonomous citizen scientists.

4.4 Relationship Between Authorities, Citizens and Researchers in the Lower Danube Basin

The analysis of the programs and activities, with public participation, that were undertaken in the Lower Danube Basin (see Sect. 3) shows the fragmented relationship between the involved parties (Fig. 4). So far, only NGOs maintained a long-term relationship with citizens and were the link between citizens, authorities and water managers. However, no direct relationship was observed between citizens and authorities, meaning that there is no evident support in heeding citizens' concerns, solving environmental issues or in developing autonomous research programs. As discussed in Sect. 4.2, European policy-makers have established means of communicating with citizens through Observers (Fig. 4), but they have limited power of decision. In addition, the policy-makers launched multiple public consultation sessions when drafting regulations. Despite these efforts, it is unclear if the messages of the policy-makers and citizens reach one another, especially in the Lower Danube Basin (see Sect. 4.3).

A critical communication gap was observed between professional scientists and citizens (Fig. 4). The decrease in consistent and predictable research funding in Romania [87] may explain the lack of involvement of professional scientists. In addition, the academic environment, in general, may not favour citizen science programs. Researchers involved in such projects may not be able to have a high publication productivity due to limited time, as they will be mostly engaging with educating citizen scientists [69]. However, professional scientists could bring the tools and expertise much needed in communities and may act as a bridge between the involved parties. They may also help citizens to identify environmental problems and encourage them to report these issues and claim their right into decision-making.

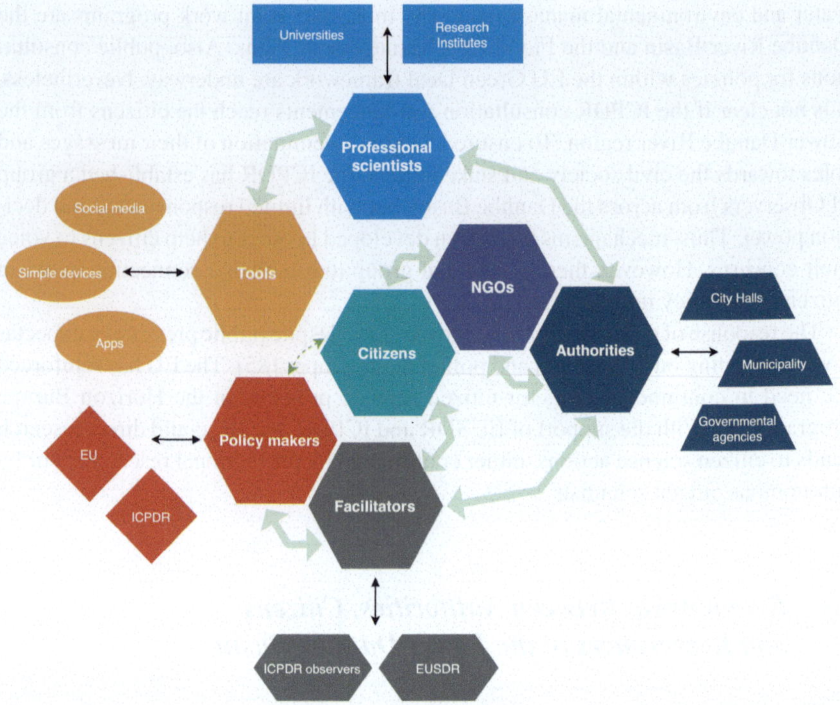

Fig. 4 Relationships between groups participating in clean-up and monitoring activities, in the Lower Danube Basin

The fragmented relationship between involved parties, citizens, policy-makers, local authorities and scientists explains the limited public participation in science projects. A survey made by Vann-Sander et al. [88] on 200 respondents, including the above-mentioned groups, regarding citizen science projects, highlighted the fragmented relationship between them, not only relating to dialogue but also to how citizen science projects are perceived and how science is disseminated. Vann-Sander et al. [88] found a level of distrust of participants over science and scientists, potentially due to the lack of community involvement and communication between parties. Participants stated that scientists are elitists, snobs and theocratic and that "society doesn't trust science". The participants also believed that science alone cannot influence decision makers. It was also the perception that policy makers and managers are interested in the citizen science only if it satisfies their goals [88]. Moreover, in some cases, formal research institutions showed a reserved attitude towards citizen scientists or even refused to collaborate invoking the uselessness of measurements or the lack of seriousness on citizens' side [24]. There was also distrust from the citizens scientists towards the involvement of the municipality [24].

A similar lack of trust between parties was found by Teodosiu et al. [29], who analysed how public participation functions in Romania. In most water management planning, the public involvement was passive and when the public actively engaged in the planning process, the authorities were reluctant to use the results. The authorities often have a top-down approach in decision making and consider that the public lacks the needed expertise. In addition, the public is sceptic to put forward their ideas when it sees how difficult it is to integrate them into formal decision making.

The political context of the past decades, which shaped the culture, society and economy, in the Lower Danube Basin, may have widened the communication gap between these groups. Communism led to extreme political inequalities, top-down approach, one-sided decisions, and excluded choices and alternatives. Suggestions, proposals, ideas and initiatives were discouraged as they could challenge the authority of the state [89]. All these severely eroded individual freedom, responsibility and influence of citizens on political development. Moreover, after the 1990s, the lack of relevant stakeholders' experience and distrust in civil societies and participatory processes limited the stakeholders' engagement in such actions [90]. Furthermore, considering the lack or slow economic and educational development in some regions of the Lower Danube Basin (see Sect. 4.2), some citizens are unlikely to report or solve their issues. According to Ottinger [26], policy makers are more likely to dismiss the environmental issues alerted by marginalized communities. The Lower Danube River region has one of the highest numbers of marginalized rural and urban communities [75]. Whilst governments mostly ignore environmental issues or downplay problems, despite public concerns, the citizens may empower themselves to monitor the environment in order to highlight the issues [91]. Community-driven data collection or open data platforms may provide marginalized communities with the needed quantitative information to increase their credibility when meeting policy makers [26]. In addition, these platforms may act as "social boundary objects" for citizens to voice their concerns and bond them to a common cause, further strengthening their position in relation to decision makers [91].

Trust and effective collaboration are especially needed in water management. Water resources are mostly taken for granted and do not become a political priority until they are severely reduced [92]. In addition, local users of water resources often do not have access to information on water planning and do not have the right to participate in decision-making processes that affect these waters [93]. There is a clear need to involve the society to solve water problems, especially considering the growing disconnection between the water environment and citizens [92]. Participation of the public and stakeholders in river basin management is increasingly promoted because it is expected to improve resource management and enable participants to engage freely and equally in management (support democratic processes) [93].

5 Recommendation

Despite the global success in involving citizens in science, either through government funded projects or independent initiations, the approach has not flourished in the Lower Danube Basin. Several opportunities arise by building on past citizen science projects and by adapting the available tools to this region. The following recommendations can be made to develop successful citizen science programs in the Lower Danube Basin:

- The first step can be taken by professional scientists to involve citizens in short temporal monitoring studies as part of their environmental research projects. Only a handful of participants with high motivation is needed to sustain a project. Prior to starting the project, a survey can be undertaken on the citizens from the Lower Danube Basin to better understand what would motivate and secure commitment during research activities.
- Later, professional scientists can join NGOs in their large-scale public participation activities, allowing them access to a large pool of volunteers and to NGOs' extensive organizational expertise.
- Researchers may use available tools specifically developed for citizen scientists, such as smartphone applications, online platforms, sensors or testing kits (Fig. 1). They can also use social media for gathering volunteers, collecting data or communicating results. Citizens can assess and measure water colour and clarity, nitrate, phosphate, pH, turbidity, water body geomorphic features and dynamics, pollution spills, water level or precipitation.
- The EU, the ICPDR and governments can step in to provide consistent and continuous financial support to large citizen science programs in the Danube Basin. In particular, a partnership between governments from the Lower Danube Basin may be effective in ensuring continuous development of citizen scientists observatories across the basin and in maintaining good ecological status. Funding can be gradually increased based on results and level of involvement, with the final goal of supporting long-term programs. However, access to funding and commitment of research institutions must be harmonized across the entire Danube Basin to ensure consistent quality and quantity of citizen generated data.
- Once communication between citizens, scientists and policy-makers is established, the authorities must be involved in the programs to guarantee that actions are implemented at local and regional levels. Policy-makers may intervene to establish an effective contact between involved parties. In addition, NGOs may help professional researchers to develop successful communication routes with the authorities, considering the NGOs' extensive expertise in engaging authorities in environmental activities.
- Also, dedicated citizens may organize themselves into an autonomous citizen science institution, which is initiated and run by them [24]. While academics are motivated to develop citizen science projects by their passion and desire to transmit science to a broader audience [69], citizens may be attracted by the core of research questions. Thus, they may have more power of decision in terms of study sites, group interaction, reporting, access to funding and communication with authorities and policy makers.

6 Conclusion

This review shows that the involvement of citizens in environmental projects, concerning the Lower Danube Basin is scarce and infrequent. The main reasons were identified as the fragmented relationship between scientists, authorities and citizens, lack of access to resources, limited funding for science and education, the urban–rural development gap and social exclusion. There are no quick solutions to improve the situation. However, gradual steps can be taken to move the wheel of growth through citizen science, as shown in Sect. 5. Water quality monitoring of the Lower Danube River by citizen scientists cannot replace traditional monitoring strategies made by professional scientists, such as the Trans National Monitoring Network or the Joint Danube Survey [21], but if a citizen science program is initiated and executed under close professional supervision, in terms of accuracy, quality control and spatial and temporal resolution, the resulting data may complement traditional records.

Citizen science represents a means of obtaining high frequency data with large geographical coverage, at scales that cannot be achieved by professionals alone. Nevertheless, citizen science also represents education, communication, shared experiences and social inclusion. This approach has the power to change the social, economic and environmental scenery in the Lower Danube Basin. Citizens will have a proactive attitude towards protecting the Danube River and the environment, in general, which will provide them with a sense of inclusion in Europe.

Acknowledgements The authors acknowledge the support of the Ministry of Research and Innovation, CNCS—UEFISCDI, project number PN-III-P1-1.1-TE-2019-0789, within PNCDI III, and by the Core Program OPTRONICA and project number 152/2016, SMIS 108109.

Author Contributions All authors contributed equally to this work.

References

1. Miller-Rushing A, Primack R, Bonney R (2012) The history of public participation in ecological research. Front Ecol Environ 10:285–290. https://doi.org/10.1890/110278
2. Earp HS, Liconti A (2020) Science for the future: The use of citizen science in marine research and conservation. In: Jungblut S, Liebich V, Bode-Dalby M (eds) YOUMARES 9—The oceans: Our research, our future: Proceedings of the 2018 conference for YOUng MArine RESearcher in Oldenburg, Germany. Springer International Publishing, Cham, pp 1–19
3. Kobori H, Dickinson JL, Washitani I, Sakurai R, Amano T, Komatsu N, Kitamura W, Takagawa S, Koyama K, Ogawara T, Miller-Rushing AJ (2016) Citizen science: a new approach to advance ecology, education, and conservation. Ecol Res 31:1–19. https://doi.org/10.1007/s11284-015-1314-y
4. Havens K, Henderson S (2013) Citizen science takes root. Am Sci 101:378–385. https://doi.org/10.1511/2013.104.378
5. Johnson BA, Mader AD, Dasgupta R, Kumar P (2020) Citizen science and invasive alien species: An analysis of citizen science initiatives using information and communications technology (ICT) to collect invasive alien species observations. Glob Ecol Conserv 21(e00812):1–14. https://doi.org/10.1016/j.gecco.2019.e00812

6. Castilla EP, Cunha DGF, Lee FWF, Loiselle S, Ho KC, Hall C (2015) Quantification of phytoplankton bloom dynamics by citizen scientists in urban and peri-urban environments. Environ Monit Assess 187(690):1–11. https://doi.org/10.1007/s10661-015-4912-9
7. Brintrup K, Amigo C, Fernández J, Hernández A, Pérez F, Félez-Bernal J, Butturini A, Saez-Carrillo K, Yevenes MA, Figueroa R (2019) Comparison of organic matter in intermittent and perennial rivers of Mediterranean Chile with the support of citizen science. Rev Chil Hist Nat 92:1–10. https://doi.org/10.1186/s40693-019-0083-3
8. Jamieson M, Elson M, Carruthers R, Ordens CM (2020) The contribution of citizen science in managing and monitoring groundwater systems impacted by coal seam gas production: An example from the Surat Basin in Australia's Great Artesian Basin. Hydrogeol J 28:439–459. https://doi.org/10.1007/s10040-019-02050-8
9. Hesley D, Burdeno D, Drury C, Schopmeyer S, Lirman D (2017) Citizen science benefits coral reef restoration activities. J Nat Conserv 40:94–99. https://doi.org/10.1016/j.jnc.2017.09.001
10. Hyder K, Wright S, Kirby M, Brant J (2017) The role of citizen science in monitoring small-scale pollution events. Mar Pollut Bull 120:51–57. https://doi.org/10.1016/j.marpolbul.2017.04.038
11. Gray S, Jordan R, Crall A, Newman G, Hmelo-Silver C, Huang J, Novak W, Mellor D, Frensley T, Prysby M, Singer A (2017) Combining participatory modelling and citizen science to support volunteer conservation action. Biol Conserv 208:76–86. https://doi.org/10.1016/j.biocon.2016.07.037
12. McKinley DC, Miller-Rushing AJ, Ballard HL, Bonney R, Brown H, Cook-Patton SC, Evans DM, French RA, Parrish JK, Phillips TB, Ryan SF, Shanley LE, Shirk JL, Stepenuck KF, Weltzin JF, Wiggins A, Boyle OD, Brigges RD, Chapin SF, Hewitt DA, Preuss PW, Soukup MA (2017) Citizen science can improve conservation science, natural resource management, and environmental protection. Biol Conserv 208:15–28. https://doi.org/10.1016/j.biocon.2016.05.015
13. Martay B, Pearce-Higgins JW, Harris SJ, Gillings S (2018) Monitoring landscape-scale environmental changes with citizen scientists: Twenty years of land use change in Great Britain. J Nat Conserv 44:33–42. https://doi.org/10.1016/j.jnc.2018.03.001
14. Bremer S, Haque MM, Bin AS, Kvamme S (2019) 'My new routine': Assessing the impact of citizen science on climate adaptation in Bangladesh. Environ Sci Policy 94:245–257. https://doi.org/10.1016/j.envsci.2018.12.029
15. Che W, Frey HC, Fung JCH, Ning Z, Qu H, Lo HK, Chen L, Wong TW, Wong MKM, Lee OCW, Carruthers D, Cheung F, Chan JWM, Yeung DW, Fung YH, Zhang X, Stocker J, Hood C, Hohenberger TL, Leung KW, Louie PYK, Li ATY, Sun L, Wei P, Li Z, Zhang Y, Wang M, Shen Q, Huang W, Lee E, Patwary A, Lei X, Cheng S, Hossain MS, Tang KTJ, Lao XQ, Leung R, Chan D, Li Y, Yuan Z, Lau AKH (2020) PRAISE-HK: A personalized real-time air quality informatics system for citizen participation in exposure and health risk management. Sustain Cities Soc 54(101986):1–7. https://doi.org/10.1016/j.scs.2019.101986
16. West SE, Büker P, Ashmore M, Njoroge G, Welden N, Muhoza C, Osano P, Maku J, njoroge P, Apondo W, (2020) Particulate matter pollution in an informal settlement in Nairobi: Using citizen science to make the invisible visible. Appl Geogr 114(102133):1–12. https://doi.org/10.1016/j.apgeog.2019.102133
17. Schröter M, Kraemer R, Mantel M, Kabisch N, Hecker S, Richter A, Neumeier V, Bonn A (2017) Citizen science for assessing ecosystem services: Status, challenges and opportunities. Ecosyst Serv 28:80–94. https://doi.org/10.1016/j.ecoser.2017.09.017
18. Njue N, Stenfert Kroese J, Gräf J, Jacobs SR, Weeser B, Breuer L, Rufino MC (2019) Citizen science in hydrological monitoring and ecosystem services management: State of the art and future prospects. Sci Total Environ 693(133531):1–16. https://doi.org/10.1016/j.scitotenv.2019.07.337
19. Michelsen N, Dirks H, Schulz S, Kempe S, Al-Saud M, Schüth C (2016) YouTube as a crowd-generated water level archive. Sci Total Environ 568:189–195. https://doi.org/10.1016/j.scitotenv.2016.05.211

20. Gasparotti C (2014) The main factors of water pollution in Danube River basin. EuroEconomica 33:1–15
21. Chapman DV, Bradley C, Gettel GM, Hatvani IG, Hein T, Kovacs J, Liska I, Oliver DM, Tanos P, Trasy B, Varbiro G (2016) Developments in water quality monitoring and management in large river catchments using the Danube River as an example. Environ Sci Policy 64:141–154. https://doi.org/10.1016/j.envsci.2016.06.015
22. Conrad CC, Hilchey KG (2011) A review of citizen science and community-based environmental monitoring: Issues and opportunities. Environ Monit Assess 176:273–291. https://doi.org/10.1007/s10661-010-1582-5
23. Lehtiniemi M, Outinen O, Puntila-Dodd R (2020) Citizen science provides added value in the monitoring for coastal non-indigenous species. J Environ Manage 267(110608):1–8. https://doi.org/10.1016/j.jenvman.2020.110608
24. Wildschut D, Zijp H (2020) The discoveries of citizens running around. Clim Risk Manag 28(100225):1–11. https://doi.org/10.1016/j.crm.2020.100225
25. Freitag A (2016) A typology for strategies to connect citizen science and management. Environ Monit Assess 188(519):1–14. https://doi.org/10.1007/s10661-016-5513-y
26. Ottinger G (2017) Making sense of citizen science: Stories as a hermeneutic resource. Energy Res Soc Sci 31:41–49. https://doi.org/10.1016/j.erss.2017.06.014
27. Becken S, Connolly RM, Chen J, Stantic B (2019) A hybrid is born: Integrating collective sensing, citizen science and professional monitoring of the environment. Ecol Inform 52:35–45. https://doi.org/10.1016/j.ecoinf.2019.05.001
28. Pretty JN (1995) Participatory learning for sustainable agriculture. World Dev 23:1247–1263. https://doi.org/10.1016/0305-750X(95)00046-F
29. Teodosiu C, Barjoveanu G, Vinke-de Kruijf J (2013) Public participation in water resources management in romania: Issues, expectations and actual involvement. Environ Eng Manag J 12:1051–1063. https://doi.org/10.30638/eemj.2013.129
30. Kiessling T, Knickmeier K, Kruse K, Brennecke D, Nauendorf A, Thiel M (2019) Plastic Pirates sample litter at rivers in Germany—Riverside litter and litter sources estimated by schoolchildren. Environ Pollut 245:545–557. https://doi.org/10.1016/j.envpol.2018.11.025
31. Let's Do It Danube! (2020) Let's Do It, Danube! https://letsdoitromania.ro/9000-voluntari-implicati-in-campania-lets-do-it-danube/. Accessed 13 Oct 2020
32. Clean Danube (2020) Pentru o Dunăre cu apele curate. http://www.dunarea-curata.ro/index.php. Accessed 9 Oct 2020
33. FreshWater Watch (2020) FreshWater Watch, Earthwatch Europe. https://freshwaterwatch.thewaterhub.org/our-data/explore-our-data. Accessed 8 Oct 2020
34. EarthEcho (2020) EarthEcho water challenge. https://www.monitorwater.org/. Accessed 9 Oct 2020
35. Breuer L, Hiery N, Kraft P, Bach M, Aubert AH, Frede HG (2015) HydroCrowd: A citizen science snapshot to assess the spatial control of nitrogen solutes in surface waters. Sci Rep 5:1–10. https://doi.org/10.1038/srep16503
36. SciShops (2020) SciShops.eu. https://www.scishops.eu/measuring-the-quality-of-the-danube-bay-zoltan-science-shop-joins-the-drinkable-rivers-initiative/. Accessed 9 Oct 2020
37. Scent (2020) The Danube Delta, Romania—Scent. https://scent-project.eu/the-danube-delta-romania. Accessed 12 Oct 2020
38. Kayaking the Continent (2020) Kayaking the continent. https://kayakingthecontinent.com/. Accessed 9 Oct 2020
39. Societatea Ornitologica Romana (2020) International waterBird count—SOR. https://www.sor.ro/proiect/international-waterbird-count/statistica/. Accessed 12 Oct 2020
40. Tian X, ten Veldhuis MC, Schleiss M, Bouwens C, van de Giesen N (2019) Critical rainfall thresholds for urban pluvial flooding inferred from citizen observations. Sci Total Environ 689:258–268. https://doi.org/10.1016/j.scitotenv.2019.06.355
41. Poisson AC, McCullough IM, Cheruvelil KS, Elliott KC, Latimore JA, Soranno PA (2020) Quantifying the contribution of citizen science to broad-scale ecological databases. Front Ecol Environ 18:19–26. https://doi.org/10.1002/fee.2128

42. Abbott BW, Moatar F, Gauthier O, Fovet O, Antoine V, Ragueneau O (2018) Trends and seasonality of river nutrients in agricultural catchments: 18 years of weekly citizen science in France. Sci Total Environ 624:845–858. https://doi.org/10.1016/j.scitotenv.2017.12.176
43. Longo A, Zappatore M, Bochicchio MA (2020) Apollon: Towards a citizen science methodology for urban environmental monitoring. Futur Gener Comput Syst 112:899–912. https://doi.org/10.1016/j.future.2020.06.041
44. Jollymore A, Haines MJ, Satterfield T, Johnson MS (2017) Citizen science for water quality monitoring: Data implications of citizen perspectives. J Environ Manage 200:456–467. https://doi.org/10.1016/j.jenvman.2017.05.083
45. Braz Sousa L, Fricker SR, Doherty SS, Webb CE, Baldock KL, Williams CR (2020) Citizen science and smartphone e-entomology enables low-cost upscaling of mosquito surveillance. Sci Total Environ 704(135349):1–10. https://doi.org/10.1016/j.scitotenv.2019.135349
46. Chiu CC, Liao CP, Kuo TC, Huang HW (2020) Using citizen science to investigate the spatial-temporal distribution of floating marine litter in the waters around Taiwan. Mar Pollut Bull 157(111301):1–12. https://doi.org/10.1016/j.marpolbul.2020.111301
47. Freihardt J (2020) Can Citizen Science using social media inform sanitation planning? J Environ Manage 259(110053):1–9. https://doi.org/10.1016/j.jenvman.2019.110053
48. EurActiv (2020) EurActiv. https://www.euractiv.ro/infosociety/aproape-zece-milioane-de-utilizatori-facebook-in-romania-13323. Accessed 28 Oct 2020
49. Abrihann R (2017) Utilizarea smartphone-urilor in Romania a crescut de peste 4 ori in ultimii ani. Cum folosesc romanii telefoanele inteligente. In: Startup cafe. https://www.startupcafe.ro/stiri-hitech-21640812-utilizarea-smartphone-urilor-romania-crescut-peste-4-ori-ultimii-ani-cum-folosesc-romanii-telefoanele-inteligente.htm. Accessed 16 Dec 2020
50. Graham EA, Henderson S, Schloss A (2011) Using mobile phones to engage citizen scientists in research. Eos (Washington DC) 92:313–315. https://doi.org/10.1029/2011EO380002
51. Fehri R, Khlifi S, Vanclooster M (2020) Testing a citizen science water monitoring approach in Tunisia. Environ Sci Policy 104:67–72. https://doi.org/10.1016/j.envsci.2019.11.009
52. Shupe SM (2017) High resolution stream water quality assessment in the Vancouver, British Columbia region: a citizen science study. Sci Total Environ 603–604:745–759. https://doi.org/10.1016/j.scitotenv.2017.02.195
53. Scott AB, Frost PC (2017) Monitoring water quality in Toronto's urban stormwater ponds: Assessing participation rates and data quality of water sampling by citizen scientists in the FreshWater Watch. Sci Total Environ 592:738–744. https://doi.org/10.1016/j.scitotenv.2017.01.201
54. Quinlivan L, Chapman D, Sullivan T (2019) Validating citizen science monitoring of ambient water quality for the United Nations sustainable development goals. Sci Total Environ 699(134255):1–9. https://doi.org/10.1016/j.scitotenv.2019.134255
55. Lévesque D, Cattaneo A, Deschamps G, Hudon C (2017) In the eye of the beholder: Assessing the water quality of shoreline parks around the Island of Montreal through citizen science. Sci Total Environ 579:978–988. https://doi.org/10.1016/j.scitotenv.2016.10.175
56. Loiselle SA, Frost PC, Turak E, Thornhill I (2017) Citizen scientists supporting environmental research priorities. Sci Total Environ 598:937. https://doi.org/10.1016/j.scitotenv.2017.03.142
57. McGoff E, Dunn F, Cachazo LM, Williams P, Biggs J, Nicolet P, Ewald NC (2017) Finding clean water habitats in urban landscapes: professional researcher vs citizen science approaches. Sci Total Environ 581–582:105–116. https://doi.org/10.1016/j.scitotenv.2016.11.215
58. Thornhill I, Ho JG, Zhang Y, Li H, Ho KC, Miguel-Chinchilla L, Loiselle SA (2017) Prioritising local action for water quality improvement using citizen science; a study across three major metropolitan areas of China. Sci Total Environ 584–585:1268–1281. https://doi.org/10.1016/j.scitotenv.2017.01.200
59. Gurnell AM, England J, Shuker L, Wharton G (2019) The contribution of citizen science volunteers to river monitoring and management: International and national perspectives and the example of the MoRPh survey. River Res Appl 35:1359–1373. https://doi.org/10.1002/rra.3483

60. Bardaji R, Sánchez AM, Simon C, Wernand MR, Piera J (2016) Estimating the underwater diffuse attenuation coefficient with a low-cost instrument: The KdUINO DIY buoy. Sensors (Switzerland) 16:1–15. https://doi.org/10.3390/s16030373
61. Storer J, Chao J, Torelli A, Ostrowski A (2016) KnoWare: A system for citizen-based environmental monitoring. Informing Sci 19:125–139. https://doi.org/10.28945/3500
62. Weeser B, Stenfert Kroese J, Jacobs SR, Njue N, Kemboi Z, Ran A, Rufino MC, Breuer L (2018) Citizen science pioneers in Kenya—A crowdsourced approach for hydrological monitoring. Sci Total Environ 631–632:1590–1599. https://doi.org/10.1016/j.scitotenv.2018.03.130
63. Etter S, Strobl B, van Meerveld I, Seibert J (2020) Quality and timing of crowd-based water level class observations. Hydrol Process 34:4365–4378. https://doi.org/10.1002/hyp.13864
64. Rambonnet L, Vink SC, Land-Zandstra AM, Bosker T (2019) Making citizen science count: Best practices and challenges of citizen science projects on plastics in aquatic environments. Mar Pollut Bull 145:271–277. https://doi.org/10.1016/j.marpolbul.2019.05.056
65. San Llorente Capdevila A, Kokimova A, Sinha Ray S, Avellán T, Kim J, Kirschke S (2020) Success factors for citizen science projects in water quality monitoring. Sci Total Environ 728(137843):1–17. https://doi.org/10.1016/j.scitotenv.2020.137843
66. Redmon JH, Levine KE, Aceituno AM, Litzenberger K, MacDonald Gibson J (2020) Lead in drinking water at North Carolina childcare centers: Piloting a citizen science-based testing strategy. Environ Res 183(109126):1–9. https://doi.org/10.1016/j.envres.2020.109126
67. Rutten M, Minkman E, van der Sanden M (2017) How to get and keep citizens involved in mobile crowd sensing for water management? A review of key success factors and motivational aspects. Wiley Interdiscip Rev Water 4(e1218):1–10. https://doi.org/10.1002/wat2.1218
68. Cappa F, Laut J, Porfiri M, Giustiniano L (2018) Bring them aboard: Rewarding participation in technology-mediated citizen science projects. Comput Human Behav 89:246–257. https://doi.org/10.1016/j.chb.2018.08.017
69. Sauermann H, Vohland K, Antoniou V, Balázs B, Göbel C, Karatzas K, Mooney P, Perelló J, Ponti M, Samson R, Winter S (2020) Citizen science and sustainability transitions. Res Policy 49(103978):1–16. https://doi.org/10.1016/j.respol.2020.103978
70. Alam A, Murthi M, Yemtsov R, Murrugarra E, Dudwick N, Hamilton E, Tiongson E (2005) Growth, poverty, and inequality. The World Bank, Washington DC, p 326
71. World Bank (2020) Financial inclusion in Romania: issues and opportunities. World Bank, Washington DC
72. Czakó K, Fekete D, Poreisz V (2014) Economic differences of countries by the river Danube. Procedia Econ Financ 9:163–175. https://doi.org/10.1016/s2212-5671(14)00017-3
73. European Commission—EAC (2019) Education and training monitor 2019—Romania. https://doi.org/10.2766/743745
74. Eurostat (2020) Europe 2020 indicators—R&D and innovation—Statistics explained. https://ec.europa.eu/eurostat/statistics-explained/index.php?title=Europe_2020_indicators_-_R%26D_and_innovation&oldid=289290. Accessed 9 Dec 2020
75. World Bank (2020) World development indicators | DataBank. https://databank.worldbank.org/reports.aspx?source=2&series=GB.XPD.RSDV.GD.ZS. Accessed 8 Dec 2020
76. Acharya KP, Pathak S (2019) Applied research in low-income countries: Why and how? Front Res Metrics Anal 4:1–9. https://doi.org/10.3389/frma.2019.00003
77. Irvine K, Weigelhofer G, Popescu I, Pfeiffer E, Păun A, Drobot R, Gettel G, Staska B, Stanica A, Hein T, Habersack H (2016) Educating for action: Aligning skills with policies for sustainable development in the Danube river basin. Sci Total Environ 543:765–777. https://doi.org/10.1016/j.scitotenv.2015.09.072
78. Fraisl D, Campbell J, See L, When U, Wardlaw J, Gold M, Moorthy I, Arias R, Piera J, Oliver JL, Masó J, Penker M, Fritz S (2020) Mapping citizen science contributions to the UN sustainable development goals. Sustain Sci 15:1735–1751. https://doi.org/10.1007/s11625-020-00833-7
79. Huddart JEA, Thompson MSA, Woodward G, Brooks SJ (2016) Citizen science: From detecting pollution to evaluating ecological restoration. WIREs Water 3:287–300. https://doi.org/10.1002/wat2.1138

80. Turşie C (2015) Macro-regional strategies of European integration: What can the Danube region learn from the Baltic Sea region? Procedia—Soc Behav Sci 183:1–10. https://doi.org/10.1016/j.sbspro.2015.04.839
81. Danube Region Strategy (2020) EUSDR. https://danube-region.eu/. Accessed 14 Dec 2020
82. Bănăduc D, Rey S, Trichkova T, Lenhardt M, Curtean-Bănăduc A (2016) The Lower Danube River-Danube Delta-North West Black Sea: A pivotal area of major interest for the past, present and future of its fish fauna—A short review. Sci Total Environ 545–546:137–151
83. ICPDR (2020) TNMN—TransNational Monitoring Network. http://www.icpdr.org/main/activities-projects/tnmn-transnational-monitoring-network. Accessed 22 Dec 2020
84. Joint Danube Survey (2019) Joint Danube survey. http://www.danubesurvey.org/jds4/about. Accessed 22 Dec 2020
85. Lisjak J, Schade S, Kotsev A (2017) Closing data gaps with citizen science? Findings from the Danube Region. ISPRS Int J Geo-Information 6(9):277(1–21). https://doi.org/10.3390/ijgi6090277
86. Warin C, Delaney N (2020) Citizen science and citizen engagement. Publications Office of the European Union, Luxembourg, p 36
87. Gusilov E (2018) Research and innovation in Romania, a land of two realities. In: Energy Ind. Rev. https://energyindustryreview.com/opinion/research-innovation-romania-land-two-realities/. Accessed 7 Dec 2020
88. Vann-Sander S, Clifton J, Harvey E (2016) Can citizen science work? Perceptions of the role and utility of citizen science in a marine policy and management context. Mar Policy 72:82–93. https://doi.org/10.1016/j.marpol.2016.06.026
89. Li K-W (2017) Why has socialism failed? In: Li K-WBT-RC in GED (ed). Chapter 19, Academic Press, pp 409–418
90. Stringer LC, Scrieciu SS, Reed MS (2009) Biodiversity, land degradation, and climate change: Participatory planning in Romania. Appl Geogr 29:77–90. https://doi.org/10.1016/j.apgeog.2008.07.008
91. Carton L, Ache P (2017) Citizen-sensor-networks to confront government decision-makers: Two lessons from the Netherlands. J Environ Manage 196:234–251. https://doi.org/10.1016/j.jenvman.2017.02.044
92. Gawlik BM, Głowacka N, Feldman DL, Elelman R (2018) The scientist, the politician, the artist and the citizen: how water united them. Environ Sci Eur 30(12):1–12. https://doi.org/10.1186/s12302-018-0141-5
93. Razzaque J (2009) Public participation in water governance. In: Dallapenna JW, Gupta J (eds) The evolution of the law and politics of water. Springer, Netherlands, pp 355–372

Stakeholders' Interests and Participation in the Sustainable Use of the Lakes Along the Danube Floodplain. A Romanian Sector as Case Study

Oana Mititelu-Ionuş, Mihaela Licurici, Cristina Şoşea, and Cristina Georgiana Zanfir

Abstract Lakes and wetlands within large river floodplains represent some of the most endangered ecosystems at global scale. In this framework, raising awareness and promoting involvement of stakeholders can substantially contribute to the protection and sustainable capitalization of the lakes along the Lower Danube Floodplain.

The present chapter aims to analyse the stakeholders' interests and participation in the sustainable use of the lakes in the Lower Danube Floodplain, the Drobeta-Turnu Severin-Bechet sector, by using an online survey conducted from August to October 2020. The questionnaire received responses from 47 Romanian stakeholders.

The different approaches used during the survey aim at better understanding the degree of knowledge and participation in the sustainable capitalization of the lakes and could help in making future decisions. The analysis of the answers shows the following main aspects: 34% of the respondents consider themselves highly informed about the state and management of lakes in the Danube Floodplain, while the most widespread manner of gaining knowledge is represented by personal theoretical documentation (44.7%); there is a significant agreement on the negative dynamics of the floodplain lakes over the last two decades, with 78.7% of the respondents agreeing or fully agreeing on this issue; 57.4% of the involved stakeholders consider the declining state of the lakes as a consequence of human activities, while 31.9% of them attribute it to a mosaic of natural and anthropogenic triggering factors.

The answers given in connection with the issues addressed provide both relevance to the case study and the potential for generalization for floodplains of large rivers.

O. Mititelu-Ionuş · M. Licurici (✉) · C. Şoşea
Faculty of Science, Geography Department, University of Craiova, A. I. Cuza Street, No. 13, 200585 Craiova, Romania
e-mail: mihaela.licurici@edu.ucv.ro

O. Mititelu-Ionuş
e-mail: oana.mititelu@edu.ucv.ro

C. Şoşea
e-mail: cristina.sosea@edu.ucv.ro

C. G. Zanfir
Faculty of Geography, University of Bucharest, Nicolae Bălcescu Blvd, No. 1, 010041 Bucharest, Romania
e-mail: cristina.zanfir@drd.unibuc.ro

A. Negm et al. (eds.), *The Lower Danube River*, Earth and Environmental Sciences Library, https://doi.org/10.1007/978-3-031-03865-5_19

In this sense, assessing stakeholders' involvement can contribute to the sustainable use of lakes by paying attention to the increasing involvement of public institutions with decision making power in the sustainable use of floodplain lakes.

Keywords Lakes · Management · Stakeholder · Online Survey · Danube Floodplain · Romania

1 Introduction

River floodplains are multifarious, rampant ecosystems, providing a complex mosaic of freshwater habitats that are characterized by their spatial–temporal heterogeneity and a certain degree of hydrological connectivity with the main channel [1]. Over time, floodplain ecosystems have been heavily modified by anthropogenic activity due to river regulation measures and techniques, which, on the one hand aimed to interrupt the periodicity of flooding disturbance, but also to capitalize the agricultural potential of lands.

A tremendous reduction of floodplain area has occurred in the last 100 years and this loss continues because of pressures such as land use change, river regulation and dam construction [2]. The Danube River Basin makes no exception as the overall extent of its floodplain has been reduced, with the highest losses occurring in the upper and lower sectors of the Danube.

Stakeholders' involvement in environmental decision-making has become an increasingly important aspect [3]. Thus, the European Water Framework Directive [4] imposes public consultation during the development of watershed management plans.

In this context, this chapter proposes a comprehensive framework to establish and implement stakeholder's interests and participation in sustainable use of the Danube Floodplain lakes (Romania), from stakeholder identification to their level of involvement, and provides new and original information on this topic, derived from own research.

2 Concepts and Approaches for Stakeholder Analysis

2.1 Stakeholders' Identification and Structuration

Understanding the perceptions and the needs of diverse stakeholders is crucial for efficient conservation policy and has become a requirement for improving environmental management around the world [5].

To implement a study on stakeholder analysis, these important actors must be identified and categorized, in order to understand the power relations between them and their specific interest and participation (Fig. 1). Regarding stakeholder identification,

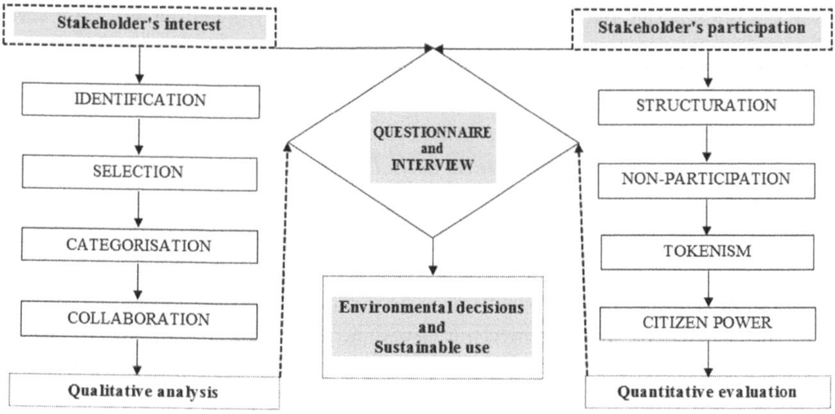

Fig. 1 Conceptual framework for stakeholder's analysis and evaluation in environmental management, adapted from [13–16]

Selman [6] distinguishes between stakeholders who have an economic interest and those motivated by principles or values. Grimble and Wellard [7] define stakeholder analysis as "holistic approach or procedure for gaining an understanding of a system and assessing the impact of changes to that system, by means of identifying the key actors or stakeholders and assessing their respective interests in the system". Recent studies [8] define the same terms as"identification of individuals, groups or organizations that have a specific interest and are likely to affect or be affected by proposed interventions". Besides all this, what constitutes a "stakeholder" or "stakeholders" is the key for stakeholder analysis, as it pre-determines who might and might not be identified as being important enough to be included in the analysis [9].

Numerous definitions have been published on what can be classed as a stakeholder, which is why a clear explanation of the term is essential. Therefore, a variety of methods for identifying stakeholders are available in the literature. One of the best-known methods of identifying stakeholders starts with brainstorming and has been presented and used by King et al. [10], as well as by Stanghellini and Collentine [11]. The *snowball* technique is based on a list of stakeholders that is submitted to one of the identified stakeholders, soliciting opinion and adding further stakeholders. Moreover, techniques such as interviews or analyses of reports and minutes increase the probability of an in-depth understanding of the process [12].

Once the stakeholders' relative importance and influence have been assessed, they can be mapped in an importance-influence matrix, also called the"interest-power matrix" [17, 18]. Van Asselt et al. [19] also drew stakeholder maps through which links, connections and relationship can be visualized. The way in which the stakeholders are organized and how decisions are made will influence the ways in which relationships between environment and community can be manipulated. As underlined by the literature review [20], the identification of the key stakeholders and of certain characteristics (such as where they come from and what is their level of involvement) represents an important step.

2.2 Stakeholders' Involvement and Participation

Over time, there have been controversies between two terms used in the field literature, namely: public participation and stakeholders. Public participation is defined, following the World Bank [21] as "a process through which stakeholders influence and share control over development initiatives and the decision and resources which affect them". On the other hand, Grimble and Wellard [7] defined stakeholders as "any group of people organised, who share a common interest or stake in a particular issue or system". Junk [22] showed that public consultations can provide institutional access to less resourceful stakeholders, e.g. NGOs, thus ensuring the stronger participation of specific local or sectorial interests.

Recent studies address stakeholder importance and their influence as a key aspect: Luyet et al. [16] propose a comprehensive framework to implement stakeholder participation in environmental projects, from stakeholder identification to evaluation. Regarding the participation, the main contribution related to stakeholder structuring and degree of involvement was detailed into three main groups: non-participation (manipulation and therapy), tokenism (informing, consultation, placation) and citizen power (partnership, delegated power and citizen control) [13] (Fig. 1).

Stakeholders' interests can be deduced and analyzed by interviews and questionnaires [23]. Therefore, in order to determine stakeholder's level of involvement, several people, such as representatives, stakeholders or experts, should be asked to complete a questionnaire or be interviewed. The questionnaire as a method of analysis has proven to be a useful tool for stakeholder views. For example, Van Dam and Junginger [24] use a stakeholder questionnaire to analyse how European stakeholders perceive various topics related to the development and implementation of sustainability criteria for biomass and bioenergy.

The "participatory modelling" concept presented by Jonsson and Alkan-Olsson [25] is aiming at serving as a platform for facilitating stakeholder participation in environmental planning at the local/regional level. The participatory environmental modelling process is based on three categories: factors influencing the willingness of stakeholders to invest time and resource in participation in the process, factors influencing the tendency of stakeholders to accept 'expert' results as legitimate descriptions of local ecosystems, and factors defining the 'room of action' for stakeholders to implement remedies suggested by model results. The type and intensity of relationships between stakeholder groups is often related to the degree of their influence and power [21, 26].

Multi-stakeholder processes represent forms of cross-sector collaboration [27]. Participation of all stakeholders is crucial for successful sustainable management of natural resources and a multi-stakeholder approach enables stakeholders to share roles and responsibilities (Fig. 1). Luyet et al. [16] use a framework with five degrees of stakeholder's participation in environmental projects: information, consultation, collaboration, co-decision and empowerment.

Regarding the decision makers, Vroom [28] proposes a systematic method to determine a specific level of involvement for each stakeholder; this model has sequentially seven questions regarding: decision significance, importance of commitment, leader's expertise, likelihood of commitment, group support for objectives, group expertise and team competence.

2.3 *Stakeholders and Water Management*

Successful management of environmental quality via regulation by state and local governments depends on citizen and stakeholder support [29].

In water management studies, stakeholders can be defined as the people who either (i) will be potentially affected by the management, (ii) will be involved by one way or another in the implementation of management activities, or (iii) who are likely to support or oppose the research or development project or the policy at stake [15]. Stakeholder interests and participation in the sustainable use of lakes and wetlands has been the subject for numerous recent studies conducted in different regions of the world:

- the case of the Lake Fundudzi catchment in Limpopo Province, South Africa [30] involved two different processes: a stakeholder analysis involving collection of qualitative data from stakeholders and participants (phase one); development of case studies of participation in communal wetland management using a case story approach, which also drew on qualitative data (phase two);
- the ecological-economic analysis of wetlands provides, after Turner et al. [31], two aspects: scientific integration for management and policy. Sustainable wetland management from floodplains requires an understanding of stakeholders' perceptions of the ecosystem and its management;
- Bosma et al. [32] conducted a study on how do individuals and groups perceive wetland functioning. The case study the Rushebeya-Kanyabaha wetland in Uganda uses fuzzy cognitive mapping in order to assess opposing interests between different stakeholders;
- Smrekar et al. [27] propose a methodology to engage relevant stakeholders in wetland governance in Ljubljansko Barje Nature Park (Slovenia) i.e. all people, groups and institutions, willing to embark on the process of improving the governance of the park;
- Christopoulou and Tsachalidis [33] conducted a survey of local residents' attitude concerning the conservation policies for protected areas (wetlands) in Greece. In this regard, questionnaires were distributed in 32 communities neighbouring four Ramsar wetlands in Northern Greece regarding the ways of management and exploitation of the wetlands and their sociological features;

- Cohen-Shacham et al. [34] conducted a stakeholder analysis based on semi-structured interviews. Their study showed that applying the ecosystem services concept facilitates to point towards sustainable solutions for conflicting stakeholder interests.

It is generally recognized that not all stakeholders have the same influence (Table 1). Mayers [35] classifies stakeholders according to power in two categories: stakeholders who may be directly impacted positively or negatively by the

Table 1 Main stakeholder types and their role in floodplain lakes management

Stakeholders: types and examples [27, 41, 42]	**General role in lake management**	**Involvement levels** [43, 44]			
Policy Public bodies with decisional functions, administrative competences, territorial jurisdiction over the lakes or over the floodplain area municipalities, local or regional authorities, government institutions.	Decision-makers, managers, providers of financial resources initiate and actively participate in the ***development, authorization and implementation*** of measures, rules, plans, projects, programmes concerning lake and floodplain management, and associated use and protection of ecosystem services at the local, regional and national scale. They are ***key-players*** with high influence/power in lake management and could benefit from the ***coordination*** of the multi-stakeholder participatory process.	***Initiating and co-ordinating***	***Co-knowing*** (stakeholders are regarded as a source of knowledge)	***Co-thinking*** (stakeholders who do not play an active role in floodplain lake management, are informed about its state and progress)	***Co-operating*** (stakeholders actually participate and actively contribute to lake management development and implementation)
Science and education Experts and professionals from universities or local schools, from research or development institutions, museums.	Multi-stakeholder involvement requires that the scientific community ***generate scientifically valid approaches, methods, and tools*** for lake management (inter- and transdisciplinary knowledge that is socially, economically, and environmentally appliable), ***elaborate*** and ***refine*** existing understandings of knowledge and the methods of its production and transfer [41], and help ***disseminating*** knowledge. Their involvement and consultation are required.				
Business Individual or collective private entities with economic profile, representing the main activities of the economic sector present on or near the floodplain lakes.	As user of floodplain resources, industry is usually a significant influencer of their state, with low official involvement in decision making. In a knowledge-based and participatory society, industry involvement in lake management enables a better assessment of specific needs and ecological footprint, ***innovation in the field of water resources use***, could contribute to ***a higher acceptance of the appropriate regulatory context*** and, finally, to a diversified and sustainable economic growth.				
Society Associations representing the interests of citizens, civil society organizations, NGOs, individuals from the local communities, occasional users of floodplain resources.	Public involvement is ***necessary*** (it can help solving major challenges concerning floodplain lakes), ***practical*** (it helps gaining field data and understanding social needs related to the floodplain, enables more welcome management directions and legitimizes them), and ***ethically just*** (the taxpayer can gain greater access to and influence over the management process and its results) [41]. These stakeholders are influenced by the lake management framework, but do not have enough power over decisions.				

process; stakeholders who may be indirectly affected by the outcomes of a proposed intervention.

Different stakeholders derive different benefits. Therefore, it is critical to explicitly integrate these multiple perspectives in policy-making [36].

Clausen et al. [9] define three categories of stakeholders, depending on the benefits and the services offered: primary stakeholders (industry and consumers); secondary stakeholders (NGOs, researchers and local authorities); tertiary stakeholders (governmental bodies, national authorities and international authorities). Besides this classification, it is generally acknowledged that the community involvement and participation in the management of natural resources is a condition of their sustainable use [15]. Moreover, it is relevant to link the scientific and technical knowledge of experts with the viewpoints expressed by local actors [37, 38].

Janse van Rensberg [39] also remarked that analysis of stakeholders' participation involves analysis of complex social behaviour or action. The degree of stakeholder's involvement is a critical point in stakeholder participation because it influences all the processes. The primary stakeholders have high importance but varying degrees of influence, whilst the major European institutions have varying degrees of importance and high influence [40].

3 Status Changes of the Lakes in the Study Area and Protection Strategies

3.1 Study Area: The Danube Floodplain, the Drobeta-Turnu Severin—Bechet Sector

Floodplains provide a multitude of ecosystem services, particularly including the aquatic surfaces and the landscape induced by them [2]. The Romanian Danube Floodplain covers a surface of about 530.5 thousand hectares between Gruia (downstream Iron Gates II) – kilometre 851 and Isaccea – kilometre 108 [45]. Within major river floodplains, lakes and wetlands are very important for the communities who live around them; thus, the values that people attribute to lakes and the effects of their management strategies are critical. Along the Lower Danube Floodplain, a tremendous lake reduction of floodplain lakes has occurred in the last 50 years and this loss continues due to pressures such as land use change, river regulation, and dam construction [46].

The study area corresponds to the Lower Danube Floodplain located between Drobeta-Turnu Severin and Bechet urban settlements, and is bounded by administrative criteria (within the South Oltenia Region), spanning a total of 29 territorial-administrative units (LAU level 2, according to Eurostat). Taking into account the external points of riparian LAU 2 in this sector, the river floodplain extension covers a length of about 260 kms, from fluvial kilometre 937 to fluvial kilometre 677 and covers an area of about 83,000 hectares [47]. In the sector under study, the Danube

Floodplain displays great transversal variability, both in terms of width and anthropogenic changes undergone by the unit in a quasi-natural state. Thus, in the sector located upstream of Calafat town, when it is observable, the floodplain reaches only neglectable widths and it extends more significantly near former clogged branches of the Danube, where lakes and ponds locally appear. The maximum width of the floodplain characterizes the sector downstream of Calafat, where this unit exceeds 10 km near the Bistreț Lake. As a result of hydrotechnical works on the Danube and on its adjacent space, the area under the former direct hydrodynamic influence of the river, the connections between watercourse and its floodplain in quasi-natural state, as well as the micromorphology of the wide floodplain in the sector located downstream of Calafat were significantly modified through longitudinal and partitioning dams and through a locally dense network of canals (there is a succession of dammed precincts with such characteristics eastwards of Ghidici settlement). The deeply rural character of the study area is obvious, as it is dominated by rural local administrative units (26), to which there are to be added three urban settlements: Drobeta-Turnu Severin, Calafat and Bechet, the latter lacking municipal status. Few settlements are located in the floodplain proper, while a chain of villages of average and large demographic size (as classified in the Romanian literature) border the northern floodplain – terrace contact [47].

The predominant morphological character of the sector under study is the presence of the floodplain and the Danube terraces, supplemented by the specific aeolian relief [48]. Roughly 60% of the study area corresponds to the floodplain of the Danube and of its main tributaries (the Topolnița, the Blahnița, the Drincea, the Desnățui, the Jiu, and the Jieț), while the rest belongs to the terrace plains [49].

Distinct taxonomic groups of the lakes in the Danube floodplain are mainly imposed by the water supply specificity, namely their area increasing with depth during the high-level waters of the river. More precisely, there can be identified the following categories of lakes in the sector under scrutiny [47, 48]:

- lakes whose existence is determined by permanent connection with the main river, either under the effect of high waters (floods or overflows) or by secondary arms, ravines, old abandoned arms, or by underground flow when the waters of the main river have high levels (Ciuperceni lake and Bistreț lake);
- lakes that appeared on site of former meanders or abandoned branches of the Danube (e.g. Gârla Mare, Maglavit and Golenţi lakes) (Fig. 2).

The relief of the study area is the result of the action exerted by three categories of factors: the fluvial processes that generate and modify the floodplain unit, the aeolian processes that have important poignancy within this sector and the human factor, whose influence has been very significant during the last decades [50].

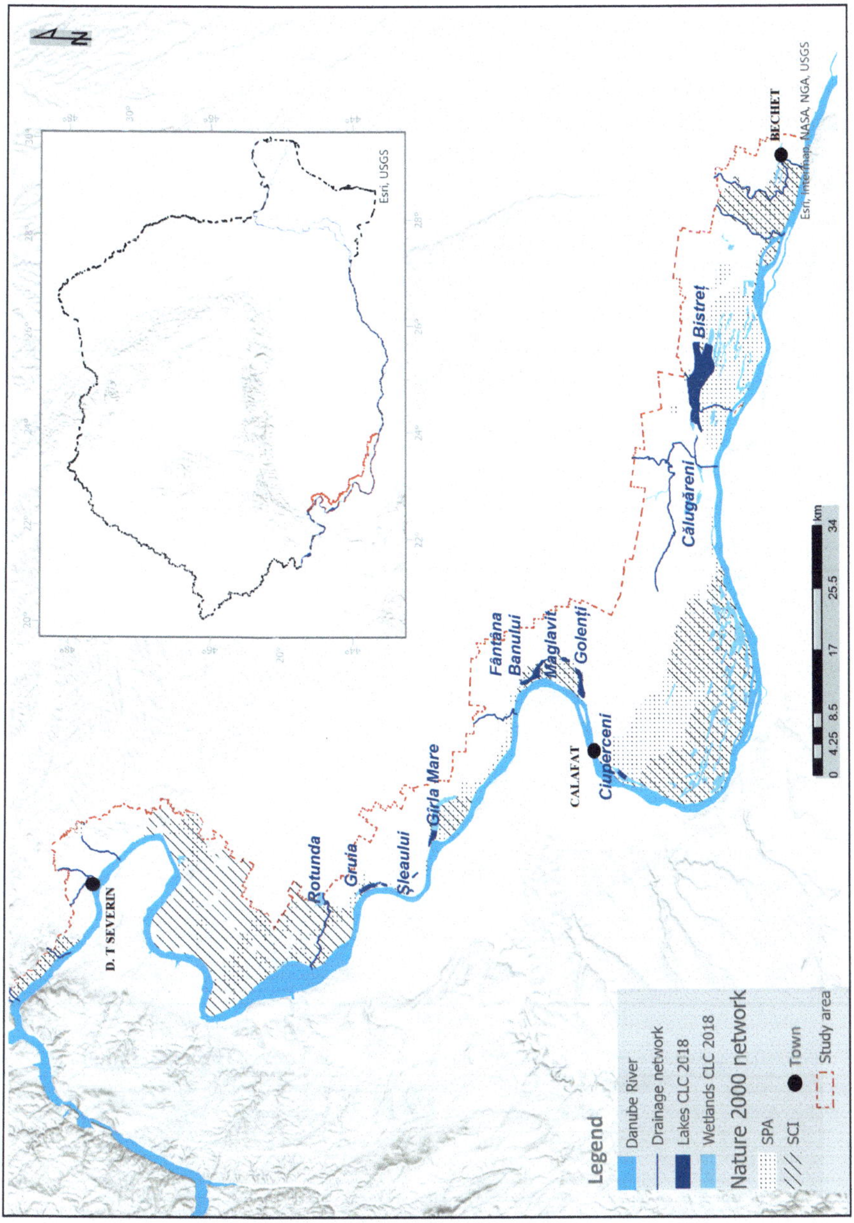

Fig. 2 Study area: location, lakes and protected areas. source of the processed data: [51, 52]

3.2 Protection and Conservation of Aquatic Ecosystems

Wetlands' conversion into arable lands was meant to generate economic development through the practice of an intensive agriculture, but their most important ecological functions are nowadays being rediscovered as they represent habitats for numerous endangered or vulnerable species as well as hydrologic polders, and areas that could mitigate floods, with the latter role becoming ever more critical in the present context of climate change. Hence, as a consequence of more and more frequent meteorological and hydrological phenomena, the tendency nowadays is to emphasize the protective benefits of natural areas such as the Danube Floodplain, giving a particular importance to stakeholders' opinions and supporting the development of projects that acknowledge natural areas' fundamental ecological functions [53].

Being considered genuine biodiversity hotspots, both permanent and temporary aquatic habitats within river floodplains stand out and have great importance for species richness in riverine landscapes as their diversity is often higher than in the fringing landscape [1].

Most of the analysed territory is SCI and/or SPA under Natura 2000 and comprises a mosaic of aquatic biotopes, which explains a high biodiversity including species and habitats of great importance [54]:

5 Sites of Community Importance:

- Maglavit, ROSPA0074 (Fig. 3);
- Calafat-Ciuperceni-Dunăre, ROSPA0013;
- Bistreț, ROSPA0010 (Fig. 4);
- Confluența Jiu—Dunăre, ROSPA0023;
- Coridorul Jiului, ROSCI0045) and

4 Special Protection Areas for birds:

Fig. 3 Golenți lake (Photo by Mititelu, May 2017)

Fig. 4 Bistreț lake (Photo by Licurici, June 2018)

Fig. 5 Gruia lake (Photo by Licurici, July 2009)

- Blahnița, ROSPA0011;
- Dunărea la Gârla Mare—Maglavit, ROSCI0299 (Fig. 5);
- Gruia—Gârla Mare, ROSPA0046;
- Ciuperceni—Desa, ROSCI0039) (see Fig. 2).

There are some differences in the values attributed to wetlands by the local population, as compared to those of the scientific community [55]. In spite of the potential to establish a synergetic relationship between community and preservation, within the protected areas there are often conflicts between local economic interests and biodiversity conservation [56, 57].

3.3 Overall Context of Hydrological and Land Use Changes. Floodplain Restoration

Although there is a wide appreciation of the ecosystem services provided by floodplains and their vital role in riverine landscapes, there have been dramatic losses of floodplain habitat due to land reclamation and channel engineering, resulting in a functional degradation of these systems worldwide [2]. The reduction and degradation of floodplain systems has diminished their capacity for water retention, thus enhancing flood risks, while other key floodplain functions and services, such as groundwater replenishment, nutrient storage and water purification have also declined in effectiveness [58]. These problems are particularly important for the Danube Floodplain.

In Romania, following the major social, economic and political shift that all South-Eastern countries have undergone after 1990, there has also been a significant change in the perspective of ecological management of its fragile ecosystems, such as those in the Danube Floodplain.

During the communist period, the Danube Floodplain sector comprised between Drobeta-Turnu Severin and Bechet was fundamentally impacted by the hydro-ameliorative extensive works of drainage and damming, especially between 1960–1980. Since the main motivation was an economic one, the management and decisional role exclusively pertained at that time to the central public administration, without the population having a consultative or decision-making role of any kind, except for the specialists working in the institutions involved in scientific research targeting this particular area. While the protection of lakes and wetlands belonging to the Danube Floodplain has become a priority in the context of necessary compliance with the European standards, the decisional aspect has been significantly improved in recent decades, but a consensus between the existing stakeholders still fails to emerge [53].

Hence, the increasing demands for extending settlements and agricultural lands have resulted in large-scale river regulation measures for flood protection. According to ICPDR [59], about 39% or 1,111 river km of the entire Danube are impounded by a total of 78 dams. As a result, more than 68% of the active floodplains of the Danube River, which are in frequent exchange with the main river channel, have already been lost, yet an important part of this active floodplain surfaces resides within the Lower Danube, in Romania. Ecological restoration of the floodplain area and economic interests of the local population has to come to terms.

According to international research studies [2], the potential area for floodplain restoration based on land use and hydro-morphological characteristics amounts to 8102 sq km for the whole Danube River, of which estimated 75% have a high restoration potential. The problem is represented by the fact that floodplain restoration is profoundly limited by stakeholders' needs, education level and awareness, acceptance and resource availability.

Since the Romanian South-Western sector of the Danube Floodplain has traditionally been known for its sandy soils and droughty climate, while being a subject area for intense human interventions, several research studies suggested that the aquatic areas of the lakes, main watercourses and wetlands within the Danube Floodplain have suffered one of the most unfavourable dynamics during the last decades. Thus, several anthropogenic changes associated with land use dynamics negatively impacted and supplementary contributed to aridization and ecological destabilization of the Danube floodplain: changes in property type and excessive fragmentation of agricultural land, predominance of subsistence individual farms, poor development of mechanization and fertilization, abandonment or destruction of the irrigation systems, deforestation, a significant increase in built-up and agricultural areas such as pastures and natural hay-field areas extended to the detriment of arable land, orchards or currently abandoned vineyards considered due to poor maintenance and low productivity [60, 61]. Moreover, further negative consequences are nowadays induced by the impact of frequent extreme climatic phenomena. Several recent research studies monitoring the hydro-morphological parameters of the floodplain lakes suggested that overall ecological conditions of the lakes are negatively impacted during recent years so appropriate measures should contribute to restoring the natural hydrological balance in the area [62–64]). In this respect, both aggressive anthropic activities that substantially changed the regime of the Danube Floodplain (the Danube dams—Iron Gates I and II, interior channels within the floodplain) and the negative consequences of more and more visible climatic changes, with a local component of rather frequent periods of prolonged drought, all contributed to the critical situation in matters of hydrological stability and ecological sustainability [65–67].

Nevertheless, restoration is often faced with conflicting socio-economic demands regarding the use of both the river channel and its floodplains. Settlements, agriculture, forestry, hydropower generation, navigation, the economic development of the catchment, but also nature protection are among the interests that commonly oppose floodplains being allowed to regain their natural flooding dynamics [68, 69]. As a consequence, project managers have to deal with various stakeholders and institutions of different legal status, originating in different sectors and, thus, often with differing and sometimes conflicting priorities. Tackling these problems may not only delay, but more often restrict or even prevent large-scale restoration measures in floodplains. However, traditional river engineering measures for providing adequate fairway parameters, such as dredging, diking or damming, are incompatible with the conservation or restoration of ecologically sensitive river stretches and floodplains [2].

Unfortunately, contemporary socio-demographical risks such as population ageing, structural changes in the workforce, the migration of the active population, as well as cultural and economic disruptions create a dominant impression in the studied sector and further contribute to delaying sustainable development of the

Danube Floodplain. As stated in the following, given the current ecological status of the ecosystem, direct involvement of stakeholders is still low due to lack of information and insufficient understanding of long-term benefits that could derive as a result of aquatic ecosystems sustainable development.

Given this context, freshwater management should no longer be just about responding to the adverse consequences and implications of quasi-present climate changes, but more about identifying solutions, opportunities and mechanisms to mitigate source problems and at the same time, to contribute to the well-being, safety and long-term resilience and development of riparian communities.

Due to the biological diversity of the lacustrine protected area, and to the spatial–temporal evolution of aquatic surfaces within the Romanian Danube Floodplain, we considered necessary a stakeholder analysis at local and regional level in order to establish their involvement and environmental sustainability.

4 A Survey on Stakeholders' Interests and Participation in the Sustainable Use of the Lower Danube Floodplain Lakes, in Drobeta-Turnu Severin—Bechet Sector

4.1 Profile of Selected Stakeholders

The identification of the most significant stakeholder types for the present study started from a broad documentation concerning the persons, groups and institutions with particular experience and interests in the Danube Floodplain and particularly in the floodplain lakes; these stakeholders could influence the interventions within the area or are directly influenced by such implemented decisions [8]. The stakeholder identification and analysis represent a useful approach to assess the stakes of interested participants in a system in more detail [7, 27].

For the present study, the sample of respondents is represented by 47 persons who fall into four major classes of stakeholders with various motivations, interests, perspectives and opportunities to influence the sustainable use of the lakes in the Danube Floodplain. Starting from the *quadruple helix* approach [41, 42, 70], the interviewees were included in the following areas of interest and influence: public institutions with decision/administrative functions (23.4% of the total respondents), education and/or research institutions (42.6%), organizations with economic profile (4.3%) and civil society (14.9%) (Fig. 6). A distinct category (14.9%) groups occasional users of floodplain resources and representatives of other institutions with tangential interests in the study area, as well as institutions that issue or disseminate important information for the sustainable capitalization of lakes within the study area (Oltenia Regional Meteorological Center, Oltenia Museum in Craiova, etc.).

The first category includes representatives of the institutions that hold competence and territorial jurisdiction within the study area, being characterized as directly interested stakeholders, with significant decision-making power (authorities of the

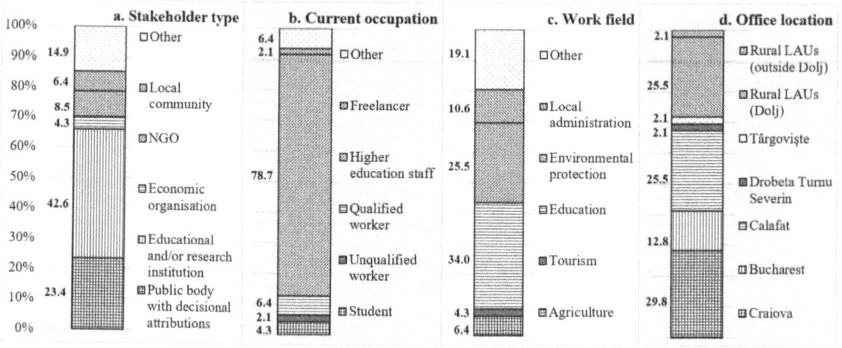

Fig. 6 The main characteristics of stakeholders' categorization. *Note* Due to rounding, some totals are slightly different from 100%

local administrative units/LAU2 located along the Danube, the National Agency of Natural Protected Areas, the National Administration *Romanian Waters* - ABA Jiu). The second category encompasses knowledge providers, being dominated by professionals from schools and universities, research centres and institutes that conduct their activities near the study area or have carried out projects concerning the floodplain (local schools, the Universities of Craiova, Bucharest, and Târgoviște, the Geography Institute of the Romanian Academy). Although it has no decision-making power, this category plays a very important part in developing and supplying information and models for sustainable development and, thus, it should be regularly involved and consulted. The last two categories include representatives of economic bodies with agricultural and fishery profile, as well as members of local communities and representatives of NGOs with activities in the study area. Through the activities they conduct, these categories can significantly influence the ecological state of the lakes within the Danube Floodplain, which means that raising their level of information, motivation and involvement in decision-making can have beneficial influences for sustainable development.

Because in addition to the institutional framework, the level of information and individual motivation for involvement in the sustainable development of the area can be significantly influenced by some socio-demographic characteristics, the sample structure included respondents from 19 to 70 years of age; although diverse, the sample is dominated by the 40–50 years contingent, which accounts for 42.6% of the total and is represented by active people, with experience in their fields of activity. Most of the selected persons have completed their university or postgraduate education (53.2% and, respectively, 36.2%), which is also reflected in their occupational profile dominated by higher education staff (78.7%) and skilled workers (6.4%). A drawback related to the sample on which the survey was conducted concerns the relatively low participation of respondents from local authorities and members of local communities (which are direct beneficiaries in the process of sustainable capitalization of lakes), as compared to other stakeholders (the academic or research community, for example).

4.2 General Framework of the Survey

4.2.1 Issues Addressed by the Questionnaire

In spite of the existing debate regarding the relative usefulness of data-gathering through interviews [71], some qualitative information cannot be gathered in any other way [34]. The subjective perspective gained during the interviews proved very important here to better understand the factors and considerations underlying the management of the study area.

The present study is based on a survey that involved the most important types of stakeholders with various interests in the Danube Floodplain, the Drobeta-Turnu Severin - Bechet sector, respectively. The survey was conducted by using a standardized questionnaire to test the degree of knowledge and participation in the sustainable capitalization of lakes within the case study area. The questionnaire was applied online, between August and October 2020, to a number of 47 respondents.

The questionnaire included a series of 25 questions, 18 of which specifically dealt with the above-mentioned issue, representing the core of the survey, while 7 questions aimed at better understanding the profile of the stakeholders who participated in the study. The core questions were designed to initially assess the general level of knowledge concerning the current state of the lakes in the study area, the main possibilities for information and the most important features of the relationships and possible collaborations between different actors interested in using these lakes. The second section of the questionnaire concerns the institutional or personal experience in activities that included the lakes in the area, as well as the perceived effects of these activities. The last section of the questionnaire involves certain qualitative assessments on lake dynamics, capitalization and management, requesting information on the most important obstacles, as well as opportunities perceived by respondents in relation to the sustainable use of the lakes within the analysed Danube Floodplain sector.

4.2.2 Types of Questions

Four main approaches were used for the design of the questions situated at the core of the survey, so that, on the one hand, the answers could clearly highlight the opinions and values of the respondents, and, on the other hand, the method would correlate with a low rejection rate. Besides the questions characterized by closed answer (nine grids with a single allowed answer and a grid with multiple possible options, respectively), there are also three questions with free answer requirement regarding particular examples of information concerning sustainably capitalized lakes or water bodies degraded as a result of unsustainable practices, as well as regarding the type and degree of personal involvement. Given the diversity of stakeholders and of their interests, which are sometimes contradictory, during the last part of the questionnaire the subjects were asked to assess the most important issues that prevent the sustainable capitalization of lakes in the case study area.

In order to obtain multiple evaluations from the respondents, alongside with the optimization of the questionnaire, two scale questions and three evaluation matrix questions were included.

4.3 Analysis of the Results

4.3.1 Information Level and Knowledge Opportunities for Stakeholders

Only a third of the respondents (namely 34% of the total) consider themselves highly informed about the state and management of lakes in the Danube Floodplain, Drobeta-Turnu Severin - Bechet sector, the dominant share being held by people not at all informed or with little knowledge about the matter (38.4%). From the viewpoint of the respondents involved in the survey, the most significant manners of information are represented by personal theoretical documentation, using written and/or video sources (44.7%), or by direct knowledge acquisition within the study area (25.5%) (Fig. 7).

In this framework, the irregular character of the collaboration between the stakeholders interested in the sustainable capitalization of the lakes within the studied area is outlined, given the fact that even knowledge dissemination between different institutional and/or social partners rarely takes place or does not take place at all (the two categories mentioned accounting for 61.7% of the total responses). However, there is interest in the use of participatory information formulas at stakeholder level, with

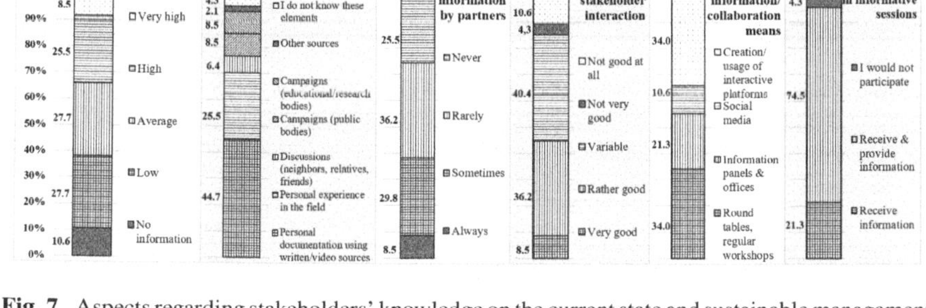

Fig. 7 Aspects regarding stakeholders' knowledge on the current state and sustainable management of lakes in the Danube Floodplain. *Note* Due to rounding, some totals are slightly different from 100%

the survey highlighting both the importance of traditional means, such as regular workshops/round tables (34% of the responses), and the use of less conventional means, such as specialized interactive platforms (34%). 74.5% of the respondents stated that they would participate in briefings and debates on the current state and management of the lakes within the Danube Floodplain, being motivated both to receive and to contribute information, while only 21.3% of them would like only to be informed, and 4.3% would not participate.

4.3.2 Stakeholders' Participation

In addition to a sound scientific background, the process of making the most appropriate integrated decisions regarding the sustainable management of the lakes in the Danube Floodplain—part of a complex natural and anthropogenic mosaic—requires the experience and knowledge of stakeholders directly involved in local activities. About half of the participants in the survey (i.e. 51.1%) share this characteristic, the manifold types of activities in which they have been involved emphasizing, on the one hand, the diversity of actors with interests in the area, and, on the other hand, the complexity of the relationships between factors that must be considered for the sustainable management of the area in general and of the floodplain lakes in particular.

The activities directly related to the management of lakes and protected areas within the study area account for 12.5% of the responses offered by the involved stakeholders (Fig. 8), to which is added a significant volume of interdisciplinary research, often accompanied by monitoring activities (58.3%). The manifold areas of expertise characteristic to stakeholders that might be involved in decision-making on this area is underlined by the types of studies they have conducted, weather specialized (addressing hydrological analyses and impact assessments of hydrotechnical constructions, research—monitoring of floodplain fauna, especially of the avifauna for which the lakes play an important part, studies on the meteorological-climatic

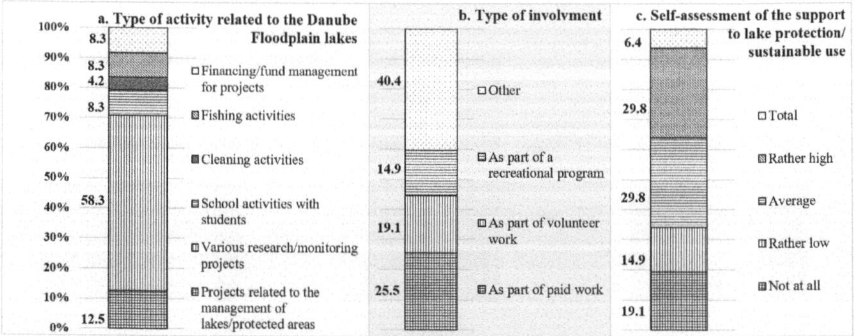

Fig. 8 Stakeholders' participation in activities involving lakes in the Danube Floodplain. *Note* Due to rounding, some totals are slightly different from 100%

parameters, etc.), but especially the integrative ones, which aimed at evaluations of the ecosystem services, studies on the possibilities for wetland restauration and biodiversity conservation within the Danube Floodplain, or on the mitigation of the effects induced by natural and anthropogenic hazards in an area with significant natural potential, but marked by socio-economic vulnerabilities. Stakeholders' interest in this area is highlighted both by the typological diversity of activities conducted in the area and by the voluntary involvement of respondents (34% of them stated that they included the lakes in the Danube Floodplain in their volunteer or recreational programs) (Fig. 8).

Although smaller individual projects can have positive effects on the state of the lakes within the study area (36.2% of respondents consider that they have fully or largely managed to support their protection and sustainable capitalization), there is also perceived a rather high failure rate associated with these activities (19.1% of the answers pointing to a total lack of support towards lake protection). Capitalizing on the expertise supplied by research and education stakeholders, as well as considering the interests of civil society stakeholders in the decision-making process and implemented projects, along with monitoring the effects of these decisions could help reduce the failure rate. This outcome could be attained as public participation affords stakeholders the opportunity to influence decisions that affect their lives and it could explain one of the reasons why, when soundly implemented as a process and not as a single event, it becomes dependable and increasingly important in environmental decision-making [3, 27].

4.3.3 Stakeholders' Perspectives on the Sustainable Capitalization of Lakes

In order to highlight stakeholders' perception on the dynamics and current state of the lakes in the studied area of the Danube Floodplain, as well as on the main factors influencing the possibilities of sustainable capitalization, both evaluation scales and questions with open answer were used.

The analysis of the answers provided during survey highlights the significant agreement on the negative dynamics of the floodplain lakes over the last two decades, both in terms of extension (78.7% of the respondents agreeing or fully agreeing on this issue) and from the viewpoint of an increased pollution level (70.2% of the respondents) (Fig. 9). Under these conditions, the main acknowledged causes for lake deterioration are primarily anthropogenic (57.4% of the interviewed stakeholders considering the declining state of the lakes as a consequence of illegal waste disposal and/or chemical spills, of the excessive use of local resources and of the inappropriate tourism practices) or complex (a combination of natural and anthropogenic factors is highlighted by 31.9% of respondents as triggers of prolonged droughts and floods that negatively affect the state of these lake ecosystems). Moreover, lake degradation is perceived as an important endangerment factor for certain wildlife species and attracts significant losses in ecosystem services.

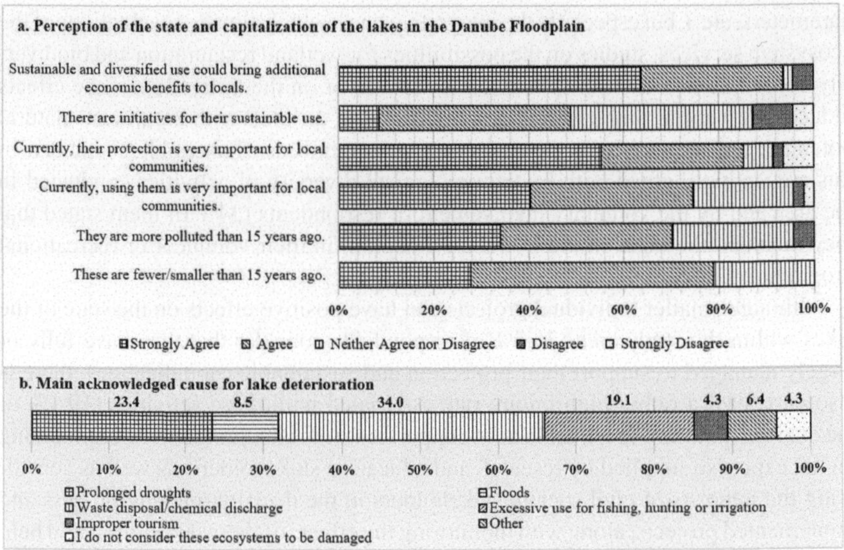

Fig. 9 Aspects of stakeholders' perception related to past and present state of the Danube Floodplain lakes

A significant part of the respondents perceives lake degradation as a complex result of the increasing anthropogenic pressure on the floodplain environment, but also of a poor management in the absence of a coherent, integrative, properly financeable strategy and of adequate informing of stakeholders with local interests. Although there is an almost general agreement that the sustainable and diversified capitalization of the floodplain lakes could bring additional economic benefits to local communities (93.6% of responses), less than half of the respondents agree that there are current initiatives to trigger this outcome (Fig. 9). The examples provided by respondents on sustainably capitalized or, on the contrary, degraded lakes suggest that the mere inclusion of these water bodies in the European Ecological Network Natura 2000—although recognized—is not enough, requiring, among other things, managerial flexibility in the selection of developed and implemented measures, depending on the specific local conditions of each lake and supported by sound knowledge (which can be achieved based on the expertise and information supplied by different stakeholders).

The diversity of opinions on the future directions for sustainable capitalization of the lakes—and, in a broader sense, of the Danube Floodplain—highlights the importance of informing and co-involving stakeholders with different capabilities and interests in the area so that the decisions made for the restoration or conservation of the natural capital could balance those concerning the development and diversification of economic capitalization for the benefit of local communities. In this framework, the main direction acknowledged by 40.4% of the respondents regards the rehabilitation activities, the adequate restoration of connectivity between river and floodplain, as it would have multiple environmental and socio-economic benefits (recovery of specific floodplain habitats, restoration of aquifers, flood risk mitigation, etc.). An important direction for the sustainable development of the region concerns the capitalization of the tourist potential of the floodplain lakes (14.9% of the respondents). On the other hand, the fact that an identic part of the answers suggests the necessity of conservative measures put in place for the aquatic biodiversity, as well as the recognition of improper tourism as a contributor to lake deterioration and stress on the aquatic bird species within the area underline that stakeholders need to be bought at the same discussion table in order to obtain a coherent and realistic management perspective. As the use of the natural capital belonging the floodplain is conducted by different stakeholders, its protection should also be multidirectional, a context in which disseminating knowledge and raising awareness and interest in the local communities play a significant part, as recognized by 29.8% of respondents.

The main problematic issues highlighted as priorities for a sustainable management of floodplain lakes require the involvement of stakeholders with different profiles and fields of activity, as their solving is not possible unidirectionally. Thus, although about 90% of respondents agree or strongly agree that there should be a greater involvement of public institutions (weather local, regional, or central ones), with decision-making power, in the sustainable capitalization of lakes (Fig. 10), this type of stakeholder does not always have the necessary resources or expertise for developing the best strategies. Local and regional decision-makers could also play an important part as coordinators of multi-stakeholder activities aimed at clarifying

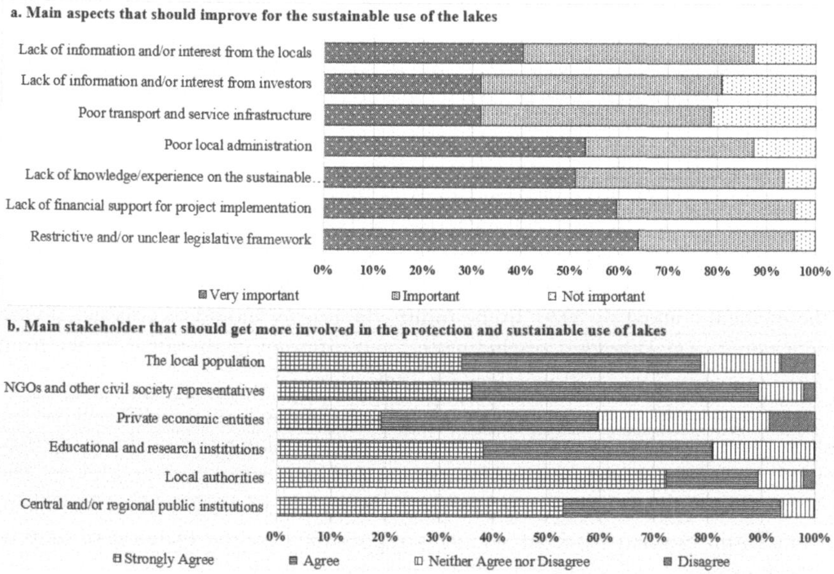

Fig. 10 Stakeholders' perspectives on the future state of the Danube Floodplain lakes—protection and sustainable capitalization

the legislative framework, outlining a solid, up-to-date database on local natural and human resources, as well as flows between them, defining a coherent strategy, with concrete financing possibilities, taking over, developing and communicating the most suitable examples of good practice for the local communities in a form adapted to their particular conditions, etc.

The analysis of the answers obtained by applying the present questionnaire confirms the diversity of stakeholder profiles with interests in the researched Danube Floodplain, of their perspectives on the area, and the fact that, in this framework, the development of final decisions only by the authorities is not beneficial. The study suggests that there should be a frequent multi-stakeholder participatory communication, which is as important as obtaining information [27, 72]. In this manner, decisions will have wider support, while reiterated participatory communication during their implementation will require feedback from final beneficiaries (representatives of the civil society, for example) and managerial flexibility.

4.4 Discussion and Recommendation

The analysis of stakeholders' interests and participation in the sustainable use of the lakes along the Danube Floodplain has a restrictive part, namely the online research was conducted in the identification process. Through the use of the survey and document analysis the number of internal stakeholders interviewed was limited. For example, difficulties were encountered in the willingness of local authorities to respond to the questionnaire. In this sense, certain limitations of the research were induced by the fact that the local authorities or other local stakeholders are not used to communicate electronically; as such, in multiple cases, either no email address is publicly available or the completion of the questionnaire was delayed. In the case of the economic sector and the regional authorities, there were difficulties in communication and completion of the questionnaire because of the conflict of interests regarding the custody of protected areas and implicitly their management.

The opportunity to further interview stakeholders after or during the completion of the questionnaire and to add a face-to-face component of the survey was hindered by the medical context of 2020. In the future, interviews should be conducted with a wider range of stakeholders, including those at different institutional levels, from the local municipality to regional decision-makers.

Nevertheless, the answers provided by the participants in the present survey confirm the authors' judgement regarding the negative dynamics of the lakes over the last two decades, both in terms of the extent and increase of pollution levels. The main cause recognized for the degradation of lakes within the Danube Floodplain, Drobeta-Turnu Severin—Bechet sector, is represented by the increase of anthropogenic presence, to which is added the faulty management in the context of Natura 2000 European Ecological Network and of a lacking coherent strategy. The answers to the last section of the survey, namely the information provided by stakeholders on their relationships and on the most important obstacles and opportunities perceived in the sustainable capitalization of lakes within the study sector of the Danube Floodplain, surprised and raised some questions. Thus, most respondents agree or strongly agree that there should be greater involvement of public institutions in the sustainable use of lakes.

Starting from these responses provided by stakeholders, it is recommended to organize on-site discussions with the active participation of public authorities, administrators of protected areas and landowners or persons/groups that conduct activities impacting the lakes. Moreover, an online platform can offer the possibility to all interested stakeholders to express their opinions supported by arguments. The development and use of such a platform are all the more important as it often happens that protected area administrators or locals in the analysed areas know more about the real situation than specialists from universities, museums, institutes of research, etc.

5 Conclusion

The choice of the approached issue started from the premise that stakeholder input and involvement may facilitate the introduction and implementation of European sustainability standards within the management plans of the lacustrine protected areas located in the Danube Floodplain. Stakeholders' classification and selection (public institutions with decision/administrative functions, education and/or research institutions, organizations with economic profile and civil society, as well as other institutions with tangential interests in the study area) was an important first step in establishing the degree of involvement and the role in the sustainable management of lakes within the Danube Floodplain, the Drobeta-Turnu Severin - Bechet sector. The qualitative analysis and the quantitative evaluation of the stakeholders were corroborated according to the field literature through the standardized questionnaire, which aimed at testing the degree of awareness and involvement in the sustainable use of the lakes in the case study area.

The relatively low participation of the respondents from local authorities and of local community members, as compared to other stakeholders, is suggestive because the results of the analysis show that the two types of stakeholders should play more active and significant parts in participatory communication and decision-making. The actions must be scientifically substantiated but, in the framework of possible gaps between theory and practice, it is necessary to involve direct beneficiaries, who can provide permanent feedback on local particularities.

The questions clarified aspects regarding the initial assessment of the general knowledge concerning the current lake status within the study area, the institutional or personal experience, and the qualitative assessments on lake dynamics, capitalization and management. The results confirmed the diversity of profiles associated with stakeholders within the Danube Floodplain, their perspectives on the sector under study, as well as the fact that, in this framework, the development of final decisions only by the authorities is not beneficial. Overall, the research ascertains the need for participatory multi-stakeholder communication for the sustainable use of the Lower

Danube Floodplain lakes. Thus, in order to continue the participatory communication during the implementation of future decisions, it is proposed to encourage and ensure feedback provision from the final beneficiaries (for example, from the representatives of civil society).

Finally, the study underlines the need to provide the local and regional public institutions with decision-making power with a clearer legislative framework corresponding to an adaptive management, in addition to material resources and expertise in decision-making.

Acknowledgements The authors address special thanks to the involved stakeholders for filling in the questionnaire and for the invaluable information that they shared.

Author Contributions All authors contributed equally to this work.

References

1. Pander J, Mueller M, Geist J (2018) Habitat diversity and connectivity govern the conservation value of restored aquatic floodplain habitats. Biol Conserv 217:1–10. https://doi.org/10.1016/j.biocon.2017.10.024
2. Hein T, Schwarz U, Habersack H, Nichersu I, Preiner S, Willby N, Weigelhofer G (2016) Current status and restoration options for floodplains along the Danube River. Sci Total Environ 543:778–790. https://doi.org/10.1016/j.scitotenv.2015.09.073
3. Richardson BJ, Razzaque J (2006) Public participation in environmental decision making. Environmental law for sustainability 6:165–194
4. Directive 2000/60/EC of October 23 2000 of the European Parliament and of the Council establishing a framework for community action in the field of water policy Off J Eur Comm, L327 (2000), pp 1–72
5. Bouahim S, Rhazi L, Ernoul L, Mathevet R, Amami B, Er-Riyahi S, Muller SD, Grillas P (2015) Combining vulnerability analysis and perceptions of ecosystem services in sensitive landscapes: A case from western Moroccan temporary wetlands. J Nat Conserv 27(1–9). https://doi.org/10.1016/j.jnc.2015.05.003
6. Selman P (2004) Community participation in the planning and management of cultural landscape. J Environ Plan Manag 47:365–392. https://doi.org/10.1080/0964056042000216519
7. Grimble R, Wellard K (1997) Stakeholder's methodologies in natural resources management: A review of principles, contexts, experiences and opportunities. Agric Syst 55(2):173–193. https://doi.org/10.1016/S0308-521X(97)00006-1
8. Gevers I, Koopmanschap E (2012) Enhancing the wise use of wetlands: a framework for capacity development. UR Centre for Development Innovation, Wageningen. http://wetlands.un-ihe.org/sites/wetlands.un-ihe.org/files/cop11-doc34-e-capacity.pdf
9. Clausen LPW, Hansen OFH, Oturai NB et al (2020) Stakeholder analysis with regard to a recent European restriction proposal on microplastics. PLoS ONE 15(6):e0235062. https://doi.org/10.1371/journal.pone.0235062
10. King CS, Feltey KM, Susel, BO'N, (1998) The question of participation: Toward authentic public participation in public administration. Public Adm Rev 58(4):317–326. https://doi.org/10.2307/977561
11. Stanghellini PS, Collentine D (2008) Stakeholder discourse and water management. Implementation of the participatory model CATCH in a Northern Italian alpine sub-catchment. Hydrol Earth Syst Sci 12(1):317–331. https://doi.org/10.5194/hess-12-317-2008

12. Blackstock KL, Kelly G, Horsey B (2007) Developing and applying a framework to evaluate participatory research for sustainability. Ecol econ 60(4):726–742. https://doi.org/10.1016/j.ecolecon.2006.05.014
13. Arnstein SR (1969) A ladder of citizen participation. J Am Inst Plann 35(4):216–224. https://doi.org/10.1080/01944366908977225
14. Olsson JA, Berg K (2005) Local stakeholders' acceptance of model-generated data used as a communication tool in water management: The Rönneå study. AMBIO: A J Hum Environ 34(7):507–512. https://doi.org/10.1579/0044-7447-34.7.507
15. Darradi Y, Grelot F, Morardet S (2006) Analysing stakeholders for sustainable wetland management in the Limpopo River basin: the case of Ga-Mampa wetland, South Africa. In Communication to the 7th WATERNET/WARFSA/GWP-SA Symposium "Mainstreaming IWRM in the Development Process". Lilongwe, Malawi
16. Luyet V, Schlaepfer R, Parlange MB, Buttler A (2012) A framework to implement stakeholder participation in environmental projects. J Environ Manage 111:213–219. https://doi.org/10.1016/j.jenvman.2012.06.026
17. Reed MS, Graves A, Dandy N et al (2009) Who's in and why? A typology of stakeholder analysis methods for natural resource management. J Environ Manage 90(5):1933–1949. https://doi.org/10.1016/j.jenvman.2009.01.001
18. Bryson JM (2003) What to do when stakeholders matter: A guide to stakeholder identification and analysis techniques. A paper presented at the London School of Economics and Political Science
19. Van Asselt M, Mellors J, Rijkens-Klomp N, Greeuw S, Molendijk K, Jelle Beers P, Van Notten P (2001) Building blocks for participation in integrated assessment: A review of participatory methods. Project FIRMA. ICIS, Maastricht. https://www.ircwash.org/sites/default/files/152-01BU-17379.pdf
20. Morris J, Baddache F (2012) Back to basics: How to make stakeholder engagement meaningful for your company. BSR publishes, London. Available at https://www.bsr.org/
21. World Bank (2003) Social analysis sourcebook: Incorporating social dimensions into bank-supported projects. Social Development Department, Washington, DC
22. Junk WM (2016) Two logics of NGO advocacy: understanding inside and outside lobbying on EU environmental policies. J Eur Public Policy 23(2):236–254. https://doi.org/10.1080/13501763.2015.1041416
23. Schmeer K (1999) Stakeholder analysis guidelines. Policy toolkit for strengthening health sector reform. http://msalamkhan.buet.ac.bd/teaching_msk_files/Stakeholder%20analysis%20guidelines%20.pdf
24. Van Dam J, Junginger M (2011) Striving to further harmonization of sustainability criteria for bioenergy in Europe: Recommendations from a stakeholder questionnaire. Energy Policy 39(7):4051–4066. https://doi.org/10.1016/j.enpol.2011.03.022
25. Jonsson A, Olsson JA (2005) Participatory Modelling: (How) can computer generated information affect the "room of action" of local stakeholders? In: Kulturstudier i Sverige. Nationell forskarkonferens 015:507–526, Linköping University
26. Cuppen E, Breukers S, Hisschemöller M et al (2010) Q methodology to select participants for a stakeholder dialogue on energy options from biomass in the Netherlands. Ecol Econ 69(3):579–591. https://doi.org/10.1016/j.ecolecon.2009.09.005
27. Smrekar A, Polajnar Horvat K, Ribeiro D (2020) Stakeholder analysis for (Mediterranean) Wetland Governance: The Case of Ljubljansko Barje Nature Park, Slovenia. In Nared J, Bole D (eds) Participatory research and planning in practice, The Urban Book Series, Springer Open, pp 169–184. https://doi.org/10.1007/978-3-030-28014-7_11
28. Vroom VH (2003) Educating managers for decision making and leadership. Manag Decis 41(10):968–978. https://doi.org/10.1108/00251740310509490
29. Maguire LA, Lind EA (2003) Public participation in environmental decisions: stakeholders, authorities and procedural justice. Int J Glob Environ Issues 3(2):133–148. https://doi.org/10.1504/IJGENVI.2003.003861

30. Silima V (2007) A review of stakeholder interests and participation in the sustainable use of communal wetlands: The case of the Lake Fundudzi catchment in Limpopo Province, South Africa, Doctoral dissertation, Rhodes University
31. Turner RK, Van Den Bergh JC, Söderqvist T, Barendregt A, Van Der Straaten J, Maltby E, Van Ierland EC (2000) Ecological-economic analysis of wetlands: scientific integration for management and policy. Ecol Econ 35(1):7–23. https://doi.org/10.1016/S0921-8009(00)00164-6
32. Bosma C, Glenk K, Novo P (2017) How do individuals and groups perceive wetland functioning? Fuzzy cognitive mapping of wetland perceptions in Uganda. Land Use Policy 60:181–196. https://doi.org/10.1016/j.landusepol.2016.10.010
33. Christopoulou OG, Tsachalidis E (2004) Conservation policies for protected areas (wetlands) in Greece: A survey of local residents' attitude. Water Air Soil Pollut Focus 4(4–5):445. https://doi.org/10.1023/B:WAFO.0000044817.88422.64
34. Cohen-Shacham E, Dayan T, de Groot R, Beltrame C, Guillet F, Feitelson E (2015) Using the ecosystem services concept to analyse stakeholder involvement in wetland management. Wetl Ecol Manag 23(2):241–256. https://doi.org/10.1007/s11273-014-9375-1
35. Mayers J (2005) Stakeholder power analysis. International Institute for Environment and Development, London. https://doi.org/10.13140/RG.2.2.22745.57446
36. Kumar R, McInnes RJ, Everard M, Gardner RC, Kulindwa KAA, Wittmer H, Infante Mata D (2017) Integrating multiple wetland values into decision-making. Ramsar Policy Brief No. 2. Gland, Switzerland: Ramsar Convention Secretariat, www.ramsar.org/sites/default/files/documents/library/rpb_values_of_wetlands_e.pdf
37. Garin P, Rinaudo JD, Rulhman J (2001) Linking expert evaluation with public consultation to design water policy at the watershed level. Proceedings of the World Water Congress, Berlin: WWC, 15–19 October
38. Rinaudo JD, Garin P (2002) Participation du public et planification de la gestion de l'eau: Nouveaux enjeux et éléments de méthode. Proceedings of the Colloque SHF "Directive Cadre et eaux souterraines". Paris, Mars 2002
39. Janse van Rensburg EJ (2001) An Introduction to research. Grahams town, Rhodes University, Environmental Education Unit, Research methods short course notes
40. Dür A, De Bièvre D (2007) Inclusion without influence? NGOs in European trade policy. J Publ Policy 27(1):79–101. https://doi.org/10.1017/S0143814X0700061X
41. Cavallini S, Soldi R, Friedl J, Volpe M (2016) Using the quadruple helix approach to accelerate the transfer of research and innovation results to regional growth. European Union. https://doi.org/10.2863/408040
42. Schütz F, Heidingsfelder ML, Schraudner M (2019) Co-shaping the future in Quadruple Helix innovation systems: Uncovering public preferences toward participatory research and innovation. She Ji: The Journal of Design, Economics, and Innovation 5(2):128–146. https://doi.org/10.1016/j.sheji.2019.04.002
43. European Commission (2003) Common implementation strategy for the water framework directive (2000/60/CE), Guidance Document no. 8. https://ec.europa.eu/environment/water/water-framework/facts_figures/guidancedocsen.htm
44. Lupo Stanghellini PS (2010) Stakeholder involvement in water management: The role of the stakeholder analysis within participatory processes. Water Policy 12(2010):675–694. https://doi.org/10.2166/wp.2010.004
45. Planul de Management al Riscului la Inundații—Fluviul Dunărea (2015) http://www.inhga.ro/documents/10184/121027/12+PMRI+Dunare.pdf/a85c0b75-dfb7-4197-a457-a03fd64b6556. Accessed June 10th 2020
46. Pleniceanu V (2003) Lacuri şi zone umede. Editura Universitaria, Craiova
47. Licurici M (2011) Human-Induced Environmental Changes and Floodplain Restoration Necessity along the Danube, on the Drobeta-Turnu Severin—Bechet Sector. Forum geografic. Studii şi cercetări de geografie şi protecţia mediului 10(2):350–363. https://doi.org/10.5775/fg.2067-4635.2011.027.d

48. Geografia văii Dunării românești (1969) vol. 1. Academia Republicii Socialiste România, București
49. Cotեț P (1957) Câmpia Olteniei. Editura Științifică, București
50. Tomescu V (1998) Lunca Dunării—sectorul oltean. Editura Sitech, Craiova
51. EEA (2000, 2018). Corine land cover project. https://land.copernicus.eu/pan-european/corine-land-cover. Accessed 5th April 2020
52. Ministry of Environment, Water and Forests, Romanian Government. Protected Areas Section. www.mmediu.ro/categorie/arii-naturale-protejate/208. Accessed April 5th 2020
53. Vijulie I, Preda M, Lequeux-Dincă AI, Cuculici R, Matei E, Mareci A, Manea G, Tudoricu A (2019) Economic productivity vs ecological protection in Danube Floodplain. Case study: Danube's sector between Olt and Vedea. Sustainability 11(22):6391. https://doi.org/10.3390/su11226391
54. European Environment Agency (2019). Natura 2000 network viewer. https://natura2000.eea.europa.eu/. Accessed April 15th 2020
55. Franco D, Luiselli L (2014) Shared ecological knowledge and wetland values: A case study. Land Use Policy 41:526–532. https://doi.org/10.1016/j.landusepol.2014.04.007
56. Salafsky N, Wollenberg E (2000) Linking livelihoods and conservation: a conceptual framework and scale for assessing the integration of human needs and biodiversity. World Dev 28(8):1421–1438. https://doi.org/10.1016/S0305-750X(00)00031-0
57. Licurici M (2010) Environmental protection and sustainable development in the Danube Floodplain, the Drobeta Turnu-Severin-Bechet sector. Annals of the University of Craiova. Series Geography XIII:55–73. http://analegeo.ro/wp-content/uploads/2010/12/4.-LICURICI-Mihaela.pdf
58. Habersack H, Schober B, Hauer C (2015) Floodplain evaluation matrix (FEM): An interdisciplinary method for evaluating river floodplains in the context of integrated flood risk management. Natural Hazards: J Int Soc Prev Mitig Nat Hazards, Springer 75(1):5–32. https://doi.org/10.1007/s11069-013-0842-4
59. ICPDR (2009) Danube River Basin Management Plan (DRBMP) according to EU Water Framework Directive (WFD), Vienna. https://www.icpdr.org/flowpaper/app/#page=16. Accessed May 18th 2020
60. Dumitrașcu M (2006) Modificări ale peisajelor în Câmpia Olteniei (Landscape changes in Oltenia Plain). Editura Academiei Române, București
61. Prăvălie R (2013a) Climate issues on aridity trends of Southern Oltenia in the last five decades. Geographia Technica (1):70–79. http://www.technicalgeography.org/pdf/1_2013/08_1_2013.pdf
62. Mititelu-Ionuș O, Licurici M (2018) Flood risk management profile. A case-study of structural measures and communication tools development. pp 285–296. In Gâștescu P, Bretcan P (edit, 2018), Water resources and wetlands, 4th International Conference Water Resources and Wetlands, Tulcea, Romania
63. Dumitrescu V, Carsmariu A (2014) The Lower Danube Green Corridor. Contributions to the analysis of the restoration opportunities for certain surveyed wetlands in the Danube Floodplain, diversified by types of potential. pp 166–175. In: Gâștescu P, Bretcan P (eds) Water resources and wetlands, 2nd International Conference—Water resources and wetlands. Tulcea (Romania). http://www.limnology.ro/water2014/proceedings.html, 166–175
64. Neculau G, Stan F-I, Adler M-J (2016) The role of evaporation in evaluating the water reserve of Romanian lakes—A case study, Lucrările Seminarului Geografic "Dimitrie Cantemir" 41:5–15
65. Prăvălie R (2013) Considerations about the reduction of marshes in Southern Oltenia. Geographia Technica 8(2):71–79. http://technicalgeography.org/pdf/2_2013/07_pravalie.pdf
66. Drăgoi IJ (2002) The interdependence of climatic and hydrological factors within the Danube alluvial plain Drobeta Turnu-Severin – Corabia sector. Forum Geografic. Studii și cercetări de geografie și protecția mediului 1:83–90
67. Prăvălie R, Peptenatu D, Sîrodoev I (2013) The impact of climate change on the dynamics of agricultural systems in south-western Romania. Carpathian J Earth Environ Sci 8(3):175–186. http://www.cjees.ro/viewTopic.php?topicId=358

68. Moss B (2008) The water framework directive: Total environment or political compromise? Sci Total Environ 400(1–3):32–41. https://doi.org/10.1016/j.scitotenv.2008.04.029
69. Buijse AD, Coops H, Staras M, Jans LH, Van Geest GJ, Grift RE, Ibelings BW, Oosterberg W, Roozen FCJM (2002) Restoration strategies for river floodplains along large lowland rivers in Europe. Fresh Biol 47:889–907. https://doi.org/10.1046/j.1365-2427.2002.00915.x
70. Nordberg K, Mariussen Å, Virkkala S (2020) Community-driven social innovation and quadruple helix coordination in rural development. Case study on LEADER group Aktion Österbotten. J Rural Stud 79:157–168. https://doi.org/10.1016/j.jrurstud.2020.08.001
71. Cast A, MacDonald DH, Grandgirard A, Kalivas T, Strathearn M, Sanderson M, Bryan B, Frahm D (2008) South Australian Murray-Darling basin environmental values report. CSIRO, Water for a Healthy Country National Research Flagship. www.clw.csiro.au/publications/waterforahealthycountry/2008/wfhc-MDB-Environmental-Values.pdf
72. Cornish L, Dunn A (2009) Creating knowledge for action: The case for participatory communication in research. Dev Pract 19(4/5):665–677. http://www.jstor.org/stable/27752104

The manufacturer's authorised representative in the EU is Springer Nature Customer Service Centre GmbH, Europaplatz 3, 69115 Heidelberg, Germany. If you have any concerns regarding our products, please contact ProductSafety@springernature.com

Printed and bound by CPI Group (UK) Ltd, Croydon, CR0 4YY
30/09/2026
02225496-0016

FRANK JUNGKURTH

Konzernleitung bei der GmbH

Konzern, Konzernrecht und Konzernfinanzierung

Herausgegeben von

Walther Hadding und Uwe H. Schneider

TEIL X

Konzernleitung bei der GmbH

Die Pflichten des Geschäftsführers

Von

Frank Jungkurth

Duncker & Humblot · Berlin

Die Deutsche Bibliothek – CIP-Einheitsaufnahme

Jungkurth, Frank:
Konzernleitung bei der GmbH : die Pflichten des Geschäftsführers /
Frank Jungkurth. – Berlin : Duncker und Humblot, 2000
(Konzern, Konzernrecht und Konzernfinanzierung ; Teil 10)
(Untersuchungen über das Spar-, Giro- und Kreditwesen : Abt. B ;
Rechtswissenschaft ; Bd. 126)
Zugl.: Darmstadt, Techn. Univ., Diss., 1998
ISBN 3-428-10146-4

D 17

Fotoprint: Werner Hildebrand, Berlin
Printed in Germany

ISSN 0720-7352
ISBN 3-428-10146-4

Gedruckt auf alterungsbeständigem (säurefreiem) Papier
entsprechend ISO 9706 ♾

Vorwort

Die hier veröffentlichte Arbeit lag im Wintersemester 1997/98 dem Fachbereich 1 Rechts- und Wirtschaftswissenschaften der Technischen Universität Darmstadt als Dissertation vor. Rechtsprechung und Literatur sind bis Juni 1999 berücksichtigt.

Besonderen Dank schulde ich meinem verehrten wissenschaftlichen Lehrer Herrn Professor Dr. *Uwe H. Schneider*. Er hat das Thema angeregt und die Anfertigung der Arbeit in jeder Hinsicht unterstützt. Mein Dank gilt auch Herrn Professor Dr. *Walther Hadding*, der die Aufnahme in die Schriftenreihe befürwortet hat. Ferner möchte ich an dieser Stelle Herrn Professor Dr. *Paul Hofmann* danken. Er hat nicht nur wertvolle Hinweise für meine Arbeit gegeben, sondern auch als Korreferent die zügige Abwicklung meines Promotionsverfahrens ermöglicht.

Darmstadt *Frank Jungkurth*

Geleitwort

1. Einzelne Lebenssachverhalte sind durch gesetzliche Vorschriften mit einem dichten Netz von Regelungen überzogen. Die letzten Lücken sind in der Lehre breitflächig diskutiert - und geschlossen. Und dabei kann man auch davon ausgehen, daß das Gesetz und die Lehre die Lebenswirklichkeit zutreffend erfassen, was nicht ausschließt, daß sich neue Sachverhalte ergeben, an die man nicht gedacht hat.

Für den Konzern und im Konzernrecht ist alles anders. Gesetzlich geregelt sind nur einzelne Problembereiche. Im übrigen ist die Normierung der Rechtsprechung und der Lehre überlassen. Die höchstrichterlichen Entscheidungen handeln aber nur punktuell von einzelnen zufällig den Gerichten vorgetragenen Sachverhalten. Die Lehre schart sich um diese Entscheidungen, vieles wird nicht beleuchtet, und der Abstand zwischen der Konzernwirklichkeit und dem Konzernrecht ist ganz und gar erstaunlich. Ein Weiteres kommt hinzu: Die Konzerne sind heute vielfach grenzüberschreitend angelegt. Das Konzernrecht ist nach wie vor national, das europäische Konzernrecht ist nicht ausgebildet, die Entwürfe sind Legion, die aufgegriffenen Problembereiche gleichfalls mehr oder weniger von Zufällen bestimmt.

2. Vor diesem Hintergrund ist die Arbeit von *Jungkurth* zu sehen. Es handelt sich um eine im Wintersemester 1997/98 dem Fachbereich Rechts- und Wirtschaftswissenschaften der Technischen Universität Darmstadt vorgelegte Dissertation. Sie ist in der Blickrichtung zwar nicht ganz neu. Ein paar wenige Vorarbeiten liegen schon vor. Sie weitet aber den Blick auf den Konzern aus der Sicht des herrschenden Unternehmens, und zwar am Beispiel der GmbH. Und sie konzentriert sich hierbei auf die Rechts- und Pflichtenstellung des Geschäftsführers der konzernleitenden GmbH.

Im ersten Teil seiner Arbeit behandelt *Jungkurth* die Pflichten des Geschäftsführers gegenüber der herrschenden GmbH. Danach hat der Geschäftsführer insbesondere vor der Konzerngründung die Zustimmung der Gesellschafter einzuholen, er hat den Konzern ordnungsgemäß zu leiten, er hat bestimmte konzernbezogene Maßnahmen den Gesellschaftern zur Entscheidung vorzulegen, er darf keine Geschäftschancen der herrschenden GmbH zu eigenen Zwekken ausnutzen, und er darf der Gesellschaft keinen Wettbewerb machen. Schließlich hat der Geschäftsführer auch für das rechtmäßige Verhalten der herrschenden GmbH nach außen zu sorgen. Diese Pflicht stellt den Übergang zum zweiten Teil der Arbeit dar, nämlich den Pflichten der herrschenden GmbH gegenüber den beherrschten Gesellschaften.

Diesen zweiten Teil untergliedert *Jungkurth* in rechtsformspezifische Pflichten und rechtsformunabhängige Pflichten der herrschenden GmbH. Bei den rechtsformspezifischen Pflichten unterscheidet er zwischen solchen gegenüber der beherrschten Aktiengesellschaft und solchen gegenüber der beherrschten GmbH. Als rechtsformunabhängige Pflicht der herrschenden GmbH arbeitet *Jungkurth* die Pflicht heraus, die Konzernleitung mit der Sorgfalt eines ordentlichen und gewissenhaften Geschäftsleiters auszuüben. Er stützt diese Pflicht mit guten Argumenten auf eine Analogie zu §§ 76, 93 AktG, § 43 GmbHG, § 34 GenG.

Nachdem Jungkurth im zweiten Teil die durch das Recht der beherrschten Gesellschaften gezogenen Grenzen für die Konzernleitung aufgezeigt hat, weist er im dritten Teil nach, daß eine Konzernintegration auch in Anbetracht dieser Grenzen zulässig ist.

Gegenstand des 4. Teils sind die Pflichten des Geschäftsführers der herrschenden GmbH gegenüber den beherrschten Gesellschaften. Hier stellt *Jungkurth* insbesondere fest, daß auch der Geschäftsführer der herrschenden GmbH gegenüber den beherrschten Gesellschaften analog §§ 76, 93 AktG, § 43 GmbHG, § 34 GenG verpflichtet sei, die Konzernleitung mit der Sorgfalt eines ordentlichen und gewissenhaften Geschäftsleiters auszuüben. Bei der Frage, ob der Geschäftsführer gegenüber den beherrschten Gesellschaften auch dem Verbot unterliege, Geschäftschancen der Gesellschaft an sich zu ziehen bzw. zu ihr in Wettbewerb zu treten, sei dagegen nach dem Zentralisierungsgrad des Konzerns zu unterscheiden.

Im fünften Teil seiner Arbeit untersucht Jungkurth die konzerndimensionalen Pflichten der herrschenden GmbH gegenüber der Allgemeinheit, für deren Erfüllung der Geschäftsführer Sorge zu tragen hat. Und im sechsten Teil der Arbeit erörtert er schließlich die Frage, ob den Geschäftsführer auch persönlich konzerndimensionale Pflichten gegenüber der Allgemeinheit treffen.

Jungkurth gebührt das Verdienst, ein Thema aufgearbeitet zu haben, das rechtstatsächlich wenig erforscht ist, gesetzlich nicht geregelt und das in der Lehre bis zum Zeitpunkt der Vorlage dieser Arbeit im Wintersemester 1997/98 vernachlässigt wurde. In Anbetracht der zunehmenden Anzahl von Unternehmensverflechtungen ist die Arbeit ein längst überfälliger Beitrag in der konzernrechtlichen Diskussion. Sie zeigt, daß sich über die methodisch sorgfältige Beantwortung konzernrechtlicher Einzelfragen eine widerspruchsfreie Gesamtkonzeption entwickeln läßt. Und sie macht deutlich, daß sich im Konzern viele Fragen völlig neu stellen.

Darmstadt/Mainz im August 1998 Prof. Dr. Uwe H. Schneider

Inhaltsverzeichnis

Abkürzungsverzeichnis

AG	Aktiengesellschaft
AktG	Aktiengesetz
Anh.	Anhang
Aufl.	Auflage
Begr. RegE	Begründung Regierungsentwurf
BB	Betriebsberater
BBodSchG	Bundes-Bodenschutzgesetz
Bd.	Band
BGB	Bürgerliches Gesetzbuch
BGBl.	Bundesgesetzblatt
BGH	Bundesgerichtshof
BGHZ	Entscheidungen des Bundesgerichtshofes in Zivilsachen
BImSchG	Bundes-Immissionsschutzgesetz
BImSchV	Bundes-Immissionsschutzverordnung
BT-Drucksache	Bundestagsdrucksache
bzw.	beziehungsweise
DB	Der Betrieb
Diss.	Dissertation
DStR	Deutsches Steuerrecht
DWiR	Deutsche Zeitschrift für Wirtschaftsrecht
erw.	erweiterte
FG	Festgabe
FS	Festschrift
GenG	Gesetz betreffend die Erwerbs- und Wirtschaftsgenossenschaften
GesZwecks	Gesellschaftszwecks
GesInteresse	Gesellschaftsinteresse
GmbH	Gesellschaft mit beschränkter Haftung

GmbHG	Gesetz betreffend die Gesellschaften mit beschränkter Haftung
GmbHR	GmbH-Rundschau
GoA	Geschäftsführung ohne Auftrag
GRUR	Gewerblicher Rechtsschutz und Urheberrecht
Habil.-Schr.	Habilitationsschrift
Halbs.	Halbsatz
HGB	Handelsgesetzbuch
h. M.	herrschende Meinung
Hrsg.	Herausgeber
i. S. d.	im Sinne des
i. S. v.	im Sinne von
i. V. m.	in Verbindung mit
jur.	juristische
JZ	Juristenzeitung
KG	Kommanditgesellschaft
KWG	Gesetz über das Kreditwesen
LG	Landgericht
mbH	mit beschränkter Haftung
neubearb.	neubearbeitete
NJW	Neue Juristische Wochenschrift
OHG	Offene Handelsgesellschaft
OLG	Oberlandesgericht
ProdHaftG	Produkthaftungsgesetz
rer. pol.	rerum politicarum
RG	Reichsgericht
RGBl.	Reichtsgesetzblatt
RGZ	Entscheidungen des Reichsgerichts in Zivilsachen
S.	Seite
überarb.	überarbeitete
u. a.	unter anderem
UmweltHaftG	Umwelthaftungsgesetz
UmwG	Umwandlungsgesetz

Uni.	Universität
Urt.	Urteil
v.	vom
Var.	Variante
WiB	Wirtschaftsrechtliche Beratung
WPg	Die Wirtschaftsprüfung
WM	Wertpapier-Mitteilungen Teil IV
z. B.	zum Beispiel
ZGR	Zeitschrift für Unternehmens- und Gesellschaftsrecht
ZHR	Zeitschrift für das gesamte Handelsrecht und Wirtschaftsrecht
ZIP	Zeitschrift für Wirtschaftsrecht
zugl.	zugleich

Einleitung

Die Rechtsform der GmbH erfreut sich zunehmender Beliebtheit. So stieg der GmbH-Bestand 1996 auf 770.000 Gesellschaften an.[1] Gleichzeitig läßt sich eine steigende Tendenz zur Konzernbildung erkennen.[2] Schätzungen besagen, daß bis zu 50 % der Gesellschaften mbH in irgendeiner Weise mit anderen Unternehmen konzernverbunden sind.[3] Die Motive für die Konzernbildung sind vielschichtig. Bezweckt werden können u. a. die Auslagerung von Haftungsrisiken, die Abgrenzung von Führungsbereichen, die Gewinnung größerer Flexibilität, das Erreichen einer größeren Marktnähe und die Verwirklichung positiver Synergieeffekte.[4]

Die Aufarbeitung der mit einer Konzernierung verbundenen Rechtsprobleme steht noch am Anfang. Ihr Schwerpunkt lag bisher bei dem Schutz der abhängigen Gesellschaft, ihrer Minderheitsgesellschafter und ihrer Gläubiger. Hiermit folgte die Diskussion der Schutzrichtung des im Aktiengesetz für die abhängige Aktiengesellschaft konzipierten Konzernrechts. Im neueren Schrifttum gewinnt jedoch zunehmend eine andere Betrachtungsweise an Raum. Angestoßen von zahlreichen Arbeiten von *M. Lutter*[5] und *Uwe H. Schneider*[6] setzte sich die Auffassung durch, daß der Schutz der Obergesellschaft, ihrer Gesellschafter und ihrer Gläubiger von mindestens ebenso großer Bedeutung ist wie der von dem Aktienkonzernrecht angestrebte Schutz. Ihre höchstrichterliche Bestätigung fand diese Auffassung in dem sogenannten „Holzmüller"-Urteil des *Bundesgerichtshofs*[7], wonach der Vorstand einer Aktiengesellschaft für die Ausgliede-

---

[1] *H. Hansen*, GmbH-Bestand, 1997, S. 205.

[2] *F. Hoffmann*, Konzern, 1993, S. 63.

[3] *V. Emmerich*/*J. Sonnenschein*, Konzernrecht, 1997, S. 4. Siehe auch die empirische Untersuchung von *C. U. Binder*, Beteiligungsstrategien, 1994, S. 391 ff.

[4] *E. Scheffler*, Konzernmanagement, 1992, S. 3 f.; siehe auch *H. Kalfass*, Konzernbildung, 1991, S. 24 ff.

[5] Grundlegend *M. Lutter*, Rechte, 1973, S. 1 ff.; ders., Binnenstruktur, 1974, S. 347 ff.; ders., Teilfusionen, 1974, S. 199 ff.; ders., Organzuständigkeiten, 1985, S. 825 ff., und zahlreiche weitere Arbeiten.

[6] *Uwe H. Schneider*, Konzernleitung, 1981, S. 249 ff.; ders., Konzernfinanzierung, 1984, S. 497 ff.; *ders.*, Konzerngründung, 1986, S. 121 ff. und öfters.

[7] *BGH*, Urt. v. 25. 2. 1982 - II ZR 174/80, in: BGHZ 83, S. 122, 131.

rung eines wesentlichen Teils des Betriebsvermögens auf eine Tochtergesellschaft die Zustimmung der Hauptversammlung einholen muß.[8]

Ausgangspunkt dieser neuen Betrachtungsweise ist aber nicht nur der Schutz der Obergesellschaft, ihrer Minderheitsgesellschafter und Gläubiger. Ihr Ansatz ist viel umfassender.[9] Sie versteht den Konzern als Organisationsform zweckgerichteten unternehmerischen Handelns, der sich in seinen Grundaspekten von anderen Formen partnerschaftlicher Kooperation nicht unterscheide. Auch im Konzern stellten sich die Fragen, wer für seine Gründung zuständig sei, wer ihn leite, nach welchen Regeln die Finanzierung und die Willensbildung erfolge, wie Dritte vor konzernspezifischen Gefahren und Minderheiten vor Ausbeutung geschützt würden und wie und nach welchen Regeln die Auflösung stattfinde.[10] Die Antworten hierauf seien in den Regelungsprinzipien des privaten Organisationsrechts zu finden, die als Rechtsprinzip auf den ungeregelten Sachverhalt des Konzerns anzuwenden seien.[11] Dieser institutionelle Ansatz denke in System und Systemrationalität und führe zu der Erkenntnis, daß der Konzern ebenso wie die Aktiengesellschaft oder GmbH ein für die grundlegenden Entscheidungen im Konzern zuständiges „Grundorgan" habe, nämlich die Haupt- bzw. Gesellschafterversammlung der Obergesellschaft. Und sie kommt zu dem Ergebnis, daß der Konzern ein Leitungsorgan habe, nämlich den Vorstand bzw. die Geschäftsführer der herrschenden Gesellschaft.[12]

Aufgabe dieser Arbeit soll es sein, einen weiteren Beitrag zur Entwicklung eines Konzernverfassungsrechts zu leisten. Untersucht werden sollen die Pflichten des Leitungsorgans eines Konzerns mit einer GmbH als Konzernspitze. Gegenstand dieser Arbeit sind mithin die Pflichten des Geschäftsführers einer herrschenden GmbH im Konzern.

Dabei wird gezeigt werden, daß die Pflichten des Geschäftsführers wesentlich von dem Verständnis des Konzerns als einem rechtlich gegliederten Unternehmen abhängen. Das hat zum einen zur Folge, daß die Pflichten des Geschäftsführers gerade im Hinblick auf die Funktion der GmbH als Leiterin dieses Unternehmens zu bestimmen sind. Wichtigste Pflicht des Geschäftsführers gegenüber seiner Gesellschaft ist denn auch die Pflicht zur Konzernleitung. Und das hat zum anderen zur Folge, daß die Pflichten des Geschäftsführers und

---

8 Ablehnend: *K. Beusch*, Aktiengesellschaft, 1984, S. 5 ff.; *K.-P. Martens*, Entscheidungsautonomie, 1983, S. 380 ff.

9 *Uwe H. Schneider*, Konzern, 1991, S. 568 f.

10 *M. Lutter*, Organzuständigkeiten, 1985, S. 827 f.; ders., Konzernrecht, 1996, S. 110 ff.; ausführlich *Uwe H. Schneider*, Konzern, 1991, S. 568 ff.

11 *M. Lutter*, Organzuständigkeiten, 1985, S. 833. Dagegen *P. O. Mülbert*, Aktiengesellschaft, 1996, S.25 ff.

12 *M. Lutter*, Organzuständigkeiten, 1985, S. 832 f.; Ablehnend: *D. Joost*, Holzmüller, 1999, S. 176; *B. Sonntag*, Konzernbildungskontrolle, 1990, S. 33.

insbesondere seine Pflicht zur Konzernleitung von den Pflichten der herrschenden GmbH gegenüber den beherrschten Gesellschaften begrenzt werden. Denn der Geschäftsführer hat dafür zu sorgen, daß sich die herrschende GmbH gegenüber den beherrschten Gesellschaften rechtmäßig verhält. Diese Pflichten der herrschenden GmbH gegenüber den beherrschten Gesellschaften hängen wiederum davon ob, welche Rechtsform die beherrschte Gesellschaft hat.

Der Gang der Untersuchung gestaltet sich damit wie folgt. Im 1. Teil werden die konzernspezifischen Pflichten des Geschäftsführers gegenüber seiner eigenen Gesellschaft dargestellt. Schwerpunkt wird die Pflicht des Geschäftsführers zur Konzernleitung sein. Wie bereits angedeutet, zeichnet sich die Konzernleitung dadurch aus, daß sie durch das Recht der beherrschten Gesellschaften begrenzt wird. Mit anderen Worten: Der Geschäftsführer muß bei der Konzernleitung die Pflichten der herrschenden GmbH gegenüber den beherrschten Gesellschaften beachten. Diese Pflichten sind Gegenstand des 2. Teils der Arbeit. Dabei wird die Darstellung auf die Pflichten der herrschenden GmbH gegenüber einer beherrschten Aktiengesellschaft und einer beherrschten GmbH beschränkt, da die beherrschten Gesellschaften in der Praxis zumeist eine dieser Rechtsformen haben. Im 3. Teil soll sodann untersucht werden, ob eine Konzernleitung in Anbetracht der durch das Recht der beherrschten Gesellschaften gezogenen Grenzen überhaupt möglich ist. Im 4. Teil werden die Pflichten des Geschäftsführers der herrschenden GmbH gegenüber den beherrschten Gesellschaften aufgezeigt. Gegenstand des 5. und 6. Teils sind schließlich die konzernbezogenen Pflichten der herrschenden GmbH bzw. ihres Geschäftsführers gegenüber der Allgemeinheit.

1. Teil: Die Pflichten des Geschäftsführers gegenüber der herrschenden GmbH

Die Pflichten des Geschäftsführers gegenüber der herrschenden GmbH ergeben sich zum einen aus dem Gesetz. Zum anderen können sie aber auch in der Satzung der herrschenden GmbH, in der Geschäftsordnung für die Geschäftsführer und in deren Anstellungsverträgen geregelt sein. Gegenstand dieser Arbeit sollen indes nur diejenigen Pflichten sein, die sich aus dem Gesetz ergeben.

A. Die gesetzliche Ausgangslage

Die Gesellschafter haben die Gesellschaft errichtet, damit die Gesellschaft den von ihnen vorgegebenen Zweck verfolgt. Für die Gesellschaft bedeutet dies, daß sie im Rahmen des Unternehmensgegenstandes alle ihr zur Verfügung stehenden Ressourcen dem Gesellschaftszweck dienstbar zu machen hat. Hiermit wird sie zur Trägerin und Leiterin eines Unternehmens.

Da die Gesellschaft nicht selbst handeln und ihr Unternehmen leiten kann, werden ihre Organe für sie tätig.[1] Im Gegensatz zur Aktiengesellschaft liegt die Unternehmensleitung in der GmbH nicht bei einem, sondern bei zwei Organen: den Gesellschaftern und dem Geschäftsführer.[2] Beide Organe sind an der Unternehmensleitung beteiligt, freilich mit unterschiedlichen Rechten und Pflichten.

Die wichtigsten Befugnisse der Gesellschafter sind in § 46 GmbHG genannt. Besonders zu nennen ist ihre Zuständigkeit für die Feststellung des Jahresabschlusses und die Bestimmung über die Verwendung des Ergebnisses, ihre Zuständigkeit für die Bestellung und Abberufung von Geschäftsführern sowie deren Entlastung und ihre Zuständigkeit für die Festlegung von Maßregeln zur Prüfung und Überwachung der Geschäftsführung. Hinzuweisen ist schließlich auf § 37 Abs. 1 GmbHG, woraus sich ergibt, daß die Gesellschafter dem Geschäftsführer hinsichtlich der Geschäftsführung verbindliche Weisungen erteilen können.[3]

---

1 *H. Wiedemann*, Gesellschaftsrecht, 1980, S. 212.

2 *M. Lutter/P. Hommelhoff*, GmbHG, 1995, § 37 Rdnr. 3.

3 *H.-G. Koppensteiner*, Rowedder, 1997, § 37 Rdnr. 25.

Damit ist gleichzeitig die Hauptpflicht des Geschäftsführers angesprochen. Seine Aufgabe ist es, die Geschäfte der Gesellschaft zu führen. Das wird aufgrund seiner vom Gesetz gewählten Bezeichnung hinreichend deutlich.[4] Gegenstand der Geschäftsführung sind alle zur Verfolgung des Gesellschaftszwecks erforderlichen Entscheidungen über den Einsatz und die Koordinierung der Unternehmensressourcen.[5] Dabei hat der Geschäftsführer gemäß § 43 Abs. 1 GmbHG die Sorgfalt eines ordentlichen Geschäftsmannes anzuwenden. Das bedeutet, daß er im Rahmen des Gesetzes, des Gesellschaftsvertrages, der für die Geschäftsführung verbindlichen Beschlüsse anderer Organe der Gesellschaft und unter der gebotenen Berücksichtigung der Interessen der Öffentlichkeit und der Arbeitnehmer an dem von der Gesellschaft betriebenen Unternehmen die Zwecke der Gesellschaft zu fördern, ihren Vorteil zu wahren und Schaden von ihr abzuwenden hat.[6]

Im folgenden soll nun untersucht werden, wie diese Geschäftsführungspflicht konzernspezifisch zu konkretisieren ist. An erster Stelle stehen dabei die Pflichten des Geschäftsführers bei der Konzerngründung.

B. Die Pflichten bei der Konzerngründung

In der Literatur werden die Pflichten des Geschäftsführers einer GmbH bei der Gründung eines Konzerns ebenso wie die entsprechenden Pflichten des Vorstands einer Aktiengesellschaft unter dem Stichwort „Konzernbildungskontrolle" diskutiert.[7] Kernpunkt ist die Frage, ob und inwieweit es zum Schutz der Gesellschafter der Konzernobergesellschaft erforderlich ist, daß die Konzerngründung von ihnen kontrolliert wird.

Dabei stellt sich das Problem der Konzernbildungskontrolle bei der GmbH nicht in derselben Schärfe wie bei der Aktiengesellschaft.[8] Denn während die Gesellschafter der GmbH jederzeit mittels ihres Weisungsrechts die Geschäftsführung an sich ziehen und bis hinunter in die letzten Verästelungen regieren können, kann die Hauptversammlung der Aktiengesellschaft ihre Kompetenzen

---

4 *W. Zöllner*, Baumbach/Hueck, 1996, § 35 Rdnr. 16.

5 *W. Zöllner*, Baumbach/Hueck, 1996, § 35 Rdnr. 17; *H.-J. Mertens*, Hachenburg, 1996, § 37 Rdnr. 3; *M. Lutter/P. Hommelhoff*, GmbHG, 1995, § 37 Rdnr. 3

6 *H.-J. Mertens*, Hachenburg, 1996, § 43 Rdnr. 19; *C. T. Ebenroth/K. W. Lange*, Sorgfaltspflichten, 1992, S. 70.

7 Siehe dazu *B. Sonntag*, Konzernbildungskontrolle, 1990, für die GmbH; *H. W. Wahlers*, Konzernbildungskontrolle, 1995, für die Aktiengesellschaft; *Th. Liebscher*, Konzernbildungskontrolle, 1995, für die Aktiengesellschaft, Personengesellschaften und die GmbH.

8 *H.-G. Koppensteiner*, Rowedder, 1997, Anh. § 52 Rdnr. 33; *M. Lutter*, Organzuständigkeiten, 1985, S. 835.

nicht selbst bestimmen. Insbesondere hat sie grundsätzlich keinen Einfluß auf die Geschäftsführung und leitet der Vorstand die Gesellschaft unter eigener Verantwortung, § 76 Abs. 1 AktG. Von der Konzernbildungskontrolle bei der Aktiengesellschaft soll hier jedoch nicht gehandelt werden.

An dieser Stelle interessieren vielmehr die Pflichten des Geschäftsführers bei der Konzerngründung, in denen sich die Konzernbildungskontrolle bei der GmbH widerspiegelt. Vorher soll jedoch die rechtstatsächliche Ausgangslage bei der Konzerngründung dargestellt werden.

I. Die rechtstatsächliche Ausgangslage

Die Gründung eines Konzerns ist ein mehrstufiger Prozeß.[9]

Zunächst erwirbt die GmbH eine Beteiligung an der zukünftigen Konzerngesellschaft. Die Beteiligung erreicht schrittweise oder sofort eine Größe, die es der GmbH ermöglicht, auf die Gesellschaft herrschenden Einfluß auszuüben, § 17 Abs. 1 AktG. Während bei einer Mehrheitsbeteiligung das Vorliegen eines solchen Abhängigkeitsverhältnisses gemäß § 17 Abs. 2 AktG vermutet wird, kann auch eine geringere Beteiligung die Möglichkeit einer beherrschenden Einflußnahme eröffnen. Dies ist z. B. der Fall, wenn durch Stimmbindungsverträge oder Treuhandabreden die Mitwirkung von Dritten sichergestellt ist.[10] Ist die Untergesellschaft eine GmbH, kann der beherrschende Einfluß auch durch ein in der Satzung verankertes Weisungsrecht eines einzelnen Gesellschafters ermöglicht werden.[11] Neben einer GmbH können auch eine Aktiengesellschaft oder eine Personengesellschaft in einem Abhängigkeitsverhältnis stehen. Eine Genossenschaft und ein Verein kommen dagegen nur ausnahmsweise als abhängige Unternehmen in Betracht, eine Stiftung scheidet als abhängiges Unternehmen gänzlich aus.[12] Wegen ihrer praktischen Relevanz sollen im folgenden nur die Konzernierung der Aktiengesellschaft und der GmbH betrachtet werden.

An die Stelle des Erwerbs einer Mehrheitsbeteiligung an einer bestehenden Gesellschaft kann auch die Gründung einer neuen Gesellschaft treten. Dieser Weg wird insbesondere dann gewählt, wenn die GmbH einen Unternehmensteil ausgliedern will.

---

[9] Ausführlicher hierzu *M. Lutter/W. Timm*, Präventivschutz, 1982, S. 412 f.; dazu kritisch aus betriebswirtschaftlicher Sicht *M. R. Theisen*, Konzern, 1991, S. 137.

[10] *U. Hüffer*, AktG, 1997, § 17 Rdnr. 6; *H.-G. Koppensteiner*, Kölner Kommentar, 1986, § 17 Rdnr. 40; *V. Emmerich/J. Sonnenschein*, Konzernrecht, 1997, S. 44.

[11] *V. Emmerich*, Scholz, 1993, Anhang Konzernrecht, Rdnr. 61.

[12] *V. Emmerich/J. Sonnenschein*, Konzernrecht, 1997, S. 449, 458, 462 f.

Die mit dem Erwerb einer ausreichend großen Beteiligung geschaffene Möglichkeit der Einflußnahme kann die GmbH sodann zur Konzernherrschaft ausbauen. Insbesondere kann sie die Konzernleitung in unterschiedlicher Intensität ausüben.

Im einfachen faktischen Konzern beschränkt sich die GmbH darauf, diejenigen Entscheidungen zu treffen, die für eine Integration der beherrschten Gesellschaft in den Konzern erforderlich sind.[13] Im übrigen läßt sie dem beherrschten Unternehmen freie Hand. Im qualifizierten faktischen Konzern hingegen nimmt die GmbH umfassend und breitflächig Einfluß auf die Untergesellschaft und führt sie gleichsam wie eine unselbständige Betriebsabteilung.[14]

Vom faktischen Konzern schließlich zu unterscheiden ist der Vertragskonzern. So kann die GmbH ihre Leitungsmacht vertraglich absichern, indem sie mit der Untergesellschaft einen Beherrschungsvertrag schließt. Zum einen gestattet er der GmbH, dem Verwaltungsorgan des Konzernunternehmens direkt Weisungen zu erteilen.[15] Zum anderen ist die GmbH nun berechtigt, das Unternehmensinteresse des Konzernunternehmens auf ihre eigenen Belange und auf das Konzerninteresse auszurichten.[16] Dies ergibt sich für die konzernierte Aktiengesellschaft unmittelbar aus § 308 Abs. 1 AktG und für die konzernierte GmbH aus einer Analogie zu dieser Vorschrift[17].

Die intensivste Form der Beherrschung vermittelt schließlich die Eingliederung gemäß §§ 319 ff. AktG. Während im Vertragskonzern die Zulässigkeit von Weisungen des herrschenden Unternehmens davon abhängt, daß die Weisungen den Belangen des herrschenden Unternehmens oder der mit ihm konzernverbundenen Gesellschaften dienen (§ 308 Abs. 1 Satz 2 AktG), gibt es eine solche Beschränkung des Weisungsrechts im Eingliederungskonzern nicht.[18] Die Ein-

---

13 Siehe dazu unten C. III.

14 Zum Vorstehenden siehe die Übersicht bei *E. Scheffler*, Konzernmanagement, 1992, S. 13.

15 *W. Zöllner*, Beherrschungsverträge, 1992, S. 175.

16 *B. Kropff*, Beherrschungsvertrag, 1993, S. 518.

17 *H. Altmeppen*, GmbHG, 1997, Anh. § 13 Rdnr. 43; *P. Ulmer*, Hachenburg, 1994, Anh. § 77 Rdnr. 218.

18 *Begr. RegE.* zu § 323 AktG, in: *B. Kropff*, AktG, 1965, S. 427; *B. Grunewald*, Geßler, 1991, § 323, Rdnr. 2. Allerdings dürfte sich der Vorstand der herrschenden Aktiengesellschaft im Verhältnis zu dieser pflichtwidrig verhalten, wenn er der eingegliederten Gesellschaft eine Weisung erteilt, die weder den Belangen der herrschenden Aktiengesellschaft noch den Belangen der übrigen Konzerngesellschaften dienen. So richtig: *B. Grunewald*, Geßler, 1991, § 323, Rdnr. 2, *H.-G. Koppensteiner*, Kölner Kommentar, 1987, § 323 Rdnr. 2.

gliederung interessiert hier jedoch nicht, da sie auf beiden Seiten eine inländische Aktiengesellschaft voraussetzt.[19]

All diese Maßnahmen der Konzerngründung werden von dem Geschäftsführer als dem gesetzlichen Vertreter der herrschenden GmbH (§ 35 Abs. 1 GmbHG) gegenüber den Konzerngesellschaften vorgenommen. Das darf er allerdings nur dann, wenn sie von seiner Geschäftsführungsbefugnis gedeckt sind oder, anders gewendet, die Grenzen seiner Geschäftsführungsbefugnis nicht überschreiten.

II. Die gesetzliche Ausgangslage

Gemäß § 37 Abs. 1 GmbHG sind die Geschäftsführer „der Gesellschaft gegenüber verpflichtet, die Beschränkungen einzuhalten, welche für den Umfang ihrer Befugnis, die Gesellschaft zu vertreten, durch den Gesellschaftsvertrag oder, soweit dieser nicht ein anderes bestimmt, durch die Beschlüsse der Gesellschafter festgesetzt sind". Entgegen ihrem Wortlaut behandelt diese Vorschrift nicht die Vertretungsbefugnis des Geschäftsführers im Außenverhältnis. Diese Vertretungsbefugnis ist gemäß § 37 Abs. 2 GmbHG unbeschränkbar. Vielmehr ist Gegenstand von § 37 Abs. 1 GmbHG die Beschränkung des Geschäftsführerhandelns im Innenverhältnis.[20] Danach ist der Geschäftsführer insbesondere verpflichtet, im Rahmen seiner Geschäftsführung diejenigen Grenzen einzuhalten, die ihm durch den Gesellschaftsvertrag gezogen sind.

Von diesen Grenzen interessieren im Rahmen der Konzerngründung die Pflicht des Geschäftsführers zur Beachtung des Verbandszwecks und des Unternehmensgegenstandes der Gesellschaft, die Pflicht zur Wahrung der Mitgliedschaftsrechte der Gesellschafter sowie seine Vorlagepflichten.

III. Die Pflicht zur Beachtung des Verbandszwecks

1. Die gesetzliche Ausgangslage

Die erste Grenze der Geschäftsführungsbefugnis des Geschäftsführers bildet seine Pflicht zur Beachtung des Verbandszwecks der Gesellschaft. Zum einen konstituiert der Verbandszweck die Gesellschaft, da das geltende Recht den Personenzusammenschluß nicht um seiner selbst willen, sondern nur als

---

[19] *U. Hüffer*, AktG, 1997, § 319 Rdnr. 4; *H.-G. Koppensteiner*, Kölner Kommentar, 1987, Vorb. § 319 Rdnr. 5, 6.

[20] *W. Zöllner*, Baumbach/Hueck, 1996, § 37 Rdnr. 1; *M. Lutter/P. Hommelhoff*, GmbHG, 1995, § 37 Rdnr. 1.

Zweckgebilde anerkennt.[21] Darüber hinaus hat er aber auch funktionelle Wirkung, indem er die Rechtsverhältnisse zwischen Verband und Mitgliedern und diejenigen zwischen den Mitgliedern untereinander inhaltlich prägt: „der vereinbarte Zweck ist das Lebensgesetz des Verbandes“.[22]

Als unbewegliche Schranke der Mehrheitsherrschaft umfaßt er nur denjenigen Teil der Satzung, der die obersten Leitsätze für die Verbandstätigkeit zum Ausdruck bringt und mit dessen Abänderung schlechterdings kein Mitglied rechnen kann. Diese Erkenntnis stützt sich für den Verein auf § 33 Abs. 1 Satz 2 1. Halbsatz BGB,[23] der die Änderung des Vereinszwecks von der Zustimmung aller Mitglieder abhängig macht. Die Regelung des § 33 Abs. 1 Satz 2 1. Halbsatz BGB ist aber nicht auf den Verein beschränkt. Vielmehr normiert sie mit dem Einstimmigkeitserfordernis ein allgemeines Prinzip des Verbandsrechts,[24] das auch für die GmbH Geltung beansprucht.[25]

Der Verbandszweck muß nicht wie z. B. der Unternehmensgegenstand der Gesellschaft (§ 3 Abs. 1 Nr. 2 GmbHG) in der Satzung ausdrücklich genannt sein. Enthält die Satzung keine ausdrückliche Aussage zum Verbandszweck, so ist dieser durch Auslegung der Satzung zu ermitteln. Mangels anderer Anhaltspunkte wird er in der Regel auf Gewinnerzielung gerichtet sein.[26] Neben der Gewinnerzielung kommen aber auch die bloße Kostendeckung, die Bedarfsdekkung und eine geistig-ideele Zielsetzung in Betracht.[27]

2. Die Lage im Konzern

Den Verbandszweck seiner Gesellschaft hat der Geschäftsführer bei der Konzerngründung zu beachten. Das bedeutet, daß er einen Konzern nur mit solchen Gesellschaften gründen darf, deren Verbandszweck entweder mit dem der herrschenden GmbH identisch ist oder diesen unterstützt.[28]

---

21 *K. Schmidt*, Gesellschaftsrecht, 1997, S. 62.

22 *H. Wiedemann*, Gesellschaftsrecht, 1980, S. 10

23 *BGH*, Beschl. v. 11.11.1985 - II ZB 5/85, in: BGHZ 96, 245, 251 f.

24 *K. Schmidt*, Gesellschaftsrecht, 1997, S. 68.

25 *M. Lutter/P. Hommelhoff*, GmbHG, 1995, § 53 Rdnr. 21; *G. H. Roth*, GmbHG, 1997, § 53 Rdnr. 30.

26 *P. Ulmer*, Hachenburg, 1989, § 1 Rdnr. 9

27 *H. Wiedemann*, Gesellschaftsrecht, 1980, S. 326; ders., Grundfragen, 1975, S. 412 ff.

28 In der Literatur wird die Bedeutung des Verbandszwecks für die Konzerngründung - soweit ersichtlich - nicht erörtert. Vielmehr konzentriert sich die Diskussion auf die Bedeutung des Unternehmensgegenstandes. Dazu sogleich.

So ist es z. B. mit dem Verbandszweck einer gemeinnützigen GmbH vereinbar, wenn ihr Geschäftsführer eine Tochtergesellschaft gründet, deren Zweck auf Gewinnerzielung gerichtet ist. Denn der Verbandszweck einer gemeinnützigen GmbH gebietet nur, daß die von ihr erwirtschafteten Gewinne dem gemeinnützigen Zweck zugeführt werden. Über die Art und Weise der Gewinnerzielung, ob nun unmittelbar oder mittelbar über eine Tochtergesellschaft, sagt der Verbandszweck dagegen nichts aus.

Aber auch der umgekehrte Fall, nämlich die Gründung einer gemeinnützigen Gesellschaft durch eine auf Gewinnerzielung gerichtete GmbH kann zulässig sein. So darf der Geschäftsführer einer GmbH ebenso wie der Vorstand einer Aktiengesellschaft Spenden machen, die in einem gewissen örtlichen und gegenständlichen Bezug zum Unternehmen stehen und deren Umfang sich in einem angemessenen Verhältnis zum Gewinn der Gesellschaft und ihren Investitionserfordernissen bewegt.[29] Dies wird zwar von einer Mindermeinung bestritten.[30] Dabei wird jedoch verkannt, daß soziale Aktivitäten zumindest mittelbar zur Gewinnerzielung beitragen.[31] Denn Voraussetzung für den wirtschaftlichen Erfolg der Gesellschaft ist auch ihre soziale Akzeptanz. Diese hängt aber in der Regel davon ab, daß die Gesellschaft „als good corporate citizen auftritt, ein vielfältiges Netz sozialer und politischer Kontakte aufbaut und sich an sozialen, politischen und kulturellen Aktivitäten beteiligt“.[32]

Folgt man dem, so verstößt es nicht gegen den Zweck der Gewinnerzielung, wenn der Geschäftsführer eine gemeinnützige GmbH gründet und über diese Spenden vornimmt. Erst wenn die dafür verwendeten Mittel den zulässigen Spendenrahmen überschreiten, wird der auf Gewinnerzielung gerichtete Verbandszweck verletzt.[33]

IV. Die Pflicht zur Beachtung des Unternehmensgegenstands

Eine weitere Grenze der Geschäftsführungsbefugnis ist der in der Satzung festgelegte Unternehmensgegenstand.

---

29 Für die GmbH: *Uwe H. Schneider*, Scholz, 1993, § 43 Rdnr. 66 f. Für die AG: *BGH*, Urt. v. 24.1.1957 - II ZR 208/55, in: BGHZ 23, S. 150, 157; *H.-J. Mertens*, Aktivitäten, 1987, S. 353; *P. Kulitz*, Unternehmerspenden, 1983, S. 159. Siehe auch *J. Tieves*, Unternehmensgegenstand, 1998, S. 216.

30 *H. Meilicke*, 1959, S. 412.

31 *H.-J. Mertens*, Aktivitäten, 1987, S. 353. Ebenso *H.-P. Westermann*, Verantwortung, 1990, S. 773 f.

32 *H.-J. Mertens*, Aktivitäten, 1987, S. 353. Ebenso *H.-P. Westermann*, Verantwortung, 1990, S. 773 f.

33 Nach *Uwe H. Schneider*, Scholz, 1993, § 43 Rdnr. 67, ist es in der Regel unbedenklich, wenn der Geschäftsführer 2 % des Bilanzgewinns für Spenden verwendet.

Der Unternehmensgegenstand bezeichnet den Tätigkeitsrahmen, innerhalb dessen der Verbandszweck zu verwirklichen ist.[34] Darüber hinaus wird die Meinung vertreten, daß er eine „rechtliche und betriebswirtschaftliche Organisationsmaxime“ enthalte.[35] Nur wenn die Konzerngründung im Unternehmensgegenstand ausdrücklich erlaubt sei, dürfe danach die Zielverwirklichung auch in Konzernunternehmen angestrebt werden.

1. Der Unternehmensgegenstand als Tätigkeitsbeschreibung

Gemäß § 3 Abs. 1 Nr. 2 GmbHG muß der Unternehmensgegenstand in der Satzung so weit individualisiert werden, daß der Schwerpunkt der Geschäftstätigkeit für die beteiligten Wirtschaftskreise hinreichend erkennbar wird.[36] Seine Festlegung in der Satzung hat mehrere Funktionen. Erstens begrenzen die Gesellschafter hiermit die Geschäftsführungsbefugnis des Geschäftsführers im Innenverhältnis.[37] Ist Unternehmensgegenstand einer Gesellschaft die Herstellung von Computern, so darf der Geschäftsführer die Produktion nicht um den Bau von Landmaschinen erweitern.[38] Zweitens dient der in das Handelsregister einzutragende Unternehmensgegenstand der Information der beteiligten Verkehrskreise über den Tätigkeitsbereich der Gesellschaft und ermöglicht drittens dem Registergericht die Prüfung, ob die Gesellschaft einen erlaubten Zweck i. S. v. § 1 GmbHG verfolgt.[39]

Auch im Konzern beschränkt der Unternehmensgegenstand die Geschäftsführungsbefugnis des Geschäftsführers. Er begrenzt den Beteiligungserwerb und die Konzernierung grundsätzlich auf solche Unternehmen, deren Unternehmensgegenstand zumindest im Kernbereich mit dem der herrschenden GmbH übereinstimmt[40] oder diesen unterstützt. Denn durch die Aufnahme der

---

[34] *K. Schmidt*, Gesellschaftsrecht, 1997, S. 69; *H. Wiedemann*, Gesellschaftsrecht, 1980, S. 328 f. Ausführlich *J. Tieves*, Unternehmensgegenstand, 1998, S. 11 ff.

[35] *W. Gross*, Zuständigkeit, 1994, S. 269.

[36] *BGH*, Beschl. v. 3.11.1980 - II ZB 1/79, in: BB 1981, S. 450; *V. Emmerich*, Scholz, 1993, § 3 Rdnr. 13; *P. Ulmer*, Hachenburg, 1989, § 3 Rdnr. 21.

[37] *H. E. Brandner*, Geschäftsführungsbefugnis, 1994, S. 45. Für den Vorstand der AG: *W. Timm*, Konzernspitze, 1980, S. 21; *M. H. Kessler*, Leitungsmacht, 1995, S. 66; *J. Tieves*, Unternehmensgegenstand, 1998, S. 268.

[38] Für die entsprechenden Pflichten des Vorstandes einer Aktiengesellschaft siehe *W. Timm*, Konzernspitze, 1980, S. 21.

[39] *V. Emmerich*, Scholz, 1993, § 3 Rdnr. 12; *P. Ulmer*, Hachenburg, 1989, § 3 Rdnr. 22.

[40] *B. Sonntag*, Konzernbildungskontrolle, 1990, S. 212 f.; *J. Tieves*, Unternehmensgegenstand, 1998, S. 276; *M. Lutter*, Organzuständigkeiten, 1985, S. 846; *Uwe H. Schneider*, Konzernverfassung, 1986, S. 1995; *E. Rehbinder*, Tochtergesellschaften, 1982, S. 430; *A. Zitzmann*, Vorlagepflichten, 1991, S. 55 f.; *H.-G. Koppensteiner*, Ro-

Konzernleitung wird die Tätigkeit der Konzernunternehmen zum integralen Bestandteil des sachlichen Tätigkeitsbereichs der herrschenden Gesellschaft. Dieser wird aber durch ihren Unternehmensgegenstand definiert. Ist also Unternehmensgegenstand, um beim obigen Beispiel zu bleiben, die Herstellung von Computern, so darf die herrschende Gesellschaft weder selbst Landmaschinen produzieren, noch darf sie eine Gesellschaft unter ihre Konzernleitung stellen, die in dieser Branche tätig ist.

Zulässig ist es dagegen, wenn die Gesellschaft betriebliche Teilfunktionen auf eine beherrschte Gesellschaft auslagert und diese nur konzerninterne Leistungen erbringt.[41]

Gliedert z. B. ein Automobilhersteller seine Datenverarbeitung auf eine Tochtergesellschaft aus, so besteht deren Unternehmensgegenstand in der Datenverarbeitung und nicht in der Herstellung von Autos. Trotzdem hat die Tochtergesellschaft nur eine die Automobilherstellung unterstützende Funktion, so daß die Ausgliederung noch vom Unternehmensgegenstand der Konzernobergesellschaft gedeckt ist. Anders gewendet muß also der Unternehmensgegenstand der beherrschten Gesellschaft im Funktionsbereich des Unternehmensgegenstandes der herrschenden GmbH liegen.

Die Beschränkung der Tochtergesellschaft auf konzerninterne Leistungen ergibt sich daraus, daß mit dem Unternehmensgegenstand gleichzeitig die Branche festgelegt wird, in der die Gesellschaft tätig sein soll. Hiermit wollen die Gesellschafter auch ihr Risiko begrenzen.[42] Würde etwa in dem obengenannten Beispiel die Tochtergesellschaft ihre Leistungen auch konzernfremden Unternehmen anbieten, so würde sie in Konkurrenz mit anderen EDV-Anbietern treten. Mithin würde sie in der EDV-Branche tätig, die gänzlichen anderen Risiken unterliegt als die Automobilindustrie. Die Gesellschafter haben aber mit der Festlegung des Unternehmensgegenstandes „Automobilherstellung" sich dafür entschieden, nur die Risiken der Automobilindustrie auf sich zu nehmen.

Vom Beteiligungserwerb zum Zwecke der Konzerngründung zu unterscheiden ist der kapitalistische Beteiligungserwerb. Er dient der bloßen Finanzanlage. Auch auf seine Zulässigkeit soll an dieser Stelle kurz eingegangen werden.[43]

---

wedder, 1997, Anh. § 52 Rdnr. 35; *H. J. Priester*, Vertragskonzern, 1986, S. 171; *W. Gross*, Zuständigkeit, 1994, S. 268; *H. W. Wahlers*, Konzernbildungskontrolle, 1995, S. 138 f. (zur Aktiengesellschaft); zu der Frage, inwieweit die Unternehmensgegenstände der Konzern-Gesellschaften übereinstimmen müssen, siehe *W. Timm*, Konzernspitze, 1980, S. 101 f.

41 Im Ergebnis ebenso *J. Tieves*, Unternehmensgegenstand, 1998, S. 212.

42 *Uwe H. Schneider*, Scholz, 1993, § 43 Rdnr. 58.

43 Ausführlich dazu *J. Tieves*, Unternehmensgegenstand, 1998, S. 421 ff.

Der kapitalistische Beteiligungserwerb ist zum einen dann zulässig, wenn er im Unternehmensgegenstand der Gesellschaft ausdrücklich vorgesehen ist.[44] Zum anderen kann sich seine Zulässigkeit aber auch ohne ausdrückliche Erwähnung in der Satzung ergeben. So ist jede Gesellschaft gezwungen, zur Erfüllung ihrer gegenwärtigen und absehbaren Verbindlichkeiten eine Liquiditätsreserve vorzuhalten. Außerdem gibt es Situationen, in denen eine Reinvestition erwirtschafteter Gewinne z. B. wegen einer Konjunkturflaute betriebswirtschaftlich nicht angezeigt ist. Schließlich kann ein Bedürfnis nach Absicherung von Zulieferer- und Abnehmerbeziehungen bestehen. Dies muß bei Auslegung des Unternehmensgegenstandes berücksichtigt werden.

Deshalb verstößt es nicht gegen den Unternehmensgegenstand, wenn der Geschäftsführer liquide Mittel im Rahmen einer vernünftigen Risikostreuung in Beteiligungen anlegt, um eine Liquiditätsreserve vorzuhalten oder um eine Investitionspause zu überbrücken.[45] Ebenso ist es mit dem Unternehmensgegenstand vereinbar, Zulieferer- und Abnehmerbeziehungen zur Absicherung mit Beteiligungen zu unterlegen. Voraussetzung ist allerdings in jedem Fall, daß der Beteiligungserwerb über eine bloße Annextätigkeit zum eigentlichen Unternehmensgegenstand nicht hinausgeht. Mit anderen Worten darf der Umfang der Beteiligungen im Verhältnis zum übrigen Vermögen der GmbH nur eine untergeordnete Rolle einnehmen. Ferner müssen die zum Vorhalten einer Liquiditätsreserve oder zur Überbrückung einer Investitionspause erworbenen Beteiligungen schnell und unauffällig wieder liquidiert werden können.[46]

Nicht erforderlich ist dagegen, daß der Unternehmensgegenstand der Beteiligungsgesellschaft mit dem der GmbH übereinstimmt.[47] Aus dem Unternehmensgegenstand läßt sich eine solche Voraussetzung nicht entnehmen. Denn dieser definiert nur den Schwerpunkt der Tätigkeit der herrschenden GmbH. Hierzu gehört der kapitalistische Beteiligungserwerb gerade nicht. Läßt sich aus dem Unternehmensgegenstand aber keine Beschränkung des kapitalistischen

---

[44] *B. Sonntag*, Konzernbildungskontrolle, 1990, S. 218. Für die Aktiengesellschaft: *P. Hommelhoff*, Konzernleitungspflicht, 1982, S. 46 f.; *M. Lutter*, Konzernrecht, 1996, S. 110; *J. Semler*, Leitung, 1996, Rdnr. 277; *W. Gross*, Zuständigkeit, 1994, S. 268; *H. W. Wahlers*, Konzernbildungskontrolle, 1995, S. 141.

[45] *P. Hommelhoff*, Konzernleitungspflicht, 1982, S. 48 f.; *E. Rehbinder*, Konzernleitungspflicht, 1983, S. 467; *Uwe H. Schneider*, Konzerngründung, 1986, S. 124; *W. Gross*, Zuständigkeit, 1994, S. 268; *W. Timm*, Konzernspitze, 1980, S. 97; *B. Sonntag*, Konzernbildungskontrolle, 1990, S. 218; *H. W. Wahlers*, Konzernbildungskontrolle, 1995, S. 141; *J. Tieves*, Unternehmensgegenstand, 1998, S. 278.

[46] Zur entsprechenden Lage bei der AG siehe *W. Gross*, Zuständigkeit, 1994, S. 268. Ausführlich *J. Tieves*, Unternehmensgegenstand, 1998, S. 429 ff.

[47] So aber *V. Emmerich*, Scholz, 1993, Anh. Konzernrecht, Rdnr. 119, 120. Im Ergebnis wie hier: *Uwe H. Schneider*, Personengesellschaft, 1979, S. 516; *W. Gross*, Zuständigkeit, 1994, S. 268; *J. Tieves*, Unternehmensgegenstand, 1998, S. 435.

Beteiligungserwerbs auf eine bestimmte Branche entnehmen, so steht es im unternehmerischen Ermessen des Geschäftsführers, welche Beteiligungen er erwirbt. Freilich ist er hierbei an die Sorgfalt eines ordentlichen Geschäftsmannes gebunden, § 43 Abs. 1 GmbHG.

2. Der Unternehmensgegenstand als rechtliche und betriebswirtschaftliche Organisationsmaxime?

Nach einer in der Literatur vertretenen Ansicht schreibt der Unternehmensgegenstand auch die organisatorische Struktur vor, in der die sachbezogene unternehmerische Tätigkeit erfolgen solle. Er regele die Frage, ob die Gesellschaft den erstrebten Erfolg unmittelbar oder auch mittelbar verwirklichen könne, nämlich durch Gründung von Tochtergesellschaften und Beteiligungserwerb.[48] Gestatte er nicht ausdrücklich die Konzerngründung, so sei die Zielverwirklichung „autark" anzustreben, „d. h. nicht durch die Verwaltung fremder Unternehmen, sondern als eigenunternehmerische Tätigkeit".[49]

Hierfür spreche erstens der Schutzzweck von § 3 Abs. 1 Nr. 2 GmbHG bzw. § 23 Abs. 3 Nr. 2 i. V. m. § 179 Abs. 2 Satz 2 AktG. Dieser gebiete es, im Unternehmensgegenstand nicht nur die Branche anzugeben, in der die Gesellschaft tätig werde, sondern auch die Art und Weise ihrer Tätigkeit. Eigene Produktion sei aber grundsätzlich etwas anderes als Geschäftsführung für oder im Hinblick auf andere Gesellschaften.[50] Insbesondere sei Gegenstand einer Holding nur noch „*der Erwerb und die Verwaltung von Beteiligungen*, d.h. vornehmlich die Wahrnehmung bestimmter *Dienstleistungsaufgaben*", nicht aber z. B. die auf eine Tochtergesellschaft ausgegliederte Produktion von Computern.[51]

---

48 Gegen diesen Ansatzpunkt *Uwe H. Schneider*, Konzerngründung, 1986, S. 126: „Das Problem „Konzerngründung" läßt sich aber nicht von der Zweckänderung und von der Gegenstandsänderung her entwickeln". Ebenso *H. Wiedemann*, Unternehmensgruppe, 1988, S. 54, für die Ausgliederung und den Erwerb einer Tochtergesellschaft durch eine AG. Ablehnend auch *A. von Riegen*, Gesellschafterschutz, 1999, S. 42 ff.; *P. O. Mülbert*, Aktiengesellschaft, 1996, S.378 f.

49 *W. Timm*, Konzernspitze, 1980, S. 92, für die AG. Für die GmbH: *H.-G. Koppensteiner*, Rowedder, 1997, Anh. § 52 Rdnr. 35; *B. Sonntag*, Konzernbildungskontrolle, 1990, S. 214 f.; *A. Zitzmann*, Vorlagepflichten, 1991, S. 56 f. Für die AG außerdem: *W. Gross*, Zuständigkeit, 1994, S. 269; *M. Lutter*, Organzuständigkeiten, 1985, S. 847; *E. Rehbinder*, Tochtergesellschaften, 1982, S. 432 f.; *H. W. Wahlers*, Konzernbildungskontrolle, 1995, S. 148 ff. A. M. *H.-J. Mertens*, Kölner Kommentar, 1988, § 82 Rdnr. 15, für die AG.

50 *W. Timm*, Konzernspitze 1980, S. 89 (für die AG).

51 *W. Timm*, Hauptversammlungskompetenzen, 1980, S. 180, für die AG (Hervorhebung im Original).

Zweitens würde der Handlungsspielraum der Geschäftsführung eingeengt, da sie nun auch auf die Belange der abhängigen Gesellschaft Rücksicht nehmen müsse. Insbesondere werde ihre Autonomie beim Erwerb mitbestimmungspflichtiger Gesellschaften durch die Mitsprache der Arbeitnehmer eingeschränkt.[52]

Drittens schließlich steige die Gefahr, daß die Geschäftsleitung ihre Kompetenzen zum Nachteil der Gesellschafter mißbrauche, wenn sie unternehmerisches Handeln partiell aus der Gesellschaft heraus verlagere.[53]

Diese Argumente überzeugen nicht.

Zunächst ist es kaum nachvollziehbar, hinter dem Begriff „Gegenstand des Unternehmens" mehr zu vermuten als eine bloße Beschreibung des wirtschaftlichen Bereichs, in dem die Gesellschaft tätig wird. Auch ist das Argument nicht stichhaltig, daß eine Holding nur noch Dienstleistungen an die Konzerngesellschaften erbringe und damit sogar einen anderen sachlichen Unternehmensgegenstand habe als vor Ausgliederung ihres operativen Geschäfts. Es ist gerade „Wesensmerkmal" des Konzerns, daß er eine wirtschaftliche Einheit darstellt.[54] Diese wirtschaftliche Einheit wird aber nicht durch schuldrechtliche Beziehungen, wie sie die Formulierung „Wahrnehmung bestimmter Dienstleistungsaufgaben" nahelegt, hergestellt, sondern allein organisationsrechtlich durch die Ausübung gesellschaftsrechtlich vermittelter Macht.[55] Die herrschende GmbH erbringt den abhängigen Gesellschaften keine Dienstleistungen, sondern realisiert im arbeitsteiligen Zusammenwirken mit ihnen ihren eigenen Unternehmensgegenstand.[56] Dabei macht es keinen Unterschied, in welchem Konzernunternehmen welche Unternehmensfunktion angesiedelt ist. Die Holding kann sich auf reine Managementaufgaben beschränken, während den Tochtergesellschaften Produktion, Vertrieb etc. zugewiesen sind. Entscheidend ist nur, daß alle Konzerngesellschaften zur Verwirklichung des Unternehmensgegenstandes der Konzernobergesellschaft beitragen. Denn der Konzern ist eine Organisationsform zweckgerichteten unternehmerischen Handelns, in der in arbeitsteiliger Weise der Unternehmensgegenstand der Konzernobergesellschaft verwirklicht wird. Im Unternehmensgegenstand der Konzernobergesellschaft muß sich also

---

52 *W. Timm*, Konzernspitze, 1980, S. 89 f. (für die AG)

53 *W. Timm*, Konzernspitze, 1980, S. 90 (für die AG).

54 *H.-G. Koppensteiner*, Kölner Kommentar, 1986, § 18 Rdnr. 15; *U. Hüffer*, AktG, 1997, § 18 Rdnr. 10; *A. Hueck/G. Hueck*, Baumbach/ Hueck, AktG, 1968, Üb vor § 329 Rdnr. 2.

55 Siehe hierzu *U. Hüffer*, AktG, 1997, § 17 Rdnr. 8 m.w.N.

56 Ebenso *P. Hommelhoff*, Konzernleitungspflicht, 1982, S. 269 ff.; *J. Tieves*, Unternehmensgegenstand, 1998, S. 480.

der Tätigkeitsschwerpunkt des Konzerns abbilden, der im Rahmen einer konzerndimensionalen Betrachtung zu ermitteln ist.

Damit schützt der Unternehmensgegenstand die gegenwärtigen und zukünftigen Gesellschafter lediglich davor, daß das in die Gesellschaft eingebrachte Kapital nicht in einer anderen Branche als der von ihnen gewählten investiert wird.[57] Ein anderer Schutzzweck des Unternehmensgegenstandes kann dem Wortlaut des § 3 Abs. 1 Nr. 2 GmbHG nicht entnommen werden.

Auch ist es nicht erforderlich, den Zweck von § 3 Abs. 1 Nr. 2 GmbHG auf den Schutz der Gesellschafter vor der Hereinnahme divergierender Interessen von Arbeitnehmern und außenstehenden Gesellschaftern zu erweitern. Ebensowenig ist es eine Frage des Unternehmensgegenstandes, ob sich die Geschäftsführer durch die Auslagerung unternehmerischen Handelns der Kontrolle der Gesellschafter entziehen dürfen. Charakteristisch für beide Fälle ist nicht, daß die Gesellschaft durch die Konzernierung ihren Unternehmensgegenstand verändert - dieser bleibt gleich -, charakteristisch ist vielmehr, daß sich Verwerfungen in der Zuständigkeitsordnung der Konzernobergesellschaft einstellen. Nicht nur der Geschäftsführer muß nun die Interessen der Arbeitnehmer und außenstehenden Gesellschafter der Tochtergesellschaft beachten, sondern auch die Gesellschafter bei Ausübung ihres Weisungsrechts. Und ob der Geschäftsführer sich teilweise der Kontrolle der Gesellschafter entziehen darf, haben diese zu entscheiden und nicht er selbst. Damit sind die Rechte der Gesellschafter und die gesellschaftsinterne Zuständigkeitsordnung der sachgerechte Ansatzpunkt, um die Frage nach der Zulässigkeit einer Konzernierung zu beantworten, nicht aber der Unternehmensgegenstand.

3. Ergebnis

Nach alledem kann als Ergebnis festgehalten werden, daß eine Konzerngründung vom Unternehmensgegenstand der herrschenden GmbH gedeckt ist, wenn

- der Unternehmensgegenstand der beherrschten Gesellschaft mit dem Unternehmensgegenstand der herrschenden GmbH im Kern identisch ist oder
- sich der Unternehmensgegenstand der beherrschten Gesellschaft im Funktionsbereich des Unternehmensgegenstandes der herrschenden GmbH bewegt und die Tochtergesellschaft ausschließlich konzerninterne Leistungen erbringt.

---

[57] Ebenso für die Aktiengesellschaft *M. H. Kessler*, Leitungsmacht, 1995, S. 67.

V. Die Pflicht zur Wahrung der Mitgliedschaftsrechte der Gesellschafter

1. Die gesetzliche Ausgangslage

Eine weitere Grenze der Geschäftsführungsbefugnis des Geschäftsführers ist seine Pflicht zur Wahrung der Gesellschafterrechte. Sie kommt zwar auch in den anschließend zu erörternden Vorlagepflichten des Geschäftsführers zum Ausdruck. Danach hat der Geschäftsführer den Gesellschaftern bestimmte Geschäftsführungsmaßnahmen zur Entscheidung vorzulegen. Trotzdem kommt der Pflicht zur Wahrung der Gesellschafterrechte im Rahmen der Konzerngründung eine eigenständige Bedeutung zu. So ist es Zweck der Vorlagepflichten, den Gesellschaftern Gelegenheit zu geben, ihre Mitverwaltungsrechte auszuüben. Bei der Konzerngründung würde dagegen eine Vorlagepflicht nicht zuerst den Zweck haben, die Gesellschaftern die *Ausübung* ihrer Mitverwaltungsrechte zu ermöglichen. Vielmehr würde sie ihnen vor allem und zuerst Gelegenheit geben, eine dauernde Schmälerung ihrer Mitgliedschaftsrechte zu verhindern.

Das wird deutlich, wenn man erkennt, daß die Konzerngründung, wie im folgenden gezeigt wird, den Gesellschaftern die Wahrnehmung ihrer Mitgliedschaftsrechte erschwert. Und von diesen Auswirkungen geht auch die herrschende Meinung aus, wenn sie die Konzerngründung von einer entsprechenden Satzungsregelung abhängig macht.[58] Denn erst mit dieser Satzungsregelung, die im folgenden als Konzernklausel bezeichnet wird, bringen die Gesellschafter zum Ausdruck, daß sie mit dem durch die Konzerngründung erfolgenden Eingriff in ihre Mitgliedschaftsrechte einverstanden sind.

Ohne eine solche Konzernklausel ergibt dagegen eine Auslegung der Satzung, daß der Geschäftsführer das Unternehmen der GmbH so zu führen hat, daß den Mitgliedschaftsrechten der Gesellschafter die größtmögliche Wirksamkeit zukommt. Dem würde es aber widersprechen, wenn der Geschäftsführer unternehmerisches Handeln auf andere Gesellschaften auslagern würde. Denn hierdurch würde die Wahrnehmung der Mitgliedschaftsrechte erschwert.

Der weitere Gang der Untersuchung gestaltet sich damit wie folgt: Zunächst werden die Auswirkungen der Konzerngründung auf die Gesellschafterrechte aufgezeigt. Sodann ist der Frage nachzugehen, in welcher Form und mit welcher Mehrheit die Gesellschafter den mit der Konzerngründung verbundenen Einschnitt in ihre Mitgliedschaftsrechte zustimmen können. Es wird gezeigt werden, daß eine solche Gestattung nur durch Aufnahme einer sogenannten Kon-

---

[58] *Uwe H. Schneider*, Konzerngründung, 1986, S. 128; *H.-G. Koppensteiner*, Rowedder, 1997, Anh. nach § 52 Rdnr. 35; *B. Sonntag*, Konzernbildungskontrolle, 1990, S. 211 ff.; *J. Tieves*, Unternehmensgegenstand, 1998, S. 495 f.

zernklausel in die Satzung erfolgen kann. Schließlich soll untersucht werden, welche Reichweite eine für die Praxis typische Konzernklausel hat.

2. Die Mediatisierung der Mitgliedschaftsrechte der Gesellschafter

a) Die Mediatisierung der Mitgliedschaftsrechte

Durch die Gründung eines Konzerns werden die Rechte und Zuständigkeiten der Gesellschafter in vielfältiger Weise berührt.[59] Ihre Kontrolle über das in der Tochtergesellschaft investierte Gesellschaftsvermögen wird mediatisiert. Entscheidungen, die vorher die Gesellschafter getroffen haben, trifft nun der Geschäftsführer bzw. Vorstand, der die Gesellschaft in der Tochtergesellschaft vertritt. *H. Wiedemann*[60] vergleicht deshalb die Mediatisierung des Gesellschaftsvermögens durch Anlage in einer Tochtergesellschaft mit der Abtretung von Mitgliedschaftsrechten an einen Treuhandgesellschafter: Gesellschaftsrechtliche Befugnisse würden durch die Konzernierung in bloß schuldrechtlich-treuhänderisch gebundene zurückgestuft.

Diesen Vergleich zog *H. Wiedemann* für die Aktiengesellschaft. Er gilt aber auch für die GmbH. Zwar ist dort der Einschnitt in die Rechte der Gesellschafter nicht so tief wie bei der Aktiengesellschaft. Denn es wird gezeigt werden, daß der Geschäftsführer im Rahmen der Konzernleitung bestimmte Maßnahmen, die in der beherrschten Gesellschaft vorgenommen werden sollen, den Gesellschaftern der herrschenden GmbH zur Entscheidung vorzulegen hat. Hält sich der Geschäftsführer aber nicht an diese Pflicht, so werden in der beherrschten Gesellschaft unter Umständen Tatsachen geschaffen, die nicht mehr rückgängig gemacht werden können.[61] Zu denken ist etwa an die Feststellung des Jahresabschlusses und Beschlußfassung über die Gewinnverwendung in einer Tochter-GmbH. Diese Maßnahme hat der Geschäftsführer den Gesellschaftern der herrschenden GmbH zur Entscheidung vorzulegen.[62] Mißachtet er diese Pflicht und stimmt er z. B. für die vollständige Ausschüttung des Gewinns, so können die Gesellschafter der herrschenden GmbH jedenfalls dann hieran nichts

---

[59] Ebenso für die AG *P. O. Mülbert*, Aktiengesellschaft, 1996, S. 363; *J. Tieves*, Unternehmensgegenstand, 1998, S. 482 ff.; *M. Habersack*, Aktienkonzernrecht, 1998, vor § 311 Rdnr. 13. Für AG und GmbH *A. von Riegen*, Gesellschafterschutz, 1999, S. 19 ff. und S. 110 ff. A. M. *W. Gross*, Zuständigkeit, 1994, S. 273 ff; *D. Joost*, Holzmüller, 1999, S. 170, für den Fall der Ausgliederung: „Nicht die Rechte werden verletzt bzw. gefährdet, sondern der Rechtsgegenstand, auf den sich die Mitgliedschaftsrechte der Aktionäre beziehen, erfährt eine Veränderung".

[60] *H. Wiedemann*, Unternehmensgruppe, 1988, S. 53.

[61] *Uwe H. Schneider*, Konzerngründung, 1986, S. 127.

[62] Siehe unten 1. Teil C. II. 1. b) (2).

mehr ändern, wenn an der Tochter-GmbH Minderheitsgesellschafter beteiligt sind. Denn mit Feststellung des Jahresabschlusses entstehen die Ansprüche der Gesellschafter auf Auszahlung ihrer Gewinnanteile, und mit dem Ergebnisverwendungsbeschluß werden ihre Auszahlungsansprüche der Höhe nach festgelegt und fällig gestellt.[63] Ein einmal entstandener und fällig gewordener Ausschüttungsanspruch kann den Minderheitsgesellschafern aber nicht mehr genommen werden.

Freilich dürfte die Gefahr einer Mißachtung der Vorlagepflichten durch den Geschäftsführer eher gering sein. Es bleibt jedoch der Effekt einer Mediatisierung der Gesellschafterrechte bei der herrschenden GmbH. Schon er rechtfertigt es, die Konzerngründung an die Zustimmung der Gesellschafter der herrschenden GmbH zu binden. Das wird deutlich, wenn man das Ausmaß der Mediatisierung näher betrachtet.

So haben die Gesellschafter der GmbH gegenüber dem Geschäftsführer ein Weisungsrecht. Dieses Weisungsrecht der Gesellschafterversammlung wird durch die Konzernierung mediatisiert.[64] Konzentriert sich die unternehmerische Tätigkeit nur in einer Gesellschaft, können die Gesellschafter über ihr Weisungsrecht auf die Leitung des Unternehmens unmittelbar Einfluß nehmen. Wird das Unternehmen aber (auch) über eine Tochtergesellschaft geführt, wird die Realisierung der Weisung erschwert. Denn nun muß der angewiesene Geschäftsführer als Vertreter der GmbH in der Tochtergesellschaft erst gemäß § 50 Abs. 1 GmbHG die Einberufung der Gesellschafterversammlung beantragen und dann einen entsprechenden Weisungsbeschluß herbeiführen. Da zwischen der Einladung der Gesellschafter und dem Stattfinden der Gesellschafterversammlung gemäß § 51 Abs. 1 Satz 2 GmbHG mindestens eine Woche liegen muß, kann die von den Gesellschaftern der herrschenden GmbH ausgehende Weisung schon gegenstandslos geworden sein.[65]

Selbst wenn die Geschäftsführer beider Gesellschaften personengleich sind, ändert sich an diesem Befund nichts. Denn auch dann muß erst die Gesellschafterversammlung der Tochtergesellschaft einberufen werden, um den außenstehenden Gesellschaftern Gelegenheit zur Stellungnahme zu geben. Andernfalls würde diesen die einzige Möglichkeit genommen, sich gegen die Weisung zu wehren. Denn erstens haben sie keine Kenntnis von der Weisung und zweitens existiert kein Beschluß, den sie anfechten können.

---

[63] *M. Lutter/P. Hommelhoff*, GmbHG, 1995, § 29 Rdnr. 4.

[64] Vgl. auch *B. Sonntag*, Konzernbildungskontrolle, 1990, S. 208.

[65] So auch für den Fall der Ausgliederung *A. Zitzmann*, Vorlagepflichten, 1991, S. 61.

Des weiteren ändern sich im Konzern auch die rechtlichen Rahmenbedingungen der Geschäftsführung,[66] was ebenfalls Auswirkungen auf das Weisungsrecht der Gesellschafter hat. So hat der Geschäftsführer bei der Konzernleitung nicht mehr nur die Interessen der eigenen Gesellschaft, sondern auch die der betroffenen Konzerngesellschaften zu beachten. Gegenüber der faktischen beherrschten Aktiengesellschaft (§§ 311, 317 AktG) und sogar gegenüber der vertraglich beherrschten Gesellschaft (§ 308 Abs. 1 Satz 2 AktG) gibt es gewichtige Schranken des Einflusses, gegenüber dem eigenen Unternehmen der herrschenden GmbH und seinen verschiedenen Betrieben aber nicht. Diese Schranken des Einflusses müssen nicht nur die Geschäftsführer bei der Konzernleitung beachten, sondern auch die Gesellschafter bei Ausübung ihres Weisungsrechts.

Der Einschnitt in das Weisungsrecht der Gesellschafterversammlung wird verstärkt durch die Auswirkungen der Konzernierung auf das Auskunfts- und Einsichtsrecht der Gesellschafter gemäß § 51a GmbHG. Während sich das Auskunftsrecht unstreitig auch auf die Beziehungen der Gesellschaft zu verbundenen Unternehmen erstreckt,[67] bezieht sich das Einsichtsrecht nur auf die Bücher und Schriften der eigenen Gesellschaft. In die der Tochtergesellschaft kann der Gesellschafter der herrschenden GmbH dagegen keine Einsicht nehmen.[68] Eine Ausnahme besteht nur dann, wenn die Berufung auf die „förmliche Verschiedenheit“ von Mutter- und Tochtergesellschaft gegen den Grundsatz von Treu und Glauben verstößt. So hat z. B. der Gesellschafter des herrschenden Gesellschaft gegen seine Gesellschaft einen Anspruch auf Einsicht in die Schriften und Bücher der Tochtergesellschaft, wenn diese zu 100 % der herrschenden Gesellschaft gehört und von dieser wie eine Betriebsabteilung geführt wird.[69]

Weiterhin ist die Mediatisierung der in § 46 Nr. 1 GmbHG genannten Organrechte zu nennen. So ist gemäß § 46 Nr. 1 GmbHG die Gesellschafterversammlung für die Feststellung des Jahresabschlusses zuständig. Nur wenn dieser einen Jahresüberschuß oder einen Bilanzgewinn ausweist, kann an die Gesellschafter gemäß § 29 Abs. 1 GmbHG eine Gewinnausschüttung erfolgen. Zum durchsetzbaren Gewinnanspruch wandelt sich das allgemeine Gewinnbezugsrecht der Gesellschafter jedoch erst dann um, wenn die Gesellschafterversamm-

---

[66] Für die Aktiengesellschaft siehe *M. Lutter*, Organzuständigkeiten, 1985, S. 838 f.

[67] *M. Lutter/P. Hommelhoff*, GmbHG, 1995, § 51a Rdnr. 10; *Uwe H. Schneider*, Konzernleitung, 1981, S. 252.

[68] *M. Lutter/P. Hommelhoff*, GmbHG, 1995, § 51a Rdnr. 14; *W. Zöllner*, Baumbach/Hueck, 1996, § 51a Rdnr. 15.

[69] *BGH*, Urt. v. 8.7.1957 - II ZR 54/56, in: BGHZ 25, 115, 117 ff.

lung gemäß § 46 Nr. 1 GmbHG einen entsprechenden Beschluß über die Ergebnisverwendung faßt.[70]

Die vollständige Bedeutung der Zuständigkeit der Gesellschafterversammlung für die Feststellung des Jahresabschlusses erschließt sich aber erst dann, wenn man berücksichtigt, daß nicht die Gesellschafter, sondern die Geschäftsführer den Jahresabschluß aufstellen, § 42a Abs. 1 Satz 1 GmbHG. Wäre diese Aufstellung für die Gesellschafter verbindlich, hätten die Geschäftsführer es in der Hand, durch die Ausübung von Gestaltungsmöglichkeiten bei der Bilanzierung und Bewertung einzelner Bilanzpositionen die Höhe des Jahresüberschusses maßgeblich zu beeinflussen.[71]

Während schon für die unabhängige GmbH die Frage streitig ist, bis zu welcher Grenze der Mehrheitsgesellschafter durch die Ausübung von Bilanzierungs- und Bewertungswahlrechten bei Feststellung des Jahresabschlusses über die Gewinnausschüttung an den Minderheitsgesellschafter entscheiden kann,[72] stellt sich diese Frage im Konzern völlig neu. Denn nun hat es nicht (nur) der Mehrheitsgesellschafter, sondern (auch) der Geschäftsführer in der Hand, über den an die Gesellschafter auszuschüttenden Gewinn zu bestimmen. Dessen Höhe hängt nämlich von dem Gewinn ab, den die Tochtergesellschaft an die Muttergesellschaft ausschüttet, und in der Gesellschafterversammlung der Tochtergesellschaft sitzt der Geschäftsführer der herrschenden GmbH und beschließt über die Feststellung des Jahresergebnisses und damit über die Höhe des ausschüttbaren Gewinns.[73] Gleiches gilt für den Beschluß über die Ergebnisverwendung.[74] Damit besteht für die Gesellschafter der herrschenden GmbH die Gefahr des sogenannten „Tresoreffekts": Gewinne werden in den Konzernunternehmen thesauriert und erreichen nicht die herrschende GmbH.[75]

Besonders an dieser Stelle zeigt sich die Bedeutung der Vorlagepflichten des Geschäftsführers im Konzern. So wird gezeigt werden, daß der Geschäftsführer in bestimmten Fällen verpflichtet ist, die Frage der Feststellung und Verwen-

---

70 *V. Emmerich*, Scholz, 1993, § 29 Rdnr. 42; *H.-G. Koppensteiner*, Rowedder, 1997, § 46 Rdnr. 8; *H. Sudhoff/M. Sudhoff*, Rechte, 1994, S. 84.

71 Zur Bilanzpolitik im Jahresabschluß siehe *Th. Mueller-Thuns*, Gewinnbezugsrecht, 1989, S. 14 ff., sowie *G. Wöhe*, Steuerbilanz, 1990, S. 267 ff.

72 Siehe hierzu *Th. Mueller-Thuns*, Gewinnbezugsrecht, 1989, S. 54 ff.; *K. Schmidt*, Scholz, 1995, § 46 Rdnr. 39.

73 Siehe hierzu auch *Uwe H. Schneider*, Konzernverfassung, 1986, S. 1997.

74 Zum Schutz der Minderheit bei Beschlüssen über die Ergebnisverwendung in der unabhängigen GmbH siehe *H.-G. Koppensteiner*, Rowedder, 1997, § 46 Rdnr. 9.

75 *Uwe H. Schneider*, Konzernfinanzierung, 1984, S. 502. Für die AG: *H. Götz*, Sicherung, 1984, S. 86; *M. Lutter*, Binnenstruktur, 1974, S. 352 f.; *H. P. Westermann*, Organzuständigkeit, 1984, S. 373; *H. W. Wahlers*, Konzernbildungskontrolle, 1995, S. 75 ff.

dung des Gewinns in der beherrschten Gesellschaft der Gesellschafterversammlung der herrschenden GmbH zur Entscheidung vorzulegen ist. Die Gefahr des Tresoreffekts wird dadurch gemindert.

Diese Ausführungen zeigen, daß durch die Konzernierung die organschaftlichen Befugnisse der Gesellschafter, über die Feststellung des Jahresabschlusses und die Ergebnisverwendung zu bestimmen, mediatisiert werden.

Vergleichbare Auswirkungen hat die Konzerngründung auf das Recht der Gesellschafter, gemäß § 46 Nr. 5 GmbHG die Geschäftsführer zu bestellen und abzuberufen sowie über ihre Entlastung zu beschließen. Denn als Vertreter der herrschenden GmbH entscheidet der Geschäftsführer, wer in der Tochtergesellschaft zum Geschäftsführer bestellt oder in den Aufsichtsrat und damit in den Vorstand gewählt wird.

Schließlich eröffnet die Gründung einer Tochtergesellschaft dem Geschäftsführer die Möglichkeit, in diese neue Gesellschafter aufzunehmen.[76] Während in der herrschenden GmbH die Gesellschafter über eine Kapitalerhöhung und damit über die Zulassung neuer Gesellschafter entscheiden, trifft in der Tochtergesellschaft der Geschäftsführer als gesetzlicher Vertreter der herrschenden GmbH die maßgeblichen Entscheidungen.[77]

b) Kein Unterschied zwischen internem und externem Unternehmenswachstum

Die vorstehenden Ausführungen haben gezeigt, daß durch die Konzernbildung die Rechte der Gesellschafter der herrschenden GmbH mediatisiert werden. Dabei macht es prinzipiell keinen Unterschied, ob es sich bei der Konzernbildung um ein internes oder um ein externes Wachstum des von der GmbH betriebenen Unternehmens handelt.[78] Gliedert die GmbH einen Unternehmensteil auf eine Tochtergesellschaft aus und schafft so neue eigene Kapazitäten (internes Wachstum[79]), werden die Rechte der Gesellschafter sofort mediatisiert. Aber auch dann, wenn die GmbH eine Beteiligung an einer bereits bestehenden Gesellschaft erwirbt, sich also mit einem unternehmensfremden System

---

[76] Ebenso für den Fall der Ausgliederung *B. Sonntag*, Konzernbildungskontrolle, 1990, S. 208.

[77] Besonders bei der Aktiengesellschaft besteht deshalb die Gefahr, daß der Vorstand Mitgliedschaftsrechte an den "Ästen" schafft und Dritten überläßt (*M. Lutter*, Binnenstruktur, 1974, S. 359).

[78] *B. Sonntag*, Konzernbildungskontrolle, 1990, S. 219; A. M. *M. R. Theisen*, Konzern, 1991, S. 159.

[79] Siehe hierzu *M. R. Theisen*, Konzern, 1991, S. 133 f.

(externes Wachstum[80]) verbindet, werden die Rechte der Gesellschafter berührt. Denn das in eine andere Gesellschaft investierte Kapital und die damit finanzierte unternehmerische Tätigkeit unterliegen einer viel geringeren Kontrolle und Einwirkungsmöglichkeit seitens der Gesellschafter als das Vermögen der eigenen GmbH.[81] Auch besteht die Gefahr, daß der Geschäftsführer auf eine Erweiterung der Geschäftstätigkeit der eigenen GmbH verzichtet und damit potentielle Rechte der Gesellschafter abschneidet. So macht es im Ergebnis keinen Unterschied, ob der Geschäftsführer zunächst die Produktion in der eigenen Gesellschaft um einen Produktionszweig erweitert und diesen dann auf eine Tochtergesellschaft ausgliedert oder ob er sofort eine Tochtergesellschaft gründet und dort das neue Produkt herstellen läßt oder aber sich an einer bereits bestehenden Gesellschaft beteiligt und in dieser seine Pläne verwirklicht. In allen Fällen werden den Gesellschaftern zumindest potentielle Einflußmöglichkeiten entzogen.[82]

c) Kein Unterschied zwischen der Gründung kleiner und großer Tochtergesellschaften

Schließlich werden die Rechte der Gesellschafter auch dann mediatisiert, wenn die Tochtergesellschaft im Verhältnis zur Muttergesellschaft nur einen unbedeutenden Umfang hat. Auch dort kann der Geschäftsführer wirksam neue Gesellschafter aufnehmen. Auch dort übt er die Gewinnbezugsrechte für die herrschende GmbH aus. Insbesondere kann sich die Tochtergesellschaft jederzeit verändern und an Bedeutung im Gesamtkonzern hinzugewinnen. Deshalb werden auch durch die Gründung oder den Erwerb einer nur kleinen Tochtergesellschaft die Rechte der Gesellschafter berührt.[83]

3. Die Legitimation durch eine Konzernklausel

Die bisherigen Ausführungen haben gezeigt, daß eine Konzerngründung nicht mehr von der Geschäftsführungsbefugnis des Geschäftsführers gedeckt ist. Freilich ist der Gesellschaft hiermit nicht der Weg zu einer Konzerngründung versperrt. Vielmehr kann die Gesellschafterversammlung als oberstes Organ der GmbH die Voraussetzungen für eine Konzerngründung schaffen.

---

80 Siehe hierzu *M. R. Theisen*, Konzern, 1991, S. 134.

81 *B. Sonntag*, Konzernbildungskontrolle, 1990, S. 219.

82 Vgl. auch *A. Zitzmann*, Vorlagepflichten, 1991, S. 60 f., der für die Ausgliederung eine besondere Satzungsbestimmung für erforderlich hält. Für die Aktiengesellschaft siehe *H. W. Wahlers*, Konzernbildungskontrolle, 1995, S. 94 ff.

83 Im Ergebnis ebenso *Uwe H. Schneider*, Konzerngründung, 1986, S. 129.

Erforderlich hierfür ist die Aufnahme einer sogenannten Konzernklausel in die Satzung.[84] Ein einfacher Gesellschafterbeschluß reicht dagegen nicht aus. Das ergibt sich aus dem Begriff der Satzung. „Satzung im materiellen Sinn ist die verbandsautonom geschaffene normative Grundordnung eines Verbandes, durch die die gesetzlich vorgesehene Grundordnung ergänzt bzw. im Rahmen des Verbandsautonomiespielraums abgeändert wird".[85] Eine solche Abänderung der gesellschaftlichen Grundordnung erfolgt aber gerade durch die Mediatisierung des Weisungsrechts, des Auskunfts- und Einsichtsrechts, des Rechts zur Feststellung des Jahresabschlusses und zur Bestimmung über die Gewinnverwendung.

Ausreichend für die Satzungsänderung ist eine Dreiviertelmehrheit in der Gesellschafterversammlung.[86] Für ein Einstimmigkeitserfordernis finden sich keine Anhaltspunkte im Gesetz. Auch spricht gegen ein Einstimmigkeitserfordernis das Bedürfnis der Gesellschafter nach flexibler Entscheidungsfindung, die eine schnelle Anpassung an das sich ständig ändernde Unternehmensumfeld ermöglicht.[87]

4. Inhalt und Reichweite einer Konzernklausel

Es soll nun untersucht werden, welchen Inhalt eine Konzernklausel haben muß, um die statutarische Grundlage für eine Konzerngründung zu schaffen. Ausgangspunkt soll folgende in der Praxis übliche Satzungsklausel sein:

> „Die Gesellschaft ist berechtigt, Beteiligungen zu erwerben und Tochtergesellschaften zu gründen".

Uwe H. Schneider[88] sieht in einer solchen Klausel eine bloße Absichtserklärung, die an der Notwendigkeit einer Satzungsänderung nichts ändere. Seiner Meinung nach müsse der Unternehmensgegenstand zumindest den Zusatz enthalten: „Die Gesellschaft ist zugleich konzernleitende Holding des X-Konzerns".

Einen Schritt weiter geht *B. Sonntag*[89]. Nach seiner Ansicht gestattet die eingangs erwähnte Klausel immerhin den Beteiligungserwerb und die Gründung von Tochtergesellschaften. Dagegen sei für die Aufnahme der Konzernleitung

84 *B. Sonntag*, Konzernbildungskontrolle, 1990, S. 232.
85 *W. Zöllner*, Baumbach/Hueck, 1996, § 53 Rdnr. 3.
86 *B. Sonntag*, Konzernbildungskontrolle, 1990, S. 232.
87 *Uwe H. Schneider*, Konzerngründung, 1986, S. 128 f.
88 *Uwe H. Schneider*, Konzerngründung, 1986, S. 134.
89 *B. Sonntag*, Konzernbildungskontrolle, 1990, S. 233.

eine weitere Satzungsänderung in der von *Uwe H. Schneider* vorgeschlagenen Weise erforderlich.

Nach *W. Timm*[90] schließlich erlaubt eine solche Klausel zwar die Aufnahme einheitlicher Leitung, jedoch nur als Annextätigkeit zum Unternehmensgegenstand. Die Umstrukturierung der Gesellschaft zu einer reinen Holding sei von ihr nicht gedeckt.

Es ist wenig überzeugend, einer Satzungsklausel lediglich den Charakter einer Absichtserklärung beizumessen. Dies widerspricht in der Regel dem Willen der Gesellschafter, die die Grundordnung der Gesellschaft rechtsverbindlich festlegen wollen.

Auch die Meinung von *B. Sonntag* ist nicht überzeugend. Hat die Gesellschaft erst einmal die Beteiligungen erworben, so haben ihre Gesellschafter auch ein Interesse daran, daß der Geschäftsführer die einheitliche Leitung ausübt und damit so weit wie möglich die Interessen der herrschenden GmbH zur Geltung bringt. Deshalb macht es keinen Sinn, dem Geschäftsführer nur den Beteiligungserwerb, nicht aber die Ausübung einheitlicher Leitung zu erlauben.

Daher entspricht es einer aus Sicht der beteiligten Verkehrskreise sachgerechten Auslegung der obengenannten Satzungsklausel, daß sie die statutarischen Voraussetzungen nicht nur für den Beteiligungserwerb und die Gründung einer Tochtergesellschaft, sondern auch für die Aufnahme einheitlicher Leitung schafft. *W. Timm* ist allerdings zuzugeben, daß zum Schutze der Gesellschafter die Klausel restriktiv ausgelegt werden muß und nur solche Konzerngründungen umfaßt, die sich als Annex zur unternehmerischen Tätigkeit in der eigenen Gesellschaft darstellt. Soll die Konzernobergesellschaft dagegen in eine reine Holding umstrukturiert werden können, muß die Satzung eine, wie *Uwe H. Schneider*[91] formuliert, qualifizierte Konzernklausel mit etwa folgendem Wortlaut enthalten:

„Die Gesellschaft ist konzernleitende Holding des X-Konzerns".

Ist danach dem Geschäftsführer eine Konzerngründung gestattet, stellt sich für ihn schließlich die Frage, ob er eine Mehrheitsbeteiligung bzw. Beteiligung mit unternehmerischer Leitung schrittweise aufbauen darf oder ob er sie als Paket erwerben muß.[92]

Gegen den ersten Weg sprechen insofern Bedenken, als die GmbH zumindest zeitweise eine Minderheitsbeteiligung halten würde, die keinen unternehmerischen Einfluß auf die Beteiligungsgesellschaft vermittelt. Grundsätzlich

---

90 *W. Timm*, Konzernspitze, 1980, S. 94.

91 *Uwe H. Schneider*, Konzerngründung, 1986, S. 134.

92 Zur Lage bei der AG siehe *W. Gross*, Zuständigkeit, 1994, S. 270.

sind aber alle Gesellschaftsmittel im Rahmen des Unternehmensgegenstandes einzusetzen. Darüber hinaus ist der Mediatisierungseffekt bei einer Minderheitsbeteiligung noch größer als bei einer Beteiligung mit unternehmerischer Leitung. Insbesondere haben die Gesellschafter keinen Einfluß auf die Gewinnverwendung in der Beteiligungsgesellschaft. Trotzdem muß dem Geschäftsführer die Möglichkeit gegeben werden, die Mehrheitsbeteiligung auch schrittweise aufzubauen. Wäre nur der andere Weg gangbar, würde die Konzerngründung erheblich erschwert. Dies widerspräche aber dem in der Konzernklausel zum Ausdruck gebrachten Interesse der Gesellschafter an einer flexiblen Unternehmensorganisation. Ein Regulativ bietet in diesem Zusammenhang der Unternehmensgegenstand. Er definiert den Schwerpunkt der Geschäftstätigkeit der GmbH und verhindert so, daß der Geschäftsführer wesentliche Mittel der GmbH in Beteiligungen investiert, die er nicht zur unternehmerischen Einflußnahme nutzen kann.[93]

5. Ergebnis

Nach alledem kann festgehalten werden, daß die Gründung eines Konzerns einer ausdrücklichen Gestattung in der Satzung bedarf. Diese Gestattung kann ein unterschiedliches Ausmaß erreichen. Am weitesten geht die sog. qualifizierte Konzernklausel, die die satzungsmäßigen Grundlagen für die Umwandlung der Gesellschaft in eine reine Holding schafft.

VI. Die Vorlagepflichten

1. Abgrenzung

Mit Aufnahme der Konzernklausel in die Satzung haben die Gesellschafter lediglich einer Mediatisierung ihrer Mitgliedschaftsrechte zugestimmt, die als Folge der Konzerngründung eintritt. Das bedeutet, daß für die Gründung eines Konzerns nicht mehr eine Satzungsänderung mit der erforderlichen Dreiviertelmehrheit notwendig ist. Nicht beantwortet ist damit jedoch die Frage, ob der Geschäftsführer nicht trotzdem für die einzelne Konzerngründungsmaßnahme die Zustimmung der Gesellschafter einholen muß. Freilich wird man eine solche Zustimmung zu einer konkreten Konzerngründungsmaßnahme annehmen können, wenn die Aufnahme der Konzernklausel in die Satzung gerade aus Anlaß dieser Maßnahme erfolgt. Dann fallen die Zustimmung zur Mediatisierung der Gesellschafterrechte und zu der konkreten Konzerngründungsmaßnahme zu-

---

93 *Uwe H. Schneider*, Konzerngründung, 1986, S. 124 f.

sammen. Häufig wird jedoch die Konzernklausel zu einem Zeitpunkt - meistens bei Gründung der Gesellschaft - in die Satzung aufgenommen, in dem eine Konzerngründung noch nicht geplant ist. In diesem Fall stellt sich für den Geschäftsführer die Frage, ob er später erfolgende Konzerngründungsmaßnahmen den Gesellschaftern zur Entscheidung vorzulegen hat.

Eine solche Vorlagepflicht kann sich ergeben, weil es sich bei der Konzerngründung um eine ungewöhnliche Maßnahme handelt.

2. *Ungewöhnliche Maßnahmen*

a) Die gesetzliche Ausgangslage

Einigkeit besteht darüber, daß den Geschäftsführer auch außerhalb der im Gesetz ausdrücklich genannten Fälle Vorlagepflichten treffen. Umstritten ist indes, nach welchen Kriterien diese Vorlagepflichten zu bestimmen sind.

Eine Ansicht bezieht sich auf § 49 Abs. 2 GmbHG und hält eine Vorlagepflicht des Geschäftsführers nur dann für geboten, wenn dies im Interesse der Gesellschaft erforderlich erscheine. Das sei nicht schon bei vom Umfang oder Gegenstand her ungewöhnlichen Geschäften der Fall, sondern nur, wenn sie substantiell in die Rechte der Gesellschafter eingriffen[94] oder mit einem außerordentlichen Risiko für diese verbunden seien[95]. Als Beispiel für einen substantiellen Eingriff in die Rechte der Gesellschafter wird die Ausgliederung wesentlicher Unternehmensteile, etwa in dem durch die Holzmüller-Entscheidung des *Bundesgerichtshofs*[96] umrissenen Rahmen genannt.[97] Es wird aber auch vertreten, daß § 49 Abs. 2 GmbHG nur bei bestandsgefährdenden Maßnahmen Platz greife.[98]

Eine andere Meinung setzt die Schwelle für die Vorlagepflicht des Geschäftsführers dagegen etwas tiefer an. Danach ist der Geschäftsführer schon dann zur Vorlage verpflichtet, wenn das Geschäft ungewöhnlich ist und über

---

[94] *W. Zöllner*, Baumbach/Hueck, 1996, § 37 Rdnr. 6e. Ebenso *H.-J. Mertens*, Hachenburg, 1996, § 37 Rdnr. 10, jedoch ohne ausdrückliche Bezugnahme auf § 49 Abs. 2 GmbHG.

[95] *H.-J. Mertens*, Hachenburg, 1996, § 37 Rdnr. 10, jedoch ohne ausdrückliche Bezugnahme auf § 49 Abs. 2 GmbHG.

[96] *BGH*, Urt. v. 25.02.1982 - II ZR 174/80, in: BGHZ 83, S. 122, 131f.

[97] *W. Zöllner*, Baumbach/Hueck, 1996, § 37 Rdnr. 6e; *H.-J. Priester*, Holzmüller, 1999, S. 198; *H.-J. Mertens*, Hachenburg, 1996, § 37 Rdnr. 10, jedoch ohne ausdrückliche Bezugnahme auf § 49 Abs. 2 GmbHG.

[98] *A. Zitzmann*, Vorlagepflichten, 1991, S. 159.

den normalen Geschäftsbetrieb hinausgeht.[99] Dabei wird zur Abgrenzung zwischen gewöhnlichen und ungewöhnlichen Geschäften auf das Material zu § 116 Abs. 2 HGB verwiesen.[100]

Eine Entscheidung dieses Meinungsstreits muß auf der Grundlage der gesetzlichen Ausgestaltung der GmbH-Verfassung erfolgen. Das Gesetz weist den Gesellschaftern in § 46 GmbHG wichtige unternehmensbezogene Entscheidungen zu. Außerdem können sie dem Geschäftsführer jederzeit Weisungen erteilen und die Geschäftsführung sogar ganz an sich ziehen. Dann müssen sie aber auch bei ungewöhnliche Geschäften ebenso mitwirken können wie ein Kommanditist gemäß § 164 HGB, der von der Geschäftsführung sogar ausgeschlossen ist. Das können sie aber nur, wenn der Geschäftsführer sie mit der geplanten Maßnahme befaßt. Daher trifft den Geschäftsführer eine Vorlagepflicht nicht nur bei Geschäften, die substantiell in die Rechte der Gesellschafter eingreifen oder mit einem besonderen Risiko für sie verbunden sind, sondern bei allen Geschäften, die über den gewöhnlichen Geschäftsbetrieb hinausgehen.

Im Rahmen der Konzerngründung ist das zum einen bei allen Konzernstrukturmaßnahmen der Fall. Zum anderen ist aber auch der Abschluß eines Beherrschungsvertrages ungewöhnlich, da er mit einem besonderen Risiko für die Gesellschaft verbunden sind.

b) Konzernstrukturmaßnahmen

Folgt man der hier vertretenen Auffassung, daß der Konzern eine Organisationsform zweckgerichteten unternehmerischen Handelns für ein wirtschaftlich einheitliches Unternehmen ist und daß dieses eine Unternehmen ein Grundorgan haben muß, nämlich die Gesellschafterversammlung der Konzernobergesellschaft, und ein Leitungsorgan, nämlich den Geschäftsführer der Konzernobergesellschaft, so ist die Antwort durch die Regelungsprinzipien unseres privaten Organisationsrechts vorgezeichnet. Die Gesellschafterversammlung ist für Konzernstrukturentscheidungen, der Geschäftsführer für Konzernleitungsentscheidungen zuständig.[101]

---

[99] *H.-G. Koppensteiner*, Rowedder, 1997, § 37 Rdnr. 10; *U. Hüffer*, Hachenburg, 1990, § 49 Rdnr. 20; *K. Schmidt*, Scholz, 1995, § 49 Rdnr. 20; *Uwe H. Schneider*, Scholz, GmbHG, 1993, § 37 Rdnr. 15; *M. Lutter/R. Leinekugel*, Kompetenzen, 1998, S. 231; *Hanseatisches OLG Hamburg*, Urt. v. 28.6.1991 - 11 U 148/90, in: GmbHR 1992, S. 43, 46.

[100] *H.-G. Koppensteiner*, Rowedder, 1997, § 37 Rdnr. 11.

[101] Für die AG *M. Lutter*, Organzuständigkeiten, 1985, S. 832. Im Ergebnis ebenso *H. Wiedemann*, Unternehmensgruppe, 1988, S. 55.

Wie Struktur- und Leitungsmaßnahmen abzugrenzen sind, ist freilich schwierig. Für die Aktiengesellschaft wird die Ansicht vertreten, daß bereits jede Auslagerung unternehmerischer Tätigkeit unabhängig von ihrem Umfang die Pflicht des Vorstands zur Vorlage an die Hauptversammlung auslöse.[102] Dagegen spricht jedoch sowohl bei der Aktiengesellschaft als auch bei der GmbH, daß die Gesellschafter durch Aufnahme einer Konzernklausel in die Satzung der Auslagerung unternehmerischer Tätigkeit auf eine Tochtergesellschaft bereits grundsätzlich zugestimmt haben. Deshalb sind Bagatellfälle von der Vorlagepflicht auszunehmen.

Auf der anderen Seite wäre es zu eng, die Vorlagepflicht des Geschäftsführers auf die Auslagerung eines Teilbetriebes im steuerrechtlichen Sinne, der für das von der Gesellschaft betriebene Unternehmen von erheblichem Gewicht ist, zu beschränken.[103] Denn dies würde bedeuten, daß nur bei substantiellen Eingriffen in die Rechte der Gesellschafter eine Vorlagepflicht entstehen würde. Oben wurde jedoch bereits festgestellt, daß die Schwelle für eine Vorlagepflicht deutlich niedriger anzusetzen ist.

Nach alledem läßt sich nur im Rahmen einer Einzelfallbetrachtung bestimmen, wann eine Strukturmaßnahme vorliegt. Maßgebliche Kriterien sind hierbei der Wert der zu übertragenden Vermögensgegenstände, die Ertragsstärke des auszulagernden Betriebsteils, dessen Nähe zum Kerngeschäft sowie dessen Einbindung in die langfristige Unternehmensstrategie der auslagernden Gesellschaft.[104]

So dürfte die Grenzlinie zur Strukturmaßnahme überschritten sein, wenn durch die Konzerngründung etwa 20-25 % der Aktivseite der Bilanz der herrschenden Gesellschaft betroffen sind.[105] Das gilt allerdings nur für konzerninterne Umstrukturierungen. Anders verhält es sich dagegen bei der Gründung einer Tochtergesellschaft mit einem dritten Partner. Übernimmt dieser unternehmerische Verantwortung, so findet eine noch stärkere Verkürzung der Rechte der Gesellschafter der herrschenden GmbH statt. Deshalb ist hier die Schwelle für eine Vorlagepflicht des Geschäftsführers deutlich geringer anzusetzen.[106]

---

[102] *D. Joost*, Holzmüller, 1999, S. 184.

[103] So für die AG *H.-J. Priester*, Ausgliederung, 1999, S. 195 f.

[104] *A. von Riegen*, Gesellschafterschutz, 1999, S. 102 ff., 121, der allerdings die Vorlagepflicht der Geschäftsführer bzw. Zuständigkeit der Gesellschafterversammlung mit einer analogen Anwendung der §§ 125 S. 1, 123 Abs. 3, 13 Abs. 1, 50 Abs. 1 UmwG begründet.

[105] *M. Lutter*, Organzuständigkeiten, 1985, S. 850. Ablehnend *M. H. Kessler*, Leitungsmacht, 1995, S. 69, 70.

[106] Für die AG: *M. Lutter*, Grundlagenbeschlüsse, 1988, S. 179 f. (durch die Konzerngründung 10 % des Umsatzes betroffen); *E. Geßler*, Hauptversammlungszuständig-

c) Abschluß eines Beherrschungsvertrages

Der Abschluß eines Beherrschungsvertrages hat keine strukturelle Wirkung, wenn die beherrschte Gesellschaft bereits faktisch in den Konzern eingebunden ist. Trotzdem stellt er eine ungewöhnliche Maßnahme dar. Denn infolge der Verlustübernahmepflicht der herrschenden GmbH analog § 302 AktG ist er für die Gesellschaft mit einem besonderen Risiko verbunden. Deshalb darf der Geschäftsführer einen Beherrschungsvertrag nicht ohne Zustimmung der Gesellschafter abschließen.[107]

Das ist heute anerkannt. Allerdings wird das Zustimmungserfordernis der Gesellschafter von der Rechtsprechung[108] und einem Großteil der Literatur[109] auf eine Analogie zu § 293 Abs. 2 AktG gestützt. Dort ist für eine Konzernobergesellschaft in der Rechtsform einer Aktiengesellschaft oder KGaA geregelt, daß deren Hauptversammlung dem Abschluß eines Beherrschungsvertrages mit einer Aktiengesellschaft als Konzernuntergesellschaft mit qualifizierter Mehrheit zustimmen muß, wenn er wirksam werden soll.

Folgt man dieser Meinung, so müßten auch die Gesellschafter einer herrschenden GmbH nicht mit einfacher Mehrheit, sondern mit einer Dreiviertelmehrheit dem Abschluß eines Beherrschungsvertrages zustimmen. Überdies

---

keiten, S. 787 (10 % vom Grundkapital oder Eigenkapital betroffen); *P. O. Mülbert*, Aktiengesellschaft, 1996, S.436 f. (Unternehmenswert der Untergesellschaft beträgt mindestens 10 % des Gesamtwertes der Unternehmensgruppe). Für AG und GmbH *J. Tieves*, Unternehmensgegenstand, 1998, S. 500, 521 (10 % des Gesellschaftsvermögens betroffen). Hauptversammlungszuständigkeit bejaht von *LG Frankfurt/M*, Urt. v. 10.3.1993 - 3/14 O 25/92, in: ZIP 1993, S. 830, 832 (50 % der Umsatzerlöse, 41 % des gezeichneten Kapitals betroffen). Hauptversammlungszuständigkeit verneint von *OLG Köln*, Urt. v. 24.11.1992 - 22 U 72/92, in: Die AG 1993, S. 86, 88 (8,25 % des Gesamtbeitragsaufkommens des Konzerns betroffen); *LG Düsseldorf*, Urt. v. 13.12.1997 - 31 O 133/96 („.W. Rau Neusser Öl und Fett AG"), in: Die AG 1999, S. 94, 95 (weniger als 50 % der Aktiva betroffen und Kernbereich der Unternehmenstätigkeit nicht berührt). Für die GmbH siehe auch *M. Henssler*, Minderheitenschutz, 1998, S. 216, nach dem die Zustimmung der Gesellschafterversammmlung erst dann erforderlich ist, wenn der Anteil der ausgegliederten Unternehmensbereiche am Jahresergebnis der ausgründenden Gesellschaft mindestens 50 % beträgt.

[107] *BGH*, Beschl. v. 24.10.1988 - II ZB 7/88, in: BGHZ 105, 324, 333 („Supermarkt"); bestätigt durch *BGH*, Beschl. v. 30.1.1992 - II ZB 15/91, in: GmbHR 1992, S. 253, 254 („Siemens"); *V. Emmerich*, Scholz, 1993, Anh. Konzernrecht Rdnr. 263; *H.-G. Koppensteiner*, Rowedder, 1997, Anh. § 52 Rdnr. 47; *M. Lutter/P. Hommelhoff*, GmbHG, 1995, Anh. § 13 Rdnr. 39.

[108] *BGH*, Beschl. v. 24.10.1988 - II ZB 7/88, in: BGHZ 105, 324, 333 („Supermarkt"); bestätigt durch *BGH*, Beschl. v. 30.1.1992 - II ZB 15/91, in: GmbHR 1992, S. 253, 254 („Siemens").

[109] *V. Emmerich*, Scholz, 1993, Anh. Konzernrecht Rdnr. 263; *ders.*, Aktienkonzernrecht, 1998, § 293 Rdnr. 9; *H.-G. Koppensteiner*, Rowedder, 1997, Anh. § 52 Rdnr. 47; *M. Lutter/P. Hommelhoff*, GmbHG, 1995, Anh. § 13 Rdnr. 39.

hätte diese Zustimmung nicht nur Wirkung im Innenverhältnis, sondern wäre auch Voraussetzung für die Wirksamkeit des Beherrschungsvertrages im Außenverhältnis. Indes ist eine analoge Anwendung von § 293 Abs. 2 AktG abzulehnen. Hier wie auch bei allen übrigen ungewöhnlichen Maßnahmen reicht vielmehr eine einfach Beschlußmehrheit aus, um den Abschluß des Beherrschungsvertrages zu billigen. Denn für eine analoge Anwendung von § 293 Abs. 2 AktG fehlt die Vergleichbarkeit des geregelten und des zu regelnden Sachverhalts.[110]

Richtigerweise ist nämlich Grund für das Zustimmungserfordernis der Hauptversammlung der herrschenden Aktiengesellschaft nicht das besondere Risiko, das sie infolge der Pflicht zum Verlustausgleich (§ 302 AktG), zur Sicherheitsleistung (§ 303 AktG) und zur Ausgleichszahlung (§ 304 AktG) trifft. Wäre dies der Fall, so müßte auch für andere riskante Geschäfte eine Zustimmungspflicht der Hauptversammlung erwogen werden. Insbesondere müßte dann auch der Abschluß eines Betriebspacht- oder Betriebsüberlassungsvertrags (§ 292 Abs. 1 Nr. 3 AktG) an die Zustimmung der Hauptversammlung gebunden sein. Denn auch hier besteht gemäß § 302 Abs. 2 AktG eine Verlustausgleichspflicht, soweit die vereinbarte Gegenleistung nicht angemessen ist, d. h. nicht ausreicht, um den Ertragswert des verpachteten oder überlassenen Unternehmens langfristig zu erhalten.[111] Vielmehr rechtfertigt sich das qualifizierte Zustimmungserfordernis aus der allein bei einem Beherrschungs- und Gewinnabführungsvertrag entstehenden Verpflichtung zur Abfindung in Aktien der herrschenden Gesellschaft, § 305 AktG.[112] Denn diese Aktien können in der Regel nur durch die Aktionäre, notfalls im Wege einer Kapitalerhöhung, aufgebracht werden. Wird aber durch den Beherrschungsvertrag in die Rechte der Aktionäre eingegriffen, so müssen diese auch ein Mitwirkungsrecht bei seinem Abschluß haben.

Eine solcher Eingriff in die Gesellschafterrechte findet bei der herrschenden GmbH dagegen nicht statt. Denn hier ist § 305 AktG, soweit er eine Abfindung in Aktien vorsieht, nicht analog anwendbar.[113] Damit fehlt aber der Grund, der

---

[110] Zur Analogie als Methode richterlicher Rechtsfortbildung siehe *K. Larenz*, Methodenlehre, 1991, S. 381 ff.

[111] *U. Hüffer*, AktG, 1997, § 302 Rdnr. 24.

[112] *W. Flume*, Gewinnübernahmevertrag, 1989, S. 667; *A. Zitzmann*, Vorlagepflichten, 1991, S. 166; *E. Vetter*, Unternehmensvertrag, 1993, S. 171. A. M. *P. O. Mülbert*, Aktiengesellschaft, 1996, S. 183.

[113] *BGH*, Urt. v. 24.10.1988 - II ZB 7/88, in: GmbHR 1989, S. 25, 28 f. Differenzierend: *M. Lutter/P. Hommelhoff*, GmbHG, 1995, Anh. § 13 Rdnr. 46.

eine analoge Anwendung von § 293 Abs. 2 AktG im Recht der GmbH rechtfertigen könnte.[114]

Mithin müssen die Gesellschafter der herrschenden GmbH nur mit einfacher Mehrheit dem Abschluß des Beherrschungsvertrages zustimmen, um ihn wirksam werden zu lassen.[115]

C. Die Pflichten bei der Konzernleitung

I. Die Pflicht zur Konzernleitung

1. Die Grundlage der Konzernleitungspflicht

Oben wurde festgestellt, daß die Gesellschaft im Rahmen des Unternehmensgegenstandes alle ihr zur Verfügung stehenden Ressourcen dem Gesellschaftszweck dienstbar zu machen hat. Zu diesen Ressourcen gehören auch die von ihr gehaltenen Beteiligungen. Sieht ihr Unternehmensgegenstand keine Vermögensverwaltung vor, kann sie sich nicht darauf beschränken, lediglich die Rendite zu kassieren und ansonsten keinen Einfluß auf das unternehmerische Geschehen in den beherrschten Gesellschaften zu nehmen.[116] Vielmehr muß sie dann mit den gehaltenen Beteiligungen ihren Gesellschaftszweck aktiv fördern. Das bedeutet, daß sie die beherrschten Gesellschaften in einen Konzern einzubeziehen und diesen zu leiten hat.[117]

Mithin müssen auch die Kompetenzen ihrer Organe konzernspezifisch fortgeschrieben werden. Für den Geschäftsführer hat das zur Folge, daß er gegenüber der Gesellschaft zur Konzernleitung verpflichtet ist.[118] Insbesondere besteht diese Pflicht unabhängig davon, ob die Konzernleitung erforderlich ist, um

---

[114] Wie hier *W. Flume*, Gewinnübernahmevertrag, 1989, S. 667; *A. Zitzmann*, Vorlagepflichten, 1991, S. 166; *E. Vetter*, Unternehmensvertrag, 1993, S. 171.

[115] Ebenso *S. Pache*, Spätlese, 1995, S. 93; ähnlich *A. Zitzmann*, Vorlagepflichten, 1991, S. 166 f.

[116] *P. Hommelhoff*, Konzernleitungspflicht, 1982, S. 46 f.

[117] *J. Semler*, Leitung, 1996, Rdnr. 270.

[118] *H.-G. Koppensteiner*, Rowedder, 1997, § 37 Rdnr. 12; *G. Krieger*, Unternehmen, 1988, § 69 Rdnr. 32; *M. Lutter/P. Hommelhoff*, GmbHG, 1995, § 37 Rdnr. 6; *Uwe H. Schneider*, Konzernleitung, 1981, S. 253. Für die Aktiengesellschaft: *P. Hommelhoff*, Konzernleitungspflicht, 1982, S. 237; *K.-P. Martens*, Organisation, 1991, S. 531; *Th. Raiser*, Kapitalgesellschaften, 1992, § 53 Rdnr. 13; *J. Semler*, Leitung, 1996, Rdnr. 277 f.; *B. Kropff*, Konzernleitungspflicht, 1984, S. 115. Zweifelnd *E. Rehbinder*, Konzernleitungspflicht, 1983, S. 468; *T. E. Abeltshauser*, Leitungshaftung, 1998, S. 156 ff. Gegen eine Konzernleitungspflicht: *W. Zöllner*, Baumbach/Hueck, 1996, GmbH-KonzernR Rdnr. 104; *M. Henssler*, Minderheitenschutz, 1998, S. 219.

wesentliche Synergieeffekte nutzbar zu machen oder sonstige wichtige Beiträge zum Erfolg des beherrschten Unternehmens zu leisten.[119] Denn mit seiner Bestellung zum Geschäftsführer wurde dieser von den Gesellschaftern beauftragt, das Unternehmen der Gesellschaft zu leiten und alle Ressourcen, also auch ihre Beteiligungen, im Rahmen des Unternehmensgegenstandes dem Gesellschaftszweck dienstbar zu machen. Das bedeutet aber auch, daß der Geschäftsführer die wesentlichen Entscheidungen im Unternehmen nicht Dritten überlassen darf wie z. B. den Vorständen oder Geschäftsführern von abhängigen Gesellschaften, sondern selbst treffen bzw. den Gesellschaftern zur Entscheidung vorlegen muß.

2. Die Pflicht zur Konzernintegration

Wichtigste Ausprägung der Konzernleitungspflicht des Geschäftsführers ist die Pflicht zur Konzernintegration. Konzernintegrative Maßnahmen sind dadurch gekennzeichnet, daß sie die Konzerngesellschaften zu einem Unternehmen im wirtschaftlichen Sinne zusammenfügen. Der Geschäftsführer hat die abhängigen Gesellschaften auf das Konzerninteresse auszurichten und seine Durchsetzung sicherzustellen.[120] Dabei kann er sich nicht darauf beschränken, lediglich die einheitliche Leitung i. S. v. § 18 Abs. 1 Satz 1 AktG herzustellen und etwa nur für das Finanzwesen Ziele zu setzen und Planungen vorzunehmen.[121] Vielmehr muß er all diejenigen Entscheidungen fällen oder zur Vorlage für die Gesellschafter vorbereiten, die für die Vermögens-, Finanz- und Ertragslage der Unternehmung und ihre künftige Entwicklung einen besonderen Bedeutungsumfang haben und die Lage des Unternehmens gravierend beeinflussen. Außerdem muß er sich mit all denjenigen Entscheidungen befassen, die wegen ihrer breitflächigen und nachhaltigen Auswirkungen nur schwer korrigiert werden können.[122]

Legt man diese Gesichtspunkte zugrunde, ergeben sich insbesondere die folgenden konzernintegrativen Aufgaben:

- Strategische Ausrichtung des Konzerns,

---

[119] So aber für den Vorstand einer AG: *H. Götz*, Leitungssorgfalt, 1998, S. 529.

[120] In betriebswirtschaftlicher Hinsicht siehe hierzu *E. Scheffler*, Konzernmanagement, 1992, S. 33; *M. R. Theisen*, Konzern, 1991, S. 203, 221.

[121] In diese Richtung auch für den Vorstand einer AG: *J. Semler*, Leitung, 1996, Rdnr. 274.

[122] Siehe hierzu auch *E. Scheffler*, Konzernmanagement, 1992, S. 36; *K.-P. Martens*, Organisation, 1991, S. 531 f. Zum Begriff der Leitung und seiner Unterscheidung zum Begriff der Geschäftsführung jüngst *T. E. Abeltshauser*, Leitungshaftung, 1998, S. 28 ff.

- Festlegung der Konzernstruktur und -organisation,
- finanzielle Führung des Konzerns und
- Besetzung wichtiger Führungspositionen im Konzern.[123]

Soweit dem Geschäftsführer diese Aufgaben zugewiesen sind, darf er sie nicht delegieren. Denn ihre ordnungsgemäße Erfüllung setzt die Kenntnis des konzerndimensionalen Gesamtzusammenhangs voraus, und über diese verfügt nur die Konzernleitung.[124] Ihre Wahrnehmung gehört mithin zu den originären Leitungsfunktionen des Geschäftsführers.[125]

a) Strategische Ausrichtung des Konzerns

Die Geschäftsführer hat eine Konzernstrategie zu entwickeln. Hierzu gehört zunächst die Festlegung strategischer Geschäftsfelder wie die Bestimmung der Produkt- und Dienstleistungsangebote sowie der Absatzmärkte. Sodann zählt zu ihr die Bestimmung der finanziellen Ziele, nämlich die Festlegung der erstrebten Kapitalrendite, des Verschuldungsgrades und der Dividendenpolitik.[126] Schließlich hat der Geschäftsführer Maßstäbe für das Verhalten gegenüber Kapitalgebern, Mitarbeitern, Kunden, Lieferanten und anderen Bezugsgruppen zu definieren.[127]

b) Festlegung von Konzernstruktur und -organisation

Des weiteren hat der Geschäftsführer für eine effiziente Konzernstruktur und -organisation[128] zu sorgen. Er muß den Konzern betriebswirtschaftlich zweck-

---

[123] *E. Scheffler*, Konzernmanagement, 1992, S. 53 ff.; *M. Lutter/P. Hommelhoff*, GmbHG, 1995, § 37 Rdnr. 4, 6; *Uwe H. Schneider*, Konzernleitung, 1981, S. 250. Für den Vorstand einer AG: *K.-P. Martens*, Organisation, 1991, S. 531 f.; *J. Semler*, Leitung, 1996, Rdnr. 273; *H. Götz*, Leitungssorgfalt, 1998, S. 531. In betriebswirtschaftlicher Hinsicht siehe auch *M. R. Theisen*, Konzern, 1991, S. 216. Zur entsprechenden Lage im rechtlichen Einheitsunternehmen *J. Semler*, Leitung, 1996, Rdnr. 11.

[124] *E. Scheffler*, Konzernmanagement, 1992, S. 37 f.; *M. R. Theisen*, Konzern, 1991, S. 209. Im Ergebnis ebenso für die konzernfreie AG: *J. Semler*, Leitung, 1996, Rdnr. 22 ff.

[125] *M. Lutter/P. Hommelhoff*, GmbHG, 1995, § 37 Rdnr. 4.

[126] Zum Vorstehenden siehe *E. Scheffler*, Konzernmanagement, 1992, S. 53 f.

[127] *Nach E. Scheffler*, Konzernmanagement, 1992, S. 53 liegt in der Formulierung dieser Verhaltensmaßstäbe die Bestimmung der Unternehmenspolitik. In dieser Arbeit wird dagegen unter Unternehmenspolitik die Gesamtheit aller Führungsentscheidungen verstanden, die den Ablauf des Geschehens im Unternehmen langfristig ordnen, beeinflussen oder unmittelbar festlegen.

[128] Zur Organisation des Konzernvorstandes einer AG siehe *K.-P. Martens*, Organisation, 1991, S. 523 ff.

mäßig organisieren, um effiziente, zielorientierte und kontrollierte Betriebsabläufe zu gewährleisten.[129] Außerdem hat er Richtlinien aufzustellen zur einheitlichen Ausrichtung von Bilanzierung, Finanzierung, Investitionen sowie des Planungs- und Berichtswesens.[130]

c) Finanzielle Führung des Konzerns

Ein weiterer Bestandteil der Konzernleitung ist die finanzielle Führung des Konzerns. Sie umfaßt die Festlegung von Kapitalausstattung und Finanzstruktur der einzelnen Konzernunternehmen, die Koordinierung der Finanzwirtschaft, die Mitentscheidung über langfristige Kreditaufnahmen und das Konzern-Controlling.[131]

d) Besetzung wichtiger Führungspositionen

Schließlich ist es Aufgabe des Geschäftsführers, die wichtigen Führungspositionen im Konzern zu besetzen. Er hat dafür zu sorgen, daß an den „Schaltstellen" des Konzerns nur qualifizierte Manager sitzen, die einerseits die beherrschten Gesellschaften effizient führen und die andererseits die Entscheidungen der Konzernleitung schnell und zuverlässig umsetzen.

e) Delegation

Ebenso wie im rechtlichen Einheitsunternehmen hat der Geschäftsführer die übrigen Aufgaben an nachgeordnete Mitarbeiter der herrschenden GmbH zu delegieren oder den Organen und Mitarbeitern der beherrschten Gesellschaften zu überlassen.[132] Dies gebietet seine Pflicht, die der Gesellschaft zur Verfügung stehenden knappen und personellen Ressourcen zum optimalen Nutzen der Gesellschaft einzusetzen. Mit anderen Worten darf der Geschäftsführer seine

---

[129] Für die konzernfreie Lage: *C. T. Ebenroth*/*K. W. Lange*, Sorgfaltspflichten, 1992, S. 71.

[130] *E. Scheffler*, Konzernmanagement, 1992, S. 54.

[131] *E. Scheffler*, Konzernmanagement, 1992, S. 54, 55. Siehe auch *M. R. Theisen*, Konzern, 1991, S. 220.

[132] Ebenso für den Vorstand einer AG: *T. E. Abeltshauser*, Leitungshaftung, 1998, S. 45.

Arbeitskraft nicht mit der Wahrnehmung von Aufgaben vergeuden, die von nachgeordnetem Personal ebenso gut erledigt werden können.[133] Freilich hat er auch hier ein unternehmerisches Ermessen und ist die Frage nach dem Delegationsgrad auch eine Frage nach der Gestaltung der Führungshierachie.[134]

f) Konzernkontrolle

Freilich entbindet die Delegation den Geschäftsführer nicht von seiner Pflicht, die ordnungsgemäße Durchführung der delegierten Aufgaben sowie die gesamte laufende Konzernverwaltung zu überwachen.[135] Die ausdrückliche Normierung einer solchen Überwachungspflicht für die Vorstandsmitglieder einer Aktiengesellschaft sah ein Entwurf des Gesetzes zur Kontrolle und Transparenz im Unternehmensbereich (KonTraG) vor.[136] Danach sollte § 93 Abs. 1 Satz 2 f. AktG folgenden Wortlaut erhalten:

> Sie [die Vorstandsmitglieder einer Aktiengesellschaft] haben insbesondere unter Berücksichtigung der Art und Größe ihrer Gesellschaft und bei Mutterunternehmen im Sinne des § 290 des Handelsgesetzbuchs ihres Konzerns geeignete Maßnahmen zu treffen, um zu gewährleisten, daß den Fortbestand der Gesellschaft gefährdende Entwicklungen, insbesondere risikobehaftete Geschäfte, Unrichtigkeiten der Rechnungslegung und Verstöße gegen gesetzliche Vorschriften, die sich auf die Vermögens-, Finanz- und Ertragslage der Gesellschaft oder des Konzerns wesentlich auswirken, früh erkannt werden. Dazu gehört auch die Einrichtung eines Überwachungssystems mit der Aufgabe, die Einhaltung der nach Satz 2 zu treffenden Maßnahmen zu überwachen".

Dieser Entwurf ist zwar nicht vollständig umgesetzt worden. Vielmehr wurde § 91 AktG ein Abs. 2 angefügt, wonach der „Vorstand ... geeignete Maßnahmen zu treffen, insbesondere ein Überwachungssystem einzurichten [hat], damit den Fortbestand der Gesellschaft gefährdende Entwicklungen früh erkannt werden". Konzernbezogene Überwachungspflichten wurden also nicht ausdrücklich normiert. Das bedeutet indes nicht, daß sie nach dem AktG nicht bestünden. Denn erstens können auch Entwicklungen in den beherrschten Gesellschaften den Fortbestand der herrschenden Aktiengesellschaft gefährden. Zweitens hatte

---

[133] Für den Vorstand einer AG siehe *P. Hommelhoff*, Konzernleitungspflicht, 1982, S. 179.

[134] *M. R. Theisen*, Konzern, 1991, S. 209, aus betriebswirtschaftlicher Sicht.

[135] *Uwe H. Schneider*, Konzernleitung, 1981, S. 250. Für das Vorstandsmitglied einer AG: *OLG Düsseldorf*, Urt. v. 28.11.1996 - 6 U 11/95, in: ZIP 1997, S. 27, 32 („ARAG"); *H. Götz*, Leitungssorgfalt, 1998, S. 534 ff. Für die konzernfreie Lage: *BGH*, Urt. v. 20.2.1995 - II ZR 9/94, in: GmbHR 1995, S. 299, 300.

[136] BMJ-Referentenentwurf - Stand: 22. November 1996, S. 3 f.

der Vorstand schon bisher dem Vorsitzenden des Aufsichtsrates über ihm bekanntgewordene geschäftliche Vorgänge bei verbundenen Unternehmen, die auf die Gesellschaft von erheblichem Einfluß sein können, zu berichten (§ 90 Abs. 1 Satz 2 Halbsatz 2 AktG). Eine solche Pflicht macht aber nur dann Sinn, wenn ihr eine entsprechende Pflicht des Vorstands zur Konzernkontrolle korrespondiert. Und drittens schließlich müssen Vorstand und Geschäftsführer schon aufgrund ihrer Pflicht zur sorgfältigen Unternehmensleitung eine Organisation schaffen, die ihnen einen Überblick über die wirtschaftliche und finanzielle Situation ihrer Gesellschaft jederzeit ermöglicht.[137] Hierzu gehört insbesondere die Einrichtung eines konzernweiten Berichts- und Kontrollwesens, welches die Beschaffung aller überwachungsrelevanten Daten der abhängigen Unternehmen gewährleistet.[138] Zu diesen Daten gehören insbesondere die Betriebsergebnisse, die Umsätze, der Auftragseingang, die Investitionen und die Entwicklung der Liquidität.[139]

Schließlich hat der Geschäftsführer der herrschenden Gesellschaft dafür zu sorgen, daß die beherrschten Unternehmen Geschäfte mit besonderer Tragweite nicht ohne seine Zustimmung vornehmen. Hierzu hat er einen Katalog zustimmungspflichtiger Geschäfte aufzustellen und ein Verfahren einzurichten, das seine rechtzeitige Unterrichtung durch die Vorstände bzw. Geschäftsführungen der beherrschten Unternehmen gewährleistet.[140]

3. Keine Pflicht zur zentralen Konzernleitung

Die obengenannten konzernintegrativen Maßnahmen bilden einen Mindeststandard, den der Geschäftsführer im Rahmen der Konzernleitung zu verwirklichen hat. Wie er die Konzernintegration konkret vollzieht, liegt in seinem unternehmerischen Ermessen.[141] Insbesondere ist er nicht verpflichtet, den Konzern zentralistisch zu organisieren.[142] Dies gilt sowohl für den faktischen Kon-

---

[137] Für die konzernfreie GmbH: *BGH*, Urt. v. 20.2.1995 - II ZR 9/94, in: GmbHR 1995, S. 299, 300.

[138] Zur Erfüllung der in § 93 Abs. 1 S. 2 f. AktG-E normierten Pflichten im rechtlichen Einheitsunternehmen siehe *P. Scharpf*, Sorgfaltspflichten, 1997, S. 737 ff. Ausführlich zur Überwachung in einer Holding *J. Semler*, Überwachung, 1998, Rz. D 1 ff. (S. 160 ff.).

[139] *H. Götz*, Leitungssorgfalt, 1998, S. 537.

[140] *H. Götz*, Leitungssorgfalt, 1998, S. 538.

[141] Eingehend für den Vorstand einer AG: *E. Rehbinder*, Konzernleitungspflicht, 1983, S. 468.

[142] So aber *P. Hommelhoff*, Konzernleitungspflicht, 1982, S. 72. Ebenso *B. Kropff*, Konzernleitungspflicht, 1984, S. 116. Dagegen: *K.-P. Martens*, Organisation, 1991, S. 531; *H. Wiedemann*, Unternehmensgruppe, 1988, S. 76; *E. Rehbinder*, Konzernleitungspflicht, 1983, S. 467; *J. Semler*, Leitung, 1996, Rdnr. 280; *T. E. Abeltshauser*,

zern wie den Vertragskonzern. Insbesondere läßt sich eine solche Pflicht nicht aus der Pflicht zur Förderung des Gesellschaftszwecks der herrschenden Gesellschaft ableiten. Denn Maßstab für die Ordnungsmäßigkeit der Konzernführung ist nicht nur die Rechtsordnung, sondern auch das Marktergebnis.[143] Eine Pflicht zur zentralistischen Organisation würde die Flexibilität des Unternehmens am Markt aber möglicherweise erheblich beeinträchtigen und damit das Marktergebnis mindern. Das liegt aber keinesfalls im Interesse der Gesellschafter, die an einer Gewinnerzielung interessiert sind. Deshalb kann für eine Konzernintegration eine dezentrale Konzernorganisation ausreichen, die den Tochtergeschäftsleitungen einen verhältnismäßig weiten Entscheidungsspielraum beläßt.

II. Die Pflicht zur kollegialen Zusammenarbeit

Der Geschäftsführer ist zur kollegialen Zusammenarbeit mit den Gesellschaftern und den Mitgeschäftsführern verpflichtet. Das kommt in mehreren Pflichten des Geschäftsführers zum Ausdruck.

1. Die Vorlagepflichten des Geschäftsführers

a) Die gesetzliche Ausgangslage

Bereits bei Darstellung der Pflichten des Geschäftsführers im Rahmen der Konzerngründung wurde festgestellt, daß der Geschäftsführer alle ungewöhnlichen Maßnahmen den Gesellschaftern zur Entscheidung vorzulegen hat. Weiterhin wird diskutiert, ob die Gesellschafter auch für die Festlegung der Unternehmenspolitik zuständig sind. Diese Frage ist Gegenstand des Streits zwischen den Befürwortern des engen und des weiten Geschäftsführungsbegriffs.

(1) Der enge Geschäftsführungsbegriff

Nach Ansicht der Vertreter des engen Geschäftsführungsbegriffs ist der Geschäftsführer lediglich für die laufende Geschäftsführung zuständig. Hierzu gehöre die Sorge für das rechtmäßige Verhalten der Gesellschaft im Außenverhältnis, die Einrichtung der Unternehmensorganisation, die Entwicklung kurzfristiger Pläne und Taktiken, die Steuerung und Überwachung des unternehmerischen Gesamtgeschehens und die Abwicklung des Tagesgeschäfts. Hierzu

---

Leitungshaftung, 1998, S. 42 ff.; *H. Götz*, Leitungssorgfalt, 1998, S. 530; *H. Altmeppen*, Managerhaftung, 1998, S. 33; *P. O. Mülbert*, Aktiengesellschaft, 1996, S. 29; *C.-P. Feick*, Pflichten, 1999, S. 104.

[143] *E. Rehbinder*, Konzernleitungspflicht, 1983, S. 468.

gehöre jedoch nicht die Entscheidung über außergewöhnliche Maßnahmen und die Grundsätze der Unternehmenspolitik.[144] Unter Unternehmenspolitik sei dabei die Gesamtheit aller Führungsentscheidungen zu verstehen, die das Geschehen im Unternehmen langfristig ordnen, beeinflussen oder unmittelbar festlegen würden. Solche Entscheidungen müsse der Geschäftsführer den Gesellschaftern vorlegen.[145]

Dies folge aus einer Gesamtschau der §§ 46 Nr. 1, 5, 6 und 7, 37 Abs. 1, 49 Abs. 2 und 38 Abs. 1 GmbHG, in denen das Gesetz den Gesellschaftern wichtige Teilbereiche der Unternehmenspolitik zugewiesen habe. So würden die Gesellschafter über einen wesentlichen Teil der Unternehmensfinanzierung entscheiden, indem sie den Jahresabschluß feststellten und den sich aus ihm ergebenden Reingewinn verteilten. Denn hiermit ergebe sich für sie zum einen die Möglichkeit, die Bilanzpolitik zu formulieren und insbesondere über die Bildung stiller Reserven zu entscheiden. Zum anderen könnten sie durch die Bedienung und Auflösung offener Rücklagen entscheidenden Einfluß auf die Selbstfinanzierung der Gesellschaft nehmen. Auch die Personalpolitik liege in wichtigen Bereichen in ihren Händen, da sie über die Bestellung und Abberufung der Geschäftsführer sowie deren Entlastung bestimmen würden und zudem für die Bestellung von Prokuristen und Handlungsbevollmächtigten zum gesamten Geschäftsbetrieb zuständig seien. Schließlich weise ihnen § 46 Nr. 6 GmbHG noch eine umfangreiche Überwachungskompetenz zu. Aus alledem folge, daß die Gesellschafter auch für die Bestimmung der Grundsätze der Unternehmenspolitik berufen seien.[146]

Freilich bedeute dies nicht, daß der Geschäftsführer an der Formulierung der Unternehmenspolitik nicht beteiligt sei. Vielmehr sei er verpflichtet, Pläne für die Zukunft des Unternehmens zu entwickeln und den Gesellschaftern vorzuschlagen.[147]

---

[144] *Uwe H. Schneider*, Scholz, 1993, § 37 Rdnr. 3; *BGH*, Urt. v. 25.2.1991 - II ZR 76/90, in: WM 1991, S. 635, 637.

[145] *Uwe H. Schneider*, Scholz, 1993, § 37 Rdnr. 10; *C. Th. Ebenroth/K. W. Lange*, Sorgfaltspflichten, 1992, S. 71; *M. Lutter/P. Hommelhoff*, GmbHG, 1995, § 37 Rdnr. 8; *G. Miller*, GmbHG, 1987, §§ 35 - 38 Rdnr. 73.

[146] Grundlegend *P. Hommelhoff*, Unternehmensführung, 1978, 124 ff.; *Uwe H. Schneider*, Scholz, 1993, § 37 Rdnr. 10.

[147] *Uwe H. Schneider*, Scholz, 1993, § 37 Rdnr. 10; *M. Lutter/P. Hommelhoff*, GmbHG, 1995, § 37 Rdnr. 9.

(2) Der weite Geschäftsführungsbegriff

Die Anhänger des weiten Geschäftsführungsbegriffs lehnen dagegen eine Vorlagepflicht des Geschäftsführers in Fragen der Unternehmenspolitik ab und befürworten seine Zuständigkeit auch für diesen Bereich.[148] Zur Begründung führen sie zunächst die Unschärfe des Begriffs „Unternehmenspolitik" an und befürchten erhebliche Abgrenzungsschwierigkeiten.[149] Darüber hinaus wird vorgetragen, daß die Interessen der Gesellschafter ausreichend durch ihr Recht geschützt seien, die Grundsätze der Unternehmenspolitik jederzeit in der Satzung oder durch Weisung festzulegen.[150] Ferner wird argumentiert, daß die Kompetenzen der Gesellschafter, aus denen die Gegenmeinung die Zuständigkeit der Gesellschafter für die Unternehmenspolitik ableite, nicht notwendigerweise unternehmenspolitischer Natur seien und deshalb ihre Folgerung nicht überzeuge.[151] So könnten die Gesellschafter z. B. den von dem Geschäftsführer aufgestellten Jahresabschluß unverändert ihrem Feststellungsbeschluß zugrunde legen und sich damit jeglicher unternehmerischer Initiative enthalten. Die Entfaltung von unternehmerischer Initiative sei aber Voraussetzung für die Qualifizierung einer Entscheidung als unternehmenspolitische Maßnahme.[152] Schließlich wird geltend gemacht, daß der Gesetzgeber mit der GmbH eine Gesellschaftsform habe schaffen wollen, die eine rein kapitalistische Beteiligung ermöglichen solle. Damit vertrage sich aber nicht die Zuständigkeit der Gesellschafter für die Unternehmenspolitik.[153]

(3) Stellungnahme

Richtigerweise wird man eine Vorlagepflicht, die sich allein aus den unternehmenspolitischen Auswirkungen der geplanten Maßnahme ergibt, ablehnen müssen. Zwar weist das Gesetz in § 46 GmbHG wichtige Entscheidungen den Gesellschaftern zu. Jedoch läßt sich hieraus keine Pflicht des Geschäftsführers ableiten, in allen unternehmenspolitischen Fragen die Gesellschafter zu konsultieren. Hierfür besteht kein Bedürfnis.

---

[148] *H. G. Koppensteiner*, Rowedder, 1997, § 37 Rdnr. 8; *W. Zöllner*, Baumbach/Hueck, 1996, § 37 Rdnr. 6g; *A. Zitzmann*, Vorlagepflichten, 1991, S. 85.

[149] *H. G. Koppensteiner*, Rowedder, 1997, § 37 Rdnr. 8; *W. Zöllner*, Baumbach/Hueck, 1996, § 37 Rdnr. 6g; *A. Zitzmann*, Vorlagepflichten, 1991, S. 65 ff.

[150] *H. G. Koppensteiner*, Rowedder, 1997, § 37 Rdnr. 8.

[151] *A. Zitzmann*, Vorlagepflichten, 1991, S. 68 ff.

[152] *A. Zitzmann*, Vorlagepflichten, 1991, S. 69.

[153] *A. Zitzmann*, Vorlagepflichten, 1991, S. 72 ff.

Erstens wird man nicht sagen können, daß die Unternehmenspolitik sinnvollerweise nur in Abstimmung mit den von den Gesellschaftern nach § 46 GmbHG zu treffenden Entscheidungen festgelegt werden kann. Selbst wenn die Gesellschafter z. B. eine vollständige Gewinnausschüttung beschließen, wird es dem Geschäftsführer in der Regel möglich sein, die bisherige Unternehmenspolitik fortzuführen. Sollte dies im Einzelfall anders sein, so hat der Geschäftsführer die Gesellschafter hierauf hinzuweisen. Das ergibt sich schon aus seiner Pflicht, die Entscheidungen der Gesellschafter sorgfältig vorzubereiten und ihre Auswirkungen den Gesellschaftern deutlich zu machen. Zweitens können die Gesellschafter jederzeit von sich aus tätig werden und die Unternehmenspolitik durch Weisungen festlegen. Ist dies geschehen, so ist der Geschäftsführer hieran gebunden. Will der Geschäftsführer die einmal von den Gesellschaftern festgelegte Unternehmenspolitik ändern, so muß er deren Zustimmung einholen. Und drittens schließlich wird die Zustimmung der Gesellschafter für die Änderung der Unternehmenspolitik auch dann erforderlich sein, wenn die Gesellschafter sich bisher noch nicht zur Unternehmenspolitik geäußert haben. Denn in aller Regel stellt die Änderung der Unternehmenspolitik eine ungewöhnliche Maßnahme dar.[154] Diese ist aber vorlagepflichtig.

Nach alledem ist der Geschäftsführer sowohl für die laufende Geschäftsführung sowie für die Festlegung der Unternehmenspolitik zuständig. Ungewöhnliche Maßnahmen hat er dagegen den Gesellschaftern zur Entscheidung vorzulegen.

b) Die Lage im Konzern

Es stellt sich nun die Frage, ob und wie diese Vorlagepflichten konzernspezifische fortgeschrieben werden können. Anders gewendet ist zu untersuchen, ob der Geschäftsführer auch solche Maßnahmen den Gesellschaftern zur Entscheidung vorzulegen hat, die nicht in der herrschenden GmbH, sondern in den Konzerngesellschaften verwirklicht werden.

(1) Meinungsstand und Stellungnahme

Konzernbezogene Vorlagepflichten werden von *H. Wiedemann*[155] abgelehnt. Er vertritt die Ansicht, daß die Gesellschafter mit ihrer Zustimmung zur Konzerngründung die gesamte Konzernleitung in die Hände des Geschäftsführers gelegt hätten. Zur Begründung beruft er sich darauf, daß nach unserer Rechts-

---

[154] So z. B. in *BGH*, Urt. v. 25.2.1991 - II ZR 76/90, in: WM 1991, S. 635, 637.

[155] *H. Wiedemann*, Unternehmensgruppe, 1988, S. 73 f.

ordnung immer nur die Begründung von Treuhandeigentum an die Zustimmung des Rechtsinhabers gebunden sei, seine Verwaltung aber allein dem Treuhänder obliege. Genauso verhalte es sich mit der Konzerngründung und Konzernleitung. Denn mit der Konzerngründung gehe Rechtsmacht der Gesellschafter auf den Geschäftsführer über, die dieser nun für die Gesellschafter treuhänderisch verwalte. Deshalb treffe den Geschäftsführer nur bei der Konzerngründung, nicht aber bei der Konzernleitung eine Vorlagepflicht. Eine Ausnahme bestehe nur dann, wenn sich die Gesellschafter eine Zustimmung zu bestimmten Maßnahmen vorbehalten hätten.

H. Wiedemann kann indes nicht gefolgt werden. Ein Grundsatz, wonach der Treugeber von der Verwaltung des Treuguts ausgeschlossen ist, existiert nicht. Schon allein durch die Gründung einer GmbH werden die Rechte der Gesellschafter an dem von ihnen eingebrachten Kapital mediatisiert. Das Kapital wird zum Treuhandeigentum der GmbH und damit der Verwaltung des Geschäftsführers unterstellt. Trotzdem haben die Gesellschafter, wie oben aufgezeigt, gewichtige Mitverwaltungsrechte. Diese stehen ihnen zu, obwohl sie Treugeber sind. Und sie schränken die Befugnisse des Geschäftsführers ein, obwohl er der Verwalter des Treuguts ist. Nach alledem wird deutlich, daß es gerade keinen Grundsatz gibt, wonach der Treugeber von der Verwaltung des Treuguts ausgeschlossen ist.

Damit bleibt die Frage offen, ob die Mitverwaltungsrechte der Gesellschafter konzernspezifisch fortgeschrieben werden müssen und den Geschäftsführer entsprechende Vorlagepflichten treffen. Das ist zu bejahen. Denn der Konzern ist eine besondere Organisationsform für Unternehmen und Konzernleitung ist funktional Unternehmensleitung. Mit anderen Worten ist der Konzern als ein wirtschaftlich einheitliches Unternehmen zu begreifen, das von der Konzernspitze geleitet wird. Deshalb ist die Leitung des Konzerns integraler Bestandteil der Unternehmensleitung bei der Konzernspitze. Da sich die Mitverwaltungsrechte der Gesellschafter eben auf diese Unternehmensleitung beziehen, erstrecken sie sich damit gleichzeitig auf die Konzernleitung.

Somit ist nur noch der Maßstab zu klären, nach dem sich beurteilt, welche Maßnahmen den Gesellschaftern vorzulegen sind.

Zum einen könnte darauf abgestellt werden, welches Organ in der herrschenden Gesellschaft zuständig wäre, wenn die im Konzernunternehmen anstehende Maßnahme bei ihr verwirklicht würde. Der Geschäftsführer der herrschenden GmbH müßte also z. B. die Gesellschafterversammlung mit der Bestellung der Prokuristen in den Tochtergesellschaften befassen (§ 46 Nr. 7 GmbHG). Dies würde selbst dann gelten, wenn die Tochtergesellschaft im Konzern nur eine unbedeutende Funktion hätte oder nur wenige Prozentpunkte zum Gesamtumsatz beitragen würde. Ebenso müßte der Geschäftsführer den Gesellschaftern die Frage vorlegen, wie er bei der Feststellung und Verwendung des Gewinns in einer Tochtergesellschaft abstimmen soll, die z. B. nur 0,5 Prozent zum Konzerngewinn beiträgt.

Zum anderen könnte man in der Bedeutung der Maßnahme für den Gesamtkonzern das ausschlaggebende Kriterium für die Vorlagepflicht des Geschäftsführers sehen. Maßnahmen der laufenden Konzerngeschäftsführung müßte der Geschäftsführer dann nicht vorlegen.[156]

Eine Stellungnahme hat sich an der Struktur der GmbH zu orientieren. Der Gesetzgeber hat die GmbH als eine konzernfreie Gesellschaft konzipiert, die als juristische Person Trägerin eines Unternehmens ist. Innerhalb dieses Unternehmens hat er der Gesellschafterversammlung die wesentlichen unternehmerischen Entscheidungen vorbehalten. Schreibt man diese gesetzliche Grundkonzeption konzernspezifisch fort, so ist wiederum zu berücksichtigen, daß der Konzern eine besondere Organisationsform für Unternehmen und Konzernleitung funktional Unternehmensleitung ist. Der Konzern ist mithin als ein einziges Unternehmen zu behandeln, an dessen Spitze die herrschende GmbH steht. An die Stelle des Unternehmens der konzernfreien GmbH tritt nun das Konzernunternehmen. Damit ist es sachgerecht, nur solche Maßnahmen den Gesellschaftern vorzulegen, die für den gesamten Konzern von Bedeutung sind. Um welche Maßnahmen es sich dabei handelt, soll im folgenden untersucht werden.

(2) Feststellung des Jahresabschlusses und Gewinnverwendung in den Konzernunternehmen

Gemäß § 46 Nr. 1 GmbHG stellen die Gesellschafter den Jahresabschluß fest und beschließen über die Gewinnverwendung. Hiermit weist ihnen das Gesetz die Aufgabe zu, über die Selbstfinanzierung des Unternehmens der GmbH zu bestimmen. Da sich das Unternehmen der GmbH, wie oben ausgeführt, auf die Tätigkeit der beherrschten Gesellschaften bezieht, fällt auch deren Finanzierung in die Entscheidungszuständigkeit der Gesellschafter der herrschenden GmbH. Für den Geschäftsführer bedeutet dies, daß er die Frage, wie er in den Konzernunternehmen bei Feststellung von deren Jahresabschluß abstimmen soll, den Gesellschaftern grundsätzlich zur Entscheidung vorzulegen hat.[157] Ebenso verhält es sich mit der Frage der Gewinnverwendung in den Konzernunternehmen.[158]

---

[156] *Uwe H. Schneider*, Konzernleitung, 1981, S. 251.

[157] *Uwe H. Schneider*, Konzernleitung, 1981, S. 253.

[158] *Uwe H. Schneider*, Konzernleitung, 1981, S. 253. A. M. *M. Henssler*, Minderheitenschutz, 1998, S. 217 f.

Fraglich ist allerdings, ob sich diese Vorlagepflicht auch auf die Jahresabschlüsse und Ergebnisverwendungsbeschlüsse solcher Konzernunternehmen erstreckt, die nur unwesentlich zum Konzernumsatz beitragen. Bei funktioneller Betrachtungsweise müßte diese Frage verneint werden. Denn die Selbstfinanzierung des Konzerns kann über die Selbstfinanzierung der wichtigsten Tochtergesellschaften gesteuert werden. Die dort festgelegten Daten würden sodann die Entscheidungsgrundlage für das Abstimmungsverhalten des Geschäftsführers in den kleineren Tochtergesellschaften bilden.

Eine solche Argumentation würde jedoch verkennen, daß der Beschluß über die Feststellung des Jahresabschlusses und die Verwendung des Ergebnisses nicht nur der Selbstfinanzierung des von der GmbH getragenen Unternehmens dient. Vielmehr ist das Beschlußrecht der Gesellschafter vor allem und in erster Linie Ausdruck ihres mitgliedschaftlichen Gewinnbezugsrechts. Denn mit Feststellung des Jahresabschlusses entstehen die Ansprüche der Gesellschafter auf Auszahlung ihrer Gewinnanteile, und mit dem Ergebnisverwendungsbeschluß werden ihre Auszahlungsansprüche der Höhe nach festgelegt und fällig gestellt.[159] Ist aber das Beschlußrecht der Gesellschafter gemäß § 46 Nr. 1 GmbHG Ausdruck ihres mitgliedschaftlichen Gewinnbezugsrechts, so hat es eigennützigen Charakter. Für die Wahrnehmung ihrer eigennützigen Rechte sind aber die Gesellschafter selbst zuständig. Deshalb hat der Geschäftsführer auch vor der Beschlußfassung in noch so unbedeutenden Konzerngesellschaften den Gesellschaftern der herrschenden GmbH die Frage seines Abstimmungsverhaltens vorzulegen.

All das gilt schließlich auch, wenn von vornherein eine Pflicht des Geschäftsführers besteht, dafür zu sorgen, daß Gewinne der beherrschten Gesellschaften zügig ausgewiesen und an die herrschende Gesellschaft ausgeschüttet werden. Das ist der Fall, wenn die herrschende GmbH gegenüber ihren Gesellschaftern verpflichtet ist, ihren gesamten Gewinn auszuschütten.[160] Denn zu diesem Gewinn gehört auch das Ergebnis, das die beherrschten Gesellschaften erzielt haben. Auch bei Bestehen eines solchen Vollausschüttungsgebots macht es noch Sinn, daß der Geschäftsführer den Gesellschaftern die Frage vorlegt, wie er in der beherrschten Gesellschaft bei der Feststellung des Jahresabschlusses und der Verwendung des Ergebnisses abstimmen soll. Denn die Gesellschafter können selbstverständlich - wenn auch nur einstimmig - beschließen, daß Gewinne in der beherrschten Gesellschaft thesauriert werden, z. B. weil diese dringend finanzielle Mittel für Neuinvestitionen benötigt.[161]

---

159 *M. Lutter/P. Hommelhoff*, GmbHG, 1995, § 29 Rdnr. 4.

160 Dagegen *M. Henssler*, Minderheitenschutz, 1998, S. 221 f.

161 *BGH*, Urt. v. 17.2.1997 - II ZR 41/96, in: WM 1997, S. 823, 825.

Indes kann sich ein Vollausschüttungsgebot heute nur noch aus der Satzung ergeben. Ein gesetzliches Vollausschüttungsgebot besteht dagegen seit Inkrafttreten des Bilanzrichtliniengesetzes am 1.1.1986 nicht mehr. An seine Stelle ist § 29 Abs. 2 GmbHG getreten, der die Höhe der Gewinnausschüttung in das unternehmerische Ermessen der Gesellschaftermehrheit stellt.[162] Etwas anderes gilt nur für sogenannte Altgesellschaften, die bei Inkrafttreten des Bilanzrichtliniengesetzes am 1.1.1986 im Handelsregister eingetragen waren. Für diese verbleibt es gemäß § 7 Abs. 1 GmbHGÄndG bei dem Vollausschüttungsgebot. Ist also eine Altgesellschaft herrschendes Unternehmen im Konzern, so muß deren Geschäftsführer dafür sorgen, daß Gewinne von Tochtergesellschaften zügig ausgewiesen und an die herrschende Gesellschaft ausgeschüttet werden.[163]

(3) Besetzung wichtiger Führungspositionen im Konzern

Gemäß § 46 Nr. 5 GmbHG obliegt den Gesellschaftern die Bestellung und Abberufung der Geschäftsführer sowie ihre Entlastung. Darüber hinaus sind die Gesellschafter gemäß § 46 Nr. 7 GmbHG im Innenverhältnis für die Bestellung der Prokuristen und von Handlungsbevollmächtigten zum gesamten Geschäftsbetrieb zuständig. Damit weist ihnen das Gesetz die Aufgabe zu, die wichtigsten Führungspositionen im Unternehmen der GmbH zu besetzen. Funktional betrachtet bedeutet dies, daß die Gesellschafter der herrschenden GmbH auch für die Besetzung der wichtigsten Führungspositionen im Konzern zuständig sind. Sollen also die Geschäftsführer einer bedeutenden Tochtergesellschaft neu gewählt werden, so hat der Geschäftsführer die Gesellschafter mit dieser Frage zu befassen.[164] Dagegen trifft ihn z. B. keine Vorlagepflicht bei der Wahl des Geschäftsführers einer unbedeutenden Enkel-GmbH.

(4) Ungewöhnliche Konzerngeschäftsführungsmaßnahmen

Ebenso wie die Gesellschafter in der konzernfreien GmbH über ungewöhnliche Geschäftsführungsmaßnahmen entscheiden, ist ihnen im Konzern die Entscheidung über ungewöhnliche Konzerngeschäftsführungsmaßnahmen vorbehalten. So muß der Geschäftsführer der herrschenden GmbH in der Regel erst

[162] *M. Lutter/P. Hommelhoff*, GmbHG, 1995, § 29 Rdnr. 1. Freilich wird die Minderheit nicht schutzlos gestellt. Wie weit der Minderheitenschutz im Rahmen von § 29 Abs. 2 GmbHG reicht, ist allerdings umstritten. Zum Meinungsstand siehe *M. Lutter/P. Hommelhoff*, GmbHG, 1995, § 29 Rdnr. 25 ff.

[163] Ebenso zur Rechtslage vor Inkrafttreten des Bilanzrichtliniengesetzes *Uwe H. Schneider*, Konzernleitung, 1981, S. 253.

[164] Ebenso *Uwe H. Schneider*, Konzernleitung, 1981, S. 252.

die Gesellschafter befragen, bevor er die Geschäftsleitung der Tochtergesellschaft anweist, einen Großkredit aufzunehmen. Ebenso wird man z. B. von der Zuständigkeit der Gesellschafter für die Frage ausgehen müssen, ob eine Tochtergesellschaft auf ihrem Geschäftsgrundstück eine neue Fabrik bauen soll. Und als letztes Beispiel sei schließlich die Einführung eines konzernweiten Cash-Managements genannt, die ebenfalls als ungewöhnliche Geschäftsführungsmaßnahme den Gesellschaftern zur Entscheidung vorzulegen ist.[165]

(5) Sonstige Entscheidungen

Sonstige Entscheidungen sind nur dann den Gesellschaftern vorzulegen, wenn sich dies aus Gesetz, Satzung, Gesellschafterbeschluß oder einer Geschäftsordnung ergibt. Beziehen sich die Vorschriften, wie das häufig der Fall sein wird, auf Maßnahmen, die in der herrschenden GmbH verwirklicht werden sollen, so ist in funktioneller Betrachtungsweise zu ermitteln, ob sie auch für entsprechende Maßnahmen gelten, die bei beherrschten Konzernunternehmen durchgeführt werden sollen.[166]

So wurde bereits festgestellt, daß die Bestellung eines Geschäftsführers in einer Tochtergesellschaft den Gesellschaftern der herrschenden GmbH nur dann zur Entscheidung vorzulegen ist, wenn die Tochtergesellschaft im Konzern eine bedeutende Position einnimmt und die Bestellung des Geschäftsführers damit eine wichtige Personalmaßnahme ist.[167] Dagegen wird eine Klausel in der Satzung oder in der Geschäftsordnung für die Geschäftsführung, die die Veräußerung eines Grundstücks an die Zustimmung der Gesellschafter bindet, auch für Grundstückskäufe durch unbedeutende Tochtergesellschaften gelten. Denn in der Regel konkretisieren die Gesellschafter in solchen Klauseln die gesetzlichen Vorlagepflichten des Geschäftsführers und bringen z. B. zum Ausdruck, welche Maßnahmen nach ihrer Einschätzung über den gewöhnlichen Geschäftsbetrieb hinaus gehen oder den Erfolg der Gesellschaft besonders gefährden. Ist Gegenstand einer solchen Klausel die Veräußerung eines Grundstücks, so ist der Erfolg der herrschenden GmbH bei der Veräußerung durch eine Tochtergesellschaft aber ebenso gefährdet wie bei der Veräußerung durch die herrschende Gesellschaft selbst.[168] Zwar ist das Verlustrisiko auf den Wert ihrer Beteiligung an der Tochtergesellschaft beschränkt, entscheidend ist jedoch nicht das Risiko ihres Verlusts, sondern die Chance ihres Erfolgs.

---

165 *M. W. Hormuth*, Cash Management, 1998, S. 178.

166 *Uwe H. Schneider*, Konzernleitung, 1981, S. 251; ders., Scholz, 1993, § 37 Rdnr. 65 ff.

167 Siehe oben 1. Teil C. II. 1. b) (3).

168 Ebenso *Uwe H. Schneider*, Konzernleitung, 1981, S. 252.

2. Die Pflicht zur Ausführung konzernbezogener Weisungen

a) Die gesetzliche Ausgangslage

Die Gesellschafter sind berechtigt, den Geschäftsführern in allen Bereichen der Unternehmensleitung Weisungen zu erteilen, § 37 Abs. 1 GmbHG. Die Geschäftsführer sind verpflichtet, diese Weisungen auszuführen.[169] Insbesondere sind sie nicht berechtigt, eine Weisung auf ihre Zweckmäßigkeit zu prüfen und ihre Ausführung wegen Unzweckmäßigkeit abzulehnen.[170]

Auch ist das Weisungsrecht der Gesellschafter nicht auf die Vorgabe allgemeiner Richtlinien begrenzt. Vielmehr erstreckt es sich auf die Erteilung von Einzelweisungen bis hinunter in das Tagesgeschäft.[171] Dabei kann die Weisungsdichte einen solchen Umfang erreichen, daß der nach § 35 Abs. 1 GmbHG vertretungsberechtigte Geschäftsführer zum bloßen Vollzugsorgan wird.[172] Die Weisungsbefugnis erstreckt sich freilich nicht auf die Erfüllung solcher Pflichten, die dem Geschäftsführer im öffentlichen Interesse auferlegt sind. So hat der Geschäftsführer unter eigener Verantwortung das Stammkapital vor verbotenen Auszahlungen zu bewahren (§§ 30, 43 Abs. 3 GmbHG), den verbotenen Erwerb eigener Anteile zu verhindern (§§ 33, 43 Abs. 3 GmbHG), für eine ordnungsgemäße Buchführung der Gesellschaft zu sorgen (§ 41 GmbHG), den Jahresabschluß aufzustellen (§ 264 Abs. 1 HGB), die Gesellschafterversammlung einzuberufen (§ 49 Abs. 2, Abs. 3 GmbHG) und, sobald erforderlich, die Eröffnung des Insolvenzverfahrens zu beantragen (§ 64 Abs. 1 GmbHG).[173]

b) Die konzernbezogene Folgepflicht

Da zur Unternehmensleitung auch die Konzernführung gehört, erstreckt sich das Weisungsrecht der Gesellschafter und die entsprechende Folgepflicht der Geschäftsführer auch auf alle konzernbezogenen Maßnahmen. Die Gesellschafter der herrschenden GmbH können den Geschäftsführer also anweisen, bei der Bestellung eines Prokuristen in der Tochter-GmbH für eine von ihnen benannte

---

169 *Uwe H. Schneider*, Scholz, 1993, § 37 Rdnr. 30.

170 *Uwe H. Schneider*, Scholz, 1993, § 43 Rdnr. 95.

171 *H.-G. Koppensteiner*, Rowedder, 1997, § 37 Rdnr. 25.

172 *H. Konzen*, Geschäftsführung, 1989, S. 2979. A. M. *M. Lutter/P. Hommelhoff*, GmbHG, 1995, § 37 Rdnr. 18; *H. Gieseke*, Interessenkonflikte, 1996, S. 490 ff., der einen Kernbereich weisungsfreier Entscheidungskompetenzen des Geschäftsführers annimmt.

173 *M. Lutter/P. Hommelhoff*, GmbHG, 1995, § 37 Rdnr. 5; *R. Höhn*, Geschäftsleitung, 1995, S. 7 ff.

Person zu stimmen. Ebenso können sie ihm Weisungen hinsichtlich seines Abstimmungsverhaltens in der Tochter-AG bezüglich der Gewinnverwendung erteilen. Diese Beispiele ließen sich beliebig fortsetzen. Problematisch ist indes, wann eine Folgepflicht des Geschäftsführers ausnahmsweise entfällt, wann eine Weisung also unbeachtlich ist.

c) Keine Folgepflicht bei unbeachtlichen Weisungen

(1) Die gesetzliche Ausgangslage

Eine Weisung an den Geschäftsführer kann auf zwei Wegen erfolgen. Zu unterscheiden ist zwischen dem gesetzlichen Normalstatut und besonderen im Gesellschaftsvertrag eingeräumten Weisungsrechten. Nach dem gesetzlichen Normalstatut erteilen die Gesellschafter mittels Beschlusses dem Geschäftsführer Weisungen. Von diesem gesetzlichen Normalstatut können die Gesellschafter im Gesellschaftsvertrag abweichen, indem sie einem einzelnen Gesellschafter ein Weisungsrecht einräumen. Dieser Gesellschafter kann dann direkt dem Geschäftsführer Weisungen erteilen.

Diese Unterscheidung wirkt sich auf die Unbeachtlichkeit von Weisungen und den Wegfall der Folgepflicht des Geschäftsführers aus.

(a) Die Unbeachtlichkeit von Weisungen aufgrund eines Gesellschafterbeschlusses

Erfolgt die Weisung aufgrund eines Gesellschafterbeschlusses, wendet die h. M. die aktienrechtlichen Nichtigkeitsvorschriften (§§ 241, 242, 249, 250, 253, 256 AktG) sowie die aktienrechtlichen Anfechtungsregeln (§§ 243 - 248, 251, 252, 254, 255, 257 AktG) auf die GmbH analog an, soweit nicht Besonderheiten des GmbH-Rechts entgegenstehen.[174] Das ist auch gerechtfertigt. Denn ebenso wie bei der Aktiengesellschaft besteht bei der GmbH ein Bedürfnis nach Rechtssicherheit.[175] Diesem Bedürfnis wird mit der Zurückdrängung der allgemeinen Nichtigkeitsgründe auf die in §§ 241 ff. AktG genannten Tatbestände

---

[174] *BGH*, Urt. v. 8.12.1954 - II ZR 291/53, in: BGHZ 15, S. 382, 384 f. (für den früheren § 195 AktG, der dem heutigen § 241 AktG entspricht); *BGH*, Urt. v. 17.2.1997 - II ZR 41/96, in: WM 1997, S. 823, 824; *H. Konzen*, Geschäftsführung, 1989, S. 2981; *K. Schmidt*, Scholz, 1995, § 45 Rdnr. 36, der allerdings von allgemeinen Institutionen des Verbandsrechts spricht; dagegen *W. Zöllner/U. Noack*, Beschlußmängel, 1989, S. 525 ff.; ihnen folgend *T. Raiser*, Hachenburg, 1990, Anh. § 47 Rdnr. 9 ff.

[175] *A. Hueck*, Gesellschafterbeschlüsse, 1962, S. 405. *M. Lutter/P. Hommelhoff*, GmbHG, 1995, Anh. § 47 Rdnr. 1. Für die Aktiengesellschaft *A. Hueck*, Anfechtbarkeit, 1924, S. 24; *U. Hüffer*, Geßler, 1984, § 241 Rdnr. 5.

Rechnung getragen. Denn durch die eindeutige Festlegung der Nichtigkeitsgründe und ihre scharfe Abgrenzung zu den Anfechtungsgründen werden die Gesellschafter und auch die Geschäftsführer in die Lage versetzt, sicher beurteilen zu können, wann ein Gesellschafterbeschluß nur anfechtbar und damit wirksam und wann er nichtig ist.

Für die Folgepflicht des Geschäftsführers hat die Unterscheidung zwischen Anfechtbarkeit und Nichtigkeit folgende Konsequenzen.

Ist der Beschluß nichtig, so darf der Geschäftsführer die Weisung nicht ausführen.[176] Nur wenn die Nichtigkeit allein auf verfahrensrechtlichen Gründen beruht, steht es dem Geschäftsführer frei, eine entsprechende Maßnahme in eigener Verantwortung auszuführen.[177]

Ist der Beschluß anfechtbar, kommt es darauf an, ob die Anfechtungsfrist abgelaufen ist. Ist der Beschluß bereits unanfechtbar geworden, besteht eine uneingeschränkte Folgepflicht.[178] Ist der Beschluß dagegen noch anfechtbar, muß der Geschäftsführer damit rechnen, daß der Beschluß in der Folgezeit angefochten wird und damit die Weisung entfällt. Deshalb muß der Geschäftsführer nur für den Fall, daß mit einer Anfechtung nicht zu rechnen ist, die Weisung ausführen.[179] Ist dagegen Anfechtungsklage erhoben, muß er die Ausführung der Weisung bis zum Ende des Rechtsstreits unterlassen. Während dieses oft langen Zeitraums steht der Geschäftsführer, wie *W. Zöllner*[180] treffend bemerkt, zwischen der „Scylla" des Mehrheitswillens, von dem er oft abhängig ist, und der „Charybdis" einer möglichen Haftung, weil er die Weisung trotz der Anfechtungsklage ausgeführt hat und sich nun herausstellt, daß der Beschluß nichtig ist und damit keine haftungsbefreiende Wirkung entfaltet. Damit wird deutlich, wie wichtig es für den Geschäftsführer ist, daß die Anfechtbarkeits- und Nichtigkeitsgründe genau bestimmt werden.

Schließlich ist von Bedeutung, daß neben den Gesellschaftern, die gegen den Beschlußantrag gestimmt haben, unter bestimmten Voraussetzungen auch der Geschäftsführer zur Anfechtung von Gesellschafterbeschlüssen berechtigt ist. Das ist analog § 245 Nr. 5 AktG der Fall, wenn es sich um einen ausführungsbedürftigen Gesellschafterbeschluß handelt und der Geschäftsführer durch des-

---

[176] *Uwe H. Schneider*, Scholz, 1993, § 43 Rdnr. 98; *W. Zöllner*, Baumbach/Hueck, 1996, § 37 Rdnr. 12; *M. Lutter/P. Hommelhoff*, GmbHG, 1995, § 37 Rdnr. 22.

[177] *Uwe H. Schneider*, Scholz, 1993, § 43 Rdnr. 98. Krit. *M. Lutter/P. Hommelhoff*, GmbHG, 1995, § 37 Rdnr. 22.

[178] *Uwe H. Schneider*, Scholz, 1993, § 43 Rdnr. 101; *W. Zöllner*, Baumbach/Hueck, 1996, § 37, Rdnr. 12; *H.-J. Fleck*, Haftung, 1974, S. 228; *H. Konzen*, Geschäftsführung, 1989, S. 2982.

[179] *C. Th. Ebenroth/K. W. Lange*, Sorgfaltspflichten, 1992, S. 73.

[180] *W. Zöllner*, Baumbach/Hueck, GmbHG, 1996, § 37 Rdnr. 12.

sen Ausführung eine zum Schadensersatz verpflichtende Handlung begehen würde.[181]

§ 245 Nr. 4 AktG, wonach der Vorstand uneingeschränkt anfechtungsberechtigt ist, findet im GmbH-Recht dagegen keine analoge Anwendung. Denn diese Anfechtungsbefugnis ist Folge der eigenverantwortlichen Stellung des Vorstands in der Zuständigkeitsordnung der Aktiengesellschaft. Da der Vorstand die Aktiengesellschaft unter eigener Verantwortung leitet, soll er auch die Möglichkeit haben, die Rechtmäßigkeit der Hauptversammlungsbeschlüsse gerichtlich überprüfen zu lassen. Der Geschäftsführer der GmbH unterliegt dagegen den Weisungen der Gesellschafterversammlung als oberstem Organ der Gesellschaft. In diese Hierachie paßt eine uneingeschränkte Anfechtungsbefugnis des Geschäftsführers aber nicht.[182]

Ist ein Beschluß nichtig, so kann jeder eine Nichtigkeitsfeststellungsklage erheben, wenn er ein Rechtsschutzbedürfnis hat. Dies ist zu bejahen, wenn der Beschluß für den Betreffenden unmittelbar erheblich ist.[183]

(b) Die Unbeachtlichkeit von Weisungen aufgrund eines gesellschaftsvertraglich eingeräumten Weisungsrechts

Eine Unterscheidung von Anfechtbarkeit und Nichtigkeit findet bei Weisungen einzelner Gesellschafter nicht statt. Sie wäre auch kaum praktikabel, weil die übrigen Gesellschafter von den Weisungen in der Regel nicht in Kenntnis gesetzt werden und deshalb ein eventuell bestehendes Anfechtungsrecht nicht ausüben können. Damit bleibt es bei den allgemeinen Nichtigkeitsgründen, die sich aus Gesetz, Satzung sowie den guten Sitten ergeben können.[184]

(2) Nichtigkeits- und Anfechtungsgründe bei konzernspezifischen Gesellschafterbeschlüssen

Es stellt sich nun die Frage, ob und wie die für die konzernfreie GmbH in Analogie zu den aktienrechtlichen Vorschriften entwickelten Nichtigkeits- und Anfechtungsgründe konzernspezifisch fortgeschrieben werden können.

---

[181] *H.-G. Koppensteiner*, Rowedder, 1997, § 47 Rdnr. 124; *M. Lutter/P. Hommelhoff*, GmbHG, 1995, Anh. § 47 Rdnr. 64 i. V. m. Rdnr. 33. Dagegen *BGH*, Urt. v. 28.1.1980 - II ZR 84/79, in: BGHZ 76, S. 154, 159.

[182] Ähnlich *T. Raiser*, Hachenburg, 1990, Anh. § 47 Rdnr. 161.

[183] *W. Zöllner*, Baumbach/Hueck, GmbHG, 1996, Anh. § 47 Rdnr. 33.

[184] *H. Konzen*, Geschäftsführung, 1989, S. 2981.

(a) Der konzernspezifische Gesellschafterbeschluß

Charakteristisch für konzernspezifische Beschlüsse der herrschenden GmbH ist ihr Bezug auf die abhängige Gesellschaft sowie die Notwendigkeit ihrer Umsetzung durch ihren Geschäftsführer. Entweder sind sie auf die Herbeiführung eines Gesellschafterbeschlusses in der abhängigen Gesellschaft gerichtet oder sie weisen den Geschäftsführer an, unmittelbar auf die Geschäftsleitung der abhängigen Gesellschaft Einfluß zu nehmen. Ihre konzernspezifische Gefahr besteht darin, daß die abhängige Gesellschaft und ihre Minderheitsgesellschafter im Konzerninteresse geschädigt werden, was jedenfalls im faktischen Konzern unzulässig ist.

Folge dieser konzernspezifischen Gefahr ist es, daß sich für die Gesellschafterbeschlüsse in der herrschenden GmbH die Bedeutung der Beschlußmängel von den Anfechtungsgründen auf die Nichtigkeitsgründe verlagert. Denn in der Regel haben die Gesellschafter kein Interesse daran, konzernbezogene Beschlüsse allein deshalb anzufechten, weil ihre Ausführung zu einem Schaden bei einer beherrschten Gesellschaft führt. Denn diesem Schaden entspricht ein Vorteil bei der herrschenden GmbH. Lediglich der Geschäftsführer wird, sofern der Gesellschafterbeschluß nicht schon nichtig ist, Anfechtungsklage erheben, wenn er durch die Ausführung des Beschlusses eine zum Schadensersatz verpflichtende Handlung, eine Straftat oder eine Ordnungswidrigkeit begehen würde. Ist dies nicht der Fall, werden die Interessen des Geschäftsführers durch einen bloß anfechtbaren Beschluß nicht berührt. Insbesondere macht er sich gegenüber seiner Gesellschaft grundsätzlich nicht schadensersatzpflichtig, wenn er einen einstimmigen bzw. nicht mehr anfechtbaren Gesellschafterbeschluß befolgt.

Etwas anderes kann jedoch dann gelten, wenn der Beschluß nichtig ist. Zwar kann sich der Geschäftsführer gegenüber einer Schadensersatzklage der Gesellschaft auf den Einwand der unzulässigen Rechtsausübung berufen, wenn alle Gesellschafter dem Beschluß zugestimmt haben.[185] Dieser Einwand ist jedoch ausgeschlossen, wenn der Geschäftsführer gegen zwingende Gläubigerschutzvorschriften verstoßen hat und der Schadensersatz zur Befriedigung der Gläubiger erforderlich ist. Denn insoweit trifft ihn eine besondere Garantiepflicht, die in § 43 Abs. 3 Satz 3 GmbHG zum Ausdruck kommt, auf die dort genannten Fälle aber nicht beschränkt ist.[186]

---

185 *H. J. Mertens*, Hachenburg, 1996, § 43 Rdnr. 78; *Uwe H. Schneider*, Scholz, 1993, § 43 Rdnr. 105; *H.-G. Koppensteiner*, Rowedder, 1997, § 43 Rdnr. 31.

186 *Uwe H. Schneider*, Scholz, 1993, § 43 Rdnr. 107; *H. J. Mertens*, Hachenburg, 1996, § 43 Rdnr. 78.

All dies macht deutlich, wie wichtig aus der Sicht des Geschäftsführers die genaue Umschreibung der Nichtigkeitsgründe ist.

(b) Konzernspezifische Nichtigkeitsgründe

Wie bereits ausgeführt, ergeben sich die Nichtigkeitsgründe für Gesellschafterbeschlüsse in der GmbH aus §§ 241 ff. AktG analog, soweit dem nicht Besonderheiten aus dem GmbH-Recht entgegenstehen. Für konzernspezifische Nichtigkeitsgründe kommen indes nur § 241 Nr. 3 Var. 2 und 3 sowie § 241 Nr. 4 AktG in Betracht.

(aa) Verstoß gegen gläubigerschützende Vorschriften

Analog § 241 Nr. 3 Var. 2 AktG ist ein Gesellschafterbeschluß nichtig, wenn er durch seinen Inhalt, d. h. nicht nur durch die Art seines Zustandekommens,[187] Vorschriften verletzt, die ausschließlich oder überwiegend zum Schutze der Gläubiger gegeben sind. Unter „überwiegend" versteht man, daß die Vorschrift wesentliche Bedeutung für den Gläubigerschutz hat, daß „sie zu seinen Grundpfeilern gehört".[188] Hierzu zählen jedenfalls die der Kapitalerhaltung dienenden Normen,[189] bei der GmbH insbesondere § 30 Abs. 1 GmbHG.[190] So ist z. B. die entgeltliche Einziehung eines Gesellschaftsanteils nichtig, wenn durch die Ausschüttung der Gegenleistung das zur Erhaltung des Stammkapitals erforderliche Vermögen angegriffen werden müßte.[191] Freilich verstößt hier nicht der Beschluß selbst, sondern erst seine Ausführung, nämlich die Erbringung der Gegenleistung gegen § 30 Abs. 1 GmbHG. Das ist jedoch für die Beurteilung der Nichtigkeit unerheblich. Denn es reicht aus, daß der Beschluß auf einen solchen Verstoß gerichtet ist. Mit anderen Worten ist es nicht erforderlich, daß wie bei einem Beschluß zur Feststellung des Jahresabschlusses, bei dem die Bilanz eine

---

[187] So schon *A.. Hueck*, Anfechtbarkeit, 1924, S. 65, für die Rechtslage vor dem AktG 1937, in dem erstmals Bestimmungen über dic Nichtigkeit von Hauptversammlungsbeschlüssen aufgenommen wurden; dazu *U. Hüffer*, Geßler, 1984, Vor § 241 Rdnr. 6 - 8.

[188] *W. Zöllner*, Schranken, 1963, S. 378; ebenso *Th. Raiser*, Hachenburg, 1990, Anh. § 47, Rdnr. 47.

[189] *U. Hüffer*, AktG, 1997, § 241 Rdnr. 17.

[190] *Th. Raiser*, Hachenburg, 1990, Anh. § 47, Rdnr. 48; *K. Schmidt*, Scholz, 1995, § 45 Rdnr. 74.

[191] *RG*, Urt. v. 24.11.1933 - II 113/33, in: RGZ 142, S. 286, 290 f.

Überbewertung der Aktiven enthält,[192] der Beschluß selbst gegen die gläubigerschützende Vorschrift verstößt.

Im Konzern wird § 30 Abs. 1 GmbHG dann relevant, wenn die Tochtergesellschaft den Gesellschaftern der Muttergesellschaft eine Leistung erbringt, ohne daß diese Leistung angemessen vergütet wird. Mindert sich nämlich aufgrund dieses Vorgangs der Buchwert der Tochterbeteiligung in der Bilanz der Muttergesellschaft und entsteht dadurch bei dieser eine Unterbilanz, so liegt eine Auszahlung aus dem zur Erhaltung des Stammkapitals der Muttergesellschaft erforderlichen Vermögen vor.[193]

Ein Beschluß der Gesellschafter der Muttergesellschaft, der deren Geschäftsführer anweist, in der Tochtergesellschaft eine Auszahlung an einen Gesellschafter der GmbH ohne angemessene Gegenleistung zu bewirken, ist also nichtig, wenn durch die Auszahlung bei der Muttergesellschaft eine Unterbilanz entsteht.

(bb) Verstoß gegen Vorschriften, die „sonst im öffentlichen Interesse gegeben sind"

Analog § 241 Nr. 2 Var. 3 AktG sind Gesellschafterbeschlüsse nichtig, wenn sie durch ihren Inhalt gegen Vorschriften verstoßen, die „sonst im öffentlichen Interesse gegeben sind". Aus dem Umstand, daß die gläubigerschützenden Vorschriften nur ein Beispiel für die sonst im öffentlichen Interesse gegebenen Vorschriften bilden, folgt, daß auch letztere „ausschließlich und überwiegend" im öffentlichen Interesse erlassen worden sein müssen.[194] Auch hier muß also das öffentliche Interesse für die Bestimmung des Normzwecks eine wesentliche Rolle spielen.[195] Das ist dann der Fall, wenn das öffentliche Interesse die strenge Nichtigkeitsfolge gebietet. Diese Voraussetzung ist erfüllt, wenn es im öffentlichen Interesse nicht hingenommen werden kann, daß der rechtswidrige Beschluß vorläufig wirksam ist und nur durch eine Anfechtungsklage beseitigt werden kann.[196] Danach können Beschlüsse nichtig sein, die beispielsweise gegen Vorschriften des Strafrechts und Ordnungswidrigkeitenrechts oder gegen zwingende Vorschriften des Verwaltungs-, Steuer- oder Sozialrechts versto-

---

192 Dazu *BGH*, Urt. v. 24.6.1982 - III ZR 19/81, in: ZIP 1982, S. 1075, 1080.

193 *Uwe H. Schneider*, Konzernfinanzierung, 1984, S. 516.

194 *U. Hüffer*, Geßler, 1984, § 241 Rdnr. 45.

195 *U. Hüffer*, Geßler, 1984, § 241 Rdnr. 45.

196 *K. Schmidt*, Scholz, 1995, § 45 Rdnr. 75.

ßen.[197] So ist z. B. ein Gesellschafterbeschluß nichtig, der den Geschäftsführer anweist, keine Sozialabgaben abzuführen.

Wie oben ausgeführt, besteht die konzernspezifische Gefahr von Gesellschafterbeschlüssen in der herrschenden GmbH darin, daß sie den Geschäftsführer anweisen, auf die Geschäftsführung der abhängigen Gesellschaft zu deren Nachteil Einfluß zu nehmen. Folgt der Geschäftsführer der abhängigen Gesellschaft der Veranlassung des Geschäftsführers der herrschenden GmbH und ist er sich des schädigenden Charakters der von ihm sodann ausgeführten Maßnahme bewußt, so begeht er eine Untreue gemäß § 266 StGB. Daraus folgt zugleich, daß der ebenfalls vorsätzlich handelnde Geschäftsführer der herrschenden GmbH sich als Anstifter gemäß §§ 266, 26 StGB strafbar gemacht hat. Damit wird deutlich, daß der entsprechende Gesellschafterbeschluß in der herrschenden GmbH auf die Begehung einer Straftat gerichtet ist. Die Wirksamkeit eines solchen Beschlusses kann aber nicht hingenommen werden. Damit ist er analog § 241 Nr. 3 Var. 3 AktG nichtig.

Dieses Beispiel veranschaulicht, daß ein Gesellschafterbeschluß immer dann nichtig ist, wenn er den Geschäftsführer dazu anweist, den Geschäftsführer oder Vorstand der beherrschten Gesellschaft zu einer Straftat anzustiften.

Gleiches muß aber auch dann gelten, wenn der Geschäftsführer der herrschenden GmbH sich durch Weitergabe der Weisung keiner Straftat schuldig macht, die Geschäftsleitung der beherrschten Gesellschaft durch Ausführung der Weisung jedoch gegen Vorschriften verstößt, die im überwiegenden öffentlichen Interesse gegeben sind. Auch dann kann die Wirksamkeit des Gesellschafterbeschlusses in der herrschenden GmbH nicht hingenommen werden. Dies gilt um so mehr, als kein Gesellschafter der herrschenden GmbH an einer Anfechtung interessiert sein wird, da er nicht selbst betroffen ist.

Hieraus folgt insbesondere, daß ein Gesellschafterbeschluß auch dann nichtig ist, wenn er eine Auszahlung aus dem zur Erhaltung des Stammkapitals erforderlichen Vermögen der beherrschten GmbH zum Gegenstand hat. In diesem Zusammenhang ist § 30 GmbHG zwar keine Vorschrift zum Schutze der „Gläubiger der Gesellschaft“, also der Gläubiger der herrschenden GmbH, er ist aber eine Vorschrift, die „sonst im öffentlichen Interesse gegeben“ ist.

Gleiches gilt für die Vorschriften des § 309 Abs. 1 und § 323 Abs. 1 i. V. m. § 309 Abs. 1 AktG, die im Vertrags- und Eingliederungskonzern mit einer beherrschten Aktiengesellschaft die Erteilung von Weisungen an die Sorgfalt eines ordentlichen und gewissenhaften Geschäftsleiters binden. Ihr wesentlicher Zweck besteht neben dem Minderheitenschutz auch im Gläubigerschutz.[198] Für

---

[197] *T. Raiser*, Hachenburg, 1990, Anh. § 47, Rdnr. 50.

[198] Für § 311 AktG *U. Hüffer*, AktG, 1997, § 311 Rdnr. 1.

§§ 309 Abs. 1, 323 Abs. 1 AktG wird das zum einen darin deutlich, daß die genannten Vorschriften zwingend sind, von ihnen im Beherrschungs- bzw. Eingliederungsvertrag also nicht abgewichen werden darf.[199] Zum anderen stellen die von ihnen statuierten Pflichten einen Ausgleich dar für die Lockerung der Vermögensbindung (§§ 291 Abs. 3 AktG, 323 Abs. 2 AktG). Die Vermögensbindung dient aber dem Gläubigerschutz. Des weiteren zeigt sich der gläubigerschützende Charakter der genannten Vorschriften darin, daß der aus ihrer Verletzung sich ergebende Schadensersatzanspruch der Gesellschaft auch von den Gesellschaftsgläubigern geltend gemacht werden kann, soweit diese von der Gesellschaft keine Befriedigung erlangen können, § 309 Abs. 4 Satz 3 AktG bzw. § 323 Abs. 1 i. V. m. § 309 Abs. 4 Satz 3 AktG. Schließlich ist auch ein Verzicht oder ein Vergleich der Gesellschaft gegenüber den Gläubigern unwirksam, § 309 Abs. 4 Satz 4 AktG und § 323 Abs. 1 i. V. m. § 309 Abs. 4 Satz 4 AktG.

Dieselben Erwägungen gelten für § 311 AktG. Danach darf das herrschende Unternehmen der abhängigen Aktiengesellschaft im faktischen Konzern nur dann einen Nachteil zufügen, wenn es bereit und in der Lage ist, den Nachteil bis zum Ende des Geschäftsjahres auszugleichen. Andernfalls hat es eine nachteilige Einflußnahme zu unterlassen. Diese Pflicht dient auch dem Schutz der Gläubiger, da durch die Zulassung eines zeitlich gestreckten Nachteilsausgleichs die Vermögensbindung in der Aktiengesellschaft zeitweise gelockert wird.[200] Schließlich läßt sich der Zweck des Gläubigerschutzes auch aus § 317 Abs. 4 i. V. m. § 309 Abs. 4 Satz 3 AktG entnehmen. Danach können die Gläubiger ebenso wie im Vertrags- und Eingliederungskonzern einen Schadensersatzanspruch der Gesellschaft wegen Verstoßes gegen § 311 Abs. 1 AktG selbst geltend machen, soweit es für ihre Befriedigung erforderlich ist.

Dagegen stellt die Verletzung der mitgliedschaftlichen Förderpflicht allein keinen Verstoß gegen eine im öffentlichen Interesse gegebene Vorschrift da. Denn im öffentlichen Interesse gegebene Vorschriften meint nur gesetzliche Vorschriften.[201] Entstehungsgrund der mitgliedschaftlichen Förderpflicht ist aber die Satzung und damit eine privatautonome Vereinbarung. Entsprechendes gilt für die Treupflicht der herrschenden GmbH gegenüber den mittelbar beherrschten Gesellschaften. Auch hier ergibt sich die von § 242 BGB vorausgesetzte rechtliche Sonderverbindung aus einer privatautonomen Gestaltung.

---

199 Für § 309 AktG *U. Hüffer*, AktG, 1997, § 309 Rdnr. 1.

200 *U. Hüffer*, AktG, 1997, § 311 Rdnr. 49; *M. Lutter*, Kölner Kommentar, 1988, § 57 Rdnr. 80; *H.-G. Koppensteiner*, Kölner Kommentar, 1987, § 311 Rdnr. 107 ff.; *W. Hefermehl/E. Bungeroth*, Geßler, 1983, § 57 Rdnr. 64. A. M. *H. Altmeppen*, Managerhaftung, 1998, S. 58; *A. Cahn*, Kapitalerhaltung, 1998, S. 65 f.

201 Ebenso zu dem Begriff der Verbotsgesetze im Sinne von § 134 BGB *R. Sack*, Staudinger, 1996, § 134 Rdnr. 16.

(cc) Verstoß gegen die guten Sitten

Analog § 241 Nr. 4 AktG ist schließlich ein Gesellschafterbeschluß nichtig, wenn er „durch seinen Inhalt gegen die guten Sitten verstößt“. Indem diese Vorschrift fordert, daß der Beschluß gerade durch seinen Inhalt gegen die guten Sitten verstößt, ist ihr Anwendungsbereich enger als der von § 138 BGB. Erforderlich ist nämlich, daß der Beschluß „für sich allein genommen“, d. h. ohne Rücksicht auf Beweggrund, Zweck oder Art des Zustandekommens gegen die guten Sitten verstößt.[202] Hiervon hat der Bundesgerichtshof[203] im Anschluß an das Reichsgericht[204] allerdings für den Fall eine Ausnahme gemacht, daß der Beschluß seinem inneren Gehalt nach in einer sittenwidrigen Schädigung nicht anfechtungsberechtigter Personen besteht. Das ist mit der h. M.[205] zu begrüßen. Denn andernfalls hätten es die Gesellschafter in der Hand, durch eine geeignete Formulierung des Gesellschafterbeschlusses die Vorschrift des § 241 Nr. 4 AktG zu umgehen und die Stellung nicht anfechtungsberechtigter Dritter zu verschlechtern. Als Beispiel für einen zwar sittlich indifferenten, aber trotzdem in diesem Sinne sittenwidrigen Gesellschafterbeschluß wird in der Literatur[206] ein Beschluß genannt, mit dem die Gesellschafter das Gesellschaftsvermögen bis zur Grenze des Stammkapitals ausschütten, um den Gesellschaftsgläubigern die Zwangsvollstreckung zu erschweren.

Für die Beurteilung konzernbezogener Beschlüsse der herrschenden GmbH spielt das genannte Urteil des Bundesgerichtshofs eine wichtige Rolle. Denn spezifisches Merkmal konzernbezogener Beschlüsse ist gerade ihre Ausrichtung auf die beherrschten Gesellschaften, die ihrerseits die Beschlüsse nicht anfechten können. Deshalb muß sich die Sittenwidrigkeit solcher Beschlüsse nicht allein aus dem Inhalt selbst ergeben, sondern kann auch aus den Motiven der Gesellschafter folgen.

Überträgt man die obengenannte Rechtsprechung auf den Konzern, so ergibt sich, daß Beschlüsse der herrschenden GmbH immer dann sittenwidrig sind, wenn durch ihre Ausführung eine beherrschte Gesellschaft geschädigt wird und dies den beschlußfassenden Gesellschaftern zumindest bewußt ist.

---

[202] *RG*, Urt. v. 11.1.1927 - II 178/26, in: RGZ 115, S. 378, 383; *T. Raiser*, Hachenburg, 1990, Anh. § 47, Rdnr. 53.

[203] *BGH*, Urt. v. 8.12.1954 - II ZR 291/53, in: BGHZ 15, S. 382, 385 f., für den früheren § 195 Nr. 4 AktG, der § 241 Nr. 4 AktG entspricht.

[204] *RG*, Urt. v. 7.6.1939 - II 199/38, in: RGZ 161, S. 129, 144.

[205] *U. Hüffer*, Geßler, 1984, § 241 Rdnr. 56; *W. Zöllner*, Baumbach/Hueck, 1996, Anh. § 47 Rdnr. 25; *K. Schmidt*, Scholz, 1995, § 45 Rdnr. 76; *H.-G. Koppensteiner*, Rowedder, 1997, § 47 Rdnr. 88; *M. Lutter/P. Hommelhoff*, GmbHG, 1995, Anh. § 47 Rdnr. 21.

[206] *M. Lutter/P. Hommelhoff*, GmbHG, 1995, Anh. § 47 Rdnr. 21.

Handeln die Gesellschafter dagegen fahrlässig, ist eine Sittenwidrigkeit zu verneinen. Das folgt in der bloßen Abhängigkeitslage mit einer GmbH als Tochtergesellschaft schon aus dem Gedanken, daß die herrschende GmbH in diesem Fall, wie noch zu zeigen ist, nicht ihre mitgliedschaftliche Förderpflicht gegenüber der Tochter-GmbH verletzt. Aber auch in den übrigen Fällen ist eine Wirksamkeit des Gesellschafterbeschlusses mit unserer Rechts- und Sittenordnung noch vereinbar. Denn Bestandteil unserer Rechtsordnung ist auch die Verbandsautonomie. Danach soll die juristischen Person durch ihre Organe selbst über ihre Willensbildung bestimmen können. Dem würde es aber widersprechen, wenn verbandsfremde Personen die Willensbildung der Gesellschaft schon mit dem Vorwurf bloßer Sorgfaltswidrigkeit „torpedieren" könnten. Überdies würde durch die Zulassung dieses Vorwurfs auch nicht der mit der Unterscheidung zwischen Anfechtbarkeit und Nichtigkeit bezweckten Rechtssicherheit in angemessener Weise Rechnung getragen. Denn oft ist nicht eindeutig ersichtlich, ob die Durchführung einer Weisung kaufmännischer Sorgfalt oder der im Verkehr erforderlichen Sorgfalt widerspricht.[207] Schließlich sind die beherrschten Gesellschaften auch nicht der herrschenden GmbH wehrlos preisgegeben. So leitet der Vorstand der abhängigen Aktiengesellschaft die Gesellschaft unter eigener Verantwortung, § 76 Abs. 1 AktG. Und der Minderheitsgesellschafter der abhängigen GmbH kann einen von der herrschenden GmbH herbeigeführten Gesellschafterbeschluß anfechten. Der Geschäftsführer ist schließlich verpflichtet, den Anweisungen der herrschenden GmbH außerhalb der Gesellschafterversammlung keine Folge zu leisten, sofern diese sorgfaltswidrig sind.

Nach alledem sind bloß sorgfaltswidrige Weisungsbeschlüsse, durch deren Ausführung eine beherrschte Gesellschaft geschädigt wird, grundsätzlich wirksam.

Das gilt auch dann, wenn sich der Geschäftsführer durch die Ausführung der Weisung gegenüber der beherrschten Gesellschaft persönlich schadensersatzpflichtig machen würde. Denn in diesem Fall ist es dem Geschäftsführer zumutbar, den Beschluß analog § 245 Nr. 5 AktG anzufechten.

(c) Konzernspezifische Anfechtungsgründe

Im Gegensatz zu den Nichtigkeitsgründen gibt es „*keinen numerus clausus*" der Anfechtungsgründe.[208] Allerdings haben sich in der Literatur Fallgruppen zu den Anfechtungsgründen herausgebildet:

---

207 Zum Unterschied zwischen diesen Sorgfaltsmaßstäben *P. Hofmann*, Handelsrecht, 1996, S. 175; *K. J. Hopt*, HGB, 1995, § 347 Rdnr. 1.

208 *K. Schmidt*, Scholz, 1995, § 45 Rdnr. 93 (Hervorhebung im Original).

- Verstoß gegen das Gesetz,
- Verstoß gegen die Satzung,
- Sittenwidrigkeit,
- Erstreben von Sondervorteilen,
- Verstoß gegen den Gleichbehandlungsgrundsatz,
- Verstoß gegen die mitgliedschaftliche Förderpflicht des Gesellschafters gegenüber der Gesellschaft und
- Verstoß gegen die Treupflicht des Gesellschafters gegenüber den Mitgesellschaftern.[209]

Wichtigster Anfechtungsgrund bei konzernbezogenen Beschlüssen dürfte der der Sittenwidrigkeit sein. Ihm kommt gegenüber den Nichtigkeitsgründen dann eigenständige Bedeutung zu, wenn die Gesellschafter keine Schädigungsabsicht haben, der Geschäftsführer sich aber durch Ausführung des Gesellschafterbeschlusses persönlich schadensersatzpflichtig machen würde. In diesem Fall wäre es mit unserer Rechtsordnung unvereinbar, den Geschäftsführer an seiner Folgepflicht festzuhalten und ihm ein Anfechtungsrecht zu verwehren. Vielmehr ist es ihm dann analog § 245 Nr. 5 AktG gestattet, den Gesellschafterbeschluß anzufechten.

Die übrigen Anfechtungsgründe sind dagegen für konzernspezifische Beschlüsse der herrschenden GmbH nicht einschlägig. So liegt kein Verstoß gegen die Satzung der herrschenden GmbH vor, wenn die Gesellschafter der herrschenden GmbH beschließen, daß eine beherrschte Gesellschaft zu einer Konzernumlage an die herrschende GmbH veranlaßt wird, ohne eine Gegenleistung zu erhalten. Denn aus der Satzung der herrschenden GmbH ergibt sich, wie noch zu zeigen sein wird, für deren Gesellschafter nur die mitgliedschaftliche Pflicht, stets das Interesse der herrschenden GmbH „im Auge zu haben". Dieses wird aber nicht verletzt, wenn die herrschende GmbH Empfängerin der Konzernumlage ist. Auch der Gleichbehandlungsgrundsatz wird durch einen solchen Beschluß nicht berührt, da er nur zur Gleichbehandlung der Gesellschafter der herrschenden GmbH verpflichtet. Ferner erstrebt der dem Beschluß zustimmende Gesellschafter keinen Sondervorteil auf Kosten der Gesellschaft. Denn nicht ihm, sondern der Gesellschaft wird ein Vorteil zugewendet. Und schließlich verstoßen die zustimmenden Gesellschafter auch nicht gegen ihre Treupflicht gegenüber ihren Mitgesellschaftern. Denn diese werden nicht geschädigt.

---

[209] *W. Zöllner*, Baumbach/Hueck, 1996, Anh. § 47 Rdnr. 43.

(3) Konzernspezifische Nichtigkeitsgründe von Weisungen aufgrund des Gesellschaftsvertrages

Weisungen aufgrund des Gesellschaftsvertrages sind entweder wirksam oder nichtig. Die Anfechtbarkeitsvorschriften sind auf sie nicht anwendbar, da erstens die übrigen Gesellschafter in der Regel von ihnen keine Kenntnis erhalten und deshalb zweitens kein Vertrauensschutz erforderlich ist. Die Nichtigkeit solcher Weisungen richtet sich deshalb nach den allgemeinen zivilrechtlichen Vorschriften, insbesondere nach §§ 134, 138 BGB.

Gemäß § 134 BGB ist eine Weisung nichtig, wenn sie gegen ein gesetzliches Verbot verstößt und sich nicht aus dem Gesetz ein anderes ergibt. Ein Verstoß gegen ein gesetzliches Verbot führt also nicht immer zur Nichtigkeit, sondern nur dann, wenn Wortlaut, Sinn und Zweck des einzelnen Verbotsgesetzes die Nichtigkeit fordern.[210]

Einen ersten Anhaltspunkt für die Ermittlung einschlägiger Verbotsgesetze für die Beurteilung gesellschaftsvertraglicher Weisungen bietet § 241 AktG. Diese Vorschrift schränkt, wie bereits erwähnt, die allgemeinen Nichtigkeitsgründe für Gesellschafterbeschlüsse im Interesse der Rechtssicherheit ein. Hieraus kann geschlossen werden, daß die dort genannten Vorschriften Verbotsgesetze i. S. v. § 134 BGB sind.[211] Verstößt also die Ausführung einer gesellschaftsvertraglichen Weisung gegen Vorschriften, die ausschließlich oder überwiegend zum Schutze der Gläubiger oder sonst im öffentlichen Interesse gegeben sind, so ist die Weisung gemäß § 134 BGB nichtig. Verbotsgesetze sind also insbesondere die §§ 30 ff. GmbHG sowie Vorschriften des Straf- und Ordnungswidrigkeitenrechts, des Sozial- und Steuerrechts. Auch die §§ 309 Abs. 1, 323 Abs. 1 i. V. m. 309 Abs. 1 AktG gehören hierher, da sie dem herrschenden Unternehmen bzw. seinen gesetzlichen Vertretern verbieten, ohne Anwendung der Sorgfalt eines ordentlichen und gewissenhaften Geschäftsleiters auf die beherrschte Aktiengesellschaft Einfluß zu nehmen. Zu nennen ist schließlich § 311 AktG, der dem herrschenden Unternehmen verbietet, der abhängigen Aktiengesellschaft einen Nachteil zuzufügen, ohne bereit und in der Lage zu sein, den Nachteil am Ende des Geschäftsjahres auszugleichen.

Keine Verbotsgesetze statuieren dagegen die Satzung oder § 242 BGB als Rechtsgrundlage für die mitgliedschaftliche Förderpflicht bzw. Treupflicht der herrschenden GmbH gegenüber den beherrschten Gesellschaften. Denn Gesetz

---

[210] *BGH*, Urt. v. 5.5.1992 - X ZR 134/90, in: BGHZ 118, S. 182, 188; *O. Jauernig*, Jauernig, 1994, § 134 Anm. 3. b).

[211] Ebenso für die in § 241 Nr. 3 AktG genannten Vorschriften *U. Noack*, Beschlüsse, 1989, S. 38.

im Sinne von § 134 BGB meint nur hoheitliche Rechtsnormen.[212] Die Satzung der Gesellschaft sowie die in § 242 BGB vorausgesetzte Sonderverbindung sind jedoch Ausdruck privatautonomer Gestaltung.

Die Sittenwidrigkeit kommt als Nichtigkeitsgrund gemäß § 138 Abs. 1 BGB zunächst dann in Betracht, wenn durch Erfüllung der Weisung die beherrschte Gesellschaft „sehenden Auges" in den Konkurs[213] geführt würde und damit zwangsläufig Gläubiger geschädigt würden.[214] Sie ist aber z. B. auch dann zu bejahen, wenn sich der Geschäftsführer der herrschenden GmbH durch Ausführung der Weisung gegenüber der beherrschten Gesellschaft schadensersatzpflichtig machen würde.

3. Die Pflicht zur Berichterstattung im Konzern

Als weitere Ausprägung der Pflicht zur kollegialen Zusammenarbeit des Geschäftsführers mit den Gesellschaftern wird die Pflicht zur Berichterstattung diskutiert. Sie kann sich aus dem mitgliedschaftlichen Informationsrecht des einzelnen Gesellschafters gemäß § 51a GmbHG sowie aus einer allgemeinen Berichtspflicht ergeben.

a) Das mitgliedschaftliche Informationsrecht des einzelnen Gesellschafters gemäß § 51a GmbHG

(1) Die gesetzliche Ausgangslage

Gemäß § 51a Abs. 1 GmbHG haben die Geschäftsführer jedem Gesellschafter auf Verlangen unverzüglich Auskunft über die Angelegenheiten der Gesellschaft zu geben und die Einsicht der Bücher und Schriften zu gestatten. Hiermit wird dem einzelnen Gesellschafter ein individuelles Informationsrecht gewährt, das ihm als eigennütziges Mitgliedschaftsrecht[215] gemäß § 51a Abs. 3 GmbHG nicht entzogen werden kann. Sein Zweck besteht darin, dem Gesellschafter eine sachgerechte Ausübung seiner Rechte zu ermöglichen.[216] Insbesondere soll jeder Gesellschafter in die Lage versetzt werden, in Ausübung seiner Gesell-

---

[212] *R. Sack*, Staudinger, 1996, § 134 Rdnr. 16.

[213] Jetzt Insolvenz.

[214] *OLG Frankfurt a. M.*, Urt. v. 7.2.1997 - 24 U 88/95, in: GmbHR 1997, S. 346, 348.

[215] *M. Lutter/P. Hommelhoff*, GmbHG, 1995, § 51a, Rdnr. 1; *M. Lutter*, Informationsrecht, 1982, S. 3.

[216] BT-Drucksache 8/1347 zu § 51 a.

schafterrechte die Tätigkeit der Geschäftsführer zu kontrollieren.[217] Allerdings besteht weitgehend Einigkeit, daß § 51a GmbHG legislatorisch mißlungen ist.[218] Denn nach dem Wortlaut der Vorschrift kann jeder einzelne Gesellschafter zu jeder Zeit ohne besondere Voraussetzungen und ohne Abstimmung mit den anderen Gesellschaftern von der Geschäftsführung Informationen verlangen und gleichzeitig Einsicht in alle Unterlagen nehmen, auch wenn sie Mitarbeiter, Kunden und Lieferanten betreffen.[219] Damit kann zum einen die Funktionsfähigkeit der Geschäftsführung erheblich beeinträchtigt werden.[220] Zum anderen besteht die Gefahr, daß Vertraulichkeitsinteressen von Arbeitnehmern und Gläubigern der Gesellschaft verletzt werden, wenn der Geschäftsführer gesellschaftsinterne Informationen über sie offenlegt.[221] Deshalb ist den Stimmen in der Literatur zu folgen, die das Informationsrecht des Gesellschafters begrenzen wollen. Voraussetzung für seine Ausübung ist danach ein allgemeines Informationsbedürfnis[222] bzw. die Zweckdienlichkeit der begehrten Auskunft für die Ausübung der Gesellschafterrechte[223].

(2) Das Problem

Die Frage einer Begrenzung des Informationsrechts stellt sich in der konzernfreien Gesellschaft nur deshalb, weil zu den Angelegenheiten der Gesellschaft alles gehört, „was überhaupt Bezug zur Gesellschaft hat.“[224] Lehnt man eine Begrenzung des Informationsrechts ab, könnte mithin der Gesellschafter über jeden noch so unbedeutenden geschäftlichen Vorgang, wie etwa über das Gehalt des Pförtners, Auskunft verlangen. Anders verhält es sich dagegen bei einem Auskunftsbegehren, das sich auf einen Vorgang in einer beherrschten Gesellschaft bezieht. Denn hier ist schon problematisch, ob überhaupt eine

---

[217] BT-Drucksache 8/1347 zu § 51 a.

[218] *K. Schmidt*, Scholz, 1995, § 51a Rdnr. 7; *H.-J. Mertens*, § 51 a Abs. 1 GmbHG, 1984, S. 557; *M. Lutter*, Informationsrechte, 1982, S. 9.

[219] *M. Lutter*, Informationsrecht, 1982, S. 2.

[220] *H.-J. Mertens*, § 51 a Abs. 1 GmbHG, 1984, S. 561 f.

[221] *H.-J. Mertens*, § 51 a Abs. 1 GmbHG, 1984, S. 562 f.

[222] *K. Schmidt*, Scholz, 1995, § 51a Rdnr. 8. Dagegen: *M. Kort*, Informationsrecht, 1987, S. 49; *J. Tietze*, Informationsrechte, 1985, S. 17 f.

[223] *B. Grunewald*, Einsichtsrecht, 1982, S. 217; *H.-J. Mertens*, § 51 a Abs. 1 GmbHG, 1984, S. 568 f.; jeweils mit Unterschieden im Details. Siehe auch *C. Wilde*, Informationsrechte, 1998, S. 424.

[224] *B. Grunewald*, Einsichtsrecht, 1982, S. 214; *K. Schmidt*, Scholz, 1995, § 51a Rdnr. 19; *M. Lutter/P. Hommelhoff*, GmbHG, 1995, § 51a Rdnr. 5; *W. Zöllner*, Baumbach/Hueck 1996, § 51a Rdnr. 10; *J. Tietze*, Informationsrechte, 1985, S. 9.

Angelegenheit der herrschenden Gesellschaft vorliegt mit der Folge einer Auskunftspflicht ihres Geschäftsführers.

(3) Meinungsstand

Unstreitig gehören zu den Angelegenheiten der herrschenden GmbH ihre rechtlichen und geschäftlichen Beziehungen *zu* den verbundenen Unternehmen.[225] Umstritten ist dagegen, unter welchen Voraussetzungen auch die Vorgänge *bei* den verbundenen Unternehmen vom Informationsrecht des Gesellschafters umfaßt sind.

Zu diesem Problem werden zwei Ansichten vertreten.

Nach einer Ansicht kann ein Gesellschafter über die Angelegenheiten von Tochtergesellschaften in gleichem Umfang Auskunft verlangen wie über die Angelegenheiten seiner eigenen Gesellschaft.[226] Andernfalls könne die Gesellschaft das Auskunftsrecht dadurch aushöhlen, daß sie Unternehmenstätigkeiten auf Tochtergesellschaften ausgliedere.[227]

Nach einer anderen Ansicht erstreckt sich das Informationsrecht des Gesellschafters dagegen nur auf diejenigen Vorgänge in der beherrschten Gesellschaft, die in ihrer Bedeutung und ihren Auswirkungen für die herrschende GmbH von objektiver Wichtigkeit sind.[228] Zur Begründung wird angeführt, daß das Auskunftsrecht des GmbH-Gesellschafters konzerndimensional keinen weiteren Umfang haben solle als das Auskunftsrecht des Aktionärs gemäß § 131 Abs. 1 AktG.[229] Aus der Regierungsbegründung zu § 131 Abs. 1 Satz 2 AktG gehe aber hervor, daß der Aktionär nur über solche geschäftlichen Vorgänge in dem verbundenen Unternehmen Auskunft verlangen könne, die die Lage der Gesellschaft oder ihre Beziehungen zu dem verbundenen Unternehmen beeinflussen könnten.[230] Außerdem könne das Auskunftsrecht des GmbH-Gesellschafters nicht weiter gehen als das Auskunftsrecht des Aufsichtsrats bei der

---

225 *U. Hüffer*, Hachenburg, 1990, § 51a Rdnr. 25.

226 *M. Lutter/P. Hommelhoff*, GmbHG, 1995, § 51a Rdnr. 11; *Uwe H. Schneider*, Konzerleitung, 1981, S. 252; *B. Grunewald*, Einsichtsrecht, 1982, S. 234. *OLG Hamm*, Beschl. v. 6.2.1985 - 8 W 52/85, in: DB 1986, S. 580.

227 *M. Lutter/P. Hommelhoff*, GmbHG, 1995, § 51a Rdnr. 11.

228 *OLG Köln*, Beschl. v. 26.4.1985 - 24 W 54/84, in: WM 1986, S. 37, 39. Ebenso *W. Zöllner*, Baumbach/Hueck 1996, § 51a Rdnr. 10a; *U. Hüffer*, Hachenburg, 1990, § 51a Rdnr. 28; *M. Kort*, Informationsrecht, 1987, S. 53 ff.; *C. v. Bitter*, Informationsrecht, 1981, S. 827; *K. Schmidt*, Scholz, 1995, § 51a Rdnr. 20; wohl auch *J. Tietze*, Informationsrechte, 1985, S. 14 f.

229 *M. Kort*, Informationsrecht, 1987, S. 52 f.

230 *U. Hüffer*, Hachenburg, 1990, § 51a Rdnr. 25.

Aktiengesellschaft. Der Aufsichtsrat dürfe aber gemäß § 90 Abs. 3 Satz 1 Var. 3 AktG nur über solche Vorgänge bei der Tochtergesellschaft Auskunft verlangen, die für die Lage der Gesellschaft von erheblicher Bedeutung sein könnten. Ebenso verhalte es sich mit der Pflicht zur Sonderberichterstattung des Vorstands gemäß § 90 Abs. 1 Satz 2 2. Halbs. AktG.[231]

(4) Stellungnahme

Richtigerweise läßt sich nur aus Sinn und Zweck von § 51a GmbHG eine Antwort auf die Frage finden, wann ein Vorgang bei einer beherrschten Gesellschaft eine Angelegenheit der herrschenden Gesellschaft ist. In diesem Zusammenhang wurde in der Literatur zutreffend auf die lediglich dienende Funktion von § 51a GmbHG hingewiesen. Mit anderen Worten soll § 51a GmbHG den Gesellschafter bei der Ausübung seiner Gesellschafterrechte unterstützen, seine Position in der GmbH aber nicht erweitern.[232] Nur soweit die Kenntnis über Vorgänge bei der beherrschten Gesellschaft für die Ausübung der Gesellschafterrechte in der herrschenden GmbH dienlich ist, besteht also ein entsprechendes Auskunftsrecht des Gesellschafters.

Ein wesentliches Recht des Gesellschafters besteht in seinem Recht auf Teilnahme an der Gesellschafterversammlung und Mitwirkung bei der Beschlußfassung. Da die Gesellschafterversammlung für die Überwachung des Geschäftsführers und dessen Entlastung zuständig ist, dient damit das Auskunftsrecht auch der Kontrolle des Geschäftsführers.[233] Und es dient ferner der Vorbereitung unternehmerischer Entscheidungen, da auch diese von den Gesellschaftern in der Gesellschafterversammlung getroffen werden.[234]

Auf den Konzern gewendet bedeutet dies, daß das Auskunftsrecht des Gesellschafters nur soweit reicht, wie die Konzernleitung ausgeübt wird. Mit anderen Worten bemißt sich die Intensität des Auskunftsrechts des Gesellschafters und damit der Auskunftspflicht des Geschäftsführers nach dem Grad der Konzernierung.[235] Während danach im zentralen Konzern nahezu alle Angelegenheiten der beherrschten Gesellschaft auch Angelegenheiten der herrschenden GmbH sind,[236] ist das Auskunftsrecht im dezentralen Konzern weniger stark ausgeprägt. Führt die herrschende GmbH ihre Tochtergesellschaft also wie eine

---

231 *M. Kort*, Informationsrecht, 1987, S. 62 f.

232 *B. Grunewald*, Einsichtsrecht, 1982, S. 217.

233 *M. Lutter*, Informationsrecht, 1982, S. 6.

234 *M. Lutter*, Informationsrecht, 1982, S. 6.

235 In diese Richtung auch *U. Hüffer*, Hachenburg, 1990, § 51a Rdnr. 29; *M. Kort*, Informationsrecht, 1987, S. 48.

236 Enger wohl *M. Kort*, Informationsrecht, 1987, S. 56.

unselbständige Betriebsabteilung und bestimmt deren Geschicke bis hinunter in das Tagesgeschäft, so kann der Gesellschafter z. B. Auskunft über Einkaufspreise, Lohnlisten und Geschäftsbeziehungen der Tochtergesellschaft zu Dritten verlangen. Legt die herrschende GmbH dagegen nur die Unternehmenspolitik der Tochtergesellschaft fest, so kann der Gesellschafter nur Auskunft über die dafür wesentlichen Daten beanspruchen. Da der Geschäftsführer der herrschenden GmbH aber auf jeden Fall zur Überwachung der Konzerngesellschaften verpflichtet ist, kann der Gesellschafter auch über solche Vorgänge in den Tochtergesellschaften Auskunft verlangen, die möglicherweise auf eine Verletzung der Überwachungspflicht schließen lassen.

All das steht freilich unter dem Vorbehalt, daß der Geschäftsführer entweder bereits im Besitz der Informationen ist oder ihm die Informationsbeschaffung rechtlich möglich ist.[237] Letzteres dürfte allerdings regelmäßig keine größeren Probleme bereiten.[238]

Erstens hat die herrschende GmbH ihrerseits gegenüber der Tochtergesellschaft ein Auskunftsrecht nach § 51a GmbHG bzw. § 131 AktG. Zweitens wurde für den faktischen Aktienkonzern festgestellt, daß das herrschende Unternehmen auch direkt auf den Vorstand der beherrschten Aktiengesellschaft Einfluß nehmen darf. Der Geschäftsführer der herrschenden GmbH muß also zunächst versuchen, den Vorstand der beherrschten Aktiengesellschaft ohne Einschaltung der Hauptversammlung zur gewünschten Auskunft zu veranlassen. Erst wenn dieser Versuch fehlschlägt, kann er sich auf die Geltendmachung des Auskunftsrechts gemäß § 131 Abs. 1 AktG beschränken, das freilich an die Hauptversammlung und die Tagesordnung gebunden ist. Drittens kann der Geschäftsführer im faktischen GmbH-Konzern die Auskunft auch durch Herbeiführung eines entsprechenden Weisungsbeschlusses in der beherrschten Gesellschaft erlangen. Eine direkte Einflußnahme auf den Geschäftsführer der beherrschten Gesellschaft ist dagegen unzulässig und muß deshalb nicht versucht werden. Viertens schließlich kann der Geschäftsführer im Vertragskonzern Geschäftsführer bzw. Vorstand der beherrschten Gesellschaft unmittelbar zur Auskunftserteilung anweisen.

Ist es dem Geschäftsführer dagegen nicht möglich, aufgrund der legitimen Wahrnehmung der Beteiligungsrechte der herrschenden GmbH, sich die Informationen zu beschaffen, so besteht auch kein entsprechendes Auskunftsrecht des Gesellschafters.[239]

---

[237] *M. Kort*, Informationsrecht, 1987, S. 58 f., 71.

[238] Ebenso *B. Grunewald*, Einsichtsrecht, 1982, S. 234 f.

[239] *M. Kort*, Informationsrecht, 1987, S. 58 f., 71.

b) Die allgemeine Berichtspflicht des Geschäftsführers

Eine allgemeine Berichtspflicht unterscheidet sich von der Pflicht zur Informationserteilung an den einzelnen Gesellschafter insbesondere dadurch, daß sie unabhängig von einen konkreten Auskunftsverlangen besteht.

(1) Die gesetzliche Ausgangslage

Eine Pflicht des Geschäftsführers, ungefragt über Vorgänge in der Gesellschaft zu berichten, besteht zunächst in den ausdrücklich im Gesetz genannten Fällen. Hierzu gehört die Pflicht des Geschäftsführers gemäß § 42a Abs. 1 Satz 1 GmbHG zur Vorlage des Jahresabschlusses und des Lageberichts. Gegebenenfalls sind diese um den Prüfungsbericht des Abschlußprüfers und des Aufsichtsrats zu ergänzen, § 42a Abs. 1 Satz 2 u. 3 GmbHG. Eine Ausnahme besteht lediglich für die kleine GmbH i. S. v. § 267 Abs. 1 HGB. Sie ist gemäß § 264 Abs. 1 Satz 3 HGB von der Pflicht zur Aufstellung eines Lageberichts befreit. Freilich kann der Gesellschaftsvertrag auch für die kleine GmbH die Aufstellung eines Lageberichts vorsehen.[240] Ferner ergibt sich aus § 49 Abs. 3 GmbHG, daß der Geschäftsführer die Gesellschafter über den Verlust der Hälfte des Stammkapitals berichten muß, sofern sich dies aus der Jahresbilanz oder einer im Laufe des Geschäftsjahres aufgestellten Bilanz ergibt. Eine entsprechende Berichtspflicht besteht aber auch dann, wenn der Geschäftsführer auf andere Weise von dem Verlust der Hälfte des Stammkapitals Kenntnis erlangt.[241] Schließlich ist der Geschäftsführer zur Rechenschaftslegung verpflichtet, um entlastet zu werden.[242]

Eine allgemeine Berichtspflicht des Geschäftsführers ähnlich der des Vorstands gegenüber dem Aufsichtsrat bei der Aktiengesellschaft existiert dagegen nicht.[243] Denn während der Aufsichtsrat zur Kontrolle des Vorstands verpflichtet ist, sind die Gesellschafter zum unternehmerischen Desinteresse berechtigt. Sie sind also gerade nicht verpflichtet, die Geschäftsführung zu kontrollieren. Ihnen steht lediglich eine entsprechende Befugnis zu, von der sie aber keinen Gebrauch machen müssen. Machen sie hiervon keinen Gebrauch, so wäre es sinnlos, die Geschäftsführer zu unnötigen Auskünften zu verpflichten. Damit ist

---

[240] *J. Schulze-Osterloh*, Baumbach/Hueck 1996, § 42 Rdnr. 25; *M. Lutter/P. Hommelhoff*, GmbHG, 1995, § 42 Rdnr. 6.

[241] *U. Hüffer*, Hachenburg, 1990, § 49 Rdnr. 22.

[242] *W. Zöllner*, Baumbach/Hueck, 1996, § 46 Rdnr. 27.

[243] *W. Zöllner*, Baumbach/Hueck, 1996, 51a Rdnr. 42; *H.-J. Mertens*, Hachenburg, 1996, § 43 Rdnr. 29. A. M. *Uwe H. Schneider*, Scholz, 1993, § 43 Rdnr. 114; *B. Grunewald*, Einsichtsrecht, 1982, S. 225 f.

der Geschäftsführer nur zu solchen Informationen verpflichtet, die die Gesellschafter benötigen, um die in ihren Zuständigkeitsbereich fallenden Entscheidungen zu treffen, wie etwa bei Feststellung des Jahresabschlusses und Verwendung des Ergebnisses. Freilich können die Gesellschafter ihren Zuständigkeitsbereich auch selbst bestimmen. Das tun sie, indem sie bestimmte Entscheidungen an sich ziehen. Ist dies geschehen, so muß der Geschäftsführer auch ungefragt über solche Vorgänge berichten, die für die von den Gesellschaftern getroffenen Entscheidungen von Bedeutung sind. Haben die Gesellschafter den Geschäftsführer z. B. angewiesen, ein bestimmtes Grundstück zu kaufen und stellt sich vor Abschluß des Kaufvertrages heraus, daß das Grundstück kontaminiert ist, so muß der Geschäftsführer den Gesellschaftern hierüber ungefragt berichten.

Sieht dagegen ein Gesellschafterbeschluß eine allgemeine Berichtspflicht des Geschäftsführers vor, so wird man zu deren Konkretisierung auf § 90 AktG zurückgreifen können. Danach muß der Geschäftsführer mindestens einmal jährlich über die beabsichtigte Geschäftspolitik und andere grundsätzliche Fragen der Unternehmensplanung berichten. Außerdem muß er die Gesellschafter mindestens einmal jährlich über die Rentabilität der Gesellschaft informieren. Des weiteren ist mindestens vierteljährlich über den Gang der Geschäfte, insbesondere den Umsatz, zu berichten. Schließlich sind die Gesellschafter über solche Geschäfte in Kenntnis zu setzen, die für die Rentabilität oder Liquidität der Gesellschaft von erheblicher Bedeutung sein können. Im letzteren Fall müssen die Geschäftsführer in der Regel aber auch ohne die Anordnung einer allgemeinen Berichtspflicht über solche Geschäfte berichten. Denn hierbei wird es sich regelmäßig um außergewöhnliche Geschäfte handeln, die ohnehin in den Entscheidungsbereich der Gesellschafter fallen.

(2) Die Lage im Konzern

Nach alledem besteht auch im Konzern keine allgemeine Berichtspflicht. Nur soweit konzernbezogene Vorgänge zur Vorbereitung der Entscheidungen der Gesellschafter dienen oder bereits getroffene Entscheidungen berühren, muß der Geschäftsführer über sie ungefragt berichten. Haben die Gesellschafter etwa die Konzernpolitik festgelegt und haben sich die Konjunkturdaten seitdem wesentlich geändert, so muß der Geschäftsführer die Gesellschafter hierüber informieren.

Sieht dagegen die Satzung eine allgemeine Berichtspflicht vor, so ist diese im Wege einer konzerndimensionalen Betrachtung zu konkretisieren. Ausgangspunkt ist dabei wiederum die Feststellung, daß der Konzern ein wirtschaftlich einheitliches Unternehmen ist. Deshalb hat der Geschäftsführer nur über solche Vorgänge ungefragt zu berichten, die für den Gesamtkonzern von

Bedeutung sind.[244] Das sind in Anlehnung an § 90 AktG insbesondere die Konzerngeschäftspolitik, die Rentabilität des Konzerns, der Gang der Geschäfte in den wichtigsten Konzerngesellschaften und Geschäfte, die auf den Gesamtkonzern einen erheblichen Einfluß haben.

4. Die Pflichten des Geschäftsführers in einer mehrgliedrigen Geschäftsführung

Hat die Gesellschaft mehrere Geschäftsführer, so gilt im Außenverhältnis als gesetzliche Regel Gesamtvertretung, § 35 Abs. 2 GmbHG. Ob spiegelbildlich im Innenverhältnis Gesamtgeschäftsführung gilt, ist im Gesetz nicht ausdrücklich geregelt. Doch ist analog § 77 AktG davon auszugehen, daß auch in der GmbH mehrere Geschäftsführer nur gemeinsam geschäftsführungsberechtigt sind.[245]

Freilich ist dies nur eine dispositive gesetzliche Regel. In der Praxis wird den einzelnen Geschäftsführern oft durch Satzung, Geschäftsordnung, Anstellungsvertrag oder Gesellschafterbeschluß ein Ressort zugewiesen, das sie selbständig zu leiten haben. Das gilt insbesondere für den Konzern. So gibt es im funktional organisierten Konzern etwa die Ressorts Finanzen, Personal, Forschung und Entwicklung, Einkauf/Logistik, während bei einer Geschäftsbereichsorganisation in der Regel ein Ressort für die Finanzen und jeweils ein Ressort für einen Geschäftsbereich eingerichtet wird.[246]

Liegt eine Geschäftsaufteilung vor, so ist klar zu unterscheiden zwischen den Geschäftsführungsaufgaben der Geschäftsführung als Ganzes und den Ressortaufgaben, die das einzelne Mitglied der Geschäftsführung in seiner Eigenschaft als Leiter eines Ressorts zu erfüllen hat.[247] Während der Geschäftsführer als Ressortleiter die Handlungsverantwortung für den ihm zugewiesenen Aufgabenbereich trägt, trifft die Geschäftsführung als Ganzes nur die Führungsverantwortung. Letztere besteht hauptsächlich in der Kontrolle der einzelnen Ressortleiter.[248] Die Ressortleiter dagegen haben dafür zu sorgen, daß die in der herrschenden GmbH beschlossenen Konzernpläne, soweit sie ihr Ressort berühren, in den beherrschten Gesellschaften umgesetzt werden. Der Ressortleiter für Finanzen hat beispielsweise zu kontrollieren, ob die Tochtergesellschaften den

---

[244] Im Ergebnis ebenso für die Aktiengesellschaft wohl auch *M. Lutter/G. Krieger*, Aufsichtsrat, 1993, Rdnr. 81.

[245] *Uwe H. Schneider*, Scholz, 1993, § 37 Rdnr. 21 m. w. N.

[246] *E. Scheffler*, Konzernmanagement, 1992, S. 24 f.

[247] *R. Höhn*, Geschäftsleitung, 1995, S. 19.

[248] *R. Höhn*, Geschäftsleitung, 1995, S. 20, 138 ff. Siehe auch *J. Semler*, Leitung, 1996, Rdnr. 19, 29 f.; *H. Altmeppen*, GmbHG, 1997, § 43 Rdnr. 8.

vorgegebenen Verschuldungsgrad einhalten. Der Geschäftsbereichsleiter hat z. B. dafür zu sorgen, daß sich die beherrschten Gesellschaften seines Geschäftsbereichs an die festgelegte Konzernstrategie halten.

III. Die Pflicht zu loyalem Verhalten

1. Unbestimmter Rechtsbegriff

Der Geschäftsführer hat in den Angelegenheiten der Gesellschaft die Sorgfalt eines ordentlichen Geschäftsmannes anzuwenden, § 43 Abs. 1 GmbHG. Verlangt wird von ihm, daß er die geschäftlichen Belange der Gesellschaft nachhaltig und vorbehaltlos fördert. Spiegelbildlich zu diesen aktiven Förderpflichten hat er alles zu unterlassen, was die Verwirklichung des Gesellschaftszwecks gefährdet und das in ihn gesetzte Vertrauen enttäuscht.[249] Hierbei handelt es sich um im Einzelfall zu konkretisierende Treupflichten, die als passive Förderpflichten weit über die allgemeinen Pflichten aus § 242 BGB hinausgehen.[250] Insbesondere sind sie stärker als die Treupflichten der Gesellschafter, für die der Sorgfaltsmaßstab des § 43 Abs. 1 GmbHG nicht gilt. Für die Geschäftsführer wird man deshalb anders als für die Gesellschafter den Grundsatz aufstellen können, daß sie ihre Privatinteressen stets hinter die Interessen der Gesellschaft zurückzustellen haben. Das bedarf der Konkretisierung. Exemplarisch betrachtet werden sollen das Verbot, Geschäftschancen der GmbH im eigenen Interesse auszunutzen, und das Wettbewerbsverbot.

2. Verbot des Ausnutzens von Geschäftschancen der GmbH

a) Die gesetzliche Ausgangslage

Zur nachhaltigen und vorbehaltlosen Förderung des Gesellschaftszwecks gehört, daß der Geschäftsführer Geschäftschancen der Gesellschaft für diese wahrnimmt und nicht für sich selbst ausnutzt. Hierauf hat namentlich *W. Timm*[251] in einem rechtsvergleichenden Aufsatz zur US-amerikanischen Doktrin der „corporate opportunity" hingewiesen. Danach ist es den Verwaltungsmitgliedern einer corporation verboten, Geschäftschancen der corporation an sich

---

[249] *V. Röhricht*, Wettbewerbsverbot, 1992, S. 767; *Uwe H. Schneider*, Scholz, 1993, § 43 Rdnr. 121.

[250] *W. Timm*, Wettbewerbsverbot, 1981, S. 179; *V. Röhricht*, Wettbewerbsverbot, 1992, S. 767.

[251] *W. Timm*, Wettbewerbsverbot, 1981, S. 177.

zu ziehen.[252] Gleiches gilt für die Geschäftsführer einer GmbH kraft ihrer Pflicht zur umfassenden Förderung der Interessen der Gesellschaft.

Das bedeutet für den Geschäftsführer, daß er bei jeder Geschäftschance unter Zurückstellung seiner eigenen Interessen prüfen muß, ob sie für die Gesellschaft von Interesse ist.[253] Dabei ist es unerheblich, wann und bei welcher Gelegenheit er von der Erwerbschance erfahren hat. Insbesondere kommt es nicht darauf an, ob er von der Geschäftschance geschäftlich oder privat Kenntnis erlangt hat.[254] Ein Geschäftsführer ist immer im Dienst.[255]

Wann eine Geschäftschance für die Gesellschaft von Interesse ist, hängt zum einen davon ab, ob sich ihre Zuordnung zur Gesellschaft schon in irgendeiner Weise manifestiert hat. So können die Gesellschafter einen Beschluß gefaßt haben, die Geschäftschance wahrzunehmen,[256] so kann die Gesellschaft schon Interesse an dem Geschäft geäußert haben,[257] sie kann schon in Vertragsverhandlungen eingetreten sein oder ihr kann schon ein Vertragsangebot unterbreitet worden sein.[258] Schließlich gehört hierher auch der Fall, daß dem Geschäftsführer nur mit Rücksicht auf seine Stellung in der Gesellschaft die Geschäftschance angetragen wurde.[259]

Liegt keine der obengenannten Fallgruppen vor, hängt die Zuordnung der Geschäftschance zum anderen davon ab, ob sie in den üblichen oder nach der einvernehmlichen Planung der Gesellschafter angestrebten Tätigkeitsbereich der Gesellschaft fällt[260] oder ob ihre Wahrnehmung für die Gesellschaft wesentlich ist[261].

Voraussetzung bei alledem ist jedoch, daß die Geschäftschance für die Gesellschaft überhaupt in Betracht kommt.[262] Das ist nicht schon dann zu vernei-

---

[252] Eingehend *J. Weisser*, Opportunities, 1991, S. 29 ff.; *N. Polley*, Wettbewerbsverbot, 1993, S. 32 ff.

[253] *V. Röhricht*, Wettbewerbsverbot, 1992, S. 770.

[254] *BGH*, Urt. v. 23.9.1985 - II ZR 246/84, in: ZIP 1985, S. 1484, 1485; *V. Röhricht*, Wettbewerbsverbot, 1992, S. 771; *H. Altmeppen*, GmbHG, 1997, § 43 Rdnr. 12.

[255] *Uwe H. Schneider*, Scholz, 1993, § 43 Rdnr. 145.

[256] *F. Kübler/J. Waltermann*, Geschäftschancen, 1991, S. 168.

[257] *BGH*, Urt. v. 24.11.1975 - II ZR 104/73, in: WM 1976, S. 77.

[258] *Uwe H. Schneider*, Scholz, 1993, § 43 Rdnr. 145.

[259] *BGH*, Urt. v. 8.5.1967 - II ZR 126/65, in: GmbHR 1968, S. 141, 142 mit Anm. *Schaudwet*; ähnlich auch *BGH*, Urt. v. 10.2.1977 - II ZR 79/75, in: GmbHR 1977, S. 129, 130.

[260] *F. Kübler/J. Waltermann*, Geschäftschancen, 1991, S. 170.

[261] *W. Timm*, Wettbewerbsverbot, 1981, S. 181.

[262] *BGH*, Urt. v. 10.2.1977 - II ZR 79/75, in: GmbHR 1977, S. 129, 130; *V. Röhricht*, Wettbewerbsverbot, 1992, S. 770.

nen, wenn der Gesellschaft die für ihre Wahrnehmung erforderlichen finanziellen Mittel fehlen. Vielmehr hat der Geschäftsführer in diesem Fall noch zu prüfen, ob nicht geeignete Finanzierungsmöglichkeiten bestehen, etwa durch eine Kreditaufnahme, die Einforderung von Nachschüssen, eine Kapitalerhöhung oder die Gewinnung eines stillen Gesellschafters.[263] Erst wenn diese Prüfung ergibt, daß die Gesellschaft die Geschäftschance nicht wahrnehmen kann, ist für den Geschäftsführer eine Hürde auf dem Weg zur eigennützigen Wahrnehmung der Geschäftschance beseitigt.

Es stellt sich dann noch die Frage, ob der Geschäftsführer nicht aufgrund eines Wettbewerbsverbots an der Wahrnehmung der Geschäftschance gehindert ist. Das soll jedoch erst weiter unten untersucht werden.

b) Die Lage im Konzern

Im Konzern ist der Geschäftsführer gegenüber seiner Gesellschaft zur Konzernleitung verpflichtet. Das bedeutet auch, daß er im Interesse seiner Gesellschaft die Interessen der beherrschten Gesellschaften mit der Sorgfalt eines ordentlichen Geschäftsmannes zu fördern hat. Sind daher die obengenannten Voraussetzungen bei einer Tochtergesellschaft erfüllt, so liegt es nahe, daß der Geschäftsführer die Erwerbschance nicht an sich ziehen darf.[264] Denn schließt die Tochtergesellschaft ein günstiges Geschäft ab, so wird dadurch auch der Gesellschaftszweck der herrschenden GmbH gefördert. Hierauf ist der Geschäftsführer aber verpflichtet.

Trotzdem wird man zu unterscheiden haben. Hat der Geschäftsführer im Rahmen der Konzernleitung von einer Geschäftschance der beherrschten Gesellschaft Kenntnis erlangt, so darf er diese nicht zu eigenen Zweck ausnutzen. Denn hiermit würde er gesellschaftsinterne Informationen mißbräuchlich verwenden. Das ist ihm aber sowohl im dezentralen wie zentralen Konzern verboten.

Im übrigen ist nach der Intensität der Konzernleitung zu unterscheiden. Denn die Treupflichten des Geschäftsführers sind, wie oben festgestellt[265], Spiegelbild seiner aktiven Förderpflichten. Je stärker also seine aktiven Förderpflichten ausgeprägt sind, desto stärker sind auch seine Loyalitätspflichten.

Folgt man dem, so verletzt der Geschäftsführer im dezentralen Konzern nicht seine Loyalitätspflichten, wenn er eine Geschäftschance an sich zieht, die nach

---

263 *V. Röhricht*, Wettbewerbsverbot, 1992, S. 770.

264 *Uwe H. Schneider*, Scholz, 1993, § 43 Rdnr. 144.

265 Siehe oben 1. Teil C. III. 1.

den obengenannten Kriterien einer beherrschten Gesellschaft zuzuordnen ist,[266] aber nicht zu ihrem üblichen Tätigkeitsbereich gehört.[267] Denn im dezentralen Konzern legt der Geschäftsführer nur die Rahmenbedingungen und Unternehmensziele fest, die die beherrschten Gesellschaften zu beachten haben. Das operative Geschäft und damit auch die Suche nach Geschäftschancen obliegt dagegen der Verwaltung der beherrschten Gesellschaften. Um die Suche nach Geschäftschancen für die beherrschten Gesellschaften muß sich der Geschäftsführer der herrschenden Gesellschaft also nicht bemühen. Dann ist er aber auch nicht verpflichtet, bei jedem eigennützigen Abschluß eines Geschäfts zu prüfen, ob das Geschäft für eine beherrschte Gesellschaft von Interesse ist.

Anders verhält es sich dagegen im zentralen Konzern. Hier ist die Konzernleitungspflicht des Geschäftsführers am stärksten ausgeprägt. Insbesondere hat er das operative Geschäft an sich gezogen. Deshalb ist es ihm zumutbar, bei jeder einzelnen Geschäftschance zu prüfen, ob sie der beherrschten Gesellschaft nach den obengenannten Kriterien zuzurechnen ist. Ist eine Zurechnung zu bejahen, darf der Geschäftsführer die Geschäftschance nicht wahrnehmen.

3. *Wettbewerbsverbot*

a) Die gesetzliche Ausgangslage

Das Wettbewerbsverbot des Geschäftsführers ist im Gegensatz zum Wettbewerbsverbot des Vorstands einer Aktiengesellschaft (§ 88 AktG) nicht gesetzlich geregelt. Es besteht jedoch Einigkeit darüber, daß auch der Geschäftsführer einer GmbH einem Wettbewerbsverbot unterliegt.[268] Zweck dieses Wettbewerbsverbots ist es, eine abstrakte Gefährdung der Gesellschaftsinteressen bereits im Vorfeld zu verhindern. Deshalb ist dem Geschäftsführer im Geschäftszweig der Gesellschaft jedwede Teilnahme am geschäftlichen Verkehr untersagt, gleichgültig ob er dabei im eigenen oder fremden Namen, für eigene oder fremde Rechnung handelt.[269] Ebensowenig darf er im Geschäftszweig seiner Gesellschaft in anderen Unternehmen als Geschäftsführer, Vorstandsmitglied oder in einer sonstigen Funktion tätig werden, die ihm Einfluß auf deren unternehmerischen Entscheidungen eröffnet. Schließlich ist ihm auch das Halten

---

[266] A. M. *V. Röhricht*, Wettbewerbsverbot, 1992, S. 770; wohl auch *Uwe H. Schneider*, Scholz, 1993, § 43 Rdnr. 144.

[267] Gehört das Geschäft zum üblichen Tätigkeitsbereich der Gesellschaft, so greift das Wettbewerbsverbot ein. Dazu sogleich.

[268] Statt aller *H. Altmeppen*, GmbHG, 1997, § 43 Rdnr. 12.

[269] *V. Röhricht*, Wettbewerbsverbot, 1992, S. 768; *W. Zöllner*, Baumbach/ Hueck, 1996, § 35 Rdnr. 22.

einer bloßen Beteiligung an einem Konkurrenzunternehmen untersagt, wenn sie ihm eine Einflußmöglichkeit auf die Entscheidungen in dem Unternehmen eröffnet. Vom Wettbewerbsverbot nicht erfaßt werden dagegen Geschäfte, die nicht auf Gewinnerzielung gerichtet sind, sondern dem privaten Bedarf dienen.[270]

Die Bestimmung des Geschäftszweigs der Gesellschaft muß von dem Zweck des Wettbewerbsverbots her erfolgen. Dieser besteht darin, die Pflicht des Geschäftsführers zur umfassenden und vorbehaltlosen Förderung der Gesellschaftsinteressen zu flankieren. Diesem Zweck wird aber nur ein Wettbewerbsverbot gerecht, daß dem Geschäftsführer alle außergesellschaftlichen Geschäftstätigkeiten untersagt, die mit einer abstrakten Gefährdung der ihm anvertrauten Geschäftsinteressen allgemein oder im Einzelfall verbunden sind.[271]

Danach ist dem Geschäftsführer ein Geschäft verboten, das in den aktuellen Tätigkeitsbereich der Gesellschaft fällt. Dabei ist es unerheblich, ob dieser Tätigkeitsbereich vom Unternehmensgegenstand der Gesellschaft gedeckt ist oder darüber hinaus geht.[272] Das gilt selbst dann, wenn die Gesellschaft erst nach Amtsantritt des Geschäftsführers ihren Tätigkeitsbereich über ihren Unternehmensgegenstand hinaus ausgedehnt hat. Denn dem Geschäftsführer ist auch in diesem Fall zumutbar, sein Geschäftsführeramt niederzulegen, wenn er eine Konkurrenztätigkeit aufnehmen will.[273]

Füllt dagegen der aktuelle Tätigkeitsbereich der Gesellschaft ihren Unternehmensgegenstand nicht vollständig aus und handelt es sich dabei auch nicht um einen nur vorübergehenden Zustand, so ist umstritten, ob der aktuelle Tätigkeitsbereich[274] oder der Unternehmensgegenstand[275] für die Bestimmung des Geschäftszweigs ausschlaggebend ist.

Zugunsten der Maßgeblichkeit des Unternehmensgegenstandes wird angeführt, daß die Gesellschaft die uneingeschränkte Freiheit besitzen müsse, ihre Geschäftstätigkeit jederzeit auf ihren vollen Unternehmensgegenstand auszudehnen. Diese Freiheit sei aber gefährdet, wenn ihr Geschäftsführer in dem

---

[270] *K.-P. Martens*, Schlegelberger, 1992, § 112 Rdnr. 9 (zu dem Wettbewerbsverbot des Gesellschafters einer oHG)

[271] *V. Röhricht*, Wettbewerbsverbot, 1992, S. 769.

[272] Ebenso *J. Tieves*, Unternehmensgegenstand, 1998, S. 298.

[273] *V. Röhricht*, Wettbewerbsverbot, 1992, S. 769.

[274] So für den Vorstand der AG: *W. Hefermehl*, Geßler, 1973, § 88 Rdnr. 11; *H.-J. Mertens*, Kölner Kommentar, 1988, § 88 Rdnr. 9; *U. Hüffer*, AktG, 1997, § 88 Rdnr. 3.

[275] So *V. Röhricht*, Wettbewerbsverbot, 1992, S. 769; *J. Tieves*, Unternehmensgegenstand, 1998, S. 296.

angestrebten Geschäftsbereich bereits tätig sei und damit ein Interesse daran habe, eine Ausdehnung der Geschäftstätigkeit der Gesellschaft zu verhindern.[276]

Dem kann indes nicht gefolgt werden. Denn bei Bemessung der Reichweite des Wettbewerbsverbots ist § 1 GWB zu beachten. Die dort geschützte Wettbewerbsfreiheit läßt nämlich nur solche gesellschaftsrechtlichen Beschränkungen der unternehmerischen Betätigungsfreiheit zu, die erforderlich sind, um die Funktionsfähigkeit eines dem gesetzlichen Typus entsprechenden Verbandes zu gewährleisten.[277] Deshalb mag es zwar richtig sein, daß es zu einer Interessenkollision kommt, wenn die Gesellschaft ihre Tätigkeit auf einen Bereich ausdehnt, auf dem ihr Geschäftsführer ein Konkurrenzunternehmen betreibt. Diese Gefahr erreicht jedoch regelmäßig nicht ein Ausmaß, das die Funktionsfähigkeit der Gesellschaft beeinträchtigen könnte.

Nach alledem ist der tatsächliche Tätigkeitsbereich der Gesellschaft für die Reichweite des Wettbewerbsverbots ihres Geschäftsführers maßgeblich.

b) Abgrenzung zur Geschäftschancenlehre

All das deutet darauf hin, daß es sich bei dem Wettbewerbsverbot nur um einen Unterfall der Geschäftschancenlehre handelt.[278] Namentlich *W. Timm*[279] vertritt die Ansicht, daß der Gesellschaft auch solche Geschäfte als Geschäftschancen zuzuordnen seien, deren Wahrnehmung dem Geschäftsführer kraft eines ausdrücklichen Wettbewerbsverbots untersagt sei. Indes sind die Geschäftschancenlehre und das Wettbewerbsverbot sorgfältig voneinander zu unterscheiden.

Zunächst fällt es schwer, ein Geschäft selbst dann als Geschäftschance der Gesellschaft zu bezeichnen, wenn die Gesellschaft auch nicht entfernt in der

---

276 *V. Röhricht*, Wettbewerbsverbot, 1992, S. 769.

277 *H.-J. Mertens/A. Cahn*, Wettbewerbsverbot, 1991, S. 555; *BGH*, Urt. v. 5.12.1983 - II ZR 242/82 („Werbeagentur"), in: BGHZ 89, S. 162, 169; *BGH*, Urt. v. 3.5.1988 - KZR 17/87, in: BGHZ 104, 246, 251; *L. Lawall*, Wettbewerbsverbot, 1996, S. 4; eingehend *M. Ivens*, Konkurrenzverbot, 1987, S. 200 ff. (alle im Zusammenhang mit dem Wettbewerbsverbot des Gesellschafters einer GmbH).

278 In diesem Sinne *V. Röhricht*, Wettbewerbsverbot, 1992, S. 768; *J. Weisser*, Opportunities, 1991, S. 148; *W. Timm*, Wettbewerbsverbot, 1981, S. 177; *H. Merkt*, Unternehmensleitung, 1995, S. 434 u. 449.

279 *W. Timm*, Wettbewerbsverbot, 1981, S. 181.

Lage ist, das Geschäft abzuschließen. Trotzdem kann dem Geschäftsführer die eigennützige Wahrnehmung der Geschäftschance verboten sein, nämlich aufgrund des Wettbewerbsverbots.

Weiterhin wird man bei der Freigabe einer Geschäftschance durch die Gesellschafterversammlung zu unterscheiden haben, ob die Geschäftschance zum üblichen Tätigkeitsbereich der Gesellschaft gehört und damit auch das Wettbewerbsverbot berührt ist oder aber das Wettbewerbsverbot nicht betroffen ist. Im letzteren Fall ist die Gesellschaft verpflichtet, die Geschäftschance freizugeben, wenn sie das Geschäft nicht wahrnehmen will. Ein sachlicher Grund, die Freigabe zu verweigern, besteht dann nicht. Wird hingegen auch das Wettbewerbsverbot berührt, so kann die Freigabe verweigert werden, wenn wenigstens eine abstrakte Gefährdung der Gesellschaftsinteressen zu befürchten ist. Das ist z. B. der Fall, wenn der Geschäftsführer ein Konkurrenzunternehmen erwerben will, an dem die Gesellschaft nicht interessiert ist. Denn nach dem Erwerb besteht die Gefahr, daß der Geschäftsführer gesellschaftsinterne Informationen zu Gunsten seines Konkurrenzunternehmens mißbraucht oder sonst schädlichen Einfluß auf die Gesellschaft nimmt.

Nach alledem ist zwischen dem Verbot der Ausnutzung von Geschäftschancen der Gesellschaft auf der einen Seite und dem Wettbewerbsverbot auf der anderen Seite zu unterscheiden. Folgt man der hier vertretenen Auffassung, so ist ein Geschäft der Gesellschaft nur dann als Geschäftschance zuzurechnen, wenn die Gesellschaft an dem Geschäft bereits Interesse gezeigt hat oder aber wenn das Geschäft für sie nützlich ist und seine Wahrnehmung für sie, wenn auch nur entfernt, in Betracht kommt. Das Wettbewerbsverbot untersagt dagegen jedwede geschäftliche Tätigkeit im Geschäftszweig der Gesellschaft.

Damit geht der Zweck des Wettbewerbsverbots weiter als der der Geschäftschancenlehre. So stellt nach der hier vertretenen Auffassung die Geschäftschancenlehre lediglich sicher, daß die sich der Gesellschaft konkret bietenden Geschäftschancen auch wahrgenommen werden. Das ist auch ein Zweck des Wettbewerbsverbots. Darüber hinaus will es aber schon im Vorfeld eine abstrakte Gefährdung der Gesellschaftsinteressen verhindern.

c) Die Lage im Konzern

Im Konzern ist der Geschäftsführer zur Konzernleitung verpflichtet. Da Konzernleitung funktionell Unternehmensleitung ist, erstreckt sich der Geschäftsbereich der herrschenden Gesellschaft auch auf den Geschäftsbereich der beherrschten Gesellschaften. In demselben Ausmaß erweitert sich das Wettbewerbsverbot des Geschäftsführers. Mit anderen Worten darf der Geschäftsführer auch im aktuellen Tätigkeitsbereich der beherrschten Gesellschaften keine Geschäfte machen.[280]

Eine Differenzierung zwischen der Lage im zentralen und dezentralen Konzern ist beim Wettbewerbsverbot dagegen nicht gerechtfertigt. Denn zum einen muß der Geschäftsführer auch im dezentralen Konzern wissen, in welchem Geschäftsbereich die beherrschten Gesellschaften tätig sind. Und zum anderen besteht auch im dezentralen Konzern die Gefahr, daß der Geschäftsführer zugunsten eines Konkurrenzunternehmens schädlichen Einfluß auf die beherrschten Gesellschaften nimmt und damit auch der herrschenden GmbH Schaden zufügt.

Das alles gilt schließlich auch in der bloßen Abhängigkeitslage. Zwar erstreckt sich hier der Geschäftsbereich der herrschenden GmbH nicht auf den der beherrschten Gesellschaft, da keine Konzernleitung ausgeübt wird. Jedoch besteht auch hier die latente Gefahr, daß der Geschäftsführer zugunsten eines Konkurrenzunternehmens schädlichen Einfluß auf die abhängige Gesellschaft nimmt und damit den Wert des Beteiligungsbesitzes der herrschenden GmbH mindert.[281]

IV. Die Haftung des Geschäftsführers

1. Die zentralen Haftungsgrundlagen

Zentrale Haftungsnorm für den Geschäftsführer ist § 43 Abs. 2 GmbHG. Während er nach § 43 Abs. 1 GmbHG in den Angelegenheiten der Gesellschaft die Sorgfalt eines ordentlichen Geschäftsmannes anzuwenden hat, haftet er ihr gemäß § 43 Abs. 2 GmbHG auf Schadensersatz, wenn er seine Obliegenheiten verletzt. Damit begründet § 43 Abs. 2 GmbHG als organisationsrechtlicher Haftungstatbestand[282] sowohl eine Verhaltenshaftung - der Geschäftsführer haftet nur für Pflichtverletzungen - als auch eine Verschuldenshaftung[283]. Die Verhaltenspflichten selbst ergeben sich jedoch nicht aus § 43 Abs. 1 GmbHG, sondern aus der Geschäftsführungsaufgabe, aus Treupflichtüberlegungen und aus Sonderregelungen.[284] Der bei Erfüllung dieser Pflichten gemäß § 43 Abs. 1

---

[280] *Uwe H. Schneider*, Scholz, 1993, § 43 Rdnr. 128; *J. Tieves*, Unternehmensgegenstand, 1998, S. 299. Für den Vorstand einer AG: *H.-J. Mertens*, Kölner Kommentar, 1988, § 88 Rdnr. 9.

[281] Im Ergebnis ebenso V. Röhricht, Wettbewerbsverbot, 1992, S. 770. A. M. für den Vorstand der AG: *H.-J. Mertens*, Kölner Kommentar, 1988, § 88 Rdnr. 9.

[282] *Uwe H. Schneider*, Scholz, 1993, § 43 Rdnr. 12 m.w.N.

[283] *Uwe H. Schneider*, Scholz, 1993, § 43 Rdnr. 7.

[284] *W. Zöllner*, Baumbach/Hueck, 1996, § 43 Rdnr. 11.

GmbHG zu beachtende Sorgfaltsmaßstab richtet sich dabei nach dem Standard einer Person in der verantwortlichen leitenden Stellung des Verwalters eines fremden Vermögens oder als eines selbständigen, treuhänderischen Verwalters fremder Vermögensinteressen, wobei die Art und Größe des Unternehmens zu berücksichtigen ist.[285]

Weiterhin haftet der Geschäftsführer aus einer Verletzung seines Anstellungsvertrages auf Schadensersatz, wenn er seine Geschäftsführungspflichten verletzt. Zwar ist der Anstellungsvertrag des Geschäftsführers von seiner Organstellung zu unterscheiden.[286] So umfaßt die Organstellung die organschaftlichen Rechte und Pflichten des Geschäftsführers sowie den Umfang seiner Zuständigkeiten in der gesellschaftsinternen Zuständigkeitsordnung. Das Anstellungsverhältnis regelt demgegenüber die persönlichen Beziehungen des Geschäftsführers zur Gesellschaft und legt z. B. das Gehalt oder die Anzahl der Urlaubstage fest. Der Anstellungsvertrag begründet aber mangels anderer Anhaltspunkte auch die Pflicht des Geschäftsführers, die Gesellschaft mit der Sorgfalt eines ordentlichen Geschäftsmannes zu führen. Das ergibt sich entweder aus einer ausdrücklichen Regelung im Anstellungsvertrag oder aber konkludent im Wege der Auslegung. Schließlich kann er auch die Leitungspflichten des Geschäftsführers konkretisieren. Auch in den Anstellungsvertrag kann also aufgenommen werden, welche konzernbezogenen Maßnahmen der Geschäftsführer der Gesellschafterversammlung vorzulegen hat. Dogmatisch läßt sich eine solche Regelung als Einzelweisung der Gesellschafter gegenüber dem Geschäftsführer sehen, die den Geschäftsführer auch organisationsrechtlich bindet.[287]

Verletzt der Geschäftsführer nun seine gegenüber der GmbH obliegenden Pflichten zur ordnungsgemäßen Konzernleitung und entsteht der Gesellschaft hieraus ein Schaden, so macht er sich sowohl nach § 43 Abs. 2 GmbHG als auch wegen Verletzung seines Anstellungsvertrages schadensersatzpflichtig.

Dieses „Damoklesschwert“ einer drohenden Schadensersatzhaftung wird jedoch oft die Risikobereitschaft des Geschäftsführers mindern. Deshalb stellt sich die Frage, ob der Haftungsmaßstab des Geschäftsführers vertraglich abgemildert werden kann.

---

[285] *W. Zöllner*, Baumbach/Hueck, 1996, § 43 Rdnr. 9 m. w. N. Ausführlich *T. E. Abeltshauser*, Leitungshaftung, 1998, S. 139 ff.

[286] *Uwe H. Schneider*, Anstellungsvertrag, 1993, S. 12.

[287] *Uwe H. Schneider*, Anstellungsvertrag, 1993, S. 16.

2. Haftungsmilderungen wegen der Verletzung konzernspezifischer Pflichten

a) Bedeutung

Im Bereich der Unternehmensleitung ist der Geschäftsführer erheblichen Haftungsrisiken ausgesetzt. Insbesondere gelten für ihn nicht die vom Bundesarbeitsgericht entwickelten Haftungserleichterungen für Arbeitnehmer.[288] Denn die Haftung des Geschäftsführers beruht nicht auf seiner Stellung als Arbeitnehmer, sondern ergibt sich aus seiner gesellschaftsrechtlichen Organfunktion.[289]

Die Haftungsrisiken des Geschäftsführers erhöhen sich im Konzern. Denn seine Pflicht zur Führung der Geschäfte der herrschenden GmbH erweitert sich zu einer Konzernleitungspflicht. Zwar hält sich das Risikogefälle gegenüber der konzernfreien Lage in Grenzen, wenn die herrschende GmbH nur einen Betriebsteil ausgliedert. Denn dann ändert sich nichts an der Größe des zu leitenden Unternehmens. Anders verhält es sich jedoch, wenn die herrschende GmbH sich an einer bereits bestehenden Gesellschaft in abhängigkeitsbegründender Weise beteiligt. Denn nun ist der Geschäftsführer nicht nur für die sorgfältige Geschäftsleitung in der herrschenden GmbH verantwortlich, sondern er muß zusätzlich das Unternehmen der Tochtergesellschaft im Rahmen der rechtlichen Grenzen führen und überwachen.

Gegen die damit verbundenen Haftungsrisiken kann sich der Geschäftsführer durch eine Vermögensschaden-Versicherung absichern.[290] Dem soll hier jedoch nicht weiter nachgegangen werden. Vielmehr soll untersucht werden, ob die Haftung des Geschäftsführers für sorgfaltswidrige Geschäftsleitung vertraglich auf Vorsatz oder grobe Fahrlässigkeit beschränkt werden kann.

b) Meinungsstand

Die Rechtsprechung[291] und ein großer Teil der Literatur[292] lehnen eine vertragliche Haftungsmilderung uneingeschränkt ab. Denn einerseits diene der

---

[288] Ganz h. M.: *BGH*, Urt. v. 27.2.1975 - II ZR 112/72, in: WM 1975, S. 467, 469; *M. Lutter/P. Hommelhoff*, GmbHG, 1995, § 43 Rdnr. 2; *B. Schaub*, Haftung, 1992, S. 987; *W. Zöllner*, Baumbach/Hueck, 1988, § 43 Rdnr. 9; *U. Haas*, Geschäftsführerhaftung, 1997, S. 294; . Differenzierend *Uwe H. Schneider*, Haftungsmilderung, 1984, S. 804.

[289] *B. Schaub*, Haftung, 1992, S. 987.

[290] Dazu und insbesondere zur Vermögensschaden-Haftpflichtversicherung siehe *Uwe H. Schneider/H. Ihlas*, Vermögensschaden-Haftpflichtversicherung, 1994, S. 1124 ff.

[291] *BFH*, Urt. v. 26.4.1984 - V R 128/79, in: BB 1984, S. 1992.

Sorgfaltsmaßstab des § 43 Abs. 1 GmbHG nicht nur dem Schutz der Gesellschaft, sondern auch ihrer Gläubiger, denen nur die beschränkte Masse des Gesellschaftsvermögens hafte.[293] Andererseits spreche auch die Stellung des Geschäftsführers als Verwalter fremden Vermögens sowie das Bedürfnis nach einem umfassenden Minderheitenschutz gegen die Zulässigkeit einer rechtsgeschäftlichen Haftungsmilderung.[294]

Dagegen stellt *H. J. Fleck*[295] darauf ab, ob die Gesellschafter den Geschäftsführer nachträglich von seiner Haftung freistellen können. Sei dies zu bejahen, müsse auch im voraus eine rechtsgeschäftliche Haftungsmilderung zulässig sein. Sei dies zu verneinen, sei eine Haftungsmilderung unzulässig. Damit scheide eine vertragliche Haftungsmilderung insbesondere für die Verletzung der in § 43 Abs. 3 GmbHG genannten Pflichten aus. Das sei die Pflicht, an die Gesellschafter keine Auszahlungen aus dem gebundenen Vermögen zu leisten, und die Pflicht, eigene Geschäftsanteile nur unter den Voraussetzungen des § 33 GmbHG zu erwerben. Beide Pflichten dienten dem Gläubigerschutz. Für die Verletzung der Pflicht zur ordnungsgemäßen Unternehmensleitung könne dagegen eine Haftungsmilderung vereinbart werden.

Ebenso wie *H.-J. Fleck* bejaht auch *Uwe H. Schneider*[296] die Zulässigkeit einer Haftungsmilderung für die Verletzung der Pflicht zur ordnungsgemäßen Unternehmensleitung. Denn wenn die Gesellschafter den Geschäftsführer mit haftungsbefreiender Wirkung zu unzweckmäßigen Maßnahmen anweisen könnten, müßten sie auch in der Lage sein, seine Haftung im voraus vertraglich abzumildern. Hieraus folge umgekehrt, daß eine Haftungsmilderung unzulässig sei, wenn eine Weisung keine Folgepflicht des Geschäftsführers auslöse und damit auch keine Haftungsbefreiung eintrete. Das sei bei allen Weisungen der Fall, deren Ausführung gegen im öffentlichen Interesse gegebene Vorschriften oder gegen die guten Sitten verstoße. Im Ergebnis zieht *Uwe H. Schneider* damit den Kreis zulässiger Haftungsmilderungen enger als *H.-J. Fleck*. Insbesondere scheide eine Haftungsmilderung nicht nur für die Verletzung von Pflichten zur Kapitalerhaltung, sondern auch für die Verletzung von Pflichten zur Kapitalaufbringung, Risikotrennung und Liquiditätssicherung aus.[297]

---

292 *C. Th. Ebenroth/K. W. Lange*, Sorgfaltspflichten, 1992, S. 76; *M. Lutter/P. Hommelhoff*, GmbHG, 1995, § 43 Rdnr. 2; *B. Bastuck*, Enthaftung, 1986, S. 91 ff.; *M. Heisse*, Geschäftsführerhaftung, 1988, S. 122 f.; *U. Haas*, Geschäftsführerhaftung, 1997, S. 298.

293 *BFH*, Urt. v. 26.4.1984 - V R 128/79, in: BB 1984, S. 1992.

294 *C. Th. Ebenroth/K. W. Lange*, Sorgfaltspflichten, 1992, S. 76; *M. Lutter/P. Hommelhoff*, GmbHG, 1995, § 43 Rdnr. 2.

295 *H.-J. Fleck*, Haftung, 1974, S. 229.

296 *Uwe H. Schneider*, Haftungsmilderung, 1984, S. 811 f.

297 *Uwe H. Schneider*, Haftungsmilderung, 1984, S. 809 f.

c) Stellungnahme

Im Recht der GmbH gilt der Grundsatz der Gestaltungsfreiheit. Deshalb bedarf es einer besonderen Rechtfertigung, den Gesellschaftern die Möglichkeit zu verwehren, mit ihrem Geschäftsführer eine Haftungsmilderung zu vereinbaren. Diese Rechtfertigung kann sich nur aus den Interessen der Minderheitsgesellschafter oder der Gläubiger ergeben.

Die Minderheitsgesellschafter lassen sich dadurch schützen, daß man die Vereinbarung der Haftungsmilderung an ihre Zustimmung bindet. Zwar läßt sich nicht absehen, in welchen Fällen die Haftungsmilderung eingreift. Jedoch gebietet es der Grundsatz der Privatautonomie, daß die Gesellschafter und nicht die Rechtsordnung die Entscheidung treffen, ob sie die damit verbundenen Risiken eingehen wollen. Eine Grenze bildet in diesem Zusammenhang nur § 276 Abs. 2 BGB, wonach die Haftung für Vorsatz nicht ausgeschlossen werden kann.[298]

Somit kann die Unzulässigkeit einer Haftungsmilderung nur mit den Interessen der Gläubiger begründet werden. Hierzu heißt es zwar in der amtlichen Begründung[299] zu § 44 des Entwurfs eines Gesetzes betreffend die Gesellschaften mit beschränkter Haftung von 1891:

> „Ein geringerer Maßstab darf an die Verantwortlichkeit derselben [Haftung des Geschäftsführers] nicht gelegt werden, zumal es sich dabei nicht blos um die Interessen der Gesellschafter, sondern auch um diejenigen der Gesellschaftsgläubiger handelt".

Das steht jedoch in Widerspruch mit dem grundsätzlich umfassenden Weisungsrecht der Gesellschafter und den Wertungen des § 43 Abs. 3 GmbHG.

So ist eine nachteilige Geschäftsführungsmaßnahme nicht pflichtwidrig, wenn sie auf einer Weisung der Gesellschafter beruht. Will man den Gesellschaftern nicht das Recht zugestehen, über die Interessen der Gläubiger zu disponieren, so kann dies nur bedeuten, daß der Sorgfaltsmaßstab des § 43 Abs. 1 GmbHG grundsätzlich nicht dem Schutz der Gläubiger dient.

Und das wird auch bestätigt durch § 43 Abs. 3 GmbHG. Diese Vorschrift macht von dem soeben genannten Grundsatz nur dann eine Ausnahme, wenn der Geschäftsführer die Vorschriften der §§ 30, 33 GmbHG verletzt hat. Ist in diesen Fällen ein Schadensersatz zur Befriedigung der Gläubiger erforderlich, wird ein Schadensersatzanspruch der Gesellschaft selbst dann nicht ausgeschlossen, wenn der Geschäftsführer aufgrund einer Weisung der Gesellschafter gehandelt hat oder die Gesellschafter auf den Schadensersatzanspruch verzich-

---

[298] Ebenso *H. Altmeppen*, GmbHG, 1997, § 43 Rdnr. 50.

[299] Amtliche Ausgabe des Entwurfs eines GmbHG nebst Begründung und Anlagen, Berlin: Franz Vahlen, 1891, S. 94.

tet haben. Hiermit hat der Gesetzgeber zum Ausdruck gebracht, daß der Geschäftsführer mit der Sorgfalt eines gewissenhaften Geschäftsmannes den Haftungsfond der Gläubiger vor den in §§ 30, 33 GmbHG genannten Schmälerungen zu schützen hat. Das hat auch zur Folge, daß die Gesellschafter im Hinblick auf die in § 43 Abs. 3 GmbHG genannten Pflichten mit dem Geschäftsführer keine Haftungsmilderung vereinbaren können. Im übrigen steht es ihnen jedoch frei, einstimmig die Haftung des Geschäftsführers bis an die Grenze des Vorsatzes im voraus auszuschließen.

Richtiger Regelungsort für den Haftungsausschluß ist dabei die Satzung.[300] Denn bei dem Haftungsausschluß handelt es sich um eine generelle, d. h. über den Einzelfall hinausgehende Regelung, die die normative Grundordnung der Gesellschaft dauerhaft abändert. Eine solche Abänderung kann aber nur durch Aufnahme des Haftungsausschlusses in die Satzung geschehen.

Als Ergebnis läßt sich somit festhalten, daß die Gesellschafter die Haftung des Geschäftsführers für Fahrlässigkeit ausschließen können. Dieser Haftungsausschluß läßt sich jedoch nicht auf die Verletzung der in §§ 30, 33 GmbHG genannten Pflichten erstrecken.

V. Überleitung: Die Pflicht zur Sorge für das rechtmäßige Verhalten der Gesellschaft nach außen

An letzter Stelle schließlich soll die Pflicht des Geschäftsführers zur Sorge für das rechtmäßige Verhalten der Gesellschaft nach außen genannt werden.[301] Sie steht deshalb an letzter Stelle des 1. Teils, weil sie den Übergang bildet zum 2. Teil. Dort werden nämlich die Pflichten der herrschenden GmbH gegenüber den beherrschten Gesellschaften untersucht. Für die Erfüllung dieser Pflichten hat der Geschäftsführer Sorge zu tragen.[302] Das bedeutet zunächst, daß er alle

---

[300] *H.-G. Koppensteiner*, Rowedder, 1997, § 43 Rdnr. 4; *W. Zöllner*, Baumbach/Hueck, 1996, § 43 Rdnr. 1d.

[301] *H.-J. Mertens*, Hachenburg, 1996, § 43 Rdnr. 20; *H.-G. Koppensteiner*, Rowedder, 1997, § 43 Rdnr.9; *Uwe H. Schneider*, Scholz, 1993, § 43 Rdnr. 51.

[302] Eine Ausnahme besteht nur dann, wenn eine Verletzung dieser Pflichten im Gesellschaftsinteresse liegt und die Einheit der Rechtsordnung nicht auch die Annahme einer Pflichtwidrigkeit des Geschäftsführers gegenüber der Gesellschaft gebietet. So handelt etwa der Geschäftsführer gegenüber seiner Gesellschaft nicht pflichtwidrig, wenn er einen Vertrag der Gesellschaft nicht erfüllt, weil ihm ein Geschäft angeboten wurde, das selbst in Anbetracht des zu leistenden Schadensersatzes und anderer etwaiger Nachteile für die Gesellschaft günstiger ist als die Durchführung des bereits abgeschlossenen Vertrages (*M. Heisse*, Geschäftsführerhaftung, 1988, S. 17; in diesem Sinne auch *Uwe H. Schneider*, Scholz, 1993, § 43 Rdnr. 53). Dagegen handelt er pflichtwidrig, wenn er z. B. veranlaßt, Altöl im Erdreich versickern zu lassen. Denn hierbei handelt es sich um eine Straftat (§ 324a StGB). In diesem Fall gebietet es aber der Grundsatz der

mitgliedschaftlichen Pflichten und Treupflichten der herrschenden GmbH gegenüber den beherrschten Gesellschaften im Rahmen der Konzernleitung zu beachten hat. Tut er dies nicht und entsteht der beherrschten Gesellschaft hierdurch ein Schaden, so macht sich die herrschende GmbH gegenüber der beherrschten Gesellschaft schadensersatzpflichtig. Denn das Verhalten ihres Geschäftsführers wird ihr analog § 31 BGB zugerechnet.[303] Gleiches gilt aber auch im außervertraglichen Bereich. Begeht der Geschäftsführer im Rahmen der Geschäftsführung eine unerlaubte Handlung und macht sich gegenüber einem Dritten schadenersatzpflichtig, ordnet § 31 BGB analog eine Mithaftung der Gesellschaft an.[304] Eine Schadensersatzpflicht der Gesellschaft entsteht aber auch dann, wenn der Geschäftsführer nicht eigene deliktische Pflichten, sondern solche der Gesellschaft verletzt.[305] Das folgt zwar nicht aus § 31 BGB, der eine persönliche Schadensersatzpflicht des Geschäftsführers voraussetzt. Das folgt aber aus dem allgemeinen Rechtsgedanken, der in § 31 BGB zum Ausdruck kommt. Dieser besteht darin, der juristischen Person nicht nur alle Vorteile, sondern auch alle Nachteile organschaftlichen Handelns aufzuerlegen.[306] Diesem Rechtsgedanken wird aber nur eine Betrachtungsweise gerecht, die die juristische Person vermöge ihrer Organe als Handlungs- und Willensträgerin begreift, das Handeln der Organe ihr also unmittelbar als Eigenhandeln zurechnet (Organtheorie).[307]

Einheit der Rechtsordnung, auch eine Pflichtwidrigkeit gegenüber der Gesellschaft anzunehmen (*M. Heisse*, Geschäftsführerhaftung, 1988, S. 19).

303 *K. Schmidt*, Gesellschaftsrecht, 1997, S. 285 f. A. M. *W. Flume*, Juristische Person, 1983, S. 395 ff. (Zurechnung nach § 278 BGB).

304 *M. Lutter*, Haftung, 1993, S. 466 f.; *H. Altmeppen*, Haftung, 1995, S. 887.

305 *D. Eckardt*, Organhaftung, 1996, S. 69; *B. Grunewald*, Haftung, 1993, S. 458 f.; *A. Sandberger*, Außenhaftung, 1997, S. 143. Siehe auch *D. Kleindiek*, Deliktshaftung, 1997, S. 232, der nach einer Analyse der Rechtsprechung vor Inkrafttreten des BGB zu dem Ergebnis kommt, daß nach dieser Rechtsprechung die deliktische Verantwortlichkeit der juristischen Person konstruktiv nicht die Eigenhaftung ihrer Organwalter im Verhältnis zum geschädigten Dritten voraussetzt. In dieselbe Richtung wie hier auch *C.-W. Canaris*, Schuldrecht, 1994, S. 421 m. Fn. 183. Dagegen: *H. Altmeppen*, Haftung, 1995, S. 888.

306 So heißt es in den Motiven zu § 31 BGB (*B. Mudgan* (Hrsg.), Materialien, 1899, S. 409): „Zur Begründung dieser Haftbarmachung [des Vereins] hat man mit Recht darauf verwiesen, daß, wenn die Körperschaft durch die Vertretung die Möglichkeit gewinne, im Rechtsverkehre handelnd aufzutreten, ihr auch angesonnen werden müsse, die Nachtheile zu tragen, welche die künstlich gewährte Vertretung mit sich bringe, ohne daß sie in der Lage sei, Dritte auf den häufig unergiebigen Weg der Belangung des Vertreters zu verweisen“. Zur Entwicklung dieses Rechtsgedankens in der Rechtsprechung vor Inkrafttreten des BGB siehe *D. Kleindiek*, Deliktshaftung, 1997, S. 218 ff.

307 *K. Schmidt*, Gesellschaftsrecht, 1997, S. 258 ff. Eine kritische Gegenüberstellung der Organtheorie *Otto von Gierkes* und der Vertretertheorie *Friedrich Carl von Savignys* nimmt *D. Kleindiek*, Deliktshaftung, 1997, S. 151 ff. vor.

Das alles verdeutlicht, welche Konsequenzen es hat, wenn der Geschäftsführer seine Pflicht gegenüber der herrschenden GmbH, für deren rechtmäßiges Verhalten zu sorgen, verletzt.

Im folgenden wird ein Teilausschnitt dieser Pflicht näher untersucht. Gegenstand des folgenden Teils sind die Pflichten der herrschenden GmbH gegenüber den beherrschten Gesellschaften. Diese Pflichten hat der Geschäftsführer aufgrund seiner Pflicht zur Sorge für das rechtmäßige Verhalten der herrschenden GmbH bei der Konzernleitung zu beachten.

2. Teil: Die Pflichten der herrschenden GmbH gegenüber den beherrschten Gesellschaften

A. Bedeutung

Die Pflichten der herrschenden GmbH gegenüber den beherrschten Gesellschaften bilden eine wesentliche Rahmenbedingung für die Konzernleitung bei der herrschenden GmbH. Wegen dieser Rahmenbedingungen besteht ein fundamentaler Unterschied zwischen der Leitung eines rechtlich Einheitsunternehmens und der Leitung eines rechtlich gegliederten Unternehmens, der Führung eines Konzerns. In ihnen spiegelt sich der Konflikt wider zwischen rechtlicher Vielheit und wirtschaftlicher Einheit. Er ist im Bereich der Konzernleitung so stark ausgeprägt, daß *M. R. Theisen* von einem Problem „existentieller Natur" spricht.[1]

Nachdem im ersten Teil der Arbeit mit der Pflicht zur Konzernintegration der Aspekt der wirtschaftlichen Einheit herausgestellt wurde, soll nun der Aspekt der rechtlichen Vielheit näher untersucht werden. Es sollen mithin die Pflichten der herrschenden GmbH gegenüber den beherrschten Gesellschaften aufgezeigt werden. Dabei wird zwischen rechtsformspezifischen Pflichten und rechtsformunabhängigen Pflichten der herrschenden GmbH unterschieden.

B. Rechtsformspezifische Pflichten

I. Faktischer Aktienkonzern

1. Die gesetzliche Ausgangslage

Im faktischen Aktienkonzern wird die Einflußnahme des herrschenden Unternehmens auf die abhängige Aktiengesellschaft durch §§ 311 ff. AktG geregelt. Gemäß § 311 Abs. 1 AktG ist jede nachteilige Einflußnahme des herrschenden Unternehmens auf die abhängige Aktiengesellschaft verboten, es sei denn, daß die Nachteile bis zum Ende des Geschäftsjahres ausgeglichen werden oder ein entsprechender Rechtsanspruch gewährt wird (siehe § 311 Abs. 2

---

1 *M. R. Theisen*, Konzern, 1991, S. 203.

AktG). Ein Nachteil liegt vor, wenn der ordentliche und gewissenhafte Geschäftsleiter einer unabhängigen Gesellschaft durch Eingehen des Rechtsgeschäfts oder Ausführung der Maßnahme seine Pflichten verletzen würde.[2] Erfolgt kein Nachteilsausgleich, sind das herrschende Unternehmen und seine gesetzlichen Vertreter zum Schadensersatz verpflichtet, § 317 AktG.

Mit dieser Regelung sollen zum einen die Minderheitsgesellschafter und Gesellschaftsgläubiger geschützt werden.[3] Diese bedürfen deshalb eines besonderen Schutzes, weil im Konzern der Gleichlauf zwischen den Interessen des herrschenden Gesellschafters und der Gesellschaft gefährdet ist. Mit anderen Worten besteht die Gefahr, daß das herrschende Unternehmen seinen anderweitigen unternehmerischen Interessen den Vorzug gibt vor dem Eigeninteresse der beherrschten Gesellschaft.

Zum anderen wird das herrschende Unternehmen durch den zeitlich gestreckten Nachteilsausgleich aber auch privilegiert.[4] Hiermit wollte der Gesetzgeber allerdings nicht die aktienrechtliche Zuständigkeitsordnung durchbrechen, nach der der Vorstand die Gesellschaft unter eigener Verantwortung leitet. Vielmehr ging er davon aus und nahm es in Kauf, daß der Unternehmensaktionär ebenso wie der private Großaktionär auf die Unternehmensleitung legal Einfluß nimmt.[5] Diesen Einfluß wollte er angesichts der bestehenden Konzernverbindungen nicht verbieten, sondern kanalisieren.[6] Denn er war der Ansicht, daß das primär als Organisationsrecht ausgestaltete Aktienrecht nicht der richtige Ort sei für die Unterscheidung zwischen wettbewerbspolitisch erwünschten und unerwünschten Unternehmensverbindungen.[7] Deshalb schuf er das System des gestreckten Nachteilsausgleichs, das die weitere Durchführung bestehender und

---

[2] *H.-G. Koppensteiner*, Kölner Kommentar, 1987, § 311 Rdnr. 22; *B. Kropff* , Geßler, 1975, § 311 Rdnr. 37.

[3] *U. Hüffer*, AktG, 1997, § 311 Rdnr. 1.

[4] *V. Emmerich/J. Sonnenschein*, Konzernrecht, 1997, S. 334; *M. Habersack*, Aktienkonzernrecht, 1998, vor § 311 Rdnr. 2.

[5] *L. Strohn*, Verfassung, 1977, S. 5 f., unter Bezug auf den Regierungsentwurf zu §§ 311 ff. AktG.

[6] So führte der Abgeordnete Dr. *Wilhelmi* als Berichterstatter des Rechtsausschusses vor dem Plenum des Deutschen Bundestages aus, daß der zeitlich gestreckte Nachteilsausgleich „es ... der Wirtschaft ermöglichen [soll], den sog. faktischen Konzern mit relativ selbständigen Untereinheiten beizubehalten" (Deutscher Bundestag - 4. Wahlperiode - Stenographische Berichte Band 5, 187. Sitzung, S. 9405). Wie hier *B. Kropff*, Geßler, 1975, § 311 Rdnr. 10, 12.

[7] Begründung RegE zum Dritten Buch des AktG bei *B. Kropff*, AktG, 1965, S. 374; *L. Strohn*, Verfassung, 1977, S. 10.

natürlich auch zukünftiger Unternehmensverbindungen ermöglichen sollte. Hiermit hat er die Zulässigkeit des faktischen Konzerns anerkannt.[8]

Freilich bedeutet die Privilegierung des herrschenden Unternehmens nicht, daß die abhängige Aktiengesellschaft den Aufforderungen des herrschenden Unternehmens Folge zu leisten hat.[9] Vielmehr steht es im Ermessen des Tochtervorstands, ob er den Willen des herrschenden Unternehmens beachten will. Berücksichtigt er diesen, muß er sich allerdings vergewissern, daß etwaige Nachteile ausgeglichen werden.[10] Das kann insbesondere durch eine geeignete Vereinbarung mit dem herrschenden Unternehmen geschehen, die noch vor der Benachteiligung erfolgt.[11]

Damit ergibt sich folgender Befund: Die herrschende GmbH darf auf den Vorstand einer abhängigen Aktiengesellschaft Einfluß ausüben. Sie darf ihn sogar zu Maßnahmen veranlassen, die für die abhängige Gesellschaft nachteilig sind, soweit der Nachteil nur ausgeglichen oder ein entsprechender Rechtsanspruch gewährt wird.

Im folgenden sollen nun die ungeschriebenen Voraussetzungen der §§ 311 ff. AktG untersucht werden. Sie beziehen sich sämtlich auf das Privilegierungselement von § 311 AktG. Mit anderen Worten greifen zwar bei Veranlassung einer nachteiligen Maßnahme das Verbot der Nachteilszufügung und der Schadensersatzanspruch nach § 317 AktG ohne Hinzutreten weiterer Voraussetzungen ein; ob das herrschende Unternehmen den Schadensersatzanspruch durch einen zeitlich gestreckten Nachteilsausgleich abwenden darf, ist jedoch an weitere Voraussetzungen gebunden.

2. Das Konzerninteresse

Erste ungeschriebene Voraussetzung für die Privilegierung des herrschenden Unternehmens ist die Einhaltung des Konzerninteresses. Das herrschende Unternehmen darf die abhängige Aktiengesellschaft nur dann zu einer nachteiligen Maßnahme veranlassen, wenn deren Durchführung den Belangen des herr-

---

8 Auf den Streit zur Zulässigkeit des faktischen Konzerns soll an dieser Stelle nicht eingegangen werden. Siehe dazu *H.-G. Koppensteiner*, Kölner Kommentar, 1987, Vorb. § 311 Rdnr. 5 ff. Vielmehr wird der h. M. gefolgt, die seine Zulässigkeit bejaht. Dazu statt aller *U. Hüffer*, AktG, 1997, § 311 Rdnr. 6. A. M. insbesondere *U. Bälz*, Unternehmen, 1992, S. 291 f.; *ders.*, Einheit, 1974, S. 308; *E. Geßler*, Leitungsmacht, 1974, S. 147; *D. Reuter*, Plädoyer, 1996, S. 283 f.

9 *H.-G. Koppensteiner*, Kölner Kommentar, 1987, § 311 Rdnr. 90; *B. Kropff*, Geßler, 1975, § 311 Rdnr. 29.

10 *U. Hüffer*, AktG, 1997, § 311 Rdnr. 48; *B. Kropff*, Geßler, 1975, § 311 Rdnr. 56 - 63; *H. Altmeppen*, Vermögensbindung, 1996, S. 696.

11 *H. Altmeppen*, Vermögensbindung, 1996, S. 696.

schenden Unternehmens oder der mit ihm und der Gesellschaft konzernverbundenen Unternehmen dient.[12]

Das folgt zum einen aus dem Grund der Privilegierung.[13] Der Gesetzgeber hat den zeitlich gestreckten Nachteilsausgleich zugelassen, um bestehende Unternehmensverbindungen nicht zu bekämpfen. Nur soweit das Handeln des herrschenden Unternehmens einer solchen Unternehmensverbindung und damit dem Konzerninteresse dient, soll die Privilegierung also reichen. Zum anderen wäre es ein Widerspruch, das herrschende Unternehmen nur im Vertragskonzern (§ 308 Abs. 1 Satz 2 AktG) und nicht auch im faktischen Konzern an das Konzerninteresse zu binden.[14]

3. Die Satzung der abhängigen Aktiengesellschaft

Weiterhin darf das herrschende Unternehmen den Vorstand der abhängigen Aktiengesellschaft nicht zu Maßnahmen veranlassen, deren Durchführung gegen die Satzung der abhängigen Aktiengesellschaft verstoßen würde. Denn die Privilegierung des herrschenden Unternehmens bezieht sich nur auf solche Maßnahmen, die in den Zuständigkeitsbereich des Vorstands der abhängigen Gesellschaft fallen.[15] Die Änderung der Satzung gehört aber in den Kompetenzbereich der Hauptversammlung (§§ 119 Abs. 1 Nr. 5, 179 Abs. 1 AktG).

Hieraus folgt insbesondere, daß das herrschende Unternehmen den Unternehmensgegenstand der beherrschten Gesellschaft zu beachten hat. Besteht dieser in der Herstellung von Computern, so darf das herrschende Unternehmen den Vorstand der abhängigen Gesellschaft nicht zur Umstellung der Produktion auf Autos veranlassen. Ebensowenig darf es die Gesellschaft in eine reine Vertriebsgesellschaft umwandeln.[16]

4. Autonomieschädigende Veranlassungen

Umstritten ist, ob die Grenzen zulässiger Einflußnahme dann überschritten werden, wenn durch die Veranlassungen des herrschenden Unternehmens die

---

[12] *B. Kropff*, Geßler, 1975, § 311 Rdnr. 34; *L. Strohn*, Verfassung, 1977, S. 64; *W. Flume*, Juristische Person, 1983, S. 122.

[13] *L. Strohn*, Verfassung, 1977, S. 64.

[14] *B. Kropff*, Geßler, 1975, § 311 Rdnr. 34.

[15] *L. Strohn*, Verfassung, 1977, S. 54.

[16] *L. Strohn*, Verfassung, 1977, S. 54 f. Ebenso *J. Semler*, Leitung, 1996, Rdnr. 321, allerdings mit der Begründung, daß der Kernbereich des Integritätsinteresses der beherrschten Gesellschaft verletzt werde.

Autonomie der beherrschten Gesellschaft beseitigt und die beherrschte Gesellschaft vollständig auf das Konzerninteresse ausgerichtet wird.

Nach einer Ansicht sind autonomieschädigende Veranlassungen verboten. Denn sie würden das finanzielle Ausgleichssystem der §§ 311 ff. AktG außer Kraft setzen. Habe nämlich die Gesellschaft erst einmal ihre Autonomie verloren, so könne bei nachfolgenden Veranlassungen nicht mehr festgestellt werden, wie sich der Vorstand einer als unabhängig gedachten Gesellschaft verhalten würde. Damit entfalle aber der Maßstab, der für die Nachteilsfeststellung erforderlich sei.[17] Hinzu komme, daß der mit dem Autonomieverlust verbundene Nachteil nicht quantifiziert werden könne. Denn die zur Wiederherstellung der selbständigen Existenzfähigkeit erforderlichen Kosten ließen sich im voraus kaum beziffern.[18] Die Zufügung nicht quantifizierbarer Nachteile sei aber unzulässig.[19]

Dagegen ist nach der Meinung von *H.-G. Koppensteiner* eine existenzgefährdende Veranlassung für die beherrschte Gesellschaft zwar meistens nachteilig,[20] unzulässig sei sie jedoch nur dann, wenn der Nachteil nicht ausgleichbar sei[21]. Bei der Frage der Ausgleichbarkeit räumt *H.-G. Koppensteiner* dem herrschenden Unternehmen jedoch einen großen Spielraum ein. Selbst wenn ein Nachteil ex ante nicht quantifiziert werden könne, sei ein Ausgleich in der Form möglich, daß sich das herrschende Unternehmen zur Kompensation all derjenigen Nachteile verpflichte, die bei Beendigung der Geschäftsbeziehungen bestünden. Es reiche also aus, wenn die Nachteile erst zu diesem Zeitpunkt ermittelt würden. Als Beispiel nennt er die Aufgabe der eigenen Vertriebsorganisation auf Weisung des herrschenden Unternehmens. Den damit verbundenen Nachteil könne das herrschende Unternehmen durch seine Verpflichtung ausgleichen, bei Beendigung des Konzernverhältnisses der beherrschten Gesellschaft Personal, Sachmittel und know-how für die Wiedererlangung ihrer Marktstellung zur Verfügung zu stellen.[22] Auch werde nicht während des Bestehens des Konzernverhältnisses die Funktionsfähigkeit der §§ 311 ff. AktG außer Kraft gesetzt. Allerdings könne sich eine Verschiebung des Maßstabs ergeben, anhand dessen die Nachteile festgestellt würden. So könne es in dem obengenannten Beispiel durchaus mit der Sorgfalt des Leiters einer unabhängigen Gesellschaft vereinbar sein, wenn dieser in Konsequenz der wirtschaftlichen

---

17 *B. Kropff*, Geßler, 1975, § 311 Rdnr. 39; *M. Lutter*, Unternehmensgruppe, 1982, S. 264 f.; *W. Flume*, Juristische Person, 1983, S. 123.

18 *L. Strohn*, Verfassung, 1977, S. 88 f.

19 *L. Strohn*, Verfassung, 1977, S. 66.

20 *H.-G. Koppensteiner*, Kölner Kommentar, 1987, § 311 Rdnr. 45.

21 *H.-G. Koppensteiner*, Kölner Kommentar, 1987, § 311 Rdnr. 48.

22 *H.-G. Koppensteiner*, Kölner Kommentar, 1987, § 311 Rdnr. 86, 87.

Ausrichtung der beherrschten Gesellschaft auf das Konzerninteresse Weisungen der Konzernspitze folge, die die Marktgängigkeit der Erzeugnisse verschlechterten.[23]

Schließlich bejaht *C. E. Decher* die Zulässigkeit eines qualifizierten faktischen Konzerns, der sich durch einen Autonomieverlust der beherrschten Gesellschaft gerade auszeichne.[24] Zur Begründung führt er an, daß der Gesetzgeber von 1965 die Besonderheiten des qualifizierten faktischen Konzerns nicht erkannt und damit nicht geregelt habe. Deshalb könne aus den Vorschriften der §§ 311 ff. AktG weder eine negative noch eine positive Wertung für den qualifizierten faktischen Konzern entnommen werden und bestehe eine Regelungslücke.[25] Beim Ausfüllen dieser Regelungslücke müsse berücksichtigt werden, daß eine zentrale Konzernleitung für die beherrschte Gesellschaft nicht schlechthin mit Nachteilen verbunden sei, sondern im Gegenteil für sie auch von Vorteil sein könne.[26] Deshalb sei die Gründung eines qualifizierten faktischen Konzerns zulässig. Eine Zustimmung der Minderheitsgesellschafter sei nicht erforderlich, da im Gegensatz zum Vertragskonzern weder die Kapitalerhaltungsregeln noch die Gewinnverwendungsregeln außer Kraft gesetzt würden.[27] Für den Schutz der Minderheitsgesellschafter sei es ausreichend, daß ihnen analog §§ 304, 305 AktG eine Ausgleichszahlung oder eine Abfindung gewährt werden müsse.[28] Die Gläubiger der abhängigen Gesellschaft würden schließlich durch einen Verlustausgleich analog §§ 302, 303 AktG geschützt.[29]

Der Ansicht von *C. E. Decher* kann nicht gefolgt werden. Der Gesetzgeber hat die Gefahren einer qualifizierten Konzernherrschaft erkannt und er hat sie geregelt, nämlich in den §§ 291 ff. AktG. Denn dort hat er zum Ausdruck gebracht, daß eine qualifizierte Konzernherrschaft nur im Vertragskonzern mit seinen Schutzvorschriften für Minderheitsgesellschafter und Gläubiger ausgeübt werden darf. Insbesondere hat er die Zulässigkeit einer qualifizierten Konzernherrschaft an die Zustimmung einer Mehrheit der Gesellschafter, die mindestens drei Viertel des bei der Beschlußfassung über den Unternehmensvertrag vetretenen Grundkapitals umfassen muß, gebunden, § 293 Abs. 1 AktG. Dieses Zustimmungserfordernis würde aber umgangen, wenn man eine qualifizierte Konzernherrschaft auch im faktischen Konzern zulassen würde, in dem das herrschende Unternehmen vielleicht noch nicht einmal über 50 % der Stimmen in

---

23 *H.-G. Koppensteiner*, Kölner Kommentar, 1987, § 311 Rdnr. 46.

24 *C. E. Decher*, Verflechtungen, 1990, S. 109 ff.

25 *C. E. Decher*, Verflechtungen, 1990, S. 109.

26 *C. E. Decher*, Verflechtungen, 1990, S. 110, 111.

27 *C. E. Decher*, Verflechtungen, 1990, S. 116.

28 *C. E. Decher*, Verflechtungen, 1990, S. 117 ff.

29 *C. E. Decher*, Verflechtungen, 1990, S. 119 ff.

der beherrschten Gesellschaft verfügt. Des weiteren greift *C. E. Decher* zu kurz, wenn er Sinn und Zweck des Zustimmungserfordernisse allein darin sieht, die Außerkraftsetzung der Kapitalerhaltungs- und Gewinnverwendungsregeln zu rechtfertigen. Zum einen wird man nicht ernsthaft bestreiten können, daß im qualifizierten faktischen Konzern zumindest die Gefahr einer Einlagenrückgewähr oder verdeckten Gewinnausschüttung besteht. Zum anderen wollte der Gesetzgeber mit dem Zustimmungserfordernis nicht nur dieser Gefahr entgegen wirken, sondern auch den Minderheitsaktionären einen Einfluß auf den Inhalt des Vertrages und insbesondere die Höhe der Ausgleichszahlungen und Abfindungen verschaffen. Schließlich sieht das Gesetz eine umfassende Prüfung des Beherrschungsvertrages durch sachverständige Prüfer (§§ 293c - 293e AktG), ein Informations- und Auskunftsrecht der Aktionäre vor Beschlußfassung über den Unternehmensvertrag (§§ 293f, 293g AktG) sowie seine Eintragung in das Handelsregister (§ 294 AktG) vor. All diese Vorschriften zum Schutze der Minderheitsaktionäre und Gläubiger der beherrschten Gesellschaft könnte das herrschende Unternehmen umgehen, wenn man die Zulässigkeit eines qualifizierten faktischen Konzerns bejahen würde. Das kann nicht sein. Deshalb ist die Ausübung einer qualifizierten Konzernherrschaft im faktischen Konzern jedenfalls dann unzulässig, wenn in der beherrschten Gesellschaft Minderheitsgesellschafter vorhanden sind und diese der Ausübung der qualifizierten Konzernherrschaft nicht zugestimmt haben.[30] Haben die Minderheitsgesellschafter zugestimmt oder handelt es sich um eine Einmann-Gesellschaft, so ändert dies freilich nichts an dem Entstehen der Ausfallhaftung der herrschenden GmbH analog § 303 AktG zugunsten der Gläubiger der beherrschten Gesellschaft.[31]

Ebenso können autonomiegefährdende Maßnahmen nicht mit dem Argument ihrer Ausgleichbarkeit gerechtfertigt werden. Denn wird einer Gesellschaft eine für ihre selbständige Lebensfähigkeit wesentliche Unternehmensfunktion entzogen und wird sie dadurch vollständig auf das Konzerninteresse ausgerichtet, läßt sich bei nachfolgenden Veranlassungen nicht mehr ermitteln, wie sich der Vorstand einer als unabhängig gedachten Gesellschaft verhalten hätte. Das ist aber erforderlich, um die Funktionsfähigkeit der §§ 311 ff. AktG aufrechtzuerhalten. Zwar ist *H.-G. Koppensteiner* zuzugeben, daß bei Bestimmung des Vergleichsmaßstabs von § 311 AktG die konkrete wirtschaftliche Lage der abhängigen Gesellschaft zu berücksichtigen ist.[32] Denn § 311 AktG will die Gesellschaft nur vor Verlusten schützen, in denen sich das spezifische zusätzliche Risiko ihrer Abhängigkeit realisiert, nicht aber vor solchen Verlusten, die auch ein wirt-

[30] *V. Emmerich*, Scholz, 1993, Anh. Konzernrecht, Rdnr. 210.

[31] *BGH*, Urt. v. 16.9.1985 - II ZR 275/84, in: BGHZ 95, S. 330, 346 f., für die Einmann-GmbH.

[32] *H.-G. Koppensteiner*, Kölner Kommentar, 1987, § 311 Rdnr. 25.

schaftlich unabhängiges Unternehmen hinnehmen muß.[33] Das hat z. B. zur Konsequenz, daß sich der Handlungsspielraum des ordentlichen Geschäftsleiters verändert, wenn die Gesellschaft in wirtschaftliche Schwierigkeiten gerät oder in eine wirtschaftliche Abhängigkeit zu einem Lieferanten kommt. Handelt es sich aber um eine wirtschaftliche Abhängigkeit gegenüber dem herrschenden Unternehmen und wurde diese wirtschaftliche Abhängigkeit kraft der rechtlichen Abhängigkeit des beherrschten Unternehmens herbeigeführt, so darf dies keinen Einfluß auf den Verhaltensmaßstab des Vorstands der abhängigen Gesellschaft haben. Denn sonst hätte es das herrschende Unternehmen in der Hand, über diesen Verhaltensmaßstab zu disponieren. Das aber würde gegen Sinn und Zweck der §§ 311 ff. AktG verstoßen. Dieser besteht nämlich darin, die außenstehenden Aktionäre und die Gesellschaftsgläubiger zu schützen, indem die Gesellschaft zumindest wirtschaftlich so gestellt wird, als ob sie ihr Unternehmen weiterhin unabhängig im gemeinsamen Interesse aller Aktionäre und nicht nur im Interesse des herrschenden Unternehmens betriebe.[34]

Damit kann als Zwischenergebnis festgehalten werden, daß autonomieschädigende Maßnahmen die Funktionsfähigkeit der §§ 311 ff. AktG außer Kraft setzen und damit unzulässig sind. Folglich darf der Geschäftsführer der herrschenden GmbH keine Maßnahmen veranlassen, die zu einem Autonomieverlust der abhängigen Aktiengesellschaft führen. Verstößt er gegen diese Pflicht und entsteht der abhängigen Gesellschaft, wie es spätestens nach Beendigung des Konzernverhältnisses der Fall sein wird, ein Schaden, so sind die herrschende GmbH gemäß § 317 Abs. 1 AktG und ihr Geschäftsführer gemäß § 317 Abs. 3 AktG gegenüber der abhängigen Gesellschaft zum Schadensersatz verpflichtet. Dabei spielt es keine Rolle, daß der zugefügte Nachteil schlechthin, also nicht nur wegen unterbliebenen Ausgleichs rechtswidrig ist. Denn ebenso wie § 317 AktG Fälle erfaßt, bei denen ein Ausgleich aus rechtlichen oder tatsächlichen Gründen nicht möglich ist,[35] gilt er auch für Veranlassungen, mit denen der Gesellschaft nicht nur ein Nachteil zugefügt, sondern ihre Autonomie vollständig aufgehoben wird.

In der Praxis lassen sich diese autonomiegefährdenden Veranlassungen in zwei Fallgruppen aufteilen. Die erste ist dadurch gekennzeichnet, daß das herrschende Unternehmen laufend in die Geschäftsführung der beherrschten Gesellschaft eingreift, sie wie eine Betriebsabteilung führt und ausschließlich den Interessen des Gesamtkonzerns dienstbar macht. In diesem Fall kann die Fülle der Eingriffe nicht mehr festgestellt und es können ihre Folgen nicht mehr be-

---

[33] *V. Emmerich/J. Sonnenschein*, Konzernrecht, 1997, S. 341; *M. Habersack*, Aktienkonzernrecht, 1998, § 311 Rdnr. 26.

[34] *V. Emmerich/J. Sonnenschein*, Konzernrecht, 1997, S. 341.

[35] *H.-G. Koppensteiner*, Kölner Kommentar, 1987, § 317 Rdnr. 6.

wertet werden. Insbesondere ist es unmöglich, von einer hypothetischen Autonomie der Gesellschaft her zu denken. Damit versagt das auf Einzelausgleich beruhende System der §§ 311 ff. AktG und ist ein entsprechendes Verhalten der Konzernleitung verboten.[36]

In der anderen Fallgruppe wird die Autonomie der Gesellschaft dadurch gefährdet, daß ihr wesentliche für ein selbständiges Auftreten am Markt erforderliche Unternehmensfunktionen genommen werden.[37] Deshalb sind insbesondere die Zentralisierung betrieblicher Funktionen wie Forschung, Entwicklung, Einkauf und Verkauf unzulässig.[38]

Eine andere Beurteilung ergibt sich jedoch, wenn die Gesellschaft ohne die Konzerneinbindung nicht überleben könnte. Nach der Aussage von Praktikern ist dies der häufigste Fall, in der wesentliche Unternehmensfunktionen ausgegliedert werden.[39] Folgt der Vorstand der Veranlassung des herrschenden Unternehmens und läßt sich z. B. auf eine Zentralisierung des Einkaufs bei einer Schwestergesellschaft ein, so handelt er nicht pflichtwidrig. Denn auch der ordentliche und gewissenhafter Geschäftsleiter einer unabhängigen Gesellschaft, die sich in derselben wirtschaftlichen Lage wie die abhängige Gesellschaft befindet, hätte so gehandelt, wenn die Ausgliederung des Einkaufs die einzige Möglichkeit gewesen wäre, die Gesellschaft zu retten.[40]

5. Konzernintegrative Veranlassungen

Die Problematik konzernintegrativer Veranlassungen besteht darin, daß ihre Folgen in der Regel nicht vorhersehbar sind. Grundlage der Nachteilsermittlung ist aber eine ex-tunc-Prognose.[41] Ein gutes Beispiel ist die Ausweitung der Autoelektrik bei AEG auf Veranlassung der Daimler-Benz AG Anfang der 90er Jahre. Hiervon versprachen sich beide Gesellschaften synergetische Vorteile. Zu Recht stellte *M. Hoffmann-Becking*[42] 1989 fest, daß man nicht prognostizie-

---

[36] *M. Lutter*, Unternehmensgruppe, 1982, S. 264 f.

[37] *M. Hoffmann-Becking*, AG-Konzern, 1989, S. 78 f.; *B. Kropff*, Geßler, 1975, § 311 Rdnr. 39; *G. Krieger*, Unternehmen, 1988, § 69 Rdnr. 67.

[38] In diese Richtung auch *J. Semler*, Leitung, 1996, Rdnr. 318, 321. Eingehend *L. Strohn*, Verfassung, 1977, S. 108 ff., unter dem Aspekt, ob die Autonomie der Gesellschaft nach ihrem Ausscheiden aus dem Konzernverbund wiederhergestellt werden kann.

[39] *P. Hommelhoff*, Gutachten, 1992, S. G 24.

[40] Ebenso *L. Strohn*, Verfassung, 1977, S. 77.

[41] *B. Kropff*, Geßler, 1975, § 311 Rdnr. 110; *U. Hüffer*, AktG, 1997, § 311 Rdnr. 28.

[42] *M. Hoffmann-Becking*, AG-Konzern, 1989, S. 79.

ren könne, ob diese strategische Ausrichtung sich zum Vor- oder Nachteil für die AEG entwickelt. Heute ist bekannt, daß Edzard Reuter mit seinem Konzept des integrierten Technologiekonzerns weithin gescheitert ist und insbesondere die AEG liquidiert wurde.

P. Hommelhoff[43] ist der Auffassung, daß sich der Tochtervorstand einer als unabhängig gedachten Gesellschaft auf solche unkalkulierbaren Maßnahmen nicht einlassen dürfe. Trotzdem verneint er nicht die Zulässigkeit solcher Veranlassungen. Vielmehr nimmt er konzernintegrative Maßnahmen aus dem Anwendungsbereich der §§ 311 ff. AktG heraus[44] und unterstellt sie eigenen Voraussetzungen[45].

Zur Begründung führt er an, daß der Gesetzgeber einerseits in den §§ 311 ff. AktG die Ausübung von Konzernleitungsmacht in Form eines dezentral strukturierten faktischen Konzern für zulässig erklärt habe.[46] Dies gehe sowohl aus den Intentionen des Gesetzgebers hervor, der der Wirtschaft die Beibehaltung relativ selbständiger Untergesellschaften habe ermöglichen wollen,[47] als auch aus gesetzlichen Bestimmungen, die die grundsätzliche Zulässigkeit des faktischen Konzerns signalisierten (z. B. §§ 18 Abs. 1, 100 Abs. 2 Satz 2 AktG, 76 Abs. 4 Satz 1 BetrVG 1952)[48]. Ein faktischer Konzern setze aber eine Ausrichtung der Unternehmenspolitik der abhängigen Gesellschaft auf die Konzernspitze voraus.[49]

Andererseits habe der Gesetzgeber durch Installierung eines vermögensmäßig orientierten, auf den Ausgleich quantifizierbarer Nachteile zugeschnittenen Ausgleichssystems einen wirksamen Außenseiterschutz realisieren wollen.[50]

Diese beiden Entscheidungen des Gesetzgebers, die in ihren Wirkungen diametral entgegengesetzt seien, müßten bei Bestimmung der Grenzen der Konzernleitungsmacht „harmonisiert" werden.[51] So dürfe die Konzernleitungsmacht nur insoweit ausgeübt werden, wie das Ausgleichssystem der §§ 311 ff. AktG nicht in seiner Funktionsfähigkeit beeinträchtigt werde. Dabei gebühre dem Interesse des Gesamtkonzerns nur dann Vorrang vor dem Eigeninteresse der faktisch konzernierten Tochtergesellschaft, sofern dies erforderlich sei, um

---

[43] *P. Hommelhoff*, Konzernleitungspflicht, 1982, S. 132.
[44] *P. Hommelhoff*, Konzernleitungspflicht, 1982, S. 133.
[45] *P. Hommelhoff*, Konzernleitungspflicht, 1982, S. 136 ff.
[46] *P. Hommelhoff*, Konzernleitungspflicht, 1982, S. 130.
[47] *P. Hommelhoff*, Konzernleitungspflicht, 1982, S. 130.
[48] *P. Hommelhoff*, Konzernleitungspflicht, 1982, S. 110 ff.
[49] *P. Hommelhoff*, Konzernleitungspflicht, 1982, S. 131.
[50] *P. Hommelhoff*, Konzernleitungspflicht, 1982, S. 137.
[51] Ebenso *M. Hoffmann-Becking*, AG-Konzern, 1989, S. 77.

überhaupt einen dezentralen faktischen Konzern bilden zu können.[52] Konzernintegrative Maßnahmen müßten somit der Tochtergesellschaft den Status eines selbständigen Unternehmensträgers belassen, so daß deren Eigeninteresse als Maßstab für die Auswirkungen von Veranlassungen des herrschenden Unternehmens gewahrt bleibe und damit die Funktionsfähigkeit der §§ 311 ff. AktG gewährleistet sei. Jedoch sei auch die Aufhebung der Selbständigkeit der Tochtergesellschaft zulässig, wenn anders ein faktischer Konzern nicht entstehen könne.[53] Befolge der Tochtervorstand solche konzernintegrativen Maßnahmen nicht, könne er wegen grober Pflichtverletzung vorzeitig abberufen werden (§ 84 Abs. 3 AktG).[54]

P. Hommelhoff kann schon im Ausgangspunkt nicht gefolgt werden. Auch der Vorstand einer unabhängigen Gesellschaft muß Entscheidungen treffen, deren Folgen nicht vorhersehbar sind. Hierbei hat er ein breites unternehmerisches Ermessen.[55] Erst wenn er dieses Ermessen überschreitet, handelt er pflichtwidrig. Auf den Konzern gewendet bedeutet dies, daß eine Veranlassung nicht nachteilhaft ist, wenn den entstehenden Risiken entsprechende Chancen gegenüberstehen.[56] Die Ausweitung der Autoelektrik von AEG war also schon nicht nachteilhaft, wenn der ordentliche und gewissenhafte Geschäftsleiter einer als unabhängig gedachten Gesellschaft nach Abwägung der Chancen und Risiken sich auf diese Maßnahme eingelassen hätte. Die Frage des Nachteilsausgleichs hätte sich dann überhaupt nicht gestellt. Nur wenn der ordentliche und gewissenhafte Geschäftsleiter einer unabhängigen Gesellschaft der Veranlassung der Daimler-Benz AG nicht nachgekommen wäre, läge ein Nachteil vor. Dieser Nachteil wäre in der Tat wegen seiner unabsehbaren Folgen nicht quantifizierbar und müßte deshalb unterbleiben.

Dies gilt entgegen der Ansicht von *P. Hommelhoff* selbst dann, wenn anders ein faktischer Konzern nicht entstehen kann. So verdient zunächst seine These keine Zustimmung, daß das Eigeninteresse der abhängigen Gesellschaft unter Umständen hinter das Konzerninteresse zurücktreten müsse. Richtig ist, daß das herrschende Unternehmen Einfluß auf die abhängige Aktiengesellschaft nehmen darf. So schützen §§ 311 ff. AktG auch nicht den Eigenwillen,[57] sondern nur das Eigeninteresse der abhängigen Gesellschaft. Dabei handelt es sich um Vermö-

---

52 *P. Hommelhoff*, Konzernleitungspflicht, 1982, S. 137.

53 *P. Hommelhoff*, Konzernleitungspflicht, 1982, S. 138.

54 *P. Hommelhoff*, Konzernleitungspflicht, 1982, S. 146.

55 So jüngst für den Vorstand einer AG: BGH, Urt. v. 21.4.1997 - II ZR 175/95, in: WM 1997, S. 970, 973. Dazu *P. Kindler*, Ermessen, 1998, S. 101 ff.

56 Im Ergebnis ebenso *L. Strohn*, Verfassung, 1977, S. 91 f.

57 A. M. *P. Hommelhoff*, Konzernleitungspflicht, 1982, S. 115.

gensinteressen.[58] Insofern soll die Gesellschaft nicht schlechter stehen als wenn sie unabhängig wäre.[59] So heißt es denn auch ausdrücklich in der *Begründung zum Regierungsentwurf* zu § 311 Abs. 1 AktG, daß bei Fehlen eines Beherrschungsvertrages weder die Interessen des herrschenden Unternehmens noch die Interessen des Konzerns eine Beeinträchtigung der Interessen der abhängigen Gesellschaft rechtfertigen würden. Diese Interessen „sind unabhängig von ihrer Größe für das Recht gleichwertig. Kein Gesichtspunkt unserer Rechts- und Wirtschaftsordnung gestattet es, den Vermögensinteressen eines Konzerns nur deswegen den Vorrang einzuräumen, weil sie quantitativ größer sind".[60] Hieraus wird deutlich, daß der Außenseiterschutz zumindest Primärzweck der vom Regierungsentwurf intendierten Regelung war. Hieran ändert auch nichts der erst in der letzten Phase der Ausschußberatungen eingefügte Abs. 2 von § 311 AktG. Wie aus dem Ausschußbericht[61] hervorgeht, werden nach Ansicht der Mehrheit des Ausschusses „die Interessen, die §§ 311 ff. wahren wollen", auch durch einen zeitlich gestreckten Nachteilsausgleich ausreichend geschützt. Mangels gegenteiliger Feststellungen im Ausschußbericht muß davon ausgegangen werden, daß mit dieser Formulierung die „Interessen" gemeint sind, die in der Regierungsvorlage definiert wurden.[62]

Damit läßt sich aus der Entstehungsgeschichte der §§ 311 ff. AktG selbst dann kein Vorrang des Konzerninteresses vor dem Eigeninteresse der Tochtergesellschaft ableiten, wenn anders ein Konzern nicht entstehen kann.

Ein anderes Ergebnis läßt sich auch nicht aus den übrigen von *P. Hommelhoff* angeführten Vorschriften (§§ 18 Abs. 1, 100 Abs. 2 Satz 1 Nr. 1 AktG, 76 Abs. 4 Satz 1 BetrVG 1952) ableiten, die sich mit dem faktischen Konzern beschäftigen. Denn erstens führt die Gründung eines faktischen Konzerns, wie noch zu zeigen sein wird, nicht notwendig zu einer Schädigung des Eigeninteresses der abhängigen Gesellschaft.[63] Und zweitens würde die Interpretation der zitierten Vorschriften überspannt, wollte man aus ihnen ein konzernfreundliche

---

[58] Weitergehend *L. Strohn*, Verfassung, 1977, S. 18 f., nach dem die §§ 311 ff. AktG auch den „Kernbestand des Integritätsinteresses" der abhängigen Gesellschaft schützen wollen.

[59] Zum Vorstehenden siehe *H.-G. Koppensteiner*, Kölner Kommentar, 1987, Vorb. § 311 Rdnr. 4.

[60] *Begr. RegE.* zu § 311 Abs. 1 AktG, in: *B. Kropff*, AktG, 1965, S. 407. Ebenso *B. Kropff*, Geßler, 1975, § 311 Rdnr. 21; *E.-J. Mestmäcker*, Systematik, 1967, S. 131, 143; *H.-G. Koppensteiner*, Kölner Kommentar, 1987, Vorb. § 311 Rdnr. 9; *H. Würdinger*, Großkommentar, 1971, § 311 Anm. 5.

[61] *Ausschußbericht* zu § 311 Abs. 2 AktG, in: *B. Kropff*, AktG, 1965, S. 409.

[62] *H.-G. Koppensteiner*, Kölner Kommentar, 1987, Vorb. § 311 Rdnr. 9.

[63] *B. Kropff*, Geßler, 1975, § 311 Rdnr. 68 f. Eingehend *J. Semler*, Leitung, 1996, Rdnr. 285 ff.

Wertung des Gesetzgebers dahingehend entnehmen, daß unter Umständen sogar eine Schädigung des Eigeninteresses der abhängigen Gesellschaft zulässig wäre.[64]

Schließlich hätte die Ansicht von *P. Hommelhoff* zur Konsequenz, daß das herrschende Unternehmen im faktischen Konzern, wo es unter Umständen nicht alle nachteiligen Maßnahmen ausgleichen muß, besser stünde als im Vertragskonzern, in dem die Vermögensinteressen der außenstehenden Aktionäre durch eine Abfindung oder eine Dividendengarantie gewahrt bleiben. Dann wäre die Regelung der §§ 300 ff. AktG aber sinnlos, da kein herrschendes Unternehmen einen Beherrschungs- oder Gewinnabführungsvertrag abschließen würde, wenn wirtschaftlich gleiche Ergebnisse „billiger" erzielt werden könnten.[65]

Nach alledem bleibt festzuhalten, daß das Eigeninteresse der abhängigen Gesellschaft vor dem Konzerninteresse absoluten Vorrang hat.[66] Das herrschende Unternehmen darf also nur solchen Einfluß auf die abhängige Gesellschaft ausüben, dessen Auswirkungen für die Gesellschaft nicht nachteilig sind oder dessen Nachteile rechtzeitig ausgeglichen werden.

6. Wettbewerbsverbot

Für den herrschenden Gesellschafter einer GmbH wird die Ansicht vertreten, daß er aufgrund seiner mitgliedschaftlichen Förderpflicht analog § 112 Abs. 1 HGB einem Wettbewerbsverbot unterliege. Das wird zum einem mit der gesteigerten Gefahr der mißbräuchlichen Verwendung gesellschaftsinterner Informationen durch den Gesellschafter zugunsten seines Wettbewerbsunternehmens begründet. Zum anderen bestehe die Gefahr, daß der Gesellschafter die Gesellschaft zum Vorteil seines Wettbewerbsunternehmens nachteilig beeinflusse.[67]

Diese Gefahren bestehen auch im faktischen Aktienkonzern. Zwar hat der herrschende Aktionär nicht die Möglichkeit, in der Hauptversammlung einen Beschluß herbeizuführen, der den Vorstand zur Vornahme einer bestimmten Geschäftsführungsmaßnahme anweist. Jedoch kann er faktisch Einfluß auf den Vorstand ausüben, wovon ja auch § 311 Abs. 1 AktG ausgeht. Dieser faktische, von der Hauptversammlung unbemerkte Einfluß des herrschenden Aktionärs kann aber für die beherrschte Gesellschaft noch größere Gefahren hervorrufen

---

[64] Ebenso *H.-G. Koppensteiner*, Kölner Kommentar, 1987, Vorb. § 311 Rdnr. 13 f.

[65] *H.-G. Koppensteiner*, Kölner Kommentar, 1987, Vorb. § 311 Rdnr. 12. Auf diesen Konflikt zwischen den Vorschriften zum Vertragskonzern und zum faktischen Konzern weist auch *L. Strohn*, Verfassung, 1977, S. 2, 16, hin.

[66] Im Ergebnis ebenso *U. Hüffer*, AktG, 1997, § 311 Rdnr. 8; *J. Neuhaus*, Grenzen, 1970, S. 1914; *M. Habersack*, Aktienkonzernrecht, 1998, vor § 311 Rdnr. 7.

[67] Siehe unten II. 10.

als die Herbeiführung von Weisungsbeschlüssen bei der GmbH, deren Durchführung im Falle der Rechtswidrigkeit durch eine Anfechtung verhindert werden kann.

So wird denn auch in der Literatur die Ansicht vertreten, daß das herrschende Unternehmen gegenüber der beherrschten Aktiengesellschaft einem Wettbewerbsverbot unterliege.[68] Und auch für Aufsichtsratsmitglieder wurde jüngst de lege ferenda ein Wettbewerbsverbot gefordert.[69] Gegen ein Wettbewerbsverbot für Aufsichtsratsmitglieder wird jedoch eingewandt, daß es mit den Wertungen des aktienrechtlichen Konzernrechts nicht vereinbar sei, da es die Gründung faktischer Konzern erschwere.[70] Dem ist zuzustimmen. Insbesondere sprechen diese Überlegungen auch gegen ein Wettbewerbsverbot des herrschenden Aktionärs.

So ist nach § 311 Abs. 1 AktG eine konkrete Nachteilszufügung zulässig, soweit der Nachteil nur rechtzeitig ausgeglichen wird. Mit dieser Privilegierung wäre aber ein präventives Wettbewerbsverbot, das solche Nachteilszufügungen schon im Vorfeld verhindern soll, unvereinbar.[71] Überdies würde damit die Absicht des Gesetzgebers konterkariert, der mit der Zulassung des gestreckten Nachteilsausgleichs in § 311 AktG die weitere Durchführung bestehender und zukünftiger Unternehmensverbindungen ermöglichen wollte. Unternehmensverbindungen bestehen aber typischerweise zwischen Wettbewerbern.

Nach alledem darf ein herrschender Aktionär ein Wettbewerbsunternehmen betreiben, und er kann sich in den Aufsichtsrat der Gesellschaft wählen lassen, ohne einem Wettbewerbsverbot unterworfen zu werden.

---

68 *M. Habersack*, Aktienkonzernrecht, 1998, vor § 311 Rdnr. 11. Dagegen jedoch *Uwe H. Schneider*, Wettbewerbsverbot, 1995, S. 367. Siehe auch *M. Lutter*, Treupflichten, 1998, S. 173, nach dem ein Mehrheitsaktionär nicht gegen seine Treupflicht verstößt, wenn er seine Aktien an einen Wettbewerber der Gesellschaft veräußert.

69 Gesetzentwurf des *Bundesrates* zur Steigerung der Effizienz von Aufsichtsräten und zur Begrenzung der Machtkonzentration bei Kreditinstituten infolge von Unternehmensbeteiligungen, BT-Drucksache 13/9716 v. 29.01.1998, Anlage 1, S. 2. Für ein Wettbewerbsverbot de lege lata jüngst *T. E. Abeltshauser*, Leitungshaftung, 1998, S. 364.

70 *Uwe H. Schneider*, Wettbewerbsverbot, 1995, S. 368.

71 Ausführlich zum Verhältnis zwischen § 311 AktG und der Treupflicht des herrschenden Unternehmens, von der das Wettbewerbsverbot eine Ausprägung ist, *W. Zöllner*, Aktienkonzernrecht, 1998, S. 239 ff. Nach *W. Zöllner* (S. 241) kann allerdings die Treupflicht dem herrschenden Unternehmen weitergehende Pflichten auferlegen als die §§ 311 ff. AktG.

7. Das Verbot der Ausnutzung von Geschäftschancen der abhängigen Gesellschaft

Weiterhin wird für das Recht der GmbH die Ansicht vertreten, daß der Gesellschafter einer personalistisch strukturierten GmbH keine Geschäftschancen der Gesellschaft an sich ziehen dürfe. Gleiches gelte für den Gesellschafter einer kapitalistisch strukturierten GmbH, wenn er in der Geschäftsführung tätig sei.

An dieser Stelle soll noch nicht erörtert werden, wann eine Geschäftschance der Gesellschaft zuzurechnen ist. Diese Diskussion soll dort geführt werden, wo sie auch in der Literatur stattfindet, nämlich bei der mitgliedschaftlichen Förderpflicht des Gesellschafters der GmbH. Dort wird gezeigt werden, daß ein Gesellschafter nur solche Geschäftschancen nicht an sich ziehen darf, die ihm aus dem Verband heraus vermittelt wurden. Das ist der Fall, wenn er aufgrund seiner Gesellschafterstellung von der Geschäftschance Kenntnis erlangt hat.[72]

Gleiches gilt für den Aktionär. Auch dieser unterliegt einer mitgliedschaftlichen Förderpflicht bzw. Treupflicht, die ihm eine Schädigung der Gesellschaft verbietet.[73] Insbesondere steht das Verbot der Ausnutzung von Geschäftschancen der Gesellschaft auch nicht der Privilegierung des herrschenden Unternehmens durch § 311 Abs. 1 AktG entgegen. Denn will das herrschende Unternehmen eine Geschäftschance der beherrschten Aktiengesellschaft an sich ziehen, so kann es deren Vorstand veranlassen, auf die Geschäftschance zu verzichten. Mit dem Ausgleich des dadurch entstandenen Nachteils kann das herrschende Unternehmen sodann bis zum Ende des Geschäftsjahres abwarten. Der Vorstand der beherrschten Aktiengesellschaft ist verpflichtet, den Verzicht in den Abhängigkeitsbericht aufzunehmen (§ 312 Abs. 1 Satz 2 AktG).

II. Einstufiger faktischer GmbH-Konzern mit Minderheitsgesellschaftern

1. Die gesetzliche Ausgangslage

Während für die Einflußnahme des herrschenden Unternehmens auf die Geschäftsführung der beherrschten Aktiengesellschaft spezielle Regelungen gelten, ist dies bei einer beherrschten GmbH nicht der Fall. Wie bei ihr die Grenzen der Konzernleitung zu ziehen sind, ist umstritten.

---

72 Siehe unten II. 9.

73 Statt aller *U. Hüffer*, AktG, 1997, § 53a Rdnr. 13 ff.

2. *Meinungsstand*

Einigkeit besteht allerdings darüber, daß der herrschende Gesellschafter einer GmbH nicht analog § 311 Abs. 1 AktG privilegiert wird.[74] Die Herbeiführung eines für die Gesellschaft nachteiligen Gesellschafterbeschlusses kann also nicht damit gerechtfertigt werden, daß der herrschende Gesellschafter bereit ist, den Nachteil bis zum Ende des Geschäftsjahres auszugleichen.

Zum einen scheitert eine analoge Anwendung der §§ 311 ff. AktG daran, daß der Geschäftsführer einer GmbH gegenüber dem herrschenden Unternehmen eine viel schwächere Stellung hat als der Vorstand einer Aktiengesellschaft. Im Gegensatz zum diesem ist er weisungsabhängig und kann jederzeit auch ohne wichtigen Grund abberufen werden. Damit entfällt ein wesentliches Funktionselement der §§ 311 ff. AktG, die davon ausgehen, daß nur ein eigenverantwortlicher Unternehmensleiter gegenüber dem herrschenden Unternehmen in der Lage ist, dessen Einflußnahme effektiv auf ihre Nachteiligkeit zu überprüfen und auf einen angemessen Ausgleich zu bestehen.[75] Auch kennt das GmbH-Recht keine Pflicht zur Erstellung eines Abhängigkeitsbericht, ohne den das finanzielle Ausgleichssystem der §§ 311 ff. AktG nicht sachgerecht durchführbar ist. Selbst wenn man entgegen der h. M.[76] eine solche Pflicht im Wege der Analogie herleiten wollte[77], würde es doch an der obligatorischen Überprüfung des Abhängigkeitsberichts durch den Abschlußprüfer (§ 313 AktG) und den Aufsichtsrat (§ 314 AktG) fehlen.

Zum anderen hätte eine analoge Anwendung der §§ 311 ff. AktG für die Gesellschafter einer GmbH weit einschneidendere Folgen als für die Gesellschafter einer Aktiengesellschaft. Während die Aktionäre meist bloße Kapitalanleger sind, deren Interessen sich vornehmlich auf die Erzielung einer Dividende richten, weisen die Gesellschafter einer GmbH mit ihren ausgeprägten Minderheitsrechten (§§ 50, 51a/b, 61, 66 GmbHG) eher unternehmerische Züge auf. Hinzu

---

[74] *BGH*, Urt. v. 5.6.1975 - II ZR 23/74 („ITT"), in: BGHZ 65, 15, 18 f.; *H.-G. Koppensteiner*, Rowedder, 1997, Anh § 52 Rdnr. 53; *M. Lutter/P. Hommelhoff*, GmbHG, 1995, Anh § 13 Rdnr. 12; *V. Emmerich/J. Sonnenschein*, Konzernrecht, 1997, S. 384; *W. Stimpel*, Innenhaftung, 1986, S. 117; *P. Hommelhoff*, Konzernleitungspflicht, 1982, S. 252 f.; *P. Ulmer*, Gläubigerschutz, 1984, S. 412; *P. Limmer*, Haftungsverfassung, 1992, S. 35; *D. Kleindiek*, Strukturvielfalt, 1991, S. 25. A. M. *U. Bälz*, Unternehmen, 1992, S. 293 f., 294, 296 ff.; *H. Rowedder*, Bestandsschutz, 1986, S. 29. Differenzierend *G. Krieger*, TBB, 1994, nach dem Nachteile so rechtzeitig zu kompensieren sind, daß sie nicht bilanzwirksam werden.

[75] *P. Hommelhoff*, Konzernleitungspflicht, 1982, S. 252; *U. Hüffer*, AktG, 1997, § 311 Rdnr. 51.

[76] *M. Lutter*, Haftung, 1985, S. 1428; *M. Paschke*, Durchgriffshaftung, 1988, S. 202 f.; *H.-D. Assmann*, Gläubigerschutz, 1986, S. 928.

[77] So *H. Rowedder*, Bestandsschutz, 1986, S. 33.

kommt oftmals eine personale Bindung des in der GmbH tätigen Gesellschafters, für den eine Konzernierung nicht nur einen finanzielle, sondern auch eine existenzielle Bedrohung bedeuten kann. Hiergegen kann ihn eine analoge Anwendung der §§ 311 ff. AktG jedoch nicht ausreichend schützen.[78]

Das alles bezieht sich freilich nur auf die Privilegierung des herrschenden Unternehmens. Im übrigen wird eine analoge Anwendung aktienrechtlicher Vorschriften jedoch vereinzelt bejaht.

So vertritt *P. Limmer*[79] die Auffassung, daß aufgrund konzernrechtlicher Leitungsmacht ein Sonderrechtsverhältnis zwischen dem herrschenden Unternehmen und der abhängigen GmbH entstehe. Zur Begründung führt er an, daß der Gesetzgeber mit dem Konzernrecht des AktG ein spezifisches Sonderrecht für die abhängige Aktiengesellschaft geschaffen habe, das das Recht der unabhängigen Aktiengesellschaft ergänze. Mit diesem Sonderrecht entstehe ein Schutzverhältnis, das allein an die konzernspezifische Kategorie der unternehmerischen Leitungsmacht anknüpfe und neben die mitgliedschaftlichen Schutzmechanismen trete.[80] Da nun die Beteiligten in der abhängigen GmbH ebenso schutzbedürftig seien wie die Beteiligten in der abhängigen Aktiengesellschaft, müsse ihnen der gleiche zusätzliche Schutz zukommen.[81] Freilich bedeute dies nicht, daß alle Vorschriften des Aktienkonzernrechts unbesehen auf die GmbH übertragen werden könnten. Vielmehr müßten aus ihnen allgemeine Rechtsprinzipien herausgearbeitet werden, die auch für die GmbH Geltung beanspruchen würden.[82] Ein solches allgemeines Rechtsprinzip sei insbesondere das allgemeine Schädigungsverbot, das in §§ 311, 317 AktG seinen gesetzlichen Ausdruck gefunden habe.[83] Dagegen scheitere die Übertragung des Privilegierungselements dieser Vorschriften an der unterschiedlichen Organisationsstruktur der GmbH.[84] Das herrschende Unternehmen dürfe der abhängigen GmbH also keinen Nachteil zufügen.

Insgesamt legt *P. Limmer* dem herrschenden Unternehmensgesellschafter damit die Sorgfaltspflichten eines ordentlichen und gewissenhaften Geschäfts-

---

78 *M. Lutter/P. Hommelhoff*, GmbHG, 1995, Anh. § 13 Rdnr. 8 f.; *U. Hüffer*, AktG, 1997, § 311 Rdnr. 51; *G. Hueck* u. a., Arbeitskreis GmbH-Reform, 1972, S. 48 f.

79 *P. Limmer*, Haftungsverfassung, 1992, S. 64 ff.

80 *P. Limmer*, Haftungsverfassung, 1992, S. 67.

81 *P. Limmer*, Haftungsverfassung, 1992, S. 67.

82 *P. Limmer*, Haftungsverfassung, 1992, S. 85, 87.

83 *P. Limmer*, Haftungsverfassung, 1992, S. 107 f.; im Ergebnis ebenso *Uwe H. Schneider*, Konzernbildung, 1980, S. 530 f.; *B. Kropff*, Beherrschungsvertrag, 1993, S. 536 ff.; *H. Altmeppen*, Managerhaftung, 1998, S.81, 86.

84 *P. Limmer*, Haftungsverfassung, 1992, S. 108; *B. Kropff*, Beherrschungsvertrag, 1993, S. 539.

leiters auf. Denn der Nachteilsbegriff zeichnet sich gerade dadurch aus, daß ein ordentlicher und gewissenhafter Geschäftsleiter die entsprechende Maßnahme nicht vorgenommen hätte.

Im Gegensatz zu *P. Limmer* will *P. Krebs*[85] die abhängige GmbH nicht nur innerhalb, sondern auch außerhalb des Konzerns vor sorgfaltswidriger Einflußnahme ihrer Gesellschafter schützen. Hierzu beruft er sich auf das Prinzip einer Haftung für sorgfaltswidrige Leitung. Zur Begründung stützt er sich auf eine Gesamtanalogie zu den §§ 309 Abs. 2, 317 Abs. 3, 323 Abs. 1 Satz 2 i. V. m. 309 Abs. 2 AktG, die eine Haftung der gesetzlichen Vertreter des herrschenden Unternehmens gegenüber der abhängigen Aktiengesellschaft für sorgfaltswidrige Weisungen/Veranlassungen statuierten. *P. Krebs*[86] will aus diesen Vorschriften im Wege einer stufenweisen Abstraktion ein allgemeines Prinzip herleiten, wonach derjenige, der eine Gesellschaft leite, gegenüber dieser für sorgfaltswidrige Leitung hafte. Dabei unterscheidet er zwischen legaler und illegaler Leitung.[87] Während jede Ausübung von legaler Leitungsmacht Leitung i. S. d. von ihm vertretenen Prinzips sei, falle die Ausübung von illegaler Leitungsmacht nur dann hierunter, wenn der legal Leitende strukturell nicht mehr in der Lage sei, illegale Leitungsversuche abzuwehren. Letzteres sei insbesondere dann der Fall, wenn der illegal Leitende die gesamte Geschäftsführung an sich gezogen habe.

Für eine herrschende GmbH im Konzern hätte die Anwendung dieses Prinzips folgende Konsequenzen. Da sie über das Weisungsrecht der Gesellschafterversammlung in zulässiger Weise Einfluß auf die Geschäftsführung ihrer Tochtergesellschaft nimmt, ist sie gegenüber dieser legal Leitende. In dieser Eigenschaft würde sie insbesondere für eine sorgfaltswidrige Stimmrechtsabgabe bei Fassung eines Weisungsbeschlusses haften.[88] In den mittelbar beherrschten Gesellschaften kann die herrschende GmbH dagegen nur illegale Leitungsmacht ausüben. Deshalb würde sie gegenüber diesen nur dann für sorgfaltswidrige Leitung einstehen müssen, wenn sie die illegale Leitungsmacht umfassend ausüben würde.

Die h. M. und die Rechtsprechung schließlich suchen einen Schutz der abhängigen GmbH über die Treupflicht des herrschenden Gesellschafters gegenüber seiner Tochtergesellschaft und seinen Mitgesellschaftern.[89]

---

[85] *P. Krebs*, Geschäftsführungshaftung, 1991, S. 232 ff.

[86] *P. Krebs*, Geschäftsführungshaftung, 1991, S. 232 ff.

[87] *P. Krebs*, Geschäftsführungshaftung, 1991, S. 174 ff.

[88] *P. Krebs*, Geschäftsführungshaftung, 1991, S. 252 ff.

[89] Siehe etwa *G. Hueck*, Baumbach/Hueck, 1996, § 13 Rdnr. 21 mit zahlreichen Nachweisen.

3. Stellungnahme

Die Ansichten von *P. Limmer* und *P. Krebs* vermögen nicht zu überzeugen.[90] Die von ihnen herangezogenen aktienrechtlichen Vorschriften können eine Haftung des herrschenden Gesellschafters einer GmbH für sorgfaltswidrige Leitung nicht begründen. Dagegen sprechen die Strukturunterschiede zwischen Aktiengesellschaft und GmbH.

So leitet der Vorstand die Aktiengesellschaft unter eigener Verantwortung, § 76 Abs. 1 AktG. Zwar geht das Gesetz davon aus, daß der private Großaktionär ebenso wie der Unternehmensaktionär auf die Unternehmensleitung Einfluß nimmt. Aber erstens besteht keine Folgepflicht des Vorstands und zweitens bleibt dieser bei seiner Entscheidung, ob er die von dem herrschenden Gesellschafter gewollte Maßnahme ausführt, an die Sorgfalt eines ordentlichen und gewissenhaften Geschäftsleiters gebunden. Der Durchsetzung dieses Sorgfaltsmaßstabes bei allen unternehmerischen Entscheidungen dienen die §§ 311 ff. AktG. Mit anderen Worten begründen die §§ 311 ff. AktG kein Schädigungsverbot, sondern knüpfen an ein bestehendes Schädigungsverbot an und wollen (nur) dessen effektive Einhaltung gewährleisten.

Folgt man dem, so ist eine analoge Anwendung der zitierten aktienrechtlichen Vorschriften nicht gerechtfertigt.[91] Denn wesentliches Strukturmerkmal der GmbH ist das Recht der Gesellschafterversammlung zur Erteilung von Weisungen an den Geschäftsführer. Anders als bei der Aktiengesellschaft liegt die Verwaltung des Gesellschaftsvermögens also nicht allein in den Händen des Geschäftsführers, sondern auch in den Händen der Gesellschafter. Für deren Einflußnahme hat das GmbHG aber keinen ausdrücklichen Sorgfaltsmaßstab normiert. Das schließt es freilich nicht aus, daß die Gesellschafter trotzdem Sorgfaltspflichten bei einer Einflußnahme auf die Gesellschaft treffen. Diese ergeben sich jedoch nicht aus einer analogen Anwendung aktienrechtlicher Vorschriften, sondern lassen sich nur aus den Wertungen des GmbHG herleiten. Ist aber erst einmal eine Pflicht des Gesellschafters zur Unterlassung nachteiliger Einflußnahme gefunden, so bedarf es, wie noch zu zeigen sein wird, nicht mehr der Anwendung aktienrechtlicher Vorschriften, um auch eine Haftung des Gesellschafters wegen Verletzung dieser Pflicht zu begründen.

Damit bleibt nur noch die Treupflicht des herrschenden Gesellschafters gegenüber seiner Tochter-GmbH und den Mitgesellschaftern übrig. Gegenüber der analogen Anwendung aktienrechtlicher Vorschriften hat sie den Vorteil, daß ihre Konkretisierung die Wertungen des GmbH-Gesetzes in sich aufnehmen und

---

[90] So auch zur Ansicht von *P. Krebs H. Ziemons*, Haftung, 1996, S. 59 ff.

[91] Ebenso *H. Altmeppen*, GmbHG, 1997, Anh. § 13 Rdnr. 114.

damit voll zur Geltung bringen kann. Entgegen der Terminologie der h. M. sollen die Pflichten des herrschenden Gesellschafters gegenüber seiner Gesellschaft jedoch nicht als Treupflichten, sondern als mitgliedschaftliche Förderpflichten bezeichnet werden. Hierdurch wird zum einen der Unterschied zwischen den Pflichten des herrschenden Gesellschafters gegenüber der Gesellschaft auf der einen Seite und seinen Treupflichten gegenüber seinen Mitgesellschaftern auf der anderen Seite auch sprachlich zum Ausdruck gebracht. Und zum anderen weist die Bezeichnung als mitgliedschaftliche Förderpflicht auf ihre Rechtsgrundlage hin.

4. Die mitgliedschaftliche Förderpflicht und die Treupflicht

Rechtsgrundlage der mitgliedschaftlichen Förderpflicht ist § 705 BGB. Dort heißt es, daß sich die Gesellschafter einer Gesellschaft bürgerlichen Rechts im Gesellschaftsvertrag verpflichten, „die Erreichung eines gemeinsamen Zweckes in der durch den Vertrag bestimmten Weise zu fördern". Das gilt jedoch nicht nur für die Gesellschafter einer Gesellschaft bürgerlichen Rechts, sondern kraft Verweisung auch für die Gesellschafter einer OHG (§ 105 Abs. 2 HGB) und Kommanditgesellschaft (§§ 161 Abs. 2, 105 Abs. 2 HGB). Darüber hinaus gilt dies aber auch für die Mitglieder von Kapitalgesellschaften und allen sonstigen privatrechtlichen Personenverbänden. Denn auch diese Verbände sind, abgesehen von der hybriden Form der Einmann-Gesellschaft, private Zweckgemeinschaften. Der Unterschied zwischen beiden Verbandstypen besteht lediglich darin, daß die Mitglieder einer Personengesellschaft in der Regel enger an den Gesellschaftszweck gebunden sind als die Mitglieder einer Kapitalgesellschaft, einer Genossenschaft oder eines Vereins[92] und daß die Förderpflicht primär gegenüber der juristischen Person und nicht den Gesellschaftern besteht[93]. Damit läßt sich aus § 705 BGB für das gesamte Recht der privaten Verbände die Grundaussage entnehmen, daß Inhalt jeder Mitgliedschaft die Pflicht zur Förderung des gemeinsamen Zwecks ist.[94]

Für den Gesellschafter einer GmbH ist diese allgemeine Förderpflicht spätestens seit der ITT-Entscheidung des *Bundesgerichtshofs*[95] im Ergebnis aner-

---

[92] *M. Lutter*, Mitgliedschaft, 1980, S. 95.

[93] *M. Winter*, Treubindungen, 1988, S. 86.

[94] *M. Lutter*, Mitgliedschaft, 1980, S. 102 f.; *ders.*, Treupflichten, 1998, S. 179; *H. Ziemons*, Haftung, 1996, S. 80. Speziell für die GmbH *P. Limmer*, Haftungsverfassung, 1992, S. 52; *M. Ivens*, Konkurrenzverbot, 1987, S. 146. Siehe auch *T. Raiser*, Hachenburg, 1989, § 14 Rdnr. 52.

[95] *BGH*, Urt. v. 5.6.75 - II ZR 23/74, in: BGHZ 65, S. 15, 18.

kannt.[96] Insbesondere seit der Girmes-Entscheidung des *Bundesgerichtshofs*[97] dürfte feststehen, daß nicht nur Gesellschafter einer personalistisch, sondern auch einer kapitalistisch strukturierten Gesellschaft einer mitgliedschaftlichen Förderpflicht unterliegen.[98] Freilich wird sie dort wie auch zumeist in der Literatur[99] nicht als Förderpflicht, sondern als Treupflicht bezeichnet.[100] Dies rührt daher, daß ihre Rechtsgrundlage nicht in § 705 BGB, sondern in § 242 BGB gesehen wird.[101] Auch wenn hiermit in der Sache keine unterschiedlichen Ergebnisse verbunden sind, so trägt doch eine dogmatische Rückbindung der Pflichten des Gesellschafters an § 705 BGB und ihre Bezeichnung als Förderpflichten zu einem besseren Verständnis der Pflichten des Gesellschafters bei.[102] Denn erstens gehen die mitgliedschaftlichen Förderpflichten oft weiter als die Sorgfaltspflichten und Mißbrauchsverbote, wie sie im Rahmen von § 242 BGB bekannt sind.[103] Zweitens weist ihre Herleitung aus § 705 BGB auf ihren spezifisch gesellschaftsrechtlichen Bezug hin und gibt zugleich die Richtung an, in der die für ihre Konkretisierung ausschlaggebenden gesetzlichen Wertungen zu suchen sind.[104] Und drittens schließlich kann gegen § 705 BGB nicht eingewandt werden, daß sich aus dieser Vorschrift nur aktive Förderpflichten ableiten ließen. Vielmehr können sich aus ihr auch passive Förderpflichten, nämlich Unterlassungs- und Loyalitätspflichten ergeben. Denn materiell macht es keinen Unterschied, ob ein Gesellschafter an einer Maßnahme der Gesellschaft aktiv mitzuwirken hat oder aber eine schädliche Einflußnahme auf die Gesellschaft zu unterlassen hat.[105]

Freilich ist mit der mitgliedschaftlichen Förderpflicht nur ein Rahmen gefunden, der mit Inhalt gefüllt werden muß. Dieser hängt nicht nur von der Rechtsform der Gesellschaft sowie von Art und Inhalt des gemeinsamen Zieles ab. Er

---

[96] Statt aller *G. Hueck*, Baumbach/Hueck, 1996, § 13 Rdnr. 21 mit zahlreichen Nachweisen.

[97] *BGH*, Urt. v. 20.3.1995 - II ZR 205/94, in: WM 1995, S. 882, 884 f. („Girmes").

[98] *T. Raiser*, Hachenburg, 1989, § 14 Rdnr. 53.

[99] Siehe nur die Monographie von *M. Winter*, Treubindungen, 1988, S. 1 ff.

[100] Zur Entwicklung der Treupflicht in Literatur und Rechtsprechung siehe *U. Immenga*, Treuepflichten, 1992, S. 189 ff.

[101] *H. Rowedder*, GmbHG, 1997, § 13 Rdnr. 13. Siehe aber auch schon *W. Zöllner*, Schranken, 1963, S. 335 ff. Für dic AG: *J. Hennrichs*, Treupflichten, 1995, S. 228 ff.

[102] *M. Lutter*, Mitgliedschaft, 1980, S. 103 ff.

[103] *U. Immenga*, Treuepflichten, 1992, S. 190.

[104] *P. Limmer*, Haftungsverfassung, 1992, S. 52.

[105] *M. Lutter*, Mitgliedschaft, 1980, S. 110.

wird auch bestimmt von der Realstruktur des betreffenden Verbandes.[106] Je stärker dessen personalistische Züge sind, desto ausgeprägter sind die Förderpflichten seiner Gesellschafter.

M. Lutter[107] hat dies wie folgt begründet:

„Richtig ist demgegenüber, die Rechtsformen als *Modelle* einer typgerechten Ordnung zu sehen. So betrachtet ist das Recht unserer privaten Verbände über alle Rechtsformen hinweg eine Einheit mit modellhaften Einzelausprägungen. Entspricht daher die Realstruktur eines Verbandes nicht der von ihr gewählten Rechtsform, so ist das innerhalb sehr weiter Grenzen nicht etwa rechtswidrig. Geht man aber von der hier postulierten Einheit des Korporationsrechts aus, so ist es durchaus systemgerecht, in solchen Fällen von dem Normengefüge der an sich typgerechten Rechtsform Aussagen für die betreffende Korporation zu übernehmen und auf diese zur Auslegungshilfe oder zur Lückenfüllung anzuwenden: vor allem dort also, wo dispositives Korporationsrecht zu unzureichenden Ergebnissen führt oder generelle Normen, wie eben die allgemeine Förderpflicht, zur Ausfüllung anstehen".

Für die Treupflicht der Gesellschafter untereinander kann § 705 BGB als Rechtsgrundlage jedoch nicht herangezogen werden. Dagegen spricht, daß sich die GmbH mit Eintragung in das Handelsregister gegenüber ihren Gründern verselbständigt (§ 11 Abs. 1 GmbHG). Das hat zur Folge, daß ab sofort nur noch gesellschaftsvertragliche Rechtsbeziehungen zwischen der GmbH und ihren Gesellschaftern bestehen, nicht mehr aber unter den einzelnen Gesellschaftern selbst.[108] Trotzdem hat der *Bundesgerichtshof*[109] in seiner ITT-Entscheidung unter Zustimmung der Literatur[110] Treupflichten zwischen den Gesellschaftern einer GmbH anerkannt. Und das zu Recht. Denn zwischen den Gesellschaftern besteht eine Sonderverbindung, an die § 242 BGB Sorgfaltspflichten knüpft. Diese Sonderverbindung ergibt sich aus der Tatsache, daß die Gesellschafter durch Einflußnahme auf die Geschäftsführung die gesellschaftsbezogenen Interessen ihrer Mitgesellschafter beeinträchtigen können. Diese erhöhten Einwirkungsmöglichkeiten auf fremde Interessen verlangen aber als Gegengewicht eine Pflicht zur Rücksichtnahme auf eben diese Interessen. Damit besteht zwischen den Gesellschaftern eine rechtliche Sonderverbindung,

---

106 *M. Lutter*, Mitgliedschaft, 1980, S. 105; *H. Henze*, Treupflicht, 1996, S. 492; *BGH*, Urt. v. 5.6.75 - II ZR 23/74, in: BGHZ 65, S. 15, 19 („ITT"); *BGH*, Urt. v. 1.2.88 - II ZR 75/87, in: BGHZ 103, S. 184, 195 („Linotype").

107 *M. Lutter*, Mitgliedschaft, 1980, S. 107. Ebenso z. B. *H. Henze*, Treupflichten, 1998, S. 186.

108 Siehe dazu *M. Lutter*, Mitgliedschaft, 1980, S. 97 f. A. M. *T. Raiser*, Hachenburg, 1989, § 13 Rdnr. 7, nach dem die gesellschaftsvertraglichen Beziehungen zwischen den Gesellschaftern vor Eintragung der GmbH auch nach deren Eintragung weiterbestehen. In diese Richtung auch *T. Verhoeven*, Innenrecht, 1978, Rdnr. 159.

109 *BGH*, Urt. v. 5.6.75 - II ZR 23/74, in: BGHZ 65, S. 15, 18.

110 Grundlegend *U. Immenga*, Kapitalgesellschaft, 1970, S. 261 ff.; ders., Treuepflichten, 1992, S. 193. Ablehnend *W. Flume*, Juristische Person, 1983, S. 270.

die ihr Verhältnis untereinander von dem allgemeinen „Jedermann-Pflichtverhältnis" des Deliktsrechts abhebt und dem Grundsatz von Treu und Glauben unterstellt, § 242 BGB.[111]

Freilich kommt bei einem Eingriff in gesellschaftsbezogene Interessen die Treupflicht zwischen den einzelnen Gesellschaftern untereinander nur dann zum Tragen, wenn nicht gleichzeitig die Interessen der Gesellschaft verletzt sind.[112] Denn ist letzteres der Fall, gehen die Interessen der Gesellschafter in den Interessen der Gesellschaft auf.[113] Das wird bei konzernbezogenen Maßnahmen des herrschenden Gesellschafters regelmäßig der Fall sein. Deshalb sollen an dieser Stelle nur diejenigen Grenzen der Konzernleitung untersucht werden, die durch die mitgliedschaftliche Förderpflicht der herrschenden GmbH gezogen werden.

5. Die Verfolgung von Sondervorteilen

Einigkeit besteht darüber, daß die Gesellschafter bei ihrer Einflußnahme auf ihre Geschäftsführung keine Sondervorteile zum Schaden der übrigen Gesellschafter verfolgen dürfen.[114] Das ist in § 243 Abs. 2 AktG für die Stimmrechtsausübung in der Aktiengesellschaft ausdrücklich normiert, gilt aber für alle Verbände. Denn ein solches Verhalten ist ein eklatanter Widerspruch zu der Verpflichtung der Gesellschafter, den gemeinsamen Gesellschaftszweck zu fördern. Zu Recht stellt *W. Flume*[115] daher fest, daß die Verfolgung von Sondervorteilen die 'Todsünde' der Stimmrechtsausübung ist. Gleiches gilt aber nicht nur für die Stimmrechtsausübung, sondern auch für eine Einflußnahme des Gesellschafters außerhalb der Gesellschafterversammlung.

6. Die Autonomie der Gesellschaft

Weiterhin bedeutet die Bindung an den Gesellschaftszweck, daß der Gesellschafter die Gesellschaft nicht ihrer Autonomie berauben darf. Denn mangels

---

[111] *J. Hennrichs*, Treupflichten, 1995, S. 238; *M. Winter*, Treubindungen, 1988, S. 69 ff.; *BGH*, Urt. v. 5.6.75 - II ZR 23/74, in: BGHZ 65, S. 15, 19 („ITT"); *BGH*, Urt. v. 1.2.88 - II ZR 75/87, in: BGHZ 103, S. 184, 195 („Linotype").

[112] *E. Rehbinder*, Treuepflichten, 1976, S. 391 f.; *H. Henze*, Treupflicht, 1996, S. 493.

[113] *M. Winter*, Treubindungen, 1988, S. 88; *P. Limmer*, Haftungsverfassung, 1992, S. 57. A. M. *T. Raiser*, Hachenburg, 1989, § 14 Rdnr. 54.

[114] *K. Schmidt*, Scholz, 1995, § 45 Rdnr. 108; *M. Lutter/P. Hommelhoff*, GmbHG, 1995, Anh. § 47 Rdnr. 52; *H. Konzen*, Geschäftsführung, 1989, S. 2982; *K. Vorwerk/J. Wimmers*, Treubindung, 1998, S. 720.

[115] *W. Flume*, Juristische Person, 1983, S. 213.

anderer Anhaltspunkte ist die Satzung einer GmbH so auszulegen, daß der Gesellschaftszweck selbständig, d. h. ohne gleichzeitige Unterworfenheit unter die Ziele eines anderen Unternehmens zu verfolgen ist.[116] Verstößt der Gesellschafter gegen diese Pflicht und veranlaßt er z. B. die Ausgliederung einer wesentlichen Unternehmensfunktion auf sich selbst oder eine Schwestergesellschaft,[117] so ist er verpflichtet, den Eingriff rückgängig zu machen.[118] Und das gilt selbst dann, wenn der Gesellschaft kein Vermögensschaden entstanden ist.[119]

7. *Stimmrechtsausübung und Sorgfaltspflichten*

Nimmt der Gesellschafter im Rahmen seines Stimmrechts Einfluß auf die Geschäftsführung, so ist streitig, ob er bestimmte Sorgfaltsanforderungen zu beachten hat.

Das wird von einem Teil der Literatur grundsätzlich verneint.[120] Nach einer anderen Ansicht ist eine Stimmabgabe dagegen nichtig, wenn ein Gesellschafter bei Ausübung seines Stimmrechts sein unternehmerisches Ermessen überschritten habe. Komme ein Verschulden des Gesellschafters hinzu, mache er sich außerdem schadensersatzpflichtig. Welche Maßstäbe für die Beurteilung von Nichtigkeit der Stimmabgabe und Verschulden des Gesellschafters zugrunde zu legen seien, ist freilich umstritten. Das Meinungsspektrum reicht bei Beurteilung der Nichtigkeit der Stimmabgabe von § 708 BGB[121] über § 276 BGB[122] bis zu § 43 GmbHG[123]. Entsprechend groß ist die Meinungsvielfalt zum Verschulden.[124]

---

[116] *W. Zöllner*, Baumbach/Hueck, 1996, GmbH-KonzernR Rdnr. 53; *P. Ulmer*, Hachenburg, 1989, § 1 Rdnr. 9; *BGH*, Beschl. v. 24.10.1988 - II ZB 7/88, in: BGHZ 105, 324, 331. Zur Zulässigkeit der satzungsmäßig abhängig gegründeten GmbH siehe *S. Beinert*, Konzernhaftung, 1995, S. 38 ff.

[117] Die Ausgliederung einer wesentlichen Unternehmensfunktion auf eine von der ausgliedernden Gesellschaft beherrschten Gesellschaft hat dagegen keinen Autonomieverlust zur Folge.

[118] *V. Emmerich*, Scholz, 1993, Anhang Konzernrecht Rdnr. 187.

[119] *W. Zöllner*, Baumbach/Hueck, 1996, GmbH-KonzernR Rdnr. 55.

[120] *M. Lutter/P. Hommelhoff*, GmbHG, 1995, Anh. § 47 Rdnr. 52; *Uwe H. Schneider*, Scholz, 1993, § 43 Rdnr. 16. Siehe auch *H.-G. Koppensteiner*, Rowedder, 1997, § 43 Rdnr. 58 und § 47 Rdnr. 110 („breiter Ermessensspielraum“).

[121] *H. Rowedder*, GmbHG, 1997, § 13 Rdnr. 13; *H.-G. Koppensteiner*, Rowedder, 1997, § 43 Rdnr. 59.

[122] *W. Zöllner*, Schranken, 1963, S. 326; ders., Baumbach/Hueck, 1996, Anh. § 47 Rdnr. 49.

[123] *U. Immenga*, Anfechtungsklage, 1973, S. 9; *M. Winter*, Treubindungen, 1988, S. 109; *P. Kleindiek*, Strukturvielfalt, 1991, 27; *H. Ziemons*, Haftung, 1996, S. 164. Eben-

Richtigerweise ist zu unterscheiden: Beschränkt sich der Gesellschafter auf die Wahrnehmung seiner Mitverwaltungsrechte und greift er nur vereinzelt in die Geschäftsführung ein, so treffen ihn keine Sorgfaltspflichten. Das ergibt sich aus der Stellung der Gesellschafterversammlung als oberstem Willensbildungsorgan der Gesellschaft. In dieser Eigenschaft ist sie berufen, die Verbandsautonomie der Gesellschaft zu verwirklichen. Ebenso wie die Privatautonomie berechtigt aber auch die Verbandsautonomie zu törichten Entscheidungen.[125] Damit wäre aber eine Bindung der Gesellschafter an Sorgfaltspflichten, die ihren Entscheidungsspielraum gerade einengen, unvereinbar.[126]

Freilich könnte man erwägen, den Gesellschafter einer GmbH wenigstens für die eigenübliche Sorgfalt einstehen zu lassen. Das ist bei dem Gesellschafter einer OHG gemäß §§ 105 Abs. 2 HGB, 708 BGB der Fall. Deshalb stellt sich die Frage, ob die §§ 105 Abs. 2 HGB, 708 BGB auf den Gesellschafter einer GmbH analog anwendbar sind. Das ist indes zu verneinen. Es fehlt an der Vergleichbarkeit des geregelten und des zu regelnden Sachverhalts. Denn im Unterschied zum Recht der GmbH besteht bei der OHG Selbstorganschaft. Das bedeutet, daß bei der OHG die Geschäfte der Gesellschaft von den Gesellschaftern selbst und nicht wie bei der GmbH von einem besonderen Organ, nämlich dem Geschäftsführer, geführt werden. Sind aber die Gesellschafter einer OHG gegenüber ihren Mitgesellschaftern zur Geschäftsführung verpflichtet, so treffen sie auch Sorgfaltspflichten. Die Gesellschafter einer GmbH sind dagegen nicht zur Geschäftsführung verpflichtet. Und sie üben auch keine Geschäftsführung aus, solange sie sich auf die Wahrnehmung ihrer Mitverwaltungsrechte und vereinzelte Eingriffe in die Geschäftsführung beschränken. Vielmehr verwirklichen sie dann nur die der GmbH als Körperschaft zustehende Verbandsautonomie. Deshalb ist es nicht gerechtfertigt, ihre Entscheidungen an Sorgfaltspflichten zu binden.

---

so *J. Wilhelm*, Rechtsform, 1981, S. 354 f., für diejenigen Gesellschafter, die geschäftsführend, leitend oder beherrschend die Geschicke der GmbH bestimmen.

124 Für § 43 GmbHG: *T. Raiser*, Hachenburg, 1989, § 14 Rdnr. 61; *E. Rehbinder*, Treuepflichten, 1976, S. 392. Dagegen: *LG Bremen*, Urt. v. 19.11.1997 - 4 O 1073/96, in: ZIP 1998, S. 561, 565; *U. Stein*, Organ, 1984, S. 154. Differenzierend *H.-G. Koppensteiner*, Rowedder, 1997, § 43 Rdnr. 59.

125 *W. Flume*, Juristische Person, 1983, S. 212. Ebenso *K. Schmidt*, Scholz, 1995, § 47 Rdnr. 28, 30.

126 *W. Flume*, Juristische Person, 1983, S. 189, 212. Das erkennt auch das, *OLG Düsseldorf*, Urt. v. 14.3.1996 - 6 U 119/94, in: GmbHR 1996, S. 689, 694, wenn es die gerichtliche Nachprüfbarkeit von Gesellschafterentscheidungen auf die Frage beschränkt, ob die Gesellschafter von ihrem Ermessen überhaupt Gebrauch gemacht haben, ob sie die Grenzen des Ermessens eingehalten haben und ob sie von dem Ermessenszweck entsprechend Gebrauch gemacht haben.

Dieses aus dem strukturellen Unterschied zwischen GmbH und OHG abgeleitete Ergebnis läßt sich schließlich durch materielle Erwägungen untermauern. Es war gerade Absicht des Gesetzgebers, eine körperschaftlich strukturierte Gesellschaft und nicht eine Personengesellschaft mit beschränkter Haftung zu schaffen. Dahinter stand der Gedanke, den Kreis möglicher Gesellschafter weiter zu ziehen als das bei einer OHG mit beschränkter Haftung möglich gewesen wäre.[127] Denn der Kreis potentieller Gesellschafter hätte bei einer OHG mit beschränkter Haftung durch das Erfordernis der Selbstorganschaft, also durch die Pflicht der Gesellschafter zur Geschäftsführung, begrenzt werden müssen.[128] Andernfalls wäre im Rechtsverkehr ein falscher Eindruck von der neuen Rechtsform entstanden. Nach Auffassung des Rechtsverkehrs nämlich stehen bei der OHG die Gesellschafter als individuelle Träger des Unternehmens im Vordergrund; und zwar deshalb, weil sie die Geschäfte selbst führen.[129] Die neue Rechtsform sollte dagegen auch solchen Personen die Teilnahme ermöglichen, die - ohne zur Geschäftsführung verpflichtet zu sein - die Geschicke der Gesellschaft mitbestimmen können.[130] Und all das sollte bei gleichzeitiger Gewährung einer Haftungsbeschränkung erfolgen. Diese Haftungsbeschränkung darf aber nicht dadurch umgangen werden, daß gleichsam durch die Hintertür der Sorgfaltspflichten eine Haftung der Gesellschafter für unternehmerische Fehlentscheidungen wieder eingeführt wird. Zudem bestünde die Gefahr, daß die Gesellschafter in Anbetracht einer drohenden Haftung von ihren Mitverwaltungsrechten erst gar keinen Gebrauch machen.[131]

Nach alledem verstößt eine Stimmabgabe nur dann gegen die mitgliedschaftliche Förderpflicht, wenn der Gesellschafter mit ihr einen Sondervorteil verfolgt, die Autonomie der Gesellschaft verletzt oder seine Mitgesellschafter bewußt schädigen will. Die Stimmabgabe ist in diesem Fall nichtig, und der Gesellschafterbeschluß ist anfechtbar.[132] Unter denselben Voraussetzungen haftet der Gesellschafter auf Schadensersatz. Das alles gilt allerdings nur dann, wenn sich der Gesellschafter auf die Wahrnehmung seiner Mitverwaltungsrechte beschränkt und nur vereinzelt in die Geschäftsführung eingreift. Geht seine

---

[127] Amtliche Ausgabe des Entwurfs eines GmbHG nebst Begründung und Anlagen, Berlin: Franz Vahlen, 1891, S. 32.

[128] Amtliche Ausgabe des Entwurfs eines GmbHG nebst Begründung und Anlagen, Berlin: Franz Vahlen, 1891, S. 32.

[129] Amtliche Ausgabe des Entwurfs eines GmbHG nebst Begründung und Anlagen, Berlin: Franz Vahlen, 1891, S. 34.

[130] Amtliche Ausgabe des Entwurfs eines GmbHG nebst Begründung und Anlagen, Berlin: Franz Vahlen, 1891, S. 34, 45.

[131] *M. Lutter/P. Hommelhoff*, GmbHG, 1995, § 43 Rdnr. 46; *Uwe H. Schneider*, Scholz, 1993, § 43 Rdnr. 16.

[132] *W. Zöllner*, Baumbach/Hueck, 1996, Anh. § 47 Rdnr. 54a.

Einflußnahme darüber hinaus, so ändern sich zwar nicht die Pflichten, die ihm als Gesellschafter obliegen. Es kommen aber aus einem anderen Rechtsgrund weitere Pflichten hinzu. Davon soll jedoch erst später gehandelt werden.

8. *Einflußnahme außerhalb der Gesellschafterversammlung*

Nimmt der Gesellschafter außerhalb der Gesellschafterversammlung, also nicht durch Ausübung seines Stimmrechts, Einfluß auf die Geschäftsführung, so könnte zweifelhaft sein, ob die mitgliedschaftliche Förderpflicht überhaupt berührt ist.

a) Verletzung der gesellschaftsinternen Zuständigkeitsordnung

Für eine Verletzung der mitgliedschaftlichen Förderpflicht spricht, daß sich die Gesellschafter verpflichtet haben, den gemeinsamen Zweck „in der durch den [Gesellschafts-] Vertrag bestimmten Weise zu fördern" (§ 705 BGB). Hierzu gehört auch die Einhaltung der gesellschaftsinternen Zuständigkeitsordnung.[133]

Allerdings wird im Recht der Personengesellschaften die Ansicht vertreten, daß ein geschäftsführender Gesellschafter, der schuldhaft seine Geschäftsführungsbefugnis überschreite, nicht mehr in „Erfüllung einer gesellschaftlichen Pflicht" handle, sondern als Geschäftsführer ohne Auftrag.[134] Das habe zur Folge, daß für die Ausführung des Geschäfts nicht der milde Haftungsmaßstab des § 708 BGB, sondern der strengere des § 276 BGB gelte. Ebenso könnte man bei dem GmbH-Gesellschafter argumentieren, daß seine direkte Einflußnahme auf die Geschäftsführung nicht mehr zur Erfüllung seiner mitgliedschaftlichen Förderpflicht gehöre und er deshalb wie ein Geschäftsführer ohne Auftrag hafte[135].

Indes ist schon der dargestellten Meinung zur Haftung des Personengesellschafters nicht zu folgen. Erstens ist es nicht richtig, daß die Haftung aus Geschäftsführung ohne Auftrag strenger ist als die aus Gesellschaftsvertrag. Denn bei letzterer muß sich das Verschulden nur auf die Verletzung der gesellschaftsinternen Zuständigkeitsordnung beziehen, nicht aber auf den durch die Ge-

---

133 Im Ergebnis ebenso *T. Verhoeven*, Innenrecht, 1978, Rdnr. 228, der allerdings auf den Schutz der Gläubiger und zukünftigen Gesellschafter abstellt, die wissen sollen, wer in der Gesellschaft die Leitungsmacht ausübt.

134 *A. Hueck*, OHG, 1971, S. 143. Ebenso *K. Hopt*, HGB, 1995, § 114 Rdnr. 15; *P. Ulmer*, Münchener Kommentar, 1986, § 708 Rdnr. 10.

135 So im Ergebnis *W. Flume*, Juristische Person, 1983, S. 88 f.

schäftsführung entstandenen Schaden.[136] Mit anderen Worten reicht es für einen Schadensersatzanspruch aus, daß der Gesellschafter - gemessen am Sorgfaltsmaßstab des § 708 BGB - seine Geschäftsführungsbefugnis überschritten hat.[137] Die Gesellschaft ist dann so zu stellen, wie sie stehen würde, wenn der Gesellschafter die Weisung nicht erteilt hätte, § 249 BGB. Bei dem Schadensersatzanspruch aus § 678 BGB ist hingegen noch erforderlich, daß die Geschäftsführung nicht dem mutmaßlichen Willen der übrigen Gesellschafter entsprach und der Gesellschafter dies bei Anwendung der im Verkehr erforderlichen Sorgfalt erkennen mußte.

Zweitens ließe sich schwerlich begründen, woraus der Gesellschaft gegen den geschäftsführenden Gesellschafter ein Anspruch auf Unterlassung der Verletzung der gesellschaftsinternen Zuständigkeitsordnung zustehen soll, wenn man eine Verletzung der mitgliedschaftlichen Förderpflicht ablehnt. Ein entsprechender Unterlassungsanspruch wird aber allgemein bejaht.[138]

Drittens schließlich überzeugt die dargestellte Meinung auch in ihrer Begründung nicht. Denn Inhalt der Geschäftsführungspflicht des Gesellschafters einer Personengesellschaft ist es, bei der Geschäftsführung die durch Gesetz und Gesellschaftsvertrag gezogenen Grenzen einzuhalten. Überschreitet der Gesellschafter seine Geschäftsführungsbefugnis, so ist zwar richtig, daß er nicht mehr in Erfüllung seiner Geschäftsführungspflichten handelt. Richtig ist aber auch, daß er gerade das Gegenteil tut, nämlich sie verletzt. Deshalb ist eine vertragliche Haftung zu bejahen.[139]

Nach alledem handelt auch der Gesellschafter einer GmbH nicht als Geschäftsführer ohne Auftrag, sondern verletzt seine gesellschaftsvertraglichen Pflichten, wenn er den Geschäftsführer außerhalb der Gesellschafterversammlung zu einer Maßnahme anweist. Hieraus folgt zugleich, daß der Gesellschaft ein entsprechender Unterlassungsanspruch zusteht.[140] Und hieraus folgt außer-

---

[136] Ebenso für die Verletzung der Pflicht zur Beachtung des Unternehmensgegenstandes: *J. Tieves*, Unternehmensgegenstand, 1998, S. 335 f. Ebenso für die Verletzung der gesellschaftsinternen Zuständigkeitsordnung durch den Geschäftsführer einer GmbH: *Uwe H. Schneider*, Scholz, 1993, § 43 Rdnr. 57.

[137] *BGH*, Urt. v. 4.11.1996 - II ZR 48/95, in: NJW 1997, S. 314. A. M. *U. Stein*, Organ, 1984, S. 182, die zusätzlich „ein konkretes inhaltlich treuwidriges Verhalten bei der Wahrnehmung von Geschäftsführungsaufgaben zum Schaden der Gesellschaft oder der Minderheit“ fordert.

[138] *K. Schmidt*, Gesellschaftsrecht, 1997, S. 1389 f.

[139] Wie hier *R. Fischer*, HGB, 1973, § 116 Anm. 29; *BGH*, Urt. v. 11.1.1988 - II ZR 192/87, in: WM 1988, S. 968, 970; *BGH*, Urt. v. 4.11.1996 - II ZR 48/95, in: NJW 1997, S. 314. Ebenso für den Geschäftsführer, der seine Geschäftsführungsbefugnis überschreitet: *BGH*, Urt. v. 12.6.1989 - II ZR 334/87, in: ZIP 1989, S. 1390, 1393.

[140] Im Ergebnis ebenso *W. Flume*, Juristische Person, 1983, S. 128.

dem, daß der Gesellschafter zum Schadensersatz verpflichtet ist, wenn der Gesellschaft durch seine Einflußnahme ein Schaden entsteht. Insbesondere ist auch hier kein Raum für eine Fahrlässigkeitshaftung des Gesellschafters. Denn eine Einflußnahme setzt ebenso wie eine Veranlassung im Sinne von § 311 Abs. 1 AktG schon begrifflich ein vorsätzliches Handeln voraus. Auch muß sich der Vorsatz nur auf die Verletzung der gesellschaftsinternen Zuständigkeitsordnung, nicht aber auf den entstanden Schaden beziehen.

b) GoA als Rechtfertigungsgrund

Stellt nach alledem eine Einflußnahme auf die Geschäftsführung außerhalb der Gesellschafterversammlung eine Verletzung der mitgliedschaftlichen Förderpflicht dar und nicht eine Geschäftsführung ohne Auftrag, so können trotzdem die Regeln über die Geschäftsführung ohne Auftrag im Rahmen eines Schadensersatzanspruchs noch an Bedeutung gewinnen. So schließt nach zutreffender Meinung eine berechtigte Geschäftsführung ohne Auftrag die Rechtswidrigkeit einer Pflichtverletzung aus.[141] Mit anderen Worten handelt ein Gesellschafter nicht rechtswidrig, wenn das angewiesene Geschäft dem Interesse und dem wirklichen oder mutmaßlichen Willen der Gesellschaft entspricht. Bei Bestimmung des mutmaßlichen Willens der Gesellschaft ist allerdings nicht darauf abzustellen, ob die Gesellschafterversammlung mit einfacher Mehrheit den Geschäftsführer zu dieser Maßnahme angewiesen hätte. Zu fragen ist vielmehr, ob die übrigen Gesellschafter mutmaßlich gewollt hätten, daß der Geschäftsführer ohne vorherige Fassung eines Gesellschafterbeschlusses zu dem Geschäft veranlaßt wird. Auf den mutmaßlichen Willen der übrigen Gesellschafter ist deshalb abzustellen, weil sich der herrschende Gesellschafter so verhält, als hätte er ein Sonderrecht zur Weisungserteilung. Für die Einräumung eines Sonderrechts ist aber Einstimmigkeit erforderlich.[142]

Im Ergebnis wird der Rechtfertigungsgrund der Geschäftsführung ohne Auftrag damit auf Fälle einer Notgeschäftsführung beschränkt. Denn nur in diesen Fällen wird nach dem mutmaßlichen Willen der übrigen Gesellschafter auf eine Einberufung der Gesellschafterversammlung verzichtet werden können.

---

[141] *H. Thomas*, Palandt, 1998, Einf. v. § 677 Rdnr. 5; *R. Wittmann*, Staudinger, 1994, Vorbem. zu §§ 677 ff, Rdnr. 8 m. w. N. Siehe jüngst auch *J. Tieves*, Unternehmensgegenstand, 1998, S. 338.

[142] *H. Winter*, Scholz, 1993, § 14 Rdnr. 20.

c) Keine Berufung auf rechtmäßiges Alternativverhalten

Ist die Rechtswidrigkeit zu bejahen, kann der Gesellschafter schließlich nicht einwenden, daß er den Erfolg auch rechtmäßig, nämlich durch Herbeiführung eines entsprechenden Gesellschafterbeschlusses, hätte bewirken können.[143]

Der Einwand rechtmäßigen Alternativverhaltens ist unbeachtlich, „wenn sich der Schädiger in einer die Interessen des Verletzten erheblich berührenden Weise über eine Verhaltenspflicht hinwegsetzt, die darauf abzielt, einen entsprechenden Schaden erst nach Ausschöpfung gewisser Verhandlungsmöglichkeiten oder Chancen zu seiner Abwendung oder nach Beachtung möglicherweise begründeter Einwände des Geschädigten zuzulassen; denn solche Verhaltenspflichten implizieren, daß die Entscheidung bis zu ihrer vollständigen Erfüllung noch offen ist, und versagen deshalb dem Verletzer den Beweis, daß ihre Einhaltung das Ergebnis seines Handelns nicht verändert hätte".[144]

Gerade eine solche Verhaltenspflicht wird durch den Gesellschafter verletzt, wenn er außerhalb der Gesellschafterversammlung auf die Geschäftsführung Einfluß nimmt. Denn die Willensbildung der Gesellschaft findet in der Gesellschafterversammlung statt. Hierdurch sollen alle Gesellschafter die Möglichkeit erhalten, auf die Entscheidungsfindung Einfluß zu nehmen. Sie sollen ihre Vorschläge und Einwände vorbringen können, um zu der für die Gesellschaft besten Entscheidung zu gelangen. Das Gesetz begreift damit die Willensbildung der Gesellschaft als einen offenen Diskurs, dessen Ergebnis erst mit der Beschlußfassung feststeht. Deshalb verschließt es dem herrschenden Gesellschafter den Beweis, daß er den Geschäftsführer auch im Wege der Herbeiführung eine Gesellschafterbeschlusses zu der Maßnahme angewiesen hätte.

Zusammenfassend kann damit festgestellt werden, daß der Geschäftsführer der herrschenden GmbH im faktischen GmbH-Konzern die Konzernleitung nur durch die Herbeiführung von Weisungsbeschlüssen in der Gesellschafterversammlung der beherrschten GmbH ausüben darf. Nur ganz ausnahmsweise, nämlich im Rahmen einer Notgeschäftsführung, ist er berechtigt, den Geschäftsführer auch außerhalb der Gesellschafterversammlung zu einer Maßnahme zu veranlassen. Damit werden der Konzernleitung im GmbH-Konzern engere Grenzen gezogen als im Aktienkonzern. Denn dort geht der Gesetzgeber davon aus, daß das herrschende Unternehmen unmittelbar auf die Unternehmenslei-

---

[143] Im Ergebnis ebenso *U. Stein*, Organ, 1984, S. 182. Ebenso für die Verletzung einer Vorlagepflicht durch den Geschäftsführer: *BGH*, Urt. v. 25.2.1991 - II ZR 76/90, in: WM 1991, S. 635, 636.

[144] *H.-J. Mertens*, Soergel, 1990, vor § 249 Rdnr. 164.; im Ergebnis ebenso *D. Medicus*, Staudinger, 1983, § 249 Rdnr. 114. Dagegen *W. Grunsky*, Münchener Kommentar, 1994, Vor § 249 Rdnr. 90b; *J. Tieves*, Unternehmensgegenstand, 1998, S. 345.

tung Einfluß nehmen darf. Im GmbH-Konzern ist eine solche Einflußnahme dagegen nur dann zulässig, wenn die herrschende GmbH ein Sonderrecht zur Weisungserteilung hat. Für die Einräumung dieses Sonderrechts ist die Zustimmung der Minderheitsgesellschafter erforderlich.

Liegt kein Sonderrecht vor und handelte der Gesellschafter rechtswidrig, so ist er zum Schadensersatz verpflichtet. Voraussetzung ist allerdings, daß der Gesellschaft ein Schaden entstanden ist.

d) Schadensbegriff

Im Rahmen von § 43 Abs. 2 BGB, der die Schadensersatzpflicht des Geschäftsführers betrifft, ist der Begriff des Schadens umstritten. Nach einer Meinung reicht jede objektive Beeinträchtigung des Vermögens der Gesellschaft für die Annahme eines Schadens aus.[145] Nach einer anderen Ansicht ist dagegen zusätzlich erforderlich, daß die Vermögensminderung dem Unternehmenszweck widerspricht.[146] Dieser Streit hat allerdings im Bereich von § 43 Abs. 2 GmbHG nur Bedeutung für die Beweislastverteilung. Während die Gesellschaft beweisen muß, daß ihr durch eine Handlung ihres Geschäftsführers ein Schaden entstanden ist, obliegt dem Geschäftsführer der Beweis, daß er die Sorgfalt eines ordentlichen und gewissenhaften Geschäftsleiters eingehalten hat, § 93 Abs. 2 Satz 2 AktG analog. Läßt man nun allein eine zweckwidrige Minderung des Gesellschaftsvermögens als Schaden gelten, so hat dies zur Folge, daß die Gesellschaft im Ergebnis auch beweisen muß, daß der Geschäftsführer zweckwidrig - gemeint ist sorgfaltswidrig - gehandelt hat. Im Ergebnis besteht jedoch darüber Einigkeit, daß der Geschäftsführer bei Einhaltung der Sorgfalt eines ordentlichen und gewissenhaften Geschäftsleiters nicht haftet. Denn dann liegt schon keine Pflichtverletzung vor.

Das ist bei dem Gesellschafter, der außerhalb der Gesellschafterversammlung auf die Geschäftsführung Einfluß nimmt, anders. Denn hier liegt die Pflichtverletzung bereits in der mit der Einflußnahme verbundenen Verletzung der gesellschaftsinternen Zuständigkeitsordnung. Deshalb muß hier der Schadensbegriff nach anderen Kriterien bestimmt werden.

Dabei ist davon auszugehen, daß die Gesellschafter ihre Einlagen der Gesellschaft zur Zweckverfolgung anvertraut haben. Sie haben ihre Einlagen also in dem Vertrauen darauf geleistet, daß diese im Rahmen des Unternehmensgegenstandes mit der Sorgfalt eines ordentlichen Geschäftsmannes verwaltet werden.

---

[145] *Uwe H. Schneider*, Scholz, 1993, § 43 Rdnr. 160.

[146] *H.-J. Mertens*, Hachenburg, 1996, § 43 Rdnr. 57; *W. Zöllner*, Baumbach/Hueck, 1996, § 43 Rdnr. 13a.

Das bedeutet aber auch, daß sie nicht mehr erwarten können als eben diesen zweckgemäßen Einsatz ihrer Einlagen. Folgt man dem, so kann man einen Schaden aber nur dann annehmen, wenn das Vermögen der Gesellschaft unter Verstoß gegen den Unternehmenszweck gemindert wurde. Das ist nur dann der Fall, wenn ein ordentlicher Geschäftsmann die von dem Gesellschafter veranlaßte Maßnahme nicht vorgenommen hätte.

Nach alledem kann als Ergebnis festgehalten werden, daß ein Schaden jede dem Gesellschaftszweck widersprechende Minderung des Gesellschaftsvermögens ist.

Damit unterliegt der Gesellschafter einer GmbH, der außerhalb der Gesellschafterversammlung auf die Geschäftsführung Einfluß nimmt, im Ergebnis derselben Haftung wie der Unternehmensgesellschafter einer Aktiengesellschaft. Bei dem Gesellschafter einer GmbH kommt jedoch hinzu, daß er von seinen Mitgesellschaftern auf Unterlassung in Anspruch genommen werden kann. Denn selbst wenn seine Einflußnahme zu keinem Schaden führt, verletzt er doch die Mitverwaltungsrechte der übrigen Gesellschafter. Solche Mitverwaltungsrechte haben die Gesellschafter einer Aktiengesellschaft dagegen nicht.

9. Verbot der Ausnutzung von Geschäftschancen der abhängigen GmbH

a) Die gesetzliche Ausgangslage

Für den Geschäftsführer wurde bereits ausgeführt, daß er keine Geschäftschancen seiner Gesellschaft zu eigenen Zwecken ausnutzen darf. Denn hiermit würde er seine organschaftliche Pflicht zur umfassenden und vorbehaltlosen Förderung des Gesellschaftszwecks verletzen. Als Geschäftschancen der Gesellschaft wurden dabei zum einen solche Geschäfte bezeichnet, deren Zuordnung zur Gesellschaft sich bereits in irgendeiner Weise manifestiert hat. Hierzu gehört etwa der Fall, daß die Gesellschaft bereits Interesse an dem Geschäft gezeigt hat. Zum anderen wurde eine Geschäftschance der Gesellschaft zugerechnet, wenn ihre Wahrnehmung für die Gesellschaft objektiv nützlich und finanziell darstellbar ist.

Anders als für den Geschäftsführer gibt es für den Gesellschafter einer GmbH dagegen keinen gesellschaftsrechtlichen Grundsatz, wonach er seine eigenen Interessen stets gegenüber denen der Gesellschaft zurückzustellen hat.[147] Deshalb müssen zusätzliche Voraussetzungen hinzukommen, um auch ihm die Wahrnehmung von Geschäftschancen der Gesellschaft zu untersagen.

---

[147] *BGH*, Urt. v. 9.6.1954 - II ZR 70/53, in: BGHZ 14, 25, 38; *H.-J. Mertens/A. Cahn*, Wettbewerbsverbot, 1991, S. 554.

b) Meinungsstand

Anknüpfungspunkt in der Literatur ist die Struktur der Gesellschaft bzw. die Stellung des Gesellschafters in der Gesellschaft. Vertreten wird, daß der Gesellschafter keine Geschäftschancen der Gesellschaft an sich ziehen dürfe, wenn die Gesellschaft personalistisch strukturiert sei oder - bei einer kapitalistischen Struktur - der Gesellschafter in der Geschäftsführung mitarbeite.[148] Eine andere Ansicht[149] stellt dagegen darauf ab, ob der Gesellschafter in der Gesellschaft eine herrschende Stellung einnehme. Sei dies der Fall, unterliege der Gesellschafter selbst dann einer Geschäftschancenbindung, wenn er von seinen Einflußmöglichkeiten keinen Gebrauch mache.

c) Stellungnahme

Richtigerweise ist zwischen der Lage im zentralen und dezentralen Konzern zu unterscheiden.

Legt der Gesellschafter nicht nur die Unternehmensziele fest, sondern hat er auch das operative Geschäft an sich gezogen, so treffen ihn - wie noch zu zeigen sein wird - dieselben Pflichten wie den Geschäftsführer. Damit unterliegt er aber auch dem Verbot der Ausnutzung von Geschäftschancen der Gesellschaft.

In allen anderen Fällen ist dagegen zu berücksichtigen, daß der Gesellschafter gerade nicht zur umfassenden und vorbehaltlosen Förderung des Gesellschaftsinteresses und damit auch nicht zur Suche nach Geschäftschancen der Gesellschaft verpflichtet ist. Auf diese Pflicht kann deshalb ein Verbot der Ausnutzung von Geschäftschancen der Gesellschaft nicht gestützt werden.

Folgt man dem, so bedeutet dies umgekehrt, daß der Gesellschafter grundsätzlich - d. h. vorbehaltlich eines Wettbewerbsverbots - berechtigt ist, zu eigenen Zwecken nach Geschäftschancen zu suchen. Deshalb verstößt er grundsätzlich nicht gegen seine mitgliedschaftliche Förderpflicht, wenn er eine Geschäftschance wahrnimmt, die für die Gesellschaft objektiv nützlich und finanziell darstellbar ist oder an der die Gesellschaft schon Interesse gezeigt hat. Von diesem Grundsatz ist jedoch eine gewichtige Ausnahme zu machen.

Zwar gibt es keinen gesellschaftsrechtlichen Grundsatz, wonach ein GmbH-Gesellschafter seine eigenen Interessen stets gegenüber denen der Gesellschaft zurückzustellen hat. Deshalb ist ihm die Ausnutzung von Geschäftschancen der Gesellschaft grundsätzlich gestattet. Nicht gestattet ist ihm jedoch die mißbräuchliche Ausnutzung seiner Gesellschafterstellung, sei es durch Verwertung

---

148 *W. Timm*, Wettbewerbsverbot, 1981, S. 179.

149 *J. Weisser*, Opportunities, 1991, S. 199.

gesellschaftsinterner Informationen,[150] sei es durch Verwendung von Mitteln der Gesellschaft für eigene Zwecke.[151] Denn hierdurch beutet der Gesellschafter Ressourcen aus, die nicht ihm, sondern der Gesellschaft zustehen. Das ist ihm jedoch verboten. Und das gilt unabhängig davon, ob die Gesellschaft personalistisch oder kapitalistisch strukturiert ist, ob der Gesellschafter in der Geschäftsführung mitarbeitet oder nicht und ob der Gesellschafter herrschender Gesellschafter oder Minderheitsgesellschafter ist.

Damit läßt sich feststellen, daß ein Gesellschafter immer dann seine mitgliedschaftliche Förderpflicht verletzt, wenn er eine Geschäftschance wahrnimmt, die ihm aus dem Verband heraus vermittelt wurde.[152] Hat also die herrschende GmbH von dem Geschäftsführer der Tochtergesellschaft erfahren, daß diese sich für ein Grundstück interessiert, das sie günstig erwerben kann, so darf die herrschende GmbH die Wahrnehmung dieser Geschäftschance nicht dadurch „torpedieren", daß sie das Geschäft an sich zieht. Verstößt sie gegen diese Pflicht und entsteht der Tochtergesellschaft dadurch ein Schaden, macht sie sich schadensersatzpflichtig. Dagegen reicht es nicht aus, daß der Gesellschafter unabhängig von seiner Gesellschafterstellung von der Geschäftschance erfahren hat. Das gilt auch dann, wenn die GmbH mit dem Geschäft als erste in Berührung gekommen ist oder bereits konkrete Maßnahmen zur Durchführung des Geschäfts getroffen hat.[153] Denn Voraussetzung ist immer, daß dem Gesellschafter die Geschäftschance aus dem Verband heraus vermittelt wurde, er sie also nicht aus eigener Leistung erworben hat.

d) Ergebnis

Als Ergebnis kann somit festgehalten werden, daß der herrschenden GmbH im dezentralen Konzern nur die Verwertung solcher Geschäftschancen verboten ist, von denen sie unter Ausnutzung ihrer Gesellschafterstellung Kenntnis erlangt hat. Im zentralen Konzern, in dem sie auch das operative Geschäft an sich

---

150 *BGH*, Urt. v. 8.5.1989 - II ZR 229/88, in: ZIP 1989, S. 986, 987; *BFH*, Urt. v. 11.6.1996 - I R 97/95, in: DStR 1996, S. 1769; *T. Raiser*, Hachenburg, 1989, § 14 Rdnr. 64; *H. Rowedder*, GmbHG, 1997, § 13 Rdnr. 14. Für die OHG: *P. Ulmer*, Staub, 1988, § 112 Rdnr. 3 u. 17, der in diesen Fällen für eine analoge Anwendung der §§ 112, 113 HGB plädiert.

151 *H.-J. Mertens/A. Cahn*, Wettbewerbsverbot, 1991, S. 551. A. M. *J. Weisser*, Opportunities, 1991, S. 172 f., der die Benutzung von Gesellschaftsmitteln als Kriterium für die Zurechnung einer Geschäftschance zur Gesellschaft ablehnt.

152 *M. Winter*, Treubindungen, 1988, S. 242; nicht so eng *V. Röhricht*, Wettbewerbsverbot, 1992, S. 775.

153 A. M. *J. Thiel*, Gewinnausschüttung, 1993, S. 1804; in diese Richtung wohl auch *BFH*, Urt. v. 13.11.1996 - I R 149/94, in: DStR 1997, S. 323, 325.

zieht, treffen sie dagegen - wie noch zu zeigen sein wird - dieselben Pflichten wie den Geschäftsführer. Dieser darf aber auch solche Geschäftschancen nicht wahrnehmen, deren Zuordnung zur Gesellschaft sich bereits in irgendeiner Weise manifestiert hat oder die für die Gesellschaft objektiv nützlich und finanziell darstellbar sind.

10. Wettbewerbsverbot

a) Die gesetzliche Ausgangslage

Oben wurde festgestellt, daß den Geschäftsführer gegenüber seiner Gesellschaft ein Wettbewerbsverbot trifft. Dieses Wettbewerbsverbot ergibt sich aus seiner organschaftlichen Pflicht, die Interessen der Gesellschaft uneingeschränkt und vorbehaltlos zu fördern. Dem würde aber eine außergesellschaftliche Tätigkeit des Geschäftsführers widersprechen, die mit einer abstrakten Gefährdung der ihm anvertrauten Gesellschaftsinteressen allgemein oder im Einzelfall verbunden ist.

Anders als für den Geschäftsführer gibt es für den Gesellschafter einer GmbH dagegen keinen gesellschaftsrechtlichen Grundsatz, wonach er seine eigenen Interessen stets gegenüber denen der Gesellschaft zurückzustellen hat. Das wurde bei der Frage der Geschäftschancenbindung bereits ausgeführt.[154] Aber ebenso wie dort werden auch für das Wettbewerbsverbot Ausnahmen von diesem Grundsatz diskutiert.

Ausgangspunkt ist dabei die Vorschrift des § 112 Abs. 1 HGB. Danach darf der Gesellschafter einer OHG „ohne Einwilligung der anderen Gesellschafter weder in dem Handelszweige der Gesellschaft Geschäfte machen noch an einer anderen gleichartigen Handelsgesellschaft als persönlich haftender Gesellschafter teilnehmen". Hierin wird zu Recht eine Konkretisierung der mitgliedschaftlichen Förderpflicht des Gesellschafters einer OHG gesehen.[155] Da auch der Gesellschafter einer GmbH einer mitgliedschaftlichen Förderpflicht unterliegt,

---

[154] Siehe oben II. 9. a).

[155] *M. Ivens*, Konkurrenzverbot, 1987, S. 129. A. M. *H. Wiedemann*/*H. Hirte*, Konkretisierung, 1986, S. 166 f., nach denen § 112 HGB keine mitgliedsbezogene, sondern auf das Amt als Geschäftsführer bezogene Pflicht statuiere. Dabei wird freilich verkannt, daß die Geschäftsführungspflicht des Gesellschafters einer OHG Ausdruck seiner mitgliedschaftlichen Förderpflicht ist und damit auch das Wettbewerbsverbot seine Grundlage in der mitgliedschaftlichen Förderpflicht findet. Dagegen trifft bei der GmbH den Gesellschafter grundsätzlich nur dann eine Geschäftsführungspflicht, wenn er zum Geschäftsführer bestellt wurde. Diese Pflicht ergibt sich allerdings nicht aus seiner Mitgliedschaft, sondern aus seinem Amt als Geschäftsführer.

stellt sich die Frage, ob und unter welchen Voraussetzungen § 112 Abs. 1 HGB im Recht der GmbH analog anwendbar ist.

b) Meinungsstand

Nach einer Meinung geht § 112 HGB von dem Regelfall einer Arbeits- und Haftungsgemeinschaft aus, die auf der tätigen Mitarbeit der einzelnen Gesellschafter aufbaue. Das hierauf begründete Vertrauensverhältnis werde aber gestört, wenn einer der Gesellschafter eine Konkurrenztätigkeit aufnehme. Davor wolle § 112 HGB schützen. Übertragen auf das Recht der GmbH bedeute dies, daß die Gesellschafter einer GmbH nur dann einem Wettbewerbsverbot unterworfen seien, wenn die Gesellschaft als Mitunternehmergemeinschaft auf eine vertrauensvolle Zusammenarbeit der Gesellschafter aufbaue. Nur in einer personalistisch strukturierten GmbH unterliege mithin der Gesellschafter einer GmbH einem Wettbewerbsverbot, und zwar unabhängig von der Höhe seiner Beteiligung.[156]

Der *Bundesgerichtshof*[157] und mit ihm ein Großteil der Literatur[158] stellen dagegen nicht auf das persönliche Vertrauensverhältnis unter den Gesellschaftern ab, sondern auf die Gefahren, die dadurch entstehen können, daß der Gesellschafter zugunsten seines Konkurrenzunternehmens auf die Gesellschaft nachteiligen Einfluß nimmt oder gesellschaftsinterne Informationen mißbräuchlich verwendet. Freilich unterscheidet sich diese Meinungsgruppe in den Details.

Nach einer Ansicht unterliegt grundsätzlich jeder GmbH-Gesellschafter einem Wettbewerbsverbot. Eine Ausnahme bestehe nur dann, wenn der Gesellschafter erstens keinen Einfluß auf das Tagesgeschäft nehme und zweitens nur über beschränkte Informationen über die Geschäftsführung verfüge. Denn dann

---

[156] Zum Vorstehenden siehe *H.-J. Mertens/A. Cahn*, Wettbewerbsverbot, 1991, S. 555 f.; ebenso *W. Timm*, Wettbewerbsverbot, 1981, S. 179; *M. Lutter/W. Timm*, Präventivschutz, 1982, S. 419; *T. Raiser*, Wettbewerbsverbote, 1985, S. 864 f.; *W. Zöllner*, Baumbach/Hueck 1996, GmbH-KonzernR, Rdnr. 70. Eher zurückhaltend *V. Röhricht*, Wettbewerbsverbot, 1992, S. 773.

[157] *BGH*, Urt. v. 16.2.1981 - II ZR 168/79 („Süssen"), in: BGHZ 80, S. 69, 74; *BGH*, Urt. v. 5.12.1983 - II ZR 242/82 („Werbeagentur"), in: BGHZ 89, S. 162, 165 ff.; *BGH*, Urt. v. 3.5.1988 - KZR 17/87, in: NJW 1988, S. 2737, 2738.

[158] *Uwe H. Schneider*, Scholz, 1993, § 43 Rdnr. 126b; *H. Altmeppen*, GmbHG, 1997, Anh. § 13 Rdnr. 98; *M. Winter*, Treubindungen, 1988, S. 249; *H. Wiedemann/H. Hirte*, Konkretisierung, 1986, S. 164; *S. Tiedchen*, Wettbewerbsverbote, 1993, S. 617; *C. P. Claussen/H.-M. Korth*, Wettbewerbsverbot, 1993, S. 127; *D. von der Osten*, Wettbewerbsverbot, 1989, S. 451; *V. Röhricht*, Wettbewerbsverbot, 1992, S. 772; *R. Salfeld*, Wettbewerbsverbote, 1987, S. 208 f.; *M. Ivens*, Konkurrenzverbot, 1987, S. 139 f.

sei seine Stellung mit der eines Kommanditisten vergleichbar, den gemäß § 165 HGB kein Wettbewerbsverbot treffe.[159]

Dagegen hat der *Bundesgerichtshof* bisher - soweit ersichtlich - nur zum Wettbewerbsverbot des herrschenden Gesellschafters einer GmbH und einer GmbH & Co. KG Stellung genommen. In seinem „Süssen"-Urteil[160] stellte er fest, daß die Befreiung eines herrschenden GmbH-Gesellschafters von einem gesellschaftsvertraglich vereinbarten Wettbewerbsverbot rechtsmißbräuchlich sei, wenn die Gesellschaft dadurch zu einem abhängigen Unternehmen werde. Ein Mißbrauch sei nur dann ausgeschlossen, wenn die Befreiung im Interesse der GmbH geboten sei. Zur Begründung führte er aus, daß der herrschende Gesellschafter zum einen jederzeit von seinen die Abhängigkeit begründenden Herrschaftsmöglichkeiten zum Nachteil der abhängigen Gesellschaft Gebrauch machen könne. Zum anderen fehle in vielen Fällen der objektive Maßstab für die jeweils sachgerechte Maßnahme, was die Feststellung einer Benachteiligung und eines Ausgleichs erheblich erschwere.

Während es in dem „Süssen"-Urteil um die rechtsmißbräuchliche Befreiung von einem gesellschaftsvertraglich vereinbarten Wettbewerbsverbot ging, stützte der *Bundesgerichtshof* zwei Jahre später in seinem „Werbeagentur"-Urteil[161] die Begründung des gesetzlichen Wettbewerbsverbots des beherrschenden Gesellschafters einer GmbH & Co. KG auf dieselben konzernrechtlichen Erwägungen.[162] Und er fügte hinzu, daß aufgrund der Möglichkeit des herrschenden Gesellschafters, gesellschaftsinterne Informationen zu erlangen, die Gefahr einer mißbräuchlichen Verwendung dieser Informationen bestehe.[163]

Diese Erwägungen für die Kommanditgesellschaft können auf die GmbH übertragen werden. Dafür spricht die ausdrückliche Anknüpfung an das zuvor dargestellte Urteil zur Aufhebung des statutarischen Wettbewerbsverbots in den Urteilsgründen als auch die in der Senatsrechtsprechung häufig hervorgehobene funktionale Austauschbarkeit von GmbH und GmbH & Co. KG.[164] Allerdings läßt der *Bundesgerichtshof* in seinem Werbeagentur-Urteil[165] offen, ob für die

---

159 *Uwe H. Schneider*, Scholz, 1993, § 43 Rdnr. 126b.

160 *BGH*, Urt. v. 16.2.1981 - II ZR 168/79 („Süssen"), in: BGHZ 80, S. 69, 74.

161 *BGH*, Urt. v. 5.12.1983 - II ZR 242/82 („Werbeagentur"), in: BGHZ 89, S. 162, 165 ff.

162 Zustimmend *H. Wiedemann/H. Hirte*, Konkretisierung, 1986, S. 164; *M. Winter*, Treubindungen, 1988, S. 249. Kritisch *U. Immenga*, Anmerkung, 1984, S. 578 ff.

163 *BGH*, Urt. v. 5.12.1983 - II ZR 242/82 („Werbeagentur"), in: BGHZ 89, S. 162, 166.

164 *M. Winter*, Treubindungen, 1988, S. 245.

165 *BGH*, Urt. v. 5.12.1983 - II ZR 242/82 („Werbeagentur"), in: BGHZ 89, 162, 167.

analoge Anwendung des § 112 HGB die bloße Möglichkeit eines beherrschenden Einflusses genügt[166] oder ob der Einfluß tatsächlich ausgeübt werden muß[167]. Umgekehrt hat ein Teil der Rechtsprechung aus diesem Urteil den Schluß gezogen, daß einen Minderheitsgesellschafter[168] und einen zu 50 % beteiligten Gesellschafter[169] kein Wettbewerbsverbot treffe.

c) Stellungnahme

Ebenso wie für die Frage der Geschäftschancenbindung ist auch für die Frage des Wettbewerbsverbots zwischen der Lage im zentralen und dezentralen Konzern zu unterscheiden.

Im zentralen Konzern treffen das herrschende Unternehmen - wie noch ausgeführt wird - dieselben Pflichten wie den Geschäftsführer. Deshalb unterliegt es schon aus diesem Grund einem Wettbewerbsverbot.

Problematisch ist nur, ob gleiches auch im dezentralen Konzern gilt. Hier läßt sich das Wettbewerbsverbot nicht mit der Pflicht zur vorbehaltlosen und umfassenden Förderung des Gesellschaftsinteresses begründen. Vielmehr müssen weitere Umstände hinzutreten, die ein Wettbewerbsverbot rechtfertigen können und einen Vergleich mit § 112 Abs. 1 HGB zulassen. Dabei ist insbesondere die Wertung von § 1 GWB zu beachten. Danach sind gesellschaftsrechtliche Beschränkungen der unternehmerischen Betätigungsfreiheit mit der in § 1 GWB geschützten Wettbewerbsfreiheit nur insoweit zulässig, als die Beschränkungen erforderlich sind, um die Funktionsfähigkeit eines dem gesetzlichen Typus entsprechenden Verbandes zu gewährleisten.[170] Mit anderen Worten darf das Wettbewerbsverbot nicht die Aufgabe haben, die Gesellschaft vor dem freien Wettbewerb zu schützen. Vielmehr darf sein Zweck nur darin

---

[166] So *H. Altmeppen*, GmbHG, 1997, Anh. § 13 Rdnr. 98; *M. Winter*, Treubindungen, 1988, S. 250; *S. Tiedchen*, Wettbewerbsverbote, 1993, S. 617; *B. Tillmann*, Wettbewerbsverbot, 1991, S. 27; *L. Lawall*, Wettbewerbsverbot, 1996, S. 116. Ebenso wohl auch *BGH*, Urt. v. 3.5.1988 - KZR 17/87, in: NJW 1988, S. 2737, 2738.

[167] So *C. P. Claussen/H.-M. Korth*, Wettbewerbsverbot, 1993, S. 127; *D. von der Osten*, Wettbewerbsverbot, 1989, S. 451; wohl auch *V. Röhricht*, Wettbewerbsverbot, 1992, S. 772; *R. Salfeld*, Wettbewerbsverbote, 1987, S. 208 f.; *U. Immenga*, Anmerkung, 1984, S. 580. Für die OHG: *C. Armbrüster*, Grundlagen, 1997, S. 267.

[168] *OLG Köln*, Urt. v. 22.2.1991 - 3 U 20/91, in: NJW-RR 1991, S. 1316, 1317; *Finanzgericht Köln*, Urt. v. 25.7.1991 - 13 K 5752/90, in: GmbHR 1992, S. 326, 327. Offengelassen in *BGH*, Urt. v. 3.5.1988 - KZR 17/87, in: GmbHR 1988, S. 334, 336.

[169] *OLG Karlsruhe*, Urt. v. 6.11.1998 - 15 U 179/97, in: GmbHR 1999, S. 539.

[170] *H.-J. Mertens/A. Cahn*, Wettbewerbsverbot, 1991, S. 555; *BGH*, Urt. v. 5.12.1983 - II ZR 242/82 („Werbeagentur"), in BGHZ 89, S. 162, 169; *BGH*, Urt. v. 3.5.1988 - KZR 17/87, in: BGHZ 104, 246, 251; *L. Lawall*, Wettbewerbsverbot, 1996, S. 4; eingehend *M. Ivens*, Konkurrenzverbot, 1987, S. 200 ff.

bestehen zu verhindern, „daß ein Gesellschafter das Unternehmen von innen her aushöhlt oder gar zerstört und damit einen leistungsfähigen Wettbewerber zugunsten seiner eigenen Konkurrenztätigkeit ausschaltet".[171]

Berücksichtigt man dies, so ist zunächst der Ansicht zuzustimmen, daß die Gesellschafter einer personalistisch strukturierten GmbH einem Wettbewerbsverbot unterliegen. Denn diese zeichnet sich ebenso wie der gesetzlichen Regeltyp der OHG dadurch aus, daß sie auf der persönlichen Zusammenarbeit ihrer Gesellschafter und nicht auf deren bloß kapitalistischer Beteiligung beruht. Eine persönliche Zusammenarbeit erfordert aber persönliches Vertrauen. Wird dieses durch den Wettbewerb eines Gesellschafters gestört, ist die Verwirklichung des Gesellschaftszwecks gefährdet. Deshalb ist ein Wettbewerbsverbot erforderlich, um die Funktionsfähigkeit einer personalistisch strukturierten GmbH zu erhalten.

Als Zwischenergebnis läßt sich somit feststellen, daß jeder Gesellschafter einer personalistisch strukturierten GmbH einem Wettbewerbsverbot unterliegt.

Entgegen der obengenannten Ansicht ist das Wettbewerbsverbot jedoch nicht auf diesen Fall beschränkt. Denn der tiefere Grund für den Wegfall der Vertrauensgrundlage ist die Gefahr, daß der Gesellschafter einerseits mit Rücksicht auf sein Konkurrenzunternehmen Geschäftsführungshandlungen zum Schaden der Gesellschaft vornimmt oder unterläßt und daß er andererseits die aufgrund seiner Geschäftsführungsbefugnis erlangten gesellschaftsinternen Informationen zum Nachteil der Gesellschaft und zum Vorteil seines Konkurrenzunternehmens einsetzt. Der Unterschied zwischen OHG und GmbH besteht insoweit lediglich darin, daß bei der OHG mit dem Wettbewerb des Gesellschafters in der Regel der Wegfall einer bestehenden Vertrauensgrundlage verbunden ist, während dies bei der kapitalistisch strukturierten GmbH nicht der Fall ist. Für das Ausmaß der Gefährdung des Gesellschaftszwecks hat dieser Unterschied jedoch keine Bedeutung. Mithin bezweckt § 112 HGB den Schutz der Gesellschaft vor konkurrenzbedingten Schädigungen ihrer Gesellschafter, insbesondere vor der mißbräuchlichen Verwendung gesellschaftsinterner Informationen. Insbesondere kann auch letztere Gefahr allein für ein Wettbewerbsverbot ausreichen. So wird in der Literatur zu Recht die Ansicht[172] vertreten, daß ein Wettbewerbsverbot des OHG-Gesellschafters vor § 1 GWB auch dann Bestand habe, wenn der Gesellschafter zwar nicht geschäftsführungsbefugt sei, aber umfassende Informations- und Kontrollrechte besitze. Nach alledem zeigt sich, daß in den Schutzbereich von § 112 HGB nicht nur personalistisch, sondern auch kapitalistisch strukturierte Gesellschaften fallen.

---

171 *BGH*, Urt. v. 3.5.1988 - KZR 17/87, in: GmbHR 1988, S. 334, 336.

172 *P. Ulmer*, Staub, 1988, § 112 Rdnr. 47.

Für diese stellt sich nun die Frage, wann die konkurrenzbedingte Schädigungsgefahr ihrer Gesellschafter ein derartiges Ausmaß erreicht, daß ihr zur Aufrechterhaltung der Funktionsfähigkeit der Gesellschaft mit einem Wettbewerbsverbot begegnet werden muß.

Bei einem Minderheitsgesellschafter scheidet die Gefahr einer schädlichen Einflußnahme auf die Geschäftsführung aus. Denn gegen den Willen seiner Mitgesellschafter kann er keine Weisungsbeschlüsse durchsetzen. Auch besteht kein Grund zur Befürchtung, daß der Geschäftsführer einer nachteiligen Veranlassung des Gesellschafters außerhalb der Gesellschafterversammlung Folge leisten werde. Damit verbleibt nur die Gefahr der mißbräuchlichen Verwendung gesellschaftsinterner Informationen, die ein Wettbewerbsverbot rechtfertigen könnte. Diese Informationen kann sich der Gesellschafter zum einen über sein individuelles Auskunftsrecht gemäß § 51a GmbHG beschaffen. Allerdings sind diesem Auskunftsrecht Grenzen gesetzt. So darf der Geschäftsführer die Auskunft gemäß § 51a Abs. 2 GmbHG verweigern, wenn zu besorgen ist, daß der Gesellschafter sie zum Schaden der Gesellschaft oder eines verbundenen Unternehmens ausnutzt. Die Verweigerung, die eines Beschlusses der Gesellschafterversammlung bedarf, ist insbesondere dann gerechtfertigt, wenn der Gesellschafter ein Konkurrenzunternehmen betreibt.[173] Mit § 51a GmbHG läßt sich deshalb keine für ein Wettbewerbsverbot ausreichende Schädigungsgefahr begründen.[174] Damit bleibt zum anderen das allgemeine Informationsrecht des Gesellschafters vor der Fassung von Gesellschafterbeschlüssen. Dieses steht ihm zu, soweit die Information zur Beurteilung des Entscheidungsgegenstandes erforderlich ist.[175] Auch dieses Informationsrecht begründet jedoch keine qualifizierte Schädigungsgefahr. Das zeigt seine Vergleichbarkeit mit dem Informationsrecht des Kommanditisten (§ 166 HGB), bei dem der Gesetzgeber ein Wettbewerbsverbot nicht für erforderlich hielt, § 165 HGB:[176]

Zunächst hat der Gesellschafter zur Feststellung des Jahresabschlusses (§ 46 Nr. 1 GmbHG) das Recht, die Vorlage des Jahresabschlusses zu verlangen und dessen Richtigkeit durch Einblick in die Bücher und Papiere der Gesellschaft zu überprüfen, § 51a GmbHG.[177] Dasselbe Recht hat der Kommanditist gemäß §

---

[173] *OLG Karlsruhe*, Beschluß vom 11.12.84 - 11 W 135/84, in: GmbHR 1985, S. 362, 363; *M. Lutter/P. Hommelhoff*, GmbHG, 1995, § 51a Rdnr. 21.

[174] Ebenso *M. Ivens*, Konkurrenzverbot, 1987, S. 136 f.

[175] *W. Zöllner*, Baumbach/Hueck, GmbHG, 1996, § 51a Rdnr. 39.

[176] Ebenso *M. Ivens*, Konkurrenzverbot, 1987, S. 135 f.

[177] Siehe auch § 42a Abs. 1 S. 1 GmbHG, wonach der Geschäftsführer den Jahresabschluß und den Lagebericht unverzüglich nach der Aufstellung den Gesellschaftern zum Zwecke der Feststellung des Jahresabschlusses vorzulegen hat. Von dieser Vorschrift kann jedoch im Gesellschaftsvertrag abgewichen werden (*J. Schulze-Osterloh*, Baumbach/Hueck, 1996, § 42a Rdnr. 9).

166 Abs. 1 HGB. Sodann ist für die Vornahme außergewöhnlicher Geschäfte bei der GmbH die Zustimmung der Gesellschafterversammlung und bei der Kommanditgesellschaft gemäß § 164 Satz 1 HGB die Zustimmung des Kommanditisten erforderlich. Auch insoweit ist der Informationsstand des Minderheitsgesellschafters bei der GmbH mit dem eines Kommanditisten vergleichbar. Die Unterschiede beginnen erst dann, wenn die Gesellschafterversammlung in das Tagesgeschäft eingreift. Ist dies der Fall, erlangt der GmbH-Gesellschafter mehr Informationen als der Kommanditist, dessen Mitwirkung auf die Vornahme außergewöhnlicher Geschäfte beschränkt ist. Auch dann besteht jedoch keine qualifizierte Schädigungsgefahr, die ein Wettbewerbsverbot rechtfertigen könnte. Denn die wirklich sensiblen Informationen erlangt der Gesellschafter schon durch Einblick in die Papiere und Bücher der Gesellschaft oder aber im Rahmen seiner Mitwirkung an außergewöhnlichen Entscheidungen. Bei allen übrigen Informationen besteht dagegen regelmäßig ein weniger ausgeprägtes Geheimhaltungsbedürfnis. Selbst wenn dies im Einzelfall aber ausnahmsweise anders sein sollte, läßt sich damit vor dem Hintergrund von § 1 GWB doch kein Wettbewerbsverbot für den Minderheitsgesellschafter einer GmbH rechtfertigen.

Als Zwischenergebnis kann damit festgehalten werden, daß ein Minderheitsgesellschafter keinem Wettbewerbsverbot unterliegt.

Etwas anderes gilt jedoch für den Mehrheitsgesellschafter. Dieser kann stets gegen den Willen seiner Mitgesellschafter Einfluß auf die Geschäftsführung nehmen. Dabei hat er sogar eine stärkere Stellung als der geschäftsführende Gesellschafter einer OHG. Denn dessen Geschäftsführungshandlungen können die übrigen geschäftsführenden Gesellschafter gemäß § 115 Abs. 1 HGB im Innenverhältnis widersprechen. Ein solches Widerspruchsrecht haben die Minderheitsgesellschafter einer GmbH nicht. Des weiteren ist der Mehrheitsgesellschafter einer GmbH häufig nicht auf die Mitwirkung der Gesellschafterversammlung angewiesen, sondern kann auch außerhalb der Gesellschafterversammlung auf die Geschäftsführung Einfluß nehmen. Der Geschäftsführer wird entsprechenden Veranlassungen Folge leisten, da er mit der Stimmenmehrheit des Mehrheitsgesellschafters gewählt wurde und jederzeit abberufen werden kann (§ 38 GmbHG).

Aus dieser umfassenden Einflußmöglichkeit ergibt sich aber eine qualifizierte Schädigungsgefahr für die Gesellschaft, wenn der Mehrheitsgesellschafter ein Konkurrenzunternehmen betreibt. Das hat der *Bundesgerichtshof* zutreffend festgestellt. So besteht die Gefahr, daß der Mehrheitsgesellschafter jederzeit von seinen Herrschaftsmöglichkeiten zum Nachteil der abhängigen Gesellschaft und zum Vorteil seines Konkurrenzunternehmens Gebrauch macht. Hinzu kommt, daß sich eine solche Nachteilszufügung nur schwer feststellen läßt, da in vielen Fällen der objektive Maßstab für die jeweils sachgerechte Maßnahme fehlt. Als Beispiel nennt der *Bundesgerichtshof*[178] die Beurteilung einer Rationalisierung oder Umstellung der Produktion, die Ausdehnung der Produktpalette in einem vom Unternehmensgegenstand gedeckten, bislang aber noch nicht ausgeschöpften Bereich, die mittel- oder langfristig für die Erhaltung der Wettbewerbsfähigkeit erforderlich sein könnten. Eine unabhängige Gesellschaft wird in diesen Frage risikofreudiger sein als der die Gesellschaft beherrschende Unternehmer, der über seine anderweitige unternehmerische Beteiligung das Konkurrenzprodukt vertreibt.

Schließlich besteht auch die qualifizierte Gefahr einer mißbräuchlichen Verwendung gesellschaftsinterner Informationen. Denn ebenso wie der Gesellschafter außerhalb der Gesellschafterversammlung auf die Geschäftsführung Einfluß nehmen kann, kann er sich auch umfassende Informationen über das Unternehmen der Gesellschaft besorgen. Die Funktionsfähigkeit der Gesellschaft ist aber gefährdet, wenn er diese Informationen für sein Konkurrenzunternehmen ausnutzt.

Trotzdem verneint eine Mindermeinung[179] ein Wettbewerbsverbot im qualifizierten faktischen Konzern und im Vertragskonzern. Dort würden zum einen andere Schutzmechanismen gelten und zum anderen ließen sich bei einer breitflächigen Einflußnahme des herrschenden Unternehmens einzelne Verletzungen des Wettbewerbsverbots nicht isolieren.

Dieser Ansicht ist für den qualifizierten faktischen Konzern zu widersprechen.[180] Denn dort werden die Minderheitsgesellschafter nicht anders, sondern zusätzlich, nämlich durch eine Verlustausgleichspflicht des herrschenden Unternehmens, geschützt. Mit anderen Worten hat die Verlustausgleichspflicht auf das Schicksal anderer Ansprüche keinen Einfluß. Etwas anderes gilt nur im Vertragskonzern und dort auch nur deshalb, weil sich dies, wenn auch nur mittelbar, aus dem Gesetz ergibt. So gestattet § 308 Abs. 1 Satz 2 AktG, der im GmbH-Recht analog anzuwenden ist[181], nachteilige Weisungen des herrschenden Unternehmens gegenüber dem Geschäftsführer der Tochtergesellschaft. Hierzu gehört auch die Weisung, daß die Tochter-GmbH der herrschenden Gesellschaft eine gesellschaftsinterne Information oder eine Geschäftschance überlassen soll. Hiermit ist aber ein Wettbewerbsverbot, das die Gefahr solcher Weisungen gerade verhindern soll, unvereinbar.

---

178 *BGH*, Urt. v. 16.2.1981 - II ZR 168/79 („Süssen"), in: BGHZ 80, 69, 74.

179 *J. Thiel*, Gewinnausschüttung, 1992, S. 342; dagegen: *R. Schwedhelm*, Wettbewerbsverbot, 1993, S. 247.

180 Im Ergebnis wie hier *L. Lawall*, Wettbewerbsverbot, 1996, S. 109.

181 *W. Zöllner*, Baumbach/Hueck, 1996, GmbH-KonzernR Rdnr. 49.

Folglich läßt sich als Zwischenergebnis festhalten, daß der herrschende Gesellschafter einer GmbH jedenfalls dann einem Wettbewerbsverbot unterliegt, wenn er die Herrschaft tatsächlich ausübt. Noch nicht beantwortet ist damit die Frage, ob allein die Möglichkeit der Herrschaftsausübung für die Begründung eines Wettbewerbsverbots schon genügt.

Diese Frage ist zu bejahen. Der Gesetzgeber hat in den §§ 311 ff. AktG zum Ausdruck gebracht, daß die besondere Gefahrenlage für die Gesellschaft nicht erst im Konzern, sondern bereits in der bloßen Abhängigkeit beginnt. So reicht schon eine gelegentliche Einflußnahme aus, um die abhängige Gesellschaft nachhaltig zu schädigen. Zu denken ist etwa an den Unternehmensgesellschafter, der sich bisher aus der Geschäftspolitik herausgehalten hat und nun der Erweiterung der Produktpalette nicht zustimmt, weil er eine zu starke Konkurrenz gegenüber dem anderen Unternehmen befürchtet, an dem er beteiligt ist. Mithin besteht schon in der Abhängigkeitslage eine qualifizierte Schädigungsgefahr, wenn der herrschende Gesellschafter ein Konkurrenzunternehmen betreibt.

Nach alledem unterliegt der herrschende Gesellschafter einer GmbH schon in der bloßen Abhängigkeitslage einem Wettbewerbsverbot gegenüber seiner Tochter-GmbH.

d) Der gegenständliche Bereich des Wettbewerbsverbots

Der gegenständliche Bereich des Wettbewerbsverbots ist identisch mit dem des Wettbewerbsverbots des Geschäftsführers. Das folgt daraus, daß das Wettbewerbsverbot in beiden Fällen den Zweck hat, eine schädliche Einflußnahme auf die Gesellschaft bzw. ein Mißbrauch gesellschaftsinterner Informationen schon im Vorfeld zu verhindern. Auf die Ausführungen zum gegenständlichen Bereich des Wettbewerbsverbots des Geschäftsführers kann deshalb verwiesen werden.[182]

III. Mehrstufiger faktischer GmbH-Konzern mit Minderheitsgesellschaftern

Nachdem die Grenzen der Konzernleitungspflicht im einstufigen Konzern aufgezeigt worden sind, stellt sich nun die Frage, welche Grenzen im zwei- und mehrstufigen Konzern[183] existieren. Dabei wird im folgenden nur von den durch

---

[182] Siehe 1. Teil C. III. 3.

[183] Zur Verbreitung mehrstufiger Unternehmensverbindungen mit einer abhängigen GmbH siehe *H. Görling*, Unternehmensverbindungen, 1993, S. 546 f.

das Eigeninteresse der Enkel-GmbH gezogenen Grenzen die Rede sein. Diese Ausführungen gelten aber auch, wie noch zu zeigen sein wird, für die Grenzen der Konzernleitung im drei- und mehrstufigen Konzern.

1. Die gesetzliche Ausgangslage

Die herrschende GmbH ist nicht Gesellschafterin ihrer Enkelgesellschaft und den Gesellschaften auf nachgeordneter Konzernstufe. Deshalb darf sie nicht unmittelbar, sondern nur mittelbar Einfluß auf die Enkelgesellschaft nehmen. Hierbei hat sie den vom Gesetz vorgezeichneten Weg einzuhalten. Will sie in der Enkelgesellschaft eine Maßnahme veranlassen, hat sie also zunächst in der Tochtergesellschaft einen Gesellschafterbeschluß herbeizuführen, der deren Geschäftsführer anweist, in der Enkelgesellschaft einen Weisungsbeschluß mit dem gewünschten Inhalt zu bewirken. Insbesondere ist ihr eine unmittelbare Einflußnahme auf die Enkelgesellschaft auch dann nicht erlaubt, wenn hierdurch weder die Tochtergesellschaft noch die Enkelgesellschaft geschädigt wird. Denn die Gesellschafter der Tochter-GmbH haben sich im Gesellschaftsvertrag verpflichtet, nur auf dem gesetzlich vorgesehenen Weg den Gesellschaftszweck der Tochtergesellschaft zu fördern. Hierdurch wird insbesondere die Mitwirkung der Minderheitsgesellschafter an der Entscheidungsfindung sichergestellt und ihnen die Möglichkeit gegeben, rechtswidrige Weisungen anzufechten. Das Mitverwaltungsrecht der Minderheitsgesellschafter würde aber umgangen, wenn die herrschende GmbH direkten Einfluß auf die Enkelgesellschaft nehmen würde.

2. Das Problem

Freilich wird das obengenannte Verfahren nicht immer eingehalten. Oft wählt das herrschende Unternehmen den „kleinen Dienstweg" und wendet sich direkt an den Geschäftsführer der Enkelgesellschaft, um ihn zu einem bestimmten Verhalten zu veranlassen. Entsteht der Enkelgesellschaft hierdurch ein Schaden, steht ihr die Tochtergesellschaft nicht als Schuldnerin eines Anspruchs aus der Verletzung mitgliedschaftlicher Förderpflichten zur Verfügung. Denn diese hat an der Schadensentstehung nicht mitgewirkt und eine Haftung für das Verhalten ihres Gesellschafters, der herrschenden GmbH, findet keine gesetzliche Grundlage.[184] Etwas anderes gilt allenfalls dann, wenn man die herrschende GmbH als verfassungsmäßig berufene Vertreterin der Tochtergesellschaft ansieht. Denn dann müßte das Verhalten der herrschenden GmbH

---

[184] *M. Lutter/P. Hommelhoff*, GmbHG, 1995, § 13 Rdnr. 19.

gemäß § 31 BGB der Tochtergesellschaft zugerechnet werden.[185] Das hätte zur Folge, daß die Tochtergesellschaft so behandelt würde, als ob sie selbst Einfluß auf die Enkelgesellschaft genommen hätte. In diesem Fall müßte in der Tat eine Verletzung der mitgliedschaftlichen Förderpflicht der Tochtergesellschaft gegenüber der Enkelgesellschaft bejaht werden, weil letztere außerhalb der Gesellschafterversammlung auf den Geschäftsführer Einfluß genommen hat. Aber auch dann könnte nur die Tochtergesellschaft, nicht aber die herrschende Gesellschaft auf Schadensersatz in Anspruch genommen werden. Damit würde die Enkelgesellschaft aber regelmäßig nur unzureichend geschützt. Vielmehr kann ein wirksamer Schutz nur in Form einer Pflicht der herrschenden GmbH gegenüber der Enkelgesellschaft auf Beachtung ihrer Interessen bewirkt werden. Denn damit wird das Entscheidungszentrum, von dem der Einfluß ausgeht, zusätzlich in die Pflicht genommen und so in verstärkter Weise von schädigenden Eingriffen abgehalten.

3. Die Treupflicht der herrschenden GmbH

W. Stimpel[186] versucht einen Schutz der mittelbar beherrschten GmbH dadurch zu erreichen, daß er der Mitgliedschaft des herrschenden Unternehmens in der Tochtergesellschaft Schutzwirkungen zugunsten der nachgeordneten Konzerngesellschaften und ihrer Minderheitsgesellschafter zuschreibt. Danach reiche die Treupflicht des Mehrheitsgesellschafters in der von der Tochter-GmbH beherrschten Unternehmensgruppe grundsätzlich so weit, wie sich dieser über die GmbH gesellschaftsrechtlich in der beherrschten Unternehmensgruppe durchsetzen könne. Hiermit knüpft *W. Stimpel* an die Rechtsprechung des *Bundesgerichtshofs*[187] an, wonach der Anstellungsvertrag des Geschäftsführers einer Komplementär-GmbH Schutzwirkungen zugunsten der Kommanditgesellschaft äußert, wenn die wesentliche Aufgabe der GmbH die Führung der Geschäfte der Kommanditgesellschaft ist.

Gegen die von *W. Stimpel* vorgeschlagene Konstruktion spricht jedoch, daß er die Treupflicht der Mutter gegenüber der Enkelgesellschaft aus einer Treupflicht zwischen Mutter und Tochter herleitet. Zwar bedeutet dies nicht, daß damit die Treupflicht gegenüber der Enkelgesellschaft zur Disposition der

---

185 In diese Richtung *Kammergericht Berlin*, Beschl. v. 9.7.1996 - 5 W 3253/96, in: BB 1996, S. 2426, 2427, das den Geschäftsführer einer herrschenden GmbH als verfassungsmäßig berufenen Vertreter einer Tochter-GmbH angesehen hat.

186 *W. Stimpel*, Innenhaftung, 1986, S. 119 f.; ihm folgend *M. Paschke*, Durchgriffsproblematik, 1988, S. 203; *M. Winter*, Treubindungen, 1988, S. 256 ff.; *L. Lawall*, Wettbewerbsverbot, 1996, S. 123 ff.; ablehnend *H.-D. Assman*, GmbH-Konzern, 1992, S. 709 f.

187 *BGH*, Urt. v. 12.11.1979 - II ZR 174/77, in: BGHZ 75, S. 321, 322 ff.

Tochter als eigentlichem Bezugspunkt der von der Mutter geschuldeten Treupflicht steht. Denn es ist mittlerweile anerkannt, daß sich die Schutz- und Rücksichtspflichten zugunsten Dritter nicht aus einer ergänzenden Auslegung des Hauptvertrages ergeben, sondern vom Willen der Parteien des Hauptvertrages grundsätzlich unabhängig sind und sich in der Sache als richterliche Rechtsfortbildung in Ergänzung dispositiven Rechts darstellen.[188] Jedoch gerät *W. Stimpel* in Begründungsschwierigkeiten, wenn die Tochtergesellschaft zu 100 % der herrschenden GmbH gehört. Denn gegenüber einer Tochtergesellschaft ohne Minderheitsgesellschafter besteht, wie noch zu zeigen sein wird, keine Treupflicht. Dann aber fehlt die Grundlage für die Schutz- und Rücksichtspflichten zugunsten der Enkelgesellschaft.[189] Gleichwohl ist die Enkelgesellschaft auch in diesem Fall schutzwürdig.

Einen anderen Ansatzpunkt wählt *E. Rehbinder*[190]. Er vertritt die Ansicht, daß das herrschende Unternehmen gegenüber der Enkelgesellschaft für jede nachteilige Einflußnahme analog § 43 Abs. 2 GmbHG auf Schadensersatz hafte. Dem kann in dieser Allgemeinheit jedoch nicht gefolgt werden. So ergeben sich aus § 43 Abs. 1 GmbHG selbst keine Verhaltenspflichten. Anders gewendet läßt sich aus § 43 Abs. 1 GmbHG nicht entnehmen, unter welchen Voraussetzungen der Geschäftsführer rechtswidrig handelt. § 43 Abs. 2 GmbHG setzt vielmehr ein pflichtwidriges Handeln des Geschäftsführers voraus und normiert nur einen Verschuldensmaßstab.[191] Die Verhaltenspflichten des Geschäftsführers ergeben sich dagegen aus seiner Geschäftsführungsaufgabe, aus Treupflichtüberlegungen und aus Sonderregelungen.[192] Will man daher § 43 Abs. 2 GmbHG auf das herrschende Unternehmen analog anwenden, so müßte zunächst der Nachweis geführt werden, daß das herrschende Unternehmen gegen eine Verhaltenspflicht verstoßen hat. Zwar erkennt *E. Rehbinder*[193] richtig, daß sich solche Verhaltenspflichten aus der Stellung des herrschenden Unternehmens als „faktisches Organ" der Enkelgesellschaft ergeben können. Darauf wird noch zurückzukommen sein. Eine solche Stellung nimmt das herrschende Unternehmen aber in der bloßen Abhängigkeitslage, in der es nur vereinzelt auf die Enkelgesellschaft Einfluß ausübt, nicht ein. Auch in der Abhängigkeitslage ist die Enkelgesellschaft aber schutzwürdig.

---

[188] *M. Winter*, Treubindungen, 1988, S. 256 f.

[189] Kritisch gegenüber Treupflichten mit Schutzwirkung zugunsten Dritter auch *P. Limmer*, Haftungsverfassung, 1992, S. 70 f.

[190] *E. Rehbinder*, Unternehmensverbindungen, 1977, S. 640 f.

[191] *H.-G. Koppensteiner*, Rowedder, 1997, § 43 Rdnr. 7.

[192] *W. Zöllner*, Baumbach/Hueck, 1996, § 43 Rdnr. 11.

[193] *E. Rehbinder*, Unternehmensverbindungen, 1977, S. 640.

Nach alledem bietet es sich an, die Treupflicht zwischen der herrschenden GmbH und den mittelbar beherrschten Gesellschaften direkt aus § 242 BGB herzuleiten.[194] Schon bei Begründung der Treupflichten zwischen den Gesellschaftern einer GmbH untereinander wurde ausgeführt, daß sich eine rechtliche Sonderverbindung, wie sie § 242 BGB verlangt, aus den erhöhten Einwirkungsmöglichkeiten auf fremde Interessen ergibt. Die herrschende GmbH ist nun nicht nur in der Lage, kraft ihrer Mitgliedschaft in der Tochtergesellschaft auf deren Interessen und die Interessen ihrer Minderheitsgesellschafter verstärkt einzuwirken. Sie ist auch in der Lage, über die Tochtergesellschaft auf die Enkelgesellschaft verstärkten Einfluß zu nehmen. Gerade dadurch, daß ihr Einfluß gesellschaftsrechtlich vermittelt ist, unterscheidet sie sich von konzernfremden Dritten, deren Eingriffe der Geschäftsführer der beherrschten Gesellschaft besser abwehren kann als die schädigende, im Extremfall durch eine Weisungskette vermittelte Einflußnahme des herrschenden Unternehmens. Daher besteht zwischen der herrschenden GmbH und den von ihr mittelbar beherrschten Unternehmen eine rechtliche Sonderverbindung, die gemäß § 242 BGB Treupflichten der herrschenden GmbH gegenüber den beherrschten Gesellschaften zur Folge hat.[195]

4. Der Inhalt der Treupflicht

Der Inhalt der Treupflicht der herrschenden GmbH gegenüber den mittelbar beherrschten Gesellschaften mbH ist nahezu deckungsgleich mit der mitgliedschaftlichen Förderpflicht der unmittelbar beteiligten Konzerngesellschaft.

Das bedeutet, daß die herrschende GmbH nur über die Gesellschafterin der mittelbar beherrschten Gesellschaft auf diese Einfluß nehmen darf. Verletzt sie diese Pflicht und entsteht der mittelbar beherrschten Gesellschaft hieraus ein Schaden, so ist die herrschende GmbH ihr zum Schadensersatz verpflichtet. Verschuldensmaßstab ist § 43 Abs. 1 GmbHG. Denn die herrschende GmbH

---

[194] Im Ergebnis ebenso *V. Emmerich*, Scholz, 1993, Anh. Konzernrecht, Rdnr. 170 f.; *H.-G. Koppensteiner*, Rowedder, 1997, Anh. nach § 52 Rdnr. 54; siehe auch *G. Teubner*, Politik, 1990, S. 274 f. In diese Richtung auch *BGH*, Urt. v. 5.12.1983 - II ZR 242/82 („Werbeagentur"), in: BGHZ 89, 162, 165, wonach die Alleingesellschafterin einer Holding-AG gegenüber einer GmbH & Co. KG, an deren Kommanditkapital und an deren Komplementär-GmbH die Holding-AG zu jeweils 80 % beteiligt war, einem Wettbewerbsverbot unterliege.

[195] Zwar spricht auch *P. Limmer*, Strukturhaftung, 1992, S. 64 ff., von einer Sonderverbindung zwischen herrschenden Unternehmen und abhängiger Gesellschaft. Der Unterschied zu der hier vertretenen Meinung besteht jedoch darin, daß er diese Sonderverbindung aus den konzernrechtlichen Vorschriften des Aktiengesetzes ableitet und damit die Sonderverbindung auf Unternehmensgesellschafter begrenzt.

maßt sich materiell Geschäftsführerbefugnisse an und muß deshalb wie ein Geschäftsführer haften.

Ferner darf sie die unmittelbar beteiligte Gesellschafterin nicht zu einer Verletzung ihrer mitgliedschaftlichen Förderpflicht veranlassen. Auch hierdurch würde sie ihre Treupflicht gegenüber der mittelbar beherrschten Gesellschaft verletzen.

Ebenso wenig darf sie Geschäftschancen der mittelbar beherrschten GmbH ausnutzen, die ihr aus dieser heraus vermittelt wurden. Und schließlich unterliegt sie gegenüber der mittelbar beherrschten Gesellschaft auch einem Wettbewerbsverbot.[196] Denn aufgrund ihrer gesellschaftsrechtlich vermittelten Macht besteht die qualifizierte Gefahr, daß sie Informationen der Enkelgesellschaft zugunsten ihres Konkurrenzunternehmens ausnutzt oder schädlichen Einfluß auf die Enkelgesellschaft nimmt.[197]

IV. Faktischer GmbH-Konzern ohne Minderheitsgesellschafter

1. Das Problem

Bisher wurde nur der Schutz der mehrgliedrigen GmbH untersucht. Sie zeichnet sich durch das Vorhandensein von Minderheitsgesellschaftern aus, deren Interesse am Erfolg der Gesellschaft durch die mitgliedschaftliche Förderpflicht des herrschenden Gesellschafters gegenüber der Gesellschaft geschützt wird. Da eine wirtschaftlich gesunde GmbH die beste Gewähr bietet, ihre Verbindlichkeiten begleichen zu können, werden auch die Gläubiger der beherrschten GmbH durch die mitgliedschaftliche Förderpflicht des herrschenden Unternehmens geschützt.[198] Allerdings handelt es sich hierbei nur um einen Reflexschutz.[199] Denn Inhaber des Anspruchs auf Erfüllung der mitgliedschaft-

---

[196] Ebenso *BGH*, Urt. v. 5.12.1983 - II ZR 242/82 („Werbeagentur"), in: BGHZ 89, 162, 165, für die Alleingesellschafterin einer Holding-AG gegenüber einer GmbH & Co. KG, an deren Kommanditkapital und an deren Komplementär-GmbH die Holding-AG zu jeweils 80 % beteiligt war. Wie hier auch *L. Lawall*, Wettbewerbsverbot, 1996, S. 124.

[197] Verletzt die herrschende GmbH durch ein- und dieselbe Maßnahme sowohl gegenüber ihrer Tochtergesellschaft wie ihrer Enkelgesellschaft ein Wettbewerbsverbot, so wird man die Rechtsfolgen dieses Verstoßes im Hinblick auf das Eintrittsrecht analog § 113 Abs. 1, 2. Halbs. HGB modifizieren müssen; denn nur *eine* Gesellschaft kann in das Geschäft eintreten. An dieser Stelle soll jedoch die Feststellung genügen, daß beide Gesellschaften von der herrschenden GmbH Unterlassung von Wettbewerb verlangen können.

[198] Zur Überschneidung von Gläubiger- und Minderheitenschutz siehe *P. Limmer*, Haftungsverfassung, 1992, S. 21 f.

[199] *P. Ulmer*, Gläubigerschutz, 1984, S. 393 f.

lichen Förderpflicht ist die Gesellschaft. Diese und damit ihre Gesellschafter können mithin über diesen Anspruch disponieren und im Extremfall den Reflexschutz der Gläubiger leerlaufen lassen. In besonderer Schärfe stellt sich dieses Problem in der Einmann-GmbH, in der per definitionem keine Minderheitsgesellschafter vorhanden sind. Hier scheint das Interesse des Gesellschafters mit dem Eigeninteresse der GmbH identisch zu sein, so daß eine Verletzung dieses Eigeninteresses ausgeschlossen erscheint.

2. *Meinungsstand*

Hiergegen wenden sich gewichtige Stimmen[200] in der Literatur. Sie wollen die Einmann-GmbH über die Gläubigerschutzvorschriften hinaus vor schädigenden Eingriffen ihres Gesellschafters schützen. Dabei lassen sich drei Begründungsansätze unterscheiden.

Die erste namentlich von *J. Wilhelm*[201] angeführte Meinung fordert, daß die juristische Person „ernstgenommen" werden müsse. Deshalb sei die GmbH als rechts- und schutzfähiges Subjekt mit eigenen Vermögensinteressen anzuerkennen. Das gelte auch für die Einmann-GmbH. Folge man dem, übten die Gesellschafter einer GmbH fremdbezogene Macht aus, wenn sie für sich Organmacht in Anspruch nähmen und die Geschicke der Gesellschaft leiteten. Für die Ausübung fremdbezogener Macht gelte aber der Rechtsgrundsatz der Pflichtgebundenheit. Deshalb treffe die bestimmungsmächtigen Gesellschafter die Sorgfaltspflichten eines Organs.[202]

Auch *K. Schmidt*[203] verlangt von den Gesellschaftern, die GmbH als rechts- und schutzfähiges Subjekt mit eigenen Vermögensinteressen zu respektieren. Im Gegensatz zu *J. Wilhelm* geht er aber davon aus, daß bei der Einflußnahme auf die Geschäftsführung nicht nur die bestimmungsmächtigen Gesellschafter, sondern alle Gesellschafter gegenüber der Gesellschaft verpflichtet seien, diese nicht durch planvolle Unterkapitalisierung oder Auszehrung in den Ruin zu treiben. Verstieße ein Gesellschafter gegen diese Pflicht, sei er zum Schadensersatz verpflichtet. Insbesondere scheitere die Haftung nach dem Gedanken des § 43 Abs. 3 Satz 3 GmbHG insoweit nicht an der Zustimmung der Mitgesellschafter, als der Schadensersatz zur Befriedigung der Gläubiger erforderlich sei.

---

200 *K. Schmidt*, Konkursverschleppungshaftung, 1988, S. 1505 f.; *M. Winter*, Eigeninteresse, 1994, S. 580 ff; *H.-J. Priester*, Eigensphäre, 1993, S. 521 ff.; *U. Burgard*, Anmerkung, 1995, S. 953 f.; *H. Ziemons*, Haftung, 1996, S. 97 ff.

201 *J. Wilhelm*, Rechtsform, 1981, S. 12 ff.

202 *J. Wilhelm*, Rechtsform, 1981, S. 337, 355 f.

203 *K. Schmidt*, Konkursverschleppungshaftung, 1988, S. 1505 f.

Der zweite Begründungsansatz sucht einen Schutz der Einmann-GmbH über eine erweiterte Anwendung der Gläubigerschutzvorschriften. So will *M. Winter*[204] aus der Wertung der §§ 65 ff. GmbHG für den Gesellschafter das Verbot herleiten, die Gesellschaft auf ‚kaltem Wege' zu liquidieren. Existenzgefährdende Maßnahmen seien deshalb unzulässig. *P. Ulmer*[205] schlägt eine Erweiterung des Ausschüttungsverbotes von § 30 Abs. 1 GmbHG vor. Werde der Bestand der Gesellschaft durch Entzug von Produktionsgrundlagen oder existenznotwendiger Liquidität gefährdet, so sei den damit verbundenen Risiken durch Wertberichtigungen und Rückstellungen in der nach § 30 GmbHG aufzustellenden Vermögensbilanz Rechnung zu tragen. Bei einer derart modifizierten Bilanz seien existenzgefährdende Gewinnausschüttungen kaum noch denkbar. Schließlich knüpft auch *H.-J. Fleck*[206] an die Kapitalschutzvorschriften an. Er leitet aus ihrem Zusammenhang und Zweck das Verbot ab, durch Entnahmen die Existenz der GmbH in so starkem Maße zu gefährden, daß ihr wirtschaftlicher Zusammenbruch mit hoher Wahrscheinlichkeit zu befürchten sei.

Der dritte Begründungsansatz schließlich ist konzernrechtlicher Natur. *P. Ulmer*[207] leitet ein Verbot der Existenzgefährdung aus einer Analogie zu den §§ 300 ff., 317, 322, 324 AktG her, die einen eigenständigen Gläubigerschutz gegenüber dem herrschenden Unternehmen statuierten. Die analoge Anwendung dieser Vorschriften sei geboten, weil dem Gläubigerschutz in der zu 100 % abhängigen GmbH nicht nur dasselbe, sondern sogar größeres Gewicht beizumessen sei als in der abhängigen Aktiengesellschaft. Denn im GmbH-Recht fehle es an einer eigenverantwortlichen Geschäftsführung und die Kapitalerhaltungsvorschriften seien weniger ausgeprägt.

Einen ähnliche Weg geht *P. Limmer*. Er führt aus, daß der Gläubigerschutz im Konzern in zweierlei Hinsicht beeinträchtigt werde.[208] Zum einen sei die Funktionsfähigkeit der Kapitalerhaltungsvorschriften wegen der breitflächigen Einwirkung des herrschenden Unternehmens nicht mehr gewährleistet. Zum anderen liege den Normen des Gesellschafts-, Unternehmens- und Wirtschaftsrechts das Leitbild eines wirtschaftlich rational handelnden Individuums zugrunde. Deshalb dürften auch die kreditgebenden Gläubiger davon ausgehen, daß sich das Handeln ihres Vertragspartners nach ökonomischen Prinzipien

---

204 *M. Winter*, Eigeninteresse, 1994, S. 586.

205 *P. Ulmer*, Schutz, 1988, S. 868 f. Ihm folgend *S. Beinert*, Konzernhaftung, 1995, S. 86.

206 *H.-J. Fleck*, Mißbrauch, 1990, S. 36 ff.

207 *P. Ulmer*, Gläubigerschutz, 1984, S. 418 ff. Siehe auch *H. Altmeppen*, Managerhaftung, 1998, S.80, nach dessen Ansicht „das in §§ 311, 317 AktG vorausgesetzte, selbstverständliche Schädigungsverbot" auch für die GmbH gilt, soweit das Vermögen der GmbH nicht disponibel ist.

208 *P. Limmer*, Haftungsverfassung, 1992, S. 41 ff.

richte. Da dies bei einem Konzernunternehmen nicht der Fall sei, werde das Vertrauen der Gläubiger aber enttäuscht. Zum Ausgleich dieser doppelten Interessenbeeinträchtigung plädiert *P. Limmer*[209] für „eine von den §§ 30 ff. GmbHG abstrakte, spezifisch konzernrechtliche Kapitalerhaltungspflicht", die den Bestand der Gesellschaft schützen soll.

Die Rechtsprechung[210] und mit ihr eine beachtliche Meinung in der Literatur[211] verneinen indes ein von dem Interesse der Gesellschafter unabhängiges Gesellschaftsinteresses. Zur Begründung führen sie an, daß die Gesellschafter den Gesellschaftszweck jederzeit ändern und insbesondere die Gesellschaft auflösen könnten.[212] Überdies steht nach *H. Wiedemann*[213] eine Bestandserhaltungspflicht in Widerspruch zur Haftung des herrschenden Unternehmens im qualifizierten faktischen Konzern. Während im letzteren Falle nur eine Verlustübernahmepflicht eingreife, seien die Gesellschafter im ersteren Falle verpflichtet, Arbeitnehmern und Gläubigern eine bestimmte Organisation zur Verfügung zu stellen. Schließlich läßt sich nach *W. Zöllner*[214] aus den Grenzen, die die Gläubigerschutzvorschriften dem Einmann-Gesellschafter ziehen würden, kein „generalklauselartiges Eigeninteresse der GmbH nach Art des (sinnvoll nur mit Hilfe des GesZwecks umschreibbaren) GesInteresses hypostasieren, aus dem in bester begriffsjuristischer Manier Folgerungen gezogen werden".

3. Stellungnahme

Die Frage, ob es ein vom Gesellschafter unabhängiges Gesellschaftsinteresse gibt, ist mißverständlich: Neben dem Gesellschaftsinteresse gibt es in der GmbH stets auch das Gläubigerinteresse, das von den Gesellschaftern nicht mißachtet werden darf.[215] Folgt man dem, so ist zwischen der mitgliedschaftli-

---

209 *P. Limmer*, Haftungsverfassung, 1992, S. 44, 156 ff.

210 *BGH*, Urt. v. 10.5.1993 - II ZR 74/92, in: GmbHR 1993, S. 427. Offengelassen von *LG Bremen*, Urt. v. 19.11.1997 - 4 O 1073/96, in: ZIP 1998, S. 561, 565.

211 *H.-D. Assmann*, GmbH-Konzern, 1992, S. 707; *T. Verhoeven*, Innenrecht, 1978, Rdnr. 159. Differenzierend zwischen der Lage innerhalb und außerhalb des Konzerns: *H.-G. Koppensteiner*, Rowedder, 1997, § 43 Rdnr. 60; Anh. nach § 52 Rdnr. 56.

212 *W. Zöllner*, Baumbach/Hueck, 1996, Anh. GmbH-KonzernR, Rdnr. 100; *M. Lutter*, Haftung, 1985, S. 1428 f.; *V. Röhricht*, Wettbewerbsverbot, 1992, S. 785; *M. Lutter/P. Hommelhoff*, GmbHG, 1995, § 13 Rdnr. 22; *H. Wiedemann*, Unternehmensgruppe, 1988, S. 87.

213 *H. Wiedemann*, Unternehmensgruppe, 1988, S. 87.

214 *W. Zöllner*, Baumbach/Hueck, 1996, Anh. GmbH-KonzernR, Rdnr. 100. Ebenso *P. Versteegen*, Konzernverantwortlichkeit, 1993, S. 92.

215 *H. Altmeppen*, GmbHG, 1997, § 13 Rdnr. 75.

chen Förderpflicht und den Pflichten des Gesellschafters zum Schutze der Gläubiger zu unterscheiden.

Die mitgliedschaftliche Förderpflicht ergibt sich aus der Verpflichtung mindestens zweier Gesellschafter, einen gemeinsamen Zweck zu verfolgen, § 705 BGB. Ist nur ein Gesellschafter vorhanden, scheidet ein gemeinsamer Zweck und damit auch die auf seine Förderung gerichtete Pflicht schon per definitionem aus. Mit anderen Worten wird die mitgliedschaftliche Förderpflicht der Gesellschaft nur als Zweckgemeinschaft geschuldet und schützt sie nur das Interesse der Gesellschafter an der Verfolgung eben dieses Zwecks. Hieraus folgt, daß der Einmann-Gesellschafter gegen keine mitgliedschaftliche Förderpflicht verstößt, wenn er der Gesellschaft Nachteile zufügt oder sie gar in den Ruin treibt. Und gleiches gilt, wenn die Gesellschaft zwar mehrere Gesellschafter hat, diese aber einvernehmlich handeln.

Somit läßt sich ein Verbot nachteiliger Eingriffe in die Geschäftsführung nur mit den Interessen der Gläubiger rechtfertigen. Deren Interessen werden in den §§ 30 ff. GmbHG geschützt. Danach darf insbesondere das zur Erhaltung des Stammkapitals erforderliche Vermögen nicht an die Gesellschafter ausgezahlt werden (§ 30 Abs. 1 GmbHG). Umgekehrt folgt hieraus, daß sonstige Schädigungen des Gesellschaftsvermögens grundsätzlich erlaubt sind. Und das ist auch interessengerecht, solange die Befriedigung der Gläubiger gewährleistet ist. Deshalb ist den in der Literatur vertretenen Ansichten zuzustimmen, die erst bei einer Existenzgefährdung einen über das geschriebenen Recht hinausgehenden Gläubigerschutz fordern. Denn erst wenn die Existenz der Gesellschaft gefährdet ist, ist die Befriedigung der Gläubiger nicht mehr gewährleistet. Und gleichzeitig wird hiermit der Ansatzpunkt für ein Verbot existenzgefährdender Eingriffe deutlich, nämlich die analoge Anwendung gläubigerschützender Vorschriften.

Allerdings vermag der Vorschlag von *P. Ulmer*, das herrschende Unternehmen bei existenzgefährdenden Eingriffen analog §§ 300 ff., 317, 322, 324 AktG auf Schadensersatz haften zu lassen, nicht zu überzeugen. So wurde oben[216] bereits festgestellt, daß die §§ 311 ff. AktG ein Schädigungsverbot voraussetzen und sie nur dessen effektive Einhaltung gewährleisten wollen. § 317 AktG scheidet damit als Analogiegrundlage für die Begründung eines Schädigungsverbots aus. Dasselbe gilt aber auch für die §§ 300 ff., 322, 324 AktG. Auch sie setzen gegenüber der beherrschten Aktiengesellschaft ein Schädigungsverbot voraus und machen hiervon nur deshalb eine Ausnahme, weil sich das herrschende Unternehmen im Beherrschungsvertrag bzw. Eingliederungsvertrag

---

216 Siehe B. II. 3.

zum Verlustausgleich, zur Ausfallhaftung oder zur Mithaftung für die Verbindlichkeiten der beherrschten Aktiengesellschaft verpflichtet hat.[217]

Als Zwischenergebnis kann somit festgehalten werden, daß sich aus den §§ 300 ff., 317, 322, 324 AktG keine Schadensersatzpflicht des Alleingesellschafters für ruinöse Maßnahmen ergibt.

Diese kann somit nur auf eine analoge Anwendung von §§ 30 ff. GmbHG bzw. §§ 65 ff. AktG gestützt werden. Hier überzeugt am meisten der Vorschlag von *P. Ulmer*, das Ausschüttungsverbot von § 30 Abs. 1 GmbHG zu erweitern und den mit existenzgefährdenden Maßnahmen verbundenen Risiken durch Wertberichtigungen und Rückstellungen in der nach § 30 GmbHG aufzustellenden Vermögensbilanz Rechnung zu tragen. Folgt man dem, so dürften existenzgefährdende Eingriffe in den meisten Fällen eine Auszahlung aus dem zur Erhaltung des Stammkapitals erforderlichen Vermögen zur Folge haben. Ist das aber so, so hat der Alleingesellschafter im Falle des Insolvenzverfahrens über das Vermögen der Gesellschaft das Stammkapital in Höhe der Auszahlung wieder aufzufüllen.

4. Existenzgefährdende Maßnahmen

Im Folgenden soll nun versucht werden, Fallgruppen für existenzgefährdende Maßnahmen aufzuzeigen. Wichtigste Fallgruppe dürfte der Abzug existenznotwendiger Liquidität sein. Denn dieser führt zur Zahlungsunfähigkeit der Gesellschaft und damit zur Eröffnung des Insolvenzverfahrens.[218] Dieselben Folgen kann der Entzug von Produktionsanlagen haben. Kann die Gesellschaft nämlich nicht mehr ausreichend produzieren, erwirtschaftet sie gegebenenfalls nicht mehr genug Gewinne, um ihre Verbindlichkeiten zu begleichen.[219] Aber auch die schlichte Veräußerung von Grundbesitz kann zu einer Bestandsgefährdung führen, wenn der Grundbesitz für die Produktion oder als Kreditunterlage existenznotwendig ist.[220] Zu denken ist schließlich an eine ‚Austrocknung' der

---

[217] Siehe z. B. *U. Hüffer*, AktG, 1997, § 302 Rdnr. 3: Verlustausgleichspflicht soll Lockerung der Vermögensbindung (§ 291 Abs. 3 AktG) kompensieren.

[218] Im Ergebnis ebenso *P. Ulmer*, Schutz, 1988, S. 868; *H.-J. Priester*, Eigensphäre, 1993, S. 525; *P. Limmer*, Haftungsverfassung, 1992, S. 159 f.; *T. Drygala*, Verhaltenshaftung, 1993, S. 325. Für die satzungsgemäß abhängig gegründete GmbH: *S. Beinert*, Konzernhaftung, 1995, S. 120.

[219] Im Ergebnis ebenso *P. Ulmer*, Schutz, 1988, S. 868; *H.-J. Priester*, Eigensphäre, 1993, S. 525.

[220] *H.-J. Priester*, Eigensphäre, 1993, S. 526.

Gesellschaft dadurch, daß Geschäfte bewußt an ihr vorbei auf eine andere Gesellschaft geleitet werden.[221]

V. Aktienvertragskonzern

1. Die gesetzliche Ausgangslage

Durch Abschluß eines Beherrschungsvertrages erhält die herrschende GmbH das Recht, dem Vorstand der beherrschten Aktiengesellschaft Weisungen zu erteilen. Dieser ist gemäß § 308 Abs. 2 AktG verpflichtet, den Weisungen Folge zu leisten. Damit ändert sich die Organisationsverfassung der beherrschten Aktiengesellschaft. Aber auch bei der herrschenden Gesellschaft tritt ebenso wie im faktischen Konzern eine Umformung der gesamten Zuständigkeitsordnung ein.[222] So wurde für die herrschende GmbH bereits aufgezeigt, daß die Gründung eines faktischen Konzerns mit der Zunahme der Pflichten ihres Geschäftsführers und der Mediatisierung der Weisungs- und Zustimmungsbefugnissen ihrer Gesellschafterversammlung verbunden ist. Ähnliches gilt für die herrschende Aktiengesellschaft, bei der der Einschnitt in die Rechte der Hauptversammlung freilich nicht so groß ist wie bei einer GmbH. Und all das gilt schließlich auch im Vertragskonzern.

Nach allem wird deutlich, daß der Beherrschungsvertrag die Organisation der herrschenden und beherrschten Gesellschaft ändert. Deshalb wird er zu Recht als Organisationsvertrag charakterisiert.[223]

Der Beherrschungsvertrag hat aber nicht nur Auswirkungen auf die Organisation der beherrschten Gesellschaft. Er beeinflußt auch ihren Gesellschaftszweck. Denn das herrschende Unternehmen ist berechtigt, auch nachteilige Weisungen zu erteilen, sofern diese nur im Konzerninteresse liegen (§ 308 Abs. 1 AktG). Hieraus ergibt sich aber, daß durch den Abschluß des Beherrschungsvertrages die unternehmerische Interessenausrichtung der beherrschten Gesell-

---

[221] *H.-J. Priester*, Eigensphäre, 1993, S. 526.

[222] *U. Bälz*, Unternehmen, 1992, S. 285.

[223] *E. Geßler*, Geßler, 1975, § 291 Rdnr. 24; *V. Emmerich*, Aktienkonzernrecht, 1998, § 291 Rdnr. 22, 23; *U. Hüffer*, AktG, 1997, § 291 Rdnr. 17; *Uwe H. Schneider*, Konzernbildung, 1980, S. 515. Für die GmbH: *M. Hoffmann-Becking*, Unternehmensvertrag, 1994, S. 58; *V. Emmerich*, Scholz, 1993, Anh. Konzernrecht Rdnr. 237; *M. Lutter/P. Hommelhoff*, GmbHG, 1995, Anh. § 13 Rdnr. 35; *BGH*, Beschl. v. 24.10.1988 - II ZB 7/88, in: BGHZ 105, S. 324, 331 („Supermarkt"); ausführlich *M. Kort*, Beherrschungsverträge, 1986, S. 55 ff.

schaft geändert wird.[224] Gesellschaftszweck ist nicht mehr die selbständige Gewinnerzielung, sondern die Förderung des Konzerninteresses.[225] Damit ändert sich bei Weisungen des herrschenden Unternehmens auch der Maßstab für das Handeln der Geschäftsleitung der beherrschten Gesellschaft, das sich nun nicht mehr am Gesellschaftsinteresse, sondern am Konzerninteresse zu orientieren hat.[226] Macht das herrschende Unternehmen dagegen von seinem Weisungsrecht keinen Gebrauch, muß sich der Vorstand der beherrschten Gesellschaft weiterhin am Gesellschaftsinteresse orientieren. Denn ob eine Maßnahme im Konzerninteresse liegt, kann letztlich nur das herrschende Unternehmen zuverlässig beurteilen.[227]

Nach alledem darf das herrschende Unternehmen der beherrschten Gesellschaft nachteilige Weisungen erteilen. Trotzdem ist sein Weisungsrecht nicht grenzenlos. Schranken können sich aus dem Konzerninteresse, dem Beherrschungsvertrag, aus der Satzung der beherrschten Gesellschaft sowie aus zwingenden gesetzlichen Vorschriften ergeben. Dem ist im folgenden nachzugehen.

2. Das Konzerninteresse

Das Konzerninteresse hat nicht nur Bedeutung für die beherrschte Gesellschaft, die mit Abschluß des Beherrschungsvertrages auf eben dieses Interesse ausgerichtet wird. Es hat auch Bedeutung für das herrschende Unternehmen. Denn dessen Weisungsrecht wird durch das Konzerninteresse begrenzt. So darf eine Weisung gemäß § 308 Abs. 1 Satz 2 AktG nur dann zu einem Nachteil bei der beherrschten Aktiengesellschaft führen, wenn die Nachteilszufügung im Konzerninteresse liegt. Mit anderen Worten muß sich die Weisung unmittelbar oder mittelbar auf die Vermögens- und Ertragslage des herrschenden Unternehmens oder einer anderen Konzerngesellschaft günstig auswirken.[228] Ungeschriebene Voraussetzung ist hierbei allerdings der Verhältnismäßigkeitsgrund-

---

224 *Uwe H. Schneider*, Konzernbildung, 1980, S. 516; ders., Konzernleitung, 1981, S. 251. Nach *U. Bälz*, Unternehmen, 1992, S. 301, entsteht durch Abschluß eines Beherrschungsvertrages ein „polykorporatives Unternehmen".

225 *V. Emmerich/J. Sonnenschein*, Konzernrecht, 1997, S. 410.

226 *BGH*, Urt. v. 24.10.88 - II ZB 7/88, in: BGHZ 105, 324, 331; *M. Hoffmann-Becking*, Unternehmensvertrag, 1994, S. 61. Siehe auch *U. Bälz*, Unternehmen, 1992, S. 307.

227 *E. Geßler*, Geßler, 1975, § 308 Rdnr. 75.

228 *U. Hüffer*, AktG, 1997, § 308 Rdnr. 17; *V. Emmerich*, Aktienkonzernrecht, 1998, § 308 Rdnr. 35; *H. Altmeppen*, Managerhaftung, 1998, S. 21.

satz. Danach darf der zugefügte Nachteil im Verhältnis zum Vorteil des Konzerns nicht unverhältnismäßig groß sein.[229]

3. Der Beherrschungsvertrag

Des weiteren können sich Schranken des Weisungsrechts aus dem Beherrschungsvertrag selbst ergeben. Das folgt aus § 308 Abs. 1 Satz 2 AktG („Bestimmt der Vertrag nichts anderes, ..."). Danach können z. B. nachteilige Weisungen ausgeschlossen werden. Möglich ist aber auch der Ausschluß nur bestimmter Weisungen oder des Weisungsrechts für bestimmte Einzelfragen. Schließlich können auch Formvorschriften für die Ausübung des Weisungsrechts vereinbart werden.[230] Betont werden muß aber auch, daß in der Praxis von einer Begrenzung des Weisungsrechts nur selten Gebrauch gemacht wird.[231]

4. Die Satzung der beherrschten Gesellschaft

Ferner muß das herrschende Unternehmen die Satzung der beherrschten Aktiengesellschaft beachten. Denn der Tochtervorstand darf nur zu solchen Maßnahmen veranlaßt werden, zu deren Ausführung er nach der gesetzlichen Zuständigkeitsordnung seiner eigenen Gesellschaft auch berechtigt ist.[232]

Das bedeutet insbesondere, daß sich die Weisungen im Rahmen des Unternehmensgegenstandes der beherrschten Gesellschaft halten müssen. Das herrschende Unternehmen darf die Geschäftsleitung der beherrschten Gesellschaft also nicht anweisen, eine Tätigkeit aufzunehmen, die außerhalb des Unternehmensgegenstandes dieser Gesellschaft liegt. Ebensowenig darf es die Geschäftsleitung veranlassen, die Geschäftstätigkeit aufzugeben oder die Verfolgung des Unternehmensgegenstandes teilweise einzustellen.[233]

---

229 *H. Würdinger*, Großkommentar, 1971, § 308 Anm. 13; *U. Hüffer*, AktG, 1997, § 308 Rdnr. 17; *H. Altmeppen*, Managerhaftung, 1998, S. 21.

230 *G. Krieger*, Vertragskonzern, 1988, § 70 Rdnr. 100.

231 *V. Emmerich/J. Sonnenschein*, Konzernrecht, 1997, S. 143; *M. Hoffmann-Becking*, Unternehmensvertrag, 1994, S. 60; *H. Altmeppen*, Managerhaftung, 1998, S. 5.

232 *E. Geßler*, Geßler, 1975, § 308 Rdnr. 56.

233 *V. Emmerich/J. Sonnenschein*, Konzernrecht, 1997, S. 321; *V. Emmerich*, Aktienkonzernrecht, 1998, § 308 Rdnr. 46.

5. Gesetz

Weiterhin muß sich die Weisung im Rahmen zwingender gesetzlicher Vorschriften halten.[234] Das herrschende Unternehmen darf den Vorstand der beherrschten Aktiengesellschaft z. B. nicht dazu anweisen, auf den Verlustausgleichsanspruch der Gesellschaft zu verzichten. Denn das würde gegen § 302 AktG verstoßen. Auch darf es ihn nicht dazu anweisen, mit einem anderen Unternehmen einen nach § 1 Abs. 1 GWB unerlaubten Kartellvertrag zu schließen. Schließlich sei noch auf die zahlreichen umweltrechtlichen Vorschriften hingewiesen, die im Rahmen der Unternehmensleitung von der Geschäftsführung der beherrschten Gesellschaft zu beachten sind.

6. Insbesondere existenzgefährdende Weisungen

Umstritten ist die Zulässigkeit existenzgefährdender Weisungen. Eine Existenzgefährdung ist trotz der Verlustausgleichspflicht des herrschenden Unternehmens möglich. Das gilt nicht nur für die Zeit nach Beendigung des Beherrschungsvertrages, sondern schon während seines Bestehens. So sichern die §§ 300 bis 302 AktG der Gesellschaft lediglich das bilanzielle Anfangsvermögen, das zum Zeitpunkt des Vertragsabschlusses bestand. Gegen den Abzug existenznotwendiger Liquidität gewähren sie dagegen keinen Schutz.[235] Deshalb ist es denkbar, daß eine vertraglich konzernierte Gesellschaft wegen Zahlungsunfähigkeit die Eröffnung des Insolvenzverfahrens beantragen muß. Dagegen ist die Spannbreite existenzgefährdender Weisungen, die sich erst nach Beendigung des Beherrschungsvertrages auswirken, breiter. Genannt sei hier nur die Ausgliederung des Vertriebs auf eine zentrale Vertriebsgesellschaft. Sie kann zur Folge haben, daß die Gesellschaft nach dem Ausscheiden aus dem Konzernverbund nicht mehr in der Lage ist, sich eigene Absatzmärkte zu erschließen. Und genannt sei schließlich die Beschränkung der Herstellung auf ein Produkt, von dessen alleinigen Vertrieb die Gesellschaft nicht mehr selbständig leben kann.[236]

---

234 *U. Hüffer*, AktG, 1997, § 308 Rdnr. 14; *H. Altmeppen*, Managerhaftung, 1998, S. 20.

235 *H. Brandes*, Kapitalerhaltung, 1991, S. 26; *H.-G. Koppensteiner*, Kölner Kommentar, 1987, § 302 Rdnr. 3; in diese Richtung auch *BGH*, Urt. v. 19.9.1988 - II ZR 255/87, in: BGHZ 105, S. 168, 182 ff. A. M. *H. Altmeppen*, Managerhaftung, 1998, S.8, 23; *P. Hommelhoff*, Eigenkapital-Ersatz, 1984, S. 1112 f.; *V. Emmerich/J. Sonnenschein*, Konzernrecht, 1997, S. 263.

236 Zu weiteren Beispielen siehe *V. Emmerich/J. Sonnenschein*, Konzernrecht, 1997, S. 322.

Nach herrschender Meinung[237] sind Weisungen, die die Existenz der beherrschten Gesellschaft vernichten, unzulässig.

Zur Begründung wird angeführt, daß es „nicht im Bereich legitimer Rechtsausübung liegen" könne, wenn gerade durch die Ausübung der vertraglich eingeräumten Rechte die rechtliche Existenz des Vertragspartners aufgehoben würde.[238] Dieses Argument überzeugt nicht. Denn es ist gerade zu prüfen, ob eine Existenzvernichtung rechtlich zulässig ist.

Überzeugender ist schon die Berufung auf § 299 AktG.[239] Nach dem Wortlaut dieser Vorschrift darf das herrschende Unternehmen die beherrschte Gesellschaft u. a. nicht anweisen, den Beherrschungsvertrag zu beenden. Denselben Effekt erzielen aber Weisungen, die die Existenz der beherrschten Gesellschaft vernichten und damit ebenfalls zur Beendigung des Beherrschungsvertrages führen. Auch könnte das herrschende Unternehmen sich damit seiner Ausgleichszahlungen an die beherrschte Gesellschaft und die außenstehenden Aktionäre entledigen, ohne daß letztere, wie es § 297 Abs. 2 AktG vorsieht, zustimmen. Allerdings läßt sich damit nur ein Bestandsschutz für die Zeit des Bestehens des Beherrschungsvertrages begründen. Ein Verbot für Weisungen, die sich erst nach Beendigung des Beherrschungsvertrages existenzschädigend auswirken, kann aus diesen Überlegungen nicht hergeleitet werden.

Schließlich wird noch die Ansicht vertreten, daß es kein Konzerninteresse gäbe, welches die Vernichtung der beherrschten Gesellschaft rechtfertigen könne.[240]

Das bedarf näherer Betrachtung. § 308 Abs. 1 Satz 2 AktG spricht nicht von Konzerninteresse, sondern bestimmt, daß auch Weisungen erteilt werden können, „die für die Gesellschaft nachteilig sind, wenn sie den Belangen des herrschenden Unternehmens oder der mit ihm und der Gesellschaft konzernverbundenen Unternehmen dienen". Nur unter dieser Voraussetzung hielt der Gesetzgeber den mit dem Beherrschungsvertrag verbundenen Eingriff in die Struktur der Aktiengesellschaft für hinnehmbar. Nicht unternehmerisch motivierte Ein-

---

[237] *V. Emmerich/J. Sonnenschein*, Konzernrecht, 1997, S. 321; *V. Emmerich*, Aktienkonzernrecht, 1998, § 308 Rdnr. 50; *H. Hirte*, Teilkodifikation, 1993, S. 25; *M. Hoffmann-Becking*, Unternehmensvertrag, 1994, S. 61; *OLG Düsseldorf*, Beschl. v. 7.6.1990 - 19 W 13/86, in: Die AG 1990, S. 490, 492; *BGH*, Urt. v. 5.2.1979 - II ZR 210/76, in: BB 1979, S. 1735, 1738. A. M. *H.-G. Koppensteiner*, Kölner Kommentar, 1987, § 308 Rdnr. 32.

[238] *U. Immenga*, Bestandsschutz, 1976, S. 307.

[239] *U. Immenga*, Bestandsschutz, 1976, S. 307.

[240] *V. Emmerich/J. Sonnenschein*, Konzernrecht, 1997, S. 322. A. M. *U. Immenga*, Bestandsschutz, 1976, S. 306; *H.-G. Koppensteiner*, Kölner Kommentar, 1987, § 308 Rdnr. 34.

griffe wollte er damit ausschließen.[241] Nimmt man diese Definition wörtlich, so sind durchaus Sachverhaltsgestaltungen denkbar, bei denen eine Vernichtung der beherrschten Gesellschaft den Belangen des herrschenden Unternehmens oder einer konzernverbundenen Gesellschaft dienen kann und überdies der Verhältnismäßigkeitsgrundsatz gewahrt ist. Läßt man dagegen das Interesse der beherrschten Gesellschaft in das Konzerninteresse einfließen und begreift das Konzerninteresse als Gesamtinteresse des herrschenden Unternehmens und der beherrschten Gesellschaft sowie der übrigen konzernverbundenen Unternehmen, so sind existenzschädigende Maßnahmen ausgeschlossen. Denn im Interesse einer Gesellschaft liegt niemals die Vernichtung ihrer Existenz.

Für eine solche Auslegung des Konzerninteresses spricht ein Vergleich mit der Lage im Eingliederungskonzern. Dort hat der Gesetzgeber gerade nicht das Weisungsrecht des herrschenden Unternehmens begrenzt. § 308 Abs. 1 AktG ist in § 323 Abs. 1 Satz 2 AktG nicht erwähnt. Das kann nur damit begründet werden, daß im Eingliederungskonzern anders als im Vertragskonzern keine außenstehenden Aktionäre vorhanden sind. Hieraus folgt umgekehrt, daß die Begrenzung des Weisungsrechts im Vertragskonzern dem Schutz der außenstehenden Aktionäre dienen soll. Deren Interesse ist aber auf den Fortbestand der Gesellschaft sowohl während des Bestehens und als auch nach Beendigung des Beherrschungsvertrages gerichtet.

Als Ergebnis kann somit festgehalten, daß das Interesse der außenstehenden Aktionäre in die Bestimmung des Konzerninteresses einfließt mit der Folge, daß existenzschädigende Weisungen verboten sind.

Hieraus folgt aber auch, daß existenzschädigende Weisungen zulässig sind, wenn sich die Gesellschaft zu 100 % in der Hand des herrschenden Unternehmens befindet.

Entsprechendes gilt für die beherrschte GmbH. Da auf diese § 308 Abs. 1 Satz 2 AktG analog angewandt wird, muß der Begriff des Konzerninteresses im GmbH-Vertragskonzern ebenso ausgelegt werden wie im Aktien-Vertragskonzern.

VI. GmbH-Vertragskonzern

1. Die gesetzliche Ausgangslage

Nach einhelliger Auffassung sind die aktienrechtlichen Vorschriften über den Beherrschungsvertrag mit einer beherrschten Aktiengesellschaft zum größten

---

[241] *Begr. RegE.* zu § 308 Abs. 1 und Abs. 2 AktG, in: *B. Kropff*, AktG, 1965, S. 403.

Teil auf den Beherrschungsvertrag mit einer beherrschten GmbH analog anwendbar.[242] Für die Konzernleitung ist insbesondere § 308 Abs. 1 u. 2 AktG zu erwähnen, wonach das herrschende Unternehmen das Recht erhält, dem Geschäftsführer der beherrschten GmbH Weisungen zu erteilen. Ein herrschender Gesellschafter ist also nicht mehr darauf angewiesen, zur Herbeiführung der von ihm gewünschten Maßnahmen den Weg über die Gesellschafterversammlung zu gehen. Ferner ist er nicht mehr an seine mitgliedschaftlicher Förderpflicht gebunden. Das bedeutet, daß er auch nachteilige Weisungen erteilen darf, sofern diese nur im Konzerninteresse liegen. Das hat auch zur Folge, daß er gegenüber der Gesellschaft keinem Wettbewerbsverbot mehr unterliegt.[243] Denn es wäre widersprüchlich, auf der einen Seite ein präventives Wettbewerbsverbot zu fordern, das eine schädliche Einflußnahme bereits im Vorfeld verhindern soll, auf der anderen Seite aber nachteilige Weisungen zuzulassen. Insoweit gilt dasselbe wie für den herrschenden Unternehmensgesellschafter einer Aktiengesellschaft im faktischen Aktienkonzern. Weiterhin hat der Beherrschungsvertrag zur Folge, daß die Weisungsbefugnis der Gesellschafterversammlung in Geschäftsführungsangelegenheiten ruht, solange der Beherrschungsvertrag besteht. Denn durch den Abschluß des Beherrschungsvertrages geht die Weisungskompetenz der Gesellschafterversammlung auf das herrschende Unternehmen über.[244] Im übrigen bleibt die gesetzliche Zuständigkeitsordnung aber unberührt. Insbesondere ist das herrschende Unternehmen an den Unternehmensgegenstand der beherrschten Gesellschaft gebunden.[245] Ferner muß es wie im Aktien-Vertragskonzern die allgemeinen gesetzlichen Vorschriften beachten und darf die beherrschte Gesellschaft nicht zu einem rechtswidrigen Verhalten anweisen. Umstritten ist indes, ob § 30 GmbHG im Vertragskonzern gilt.

---

[242] Statt aller *W. Zöllner*, Baumbach/Hueck, 1996, GmbH-KonzernR Rdnr. 36 ff.

[243] *F. Wassermeyer*, Wettbewerbsverbot, 1993, S. 333.

[244] *W. Zöllner*, Beherrschungsverträge, 1992, S. 178 ff., 182; *BGH*, Beschl. v. 24.10.1988 - II ZB 7/88, in: BGHZ 105, S. 324, 331 („Supermarkt"); *M. Lutter/P. Hommelhoff*, GmbHG, 1995, Anh. § 13 Rdnr. 34; *H.-G. Koppensteiner*, Rowedder, 1997, Anh. § 52 Rdnr. 78; enger *M. Kort*, Beherrschungsverträge, 1986, S. 140, nach dem das Weisungsrecht des herrschenden Unternehmens und das der Gesellschafterversammlung nebeneinander bestehen.

[245] Im Ergebnis ebenso *OLG Stuttgart*, Urt. v. 29.10.1997 - 20 U 8/97 („Dornier/Daimler Benz"), in: Die AG 1998, S. 585, 586, wonach das herrschende Unternehmen die Satzung der beherrschten Gesellschaft beachten muß und deshalb nicht deren Firma ändern darf.

2. Insbesondere § 30 GmbHG

§ 30 GmbHG verbietet Auszahlungen aus dem zur Erhaltung des Stammkapitals erforderlichen Gesellschaftsvermögen. Die Befürworter[246] eines solchen Verbots auch im Vertragskonzern begründen ihre Ansicht damit, daß es sich bei § 30 GmbHG um eine zwingende gesetzliche Vorschrift handele, über die sich das herrschende Unternehmen nicht hinwegsetzen dürfe. Auch wird auf die unterschiedliche Ausgestaltung der Kapitalerhaltung im Aktien- und GmbH-Recht hingewiesen.[247] Demgegenüber führen die Gegner[248] eines Ausschüttungsverbots an, daß die Interessen der Gläubiger der beherrschten Gesellschaft durch die unstreitig[249] bestehende Verlustausgleichspflicht des herrschenden Unternehmens analog § 302 AktG ausreichend gewahrt seien.

Die Entscheidung dieses Streits hängt von der Frage ab, ob § 291 Abs. 3 AktG im GmbH-Vertragskonzern analog anwendbar ist. § 291 Abs. 3 AktG bestimmt, daß Leistungen der Gesellschaft aufgrund eines Beherrschungs- oder eines Gewinnabführungsvertrages nicht als Verstoß gegen die §§ 57, 58 und 60 AktG gelten. Das hat Bedeutung für den Fall, daß ein Aktionär der beherrschten Gesellschaft Partei des Beherrschungsvertrages ist. Denn regelmäßig sind mit nachteiligen Weisungen verdeckte Gewinnausschüttungen verbunden. Gemäß § 57 Abs. 3 AktG darf an den Aktionär aber nur der festgestellte Bilanzgewinn ausgeschüttet werden. Würde § 57 Abs. 3 AktG im Vertragskonzern fortgelten, wäre deshalb ein Beherrschungsvertrag mit einem Aktionär der Gesellschaft praktisch nicht durchführbar. Gerade wegen dieses Umstands wird die Ansicht vertreten, daß § 291 Abs. 3 AktG nur klarstellende Wirkung habe.[250] Denn der Dispens von den Vorschriften der §§ 57, 58 und 60 AktG ergebe sich schon daraus, daß das Gesetz einen Beherrschungsvertrag zulasse, was nur dann sinnvoll sei, wenn er auch in den praktisch häufigsten Fällen, nämlich mit einem Aktionär als herrschendem Unternehmen, durchgeführt werden könne. Schließt man sich dieser Meinung an, so scheidet eine Analogie von § 291 Abs. 3 AktG im GmbH-Vertragskonzern aus. Denn dann hat diese Vorschrift keine normative Wirkung.

---

[246] *H. Brandes*, Kapitalerhaltung, 1991, S. 32 f.; *V. Emmerich*, Scholz, 1993, Anhang Konzernrecht Rdnr. 277; *P. Ulmer*, Hachenburg, 1994, Anh. § 77, Rdnr. 219; *W. Zöllner*, Baumbach/Hueck, 1996, GmbH-KonzernR Rdnr. 77.

[247] *U. Kühbacher*, Darlehen, 1993, S. 51.

[248] *H.-J. Fleck*, Mißbrauch, 1990, S. 47; *P. Hommelhoff*, Eigenkapital-Ersatz, 1984, S. 1110; *M. Lutter*, GmbH-Konzernrecht, 1986, S. 200.

[249] Statt aller *W. Zöllner*, Baumbach/Hueck, 1996, GmbH-KonzernR Rdnr. 77 m. w. N.

[250] *Begr. RegE.* zu § 291 AktG, in: *B. Kropff*, AktG, 1965, S. 377 f.; *H. Brandes*, Kapitalerhaltung, 1991, S. 26 f.

Zu fragen bleibt in diesem Fall nur, ob sich mit der gleichen Argumentation wie im Aktienrecht aus der Zulässigkeit des Beherrschungsvertrages mit einer beherrschten GmbH eine Lockerung von deren Vermögensbindung ergibt. Das ist jedoch zu verneinen, da die Aufhebung von § 30 GmbHG nicht notwendige Voraussetzung für die Durchführung eines Beherrschungsvertrages ist. Denn anders als im Aktienrecht sind im GmbH-Recht verdeckte Gewinnausschüttungen bis zur Grenze des § 30 GmbHG zulässig, sofern die übrigen Gesellschafter zustimmen.

Indes kann der Ansicht, daß § 291 Abs. 3 AktG nur klarstellende Wirkung habe, nicht gefolgt werden. Denn für die Durchführung eines Beherrschungsvertrages mit einer beherrschten Aktiengesellschaft würde es ebenso wie im GmbH-Konzern ausreichen, wenn die Vermögensbindung nur bis zur Grenze des Grundkapitals aufgehoben würde. Deshalb kommt § 291 Abs. 3 AktG mindestens insoweit normative Wirkung zu, als er das herrschende Unternehmen auch von dieser Grenze befreit. Mithin bleibt die Frage einer Analogie offen.

Gegen eine Analogie wird man nicht die unterschiedliche Ausgestaltung des Kapitalerhaltungsrechts bei der Aktiengesellschaft und der GmbH anführen können. Im Gegenteil spricht diese gerade für eine Analogie. Denn da bei der konzernfreien Aktiengesellschaft der gesetzlichen Gläubigerschutz stärker ausgeprägt ist als bei der konzernfreien GmbH, ist kein Grund ersichtlich, warum das im Vertragskonzern gerade umgekehrt sein sollte.

Nach alledem kann der Sinn und Zweck von § 291 Abs. 3 AktG nur darin bestehen, durch Aufhebung des Verbots der Einlagenrückgewähr die Rahmenbedingungen für die Durchführung des Beherrschungsvertrages zu verbessern. Hiermit trug der Gesetzgeber dem Interesse des herrschenden Unternehmens an einer effizienten Konzernleitung Rechnung. Dasselbe Interesse besteht aber auch im GmbH-Vertragskonzern. Da dort außerdem die Interessen der Minderheitsgesellschafter und Gläubiger ebenso geschützt werden wie im Aktien-Vertragskonzern, ist die Interessenlage hinsichtlich der Aufhebung des Verbots der Einlagenrückgewähr bei GmbH und Aktiengesellschaft gleich. Damit ist eine analoge Anwendung von § 291 Abs. 3 AktG im Vertragskonzern mit einer beherrschten GmbH geboten.

Folglich gilt im GmbH-Vertragskonzern nicht die Kapitalerhaltungsvorschrift des § 30 GmbHG. Ausschüttungen aus dem zur Erhaltung des Stammkapitals erforderlichen Vermögen sind damit zulässig.

C. Rechtsformunabhängige Pflichten: Die Pflicht der herrschenden GmbH zur ordnungsgemäßen Konzernleitung

I. Abgrenzung

Bisher wurden nur diejenigen Grenzen der Konzernleitungspflicht untersucht, die sich im faktischen Aktienkonzern aus §§ 311 ff. AktG und im faktischen GmbH-Konzern aus der mitgliedschaftlichen Förderpflicht bzw. der gesellschaftlichen Treupflicht der herrschenden GmbH ergeben. Ferner wurde für den Vertragskonzern aufgezeigt, daß die herrschende GmbH die beherrschte Gesellschaft zwar zu einer für diese nachteiligen Maßnahme veranlassen darf, die Maßnahme aber im Konzerninteresse liegen muß. Als weitere Grenze wird nun in der Literatur die Pflicht der herrschenden GmbH gegenüber den konzernierten Gesellschaften zur ordnungsgemäßen Konzernleitung unter Anwendung der Sorgfalt eines ordentlichen und gewissenhaften Geschäftsleiters diskutiert.

Von der Haftung nach § 317 Abs. 1 AktG unterscheidet sich eine solche Pflicht dadurch, daß die herrschende GmbH nicht nur eine Schadensersatzpflicht wegen der Veranlassung und des nicht rechtzeitig erfolgten Ausgleichs nachteiliger Maßnahmen trifft. Vielmehr obliegen ihr dann auch aktive Handlungspflichten, die von der Pflicht zur Kontrolle bis zur Pflicht zur unternehmerischen Planung und Entscheidung sowie zur Sorge für das rechtmäßige Verhalten der Konzerngesellschaften reichen können. Im faktischen GmbH-Konzern verschiebt sich der Blickwinkel von der mitgliedschaftlichen Förderpflicht und den Treupflichten mit ihrem Schwerpunkt bei den Rücksichts- und Loyalitätspflichten zu einer Konzernleitungspflicht. Darüber hinaus wird die herrschende GmbH den für einen ordentlichen Geschäftsleiter geltenden Sorgfaltspflichten unterworfen und sie muß bei Pflichtverletzungen beweisen, daß sie kein Verschulden trifft (§§ 93 Abs. 2 Satz 2 AktG, 34 Abs. 2 GenG analog). Im Vertragskonzern schließlich wäre die herrschende GmbH zwar nach wie vor berechtigt, die beherrschte Gesellschaft zu nachteiligen, aber im Konzerninteresse liegende Maßnahmen zu veranlassen. Hinzu kommen würde aber ihre Pflicht, das Unternehmen der abhängigen Gesellschaft zu leiten, wobei freilich nicht deren Interesse, sondern das Konzerninteresse Richtschnur wäre.

Die Bedeutung der Konzernleitungspflicht soll folgendes Beispiel verdeutlichen: Man stelle sich vor, daß das herrschende Unternehmen die abhängige Gesellschaft veranlaßt, sich dem konzernweiten Cash-Management[251] anzu-

[251] Zur Zulässigkeit der Einführung eines zentralen Cash Managements im Konzern ausführlich *M. W. Hormuth*, Cash Management, 1998, S. 115 ff. Siehe ferne *Uwe H. Schneider*, Cash-Management, 1998, S. 805 ff. Aus betriebswirtschaftlicher Sicht: *E. Wehlen*, Cash-Management, 1998, Rdnr. 23.1 ff.

schließen. Führt eine Schwestergesellschaft das Cash-Management durch, so muß die abhängige Gesellschaft nun ihre gesamte freie Liquidität an diese abführen. Wird die Liquidität nicht angemessen verzinst oder ist ihre Rückführung gefährdet, so hat das herrschende Unternehmen die Gesellschaft zu einer nachteiligen Maßnahme veranlaßt.[252] In der Regel wird sie zudem einen Sondervorteil verfolgen. Mithin greift sowohl § 317 Abs. 1 AktG als auch die mitgliedschaftliche Förderpflicht ein, je nachdem, ob es sich bei der beherrschten Gesellschaft um eine Aktiengesellschaft oder GmbH handelt.

Etwas anderes gilt aber dann, wenn zwar die Liquidität zunächst angemessen verzinst und ihre Rückführung auch gesichert ist, die Betreibergesellschaft aber später in die Krise gerät und die Rückführung der Liquidität damit erst zu einem späteren Zeitpunkt gefährdet ist. §§ 309 Abs. 1, 317 Abs. 1 AktG und die mitgliedschaftliche Förderpflicht können die beherrschte Gesellschaft in diesem Fall nicht schützen.[253] Denn für die Beurteilung der Sorgfaltswidrigkeit einer Weisung bzw. Nachteiligkeit einer Maßnahme ist auf die im Zeitpunkt ihrer Vornahme maßgeblichen Verhältnisse abzustellen.[254] Weitere Pflichten, insbesondere eine Pflicht zur Erteilung von Weisungen oder gar zur ordnungsgemäßen Konzernleitung, lassen sich aus diesen Vorschriften nicht entnehmen.[255] Gleiches gilt für die mitgliedschaftliche Förderpflicht. In dem entscheidenden Zeitpunkt der Veranlassung war die Krise der Schwestergesellschaft aber noch nicht absehbar. Nimmt man nun eine Konzernleitungspflicht an, so ist das herrschende Unternehmen verpflichtet, die Durchführung des Cash-Managements zu überwachen und bei den ersten Anzeichen einer Krise die erforderlichen Maßnahmen zu ergreifen.[256] Diese könnten z. B. darin bestehen, die anderen Konzernunternehmen zu veranlassen, wieder selbst das Cash-Management durchzuführen.

Damit wird deutlich, daß die Annahme einer Konzernleitungspflicht des herrschenden Unternehmens gegenüber der beherrschten Gesellschaft letztere erheblich stärker schützen würde als die bisher aufgezeigten Grenzen der Konzernleitung.

Im folgenden soll untersucht werden, ob und unter welchen Voraussetzungen eine Pflicht der herrschenden GmbH gegenüber den beherrschten Gesellschaf-

---

[252] *M. W. Hormuth*, Cash Management, 1998, S. 127 ff.

[253] A. M. *U. Kühbacher*, Darlehen, 1993, S. 60; *K. Eschenbruch*, Konzernhaftung, 1996, Rz. 3057.

[254] *U. Hüffer*, AktG, 1997, § 311 Rdnr. 28.

[255] A. M. für den Vertragskonzern wohl *H. Würdinger*, Großkommentar, 1971, § 309 Anm. 3. Grundsätzlich wie hier *H.-G. Koppensteiner*, Kölner Kommmentar, 1987, § 309 Rdnr. 3.

[256] *Uwe H. Schneider*, Cash-Management, 1998, Rdnr. 25.45.

ten zur Konzernleitung besteht. In der Literatur wird diese Frage hauptsächlich für die GmbH diskutiert. Gefragt wird, wann ein Gesellschafter oder gesellschaftsfremder Dritter ausnahmsweise als „faktisches Organ" zu behandeln ist, mit der Folge, daß ihn die Pflichten eines Geschäftsführers treffen. Aus diesem Grund soll auch hier mit der Konzernleitungspflicht der herrschenden GmbH gegenüber einer beherrschten GmbH begonnen werden. Sodann wird gezeigt werden, daß auch gegenüber einer beherrschten Aktiengesellschaft eine Konzernleitungspflicht besteht.

II. GmbH-Konzern

1. Meinungsstand

Befürwortet wird eine Konzernleitungspflicht des herrschenden Unternehmens gegenüber der beherrschten Gesellschaft von *Uwe H. Schneider*. Zur Begründung stützt er sich auf die §§ 76, 93 AktG, § 43 GmbHG, § 34 GenG, aus denen er ein allgemeines Rechtsprinzip des Organisationsrechts herleitet, das er wie folgt formuliert:

> „Wer die Leitung über ein Unternehmen übernimmt, hat sie mit der Sorgfalt eines ordentlichen und gewissenhaften Geschäftsleiters auszuüben".[257]

Hiermit knüpft *Uwe H. Schneider* an die Grundsätze zum faktischen Organ an und begründet damit für das herrschende Unternehmen Organpflichten gegenüber den beherrschten Gesellschaften.[258]

Einen ähnlichen Weg geht *J. Wilhelm*. Wie bereits ausgeführt, plädiert er dafür, die juristische Person ernstzunehmen und als selbständigen Interessenträger auch gegenüber ihren Gesellschaftern anzuerkennen.[259] Folge man dem, übten die Gesellschafter fremdbezogene Macht aus, wenn sie auf die Geschäftsführung der Gesellschaft Einfluß nähmen. Fremdbezogene Machtausübung sei im Gesellschaftsrecht aber an die Sorgfalt eines ordentlichen und gewissenhaften Geschäftsführers gebunden, wie sich aus § 43 Abs. 1 GmbHG bzw. §§ 93 Abs.

---

257 *Uwe H. Schneider*, Konzernleitung, 1981, S. 257.

258 Im Ergebnis ebenso *W. Flume*, Juristische Person, 1983, S. 86, 88 ff. In diese Richtung auch *E. Rehbinder*, Unternehmensverbindungen, 1977, S. 640 f.; *K. Schmidt*, Abhängigkeit, 1981, S. 475; *J. Schulze-Osterloh*, Betriebsaufspaltung, 1983, S. 158 f.; *G. Geitzhaus*, Verlustausgleichspflicht, 1989, S. 403; *Unternehmensrechtskommission*, Bericht, 1980, Tz. 1673; die allerdings nur eine Haftung wegen nachteiliger Einflußnahme analog § 43 GmbHG annehmen bzw. eine solche Haftung fordern. Ebenso wohl auch *M. Lutter*, Unternehmensgruppe, 1982, S. 264, und Haftung, 1985, S. 1434, der das herrschende Unternehmen für verpflichtet hält, die Regeln einer ordnungsgemäßen Konzerngeschäftsführung zu beachten.

259 *J. Wilhelm*, Rechtsform, 1981, S. 1 ff., 12 ff.

1 Satz 1, 317 Abs. 2 AktG ergebe. Deshalb müßten alle Gesellschafter, die allein oder zusammen mit anderen die Geschicke der Gesellschaft bestimmten, die Sorgfalt eines ordentlichen und gewissenhaften Geschäftsleiters anwenden.[260] Insbesondere träfen sie nicht nur Unterlassungs-, sondern auch Handlungspflichten.[261] Von *Uwe H. Schneider* unterscheidet sich *J. Wilhelm* dadurch, daß er eine Sorgfaltshaftung schon dann bejaht, wenn die bestimmungsmächtigen Gesellschafter zwar nicht die Leitung über das Unternehmen der Gesellschaft übernommen haben, den Geschäftsführer jedoch zu einer Maßnahme angewiesen haben, die dieser für sorgfaltswidrig hält und von der er deshalb abgeraten hat.[262]

Überwiegend wird eine Konzernleitungspflicht des herrschenden Unternehmens gegenüber der beherrschten Gesellschaft jedoch abgelehnt.[263] Begründet wird dies vornehmlich damit, daß eine Pflicht des Gesellschafters zur sorgfältigen Geschäftsführung ohne Grund die gesetzliche Kompetenzverteilung zwischen Geschäftsführern und Gesellschaftern durchbrechen würde. Denn diese bestehe gerade darin, daß die Gesellschafter Einfluß auf die Geschäfte der GmbH nehmen dürften, ohne sogleich mit den Organpflichten eines Geschäftsführers belastet zu werden.[264] Außerdem seien nachteilige Weisungen aufgrund entsprechender Gesellschafterbeschlüsse grundsätzlich zulässig, was sich daran zeige, daß die Geschäftsführer diese befolgen müßten und deshalb von einer Haftung befreit seien. Dann könne es aber nicht angehen, nun die Gesellschafter als „Herren der Gesellschaft" einer organschaftlichen Haftung zu unterwerfen.[265] Des weiteren regele § 43 GmbHG nur die Pflichten des Geschäftsführers als Verwalter fremder Interessen. Der Gesellschafter nehme jedoch bei der

---

[260] *J. Wilhelm*, Rechtsform, 1981, S. 336 ff., 354 ff.

[261] *J. Wilhelm*, Rechtsform, 1981, S. 356.

[262] *J. Wilhelm*, Rechtform, 1981, S. 355, Fußnote 257.

[263] *H. Konzen*, Geschäftsführung, 1989, S. 2986; *P. Ulmer*, Gläubigerschutz, 1984, S. 414; *H.-J. Mertens*, Geschäftsführungshaftung, 1979, S. 465; *H. Wiedemann*, Unternehmensgruppe, 1988, S. 84 ff.; *P. Limmer*, Haftungsverfassung, 1992, S. 77 u. 125; *W. Zöllner*, Baumbach/Hueck, 1996, § 43 Rdnr. 1b; *J. Semler*, Leitung, 1996, Rdnr. 279; *H.-G. Koppensteiner*, Rowedder, 1997, § 43 Rdnr. 58; *U. Hüffer*, AktG, 1997, § 309 Rdnr. 10, der freilich „systemimmanent" begründbare Ausnahmen zuläßt, „wenn Repräsentanten des herrschenden Unternehmens durch vorangegangene Weisung eine Lage geschaffen haben, die nunmehr weitere Weisungen erfordert". Siehe auch *H. Würdinger*, Großkommentar, 1971, § 309 Anm. 3.

[264] *P. Ulmer*, Gläubigerschutz, 1984, S. 414; *P. Limmer*, Haftungsverfassung, 1992, S. 77.

[265] *P. Ulmer*, Gläubigerschutz, 1984, S. 415; *H.-J. Mertens*, Geschäftsführungshaftung, 1979, S. 465; *H. Wiedemann*, Unternehmensgruppe, 1988, S. 84.

Einflußnahme auf die Gesellschaft auch seine eigenen Interessen wahr, weshalb § 43 GmbHG auf ihn nicht passe.[266]

Nach *U. Stein* schließlich ergeben sich Organpflichten wie die Pflicht zur Konzernleitung nur für solche Nichtorgane, die fehlerhaft bestellt wurden. Dies rechtfertige sich daraus, daß das fehlerhaft bestellte Organ von der Gesellschaft mit seinem Einverständnis als Organ gewollt sei, daß es sich aufgrund der tatsächlich erfolgten Bestellung auf die von beiden Seiten gewollte Übernahme der Organverantwortung habe einstellen können und sich mit der tatsächlichen Annahme der Bestellung darauf eingelassen habe.[267] Im übrigen bestehe im Bereich der Konzernsachverhalte[268] nur für die Minderheit im faktischen GmbH-Konzern eine Haftungslücke, und zwar bei illegalen Weisungen und illegalen Eingriffe des herrschenden Unternehmens in die Geschäftsführung.[269] Diese Haftungslücke will *U. Stein* aber nicht durch die Zuweisung der vollen Organverantwortung, sondern durch eine partielle Haftungserweiterung[270] schließen. Danach soll das herrschende Unternehmen nur verpflichtet sein, „die inhaltliche Rechtmäßigkeit schädlicher Weisungen mit der organüblichen Sorgfalt zu prüfen“[271] bzw. „bei der Wahrnehmung von Geschäftsführungsaufgaben die Sorgfalt eines Organs einzuhalten“[272]. Darüber hinausgehende Handlungspflichten seien dagegen abzulehnen.

2. *Stellungnahme*

Schon oben wurde *J. Wilhelm* widersprochen, der die Gesellschaft auch gegenüber ihren Gesellschaftern als selbständige Interessenträgerin begreift. Richtig ist zwar, daß die Gesellschaft Rechtsfähigkeit besitzt, materiell ist sie jedoch Zweckgemeinschaft ihrer Gesellschafter und sind diese die eigentlich Berechtigten des Gesellschaftsvermögens. Für die Gesellschafter bedeutet dies, daß sie mit ihrer Einflußnahme auf die Geschäftsführung immer auch Eigenverwalter des in der Gesellschaft gebundenen Vermögens sind. Diesem Element

---

266 *H.-J. Mertens*, Geschäftsführungshaftung, 1979, S. 464.

267 Zum Vorstehenden siehe *U. Stein*, Organ, 1984, S. 114 ff.

268 Eine Haftungslücke bejaht *U. Stein*, Organ, 1984, außerdem für die Haftung von solchen Personen, die anstelle des Organs für die Gesellschaft tätig seien (S. 149 ff.), für die Haftung des Hintermanns eines Strohmanngeschäftsführers (S. 152 ff.) und für die Haftung des privaten Mehrheitsaktionärs für illegale Eingriffe in dic Geschäftsführung (S. 177 f.).

269 *U. Stein*, Organ, 1984, S. 170 ff., 177 ff.

270 Zur unterschiedlichen Interessenlage bei Organersetzung und Haftungserweiterung siehe *U. Stein*, Organ, 1984, S. 73 ff.

271 *U. Stein*, Organ, 1984, S. 187.

272 *U. Stein*, Organ, 1984, S. 190.

der Ausübung eigener Interessen trägt *J. Wilhelm* nicht Rechnung, wenn er allein mit dem Grundsatz der Pflichtgebundenheit fremdbezogener Machtausübung Sorgfaltspflichten des bestimmungsmächtigen Gesellschafters begründen will. Vielmehr führt dieses Element richtigerweise dazu, daß den Gesellschafter jedenfalls bei einer vereinzelten Einflußnahme keine Handlungspflichten, sondern nur die oben aufgezeigten und mit der mitgliedschafltichen Förderpflicht begründeten Unterlassungspflichten treffen.

Umgekehrt lassen sich mit den Hinweis auf die Wahrnehmung eigener Interessen Handlungspflichten des herrschenden Gesellschafters, der nicht nur vereinzelt, sondern umfassend auf die Gesellschaft Einfluß nimmt und sie gleichsam wie ein Organ leitet, nicht ausschließen. Das wird deutlich, wenn man die Pflichten des Gesellschafter-Geschäftsführers näher betrachtet. Dieser hat gemäß § 43 Abs. 1 GmbHG unbestritten die Sorgfalt eines ordentlichen Geschäftsmannes einzuhalten, obwohl er zugleich Gesellschafter ist und damit im Rahmen der Geschäftsführung auch eigene Interessen wahrnimmt.

Damit stehen sich zwei Strukturprinzipien des GmbH-Rechts gegenüber. Auf der einen Seite soll es den Gesellschaftern gestattet sein, auf die Geschäftsführung Einfluß zu nehmen, ohne zugleich wie ein professionelles Management zu haften. Und auf der anderen Seite hat der Gesetzgeber in § 43 Abs. 1 GmbHG zum Ausdruck gebracht, daß der Geschäftsführer die Geschäfte der GmbH mit der Sorgfalt eines ordentlichen Geschäftsmannes führen soll. Diese beiden Strukturprinzipien müssen beachtet werden, wenn es darum geht, die Voraussetzungen für die Pflicht des Gesellschafters zur Geschäftsführung zu begründen. Diese Pflicht kann sich nur aus einer Analogie zu § 43 Abs. 1 GmbHG ergeben.

§ 43 Abs. 1 GmbHG will im Interesse der Gesellschafter und Gesellschaftsgläubiger gewährleisten, daß das Unternehmen der GmbH mit der Sorgfalt eines ordentlichen Geschäftsmannes geführt wird. Dieser Grundsatz wird nur dann durchbrochen, wenn die Gesellschafter auf die Geschäftsführung Einfluß nehmen. Von der Durchbrechung dieses Grundsatzes geht das Gesetz aus, wenn es einerseits in § 37 Abs. 1 GmbHG den Geschäftsführer an die Weisungen der Gesellschafter bindet, andererseits aber nur dem Geschäftsführer, nicht aber den Gesellschaftern die Sorgfalt eines ordentlichen Geschäftsmannes auferlegt. Allerdings liegt dieser Regelung die Annahme des Gesetzgebers zugrunde, daß die Gesellschafter nur ihre originären Entscheidungskompetenzen wahrnehmen und im übrigen nur vereinzelt in die Geschäftsführung eingreifen. So heißt es in der amtlichen Begründung zum Entwurf eines GmbH-Gesetzes[273]:

---

[273] Amtliche Ausgabe des Entwurfs eines GmbHG nebst Begründung und Anlagen, Berlin: Franz Vahlen, 1891, S. 34.

„... in der Mehrzahl der Fälle wird dagegen die Wahl der beschränkten Haftung nur der Ausdruck für den Willen der Theilnehmer sein, ihre Betheiligung für die Zwecke der Gesellschaft auf die Leistung der übernommenen Einlage und eine mehr oder weniger nachdrückliche Mitwirkung bei der Oberleitung und Beaufsichtigung der Geschäfte zu beschränken. Da, wo die Zahl der Teilnehmer nicht eine ganz geringe ist, liegt dies ohnehin in der Natur der Verhältnisse, und von diesem Falle, als dem regelmäßigen, ist im Gesetze auszugehen".

Liegt dieser Fall vor, ist die Gefahr einer sorgfaltswidrigen Geschäftsleitung gering.

Beschränken sich die Gesellschafter auf die Wahrnehmung ihrer originären Entscheidungskompetenzen, so nehmen sie nur über die Feststellung des Jahresabschlusses und die Verwendung des Ergebnisses, über die Einforderung von Einzahlungen auf die Stammeinlagen und die Rückzahlung von Nachschüssen (§ 46 Nr. 1 - 3 GmbHG) unmittelbar Einfluß auf die Unternehmensführung. Diese Entscheidungen haben zwar Auswirkungen auf die Finanzlage des Unternehmens, sie beeinflussen den Geschäftsführer jedoch nicht in der Wahrnehmung seiner Pflicht zur sorgfältigen Unternehmensleitung. Denn erstens sind die Gesellschafter bei der Feststellung des Jahresabschlusses an die Regeln des Bilanzrechts gebunden. Diese verhindern, daß der Gesellschaft Mittel entzogen werden, die der Geschäftsführer für eine sorgfältige Unternehmensleitung benötigt.[274] Zweitens darf nur dann an die Gesellschafter ein Gewinn ausgeschüttet werden, wenn dadurch das zur Erhaltung des Stammkapitals erforderliche Vermögen nicht angetastet wird, § 30 Abs. 1 GmbHG. Drittens entscheiden die Gesellschafter nur über die Höhe des zu thesaurierenden Gewinns, nicht aber über dessen Verwendung. Viertens darf der Geschäftsführer bei der Finanzplanung nicht davon ausgehen, daß die Gesellschafter Einzahlungen auf die Stammeinlage einfordern. Und fünftens schließlich unterliegt die Rückzahlung von Nachschüssen gemäß § 30 Abs. 2 GmbHG Beschränkungen zum Schutze der Gläubiger, so daß auch hierdurch der Gesellschaft keine Mittel entzogen werden, die für eine sorgfältige Unternehmensleitung erforderlich sind.

Nach alledem ist die sorgfältige Leitung des Unternehmens der GmbH durch ihren Geschäftsführer auch dann noch möglich, wenn die Gesellschafter nur ihre originären Entscheidungszuständigkeiten wahrnehmen.

Dasselbe gilt aber auch, wenn sie nur vereinzelt in die Geschäftsführung eingreifen und z. B. darüber entscheiden, ob ein Produktionsbereich eingestellt werden soll oder nicht. Denn der Geschäftsführer hat die Entscheidungen der Gesellschafter vorzubereiten, und die Gesellschafter werden in der Regel seinen

---

[274] Freilich können die Gesellschafter einer GmbH auch außerhalb eines Ergebnisverwendungsbeschlusses die Ausschüttung von Gesellschaftsvermögen beschließen. Aber auch dann sind sie selbstverständlich an die Kapitalerhaltungsvorschriften der §§ 30 ff. GmbHG gebunden.

Ratschlägen folgen. Auch ist der Geschäftsführer in der Folgezeit verpflichtet zu überprüfen, ob sich die Entscheidungsgrundlagen der Gesellschafter geändert haben, so daß diese ihre Entscheidungen möglicherweise revidieren wollen.

Anders verhält es sich jedoch dann, wenn die Gesellschafter nicht nur vereinzelt in die Geschäftsführung eingreifen, sondern die Zielsetzungen in den einzelnen Funktionsbereichen des Unternehmens (Beschaffung-, Produktions-, Absatzpolitik sowie Finanzierung und Personalpolitik) bestimmen. Ist dies der Fall, so ist der Geschäftsführer regelmäßig nicht mehr in der Lage, das Unternehmen der Gesellschaft ordnungsgemäß zu leiten. Denn ein wesentliches Merkmal von Leitung ist die von einer zentralen Instanz ausgehende Koordination der einzelnen Zielsetzungen des Unternehmens, gerichtet auf die Herstellung und den Absatz von Gütern und Dienstleistungen.[275] Diese Koordination ist deshalb erforderlich, weil zwischen einzelnen Tätigkeiten und Entscheidungsbereichen häufig Wechselwirkungen bestehen.[276] So hängt z. B. der Handlungsspielraum des Fertigungsbereichs von den Produktionsfaktoren ab, die die Beschaffungsabteilung bereitstellt.[277] Deshalb müssen die einzelnen Zielvorgaben aufeinander abgestimmt werden. Übernehmen aber die Gesellschafter die Zielsetzung in auch nur einem Entscheidungsbereich, so ist der Geschäftsführer nicht mehr in der Lage, seine Koordinierungsaufgabe ordnungsgemäß zu erfüllen. Ihm fehlen dann die originären Führungsmöglichkeiten.[278] Auch ist nicht zu erwarten, daß sich ein in diesem Ausmaß betätigender Gesellschafter in seinen Entscheidungen von dem Geschäftsführer wesentlich beeinflussen läßt.

Das alles gilt erst recht im Konzern. Denn in der Konzernlage orientiert das herrschende Unternehmen seine Einflußnahme nicht mehr am Unternehmensinteresse der beherrschten Gesellschaft, sondern am Konzerninteresse. Das ist auch zulässig, solange der Gesellschafter keinen Sondervorteil verfolgt. Durch die Orientierung am Konzerninteresse entsteht aber erstens ein erhöhter Koordinierungsbedarf und zweitens ein Informationsgefälle zwischen dem herrschenden Unternehmen und dem Geschäftsführer der beherrschten Gesellschaft. Das wird deutlich, wenn man sich nochmals den Inhalt der Konzernführung betrachtet. Dieser besteht nach *M. R. Theisen*[279] in der „*konzernumfassende*[n] *Planung*, *Koordinierung* und *Ausführung* des für die Konzernunternehmung entwickelten strategischen, taktischen und operativen Programms". Hieraus folgt, daß nicht nur die Unternehmensziele einer einzelnen Gesellschaft unter-

---

275 *W. Dorow*, Leitungsstruktur, 1993, Spalte 2569. Siehe auch *J. Semler*, Leitung, 1996, Rdnr. 29.

276 *H. Laux*, Koordination, 1993, Spalte 2309.

277 *H. Laux*, Koordination, 1993, Spalte 2309 („Restriktionsverbund").

278 *J. Semler*, Leitung, Rdnr. 273.

279 *M. R. Theisen*, Konzern, 1991, S. 203 (Hervorhebung im Original).

einander abgestimmt werden müssen, sondern auch eine Koordinierung mit den Unternehmenszielen der übrigen Konzernunternehmen erfolgen muß. Und da der Geschäftsführer der abhängigen Gesellschaft in der Regel keine Kenntnis von den Unternehmenszielen der übrigen Konzernunternehmen hat und auch keinen Einfluß auf ihre Festlegung nehmen kann, ist er jetzt noch weniger in der Lage, das Unternehmen der Gesellschaft ordnungsgemäß zu führen.

Aus alledem folgt, daß der Geschäftsführer nicht mehr in der Lage ist, die Gesellschaft ordnungsgemäß zu leiten, wenn die Gesellschafter auch nur in einem unternehmerischen Funktionsbereich die Unternehmensziele bestimmen. Denn diese Entscheidung äußert Wirkungen auf einen anderen Unternehmensbereich und das wirkt sich wiederum auf einen anderen Unternehmensbereich aus. Dies zeigt sich besonders deutlich bei der Finanzierung, von der die Führung aller anderen Unternehmensbereiche abhängig ist. Nicht zuletzt deshalb reicht nach h. M. die verbundweite Finanzplanung und -kontrolle zur Begründung der einheitlichen Leitung im Sinne von § 18 Abs. 1 Satz 1 AktG aus.[280] Nicht zu folgen ist allerdings einer in der Literatur vertretenen Ansicht, wonach die verbundweite Finanzplanung und -kontrolle nicht nur eine ausreichende, sondern auch eine notwendige Bedingung einheitlicher Leitung ist.[281] Denn hierbei wird verkannt, daß sich auch die Entscheidungen in anderen Unternehmensbereichen notwendigerweise auf das gesamte Unternehmen auswirken, so daß der beherrschten Gesellschaft eine selbständige Planung unmöglich gemacht wird.[282] Bestimmt z. B. das herrschende Unternehmen die Produktionspolitik der beherrschten Gesellschaften, so hat dies auch Einfluß auf den Finanzierungs-, Beschaffungs- und Absatzbereich. Damit zieht das herrschende Unternehmen aber letztlich die gesamte Leitung der beherrschten Gesellschaft an sich. Zu Recht stellt *Uwe H. Schneider* deshalb fest, daß Leitung unteilbar ist.[283]

Ist das aber der Fall, so folgt aus einer Analogie zu § 43 Abs. 1 GmbHG, daß das herrschende Unternehmen die Leitung mit der Sorgfalt eines ordentlichen Geschäftsmannes auszuüben hat. Das gebieten Sinn und Zweck von § 43 Abs. 1 GmbHG, wonach derjenige der die Leitung über das Unternehmen der Gesellschaft übernommen hat, diese mit der Sorgfalt eines ordentlichen Geschäftsmannes auszuüben hat.

---

280 *E. Geßler*, Geßler, 1973, § 18 Rdnr. 32; *V. Emmerich/J. Sonnenschein*, Konzernrecht, 1997, S. 57 ff.

281 *H.-G. Koppensteiner*, Kölner Kommentar, 1986, § 18 Rdnr. 20; *U. Hüffer*, AktG, 1997, § 18 Rdnr. 11. Dagegen *J. Semler*, Leitung, 1996, Rdnr. 274.

282 *V. Emmerich/J. Sonnenschein*, Konzernrecht, 1997, S. 58.

283 *Uwe H. Schneider*, Compliance, 1996, S. 243; ders., Pflichten, 1992, S. 497.

Freilich bedeutet dies nicht in jedem Fall, daß das herrschende Unternehmen nun dieselben Pflichten treffen wie den Geschäftsführer der beherrschten Gesellschaft. Vielmehr ist nach dem Zentralisierungsgrad des Konzerns zu unterscheiden.

Im dezentralen Konzern kann sich das herrschende Unternehmen darauf beschränken, mit kaufmännischer Sorgfalt die Konzernpolitik, die Konzernstrategie und die Konzernpläne zu entwickeln, ihre Durchführung zu überwachen und bei Bedarf zu korrigieren. Insbesondere muß es etwaige Nachteile von der Gesellschaft abwenden, wenn es im Rahmen der Überwachung feststellt, daß der beherrschten Gesellschaft aus der konzernumfassenden Koordinierung der Unternehmenspolitik Schäden drohen. Im obigen Beispiel wäre es deshalb verpflichtet, die beherrschte Gesellschaft über die Krise der Cash-Management-Betreiber-Gesellschaft zu informieren und sie zum Ausscheiden aus dem Konzern-Cash-Managenment zu veranlassen. Dagegen koordinieren die Verwaltungsorgane der beherrschten Gesellschaften die Durchführung der Konzernpläne in ihrer Gesellschaft und nehmen die übrigen Aufgaben eines sorgfältigen Unternehmensleiters wahr.

Im zentralen Konzern werden die Entscheidungsrechte in der Konzernspitze konzentriert. Die Konzernspitze gibt nicht mehr nur die Ziele in den verschiedenen Unternehmensbereichen an, sondern entscheidet auch über deren konkrete Durchführung, zieht also auch die Leitung des operativen Geschäfts an sich. So trifft sie z. B. Entscheidungen über Preise, Fertigungsverfahren, Produktionsmengen, Personaleinstellungen usw. Die beherrschte Gesellschaft wird damit gleichsam zu einer bloßen Betriebsabteilung. Umgekehrt bedeutet dies für die herrschende GmbH, daß sie vollständig in die Pflichtenstellung des Geschäftsführers einrückt. Das bedeutet insbesondere auch, daß sie alle Geschäftschancen, die die beherrschte Gesellschaft wahrnehmen könnte, dieser überlassen muß. Denn der Geschäftsführer einer GmbH ist verpflichtet, mit seiner ganzen Kraft die Interessen der Gesellschaft zu fördern. Das wurde bereits ausgeführt.

3. Insbesondere die Konzernleitungspflicht gegenüber der Einmann-GmbH

Ob im Hinblick auf eine Konzernleitungspflicht für die Einmann-GmbH dasselbe gilt wie für die mehrgliedrige GmbH, hängt von der Rechtsgrundlage der Konzernleitungspflicht ab. Das sind die Vorschriften der §§ 76, 93 AktG, § 43 GmbHG, § 34 GenG. Aus ihnen wurde der allgemeine Rechtsgrundsatz hergeleitet, daß derjenige, der die Leitung über ein Unternehmen übernimmt, sie mit der Sorgfalt eines ordentlichen und gewissenhaften Geschäftsleiters auszuüben hat. Diese Pflicht dient dem Schutz der Gesellschafter. In der Einmann-GmbH bedarf aber der Alleingesellschafter, der die Konzernleitung übernommen hat, keines Schutzes vor der sorgfaltswidrigen Ausübung dieser Konzernleitung. Deshalb würde eine Konzernleitungspflicht nur dem Interesse der Gläubiger der Gesellschaft dienen. Denn eine sorgfältige Unternehmensleitung steigert den wirtschaftlichen Erfolg der Gesellschaft und vergrößert damit auch die Befriedigungsmöglichkeiten der Gläubiger.

Hieraus folgt, daß eine Konzernleitungspflicht des Alleingesellschafters gegenüber der Einmann-GmbH nur dann zu bejahen ist, wenn die Vorschriften der §§ 76, 93 AktG, § 43 GmbHG, § 34 GenG auch dem Schutz der Gläubiger dienen. Das ist bei den §§ 76, 93 AktG[284] und bei § 34 GenG[285] unstreitig der Fall. Bei § 43 Abs. 1 GmbHG bestehen dagegen Zweifel. So wird in Rechtsprechung[286] und Literatur[287] die Auffassung vertreten, daß der Alleingesellschafter-Geschäftsführer für nachteilige Maßnahmen nicht hafte, sofern nicht Kapitalerhaltungsvorschriften verletzt worden seien. Zur Begründung beruft man sich darauf, daß der Geschäftsführer von seiner Haftung befreit sei, wenn er aufgrund einer bindenden Weisung der Gesellschafter handele. Der Alleingesellschafter-Geschäftsführer führe aber „praktisch seine eigenen Weisungen" aus.[288]

Dem kann in dieser Allgemeinheit nicht gefolgt werden. Oben[289] wurde zwar für die mehrgliedrige GmbH ausgeführt, daß der Sorgfaltsmaßstab des § 43 Abs. 1 GmbHG bis zur Grenze des Vorsatzes abgemildert werden kann. Das muß für den Alleingesellschafter-Geschäftsführer dahingehend ergänzt werden, daß für diesen auch die Grenze des Vorsatzes wegfällt. Denn in der Tat führt dieser nur seine eigenen Weisungen aus und ist deshalb nicht schutzbedürftig.

All das gilt jedoch nur außerhalb des Bereichs der Kapitalerhaltung. Soweit es dagegen um die Erfüllung von Kapitalerhaltungspflichten geht, bleibt es bei der Pflicht des Geschäftsführers zur ordnungsgemäßen Geschäftsleitung und dem Sorgfaltsmaßstab des § 43 Abs. 1 GmbHG. Nur ist Zweck dieser Pflicht nicht mehr die Erzielung eines Gewinns, sondern die Einhaltung der Kapitalerhaltungsvorschriften. Mit anderen Worten hat der Alleingesellschafter-Geschäftsführer die Gesellschaft mit der Sorgfalt eines ordentlichen Geschäftsmannes so zu führen, daß Kapitalerhaltungsvorschriften nicht verletzt werden.

---

[284] *H.-J. Mertens*, Kölner Kommentar, 1988, § 93 Rdnr. 22: Pflichtenbindung des Alleinaktionärs als Verwaltungsmitglied dient dem Schutz öffentlicher Interessen.

[285] *K. Müller*, GenG, 1996, § 34, Rdnr. 1: Haftung aus § 34 Abs. 2 GenG zwingend.

[286] *BGH*, Urt. v. 21.3.1994 - II ZR 260/92, in: GmbHR 1994, S. 459, 460.

[287] *H.-G. Koppensteiner*, Rowedder, 1997, § 43 Rdnr. 5; *W. Zöllner*, Baumbach/Hueck, GmbHG, 1996, § 43 Rdnr. 22; *Uwe H. Schneider*, Scholz, 1993, § 43 Rdnr. 108.

[288] *BGH*, Urt. v. 28.9.1992 - II ZR 299/91, in: WM 1992, S. 2053, 2055.

[289] Siehe 1. Teil C. IV. 2.

Und das gilt auch für das herrschende Unternehmen, wenn es die Leitung der Gesellschaft an sich gezogen hat. Das herrschende Unternehmen ist also verpflichtet, die Konzernleitung mit der Sorgfalt eines ordentlichen und gewissenhaften Geschäftsleiters so auszuüben, daß bei der Einmann-GmbH keine Kapitalerhaltungsvorschriften verletzt werden. Verletzt es diese Pflicht und ist der Ersatz des dadurch entstandenen Schadens zur Befriedigung der Gläubiger erforderlich, wird das herrschende Unternehmen gemäß § 43 Abs. 3 Satz 3 GmbHG nicht dadurch von seiner Ersatzpflicht befreit, daß es in Ausführung „einer eigenen Weisung" handelte. Und unter denselben Voraussetzung kann es auf den Schadensersatzanspruch nicht verzichten, § 43 Abs. 3 Satz 2 i. V. m. § 9b Abs. 1 GmbHG analog.

Spezifischer Inhalt dieser Konzernleitungspflicht ist es zunächst, daß das herrschende Unternehmen keine Zahlungen an sich selbst aus dem zur Erhaltung des Stammkapitals erforderlichen Vermögen veranlaßt. Das ist unproblematisch, da sich diese Pflicht bereits ausdrücklich aus § 43 Abs. 3 GmbHG ergibt. Den Auszahlungen aus dem zu Erhaltung des Stammkapitals erforderlichen Vermögen können, wie oben ausgeführt, existenzgefährdende Eingriffe gleichstehen. Auch solche darf das herrschende Unternehmen also nicht vornehmen. Darüber hinaus ist es aber auch verpflichtet, die Gesellschaft so zu führen, daß die Einhaltung der Kapitalerhaltungsvorschriften effektiv durchgesetzt werden kann. Mit anderen Worten darf es nicht eine Lage entstehen lassen, die bei einem späteren Insolvenzverfahren über das Vermögen der Gesellschaft die Feststellung unmöglich macht, ob Kapitalerhaltungsvorschriften verletzt wurden. Genau hier beginnt die gegenüber §§ 30, 33 GmbHG selbständige Bedeutung der Pflicht des herrschenden Unternehmens zur ordnungsgemäßen Konzernleitung. Und genau die Verletzung dieser Pflicht führt in den qualifizierten faktischen Konzern.[290]

Denn dieser zeichnet sich nach der Rechtsprechung des *Bundesgerichtshofs* dadurch aus, daß das herrschende Unternehmen „die Konzernleitungsmacht in einer Weise ausübt, die keine angemessene Rücksicht auf die eigenen Belange der abhängigen Gesellschaft nimmt, ohne daß sich der ihr insgesamt zugefügte Nachteil durch Einzelausgleichsmaßnahmen kompensieren ließe".[291] Die Belange einer Einmann-Gesellschaft sind aber die Belange ihrer Gläubiger. Denn, so der *Bundesgerichtshof*[292] weiter, im „Fall einer Einpersonengesellschaft fehlt es an einer solchen angemessenen Rücksichtnahme, wenn die Gesellschaft infolge der im Konzerninteresse ausgeübten Einwirkungen ihren Verbindlichkeiten nicht nachkommen kann". Entscheidend ist aber, daß Einzelausgleichsmaßnah-

---

[290] Im Ergebnis ebenso *U. Haas*, Geschäftsführerhaftung, 1997, S. 134 f.

[291] *BGH*, Urt. v. 29.3.1992 - II ZR 265/91 („TBB"), in: BGHZ 122, S. 123, 130.

[292] *BGH*, Urt. v. 29.3.1992 - II ZR 265/91 („TBB"), in: BGHZ 122, S. 123, 130.

men, d. h. die Erstattung nach § 30 Abs. 1 GmbHG verbotener Zahlungen, nicht mehr möglich sind. Und dies ist der Fall, weil das herrschende Unternehmen seine Pflicht verletzt hat, die Konzernleitung mit der Sorgfalt eines ordentlichen und gewissenhaften Geschäftsleiters so auszuüben, daß die effektive Einhaltung und Durchsetzung der Kapitalerhaltungsvorschriften gewährleistet ist.

Damit kann für die Konkretisierung der Konzernleitungspflicht der herrschenden GmbH gegenüber der beherrschten Einmann-GmbH auf die Rechtsprechung zum qualifizierten faktischen Konzern zurückgegriffen werden. Exemplarisch genannt sei wiederum das TBB-Urteil des *Bundesgerichtshofs*. In dem dort zugrunde liegenden Fall hatte das herrschende Unternehmen die abhängigen Gesellschaften mbH veranlaßt, die gesamtschuldnerische Haftung für die Verbindlichkeiten jeder einzelnen von ihnen gegenüber der kreditgebenden Bank zu übernehmen.[293] Den damit verbundenen Risiken sei, sofern sie sich unmittelbar auf die Vermögenslage der betroffenen Gesellschaft auswirken, durch entsprechende Rückstellungen in der Bilanz Rechnung zu tragen. Erfolge dies nicht, führe der Haftungsverbund zu einer Unübersichtlichkeit der das Vermögen des Unternehmens berührenden Einwirkungen und mache einen Einzelausgleich unmöglich.[294]

Hieraus folgt, daß der Einmann-Gesellschafter verpflichtet ist, alle konzernbezogenen Risiken angemessen zu dokumentieren und zu bilanzieren. Verletzt er diese Pflicht, ist die Durchsetzung der Kapitalerhaltungsvorschriften gefährdet.

Nach alledem ist den Stimmen in der Literatur Recht zu geben, die in der Haftung des herrschenden Unternehmens im qualifizierten faktischen Konzern einen Haftung wegen fehlerhafter Konzernleitung sehen.[295] Und ebenso ist eine Verlustausgleichspflicht bzw. Ausfallhaftung analog §§ 302, 303 AktG zu bejahen. Denn hat das herrschende Unternehmen in der Gesellschaft eine Lage verursacht, die die Kapitalerhaltungsvorschriften leerlaufen lassen, so besteht dieselbe Situation wie im aktienrechtlichen Vertragskonzern, bei dem die Vermögensbindung aufgehoben ist. Schließlich ist auch trotz der Haftung des herrschenden Unternehmens analog § 43 Abs. 3 GmbHG die für eine Analogie erforderliche Regelungslücke zu bejahen. Denn da ein Einzelausgleich nicht möglich ist, läßt sich auch nicht nachweisen, welche Lage bestehen würde, wenn der zum Schadensersatz verpflichtende Umstand nicht eingetreten wäre, das herrschende Unternehmen seine Konzernleitungspflicht also nicht verletzt hätte.

---

293 *BGH*, Urt. v. 29.3.1992 - II ZR 265/91 („TBB"), in: BGHZ 122, S. 123, 128 f.

294 *BGH*, Urt. v. 29.3.1992 - II ZR 265/91 („TBB"), in: BGHZ 122, S. 123, 132.

295 *T. Drygala*, Verhaltenshaftung, 1993, S. 319; *M. Lutter*, „TBB", 1993, S. 580 f.

III. Aktienkonzern

Schließlich ist die herrschende GmbH auch im Aktienkonzern verpflichtet, die Konzernleitung mit der Sorgfalt eines ordentlichen und gewissenhaften Geschäftsleiters auszuüben. Denn auch der Vorstand einer Aktiengesellschaft ist nicht mehr in der Lage, die Gesellschaft ordnungsgemäß zu führen, wenn das herrschende Unternehmen die Unternehmensleitung an sich gezogen hat. Dagegen spricht auch nicht, daß der Vorstand einer faktisch beherrschten Aktiengesellschaft eine Veranlassung nur dann ausführen darf, wenn er die Auswirkungen dieser Maßnahmen nicht als nachteilig einschätzt oder auf Nachteilsausgleich besteht. Deshalb wird ihm das herrschende Unternehmen seine Planung offenlegen und die relevanten Daten nennen. Ist das geschehen und hat der Vorstand die Veranlassung befolgt, so kann er den weiteren Lauf des konzerninternen Geschehens und seine eingetretenen oder voraussichtlichen Auswirkungen auf die beherrschte Gesellschaft nicht mehr so intensiv und umfassend kontrollieren wie das herrschende Unternehmen. Deshalb ist es gerechtfertigt, insoweit die Organpflichten auf das herrschende Unternehmen zu erstrecken. Das alles gilt unabhängig davon, ob der herrschende Gesellschafter Unternehmen im Sinne von §§ 15 ff. AktG ist oder nicht.

3. Teil: Die Bestätigung der Konzernleitungspflicht des Geschäftsführers gegenüber der herrschenden GmbH am Beispiel konzernintegrativer Maßnahmen

Es soll nun untersucht werden, ob der Geschäftsführer seine im 1. Teil festgestellte Konzernleitungspflicht gegenüber der herrschenden GmbH überhaupt erfüllen kann, wenn ihm durch das Recht der beherrschten Gesellschaften Grenzen gezogen sind. Das ist im Vertragskonzern zweifellos zu bejahen, da dort nachteilige Maßnahmen ohne die Pflicht zum Nachteilsausgleich oder Schadensersatz zulässig sind (§ 308 AktG). Deshalb beschränken sich die folgenden Ausführungen auf den faktischen Konzern. Denn im faktischen Konzern muß der Geschäftsführer die Pflichten der herrschenden GmbH gegenüber den beherrschten Gesellschaften beachten. Damit stellt sich für den Geschäftsführer immer wieder von neuem die Frage nach der Zulässigkeit konzernbezogener Maßnahmen.

Da der Kern der Konzernleitungspflicht des Geschäftsführers die Konzernintegration ist, sollen im folgenden konzernintegrative Maßnahmen exemplarisch auf ihre Vereinbarkeit mit den Pflichten der herrschenden GmbH gegenüber den beherrschten Gesellschaften im faktischen Konzern untersucht werden.

A. Festlegung strategischer Geschäftsfelder und die Abgabe von Ressourcen

Aufgabe des Geschäftsführers ist es, in den beherrschten Gesellschaften die von der Konzernspitze definierte Konzernstrategie durchzusetzen. Dazu muß er etwa die Konzernunternehmen anweisen, einzelne Produktlinien, Dienstleistungsbereiche oder Absatzmärkte aufzugeben und sich in anderen Bereichen stärker zu engagieren. Ebenso kann es im Konzerninteresse angezeigt sein, daß die Konzernunternehmen im Sinne einer effizienten Arbeitsteilung im Konzern Ressourcen an ein anderes Konzernunternehmen abgeben. Dabei kann es sich um eine Führungskraft handeln, um Know-how, um eine Produktlinie oder um einen ganzen Betrieb oder Geschäftsbereich. Solche Maßnahmen sind in der Regel nicht nachteilig, wenn sie mit entsprechenden Vorteilen verbunden sind. Erfolgt also die Aufgabe eines Marktes zugunsten eines ebensoviel Erfolg versprechenden anderen Marktes, wird die Entwicklung eines Produkts gegen Zuweisung eines ebenso erfolgversprechenden anderen Produkts eingestellt, und stehen schließlich der Übernahme von Risiken entsprechende Chancen gegenüber, so liegt keine nachteilige Maßnahme und damit keine Verletzung des Eigeninteresses der beherrschten Gesellschaft vor.[1]

Freilich mag man hier einwenden, daß die Auswirkungen der genannten Maßnahmen sich in aller Regel nicht hinreichend präzisieren lassen. Dies gilt jedoch nicht nur für vom herrschenden Unternehmen veranlaßte Maßnahmen, sondern überhaupt für alle unternehmerischen Führungsentscheidungen, die in Anbetracht unvorhersehbarer Entwicklungen der Konjunktur, der Nachfrage und der Finanzierungsbedingungen notwendigerweise mit mehr oder minder großen Unsicherheiten behaftet sind.[2] Auch der ordentliche und gewissenhafte Geschäftsleiter einer unabhängigen Gesellschaft trifft also Entscheidungen, deren Auswirkungen sich im Zeitpunkt ihrer Ausführung nicht hinreichend als vorteil- oder nachteilhaft abschätzen lassen. Dies eröffnet der Konzernleitung ein breites Spektrum von Maßnahmen.[3] Erst wenn die Grenzen dieses Spektrums überschritten werden, darf der Vorstand bzw. Geschäftsführer der abhängigen Gesellschaft die Weisung der Konzernleitung nicht ausführen.[4]

Dieses Spektrum ist jedenfalls dann überschritten, wenn die Gesellschaft nach ihrem Ausscheiden aus dem Konzernverbund nicht mehr selbständig lebensfähig wäre.[5] Insbesondere die Zentralisierung bedeutsamer betrieblicher Funktionen wie Forschung, Entwicklung, Einkauf und Verkauf wird zu einer solchen Existenzgefährdung führen und ist deshalb unzulässig.[6]

Eine andere Beurteilung ergibt sich jedoch, wenn die Gesellschaft kurz vor dem Bankrott steht. Stellt sie in dieser Situation die Produktion auf Veranlassung des herrschenden Unternehmens um, obwohl kein rentabler Absatz für die neue Produktion gesichert ist und die Gesellschaft bei einem wirtschaftlichen Fehlschlag nicht überleben könnte, so liegt kein Nachteil vor. Denn auch der Vorstand einer unabhängigen Gesellschaft, die sich in derselben wirtschaftlichen Lage wie die abhängige Gesellschaft befindet, handelt nicht pflichtwidrig, wenn die Produktionsumstellung die einzige Möglichkeit ist, die Gesellschaft zu retten.[7]

---

[1] *G. Krieger*, Tatbestand, 1992, S. 48.

[2] Zum Vorstehenden siehe *H.-G. Koppensteiner*, Kölner Kommentar, 1987, § 311 Rdnr. 44.

[3] Ebenso *M. Hoffmann-Becking*, AG-Konzern, 1989, S. 81.

[4] Siehe auch *H.-G. Koppensteiner*, Kölner Kommentar, 1987, § 311 Rdnr. 48.

[5] *G. Krieger*, Unternehmen, 1988, § 69 Rdnr. 67.

[6] *G. Krieger*, Tatbestand, 1992, S. 48 f., 52; *H.-G. Koppensteiner*, Kölner Kommentar, 1987, § 311 Rdnr. 45; *L. Strohn*, Verfassung, 1977, S. 90.

[7] *L. Strohn*, Verfassung, 1977, S. 77.

B. Festlegung von Konzernstruktur und Organisation

Hierunter fällt die Festlegung der Konzernorganisation[8] und der Managementstruktur mit dem Ziel einer betriebswirtschaftlich zweckmäßigen Struktur zur Sicherstellung effizienter, zielorientierter und kontrollierter Betriebsabläufe. Außerdem zählen zu diesem Aufgabenbereich die Aufstellung von Richtlinien zur einheitlichen Ausrichtung von Bilanzierung, Finanzierung, Investitionen sowie des Planungs- und Berichtswesens.[9]

Auch hier sind im Einzelfall die konkreten Auswirkungen konzernstruktureller und organisatorischer Maßnahmen daraufhin zu untersuchen, ob sie für die abhängige Gesellschaft nachteilig sind. Dies wird, wie oben gesehen, zu bejahen sei, wenn bedeutsame betriebliche Funktionen zentralisiert werden. Werden hingegen nur äußere Betriebsabläufe gestrafft oder das konzerninterne Berichtssystem verbessert, liegt kein Nachteil vor.[10]

C. Finanzielle Führung des Konzerns

Zur finanziellen Führung des Konzerns gehört die Festlegung von Kapitalausstattung und Finanzstruktur der einzelnen Konzernunternehmen, die Koordinierung der Finanzwirtschaft, die Mitentscheidung über langfristige Kreditaufnahmen und das Konzern-Controlling. Neben diesen originären Konzernführungsaufgaben kann die Konzernleitung aber auch weitere Bereiche an sich ziehen, um Synergiepotentiale auszunutzen. So kann sie insbesondere die Nachfrage der Konzernunternehmen in den Finanzmärkten bei sich oder einem anderen Konzernunternehmen bündeln, um die Finanzierungskosten für den Gesamtkonzern zu minimieren und Kapitalnutzen und Kapitalkosten zu optimieren. Des weiteren kann sie ein zentrales Cash-Management und ein zentrales Devisenmanagement sowie eine Konzernrevision und ein Risk-Management einführen.[11]

Soweit die Festlegung der Kapitalausstattung durch den Umfang des auszuschüttenden Gewinns bestimmt wird, ist die Vorschrift des § 311 AktG schon nicht einschlägig. Denn sie erfaßt nicht diejenigen Beschlüsse, die wie der Gewinnverwendungsbeschluß alle Aktionäre gleich behandeln und auch legitime Gläubigerinteressen nicht berühren. Dies gilt selbst dann, wenn die Gesellschaft

---

8 Zur Organisation des Konzernvorstandes einer AG siehe *K.-P. Martens*, Organisation, 1991, S. 523 ff.

9 *E. Scheffler*, Konzernmanagement, 1992, S. 54.

10 *G. Krieger*, Tatbestand, 1992, S. 56.

11 *E. Scheffler*, Konzernmanagement, 1992, S. 54, 55.

durch eine überhöhte Gewinnausschüttung in eine finanzielle Notlage gerät. § 311 AktG schützt die Gesellschaft nur vor abhängigkeitsspezifischen Gefahren. Eine solche wird man aber in der Abstimmung über die Gewinnverwendung nicht sehen können, da auch die Bedürfnisse eines nichtunternehmerischen Aktionärs oder das Dividendeninteresse gleichsinnig abstimmender Minderheitsaktionäre auf eine hohe Gewinnausschüttung gerichtet sein können.[12]

Die Kapitalausstattung der abhängigen Unternehmen kann aber auch über die Bestimmung von Art und Umfang der Fremdfinanzierung erfolgen. Ebenso wie bei der Festlegung des Verschuldungsgrades ist hier zu fragen, welche konkreten Maßnahmen die Entscheidung über die Kapitalausstattung zur Folge hat. Zu prüfen ist also z. B., ob der Vorstand einer als unabhängig gedachten Gesellschaft ebenso hohe Schulden wie von der Konzernleitung angewiesen aufgenommen hätte, um ein neues Werk zu bauen.

Die Festlegung der Kapitalstruktur leitet über zur Koordinierung der Finanzwirtschaft. Sie beginnt bei der Auswahl der Banken und sonstigen Kreditgebern, erstreckt sich über die Auswahl der Kreditarten und Kreditsicherheiten bis hin zur Bündelung der Nachfrage auf den Kapital- und Geldmärkten.[13] Sie ist regelmäßig mit Zinsvorteilen für alle Konzerngesellschaften verbunden und stellt für diese grundsätzlich keinen Nachteil dar. Insbesondere wird allein durch eine finanzielle Konzernführung die selbständige Existenzfähigkeit der abhängigen Gesellschaften nicht gefährdet, da sie nach ihrem Ausscheiden aus dem Konzernverbund nicht mehr Schwierigkeiten bei der Suche nach neuen Kreditgebern haben dürften als wenn sie unabhängig geblieben wären.[14]

Auch das zentrale Cash- und Devisenmangement ist für die abhängigen Unternehmen wegen des Zinsgewinns regelmäßig vorteilhaft.[15] Nachteile stellen sich erst dann ein, wenn nicht mehr genügend Liquidität vorhanden ist, um fällige Schulden zu bezahlen oder notwendige Investitionen vorzunehmen.[16] Eine weitere Einschränkung wird man im Hinblick auf eine mögliche Krise der Konzerngesellschaft, bei der das Cash-Management angesiedelt ist, vornehmen müssen. Damit im Falle von deren Zahlungsunfähigkeit die übrigen Konzerngesellschaften nicht auch in Liquiditätsschwierigkeiten geraten, wird man fordern müssen, daß den Konzerngesellschaften ein Minimum an direkten Bankkontak-

---

12 *H.-G. Koppensteiner*, Kölner Kommentar, 1987, § 311 Rdnr. 17.

13 *E. Scheffler*, Konzernmanagement, 1992, S. 122.

14 *H.-G. Koppensteiner*, Kölner Kommentar, 1987, § 311 Rdnr. 103.

15 *P. Ulmer*, Vermutungsfragen, 1992, S. 81.

16 *J. Ensthaler/M. Kreher*, Haftungspotentiale, 1996, S. 389.

ten und Kreditlinien verbleibt, auf die sie im Falle der Beendigung des Konzernverhältnisses schnell zugreifen können.[17]

D. Besetzung wichtiger Führungspositionen

Die Besetzung wichtiger Führungspositionen wirft nur dann Probleme auf, wenn es zu einer Organverflechtung zwischen dem herrschenden Unternehmen und der beherrschten Gesellschaft auf der Ebene des Geschäftsführungsorgans der beherrschten Gesellschaft kommt. Für diesen Fall bejaht eine in der Literatur vertretene Ansicht das Vorliegen eines qualifizierten faktischen Konzerns. Denn durch die Doppelmandate verdichte sich die Herrschaft der Konzernspitze zu einer dauernden und umfassenden Leitung, bei der sich einzelne Veranlassungen der Konzernleitung nicht mehr isolieren ließen. Daher versage das finanzielle Ausgleichssystem der §§ 311 ff. AktG, weshalb die Doppelmandate unzulässig seien und das herrschende Unternehmen bei einem Verstoß gegen dieses Verbot eine Verlustausgleichspflicht analog §§ 302, 303 AktG treffe.[18]

Dieser Ansicht kann nicht gefolgt werden. Ihr liegt die Annahme zugrunde, daß aufgrund der Doppelmandate zwischen den Weisungen des herrschenden Unternehmens und dem eigenen Willen des Tochtervorstands nicht mehr unterschieden werden könne. Das überzeugt nicht. Eine Person ist sehr wohl in der Lage, zwei Gesellschaften gleichzeitig zu führen und dabei jeweils deren Eigeninteressen zu wahren.[19] Erforderlich ist nur, daß die einzelnen Unternehmensbereiche klar auseinandergehalten werden, die Bücher in Ordnung sind und hinter der Führung und Verteilung der Geschäfte ein nachvollziehbares unternehmerisches Prinzip steht. Unterstützt wird dies durch Berichte aus der Praxis, wonach Unternehmen nach Einführung einer Personalunion genauso weitergeführt wur-

---

[17] *G. Krieger*, Tatbestand, 1992, S. 55; *H. P. Westermann*, Tatbestand, 1992, S. 33 f. Siehe auch *M. Hoffmann-Becking*, AG-Konzern, 1989, S. 81.

[18] *F. J. Säcker*, Problematik, 1987, S. 67 ff. Ähnlich *P. Ulmer*, Vermutungsfragen, 1992, S. 79, nach dem eine personelle Verflechtung „von oben nach unten" vergleichbar sei mit einer dauernden und umfassenden Leitung durch das herrschende Unternehmen. Diese begründe aber die Vermutung für eine Verletzung der Eigeninteressen der beherrschten Gesellschaft und führe, wenn sie nicht widerlegt werde, zu einer Haftung analog §§ 302, 303 AktG. Ähnlich *C. E. Decher*, Verflechtungen, 1990, S. 92, der freilich die Zulässigkeit eines qualifizierten faktischen Konzerns bejaht und damit wie hier zu dem Ergebnis kommt, daß Doppelmandate zulässig seien (S. 131, 151 ff.).

[19] *G. Krieger*, Tatbestand, 1992, S. 50 f.; ders., TBB, 1994, S. 386; *H. P. Westermann*, Tatbestand, 1992, S. 38; *BAG*, Beschl. v. 16.8.1995 - 7 ABR 57/94, in: ZIP 1996, S. 292, 294. Dagegen *C. E. Decher*, Verflechtungen, 1990, S. 144 f. Einschränkend *M. Hoffmann-Becking*, AG-Konzern, 1989, S. 83, nach dem die Funktionsfähigkeit der §§ 311 ff. AktG jedenfalls dann nicht gefährdet ist, wenn sich die personelle Verflechtung auf ein Mitglied des Tochtervorstands beschränke.

den wie zuvor.[20] Und anzumerken ist schließlich, daß nach der neuesten Rechtsprechung zum qualifizierten faktischen Konzern die dauernde und umfassende Ausübung der Leitungsmacht durch das herrschende Unternehmen noch nicht einmal mehr die Vermutung begründet, daß auf die Belange der abhängigen Gesellschaft keine angemessene Rücksicht genommen worden ist.[21]

Nach alledem steht es dem Geschäftsführer der herrschenden GmbH frei, sich selbst in das Verwaltungsorgan der beherrschten Gesellschaft wählen zu lassen oder diese mit Personen seines Vertrauens zu besetzen.

---

20 *G. Krieger*, Tatbestand, 1992, S. 50 f.

21 BGH, Urt. v. 29.3.1993 - II ZR 265/91 („TBB"), in: WM 1993, S. 687, 691.

4. Teil: Die Pflichten des Geschäftsführers gegenüber den beherrschten Gesellschaften

A. Die Pflicht zur ordnungsgemäßen Unternehmensleitung

I. Die gesetzliche Ausgangslage

Den Geschäftsführer treffen grundsätzlich Pflichten nur im Verhältnis zu seiner eigenen Gesellschaft. Denn Entstehungsgrund seiner Pflichten ist seine Organstellung und Organ ist er nur bei seiner eigenen Gesellschaft. Trotzdem stellt sich die Frage, ob den Geschäftsführer einer herrschenden GmbH nicht auch ausnahmsweise Pflichten im Verhältnis zu den Konzernunternehmen obliegen können.

Das ist im Vertragskonzern und im Eingliederungskonzern mit einer beherrschten Aktiengesellschaft der Fall. Dort ist der Geschäftsführer der herrschenden GmbH gegenüber der beherrschten Gesellschaft verpflichtet, bei der Erteilung von Weisungen die Sorgfalt eines ordentlichen und gewissenhaften Geschäftsleiters anzuwenden, § 309 Abs. 2 AktG bzw. § 323 Abs. 1 i. V. m. § 309 AktG. Er muß also bei der Vorbereitung und Ausübung der Weisungen kaufmännische Vorsichtsmaßnahmen und gesicherte betriebswirtschaftliche Erkenntnisse berücksichtigen sowie die sonstigen rechtlichen Rahmenbedingungen für Geschäftsführungsmaßnahmen beachten.[1] Verletzt er diese Pflicht, macht er sich gegenüber der beherrschten Gesellschaft schadensersatzpflichtig.

Im faktischen Aktienkonzern ist der Geschäftsführer zwar gegenüber der beherrschten Gesellschaft zum Schadensersatz verpflichtet, wenn die herrschende GmbH bei dieser eine nachteilige Maßnahme veranlaßt und den Nachteil nicht rechtzeitig ausgeglichen hat, § 317 Abs. 1 u. Abs. 3 AktG. Hierbei handelt es sich jedoch um eine Haftung für fremdes Verhalten, nämlich dem Verhalten der herrschenden GmbH. Nur diese trifft gemäß § 311 Abs. 1 AktG die deliktische Pflicht, die Zufügung von Nachteilen zu unterlassen.

Sowohl im Vertragskonzern wie im faktischen Konzern mit einer beherrschten Aktiengesellschaft haftet der Geschäftsführer also nur für die Erteilung von pflichtwidrigen Weisungen bzw. für nachteilige Veranlassungen. Mit anderen

---

[1] *K. Eschenbruch*, Konzernhaftung, 1996, Rz. 3041.

Worten: Voraussetzung für eine Haftung des Geschäftsführers ist immer eine aktive Einflußnahme auf die Gesellschaft. Für Unterlassungen haftet er dagegen nicht. Weder § 309 AktG noch § 317 Abs. 3 AktG statuieren aktive Handlungspflichten.

Im Vertragskonzern mit einer beherrschten GmbH gilt § 309 Abs. 1, Abs. 2 AktG entsprechend.[2] Auch hier ist der Geschäftsführer der herrschenden GmbH gegenüber der beherrschten GmbH verpflichtet, bei der Erteilung von Weisungen die Sorgfalt eines ordentlichen und gewissenhaften Geschäftsleiters anzuwenden. Verletzt er diese Pflicht, macht er sich gegenüber der beherrschten Gesellschaft schadensersatzpflichtig.

Im faktischen GmbH-Konzern gibt es dagegen keine gesetzlichen Vorschriften, die eine persönliche Haftung des Geschäftsführers der herrschenden GmbH gegenüber der beherrschten Gesellschaft ausdrücklich statuieren.

An erster Stelle soll deshalb untersucht werden, ob im Wege der Analogie persönliche Pflichten des Geschäftsführers der herrschenden GmbH gegenüber der faktisch beherrschten GmbH hergeleitet werden können. Sodann stellt sich die Frage, ob auf demselben Wege auch weitere Pflichten des Geschäftsführers gegenüber der vertraglich konzernierten GmbH und der beherrschten Aktiengesellschaft begründet werden können.

II. Meinungsstand

In Rechtsprechung und Literatur gibt es nur vereinzelte Stellungnahmen zu den Pflichten des Geschäftsführers einer herrschenden GmbH gegenüber den Konzernunternehmen. Dagegen gibt es eine breite Diskussion über die ähnlich gelagerte Frage, ob bei einer GmbH & Co. KG den Geschäftsführer der Komplementär-GmbH auch Pflichten gegenüber der Kommanditgesellschaft treffen. Diese Diskussion soll in die Betrachtung mit einbezogen werden.

Die erste Ansicht zu den Pflichten des Geschäftsführers einer Komplementär-GmbH beruft sich auf den Vertrag mit Schutzwirkung zugunsten Dritter. Vertreten wird, daß die Organstellung[3] bzw. der Anstellungsvertrag[4] des Geschäftsführers der Komplementär-GmbH Schutzwirkung zugunsten der Kommanditgesellschaft äußere, wenn die wesentliche Aufgabe der Komplementär-

---

[2] *P. Ulmer*, Hachenburg, 1994, Anh. § 77 Rdnr. 220; *B. Kropff*, Beherrschungsvertrag, 1993, S. 528 f.; *H.-G. Koppensteiner*, Rowedder, 1997, Anh. nach § 52 Rdnr. 79. Offengelassen von *V. Emmerich*, Scholz, 1993, Anhang Konzernrecht, Rdnr. 289. Dagegen *M. Kort*, Beherrschungsverträge, 1986, S. 143.

[3] *H. Brandes*, Rechtsprechung, 1987, S. 7.

[4] *BGH*, Urt. v. 12.11.1979 - II ZR 174/77, in: BGHZ 75, S. 321, 324.

GmbH in der Führung der Geschäfte der Kommanditgesellschaft bestehe. Sei dies der Fall, erstrecke sich der Schutzbereich des zwischen der GmbH und dem Geschäftsführer bestehenden Dienst- bzw. Organverhältnisses hinsichtlich einer Haftung des letzteren auch auf die Kommanditgesellschaft. Namentlich der *Bundesgerichtshof*, der die Schutzwirkung aus dem Anstellungsvertrag herleitet, begründet dies mit Sinn und Zweck des Anstellungsvertrages und dem Grundsatz von Treu und Glauben. Denn zum einen wirkten sich Fehlleistungen der Geschäftsführung zwangsläufig stets und in erster Linie zum Nachteil der Kommanditgesellschaft aus. Zum anderen sei der Kommanditgesellschaft ein Weisungsrecht gegenüber dem Geschäftsführer versagt, obwohl sie auf dessen Sorgfalt angewiesen sei. Schließlich hätte die Komplementär-GmbH die Möglichkeit, auf Ersatzansprüche gegen den Geschäftsführer zu verzichten. In dieser Lage dürften die Kommanditisten erwarten, daß die Kommanditgesellschaft bei pflichtwidriger Schädigung durch den Geschäftsführer der Komplementär-GmbH diesen unmittelbar in Anspruch nehmen dürfe.[5]

P. Krebs hat versucht, ein Prinzip der Haftung für sorgfaltswidrige Leitung nachzuweisen,[6] das auch die Haftung des Geschäftsführers der herrschenden GmbH gegenüber den Konzernunternehmen für sorgfaltswidrige Leitung umfasse[7]. Diese Haftung stützt er auf eine Gesamtanalogie zu den §§ 309 Abs. 2, 317 Abs. 3, 323 Abs. 1 Satz 2 i. V. m. § 309 Abs. 2 AktG.[8] Dabei unterscheidet er zwischen legaler Leitung, die bei der GmbH von den Geschäftsführern und Gesellschaftern ausgeübt werde, und illegaler Leitung durch nicht leitungsbefugte Personen.[9] Außerdem unterscheidet er zwischen Primärleitenden, die zur Leitung verpflichtet seien, und Sekundärleitenden, die weisungsberechtigt, aber nicht zur Leitung verpflichtet seien.[10] Jede, auch die erste und einzige Handlung in Ausführung legaler Leitungsmacht, also bei der GmbH durch den Geschäftsführer und Gesellschafter, sei Leitung.[11] Dagegen sei ein illegaler Leitungsversuch nur dann als Leitung zu qualifizieren, wenn die primär legal Leitenden, also die zur Leitung verpflichteten Personen, strukturell nicht mehr in der Lage

---

5 Zum Vorstehenden siehe *BGH*, Urt. v. 12.11.1979 - II ZR 174/77, in: BGHZ 75, S. 321, 323 f.

6 *P. Krebs*, Geschäftsführungshaftung, 1991, S. S. 232 ff.

7 *P. Krebs*, Geschäftsführungshaftung, 1991, S. 201.

8 *P. Krebs*, Geschäftsführungshaftung, 1991, S. 154 ff. Gegen einc Haftung der Vorstandsmitglieder einer herrschen AG gegenüber einer beherrschten GmbH analog § 317 Abs. 3 AktG jüngst *LG Bremen*, Urt. v. 19.11.1997 - 4 O 1073/96, in: ZIP 1998, S. 561, 563.

9 *P. Krebs*, Geschäftsführungshaftung, 1991, S. 174 ff.

10 *P. Krebs*, Geschäftsführungshaftung, 1991, S. 173 f.

11 *P. Krebs*, Geschäftsführungshaftung, 1991, S. 180.

seien, die illegalen Leitungsversuche abzuwehren.[12] Dies sei regelmäßig der Fall, wenn die legal zur Leitung Verpflichteten verdrängt würden und damit die Funktionsfähigkeit der Gesellschaft nicht mehr gewährleistet sei. Nur dann bestehe eine Leitungspflicht für illegal Leitende.[13]

Uwe H. Schneider[14] schließlich will nicht nur für die herrschende GmbH, sondern auch für ihren Geschäftsführer aus §§ 76, 93 AktG, 43 GmbHG, 34, 41 GenG die Pflicht herleiten, bei Ausübung der Konzernleitung die Sorgfalt eines ordentlichen und gewissenhaften Geschäftsleiters zu beachten.

III. Stellungnahme

Gegen die vom *Bundesgerichtshof* befürwortete Konstruktion eines Anstellungsvertrages des Geschäftsführers mit Schutzwirkung zugunsten der Kommanditgesellschaft spricht, daß allein der in der Satzung der GmbH niedergelegte Zweck, nämlich die Führung der Geschäfte der Kommanditgesellschaft, schwerlich eine Drittbegünstigung zu begründen vermag. Eine Auslegung der Satzung der GmbH unter Berücksichtigung der Interessen ihrer Gesellschafter trägt ein solches Ergebnis jedenfalls nicht. Denn die Gesellschafter sind nur daran interessiert, daß der Geschäftsführer gegenüber der GmbH haftet, ein Interesse an einer Haftung des Geschäftsführers gegenüber der Kommanditgesellschaft haben sie dagegen nicht. Weiterhin gerät der *Bundesgerichtshof* in Argumentationsschwierigkeiten, wenn ein Anstellungsvertrag überhaupt nicht geschlossen wurde. Und schließlich ist ihm entgegenzuhalten, daß primär das Organisationsrecht darüber entscheidet, welche Pflichten den Geschäftsführer treffen, der Anstellungsvertrag aber nur das aufnimmt, was durch das Gesellschaftsrecht schon vorgegeben ist.[15] Nach alledem ist die Konstruktion des *Bundesgerichtshofs* abzulehnen.

Auch die Anknüpfung der Schutzwirkung an die Organstellung des Geschäftsführers begegnet erheblichen Bedenken. Zwar hat diese Ansicht den Vorzug, daß die Haftung des Geschäftsführer von seinem Anstellungsvertrag unabhängig ist. Allerdings bietet sie für die Haftung des Geschäftsführers gegenüber den Konzernunternehmen keine angemessene Lösung. Denn dem Schutzbedürfnis der Konzernunternehmen wird nicht ausreichend Rechnung getragen, wenn man als Voraussetzung fordert, daß die Geschäftsführung in den Konzernunternehmen wesentlicher Zweck der herrschenden GmbH ist.

---

[12] *P. Krebs*, Geschäftsführungshaftung, 1991, S. 177, 180.

[13] *P. Krebs*, Geschäftsführungshaftung, 1991, S. 177, 179 f.

[14] *Uwe H. Schneider*, Konzernleitung 1981, S. 257; ders., Scholz, 1993, § 43 Rdnr. 265. Dagegen: *G. Krieger*, Haftungsrisiken, 1998, Rdnr. 4.52.

[15] Ebenso *U. Hüffer*, Organpflichten, 1981, S. 356.

Auch die Meinung von *P. Krebs* ist abzulehnen. Oben wurde ausgeführt, daß der Gesellschafter einer GmbH bei seiner regelmäßigen Mitwirkung an den Entscheidungen der Gesellschafterversammlung nicht die Sorgfalt eines ordentlichen und gewissenhaften Geschäftsleiters anzuwenden hat.[16] Vielmehr gebietet ihm seine mitgliedschaftliche Förderpflicht nur, bei seinen Entscheidungen stets das Interesse der Gesellschaft zu verfolgen.[17] Wollte man nun mit *P. Krebs* den Geschäftsführer der herrschenden GmbH gegenüber der Tochter-GmbH einer Haftung für sorgfaltswidrige Leitung unterwerfen, so würde man damit für ihn Pflichten statuieren, die weiter reichen als die Pflichten der herrschenden GmbH selbst. Das kann aber nicht sein.

Diesen Widerspruch vermeidet *Uwe H. Schneider*. Übernimmt die herrschende GmbH die Leitung über die Tochtergesellschaft, so ist nach seiner Ansicht nicht nur die GmbH, sondern auch der Geschäftsführer zur sorgfältigen Unternehmensleitung verpflichtet. Ersteres wurde oben ausführlich begründet.

Uwe H. Schneider ist zuzustimmen. Das ergibt sich aus dem Umstand, daß der Geschäftsführer diejenige natürliche Person ist, die die einheitliche Leitung ausübt. Und allein an diese tatsächliche Verhaltensweise, die der herrschenden GmbH analog § 31 GmbHG zugerechnet wird, knüpfen die §§ 76, 93 AktG, 43 GmbHG, 34, 41 GenG an.

Das wird schließlich bestätigt durch § 309 Abs. 2 AktG bzw. § 323 Abs. 1 i. V. m. § 309 AktG und § 317 Abs. 3 AktG, die gerade im Bereich der Konzernleitung persönliche Pflichten bzw. eine persönliche Haftung des Geschäftsleitungsorgans des herrschenden Unternehmens begründen. Freilich knüpfen diese Vorschriften an eine Weisung bzw. Veranlassung an, begründen also keine Pflicht zum Tätigwerden. Das läßt sich jedoch mit der Annahme des Gesetzgebers erklären, daß die Verantwortlichkeit des Geschäftsleitungsorgans der beherrschten Gesellschaft außerhalb von Weisungen des herrschenden Unternehmens unberührt bleibe. Diese Annahme stimmt jedoch dann nicht mehr, wenn das herrschende Unternehmen die Gesellschaft in einen Konzern eingegliedert hat, also ihre Einflußnahme am Konzerninteresse ausrichtet. Denn dann ist das Geschäftsleitungsorgan des beherrschten Unternehmens nicht mehr in der Lage ist, die Geschäftsleitung ordnungsgemäß auszuüben. In diesem Fall ist der Geschäftsführer der herrschenden GmbH ebenso wie die herrschende GmbH selbst gegenüber der beherrschten Gesellschaft zur ordnungsgemäßen Konzernleitung analog §§ 76, 93 AktG, 43 GmbHG, 34 GenG verpflichtet.

Das gilt im GmbH-Konzern wie im Aktienkonzern, und das gilt im faktischen Konzern wie im Vertragskonzern.

---

[16] Siehe 2. Teil B. II. 7.

[17] Siehe 2. Teil B. II. 5.

Hinsichtlich des Inhalts der Pflicht zur ordnungsgemäßen Unternehmensleitung kann schließlich auf die entsprechende Pflicht der herrschenden GmbH verwiesen werden.

B. Die Treupflichten

Es stellt sich nun die Frage, ob den Geschäftsführer der herrschenden GmbH gegenüber den beherrschten Gesellschaften auch Treupflichten treffen. Das soll am Beispiel des Wettbewerbsverbots und des Verbots der Ausnutzung von Geschäftschancen der Gesellschaft untersucht werden.

I. Entstehungsgrund der Treupflichten

Rechtsgrundlage von Treupflichten ist § 242 BGB. Einigkeit besteht darüber, daß erste Anwendungsvoraussetzung von § 242 BGB das Bestehen einer Sonderverbindung ist, die das Rechtsverhältnis der Beteiligten von den gegenüber jedermann bestehenden Pflichten heraushebt. Im Verhältnis zu seiner eigenen Gesellschaft ergab sich diese Sonderverbindung aufgrund der Organstellung des Geschäftsführers. Darauf kann im Verhältnis zu den beherrschten Gesellschaften nicht zurückgegriffen werden, da der Geschäftsführer nicht deren Organ ist.

Allerdings reichen für die Begründung einer Sonderverbindung auch verstärkte Einwirkungsmöglichkeiten auf die Interessen eines Dritten aus. Und über diese verfügt der Geschäftsführer als Organ der herrschenden GmbH. Das zeigt sich besonders deutlich im Konzern. Dieser zeichnet sich gerade dadurch aus, daß der Geschäftsführer Leitungsmacht ausübt und damit auf die Interessen der beherrschten Gesellschaft einwirkt. Aber auch in der bloßen Abhängigkeitslage ist eine Sonderverbindung zu bejahen, da diese bereits durch das Bestehen erhöhter Einwirkungsmöglichkeiten auf die Interessen der beherrschten Gesellschaft hergestellt wird.

Als Ergebnis kann somit festgehalten werden, daß den Geschäftsführer der herrschenden GmbH gegenüber den beherrschten Gesellschaften Treupflichten treffen.

Es soll nun untersucht werden, ob sich diese Treupflichten zu einem Verbot der Ausnutzung von Geschäftschancen der beherrschten Gesellschaft bzw. einem Wettbewerbsverbot konkretisieren lassen.

II. Die inhaltliche Präzisierung der Treupflichten

Zur inhaltlichen Präzisierung der Treupflichten kann an dieser Stelle auf die Gesichtspunkte zurückgegriffen werden, mit denen die Treupflichten des Geschäftsführers gegenüber seiner eigenen Gesellschaft konkretisiert wurden. Sprechen diese Gesichtspunkte auch für ein Verbot der Ausnutzung von Geschäftschancen der Gesellschaft und ein Wettbewerbsverbot im Verhältnis zu den Konzernunternehmen, so ist es auch hier geboten, entsprechende Verbote anzunehmen.

1. Das Verbot der Ausnutzung von Geschäftschancen der beherrschten Gesellschaften

Das Verbot der Ausnutzung von Geschäftschancen der Gesellschaft im Verhältnis zur eigenen Gesellschaft wurde damit begründet, daß der Geschäftsführer zur umfassenden Förderung des Gesellschaftszwecks verpflichtet ist. Damit verträgt es sich aber nicht, wenn er potentielle Geschäftschancen der Gesellschaft an sich zieht.

Gegenüber den Konzernunternehmen ist der Geschäftsführer der herrschenden GmbH dagegen nur dann zur umfassenden Förderung des Gesellschaftszwecks verpflichtet, wenn er die Geschäftsführung der Unternehmen vollständig an sich gezogen hat und damit wie ein faktisches Organ auftritt. Das ist nur dann der Fall, wenn er der beherrschten Gesellschaft nicht nur die Unternehmensziele vorgibt, sondern auch die Ausführungsentscheidungen trifft. In allen übrigen Fällen stehen den Konzernunternehmen dagegen ihre eigenen Geschäftsleitungsorgane zur Verfügung, so daß der Geschäftsführer jedenfalls nicht alle potentiellen Geschäftschancen den Konzernunternehmen überlassen muß.

Anders verhält es sich jedoch bei solchen Geschäftschancen, die ihm aus den Konzernunternehmen heraus vermittelt wurden. Das sind alle Geschäftschancen einer abhängigen Gesellschaft, von denen er im Rahmen der Konzernleitung Kenntnis erlangt hat. In diesem Fall würde es einen Mißbrauch seiner Einwirkungsmöglichkeiten darstellen, wenn er die Geschäftschancen an sich ziehen würde. Denn diese Geschäftschancen wurden von den Konzernunternehmen erwirtschaftet und stehen diesen folglich zu.

2. Wettbewerbsverbot

Das Wettbewerbsverbot des Geschäftsführers im Verhältnis zu seiner eigenen Gesellschaft beruht auf drei Säulen. Erstens ist der Geschäftsführer verpflichtet, den Gesellschaftszweck der Gesellschaft dadurch umfassend zu fördern, daß er Geschäftschancen für die Gesellschaft wahrnimmt. Diese Pflicht wird durch das Wettbewerbsverbot flankiert. Zweitens besteht die Gefahr, daß er zugunsten eines Konkurrenzunternehmens schädlichen Einfluß auf die Gesellschaft nimmt. Drittens besteht die Gefahr, daß er gesellschaftsinterne Informationen zugunsten seines Konkurrenzunternehmens mißbräuchlich verwendet. Beiden Gefahren soll ein Wettbewerbsverbot vorbeugen.

Der erste Aspekt kommt im Verhältnis zu den Konzernunternehmen nur dann in Betracht, wenn die herrschende GmbH vollständig an die Stelle des Geschäftsleitungsorgans der beherrschten Gesellschaft getreten ist und sie damit dieselben Pflichten treffen wie dieses. Ist dies der Fall, ist der Geschäftsführer gegenüber den Konzernunternehmen zur Unterlassung von Wettbewerb ebenso verpflichtet wie zur ordnungsgemäßen Geschäftsführung. Denn die erstere Pflicht ist nur eine Folge der letzteren Pflicht.

Aber auch in den Fällen, in denen dem Geschäftsführer nicht die Pflicht zur Wahrnehmung von Geschäftschancen der Gesellschaft obliegt, sondern in denen er nur die originären Führungsentscheidungen trifft, die von den Geschäftsleitungsorganen der beherrschten Gesellschaften sodann umgesetzt werden müssen, ist ein Wettbewerbsverbot zu bejahen. Denn auch dann besteht die Gefahr, daß der Geschäftsführer Informationen der Gesellschaft zugunsten seines Konkurrenzunternehmens mißbraucht oder auf andere Weise schädlichen Einfluß auf die Gesellschaft nimmt.

5. Teil: Die konzerndimensionalen Pflichten der herrschenden GmbH gegenüber der Allgemeinheit

Die herrschende GmbH treffen nicht nur Pflichten gegenüber den beherrschten Gesellschaften, sondern auch gegenüber der Allgemeinheit. Diese lassen sich in deliktsrechtliche Pflichten und öffentlich-rechtliche Pflichten unterscheiden. An dieser Stelle interessieren nur diejenigen Pflichten, die einen konzerndimensionalen Bezug haben. Sie zeichnen sich dadurch aus, daß sie eine Verantwortlichkeit des herrschenden Unternehmens für einen Vorgang bei einer beherrschten Gesellschaft begründen. So stellt sich etwa die Frage, ob das herrschende Unternehmen für den Fehler eines Produkts verantwortlich ist, das nicht von ihm, sondern von einer Tochtergesellschaft hergestellt wurde.

Für die Systematisierung der konzerndimensionalen Pflichten der herrschenden GmbH bietet es sich an, nach den Aufgreifmerkmalen zu unterscheiden, an die die einzelnen Vorschriften anknüpfen. Diese Systematisierung hat gegenüber der üblichen Unterscheidung zwischen deliktsrechtlichen und öffentlichrechtlichen Pflichten den Vorteil, daß sie Wiederholungen vermeidet. Denn es wird gezeigt werden, daß deliktsrechtliche Pflichten und öffentlich-rechtliche Pflichten oft ähnliche Aufgreifmerkmale haben.

Der weitere Gang der Untersuchung gestaltet sich damit wie folgt: Exemplarisch untersucht werden Pflichten, deren Rechtsgrundlage an die Konzernlage, an eine tatsächliche Geschehensbeherrschung oder an die Geschäftsführerstellung anknüpft.

A. Aufgreifmerkmal: Konzernlage

Vorschriften, die ausdrücklich an die Konzernlage anknüpfen, finden sich insbesondere im Handelsgesetzbuch sowie im Wirtschaftsaufsichtsrecht. So verpflichtet etwa § 290 HGB die gesetzlichen Vertreter einer an der Spitze eines Konzerns stehenden Kapitalgesellschaft, unter bestimmten Voraussetzungen einen Konzernabschluß und einen Konzernlagebericht aufzustellen. Hierdurch soll den Gesellschaftern, Gläubigern und Arbeitnehmern der Gesellschaft ein den tatsächlichen Verhältnissen entsprechendes Bild der Vermögens-, Finanz- und Ertragslage des Konzerns als einem wirtschaftlich einheitlichen Unternehmen vermittelt werden (§ 297 Abs. 2 und 3 HGB). Gemäß § 10a Abs. 8

KWG ist bei einer Institutsgruppe[1] das übergeordnete Unternehmen für eine angemessene Eigenkapitalausstattung der Institutsgruppe verantwortlich. Hierdurch soll verhindert werden, daß die Vorschriften über die Angemessenheit des haftenden Eigenkapitals nicht durch Gründung von Tochtergesellschaften umgangen werden können und auf dem der Tochtergesellschaft zugewiesenen Geschäftskapital ein neues Kreditvolumen aufgebaut wird, ohne daß tatsächlich ein zusätzliches Risikokapital in die Institutsgruppe eingebracht wird.[2] Insbesondere kehrt in § 10a Abs. 8 Satz 2 KWG dasselbe Prinzip wieder, das für die Grenzen der Konzernleitung ausschlaggebend ist. So heißt es dort ausdrücklich, daß das übergeordnete Unternehmen „zur Erfüllung seiner Verpflichtungen nach Satz 1 auf gruppenangehörige Unternehmen nur einwirken [darf], soweit dem das allgemein geltende Gesellschaftsrecht nicht entgegensteht". Ferner ist § 15 GWG zu nennen. Danach haben bestimmte Unternehmen dafür zu sorgen, daß von ihnen abhängige Unternehmen im Ausland, die mit ihnen unter einheitlicher Leitung zusammengefaßt sind, bestimmte Verpflichtungen aus dem GWG erfüllen. Schließlich sei aus dem Umweltschutzrecht noch § 4 Abs. 3 Satz 4 BBodSchG erwähnt. Gehört einer juristischen Person ein kontaminiertes Grundstück, so ist nach dieser Vorschrift zur Sanierung auch verpflichtet, wer aus gesellschaftsrechtlichem Rechtsgrund für die juristische Person einzustehen hat. Als Hauptanwendungsfall einer solchen Einstandspflicht wird die Haftung des herrschenden Unternehmens gegenüber der abhängigen Gesellschaft im qualifizierten faktischen Konzern genannt.[3]

B. Aufgreifmerkmal: Tatsächliche Geschehensbeherrschung

I. Die gesetzliche Ausgangslage

Zahlreiche Vorschriften knüpfen an eine tatsächliche Geschehensbeherrschung an und machen sie zu einer notwendigen, wenn auch nicht immer ausreichenden Voraussetzung für die Begründung von Pflichten. Solche Vorschriften finden sich sowohl im Deliktsrecht wie im öffentlichen Recht.

---

[1] Eine Institutsgruppe liegt gemäß § 10a Abs. 2 KWG vor, wenn ein Institut (Kreditinstitut oder Finanzdientstleistungsinstitut, § 1 Abs. 1 u. 2 KWG) ein anderes Institut, ein Finanzunternehmen (§ 1 Abs. 3 KWG) oder ein Unternehmen mit bankbezogenen Hilfsdiensten (§ 1 Abs. 3 c) KWG) zum Tochterunternehmen hat.

[2] *M. Schneider*, KWG, 1986, § 10a Anm. 1.

[3] *A. Turiaux/D. Knigge*, Bundes-Bodenschutzgesetz, 1999, S. 380. Siehe auch *L. Knopp/E. Albrecht*, Bundes-Bodenschutzgesetz, 1998, S. 1856; *W. Schmitz-Rode/S. Bank*, Haftung, 1999, S. 417, 419 f.

Zum Deliktsrecht gehören insbesondere die zu § 823 Abs. 1 BGB entwickelten Verkehrspflichten. Ihnen liegt der Gedanke zugrunde, daß derjenige, der eine Gefahrenquelle schafft oder andauern läßt, zumutbare Vorkehrungen zum Schutze Dritter zu treffen hat.[4] Ihre schuldhafte Verletzung ist bei Fahrlässigkeitsdelikten Voraussetzung für einen Schadensersatzanspruch nach § 823 Abs. 1 BGB. Dagegen knüpfen die Tatbestände der Gefährdungshaftung nicht an eine schuldhafte Pflichtverletzung, sondern allein an die Schaffung einer bestimmten Gefahrenlage an.[5] Auch wenn sie keine Pflichten begründen, sollen sie in die Betrachtung einbezogen werden. Zu nennen ist etwa die Haftung des Herstellers eines fehlerhaften Produkts für Körper-, Gesundheits- und Sachschäden, die durch das fehlerhafte Produkt verursacht wurden, § 1 Abs. 1 ProdHaftG. Und zu nennen ist die Haftung des Inhabers einer im Anhang 1 zum UmweltHG genannten Anlage für Körper-, Gesundheits- und Sachschäden, die durch Umwelteinwirkungen der Anlage entstanden sind, § 1 UmweltHG.

An letzter Stelle schließlich soll aus dem öffentlichen Recht noch das BImSchG genannt werden. Es legt dem Betreiber einer Anlage bestimmte Pflichten zum Schutz von Mensch und Natur auf, §§ 5, 22 BImSchG.

In der Literatur wird die Ansicht vertreten, daß sich aus diesen Vorschriften konzerndimensionale Pflichten des herrschenden Unternehmens ergeben würden.

II. Meinungsstand

So träfen das herrschende Unternehmen die zu § 823 Abs. 1 BGB entwickelten Verkehrspflichten eines Produzenten, wenn es auf den technischen Produktionsprozeß bei der Tochtergesellschaft Einfluß nehme.[6] Und unter denselben Voraussetzungen wird es als Hersteller im Sinne von § 4 Abs. 1 Satz 1 ProdHaftG angesehen.[7] Entsprechendes gelte schließlich für seine Qualifizierung als Inhaber einer Anlage im Sinne von § 1 UmweltHaftG[8] sowie als Be-

---

4 *BGH*, Urt. v. 28.4.1952 - III ZR 118/51, in: BGHZ 5, S. 378, 380 f.

5 *H. Brox*, Schuldrecht, 1997, Rdnr. 532; *C.-W. Canaris*, Schuldrecht, 1994, S. 600.

6 *W. Oehler*, Produzentenhaftung, 1990, S. 1449. Nicht so weitgehend *E. Rehbinder*, Konzernaußenrecht, 1969, S. 510.

7 *P. Hommelhoff*, Produkthaftung, 1990, S. 764.

8 *H.-P. Westermann*, Umwelthaftung, 1991, S. 238 f.; *M. Lutter*, Haftungsfragen, 1998, F 24 (S. 257 f.). Enger *C.-P. Feick*, Pflichten, 1999, S. 149, der eine „umfassende und dauernde Ausübung der operativen Leitung der Anlage durch eigene Mitarbeiter" des herrschenden Unternehmens fordert. Siehe auch *A. Hucke/H. Schröder*, Umwelthaftung, 1998, S. 2206, die bestimmte Indizien für die Inhabereigenschaft des herrschenden Unternehmens aufstellen.

treiber einer Anlage im Sinne des Bundesimmissionsschutzgesetzes[9]. Dagegen reiche z. B. die Festlegung des finanziellen Rahmens für die Produktion oder den Betrieb der Anlage zur Begründung einer konzerndimensionalen Haftung gegenüber Dritten nicht aus, soweit nur die Fertigungsverantwortung bzw. die Verantwortung für den technischen Betrieb der Anlage auf seiten der Tochtergesellschaft unberührt bleibe.[10]

Diesen Ansichten kann indes nur teilweise gefolgt werden.

III. Keine konzerndimensionale Gefährdungshaftung

Eine konzerndimensionale Gefährdungshaftung des herrschenden Unternehmens ist abzulehnen. Sie widerspricht dem in §§ 1 Abs. 1 Satz 2 AktG, 13 Abs. 2 GmbHG niedergelegten Grundsatz der Haftungstrennung. Dieser besagt, daß für Verbindlichkeiten der Gesellschaft nur das Gesellschaftsvermögen haftet. Hiermit soll dem Rechtsverkehr die Möglichkeit gegeben, Haftungsrisiken zu begrenzen. Wollte man nun das herrschende Unternehmen als Gesellschafter teilweise denselben Haftungsrisiken aussetzen wie die Gesellschaft selbst, so würde damit der Grundsatz der Haftungstrennung durchbrochen. Das wird besonders deutlich, wenn man erkennt, daß sich die in der Literatur entwickelten Grundsätze nicht auf den Unternehmensgesellschafter beschränken lassen. Vielmehr müßten sie folgerichtig auch für den privaten Mehrheitsgesellschafter gelten, der die Leitung des Unternehmens an sich gezogen hat. Denn auch der private Mehrheitsgesellschafter übt dann die tatsächliche Gewalt über die Gefahrenquellen in dem Unternehmen der GmbH bzw. Aktiengesellschaft aus. Hinzu kommt bei der GmbH, daß der Gesetzgeber diese Rechtsform gerade zur Verfügung gestellt und entsprechend ausgestaltet hat, um ihren Gesellschaftern eine Mitwirkung in der Unternehmensführung zu ermöglichen.[11] Von dieser Mitwirkung würden sie aber abgehalten, wenn sie befürchten müßten, daß eine ihrer Entscheidungen den technischen Betrieb einer Anlage oder den technischen Produktionsprozeß beeinflußt und sie damit einer Gefährdungshaftung nach § 1 UmweltHaftG bzw. § 1 Abs. 1 Satz 1 ProdHaftG ausgesetzt werden.

---

[9] *Uwe H. Schneider*, Compliance, 1996, S. 239. Enger *C.-P. Feick*, Pflichten, 1999, S. 144, nach dem das herrschende Unternehmen sich „dauernd und umfassend sowohl in das operative Geschäft der Tochter einmischen, als auch auf den technischen Betrieb der Anlage erheblichen Einfluß ausüben" müsse, um als Betreiber qualifiziert werden zu können.

[10] *H.-P. Westermann*, Umwelthaftung, 1991, S. 238 f.; *P. Hommelhoff*, Produkthaftung, 1990, S. 764. Im Ergebnis wohl auch *E. Rehbinder*, Konzernaußenrecht, 1969, S. 509.

[11] Amtliche Ausgabe des Entwurfs eines GmbHG nebst Begründung und Anlagen, Berlin: Franz Vahlen, 1891, S. 45 f.

Freilich könnte man argumentieren, daß der Gesellschafter deshalb die Gefahrenquelle beherrscht, weil aufgrund der von ihm ausgeübten Leitung der Geschäftsführer und damit auch die GmbH nicht mehr in der Lage ist, die Gefahrenquelle vollständig zu steuern. Diese Situation könnte der Gesellschafter aber dadurch vermeiden, daß er sich zum Geschäftsführer bestellen läßt und damit die Gesellschaft wieder vollständig die Gefahr beherrscht. Folge müßte sein, daß der Gesellschafter nun nicht mehr persönlich die Risiken der Gefährdungshaftung treffen. Das wäre allerdings widersprüchlich. Denn als Geschäftsführer kann der Gesellschafter noch stärker auf die Gefahrenquelle einwirken als nur aufgrund seiner Gesellschafterstellung. Deshalb müßten ihn jetzt erst recht die Haftungsrisiken aus den genannten Gefährdungstatbeständen treffen.

Nach alledem bietet sich ein anderes Kriterium für die Bestimmung der Adressaten der Gefährdungshaftung an. Es ist derjenige Hersteller eines Produkts, der das Produkt auf eigene Rechnung und im eigenen Namen herstellt. Und es ist derjenige Inhaber einer Anlage, der die Anlage auf eigene Rechnung und im eigenen Namen betreibt.[12] Und entsprechendes gilt schließlich für den Betreiberbegriff des BImSchG.[13] Für letzteren wird zwar auch darauf abgestellt, wer den bestimmenden Einfluß auf die Beschaffenheit und den Betrieb der Anlage ausübt, wer darüber entscheiden kann, ob und wie die Anlage betrieben wird.[14] In einem Unternehmen sei das derjenige, der das Unternehmen selbständig führe.[15] Für die Annahme der Selbständigkeit reiche es jedoch aus, daß das Unternehmen im eigenen Namen und auf eigene Rechnung geführt werde.[16]

Dieses Ansicht wird schließlich durch § 4 der 5. BImSchV bestätigt. Nach dieser Vorschrift kann die zuständige Behörde dem Betreiber einer genehmigungsbedürftigen Anlage gestatten, statt eines Betriebsangehörigen eine betriebsfremde Person, nämlich den Immissionsschutzbeauftragten für den Konzernbereich zum Immissionsschutzbeauftragten zu bestellen. Das begründet der Verordnungsgeber damit, daß der Immissionsschutzbeauftragte am wirksamsten in der Konzernspitze die Belange des Umweltschutzes vertreten könne. Denn dort würden die für den Immissionsschutz wesentlichen Entscheidungen getroffen.[17] Aus der Tatsache, daß der Verordnungsgeber lediglich die Möglichkeit der Bestellung eines Konzernbeauftragten für Immissionsschutz vorsieht, dies jedoch nicht zur Pflicht macht, läßt sich schließen, daß er weiterhin nur das

---

12 *P. Salje*, UmweltHG, 1993, §§ 1, 3 Rdnr. 16.

13 *H. D. Jarass*, BImSchG, 1995, § 3 Rdnr. 71.

14 *H. D. Jarass*, BImSchG, 1995, § 3 Rdnr. 70; *H. P. Westermann*, Umwelthaftung, 1991, S. 234.

15 *H. D. Jarass*, BImSchG, 1995, § 3 Rdnr. 71.

16 *H. D. Jarass*, BImSchG, 1995, § 3 Rdnr. 71.

17 BR-Drucksache 660/74.

beherrschte Unternehmen als Betreiberin ansieht und dem herrschenden Unternehmen keine immissionsschutzrechtlichen Pflichten im Hinblick auf die beherrschte Gesellschaft auferlegen will, und zwar selbst dann nicht, wenn die wesentlichen Entscheidungen für den Immissionsschutz in der Konzernspitze getroffen werden.[18]

Nach alledem ist *H. D. Jarass*[19] Recht zu geben, wenn er ausführt, daß selbst im Vertragskonzern das herrschende Unternehmen nicht als Betreiberin der von den beherrschten Gesellschaften im eigenen Namen und auf eigene Rechnung betriebenen Anlagen anzusehen ist. Und hinzuzufügen ist, daß das herrschende Unternehmen unter diesen Voraussetzungen auch nicht Herstellerin im Sinne von § 4 Abs. 1 Satz 1 ProdHaftG und Inhaberin einer Anlage nach § 1 UmweltHaftG ist.

IV. Konzerndimensionale Verkehrspflichten

1. Die gesetzliche Ausgangslage

Wie bereits erwähnt ist die Verletzung einer sogenannten Verkehrspflicht im Bereich fahrlässiger Rechtsgutsverletzungen Voraussetzung für die Erfüllung des Tatbestandes einer unerlaubten Handlung. In der Literatur werden nach ihrem Entstehungsgrund drei Arten von Verkehrspflichten unterschieden: Die Verkehrspflicht aufgrund der Bestimmungsgewalt über einen bestimmten Bereich, die Verkehrspflicht aufgrund der Übernahme einer Aufgabe und die Verkehrspflicht aufgrund vorangegangenen gefahrerhöhenden Tuns.[20]

2. Bestimmungsgewalt über einen bestimmten Bereich

a) Die gesetzliche Ausgangslage

Erster Entstehungsgrund für eine Verkehrspflicht ist die Bestimmungsgewalt über einen bestimmten Bereich. Diese Verkehrspflicht wird damit begründet, daß derjenige, der eine Gefahrenquelle schafft oder andauern läßt, zumutbare Vorkehrungen zum Schutze Dritter zu treffen hat.[21] Das bedarf jedoch der Er-

---

[18] Gerade deshalb bejaht aber *Uwe H. Schneider*, Pflichten, 1992, S. 493, eine Pflicht der Konzernobergesellschaft zur Bestellung eines „Konzernverantwortlichen“.

[19] *H. D. Jarass*, BImSchG, 1995, § 3 Rdnr. 71. A. M. für den Begriff des Anlageninhabers im Sinne von § 1 UmweltHG wohl aber *P. Salje*, UmweltHG, 1993, §§ 1, 3 Rdnr. 21

[20] Dazu *C.-W. Canaris*, Schuldrecht, 1994, S. 406 ff.

[21] *BGH*, Urt. v. 28.4.1952 - III ZR 118/51, in BGHZ 5, S. 378, 380 f.

gänzung. Denn nicht jeder, der eine Gefahrenquelle andauern läßt, verletzt eine Verkehrspflicht. Erforderlich ist vielmehr, daß er auch den Vorteil aus dieser Gefahrenquelle zieht.[22] Nur dann ist es gerechtfertigt, ihn für die unterlassene Beseitigung der Gefahrenquelle haftbar zu machen. Deshalb ist z. B. nur der Gebäudeeigentümer dafür verantwortlich, daß von seinem Gebäude keine Gefahren ausgehen, nicht aber die Mietparteien, die in dem Gebäude wohnen.[23]

Damit wird als „Gerechtigkeitskriterium" für die Verkehrspflicht aufgrund der Bestimmungsgewalt über einen bestimmten Bereich folgender Gedanke deutlich: Wer die Bestimmungsgewalt über einen bestimmten Bereich hat und darauf einen Verkehr eröffnet oder duldet, ist deshalb zur Abwehr von Gefahren verpflichtet, weil er erstens zur Beherrschung der Gefahr in der Lage ist und zweitens den Vorteil aus der Gefahrenquelle zu ziehen vermag.[24] Dieses Prinzip liegt auch den §§ 831 Abs. 1, 833, 836 f. BGB zugrunde, so daß man von einem allgemeinen Rechtsgedanken sprechen kann.[25]

b) Die Lage im Konzern

Eine konzerndimensionale Verkehrspflicht aufgrund der Bestimmungsgewalt über einen bestimmten Bereich ist aus denselben Gründen abzulehnen wie eine konzerndimensionale Gefährdungshaftung. Das herrschende Unternehmen kann also nicht allein deshalb, weil es eine Machtposition in der Gesellschaft innehat, z. B. der von der Rechtsprechung zu § 823 Abs. 1 BGB entwickelten Produzentenhaftung unterworfen werden.[26] Dafür spricht auch, daß andernfalls jeden Mehrheitsgesellschafter, gleichgültig ob Privatgesellschafter oder Unternehmensgesellschafter, Verkehrspflichten im Hinblick auf die von der Gesellschaft eröffneten Gefahrenquellen treffen müßten. Insbesondere könnte sich ein Mehrheitsgesellschafter seinen Verkehrspflichten nicht dadurch entziehen, daß er keinen Einfluß auf den technischen Betrieb der Anlage oder den Produktionsprozeß nimmt. Im Gegenteil müßte er sogar sicherstellen, daß alle erforderlichen Vorkehrungen zum Schutze Dritter getroffen werden. Das widerspricht

---

[22] *C.-W. Canaris*, Schuldrecht, 1994, S. 408. Siehe auch *M. Rottkemper*, Außenhaftung, 1996, S. 180 f., der unter Hinweis auf den Zusammenhang von Risikoverantwortlichkeit und Vorteilsziehung eine Verkehrspflicht von Organmitgliedern aufgrund der Eröffnung von der der Gesellschaft zuzurechnenden Gefahrenquellen ablehnt.

[23] Eine Ausnahme besteht freilich, wenn ein Mieter die Verkehrspflichten des Eigentümers übernommen hat.

[24] *C.-W. Canaris*, Schuldrecht, 1994, S. 408.

[25] *C.-W. Canaris*, Schuldrecht, 1994, S. 411.

[26] Im Ergebnis ebenso für die Verkehrspflichten des Geschäftsführers *M. Grünwald*, Außenhaftung, 1999, S. 114.

aber der Stellung der Gesellschafter, die grundsätzlich zum geschäftlichen Desinteresse berechtigt sind.

Deshalb gilt auch hier, daß nur derjenige für eine Gefahrenquelle verantwortlich ist, der die Gefahrenquelle auf eigenen Namen und auf eigene Rechnung geschaffen hat bzw. aufrechterhält.

3. Übernahme einer Aufgabe

a) Die gesetzliche Ausgangslage

Das zweite Kriterium für die Bestimmung des Adressaten einer Verkehrspflicht ist die Übernahme einer Aufgabe.[27] Danach ist derjenige verkehrspflichtig, der durch die tatsächliche Übernahme eines Aufgaben- und Einflußbereichs in die faktische Beherrschung einer Gefahrenquelle einrückt.[28] Denn dann darf der Rechtsverkehr darauf vertrauen, daß der Übernehmende in dem übernommen Aufgaben- und Einflußbereich die der Schadensabwendung dienenden Obliegenheiten erfüllt. Deshalb ist z. B. der Beifahrer, der es übernommen hat, bei einem Abbiegemanöver die rückwärtige Fahrbahn auf das Herannahen anderer Verkehrsteilnehmer zu beobachten, letzteren schadensersatzpflichtig, wenn es durch seine Unaufmerksamkeit zu einem Unfall kommt.[29] Und der Produktionsleiter eines Herstellers von Spannkupplungen haftet auf Schadensersatz, wenn durch eine fehlerhafte Spannkupplung ein Mensch getötet wird.[30]

b) Das Problem

Der Konzern zeichnet sich dadurch aus, daß das herrschende Unternehmen die Leitung der beherrschten Gesellschaften an sich gezogen hat. Nicht mehr die Verwaltungsorgane der beherrschten Gesellschaften, sondern das herrschende Unternehmen übt nun die Unternehmensleitung aus. Damit rückt das herrschende Unternehmen aber auch in den Aufgabenbereich der Verwaltungsorgane der beherrschten Gesellschaften ein. Deshalb sind Verkehrspflichten des herrschenden Unternehmens aufgrund der Übernahme einer Aufgabe zu beja-

---

[27] *BGH*, Urt. v. 17.1.1989 - VI ZR 186/88, in: NJW-RR 1989, S. 394, 395. Grundlegend *P. Ulmer*, Handlungspflichten, 1969, S. 163 ff. Siehe aber auch *C. v. Bar*, Verkehrspflichten, 1980, S. 120 ff.

[28] *D. Eckardt*, Organhaftung, 1996, S. 71.

[29] *BGH*, Urt. v. 10.1.1961 - VI ZR 62/60, in: VersR 1961, 233, 234.

[30] *BGH*, Urt. v. 3.6.1975 - VI ZR 192/73, in: BB 1975, S. 1031

hen, wenn auch die Verwaltungsorgane der beherrschten Gesellschaften persönliche Verkehrspflichten treffen.[31] Das ist indes umstritten.

c) Meinungsstand

Nahezu unstreitig ist nur, daß der Geschäftsführer aus Delikt haftet, wenn er eine unerlaubte Handlung entweder eigenhändig begangen hat[32] oder von ihrer Begehung durch andere wußte und die Schädigung pflichtwidrig nicht verhinderte.[33] Dabei ergebe sich eine Rechtspflicht zum Einschreiten aus der verantwortlichen Stellung des Geschäftsführers in der Gesellschaft und für die Gesellschaft.[34]

Streitig ist dagegen, ob den Geschäftsführer darüber hinaus deliktische Pflichten treffen. Dabei geht es um die Frage, ob der Geschäftsführer auch dann aus Delikt haftet, wenn zwar keine Handlung von ihm für die Rechtsgutsverletzung kausal war, er jedoch die Rechtsgutsverletzung durch eine bessere Überwachung bzw. Organisation des Unternehmens der Gesellschaft hätte verhindern können.

Ausgelöst wurde dieser Streit durch das viel beachteten Urteil des *Bundesgerichtshofs*[35] vom 5. Dezember 1989. Dort hat der *Bundesgerichtshof* den Geschäftsführer einer GmbH persönlich für eine Eigentumsverletzung haften lassen, die dadurch verursacht worden war, daß ein anderer Mitarbeiter der Gesellschaft unter Eigentumsvorbehalt gelieferte Materialien unbefugtermaßen bei einem Kunde der GmbH hatte einbauen lassen, wodurch dieser gemäß § 946 BGB Eigentümer wurde. Unbefugt war dieses Verhalten allein deshalb, weil in den AGB des Kunden ein Abtretungsverbot enthalten war und daher die Vorausabtretung der Forderungen gegen ihn, die sich der Vorbehaltslieferant im Wege eines verlängerten Eigentumsvorbehalts ausbedungen hatte, nicht wirksam werden konnte. Das Berufungsgericht wies die Klage ab, weil der Ge-

---

31 In diesem Sinne auch *P. Ulmer*, Hachenburg, 1991, § 64 Rdnr. 11, für die Konkursantragspflicht (jetzt Pflicht zur Beantragung der Eröffnung des Insolvenzverfahrens) des faktischen Geschäftsführers.

32 *H. Altmeppen*, GmbHG, 1997, § 43 Rdnr. 25; *D. Medicus*, Eigenhaftung, 1991, S. 169; *S. Recksiek*, Außenhaftung, 1998, S. 20.

33 *BGH*, Urt. v. 26.9.1985 - I ZR 86/83, in: GmbHR 1986, S. 83, 84; *M. Lutter*, Haftung, 1993, S. 468 f., 482; ders., Haftungsrisiken, 1997, S. 334; *W. Groß*, Außenhaftung, 1998, S. 563 f.; *Heil/Russenschuck*, Haftung, 1998, S. 1753. Für das Vorstandsmitglied einer AG: *OLG Karlsruhe*, Urt. v. 11.12.1996 - 6 U 56/96, in: ZIP 1997, S. 125, 126 f.; *M. Grünwald*, Außenhaftung, 1999, S. 54 f. A. M. *S. Recksiek*, Außenhaftung, 1998, S. 47 ff.

34 *M. Lutter*, Haftung, 1993, S. 468 f., 482.

35 *BGH*, Urt. v. 5.12.1989 - VI ZR 335/88, in: BGHZ 109, S. 297 ff.

schäftsführer weder am Abschluß der Verträge mit dem Vorbehaltslieferanten mitgewirkt habe noch den mit diesem vereinbarten verlängerten Eigentumsvorbehalt und das mit dem Kunden vereinbarte Abtretungsverbot gekannt habe.[36] Der *Bundesgerichtshof*[37] gab der Klage statt, weil den Geschäftsführer eine „Garantiepflicht dahin [traf], dafür zu sorgen, daß der verlängerte Eigentumsvorbehalt bei Verarbeitung der Baumaterialien nicht aufgrund eines Abtretungsverbots des Auftragsgebers ins Leere ging“. Zwar bestünde die Pflicht zur ordnungsgemäßen Unternehmensführung, zu der auch die Abwendung von Schäden Dritter gehöre, nur gegenüber der Gesellschaft. Jedoch könne den Geschäftsführer aus *besonderen Gründen* auch persönlich eine Pflicht gegenüber einem Dritten treffen. Hierzu führte der Bundesgerichtshof[38] aus:

> „Dies kann im außervertraglichen, deliktischen Bereich insbesondere wegen einer dem Geschäftsführer als Aufgabe zugewiesenen oder von ihm jedenfalls in Anspruch genommenen Garantenstellung zum Schutz fremder Schutzgüter i. S. des § 823 I BGB der Fall sein, die ihre Träger der Einflußsphäre der Gesellschaft anvertraut haben. Hier kann über die Organstellung hinaus eine mit der Zuständigkeit für die Organisation und Leitung und der daraus erwachsenden persönlichen Einflußnahme auf die Gefahrenabwehr bzw. -steuerung verbundene persönliche Verantwortung des Organs den betroffenen Außenstehenden gegenüber zum Tragen kommen“.

Voraussetzung sei allerdings, daß der Geschäftsführer zur Abwehr der Gefahrenlage gerade in seinem Aufgabenbereich gefordert sei.[39] Im zu entscheidenden Fall habe eine Gefahrenlage bestanden, die Organisationsvorkehrungen verlangt habe, um das Vorbehaltseigentum der Klägerin zu schützen. Zu solchen organisatorischen Vorkehrung sei aber zu allererst der Geschäftsführer berufen.[40]

Die Bejahung deliktischer Verkehrspflichten des Geschäftsführers durch den *Bundesgerichtshof* stößt in der Literatur zum größten Teil auf massive Kritik.[41] Befürchtet wird eine unangemessene Erhöhung des Haftungsrisikos des Ge-

---

36 So die in dem Urteil des *BGH* in BGHZ 109, S. 297, 299, mitgeteilten Gründe des Berufungsgerichts.

37 *BGH*, Urt. v. 5.12.1989 - VI ZR 335/88, in: BGHZ 109, S. 297, 302.

38 *BGH*, Urt. v. 5.12.1989 - VI ZR 335/88, in: BGHZ 109, S. 297, 303.

39 *BGH*, Urt. v. 5.12.1989 - VI ZR 335/88, in: BGHZ 109, S. 297, 304.

40 *BGH*, Urt. v. 5.12.1989 - VI ZR 335/88, in: BGHZ 109, S. 297, 304.

41 *H.-J. Mertens/G. Mertens*, Anmerkung, 1990, S. 488 ff.; *M. Lutter*, Haftung, 1993, S. 470 ff.; *D. Medicus*, Eigenhaftung, 1991, S. 160; *ders.*, Außenhaftung, 1998, S. 585; *M. Dreher*, Verantwortlichkeit, 1992, S. 34; *M. Kort*, Eigenhaftung, 1990, S. 923; *H. P. Westermann/S. Mutter*, Verantwortlichkeit, 1995, S. 189; *H.-P. Götting*, Haftung, 1994, S. 11 f.; *A. Sandberger*, Außenhaftung, 1997, S. 154 f.; *Heil/Russenschuck*, Haftung, 1998, S. 1750 f.; *M. Grünwald*, Außenhaftung, 1999, S. 139. Siehe auch *OLG Köln*, Urt. v. 20.6.1992 - 6 U 72/91, in: GmbHR 1993, S. 586 f. Differenzierend: *B. Grunewald*, Haftung, 1993, S. 454 ff.

schäftsführers.[42] Auch sei das Rechtsinstitut der beschränkten Haftung sowie die Entscheidungs- und Handlungsfähigkeit der Gesellschaftsorgane gefährdet; denn im Konkurs[43] der Gesellschaft würde sich das Ausfallrisiko (§ 13 Abs. 2 GmbHG) vom Gläubiger auf den Geschäftsführer verlagern, dessen Ersatzanspruch wertlos werde.[44] Allerdings führt diese Kritik nicht zu einer uneingeschränkten Ablehnung der Rechtsprechung des Bundesgerichtshofs. Vielmehr gibt es auch Stimmen, die eine Außenhaftung des Geschäftsführers lediglich an engere Voraussetzungen knüpfen wollen. So wird die Ansicht vertreten, daß der Geschäftsführer nur für eine grobe Verletzung des Kernbereichs der ihm obliegenden Organisations- und Leitungspflichten hafte.[45] Eine solche Verletzung liege vor, wenn der Geschäftsführer eine jedem durchschnittlich sorgfältigen Rechtsgenossen ohne weiteres erkennbare erhebliche Gefahr, deren Verwirklichung zu einem schwerwiegenden Schaden für ein vom Schutzbereich des § 823 Abs. 1 BGB erfaßten Rechtsgut führe, nicht abwende.[46]

Das Baustoff-Urteil des *Bundesgerichtshofs* wurde jedoch auch begrüßt.[47] Bejaht wird eine Garantenstellung des Geschäftsführers aus Organisationsherrschaft, die ihm nicht nur gegenüber der Gesellschaft, sondern auch gegenüber Dritten obliege. Diese Garantenstellung begründe für die Geschäftsführer Organisations- und Überwachungspflichten, die im Hinblick auf eine Begrenzung von deren Haftungsrisiko angemessen konkretisiert werden müßten.[48] Weitergehend wird sogar eine generelle Schutzpflicht von Organwaltern gegenüber Dritten innerhalb ihres Zuständigkeitsbereichs angenommen.[49] Hierdurch werde die Haftungssituation hergestellt, die das Gesetz für die traditionellen (eigenhändigen) Delikte der Organe und satzungsmäßigen Vertreter vorsehe. Außerdem sei der Organwalter nicht schutzbedürftiger als der abhängig Beschäftigte, der auch nach außen für die Verletzung von Verkehrspflichten des Organisationsträgers hafte. Schließlich stelle die Eigenhaftung des Organs „eine gewisse Kompensation zu der Möglichkeit inkorporierter Organisationsträger

---

42 *H.-J. Mertens/G. Mertens*, Anmerkung, 1990, S. 488.; *M. Dreher*, Verantwortlichkeit, 1992, S. 34; *H.-P. Götting*, Haftung, 1994, S. 11.

43 Jetzt Insolvenzverfahren.

44 *M. Lutter*, Haftung, 1993, S. 473.

45 *M. Grünwald*, Außenhaftung, 1999, S. 137 ff.

46 *M. Grünwald*, Außenhaftung, 1999, S. 140.

47 *Uwe H. Schneider*, Scholz, 1993, § 43 Rdnr. 230; *G. Brüggemeier*, Organisationshaftung, 1991, S. 65; *W. Groß*, Außenhaftung, 1998, S. 567.

48 *Uwe H. Schneider*, Scholz, 1993, § 43 Rdnr. 230; *J. Keßler*, Eigenhaftung, 1994, S. 436. Dagegen *P. Haß*, Haftung, 1994, S. 671; *U. Haas*, Geschäftsführerhaftung, 1997, S. 239 ff.

49 *G. Brüggemeier*, Organisationshaftung, 1991, S. 64 f.; *J.-U. Nölle*, Eigenhaftung, 1995, S. 98 ff. Dagegen: *D. Kleindiek*, Deliktshaftung, 1997, S. 207.

dar, die Haftung auf die umfangsmäßig limitierte Vermögensmasse der juristischen Person zu beschränken, wobei es de lege lata keine Rechtspflicht zu angemessener Kapitalausstattung gibt".[50]

d) Stellungnahme

Eine Stellungnahme zu diesem Streit muß im Lichte von § 13 Abs. 2 GmbHG erfolgen. Das dort zum Ausdruck kommende Trennungsprinzip besagt, daß nur die Gesellschaft, nicht aber die Gesellschafter oder Geschäftsführer für die Verbindlichkeiten der Gesellschaft haften sollen. Mit diesem Prinzip ist es aber nicht vereinbar, dem Geschäftsführer allein aufgrund seiner Organstellung die Verkehrspflichten der Gesellschaft aufzubürden.[51] Aus diesem Grund ist die Ansicht abzulehnen, wonach eine generelle Schutzpflicht von Organwaltern gegenüber Dritten innerhalb ihres Zuständigkeitsbereichs bestehe. Vielmehr müssen besondere Umstände, wie ja auch der *Bundesgerichtshof* betont, hinzukommen, um persönliche Verkehrspflichten für den Geschäftsführer begründen zu können. Hierbei ist der vom *Bundesgerichtshof* angesprochene Vertrauensgedanke zwar der richtige Ansatz.[52] Denn Rechtsgrund der Verkehrspflichten sind die Sicherheitserwartungen des Rechtsverkehrs. Nicht alle Sicherheitserwartungen können jedoch durch Verkehrspflichten geschützt werden. Das würde den Umfang der Verkehrspflichten sprengen und den Handlungsspielraum des einzelnen unangemessen einschränken.[53] Deshalb ist zu fordern, daß nur berechtigte Sicherheitserwartungen Verkehrspflichten auslösen. Damit erhält der Rechtsbegriff „Vertrauen" ein normatives Element.[54] Und dieses normative Element läßt es zu, die Wertung des Trennungsprinzips bei der Begründung von Verkehrspflichten angemessen zu berücksichtigen.

Folgt man dem, so läßt sich zunächst begründen, warum der Geschäftsführer verpflichtet ist, gegen eine bevorstehende Rechtsgutsbeeinträchtigung durch andere Mitarbeiter der Gesellschaft einzuschreiten, wenn er davon Kenntnis erlangt. Denn aufgrund seiner verantwortlichen Stellung in der Gesellschaft und für die Gesellschaft kann ein solches Einschreiten von ihm berechtigterweise erwartet werden. Gleiches wird man aber auch dann anzunehmen haben, wenn

---

50 *G. Brüggemeier*, Organisationshaftung, 1991, S. 64 f.

51 In diese Richtung auch *D. Eckardt*, Organhaftung, 1996, S. 74. A. M. *W. Groß*, Außenhaftung, 1998, S. 552, nach dem nur die Gesellschafter durch das in § 13 Abs. 2 GmbHG zum Ausdruck kommende Trennungsprinzip privilegiert werden.

52 Zur allgemeinen Bedeutung des Vertrauensschutzaspekts *C. v. Bar*, Verkehrspflichten, 1980, S. 117.

53 Zum Spannungsverhältnis zwischen Güterschutz und Handlungsfreiheit siehe *C.-W. Canaris*, Schuldrecht, 1994, S. 350.

54 *D. Eckardt*, Organhaftung, 1996, S. 71.

er zwar keine positive Kenntnis von der bevorstehenden Rechtsgutsbeeinträchtigung hat, sich ihm hierfür aber konkrete Anhaltspunkte aufdrängen. Auch dann können Maßnahmen zur Abwendung von Rechtsgutsbeeinträchtigungen von ihm erwartet werden.

Im übrigen wird man als Wertungskriterien für die Bestimmung von Verkehrspflichten die Schadensneigung der übernommenen Aufgabe, ihre Nähe zum tatbestandsmäßigen Verletzungserfolg und das Maß der drohenden Rechtsgutsbeeinträchtigung heranziehen können.[55] Mit anderen Worten ist um so eher eine persönliche Verkehrspflicht anzunehmen, je größer die Gefahr für ein Rechtsgut ist. Und das gilt unabhängig davon, ob ein Organmitglied die gefahrenwesentliche Aufgabe übernommen hat oder ein Mitarbeiter der Gesellschaft. Deshalb ist dem *Bundesgerichtshof*[56] zuzustimmen, wenn er den Produktionsleiter von Spannkupplungen persönliche Verkehrspflichten auferlegt. Denn dieser übt eine gefahrenwesentliche Aufgabe aus. Und gleiches würde auch dann gelten, wenn der Produktionsleiter zugleich Geschäftsführer wäre. Macht ein Geschäftsführer dagegen nur allgemeine Vorgaben für die Produktion, legt er etwa nur die Maschinenlaufzeit fest oder bestimmt deren Auslastung, nimmt er aber ansonsten keinen Einfluß auf den technischen Produktionsprozeß, treffen ihn dagegen keine persönlichen Verkehrspflichten. Insoweit ist auch der in der Literatur vertretenen Ansicht zuzustimmen, daß das herrschende Unternehmen die Verkehrspflichten eines Produzenten trifft, wenn es auf den technischen Produktionsprozeß bei der Tochtergesellschaft Einfluß nimmt.[57]

e) Die Lage im Konzern

Nach alledem zeigt sich, daß der Geschäftsführer zum Einschreiten gegen bevorstehende Rechtsgutsbeeinträchtigungen verpflichtet ist, wenn er von dieser positive Kenntnis hat oder sich ihm hierfür konkrete Anhaltspunkte aufdrängen. Gleiches muß für das herrschende Unternehmen gelten, das durch Übernahme der einheitlichen Leitung in die verantwortliche Stellung des Geschäftsführers eingerückt ist. Insbesondere hängt diese Handlungspflicht nicht von dem Zentralisierungsgrad des Konzerns ab. Denn aufgrund ihrer subjektiven Voraussetzungen werden die Pflichten des herrschenden Unternehmens auch im dezentralen Konzern nicht überspannt.

---

[55] *D. Eckardt*, Organhaftung, 1996, S. 76. In diese Richtung auch *M. Lutter/P. Hommelhoff*, GmbHG, 1995, § 43 Rdnr. 41.

[56] *BGH*, Urt. v. 3.6.1975 - VI ZR 192/73, in: BB 1975, S. 1031

[57] *W. Oehler*, Produzentenhaftung, 1990, S. 1449. Nicht so weitgehend *E. Rehbinder*, Konzernaußenrecht, 1969, S. 510.

Anders verhält es sich dagegen bei denjenigen Verkehrspflichten, die an die Schadensneigung der übernommenen Aufgabe, ihre Nähe zum tatbestandsmäßigen Verletzungserfolg und das Maß der drohenden Rechtsgutsbeeinträchtigung anknüpfen. Hier hängen die Verkehrspflichten des herrschenden Unternehmens von dem konkreten Aufgabenbereich ab, den es im Rahmen der Konzernleitung an sich gezogen hat.

4. *Vorangegangenes gefahrerhöhendes Tun*

a) Die gesetzliche Ausgangslage

Letztes Zurechnungskriterium für eine Verkehrspflicht ist schließlich die Schaffung einer erhöhten Gefahr durch vorangegangenes Tun. Ein erhöhtes Ausmaß der geschaffenen Gefahr ist deshalb erforderlich, damit nicht die Grenzen zum allgemeinen Lebensrisiko, das jeder selbst trägt, verwischt werden.[58] Eine erhöhte Gefahr liegt z. B. vor, wenn ein Produzent die Entsorgung umweltschädlichen Industrieabfalls einem Dritten überläßt. Ihn trifft dann die Pflicht zu überprüfen, ob der Dritte seine Aufgabe ordnungsgemäß erfüllt.[59] Und ebenso ist eine Gesellschaft zum Rückruf eines von ihr vertriebenen Ledersprays verpflichtet, wenn dieses gesundheitsschädlich ist. Denn durch den Vertrieb des gesundheitsschädlichen Ledersprays hat die Gesellschaft eine erhöhte Gefahrenlage geschaffen. Nicht zu folgen ist dagegen dem *Bundesgerichtshof*, nach dem in einem solchen Fall auch die Geschäftsführer der Gesellschaft aufgrund vorangegangenem, pflichtwidrigen Gefährdungsverhaltens die deliktische Pflicht zum Rückruf des Ledersprays treffe.[60] Denn diese Ansicht verstößt gegen das bereits erwähnte Trennungsprinzip, wonach grundsätzlich nur die Gesellschaft, nicht aber die Gesellschafter oder ihre Geschäftsführer Pflichten gegenüber Dritten treffen. Trotzdem kam der *Bundesgerichtshof* in dem von ihm entschiedenen Fall zum richtigen Ergebnis. Denn wie aus dem Urteil hervorgeht, hatten die Geschäftsführer konkrete Hinweise auf eine Gesundheitsschädlichkeit des Ledersprays. Dann traf sie aber auch nach der hier vertretenen Auffassung die deliktische Pflicht zum Rückruf des Ledersprays. Diese Pflicht ergab sich allerdings nicht aufgrund gefahrerhöhenden Tuns, sondern aufgrund der verantwortlichen Stellung der Geschäftsführer in der Gesellschaft.

---

[58] *C.-W. Canaris*, Schuldrecht, 1994, S. 410.

[59] *BGH*, Urt. v. 7.10.1975 - VI ZR 43/74, in: NJW 1976, S. 46, 47.

[60] *BGH*, Urt. v. 6.7.1990 - 2 StR 549/89, in: NJW 1990, S. 2560, 2563.

b) Die Lage im Konzern

Im Konzern kann die Schaffung einer besonderen Gefahrenlage noch nicht darin gesehen werden, daß das herrschende Unternehmen überhaupt eine Tochtergesellschaft gegründet bzw. in die einheitliche Leitung einbezogen hat.[61] Denn das ist ihm ja gerade gestattet. Ebenso wenig reicht es im Lichte des Trennungsprinzips aus, daß es die Tochtergesellschaft zur Herstellung eines Produkts veranlaßt hat, das sich nachträglich als gesundheitsschädlich herausstellt. Denn § 13 Abs. 2 GmbHG will den Gesellschafter auch vor solchen Risiken abschirmen. Erforderlich ist vielmehr, daß das herrschende Unternehmen in der beherrschten Gesellschaft eine Lage schafft, die es dieser von vornherein unmöglich macht, ihre Verkehrspflichten zu erfüllen. Erst wenn diese Voraussetzung erfüllt ist, liegt eine besondere Gefahrenlage vor, die persönliche Verkehrspflichten des herrschenden Unternehmens rechtfertigt.

Das kann der Fall sein, wenn das herrschende Unternehmen die beherrschte Gesellschaft mit einer Aufgabe betraut, für die sie nicht das nötige Sachverständnis hat. Das kann auch dann der Fall sein, wenn der beherrschten Gesellschaft nicht genügend finanzielle Mittel zur Verfügung stehen, um ihre Verkehrspflichten zu erfüllen. Und das kann schließlich der Fall sein, wenn es den Geschäftsführer aus seiner Position verdrängt und selbst nicht willens und in der Lage ist, die Gesellschaft ordnungsgemäß zu führen. Übt das herrschende Unternehmen dagegen die Unternehmensleitung aus, so nimmt es nur die Stelle des Geschäftsführers ein und schafft damit als Gesellschafter kein über das normale Maß hinausgehende Risiko.

5. Sorgfaltsanforderungen

a) Bedeutung

Nachdem aufgezeigt wurde, daß die herrschende GmbH nur ausnahmsweise konzerndimensionale Verkehrspflichten treffen, scheint die Frage nach deliktischen Sorgfaltspflichten nur von untergeordneter Bedeutung zu sein. Denn die deliktischen Sorgfaltspflichten geben nur den Sorgfaltsmaßstab an, der bei der Wahrnehmung der Verkehrspflichten einzuhalten ist. Trotzdem spielen die deliktischen Sorgfaltspflichten auch nach der hier vertretenen Ansicht eine wichtige Rolle. Denn das herrschende Unternehmen trifft gegenüber den beherrschten Gesellschaften die Pflicht, die Konzernleitung mit der Sorgfalt eines ordentlichen und gewissenhaften Geschäftsleiters auszuüben. Zu dieser Pflicht

---

[61] Siehe auch *M. Grünwald*, Außenhaftung, 1999, S. 106: „Die Gründung einer GmbH verursacht keine Gefahr".

gehört es aber, dafür zu sorgen, daß die beherrschten Gesellschaften ihre Verkehrspflichten erfüllen. Deshalb ist auch für die herrschende GmbH von Bedeutung, welche Sorgfaltsanforderungen an die Erfüllung von Verkehrspflichten zu stellen sind.

b) Abgrenzung

Allerdings sind die Sorgfaltsanforderungen bei der Erfüllung von Verkehrspflichten nicht identisch mit den Sorgfaltsanforderungen bei der Erfüllung der Pflicht zur ordnungsgemäßen Konzernleitung. Denn bei der Bestimmung von ersteren kommt ein Gedanke zum Tragen, der auf letztere nicht zutrifft.

Die Sorgfaltsanforderungen bei der Erfüllung der Verkehrspflichten werden in der Rechtsprechung so hoch angesetzt, daß es dem Verkehrspflichtigen nur selten gelingt, den ihm in vielen Fällen obliegenden Beweis mangelnden Verschuldens[62] zu führen.[63] Besonders deutlich wird dies im Rahmen der Produzentenhaftung. Wird jemand durch ein fehlerhaftes Produkt geschädigt und nimmt er den Produzenten nach § 823 Abs. 1 BGB oder § 831 Abs. 1 BGB auf Schadensersatz in Anspruch, so muß der Produzent zum einen nachweisen, daß ihn kein Organisationsverschulden trifft, die gesamte Produktion also lückenlos organisiert und überwacht war, und Fehlerquellen ausgeschlossen sind.[64] Zum anderen muß er aber im Rahmen von § 831 Abs. 1 Satz 2 BGB auch nachweisen, daß der organisierte Produktionsablauf keiner Störung durch individuelle Fehlleistungen von Bediensteten ausgesetzt war, oder aber alle in Frage kommenden Personen benennen und sich für jeden einzelnen von ihnen hinsichtlich Auswahl und Überwachung entlasten.[65] Damit wird dem Produzenten eine tendenziell umfassende, verschuldensunabhängige Einstandpflicht für die durch Anlagen, Betriebsmittel und Gehilfen verursachten Schäden auferlegt.[66] All dem liegt unausgesprochen der Gedanke zugrunde, daß derjenige, der die Vorteile aus einer Tätigkeit zieht, auch deren Risiken zu tragen hat.[67] Das ist aber, wie bereits festgestellt wurde, nur die beherrschte Gesellschaft. Und in der konzernfreien Lage ist es insbesondere nicht der Geschäftsführer der Gesellschaft.

---

[62] Grundlegend zur Beweislastumkehr bei der Produzentenhaftung *BGH*, Urt. v. 26.11.1968 - VI ZR 212/66 („Hühnerpest"), in: BGHZ 51, S. 91, 102 ff.

[63] *H.-J. Mertens*, Münchener Kommentar, 1986, § 823 Rdnr. 309 (für die Produzentenhaftung). Kritisch zu dieser Rechtsprechung: *C.-W. Canaris*, Schuldrecht, 1994, S. 427 f., 483.

[64] *BGH*, Urt. v. 19.6.1973 - VI ZR 178/71, in: NJW 1973, S. 1602, 1603.

[65] *BGH*, Urt. v. 19.6.1973 - VI ZR 178/71, in: NJW 1973, S. 1602, 1603.

[66] Ebenso zu anderen Beispielen aus der Rechtsprechung *C. v. Bar*, Verkehrspflichen, 1980, S. 48 f.

[67] *D. Eckardt*, Organhaftung, 1996, S. 62,

Deshalb sind an die Sorgfaltsanforderungen des Geschäftsführers bzw. des herrschenden Unternehmens im Rahmen der Konzernleitung nicht so hohe Anforderungen zu stellen wie an dic beherrschte Gesellschaft bei Erfüllung ihrer Verkehrspflichten. Mithin genügt es, wenn die im folgenden genannten Sorgfaltsanforderungen erfüllt werden.

c) Die gesetzliche Ausgangslage

Da die herrschende GmbH gegenüber den beherrschten Gesellschaften weisungsberechtigt ist oder letztere jedenfalls den Weisungen der herrschenden GmbH folgen, liegt in deliktsrechtlicher Hinsicht dieselbc Situation vor wie bei einer innerbetrieblichen Arbeitsteilung. Für diese gilt aber, daß das Haftungsrecht den tatsächlichen Arbeitsteilungen folgt. Das bedeutet, daß jeder nur für die von ihm übernommene Aufgabe verantwortlich ist. Mit anderen Worten hat jeder die von ihm übernommene Aufgabe ordnungsgemäß zu erfüllen.[68] Hierin kommt ein Grundgedanke der außervertraglichen Haftung zum Ausdruck.[69]

Hieraus folgt für die Geschäftsleitung die Allzuständigkeit und Generalverantwortung für das Geschehen im Unternehmen. Denn Aufgabe der Geschäftsleitung ist es gerade, das Unternehmen zu führen. Diese Generalverantwortung für das Geschehen im Unternehmen äußert sich zuerst und vor allem in einer Organisationspflicht. Sie läßt sich wie folgt formulieren:

> „Die Geschäftsleitung hat den Betrieb so zu organisieren, daß die mit der betrieblichen Tätigkeit verbundenen Pflichten aller Art und damit auch die Dritten gegenüber bestehenden Sorgfalts- und Gefahrabwendungspflichten erfüllt werden".[70]

Hieraus folgt zugleich, daß die Geschäftsleitung sich nicht um alles selbst kümmern muß. Vielmehr hat sie nur diejenigen Maßnahmen zu veranlassen, damit das nach Lage der Dinge zur Gefahrvermeidung Erforderliche geschieht. Danach ist für die Geschäftsleitung folgendes zumutbar und möglich:

- Zunächst kann bei einer mehrgliedrigen Geschäftsleitung eine Aufgabenverteilung auf die einzelnen Geschäftsführer erfolgen. Jeder Geschäftsführer ist sodann für seinen eigenen Aufgabenbereich und damit auch für die Erfüllung der in diesen Aufgabenbereich fallenden Pflichten verantwortlich.[71]

---

[68] *J. Schmidt-Salzer*, Produkthaftung, 1990, Rdnr. 4.168.

[69] *J. Schmidt-Salzer*, Produkthaftung, 1990, Rdnr. 4.160.

[70] *J. Schmidt-Salzer*, Produkthaftung, 1990, Rdnr. 4.180; *BGH*, Urt. v. 20.2.1995 - II ZR 9/94, in: ZIP 1995, S. 560, 561; *C. T. Ebenroth/K. W. Lange*, Sorgfaltspflichten, 1992, S. 71.

[71] *M. Grünwald*, Außenhaftung, 1999, S. 171; *D. Medicus*, Geschäftsverteilung, 1998, S. 14. Aus strafrechtlicher Sicht *OLG Frankfurt/M*, Urt. v. 9.12.1994 - 24 U 254/93, in: ZIP 1995, S. 213, 215.

Diese Aufgabenverteilung muß jedoch klar und eindeutig dokumentiert sein.[72] Denn es besteht die Gefahr, daß im Schadensfall der eine Geschäftsführer auf die Verantwortlichkeit des anderen verweist. Trotz dieser Aufgabenverteilung bleiben die Geschäftsführer jedoch verpflichtet, ihre Mitgeschäftsführer zu überwachen.[73] Es verbleibt dem Geschäftsführer die „Pflicht der Überwachung der Überwacher".[74] Allerdings dürfen an diese Überwachungspflicht keine überzogenen Anforderungen gestellt werden. Denn es ist zu bedenken, daß ein Geschäftsführer nicht auf allen Gebieten sachverständig sein kann. Deshalb reicht die Prüfung aus, ob der Mitgeschäftsführer weiterhin die Fähigkeiten besitzt, die Pflichten in seinem Zuständigkeitsbereich ordnungsgemäß zu erfüllen, und ob er seiner Aufgabe auch tatsächlich nachkommt.[75] Das alles betrifft die Ebene der Geschäftsleitung.

- Aber auch unterhalb der Geschäftsleitungsebene hat die Geschäftsleitung durch organisatorische Maßnahmen den erforderlichen Rahmen dafür zu schaffen, daß die in dem Betrieb mit Einzelaufgaben Betrauten ihre Tätigkeiten ordnungsgemäß verrichten. Dazu gehören insbesondere ausreichende Kontrollmechanismen, um die Effizienz der getroffenen Maßnahmen und ihre Realisierung und laufende Einhaltung zu überprüfen.[76]
- Schließlich hat die Geschäftsleitung einzugreifen, wenn sich innerhalb einer allgemein ausreichenden Organisation Situationen ergeben, in denen die getroffenen generellen, organisatorischen Maßnahmen nicht ausreichen und konkreter Anlaß zum Handeln besteht.[77]

---

[72] Dazu näher *Uwe H. Schneider*, Pflichten, 1992, S. 485.

[73] *H. Altmeppen*, GmbHG, 1997, § 43 Rdnr. 8; *M. Grünwald*, Außenhaftung, 1999, S. 172. Ebenso der *BFH*, Urt. v. 26.4.1984 - V R 128/79, in: GmbHR 1985, S. 30, 31 f., für die Wahrnehmung der steuerlichen Pflichten im Betrieb.

[74] *Uwe H. Schneider*, Pflichten, 1992, S. 487, zur entsprechenden Lage bei der Wahrnehmung öffentlich-rechtlicher Pflichten. In diesem Sinne auch *BGH*, Urt. v. 1.3.1993 - II ZR 81/94, in: GmbHR 1994, S. 460, 461; *BGH*, Urt. v. 15.10.1996 - VI ZR 319/95, in: BB 1996, S. 2531, 2532; *OLG Düsseldorf*, Urt. v. 12.11.1993 - 22 U 90/93, in: GmbHR 1994, S. 403.

[75] *Uwe H. Schneider*, Pflichten, 1992, S. 482.

[76] *J. Schmidt-Salzer*, Produkthaftung, 1990, Rdnr. 4.192. Zur entsprechenden Situation bei der Wahrnehmung öffentlich-rechtlicher Pflichten siehe *Uwe H. Schneider*, Unternehmensorganisation, 1993, S. 1914 f.; ders., Pflichten, 1992, S. 485 ff.

[77] *J. Schmidt-Salzer*, Produkthaftung, 1990, Rdnr. 4.192. Zur entsprechenden Lage bei der Wahrnehmung öffentlich-rechtlicher Pflichten siehe *Uwe H. Schneider*, Pflichten, 1992, S. 483; *BGH*, Urt. v. 6.7.1990 - 2 StR 549/89, in: NJW 1990, S. 2560, 2565 („Lederspray"); *BGH*, Urt. v. 21.1.1997 - VI ZR 338/95, in: GmbHR 1997, S. 305, 307.

d) Die Lage im Konzern

Das alles läßt sich auf den Konzern übertragen. Danach ist die herrschende GmbH verpflichtet, die konzerndimensionale Arbeitsteilung so zu organisieren, daß jede Konzerngesellschaft die ihr übertragenen Aufgaben ordnungsgemäß wahrnehmen kann. Diese Arbeitsteilung muß klar und eindeutig geregelt sein.

Die Organisationspflicht des herrschenden Unternehmens fordert zunächst, daß es die Konzerngesellschaft, der es eine Teilaufgabe bei der Herstellung des Produkts übertragen will, sorgfältig auswählt. Es darf nur eine solche Tochtergesellschaft mit der Aufgabe betrauen, die die Gewähr für eine ordnungsgemäße Erfüllung der Aufgabe bietet. Dabei sind die Pflichten zur Auswahl der Tochtergesellschaft um so strenger, je enger die Grenzen sind, die der Einflußnahme der herrschenden GmbH durch die Selbständigkeit der beherrschten Gesellschaft gezogen werden.[78]

Hat das herrschende Unternehmen die Tochtergesellschaft sorgfältig ausgewählt, so muß es deren Arbeitsweise nicht mehr auf „Schritt und Tritt" kontrollieren oder Sicherheitskontrollen vornehmen oder vornehmen lassen, die schon von der Tochtergesellschaft mit ihrer besonderen fachlichen Erfahrung und Einrichtung vorzunehmen sind. Entsteht allerdings durch die von der Tochtergesellschaft erbrachten Leistungen eine erhöhte Gefahrenlage, die auch von dem herrschenden Unternehmen ohne spezielle Fachkenntnisse erkannt und durch geeignete Anweisungen abgestellt werden kann, so ist es gegenüber der Tochtergesellschaft zum Einschreiten verpflichtet.

Beschränkt sich das herrschende Unternehmen dagegen nicht auf allgemeine Vorgaben, sondern nimmt es auch Einfluß auf deren Umsetzung, behandelt es insbesondere die beherrschten Gesellschaft wie eine Betriebsabteilung, so trifft es dieselben Pflichten wie das Verwaltungsorgan der beherrschten Gesellschaft. Insbesondere ist es nun gegenüber der beherrschten Gesellschaft verpflichtet, deren Betrieb so zu organisieren, daß deren Verkehrspflichten ordnungsgemäß erfüllt werden.

6. Ergebnis

Nach alledem treffen das herrschende Unternehmen konzerndimensionale Verkehrspflichten erstens dann, wenn es von einer bevorstehenden Rechtsgutsbeeinträchtigung im Konzern positive Kenntnis hat oder sich ihm konkrete

---

[78] *BGH*, Urt. v. 7.10.1975 - VI ZR 43/74, in: NJW 1976, S. 46, 47, für die Beauftragung eines selbständigen Unternehmers mit der Beseitigung gefährlicher Industrieabfälle.

Anhaltspunkte hierfür aufdrängen. Dann ist es zum Einschreiten verpflichtet. Zweitens treffen es Verkehrspflichten, wenn es eine gefahrenwesentliche Aufgabe in einer beherrschten Gesellschaft übernommen hat. In diesem Fall ist es verpflichtet, die Aufgaben ordnungsgemäß zu erfüllen, damit Dritte nicht in ihren Rechtsgütern geschädigt werden. Und drittens schließlich treffen das herrschende Unternehmen Verkehrspflichten, wenn es in der Tochtergesellschaft eine erhöhte Gefahrenlage geschaffen hat. Das ist z. B. der Fall, wenn es die Tochtergesellschaft mit einer gefahrenwesentlichen Aufgabe betraut hat, die diese von vornherein nicht ordnungsgemäß erfüllen kann.

C. Aufgreifmerkmal: Geschäftsführerstellung

Zahlreiche Pflichten knüpfen schließlich an die Geschäftsführerstellung an. Exemplarisch untersucht werden sollen die Pflicht des Geschäftsführers zur Stellung des Antrags auf Eröffnung des Insolvenzverfahren gemäß § 64 Abs. 1 GmbHG und seine steuerlichen Pflichten gemäß § 34 Abs. 1 AO.

I. Die gesetzliche Ausgangslage

Gemäß § 64 Abs. 1 GmbHG sind die Geschäftsführer einer GmbH verpflichtet, im Falle der Zahlungsunfähigkeit oder Überschuldung der Gesellschaft ohne schuldhaftes Zögern die Eröffnung des Insolvenzverfahrens zu beantragen. Gleiches gilt gemäß § 92 Abs. 2 AktG für den Vorstand einer Aktiengesellschaft. Zweck dieser Vorschriften ist es zu verhindern, den Rechtsverkehr durch Fortführung einer zahlungsunfähigen oder überschuldeten Gesellschaft zu gefährden.[79] Hiermit dienen sie insbesondere dem Schutz der Gläubiger und sind deshalb zu deren Gunsten als Schutzgesetze im Sinne von § 823 Abs. 2 BGB zu qualifizieren.[80]

Gemäß § 34 Abs. 1 AO haben die gesetzlichen Vertreter juristischer Personen, bei der GmbH also der Geschäftsführer, deren steuerliche Pflichten zu erfüllen.[81] Das sind z. B. die Pflicht zur Abgabe von Steuererklärungen, Buchführungspflichten und Auskunftspflichten. Verletzen sie diese Pflichten vorsätzlich oder grob fahrlässig, so haften sie, soweit infolge der Pflichtverletzung Ansprüche aus dem Steuerverhältnis mit der juristischen Person nicht oder nicht

---

[79] Für die GmbH: *K. Schmidt*, Scholz, 1995, § 64 Rdnr. 10.

[80] Für die GmbH: *J. Schulze-Osterloh*, Baumbach/Hueck, 1996, § 64 Rdnr. 24 m. w. N. Für die AG: *U. Hüffer*, AktG, 1997, § 92 Rdnr. 16.

[81] Dazu jüngst *T. Neusel*, Haftung, 1997, S. 1129 ff.

rechtzeitig festgesetzt oder erfüllt oder soweit infolgedessen Steuervergütungen oder Steuererstattungen ohne rechtlichen Grund gezahlt werden, § 69 AO.

II. Die Lage im Konzern

Im Konzern stellt sich die Frage, ob auch das herrschende Unternehmen Adressatin der genannten Pflichten sein kann.

Hierzu scheint § 35 AO für den Bereich des Steuerrechts eine abschließende Regelung getroffen zu haben. Danach hat derjenige, der als Verfügungsberechtiger im eigenen oder fremden Namen auftritt, die steuerlichen Pflichten eines gesetzlichen Vertreters, soweit er sie rechtlich und tatsächlich erfüllen kann. Verfügungsberechtigt ist jeder, der rechtlich und wirtschaftlich über Mittel, die einem anderen gehören, verfügen kann und als Verfügungsberechtigter nach außen hin auftritt. Dabei kann die rechtliche Verfügungsbefugnis auf einer ausdrücklichen Ermächtigung des Steuerschuldners beruhen. Sie kann sich aber auch aus den Grundsätzen der Anscheins- und Duldungsvollmacht ergeben.[82] Und sie kann schließlich bei einem herrschenden GmbH-Gesellschafter darauf beruhen, daß er maßgeblichen Einfluß auf die Geschäftsführung nehmen kann.[83]

Auf den Konzern gewendet bedeutet dies, daß das herrschende Unternehmen jedenfalls im GmbH-Konzern immer die rechtliche Möglichkeit hat, die steuerlichen Pflichten der beherrschten GmbH zu erfüllen. Normadressat von § 35 AO ist es jedoch nur dann, wenn es auch nach außen hin als Verfügungsberechtigter auftritt. Da nach außen hin immer nur natürliche Personen auftreten und es nicht darauf ankommt, ob diese im eigenen oder fremden Namen handeln, muß es sich bei dem herrschenden Unternehmen zudem um eine natürliche Person handeln. Eine GmbH ist also niemals Normadressatin von § 35 AO. Das zeigt auch ein Urteil des *Bundesfinanzhofs*[84]. Dort war der Geschäftsführer einer herrschenden GmbH für das Finanz- und Rechnungswesen der Tochtergesellschaften zuständig, leistete für diese Zahlungen, nahm sie entgegen, führte Kreditverhandlungen, unterzeichnete Zahlungsanmeldungen für das Hauptzollamt usw. Obwohl der *Bundesfinanzhof* feststellte, daß der Betreffende in seiner Eigenschaft als Geschäftsführer der herrschenden GmbH aufgetreten war, sah er

---

[82] Zum Vorstehenden siehe *B. Brockmeyer*, Klein/Orlopp, 1995, § 35 Anm. 2.

[83] *Finanzgericht Rheinland-Pfalz*, Urt. v. 15.7.1993 - 4 K 2664/92, in: EFG 1994, S. 134, 135. Siehe aber auch *BFH*, Urt. v. 27.11.1990 - VII R 20/89, in: BStBl. 1991, S. 284, 286, wonach sich die Verfügungsbefugnis einer Alleingesellschafterin daraus ergab, „daß es ihr in dieser Eigenschaft rechtlich und tatsächlich jederzeit möglich war, sich selbst zur Geschäftsführerin zu bestellen oder dafür zu sorgen, daß ein anderer Geschäftsführer bestellt wurde".

[84] *BFH*, Urt. v. 21.2.1989 - VII R 165/85, in: GmbHR, 1989, S. 520, 521.

diesen und nicht die herrschende GmbH als Verfügungsberechtigten im Sinne von § 35 AO an. Da dieser grob fahrlässig nicht dafür gesorgt hatte, daß die beherrschte Gesellschaft als Inhaberin eines offenen Zollagers gemäß § 46 Abs. 3 ZG und § 97 Abs. 1 AZO die aus dem offenen Zollager entnommenen Waren rechtzeitig anmeldete und den Zoll zahlte, haftete er gemäß § 69 AO für die Ansprüche aus dem Steuerschuldverhältnis.

An § 35 AO mag der *Bundesgerichtshof* gedacht haben, als er entscheiden mußte, ob auch eine nicht zum Geschäftsführer bestellte Person Normadressatin von § 64 Abs. 1 GmbHG sein kann. Hierzu führte er aus, daß die Pflicht zur Stellung des Konkursantrags[85] auch demjenigen obliege, der, ohne zum Geschäftsführer bestellt zu sein, die Geschäfte der GmbH tatsächlich wie ein Geschäftsführer oder Mitgeschäftsführer führe.[86] Hierfür sei erforderlich, daß

> „der Betreffende die Geschicke der Gesellschaft - und zwar nicht nur durch interne Einwirkung auf die satzungsmäßigen Geschäftsführer, sondern durch eigenes, auch nach außen hervortretendes, üblicherweise der Geschäftsführung zuzurechnendes Handeln - so maßgeblich in die Hand genommen hat, daß ihm auch die Verantwortung für die rechtzeitige Stellung des Konkursantrages zufällt".[87]

Diese Entscheidung zum sog. faktischen Geschäftsführer ist in der Literatur auf Zustimmung gestoßen.[88] Auf den Konzern gewendet hätte sie zur Folge, daß das herrschende Unternehmen nur dann die Verantwortung für die rechtzeitige Stellung des Antrags auf Eröffnung des Insolvenzverfahrens trifft, wenn es eine natürliche Person ist und einen so maßgebenden Einfluß auf den Geschäftsführer der beherrschten Gesellschaft ausübt, daß dieser nur noch als „sein Werkzeug" erscheint.[89] Andere Stimmen in der Literatur ziehen den Kreis der Normadressaten von § 64 Abs. 1 GmbHG dagegen weiter. Vertreten wird, daß auch ein beherrschender Gesellschafter wegen Konkursverschleppung[90] hafte, wenn und soweit er auf die gebotene Geschäftsführungsmaßnahme des Konkursantrags[91] negativ Einfluß genommen und dabei schuldhaft die Pflicht verletzt habe, sorgfältig zu prüfen, ob die Gesellschaft „stillgelegt" werden müsse.[92] Darüber hinaus wird die Ansicht vertreten, daß das herrschende Unternehmen unabhängig von einer konkreten Einflußnahme allein wegen Ausübung der Konzernleitung dazu verpflichtet sei, für die rechtzeitige Stellung des Kon-

---

85 Jetzt Antrag auf Eröffnung des Insolvenzverfahrens.

86 *BGH*, Urt. v. 21.3.1988 - II ZR 194/87, in: BGHZ 104, S. 44, 46.

87 *BGH*, Urt. v. 21.3.1988 - II ZR 194/87, in: BGHZ 104, S. 44, 48.

88 *M. Lutter/P. Hommelhoff*, GmbHG, 1995, § 64 Rdnr. 26; *J. Schulze-Osterloh*, Baumbach/Hueck, 1996, § 64 Rdnr. 6; *P. Ulmer*, Hachenburg, 1991, § 64 Rdnr. 12.

89 *M. Lutter/P. Hommelhoff*, GmbHG, 1995, § 64 Rdnr. 26.

90 Jetzt Verschleppung der Eröffnung des Insolvenzverfahrens.

91 Jetzt Antrag auf Eröffnung des Insolvenzverfahrens.

92 *H. Altmeppen*, GmbHG, 1997, § 64 Rdnr. 36 f.

kursantrags[93] zu sorgen.[94] Im Gegensatz zur herrschenden Meinung reicht danach auch eine dezentrale Konzernleitung für die Begründung der Pflicht aus, die Eröffnung des Konkursverfahrens bzw. - nach Inkrafttreten der Insolvenzordnung - die Eröffnung des Insolvenzverfahrens zu beantragen.

Richtigerweise ist zwischen der Pflicht zur Stellung des Antrags auf Eröffnung des Insolvenzverfahrens einerseits und der Pflicht, für die rechtzeitige Stellung dieses Antrags zu sorgen, andererseits zu unterscheiden.[95] Entsprechendes gilt für die steuerlichen Pflichten.

Die Pflicht zur Abgabe der Steuererklärungen für die GmbH, zur Buchführung, zur Auskunft gegenüber dem Finanzamt u.s.w. sowie die Pflicht zur Stellung des Antrags auf Eröffnung des Insolvenzverfahres können nur solche nicht zum Geschäftsführer bestellte Personen treffen, die zur Erfüllung dieser Pflichten auch rechtlich in der Lage sind. Das wird in § 35 AO ausdrücklich vorausgesetzt, gilt aber auch für das Insolvenzrecht. Hieraus folgt zum einen, daß nur natürliche Personen die genannten Pflichten treffen. Denn § 34 AO sowie § 64 GmbHG und § 92 Abs. 2 AktG knüpfen an die Stellung des Geschäftsführers bzw. Vorstands als gesetzlichem Vertreter der Gesellschaft an. Gesetzlicher Vertreter einer GmbH oder Aktiengesellschaft kann aber nur eine natürliche Person sein, §§ 6 Abs. 2 GmbHG, 76 Abs. 3 Satz 1 AktG. Entsprechendes muß aus Gründen der Rechtssicherheit für die Personen gelten, die an seine Stelle treten. Zum anderen muß diese Person zumindest mit Duldung des für die Geschäftsführerbestellung bzw. Vorstandsbestellung zuständige Gesellschaftsorgans nach außen als Geschäftsführer auftreten. Denn nur dann wird man aufgrund einer Duldungsvollmacht die für die Erfüllung der Pflichten erforderliche Verfügungsbefugnis bzw. Vertretungsbefugnis annehmen können.

Die Pflicht, für die rechtzeitige Erfüllung der steuerlichen Pflichten bzw. der Pflicht zur Stellung des Antrags auf Eröffnung des Insolvenzverfahrens durch die dazu berechtigten Personen zu sorgen, obliegt dagegen (auch) dem herrschenden Unternehmen. Denn der Geschäftsführer ist nicht nur deshalb Normadressat von § 34 AO bzw. § 64 GmbHG, weil er gesetzlicher Vertreter der Gesellschaft ist. Vielmehr obliegen ihm die steuerlichen Pflichten und die Pflicht zur Beantragung der Eröffnung des Insolvenzverfahrens auch aufgrund

---

93 Jetzt Antrag auf Eröffnung des Insolvenzverfahrens.

94 *Uwe H. Schneider*, Konzernleitung, 1981, S. 255; *G. H. Roth*, Haftung, 1989, S. 430. Dagegen ausdrücklich: *K. Schmidt*, Scholz, 1995, § 64 Rdnr. 9; *J. Meyer-Landrut*, GmbHG, 1987, § 64 Rdnr. 5.

95 *K. Schmidt*, Scholz, 1995, § 63 Rdnr. 37 u. § 64 Rdnr. 7; *U. Stein*, Konkursverschleppung, 1984, S. 229 ff.

seiner Funktion als Leiter des Unternehmens der Gesellschaft. Denn aufgrund dieser Funktion ist er am besten in der Lage, die steuerlichen Pflichten der Gesellschaft und die Pflicht zur Stellung des Antrags auf Eröffnung des Insolvenzverfahrens zu erfüllen. Im Konzern wird die Unternehmensleitung aber nicht mehr von dem Verwaltungsorgan der beherrschten Gesellschaft, sondern von dem herrschenden Unternehmen ausgeübt. Soll dies keinen Einschnitt in die Wirksamkeit der von § 34 AO bzw. § 64 GmbHG normierten Pflichten zur Folge haben, so ist es geboten, dem herrschenden Unternehmen die Pflicht aufzuerlegen, für die rechtzeitige Abgabe der Steuererklärung bzw. für die rechtzeitige Stellung des Antrags auf Eröffnung des Insolvenzverfahrens Sorge zu tragen. Methodisch läßt sich dies durch eine analoge Anwendung von § 69 AO bzw. § 64 Abs. 1 GmbHG i. V. m. § 823 Abs. 2 BGB bewerkstelligen. Danach haftet das herrschende Unternehmen nach § 69 AO, wenn es seine Pflicht, für die Erfüllung der steuerlichen Pflichten der beherrschten Gesellschaft Sorge zu tragen, vorsätzlich oder grob fahrlässig verletzt. Und es haftet den Gläubigern der beherrschten Gesellschaft auf Schadensersatz, wenn es schuldhaft nicht für die rechtzeitige Stellung des Antrags auf Eröffnung des Insolvenzverfahrens gesorgt hat.

Das alles gilt sowohl im zentralen wie im dezentralen Konzern. Denn auch im dezentralen Konzern hat das herrschende Unternehmen, um mit den Worten des *Bundesgerichtshofs* zu sprechen, die „Geschicke" der beherrschten Gesellschaft in die Hand genommen, indem es sie mit den „Geschicken" der anderen Konzerngesellschaften wirtschaftlich verflochten hat. Hat etwa das herrschende Unternehmen die beherrschte Gesellschaft lediglich angewiesen, an dem konzernweiten Cash-Management teilzunehmen, so liegen die Geschicke der beherrschten Gesellschaft schon allein deshalb (auch) in seiner Hand, weil nur das herrschende Unternehmen einen Überblick über die wirtschaftliche Lage der Cash-Management-Betreibergesellschaft hat. Zeichnet sich bei der Cash-Management Betreibergesellschaft eine Krise ab, so kann die bloße Weisung zum Ausscheiden aus dem konzernweiten Cash-Management genügen, um einen Übergriff der Krise auf die beherrschte Gesellschaft zu verhindern. Ist dagegen ein „Flächenbrand" unvermeidbar, so wird das herrschende Unternehmen jedenfalls früher Kenntnis bzw. fahrlässige Unkenntnis von der (drohenden) Zahlungsunfähigkeit der beherrschten Gesellschaft haben als deren Geschäftsführer. Damit findet auch eine Vorverlagerung des Zeitpunkts statt, ab dem es für masseschmälernde Zahlungen durch die beherrschte Gesellschaft haftet. Dadurch steigen aber die Befriedigungsmöglichkeiten der Gläubiger.

Schließlich werden die Pflichten des herrschenden Unternehmens durch die hier vertretene Ansicht auch nicht überspannt. Zwar ist es verpflichtet, die Stellung des Antrags auf Eröffnung des Insolvenzverfahrens zu veranlassen, wenn die beherrschte Gesellschaft zahlungsunfähig oder überschuldet ist; auf Schadensersatz haftet es jedoch nur dann, wenn es diese Pflicht schuldhaft verletzt.[96] Voraussetzung ist damit u. a. die zumindest fahrlässige Unkenntnis von der Zahlungsunfähigkeit oder Überschuldung der Gesellschaft.[97] Damit hängt der Schuldvorwurf von den Überwachungspflichten ab, die das herrschende Unternehmen treffen. Und hier ist Raum für eine Differenzierung nach dem Zentralisierungsgrad des Konzerns.

Im zentralen Konzern trifft es dieselben Pflichten wie den Geschäftsführer. Dieser ist aber verpflichtet, die wirtschaftliche Lage der Gesellschaft laufend zu beobachten und sich bei den Anzeichen einer Krise durch die Aufstellung eines Vermögensstatus einen Überblick über den Vermögensstand zu verschaffen.[98]

Im dezentralen Konzern kann sich das herrschende Unternehmen dagegen darauf beschränken, regelmäßige Berichte der Geschäftsleitung der beherrschten Gesellschaften anzufordern. Erst wenn diese Berichte auf eine Krise hindeuten oder sich aus anderen Umständen für das herrschende Unternehmen solche Anzeichen ergeben, handelt das herrschende Unternehmen schuldhaft, wenn es nicht für die rechtzeitige Stellung des Antrags auf Eröffnung des Insolvenzverfahrens sorgt.

---

96 Für den GmbH-Geschäftsführer: *K. Schmidt*, Scholz, 1995, § 64 Rdnr. 14.

97 *BGH*, Urt. v. 6.6.1994 - II ZR 292/91, in: NJW 1994, S. 2220, 2224; *H. Rowedder*, GmbHG, 1997, § 64 Rdnr. 23. A. M. *J. Schulze-Osterloh*, Baumbach/Hueck, 1996, § 64 Rdnr. 27 (positive Kenntnis).

98 *BGH*, Urt. v. 6.6.1994 - II ZR 292/91, in: NJW 1994, S. 2220, 2224.

6. Teil: Die konzernbezogenen Pflichten des Geschäftsführers gegenüber der Allgemeinheit

Oben wurde bereits festgestellt, daß der Geschäftsführer nur dann aus unerlaubter Handlung haftet, wenn er das Delikt entweder eigenhändig begangen hat oder von seiner Begehung durch andere wußte bzw. konkrete Anhaltspunkte hierfür hatte und dies pflichtwidrig nicht verhinderte.[1] Dabei ergibt sich seine Pflicht zum Einschreiten aus seiner verantwortlichen Stellung als Geschäftsführer der Gesellschaft. Und es wurde festgestellt, daß den Geschäftsführer Verkehrspflichten treffen, wenn er gefahrenwesentliche Aufgaben übernommen hat. Führt er diese Aufgaben nicht ordnungsgemäß aus und entsteht einem Dritten dadurch ein Schaden, ist er diesem aus § 823 Abs. 1 BGB zum Schadensersatz verpflichtet.

Diese Grundsätze lassen sich auf den Konzern übertragen. Weist etwa der Geschäftsführer der herrschenden GmbH den Geschäftsführer der beherrschten GmbH an, ein gesundheitsschädliches Produkt auf den Markt zu bringen, und wußte er von der Gesundheitsschädlichkeit oder hätte er davon wissen müssen, so haftet er einem geschädigten Dritten auf Schadensersatz. War die Gesundheitsschädlichkeit dagegen bei Anwendung der im Verkehr erforderlichen Sorgfalt zunächst nicht erkennbar und stellt sie sich erst später heraus, so ist der Geschäftsführer nur dann zum Rückruf des Produkts verpflichtet, wenn er von der Gesundheitsschädlichkeit Kenntnis erlangt hat oder sich ihm hierfür konkrete Anhaltspunkte boten. Unterläßt er in diesem Fall den Rückruf, so ist er einem geschädigten Dritten gemäß § 823 Abs. 1 BGB zum Schadensersatz verpflichtet. Seine Pflicht zum Tätigwerden ergibt sich dabei aus seiner verantwortlichen Stellung, die er durch Ausübung der Konzernleitung in der beherrschten Gesellschaft hat. Insbesondere reicht schon die Ausübung einer nur „lockeren" Konzernleitung aus. Denn da Voraussetzung für eine Haftung die positive Kenntnis des Geschäftsführers von der Gesundheitsschädlichkeit des Produkts ist, wird er nicht überfordert, wenn von ihm bereits im dezentralen Konzern ein Einschreiten erwartet wird.

Anders verhält es sich dagegen in den Fällen, in denen das Gesetz dem Geschäftsführer besondere Pflichten zuweist. Exemplarisch genannt sei § 64 Abs. 1 GmbHG, wonach der Geschäftsführer im Falle der Zahlungsunfähigkeit oder

---

[1] Siehe 5. Teil B. IV. 3. d).

Überschuldung der Gesellschaft Antrag auf Eröffnung des Insolvenzverfahrens zu stellen hat. Hierzu wurde bereits oben[2] ausgeführt, daß denjenigen, der die Leitung über die Gesellschaft übernommen hat, die Pflicht trifft, für die rechtzeitige Stellung des Antrags auf Eröffnung des Insolvenzverfahrens zu sorgen. Das ist im Konzern auch der Geschäftsführer der herrschenden GmbH. Denn er übt die Konzernleitung aus. Dabei schadet es nicht, daß er dies im Namen der herrschenden GmbH tut. Insoweit gelten dieselben Grundsätze wie bei der Pflicht des Geschäftsführers gegenüber den beherrschten Gesellschaften, die Konzernleitung mit der Sorgfalt eines ordentlichen und gewissenhaften Geschäftsleiters auszuüben.

Freilich haftet der Geschäftsführer nur dann wegen Verschleppung des Insolvenzverfahrens auf Schadensersatz, wenn er es schuldhaft unterlassen hat, für die Stellung des Antrags auf Eröffnung des Insolvenzverfahrens in der beherrschten Gesellschaft zu sorgen. Auch hier ist der Inhalt seiner Sorgfaltspflichten von dem Zentralisierungsgrad des Konzerns abhängig.

---

2 Siehe 5. Teil C. II.

7. Teil: Zusammenfassung

1. Die Rechtsform der GmbH erfreut sich zunehmender Beliebtheit. So stieg der GmbH-Bestand 1996 auf 770.000 Gesellschaften an. Gleichzeitig läßt sich eine steigende Tendenz zur Konzernbildung erkennen.

2. Die Aufarbeitung der mit einer Konzernierung verbundenen Rechtsprobleme steht noch am Anfang. Ihr Schwerpunkt lag bisher bei dem Schutz der abhängigen Gesellschaft, ihrer Minderheitsgesellschafter und ihrer Gläubiger. Im neueren Schrifttum gewinnt jedoch zunehmend eine neue Betrachtungsweise an Raum. Danach ist der Schutz der Obergesellschaft, ihrer Gesellschafter und ihrer Gläubiger von mindestens ebenso großer Bedeutung.

3. Ausgangspunkt dieser neuen Betrachtungsweise ist aber nicht nur der Schutz der Obergesellschaft, ihrer Minderheitsgesellschafter und Gläubiger. Ihr Ansatz ist viel umfassender. Sie versteht den Konzern als Organisationsform zweckgerichteten unternehmerischen Handelns, der sich in seinen Grundaspekten von anderen Formen partnerschaftlicher Kooperation nicht unterscheidet. Auf dieser Grundlage versucht sie, ein Konzernverfassungsrecht zu entwickeln.

4. Mit den Pflichten des Geschäftsführers einer herrschenden GmbH im Konzern soll ein weiterer Beitrag zur Entwicklung eines Konzernverfassungsrechts geleistet werden.

5. Untersucht wurden im 1. Teil die Pflichten des Geschäftsführers gegenüber seinen eigenen Gesellschaft. Dabei wurde zwischen den Pflichten bei der Konzerngründung und den Pflichten bei der Konzernleitung unterschieden.

a) Bei der Konzerngründung werden dem Geschäftsführer Grenzen durch den Verbandszweck und den Unternehmensgegenstand der herrschenden GmbH sowie durch die Mitgliedschaftsrechte der Gesellschafter der herrschenden GmbH gezogen.

Der Verbandszweck umfaßt als unbewegliche Schranke der Mehrheitsherrschaft nur denjenigen Teil der Satzung, der die obersten Leitsätze für die Verbandstätigkeit zum Ausdruck bringt und mit dessen Abänderung schlechterdings kein Mitglied rechnen kann. Er kann in der Gewinnerzielung, der bloßen Kostendeckung, der Bedarfsdeckung oder einer geistig-ideelen Zielsetzung bestehen. Der Geschäftsführer darf keine Gesellschaft in die einheitliche Leitung einbeziehen, die dem Verbandszweck der herrschenden GmbH widerspricht.

Der Unternehmensgegenstand bezeichnet den Tätigkeitsrahmen, innerhalb dessen der Verbandszweck zu verfolgen ist. Der Geschäftsführer darf zum einen solche Gesellschaften in die einheitliche Leitung einbeziehen, deren Unternehmensgegenstand im Kernbereich mit dem der herrschenden GmbH übereinstimmt. Liegt diese Voraussetzung nicht vor, so ist die Einbeziehung einer Gesellschaft in die einheitliche Leitung zum anderen dann zulässig, wenn die Gesellschaft eine betriebliche Teilfunktion für die herrschende GmbH erfüllt und sich auf konzerninterne Leistungen beschränkt.

Durch die Konzerngründung werden schließlich die Vermögens- und Mitverwaltungsrechte der Gesellschafter der herrschenden GmbH mediatisiert. Deshalb ist eine Konzerngründung nur zulässig, wenn sie in der Satzung gestattet ist. Eine entsprechende Satzungsklausel kann etwa lauten:

> „Die Gesellschaft ist berechtigt, Beteiligungen zu erwerben und Tochtergesellschaften zu gründen“.

Enthält die Satzung der GmbH eine solche Konzernklausel, so muß der Geschäftsführer eine Maßnahme der Konzerngründung trotzdem den Gesellschaftern gemäß § 49 Abs. 2 GmbHG zur Entscheidung vorlegen, wenn die Maßnahme ungewöhnlich ist. Das ist zu bejahen, wenn es sich um eine Konzernstrukturmaßnahme handelt.

Ebenso stellt der Abschluß eines Beherrschungsvertrages eine ungewöhnliche Maßnahme dar. Entgegen der h. M. müssen ihm die Gesellschafter aber nur mit einfacher Mehrheit und nicht analog § 293 Abs. 2 AktG mit qualifizierter Mehrheit zustimmen.

b) Der Geschäftsführer ist gegenüber seiner Gesellschaft zur Konzernleitung verpflichtet. Dabei hat er mit den Gesellschaftern und Mitgeschäftsführern kollegial zusammenzuarbeiten und ihn treffen Loyalitätspflichten. Schließlich ist er zur Sorge für das rechtmäßige Verhalten der Gesellschaft nach außen verpflichtet.

(1) Inhalt der Konzernleitungspflicht ist es, die beherrschten Gesellschaften in den Konzern zu integrieren. Hierzu gehören die strategische Ausrichtung des Konzerns, die Festlegung der Konzernstruktur und -organisation, die finanzielle Führung des Konzerns und die Besetzung wichtiger Führungspositionen im Konzern. Zu einer zentralistischen Konzernorganisation ist der Geschäftsführer dagegen nicht verpflichtet.

(2) Im Rahmen der kollegialen Zusammenarbeit mit den Gesellschaftern hat er bestimmte Maßnahmen den Gesellschaftern zur Entscheidung vorzulegen. Hierzu gehören die Feststellung des Jahresabschlusses und die Gewinnverwendung in den Konzernunternehmen, die Besetzung wichtiger Führungspositionen im Konzern sowie alle ungewöhnlichen Konzerngeschäftsführungsmaßnahmen.

Weiterhin ist der Geschäftsführer verpflichtet, die Weisungen der Gesellschafterversammlung auszuführen. Eine Folgepflicht besteht allerdings nicht, wenn der Weisungsbeschluß nichtig ist, z. B. weil er gegen gläubigerschützende Vorschriften verstößt, § 241 Nr. 3 Var. 2 AktG analog. Und sie besteht nicht, wenn der Beschluß wirksam angefochten wurde.

Gemäß § 51a GmbHG ist der Geschäftsführer verpflichtet, einem Gesellschafter auf dessen Verlangen Auskunft über die Angelegenheiten der Gesellschaft zu erteilen. Zu den Angelegenheiten der herrschenden Gesellschaft gehören auch Angelegenheiten der beherrschten Gesellschaft, wenn die Kenntnis von ihnen zur Ausübung der Gesellschafterrechte zweckdienlich ist. Eine allgemeine Berichtspflicht des Geschäftsführers gegenüber der Gesellschafterversammlung ähnlich der des Vorstands gegenüber dem Aufsichtsrat besteht dagegen nicht.

(3) Der Geschäftsführer ist zu loyalem Verhalten verpflichtet.

Er darf zum einen keine Geschäftschancen der herrschenden GmbH an sich ziehen. Insbesondere sind der herrschenden GmbH im Verhältnis zu ihrem Geschäftsführer auch Geschäftschancen der beherrschten Gesellschaften zuzurechnen. Dabei ist allerdings nach dem Zentralisierungsgrad des Konzerns zu unterscheiden.

Zum anderen unterliegt der Geschäftsführer einem Wettbewerbsverbot. Dieses erstreckt sich auf den Geschäftsbereich der herrschenden GmbH und der beherrschten Gesellschaften.

(4) Der Geschäftsführer haftet der herrschenden GmbH gemäß § 43 Abs. 2 GmbHG auf Schadensersatz, wenn er seine Konzernleitungspflicht verletzt. Eine vertragliche Haftungsmilderung ist bis zur Grenze des Vorsatzes zulässig. Von dieser Haftungsmilderung wird allerdings nicht die Verletzung der in §§ 30, 33 GmbHG genannten Pflichten erfaßt.

(5) Der Geschäftsführer ist verpflichtet, für das rechtmäßige Verhalten der herrschenden GmbH gegenüber den beherrschten Gesellschaften zu sorgen. Das bedeutet, daß die Pflichten der herrschenden GmbH gegenüber den beherrschten Gesellschaften der Konzernleitung Grenzen setzen.

6. Die Pflichten der herrschenden GmbH gegenüber den beherrschten Gesellschaften waren Gegenstand des 2. Teils. Dabei wurde zwischen rechtsformspezifischen und rechtsformunabhängigen Pflichten unterschieden.

a) Bei den rechtsformspezifischen Pflichten ist zwischen dem faktischen Aktienkonzern, dem einstufigen faktischen GmbH-Konzern mit Minderheitsgesellschaftern, dem mehrstufigen faktischen GmbH-Konzern mit Minderheitsgesellschaftern, dem faktischen GmbH-Konzern ohne Minderheitsgesellschafter, dem Aktien-Vertragskonzern und dem GmbH-Vertragskonzern zu unterscheiden.

Die rechtsformspezifischen Pflichten im faktischen Aktienkonzern ergeben sich aus §§ 311 ff. AktG. Danach darf die herrschende GmbH nur dann die beherrschte Gesellschaft zu einer nachteiligen Maßnahme veranlassen, wenn der Nachteil bis zum Ende des Geschäftsjahres ausgeglichen wird oder ein entsprechender Rechtsanspruch gewährt wird.

Im einstufigen faktischen GmbH-Konzern mit Minderheitsgesellschaftern hat die herrschende GmbH bei der Konzernleitung ihre mitgliedschaftliche Förderpflicht gegenüber der beherrschten Gesellschaft zu beachten. Danach darf die herrschende GmbH bei ihrer Einflußnahme auf die Geschäftsführung der beherrschten Gesellschaft keine Sondervorteile zum Schaden der übrigen Gesellschafter verfolgen. Sie darf die Autonomie der beherrschten Gesellschaft nicht aufheben. Sie darf außerhalb der gesellschaftsinternen Zuständigkeitsordnung keinen Einfluß auf den Geschäftsführer der beherrschten Gesellschaft nehmen. Weiterhin unterliegt sie schon in der Abhängigkeitslage einem Wettbewerbsverbot gegenüber der beherrschten Gesellschaft. Und sie darf schließlich keine Geschäftschancen wahrnehmen, von denen sie aufgrund ihrer Gesellschafterstellung Kenntnis erlangt hat.

Der herrschenden GmbH obliegen gegenüber einer Enkel-GmbH mit Minderheitsgesellschaftern Treupflichten. Diese haben denselben Inhalt wie die mitgliedschaftliche Förderpflicht gegenüber der Tochter-GmbH.

Gegenüber der beherrschten Einmann-GmbH ist es der herrschenden GmbH lediglich verboten, Ausschüttungen aus dem zur Erhaltung des Stammkapitals erforderlichen Vermögen zu veranlassen. Zu einer solchen Auszahlung können allerdings auch existenzgefährdende Eingriffe zugunsten eines anderen Konzernunternehmens führen.

Im Aktien-Vertragskonzern dürfen der beherrschten Gesellschaft zwar nachteilige Weisungen erteilt werden, diese müssen aber im Konzerninteresse liegen. Darüber hinaus sind existenzgefährdende Weisungen verboten.

Gleiches gilt für den GmbH-Vertragskonzern.

b) Die rechtsformunabhängige Pflicht der herrschenden GmbH gegenüber den beherrschten Gesellschaften hat zum Inhalt, die Konzernleitung mit der Sorgfalt eines ordentlichen und gewissenhaften Geschäftsleiters auszuüben. Diese Pflicht entsteht, wenn die herrschende GmbH nicht nur vereinzelt in die Geschäftsführung der beherrschten Gesellschaft eingreift, sondern die Zielsetzungen in den einzelnen Funktionsbereichen des Unternehmens (Beschaffungs-, Produktions-, Absatzpolitik sowie Finanzierung und Personalpolitik) bestimmt. Denn dann ist das Verwaltungsorgan der beherrschten Gesellschaft regelmäßig nicht mehr in der Lage, das Unternehmen der Gesellschaft ordnungsgemäß zu leiten.

Die Pflicht, die Konzernleitung mit der Sorgfalt eines ordentlichen und gewissenhaften Geschäftsleiters auszuüben, besteht in modifizierter Form auch gegenüber der Einmann-GmbH. Dort hat die herrschende GmbH allerdings nur noch die effektive Einhaltung der Kapitalerhaltungsvorschriften zu gewährleisten.

7. Im 3. Teil wurde gezeigt, daß eine Konzernintegration in Anbetracht der durch das Recht der beherrschten Gesellschaften gezogenen Grenzen möglich ist.

8. Gegenstand des 4. Teils waren die Pflichten des Geschäftsführers der herrschenden GmbH gegenüber den beherrschten Gesellschaften. Auch gegenüber diesen ist der Geschäftsführer verpflichtet, die Konzernleitung mit der Sorgfalt eines ordentlichen und gewissenhaften Geschäftsleiters auszuüben. Im zentralen Konzern ist es ihm gegenüber der beherrschten Gesellschaft außerdem verboten, potentielle Geschäftschancen der Gesellschaft an sich zu ziehen. Im dezentralen Konzern darf er hingegen nur solche Geschäftschancen der beherrschten Gesellschaft nicht ausnutzen, von denen er im Rahmen der Konzernleitung Kenntnis erlangt hat. Schließlich unterliegt der Geschäftsführer sowohl im zentralen wie im dezentralen Konzern gegenüber den beherrschten Gesellschaften einem Wettbewerbsverbot.

9. Gegenstand des 5. Teils waren die konzerndimensionalen Pflichten der herrschenden GmbH gegenüber der Allgemeinheit. Sie zeichnen sich dadurch aus, daß sie eine Verantwortlichkeit des herrschenden Unternehmens für einen Vorgang bei einer beherrschten Gesellschaft begründen. Es wurde gezeigt, daß die herrschende GmbH keiner konzerndimensionalen Gefährdungshaftung unterliegt. Aus Delikt haftet sie erstens dann, wenn sie eigenhändig eine unerlaubte Handlung beging oder von deren Begehung in einem Konzernunternehmen wußte bzw. konkrete Anhaltspunkte hierfür hatte und pflichtwidrig nicht einschritt. Die Pflicht zum Einschreiten ergibt sich dabei aus ihrer verantwortlichen Stellung im Konzern. Zweitens treffen sie deliktische Verkehrspflichten, wenn sie eine gefahrenwesentliche Aufgabe in der beherrschten Gesellschaft übernommen hat oder dort eine erhöhte Gefahrenlage für die Rechtsgüter Dritter geschaffen hat. Und drittens wurde exemplarisch festgestellt, daß sie für die rechtzeitige Stellung des Antrags auf Eröffnung des Insolvenzverfahrens bzw. für die Erfüllung der steuerlichen Pflichten in den beherrschten Gesellschaften zu sorgen hat.

10. Im 6. Teil wurden schließlich die konzernbezogenen Pflichten des Geschäftsführers gegenüber der Allgemeinheit untersucht. Für sie gelten dieselben Grundsätze wie für die konzerndimensionalen Pflichten der herrschenden GmbH gegenüber der Allgemeinheit.

Literaturverzeichnis

Abeltshauser, Thomas E. [Leitungshaftung, 1998]: Leitungshaftung im Kapitalgesellschaftsrecht - Zu den Sorgfalts- und Loyalitätspflichten von Unternehmensleitern im deutschen und im US-amerikanischen Kapitalgesellschaftsrecht, Köln u. a.: Heymann, 1998, zugl. Habil.-Schr. Univ. Gießen, 1994

Altmeppen, Holger [Haftung, 1995]: Haftung der Geschäftsleiter einer Kapitalgesellschaft für Verletzung von Verkehrssicherungspflichten, in: ZIP 1995, S. 881 - 891

– [Vermögensbindung, 1996]: Zur Vermögensbindung in der faktisch abhängigen AG, in: ZIP 1996, S. 693 - 698

– [GmbHG, 1997]: §§ 13, 43, 64 GmbHG, Konzernrecht der GmbH (Anh. § 13), in: *G. H. Roth/H. Altmeppen*, GmbHG, 1997

– [Managerhaftung, 1998]: Die Haftung des Managers im Konzern, München: Beck, 1998

Armbrüster, Christian [Grundlagen, 1997]: Grundlagen und Reichweite von Wettbewerbsverboten im Personengesellschaftsrecht, in: ZIP 1997, S. 261 - 273

Assmann, Heinz-Dieter [GmbH-Konzern, 1992]: Der faktische GmbH-Konzern, in: *M. Lutter/P. Ulmer/W. Zöllner* (Hrsg.), FS GmbHG, 1992, S. 657 - 735

Bälz, Ulrich [Einheit, 1974]: Einheit und Vielheit im Konzern, in: *F. Baur* u. a. (Hrsg.), FS Raiser, 1974, S. 287 - 338

– [Unternehmen, 1992]: Verbundene Unternehmen, in: Die AG 1992, S. 277 - 311

Bar, Christian von [Verkehrspflichten, 1980]: Verkehrspflichten: Richterliche Gefahrsteuerungsgebote im deutschen Deliktsrecht, Köln u. a.: Heymann, 1980, zugl. Habil.-Schr. Univ. Göttingen, 1979/1980

Barz, Carl Hans u. a. [Großkommentar, 4. Bd., 1975]: Aktiengesetz, Großkommentar, 4. Bd., §§ 291 - 410, Einführungsgesetz zum Aktiengesetz, 3., neu bearb. Aufl., Berlin, New York: De Gruyter, 1975

Bassenge, Peter u. a. [Palandt, 1998]: Palandt, Bürgerliches Gesetzbuch, 57., neubearb. Aufl., München: Beck, 1998

Bastuck, Burkhard [Enthaftung, 1986]: Enthaftung des Managements - Corporate Indemnification im amerikanischen und deutschen Recht, Köln u. a.: Heymann, 1986 (Abhandlungen zum deutschen und europäischen Handels- und Wirtschaftsrecht, Bd. 51), zugl. Diss. jur. Univ. Bonn, 1984/85

Baur, Jürgen F., u. a. (Hrsg.): [FS Raiser, 1974]: Funktionswandel der Privatrechtsinstitutionen - Festschrift für Ludwig Raiser zum 70. Geburtstag, Tübingen: Mohr, 1974

– u. a. (Hrsg.) [FS Steindorff, 1990]: Festschrift für Ernst Steindorff zum 70. Geburtstag am 13. März 1990, Berlin, New York: De Gruyter, 1990

Beinert, Stefanie [Konzernhaftung, 1995]: Die Konzernhaftung für die satzungsgemäß abhängig gegründete GmbH, Köln: O. Schmidt, 1995 (Rechtsfragen der Handelsgesellschaften, Heft 83), zugl. Diss. jur. Univ. Berlin, 1995

Beisse, Heinrich/Lutter, Marcus/Närger, Heribald (Hrsg.) [FS Beusch, 1993]: Festschrift für Karl Beusch zum 68. Geburtstag am 31. Oktober 1993, Berlin, New York: De Gruyter, 1993

Beusch, Karl [Aktiengesellschaft, 1984]: Die Aktiengesellschaft - eine Kommanditgesellschaft in der Gestalt einer juristischen Person?, Betrachtungen zum sogenannten Holzmüller-Urteil des Bundesgerichtshofs, in: *W. Hadding* u. a. (Hrsg.), FS Werner, 1984, S. 1 - 21

Biedenkopf, Kurt H./Coing, Helmut/Mestmäcker, Ernst-Joachim (Hrsg.) [FG Kronstein, 1967]: Das Unternehmen in der Rechtsordnung - Festgabe für Heinrich Kronstein aus Anlaß seines 70. Geburtstages am 12. September 1967, Karlsruhe: C. F. Müller, 1967

Bierich, Marcus/Hommelhoff, Peter/Kropff, Bruno (Hrsg.) [FS Semler, 1993]: Festschrift für Johannes Semler zum 70. Geburtstag am 28. April 1993, Unternehmen und Unternehmensführung im Recht, Berlin, New York: De Gruyter, 1993

Binder, Christof U. [Beteiligungsstrategien, 1994]: Beteiligungsstrategien in der Konzernpraxis: Eine empirische Untersuchung der Beteiligungshöhen in deutschen Konzernen, in: Die AG 1994, S. 391 - 396

Bitter, Walter/Bitter, Georg [Konzern, 1996]: Alles klar im qualifiziert faktischen Konzern? Oder: Die Ausweitung der BGH-Rechtsprechung durch das BAG - Anmerkungen zum Konzernhaftungsrecht aus Anlaß der Entscheidung des BAG vom 1.8.1995, in: BB 1996, S. 2153 - 2160

Böttcher, Reinhard/Hueck, Götz/Jähnke, Burkhard (Hrsg.): [FS Odersky, 1996]: Festschrift für Walter Odersky zum 65. Geburtstag am 17. Juli 1996, Berlin, New York: Walter de Gruyter, 1996

Bork, Reinhard u. a. [Staudinger, 1996]: J. von Staudingers: Kommentar zum Bürgerlichen Gesetzbuch mit Einführungsgesetz und Nebengesetzen, Erstes Buch, Allgemeiner Teil, §§ 134 - 163, Redaktor *Norbert Habermann*, 13. Bearbeitung, Berlin: Sellier - de Gruyter, 1996

Brandes, Helmut [Rechtsprechung, 1987]: Die Rechtsprechung des BGH zur GmbH & Co. KG und zur Publikumsgesellschaft, in: WM 1987, Sonderbeilage Nr. 1

– [Kapitalerhaltung, 1991]: Grundsätze der Kapitalerhaltung im Vertragskonzern, in: *R. Goerdeler* u. a. (Hrsg.), FS Kellermann, 1991, S. 25 - 35

Brandner, Hans Erich [Geschäftsführungsbefugnis, 1994]: Geschäftsführungsbefugnis, Unternehmensgegenstand und Unternehmenszweck, in: *G. Pfeiffer/G. Wiese/K. Zimmermann* (Hrsg.), FS Rowedder, 1994, S. 41 - 47

Brockmeyer, Bernhard [Klein/Orlopp, 1995]: § 35 AO, in: *F. Klein/G.Orlopp/B. Brockmeyer*, AO, 1995

Brox, Hans [Schuldrecht, 1997]: Besonderes Schuldrecht, 22. Aufl., München: Beck, 1997

Brüggemann, Dieter, u. a. [HGB, 1973]: Handelsgesetzbuch, Großkommentar, begründet von Hermann Staub, weitergeführt von Mitgliedern des Reichsgerichts, 3., neubearb. Aufl., II. Bd., 1. Halbband: §§ 105 - 144, Berlin, New York: De Gruyter, 1973

Brüggemeier, Gert [Organisationshaftung, 1991]: Organisationshaftung - Deliktsrechtliche Aspekte innerorganisatorischer Funktionsdifferenzierung, in: AcP 191 (1991), S. 33 - 68

Bundesministerium der Justiz (Hrsg.) [Bericht, 1980]: Bericht über die Verhandlungen der Unternehmensrechtskommission, Köln: O. Schmidt, 1980

Burgard, Ulrich [Anmerkung, 1995]: Anmerkung zum Urteil des BGH vom 19.9.94 - II ZR 237/93 („Freiberufler"), in: DNotZ 1995, S. 951 - 956

Caemmerer, Ernst von/Nikisch, Arthur/Zweigert, Konrad [FS Dölle, 1963]: Vom deutschen zum europäischen Recht, Festschrift für Hans Dölle, Bd. I: Deutsches Privat- und Zivilprozessrecht, Rechtsvergleichung, Tübingen: J. C. B. Mohr (Paul Siebeck), 1963

Cahn, Andreas [Kapitalerhaltung, 1998]: Kapitalerhaltung im Konzern, Köln u. a.: Heymanns, 1998, zugl. Habil.-Schr. Univ. Frankfurt am Main, 1995

Canaris, Claus-Wilhelm [Schuldrecht, 1994]: Lehrbuch des Schuldrechts, 2. Bd., Besonderer Teil, 2. Halbband, begr. von Karl Larenz, 13., völlig neuverfaßte Aufl., München: Beck, 1994

– */Schilling, Wolfgang/Ulmer, Peter* (Hrsg.) [Staub, 1989]: Staub, Großkommentar zum HGB, 4., neubearb. Aufl., 12. Lieferung: §§ 105 - 113, Berlin, New York: de Gruyter, 1989

Claussen, Carsten Peter/Korth, H.-Michael [Wettbewerbsverbot, 1993]: Das Wettbewerbsverbot des Geschäftsführers/Gesellschafters einer GmbH, in: *H. Beisse/M. Lutter/H. Närger* (Hrsg.), FS Beusch, 1993, S. 111 - 129

Crezelius, Georg, u. a. [Scholz, I. Bd., 1993]: Scholz, Kommentar zum GmbH-Gesetz mit Nebengesetzen und den Anhängen Konzernrecht sowie Umwandlung und Verschmelzung, I. Bd., §§ 1 - 44, Anh. Konzernrecht, 8., neubearb. und wesentlich erw. Aufl., Köln: O. Schmidt, 1993

– u. a. [Scholz, II. Bd., 1995]: Scholz, Kommentar zum GmbH-Gesetz mit Nebengesetzen und dem Anhang Konzernrecht, II. Bd., §§ 45 - 85, 8. neubearb. und wesentlich erw. Aufl., Köln: O. Schmidt, 1995

Decher, Christian E. [Verflechtungen, 1989]: Personelle Verflechtungen im Aktienkonzern - Loyalitätskonflikt und qualifizierter faktischer Konzern, Heidelberg: Verlag Recht und Wirtschaft, 1990 (Abhandlungen zum Arbeits- und Wirtschaftsrecht, Bd. 61), zugl. Diss. iur. Univ. Köln, 1989

Dorow, Wolfgang [Leitungsstruktur, 1993]: Leitungs- und Machtstruktur, in: *W. Wittmann* u. a. (Hrsg.), Handwörterbuch, 1993, Spalten 2568 - 2581

Dreher, Meinrad [Verantwortlichkeit, 1992]: Die persönliche Verantwortlichkeit von Geschäftsleitern nach außen und die innergesellschaftliche Aufgabenverteilung, in: ZGR 1992, S. 22 - 63

– [Treuepflichten, 1993]: Treuepflichten zwischen Aktionären und Verhaltenspflichten bei der Stimmrechtsbündelung - Gesellschaftsrechtliche und zivilrechtliche Grundlagen, in: ZHR 157 (1993), S. 150 - 171

Drygala, Tim [Verhaltenshaftung, 1993]: Verhaltenshaftung im faktischen Konzern, in: GmbHR 1993, S. 317 - 329

Ebenroth, Carsten Thomas/Lange, Knut Werner [Sorgfaltspflichten, 1992]: Sorgfaltspflichten und Haftung des Geschäftsführers einer GmbH nach § 43 GmbHG, in: GmbHR 1992, S. 69 - 76

Eckardt, Dieterich [Organhaftung, 1996]: Persönliche Organ- und Gehilfenhaftung für Verkehrspflichtverletzungen im Unternehmen, in: *A. Willingmann* u. a. (Hrsg.), Jahrbuch, 1996, S. 61 - 80

Emmerich, Volker [Konzernbildungskontrolle, 1991]: Konzernbildungskontrolle, in: Die AG 1991, S. 303 - 312

– [Scholz, 1993]: §§ 3, 29 GmbHG, GmbH-Konzernrecht, in: *G. Crezelius* u. a., Scholz, I. Bd., 1993

– [Aktienkonzernrecht, 1998]: §§ 291, 293, 308 AktG, in: *V. Emmerich/M. Habersack*, Aktienkonzernrecht, 1998

– */Sonnenschein, Jürgen* [Konzernrecht, 1997]: Konzernrecht, Das Recht der verbundenen Unternehmen bei Aktiengesellschaft, GmbH, Personengesellschaften, Genossenschaft, Verein und Stiftung, 6., völlig neubearb. Aufl., München: Beck, 1997

– */Habersack, Mathias* [Aktienkonzernrecht, 1998]: Aktienkonzernrecht, Kommentar zu den §§ 15 - 22 und 291 - 328 AktG, München: Beck 1998

Ensthaler, Jürgen/Kreher, Markus [Haftungspotentiale, 1996]: Haftungspotentiale unterschiedlicher Finanzierungsformen im Zusammenhang mit qualifizierten faktischen GmbH-Konzernen, in: BB 1996, S. 385 - 389

Eschenbruch, Klaus [Konzernhaftung, 1996]: Konzernhaftung: Haftung der Unternehmen und der Manager, Düsseldorf: Werner, 1996

Feick, Carl-Peter [Pflichten, 1999]: Die Wahrnehmung öffentlich-rechtlicher Pflichten im konzernfreien und im konzernverbundenen Unternehmen, Herdecke: GCA-Verl., 1999, zugl. Diss. rer. pol. TU Darmstadt, 1997

Fischer, Robert [HGB, 1973]: § 116 HGB, in: *D. Brüggemann* u. a., HGB, 1973

– */Möhring, Philipp/Westermann, Harry* (Hrsg.) [FS Barz, 1974]: Wirtschaftsfragen der Gegenwart - Festschrift für Carl Hans Barz zum 65. Geburtstag am 6. Dezember 1974, Berlin, New York: De Gruyter, 1974

Fleck, Hans-Joachim [Haftung, 1974]: Zur Haftung des GmbH-Geschäftsführers, in: GmbHR 1974, S. 224 - 235

– [Mißbrauch, 1990]: Mißbrauch der Vertretungsmacht oder Treubruch des mit Einverständnis aller Gesellschafter handelnden GmbH-Geschäftsführers aus zivilrechtlicher Sicht, in: ZGR 1990, S. 31 - 49

Flume, Werner [Juristische Person, 1983]: Allgemeiner Teil des Bürgerlichen Rechts, Erster Band, Zweiter Teil: Die juristische Person, Berlin u. a.: Springer, 1983 (Enzyklopädie der Rechts- und Staatswissenschaft, Abteilung Rechtswissenschaft)

– [Gewinnübernahmevertrag, 1989]: Der Gewinn- und Verlustübernahmevertrag im GmbH-Recht, in: DB 1989, S. 665 - 670

Gamm, Otto Friedrich Freiherr von/ Raisch, Peter/Tiedemann, Klaus (Hrsg.) [FS Pfeiffer, 1988]: Strafrecht, Unternehmensrecht, Anwaltsrecht - Festschrift für Gerd Pfeiffer zum Abschied aus dem Amt als Präsident des Bundesgerichtshofes, Köln u. a.: Heymann, 1988

Geitzhaus, Georg [Verlustausgleichspflicht, 1989]: Verlustausgleichspflicht als Motor der Konzernplanung (I), in: GmbHR 1989, S. 397 - 406

Geßler, Ernst [Geßler, 1973]: § 18 AktG (Stand: 1973), in: *E. Geßler* u. a., AktG, Bd. I, 1973/84

– [Leitungsmacht, 1974]: Leitungsmacht und Verantwortlichkeit im faktischen Konzern, in: *W. Hefermehl/R. Gmür/H. Brox* (Hrsg.), FS Westermann, 1974, S. 145 - 165

– [Geßler, 1975]: §§ 291, 308 AktG (Stand: Oktober 1975), in: *E. Geßler* u. a., AktG, Bd. IV, 1976/94

– [Hauptversammlungszuständigkeiten, 1985]: Einberufung und ungeschriebene Hauptversammlungszuständigkeiten, in: *M. Lutter/H.-J. Mertens/P. Ulmer* (Hrsg.), FS Stimpel, 1985, S. 771 - 789

– u. a. [AktG, Bd. I, 1973/84]: Kommentar zum Aktiengesetz, Bd. I: §§ 1 - 22 (1. Lieferung, 1973), §§ 23 - 58 (8. Lieferung, 1984), §§ 53a - 75 (7. Lieferung, 1983), München: Vahlen, 1973/84

– u. a. [AktG, Bd. II, 1973/1974]: Kommentar zum Aktiengesetz, Bd. II, §§ 76 - 147, München: Vahlen, 1973/74

– u. a. [AktG, Bd. IV, 1976/94]: Kommentar zum Aktiengesetz, Bd. IV: §§ 291 - 318 (6. Lieferung 1976, Stand: Oktober 1975), §§ 319 - 358a (13. Lieferung 1991, Stand: Juli 1991), §§ 359 - 395 (14. Lieferung 1993, Stand: Dezember 1992), §§ 396 - 410 (18. Lieferung 1994, Stand: Februar 1994), München: Vahlen, 1976 ff.

– u. a. [AktG, Bd. V, 1984/86]: Kommentar zum Aktiengesetz, Bd. V: §§ 241 - 261 (9. Lieferung 1984, Stand: Juni 1984), §§ 262 - 277 (11. Lieferung 1986, Stand: Mai 1986), §§ 278 - 290 (10. Lieferung 1986, Stand: Juli 1985), München: Vahlen, 1984 ff.

– u. a. [AktG, Bd. VI, 1976/94]: Kommentar zum Aktiengesetz, Bd. VI: §§ 291 - 318 (6. Lieferung 1976, Stand: Oktober 1975), §§ 319 - 358a (13. Lieferung 1991, Stand: Juli 1991), §§ 359 - 395 (14. Lieferung 1993, Stand: Dezember 1992), §§ 396 - 410 (18. Lieferung 1994, Stand: Februar 1994), München: Vahlen, 1976 ff.

– u. a. [Schlegelberger, 1992]: Schlegelberger, Handelsgesetzbuch, Bd. III/1. Halbbd.: §§ 105 - 160, 5., neubearb. Aufl., München: Vahlen, 1992

Gieseke, Horst [Interessenkonflikte, 1996]: Interessenkonflikte der GmbH-Geschäftsführer bei Pflichtenkollisionen, in: GmbHR 1996, S. 486 - 492

Goerdeler, Reinhard u. a. (Hrsg.) [FS Fleck, 1988]: Festschrift für Hans-Joachim Fleck zum 70. Geburtstag am 30. Januar 1988, Berlin, New York: de Gruyter, 1988 (ZGR-Sonderheft 7)

– u. a. (Hrsg.) [FS Kellermann, 1991]: Festschrift für Alfred Kellermann zum 70. Geburtstag am 29. November 1990, Berlin, New York: Walter de Gruyter, 1991 (ZGR-Sonderheft 10)

Görling, Helmut [Unternehmensverbindungen, 1993]: Die Verbreitung zwei- und mehrstufiger Unternehmensverbindungen - Ergebnisse einer empirischen Untersuchung, in: Die AG 1993, S. 538 - 547

Götting, Horst-Peter [Haftung, 1994]: Die persönliche Haftung des GmbH-Geschäftsführers für Schutzrechtsverletzungen und Wettbewerbsverstöße, in: GRUR 1994, S. 6 - 12

Götz, Heinrich [Sicherung, 1984]: Die Sicherung der Rechte der Aktionäre der Konzernobergesellschaft bei Konzernbildung und Konzernleitung - zugleich auch eine Besprechung der "Holzmüller"-Entscheidung des Bundesgerichtshofs -, in: Die AG 1984, S. 85 - 94

– [Leitungssorgfalt, 1998] Leitungssorgfalt und Leitungskontrolle der Aktiengesellschaft hinsichtlich abhängiger Unternehmen, in: ZGR 1998, S. 524 - 546

Groß, Werner [Außenhaftung, 1998]: Deliktische Außenhaftung des GmbH-Geschäftsführers, in: ZGR 1998, S. 551 - 569

Groß, Wolfgang [Zuständigkeit, 1994]: Zuständigkeit der Hauptversammlung bei Erwerb und Veräußerung von Unternehmensbeteiligungen, in: Die AG 1994, S. 266 - 276

Grunewald, Barbara [Einsichtsrecht, 1982]: Einsichts- und Auskunftsrecht des GmbH-Gesellschafters nach neuem Recht, in: ZHR 146 (1982), S. 211 - 237

– [Geßler, 1991]: § 323 AktG, in: *E. Geßler* u. a., AktG, Bd. VI, 1976/94

– [Haftung, 1993]: Die Haftung von Organmitgliedern nach Deliktsrecht, in: ZHR 157 (1993), S. 451 - 463

Grunsky, Wolfgang [Münchener Kommentar, 1994]: Vorbemerkungen zu § 249 BGB, in: *K. Rebmann/F.-J. Säcker* (Hrsg.), Münchener Kommentar, 1994

Grünwald, Michael [Außenhaftung, 1999]: Die deliktische Außenhaftung des GmbH-Geschäftsführers für Organisationsdefizite, Baden-Baden: Nomos, 1999 (Studien zum Handels-, Arbeits- und Wirtschaftsrecht, Bd. 56), zugl. Diss. jur. Univ. München, 1997/98

Haas, Ulrich [Geschäftsführerhaftung, 1997]: Geschäftsführerhaftung und Gläubigerschutz - Unternehmerische Verhaltenspflichten des GmbH-Geschäftsführers zum Schutz Dritter, München: Beck, 1997, zugl. Habil.-Schr. Univ. Regensburg, 1996

Habersack, Mathias [Aktienkonzernrecht, 1998]: § 311 AktG, in: *V. Emmerich/M. Habersack*, Aktienkonzernrecht, 1998

Hadding, Walther u. a. (Hrsg.) [FS Werner, 1984]: Festschrift für Winfried Werner zum 65. Geburtstag am 17. Oktober 1984, Handelsrecht und Wirtschaftsrecht in der Berufspraxis, Berlin, New York: De Gruyter, 1984

Hansen, Herbert [GmbH-Bestand, 1997]: Der GmbH-Bestand stieg auf 770 000 Gesellschaften an - Gleichzeitig Anmerkungen zu der Einstellung der statistischen Erhebungen bei den Kapitalgesellschaften, in: GmbHR 1997, S. 204 - 207

Haß, Peter [Haftung, 1994]: Die persönliche Haftung des GmbH-Geschäftsführers bei Wettbewerbsverstößen und Verletzung gewerblicher Schutzrechte, in: GmbHR 1994, S. 666 - 673

Havermann, Hans (Hrsg.) [FS Goerdeler, 1987]: Bilanz- und Konzernrecht, Festschrift zum 65. Geburtstag von Dr. Dr. h. c. Reinhard Goerdeler, Düsseldorf: IDW-Verlag, 1987

Hefermehl, Wolfgang [Geßler, 1973]: § 88 AktG (Stand: März 1973), in: *E. Geßler* u. a., AktG, Bd. II, 1973/1974

– */Bungeroth, Erhard* [Geßler, 1983]: § 57 AktG (Stand: 1983), in: *E. Geßler* u. a., AktG, Bd. I, 1973/84

Hefermehl, Wolfgang/Gmür, Rudolf/Brox, Hans (Hrsg.) [FS Westermann, 1974]: Festschrift für Harry Westermann zum 65. Geburtstag, Karlsruhe: C. F. Müller, 1974

Heil, Ulf/Russenschuck, Viola [Haftung, 1998]: Die persönliche Haftung des GmbH-Geschäftsführers - Im Blickpunkt: Grundsätze der sogenannten „Managerprodukthaftung", in: BB 1998, S. 1749 - 1755

Heisse, Matthias [Geschäftsführerhaftung, 1988]: Die Beschränkung der Geschäftsführerhaftung gegenüber der GmbH, Köln: O. Schmidt, 1988 (Rechtsfragen der Handelsgesellschaften, Heft 55), zugl. Diss. jur. Univ. München, 1987

Hennrichs, Joachim [Treupflichten, 1995]: Treupflichten im Aktienrecht - zugleich Überlegungen zur Konkretisierung der Generalklausel des § 242 BGB sowie zur Eigenhaftung des Stimmrechtsvertreters -, in: AcP 195 (1995), S. 221 - 273

Henssler, Martin [Minderheitenschutz, 1998]: Minderheitenschutz im faktischen GmbH-Konzern - Zugleich ein Plädoyer für die Aufwertung des Konzernabschlusses, in: *M. Lieb/U. Noack/H. P. Westermann*, FS Zöllner, 1998, S. 203 - 233

Henze, Hartwig [Treupflicht, 1996]: Die Treupflicht im Aktienrecht - Gedanken zur Rechtsprechung des Bundesgerichtshofes von „Kali und Salz" über „Linotype" und „Kochs Adler" bis zu „Girmes", in: BB 1996, S. 489 - 499

– [Treupflichten, 1998]: Treupflichten der Gesellschafter im Kapitalgesellschaftsrecht, in: ZHR 162 (1998), S. 186 - 196

Hirte, Heribert (Hrsg.) [Vertragskonzern, 1993]: Der Vertragskonzern im Gesellschaftsrecht, Köln: Verlag Kommunikationsforum Recht, Wirtschaft, Steuern, 1993 (RWS-Dokumentation, Bd. 13)

– [Teilkodifikation, 1993]: Der Vertragskonzern im Gesellschaftsrecht - eine Teilkodifikation des Rechts des qualifizierten faktischen Konzerns, in: *H. Hirte* (Hrsg.), Vertragskonzern, 1993, S. 1 - 30

Hofmann, Paul [Handelsrecht, 1996]: Handelsrecht, 9., überarb. Aufl., Neuwied, Kriftel, Berlin: Luchterhand, 1996

Höhn, Reinhard [Geschäftsleitung, 1995]: Die Geschäftsleitung der GmbH - Organisation, Führung und Verantwortung, Kontrolle durch Aufsichtsrat und Beirat, 2. überarb. und erw. Aufl., Köln: O. Schmidt, 1995

Hoffmann, Friedrich (Hrsg.) [Konzernhandbuch, 1993]: Konzernhandbuch: Recht, Steuern, Rechnungslegung, Führung, Organisation, Praxisfälle, Wiesbaden: Gabler, 1993

– [Konzern, 1993]: Der Konzern als Gegenstand betriebswirtschaftlicher Forschung, in: *F. Hoffmann* (Hrsg.), Konzernhandbuch, 1993, S. 1 - 79

Hoffmann-Becking, Michael (Hrsg.) [Münchener Handbuch, Bd. 4, 1988]: Münchener Handbuch des Gesellschaftsrechts, Bd. 4: Aktiengesellschaft, München: Beck, 1988

– [AG-Konzern, 1989] Der qualifizierte faktische AG-Konzern - Tatbestand und Abwehransprüche -, in: *P. Ulmer* (Hrsg.), Probleme, 1989, S. 68 - 86

– [Unternehmensvertrag, 1994]: Gelöste und ungelöste Fragen zum Unternehmensvertrag der GmbH, in: WiB 1994, S. 57 - 63

Hommelhoff, Peter [Unternehmensführung, 1978]: Unternehmensführung in der mitbestimmten GmbH, in: ZGR 1978, S. 119 - 155

– [Konzernleitungspflicht, 1982]: Die Konzernleitungspflicht, Köln u. a.: Heymann, 1982, zugl. Habil.-Schr. Univ. Bochum, 1980/81

– [Eigenkapital-Ersatz, 1984]: Eigenkapital-Ersatz im Konzern und in Beteiligungsverhältnissen, in: WM 1984, S. 1105 - 1118

– [Produkthaftung, 1990]: Produkthaftung im Konzern, in: ZIP 1990, S. 761 - 771

– [Gutachten, 1992]: Empfiehlt es sich, das Recht faktischer Unternehmensverbindungen neu zu regeln? Gutachten G für den 59. Deutschen Juristentag, in: *Ständige Deputation des Deutschen Juristentages* (Hrsg.), Verhandlungen, 1992, S. G 1 - G 85

– /*Stimpel, Walter*/*Ulmer, Peter* (Hrsg.) [Konzernrechtstage, 1992]: Heidelberger Konzernrechtstage: Der qualifizierte faktische GmbH-Konzern - Analyse der Rechtsprechung - Folgerungen für die Praxis, Köln: O. Schmidt, 1992 (Rechtsfragen der Handelsgesellschaften, Heft 75)

– u. a. (Hrsg.) [Entwicklungen, 1986]: Entwicklungen im GmbH-Konzernrecht, 2. Deutsch-Österreichisches Symposium zum Gesellschaftsrecht vom 21. und 22. Februar 1986 in Landshut, Berlin, New York: De Gruyter, 1986 (ZGR-Sonderheft 6)

Hopt, Klaus J. [HGB, 1995]: Handelsgesetzbuch mit GmbH & Co., Handelsklauseln, Bank- und Börsenrecht, Transportrecht (ohne Seerecht), begründet von Adolf Baumbach, fortgeführt bis zur 24. Aufl. von Konrad Duden, 29., völlig neubearb. u. erw. Aufl., München: Beck, 1995 (Beck'sche Kurzkommentare; Bd. 9)

Hormuth, Mark W. [Cash-Management, 1998]: Recht und Praxis des konzernweiten Cash Managements - Ein Beitrag zur Konzernfinanzierung, Schriftenreihe „Untersuchungen über das Spar-, Giro- und Kreditwesen, Abteilung B: Rechtswissenschaft", hrsg. von *Walther Hadding* und *Uwe H. Schneider*, Bd. 116, Konzern, Konzernrecht und Konzernfinanzierung, Teil VIII, Berlin: Duncker & Humblot, 1998, zugl. Diss. rer. pol. Technische Universität Darmstadt, 1997

Horn, Norbert/*Luig, Klaus*/*Söllner, Alfred* (Hrsg.) [FS Coing, Bd. II, 1982]: Europäisches Rechtsdenken in Geschichte und Gegenwart, Festschrift für Helmut Coing zum 70. Geburtstag, Bd. II, München: Beck, 1982

Hucke, Anja/*Schröder, Henning* [Umwelthaftung, 1998]: Umwelthaftung von Konzernen, in: DB 1998, S. 2205 - 2210

Hueck, Alfred [Anfechtbarkeit, 1924]: Anfechtbarkeit und Nichtigkeit von Gesellschafterbeschlüssen bei Aktiengesellschaften, Mannheim, Berlin, Leipzig: J. Bensheimer, 1924

– [Gesellschafterbeschlüsse, 1962]: Mangelhafte Gesellschafterbeschlüsse bei der GmbH, in *H. C. Nipperdey* (Hrsg.), FS Molitor, 1962, S. 401 - 427

– [OHG, 1971]: Das Recht der offenen Handelsgesellschaft, 4., neubearb. Auf., Berlin, New York: De Gruyter, 1971

– */Hueck, Götz* [Baumbach/Hueck, AktG, 1968]: Baumbach/Hueck, Aktiengesetz, begr. von Adolf Baumbach, bearb. unter Mitwirkung von Joachim Schulze, 13., völlig neu bearb. Aufl., München: Beck, 1968

Hueck, Götz [Baumbach/Hueck, 1996]: § 13 GmbHG, in: *G. Hueck/J. Schulze- Osterloh/W. Zöllner*, Baumbach/Hueck, 1996

Hueck, Götz/Schulze-Osterloh, Joachim/Zöllner, Wolfgang [Baumbach/ Hueck, GmbHG, 1996]: Baumbach/Hueck, GmbH-Gesetz, Gesetz betreffend die Gesellschaften mit beschränkter Haftung, begr. von Adolf Baumbach, fortgeführt von Alfred Hueck, 16., erw. und völlig überarb. Aufl., München: Beck, 1996

– u. a. [Arbeitskreis GmbH-Reform, 1972]: Arbeitskreis GmbH-Reform, Thesen und Vorschläge zur GmbH-Reform, Band II, Heidelberg: Verlag Recht und Wirtschaft, 1972

Hüffer, Uwe [Organpflichten, 1981]: Organpflichten und Haftung in der Publikumspersonengesellschaft - Besprechung der Entscheidungen BGH, BB 1980, 546 und BGHZ 75, 321 -, in: ZGR 1981, S. 348 - 363

– [Geßler, 1984]: § 241 AktG (Stand: Juni 1984), in: *E. Geßler* u. a., AktG, Bd. V, 1984/86

– [Hachenburg, 1990]: § 45 GmbHG (Stand: 1.11.1990), in: *P. Ulmer* (Hrsg.), Hachenburg, 1989/96

– [AktG, 1997], Aktiengesetz, 3., neubearb. Aufl., München: Beck, 1997

Immenga, Ulrich [Kapitalgesellschaft, 1970]: Die personalistische Kapitalgesellschaft - Eine rechtsvergleichende Untersuchung nach deutschem GmbH-Recht und dem Recht der Corporations in den Vereinigten Staaten -, Bad Homburg v.d.H.: Athenäum Verlag, 1970 (Wirtschaftsrecht und Wirtschaftspolitik, Bd. 15), zugl. Habil.-Schr. Univ. Bielefeld, 1970

– [Anfechtungsklage, 1973]: Die Problematik der Anfechtungsklage im GmbH-Recht - Ein Beitrag zum Minderheitenschutz in der Gesellschafterversammlung, in: GmbHR 1973, S. 5 - 11

– [Bestandsschutz, 1976]: Bestandsschutz der beherrschten Gesellschaft im Vertragskonzern?, in: ZHR 140 (1976), S. 301 - 316

– [Anmerkung, 1984]: Anmerkung zu BGH, Urt. v. 5.12.1983 - II ZR 242/82, in: JZ 1984, S. 578 - 580

– [Treuepflichten, 1992]: Bindung von Rechtsmacht durch Treuepflichten, in: *M. Lutter/P. Ulmer/W. Zöllner* (Hrsg.), FS GmbHG, 1992, S. 189 - 208

– /*Möschel, Wernhard*/*Reuter, Dieter* (Hrsg.) [FS Mestmäcker, 1996]: Festschrift für Ernst-Joachim Mestmäcker zum siebzigsten Geburtstag, Baden-Baden: Nomos, 1996

Ivens, Michael [Konkurrenzverbot, 1987]: Das Konkurrenzverbot des GmbH-Gesellschafters - Verbotsklauseln, Treuebindung, Kartellrecht, Hamburg: W. Mauke Söhne, 1987

Jarass, Hans D. [BImSchG, 1995]: Bundes-Immissionsschutzgesetz (BImSchG), Kommentar, 3., überarb. Aufl., München: Beck, 1995

Jauernig, Othmar (Hrsg.) [BGB, 1994]: Bürgerliches Gesetzbuch mit Gesetz zur Regelung des Rechts der Allgemeinen Geschäftsbedingungen, 7., neubearb. Aufl., München: Beck, 1994

– [Jauernig, 1994]: § 134 BGB, in: *O. Jauernig* (Hrsg.), BGB, 1994

Joost, Detlev [Holzmüller, 1999]: „Holzmüller 2000" vor dem Hintergrund des Umwandlungsgesetzes, in: ZHR 163 (1999), S. 164 - 186

Kalfass, Hermann H. [Konzernbildung, 1991]: Ökonomische Analyse der Konzernbildung, in: *E.-J. Mestmäcker*/*P. Behrens* (Hrsg.), Gesellschaftsrecht, 1991, S. 19 - 48

Keßler, Jürgen [Eigenhaftung, 1994]: Die deliktische Eigenhaftung des GmbH-Geschäftsführers, in: GmbHR 1994, S. 429 - 437

Kessler, Manfred H. [Leitungsmacht, 1995]: Die Leitungsmacht des Vorstandes einer Aktiengesellschaft, in: Die AG 1995, S. 61 - 76 (Teil I), 120 - 132 (Teil II)

Kindler, Peter [Ermessen, 1998]: Unternehmerisches Ermessen und Pflichtenbindung - Voraussetzungen und Geltendmachung der Vorstandshaftung in der Aktiengesellschaft, in: ZHR 1998, S. 101 - 119

Klein, Franz/*Orlopp, Gerd*/*Brockmeyer, Bernhard* [AO, 1995]: Abgabenordnung - einschließlich Steuerstrafrecht -, begründet von *Franz Klein* und *Gerd Orlopp*, 5., völlig neubearb. Auflage, München: Beck, 1995

Kleindiek, Detlev [Strukturvielfalt, 1991]: Strukturvielfalt im Personengesellschafts-Konzern - Rechtsformspezifische und rechtsformübergreifende Aspekte des Konzernrechts -, Köln u. a.: Heymann, 1991 (Abhandlungen zum deutschen und europäischen Handels- und Wirtschaftsrecht, Bd. 79), zugl. Diss. jur. Univ. Bielefeld, 1990

– [Deliktshaftung, 1997]: Deliktshaftung und juristische Person - Zugleich zur Eigenhaftung von Unternehmensleitern, Jus Privatum, Beiträge zum Privatrecht, Bd. 22, Tübingen: Mohr Siebeck, 1997, zugl. Habil.-Schr. Univ. Tübingen, 1996/97

Knopp, Lothar/*Albrecht, Eicke* [Bundes-Bodenschutzgesetz, 1998]: Das neue Bundes-Bodenschutzgesetz und Altlasten - Im Blickpunkt: Ausgewählte Fragen aus betrieblicher Sicht, in: BB 1998, S. 1853 - 1858

Konzen, Horst [Geschäftsführung, 1989]: Geschäftsführung, Weisungsrecht und Verantwortlichkeit in der GmbH und GmbH & Co KG, in: NJW 1989, S. 2977 - 2987

Koppensteiner, Hans-Georg [Kölner Kommentar, 1986]: §§ 17, 18 AktG (Stand: Januar 1986), in: *W. Zöllner* (Hrsg), Kölner Kommentar, Bd. 1, 1988

– [Kölner Kommentar, 1987]: §§ 291 - 328 AktG (Stand: Mai 1987), in: *W. Zöllner* (Hrsg.), Kölner Kommentar, Bd. 6, 1987

– [Rowedder, 1997]: §§ 37, 43, 46, 47, 51a GmbHG, GmbH-Konzernrecht, in: *H. Rowedder* u. a., GmbHG, 1997

Kort, Michael [Beherrschungsverträge, 1986]: Der Abschluß von Beherrschungs- und Gewinnabführungsverträgen im GmbH-Recht - Übergang vom faktischen zum Vertragskonzern, Köln u. a.: Heymann, 1986 (Abhandlungen zum deutschen und europäischen Handels- und Wirtschaftsrecht, Bd. 49), zugl. Diss. jur. Univ. München, 1985

– [Informationsrecht, 1987]: Das Informationsrecht des Gesellschafters der Konzernobergesellschaft, in: ZGR 1987, S. 46 - 75

– [Eigenhaftung, 1990]: Zur Eigenhaftung des GmbH-Geschäftsführers gegenüber privatrechtlichen Dritten - Zugleich Besprechung von BGH, Urteil vom 3.10.1989 - XI ZR 157/88, DB 1989 S. 2320 u. BGH, Urteil vom 5.12.89 - VI ZR 335/88, DB 1990, S. 268, in: DB 1990, S. 921 - 923

Krebs, Peter [Geschäftsführungshaftung, 1991]: Geschäftsführungshaftung bei der GmbH & Co. KG und das Prinzip der Haftung für sorgfaltswidrige Leitung, Baden-Baden: Nomos, 1991 (Studien zum Handels-, Arbeits- und Wirtschaftsrecht, Bd. 17), zugl. Diss. jur. Univ. Köln, 1991

Krieger, Gerd [Unternehmen, 1988]: Abhängige Unternehmen und faktische Konzerne, in: *M. Hoffmann-Becking* (Hrsg.), Münchener Handbuch, Bd. 4, 1988, § 69

– [Vertragskonzern, 1988]: Vertragskonzern (Beherrschungsvertrag), in: *M. Hoffmann-Becking* (Hrsg.), Münchener Handbuch, Bd. 4, 1988, § 70

– [Tatbestand, 1992]: Der Tatbestand des qualifizierten faktischen Konzerns, in: *P. Hommelhoff/W. Stimpel/P. Ulmer* (Hrsg.), Konzernrechtstage, 1992, S. 41 - 59

– [TBB, 1994]: Kann die Praxis mit TBB leben?, in: ZGR 1994, S. 375 - 394

– [Haftungsrisiken, 1998]: Die rechtlichen Haftungsrisiken im Konzern, in: in: *M. Lutter/E. Scheffler/Uwe H. Schneider* (Hrsg.), Konzernfinanzierung, 1998, Rdnr. 4.1 - 4.69

Kropff, Bruno [AktG, 1965]: Aktiengesetz, Textausgabe des Aktiengesetzes vom 6.9.1965 (Bundesgesetzbl. I S. 1089) und des Einführungsgesetzes zum Aktiengesetz vom 6.9.1965 (Bundesgesetzbl. I S. 1185) mit Begründung des Regierungsentwurfs, Bericht des Rechtsausschusses des Deutschen Bundestags, Verweisungen und

Sachverzeichnis, im Anhang: Aktiengesetz von 1937, Düsseldorf: Verlagsbuchhandlung des Instituts der Wirtschaftsprüfer GmbH, 1965

– [Geßler, 1975]: § 311 AktG (Stand: 1.10.1975), in: *E. Geßler* u. a., AktG, Bd. IV, 1976/94

– [Konzernleitungspflicht, 1984]: Zur Konzernleitungspflicht, in: ZGR 1984, S. 112 - 133

– [Beherrschungsvertrag, 1993]: Der GmbH-Beherrschungsvertrag: Voraussetzung für den Vorrang von Konzerninteressen?, in: *M. Bierich/P. Hommelhoff/B. Kropff* (Hrsg.), FS Semler, 1993, S. 517 - 540

Kübler, Friedrich/Waltermann, Jens [Geschäftschancen, 1991]: Geschäftschancen der Kommanditgesellschaft - Besprechung der Entscheidung BGH ZIP 1989, 986 ff -, in: ZGR 1991, S. 162 - 174

– */Mertens, Hans-Joachim/Werner, Winfried* (Hrsg.) [FS Heinsius, 1991]: Festschrift für Theodor Heinsius zum 65. Geburtstag am 25. September 1991, Berlin, New York: De Gruyter, 1991

Kühbacher, Uli [Darlehen, 1993]: Darlehen an Konzernunternehmen: Besicherung und Vertragsanpassung, Schriftenreihe „Untersuchungen über das Spar-, Giro- und Kreditwesen, Abteilung B: Rechtswissenschaft", hrsg. von *Walther Hadding* und *Uwe H. Schneider*, Bd. 86, Konzern, Konzernrecht und Konzernfinanzierung, Teil III, Berlin: Duncker & Humblot, 1993, zugl. Diss. rer. pol. Technische Hochschule Darmstadt, 1992

Kulitz, Peter [Unternehmerspenden, 1983]: Unternehmerspenden an politische Parteien, Berlin: Duncker & Humblot, 1983 (Beiträge zur politischen Wissenschaft, Bd. 44), zugl. Diss. jur. Univ. Tübingen, 1982/83

Larenz, Karl [Methodenlehre, 1991]: Methodenlehre der Rechtswissenschaft, 6., neu bearbeitete Auflage, Berlin u. a.: Springer, 1991

Laux, Helmut [Koordination, 1993]: Koordination in der Unternehmung, in: *W. Wittmann* u. a. (Hrsg), Handwörterbuch, 1993, Spalten 2308 - 2319

Lawall, Lars [Wettbewerbsverbot, 1996]: Das ungeschriebende Wettbewerbsverbot des GmbH-Gesellschafters, Frankfurt am Main u. a.: Lang, 1996 (Europäische Hochschulschriften: Reihe 2, Rechtswissenschaft; Bd. 1910), zugl. Diss. jur. Univ. Bayreuth, 1995

Lieb, Manfred/Noack, Ulrich/ Westermann, Harm Peter (Hrsg.) [FS Zöllner, 1998]: Festschrift für Wolfgang Zöllner zum 70. Geburtstag, Bd. 1, Köln u. a.: Heymanns, 1998

Liebscher, Thomas [Konzernbildungskontrolle, 1995]: Konzernbildungskontrolle - Rechtsformspezifische und rechtsformunabhängige Aspekte eines konzernrechtlichen Präventivschutzes im Rahmen des Konzernierungsprozesses, Berlin: Duncker und Humblot, 1995 (Schriften zum Wirtschaftsrecht, Bd. 85), zugl. Diss. jur. Mannheim, 1994

Limmer, Peter [Haftungsverfassung, 1992]: Die Haftungsverfassung des faktischen GmbH-Konzerns - Haftungsrechtliche Konsequenzen von Konzernleitung und --organisation, Köln: O. Schmidt, 1992 (Rechtsfragen der Handelsgesellschaften, Heft 70), zugl. Diss. jur. Univ. Bochum, 1991

Lutter, Marcus [Rechte, 1973]: Die Rechte der Gesellschafter beim Abschluß fusionsähnlicher Unternehmensverbindungen, in: DB 1973, Beilage Nr. 21 zu Heft Nr. 46, S. 1 - 15

– [Teilfusionen, 1974]: Teilfusionen im Gesellschaftsrecht, in: *R. Fischer/Ph. Möhring/H. Westermann*, FS Barz, 1974, S. 199 - 218

– [Binnenstruktur, 1974]: Zur Binnenstruktur des Konzerns, in *W. Hefermehl/ R. Gmür/H. Brox* (Hrsg.), FS Westermann, 1974, S. 347 - 368

– [Mitgliedschaft, 1980]: Theorie der Mitgliedschaft - Prolegomena zu einem Allgemeinen Teil des Korporationsrechts -, in: AcP 180 (1980), S. 84 - 159

– [Informationsrecht, 1982]: Zum Informationsrecht des Gesellschafters nach neuem GmbH-Recht, in: ZGR 1982, S. 1 - 14

– [Unternehmensgruppe, 1982]: Die zivilrechtliche Haftung in der Unternehmensgruppe, in: ZGR 1982, 244 - 275

– [Haftung, 1985]: Die Haftung des herrschenden Unternehmens im GmbH-Konzern - Überlegungen zur "Autokran"-Entscheidung des BGH vom 16.9.1985, ZIP 1985, 1263, in: ZIP 1985, S. 1425 - 1435

– [Organzuständigkeiten, 1985]: Organzuständigkeiten im Konzern, in: *M. Lutter/H.-J. Mertens/P. Ulmer* (Hrsg.), FS Stimpel, 1985, S. 825 - 854

– [GmbH-Konzernrecht, 1986]: Das System des GmbH-Konzernrechts, in: *P. Hommelhoff* u. a., Entwicklungen, 1986, S. 192 - 218

– [Grundlagenbeschlüsse, 1988]: Zur Vorbereitung und Durchführung von Grundlagenbeschlüssen in Aktiengesellschaften, in: *R. Goerdeler* u. a. (Hrsg.), FS Fleck, 1988, S. 169 - 190

– [Kölner Kommentar, 1988]: § 57 AktG (Stand: Juli 1998), in: *W. Zöllner* (Hrsg.), Kölner Kommentar, Bd. 1, 1988

– ["TBB", 1993]: Anmerkung zum Urteil des BGH vom 29.3.93 - II ZR 265/91 ("TBB"), in: JZ 1993, S. 580 - 582

– [Haftung, 1993]: Zur persönlichen Haftung des Geschäftsführers aus deliktischen Schäden im Unternehmen, in: ZHR 157 (1993), S. 464 - 482

– [Konzernrecht, 1996]: Konzernrecht: Schutzrecht oder Organisationsrecht?, in: *K. Reichert* u. a. (Hrsg.), FS Volhard, 1996, S. 105 - 113

– [Haftungsrisiken, 1997]: Haftungsrisiken des Geschäftsführers einer GmbH, in: GmbHR 1997, S. 329 - 335

– [Treupflichten, 1998]: Treupflichten und ihre Anwendungsprobleme, in: ZHR 1998, S. 164 - 185

– (Hrsg.) [Holding, 1998]: Holding Handbuch, Recht - Management - Steuern, 3. aktualisierte und erw. Aufl., Köln: O. Schmidt, 1998

– [Haftungsfragen, 1998]: Haftungsfragen in der Holding, in: *M. Lutter* (Hrsg.), Holding, 1998, S. 227 - 266

– /*Timm*, *Wolfram* [Präventivschutz, 1982]: Konzernrechtlicher Präventivschutz im GmbH-Recht, in: NJW 1982, S. 409 - 420

– /*Krieger*, *Gerd* [Aufsichtsrat, 1993]: Rechte und Pflichten des Aufsichtsrats, 3. vollständig neu bearb. u. erw. Aufl., Freiburg im Breisgau: Haufe, 1993

– /*Hommelhoff*, *Peter* [GmbHG, 1995] GmbH-Gesetz, Kommentar, 14., neubearb. und erw. Aufl., Köln: O. Schmidt, 1995

– /*Leinekugel*, *Rolf* [Kompetenzen, 1998]: Kompetenzen von Hauptversammlung und Gesellschafterversammlung beim Verkauf von Unternehmensteilen, in: ZIP 1998, S. 225 - 232

– /*Stimpel*, *Walter*/*Wiedemann*, *Herbert* (Hrsg.) [FS Fischer, 1979]: Festschrift für Robert Fischer, Berlin, New York: de Gruyter, 1979

– /*Mertens*, *Hans-Joachim*/*Ulmer*, *Peter* (Hrsg.) [FS Stimpel, 1985]: Festschrift für Walter Stimpel zum 68. Geburtstag am 29. November 1985, Berlin, New York: De Gruyter, 1985

– /*Ulmer*, *Peter*/*Zöllner*, *Wolfgang* (Hrsg.) [FS GmbHG, 1992]: Festschrift 100 Jahre GmbHG, Köln: O. Schmidt, 1992

– /*Scheffler*, *Eberhard*/*Schneider*, *Uwe H.* (Hrsg.) [Konzernfinanzierung, 1998]: Handbuch der Konzernfinanzierung, Köln: O. Schmidt, 1998

Martens, *Klaus-Peter* [Entscheidungsautonomie, 1983]: Die Entscheidungsautonomie des Vorstands und die "Basisdemokratie" in der Aktiengesellschaft - Anmerkungen zu BGHZ 83, S. 122 ("Holzmüller"), in: ZHR 147 (1983), S. 377 - 428

– [Organisation, 1991]: Die Organisation des Konzernvorstands, in: *F. Kübler*/*H.-J. Mertens*/*W. Werner* (Hrsg.), FS Heinsius, 1991, S. 523 - 544

– [Schlegelberger, 1992]: § 112 HGB, in: *E. Geßler* u. a., Schlegelberger, 1992

Martinek, *Michael* u. a. [Staudinger, 1995]: J. von Staudingers: Kommentar zum Bürgerlichen Gesetzbuch mit Einführungsgesetz und Nebengesetzen, Zweites Buch: Recht der Schuldverhältnisse, §§ 652 - 704, 13. Bearbeitung, Berlin: Sellier - de Gruyter, 1995

Medicus, Dieter [Staudinger, 1983]: § 249 BGB, in: *D. Medicus/K. Schmidt*, Staudinger, 1983

– [Eigenhaftung, 1991]: Zur deliktischen Eigenhaftung von Organpersonen, in: *B. Pfister/M. R. Will* (Hrsg.), FS Lorenz, 1991, S. 155 - 169

– [Geschäftsverteilung, 1998]: Die interne Geschäftsverteilung und die Außenhaftung des GmbH-Geschäftsführers, in: GmbHR 1998, S. 9 - 16

– [Außenhaftung, 1998]: Deliktische Außenhaftung der Vorstandsmitglieder und Geschäftsführer, in: ZGR 1998, S. 570 - 585

– */Schmidt, Karsten* [Staudinger, 1983]: J. von Staudingers Kommentar zum Bürgerlichen Gesetzbuch mit Einführungsgesetz und Nebengesetzen, Zweites Buch: Recht der Schuldverhältnisse, §§ 243 - 254, Redaktor: Dieter Medicus, Berlin: J. Schweitzer Verlag KG Walter de Gruyter & Co., 1983

Meilicke, Heinz [Zuwendungen, 1959]: Zuwendungen an politische Parteien aus Mitteln wirtschaftlicher Unternehmen, in: NJW 1959, S. 409 - 413

Merkt, Hanno [Unternehmensleitung, 1995]: Unternehmensleitung und Interessenkollision, in: ZHR 159 (1995), S. 423 - 453

Mertens, Hans-Joachim [Geschäftsführungshaftung, 1979]: Die Geschäftsführungshaftung in der GmbH und das ITT-Urteil, in: *M. Lutter/W. Stimpel/H. Wiedemann* (Hrsg.), FS Fischer, 1979, S. 461 - 475

– [§ 51 a Abs. 1 GmbHG, 1984]: § 51 a Abs. 1 GmbHG und die kapitalistisch strukturierte GmbH, in: *W. Hadding* u. a. (Hrsg.), FS Werner, 1984, S. 557 - 572

– [Münchener Kommentar, 1986]: Vorbemerkungen zu §§ 823 - 853, in: *K. Rebmann/F. J. Säcker*, Münchener Kommentar, 1986

– [Aktivitäten, 1987]: Zur Auslegung und zum Verhältnis von § 76 Abs. 1 und § 58 AktG im Hinblick auf uneigennützige soziale Aktivitäten der Aktiengesellschaft, in: *H. Havermann* (Hrsg.), FS Goerdeler, 1987, S. 349 - 360

– [Kölner Kommentar, 1988]: §§ 82, 88, 93 AktG (Stand: Dezember, 1988), in: *W. Zöllner* (Hrsg.), Kölner Kommentar, Bd. 2, 1989

– [Soergel, 1990]: Vorbemerkungen zu § 249 BGB, in: *J. F. Baur* u. a., Soergel, 1990

– [Hachenburg, 1996]: §§ 37, 43 GmbHG (Stand: 1.6.1996), in: *P. Ulmer* (Hrsg.), Hachenburg, 8. Lieferung, 1996

– */Mertens, Georg* [Anmerkung, 1990]: Anmerkung zum Urteil des Bundesgerichtshofs vom 5. Dezember 1989 - VI ZR 335/88, in: JZ 1990, S. 488 - 490

– */Cahn, Andreas* [Wettbewerbsverbot]: Wettbewerbsverbot und verdeckte Gewinnausschüttung im GmbH-Konzern, in: *F. Kübler/H.-J. Mertens/W. Werner* (Hrsg.), FS Heinsius, 1991, S. 545 - 574

Mestmäcker, Ernst-Joachim [Systematik, 1967]: Zur Systematik des Rechts der verbundenen Unternehmen im neuen Aktiengesetz, in: *K. H. Biedenkopf/H. Coing/E.-J. Mestmäcker* (Hrsg.), FG Kronstein, 1967, S. 129 - 150

– */Behrens, Peter* (Hrsg.) [Gesellschaftsrecht, 1991]: Das Gesellschaftsrecht der Konzerne im internationalen Vergleich - Ein Symposium des Max-Planck-Instituts für ausländisches und internationales Privatrecht in Hamburg, Baden-Baden: Nomos Verlagsgesellschaft, 1991 (Wirtschaftsrecht und Wirtschaftspolititk, Bd. 110)

Meyer-Landrut, Joachim [GmbHG, 1987]: § 64 GmbHG, in: *J. Meyer-Landrut/G. Miller/ R. J. Niehus*, GmbHG, 1987

Meyer-Landrut, Joachim/Miller, Georg/Niehus, Rudolf J. [GmbHG, 1987]: Gesetz betreffend die Gesellschaften mit beschränkter Haftung (GmbHG) einschließlich Rechnungslegung zum Einzel- sowie zum Konzernabschluß, Kommentar, unter Mitarbeit von *Willi Scholz*, Berlin, New York: De Gruyter, 1987

Miller, Georg [GmbHG, 1987]: §§ 35 - 38 GmbHG, in: *J. Meyer-Landrut/G. Miller/R. J. Niehus*, GmbHG, 1987

Mudgan, B. (Hrsg.) [Materialien, 1899]: Die gesamten Materialien zum Bürgerlichen Gesetzbuch für das Deutsche Reich, I. Bd.: Einführungsgesetz und Allgemeiner Theil, Berlin: R. v. Decker`s Verlag, 1899

Mueller-Thuns, Thomas [Gewinnbezugsrecht, 1989]: Gewinnbezugsrecht und bilanzpolitische Gestaltungsmöglichkeiten in der GmbH, Köln: O. Schmidt, 1989

Mülbert, Peter O. [Aktiengesellschaft, 1996]: Aktiengesellschaft, Unternehmensgruppe und Kapitalmarkt, 2., unveränderte Aufl., München: Beck, 1996, zugl. Habil.-Schr. Univ. München, 1994

Müller, Klaus [GenG, 1996]: Kommentar zum Gesetz betreffend die Erwerbs- und Wirtschaftsgenossenschaften, Zweiter Band (Anh. IV nach § 33 bis 42), Bielefeld: Gieseking, 1996

Neuhaus, Jürgen [Grenzen, 1970]: Die Grenzen der Konzernleitungsgewalt im faktischen Konzern und der Nachteilsbegriff des § 311 AktG 65, in: DB 1970, S. 1913 - 1919

Neusel, Tibet [Haftung, 1997]: Die persönliche Haftung des Geschäftsführers für Steuern der GmbH, in: GmbHR 1997, S. 1129 - 1134

Nipperdey, Hans Carl (Hrsg.) [FS Molitor, 1961]: Festschrift für Erich Molitor zum 75. Geburtstag, 3. Oktober 1961, München, Berlin: Beck 1962

Noack, Ulrich [Beschlüsse, 1988]: Fehlerhafte Beschlüsse in Gesellschaften und Vereinen, Köln u. a.: Heymann, 1989 (Abhandlungen zum deutschen und europäischen Handels- und Wirtschaftsrecht, Bd. 62), zugl. Diss. jur. Univ. Tübingen, 1988

Nölle, Jens-Uwe [Eigenhaftung, 1995]: Die Eigenhaftung des GmbH-Geschäftsführers für Organisationspflichtverletzungen, Regensburg: Roderer, 1995 (Theorie und Forschung; Bd. 359: Rechtswissenschaften, Bd. 28), zugl. Diss. jur. Univ. Bremen, 1995

Oehler, Wolfgang [Produzentenhaftung, 1990]: Produzentenhaftung im Konzern - Deliktsrecht und Haftungsbeschränkung, in: ZIP 1990, S. 1445 - 1455

Osten, Dinnies von der [Wettbewerbsverbot, 1989]: Das Wettbewerbsverbot von Gesellschaftern und Gesellschafter-Geschäftsführern in der GmbH, in: GmbHR 1989, S. 450 - 455

Pache, Sven [Spätlese, 1995]: Spätlese - Die Rechtsentwicklung nach dem "Supermarkt"-Beschluß des BGH, in: GmbHR 1995, S. 90 - 93

Paschke, Marian [Durchgriffsproblematik, 1988]: Rechtsfragen der Durchgriffsproblematik im mehrstufigen Unternehmensverbund, in: Die AG 1988, S. 196 - 206

Pfeiffer, Gerd/Wiese, Günther/Zimmermann, Klaus (Hrsg.) [FS Rowedder, 1994]: Festschrift für Heinz Rowedder zum 75. Geburtstag, München: Beck, 1994

Pfister, Bernhard/Will, Michael R. (Hrsg.) [FS Lorenz, 1991]: Festschrift für Werner Lorenz zum siebzigsten Geburtstag, Tübingen: Mohr, 1991

Polley, Notker [Wettbewerbsverbot, 1993]: Wettbewerbsverbot und Geschäftschancenlehre - Eine Untersuchung am Beispiel der Geschäftsleitung von US-Corporation und deutscher GmbH, Baden-Baden: Nomos, 1993 (Nomos Universitätsschriften: Recht, Bd. 96), zugl. Diss. jur. Univ. Osnabrück, 1992

Priester, Hans-Joachim [Vertragskonzern, 1986]: Bildung und Auflösung des GmbH-Vertragskonzerns, in: *P. Hommelhoff* u. a. (Hrsg.), Entwicklungen, 1986, S. 151 - 186

– [Eigensphäre, 1993]: Die eigene GmbH als fremder Dritter - Eigensphäre der Gesellschaft und Verhaltenspflichten ihrer Gesellschafter -, in: ZGR 1993, S. 512 - 533

– [Ausgliederung, 1999]: Die klassische Ausgliederung - ein Opfer des Umwandlungsgesetzes 1994?, in: ZHR 163 (1999), S. 187 - 202

Raiser, Thomas [Wettbewerbsverbote,1985]: Wettbewerbsverbote als Mittel des konzernrechtlichen Präventivschutzes, in: *M. Lutter/H.-J. Mertens/P. Ulmer* (Hrsg.), FS Stimpel, 1985, S. 855 - 865

– [Hachenburg, 1989]: §§ 13, 14 GmbHG (Stand: 1.9.1989), in: *P. Ulmer* (Hrsg.), Hachenburg, 1989/96

– [Hachenburg, 1990]: Nichtigkeit, Anfechtbarkeit und Unwirksamkeit von Gesellschafterbeschlüssen, in: *P. Ulmer* (Hrsg.), Hachenburg 1989/96, Anh. § 47

– [Kapitalgesellschaften, 1992]: Recht der Kapitalgesellschaften: ein Handbuch für Praxis und Wissenschaft, Aktiengesellschaft, Gesellschaft mit beschränkter Haftung, Kapitalgesellschaft & Co., Umwandlung, Verschmelzung, Spaltung, Konzernrecht, 2., neubearb. und wesentlich erw. Aufl., München: Vahlen, 1992

Rebmann, Kurt/Säcker, Franz Jürgen (Hrsg.) [Münchener Kommentar, 1986]: Münchener Kommentar zum Bürgerlichen Gesetzbuch, Bd. 3: Schuldrecht, Besonderer Teil, 2. Halbband (§§ 652 - 853), Redakteur: *P. Ulmer*, 2. Aufl., München: Beck, 1986

– */Säcker, Franz Jürgen* (Hrsg.) [Münchener Kommentar, 1994]: Münchener Kommentar zum Bürgerlichen Gesetzbuch, Bd. 2: Schuldrecht, Allgemeiner Teil (§§ 241 - 423), Redakteur: *H. Heinrichs*, 3. Aufl., München: Beck, 1994

Recksiek, Silke [Außenhaftung, 1998]: Deliktische Außenhaftung des GmbH-Geschäftsführers insbesondere wegen Verkehrspflichtverletzung und sogenannter Konkursverschleppung, Diss. iur. Univ. Bonn 1998

Rehbinder, Eckard [Konzernaußenrecht, 1969]: Konzernaußenrecht und allgemeines Privatrecht, eine rechtsvergleichende Untersuchung nach deutschem und amerikanischem Recht, Bad Homburg von der Höhe, Berlin, Zürich: Gehlen, 1969 (Wirtschaftsrecht und Wirtschaftspolitik, Bd. 8), zugl. Habil.-Schr. Univ. Frankfurt am Main, 1968

– [Treuepflichten, 1976]: Treuepflichten im GmbH-Konzern - Besprechung der Entscheidung BGHZ 65, 15 -, in: ZGR 1976, S. 386 - 399

– [Unternehmensverbindungen, 1977]: Gesellschaftsrechtliche Probleme mehrstufiger Unternehmensverbindungen, in: ZGR 1977, S. 581 - 649

– [Tochtergesellschaften, 1982]: Ausgründung und Erwerb von Tochtergesellschaften und Rechte der Aktionäre, in: *N. Horn/K. Luig/A. Söllner* (Hrsg.), FS Coing, 1982, S. 423 - 439

– [Konzernleitungspflicht, 1983]: Peter Hommelhoff, Die Konzernleitungspflicht. Zentrale Aspekte eines Konzernverfassungsrechts. In: ZHR 147 (1983), S. 464 - 473

Reichert, Klaus u. a. (Hrsg.) [FS Volhard]: Recht, Geist und Kunst, liber amicorum für Rüdiger Volhard, Baden Baden: Nomos, 1996

Reuter, Dieter [Plädoyer, 1996]: Ein Plädoyer für das institutionelle Rechtsdenken, in: *U. Immenga/W. Möschel/D. Reuter* (Hrsg.), FS Mestmäcker, 1996, S. 271 - 291

Riegen, Arend von [Gesellschafterschutz, 1999]: Gesellschafterschutz bei Ausgliederungen durch Einzelrechtsnachfolge, Köln u. a.: Heymanns, 1999 (Abhandlungen zum deutschen und europäischen Handels- und Wirtschaftsrecht, Bd. 113), zugl. Diss. jur. Univ. Hamburg 1998

Röhricht, Volker [Wettbewerbsverbot, 1992]: Das Wettbewerbsverbot des Gesellschafters und des Geschäftsführers, in: WPg 1992, S. 766 - 786

Roth, Günther H. [Haftung, 1989]: Die Haftung als faktischer Geschäftsführer im Konkurs der GmbH - Besprechung der Entscheidung BGHZ 104, 44 ff -, in: ZGR 1989, S. 421 - 433

– [GmbHG, 1997]: § 54 GmbHG, in: *G. H. Roth/H. Altmeppen*, GmbHG, 1997

– /*Altmeppen, Holger* [GmbHG, 1997]: Gesetz betreffend die Gesellschaft mit beschränkter Haftung (GmbHG), 3., neubearb. Aufl. München: Beck, 1997

Rottkemper, Michael [Außenhaftung, 1996]: Deliktische Außenhaftung der Leitungsorganmitglieder rechtsfähiger Körperschaften, Frankfurt a. M. u. a.: Lang, 1996 (Europäische Hochschulschriften, Reihe II: Rechtswissenschaft, Bd. 2013), zugl. Diss. jur. Univ. Marburg, 1996

Rowedder, Heinz, [Bestandsschutz, 1986]: Bestandsschutz im faktischen GmbH-Konzern, in: P. Hommelhoff u. a. (Hrsg.), Entwicklungen, 1986, S. 20 - 38

– [GmbHG, 1997]: §§ 13, 64 GmbHG, in *H. Rowedder* u. a., GmbHG, 1997

– u. a. [GmbHG, 1997]: Gesetz betreffend die Gesellschaften mit beschränkter Haftung (GmbHG), 3. neubearb. Aufl., München: Vahlen, 1997

Sack, Rolf [Staudinger, 1996]: § 134 BGB (Stand: Januar 1996), in: *R. Bork* u. a., Staudinger, 1996

Säcker, Franz Jürgen [Problematik, 1987]: Zur Problematik von Mehrfachfunktionen im Konzern, in: ZHR 151 (1987), S. 59 - 71

Salfeld, Rainer [Wettbewerbsverbote, 1987]: Wettbewerbsverbote im Gesellschaftsrecht, Frankfurt am Main, Bern, New York: Lang, 1987 (Europäische Hochschulschriften, Reihe II: Rechtswissenschaft, Bd. 602), zugl. Diss. jur. Univ. Bielefeld, 1985/86

Salje, Peter [UmweltHG, 1993]: Umwelthaftungsgesetz, Kommentar, unter Mitarbeit von *Jörg Peter*, München: Beck, 1993

Sandberger, Andreas [Außenhaftung, 1997]: Die Außenhaftung des GmbH-Geschäftsführers, Baden-Baden: Nomos, 1997 (Nomos Universitätsschriften: Recht; Bd. 277), zugl. Diss. jur. Univ. Tübingen, 1996

Scharpf, Paul [Sorgfaltspflichten, 1997]: Die Sorgfaltspflichten des Geschäftsführers einer GmbH - Pflicht zur Einrichtung eines Risikomanagement- und Überwachungssystems aufgrund der geplanten Änderung des AktG auf für den GmbH-Geschäftsführer -, in: DB 1997, S. 737 - 743

Schaub, Bernhard [Haftung, 1992]: Die Haftung des Geschäftsführers einer GmbH gegenüber der Gesellschaft und den Gesellschaftern, in: DStR 1992, S. 985 - 989 (Teil I), 1021 - 1024 (Teil II)

Scheffler, Eberhard [Konzernmanagement, 1992]: Konzernmanagement. Betriebswirtschaftliche und rechtliche Grundlagen der Konzernführungspraxis. München: Beck, 1992

Schmidt, Karsten [Abhängigkeit, 1981]: Abhängigkeit, faktischer Konzern, Nichtaktienkonzern und Divisionalisierung im Bericht der Unternehmensrechtskommission, in: ZGR 1981, S. 455 - 486

– [Konkursverschleppungshaftung, 1988]: Konkursverschleppungshaftung und Konkursverursachungshaftung - Überlegungen zum Haftungsrecht der GmbH und der GmbH & Co. -, in: ZIP 1988, S. 1497 - 1508

– [Scholz, 1995]: §§ 45, 46, 47, 49, 51a, 63, 64 GmbHG, in: *G. Crezelius* u. a., Scholz, II. Bd., 1995

– [Gesellschaftsrecht, 1997]: Gesellschaftsrecht, 3., völlig neu bearb. und erw. Aufl., Köln u. a.: Heymann, 1997

Schmidt-Salzer, *Joachim* [Produkthaftung, Band III/1, 1990]: Produkthaftung, Band III/1, 1990, Deliktsrecht, 1. Teil, 2. neubearb. u. wesentl. erw. Aufl., Heidelberg: Verlag Recht und Wirtschaft, 1990

Schmitz-Rode/Bank, Stephan [Haftung, 1999]: Die konzernrechtliche Haftung nach dem Bundes-Bodenschutzgesetz, in: DB 1999, S. 417 - 420

Schneider, *Manfred* [KWG, 1986]: KWG-Kommentar: Kreditwesengesetz mit den wichtigsten Ausführungsvorschriften - begründet von *Johannes Consbruch* und *Annemarie Möller*, in 2. Aufl. fortgeführt von *Inge Lore Bähre*, *Manfred Schneider* -, 3., völlig neubearb. Aufl., München: Beck, 1986

Schneider, *Uwe H.* [Personengesellschaft, 1979]: Die Personengesellschaft als herrschendes Unternehmen im Konzern - Ein Beitrag zum Konzernrecht der Personengesellschaften -, in: ZHR 143 (1979), S. 485 - 521

– [Konzernbildung, 1980]: Konzernbildung, Konzernleitung und Verlustausgleich im Konzernrecht der Personengesellschaften - Zugleich ein Beitrag zur Bedeutung des Gervais-Urteils für die Entwicklung des Konzernrechts -, in: ZGR 1980, S. 511 - 547

– [Konzernleitung, 1981]: Konzernleitung als Rechtsproblem, in: BB 1981, S. 249 - 259

– [Konzernfinanzierung, 1984]: Das Recht der Konzernfinanzierung, in: ZGR 1984, S. 497 - 537

– [Haftungsmilderung, 1984]: Haftungsmilderung für Vorstandsmitglieder und Geschäftsführer bei fehlerhafter Unternehmensleitung?, in: *W. Hadding* u. a. (Hrsg.), FS Werner, 1984, S. 795 - 815

– [Konzerngründung, 1986]: Die Gründung von faktischen Konzernen - Zuständigkeiten und Finanzierung, in: *P. Hommelhoff* u. a. (Hrsg.), Entwicklungen, 1986, S. 121 - 150

– [Konzernverfassung, 1986]: Die vertragliche Ausgestaltung der Konzernverfassung - Zur Anpassung von Gesellschaftsverträgen, Geschäftsordnungen, Stimmbindungsverträgen und Konsortialverträgen an die Konzernlage, in: BB 1986, S. 1993 - 1999

– [Konzern, 1991]: Der Konzern als Rechtsform für Unternehmen - Zum Regelungsgegenstand eines Konzernverfassungsrechts -, in: *E. J. Mestmäcker/P. Behrens* (Hrsg.), Gesellschaftsrecht, 1991, S. 563 - 580

– [Pflichten, 1992]: Die Wahrnehmung öffentlich-rechtlicher Pflichten durch den Geschäftsführer - Zum Grundsatz der Gesamtverantwortung bei mehrköpfiger Geschäftsführung in der konzernfreien GmbH und im Konzern, in: *M. Lutter/P. Ulmer/W. Zöllner* (Hrsg.), FS GmbHG, 1992, S. 473 - 499

– [Anstellungsvertrag, 1993]: Der Anstellungsvertrag des Geschäftsführers einer GmbH im Konzern - Überlegungen zum Vertragspartner, zur Rechtsnatur und zum Inhalt des Konzern-Anstellungsvertrags, in: GmbHR 1993, S. 10 - 21

– [Scholz, 1993]: §§ 35, 37, 43 GmbHG, in: *G. Crezelius* u. a., Scholz, I. Bd., 1993

– [Unternehmensorganisation, 1993]: Gesellschaftsrechtliche und öffentlich-rechtliche Anforderungen an eine ordnungsgemäße Unternehmensorganisation - Zur Überlagerung des Gesellschaftsrechts durch öffentlich-rechtliche Verhaltenspflichten und öffentlich-rechtliche Strukturnormen, in: DB 1993, S. 1909 - 1915

– [Wettbewerbsverbot, 1995]: Wettbewerbsverbot für Aufsichtsratsmitglieder einer Aktiengesellschaft? - 12 Thesen zu einer rechtspolitischen Diskussion, in: BB 1995, S. 365 - 370

– [Compliance, 1996]: Die Überlagerung des Konzernrechts durch öffentlich-rechtliche Strukturnormen und Organisationspflichten - Vorüberlegungen zu „Compliance im Konzern“ -, in: ZGR 1996, S. 225 - 246

– [Cash-Management, 1998]: Das Recht des konzernweiten Cash-Managements, in: *M. Lutter/E. Scheffler/Uwe H. Schneider* (Hrsg.), Konzernfinanzierung, 1998, Rdnr. 25.1 - 25.92

– */Ihlas, Horst* [Vermögensschaden-Haftpflichtversicherung, 1994]: Die Vermögensschaden-Haftpflichtversicherung des Geschäftsführers einer GmbH, in: DB 1994, S. 1123 - 1128

Schulze-Osterloh, Joachim [Betriebsaufspaltung, 1983]: Gläubiger- und Minderheitenschutz bei der steuerlichen Betriebsaufspaltung, in: ZGR 1983, S. 123 - 161

– [Baumbach/Hueck, 1996]: §§ 42, 42a, 64 GmbHG, in: *G. Hueck/J. Schulze-Osterloh/ W. Zöllner*, Baumbach/Hueck, 1996

Schwedhelm, Rolf [Wettbewerbsverbot, 1993]: Wettbewerbsverbot und Betriebsabgrenzung - Zu den zivilrechtlichen Voraussetzungen verdeckter Gewinnausschüttungen im Steuerrecht, in: DStR 1993, S. 245 - 251

Semler, Johannes [Überwachung, 1995]: Die interne Überwachung in einer Holding, in: *M. Lutter* (Hrsg.), Holding, 1995, S. 149 - 192

– [Leitung, 1996]: Leitung und Überwachung der Aktiengesellschaft: Die Leitungsaufgabe des Vorstands und die Überwachungsaufgabe des Aufsichtsrats, 2., völlig überarb. und erw. Aufl., Köln u. a.: Heymann, 1996 (Abhandlungen zum deutschen und europäischen Handels- und Wirtschaftsrecht, Bd. 31), zugl. Diss. jur. Univ. Bochum, 1980

Sonntag, Bernd [Konzernbildungskontrolle, 1990]: Konzernbildungs- und Konzernleitungskontrolle bei der GmbH, Frankfurt am Main u. a.: Lang, 1990 (Europäische Hochschulschriften: Reihe 2, Rechtswissenschaft, Bd. 968), zugl. Diss. jur. Univ. Bayreuth, 1989

Ständige Deputation des Deutschen Juristentages (Hrsg.) [Verhandlungen, 1992]: Verhandlungen des neunundfünfzigsten Deutschen Juristentages, Hannover 1992, Band I, Gutachten, München: Beck, 1992

Stein, Ursula [Organ, 1984]: Das faktische Organ, Köln u. a.: Heymann, 1984 (Abhandlungen zum deutschen und europäischen Handels- und Wirtschaftsrecht, Bd. 41), zugl. Diss. jur. Univ. Frankfurt am Main, 1983

– [Konkursverschleppung, 1984]: Die Normadressaten der §§ 64, 84 GmbHG und die Verantwortlichkeit von Nichtgeschäftsführern wegen Konkursverschleppung, in: ZHR 148 (1984), S. 207 - 237

Stimpel, Walter [Innenhaftung, 1986]: Die Rechtsprechung des Bundesgerichtshofes zur Innenhaftung des herrschenden Unternehmens im GmbH-Konzern, in: Die AG 1986, S. 117 - 123

Strohn, Lutz [Verfassung, 1977]: Die Verfassung der Aktiengesellschaft im faktischen Konzern - Zur Harmonisierung der §§ 311 ff. AktG mit den allgemeinen Bestimmungen des Aktienrechts -, Köln u. a.: Heymann, 1977 (Abhandlungen zum deutschen und europäischen Handels- und Wirtschaftsrecht, Bd. 18), zugl. Diss. jur. Univ. Köln, 1976/77

Sudhoff, Heinrich/Sudhoff, Martin [Rechte, 1994]: Rechte und Pflichten des Geschäftsführers einer GmbH und einer GmbH & Co, 14., überarb. Aufl., Köln: O. Schmidt, 1994 (Rechtsfragen der Handelsgesellschaften, Heft 4)

Teubner, Gunther [Politik, 1990]: Die ‚Politik des Gesetzes' im Recht der Konzernhaftung - Plädoyer für einen sektoralen Konzerndurchgriff, in: *J. F. Baur* u. a. (Hrsg.), FS Steindorff, 1990, S. 261 - 279

Thamm, Manfred [Haftung, 1994]: Die persönliche Haftung von Führungskräften und Mitarbeitern in Unternehmen, in: DB 1994, S. 1021 - 1023

Theisen, Manuel R. [Konzern, 1991]: Der Konzern - Betriebswirtschaftliche und rechtliche Grundlagen der Konzernunternehmung, Stuttgart: Poeschel, 1991

Thiel, Jochen [Gewinnausschüttung, 1992]: Verdeckte Gewinnausschüttung bei Verletzung des Wettbewerbsverbots durch Gesellschafter/Geschäftsführer einer GmbH - Anmerkungen zum BMF-Schreiben vom 4.2.1992, in: GmbHR 1992, S. 338 - 346

– [Gewinnausschüttung, 1993]: Die verdeckte Gewinnausschüttung im Spannungsfeld zwischen Zivil- und Steuerrecht, in: DStR 1993, S. 1801 - 1808

Thomas, Heinz [Palandt, 1998]: § 677 BGB, in: *P. Bassenge* u. a., *Palandt*, 1998

Tiedchen, Susanne [Wettbewerbsverbote, 1993]: Wettbewerbsverbote im GmbH-Konzern, in: GmbHR 1993, S. 616 - 622

Tieves, Johannes [Unternehmensgegenstand, 1998]: Der Unternehmensgegenstand der Kapitalgesesellschaft, Köln: O. Schmidt, 1998 (Rechtsfragen der Handelsgesellschaften, Bd. 93), zugl. Diss. jur. Univ. Heidelberg 1997

Tillmann, Bert [Wettbewerbsverbot, 1991]: Wettbewerbsverbot des Gesellschafter-Geschäftsführers einer GmbH - Zivil- und steuerrechtliche Aspekte, in: GmbHR 1991, S. 26 - 31

Timm, Wolfram [Konzernspitze, 1980]: Die Aktiengesellschaft als Konzernspitze - Die Zuständigkeitsordnung bei der Konzernbildung, Köln u. a.: Heymann, 1980 (Abhandlungen zum deutschen und europäischen Handels- und Wirtschaftsrecht, Bd. 31), zugl. Diss. jur. Univ. Bochum, 1979

– [Hauptversammlungskompetenzen, 1980]: Hauptversammlungskompetenzen und Aktionärsrechte in der Konzernspitze - zugleich Überlegungen zum Urteil des LG Hamburg vom 1.10.1979 -, in: Die AG 1980, S. 172 - 187

– [Wettbewerbsverbot, 1981]: Wettbewerbsverbot und "Geschäftschancen" - Lehre im Recht der GmbH, in: GmbHR 1981, S. 177 - 186

Turiaux, André/Knigge, Dagmar [Bundes-Bodenschutzgesetz, 1999]: Bundes-Bodenschutzgesetz - Altlastensanierung und Konzernhaftung - Im Blickpunkt: Der erweiterte Kreis der Sanierungsverpflichteten, in: BB 1999, S. 377 - 384

Ulmer, Peter [Handlungspflichten, 1969]: Die deliktische Haftung aus der Übernahme von Handlungspflichten - Zugleich ein Beitrag zum Verhältnis von Vertragsverletzung und unerlaubter Handlung, in: JZ 1969, S. 163 - 174

– [Gläubigerschutz, 1984]: Der Gläubigerschutz im faktischen GmbH-Konzern beim Fehlen von Minderheitsgesellschaftern, in: ZHR 148 (1984), S. 391 - 421

– [Münchener Kommentar, 1986]: § 708 BGB, in: *K. Rebmann/F. J. Säcker* (Hrsg.), Münchener Kommentar, 1986

– [Schutz, 1988]: Schutz der GmbH gegen Schädigungen zugunsten ihrer Gesellschafter? - Zur Relevanz der Rechtsprechung zu § 266 StGB für das Gesellschaftsrecht -, in: *O. F. Freiherr von Gamm/P. Raisch/K. Tiedemann* (Hrsg.), FS Pfeiffer, 1988, S. 853 - 871

– [Staub, 1988]: § 112 HGB (Stand: 1.7.1988), in: *C.-W. Canaris/W. Schilling/P. Ulmer* (Hrsg.), Staub, 1989

– (Hrsg.) [Probleme, 1989]: Probleme des Konzernrechts - Symposium zum 80. Geburtstag von Wolfgang Schilling, Heidelberg: Verlag Recht und Wirtschaft, 1989 (Abhandlungen aus dem gesamten Bürgerlichen Recht, Handelsrecht und Wirtschaftsrecht, Beihefte der ZHR, Heft 62)

– (Hrsg.) [Hachenburg, 1989/96]: Hachenburg, Gesetz betreffend die Gesellschaften mit beschränkter Haftung (GmbHG), Großkommentar, 8. neubearb. Aufl.: 1. Lieferung: Allgemeine Einleitung, §§ 1 - 14, 1989, 2. Lieferung: §§ 15 - 28, 1990, 3. Lieferung: §§ 45 - 52, 1991, 4. Lieferung: § 32a - Anh. § 34, 1991, 5. Lieferung: §§ 53 - 64, 1991, 6. Lieferung: §§ 29 - 32, 1992, 7. Lieferung: §§ 65 - 85 mit Anh. § 77, 1994, 8. Lieferung: §§ 35 - 44, Register, 1996, Berlin, New York: De Gruyter, 1989/96

– [Hachenburg, 1989]: §§ 1, 3 GmbHG (Stand: 1.9.1989), in: *P. Ulmer* (Hrsg.), Hachenburg, 1989/96

– [Hachenburg, 1991]: § 64 GmbHG (Stand: 1.7.1991), in: *P. Ulmer* (Hrsg.), Hachenburg, 1989/96

– [Hachenburg, 1994]: GmbH-Konzernrecht (Stand: 1.3.1994), in: *P. Ulmer* (Hrsg.), Hachenburg, 1989/1996, Anh. § 77

– [Vermutungsfragen, 1992]: Vermutungs- und Beweisfragen: Qualifizierungsvermutung; Kausalitätsgegenbeweis, in: *P. Hommelhoff*/*W. Stimpel*/*P. Ulmer* (Hrsg.), Konzernrechtstage, 1992, S. 65 - 88

– [Hachenburg, 1994]: GmbH-Konzernrecht (Stand: 1.3.1994), in: *P. Ulmer* (Hrsg.), Hachenburg, 1989/96, Anh. § 77

Verhoeven, *Thomas* [Innenrecht, 1978]: GmbH-Konzern-Innenrecht - Grundsatzfragen des geltenden Rechts -, Schriften aus Rechtswissenschaft und Rechtspraxis, hrsg. von *Brigitte Knobbe-Keuk*, *Manfred Lieb* u. *Harm Peter Westermann*, Bd. 1, Bonn: Stollfuß, 1978

Versteegen, *Peter* [Konzernverantwortlichkeit, 1993]: Konzernverantwortlichkeit und Haftungsprivileg, Köln u. a.: Heymann, 1993 (Abhandlungen zum deutschen und europäischen Handels- und Wirtschaftsrecht, Bd. 83), zugl. Diss. jur. Univ. Freiburg, 1993

Vetter, *Eberhard* [Unternehmensvertrag, 1993]: Die Geltung von § 293 Abs. 2 AktG beim Unternehmensvertrag zwischen herrschender AG und abhängiger GmbH - Anmerkungen zum Beschluß des BGH vom 30.1.1992, in: Die AG 1993, S. 168 - 173

Vorwerk, *Klaus*/*Wimmers*, *Jörg* [Treubindung, 1998]: Treubindung des Mehrheitsgesellschafters oder der Gesellschaftermehrheit bei Beschlußfassung in der GmbH-Gesellschafterversammlung, in: GmbHR 1998, S. 717 - 723

Wahlers, *Henning W.* [Konzernbildungskontrolle, 1995]: Konzernbildungskontrolle durch die Hauptversammlung der Obergesellschaft, Köln u. a.: Heymann, 1995 (Abhandlungen zum deutschen und europäischen Handels- und Wirtschaftsrecht, Bd. 91), zugl. Diss. jur. Univ. Köln, 1994

Wassermeyer, *Franz* [Wettbewerbsverbot, 1993]: Das Wettbewerbsverbot des Gesellschafters und des Gesellschafter-Geschäftsführers einer GmbH, in: GmbHR 1993, S. 329 - 336

Wehlen, Erhard [Cash-Management, 1998]: Das Cash-Management im Konzern, in: *M. Lutter/E. Scheffler/Uwe H. Schneider* (Hrsg.), Konzernfinanzierung, 1998, Rdnr. 23.1 - 23.82

Weisser, Johannes [Opportunities, 1991]: Corporate Opportunities: Zum Schutz der Geschäftschancen des Unternehmens im deutschen und im US-amerikanischen Recht, Köln u. a.: Heymann, 1991 (Abhandlungen zum deutschen und europäischen Handels- und Wirtschaftsrecht, Bd. 77), zugl. Diss. jur. Univ. Bonn, 1990

Westermann, Harm Peter [Organzuständigkeit, 1984]: Organzuständigkeit bei Bildung, Erweiterung und Umorganisation des Konzerns, in: ZGR 1984, S. 352 - 411

– [Verantwortung, 1990]: Gesellschaftliche Verantwortung des Unternehmens als Gesellschaftsrechtsproblem, in: ZIP 1990, S. 771 - 777

– [Umwelthaftung, 1991]: Umwelthaftung im Konzern, in: ZHR 155 (1991), S. 223 - 246

– [Tatbestand, 1992]: Der Tatbestand des qualifizierten faktischen Konzerns, in: *P. Hommelhoff/W. Stimpel/P. Ulmer* (Hrsg.), Konzernrechtstage, 1992, S. 21 - 40

– */Mutter, Stefan* [Verantwortlichkeit, 1995]: Die Verantwortlichkeit von Geschäftsführern einer GmbH gegenüber Dritten - Eine Bestandsaufnahme nach der Neuordnung durch die Rechtsprechung -, in: DWiR 1995, S. 184 - 192

Wiedemann, Herbert [Grundfragen, 1975]: Grundfragen der Unternehmensverfassung, in: ZGR 1975, S. 385 - 432

– [Gesellschaftsrecht, 1980]: Gesellschaftsrecht, Bd. I, Grundlagen, München: Beck, 1980

– [Unternehmensgruppe, 1988]: Die Unternehmensgruppe im Privatrecht - Methodische und sachliche Probleme des deutschen Konzernrechts, Tübingen: Mohr, 1988

– */Hirte, Heribert* [Konkretisierung, 1986]: Die Konkretisierung der Pflichten des herrschenden Unternehmens - Besprechung der Heumann-Ogilvy-Entscheidung BGHZ 89, 162 -, in: ZGR 1986, S. 163 - 174

Wilde, Christian [Informationsrechte, 1998]: Informationsrechte und Informationspflichten im Gefüge der Gesellschaftsorgane, in: ZGR 1998, S. 423 - 465

Wilhelm, Jan [Rechtsform, 1981]: Rechtsform und Haftung bei der juristischen Person, Köln u. a.: Heymann, 1981, zugl. Habil.-Schr. Univ. Bonn, 1978

Willingmann, Armin u. a. (Hrsg) [Jahrbuch, 1996]: Jahrbuch Junger Zivilrechtswissenschaftler 1996, Das deutsche Zivilrecht 100 Jahre nach Verkündung des BGB : Erreichtes, Übersehenes, Verfehltes, Rostocker Tagung 11. - 14. September 1996, Stuttgart u. a.: Richard Boorberg, 1996

Winter, Heinz [Scholz, 1993]: § 14 GmbHG, in: *G. Crezelius* u. a., Scholz, I. Bd., 1993

Winter, Heinz [Scholz, 1993]: § 14 GmbHG, in: *G. Crezelius* u. a., Scholz, I. Bd. 1993

Winter, *Martin* [Treubindungen, 1988]: Mitgliedschaftliche Treubindungen im GmbH-Recht - Rechtsformspezifische Aspekte eines allgemeinen gesellschaftsrechtlichen Prinzips -, München: Beck, 1988 (Schriften des Instituts für Arbeits- und Wirtschaftsrecht der Universität zu Köln, Bd. 55), zugl. Diss. jur. Univ. Heidelberg, 1986/87

– [Eigeninteresse, 1994]: Eigeninteresse und Treupflicht bei der Einmann-GmbH in der neueren BGH-Rechtsprechung, in: ZGR 1994, S. 570 - 594

Wittmann, *Roland* [Staudinger, 1994]: Vorbemerkungen zu §§ 677 ff. BGB (Stand: Juni 1994), in: *M. Martinek* u. a., Staudinger, 1995

Wittmann, *Waldemar* u. a. (Hrsg.) [Handwörterbuch, 1993]: Handwörterbuch der Betriebswirtschaft, Teilband 2: I - Q, 5., völlig neu gestaltete Aufl., Stuttgart: Schäffer-Poeschel, 1993

Wöhe, *Günter* [Handelsbilanz, 1990]: Die Handels- und Steuerbilanz, 2., völlig neu bearb. Aufl., München: Beck, 1990

Würdinger, *Hans* [Großkommentar, 1971]: §§ 308, 309, 311 AktG (Stand: Dezember 1971), in: *C. H. Barz* u. a., Großkommentar, 4. Bd., 1975

Ziemons, *Hildegard* [Haftung, 1996]: Die Haftung der Gesellschafter für Einflußnahmen auf die Geschäftsführung der GmbH, Köln u. a.: Heymann, 1996 (Abhandlungen zum deutschen und europäischen Handels- und Wirtschaftsrecht, Bd. 104), zugl. Diss. jur. Univ. Bonn, 1996

Zitzmann, *Axel* [Vorlagepflichten, 1991]: Die Vorlagepflichten des GmbH-Geschäftsführers, Heidelberg: Verlag Recht und Wirtschaft, 1991 (Abhandlungen zum Arbeits- und Wirtschaftsrecht, Bd. 64), zugl. Diss. jur. Univ. Bonn, 1990

Zöllner, *Wolfgang* [Schranken, 1963]: Die Schranken mitgliedschaftlicher Stimmrechtsmacht bei den privatrechtlichen Personenverbänden, München, Berlin: Beck, 1963 (Schriften des Instituts für Wirtschaftsrecht an der Universität Köln, Bd. 15), zugl. Habil.-Schr. Univ. München, 1960

– (Hrsg.) [Kölner Kommentar, Bd. 6, 1987]: Kölner Kommentar zum Aktiengesetz, Bd. 6, 1. Lieferung: §§ 291 - 328, Stand: Mai 1987, 2., überarb. und erw. Aufl., Köln u. a.: Heymann, 1987

– (Hrsg.) [Kölner Kommentar, Bd. 1, 1988]: Kölner Kommentar zum Aktiengesetz, Bd. 1: §§ 1 - 75, 2., neubearb. u. erw. Aufl., Köln u. a.: Heymann, 1988

– (Hrsg.) [Kölner Kommentar, Bd. 2, 1989]: Kölner Kommentar zum Aktiengesetz, Bd. 2, 1. Lieferung: §§ 76 - 94, 2., neubearb. u. erw. Aufl., Köln u. a.: Heymann, 1989

– [Beherrschungsverträge, 1992]: Inhalt und Wirkungen von Beherrschungsverträgen bei der GmbH, in: ZGR 1992, S. 173 - 202

– [Aktienkonzernrecht, 1998]: Treupflichtgesteuertes Aktienkonzernrecht, in: ZHR 162 (1998), S. 235 - 248

– /*Noack, Ulrich* [Beschlußmängel, 1989]: Die Geltendmachung von Beschlußmängeln im GmbH-Recht, in: ZGR 1989, S. 525 - 544

– [Baumbach/Hueck, 1996]: §§ 35, 37, 43, 47, 51a, 53 GmbHG, GmbH-Konzernrecht, in: *G. Hueck/J. Schulze-Osterloh/W. Zöllner*, Baumbach/Hueck, 1996

Sachwortverzeichnis

Die Zahlen verweisen auf die Seiten des Buches